D1736663

# GEOTECHNICAL ENGINEERING INVESTIGATION MANUAL

McGRAW-HILL SERIES IN GEOTECHNICAL ENGINEERING

*Legget* and *Karrow* **Handbook of Geology in Civil Engineering** (1983)
*Hunt* **Geotechnical Engineering Investigation Manual** (1984)
*Hunt* **Geotechnical Engineering Practice** (1984)

# GEOTECHNICAL ENGINEERING INVESTIGATION MANUAL

ROY E. HUNT

*Consulting Engineer*

**McGraw-Hill Book Company**

*New York St. Louis San Francisco Auckland Bogotá Hamburg London Madrid Mexico Montreal New Delhi Panama Paris São Paulo Singapore Sydney Tokyo Toronto*

Library of Congress Cataloging in Publication Data

Hunt, Roy E.
Geotechnical engineering investigation manual.

(McGraw-Hill series in geotechnical engineering)
Includes index.
1. Engineering geology—Handbooks, manuals, etc.
I. Title. II. Series.
TA705.H86 1983 624.1′51 82-22886
ISBN 0-07-031309-1

Copyright © 1984 by McGraw-Hill, Inc. All rights reserved.
Printed in the United States of America. Except as permitted under the United States Copyright Act of 1976, no part of this publication may be reproduced or distributed in any form or by any means, or stored in a data base or retrieval system, without the prior written permission of the publisher.

3 4 5 6 7 8 9 0 VNH/VNH 8 9

ISBN 0-07-031309-1

The editors for this book were Joan Zseleczky and Geraldine Fahey, the designer was Elliot Epstein, and the production supervisor was Sally Fliess. It was set in Melior by University Graphics, Inc.

Printed and bound by Von Hoffman Press, Inc.

# CONTENTS

## APPENDICES

# FOREWORD

Whoever reads this book will never approach geotechnical exploration in the same way again. Nowhere else has it been made so evident that all boundaries among engineering geology, geophysics, rock mechanics, soil mechanics, geohydrology, seismology, and a host of other disciplines are meaningless; that contributions to the solution of geotechnical problems may come from any or all of these sources; or that the practitioner who holds too narrowly to a specialty is likely to overlook knowledge that could be of the greatest benefit to him in reaching a proper judgment. If for no other reason than to appreciate the breadth of geotechnics and to escape from the trap of too narrow a perspective, this book should be read.

The book also rights an imbalance of long standing: it gives residual soils, tropically weathered soils, and the transitional materials between soil and rock the attention their widespread occurrence deserves. It does this, not because of any abstract desire of the author to present a complete picture, but because he has worked with these materials in many parts of the world, has found them no less deserving of rational treatment than transported soils in temperate zones, and sees no reason to treat them as oddities in the world of workaday geotechnics. His recognition of the prevalence and significance of colluvium is also notable.

The reader can use the book in several ways. Even the most experienced professional will find it to be a useful check list of the adequacy of his exploratory programs. The less experienced will discover a wealth of useful data as well, and will learn of procedures and resources that have not previously come to his attention.

The reader will certainly not find all he needs to solve his exploration problems. Nor will he learn the nuts and bolts of investigation. The book does not tell him in detail how to make a boring, use a sophisticated sampler, or interpret an airphoto. It does, however, give him the information required to judge the circumstances under which a certain procedure or test would be useful, to know whether the field work is being competently done, to assess the validity of the results, and to reach reasonable engineering conclusions. It also points him, through the many references and the indexes of source material, to more complete information if he needs it.

The purist in any of the geotechnical disciplines will no doubt find shortcomings, but the practical man with problems to solve will be fascinated by the vast array of useful information and varieties of approaches available to him, and will find much of what he needs at his fingertips.

*Ralph B. Peck*

# PREFACE

Geotechnical engineering is considered as the discipline of civil engineering that involves the interrelationship between the geologic environment and the works of humans. Soil mechanics and rock mechanics are subdivisions in which the mathematical aspects of analysis for the design of engineering works are defined and described as related to the geologic environment. For projects involving excavations in rock there is a close relationship with mining engineering. The base upon which the knowledge structure is built in geotechnical engineering is a thorough comprehension of the elements of the geologic environment.

In reality, therefore, geotechnical engineering consists of two major, but separate disciplines: geology and civil engineering. Both disciplines are branches of applied science, but there is a major philosophical difference between them. The geologist bases his conclusions primarily on observations and intuitive reasoning, whereas the engineer measures properties and applies mathematical relationships to reach his conclusions. The discipline of engineering geology (or geological engineering) has attempted to fill the philosophical gap, but primarily as related to the evaluation of geologic phenomena such as slope movements, earthquakes, etc., not as related to the design and construction of engineering works such as foundations and retaining structures.

This book was conceived as a vehicle to create a merger between geology and civil engineering; it is a comprehensive guide to the elements of geotechnical engineering from the aspects of investigating and defining the geologic environment for the purpose of providing criteria for the design of engineering works—whether they are in soil or rock.

The geotechnical engineer must be familiar with the many components of the geologic environment and their characteristics: rock types and rock masses, soil types and soil formations, and groundwater, as well as the phenomena generally referred to as the geologic hazards, i.e., flooding and erosion, landslides, ground heave, subsidence and collapse, and earthquakes.

In geotechnical investigations it is necessary to identify these elements and to define their spatial orientation employing various techniques of exploration. Engineering design criteria are established based on measurements of the hydraulic and mechanical properties of the component geologic materials, either in the laboratory by testing samples retrieved from the field, or in the field itself, i.e., in situ. The response of the geologic environment to changing stress fields or other transient conditions, occurring naturally or as a result of construction activity, is measured with instrumentation.

The emphasis in this text is on the identification and description of the elements of the geologic environment, the data required for analysis and design of engineering works, the physical and engineering properties of the geologic materials, and procurement of the data. Approaches to evaluations to arrive at solutions to engineering problems are described for some conditions as an aid to understanding the necessity for the data and their application; general solutions are described for those problems that can be resolved based on experience and judgment, without resorting to rigorous mathematical analysis. The analytical aspects of soil and rock mechanics as applied to the design of foundations, retaining structures, dams, pavements, tunnels and other engineering works are not included in this text, except on occasion as a brief reference to some particular aspect of analysis such as settlements, slope stability, or seepage forces.

The most serious elements of the geologic environment that impact on the works of humans are the activities of the geologic hazards, and approaches for dealing with these conditions are described in some detail. These phenomena can be considered in terms of the degree of their hazard or activity and the degree of the risk of their occurrence in terms of the consequences that might affect a person's ability to arrive at a solution to the problem which can take one of several approaches: avoid the hazard, reduce it, or eliminate it. It must be recognized that in many instances it is not possible to totally eliminate a hazardous condition and it must either be avoided or reduced to the point where the risk is tolerable.

It also must be recognized that there are many limitations to our capabilities in geotechnical engineering. It is not always possible to define all significant conditions at a given location, to obtain accurate measurements of the properties of many geologic materials, or to predict the occurrence and magnitude of slope failures, floods and earthquakes, for example. In addition, many analytical approaches are based on empirical concepts, and most are based on concepts of elastic theory which apply only in a general manner to most geologic materials. Realization of our limitations leads to conservative design and adequate safety factors to provide for the unknown.

The greater majority of the techniques and concepts presented herein are expected to be applicable for many years to come, but there are certain areas where the state of the art is changing rapidly and some parts of this work are expected to become modified with time, particularly in earthquake engineering, offshore technology, and the application of electronics to measurements of properties and instrumentation.

The reader will find the presentation somewhat different from the usual engineering text. Emphasis is on the practical approach to problem solving rather than the theoretical, with the objective of providing a medium to enable the user to rapidly retrieve key data.

The author thanks his many colleagues in the profession who offered comments on the manuscript, particularly Dr. Ralph B. Peck, as well as the many authors and publishers who provided permission to reproduce many of the tables and figures, or who provided figures themselves. *Note:* Photographs without credits were taken by the author.

*Roy E. Hunt*

# GEOTECHNICAL ENGINEERING INVESTIGATION MANUAL

# CHAPTER ONE

# INTRODUCTION

## 1.1 BASIC CONSIDERATIONS

### 1.1.1 HUMAN ACTIVITIES AND THE GEOLOGIC INTERFACE

#### The Geologic Environment

The geologic environment interfacing with human activities presents conditions that can be considered to include not only the basic components of soil, rock, and water, but also the associated phenomena referred to as the geologic hazards, i.e., flooding, erosion, landslides and other slope failures, ground subsidence and collapse, ground heave, and earthquakes.

#### Interfacing Impacts

Human activities, as related to the geologic environment, can be considered in two modes: aggressive and passive. *Aggressive* activities, such as the construction of buildings, dams, roadways, and tunnels impact for the most part on their immediate surroundings, whereas the development of mineral resources by the subsurface extraction of water, oil, gas, or coal can impact on areas substantially removed from the activity. The *passive* mode refers to geologic occurrences impacting on human works in such forms as floods, slope failures, and earthquakes, which can occur naturally or as a result of human activities. Construction or development requires a thorough assessment of the geological environment to ensure that the interfacing is compatible.

The impact of the geologic environment on human activities can be considered as direct or indirect. *Direct impact* occurs in the use of geologic materials in construction, the removal of materials in excavation, and the provision for secure support of structures, excavations, and slopes. The objectives are economical construction and satisfactory performance. *Indirect impact* occurs during extraction of materials from the surface, which can cause adjacent ground subsidence, and extraction from the subsurface, which can cause subsidence and collapse. The objective is to avoid or control these consequences of human activity. The naturally occurring geologic hazards, whose incidence can be increased by human activities, also require treatment.

#### Engineering Assessment

An adequate assessment of the geologic environment is prerequisite to formulating effective treatments for the consequences of these interfacing activities. This requires the capability to identify and describe rock and soil types, rock mass and soil formation characteristics, and groundwater conditions and to recognize and describe the potential for the phenomena of flooding, erosion, slope failures, ground subsidence and collapse, ground heave, and earthquakes. The basis for adequate assessment is thorough investigation.

For many years the services required for investigation of the geologic environment for new construction were performed by soil and foundation engineers practicing the discipline of soil mechanics. The discipline of *geotechnical engineering* has evolved to include rock mechanics and geological engineering as well as soil mechanics to provide a broader intelligence base for resolving modern problems. World population has exploded in recent decades, resulting in an increase in land use for construction works and mineral resources extraction, which often interface, an increase in the size and depth of construction works, and an increase in underground construction. These increases present the practitioner with a much wider range of problems than encountered previously.

### 1.1.2 INVESTIGATION

#### Importance

The engineering works interfacing with the geologic environment can be constructed economically, perform safely, and have nondetrimental impact on other works only if all geologic elements are accurately identified and their properties properly measured and evaluated. Reliable evaluations, however, are possible only when complete and representative data are available. Investigation, therefore, is the most important phase of any construction or development program.

#### Limitations

It must be recognized that geologic conditions can be extremely complex and variable as well as subject to changes with time, and that there are many limitations in the state of the art of geotechnical investigation. There will be occasions when it is not possible to identify all of the critical aspects of geologic conditions, regardless of the comprehensiveness of the investigation. Familiarity on the part of the geotechnical engineer with those aspects of the geologic environment that are unusually complex and variable with a low degree of predictability, as well as with the limitations of investigational methodology, should lead to incorporation of conservative measures into design and construction to avoid unsatisfactory results. The alternative may be construction delays and extra costs, or even structural collapse or other forms of failure.

### 1.1.3 OBJECTIVES

#### General

This work was produced with the objectives of providing a guide to investigation and assessment of the geologic environment for practitioners involved with engineering works for new construction or development, and a guide to general solutions of problems that may be based primarily on intuitive reasoning and experience rather than rigorous mathematical analysis. *Investigation* as used herein signifies field exploration, field and laboratory measurement of properties, and field instrumentation to monitor deformations and stresses in situ. Effective planning and execution of investigations and interpretation and evaluation of data, the elements of performing a study, require a thorough comprehension of the characteristics of the geologic materials and the geologic hazards.

#### Assessment and Problem Solving

In many cases, assessment and problem solving may be based on intuitive reasoning and judgment in certain aspects of groundwater control, slope stabilization, and the control of ground heave, collapse, and subsidence. Although some elements of mathematical analysis for evaluating foundation settlements, groundwater flow, slope stability, and earthquake engineering are presented in this work, the main objective is to provide the basis for understanding the physical phenomena involved as background for comprehending the need for the various elements of investigation. It is not an objective to provide the basis for the analysis and design of slopes, retaining structures, closed excavations, foundations, and ground improvement systems.

## 1.2 SCOPE

### 1.2.1 GENERAL

This book is divided into three basic parts:

- Part I: *Investigation Methods and Procedures* covers field exploration, field and laboratory measurements of properties, and field instrumentation.
- Part II: *Geologic Materials, Characteristics, and Recognition* covers rock masses, soil formations, and surface and subsurface water. The

emphasis is on origin, mode of occurrence or deposition as related to their significance in terrain analysis and the prediction of geologic conditions including characteristic properties, and their impact on engineered construction.

- Part III: *Geologic Hazards, Recognition, and Treatment* covers landslides and other forms of slope failure; ground subsidence, collapse and heave; and earthquakes (flooding and erosion are covered in Part II under *water*). The causes, effects, modes of occurrence, and methods of investigation and treatment are described for each phenomenon.

### 1.2.2 INVESTIGATION METHODS AND PROCEDURES (Part I)

#### Exploration (Chap. 2)

The test boring has long been the standard investigation procedure in many countries, but there are numerous other tools and methods available that can be applied to provide comprehensive information on geologic conditions.

*Data collection and terrain analysis* are particularly important where large land areas are involved, or where the project area is unfamiliar to the investigator. Included is the preliminary determination of geologic conditions through literature search, terrain analysis, and field reconnaissance. Terrain analysis is stressed throughout this volume because of its high value in establishing geologic conditions, in particular in identifying those that are hazardous and in classifying the geologic materials by origin and mode of occurrence. When these latter factors are established it is possible to reach preliminary conclusions regarding the engineering characteristics of the various formations. For moderate to large land areas, the preparation of an *engineering geology map,* on which are depicted surface and shallow subsurface conditions of rock, soil, water, and hazards, is not only useful for preliminary site planning, but is necessary for the thorough programming of the field investigation.

*Subsurface sectioning* is accomplished with geophysical methods, test borings, and various reconnaissance methods involving test pits, augers, etc. Information over large areas and areas with difficult access is obtained efficiently with geophysical techniques, which commonly include refraction seismology and electrical resistivity on land and refraction and reflection profiling on water. Other techniques employed occasionally are surveys with gravimeters, magnetometers, and ground-probing radar. There are a number of simple and economical methods of exploring to shallow depths, and a number of procedures for performing test borings on land and water. The static cone penetrometer, long used in Europe, is finding increasing application in the United States. Various types of remote-sensing equipment such as cameras and probes are available for providing a continuous log of borehole conditions. Nuclear probes provide data on in situ water contents and densities. The determination of groundwater conditions is an important aspect of investigation, and reliable data are obtained only with proper procedures.

*Recovery of samples* of the geologic materials for identification and laboratory testing is a major objective of the exploration program. A wide variety of tools is available, and their selection depends on the type of material to be sampled and the use to which the sample will be put. Sample quality depends very much on extraction techniques, whether one is sampling soils or coring rock.

#### Measurement of Properties (Chap. 3)

The basic and index properties, used for identification and correlation of engineering properties, and the engineering properties of permeability, strength, and deformation are obtained for soil and rock both in the field and in the laboratory. When representative undisturbed samples can be obtained, as from soft to stiff intact clays, properties are measured in the laboratory. There are many materials, however, from which undisturbed samples are difficult or impossible to obtain, such as clean cohesionless sands, residual soils, glacial till, and soft or heavily jointed rock masses. For these materials properties are best measured in situ. Soft to firm clays and hard fissured clays are also tested in situ.

*Permeability* measurements in soils and rock masses are best made by field tests because of mass effects of stratification and joints, but numerous correlative data exist, particularly for soils, to permit estimates that are adequate for many studies.

*Shear strength and deformation* characteristics are measured by in situ testing with either full-scale load tests on the surface, load tests in pits or tunnels, or load tests with instruments lowered into boreholes. Geophysical methods are used to obtain measures of dynamic properties, and the field data are correlated with laboratory dynamic testing data to obtain deformation moduli for design. Dynamic testing is also important for evaluating rock-mass quality.

Many types of geologic materials have characteristic engineering properties, and values are included throughout the chapter as well as elsewhere in the book. The values are useful for preliminary assessments and for the evaluation of the "reasonableness" of data obtained from field and laboratory testing. Very often, such values are obtained from back-analysis of a failure condition in the field.

### Field Instrumentation (Chap. 4)

*General*

There are many instruments available to measure, either qualitatively or quantitatively, surface and subsurface movements, strains, and in situ stresses and pressures. They range from simple to complex, from low-cost to expensive, from recoverable to expendable to permanent. The selection depends on the importance of the work, the information required, and the time available for installation and study.

*Applications*

Instruments are installed to provide information for the anticipation of failure of natural or cut slopes, measurements of the settlement and heave of structures, control of preloading operations, anticipation of objectionable ground subsidence beneath structures adjacent to excavations or over tunnels, control of the performance of earth dams and tunnels, monitoring vibrations and seismic forces, and the investigation of fault activity and surface warping. In many engineering projects, particularly braced excavations, tunnels in rock, and steep cut slopes in open-pit mines, the monitoring of deformations permits the use of economical contingent designs with low safety factors and reduced construction costs. When deformations reach dangerous levels, as predetermined by analysis, additional support can be added to arrest movements, or other procedures invoked, in accordance with contingency plans already prepared.

## 1.2.3 GEOLOGIC MATERIALS: CHARACTERISTICS AND RECOGNITION (Part II)

### Rock and Soil: Identification and Classification (Chap. 5)

*Rock types* are described, from the aspects of their identification characteristics of mineral content, fabric, and texture, as having been originally formed as igneous, sedimentary, or metamorphic rocks. Rock-mass characteristics as affected by discontinuities are introduced to permit presentation of rock-mass description and classification systems. It is the characterisitcs of the in situ rock mass with its systems of discontinuities which normally control response to changing stress fields.

*Soils* are described from the aspects of their general characteristics, mineralogy, and related engineering properties. Mineralogy is more important than often realized; not all sands are composed of the essentially indestructible quartz grains, and clay activity depends very much on mineral composition. Classification systems are presented along with procedures for identifying and describing the various soil types.

### Rock-Mass Characteristics (Chap. 6)

Rock masses, as originally formed, have characteristic forms, shapes, and structural features which are related to rock type. Tectonic activity and other phenomena deform the rock mass, resulting in folding and fracturing which provide systems of faults, joints, and other discontinuities. Differential weathering and erosion attack the mass, resulting in terrain features indicative of various conditions and provide the basis for their prediction.

*Faults* are closely associated with shallow-focus earthquakes. Their recognition and positive identification, as well as the determination of their activity, are a critical element in earthquake engineering studies. Faults also create a major weakness zone in rock masses.

*Joints*, the most common form of discontinuity, control rock-mass behavior in most situations. A description of their spatial orientations and characteristics is necessary for the solution of rock-mechanics problems, particularly in closed and open excavations and concrete dam foundations and abutments.

*Residual stresses* result in strains and rock bursts in excavations and are an important element of rock-mass conditions. Unless specifically sought by means of in situ instrumentation, their prediction is difficult although their incidence is often associated with a particular set of conditions.

*Weathering* causes rock masses to undergo disintegration and decomposition, completely changing characteristics and producing new materials ranging from altered minerals along joints, to soft and decomposed rock and finally to residual soil. Various rock types develop characteristic profiles in given climatic conditions, thus providing the basis for the prediction of soil types and their general properties when climate and rock type are known.

### Soil Formations: Geologic Classes and Characteristics (Chap. 7)

Soils are classed by origin as residual, colluvial, alluvial, etc.; subclasses are based on mode of occurrence or deposition such as fluvial, lacustrine, etc., and subdivided as stream-channel deposits, floodplains, tidal marshes, etc. Soils of various origins and modes of occurrence or deposition have characteristic terrain features of landform, drainage patterns, and vegetation which are used as indicators to provide the basis for their identification, anticipation of their structural features such as stratification, and estimation of their characteristic engineering properties. The origin and mode of occurrence most significantly affect their gradation characteristics as well as their stress history.

*Residual soil* characteristics reflect parent rock type. For example, relatively thin deposits of inactive clays are normal in most sedimentary and sialic igneous rocks; thick deposits of inactive clays are normal to foliated rocks in moist climates; active clays develop from mafic rocks and marine shales in moist climates. The landform reflects the parent rock characteristics.

*Colluvium* frequently originates from residual soils, but can originate from any soil formation on slopes. Colluvium develops characteristic land forms and leaves scars of denuded vegetation and fresh surface materials after major slope failures. Colluvial soil slopes can be expected to be unstable, and the formation may be found overlying weak alluvia.

*Alluvium* refers to channel deposits that relate to stream shape and gradient, which influence the carrying capacity of the stream. Only boulders and cobbles remain in the steep gradients of young streams; gravels and sands in the moderate gradients of mature streams; and sands, silts, and clays in the small-to-nil gradients of the pastoral zone (except during flood stages). The fluvial environment creates floodplains, terraces, and such other terrain features as back swamp deposits and oxbow lakes, all with characteristic soil types. Rivers terminate in lakes and oceans to deposit deltas or to have their sediments carried by currents for deposition along shorelines as beaches, spits, etc. or offshore in deep waters where accumulations may reach great thicknesses. The types of materials deposited reflect currents and water depths. Emergence can cause the formation of coastal plains or exposed lake beds.

*Aeolian* deposits of dunes, sand sheets, and loess all have characteristic terrain features of landforms and drainage patterns and very typical engineering properties.

*Glacial deposits* have many characteristic landforms reflecting their mode of deposition as either directly deposited from the glacier (till) or from meltwaters (stratified drift or outwash). Till, normally strong and relatively incompressible, is typically associated with a very irregular ground surface containing numerous lakes and poorly developed drainage systems. Two important glacial deposits are lacustrine (varved clays) and marine (often sensitive clays), both having identifiable modes of occurrence and characteristic properties.

*Secondary deposits* are considered in this work to include the duricrusts (laterites, ironstone, caliche, and silcrete), and permafrost and seasonal frost. The duricrusts, surficial formations within soil deposits and limited in extent, often have rocklike characteristics. They form in certain environments under a limited set of conditions.

### Water: Surface and Subsurface (Chap. 8)

*Flooding* is a major geologic hazard, and its prediction can be based on hydrological analysis or on the interpretation of geological conditions. The latter procedure can be the more reliable, especially where precipitation and flood level data are meager. *Erosion* is also a hazard wasting the land and resulting in the unnecessary filling of waterbodies with sediments. Control is an important element of earthwork construction.

*Groundwater* has various significant aspects including several modes of occurrence (static water table, perched, artesian conditions), subsurface flow, and water quality. The analysis of flow and the determination of seepage forces through embankments, in slopes, and beneath excavations is usually done by means of flow nets. Seepage forces are particularly significant in the stability of slopes, earth dams, and excavation bottoms. Groundwater control, important in the design and construction of slopes, basements, tunnels, pavements, etc. can be affected by cutoffs and barriers, dewatering, and drains. Proper filter design is an important element of drainage systems.

## 1.2.4 GEOLOGIC HAZARDS: RECOGNITION AND TREATMENT (Part III)

### Landslides and Other Slope Failures (Chap. 9)

The major forms of slope failures (falls, slides, flows, and avalanches) are related primarily to climate and geologic conditions. The elements of slope stability involve the slope inclination and height, material structure and strength, seepage forces, and runoff. Each of the particular forms of failure has its own characteristic features, the recognition of which, together with an understanding of the elements of stability, provides the basis for prediction of potential failures and for formulating treatments.

Prediction based on mathematical analysis is applicable only to slide forms of failure under certain conditions. In the prediction and treatment of slope failures, consideration is given to the degree of the hazard (basically its magnitude); slope geology and geometry; surface conditions of seepage, vegetation, and movements (degree of activity in terms of failure); and rainfall and other weather data. The selection of the treatment is based on an evaluation of the degree of hazard and the degree of risk (consequences of failure). Failure may be permitted in certain cases, or stabilization achieved by such methods as removal of materials, improvement of external and internal drainage, or retention. Some conditions cannot be stabilized from the aspects of practical considerations and should be avoided.

### Ground Subsidence, Collapse, and Heave (Chap. 10)

*Regional subsidence* is caused by earthquakes; extraction of water, oil, or gas; and subsurface mining operations. The result for the surface can be flooding, faulting, and the distortion of structures.

*Ground collapse* results from subsurface mining operations, water-table lowering in cavernous soluble rocks, or the weakening of the intergranular structure of certain soil formations. Surface structures can be lost, destroyed, or distorted.

*Subsidence in soils* results from leaching, internal erosion from piping, compression under externally applied loads, or evaporation causing desiccation and shrinkage. It can cause distortion of surface structures and the collapse of earthen embankments and slopes.

*Surface heave* occurs on a regional basis from tectonic forces, and on a local basis from stress relief during excavation and from the expansion of soil and rock which can result from swelling characteristics or frost. Expansion from swelling or frost can result in the distortion of structures and the weakening of slopes.

Each of these phenomena is characteristic of certain environments and interfacing elements and, therefore, is predictable. In many cases the most suitable treatment is their prevention.

### Earthquakes (Chap. 11)

The elements of earthquakes include the energy source or cause, the position of the source in the earth (focus and epicenter), the seismic waves generated by the source, the ground motion resulting from the waves, the characteristics of intensity and magnitude, and the attenuation or amplification of the intensity.

*Surface effects* are numerous and include faulting, dynamic soil behavior (such as subsidence and liquefaction), slope failures, tsunamis and seiches, and the shaking of structures.

*Design studies* require an evaluation of the regional seismic risk, identification of capable faults, the development of the design earthquake, and the selection of ground response factors including acceleration of gravity, frequency content, duration, and the influence of local soil conditions. Suitable treatment requires resistant design.

### 1.2.5 APPENDICES

#### The Earth and Geologic History (A)

The earth's crust and surface are undergoing constant change, usually barely perceptible, as the result of global tectonics. The nomenclature of geological history is useful for indexing formations and for correlations between widely separated geographic locations. Ages are determined by radiometric dating, for which there are a number of procedures.

#### Sources of Publications (B)

Sources for the procurement of geologic publications, maps and remote sensing imagery, and earthquake information are given.

#### Conversion Tables (C)

Conversions from English to metric to S.I. units and metric to English to S.I. units are given.

#### Symbols (D)

Symbols used in the text are summarized and identified.

#### Engineering Properties of Geologic Materials (E)

A synopsis of the tables and figures from the text providing data and correlations for rock, soil, and groundwater is presented. Rock and soil properties are keyed to index or basic properties, compression, expansion, and strength.

## 1.3 GEOTECHNICAL INVESTIGATION: OBJECTIVES, STAGES, SCOPES, PLANNING

### 1.3.1 BASIC OBJECTIVES

The basic objectives of a geotechnical investigation include the determination of the following:

1. Lateral distribution and thickness of the soil and rock strata within the zone of influence of the proposed construction or development
2. Groundwater conditions with consideration of seasonal changes and the effects of extraction due to construction or development
3. Physical and engineering properties of the soil and rock formations and groundwater quality
4. Hazardous conditions including unstable slopes, active or potentially active faults, regional seismicity, floodplains, and ground subsidence, collapse, and heave potential
5. Ground response to changing natural conditions and construction or development brought about by surface loadings from structures, unloadings by surface or subsurface excavations, or unloadings from extraction of mineral resources
6. Suitability of the geologic materials for aggregate and for the construction of pavements and embankments.

### 1.3.2 STAGES OF INVESTIGATION

Investigations are performed in a number of stages, each with a different objective, and each requiring interpretation, analysis, and evaluation.

- *Stage 1:* General geologic conditions are identified through the retrieval of existing data, interpretation of remote sensing imagery (terrain analysis), and field reconnaissance. Engineering geology maps are prepared.
- *Stage 2:* Subsurface sections are prepared from exploration data obtained by test borings and pits, geophysical methods, etc.
- *Stage 3:* Samples are recovered of the soils and rocks for identification and laboratory testing.
- *Stage 4:* Measurements of engineering properties are made in the laboratory and in situ.
- *Stage 5:* Instrumentation is installed to monitor

ground and structural response to changing field conditions.

### 1.3.3 PHASES OF INVESTIGATION

Investigations can be divided into a number of phases based on purpose, with various investigation stages in each phase. In general, phases range from feasibility, to preliminary design, to final design, to construction, to postconstruction.

#### Feasibility

A feasibility phase is imperative when the investigator has no prior experience in the project area, or the site is located in virgin or rural country, or for large important projects. The purpose is to identify potentially hazardous or unfavorable conditions such as active faults, unstable slopes, sinkholes, and deep deposits of weak soils that would cause construction to be either risky from the aspect of safety, or extraordinarily costly. Investigation is made with stage 1, and often with parts of stages 2 and 3.

The data obtained have a number of useful applications. At a high-hazard site the decision may be made to abandon it and search for one with more favorable conditions. If the site is marginal, preliminary planning can consider optimum utilization of favorable areas and the avoidance of unfavorable areas to minimize construction costs. Depending upon the nature of the project, unfavorable areas may be characterized by shallow rock, shallow water table, deposits of weak soils, or expansive or collapsible soils. On sites with no obviously severe constraints, preliminary designs and cost estimates may be prepared. The data gathered provide the basis for intelligent planning of more detailed investigation, as the selection of proper methods and tools is much dependent upon geologic conditions.

#### Preliminary Design Phase

The preliminary design phase usually proceeds when locations, dimensions, and loadings of the proposed construction have been defined. Its purpose is the comprehensive determination of the distribution of the geologic materials and their engineering characteristics, and the evaluation of hazardous or constraining conditions. Investigation includes stages 2, 3, and 4, and on many projects may be combined with the feasibility phase. On routine, straightforward projects, investigation may only involve a preliminary phase.

The data obtained are used for the selection and dimensioning of foundations, retaining structures, groundwater control, slope stabilization, and tunnel and pavement support systems. If the project involves mineral extraction, the effects on the surface are evaluated and methods formulated to avoid detrimental effects such as subsidence and collapse.

#### Final Design Phase

Sites with difficult conditions requiring the designer to change building locations or dimensions or to confirm or modify designs, primarily to realize economies, are subject to a final design phase. The purpose is to procure supplementary data, usually from additional undisturbed samples for laboratory tests or full-scale load tests to confirm or improve design criteria. Investigation requires detailed information employing stages 2, 3, and 4.

#### Construction Phase

Unforeseen geologic conditions may be encountered, common on projects with deep excavations or deep foundations, or unconventional designs may be used with contingency plans. Either case requires additional investigation during the construction phase. Investigation stages 2, 3, 4, and frequently 5 are performed. Instrumentation is installed to monitor changing groundwater levels; stability and movements of slopes and retaining structures; settlements of fills, embankments, and foundations; ground subsidence and its effect on adjacent structures; and deformations and stresses in underground support systems.

Instrumentation provides an early warning system and the data base for decisions for invoking contingency plans when deformations and stresses exceed those anticipated during design.

#### Postconstruction Phase

Instrumentation (stage 5) is good practice for heavy foundation loading or other conditions in which settlements are anticipated, for retaining structures in critical locations, and for slopes, dams, and tunnels. The purpose is to provide early warning systems for potentially troublesome conditions, and to provide data to advance the state of the art and reduce the degree of conservatism in future similar projects.

### 1.3.4 STUDY PLANNING

The basic considerations in study planning are the lateral and vertical extent to which the proposed construction or development will significantly influence the geologic environment (or be influenced by the environment).

#### Lateral Extent of Influence

The lateral extent of influence can vary from the immediate area of the project, to many meters beyond if dewatering of excavations is anticipated, to an entire valley or region where deep mineral extraction is planned, to several hundred kilometers where area seismicity is of concern.

#### Vertical Depth

For structures on or near the surface, the vertical influence is a function of several variables, including the size and loading of the structure which imposes changes on the natural stress conditions to some significant depth; unfavorable geologic conditions which require explorations to depths adequate to define suitable support for deep foundations; and excavations which require explorations to some distance below their maximum depth to determine groundwater, rock, and soil conditions both within and below the excavation zone. Deep excavations for tunnels or mines and the deep extraction of fluids require knowledge of geologic conditions between the point of excavation or extraction and the surface, and at times between even deeper points and the surface.

# PART I

# INVESTIGATION METHODS AND PROCEDURES

**CHAPTERS**



**PURPOSE AND SCOPE**

Part I describes and provides the basis for the selection of the numerous methods and procedures for:

1. Exploring the geologic environment and mapping surficial conditions including rock, soil, and water, and the geologic hazards; preparing subsurface sections; and obtaining samples of the materials for identification, classification, and laboratory testing.
2. Measurement of material properties (basic, index, hydraulic, and mechanical) in the field and laboratory.
3. Measurement and monitoring with field instrumentation, movements, deformations, and stresses occurring naturally or as a consequence of construction.

Although in practice analytical procedures and design criteria are often presented as part of an investigation, they are not included in the scope of Part I.

**SIGNIFICANCE**

The investigation phase of any study, as covered by the scope of Part I, undertaken for development, construction, or any other engineering works involving the earth, is by far the most important phase. Not only must conditions at the project site be thoroughly identified, but for many projects the regional geologic characteristics must also be determined. For all phases of investigation there are a large number of methods and devices to choose from, ranging from simple to complex, and usually several are applicable for a given subject of study.

Engineering analyses and evaluations are valid only when based on properties truly representing all of the natural materials that may influence the works. Properties of some materials are best measured in the laboratory, while others must be field tested. In some cases properties cannot be adequately defined by direct testing and the result will be designs that are conservative and too costly, unconservative and risky, or unconservative but based on contingency plans. Field instrumentation to monitor ground conditions during construction is an important element of many studies where subsurface conditions cannot be adequately defined by exploration and testing. Instrumentation is used also to obtain design data and to monitor changing natural conditions such as slope failures and fault movements.

# CHAPTER TWO

# EXPLORATION

## 2.1 INTRODUCTION

### 2.1.1 OBJECTIVES

The general objective of an exploration program is to identify all of the significant features of the geologic environment which may impact on proposed construction. Specific objectives are to:

1. Define the *lateral distribution and thickness* of soil and rock strata within the zone of influence of the proposed construction
2. Define *groundwater conditions* with consideration of seasonal changes and effects of construction or development extraction
3. Identify *geologic hazards* including unstable slopes, faults, ground subsidence and collapse, floodplains, and regional seismicity
4. Procure *samples* of the geologic materials for identification, classification, and the measurement of engineering properties
5. Perform *in situ testing* to measure engineering properties of the geologic materials *(see Chap. 3)*.

### 2.1.2 METHODOLOGY

Three general categories subdivide exploration methodology:

1. *Surface mapping of geologic conditions (see Art. 2.2)*, which requires review of reports and publications, interpretation of topographic and geographic maps and remote sensing imagery, and site reconnaissance
2. *Subsurface sectioning (see Art. 2.3)*, for which data are obtained by geophysical prospecting, test and core borings, and excavations and soundings
3. *Sampling* the geologic materials *(see Art. 2.4)* utilizing test and core borings and excavations

A general summary of exploration methods and objectives is given on Table 2.1.

## 2.2 SURFACE MAPPING

### 2.2.1 GENERAL

**Objectives**

*Data Base*

For all sites it is important to determine the general geologic conditions and identify significant development and construction constraints. For large sites (generally those other than a single building at a fixed location) it is useful to prepare a map illustrating the surficial and shallow geologic conditions.

*Preliminary Site Evaluations*

An overview of geologic conditions permits preliminary evaluations regarding the site suitability for development in consideration of major hazards or constraints; the optimum location for

**TABLE 2.1**
**EXPLORATION OBJECTIVES AND APPLICABLE METHODS**

| | General | | | | | | | | Geophysics | | | | | | |
|---|---|---|---|---|---|---|---|---|---|---|---|---|---|---|---|
| **Objectives** | **Reports and publications** | **Topographic maps** | **Imagery: Landsat, SLAR** | **Imagery: Low-altitude photos** | **Bathymetry** | **Side-scan sonar** | **Underwater TV** | **Reconnaissance** | **Seismic refraction** | **Seismic reflection** | **Electrical resistivity** | **Gravimeter** | **Magnetometer** | **Radar profiling** | **Video-pulse radar** |
| Geology: | | | | | | | | | | | | | | | |
| Regional | X | X | X | X | | | | X | | | | | | | |
| Surficial-land | X | X | X | X | | | | X | | | | | | | |
| Surficial-seafloor | | | | | X | X | X | X | | | | | | | |
| Major structures | X | X | X | X | | | | | X | X | | X | X | | |
| Faults* | X | X | X | X | | | | | X | X | | X | | X | |
| Sections: | | | | | | | | | | | | | | | |
| Deep-land | X | | | | | | | | X | | | | | | |
| Shallow-land | X | | | | | | | | X | | X | | | X | |
| Subaqueous | | | | | | | | | X | X | | | | | |
| Soft-soil depth | | | | | | | | | | | | | | | |
| Sliding masses* | | | | X | | | | X | | | | | | | |
| Rock depth | | | | | | | | | X | | X | | | X | |
| Rock-mass conditions | | | | | | | | | X | | | X | | X | X |
| Soil samples: | | | | | | | | | | | | | | | |
| Disturbed to GWL | | | | | | | | | | | | | | | |
| Representative | | | | | | | | | | | | | | | |
| Undisturbed | | | | | | | | | | | | | | | |
| Deep, offshore | | | | | | | | | | | | | | | |
| Rock cores: | | | | | | | | | | | | | | | |
| Normal depths | | | | | | | | | | | | | | | |
| Deep | | | | | | | | | | | | | | | |

*See also Instrumentation.

| Boring | | | | | | Borehole sensing | | | | | | Miscellaneous | | | | | | |
|---|---|---|---|---|---|---|---|---|---|---|---|---|---|---|---|---|---|---|
| Wash boring | Rotary drilling | Rotary probe | Continuous-flight auger | Hollow-stem auger | Wire-line drilling | Borehole cameras | Acoustical sounding | Electric well log | Radioactive probes | Ultrasonic acoustics | 3-D velocity log | Test pits/trenches | Adits | Bar soundings | Retractable plug | Continuous cone penetrometer | Hand augers | Bucket auger |
| | | | | | | | | | | | | | | | | | | |
| | | | | | | | | | | | | | | | | | | |
| | | | | | | | | | | | | | | | | | | |
| | | | | | | | | | | | | | | | | | | |
| | X | | | | X | X | | X | X | X | X | X | X | | | | | |
| | X | | | | X | | | X | X | | | | | | | | | |
| X | X | | X | X | | X | X | X | X | | | X | X | | X | X | X | X |
| | X | | | | X | | | | X | | | | | | | X | | |
| X | X | | | X | | | | | | | | | | X | X | X | | |
| X | X | | | X | | | | X | X | | | X | | | | | | X |
| | X | X | | | | | X | | | | | X | | | | | | |
| | | X | | | X | X | | X | X | X | X | | X | | | | | |
| X | X | | X | X | | | | | | | | X | | | | | X | X |
| X | X | | | X | | | | | | | | | | | X | | | |
| X | X | | | X | | | | | | | | X | | | | | | |
| | | | | | X | | | | | | | | | | | X | | |
| | | X | | | | | | | | | | | | | | | | |
| | | | | | X | | | | | | | | | | | | | |

the proposed construction in view of the constraints; and the planning of an investigation program required to determine subsurface sections, procure samples of the materials, and measure their engineering properties in situ.

### Methodology

A geologic land reconnaissance study advances through a number of steps as described briefly in Fig. 2.1, including:

- Research of reference materials and collection of available data
- Terrain analysis based on topographic maps and interpretation of remotely sensed imagery
- Preparation of a preliminary engineering geology map
- Site reconnaissance to confirm and amplify the engineering geology map, after which it is prepared in final form
- Preparation of a subsurface exploration program based on the anticipated conditions.

## 2.2.2 RESEARCH DATA

### Basic Objectives

A large amount of information often is available in the literature for a given location, and a search should be made to gather as much data as possible before initiating any exploration work, particularly when large sites are to be studied, or the site is located in a region not familiar to the design team. Information should be obtained on:

- *Bedrock geology* including major structural features such as faults
- *Surficial geology* in terms of soil types on a regional basis or, if possible, on a local basis
- *Climatic conditions,* which influence soil development, groundwater occurrence and fluctuations, erosion, flooding, slope failures, etc.
- *Regional seismicity* and *earthquake history*
- *Geologic hazards,* both regional and local, such as ground subsidence and collapse, slope failures, and floods
- *Geologic constraints,* both regional and local, such as expansive soils, weak soils, shallow rock, groundwater, etc.

### Information Sources

*Geologic texts* provide information on physiography, geomorphology, and geologic formation types and structure, although usually on a regional basis.

*Federal and state agencies* issue professional papers, bulletins, reports, and geologic maps, as do some cities such as Los Angeles. Sources of geologic information are given in Appendix B, including the U.S. Geological Survey (USGS) and the U.S. state geological departments. In addition to those for geology, agencies exist for agriculture, mining, and groundwater and issue reports, bulletins, and maps.

*Engineering Soil Surveys* have been prepared for New Jersey [Rogers (1950)[1]] and Rhode Island which are presented as reports and maps on a county basis. The maps illustrate shallow soil and rock conditions and classify the soils by origin in combination with AASHO Designation M145–49, and the prevailing or average drainage conditions.

*Technical publications* such as the *Journal of the Geotechnical Engineering Division,* ASCE; *Géotechnique,* Institute of Civil Engineers, London; the *Bulletin of the Association of Engineering Geologists,* AEG; the *Canadian Geotechnical Journal;* and the various international conferences on soil and rock mechanics, engineering geology, and earthquakes, often contain geologic information on a specific location.

*Climatic data* are obtained from the U.S. Weather Bureau or other meteorological agencies.

### Geologic Maps

Geologic maps vary in scale generally from 1:2,500,000 (U.S. map) to various scales used by state agencies to USGS quadrangle maps at 1:24,000, and vary in the type of geologic information provided. A guide to map scale conversions is given on Table 2.2.

*Bedrock geology maps* (Fig. 2.2) often give only the geologic age; the rock types are usually described in an accompanying text. There is a general correlation between geologic age and rock type; the geologic time scale and the dominant rock types in North America for the various

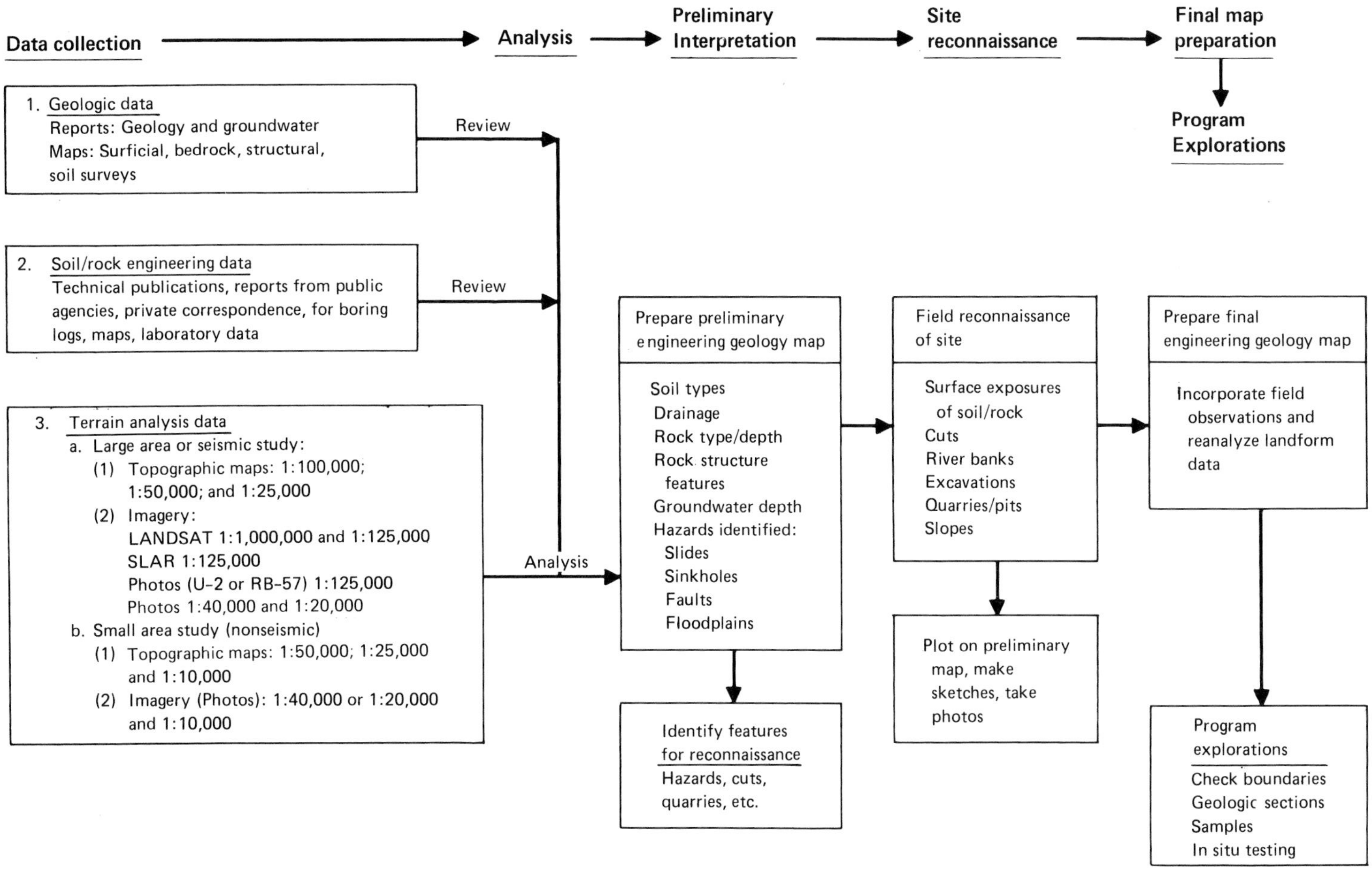

FIG. 2.1 The elements of the geologic land reconnaissance study.

## TABLE 2.2
## GUIDE TO MAP SCALES

| Scale | Feet per inch | Inches per 1000 feet | Inches per mile | Miles per inch | Meters per inch | Acres per square inch |
|---|---|---|---|---|---|---|
| 1:500 | 41.67 | 24.00 | 126.72 | 0.008 | 12.70 | 0.040 |
| 1:600 | 50.00 | 20.00 | 105.60 | 0.009 | 15.24 | 0.057 |
| 1:1000 | 83.33 | 12.00 | 63.36 | 0.016 | 25.40 | 0.159 |
| 1:1200 | 100.00 | 10.00 | 52.80 | 0.019 | 30.48 | 0.230 |
| 1:1500 | 125.00 | 8.00 | 42.24 | 0.024 | 38.10 | 0.359 |
| 1:2000 | 166.67 | 6.00 | 31.68 | 0.032 | 50.80 | 0.638 |
| 1:2400 | 200.00 | 5.00 | 26.40 | 0.038 | 60.96 | 0.918 |
| 1:2500 | 208.33 | 4.80 | 25.34 | 0.039 | 63.50 | 0.996 |
| 1:3000 | 250.00 | 4.00 | 21.12 | 0.047 | 76.20 | 1.435 |
| 1:4000 | 333.33 | 3.00 | 15.84 | 0.063 | 101.60 | 2.551 |
| 1:5000 | 416.67 | 2.40 | 12.67 | 0.079 | 127.00 | 3.986 |
| 1:6000 | 500.00 | 2.00 | 10.56 | 0.095 | 152.40 | 5.739 |
| 1:7920 | 660.00 | 1.515 | 8.00 | 0.125 | 201.17 | 10.000 |
| 1:8000 | 666.67 | 1.500 | 7.92 | 0.126 | 203.20 | 10.203 |
| 1:9600 | 800.00 | 1.250 | 6.60 | 0.152 | 243.84 | 14.692 |
| 1:10000 | 833.33 | 1.200 | 6.336 | 0.158 | 254.00 | 15.942 |
| 1:12000 | 1,000.00 | 1.000 | 5.280 | 0.189 | 304.80 | 22.957 |
| 1:15000 | 1,250.00 | 0.800 | 4.224 | 0.237 | 381.00 | 35.870 |
| 1:15840 | 1,320.00 | 0.758 | 4.000 | 0.250 | 402.34 | 40.000 |
| 1:19200 | 1,600.00 | 0.625 | 3.300 | 0.303 | 487.68 | 58.770 |
| 1:20000 | 1,666.67 | 0.600 | 3.168 | 0.316 | 508.00 | 63.769 |
| 1:21120 | 1,760.00 | 0.568 | 3.000 | 0.333 | 536.45 | 71.111 |
| 1:24000 | 2,000.00 | 0.500 | 2.40 | 0.379 | 609.60 | 91.827 |
| 1:25000 | 2,083.33 | 0.480 | 2.534 | 0.305 | 635.00 | 99.639 |
| 1:31680 | 2,640.00 | 0.379 | 2.000 | 0.500 | 804.67 | 160.000 |
| 1:48000 | 4,000.00 | 0.250 | 1.320 | 0.758 | 1,219.20 | 367.309 |
| 1:62500 | 5,208.33 | 0.192 | 1.014 | 0.986 | 1,587.50 | 622.744 |
| 1:63360 | 5,280.00 | 0.189 | 1.000 | 1.000 | 1,609.35 | 640.000 |
| 1:100000 | 8,333.33 | 0.120 | 0.634 | 1.578 | 2,540.00 | 1,594.225 |
| 1:125000 | 10,416.67 | 0.096 | 0.507 | 1.973 | 3,175.01 | 2,490.980 |
| 1:126720 | 10,560.00 | 0.095 | 0.500 | 2.000 | 3,218.69 | 2,560.000 |
| 1:250000 | 20,833.33 | 0.048 | 0.253 | 3.946 | 6,350.01 | 9,963.907 |
| 1:253440 | 21,120.00 | 0.047 | 0.250 | 4.000 | 6,437.39 | 10,244.202 |
| 1.500000 | 41,666.67 | 0.024 | 0.127 | 7.891 | 12,700.02 | 39,855.627 |
| 1.750000 | 62,500.00 | 0.016 | 0.084 | 11.837 | 19,050.04 | 89,675.161 |
| 1:1000000 | 83,333.33 | 0.012 | 0.063 | 15.783 | 25,400.05 | 159,422.507 |
| **Formula** | $\frac{\text{Scale}}{12}$ | $\frac{12.000}{\text{Scale}}$ | $\frac{63.360}{\text{Scale}}$ | $\frac{\text{Scale}}{63.360}$ | Feet per inch × 0.3046 | $\frac{(\text{Scale})^2}{43,560 \times 144}$ |

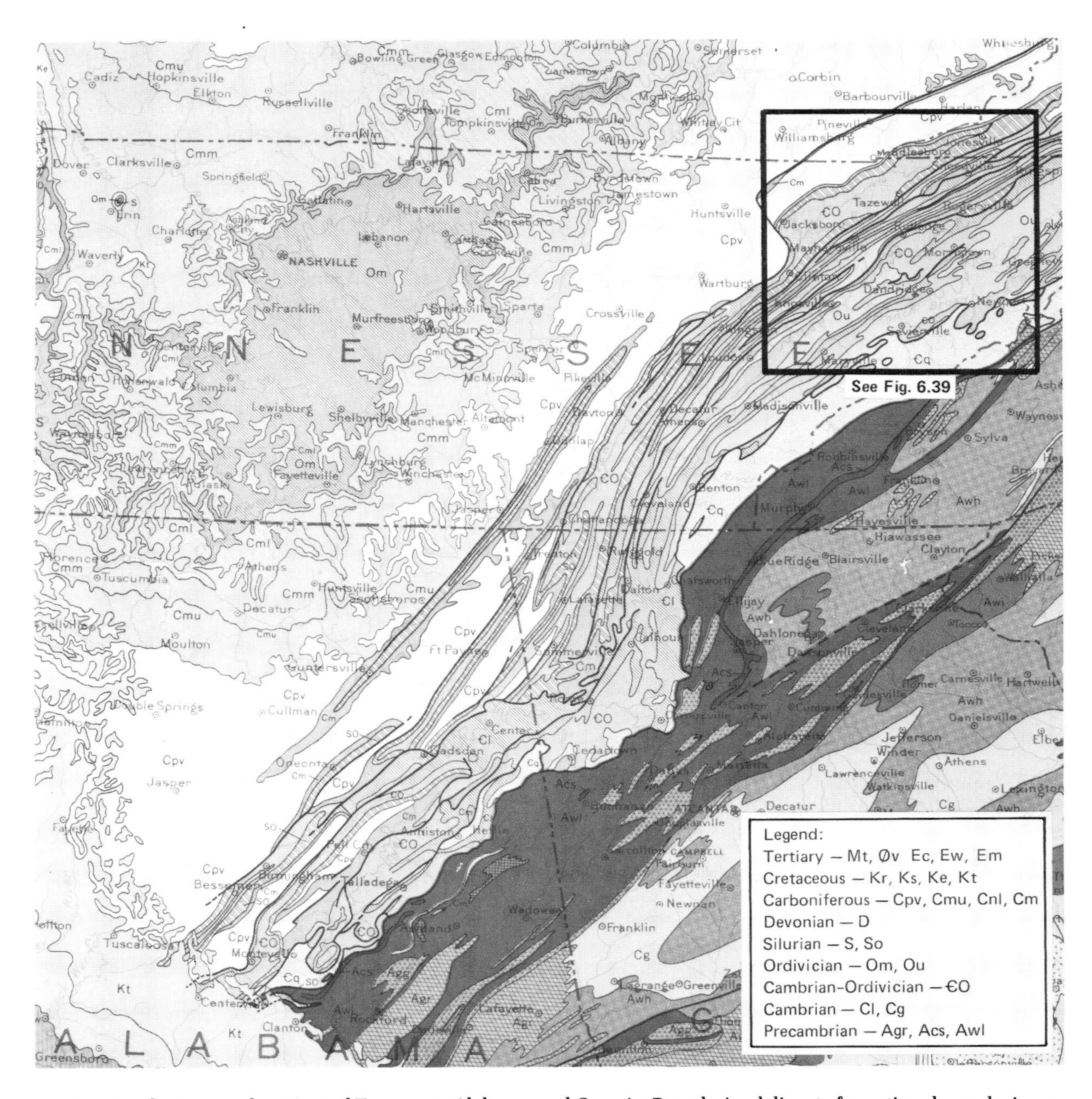


**FIG. 2.2 Geologic map of portions of Tennessee, Alabama, and Georgia. Boundaries delineate formations by geologic age, and their shapes and distribution reflect the regional rock structure and geomorphic provinces as follows: Northwest map area, generally horizontally bedded sedimentary rocks; central area, folded sedimentary rocks of the Appalachian Mountains; and southeast area, the crystalline masses of the Precambrian. See Fig. 6.39 for ERTS image of area indicated in northeast corner. In the legends accompanying geologic maps, capital letters denote geologic periods and small letters denote rock groups.** *(From Geologic Map of the United States, USGS.)*

periods are given in Appendix A. The formations for a given period are often similar on other continents. For the purpose of mapping, rocks are divided into formations, series, systems, and groups. The *formation* is the basic unit; it has recognizable contacts to enable tracing in the field and is large enough to be shown on the map. *Series* are coordinate with epochs, *systems* with periods, and *groups*, the largest division, with eras.

*Structural geology* may be shown on special maps or included on bedrock geology maps with symbols used to identify faults, folding, bedding, jointing, foliation, and cleavage. The maps often include geologic columns and sections. The common symbols used to define geologic structure are given on Fig. 6.3.

*Surficial geology maps* depict shallow or surficial soil and rock types; Fig. 7.1 is a surficial geology map of the United States presented at a much reduced scale.

"*Folios of the Geologic Atlas of the United States*" were produced by the USGS until 1945, and included maps of bedrock geology, structural geology, and surficial geology for many cities in the United States and other areas of major geologic importance.

*Soil survey maps*, produced by the Soil Conservation Service (SCS) of the U.S. Dept. of Agriculture, usually are plotted as overlays on aerial photographs at relatively large scales as shown on the example (Fig. 2.3). They are prepared on a county basis and illustrate the soil cover to a depth of about 2 m, based on pedological soil classifications *(see Art. 7.8)*, often combined with symbology describing slopes, shallow groundwater, and soil drainage conditions. Recent maps contain engineering-oriented data prepared by the Bureau of Public Roads in conjunction with the SCS. The shallow depth depicted limits their usefulness in many engineering studies.

*Tectonic maps* give regional lineations often indicative of faulting, as shown on the example given as Fig. 6.49.

*Earthquake data* may be presented as *intensity maps (see Fig. 11.3)*, as *isoseismal maps (see Fig. 11.28)*, as various forms of *seismic risk maps (see Fig. 11.15)*, or as *microzonation maps (see Art. 11.4.3)*.

*Other useful maps* published by the Geological Society of America include the glacial map of the United States and the loessial soils or windblown deposits of the United States.

### 2.2.3 TERRAIN ANALYSIS

#### General

*Significance*

Landforms and other surface characteristics are strong indicators of geologic conditions. Characteristic terrain features reveal rock type and structural forms where the rock is relatively shallow and subject to weathering and erosion *(see Art. 6.1.2)*, or represent typical soil formations in terms of their origin and mode of deposition where deposits are sufficiently thick *(see Art. 7.1.2)*.

*Objectives*

The delineation and mapping of the significant aspects of the geologic environment are the objective of terrain analysis, providing information on rock types and structures, soil types and formations, groundwater conditions, and floodplains and providing the locations of such hazards as landslides and other slope failures, of sinkholes, and other evidence of ground collapse and subsidence.

*Methodology*

Terrain analysis is based on the interpretation of features evident on topographic maps and remote-sensing imagery.

Topographic maps, such as quadrangle sheets, show landforms, drainage patterns, stream shape, and groundwater conditions, all indicators of geologic conditions. Because of their availability and usefulness they should be procured as a first step in almost any study.

There are many forms of remote-sensing imagery which show the indicators evident on the topographic maps, as well as vegetation types and density and image tone, and are useful for environmental as well as geological studies. The selection of remote-sensing imagery

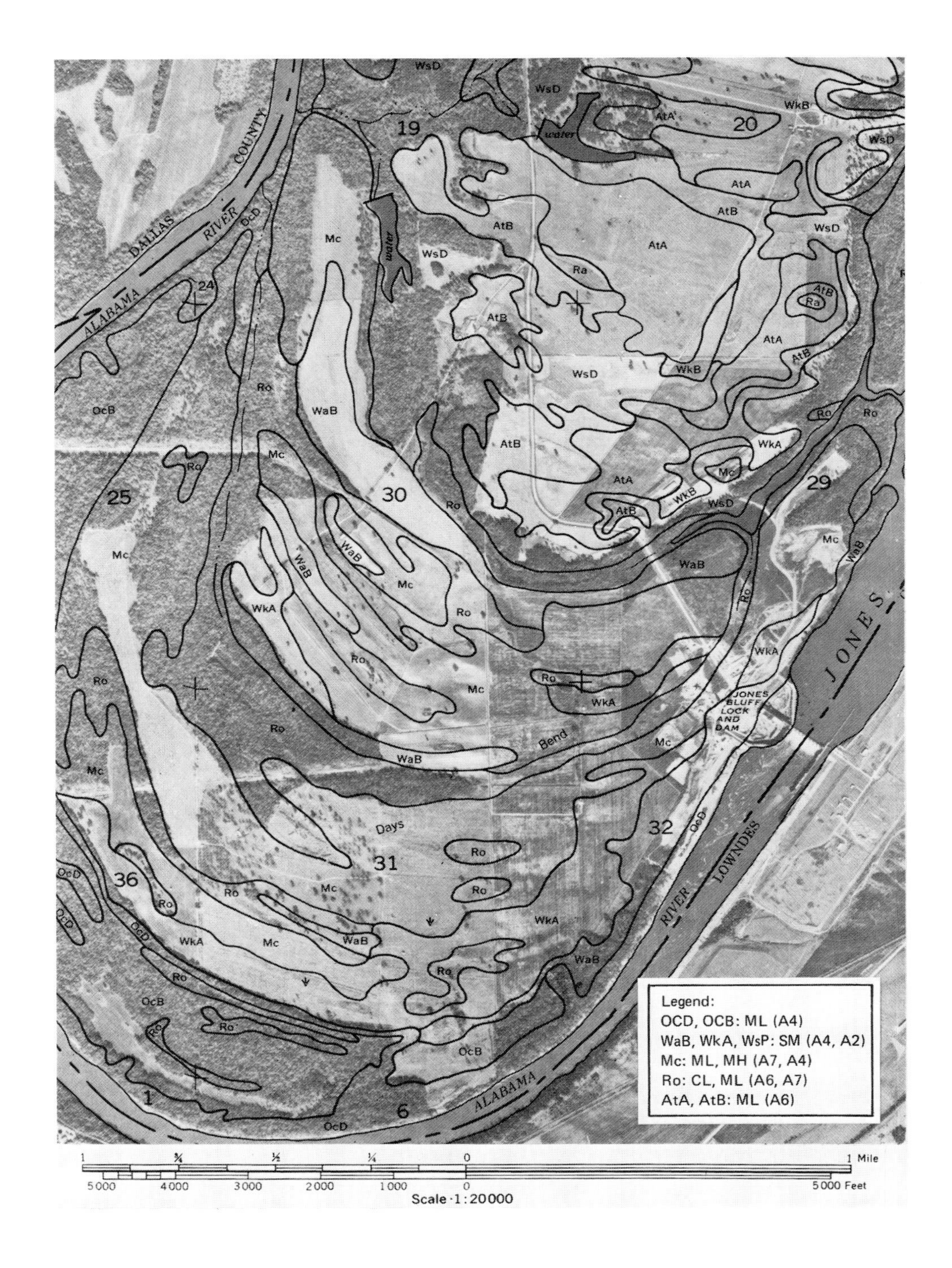


FIG. 2.3 Portion of soil map for Autauga County, Alabama, prepared by the Soil Conservation Service, USDA (1977),[2] delineating surficial conditions including soils deposited as point bars and swales in a meander of the Alabama River.

depends upon availability, the study purpose, and the land area involved. Stereoscopic examination and interpretation of aerial photographs is the basic analytical method.

On a regional basis, landform is the most important element of interpretation. Some relationships between landform, rock type, and structure are apparent on the portion of the physiographic diagram of northern New Jersey in Fig. 2.4.

### Topographic Maps and Charts

#### *General*

*Topographic maps* are available from a number of sources (*see Appendix B*) and in a variety of scales as follows:

- USGS provides maps covering a quadrangle area bounded by lines of latitude and longitude available in 7.5′ series (1:24,000), 15′ series (1:62,500), 30′ series (1:125,000), and 1° series (1:250,000) for most of the United States, although many of the larger scales are out of print.
- U.S. Army Map Service provides maps based on the U.S. military grid at scales of 1:25,000, 1:50,000, and 1:100,000.
- Other countries use scales ranging from 1:10,000 to 1:1,000,000 but coverage is often incomplete. 1:50,000 is a common scale available for many areas, even in countries not fully developed.

*River survey maps* are available from the USGS but primarily for the western United States and show the course and fall of the stream, the configuration of the valley floor, and the adjacent slopes and cultural features.

*Coastline charts,* available from the National Ocean Survey (NOS), provide information on water depths and near-shore topography.

#### *Interpretive Features of Topographic Maps*

Scales of 1:50,000 to 1:250,000 show regional landforms and drainage patterns and can indicate rock type and regional structural features such as folds and lineations, the latter often representing faults. Scales of 1:10,000 to 1:24,000 provide more local detail on features such as slopes, soil formations, and sinkholes.

*Interpretation* of geologic conditions is primarily based on landform as disclosed by contour lines and drainage patterns; many relationships between these features and geologic conditions are given in Chaps. 6 and 7, as well as in Chaps. 8 through 11.

Figures 2.5 and 2.6 provide illustrations of the relationship between landform as disclosed by contour lines and geologic conditions. Landform reflects the differential resistance to erosion of the various rock types, as shown clearly on Fig. 2.5, an area in the folded sedimentary rocks of the Appalachian Mountains. The distinctive ridge crossing the upper area is an exposure of the very resistant Shawangunk conglomerate, which causes the river to bend at its encounter with it. Southeast of the ridge the rocks are predominantly the Martinsburg shale which is less resistant to erosion and characteristically forms gently rolling hills in a somewhat irregular pattern. The Delaware River cuts across the rock structure, the original surface of which represented an uplifted peneplain, whereas the tributary streams follow the rock structure, flowing along joints or along the strike of folded beds. The broader valleys in the area generally follow the strike of the beds and are the result of relatively rapid removal of limestone by solution.

The landforms evident on Fig. 2.6 are indicative of several geologic formations; the steep-sloped, irregularly shaped form in the upper left is an area of very resistant granite gneiss at shallow depths. The Quinnipiac River is in its middle stage or floodway zone (*see Art. 7.4.1*), flowing over the remnants of a glacial lake bed and depositing fine-grained soils during flood stages. The very flat areas, generally between the railroad and the river extending through the middle portion of the map, are a sand and gravel terrace formation which at one time was the valley floor. A gravel pit is noted in the terrace formation, and areas of poor drainage (swamps and ponds) are apparent on the right-hand portion of the map. These perched water conditions above the valley floor result from the poor internal drainage of the underlying clayey glacial till.

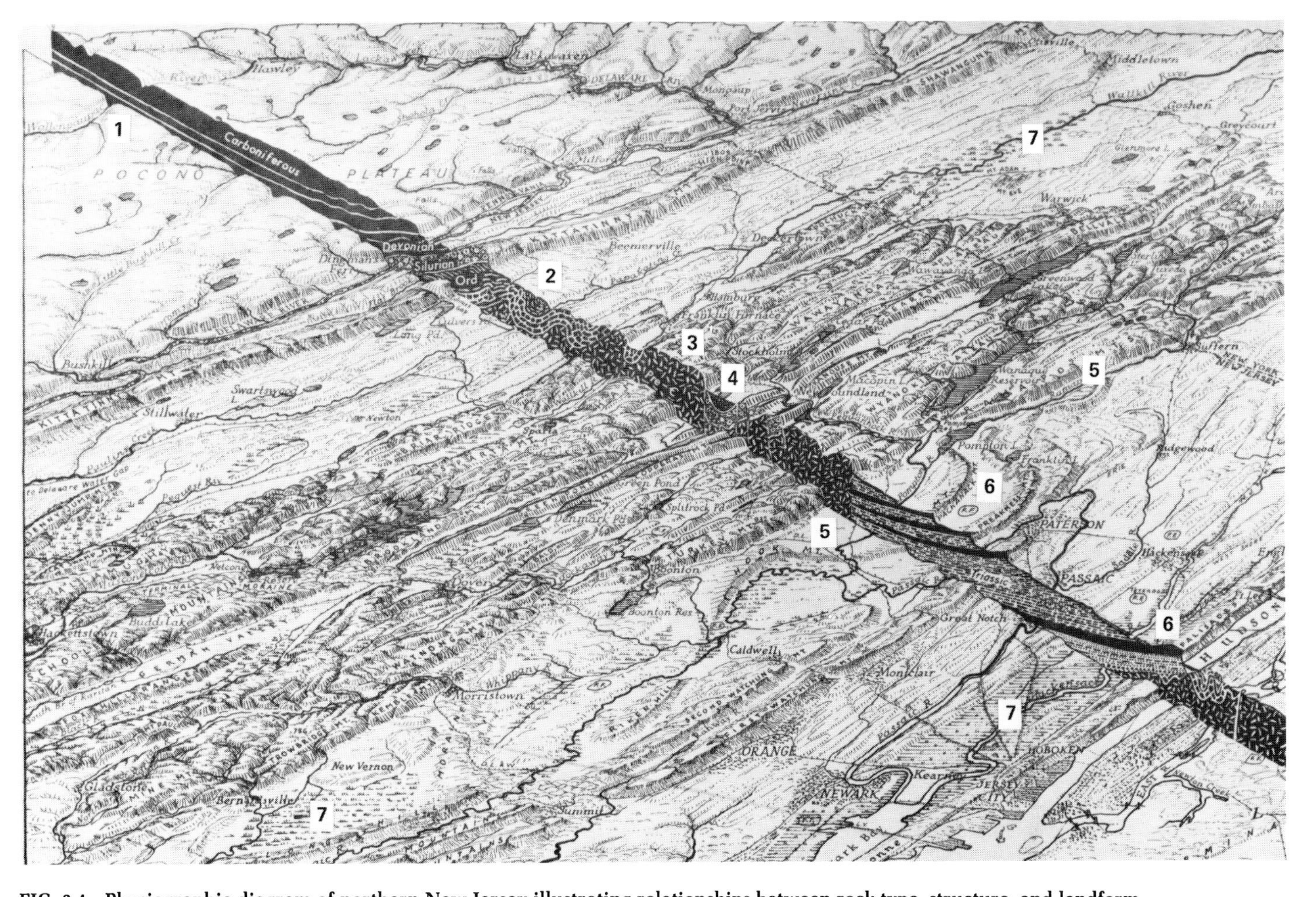


**FIG. 2.4 Physiographic diagram of northern New Jersey illustrating relationships between rock type, structure, and landform.**

1. **horizontally bedded sedimentary rocks**
2. **folded sedimentary rocks**
3. **batholith of Precambrian gneiss**
4. **graben formed by fault blocks**
5. **scarp of the Ramapo fault**
6. **sills of basalt and diabase**
7. **glacial lake beds**

**(See also ERTS image, Fig. 2.8)** *(Drawn by E. J. Raisz. Courtesy of The Geographic Press, A Division of Hammond, Inc.)*

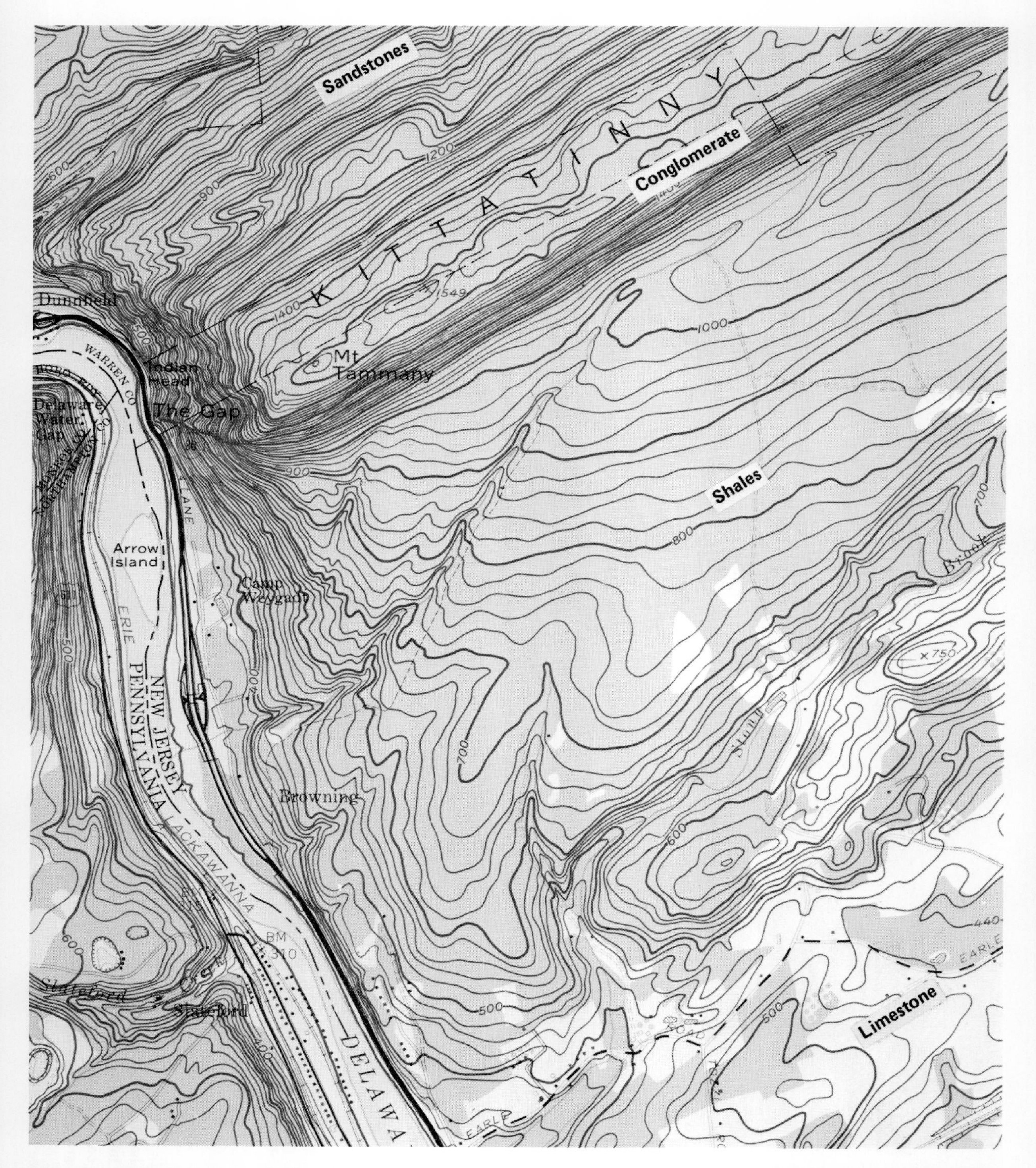


FIG. 2.5 USGS quadrangle map for the Delaware Water Gap, New Jersey-Pennsylvania (scale 1:24,000). Substantial information can be obtained from these topographic maps by interpretation of landform, drainage patterns, and water bodies depicted.

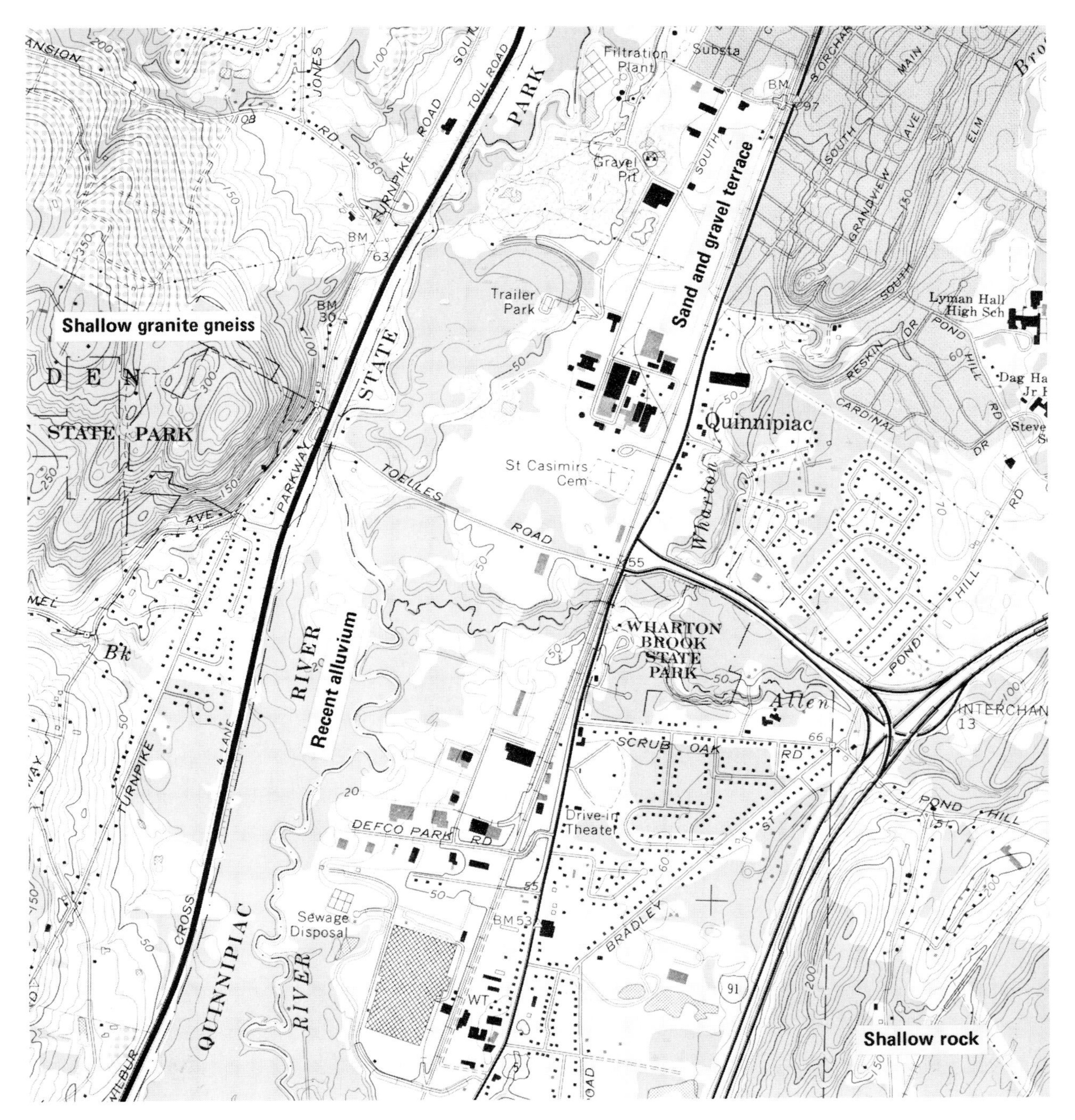


FIG. 2.6 USGS quadrangle map, Wallingford, Connecticut (scale 1:24,000). Larger-scale topographic maps provide more detailed information on terrain features.

**TABLE 2.3**
**FORMS OF REMOTE-SENSING IMAGERY**

| Imagery | Procurement platform | Image scale | Image form | Availability | Source | Example |
|---|---|---|---|---|---|---|
| Multispectral scanner (MSS)* | Unmanned satellite (ERTS-1, LANDSAT) | 1:1,000,000–1:250,000 (resolution 80m) | B&W negative or print of electronic image in four wavelength ranges (can be color-enhanced): band 4 (green), 0.5–0.6 $\mu$m, band 5 (red), 0.6–0.7 $\mu$m; band 6 (near IR), 0.7–0.8 $\mu$m; band 7 (near IR), 0.8–1.1 $\mu$m | Worldwide coverage every 18 days since July 1972. | EROS Data Center | Fig. 2.8 |
| Multispectral camera (S190A) | Manned spacecraft (SKYLAB) | 1:1,000,000–1:250,000 (resolution 30–50 m) | Negatives and prints from six film and filter combinations: B&W IR (two wavelengths), color IR, high-resolution color, B&W (two wavelengths) | Limited worldwide coverage. | EROS Data Center | None given |
| Earth terrain camera (S190B) | Manned spacecraft (SKYLAB) | 1:950,000–1:125,000 (resolution 20–30 m) | Photographs with four film and filter combinations: IR color high-resolution color, high-definition B&W, high-resolution IR color | Limited worldwide coverage. | EROS Data Center | Fig. 2.9 |
| Side-looking airborne radar (SLAR) | Aircraft | 1:300,000, 1:125,000, 1:30,000 (occasional) | B&W print of electronic image (wavelengths 1 mm–4 m) | Very limited coverage. Procurement costly. | Private companies | Fig. 2.10 |
| Thermal IR scanner | Aircraft | Varies with platform altitude | B&W or color enhanced prints† of electronic image; wavelengths 8–14 $\mu$m | Very limited coverage. Procurement relatively costly. | Private companies | None given |
| Aerial cameras | High-altitude aircraft (U-2, RB-57) | 1:125,000, 1:100,000 | B&W or color stereo pairs of aerial photos of high resolution | Limited worldwide coverage. | EROS Data Center | Fig. 2.11 |
| Aerial cameras | Moderate- to low-altitude aircraft | Varies with altitude; usually 1:60,000–1:8,000 | (*a*) B&W or color stereo pairs of aerial photos of high resolution; (*b*) B&W or color IR | (*a*) U.S. complete coverage; (*b*) worldwide coverage, extensive. | Private companies, government agencies‡ | Fig. 2.12, Fig. 2.13 |
| Multispectral cameras | Moderate- to low-altitude aircraft | Varies with platform altitude | Negatives and photos from various film and color combinations; prints or projections color-enhanced† for interpretation | Very limited coverage. | Private companies | None given |
| Side-scan sonar | Water vessels | Large scales | B&W print of electronic image | Very limited coverage. | Private companies | Fig. 2.14 Fig. 9.59 |

*Stereosat with resolution of 15 m and image pairs suitable for stereoscopic viewing and production of topographic maps at 1:100,000 may be available by 1984 [Godfrey (1979)[3]].

†Color-enhancement obtained by projection of various wavelengths through filters and combining the results (false-color imagery).

‡Photos normally used for photogrammetric mapping.

NOTE: B&W—black and white, IR—infrared.

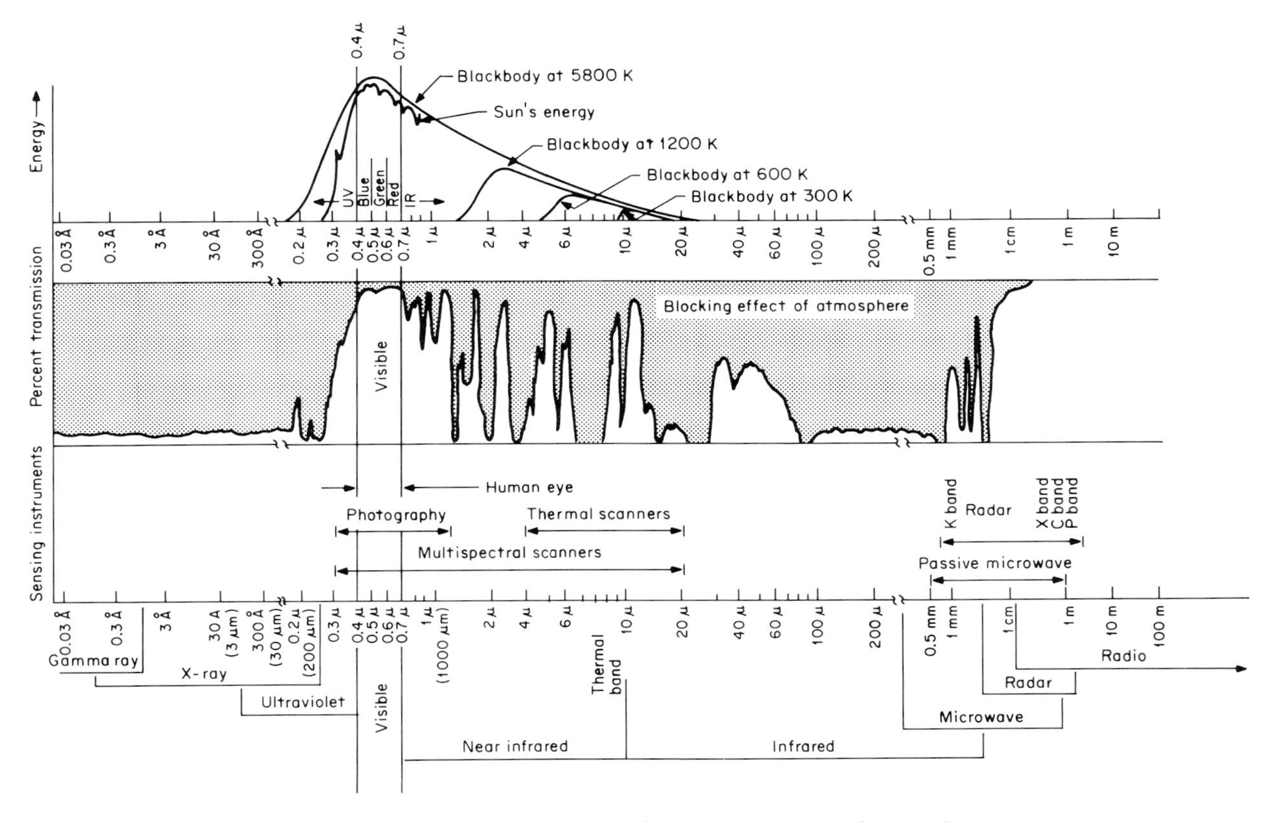


**FIG. 2.7 The electromagnetic spectrum illustrating atmospheric attenuation and general sensor categories.** [*From Way (1978).*[4] *Reprinted with permission of Dowden, Hutchinson & Ross, Inc.*]

## Remote-Sensing Imagery and Interpretation

### *General*

Imagery is procured by a number of methods in a number of forms at various scales as summarized on Table 2.3. The relationship between the various forms and the electromagnetic spectrum is given on Fig. 2.7. Some sources for obtaining imagery are given in Appendix B.

### *Imagery Examples*

The more useful types for engineering geology studies include:

- *ERTS or LANDSAT* imagery, obtained at scales of 1:1,000,000 or enlarged to 1:250,000, is the most important data source for regional geologic studies. Figure 2.8 illustrates a number of significant landforms.
- *SKYLAB* imagery, obtained by a multispectral camera or the earth terrain camera, provides images at scales similiar to ERTS or LANDSAT but at higher resolution, and can be enlarged to 1:125,000 without significant quality loss. The strong lineation of the San Andreas fault is apparent on Fig. 2.9.
- *SLAR* (side-looking airborne radar) penetrates cloud cover, and to some degree vegetation, providing low-resolution images at scales of 1:250,000 to 1:125,000, and occasionally larger (Fig. 2.10).
- *High-altitude stereo aerial photographs* provide the smallest-scale images for stereo viewing, ranging in scale from 1:125,000 to 1:100,000, yielding substantial detail on terrain features (Fig. 2.11).
- *Stereo pairs of aerial photographs* provide the

**FIG. 2.8** **ERTS image of New Jersey and eastern Pennsylvania (scale 1:1,000,000). The landforms evident are indicative of several geologic formations: (1) folded sedimentary rocks, (2) glaciated Precambrian crystalline rocks, (3) basalt sills of the Triassic, (4) another Precambrian crystalline complex including a major fault system. Comparative information is given on the physiographic diagram (Fig. 2.4) and the topographic map (Fig. 2.5).** *(Original image by NASA, reproduction by U.S. Geological Survey, EROS Data Center.)*

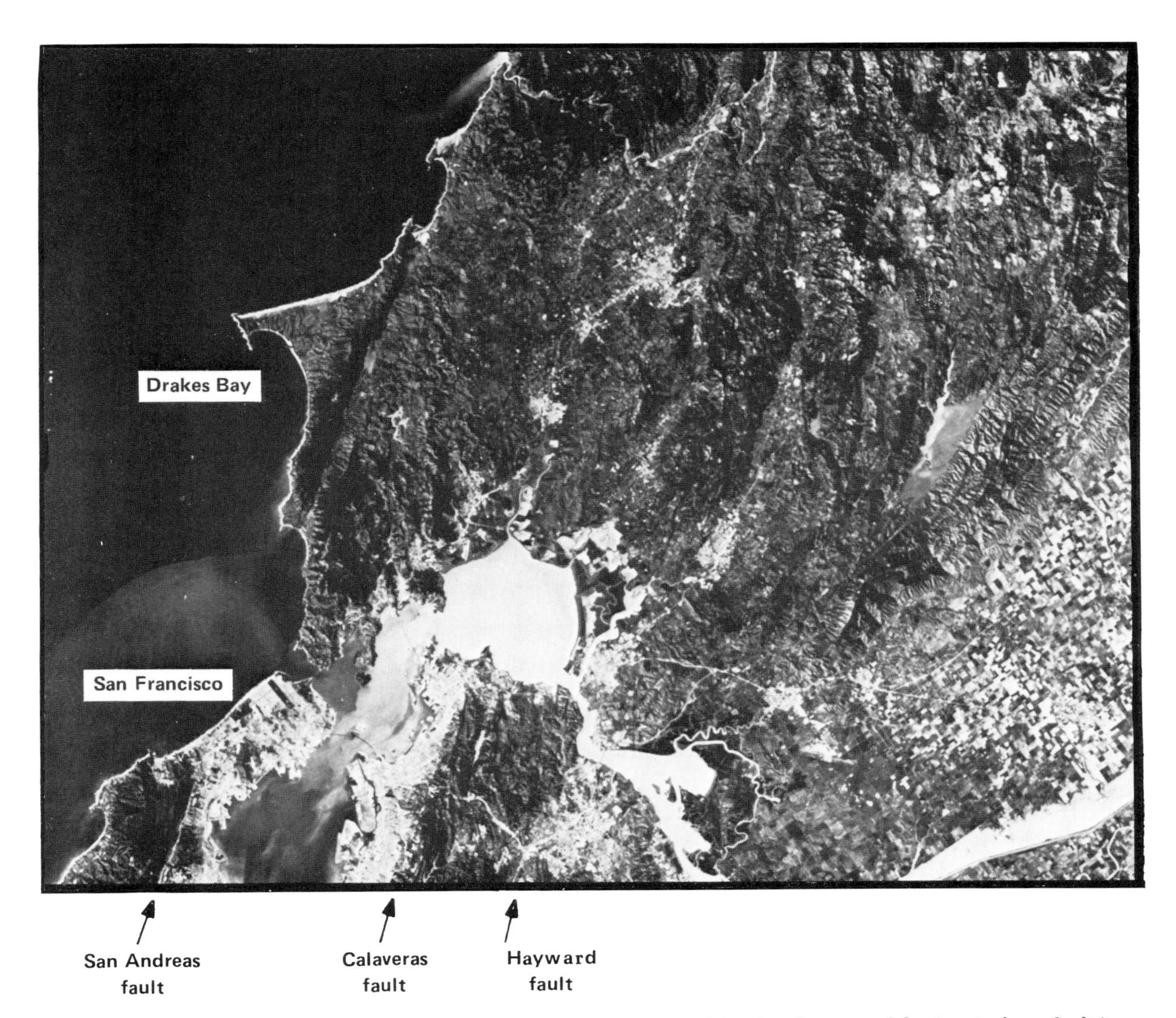


**FIG. 2.9 SKYLAB image of the Point Reyes and San Francisco area of California. The trace of the San Andreas fault is clearly evident along the coastline; the Calaveras and Hayward faults are less evident. Compare with ERTS image in Fig. 6.66 and Fig. 7.49, a topographic map.** *(Original image by NASA, reproduction by U.S. Geological Survey, EROS Data Center.)*

FIG. 2.10 SLAR (side-looking airborne radar) image of a portion of the coastline of Brazil between Macaé and Cabo S. Tome, state of Rio de Janeiro (scale 1:250,000). The shadow effect of the scanning radar is apparent. Three geologic provinces are covered: (1) Precambrian crystalline rocks, the rough-textured area in the upper left; (2) Tertiary sediments of the coastal plain, the subdued texture of the central portion; and (3) recently uplifted coastal zone showing strand lines, or old beach ridges.

**FIG. 2.11** NASA high-altitude stereo pair of an area northwest of Tucson, Arizona (scale 1:125,000). Apparent are sheet wash and sheet erosion of the "bajadas," alluvial fans of granular soils, valley fill of fine-grained soils, and the "dry wash" of the Santa Cruz River, all typical depositional forms in valleys adjacent to mountains in arid to semiarid climates. The blocked area is shown at a larger scale on a stereo pair given as Fig. 7.30. *(Original image by NASA, reproduction by U.S. Geological Survey, EROS Data Center.)*

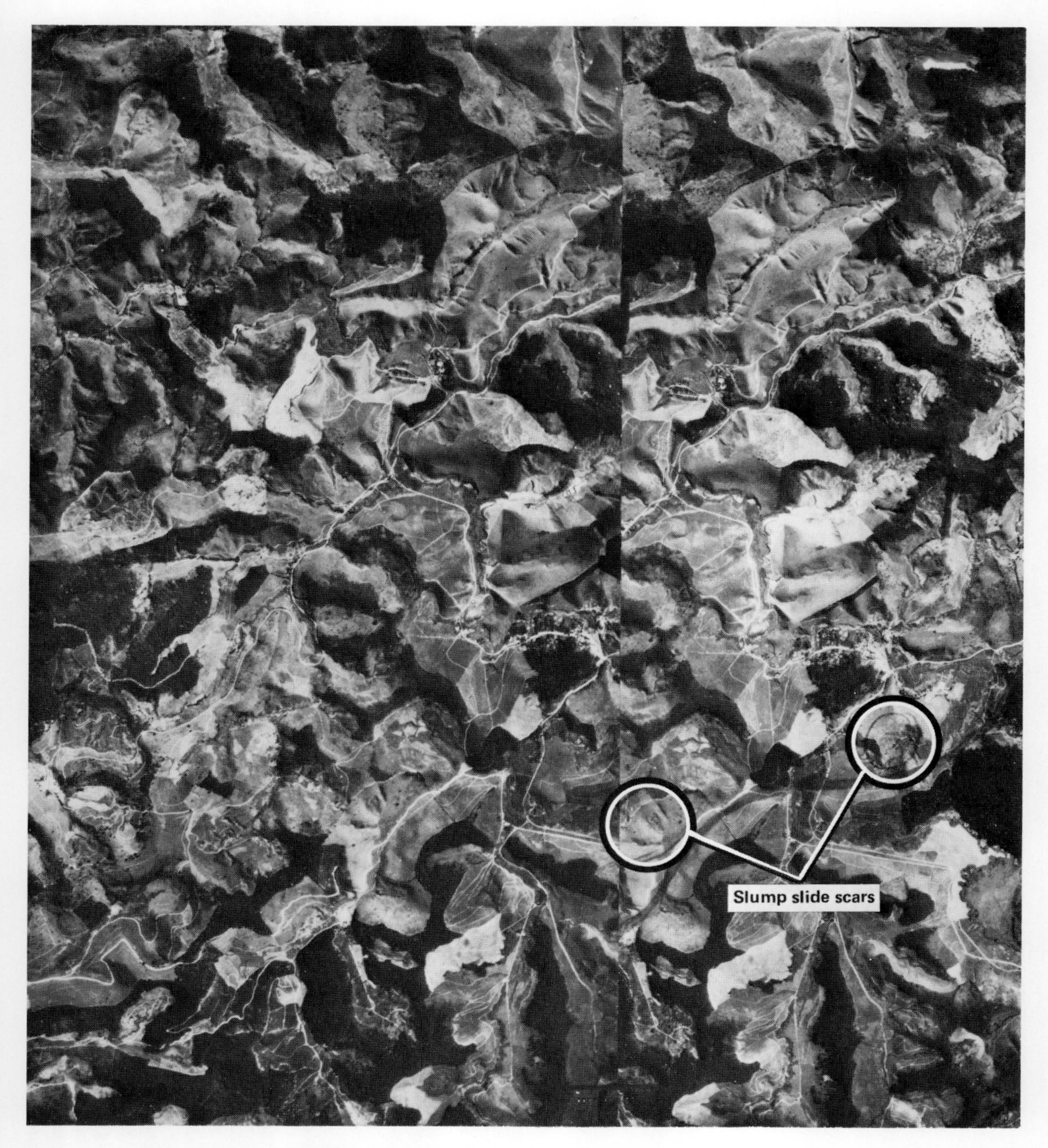


FIG. 2.12 Stereo pair of serial photos (scale 1:40,000) shows landform developing in metamorphic rocks from subtropical weathering. Severe erosion and a number of landslides are apparent, including the rotational slides shown on the larger scales given as Figs. 2.13 and 9.28.

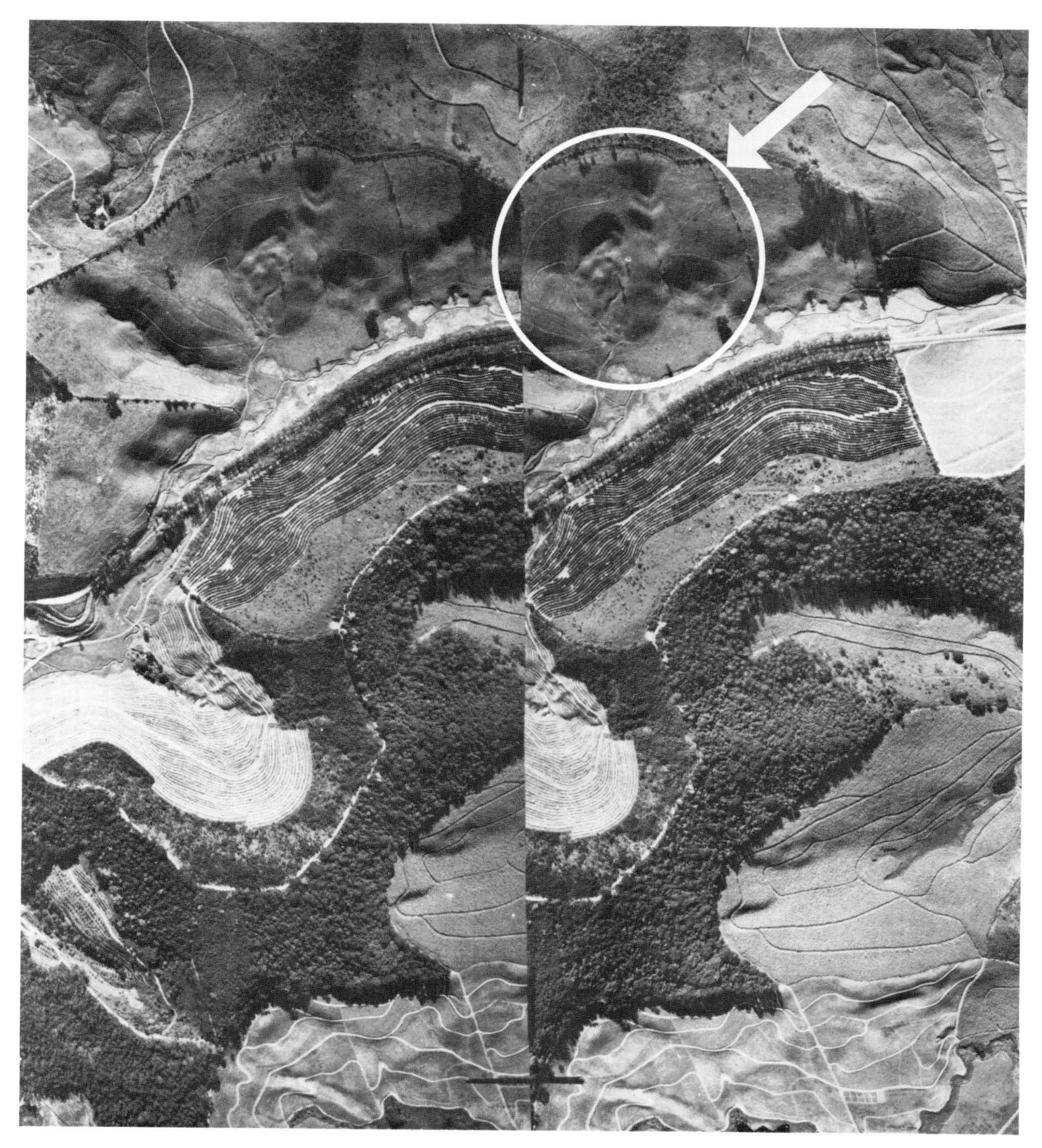

FIG. 2.13 Stereo pair of aerial photos (scale 1:8,000) of a portion of Fig. 2.12 provides substantially more detailed information on the soil conditions and their distribution. The spoon-shaped slump slide that occurred in residual soils is clearly apparent. Its rounded forms, resulting from erosion, indicate that the slide is relatively old.

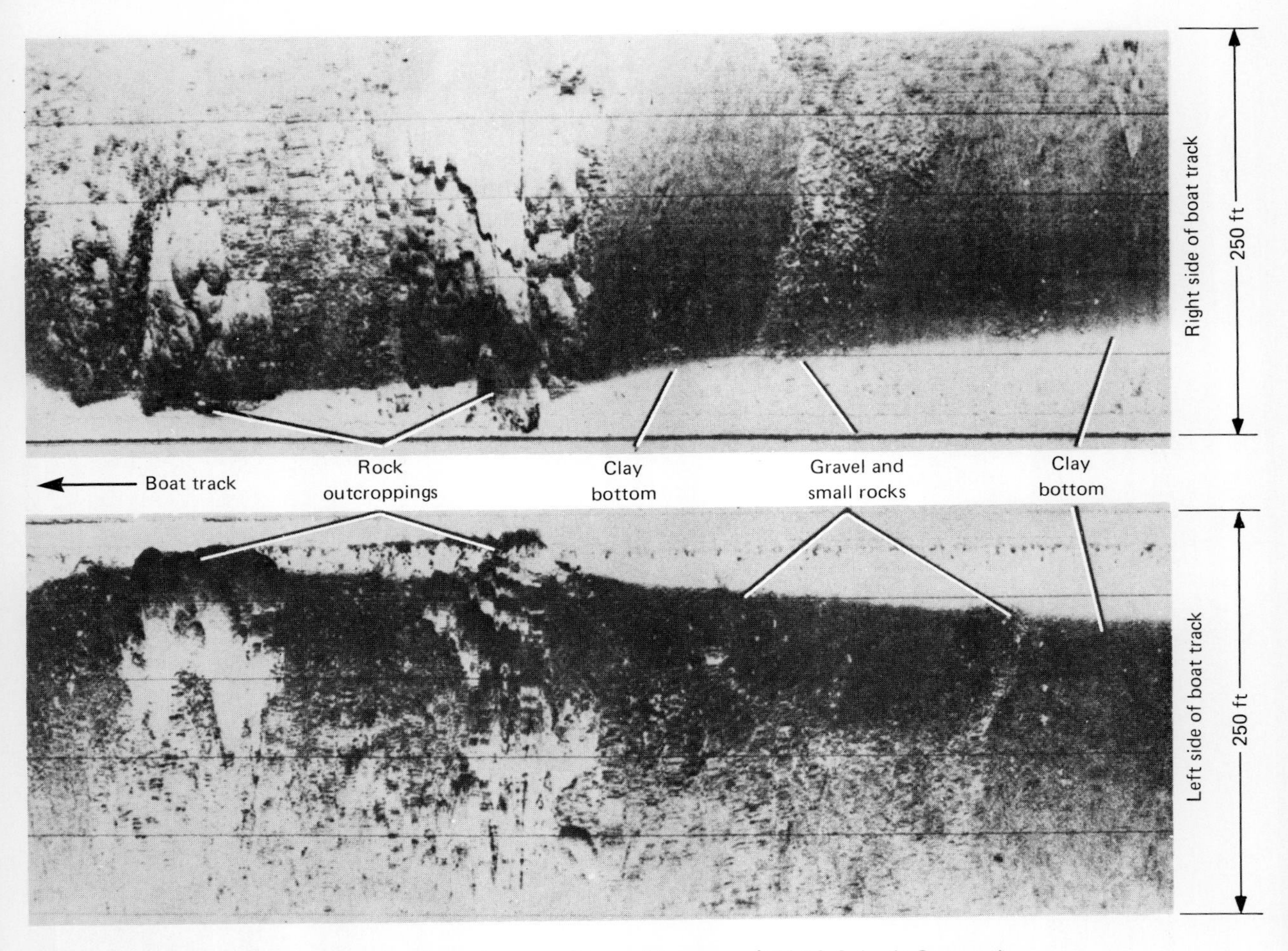


**FIG. 2.14 Side-scan sonar image reveals seafloor conditions.** *(Courtesy of EG&G Seismic Systems.)*

basis for detailed engineering geologic mapping. Detailed studies of large to small areas should be based on stereoscopic interpretation of aerial photos, preferably obtained at two scale ranges: 1:60,000 to 1:40,000 (Fig. 2.12) and 1:20,000 to 1:8,000 (Fig. 2.13), with the smaller scales providing an overview and the larger scales providing details.

- *Side-scan sonar* provides images of the seafloor, which often has features indicative of significant geologic conditions (Fig. 2.14). (*See also Fig. 9.59.*)

*Applications*

Imagery interpretation as applied to engineering geologic mapping is summarized on Table 2.4, and as applied to environmental and natural-resource studies on Table 2.5.

*Interpretative Techniques*

The elements of imagery interpretation are summarized on Table 2.6. Detailed mapping on large scales requires stereoscopic interpretation of pairs of aerial photographs (Fig. 2.15). Stereoscopic viewing of twin images of either ERTS, LANDSAT, or SLAR imagery provides pseudostereo images and aids in the interpretation of these forms. Techniques of terrain analysis and air photo interpretation are described by Way (1978),[4] Lueder (1959),[5] American Society of Photogrammetry (1960),[6] and Belcher (1948),[7] among others.

**TABLE 2.4**
**USES OF REMOTE SENSING FOR ENGINEERING GEOLOGIC MAPPING**

| Information desired | Applicable imagery |
|---|---|
| *Regional geologic mapping* and delineation of major structural features<br>(*a*) Global coverage, moderate resolution<br>(*b*) High resolution, but incomplete global coverage<br>(*c*) Useful for areas of perennial cloud cover and heavy vegetation; low resolution | Space platform and SLAR:<br>(*a*) ERTS and LANDSAT (MSS)<br>(*b*) Multispectral camera (SKYLAB), earth terrain camera (SKYLAB)<br>(*c*) Side-looking airborne radar (SLAR) |
| *Detailed mapping* of rock type, structure, soil formations, drainage, groundwater, slope failures, sinkholes, etc.:<br>(*a*) Moderately large areas<br>(*b*) Large areas, general mapping<br>(*c*) Small areas, detailed mapping | Stereo pairs of aerial photos (preferably B&W):<br>(*a*) Scale 1:100,000<br>(*b*) Scale 1:60,000–1:40,000<br>(*c*) Scale 1:20,000–1:8,000 |
| *Improved definition of surface and groundwater conditions* on large- to local-area basis, such as land-water interface, seepage, ground moisture (important for sinkhole and fault identification) | B&W or color IR photos, preferably in stereo pairs |
| *Seafloor and other underwater conditions* (rock outcrops, soils, sunken vessels, pipelines, etc.) | Side-scan sonar |

NOTES: Normal studies of large land areas, such as for highways, airports, industrial zones, new communities should be based on the interpretation of aerial photos of at least two scale ranges (1:60,000–1:40,000 and 1:20,000–1:8,000).

Studies of areas where seismicity is of concern should always begin with interpretations of ERTS/LANDSAT imagery, then be supplemented by interpretation of normal study imagery scales.

**TABLE 2.5**
**USES OF REMOTE SENSING FOR ENVIRONMENTAL AND NATURAL RESOURCE STUDIES**

| Information desired | Applicable imagery |
|---|---|
| *Regional environmental studies* of air, water and vegetation quality, flooding:<br>(*a*) On a changing or seasonal basis<br>(*b*) High resolution but incomplete global coverage | Satellite and space platforms:<br>(*a*) ERTS and LANDSAT (MSS)<br>(*b*) SKYLAB |
| *Surface and groundwater studies* (large to local areas):<br>(*a*) General<br>(*b*) Thermal gradients indicative of pollution or saltwater intrusion of surface waters<br>(*c*) Subsurface seepage | IR and multispectral imagery:<br>(*a*) B&W or color IR photos, multispectral photos<br>(*b*) Thermal IR scanner<br>(*c*) Thermal scanner |
| *Vegetation:* Forestry and crop studies (large to local areas); identify types, differentiate healthy from diseased vegetation | B&W or color IR, multispectral photos |
| *Mineral resource studies:*<br>(*a*) Based on landform analysis<br>(*b*) Based on plant indicators | Various types:<br>(*a*) ERTS, LANDSAT, SKYLAB, aerial photos<br>(*b*) Multispectral photos |

**TABLE 2.6**
**ELEMENTS OF IMAGERY INTERPRETATION**

| Imagery feature | Imagery type | Interpretation |
|---|---|---|
| Topography | ERTS/LANDSAT/ SLAR: Stereo pairs of aerial photos and topographic maps. | Rock masses as formed or subsequently deformed have characteristic landforms (Chap. 6) as do soil formations classed by mode of deposition of occurrence (Chap. 7), which in all cases depend strongly on climate.<br>Slope inclinations and heights are related to material types in terms of strength and structure.<br>Slope failures, sinkholes, erosion gullies, etc., have characteristic forms. |
| Drainage patterns and stream forms | ERTS/LANDSAT/ SLAR: Stereo pairs of aerial photos and topographic maps. | Drainage patterns on a regional and local basis reflect rock type and variations, rock structure, and where soil cover is adequately thick, the soil type (see Tables 6.2 and 7.4 for typical patterns).<br>Stream form is also related to its geologic environment *(see Art. 7.4.1).*<br>Streams, lakes, and swamps are indicators of the groundwater table, which usually follows the surface at depressed contours *(Art. 8.3.1).* |
| Gully characteristics | Stereo pairs of aerial photos (large scale). | Various soil types have characteristic gulley shapes *(see Fig. 7.2).* |
| Photo tone | B&W aerial photos. | Tone shows relative ground moisture and texture. Some general relationships are:<br>White—concrete, or free-draining soils above the water table.<br>Light gray—primarily coarse soils with some fines. Acid rocks.<br>Dull gray—slow-draining soils. Basic rocks. |

### Engineering Geology Maps

*General*

Data obtained from terrain analysis are plotted to form an engineering geology map which provides information on geologic conditions over the entire study area. When interpreted with experience, significant knowledge regarding the engineering characteristics of the formations and materials becomes available.

*Preliminary map assessment* permits conclusions regarding:

1. Abandonment of the site to avoid extremely hazardous conditions
2. Location of structures to avoid unfavorable conditions
3. General requirements for foundations and excavation
4. Formulation of the program of subsurface investigation.

*Map Preparation*

Use a topographic map as a base map, selecting a scale convenient to the study area and purpose.

Plot the data obtained from terrain analysis including delineations of the various soil and rock types, major structural features, areas of

| Imagery feature | Imagery type | Interpretation |
|---|---|---|
| | | Dark gray to black—poor draining soils, organic soils groundwater near the surface |
| Vegetation | B&W aerial photos. | Vegetation varies with climate, geologic material, and land use. Some relationships between vegetation and soil types are given in Art. 7.8.3.<br><br>Tree lines often delineate floodplain limits and fault traces. |
| Land use | B&W aerial photos. | Most significant are the locations of man-made fills, cut for roadways, borrow pits, open-pit mines, and other man-made features. Development is usually related to landform. |
| Color-enhanced (false-color) imagery | ERTS/LANDSAT:<br><br>Red normally used for near infrared. Can be presented in various colors to enhance specific features. | Filtered through red: color significance:<br><br>Vegetation—the brighter the red the healthier is the vegetation.<br><br>Water bodies—water absorbs sun's rays, clear water shows black. Silt reflects sun's rays, sedimentation shows light blue.<br><br>Urban areas—bluish-gray hues. |
| | Multispectral photos or color IR. | Various filters are used to emphasize the desired feature (vegetation type, water-body pollution, thickness of snow field, etc.) |
| | Thermal IR. | Various filters are used to emphasize a particular feature. Can delineate water gradients to 1°F. |

NOTE: In all color-enhanced imagery, ground truth is required to identify the feature related to a specific color.

shallow groundwater and rock, and locations of hazards such as sinkholes and landslides. A suggested nomenclature for the identification of various soil and rock formations is given on Table 2.7.

Preparation is first on a preliminary basis; a final map is prepared after site reconnaissance and, preferably, after at least some subsurface investigation. To aid in reconnaissance all significant cuts and other surface exposures should be noted on the map, as well as areas of questionable conditions.

*Assessment of Mapped Conditions*

SOIL FORMATIONS Soils may be geologically classified by their origin and mode of occurrence (Chap. 7), the engineering significance of which lies in the characteristic properties attendant to the various classes. Therefore, if a soil formation is classed in terms of origin and mode of occurrence, preliminary judgments can be made regarding their influence on construction.

ROCK FORMATIONS The various rock types have characteristic engineering properties and structural features, either as originally formed or as deformed by tectonic or other geologic activity, and when attacked by weathering agents they decompose to form characteristic soil types as described in Chap. 6. The identification of rock-mass features allows the formulation of prelim-

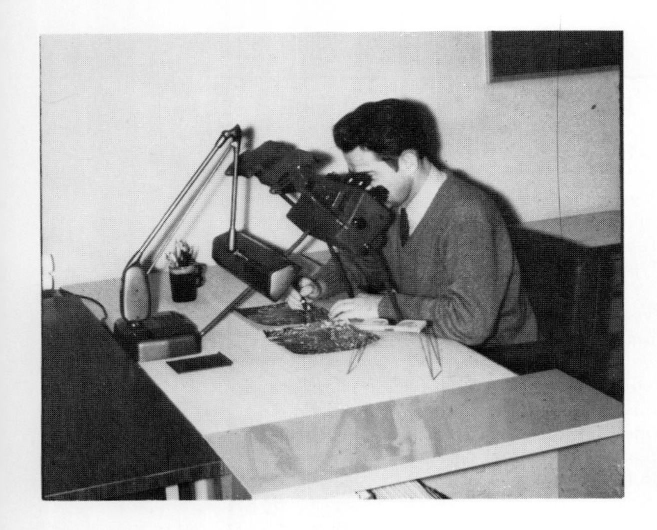

**FIG. 2.15 Stereoscopic interpretation of aerial photographs. Two types of viewers are shown.**

inary judgments regarding their influence on construction.

Three examples illustrate the approach:

1. Figure 2.16—preliminary engineering geology map for proposed new community in a region of glacial soils. Conditions may be generally interpreted for engineering purposes solely on the basis of the soil types, classed geologically as follows:
   a. *Foundation conditions:* RX, GT—good support all loads; GK—good support moderate loads; GL, GT/GL—possible suitable support for light to moderate loads; AF—probable poor support. Areas of GK, GL, GT/GL, and AR in particular require detailed investigation. (NOTE: Terms such as moderate loads or poor support require definition in the report accompanying the map.)
   b. *Excavation conditions:* RX areas will require blasting. High groundwater can be expected in areas of AF and GL.
   c. *Borrow materials:* Coarse-grained granular soils are found in GK.
   d. *Septic tanks:* High groundwater, clayey soils, or shallow rock over much of the area imposes substantial constraint to their use.
   e. *Groundwater for potable water supply:* Most feasible locations for wells are at the base of slopes in the GT and GK materials. Buried channel aquifers may exist in the valley under the GL deposits.
2. Figure 2.17—general engineering geology map prepared for an interstate highway through an area with potential slope stability problems as shown on the stereo pair (Fig. 2.18).
3. Figure 2.19—detailed engineering geology map prepared for new community proposed for an area of shallow limestone, some parts with expansive clays and others with a potential for collapse. The aerial photo (Fig. 2.20) illustrates ground conditions, and the topographic map (Fig. 2.21)illustrates the regional landform.

*Other map forms* prepared for engineering studies can include:

1. *Geologic hazard,* or *risk, maps* which delineate geologic conditions in terms of various degrees of hazard or risk such as terrain where soil liquefaction or slope failures are of concern. An example of a slope-failure hazard map prepared for a mountain roadway is given as Fig. 9.96.
2. *Geologic constraint maps (see Fig. 8.61)* which form the basis for the preparation of land-use maps *(see Fig. 8.62).*

### 2.2.4 SITE RECONNAISSANCE

#### General

All sites should be visited by an experienced professional to collect first-hand information on geology, terrain and exploration equipment access, existing structures and their condition, existing utilities, and potentially hazardous conditions. Prior examination of aerial photos will identify many of the points to be examined.

#### Reconnaissance Checklist

1. Examine exposures of soils and rocks in cuts (highways, railroads, building excavations, gravel pits, quarries, stream banks and terraces), and on the surface, and note effluent groundwater seepage.
2. Examine slopes for signs of instability (creep ridges, tilted and bent trees, tilted poles, slope seepage).

**TABLE 2.7**
**SUGGESTED MAP SYMBOLS FOR ENGINEERING GEOLOGY MAPS**

| Classification | Symbol | Modifiers based on coarseness or occurrence |
|---|---|---|
| Residual soil | R | Rm—massive |
| | | Rs—saprolite |
| | | Rc—coarse-grained (granular) |
| | | Rf—fine-grained (clayey or cohesive) |
| Colluvial soil | C | Cr—originally residual soil |
| | | Cm—originally glacio-marine soils |
| | | Cl—originally glacio-lacustrine soils |
| | | T—talus |
| Alluvial soil | A | Ao—oxbow lake |
| | | At—terrace |
| | | Ar—recent alluvium; usually silt-sand mixtures with organic matter |
| | | Ac—coarse grained; sand and gravel mixtures |
| | | Am—medium grained; sand-silt mixtures |
| | | Af—fine grained; silt-clay mixtures |
| Eolian soil | E | El—loess |
| | | Ed—dunes |
| Glacial soils | G | Gm—moraine |
| | | Gt—till |
| | | Gs—stratified drift (outwash plains) |
| | | Gk—kane |
| | | Ge—esker |
| | | Gl—lakebed |
| Organic soils | O | Om—marsh |
| | | Os—swamp |
| Man-made fill | F | |
| High water table | Hg | |

**Rock symbols**

| Igneous | Sedimentary | Metamorphic |
|---|---|---|
| gr—granite | sg—conglomerate | qz—quartzite |
| ry—rhyolite | ss—sandstone | ma—marble |
| sy—syenite | si—siltstone | hr—hornfeld |
| mo—monzonite | sh—shale | gn—gneiss |
| di—diorite | ls—limestone | sc—schist |
| ga—gabbro | ak—arkose | ph—phyllite |
| ba—basalt | do—dolomite | sl—slate |

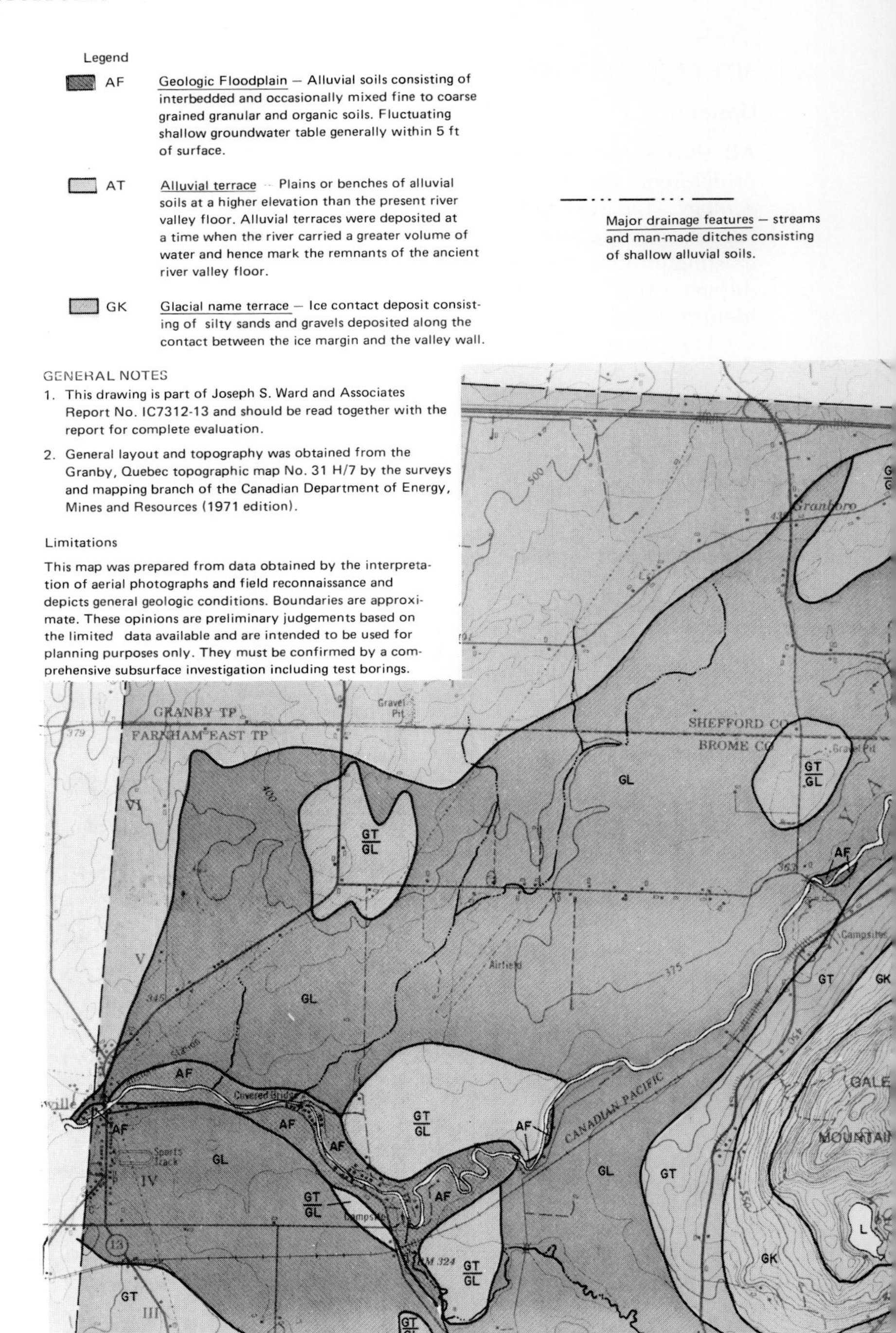


**FIG. 2.16 Preliminary engineering geology map, Bromont, Quebec, Canada. Conditions may be generally interpreted for engineering evaluations as described in the text.** *(Courtesy Joseph S. Ward and Associates.)*

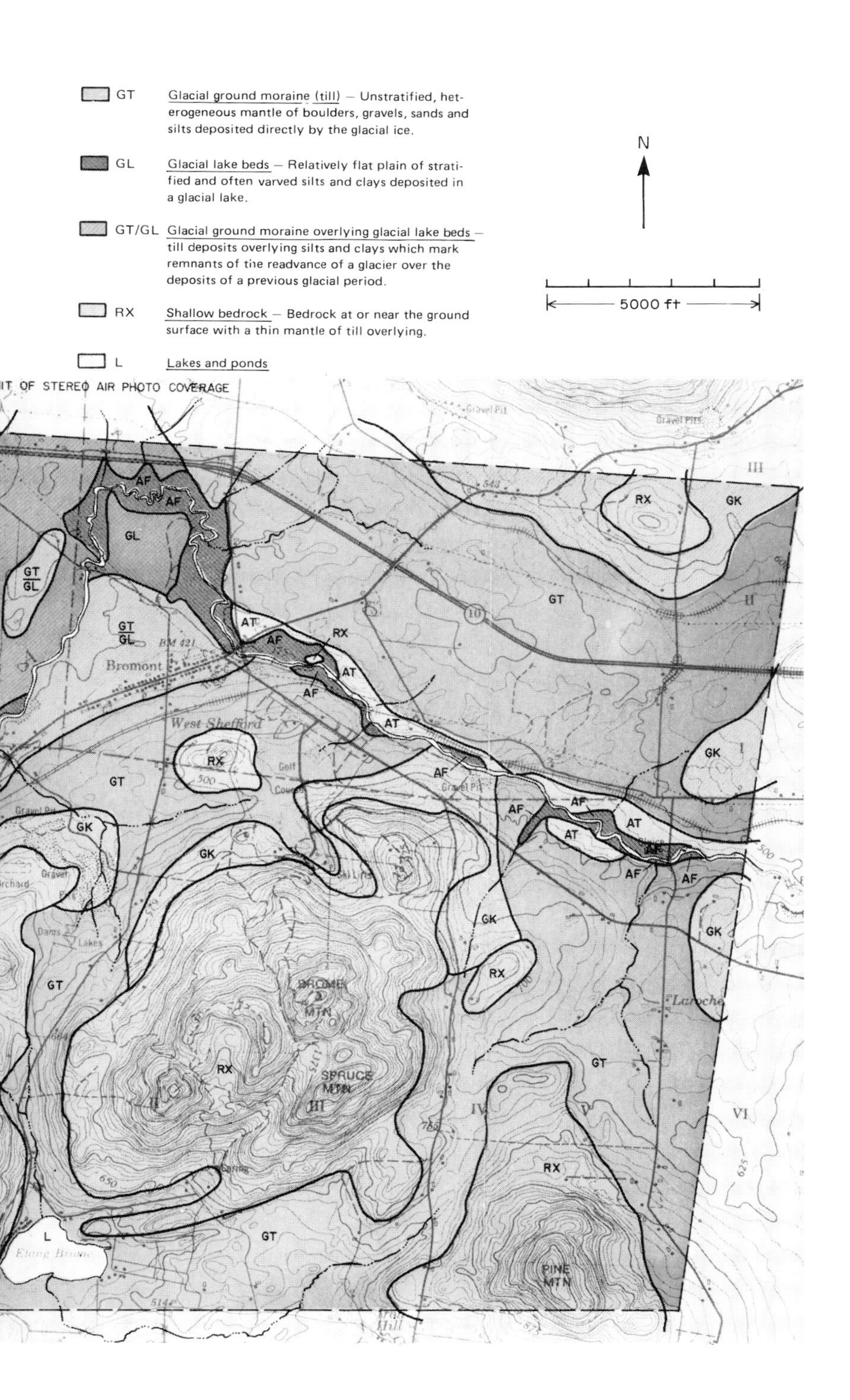
GT Glacial ground moraine (till) – Unstratified, heterogeneous mantle of boulders, gravels, sands and silts deposited directly by the glacial ice.
GL Glacial lake beds – Relatively flat plain of stratified and often varved silts and clays deposited in a glacial lake.
GT/GL Glacial ground moraine overlying glacial lake beds – till deposits overlying silts and clays which mark remnants of the readvance of a glacier over the deposits of a previous glacial period.
RX Shallow bedrock – Bedrock at or near the ground surface with a thin mantle of till overlying.
L Lakes and ponds
N
5000 ft
MIT OF STEREO AIR PHOTO COVERAGE
Bromont
West Shefford
BROME MTN
SPRUCE MTN
PINE MTN
AF
AT
GK
RX
GT
GL
L

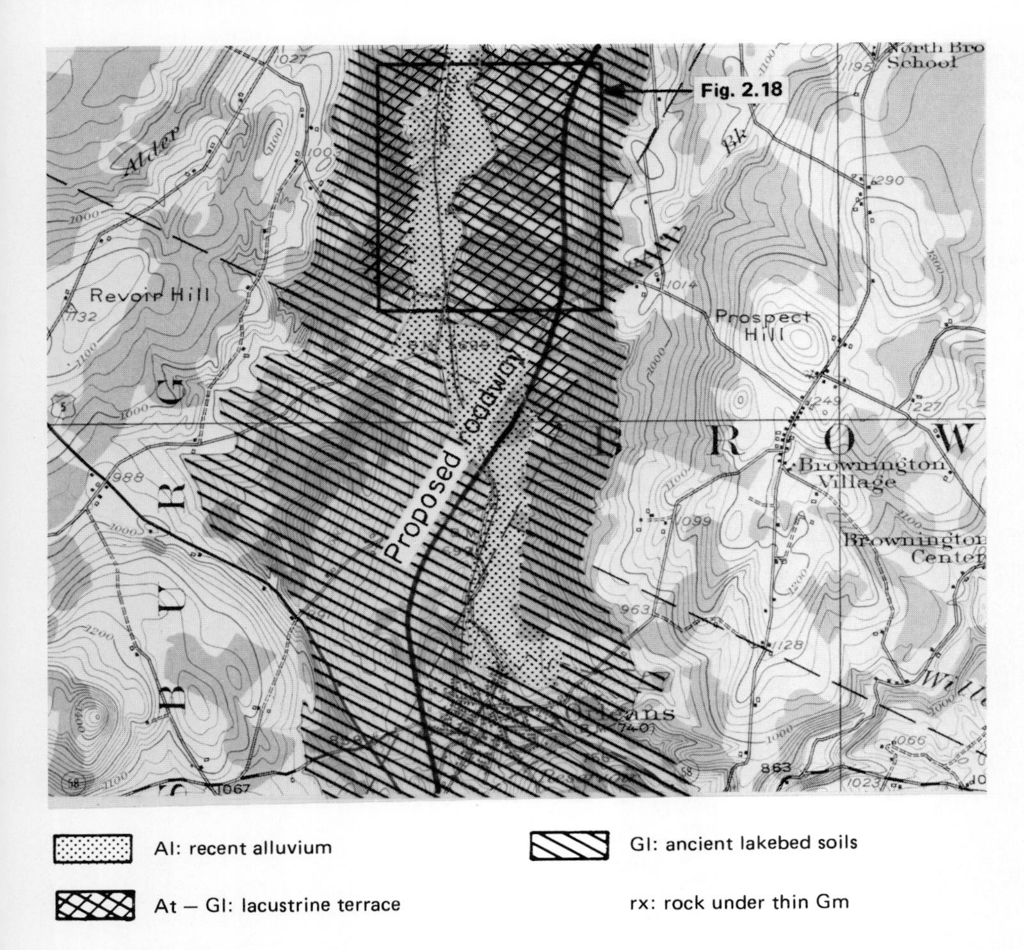


**FIG. 2.17 General engineering geology map prepared for interstate highway through a valley with glacial lacustrine soils on the slopes (Barton River Valley, Orleans, Vermont). The lower slopes in the At-GL material are subjected to active movements (see stereo pair, Fig. 2.18); the upper slopes in overconsolidated GL soils (stiff to hard varved clays) will tend to be unstable in cut. In many areas the Al soils will be highly compressible under embankment fills.**

**FIG. 2.18 Stereo pair of aerial photos showing old slide scars in glacial lakebed terrace soils (Barton River Valley, Orleans, Vermont).**

Slide scar
6 91

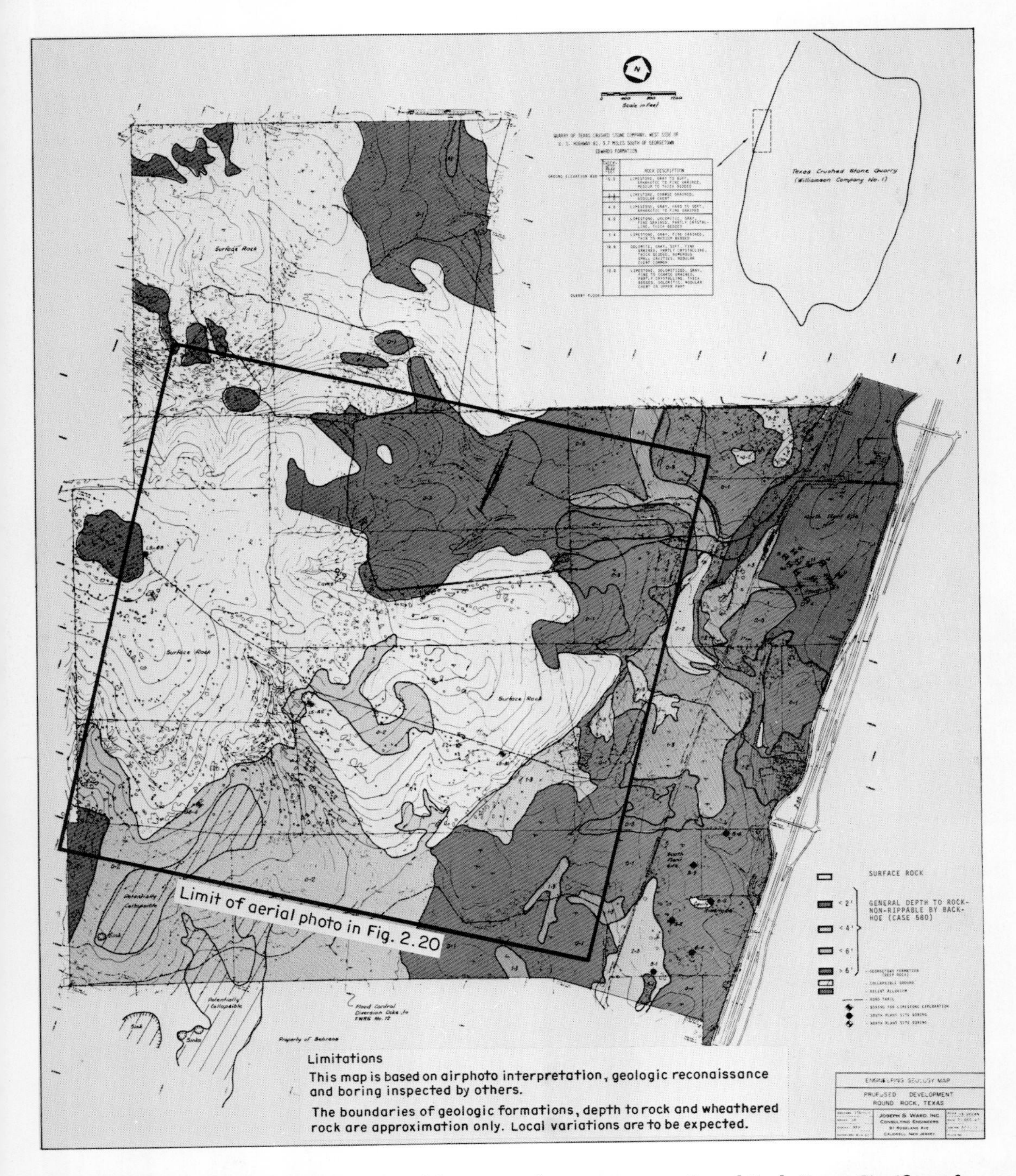


**FIG. 2.19 Engineering geology map prepared for proposed new town, near Round Rock, Texas. Significant features zoned for planning evaluation are: (*a*) surface exposures of hard limestone which will cause excavation difficulties, (*b*) depth to hard limestone where overlain by soils, often highly expansive, and (*c*) areas with sinkholes and potential for ground collapse. An aerial photo of the site is given as Fig. 2.20 and the regional topography on Fig. 2.21.** *(Map courtesy of Joseph S. Ward and Associates.)*

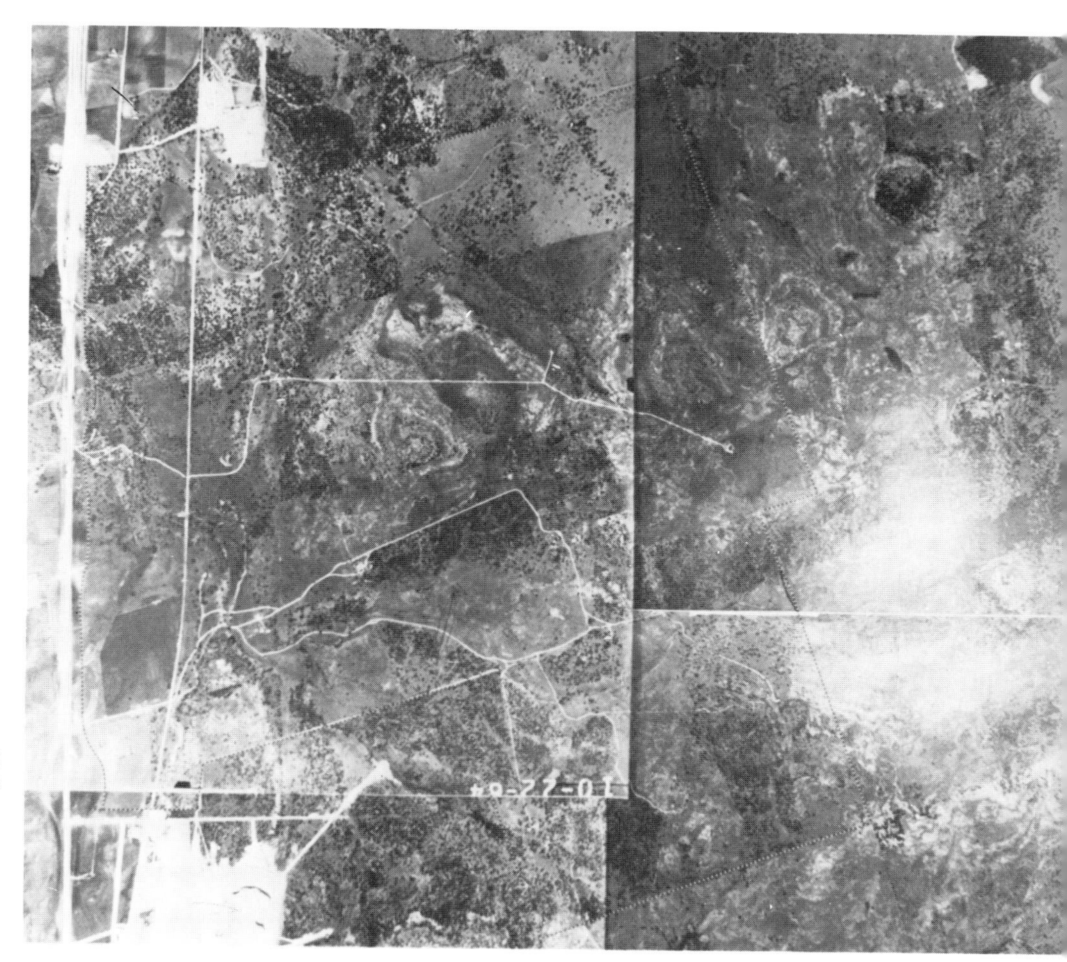

**FIG. 2.20 Aerial photo (scale 1:8,000) illustrating features of surface limestone, Round Rock, Texas, site of Fig. 2.19.** ***(See also Figs. 7.9, 10.19, and 10.20.)***

3. Examine existing structures and pavements for signs of distress.
4. Note evidence of flood levels along streams.
5. Contact local architects and engineers for information on foundations and local soil conditions.
6. Contact local well drillers for information on groundwater conditions.
7. Contact local public officials for building code data and information on foundations, soil conditions, and on-site utilities.
8. Note site conditions imposing constraints on access for exploration equipment.
9. Note present land use.

### Revise Engineering Geology Map

The information gathered during site reconnaissance is used to revise the preliminary engineering geology map where necessary.

### 2.2.5 PREPARATION OF SUBSURFACE EXPLORATION PROGRAM

Prepare the subsurface exploration program, considering the necessity to:

- Confirm the boundaries of the various geologic formations as mapped
- Obtain data for the preparation of geologic sections
- Obtain samples for identification, classification, and laboratory testing
- Obtain in situ measurements of the engineering properties of the materials

## 2.3 SUBSURFACE EXPLORATION

### 2.3.1 GENERAL

### Objectives

- To confirm or supplement the engineering

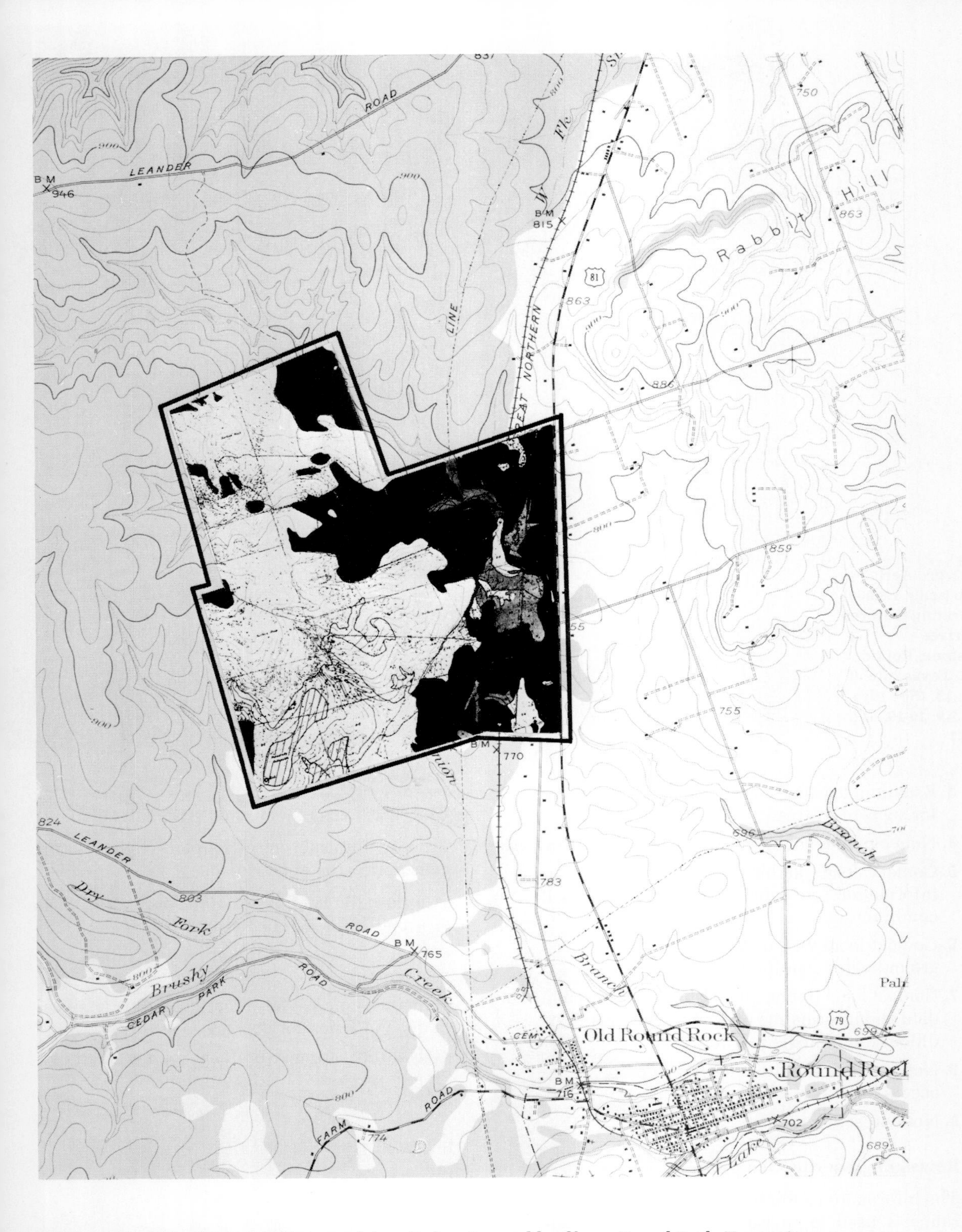


FIG. 2.21 Portion of topographic map giving site location and landform, Round Rock, Texas, site.

geology map showing shallow and surficial distributions of the various formations

- To determine the subsurface distribution of the geologic materials and groundwater conditions (this article)
- To obtain samples of the geologic materials for identification and laboratory testing *(see Art. 2.4)*
- To obtain in situ measurements of engineering properties *(see Chap. 3)*

**Exploration Method Categories**

*General Categories*

- *Direct methods* allow examination of materials, usually with the recovery of samples; examples are excavations and test borings
- *Indirect methods* provide a measure of material properties; examples are geophysical methods and the use of the cone penetrometer, which through correlations with other data allow an estimation of material type

*Specific Categories*

- *Geophysical methods* provide indirect data *(see Art. 2.3.2)*
- *Reconnaissance methods* provide direct and indirect data *(see Art. 2.3.3)*
- *Test and core borings* provide direct data *(see Art. 2.3.4)*
- *Remote borehole sensing* and *logging* provide direct and indirect data *(see Art. 2.3.5)*

**Method Selection**

*Basic Factors*

Selection is based on consideration of the study objectives and phase *(see Art. 1.3)*, the size of the study area, project type and design elements, geologic conditions, surface conditions and accessibility, and the limitations of budget and time.

The various methods in terms of their applicability to general geologic conditions are listed on Table 2.1 and described in general terms of applicability and limitations on Table 2.8.

*Key Methods*

*Geophysical methods,* particularly seismic refraction surveys, provide the quickest and often the most economical method of obtaining general information over large land areas, or in areas with difficult access, such as mountainous regions or large water bodies.

*Test pits and trenches* are rapid and economical reconnaissance methods for obtaining information on shallow soil and groundwater conditions and depth and rippability of rock and for investigating landfills of miscellaneous materials.

*Test borings* are necessary in almost all investigations for the procurement of soil and rock samples below depths reachable by test pits.

*Other methods* generally can be considered to provide information supplemental to that obtained by key methods.

2.3.2 GEOPHYSICAL METHODS

**Seismic Methods: General**

*Theoretical Basis (see also Art. 11.2.2)*

*Elastic waves,* initiated by some energy source, travel through geologic media at characteristic velocities and are refracted and reflected by material changes or travel directly through the material, finally arriving at the surface where they are detected and recorded by instruments (Fig. 2.22). There are several types of elastic waves.

*Compression or Primary (P) waves* are body waves which may propagate along the surface and into the subsurface, returning to the surface by reflection and refraction, or which may travel through the materials as direct waves. *P* waves have the highest velocities $V_p$ and arrive first at the recording instrument.

*Shear (S) waves* are also body waves propagating and traveling in a manner similar to *P* waves. *S* waves travel at velocities $V_s$ from roughly $0.58V_p$ for well-consolidated materials to $0.45V_p$ for poorly consolidated soils. They are not transmitted through water or across air gaps.

*Rayleigh (R) waves* propagate only near the surface as a disturbance whose amplitude atten-

**TABLE 2.8**
**TOOLS AND METHODS FOR SUBSURFACE SECTIONING**

| Category | Applications | Limitations |
|---|---|---|
| | GEOPHYSICAL METHODS | |
| Surface seismic refraction | Determine stratum depths and characteristic velocities, land or water. | May be unreliable unless velocities increase with depth and bedrock surface is regular. Data are indirect and represent averages. |
| Uphole, downhole, and crosshole surveys (seismic direct methods) | Obtain velocities for particular strata; dynamic properties and rock-mass quality. | Data are indirect and represent averages, and may be affected by mass characteristics. |
| Seismic reflection | Not used on land for engineering studies. Useful offshore for continuous profiling. | Does not provide velocities. Computations of depths to stratum changes requires velocity data obtained by other means. |
| Electrical resistivity | Locate saltwater boundaries, clean granular and clay strata; rock depth. | Difficult to interpret and subject to wide variations. Does not provide engineering properties. |
| Gravimeters | Detect major subsurface structures: faults, domes, intrusions, cavities. | Normally used only for cavity information for engineering studies. |
| Magnetometer | Mineral prospecting and location of large igneous masses. | Normally not used in engineering studies. |
| Radar subsurface profiling | Provides subsurface profile; used to locate buried pipe, bedrock, boulders. | In development stages. Does not provide depths or engineering properties. Shallow penetration. |
| Video-pulse radar | Used to locate faults, caverns, voids, buried pipe, general rock structure. | Same as for radar subsurface profiling. |
| Geophysical well logging—see borehole remote sensing and logging | | |
| | RECONNAISSANCE METHODS | |
| Test pits and trenches | Provide visual examination of soil stratigraphy, groundwater and rock depth, fault features. | Limited depth when machine-excavated. Deep excavation below GWL is costly when sheeting and pumping required. |
| Adits and tunnels | Examination of rock quality in situ and access for in situ testing. | Costly for small projects. Normally not used in soil. |
| Bar soundings | Pushed or driven to determine soft soil thickness or depth to shallow rock. | No samples obtained. Strata cannot be identified. |

| Category | Applications | Limitations |
|---|---|---|
| | RECONNAISSANCE METHODS | |
| Retractable-plug sampler | Determine thickness of weak soils. Small samples obtained for identification. | Best for soft soils. Samples are small. |
| Continuous cone penetrometer | Continuous penetration resistance for all but strong soils. Fast and efficient. | Samples not recovered. Cannot penetrate strong soils or rock. |
| Hand augers | Continuous profiling in granular soils above GWL and in clayey soils of firm or greater consistency, locating GWL. | Samples disturbed. Cannot penetrate below GWL in granular soils. Penetration in strong soils very difficult. |
| Bucket auger | Similar to hand auger but greater penetration in strong soils. | Not used in unstable soils or below GWL. |
| | TEST BORINGS | |
| Wash boring | Obtain soil samples primarily for identification and index testing, and SPT testing. | Slow procedure. Cannot penetrate strong soils or rock. UD sampling difficult. |
| Rotary drilling | Obtain samples of all types in soil or rock for identification and laboratory testing of index and engineering properties, and in situ testing. | Requires relatively large and costly equipment. Soil samples and rock cores normally limited to 6-in dia. |
| Rotary probes | Rapidly determined depth to rock with a rotary drill rig. | No samples are obtained. |
| Continuous flight auger | Rapid drilling and disturbed sampling in soils with cohesion and greater than soft consistency. Normal sampling possible if hole remains open. Can penetrate soft rock. | Hole collapses in soft soils, dry granular soils without cohesion, and many soils below GWL. |
| Hollow-stem flight auger | Similar to continuous flight but hollow-stem serves as casing, permitting normal soil sampling. | Cannot penetrate very strong soils, boulders, or rock. |
| Percussion drilling (cable tool) | Usually used to drill water wells. | Large cumbersome equipment. Normal sampling difficult. |
| Hammer drilling | Good penetration in boulders and cobbles. | Large cumbersome equipment. Much soil disturbance results in samples of questionable quality. |
| Wireline drilling | Fast and efficient for deep core drilling on land and offshore borings. | Equipment costly and no more efficient than normal rotary drilling for most land investigations. |

(continued)

**TABLE 2.8**
**TOOLS AND METHODS FOR SUBSURFACE SECTIONING**

| Category | Applications | Limitations |
|---|---|---|
| | BOREHOLE SENSING AND LOGGING | |
| Borehole cameras | Obtain continuous image of borehole in rock showing orientation of faults and joints. Small cavern examination possible when equipped with telephoto lens and spotlights. | Requires open hole. Images are affected by water quality. |
| Rock detector (acoustical sounding) | Differentiates boulders from bedrock and locates bedrock | Results are purely qualitative. |
| Electric well logger | Provides continuous record of resistivity from which material types can be deduced when correlated with test borings. | Generally provides qualitative information. Best used with test boring information. Limited to uncased hole. |
| Scintillometer | Measures gamma rays. Used to locate shale and clay beds and in mineral prospecting. | Data generally of limited engineering use. |
| Gamma-gamma probe | Provides continuous measure of material density. | Value limited to density measurements. |
| Neutron probe | Provides continuous measure of natural moisture content. Has been used with the density probe to locate failure zones in slopes. | Value limited to in situ moisture-content measures. |
| Ultrasonic acoustical devices | Provides continuous image of borehole wall showing fractures and other discontinuities. Can be used to compute dip. | Images are much less clear than those obtained with borehole cameras. |
| 3-D velocity logger | Provides an image of shear and compression waves for a short distance beyond borehole, and reveals fracture patterns. Used with acoustical device for rock quality and with gamma-gamma probe to obtain dynamic elastic properties. | Penetration depth beyond hole wall of a meter or so. Hole diameter must be known accurately and is measured with borehole calipers. |
| Caliper logging | Used to continuously measure and record borehole diameter. | Maximum range about 32 in. |
| Temperature logging | Continuous measures of borehole temperature after fluid has stabilized. | Normally used in deep-hole drilling for petroleum exploration and well development. |
| Tro-pari surveying | Measures borehole inclination and direction in rock. | Relatively deep boreholes in good quality rock. |

uates rapidly with depth, traveling at a velocity approximately $0.9V_s$. The recorded velocity may be less because $R$ waves travel near the surface where lower-velocity material normally occurs, and usually consist of a trail of low-frequency waves spread out over a long time interval.

*Transmission Characteristics*

In a given material the arrival time of each wave at the recording instrument depends on the travel distance between the energy source and the detector, which is in turn a function of the depth to the stratum. In a sequence of strata with successively higher velocities there is a distance between the energy source and the detector at which the refracted wave is transmitted through a higher velocity material and arrives at the detector before the direct or reflected wave. Even though the direct and reflected waves travel shorter distances, they are transmitted at lower velocities.

For land explorations to depths of less than about 300 m, seismic refraction techniques are used rather than reflection because the direct and refracted waves arrive first and tend to mask the reflected waves. Reflection seismology is normally used for deep exploration and for marine studies for profiling, but does not directly yield velocity data as do refraction and direct techniques.

*Seismic Exploration Techniques*

*Refraction* techniques are used to measure compression ($P$) wave velocities in each geologic stratum, which are indicative of type of material and location of the groundwater table, and to estimate depths to various substrata and to indicate the locations of faults and large caverns.

*Direct* techniques provide information on rock-mass characteristics, such as fracture density and degree of decomposition, and on dynamic soil and rock properties including Young's modulus, Poisson's ratio, shear modulus, and bulk modulus (see *Arts. 3.5.3 and 3.5.5*).

*Reflection* techniques are used primarily in marine investigations. They provide a pictorial record of the sea-bottom profile showing changes in strata, salt domes, faults, and marine slides. Since velocities are not directly measured, material types and depths of strata can only be inferred unless correlations are made with other data.

*Energy Sources for Wave Propagation*

*Impact source* (hammer or weight drop), used for shallow explorations on land, tends to generate disproportionately large Rayleigh surface waves, but also produces large $P$ waves, helpful for engineering studies.

*Explosives*, used for land and subaqueous studies, convert a smaller proportion of their energy

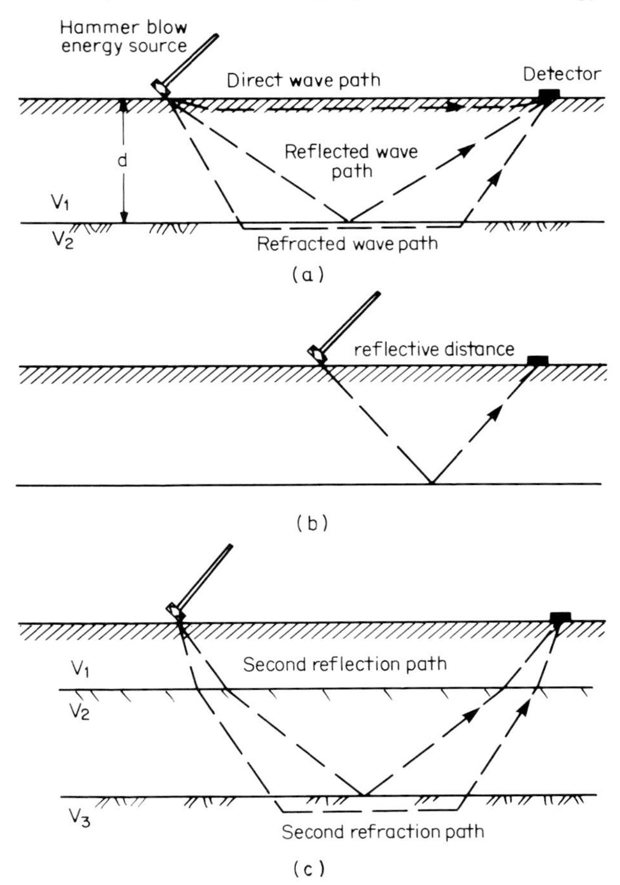


**FIG. 2.22 Transmission paths of direct, reflected, and refracted seismic waves through the shallow subsurface.**

**FIG. 2.23 Single channel refraction seismograph. Hammer striking metal plate causes seismic waves.**

into surface waves, especially when placed at substantial depths below the surface.

*High-energy spark* is used for subaqueous studies.

See also Griffiths and King (1969)[8] and Mooney (1973).[9]

### Seismic Refraction Method

*General*

Seismic refraction techniques are used to measure material velocities, from which are computed depths to changes in strata. Material types are judged from correlations with velocities.

Basic equipment includes an energy source (hammer or explosives); elastic-wave detectors (seismometers) which are geophones (electromechanical transducers) for land exploration or hydrophones (pressure-sensitive transducers) for aqueous exploration; and a recording seismograph which contains a power source, amplifiers, timing devices, and a recorder. Equipment may provide single or multiple recording channels.

The recorded elastic waveforms are presented as seismograms.

*Operational Procedures*

*Single-channel seismograph* operation employs a single geophone set into the ground a short distance from the instrument. A metal plate, located on the ground about 3 m from the instrument, is struck with a sledge hammer (Fig. 2.23). The instant of impact is recorded through a wire connecting the hammer with the instrument. The shock waves travel through the soil media and their arrival times are recorded as seismograms or as digital readouts. The plate is placed alternately at intervals of about 3 m from the geophone and struck at each location with the hammer. Single-channel units are used for shallow exploration in simple geologic conditions.

*Multichannel seismographs* (Fig. 2.24) employ 6 to 24 or more geophones set out in an array to detect the seismic waves which are transmitted and recorded simultaneously and continuously on photographic film or magnetic tape. The energy source is usually an explosive charge set in an auger hole at a shallow depth. The desired depth of energy penetration is a function of the *spread* length (distance between the shot point and the farthest geophone), which should be, in general, 3 to 4 times the desired penetration depth. A normal spread would be about 100 m to investigate depths to 25 or 30 m with geophones spaced at 10-m intervals to define the velocity curves. The geophysicist determines the spread length and the geophone spacing to suit the anticipated geological conditions. In practice a shot is usually set off at one end of the spread and then another at the opposite end (reverse profiling) to detect stratum changes and sloping rock surfaces. At times, charges are set off in the middle of the spread, or at other locations. Multichannel units are used for deep exploration and all geologic conditions.

*Seismograms*

The seismic waveforms are usually recorded on photographic paper as seismograms. In Fig. 2.25, the $P$ wave, traveling at the highest velocity, is the *first arrival* to be recorded and is easily recognized. It is used to determine the depths to the various strata on the basis of their characteristic transmission velocities.

The $S$ wave appears later in the wave train as a large pulse and often is difficult to recognize. On the figure it is observed crossing the spread at an intermediate angle from the first arrivals, indicating a lower velocity. $S$ wave velocities are used in conjunction with $P$ wave velocities to compute the dynamic properties of the transmitting media *(see Art. 3.5.3).*

The Rayleigh wave ($R$) appears as a large-amplitude, low-frequency signal arriving late on the train. On the figure it crosses the spread at an angle larger than the $S$ wave and leaves the record at about the 200-ft spread. Although easy to recognize, the beginning is essentially indeterminate and does not provide much information for engineering studies.

*Time-Distance Charts and Analysis*

A typical time-distance chart of the first arrivals obtained with a single-unit seismograph is given as Fig. 2.26.

As the distance from the geophone to the shot point is increased, eventually the shock waves will have sufficient time to reach the interface between media of lower and higher velocities, to be refracted and travel along the interface at the higher velocity, and to arrive before the direct and reflected waves traveling through the shallower, lower-velocity material.

The travel times of the first arrivals are plotted versus the distance from the geophone and the velocity of the various media determined from the slopes of the lines connecting the plotted points as shown on the figure.

Various formulas are available for computing the depth to the interfaces of the various layers, varying from simple to complex depending on the number of layers involved and the dip of the beds. The formulas for computing the depths of

**FIG. 2.24 A 24-channel SIE refraction seismograph and geophones before studies at a nuclear power plant site. Only 12 channels are set up in the photo.**

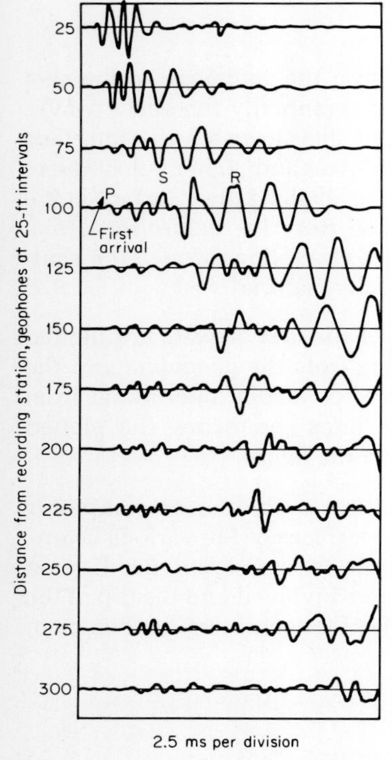


**FIG. 2.25** **(*Left*) Seismograms of waveforms recorded at 25-ft intervals as caused by a 300-lb weight drop.**[9]

**FIG. 2.26** **(*Below*) Time-distance graph and the solution to a three-layer problem.**

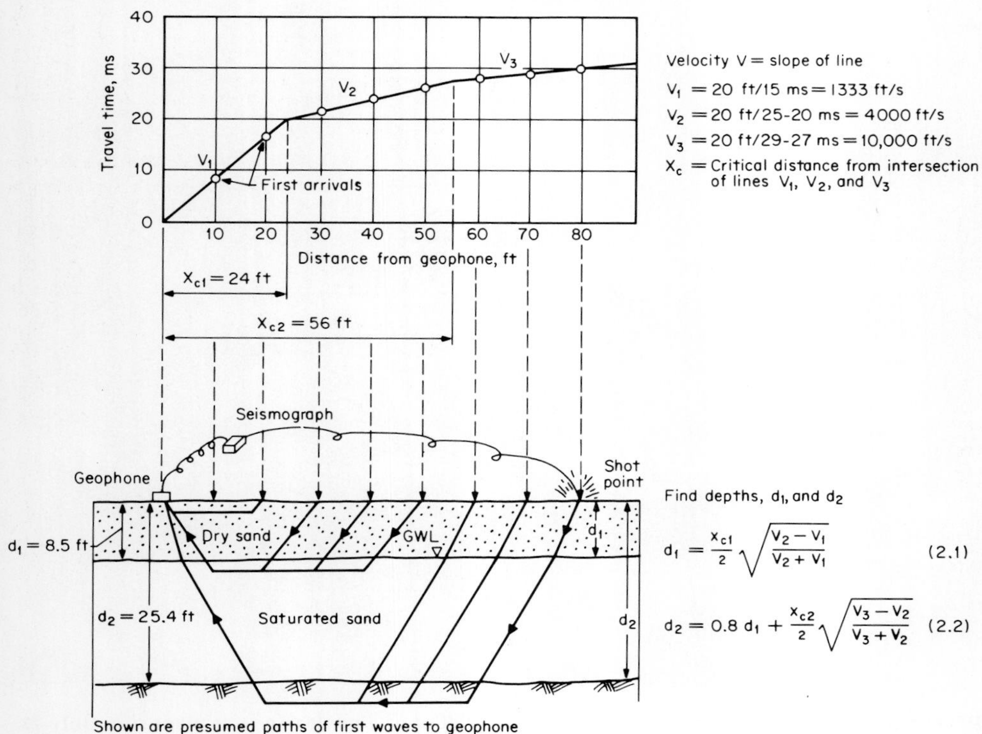


Velocity V = slope of line

$V_1$ = 20 ft/15 ms = 1333 ft/s

$V_2$ = 20 ft/25-20 ms = 4000 ft/s

$V_3$ = 20 ft/29-27 ms = 10,000 ft/s

$X_c$ = Critical distance from intersection of lines $V_1$, $V_2$, and $V_3$

Find depths, $d_1$, and $d_2$

$$d_1 = \frac{X_{c1}}{2}\sqrt{\frac{V_2 - V_1}{V_2 + V_1}} \tag{2.1}$$

$$d_2 = 0.8\,d_1 + \frac{X_{c2}}{2}\sqrt{\frac{V_3 - V_2}{V_3 + V_2}} \tag{2.2}$$

the relatively simple three-layer problem are given on Fig. 2.26 (Eqs. 2.1 and 2.2). For interpretation some information on topography must be available.

Actual seismograms for three shots along the same spread (each end and the middle) are given on Fig. 2.27, and the time-distance plots on Fig. 2.28. The example, a three-layer problem in a residual soil profile, is from a continuous profiling study for a railroad. The resulting subsurface section is given as Fig. 2.29.

*Limitations*

- Softer, lower-velocity material will be masked by an overlying denser, higher-velocity material and cannot be directly disclosed.
- A stratum with a thickness of less than about one-fourth the depth from the ground surface to the top of the stratum cannot be distinguished.
- Erratic or "average" results are obtained in boulder formations, areas of irregular bedrock surfaces, or rock with thin, hard layers dipping in softer rock.
- Well-defined stratum interfaces are not obtained where velocity increases gradually with depth, as in residual soil grading to weathered to sound rock.
- In frozen ground the shot point and geophones must be below the frozen zone because the shock waves travel much faster through frost than through the underlying layers.
- Application in urban areas is limited because of utility lines, pavements, foundations, and extraneous noise sources.

*Applications*

The method is most suitable as an exploration tool where there are media having densities that increase distinctly with depth and fairly planar interface surfaces. In such instances it can economically and efficiently provide a general profile of geologic conditions (Fig. 2.29).

Material types are estimated from computed $P$ wave velocities. Typical velocities for many types of materials are given on Table 2.9, and for weathered and fractured igneous and metamorphic rocks on Table 2.10.

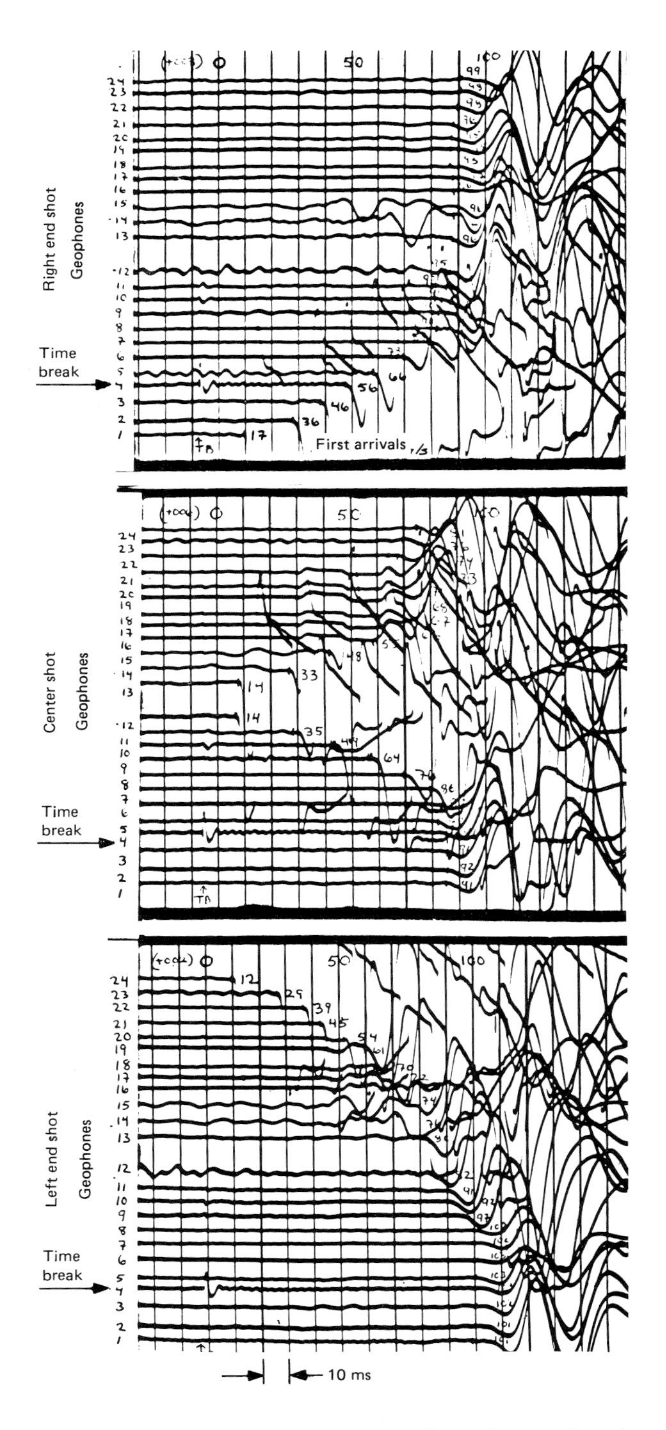


**FIG. 2.27 Seismograms for one 24-channel spread with three shot points.**

Velocity data are used also to estimate rock rippability *(see Table 3.7)*.

## Seismic Direct Methods

*Applications*

Seismic direct methods are used to obtain data on the dynamic properties of rocks and soils *(see Chap. 3)*, and to evaluate rock-mass quality *(see Art. 5.2.7)*. See also Auld (1977),[10] Ballard (1976),[11] and Dobecki (1979).[12]

*Techniques (Fig. 2.30)*

UPHOLE SURVEYS The geophones are laid out on the surface in an array, and the energy source is set off in an uncased mud-filled borehole at successively decreasing depths starting at the bottom of the hole. The energy source is usually either explosives or a mechanical pulse instrument composed of a stationary part and a hammer.

DOWNHOLE SURVEYS The energy source is located on the surface and the detectors incorporated in a sonde which is raised or lowered in the borehole to give either a continuous or intermittent log of adjacent materials.

CROSSHOLE SURVEY The energy source is located in a center test boring and the detectors are placed at the same depth as the energy source in a number of surrounding boreholes.

*Advantages over Refraction Surveys*

In the uphole and downhole methods the influence of reflection and refraction from the layers surrounding the layer of interest is substantially reduced.

In the crosshole method the influence of surrounding layers is eliminated (unless they are dipping steeply) and the seismic velocities are measured directly for a particular stratum. It is the dominant technique.

## Seismic Reflection Method

*Application*

Seismic reflection surveys obtain a schematic

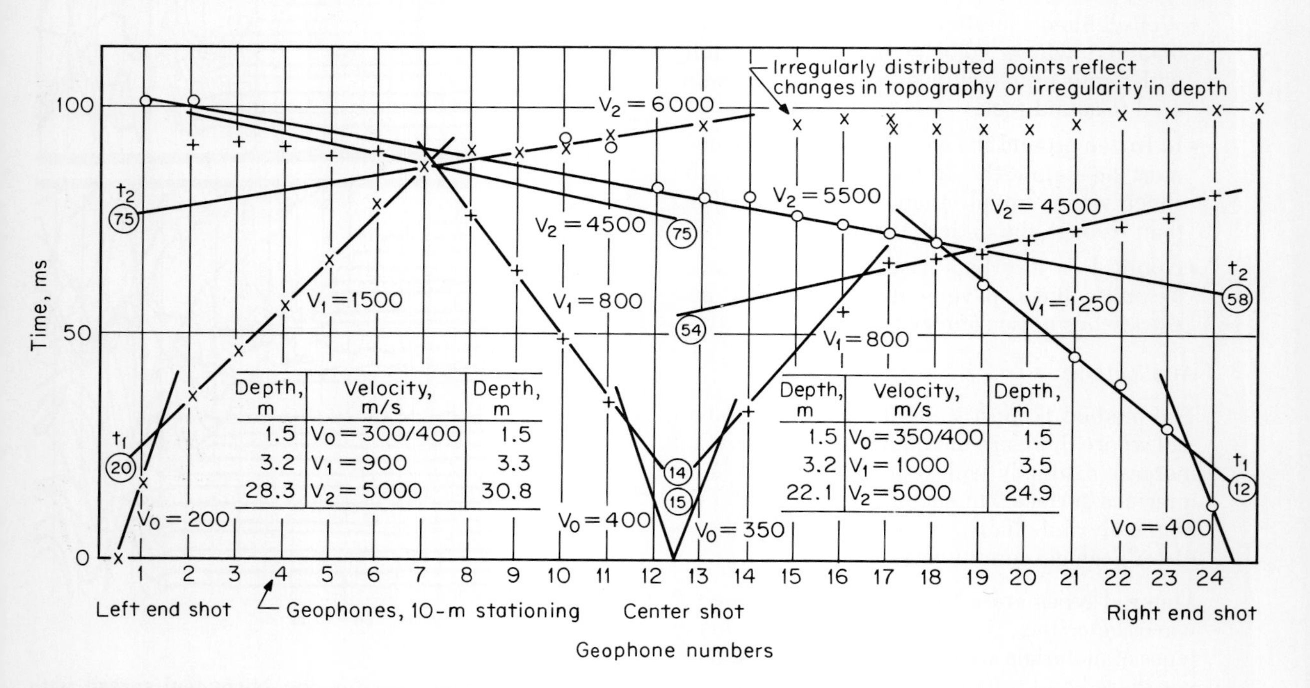


**FIG. 2.28 Time-distance graphs for the seismograms of Fig. 2.27.**

representation of the subsurface in terms of time and, because of their very rapid accumulation of data over large areas, are used in engineering studies primarily offshore.

For marine surveys reflection methods are much more rapid than refraction surveys and, when obtained from a moving vessel, provide a continuous image of subseafloor conditions.

*Operational Procedures*

*Continuous marine profiling* is usually performed with an electric-sonic energy source generating continuous short-duration pulses, while towed behind a vessel, in conjunction with a hydrophone which detects the original pulses and reflected echo signals. The output of the hydrophone is amplified and passed to a recorder which transcribes each spark event to sensitized paper, resulting directly in a pictorial section of the seafloor as shown on Fig. 2.31.

*Equipment* in use is generally of two types:

- "Boomers" operate in water depths from 3 to 200 m and provide high resolution (range 15 to 30 cm) but moderate penetration (about 100 m). Systems are available to provide various energies and frequencies; the selection depends on the penetration and resolution desired.
- "Sparkers" operate in water depths generally from 10 to 600 m with resolution capabilities of about 15 to 25 m and with penetration depths of about 1200 m or more, depending on the energy selection.

*Interpretive Information Obtained*

A pictorial section beneath the seafloor is obtained, showing the general stratigraphy, and such features as slumps and gas pockets (Fig. 2.32) when the resolution is high.

Deeper penetration methods, such as sparker surveys, can indicate major geologic structures such as faults and salt domes (Fig. 2.33) and massive submarine slides *(see Fig. 9.56)*.

Velocities cannot be calculated with reliability since distances are not accurately known, and therefore material types and stratum depths

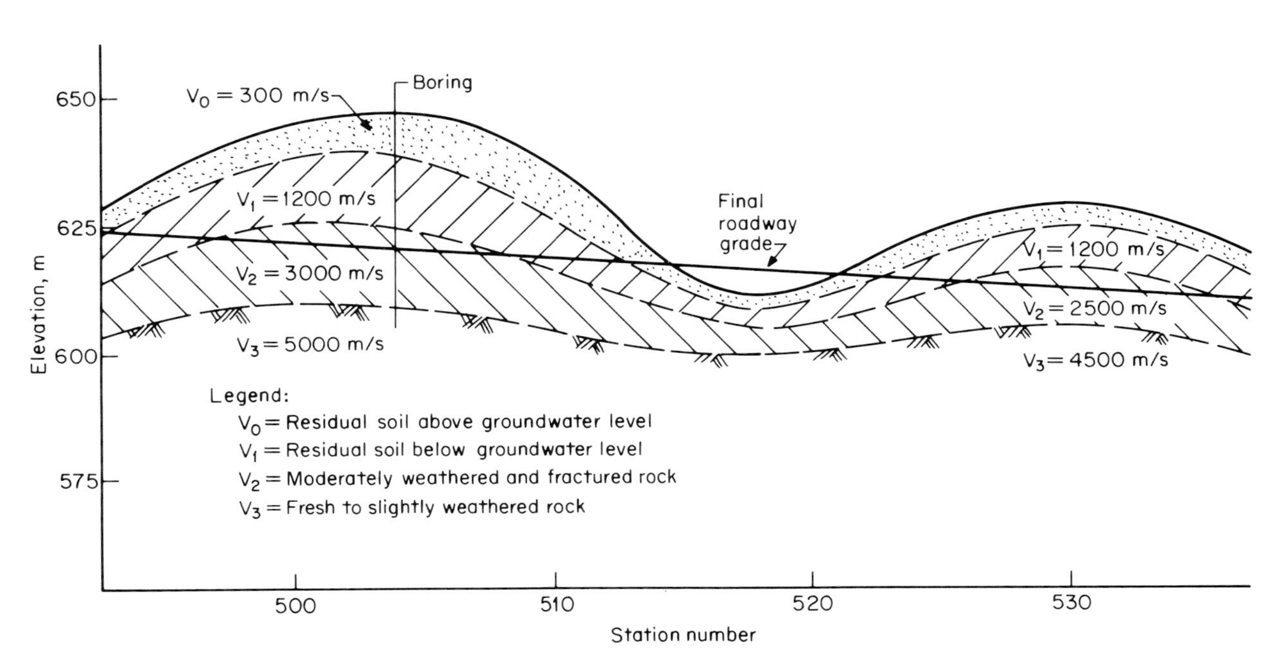


**FIG. 2.29 Example of a subsurface section prepared from a continuous refraction profile. Velocities are correlated with rippability as shown on Table 3.7.**

**TABLE 2.9**
**TYPICAL COMPRESSION-WAVE VELOCITIES IN SOILS AND ROCKS**

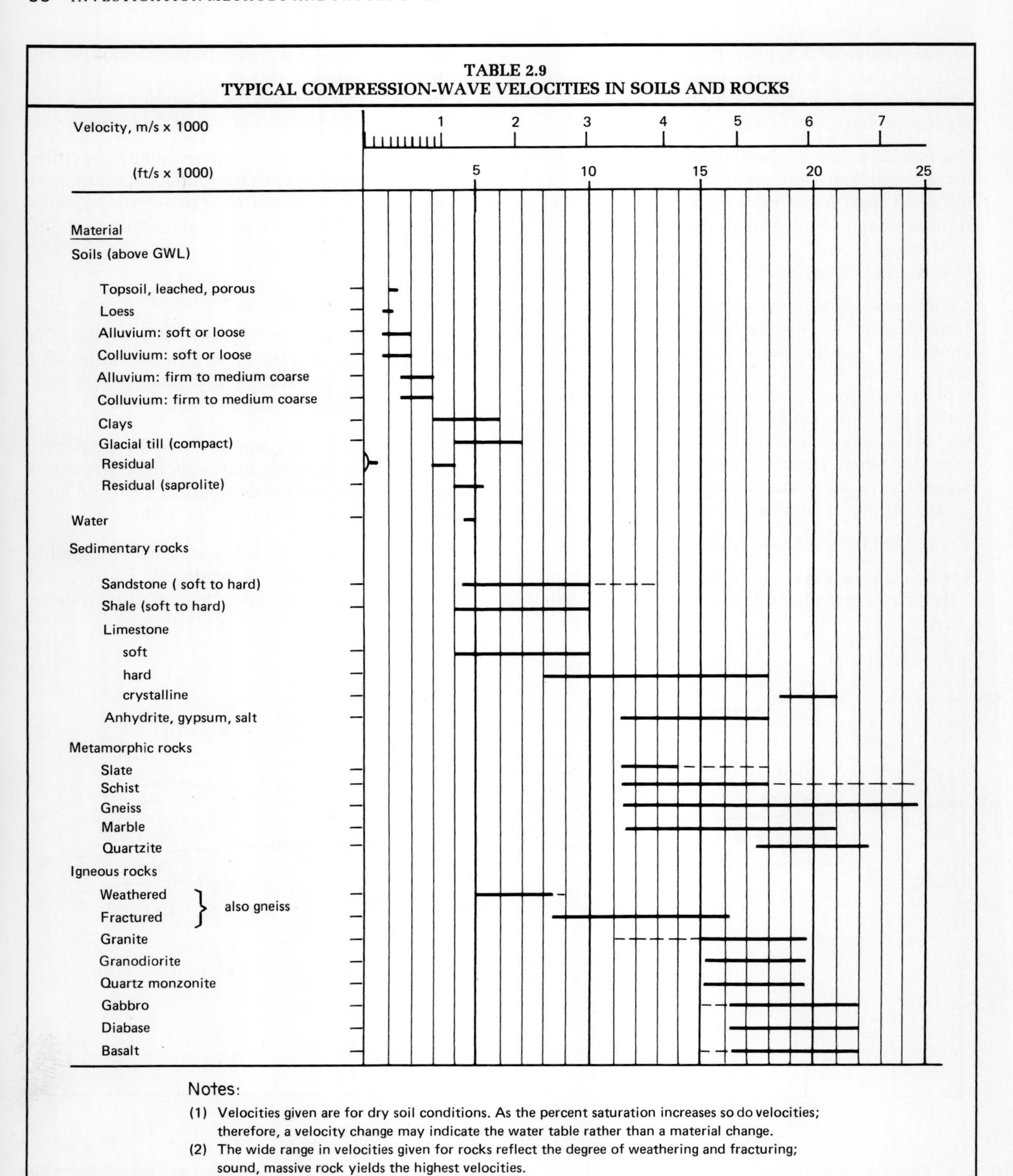


Notes:

(1) Velocities given are for dry soil conditions. As the percent saturation increases so do velocities; therefore, a velocity change may indicate the water table rather than a material change.

(2) The wide range in velocities given for rocks reflect the degree of weathering and fracturing; sound, massive rock yields the highest velocities.

**TABLE 2.10**
**TYPICAL *P*-WAVE VELOCITIES OF WEATHERED AND FRACTURED IGNEOUS AND METAMORPHIC ROCKS**

| Material | Grade* | $V_p$, m/s |
|---|---|---|
| Fresh, sound rock | F | 5000+ |
| Slightly weathered and/or widely spaced fractures | WS | 5000–4000 |
| Moderately weathered and/or moderately close fractures | WM | 4000–3000 |
| Strongly weathered and/or close fractures | WH | 3000–2000 |
| Very strongly weathered (saprolite) and/or crushed | WC | 2000–1200† |
| Residual soil (unstructured saprolite), strong | RS | 1200–600† |
| Residual soil, weak, dry | RS | 600–300† |

*See Table 5.21.

†$V_p$ (water) ≈ 1500 m/s, $V_p$ (min) of saturated soil ≈ 900 m/s.

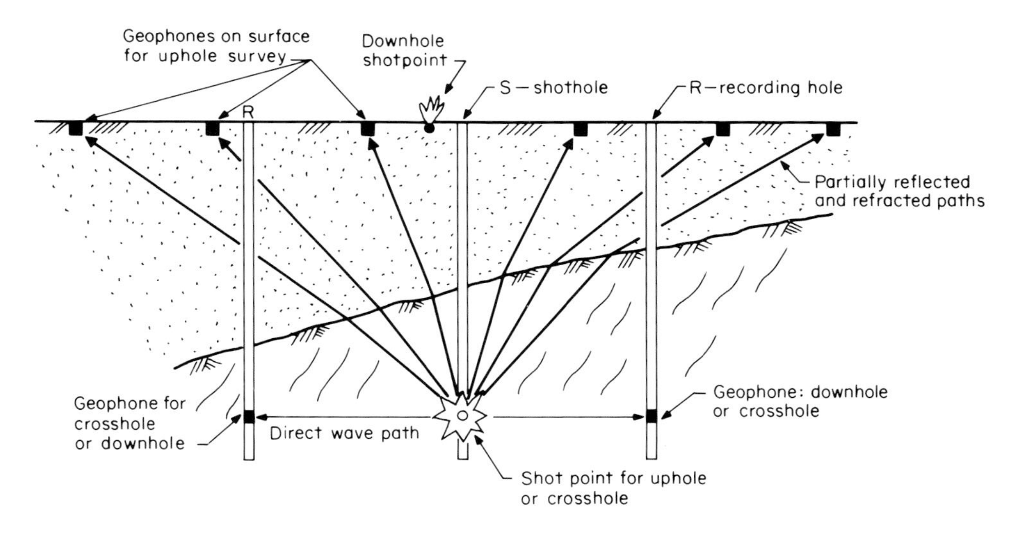


**FIG. 2.30 Direct seismic methods to measure dynamic properties of soils and rocks, and assess rock mass quality. A single borehole is used in the uphole or downhole survey; array of usually four borings is used in the crosshole survey. In uphole surveys the geophones should be set on rock, if possible, to obtain measurements of rock quality.**

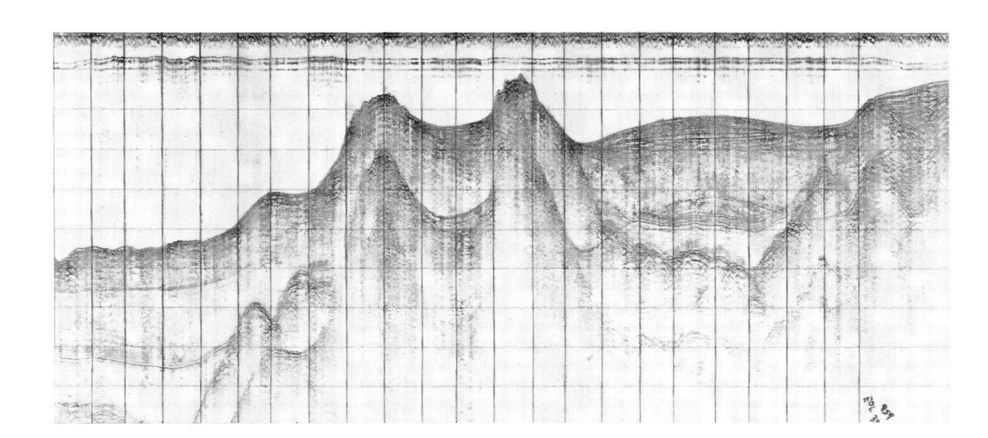

**FIG. 2.31 Example of a Boomer survey record.** *(Courtesy of EG&G.)*

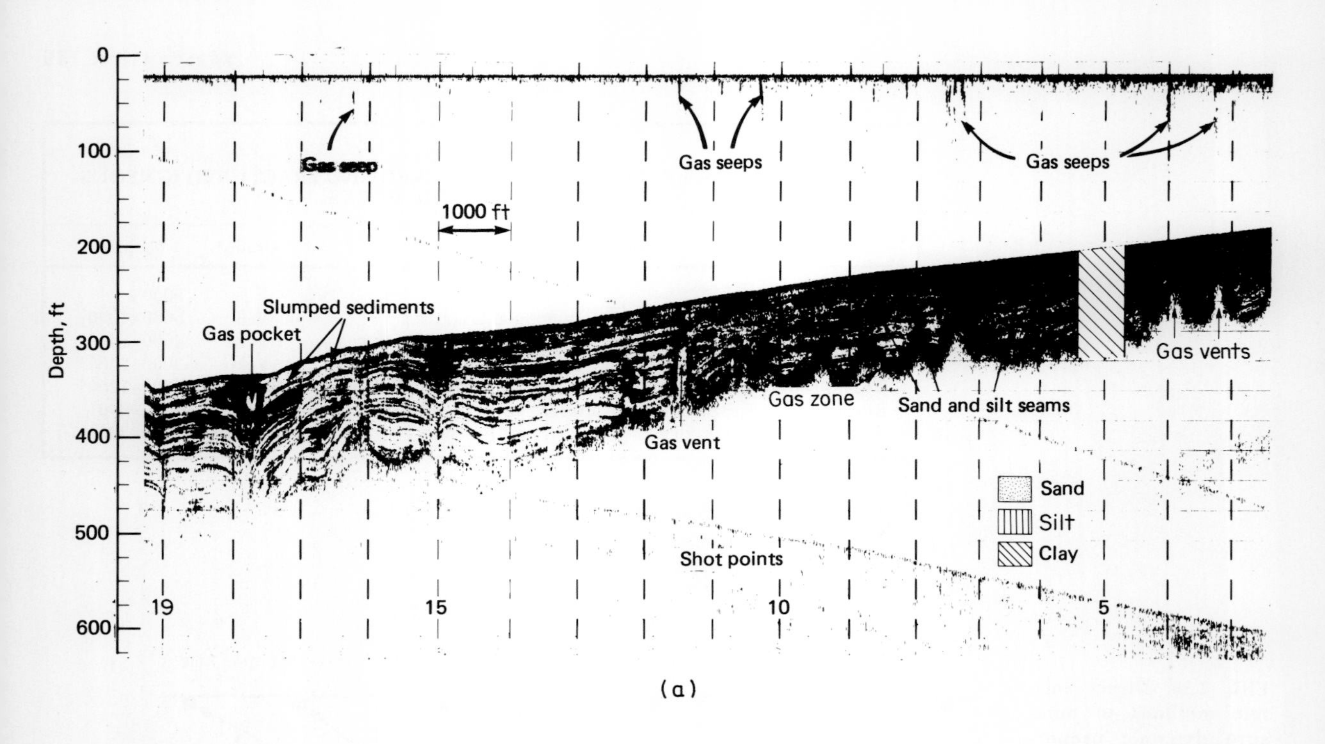


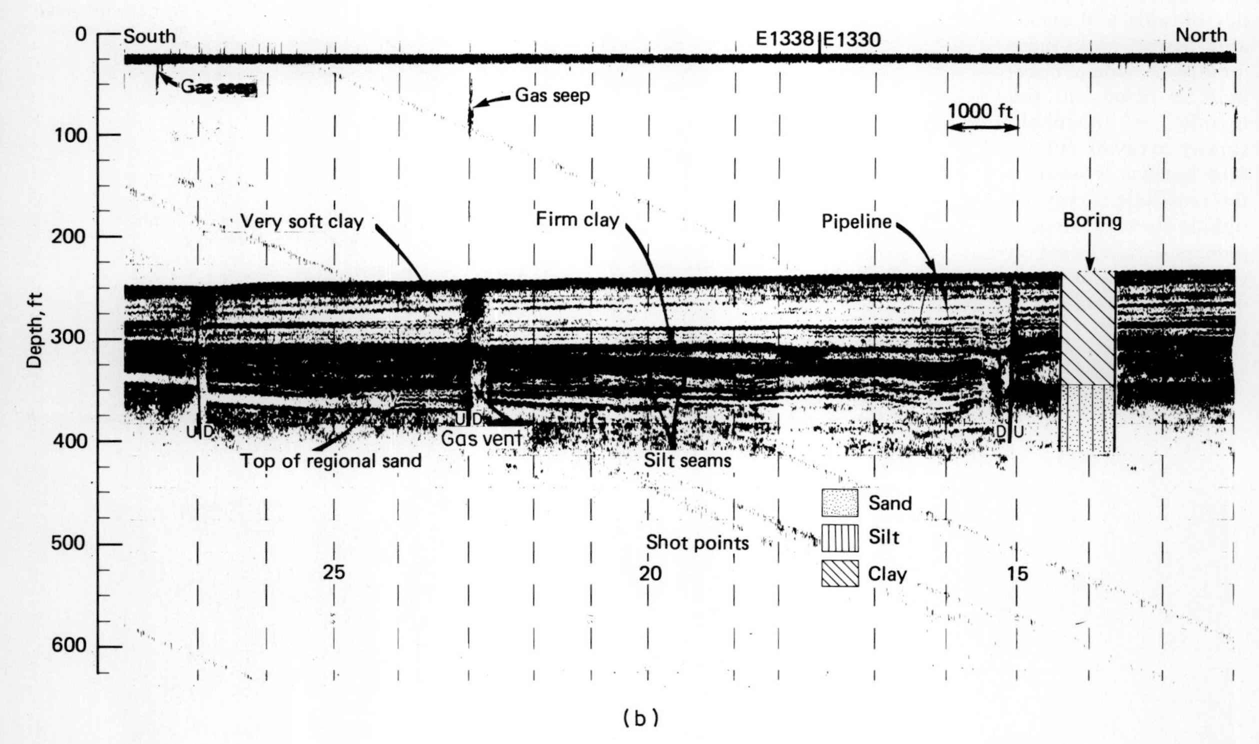


**FIG. 2.32** Examples of Boomer profiles from the Gulf of Mexico. Note gas pockets and vents.

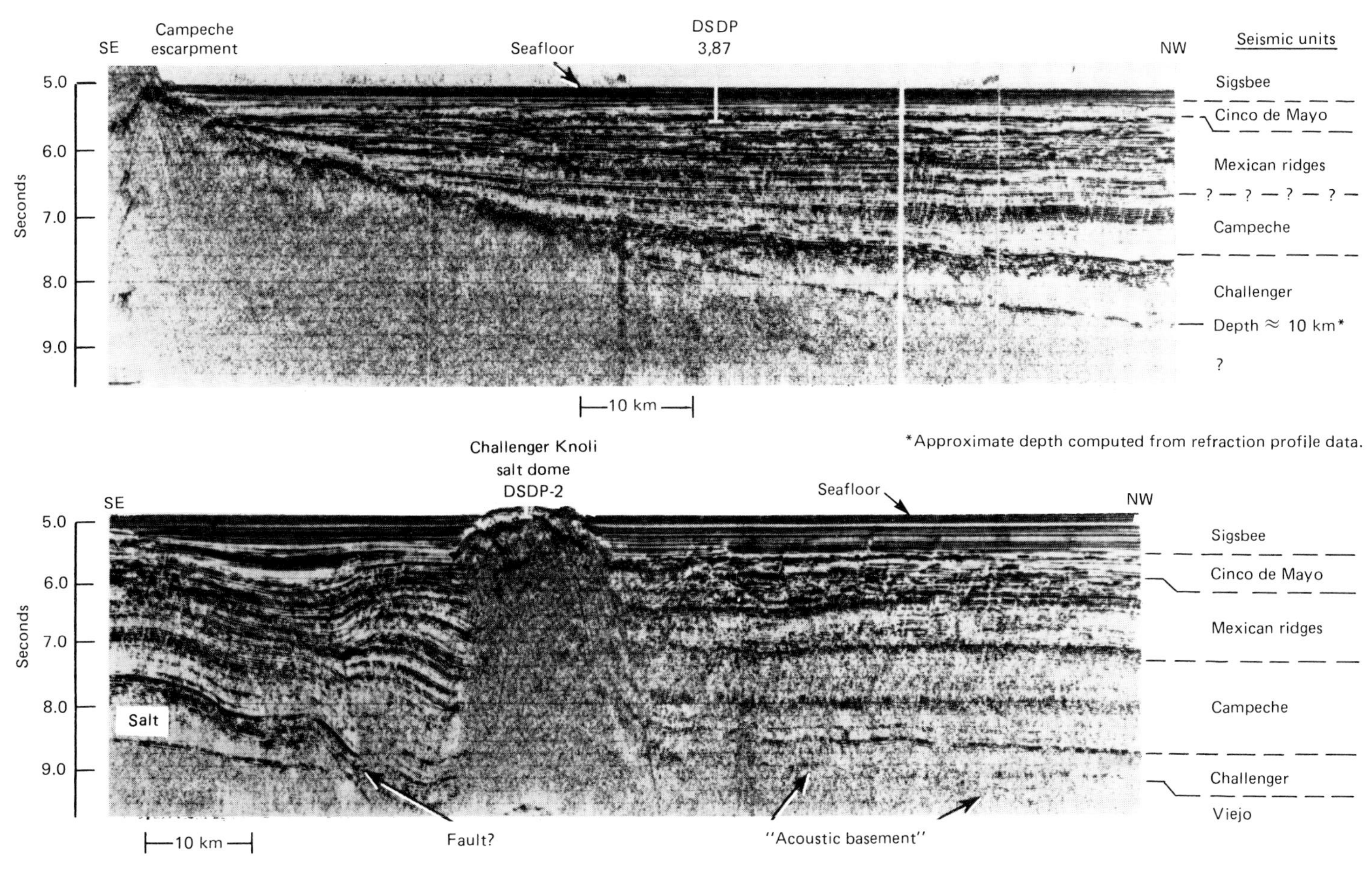


**FIG. 2.33 Part of multichannel seismic reflection section, abyssal western Gulf of Mexico.** [*From Ladd, et al. (1976).*[13]]

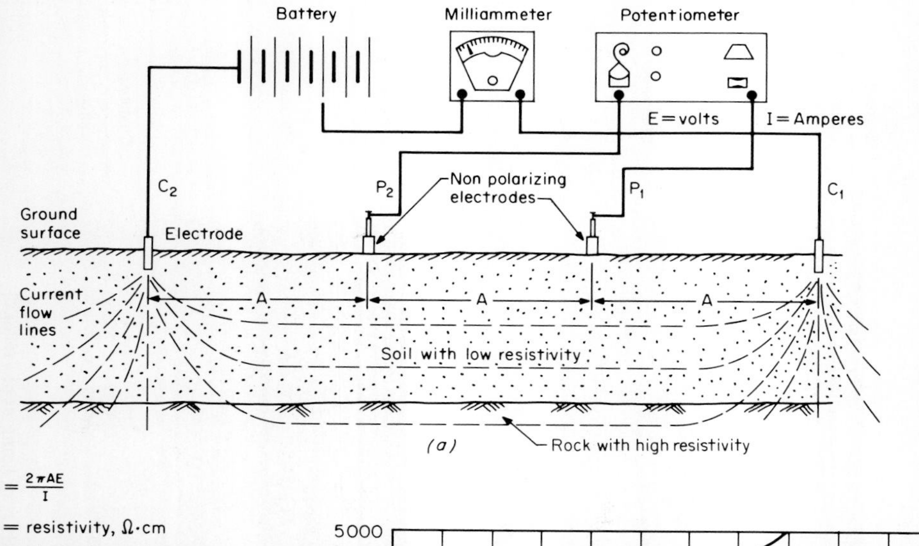


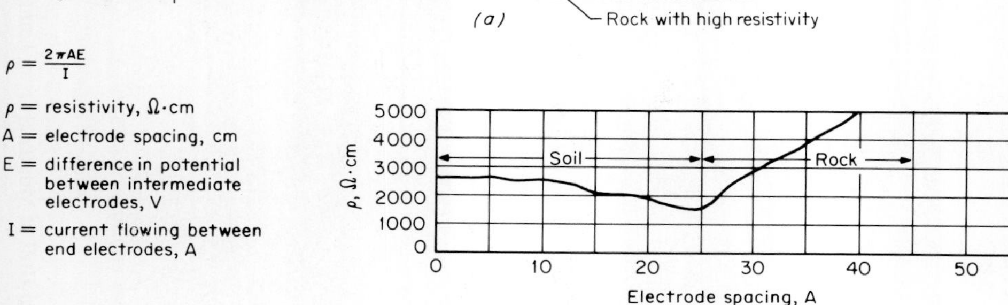


**FIG. 2.34** **(a) Components of the electrical resistivity apparatus and the common four-electrode configuration of the Wenner array. (b) typical resistivity curve.** [*From ASTM (1951).*[14] *Adapted with permission of the American Society for Testing and Materials.*]

cannot be evaluated as they can with refraction methods. Depths are estimated by assuming a water velocity of 1500 m/s but variations in strata impedance affect the thickness scale. Test borings or refraction studies are necessary for depth and material-type determinations.

## Electrical Resistivity Methods

*Applications*

- Differentiation between clean granular materials and clay layers for borrow-material location
- Measurement of the thickness of organic deposits in areas with difficult access
- Measurement of the depth to a potential failure surface in "quick" clays in which the salt content, and therefore the resistivity, is characteristically different near the potential failure surface *(see Art. 9.5.2)*
- Location of subsurface saltwater boundaries
- Identification of variations in groundwater quality in homogeneous granular deposits, as may be caused by chemical wastes leaking from a storage basin
- Measurement of depth to bedrock
- Location of solution cavities in limestone (not always successful)

NOTE: Measurements in all cases given above are often only approximations.

*Theoretical Basis*

Various subsurface materials have characteristic conductances for direct currents of electricity. Electrolytic action, made possible by the presence of moisture and dissolved salts within the soil and rock formations, permits passage of current between electrodes placed in the surface soils. In general, conductance is good in such materials as moist clays and silts, and poor in such materials as dry loose sands, gravels, and sound rocks.

*Resistivity* refers to the resistance to current flow developed in geologic materials and is expressed as ohm-centimeter squared per cen-

**TABLE 2.11**
**TYPICAL RESISTIVITY VALUES FOR GEOLOGIC MATERIALS***

| Material | Resistivity | |
|---|---|---|
| | **Ohm-feet** | **Ohm-meters** |
| Clayey soils: wet to moist | 5–10 | 1.5–3.0 |
| Silty clay and silty soils: wet to moist | 10–50 | 3–15 |
| Silty and sandy soils: moist to dry | 50–500 | 15–150 |
| Bedrock: well-fractured to slightly fractured with moist soil-filled cracks | 500–1000 | 150–300 |
| Sand and gravel with silt | About 1000 | About 300 |
| Sand and gravel with silt layers | 1000–8000 | 300–2400 |
| Bedrock: slightly fractured with dry soil-filled cracks | 1000–8000 | 300–2400 |
| Sand and gravel deposits: coarse and dry | >8000 | >2400 |
| Bedrock: massive and hard | >8000 | >2400 |
| Freshwater | 67–200 | 20–60 |
| Seawater | 0.6–0.8 | 0.18–0.24 |

*From Soiltest, Inc.

NOTES: 1. In soils resistivity is controlled more by water content than by soil minerals.
2. The resistivity of the pore or cleft water is related to the number and type of dissolved ions and the water temperature.

timeter, or simply as ohm-centimeters or ohm-feet. Some typical values of resistivity for various geologic materials are given on Table 2.11.

*Apparatus*

The electrical resistivity apparatus consists of a battery as energy source, a milliammeter, a potentiometer, and electrodes (Fig. 2.34). There are two basic electrode configurations.

WENNER ARRAY Commonly used in the United States, it employs four equally spaced electrodes as shown on Fig. 2.34.

SCHLUMBERGER ARRAY Commonly used in Europe, it is similar to the Wenner, except that the spacing of the two center electrodes is made a small portion of that between the other two.

*Operational Procedures*

With a battery as a direct-current source, a current flow is established between the two outer electrodes. The current drop is detected by the two inner electrodes and recorded on the potentiometer, and the "apparent" resistivity (Wenner array) is computed from the expression:

$$\rho = \frac{2\pi AE}{I} \tag{2.3}$$

where $\rho$ = soil resistivity
$A$ = distance between electrodes in centimeters
$E$ = differential potential between intermediate electrodes in volts
$I$ = current flowing between end electrodes in amperes.

"Apparent" resistivity signifies an average value resulting from layering effects.

In *vertical profiling* the electrode spacing is increased while the resistivity changes are recorded, and a curve of resistivity vs. electrode spacing is drawn as shown on Fig. 2.34. As the value of resistivity obtained is largely dependent upon material resistivities to a depth equal to the electrode spacing $A$, material changes can be inferred from the change in slope of the curve. For any given depth, in terms of electrode spacing, the *lateral variations* in resistivity are measured by moving the rear electrode to a front position and marching the array laterally. Subsurface conditions are inferred from the variations in vertical and lateral values. In multilayered systems, interpretations must be correlated with test borings.

To interpret vertical profiling, a set of empirical curves (Wetzel-Mooney curves) is used to estimate the depth to an interface and the resistivity. The curves, log-log plots of apparent resistivity vs. electrode separation, are matched to the curves drawn from the field data.

*Limitations*

Since resistivity is a function of water content and soluble salts, materials with widely differing engineering properties can have the same resistivity. Therefore, correlations from one location to another may not be possible. Differentiation between strata may not be possible where the overlying material has an extremely high resistance.

Water-table location often limits the depth for practical study because conductivity rises sharply in saturated materials and makes differentiation between horizons impossible.

Because of the difficulties of relating measured resistivity values to specific soil or rock types, subsurface conditions are usually inferred from the vertical and lateral variations in the measured values. In multilayered systems, interpretations must be confirmed by correlations with test boring data, and in general, electrical resistivity should be considered as a preliminary exploration method.

**Magnetometer Surveys**

*Applications*

Magnetometer surveys are used for detection of magnetic ore bodies or rocks that are strongly magnetic, such as the crystalline types as differentiated from sedimentary types. They are seldom used for engineering studies.

*Theoretical Basis*

Many rocks contain small but significant quantities of ferromagnetic minerals which vary with rock type. The weak magnetization modifies the earth's magnetic field to an extent that can be detected by sensitive instruments.

*Operational Procedure*

Magnetometers provide the measurements, and when towed behind aircraft they can cover large areas and provide appreciable data in a relatively short time.

*Data Presentation*

Contour maps are prepared showing lines of equal values which are qualitatively evaluated to identify anomalies indicative of ore bodies or rock type changes.

**Gravimeter Surveys**

*Applications*

Gravimetric surveys in their normal geologic application are used for detection of major subsurface structures such as faults, domes, anticlines, and intrusions. Gravimetric surveys have been used during engineering studies to detect cavities in limestone.

*Theoretical Basis*

Major geologic structures impose a disturbance on the earth's gravitational field. That part of the difference between measured gravity and theoretical gravity which is a result purely of lateral variations in *material density* is known as the Bouguer anomaly. Other factors affecting gravity are latitude, altitude, and topography. They require consideration during gravitational measurements to obtain the quantity representing the Bouguer anomaly.

*Gravimeters* consist of spring-supported pendulums similar in design to a long-period seismograph *(see Art. 11.2.3)*.

*Data Presentation*

Isogal maps are prepared showing contours of similar values given in milligals (mgal) to illustrate the gravity anomalies. (NOTE: 1 mgal = 0.001 gal; 1 gal = an acceleration of gravity = 1 cm/s$^2$.)

**TABLE 2.12**
**MATERIAL TRANSPARENCY TO GROUND-PROBING RADAR***

| Transparency | Material |
|---|---|
| High | Glacial ice and rock salt |
| Good | Metamorphic rocks, limestone, and dune sands |
| Medium | Most coarse-grained soils, bituminous coals, oil shales, concrete above grade, lakes, and river water |
| Poor | Wet clays and shales, seawater |
| Opaque | Metallic objects |

*From Cook (1974).[16]

*Cavity Exploration in Soluble Rock (Art. 10.4.4)*

The bulk density of limestones is about 2.6 g/cm$^3$, and that of soils generally ranges from 1.6 to 2.0 g/cm$^3$. In karst regions a gravity-low anomaly may indicate an empty cavity, a cavity filled with low-density material, or a change in soil or groundwater conditions. Microgravimetric instruments have been developed in recent years which permit a precision of 0.01 mgal (10 $\mu$gal) or better [Greenfield (1979)[15]], equivalent to a change in soil thickness of 24 cm for a density contrast between soil and rock of 1.0 g/cm$^3$. A detected anomaly is then explored with test borings.

## Ground-Probing Radar

*Applications*

Although still in the development stage, ground-probing radars appear to have important potential as a rapid method of subsurface profiling. They are designed to probe solids relatively opaque to radar waves, such as pavement reinforcing rods and base, as well as subpavement voids, buried pipes, the bedrock surface and overlying boulders, caverns, tunnels, clay zones, faults, and ore bodies.

*Theoretical Basis*

Energy is emitted in the radio portion of the electromagnetic spectrum (*see Fig. 2.7*), of which some portion is reflected back to the radar equipment. Various materials have differing degrees of transparency to radar penetration [Cook (1974)[16]], as given on Table 2.12.

*Techniques*

*Single-pulse radar wave* is applied directly to the ground surface, or underground in mines, tunnels, and boreholes. The radar wave is returned and registered as a video pulse, which appears similar to a seismic refraction wave. Irregularities on the wave train indicate a reflector such as a saturated clay-filled fault zone in crystalline rocks or a cavern or void [Moffatt (1974)[17]].

*Continuous subsurface profiling by impulse radar,* also termed electromagnetic subsurface profiling (ESP), provides a registration of a continuously reflected radar pulse similar to seismic reflection images. A sled-mounted antenna is towed behind a small vehicle or boat containing the ESP system. It has been used since 1970 [Morey (1974)[18]] to locate buried sewer lines and cables, evaluate pavement condition, detect voids beneath floors, and profile the bottoms of rivers and lakes. The ESP technique provides clear images in low-conductivity materials such as sand, freshwater, or rock, and poor results in high-conductivity materials such as wet clay because the penetration depth is limited by the strong attenuation of the signal.

### 2.3.3 RECONNAISSANCE METHODS

## General

Reconnaissance methods of exploration are divided into two general groups as follows:

- *Large excavations* allow close examination of

the geologic materials and include test pits, test trenches, and large-diameter holes which can be made relatively rapidly and cheaply; and adits and tunnels, which, although costly to excavate, are valuable for investigating rock-mass conditions.

- *Hand tools and soundings* provide low-cost and rapid means of performing preliminary explorations. Samples are recovered with the hand auger and 1-inch retractable plug sampler, and probings with bars provide indications of penetration resistance to shallow depths. The cone penetrometer test (CPT) *(see Art. 2.3.4.)* and the standard penetration test (SPT) *(see Art. 2.4.2)* are used also as reconnaissance methods for preliminary explorations.

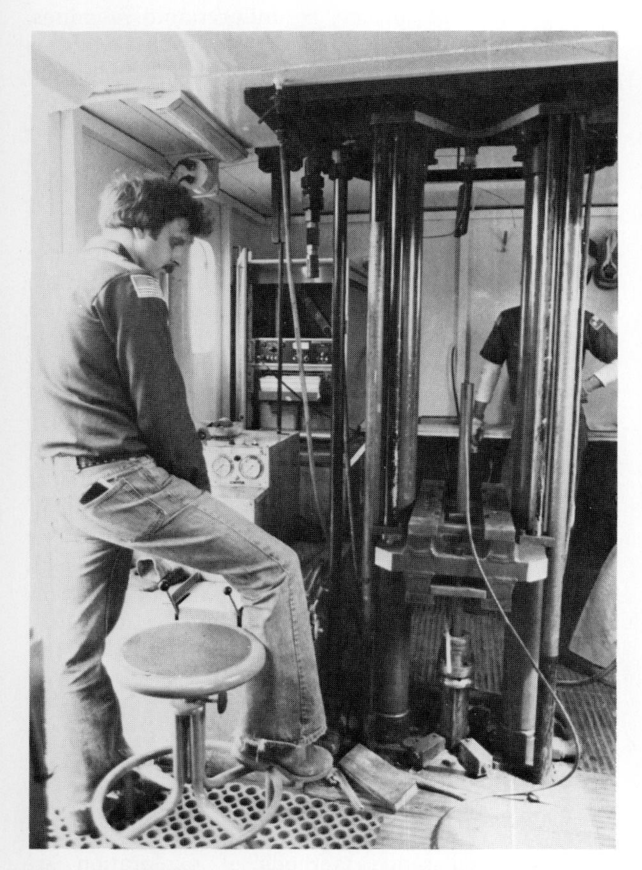

**FIG. 2.35 Interior of Fugro truck-mounted penetrometer rig with 15-ton thrust capability.** *(Courtesy of Fugro Gulf, Inc.)*

**Methods**

- Test pit or trench excavation—see Table 2.13.
- Large-diameter holes—see Table 2.13.
- Adits and tunnels—see Table 2.13.
- Bar soundings—see Table 2.13.
- Hand auger or posthole digger—see Table 2.13.
- One-inch retractable plug sampler—see Table 2.13.
- Continuous penetrometer test—see Table 2.13 and Art. 2.3.4.
- Standard penetration test—see Table 2.13 and Art. 2.4.2.

### 2.3.4 CONTINUOUS CONE PENETROMETER TEST (CPT) (ASTM D-3441)

**General**

*Operation*

Rods with a cone tip are forced into the ground while the resistance to penetration is recorded continuously. No samples are recovered. The subsurface profile is inferred from correlations. Commonly used in Europe, the method is finding increased acceptance in the United States.

*Types of Force Application*

- *Dynamic*, wherein the rods are driven by a hammer
- *Static*, wherein the rods are pushed as deadweights are applied
- *Quasi-static*, wherein the rods are pushed hydraulically by reaction against a machine. The apparatus can be mounted in a truck (Fig. 2.35) or in a light portable unit (Fig. 2.36) which is anchored to the ground to provide the reaction. Often referred to as the static cone penetrometer test, the quasi-static method is currently the most advanced and the most commonly used.

*Applications*

The method permits rapid and economical exploration of thick deposits of weak to moderately strong soils and provides detailed information on soil stratification.

Measurements are provided of engineering properties *(see Art. 3.4.5)*. In relatively permea-

**TABLE 2.13**
**RECONNAISSANCE METHODS OF EXPLORATION**

| Method | Applications | Procedure | Limitations |
|---|---|---|---|
| Test pit or trench excavation | Detailed examination of soil strata.<br>Observation of groundwater seepage.<br>Identification of GWL.<br>Recovery of disturbed or undisturbed samples above GWL and in situ density tests.<br>Examination of fault zones.<br>Examination of miscellaneous and rubble fills.<br>Identification of rock surface and evaluation of rippability.<br>Borrow material investigations. | Excavation by backhoe, bulldozer, or by hand.<br>Can be extended below GWL by sheeting and pumping if soils have at least some cohesion. | Usually limited in depth by water table (GWL), rock depth, or reach of equipment.<br>Can be dangerous if left unsheeted and depths are over 4 to 5 ft. |
| Large-diameter holes | Detailed examination of strong cohesive soil strata and location of slickensides and other details affecting stability and seepage. | Holes 60 to 100 cm in diameter excavated by rotating large auger bucket (Fig. 2.54), or excavated by hand. | Strong cohesive soils with no danger of collapse.<br>Rock penetration limited except by calyx drilling *(Art. 2.4.1)*. |
| Adits and tunnels | Used in rock masses for preparation of detailed geological sections and in situ testing; primarily for large dams and tunnels. | Excavation by rock tunneling methods. | Very costly.<br>Rock masses that do not require lining for small diameter tunnels are left open for relatively short time intervals. |
| Bar soundings | To determine thickness of shallow stratum of soft soils. | A metal bar is driven or pushed into ground. | No samples obtained.<br>Penetration limited to relatively weak soils such as organics or soft clays. |
| Hand auger or posthole digger | Recovery of disturbed samples and determination of soil profile to shallow depths.<br>Locate GWL (hole usually collapses in soils with little to slight cohesion). | Rotation of a small-diameter auger into the ground by hand. | Above GWL in clay soils or granular soils with at least apparent cohesion.<br>Below GWL in cohesive soils with adequate strength to prevent collapse.<br>Penetration in dense sands and gravels or slightly plastic clays can be very difficult. |
| One-inch retractable-plug sampler | Blows from driving give qualitative measure of penetration resistance to depths of 30 m in soft clays.<br>One-inch diameter samples can be retrieved up to 1 m in length. | Small-diameter casing is driven into ground by 30-lb slip hammer dropped 12 in.<br>Samples are obtained by retracting driving plug and driving or pressing the casing forward. | The entire rod string must be removed to recover sample.<br>Penetration depth in strong soils limited.<br>Small-diameter samples. |
| Continuous cone penetrometer (CPT) (see following and *Art. 3.4.5)* | Continuous penetration resistance including side friction and point resistance for all but very strong soils on land or water. | Probe is jacked against a reaction for continuous penetration. | No samples are recovered.<br>Penetration generally limited to a meter or so in 100 blow material (SPT). |
| Standard penetration test (SPT) *(see Art. 3.4.5)* | Recovery of disturbed samples and determination of soil profile.<br>Locate GWL. | Split-barrel sampler driven into ground by 140-lb hammer dropped 30 in. | Penetration limited to soils and soft rocks. Not suitable for boulders and hard rocks. |

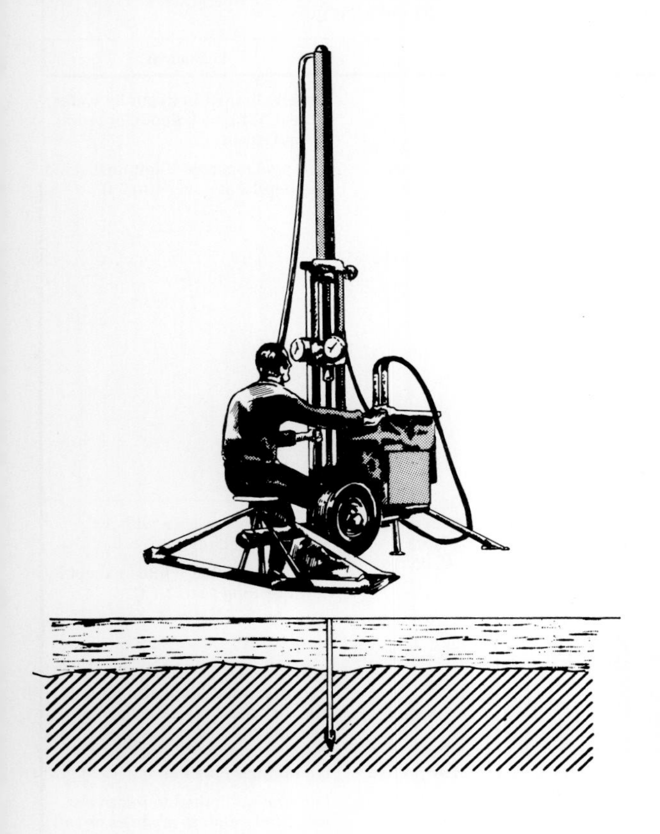

FIG. 2.36 **Motorized Dutch cone penetrometer machine with 10-ton capacity.** *(Courtesy of Joseph S. Ward & Associates.)*

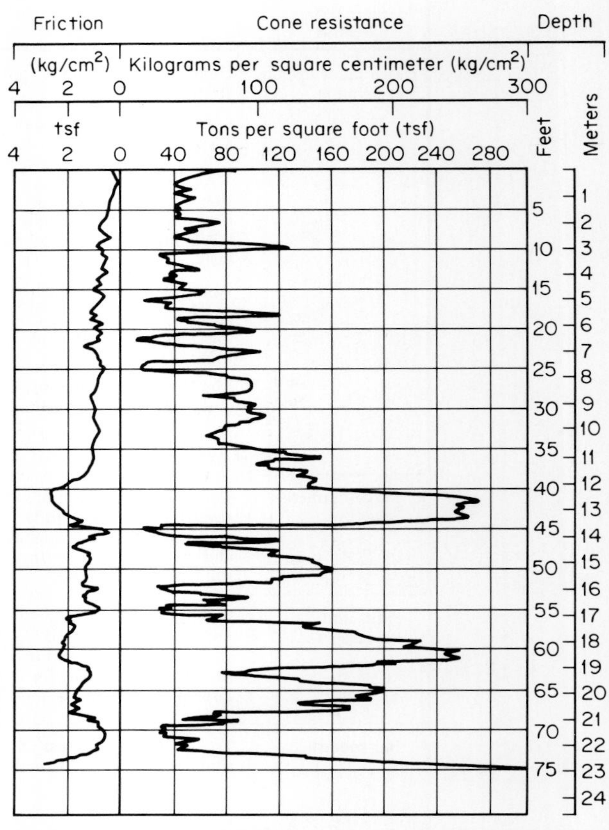


FIG. 2.37 **CPT sounding log shows point and shaft friction resistance.**

ble soils, such as fine and coarser sands, pore-pressure effects during penetration at standard rates often have negligible influence, and the CPT measures approximately fully drained behavior. In homogeneous, plastic clays the CPT measures approximately fully undrained behavior. Mixed soils produce in-between behavior.

## Cone Apparatus

*Simple types,* such as the Dutch mantle cone (not shown), consist of a small-diameter rod tipped with a cone having standard dimensions of 1.4 in (36 mm) base diameter (area = 10 cm²) and 60° inclination from the point. The simple types measure combined penetration resistance from both shaft and point.

*Advanced types,* developed by engineers in Sweden and Holland, incorporate a friction jacket above the standard cone tip permitting the separation of end resistance $q_c$ from shaft friction $f_s$ during penetration as shown on Fig. 2.37. Two types are currently in use: the Begemann friction jacket mechanical cone (Figs. 2.38 and 2.39) and the Fugro electric friction jacket cone (Fig. 2.40).

## Operations

*Begemann Friction Jacket Cone*

A hydraulic thrust is applied to the rods from the machine, which also provides reaction. The outer rod, shaft, and cone tip (Fig. 2.39) are pushed to test depth; the inner rod is then pushed with a slow thrust (2 cm/s or less) to advance the cone 1.5 in while the required force is measured by the cone-tip resistance $q_c$. The cone and friction jacket are then advanced and

**FIG. 2.38 Begemann's friction jacket cone.**

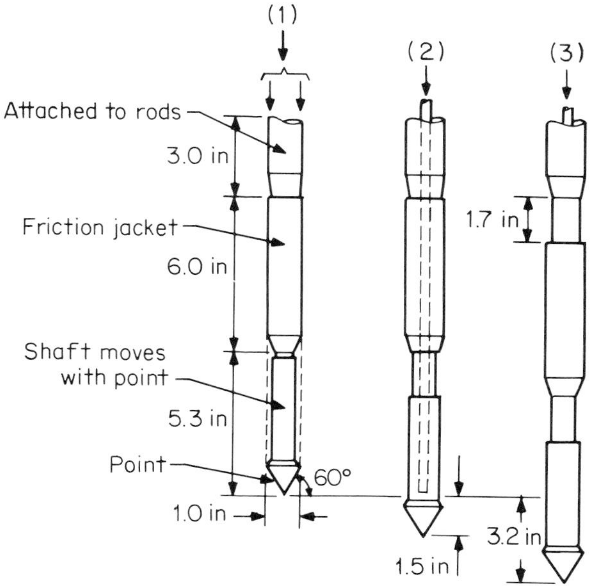

(1)
(2)
(3)
Attached to rods
3.0 in
Friction jacket
6.0 in
Shaft moves with point
5.3 in
Point
60°
1.0 in
1.7 in
3.2 in
1.5 in

**FIG. 2.39 The static cone penetrometer with the Begemann friction jacket.** [*From Alperstein and Leifer (1976).*[19]]

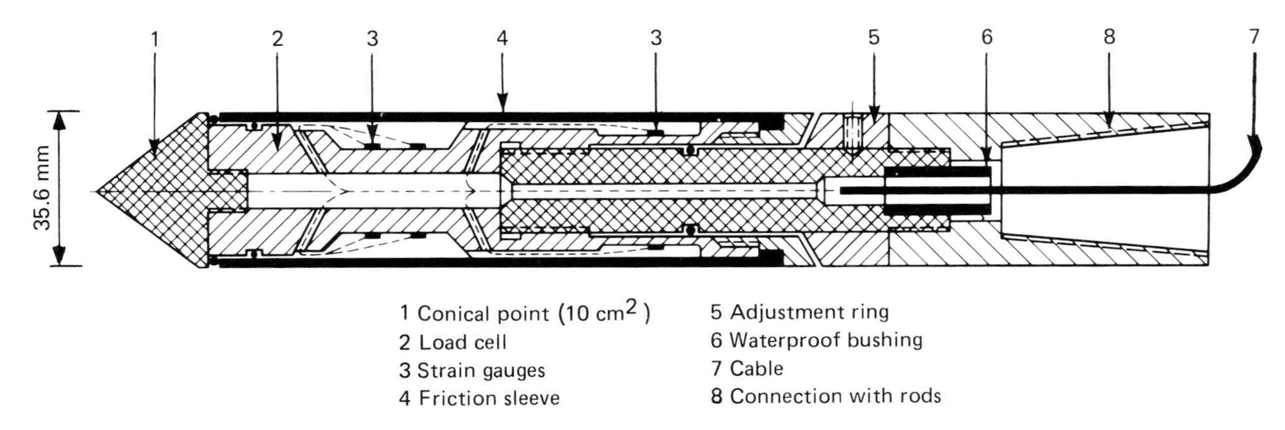


**FIG. 2.40 The Fugro electric friction cone penetrometer tip.** *(Courtesy of Fugro Gulf Inc.)*

the total resistance provided by the cone tip and the shaft (sleeve) friction $f_s$ is measured; $f_s$ is equal to the total force minus $q_c$.

The cone tip should not remain at depth for extended time intervals, since $f_s$ builds up from thixotropic effects.

If the entire apparatus is in good working order the CPT is much less subject to operational factors than is the SPT *(see Art. 3.4.5).*

*Fugro Electric Friction Cone*

Advanced by hydraulic thrust, the Fugro cone (Fig. 2.40) employs load cells and strain gages which measure electronically both tip resistance and local sleeve friction simultaneously. The results are recorded by chart plotters at the surface with an accuracy of measurement of approximately $\pm 1\%$.

A special Fugro penetrometer with a porous

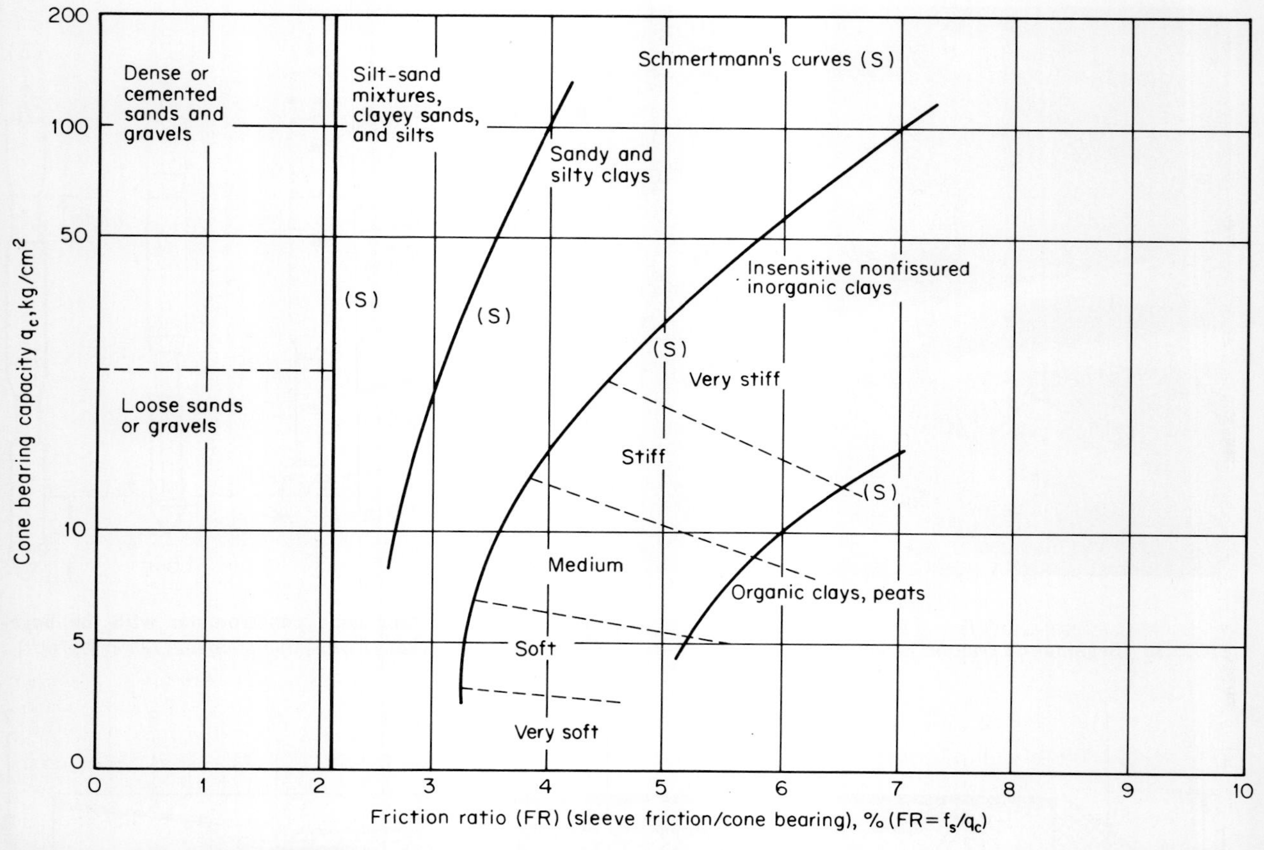


**FIG. 2.41 Soil classification from cone penetrometer tests with the Begemann cone.** [*After Alperstein and Leifer (1976),*[19] *Sanglerat (1972),*[20] *and Schmertmann (1970).*[21]]

stone element in the tip and a built-in electric transducer has been developed to measure pore pressures. To obtain pore-pressure data, penetration is stopped at the desired depth and readings are made until the pore pressure generated by penetration has dissipated.

See also ASTM D3441, Sanglerat (1972),[20] and Schmertmann (1977).[21]

*Penetration Capacity*

Ten-ton equipment using friction cones can penetrate up to a few meters into granular soils with SPT values as great as about 100 at depths of about 8 m. Layers of very dense or cemented sand often require forces that exceed the thrust or reaction capability of the CPT equipment. Significant amounts of gravel can render CPT values very erratic and difficult to interpret quantitatively. The presence of cobbles can stop penetration and damage equipment. Limited penetration is often possible in some soft or weathered rocks.

The Dutch mantle cone is rugged and has a static thrust capacity of up to 17.5 tons, depending upon the size of the reaction machine used, and therefore has a greater penetration capacity than do the friction types.

## Classification of Materials

*Correlations*

CPT values are influenced by soil type and gradation, compactness, and consistency, which also affect the relationship of $q_c$ to $f_s$. Correlations have been developed between cone-tip resistance ($q_c$, also referred to as cone-bearing capacity) and the friction ratio (FR, in which FR $= f_s/q_c$) to provide a guide to soil classification as given on Fig. 2.41, since samples are not recovered for examination. The groundwater table does not appear to have a direct effect on CPT values. An example of a CPT soil profile is given on Fig. 2.42; borings were made initially, followed by the placement of hydraulic fill and the CPT tests.

*Limitations*

FR values may be misleading because of a number of factors, as follows:

- Soil sensitivity *(see Art. 3.4.2)*
- Thin stratification (the Begemann tip travels 8 cm)
- Soil bearing on the bottom bevel of the Begemann friction sleeve, which accounts for one-half to one-third the $f_s$ value in sands, but is not significant in clays
- Temporary pore-pressure effects in fine-grained granular soils
- Soil mineral compositions such as mica flakes and crushable shell fragments
- Coarse particles such as gravel or cobbles

## Operations Offshore

*General*

Offshore exploration, such as for oil-production platforms, usually requires deep penetration, and the CPT is usually operated in conjunction with wire-line drilling techniques *(see Art. 2.3.5)*, with equipment mounted on large vessels such as shown in Fig. 2.43. The major problem, maintaining adequate thrust reaction from a vessel subjected to sea swells, can be provided by a heave compensator (Fig. 2.44). Thrust reaction can be provided by weighted frames set on the seafloor or from the drill string.

*Seafloor Reaction Systems*

Both the Dutch company Fugro and McClelland Engineers of Houston, Texas, have developed underwater cone penetrometer rigs that operate from the seafloor. The Fugro system, called "Seacalf," operates in water depths to 300 m. A hydraulic jacking system mounted in a ballasted frame is lowered to the seabed (Fig. 2.45) to provide a reaction force up to 20 tons. The jack pushes a string of steel rods, on which the electric friction cone is mounted, at a constant rate of penetration. Penetration depth depends on

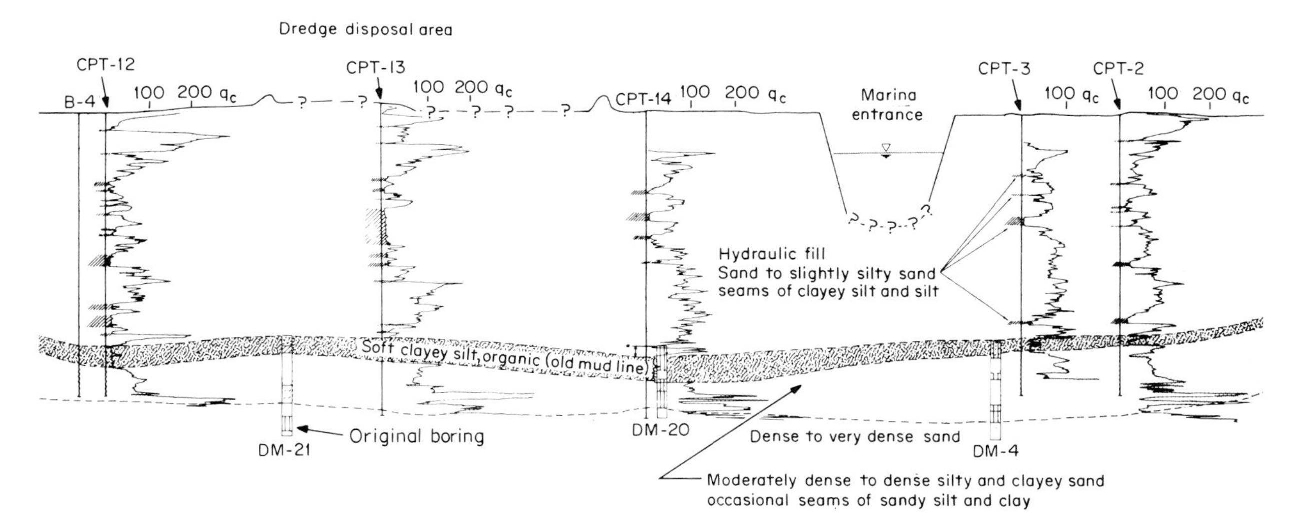


**FIG. 2.42 Soil profile as determined from cone penetrometer tests. Borings were made first, fill was placed hydraulically, and the CPT test was made.**

**FIG. 2.43 Drill ship "Surveyor" owned and operated by Heerema, Marine Contractors, Fribourg, Switzerland.**

**FIG. 2.44 Preparing for drilling and testing with wireline equipment onboard the "Surveyor."**

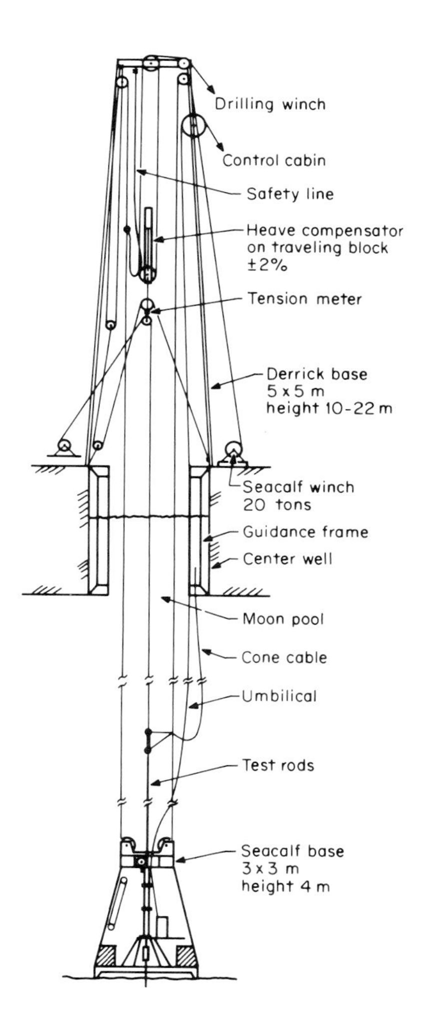


**FIG. 2.45** ***(Left)*** **Fugro's Seacalf system for offshore CPT.**

**FIG. 2.46** ***(Below)*** **Fugro's Wison Mark III system for offshore CPT.**

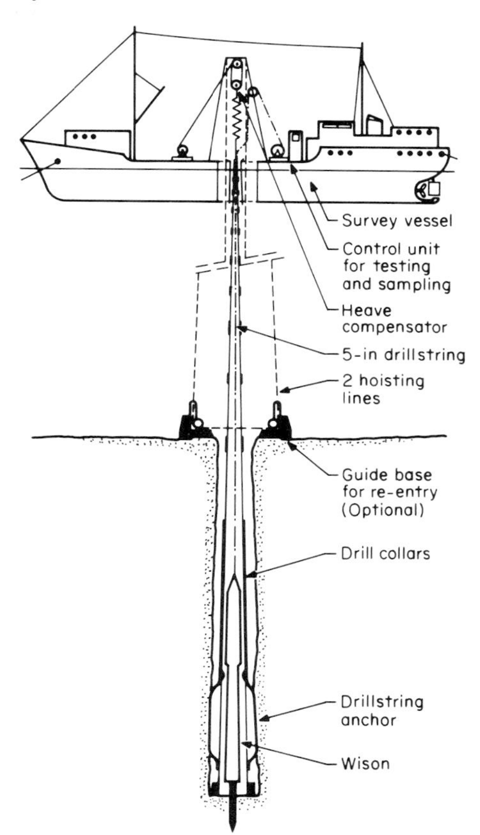


soil conditions and typically ranges from about 20 m in dense sands and hard gravelly clays to between 30 and 40 m in soft, normally consolidated clays.

*Downhole Reaction System*

Where deeper penetration is required than is possible with a seabed apparatus, Fugro employs a system termed the Wison Mark III, which permits CPT testing from the borehole bottom. The hole is advanced to the test depth and a hydraulic jacking unit is lowered inside the drill string on an umbilical cable. Upon reaching the drill bit, it latches automatically into a seating assembly.

Pressure is applied (to 8 tons maximum) to the internal hydraulic piston, which forces the cone into the soil at a constant penetration rate to a maximum depth of 1.5 or 3 m, depending upon the equipment used. If insufficient reaction force is available from the drill string for the required thrust, then a drill string anchor is employed (Fig. 2.46) which consists of an inflatable rough-hole packer (Fig. 2.47). The packer fixes the drill string in the borehole and provides a reaction force up to 20 tons during a test. After

**FIG. 2.47 Borehole packer inflated for testing prior to installation; provides a reaction force to 20 tons.**

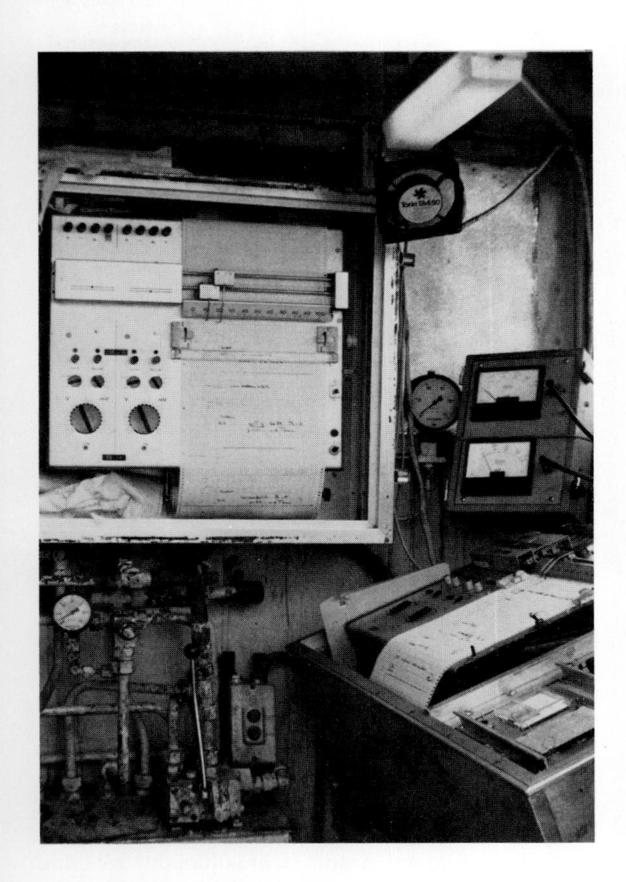

**FIG. 2.48 Fugro instrumentation and control cab for CPT and Wipsampling. Chart on left records point and friction resistance vs. depth; that on right records point and friction resistance vs. time. Instrumentation is connected to the cone via the umbilical cable.**

each test the hole is drilled to the next test depth. Present equipment is used to depths of 450 m below deck level.

Point and shaft resistance are recorded on charts giving both resistance vs. depth and vs. time. The interior of the control and instrumentation cab is shown on Fig. 2.48; an example of a cone-tip resistance log is given as Fig. 2.49.

## 2.3.5 TEST AND CORE BORINGS

### General

Test and core borings are made to:

- Obtain samples of the geologic materials for examination, classification, and laboratory testing
- Permit in situ measurements of the physical and engineering properties of the materials
- Obtain information on groundwater conditions

They may be classed according to sampling operations:

- *Wash sample borings* are made to recover completely disturbed samples for general classification only
- *Sample borings* are made to recover partially disturbed samples (SPT) or undisturbed samples (UD)
- *Core borings* are made to recover rock cores
- *Rotary probes* recover only rock cuttings and are made to provide a rapid determination of the bedrock depth

## Operational Elements

The execution of a boring requires fragmentation of materials, removal of the materials from the hole, and stabilization of the hole walls to prevent collapse.

*Fragmentation*

Materials in the hole are fragmented for removal by:

- Circulating water in loose sands or soft clays and organic soils
- Chopping while twisting a bit by hand (wash boring), or rotary drilling or augering in moderately strong soils
- Blasting or rotary drilling with a rock bit for "floating" boulders
- Rotary or percussion drilling in rock

*Material Removal*

Materials are removed to form the hole by:

- *Dry methods* used in cohesive soils, employing continuous-flight augers above the water table and the hollow-stem auger below groundwater level (GWL)
- *Circulating fluids* from a point in the hole bottom, which is the more common procedure, accomplished by either:

  Clean water used with casing

  Mud slurry formed either naturally or with additives used without casing

  Air pressure in highly fractured or cavernous rock above the GWL where circulating water is lost and does not return to the surface

*Stabilization*

Some form of stabilization is often needed to prevent hole collapse. *None is required* in strong soils above GWL or with hollow-stem augers, which serve as casing.

*Casing,* used in sands and gravels above the water table, in most soils below GWL, and normally in very soft soils, is usually installed by driving with a 300-lb hammer, although 140-lb hammers are often used with lighter tripod rigs. Casing has a number of disadvantages:

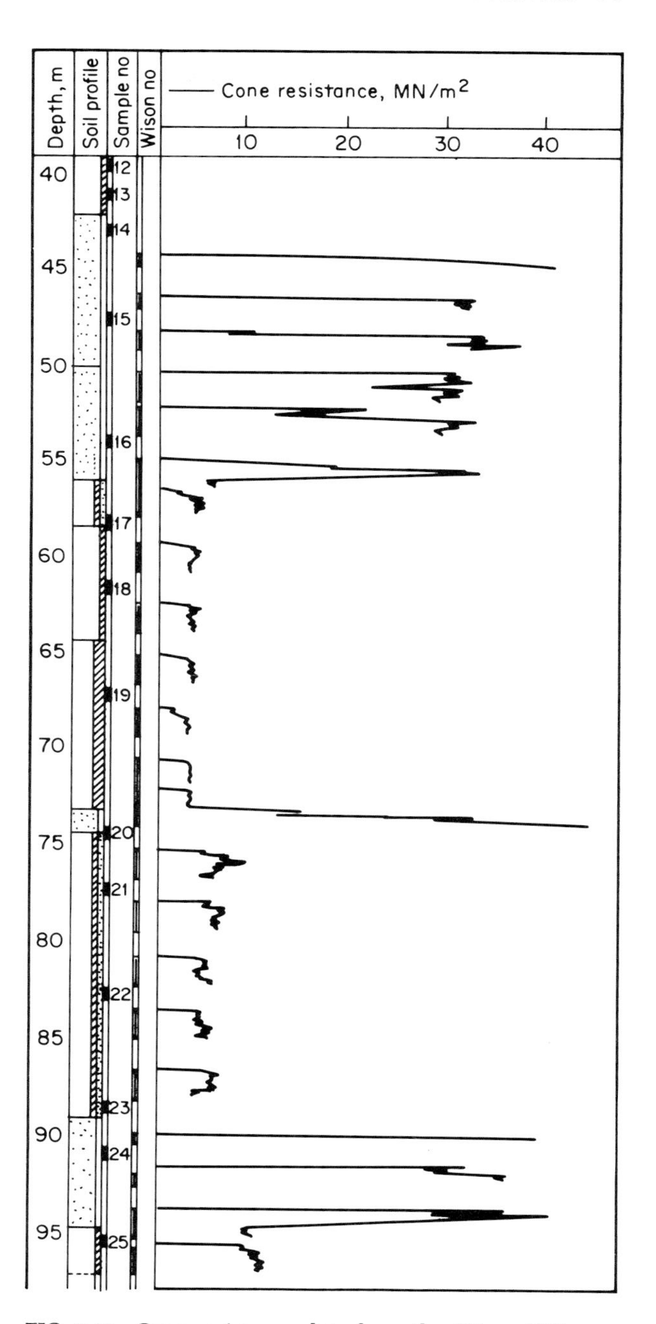


**FIG. 2.49 Cone resistance data from the Wison CPT performed with strokes of 1.5 or 3 m at sequential depths.** *(Courtesy of Fugro Gulf, Inc.)*

**TABLE 2.14**
**COMPARISON OF VARIOUS DRILLING MACHINES AND METHODS**

| Machine | Application | Drilling method | Advantages | Limitations |
|---|---|---|---|---|
| *Tripod* with block and tackle (Fig. 2.50) or motor-driven winch | Procure soil samples.<br>Exploratory borings for preliminary studies.<br>Holes for some types of in situ testing. | Hole advanced by chopping while twisting rods and washing with pump-circulated water.<br>Commonly called wash-boring method. | Requires only minimum-skill labor.<br>Almost any location accessible to the light, portable equipment. | Slow operation, especially below 10 m.<br>Penetration difficult in strong soils and impossible in rock.<br>Difficult to remove gravel from casing; leads to poor samples.<br>UD sampling difficult except in very soft soils because of lack of reaction. |
| *Rotary drills*<br>Skid-mounted (Fig. 2.51)<br>Truck-mounted (Fig. 2.52)<br>Trailer-mounted<br>Track-mounted | Procure all types of soil and rock samples.<br>Make hole for many types of in situ testing.<br>Drilling inclined holes in soil or rock for horizontal drains or anchors. | Hole advanced by cutting bit on end of power-driven rotating drill rod to which pressure is applied hydraulically.<br>Hole normally retained by mud slurry. | Relatively rapid<br>Can penetrate all types of materials.<br>Suitable for all types of samples. | Equipment access in swampy or rugged terrain difficult.<br>Requires trail or road.<br>Requires level platform for drilling.<br>Efficiency of drilling varies with rig size. |
| *Continuous-flight auger* (Fig. 2.53) | Drill small- to moderate-size hole for continuous but disturbed samples.<br>Other samples possible.<br>Normally used in cohesive soils with adequate strength to prevent open hole collapse. | Rotating continuous flights of helical augers.<br>Removal of all flights allows soil examination. | Rapid procedure for exploratory boring in strong cohesive soils and soft rock.<br>SPT sampling possible when hole remains open after auger removal. | Hole collapses when auger withdrawn from weak cohesive or cohesionless granular soils, thereby limiting depth, usually to near water table.<br>Auger samples disturbed.<br>Sampling methods limited. Requires rig samples and soil stratum examination modification. |
| *Hollow-stem auger* | Drill small- to moderate-size holes for soil sampling. | Similar to continuous-flight auger except hollow stem is screwed into ground to act as casing. | Rapid method in weak to moderately strong soils.<br>SPT and UD sampling possible. | Penetration in strong soils to significant depths or through gravel layers difficult, and not possible through boulders and rock.<br>Considerable disturbance may occur from auger bit. |

| | | | | |
|---|---|---|---|---|
| *Large-diamater augers* (Fig. 2.54)<br>Bucket auger<br>Disk auger<br>Helical auger | Drill large-diameter holes (to 4 ft) for disturbed samples and soil strata examination in cohesive soils where hole remains open. | Rotating large-diameter auger cuts soil to form hole. | Rapid method.<br>Enables close examination of subsurface soil conditions. | Depth limited by groundwater and rock conditions.<br>Large machine requires easy access.<br>Not suitable in cohesionless soils, soft clays, or organic soils.<br>Samples disturbed. |
| *Percussion drills* (cable-tool or churn-drilling) | Commonly used to drill water wells.<br>Recovers "wash" samples in bailers.<br>Define rock depth. | Heavy bits are raised and dropped to break up materials and form a slurry which is removed by bailers or sand pumps. Casing retains the hole. | Relatively economical method of making large-diameter holes through any material [up to 2 ft (60 cm)]. | Equipment large and cumbersome.<br>Slow progress in strong soils and rock.<br>Disturbance around bit from high-energy impact seriously affects SPT values.<br>Rock coring and UD sampling not possible. |
| *Hammer drills* | Water wells.<br>Exploratory holes through cobbles and boulders. | Similar to percussion. Diesel pile-driving hammer used to drive double-wall casing while circulating air through annulus to blow cuttings from inner barrel. | Relatively rapid penetration through cobbles and boulders. | Similar to percussion drills, except progress is much more rapid. |
| *Pneumatic percussion drill* | Drilling holes for:<br>Rock anchors<br>Blasting<br>Grout nipples | Percussion rock bit chips and crushes rock with hammer blows as bit rotates. Chips removed by air pressure. | Rapid procedure for making small-diameter holes in hard rock. | Samples are only small chips. Not used for sampling.<br>Possible to lose entire drill stem in loose, fractured rock, clay seams, wet shale, etc.<br>Best use is hard massive rock. |
| *Impact drill* | Rapid drilling of exploratory hole in rock (one case: 640-ft hole, 6.5-in-dia., drilled in 24 hr.[22]) | Pneumatically energized tungsten carbide bit hammers hard rock at as high a rate as 700 blows per minute. | Very rapid penetration in rock masses. Could be used to drill pilot holes to substantial depths for tunnel studies, then core rock in critical depths with rotary methods. | Limited to rock masses.<br>No sample recovery.<br>Danger of hole closure in loose fractured or seamy rock zones. (Could be corrected by cement injection and redrilling). |

**FIG. 2.50 The wash boring method. The hole is advanced by hand by twisting a bladed bit into the soil as water under pressure removes cuttings from the hole. In the photo a 140-lb hammer is being positioned before driving an SPT sample.**

- Installation is slow in strong soils and casing recovery often difficult.
- Sampling stratum changes is prevented unless they occur at the end of a driven section, and in situ testing is limited.
- Obstacles such as boulders cannot be penetrated and require removal by blasting or drilling. The latter results in a reduced hole diameter which restricts sampling methods unless larger-diameter casing is installed before drilling.
- Loose granular soils below GWL tend to rise in casing during soil removal, resulting in plugged casing and loosened soils below.
- Removal of gravel particles is difficult and requires chopping to reduce particle sizes.
- Casing plugged with sand or gravel prevents sampler penetration adequate for recovery of undisturbed samples and representative SPT values. (The length of drill rods and sampling tool must be measured carefully to ensure that the sampler rests on the bottom slightly below the casing.)

*Mud Slurry*, formed naturally by mixing of clayey soils during drilling, or by the addition of bentonite, is a fast and efficient method suitable for all forms of sampling and in situ testing. There are several disadvantages:

- Hole closure may occur in soft soils.
- Relatively large pumps are required to circulate the slurry, particularly when boring depths exceed 10 m.
- Mud-cased hole does not permit accurate water level readings, unless biodegradable muds such as "Revert" are used. Such muds incorporate an organic substance that degrades in a period of 24 to 48 hours, allowing GWL measurements.
- Excessive wear on pumps and other circulating equipment occurs unless sand particles are removed in settling pits.
- Mud may penetrate some soils and contaminate samples.

*Grouting* is used where closure or hole collapse occurs in fractured or seamy zones in rock masses. Cement grout is injected into the hole in the collapsing zone and then the hole is redrilled.

*Boring Inclination*

*Vertical borings* are normal in soil formations and most rock conditions during investigation.

*Angle borings* often are drilled in rock masses to explore for joints, faults, or solution cavities, or for the installation of anchors in soil or rock.

*Horizontal borings* are drilled to explore for tunnels or the installation of rock bolts, instrumentation, or horizontal drains. Maintaining a straight horizontal boring is virtually impossible with current methods. At the start of the boring, gravity tends to pull the drill bit downward; then as penetration increases, gravity acts on the heavy drill string and the bit may tend to drift upward. Rock quality variations will also cause inclination changes.

**Drilling Machines and Other Equipment**

The basic components required for test or core borings include a drilling machine, casing, drill rods, drilling bits, and sampling tools.

**FIG. 2.51 A skid-mounted rotary drilling machine which advances the hole in soil or rock with a cutting bit on the end of a power-driven rotating drill rod to which pressure is applied by a hydraulic ram. In the photo the drill rods are being lowered into the hole before the hole is advanced.**

**FIG. 2.52 Rotary drilling with a truck-mounted Damco drill rig using a mud slurry to prevent hole collapse. Rope on the cathead is used to lift rods to drive the SPT sampler, which in the photo is being removed from the hole. A hydraulic piston applies pressure to rods during rock coring, or during pressing of undisturbed samples. The table supporting the hydraulic works is retractable to allow driving of SPT samples or casing.**

*Drilling machines* consisting of a power source, a mast for lifting apparatus, and a pump for circulating water or mud (or a compressor for air drilling), to lower, rotate, and raise the drilling tools to advance the hole and obtain samples. There are a large number of drilling machines and hole-making methods which are summarized on Table 2.14 in terms of application, method, advantages and limitations. Test borings for obtaining representative or undisturbed samples under all conditions are normally made by rotary drills, and under certain conditions, with the tripod, block and tackle, continuous flight auger, or hollow-stem auger. Exploratory holes in which only disturbed samples are obtained are made with large-diameter augers in clays or by percussion or hammer drilling in all types of materials. Holes are made rapidly in rock, without core recovery, by pneumatic percussion and impact drills. Some of the more common machines are illustrated on Figs. 2.50 through 2.54.

*Casing* is used to retain the hole in the normal test boring operation, with tripods or rotary machines, at the beginning of the hole and for the cases described under *Stabilization*. Boring cost is related directly to casing and hole size. Standard sizes and dimensions are given on Table 2.15; casing is designed for telescoping. There are several types of casing as follows:

- *Standard drive pipe* has couplings larger than the outside pipe diameter and is used for heavy-duty driving.
- *Flush-coupled* casing has couplings with the same diameter as the outside pipe diameter and for a given diameter is lighter and easier to drive, because of the smooth outside surface, than standard pipe, although more costly to purchase.
- *Flush-jointed* casing has no couplings and is even lighter in weight than flush-coupled casing, but it is not as rugged as the other types and should not be driven.

*Drill rods* connect the drilling machine to the drill bits or sampler during the normal test or

**FIG. 2.53 A Penn-Drill power auger machine drilling soft limestone with a solid-stem continuous-flight auger. Drilling can also be performed with a hollow-stem auger from which samples can be retrieved with normal sampling equipment. Rig in photo has been adapted for rock coring and pushing undisturbed samples as shown by the hydraulic works at right.**

core boring operation with rotary machines (or tripods for soil borings). Standard sizes are given on Table 2.15. Selection is a function of anticipated boring depth, sampler types, and rock-core diameter but must be related to machine capacity. The more common diameters are as follows:

- "A" rod is normally used in wash boring or shallow-depth rotary drilling to take SPT samples. In developing countries standard 1-in pipe often is substituted for wash borings because of its low weight and ready availability.
- "B" rod is often used for shallow rotary core drilling, especially with light drilling machines.
- "N" rod is the normal rod size for use with large machines for all sampling and coring operations; it is especially necessary for deep core drilling (over 20 m).
- "H" rod is used in deep core borings in fractured rock since it is heavier and stiffer than "N" rod and will permit better core recoveries.

*Drilling bits* are used to cut soil or rock; some common types are shown on Fig. 2.55. *Chopping bits* (fishtail, offset, etc.—not shown) are used for wash borings. *Drag bits* (fishtail or bladed bits) are used for rotary soil boring. They are provided with passages or jets through which is pumped the drilling fluid that serves to clean the cutting blades; the jets must be designed to prevent the fluid stream from directly impinging on the hole walls and creating cavities, or from directing the stream straight downward and disturbing the soil at sampling depth. Low pump pressure is always required at sampling depth to avoid cavities and soil disturbance. *Rock bits* (tricone, roller bit, or "Hughes" bit) are used for rock drilling. *Core bits* (tungsten carbide teeth or diamonds) are used for rock coring. Sizes are given on Table 2.15. Their design features are described in Art. 2.4.5 since they are sampling tools which also advance the hole.

*Sampling tools* are described in Art. 2.4.

### Standard Boring Procedures

1. Take surface sample. (In some cases samples are taken continuously from the surface to some depth. Sampling procedures are described in Art. 2.4.)
2. Drive "starter" casing to 5-ft penetration (1-m penetration in metric countries).
3. Fragment and remove soil from casing to a depth of about 4 in (10 cm) below the casing to remove material at sampling depth which will be disturbed by the force of the plugged casing during driving. Completely disturbed "wash" samples (cuttings) may be collected at the casing head for approximate material classification.
4. Take sample.

(a)

FIG. 2.54 (*a*) Heavy-duty auger machine excavating with a large-diameter barrel bucket. Generally suitable only in soils with cohesion where the hole remains open without support. (*b*) Types of large augers. [*From USBR* (1974)[22]]

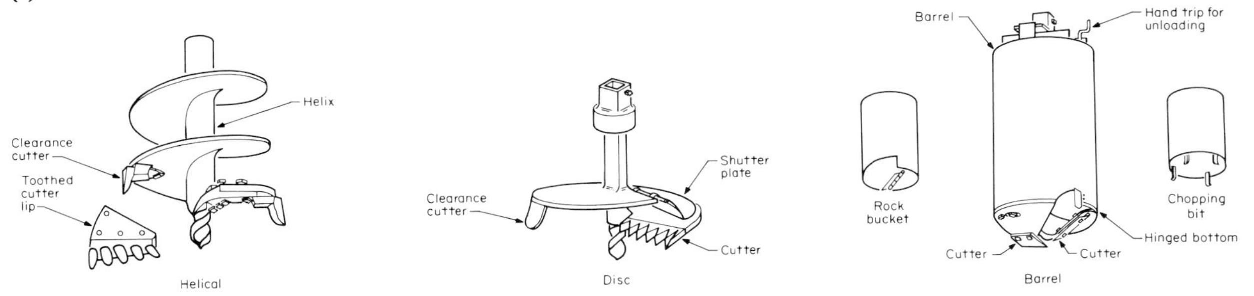


(b)

FIG. 2.55 Several types of drilling bits. From left to right: three-bladed soil-cutting bits, tricone roller rock-cutting bits, carboloy-tooth bits for cutting soft rock, and diamond rock-coring bit.

**TABLE 2.15**
**STANDARD SIZES OF DRILL TOOLS***

| Size | OD | | ID | | Weight | | Coupling OD | |
|---|---|---|---|---|---|---|---|---|
| | in | mm | in | mm | lb/ft | kg/m | in | mm |
| DRILL RODS—FLUSH COUPLED | | | | | | | | |
| E† | 1 5/16 | 33.3 | 7/8 | 22.2 | 2.7 | 4.0 | 7/16 | 11.1 |
| A† | 1 5/8 | 41.3 | 1 1/8 | 28.5 | 3.7 | 5.7 | 9/16 | 14.3 |
| B† | 1 7/8 | 47.6 | 1 1/4 | 31.7 | 5.0 | 7.0 | 5/8 | 15.9 |
| N† | 2 3/8 | 60.3 | 2 | 50.8 | 5.2 | 7.5 | 1 | 29.4 |
| EW‡ | 1 3/8 | 34.9 | 15/16 | 23.8 | 3.1 | 4.7 | 7/16 | 11.1 |
| AW‡ | 1 3/4 | 44.4 | 1 1/4 | 31.8 | 4.2 | 6.5 | 5/8 | 15.9 |
| BW‡ | 2 1/8 | 54.0 | 1 3/4 | 44.5 | 4.3 | 6.7 | 3/4 | 19.3 |
| NW‡ | 2 5/8 | 66.7 | 2 1/4 | 57.1 | 5.5 | 8.4 | 1 3/8 | 34.9 |
| HW‡ | 3 1/2 | 88.9 | 3 1/16 | 77.8 | 7.7 | 11.5 | 2 3/8 | 60.3 |
| CASING—FLUSH JOINTED | | | | | | | | |
| EW | 1 13/16 | 43.0 | 1 1/2 | 38.1 | 2.76 | 4.2 | | |
| AW | 2 1/4 | 57.2 | 1 29/32 | 48.4 | 3.80 | 5.8 | | |
| BW | 2 7/8 | 73.9 | 2 3/8 | 60.3 | 7.00 | 10.6 | | |
| NW | 3 1/2 | 88.9 | 3 | 76.2 | 8.69 | 13.2 | | |
| HW | 4 1/2 | 114.3 | 4 | 101.6 | 11.35 | 16.9 | | |
| PW | 5 1/2 | 139.7 | 4 7/8 | 127.0 | 15.35 | 22.8 | | |
| SW | 6 7/8 | 168.3 | 6 1/32 | 152.4 | 19.49 | 29.0 | | |
| UW | 7 5/8 | 193.7 | 7 | 177.8 | 23.47 | 34.9 | | |
| ZW | 8 5/8 | 219.1 | 8 3/32 | 203.2 | 27.80 | 41.4 | | |
| CASING—FLUSH COUPLED | | | | | | | | |
| EX | 1 13/16 | 46.0 | 1 5/8 | 41.3 | 1.80 | 2.7 | 1 1/2 | 33.1 |
| AX | 2 1/4 | 57.2 | 2 | 50.8 | 2.90 | 4.4 | 1 29/32 | 48.4 |
| BX | 2 7/8 | 73.0 | 2 9/16 | 65.1 | 5.90 | 8.8 | 2 3/8 | 69.3 |
| NX | 3 1/2 | 83.9 | 5 3/16 | 81.0 | 7.80 | 11.8 | 3 | 76.2 |
| HX | 4 1/2 | 114.3 | 4 1/8 | 104.8 | 8.65 | 13.6 | 3 15/16 | 100.0 |
| CASING—STANDARD DRIVE PIPE | | | | | | | | |
| Size in. | OD in | OD mm | ID in | ID mm | Weight lb/ft | Weight kg/m | Coupling OD in | Coupling OD mm |
| 2 | 2 3/8 | 60.3 | 2 1/16 | 52.4 | 5.5 | 8.3 | 2 7/8 | 73.0 |
| 2 1/2 | 2 7/8 | 73.0 | 2 15/32 | 62.7 | 9.0 | 13.6 | 3 3/8 | 85.7 |
| 3 | 3 1/2 | 88.9 | 3 1/16 | 77.8 | 11.5 | 17.4 | 4 | 101.6 |
| 3 1/2 | 4 | 101.6 | 3 9/16 | 90.5 | 15.5 | 23.4 | 4 5/8 | 117.3 |
| 4 | 4 1/2 | 114.3 | 4 1/32 | 102.4 | 18.0 | 27.2 | 5 3/16 | 131.8 |
| CASING—EXTRA HEAVY DRIVE PIPE | | | | | | | | |
| 2 | 2 3/8 | 60.3 | 1 15/16 | 49.2 | 5.0 | 7.6 | 2 7/32 | 56.4 |
| 2 1/2 | 2 7/8 | 73.0 | 2 21/64 | 59.1 | 7.7 | 11.6 | 2 5/8 | 66.7 |
| 3 | 3 1/2 | 88.9 | 2 29/32 | 73.8 | 10.2 | 15.4 | 3 1/4 | 82.5 |
| 3 1/2 | 4 | 101.6 | 3 23/64 | 85.3 | 12.5 | 18.9 | 3 3/4 | 95.3 |
| 4 | 4 1/2 | 114.3 | 3 53/64 | 97.2 | 15.0 | 22.7 | 4 1/4 | 107.8 |

*From Diamond Core Drill Manufacturers Association (DCDMA).

†Original diamond core drill tool designations.

‡Current DCDMA standards.

| DIAMOND CORE BITS | | | | |
|---|---|---|---|---|
| **DCDMA standards** | **Core diam. (bit ID)** | | **Hole diam. (reaming shell OD)** | |
| **Size** | **in** | **mm** | **in** | **mm** |
| EWX and EWM | 0.845 | 21.5 | 1.485 | 37.7 |
| AWX and AWM | 1.185 | 30.0 | 1.890 | 48.0 |
| BWX and BWM | 1.655 | 42.0 | 2.360 | 59.9 |
| NWX and NWM | 2.155 | 54.7 | 2.930 | 75.7 |
| 2¾ in, 3⅞ in | 2.690 | 68.3 | 3.875 | 98.4 |
| 4 in, 5½ in | 3.970 | 100.8 | 5.495 | 139.6 |
| 8 in, 7¾ in | 5.970 | 151.6 | 7.755 | 196.8 |
| **Wireline Size** | | | | |
| AQ | 1 1/16 | 27.0 | 1 57/64 | 48.0 |
| BQ | 1 7/16 | 36.5 | 2 23/64 | 60.0 |
| NQ | 1⅞ | 47.6 | 2 63/64 | 75.8 |
| HQ | 2½ | 63.5 | 3 25/32 | 96.0 |
| PQ | 3 11/32 | 85.0 | 4 53/64 | 122.6 |

5. Advance hole through the next interval of 5 ft (or 1 m) by either driving casing and removing the soils, as in step 3, or using a mud slurry to retain the hole, and take sample.
6. Continue sequences until prescribed final boring depth is reached. [In deep borings, or supplemental boring programs, sampling intervals are often increased to 10 or even 20 ft (3 to 6 m).]
7. When rock is encountered, set casing to the rock surface to permit coring with clean water to keep the bit cool and clean and prevent clogging.
8. Record casing lengths; measure and record drill rod and bit length and drill rod and sample tool length each time the hole is entered to ensure that the hole bottom is reached, the hole is not collapsing if uncased, and the sampler is at the required depth *below* the casing or final boring depth prior to sampling.

### Wireline Drilling

*General*

Wireline drilling eliminates the necessity of removing a string of drill rods for sampling and coring and therefore is a very efficient method for deep core drilling on land or offshore (see *Figs. 2.43 and 2.44*). The coring device is integral with the drill rods, which also serve as casing. It is not normally necessary to remove the casing except when making bit changes. The drill string is 4- to 6-in pipe with a bit at the end.

*Procedure*

The drill string is rotated as the drilling fluid is pumped down through it. (In offshore drilling the mud, mixed onboard, is normally not recirculated, but rather flows up through the hole onto the seafloor.)

Soil sampling, rock coring, and in situ testing are carried out from the inner barrel assembly (see *Fig. 2.84*). Core samples are retrieved by removal of the inner barrel assembly from the core barrel portion of the casing drill rods. An "overshot" or retriever is lowered by the wireline through the drill rod to release a locking mechanism in the inner barrel head. The inner barrel with the core is then lifted with the wireline to the surface, the core removed, and the barrel returned to the bottom. In deep holes it is necessary to pump the inner barrel into place with fluid pressure. Wireline core diameters are given on Table 2.15.

## Subaqueous Drilling

See Art. 2.4.4 for various types of platforms for the drilling equipment and applications for subaqueous test borings.

## Planning and Executing a Test Boring Program

*Equipment Selection*

Consideration is given to the study phase, terrain features and accessibility, geologic conditions, boring depths, and the sample types required when a test boring program is planned.

*Boring Spacing*

In the feasibility and preliminary studies borings are located to explore surface boundaries and stratigraphy as depicted on the engineering geology map, with provisions for additional borings as may be required for increased definition. Grid systems may be appropriate in uniform conditions, and depending on the study area size may range in spacing from 30 to 100 m.

Final study programs depend upon the project type as follows:

- Structures (buildings, industrial plants, etc.) in urban areas usually are required by code to be investigated by borings at spacings which provide at least one boring for a given building area. In other than code-controlled areas, boring layout depends on building configuration, and spacing is generally about 15 to 30 m depending on the uniformity of geologic conditions, the importance of the structure, and the foundation type.
- Dams are usually investigated on a grid of about 15 to 30 m spacing.
- Highway and railroad study programs depend on the adequacy of data obtained during the mapping phase as supplemented by geophysical and reconnaissance studies (excavations, augers, probes, etc), unless specified otherwise by a highway department or other owner. A minimum program requires borings in major cuts and fills and at tunnel portals and all structure locations.
- In all cases flexibility must be maintained and the program closely supervised to permit investigation of irregular or unforeseen conditions as they appear.

*Boring Depth*

*Excavations* (open cuts for buildings, highways, subways, etc.; closed excavations for tunnels, caverns, mines) require borings adequately deep to explore to at least a short depth below final grade, or deeper if conditions are unfavorable, and to determine piezometric levels which may be artesian. The latter condition may require borings substantially below final grade.

*Foundations* for buildings and other structures require boring depths programmed and controlled to satisfy several conditions. As a general rule, borings must explore the entire zone of significant stress (about 1½ to 2 times the minimum width of the loaded area) in which *deformable material exists.* (Note that significant stress can refer to that imposed by a controlled fill and the floor it supports rather than the foundations bearing in the fill, or can refer to stresses imposed at some depth along a pile group rather than at basement or floor level.)

The primary objective is to locate suitable bearing for some type of foundation and to have knowledge of the materials beneath the bearing stratum. Drilling to "refusal" is never a satisfactory procedure unless the materials providing refusal to penetration (often inadequately defined by an SPT value) and those underlying have been previously explored and adequately defined. In rock, penetration must be adequate to differentiate boulders from bedrock.

*Specifications*

Boring type (exploratory, undisturbed, core boring), spacing and depth, sample type and intervals, sample preservation and shipment, groundwater depths, and often drilling procedures are covered in specifications provided to the boring contractor. Standard boring specifications require modification to suit a particular project.

*Inspection*

During all phases of the boring program inspection should be provided to ensure that the intent of the specifications is properly interpreted and executed, and that the desired results are achieved.

Functions of the inspector are generally as follows:

1. Enforce the specifications.
2. Maintain liaison with the designer and modify the program as necessary (add or delete borings; change types, depths and intervals of sampling; etc.).
3. Ensure complete and reliable drilling information (accurate reporting of depths, proper drilling and sampling techniques).
4. Accurately identify all geologic conditions encountered and prepare reports and field logs including all pertinent information (*see Art. 2.4.7*).

Some conditions where experienced geologic interpretations are necessary include:

- Differentiating a fill from a natural deposit. Some granular fills may appear to be natural, when in reality they overlay a thin organic stratum, or contain zones of trash and rubbish. Mistakenly identifying a deposit as fill can result in unnecessarily costly foundations when local codes prohibit foundations on fill.
- Judging the recovered sample material to be wash remaining in the casing rather than undisturbed soil.
- Determining groundwater conditions.
- Differentiating boulders from bedrock on the basis of rock identification (boulders may be of a rock type different from the underlying bedrock, especially in glaciated terrain).

### *Supplemental Information*

Many types of devices are available for use in boreholes to remotely sense and log various subsurface conditions which should be considered in the planning of any subsurface exploration program, as described in Art. 2.3.6.

## 2.3.6 BOREHOLE REMOTE SENSING AND LOGGING

A number of devices and instruments can be lowered into boreholes to obtain a variety of information. They are particularly useful for investigating geologic conditions in materials from which it is difficult or impossible to obtain undisturbed samples, such as cohesionless granular soils and badly fractured rock masses.

Applications, equipment, operation and limitations of various devices for remotely sensing and logging boreholes are summarized on Table 2.16; additional information is provided on Table 2.17.

*Borehole cameras* (TV and photographic) furnish images of borehole walls in fractured rock masses.

*Seisviewer*, an ultrasonic acoustical device, also is used to obtain images of borehole walls in fractured rock masses.

*The 3-D velocity probe*, which ranges up to 15 ft in length, 3 in in diameter, and about 150 lb in weight, is lowered into the borehole to measure compression and shear-wave velocities which provide information on fracture patterns in rock masses and, with the gamma-gamma probe for density measurements, provide the basis for computing dynamic properties.

In operation, a magnetic field is produced by alternating current in a coil wrapped around a rod with magnetostrictive properties. This field produces a sonic pulse that penetrates into the surrounding rock about 1 m (Fig. 2.56). The

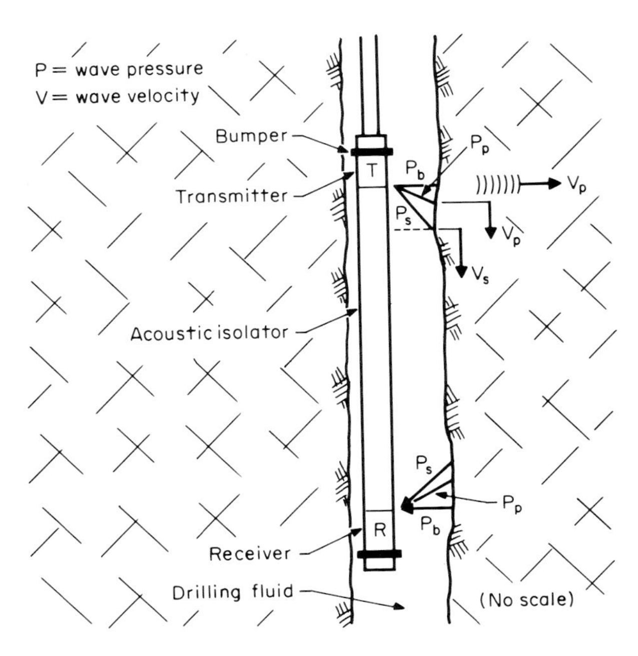


**FIG. 2.56 Schematic of the principles of the 3-D velocity probe.** [*From Myung and Baltosser (1972).*[24]]

**TABLE 2.16**
**BOREHOLE REMOTE-SENSING AND LOGGING METHODS**

| Device | Applications | Equipment | Operation | Limitations |
|---|---|---|---|---|
| Borehole film camera | In rock masses provides continuous undistorted record, with depth and orientation controlled, of all geologic planes in proper polar coordinates. Records fractures to 0.01 in. | Camera with conical or rotating reflecting mirror, transmitter, illumination device, and compass contained in assembly with length of 33 in and diameter of 2.75 in. | Camera lowered into dry or water-filled hole of NX diameter. | In NX hole, depth limited to about 150 m. Images affected by water quality. Lens has limited depth of focus and cannot "see" depth of openings beyond more than a few centimeters. |
| Borehole TV camera | Examination of boreholes in rock. Some types allow examination of voids such as caverns. | TV camera. Some can be fitted with zoom lens and powerful miniature floodlight. Record can be taped. | Camera lowered into dry or water-filled hole. | Resolution less than photographic image. |
| Seisviewer | To procure image of borehole walls showing fractures and discontinuities in rock masses. | Ultrasonic acoustical device mounted in probe produces images of the entire hole. | Probe lowered into dry or water-filled hole. | Rock masses. |
| 3-D velocity probe (3-D sonic logger) (Fig. 2.56) | To procure images of sonic compression and shear-wave amplitudes and rock fracture patterns. $V_p$ and $V_s$ used to compute dynamic properties. | A sonde from which a sonic pulse is generated to travel into the rock mass (see text). | Sonde is lowered into borehole. Arrival times of compression and shear waves are transmitted to surface, amplified, and recorded. | Fracture patterns in rock masses revealed to a depth of about 1 m. |
| Mechanical calipers | Continuous measurement of borehole diameter to differentiate soft from hard rock and swelling zones. Useful for underground excavations in rock. | Mechanical caliper connected to surface recorder. | Lowered into hole and spread mechanically as it is raised. Soft rock gives large diameters from drilling operation. | In rock masses, measures to about 32 in maximum. |
| Recording thermometers | To procure information on groundwater flow and evaluate grouting conditions. | Recording thermometer. | Lowered into borehole after fluid temperature stabilizes. | Rock masses and water-filled holes. |

| | | | | |
|---|---|---|---|---|
| Electric well log | In all materials to obtain continuous record of relative material characteristics and groundwater conditions from ground surface. | Devices to record apparent resistivities and spontaneous potential (natural) generated in borehole. Similar to electrical resistivity measurements from surface. | Instrument lowered into mud or water-filled hole. Dry-hole devices available. | Data are essentially qualitative for correlations with sample or core borings. |
| Gamma-gamma probe | To measure material density in situ. Particularly useful in cohesionless soils and fractured rock masses.* | Nuclear probe measures back-scatter of gamma rays emitted from a source in the probe. | Probe lowered into cased or mud-filled hole. Back-scatter must be calibrated for the casing or mud. | Measurements of in situ density. |
| Neutron-gamma probe | To measure material moisture content in situ.* | Nuclear probe measures back-scatter of gamma and neutron rays (resulting from bombardment by fast neutrons), which gives measure of hydrogen content of materials. | Probe lowered into cased or mud-filled hole. Back-scatter must be calibrated for the casing or mud. Hydrogen content is correlated with the water or hydrocarbon content. | Measurements of in situ water content. |
| Scintillometer | Locate shales or clay zones. Used primarily for petroleum exploration. | Nuclear probe measures gamma rays emitted naturally from the mass. | Probe lowered into borehole. Shales and clays have a high emission intensity compared with sands. | Qualitative assessments of shale or clay formations. |
| Rock detector (acoustic-sounding technique) (Fig. 2.57) | To differentiate boulders from bedrock. | Microphone transmitter set into bedrock and series of holes drilled at various locations to produce noise which is monitored with earphones and observed and recorded by oscilloscope. | One observer listens to drilling sounds while the other records drilling depth for each significant change. Characteristic sounds enable differentiation between soil, boulders, and rock. | Area about 60–100 m from each geophone can be investigated. In overburden, drill bits tend to clog in clays and hole collapses below GWL. Best conditions are shallow deposits of dry, slightly cohesive granular soils. |
| Tro-pari | Measure borehole inclination and direction (see text and Table 2.8). | | | |

*Nuclear probes (gamma and neutron) have been used to monitor changes in moving slopes [Cotecchia (1978)[25]] and even to locate the failure zone in a uniform deposit which was evidenced by a sudden change in density and moisture on a relatively uniform log.

NOTE: See Table 2.17 for data on borehole logging devices used by the U.S. Army Corps of Engineers.

TABLE 2.17
BOREHOLE LOGGING DEVICES USED BY U.S. CORPS OF ENGINEERS

| Method | Tool | | Boring | | | | | | | Normal logging speed, ft/min |
|---|---|---|---|---|---|---|---|---|---|---|
| | | | Diameter, in | | Operating medium | | | Hole condition | | |
| | Diameter, in | Length, ft | Min. | Max. | Air | $H_2O$ | Mud | Cased | Uncased | |
| **Electrical** | | | | | | | | | | |
| Spontaneous potential | 1⅜ | 4 | 2 | 12 | | x | x | | x | 15 |
| Single-point resistivity | 1⅜ | 4 | 2 | 12 | | x | x | | x | 15 |
| Multielectrode resistivity | 1⅜ | 31 (flex.) | 2 | 12 | | x | x | | x | 15 |
| **Radiation** | | | | | | | | | | |
| Natural gamma | 1⅜ | 6¼ | 2 | 8 | x | x | x | x | x | 10 |
| Gamma-gamma | 1⅜ | 9 | 2 | 8 | x | x | x | x | x* | 5 |
| Neutron | 1⅜ | 8½ | 2 | 8 | x | x | x | x | x* | 5 |
| **Sonic** | | | | | | | | | | |
| 3-D velocity | 2¼ | 15 | 3 | 8 | | x | x | | x | 15 |
| Televiewer | 2 | 16½ | 3 | 10 | | x | x | x | x* | 4 |
| **Fluid** | | | | | | | | | | |
| Temperature | 1⅜ | 4¼ | 2 | Inf. | | x | x | x | x | 3 |
| Resistivity | 1⅜ | 4¼ | 2 | Inf. | | x | x | x | x | 3 |
| Tracejector | 1⅝ | 6 | 2 | Inf. | | x | | x | x | |
| Sampler | 2½ | 8 | 3 | Inf. | | x | x | x | x | |
| **Optical** | | | | | | | | | | |
| Television | 4¾† | 3¼§ | 5 | 16¶ | x | x | | | x | 1 |
| Borescope | 2 | 80 | 2¼ | 4 | x | x | | | x | Slow |
| NX camera | 2¾ | 2½ | 3 | 4 | x | x | | | x | 10 |
| **Mechanical** | | | | | | | | | | |
| Caliper | 1⅝ | 7½ | 2 | 32 | x | x | x | | x | 10 |
| Surveyor | 3‡ | 2 | 4 | Cntr. | x | x | x | | x | 1 |

*Extreme roughness may affect results.
†Radial lens 3 in without rotation device.
‡Modified to operate on TV cable.
§13 ft with gyroscope and rotation device.
¶Maximum of 20 ft with telephoto lens
NOTE: Data from Underwood (1974).[26]

arriving pulse deforms cylindrical piezometric crystals used as receivers to produce the electricity necessary to record the energy arrival. The receivers are located 3 to 6 ft from the transmitter along the probe, and are isolated from the transmitter to prevent sonic energy from traveling along the probe.

The arrival times are transmitted to the surface, amplified, and recorded, to provide an image similar to that produced by a recording seismograph. The 3-D sonic logger was so-named because the records show the amplitude and arrival times of the sonic energy for a given travel distance (between the transmitter and receiver). For velocity computations the hole diameter must be known and is measured with mechanical calipers.

*Mechanical calipers* are used to measure hole diameter in rock masses.

*Recording thermometers* are used to measure fluid temperatures in rock masses.

The *electric well logger* is used in soil and rock masses for continuous measurements of resistivity.

The *gamma-gamma probe* is used in soil and rock masses to obtain continuous measurements of in situ densities.

The *neutron-gamma probe* is used in soil and rock masses to obtain continuous measurements of in situ moisture contents.

The *scintillometer* is used to locate shales and clay zones in soil and rock masses.

The *rock detector* is an acoustical sounding device used to differentiate boulders and other obstructions from bedrock. A geophone is set into bedrock and connected to an amplifier, headphone, and oscilloscope. A series of holes is drilled with a wagon drill, or other drilling machine (Fig. 2.57), while the observer listens to the volume and nature of the generated sounds.

The *Tro-pari surveying instrument* is used to measure borehole inclination and direction in rock (*see Table 2.8*).

## 2.3.7 GROUNDWATER AND SEEPAGE DETECTION

### General Groundwater Conditions

Figure 2.58 illustrates various groundwater conditions, which are discussed in detail in Art. 8.3.1 and are summarized as follows:

- *Static water table or level* (GWL) is located at the depth below which the ground is continuously saturated and at which the pressure in the water is atmospheric.
- *Perched water table* can also be a measured groundwater level, and represents a saturated zone overlying an impervious stratum below which the ground is not saturated.
- *Artesian conditions* result from groundwater under a head in a confined stratum which is greater than the static water-table head and can result in free flow at the surface when the confined stratum is penetrated.
- *Variations* in conditions occur with time, and are affected by seasonal conditions, tidal fluctuations, flooding, and pumping. Variations also occur with physical conditions in terms of soil type and density, ground contours, surface drainage, rock-mass discontinuities, etc.

### Determining Conditions

*Terrain analysis* provides general information on water-table location.

*Geophysical methods* (seismic refraction and electrical resistivity) provide indirect measures of the approximate depth to groundwater.

*Reconnaissance methods* employing excavations locate the groundwater level in a positive manner but with a short interval of time. Probable variations with time must always be considered during site analysis. Test pits are the best method for measuring short-term water levels and seepage rates. Auger borings also may clearly reveal the water table, especially in slightly cohesive soils. In cohesionless soils the hole will tend to collapse within a few centimeters of the groundwater level but direct measurement is not usually possible.

#### *Test Borings*

Moisture condition of drive samples (SPT) may provide an indication of groundwater level. Samples will range from dry to moist to wet as the groundwater level is approached.

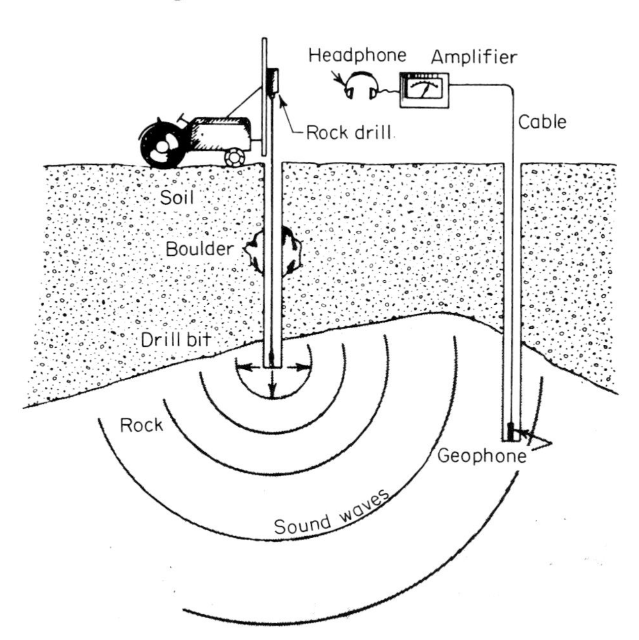


**FIG. 2.57 The elements of the rock indicator, or acoustic sounding technique, to differentiate boulders from bedrock.**

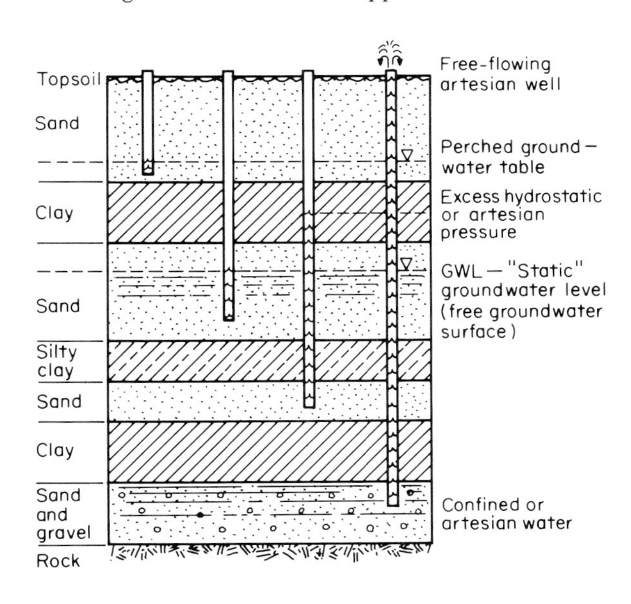


**FIG. 2.58 Various groundwater conditions.**

In cased borings the water level is generally determined by pumping or bailing water from the hole and permitting stabilization for 24 hr. This method is reliable in uniform sand strata or other pervious materials, but differentiating perched from static conditions is difficult. Site stratigraphy provides some clues for judgment. The method is unreliable when the casing terminates in an impervious stratum which blocks the entrance of water. The casing should be raised until it terminates in a permeable stratum.

If the casing ends in an aquifer with an artesian head, water may flow from the casing when pumping ceases during drilling operations, or it may rise to a point above the estimated GWL.

Casing water levels should be noted periodically during boring operations and for a period thereafter, and water loss as well as artesian and static conditions should be noted.

Boring with a mud slurry will not provide reliable water level readings unless the hole is flushed with clean water or biodegradable mud is used for boring. After flushing, most holes in permeable soils collapse near or slightly above the perched or static water level.

*Borehole remote-sensing probes,* such as the electric well logger and the neutron probe, provide good indications of perched or static conditions.

*Piezometers (see Art. 4.4.2)* provide the most accurate method of measuring groundwater conditions, and are especially valuable in recording changes with time. Measurements made during pumping tests performed at various levels are useful in differentiating perched from static conditions.

### Seepage Detection

*Conditions of engineering significance* include:

- Flow through, around, and beneath earth dam embankments
- Flow beneath and around concrete dams
- Groundwater pollution as caused by flow from sanitary landfills, mine tailings storage areas, chemical waste ponds, etc.

*Detection Methods*

Nonradioactive tracers include fluorescent and nonfluorescent dyes. Radioactive tracers include bromine 82 and iodine 131 which are readily detectable with a Geiger-Muller counter. Temperature can be sensed with a thermistor attached to the top of an insulated aluminum-tipped probe inserted into the ground. Acoustical emission monitoring *(see Art. 4.3.5)* may detect large flows.

## 2.4 RECOVERY OF SAMPLES AND CORES

### 2.4.1 GENERAL

#### Objectives

Samples of the geologic materials are recovered to allow detailed examination for identification and classification, and to provide specimens for laboratory testing to obtain data on their physical and engineering properties.

#### Sample Classes Based on Quality

*Totally disturbed* samples are characterized by complete destruction of fabric and structure and the mixing of materials, such as occurs in wash and auger samples.

*Representative samples* are partially deformed. The engineering properties (strength, compressibility, and permeability) are changed, but the original fabric and structure vary from unchanged to distorted, but still apparent. Such distortion occurs with split-barrel samples.

*Undisturbed* samples may have slight deformation around their perimeter, but for the most part the engineering properties are unchanged. Such results are obtained with tube or block samples.

#### Sampler Selection

A number of factors are considered in the selection of samplers, including:

- Sample use, which varies with disturbance from general determination of material (wash sampler), to examination of material and fabric and in situ testing (split-barrel sampler), to performing laboratory index tests (split-barrel

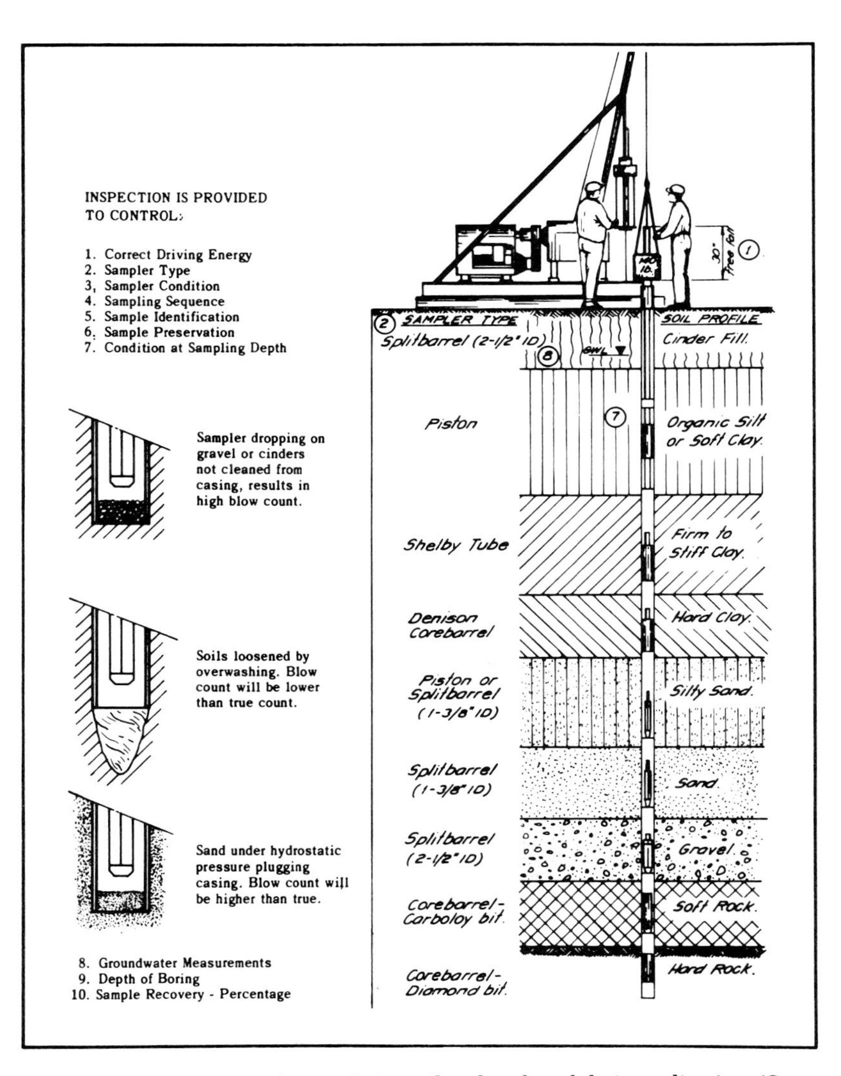


**FIG. 2.59 Common sampling tools for soil and rock and their application.** *(Courtesy of Joseph S. Ward & Associates.)*

sampler), to carrying out laboratory engineering-properties tests (undisturbed samples)

- Soil type, since some samplers are suited only for particular conditions
- Rock conditions, since various combinations of rock bits and core barrels are used, depending on rock type and quality and the amount of recovery required
- Surface conditions, which vary from land or quiet water to shallow or deep water with moderate to heavy swells

Some common sampling tools and their application to various subsurface conditions are illustrated on Fig. 2.59. A general summary of most tools available with respect to type of sample required and ground conditions is contained in Table 2.18. The various tools and methods and their application and limitations are described on Table 2.19.

**TABLE 2.18**
**SOIL AND ROCK SAMPLING METHODS AND GENERAL APPLICATION CONDITIONS**

| Samples required and/or material to be sampled | Soils | | | | | | | | | | | | Shallow subaqueous | | | | | Rock | | | | | | | |
|---|---|---|---|---|---|---|---|---|---|---|---|---|---|---|---|---|---|---|---|---|---|---|---|---|---|
| | Wash sample | Auger sample | Retractable plug | Block sample | Split barrel | Shelby tube | Stationary piston | Osterberg piston | Shear-pin piston | Swedish foil sampler | Denison sampler | Pitcher sampler | Free-fall gravity tube | Harpoon-type gravity tube | Explosive coring tube | Gas-operated piston | Vibracore | Single-tube core barrel | Double-tube core barrel | Double-tube swivel type | Double-tube series"M" | Wire-line core barrel | Oriented core barrel | Integral coring | Calyx or shot coring |
| Disturbed samples: Above GWL | X | X | | | | | | | | | | | | | | | | | | | | | | | |
| Representative samples: Soft soils | | | X | | X | | | | | | | | | | | | | | | | | | | | |
| All soils | | | | | X | | | | | | | | | | | | | | | | | | | | |
| Undisturbed samples: Above GWL in cohesive soils | | | | X | | X | | X | X | | X | X | | | | | | | | | | | | | |
| Soft to firm clays and silts | | | | | | | X | X | X | X | | | | | | | | | | | | | | | |
| Cohesive soils, except hard | | | | | | X | | X | X | | | | | | | | | | | | | | | | |
| For natural and sand density | | | | | | | | | X | | | | | | | | | | | | | | | | |
| Hard soils—residual, till, clay | | | | | | | | | | | X | X | | | | | | | | | | | | | |
| Soils alternating hard/soft | | | | | | | | | | | | X | | | | | | | | | | | | | |
| Underwater samples: Representative: Most soils to 5 m | | | | | | | | | | | | | X | | X | | | | | | | | | | |
| Muds and silts to 12 m | | | | | | | | | | | | | X | X | | X | X | | | | | | | | |
| Stiff to hard soils | | | | | | | | | | | | | | | X | | | | | | | | | | |
| Undisturbed: Soft to firm soils | | | | | | | | | | | | | | | | X | X | | | | | | | | |
| Rock coring: Good quality rock | | | | | | | | | | | | | | | | | | X | | | | | | | |
| Good to medium quality rock | | | | | | | | | | | | | | | | | | | X | | | | | | |
| Poor quality rock | | | | | | | | | | | | | | | | | | | | X | X | | | X | |
| Deep borings | | | | | | | | | | | | | | | | | | | | | | X | | | |
| For core orientation | | | | | | | | | | | | | | | | | | | | | | | X | X | |
| For diameters to 36 in or greater | | | | | | | | | | | | | | | | | | | | | | | | | X |

### 2.4.2 TEST BORING SOIL SAMPLING

**Types Commonly Used**

For *representative samples* for identification and index tests, the *split-barrel sampler* (SS) is used in all soil types.

For *undisturbed samples* for engineering properties tests, *thin-wall tubes* are used in soft to firm clays and *coring samplers* are used in stiff to hard cohesive soils.

**Required Boring Diameters**

Wash or exploratory borings for split-barrel sampling are normally of 2½ in diameter (casing ID). Undisturbed sample borings are normally of 4 in diameter but may be larger to improve sample quality. Core borings vary from 2 in to larger, normally with NX core taken in a hole started with 4-in diameter casing.

**Sampling Interval**

Samples are normally prescribed for 5-ft (1-m) intervals and a change in strata. Samples should also be taken at the surface to record the topsoil thickness, and continuously from the surface to below the depth of shallow foundations to assure information at footing depth. (Five-foot intervals often do not provide information at shallow footing elevations.)

Continuous sampling is also important through miscellaneous fills, which vary widely in materials and which often overlie a layer of organic soil which may be thin but significant, and through formations with highly variable strata. In deep borings sampling depths are often changed to 10- or 20-ft intervals after several normally sampled borings are completed and general subsurface conditions defined.

**Factors Affecting Sample Quality**

*Sampler Wall Thickness*

Large outside diameter relative to the inside diameter causes deformation by material displacement.

*Sampler Conditions*

- Dull, bent, or otherwise deformed cutting edges on the sampler cause sample deformation.
- Inside friction, increased by rust, dirt, or, in the case of tubes, omission of lacquer, causes distortion which is evidenced by a turning downward of layers, resulting in conical shapes in extreme cases. Very slight edge disturbance is illustrated on Fig. 7.93.

*Boring Operations*

Dynamic forces caused by driving casing can loosen dense granular soils or densify loose granular soils.

Overwashing, jetting, and high fluid pressures also loosen granular soils or soften cohesive materials.

Coarse materials often remain in the hole after washing, particularly in cased borings. These "cuttings" should be removed by driving a split-barrel sampler, by pushing a Shelby tube, or with a cleanout auger. Contamination is common after boring through gravel layers or miscellaneous fills containing cinders, etc.

Hole squeezing may occur in soft clays if the drilling mud is too thin.

Plastic clays may remain along the casing walls if cleaning is not thorough.

Hollow stem augers can cause severe disturbance depending on the rate of advance and rotation, and the choice of teeth on the bit.

*Sampler Insertion*

All ball check valves and other mechanisms should be working properly before the sampler is lowered into the hole. The sampler should be lowered to the bottom immediately after the hole is cleaned.

Measurements should be made carefully of the total length of the sampler and rods to ensure that the sampler is resting on the bottom elevation to which the hole was cleaned and to avoid sampling cuttings.

Since complete cleaning usually is not practical, split-barrel samplers are seated under the rod weight and often tapped lightly with the drive hammer; piston samples are forced gently through the zone of soft cuttings.

**TABLE 2.19**
**SAMPLING TOOLS AND METHODS**

| Category-method/tool | Application | Limitations |
|---|---|---|
| | RECONNAISSANCE | |
| Wash sample | Indication of material type only. | Completely mixed, altered, segregated. |
| Auger sample | Material identification. | Completely disturbed. |
| Retractable-plug sampler | Material identification. | Slight disturbance, very small sample of soft soils. |
| Block samples | Large undisturbed samples of cohesive materials. | Taken from test pits, cohesive soils only. |
| | TEST BORING SAMPLING (SOILS) | |
| Split barrel (spoon) | Disturbed samples in soils suitable for identification and lab index tests. | Samples not suitable for engineering properties testing. Sampling impossible in very coarse granular soils. |
| Shelby tube | Undisturbed samples in firm to stiff cohesive soil. Can be driven into hard soils. | Will not retrieve very soft or clean granular soils. |
| Standard stationary piston | Undisturbed samples in soft to firm clays and silts. | Will not penetrate compact sands, stiff clays, and other strong soils. Will not retrieve sands. Can be overpushed. |
| Osterberg piston sampler | Undisturbed samples in all soils with cohesion except very strong. Less successful in clean sands. | Usually cannot penetrate strong residual soil and glacial till. Some disturbance in sands and often loss of sample. User can not observe amount of partial penetration. |
| Shear-pin piston (Greer and McClelland) | Undisturbed samples in all soils with cohesion except very strong. Often recovers samples in sands and can be used to determine natural density. | Usually cannot penetrate strong residual soil or glacial till. Disturbance in sands. Cannot observe amount of partial penetration. |
| Swedish foil sampler | Continuous undisturbed samples in soft to firm cohesive soils. | Gravel and shells will rupture foil. Cannot penetrate strong soils. |
| Denison sampler | Undisturbed samples in strong cohesive soils such as residual soils, glacial till, soft rock. | Not suitable in clean granular soils, and soft to firm clays. |
| Pitcher sampler | Similar to Denison above. Superior in alternating soft to hard layers. Can be used in firm clays. | Similar to Denison above. |
| | SUBAQUEOUS SAMPLING WITHOUT TEST BORING | |
| Free-fall gravity coring tube | Samples firm to stiff clays, sands and fine gravel in water depths to 4000 m. | Maximum length of penetration about 5 m in soft soils, 3 m in firm soils. |
| Harpoon-type gravity sampler | Samples river bottom muds and silts to depths of about 3 m. | Penetration limited to few meters in soft soils. |

| Category-method/tool | Application | Limitations |
|---|---|---|
| Explosive coring tube (Piggot tube) | Small-diameter samples of stiff to hard ocean bottom soils to water depths of 6000 m. | Sample diameters only 1⅞ in. Penetration only to 3 m below seafloor. |
| Gas-operated free-fall piston (NGI) | Good quality samples up to 10 m depth from seafloor. | Penetration limited to 10 m below seafloor. |
| Vibracore | Undisturbed samples of soft to firm bottom sediments, 3½ in dia. to depths of 12 m. | Limited to soft to firm soils and maximum penetration of 12 m. Water depth limited to 60 m. |
| | SUBAQUEOUS SAMPLING WITH TEST BORING | |
| Wireline drive sample | Disturbed samples in soils. | Penetration length during driving not known. |
| Wireline push samples | Relatively undisturbed samples may be obtained in cohesive materials. | Often poor or no recovery in clean granular soils. |
| | ROCK CORING | |
| Single-tube core barrel | Coring hard homogeneous rock where high recovery not necessary. | Circulating water erodes soft, weathered, or fractured rock. |
| Double-tube core barrel | Coring most rock types where high recovery not necessary, and rock not highly fractured or soft. | Recovery often low in soft or fractured rocks. |
| Double-tube swivel-type core barrel | Coring all rock types where high recovery is necessary. | Usually not needed in good-quality rock. Barrel more complicated than single or double tube. |
| Double-tube swivel-type, series M | Superior to double-tube swivel-type core barrel, above. Particularly useful to obtain high recovery in friable, highly fractured rock. | Not needed in good quality rock. Barrel more costly and complicated than others above. |
| Wireline core barrel | Deep hole drilling in rock or offshore because of substantial reduction of in-out times for tools. | No more efficient than normal drilling to depths of about 30 m. |
| Oriented core barrel | Determination of orientation of geologic structures. | Procedure is slow and costly. Requires full recovery. |
| Integral coring method | Recover cores and determine orientation in poor-quality rock with cavities, numerous fractures, and shear zones. | Slow and costly procedure. |
| Calyx or shot coring | Obtain cores in medium- to good-quality rock to 2-m dia. | Slow and costly. Difficult in soft or seamy rock. |

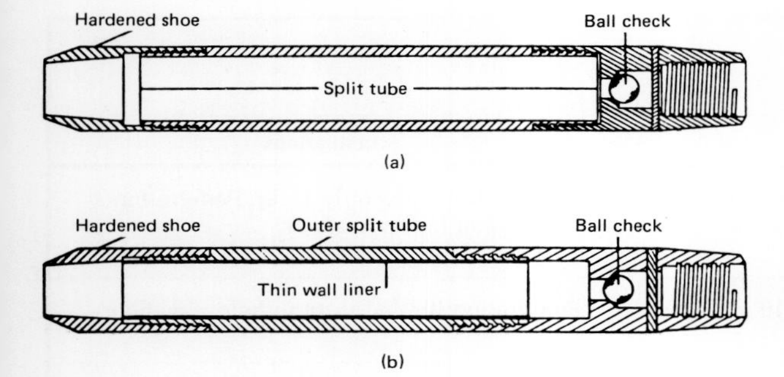


**FIG. 2.60 The split-barrel sampler or split spoon: (*a*) without liner; (*b*) with liner.** *(Courtesy of Sprague and Henwood, Inc.)*

*Soil Factors*

- Soft to firm clays generally provide the best "undisturbed" samples, except for "quick" clays which are easily disturbed.
- Air or gas dissolved in pore water and released during sampling and storage can reduce shear strength.
- Heavily overconsolidated clays may be subject to the opening of fissures from stress release during boring and sampling, thereby substantially reducing strength.
- Gravel particles in a clay matrix will cause disturbance.
- Cohesionless granular soils cannot be sampled "undisturbed" in the present state of the art.
- Disturbance in cohesive materials usually results in a decrease in shear strength and an increase in compressibility.

*Sample Preservation, Shipment, Storage, and Extraction*

See Art. 2.4.6.

*Reference*

Sample disturbance and its effect on engineering properties are described in detail by Broms (1980).[27]

**Split-Barrel Sampler (Split Spoon)**

*Purpose*

Split-barrel samplers are used to obtain representative samples suitable for field examination of soil texture and fabric and for laboratory tests, including measurements of grain-size distribution, specific gravity, and plasticity index, which require retaining the entire sample in a large jar.

*Sampler Description*

Split-barrel samplers are available without and with liners; the components are shown on Fig. 2.60. OD ranges from 2 to 4½ in; a common OD is 2 in with ¼-in wall thickness (1½-in sample). Larger diameters are used for sampling gravelly soils. Lengths are either 18 or 24 in.

A ball check valve prevents drill pipe fluid from pushing the sample out during retrieval. To prevent sample spillage during retrieval, flap valves can be installed in the shoe for loose sands, or a leaf-spring core retainer (basket) can be installed for very soft clays and fine cohesionless soils. Upon retrieval, the barrel between the head and the shoe is split open (Fig. 2.61), the sample is examined and described, removed, and stored.

Brass liners are used for procuring drive samples of strong cohesive soils for laboratory direct-shear testing in some sampler types.

*Sampling Procedure*

The sampler is installed on the hole bottom, then driven into the soil with a hammer (normally 140 lb) falling on the drill rods. The number of blows required for a given weight and drop height, and a given penetration, are recorded to provide a measure of soil compactness or consistency as described in Art. 3.4.5.

### Thin-Wall Tube Samplers

*Purpose*

Thin-wall tube samplers are used to obtain undisturbed samples of soft to stiff cohesive soils for laboratory testing of strength, compressibility, and permeability.

*Thin-Wall Tubes*

MATERIALS Cold-drawn, seamless steel tubing (trade name "Shelby tube") is used for most soil materials, or brass tubes are used for organic soils where corrosion resistance is required. Wall thickness is usually 18 gauge; heavier gauges are available.

Lacquer coating can provide corrosion protection and reduce inside frictional resistance and sample disturbance.

TUBE DIAMETERS AND LENGTHS Ranges are 2 to 6 in in diameter, 24 to 30 in in length. Tubes 2 in in diameter are used in 2½-in exploratory borings, but samples have a large ratio of perimeter disturbance to area and are considered too small for reliable laboratory engineering-property testing.

Tubes 3 in (2.87″) in diameter are generally considered the standard type for laboratory test samples. The tube should be provided with a cutting edge drawn in to provide about 0.04 in inside clearance (or 0.5 to 1.5% less than the tube ID), which permits the sample to expand slightly upon entering the tube, thereby relieving sample friction along the walls and reducing disturbance.

Tubes 4 to 6 in in diameter reduce disturbance but require more costly borings. A 5-in tube yields four samples of 1-in diameter from the same depth for triaxial testing.

SAMPLING WITH TUBES Thin-wall tubes are normally pressed into the soil by hydraulically applied force. After pressing, the sample is left to rest in the ground for 2 to 3 min to permit slight expansion and an increase in wall friction to aid in retrieval. The rods and sampler are rotated clockwise about two revolutions to free the sampler by shearing the soil at the sampler bottom. The sample is withdrawn slowly from the hole with an even pull and no jerking. In soft soils and loose granular soils the sampler bottom is capped just before it emerges from the casing fluid to prevent loss.

TYPES OF THIN-WALL TUBE SAMPLERS

- Shelby tube sampler
- Stationary piston samplers
  - Standard stationary piston
  - Osterberg hydraulic piston sampler
  - Greer and McClelland shear-pin piston

*Shelby Tube Sampling*

THE SAMPLER A thin-wall tube is fitted to a head assembly (Fig. 2.62) which is attached to drill rod. The O ring provides a seal between the head and the tube, and a ball check valve prevents water in the rods from flushing out the sample during retrieval. Application is most satisfactory in firm to hard cohesive soils.

**FIG. 2.61 Split-barrel sample of sand.**

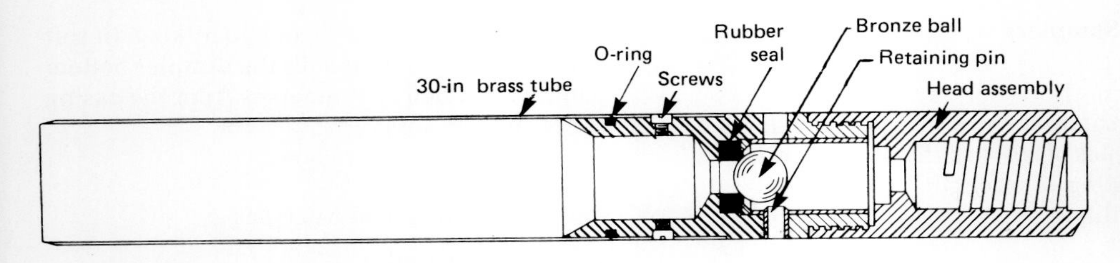


**FIG. 2.62 Thin-wall "Shelby tube" sampler.** *(Courtesy of Sprague and Henwood, Inc.)*

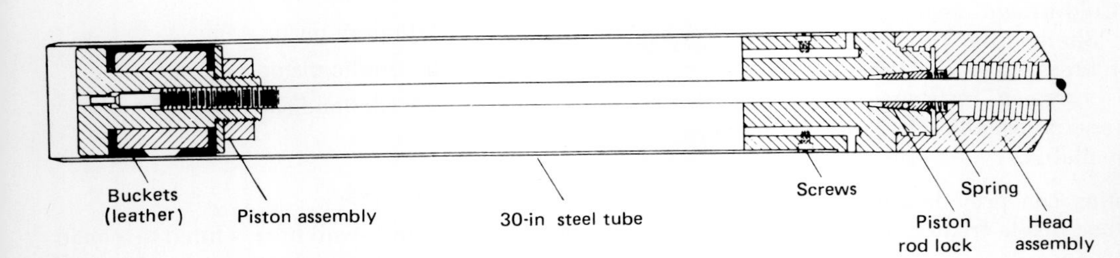


**FIG. 2.63 Stationary piston sampler.** *(Courtesy of Sprague and Henwood, Inc.)*

SAMPLING OPERATION In firm to stiff soils, the tube is pushed into the soil by a steady thrust of the hydraulic system on a rotary drilling machine using the machine weight as a reaction. Care is required that the sampler is not pressed a distance greater than its length.

Soft soils are difficult to sample and retain because they have insufficient strength to push the column of fluid in the tube past the ball check valve.

In stiff to hard cohesive soils, samples are often taken by driving heavy-gauge tubes (*see Fig. 7.80* for a driven sample of glacial till).

*Standard Stationary Piston Sampler*

THE SAMPLER A thin-wall tube is attached to a head assembly. The tube contains a piston (Fig. 2.63), which is connected to a rod passing through the drill rod to the surface. The piston when at the bottom of the tube prevents soil from entering the tube as it is lowered into the hole and permits seating through soft cuttings. Sampler is used to retrieve very soft to firm cohesive soils.

SAMPLING OPERATION The rod connected to the piston is held fixed at the surface while the hydraulic system on the drilling machine presses the tube past the piston into the soil. With light rigs the reaction can be increased by using earth anchors. In properly fitted piston samplers a strong vacuum is created to hold the sample in the tubes during withdrawal from the hole.

*Osterberg Hydraulic Piston Sampler*

THE SAMPLER A thin-wall tube contains an actuating and a fixed piston (Fig. 2.64). An opening in the head assembly permits applying fluid pressure to the actuating piston at the tube top. The sampler is used for very soft to firm cohesive soils.

SAMPLING OPERATION Fluid pressure is applied to the actuating piston, which presses the tube past the fixed piston into the soil. The actuating piston eliminates the cumbersome rods of the standard stationary piston as well as the possibility of overpushing.

*Greer and McClelland Shear-Pin Piston*

This device is similar to the Osterberg sampler except that the tube is attached to the piston with shear pins, which permit the fluid pressure to build to a high value before it "shoots" the piston when the pins shear. The sampler can be used in soft to stiff cohesive soils and loose sands. In loose sands disturbance is unavoidable. The "apparent" density, however, can be determined by measuring the weight of the total sample in the tube and assuming the volume calculated from the tube diameter and the stroke length. With the shear-pin piston, the sampling tube will almost always be fully extended because of the high thrust obtained.

## Double-Tube Soil Core Barrels

*Purpose*

Double-tube soil core barrels are used to obtain undisturbed samples in stiff to hard cohesive soils, saprolite, and soft rock.

*Denison Core Barrel*

THE SAMPLER A rotating outer barrel and bit contain a fixed inner barrel with a liner, as shown on Fig. 2.65. The cutting shoe on the inner barrel can extend below the cutting bit. Liners range from 28-gauge galvanized steel to brass and other materials such as phenolic-resin-impregnated paper with a 1/16-in wall. Various bits are available for cutting materials of varying hardness, to obtain samples ranging in diameter

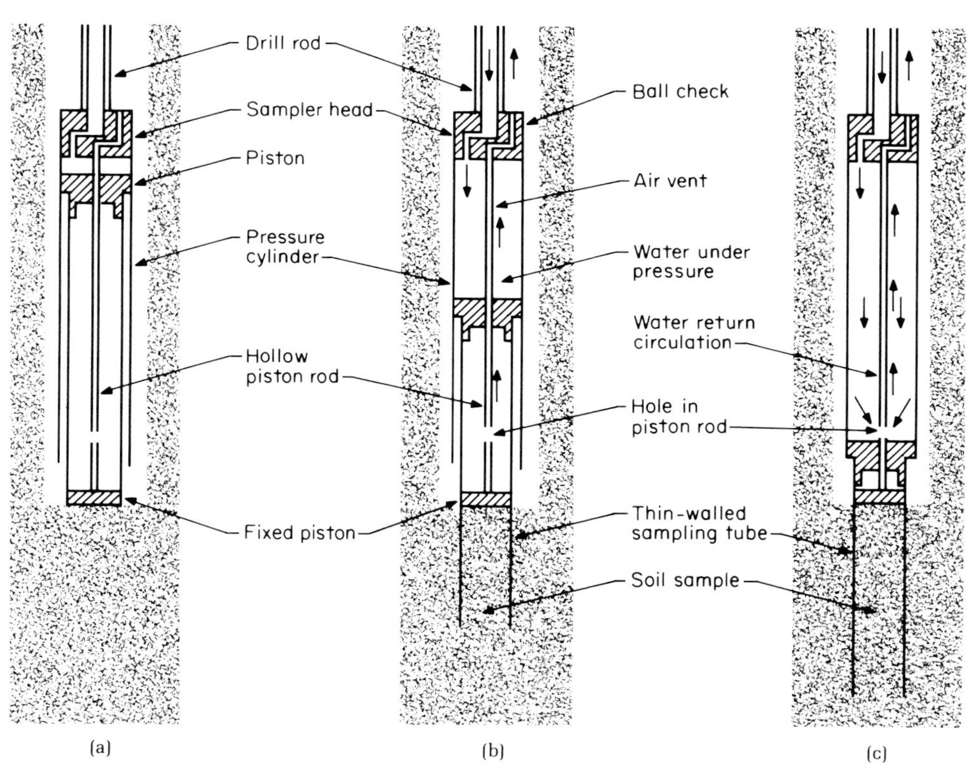


**FIG. 2.64 Operation of the Osterberg piston sampler. (a) Sampler is set on cleaned bottom of borehole. (b) Hydraulic pressure propels sampling tube into the soil. (c) Pressure is released through hole in piston rod.** *[From ENR (1952).[28] Reprinted with permission of McGraw-Hill Book Company.]*

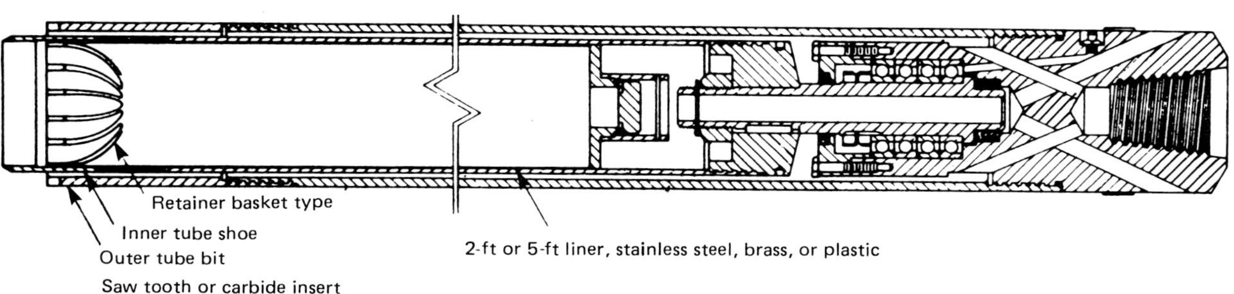


**FIG. 2.65 Denison core barrel.** *(Courtesy of Sprague and Henwood, Inc.)*

**FIG. 2.66 Removal of the cutting bit from the Denison core barrel.**

from $2\frac{3}{8}$ to $6\frac{5}{16}$ in. Sample tubes range from 2 to 5 ft in length.

SAMPLING OPERATION During drilling, pressure is applied by the hydraulic feed mechanism on the drill rig to the inner barrel while the bit on the outer barrel cuts away the soil. The sampler is retrieved from the hole, the cutting bit removed from the barrel (Fig. 2.66), and the thin liners, retained in the inner tube by wall friction or a basket-type retainer, are removed.

The extension of the cutting shoe below the cutting bit is adjustable. The maximum extension is used in relatively soft or loose materials, whereas in hard materials the shoe is maintained flush with the bit.

*Pitcher Sampler*

The operation of the Pitcher sampler is similar to that of the Denison core barrel except that the inner barrel is spring-loaded and thus provides for the automatic adjustment of the distance by which the cutting edge of the barrel leads the coring bit (Fig. 2.67). Because of adjusting-spring pressure, the Pitcher sampler is particularly suited to sampling deposits consisting of alternating soft and hard layers.

2.4.3 MISCELLANEOUS SOIL-SAMPLING METHODS

### Wash Samples

Completely disturbed cuttings from the hole advance operation carried to the surface by the wash fluid and caught in small sieves or by hand are termed washed samples. Value is limited to providing only an indication of the type of material being penetrated.

### Auger Samples

Completely disturbed cuttings from the penetration of posthole diggers (hand augers, *see Table 2.13*), continuous flight augers, or large diameter augers are brought to the surface when the auger is removed from the hole. In cohesive soils they are useful for soil identification, moisture content, and plasticity index tests.

### Retractable Plug Sampler

One-inch-diameter tubes containing slightly disturbed samples are obtained in soft to firm organic and cohesive soils suitable for soil iden-

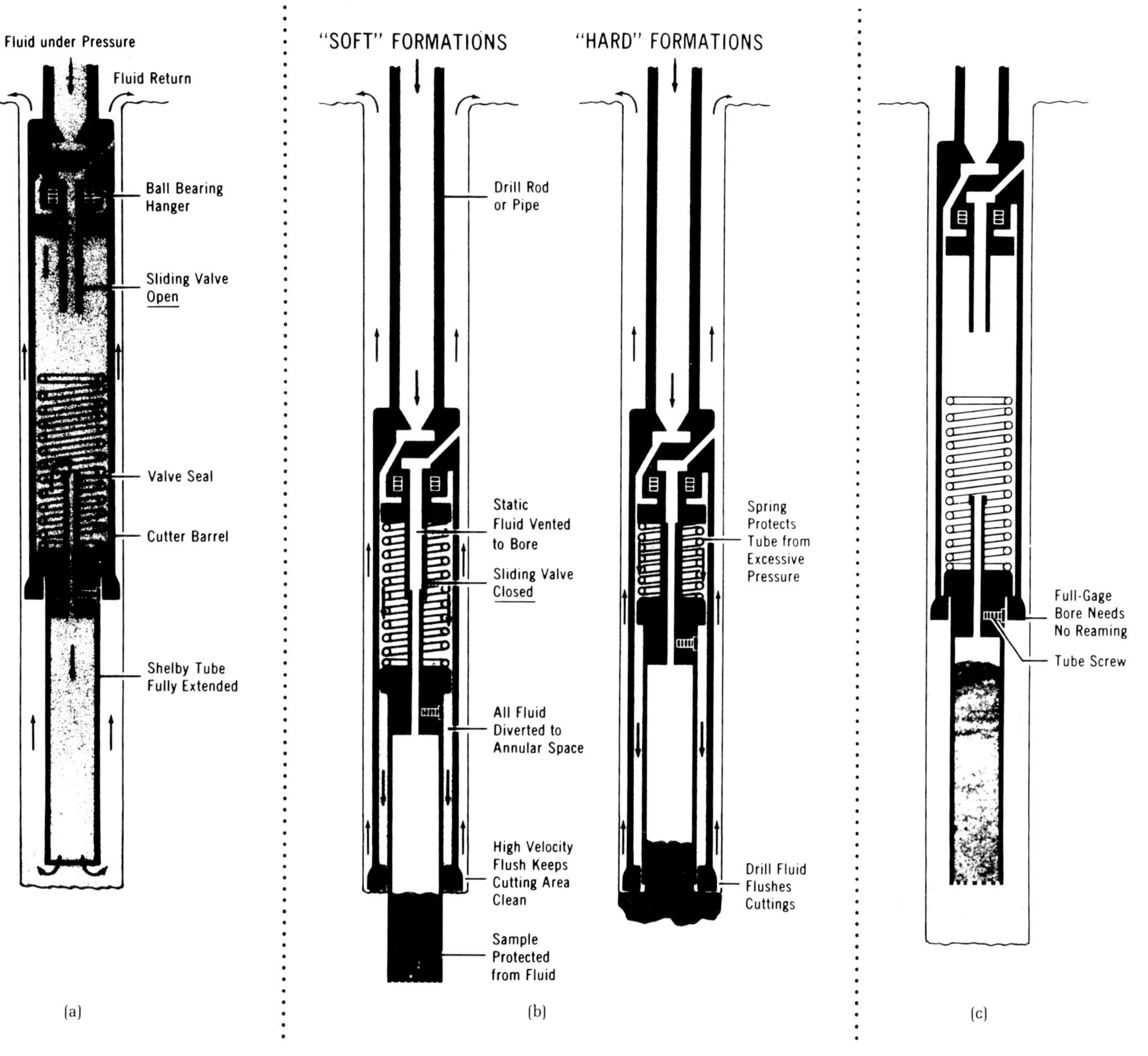


**FIG. 2.67 Operation of the Pitcher sampler: (*a*) down the hole; (*b*) sampling; (*c*) recovery.** *(Courtesy of Mobile Drilling Inc.)*

**FIG. 2.68 Float-mounted tripod rig; casing is being driven prior to SPT exploratory sampling.**

**FIG. 2.69 Barge-mounted rotary drill rig operating in the Hudson River for the third tube of the Lincoln Tunnel, New York City, a location with strong currents and heavy boat traffic.**

tification and moisture content measurements (see *Table 2.13*).

**Test Pit Block Samples**

High-quality undisturbed block or "chunk" samples in stiff to hard cohesive soils, or small cylinder samples in softer soils with some cohesion, are taken from test pits.

Strong cohesive soils are sampled by carefully hand-cutting a block from the pit walls. The sample is trimmed by knife and encased in paraffin on the exposed sides. The block is cut loose from the pit, overturned and the remaining side coated, and then sealed in a box for shipment to the laboratory. Very large samples are possible.

Weaker soils with come cohesion are carefully hand-trimmed into a small cylinder and sealed. The method has been used to obtain samples for density tests in partially saturated silty or slightly clayey sands.

**Swedish Foil Sampler**

High-quality continuous samples in soft, sensitive cohesive soils, useful in locating the shear zone in a slope failure problem, are possible with the Swedish foil sampler. A sampling tube, usually 8 ft in length, is pushed into the soil by a special drill rig as a reaction. To eliminate friction between the sample and the tube walls, thin steel strips, or foils, unroll to follow along the sampler walls as the sampler penetrates the soil.

2.4.4 SUBAQUEOUS SAMPLING

**Categories**

Sample procurement in subaqueous conditions can be placed in one of four general categories on the basis of the sampling technique:

1. Normal cased-boring methods
2. Wireline drilling techniques
3. Sampling to shallow depths below the bottom without drill rigs and casing
4. Sample recovery from deep borings in offshore sands is often difficult. Borehole remote-sensing and logging methods (see *Art. 2.3.6*), such as the electric well logger and nuclear probes, should be considered since they provide important supplemental data.

**FIG. 2.70 Rotary drilling from a self-propelled jack-up suitable for water depths to 30 m.** *(Courtesy of Otis Engineering Corp.)*

**Normal Cased-Boring Methods**

*General*

Normal cased borings require a stable platform for mounting the boring equipment and procurement of samples. The up and down movements from swells severely affect drilling and sampling operations as bits and samplers are removed from contact with the hole bottom. Tidal effects require careful considerations in depth measurements.

*Platforms*

*Floats* (Fig. 2.68) or *barges* (Fig. 2.69) are used in shallow water, generally less than 15 m deep, with slight swells. Penetration depths are in moderate ranges, depending on drill rig capacity.

*Large barges* or *jack-up platforms* (Fig. 2.70) are used in water to depths in the order of 30 m with slight to moderate swells. Penetration depths below bottom are moderate depending upon the drilling equipment.

*Submersibles* (Fig. 2.71) containing a small rotary

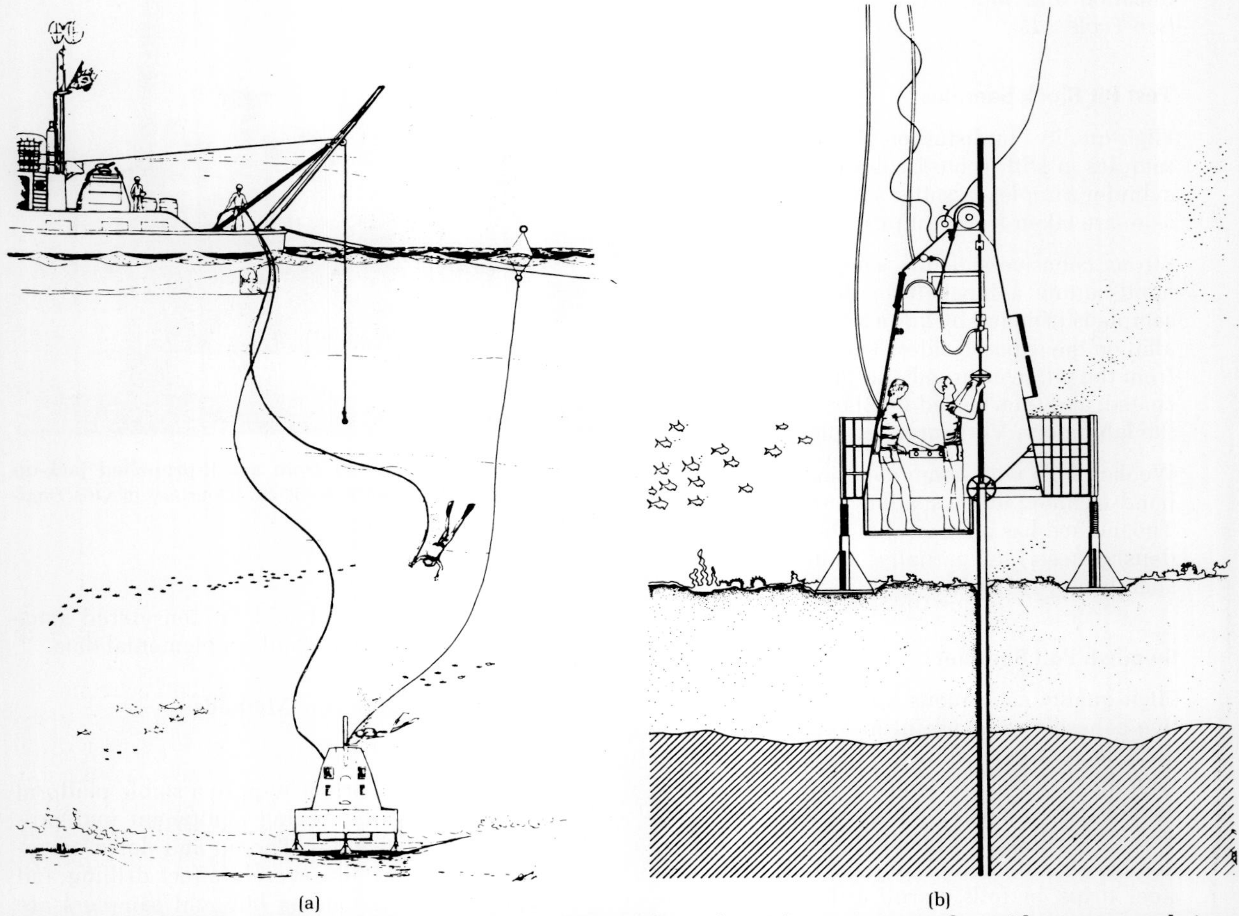


**FIG. 2.71 The seafloor drilling bell developed by ENGESUB of Brazil. (*a*) The support vessel provides compressed air to the bell. (*b*) Operation in the bell interior performed by divers.**

rig are suitable at present in water depths to about 30 m and moderate to heavy swells. The example shown has a maximum boring penetration of about 20 m and is suitable for split-barrel and undisturbed sampling.

### Wireline Drilling Methods

*General*

Wireline drilling techniques are used in deep water, are capable of much deeper penetration depths than are normal cased borings, and can tolerate much more severe sea conditions than can cased borings.

*Platforms*

*Large barges* or *moderately large ships* are used in relatively calm water and water depths over 15 m where deep penetration of the seafloor is required.

*Large drill ships* (Fig. 2.43) or *jack-up platforms* are used in deep water with heavy swells where deep penetration below the seafloor is required.

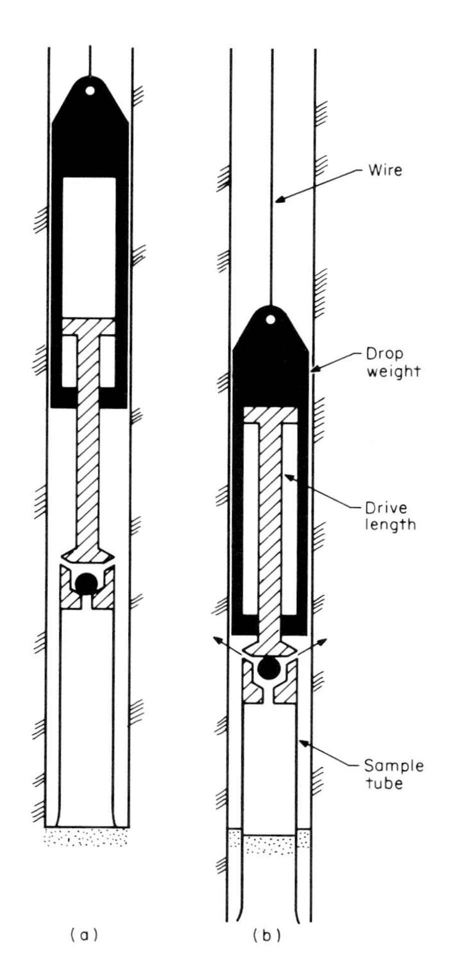


**FIG. 2.72 Wireline drive sampler; (*a*) before driving and (*b*) after driving.**

*Samplers*

DRIVE SAMPLERS (Fig. 2.72.) Either split-barrel samplers or tubes are driven with an 80-kg hammer dropped 3 m by release of a wire-hoisting drum. Penetration is only approximated by measuring sample recovery since the sampler is attached to the wire, not to drill rods. Recovery is related to the blow count for a rough estimate of relative density. Tube samples recovered in deep water at substantial depths below the seafloor in stiff clays will undergo significant strength decrease from stress release upon extraction from the seabed and extrusion in the shipboard laboratory.

PUSHED-TUBE SAMPLERS The Fugro Wipsampler (Fig. 2.73) is used to recover undisturbed samples from cohesive seabed soils. Lowered on an umbilical cable to the bottom of the wireline drill string, it latches into a special sub with an internal groove. The packer, serving as an anchor for reaction (Fig. 2.47), is inflated by drilling mud under pressure, and the sampler is pushed hydraulically. Penetration and hydraulic pressure are monitored in the control cab (Fig. 2.48). After sampling the anchor is deflated and hole advance continues.

### Sampling to Shallow Penetration without Drill Rigs and Casing

*General*

Various devices and methods are available for sampling shallow seafloor conditions without the necessity of mounting a drill rig on a platform and maintaining a fixed position for extended time intervals. Sampling procedure involves operating the equipment from the side of a vessel equipped with a crane. Sampling is generally not feasible in strong materials or to bottom penetration depths greater than about 12 m, depending upon the device used.

*Sampling Methods and Devices*

The application, description, and penetration depths of the various devices and methods are summarized on Table 2.20. The devices include the following:

- Petersen dredge
- Harpoon-type gravity corer (Fig. 2.74)
- Free-fall gravity corer (Fig. 2.75)
- Piston gravity corer
- Piggot explosive coring tube
- NGI (Norwegian Geotechnical Institute) gas-operated free-fall piston (Fig. 2.76)
- Vibracore (Fig. 2.77)

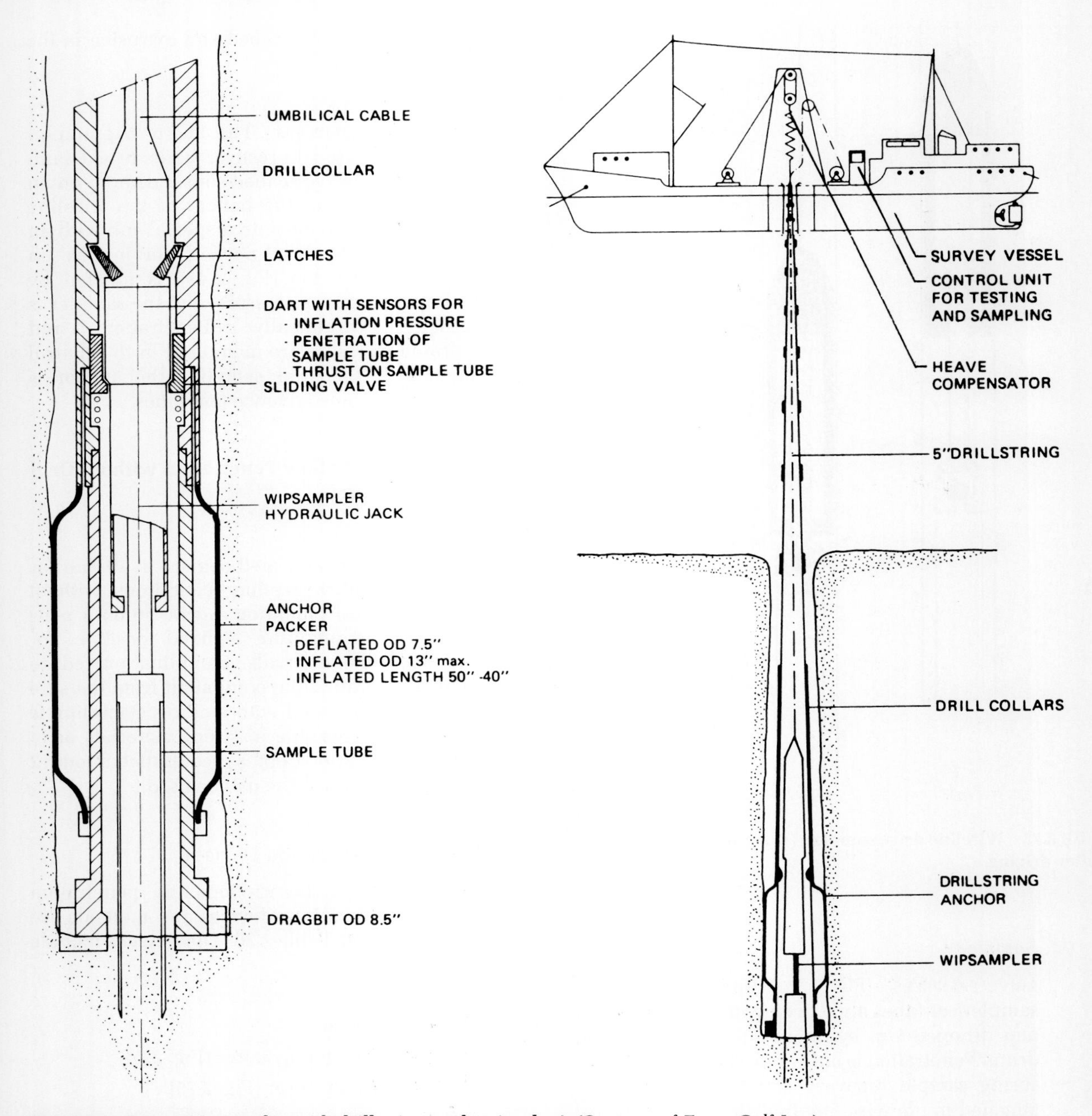


**FIG. 2.73 Fugro Wipsampler with drillstring anchor (packer).** *(Courtesy of Fugro Gulf Inc.)*

**TABLE 2.20**
**SUBAQUEOUS SOIL SAMPLING WITHOUT DRILL RIGS AND CASING**

| Device | Application | Description | Penetration depth | Comments |
|---|---|---|---|---|
| Petersen dredge | Large, relatively intact "grab" samples of seafloor. | Clam-shell type grab weighing about 100 lb with capacity about 0.4 $ft^3$. | To about 4 in. | Effective in water depths to 200 ft. More with additional weight. |
| Harpoon-type gravity corer (Fig. 2.74) | Cores from 1.5- to 6-in dia. in soft to firm soils. | Vaned weight connected to coring tube dropped directly from boat. Tube contains liners and core retainer. | To about 30 ft. | Maximum water depth depends only on weight. UD sampling possible with short, large-diameter barrels. |
| Free-fall gravity corer (Fig. 2.75) | Cores 1.5- to 6-in dia. in soft to firm soils. | Device suspended on wire rope over vessel side at height above seafloor about 15 ft and then released. | Soft soils to about 17 ft. Firm soils to about 10 ft. | As above for harpoon type. |
| Piston gravity corer (Ewing gravity corer) | 2.5-in sample in soft to firm soils. | Similar to free-fall corer except that coring tube contains a piston that remains stationary on the seafloor during sampling. | Standard core barrel 10 ft; additional 10 ft sections can be added. | Can obtain high-quality UD samples. |
| Piggot explosive coring tube | Cores of soft to hard bottom sediments. | Similar to gravity corer. Drive weight serves as gun barrel and coring tube as projectile. When tube meets resistance of seafloor, weighted gun barrel slides over trigger mechanism to fire a cartridge. The exploding gas drives tube into bottom sediments. | Cores to 1⅞ in and to 10 ft length have been recovered in stiff to hard materials. | Has been used successfully in 20,000 ft of water. |
| Norwegian Geotechnical Institute gas-operated piston (Fig. 2.76) | Good-quality samples in soft clays. | Similar to the Osterberg piston sampler except that the piston on the sampling tube is activated by gas pressure. | About 35 ft. | |
| Vibracore (Fig. 2.77) | High-quality samples in soft to firm sediments. Dia. 3½ in. | Apparatus is set on seafloor. Air pressure from the vessel activates an air-powered mechanical vibrator to cause penetration of the tube, which contains a plastic liner to retain the core. | Length of 20 and 40 ft. Rate of penetration varies with material strength. Samples a 20-ft core in soft soils in 2 min. | Maximum water depth about 200 ft. |

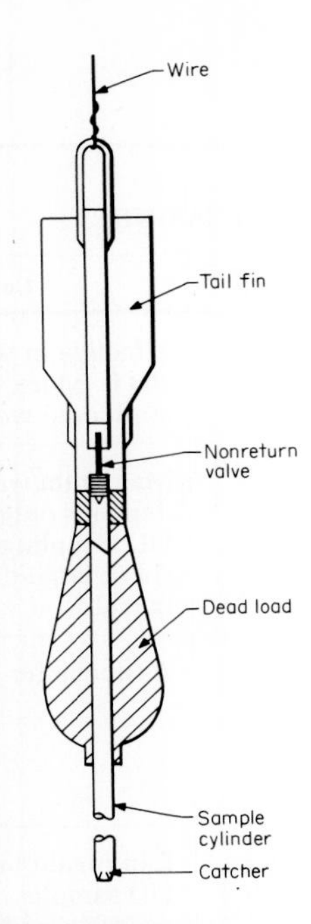


**FIG. 2.74 Harpoon-type gravity corer.**

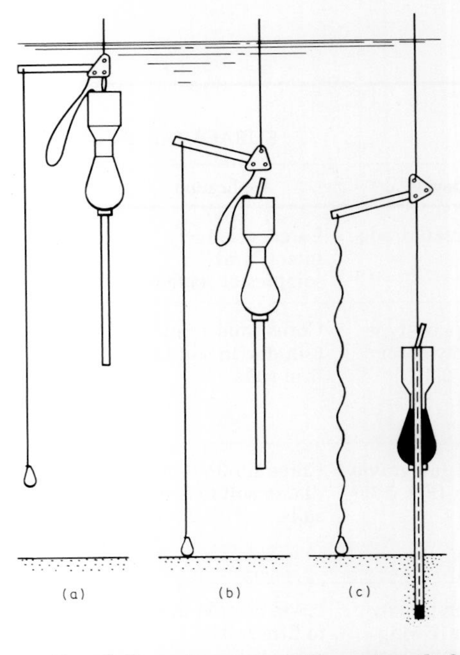


**FIG. 2.75 Free-fall gravity corer: (*a*) lowering; (*b*) free fall; (*c*) sampling.**

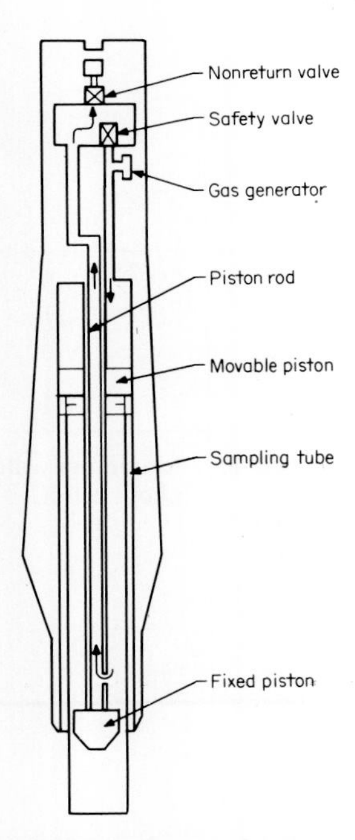


**FIG. 2.76 NGI gas-operated free-fall piston sampler.** [*From Eide (1974).*[29]]

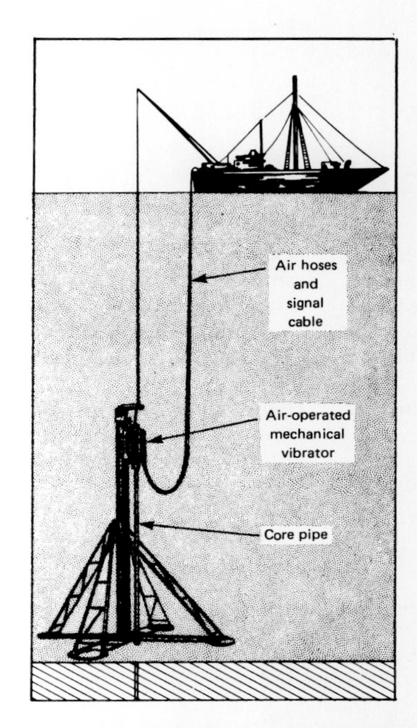


**FIG. 2.77 The Vibracore lowered to the seafloor.** (*Courtesy of Ocean/Seismic/Survey, Inc.*)

## 2.4.5 ROCK CORING

### Objectives

Rock coring is intended to obtain intact cores and a high percentage of core recovery.

### Equipment

Rotary drilling machine, drill rods, a core barrel to receive the core, and a cutting bit are needed.

### Operations

*General*

The core barrel is rotated under pressure from the drill rig applied directly to it while water flows through the head, down the barrel, out through the waterways in the bit, and up through the rock hole and casing (in soil) to return to the surface.

*Coring Start*

When rock is first encountered in a borehole, the initial core runs are usually short because of the possibility that the upper rock will be soft and fractured. As rock quality improves, longer core runs are made.

*Speed of Barrel Rotation*

Generally between 50 and 1750 rev/min, rotation speed is a function of the bit diameter and rock quality. Slow speeds are used in soft or badly fractured rocks, and high speeds are used in sound hard rocks. If large vibrations and "chatter" of the drill stem occurs, the speed should be reduced or core recovery and quality will be severely affected.

*Bit Pressure*

Pressure is also modified to suit conditions. Low bit pressure is used in soft rocks and high pressure is used in hard rocks. When vibrations and "chatter" occur, the pressure, which is imposed hydraulically, should be reduced.

*Drilling Fluid Pressure*

Fluid pressure should be the minimum required to return the cuttings adequately to the surface to avoid erosion of borehole walls.

*Lack of Drilling Fluid Return*

If there is no fluid return, drilling should immediately stop and the core barrel returned to the surface to avoid overheating the bit, which would result in bit damage (loss of diamonds) and possible jamming in the hole. Lack of fluid can result from:

- Blockage of the core barrel, which occurs in clayey zones.
- Loss in caverns, large cavities, or highly fractured zones. In Fig. 2.78, which shows coring in a limestone with highly fractured zones above the water table, a light drilling mud is being used to minimize fluid loss (note the mud "pit").

*Core Grinding*

Continued drilling after a broken core has blocked entry into the core barrel results in core grinding. Indications of blockage may be heavy rod vibrations, a marked decrease in penetration rate accompanied by an increase in engine speed, return fluid more heavily laden with cuttings than normal, and a rise in circulation fluid pressure.

*Core Retrieval*

The core barrel is retrieved from the ground. The core is removed from the barrel (Fig. 2.79) and laid out in wooden boxes exactly as recovered (Fig. 2.80). Wooden spacers are placed to divide each run. The depths are noted, the core is examined, and a detailed log is prepared.

### Core Barrels

*General*

The selection of a core barrel is based on the condition of the rock to be cored and the amount and quality of core required.

*Lengths*

Core barrels vary in length from 2 to 20 ft, with 5 and 10 ft being the most common.

FIG. 2.78 *(Above left)* Core drilling for quarry sites in the Negev Desert of Israel with a Failing Holemaster. Light drilling mud is necessary in the fractured limestone above the water table to prevent loss of drilling fluid.

FIG. 2.79 *(Left)* Removal of HX diameter limestone core from the inner barrel of a double-tube swivel-type core barrel.

FIG. 2.80 *(Above right)* Core recovery of 100% in hard, sound limestone; very poor recovery in shaly, clayey, and heavily fractured zones.

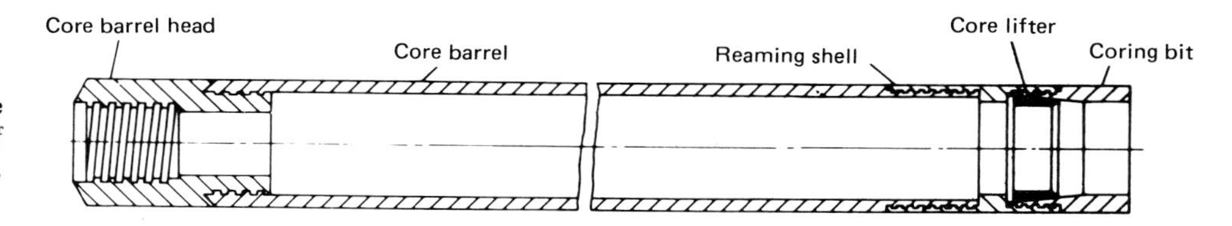


**FIG. 2.81 Single-tube core barrel.** *(Courtesy of Sprague and Henwood, Inc.)*

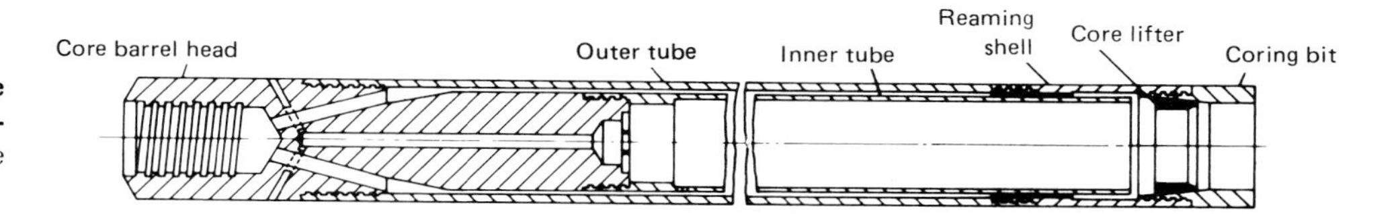


**FIG. 2.82 Rigid-type double-tube core barrel.** *(Courtesy of Sprague and Henwood, Inc.)*

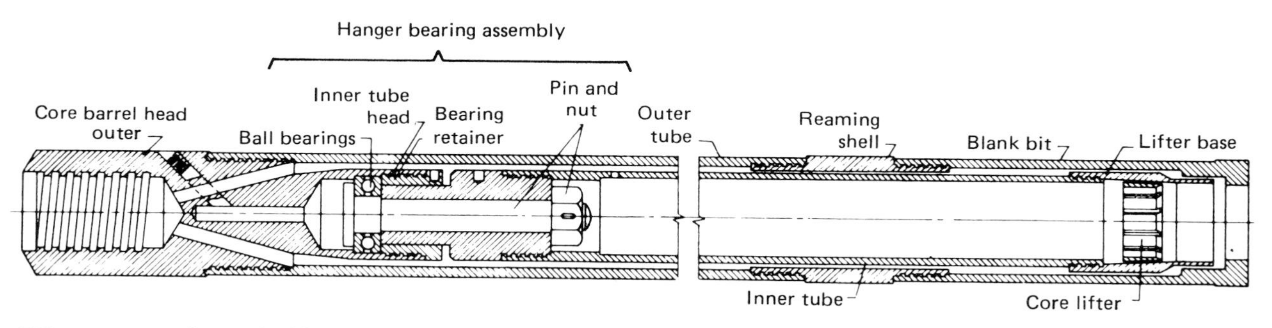


**FIG. 2.83 Swivel-type double-tube core barrel, series M.** *(Courtesy of Sprague and Henwood, Inc.)*

*Types*

Table 2.21 provides summary descriptions of suitable rock conditions for optimum application, descriptions of barrel operation, and general comments. The types include:

- Single-tube core barrel (Fig. 2.81)
- Double-tube rigid core barrel (Fig. 2.82)
- Swivel-type double-tube core barrel, two types:
  - Conventional
  - Series M (Fig. 2.83)
- Wireline core barrel (Fig. 2.84)
- Oriented core barrel

### Coring Bits

*General*

*Types* of coring bits are based on the cutting material, i.e., sawtooth, carbide inserts, diamonds.

*Waterways* are required in the bits for cooling. Conventional waterways are passages cut into the bit face; they result in enlarged hole diameter in soft rock. Bottom-discharge bits should be used for coring soft rock or rock with soil-filled fractures. Discharge occurs behind a metal skirt separating the core from the discharging fluid, providing protection from erosion.

*Common bit sizes and core diameters* are given on Table 2.15. The smaller diameters are used in exploratory borings for rock identification or in good-quality rock, but when maximum core recovery is required in all rock types, NX cores or larger are obtained. In seamy and fractured rock, core recovery improves with the larger diameters, and HX size is commonly used.

*Reaming shells,* slightly larger than the core barrel diameter and set with diamonds or carbide insert strips, ream the hole, maintaining its gauge and reducing bit wear.

**TABLE 2.21**
**TYPES OF ROCK CORE BARRELS**

| Core barrel | Suitable rock conditions | Operation | Comments |
|---|---|---|---|
| *Single-tube* (Fig. 2.81) | Hard homogeneous rock which resists erosion. | Water flows directly around the core. Uses split-ring core catcher. | Simple and rugged. Severe core loss in soft or fractured rock. |
| *Double-tube, rigid type* (Fig. 2.82) | Medium to hard rock, sound to moderately fractured. Somewhat erosion-resistant. | Inner barrel attached to head and rotates with outer barrel as water flows through annular space. | Water makes contact with core only in reamer shell and bit area, reducing core erosion. Holes in inner tube may allow small flow around core. |
| *Double-tube, swivel type* (conventional series) | Fractured formations of average rock hardness not excessively susceptible to erosion. | Inner barrel remains stationary while outer barrel and bit rotate. Inner barrel terminates above core lifter. | Torsional forces on core are eliminated minimizing breakage. Core lifter may tilt and block entrance to inner barrel, or may rotate with the bit causing grinding of the core. |
| *Double-tube, swivel type* (series M) (Fig. 2.83) | Badly fractured, soft, or friable rock easily eroded. | Similar to conventional series except that core lifter is attached to inner barrel and remains oriented. Inner barrel is extended to the bit face. | Superior to the conventional series. Blocking and grinding minimized. Erosion minimized by extended inner barrel. |
| *Wireline core barrel* (Fig. 2.84) | Deep core drilling in all rock conditions. | See Art. 2.3.4. | Retriever attached to wireline retrieves inner barrel and core without necessity of removing core bit and drill tools from the hole. |
| *Oriented core barrel* | Determine orientation of rock structure. | Similar to conventional core barrels. Orienting barrel has three triangular hardened scribes mounted in the inner barrel shoe which cuts grooves in the core. A scribe is aligned with a lug on a survey instrument mounted in a nonmagnetic drill collar. The instrument contains a compass-angle device, multishot camera, and a clock mechanism. After 30 cm or so of coring, the advance is stopped and a photograph taken of the compass, clock, and lug. Geologic orientation is obtained by correlation between photographs of the core grooves and the compass photograph. | |

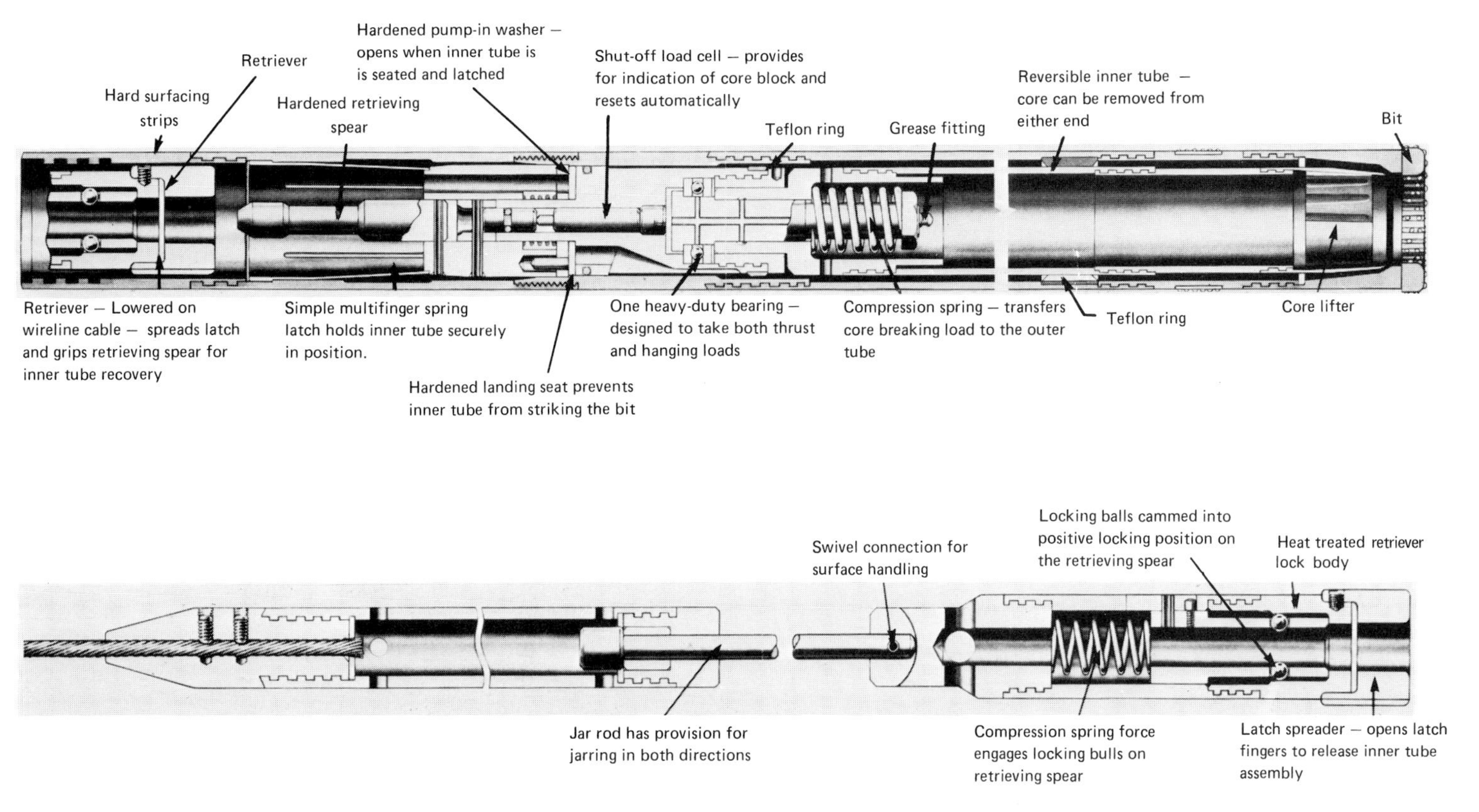


**FIG. 2.84 Wireline core barrel and retriever assembly.** *(Courtesy of Sprague and Henwood, Inc.)*

**FIG. 2.85 Massive, hard granite gneiss at the mouth of a water tunnel. Note diorite dike and seepage from joints. Core recovery in such materials should be high.**

**FIG. 2.86 Jointed granite gneiss and crushed rock zone at other end of tunnel of Fig. 2.85, about 400 m distant. Core recovery is shown on Fig. 2.87.**

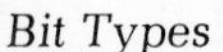

*Bit Types*

*Sawtooth bits* are the lowest in cost and have a series of teeth cut in the bit which are faced with tungsten carbide. They are used primarily to core overburden and very soft rock.

*Carbide insert bits* (see *Fig. 2.55*) have tungsten carbide teeth set in a metal matrix and are used in soft to medium-hard rocks.

*Diamond bits* (see *Fig. 2.55*) are the most common type, producing high-quality cores in all rock types from soft to hard. Coring is more rapid and smaller and longer cores are retrieved than with other bit types. The diamonds are either surface-set in a metal matrix, or the metal matrix is impregnated throughout with diamond chips. There are various designs for cutting various rock types, differing in quality, size, and spacing of the diamonds, matrix composition, face contours, and the number and locations of the waterways.

### Core Recovery and RQD

*Reporting Methods*

*Percent core recovery* is the standard reporting method wherein core recovery is given as a percentage of total length cored. *RQD (rock quality designation)* was proposed by Deere (1963)[30] as a method for classifying core recovery to reflect the fracturing and alteration of rock masses. For RQD determination the core should be at least 50 mm in diameter (NX) and recovery with double-tube swivel-type barrels is preferred.

RQD is obtained by summing the total length of core recovered, but counting only those pieces of hard, sound core which are 10 cm (4 in) in length or longer, and taking that total length as a percentage of the total length cored. If the core is broken by handling or drilling, as evidenced by fresh breaks in the core (often perpendicular to the core), the pieces are fitted together and counted as one piece.

*Causes of Low Recovery*

ROCK CONDITIONS Fractured or decomposed rock and soft clayey seams cause low recovery, for example, as shown on Fig. 2.80. Rock quality can vary substantially for a given location and rock types as illustrated in Figs. 2.85 and 2.86, which show a formation of granite gneiss, varying from sound and massive to jointed and seamy. Core recovery in the heavily jointed zone is illustrated on Fig. 2.87.

CORING EQUIPMENT Worn bits, improper rod sizes (too light), improper core barrel and bit, and inadequate drilling machine size all result in low recovery. In one case in the author's experience, coring to depths of 30 to 50 m in a weathered to sound gneiss with light drill rigs, light rods (A), and NX double-tube core barrels resulted in 40 to 70% recoveries and 20 to 30% RQD values. When the same drillers redrilled the holes within 1 m distance using heavier machines, N rod and HX core barrels, recovery increased to 90 to 100% and RQDs to 70 to 80%, even in highly decomposed rock zones, layers of hard clay, and seams of soft clay within the rock mass.

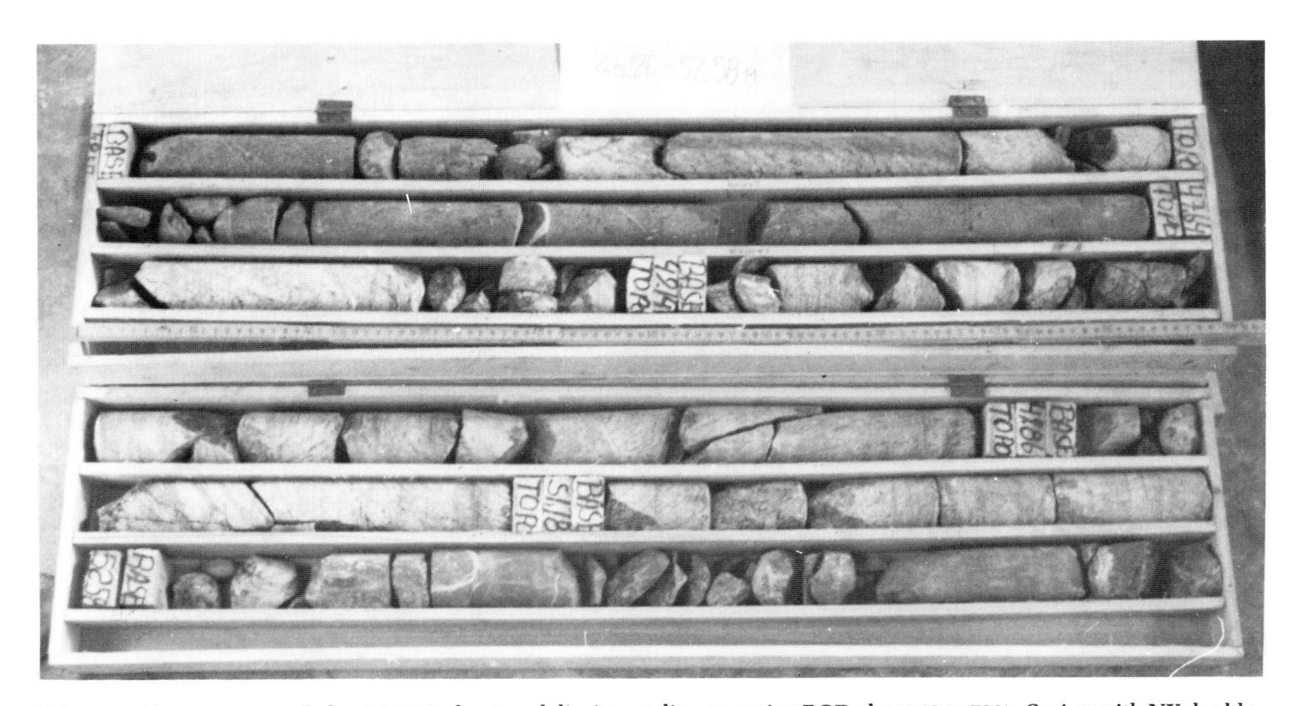

**FIG. 2.87 Core recovery of about 90% in fractured diorite grading to gneiss; RQD about 40 to 70%. Coring with NX double-tube swivel-type barrel.**

CORING PROCEDURE Inadequate drilling fluid quantities, too much fluid pressure, improper drill rod pressure, or improper rotation speed all affect core recovery.

### Integral Coring Method

*Purpose*

Integral coring is used to obtain representative cores in rock masses in which recovery is difficult with normal techniques and to reveal defects and discontinuities such as joint openings and fillings, shear zones, and cavities. The method, developed by Dr. Manual Rocha of Laboratorio Nacional de Engenheira Civil (LNEC) of Lisbon, can produce cores of 100% recovery with the orientation known. Defect orientation is an important factor in rock-mass-stability analysis.

*Technique*

1. An NX-diameter hole or larger is drilled to where integral coring is to begin.
2. A second, smaller hole (nominally about 2.6 mm in diameter) is drilled coaxially with the first through the desired core depth, although usually not exceeding 1.5 m in depth.
3. A notched pipe is lowered into the hole and bonded to the rock mass with cement or epoxy resin grout, which leaves the pipe through perforations.
4. After the grout has set, a core is recovered by overcoring around the pipe and through the cemented mass.
5. During installation of the pipe, the notch positions are carefully controlled by a special adapter and recorded so that when the core is retrieved the orientation of the fractures and shear zones in the rock mass are known.

### Large-Diameter Cores by Calyx or Shot Drilling

*Purpose*

Calyx or shot drilling is intended to allow borehole inspection in rock masses in holes up to 2 m in diameter.

*Method*

Calyx coring uses chilled shot as a cutting medium. The shot is fed with water and lodges around and partially embeds in a bit of soft steel. The flow of freshwater is regulated carefully to remove the cuttings but not the shot. The cores are recovered by a special corelifter barrel, wedge pins, or mucking after removal of the core barrel. An example of a 36-in-diameter core taken in shale in which slickensides are apparent is shown in Fig. 6.56.

*Limitations*

The method is limited to rock of adequate hardness to resist erosion by the wash water and to vertical or nearly vertical holes.

## 2.4.6 SAMPLE AND CORE TREATMENT

### Upon Retrieval

The sampler is dismantled carefully to avoid shocks and blows (in soils), obvious cuttings are removed, and the recovery is recorded (RQD is also recorded in rock masses). The sample is immediately described and logged. It is not allowed to dry out, since consistency of cohesive soils changes and details of stratification become obscured. The sample is then preserved and protected from excessive heat and freezing.

### Preservation, Shipment, and Storage

*Split-Barrel Samples*

Carefully place intact uncontaminated short cores in wide-mouth jars of sufficient size (16 oz) to store 12 in of sample. The samples, which may be used for laboratory examination, should not be mashed or pushed into any container; such action would result in complete loss of fabric and structure. The jar caps should contain a rubber seal, be closed tightly, and be waxed to prevent moisture loss. Liner samples are preserved as thin-wall tube samples.

*Thin-Wall Tube Samples*

Remove all cuttings from the sample top with a small auger and fill the top with a mixture of paraffin and a microcrystalline wax such as Petrowax, applied at a temperature close to the congealing point. Normal paraffin is subjected to excessive shrinkage during cooling and should not be used or an ineffective moisture seal will result. The top is capped, taped, and waxed.

Invert the tube, remove a small amount of soil from the bottom and fill the tube with wax. Cap, tape, and wax the bottom.

Tubes should be shipped upright, if possible, in containers separating the tubes from each other and packed with straw.

In the laboratory, tubes that are not to be immediately tested are stored in rooms with controlled humidity to prevent long-term drying. Soil properties can change with time; therefore, for best results, samples should be tested as soon as they are received in the laboratory.

*Rock Cores*

Rock cores are stored in specially made boxes (see *Fig. 2.87*) in which wooden spacers are placed along the core to identify the depth of run.

**Extrusion of Undisturbed (UD)** Samples

*Thin-Wall Pushed Samples*

These should be extruded from the tube in the laboratory in the same direction as the sample entered the tube, with the tube held vertical, as shown on Fig. 2.88. This procedure avoids the effects on sample quality of reverse wall friction and of the sample's passing the cutting edge of the tube.

*Thin-Wall Cored Samples*

Because they contain strong cohesive soils, wall friction in cored soil samples is usually too high to permit extrusion from the entire tube without causing severe disturbance. Removal normally requires cutting the tube into sections and then extruding the shorter lengths.

*Field Extrusion*

Some practitioners extrude the sample in the field, cutting off 6-in sections, wrapping them in aluminum foil, and surrounding them with wax in a carton. The procedure simplifies transport but leads to additional field and laboratory handling which may result in disturbance of easily remolded soils.

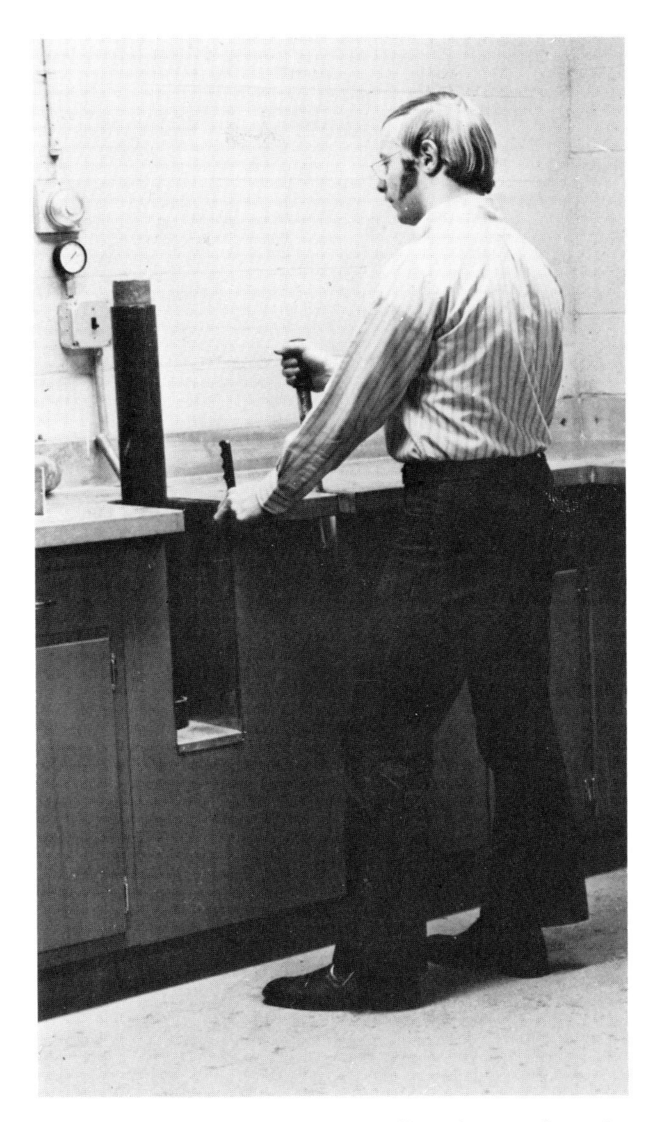

**FIG. 2.88 Vertical extrusion of Shelby tube sample in the laboratory. Sample is extruded in the same direction as taken in the field to minimize disturbance.** *(Courtesy of Joseph S. Ward & Associates.)*

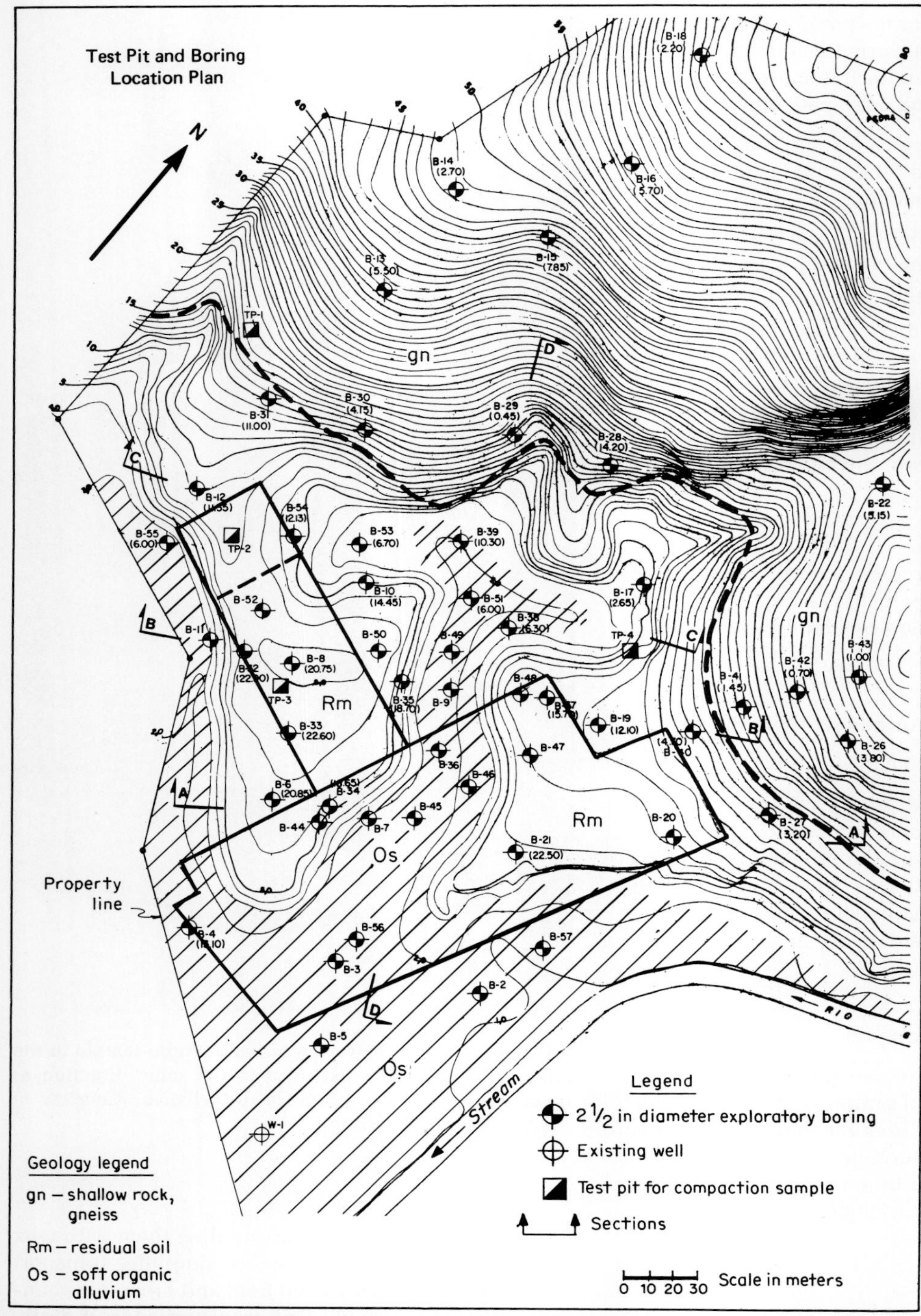


**FIG. 2.89 Example of test pit and boring location plan using topographic map and engineering geology map as base map.** *(Courtesy of Joseph S. Ward & Associates.)*

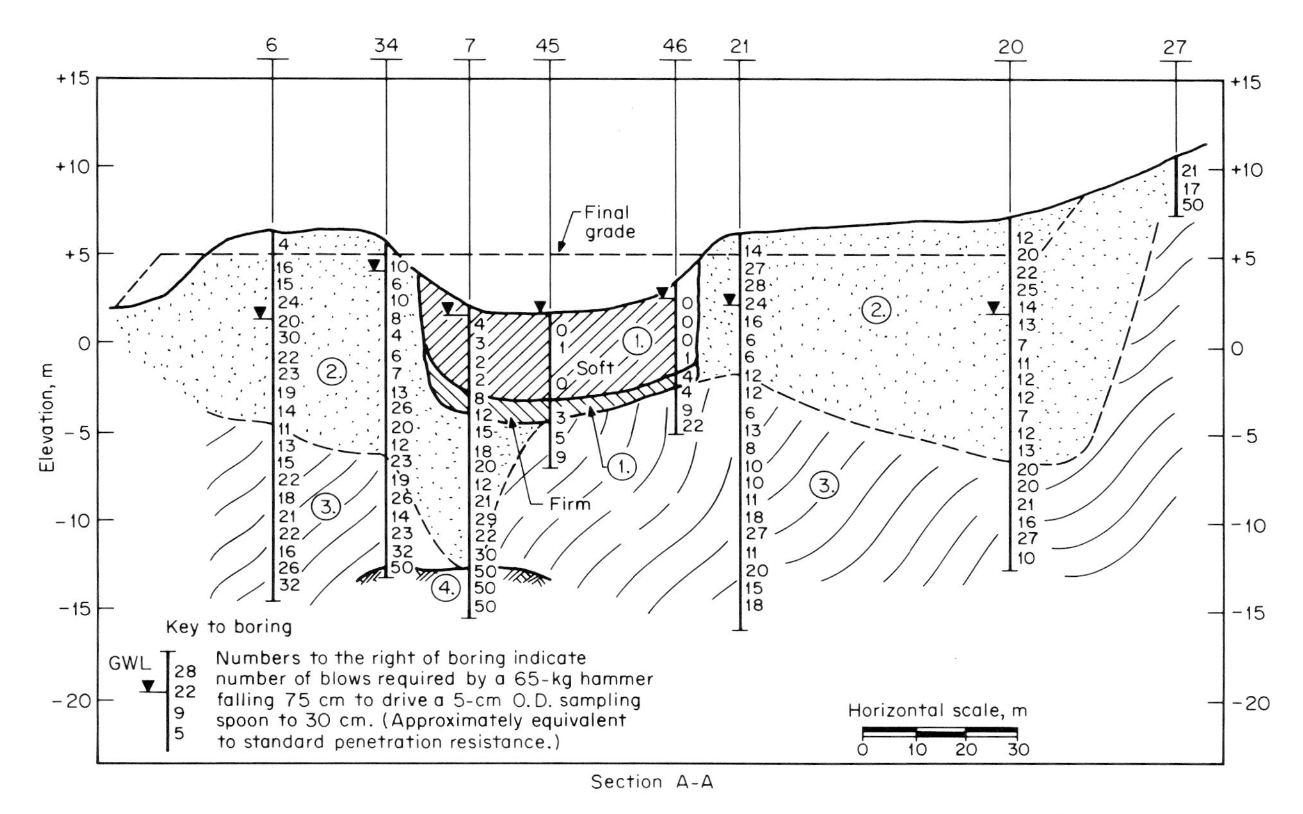


**FIG. 2.90 Typical geologic section across site shown on Fig. 2.89. Stratum descriptions: (1) recent alluvium and marine deposits consisting of interbedded organic silts, sands, and clays; (2) residual soil: silt, clay, and sand mixtures; (3) micaceous saprolite: highly decomposed gneiss retaining relict structure; (4) weathered and partially decomposed gneiss.**

## 2.4.7 DATA PRESENTATION

### Basic Elements

*Location Plan*

Locations of all explorations should be shown accurately on a plan. As a base map it is helpful to use a topographic map that also shows the surficial geology as in Fig. 2.89. A map providing the general site location is also useful, especially for future reference to local geologic conditions. Many reports lack an accurate description of the site location.

*Geologic Sections*

Data from the various exploration methods are used as a basis for typical geologic sections to illustrate the more significant geologic conditions, as in Fig. 2.90. The objective is to illustrate clearly the problems of the geologic environment influencing design and construction.

For engineering evaluations, it is often useful to prepare large-scale sections on which are plotted all of the key engineering property data as measured in the field and in the laboratory.

Fence diagrams, or three-dimensional sections, are helpful for sites with complex geology.

*Logs*

The results of test and core borings, test pits, and other reconnaissance methods are presented on logs which include all pertinent information.

## Boring Logs

### General

Logs are prepared to provide complete documentation on the drilling, sampling, and coring operations and on the materials and other aspects of the subsurface encountered, including groundwater conditions.

Logs provide the basis for analysis and design, and therefore complete documentation and clear and precise presentation of all data are necessary. Normally two sets of logs are required: *field logs* and *report logs*, each serving a different purpose.

### Field Log

PURPOSE A field log is intended to record all of the basic data and significant information regarding the boring operation.

INFORMATION INCLUDED Typical contents are indicated on the example given as Fig. 2.91, which is quite detailed, including the sample description and remarks on the drilling operations. The field log is designed to describe each sample in detail as well as other conditions encountered. All of the information is necessary for the engineer to evaluate the validity of the data obtained, but it is not necessary for design analysis. (For material descriptions *see Arts. 5.2.7 and 5.3.6.*)

*A preliminary report log* is begun by the field inspector as the field boring log is prepared.

### Report Log

PURPOSE A report log is intended to record the boring data needed for design analysis as well as some laboratory identification test data and a notation of the various tests performed. The report log also allows changes to be made in the material description column so that the descriptions agree with gradation and plasticity test results from the laboratory.

INFORMATION INCLUDED The examples given as Fig. 2.92 (test boring report log) for a soil and rock borehole and Fig. 2.93 (core boring report log) for rock core borings illustrate the basic information required for report logs.

Field boring log No. 1 Type UD-C Coordinates: N1025630 E78765.50 Elevation: +5.20 Datum: MSL Sheet 1 of 1

Project: New Plant Building Location: Uptown, N.J. Client: Widget Inc Project No. 78-102

Contractor: Drill Good Inc. Driller: W. Dowell Inspector: J. Eyesharp Start: 7.21.78

Drill rig type: Strong 1200 Rod diameter: N Casing (Dia. / type / wgt. / drop): 4-Flush Jt 300 lb 24 in Sampler (Dia. / wgt. / drop): 2 ft 140 lb 30 in Finish: 7.21.78

GWL: At start 4.2 ft, end of Boring 3.0 ft, 24 hrs 4.5 ft

| Casing blows per foot | Samples | | | | Blows per 6 in. | | | | Pen, in | Rec, in | Sample description | Remarks on Operations |
|---|---|---|---|---|---|---|---|---|---|---|---|---|
| | Depth, ft | No. | Type | Diameter | 6 | 12 | 18 | 24 | | | | |
| 3/6 | 0-1.5 | 1 | SS | 2 in | 2 | 4 | 4 | | 18 | 9 | Topsoil, DfS, l·$ to 5 in, then | 7/21 — 8 a.m.-11 a.m. set rig over |
| Mud drilled | | | | | | | | | | | BC-fS, t·$, dry, stratified | hole and prepare for boring |
| below 2 ft | 1.5-3.0 | 2 | SS | 2 in | 3 | 5 | 7 | | 18 | 12 | BC-fs t·$, moist, stratified, | S-1 taken from surface, then |
| to 21.0 ft | 3.0-4.5 | 3 | SS | 2 in | 1 | 2 | 4 | | 18 | 0 | Lost sample | 2 ft section 4 in casing set |
| then 4 in | 4.5-6.0 | 4 | SS | 2 in | 1 | 2 | 2 | | 18 | 10 | Top of sample: 3 in Bm-fs, l-fs soft | and mixed mud |
| casing | | | | | | | | | | | Bot. " " : DO$ w/shell fgmts. | S-3 Sampling spoon lacked |
| stalled | 6.0-8.0 | 5 | P | 3 in | Push-100 psi | | | | 24 | 24 | Top of sple: " " , soft | core catcher |
| to 20 ft | | | | | | | | | | | Bot. of sple: " , more plastic | Stratum change at 5.5 ft |
| for rock | 8.0-9.5 | 6 | SS | 2 in | 0 | 1 | 1 | | 18 | 7 | DO$, clayey, w/shell fgmts. | " " at 10.5 ft |
| coring | 9.5-11.0 | 7 | SS | 2 in | 2 | 4 | 9 | | 18 | 10 | Top of sple: 4 in DO$ w/shell fgmts. | Below 17.0 ft hard drilling |
| | | | | | | | | | | | Bot. of sple: 6 in residual soil: | Refusal to s.s at 20.1 ft |
| | | | | | | | | | | | multicolored c-fS l·$ plastic | Cleaned hole to 21.0 ft, drove |
| | | | | | | | | | | | micaceous, compact | casing to 21.0 ft; starting |
| | 11.0-13.0 | 8 | D | 2.5 in | Drilled | | | | 24 | 20 | Same as S-7, compact | coring at 3 p.m. w/ DS barrel |
| | 13.0-14.5 | 9 | SS | 2 in | 7 | 8 | 12 | | 18 | 10 | " " " | series "M" @ NX diamond bit w/ |
| | 15.0-16.5 | 10 | SS | 2 in | 7 | 10 | 17 | | 18 | 11 | Grading to saprolite, rock | bottom discharge. |
| | | | | | | | | | | | structure apparent, mica and | |
| | | | | | | | | | | | quartz in sand sizes, silty | Boring completed: 6 P.M. 7/21 |
| | 20.0-20.1 | 11 | SS | 2 in | 50/2 in | | | | 2 | 2 | Gneiss Rock: Soft, L.G, mod. dec. | GWL at 6 P.M.-Depth = 3.0 ft |
| | 21.0-26.0 | 12 | DS | NX in | 125 rpm @ 12 lb/in² | | | | 60 | 51 | GN, LG., soft-hard, mod. dec. | Hole Flushed; 5 ft casing left in hole |
| | | | | | (rate = 1 ft/6 in) | | | | | | closely jointed, RQD = 60% | GWL at 6 P.M. 7.22 at d = 4.5 ft |
| | 26.0-31.0 | 13 | DS | NX in | 125 rpm @ 250 lb/in² | | | | 60 | 57 | GN, LG., V.Hd, slightly dec., mod. jointed | Field log boring No. 1 |
| | | | | | (rate = 1 ft/12 in) | | | | | | RQD = 75% | Sheet 1 of 1 |

FIG. 2.91 Example of a field test boring report for soil and rock drilling.

| | | |
|---|---|---|
| Happy Engineering Co | Test boring report log | UD-C (1) Boring No. 1 |
| Project: New Plant Bldg. | Location: Uptown, N.J. | Sheet 1 of 1 |
| Client: Widget Inc. | Coordinates: N102563.02, E78 765.50 | Project No. 78-102 |
| Contractor: Drill-Good Inc. | Driller: W. Dowell | Inspector: J. Eyesharp |
| Date started: 7-21-78 | Date finished: 7-21-78 Surface Elevation: 15.20 ft | Datum: MSL |

| Depth, ft | Casing blows | Sample No. | N(3) | Sampler (2) | Recovery, % | Symbol | Strat. Elev. | Material and stratum description | UCS or RQD | Moisture (4) | Laboratory tests NMC | LL | PL | Other (5) |
|---|---|---|---|---|---|---|---|---|---|---|---|---|---|---|
| | 3 6 | 1 | 8 | SS | 50 | | 4.80 | Topsoil; dark grey silty fine sand | | D | | | | |
| | Mud | 2 | 12 | SS | 75 | | | Brown c-f sand, trace silt, stratified, loose | SM | m | | | | G |
| | | 3 | 6 | SS | 0 | | | | | | | | | |
| 5 | | 4 | 4 | SS | 60 | | -0.30 | | SM | ▼ W | | | | |
| | | 5 | P 100 | P | 100 | | | Dark grey organic silt and clay, w/shell fragments, soft, | OL | W | 110 | 120 | 70 | C, V-lab |
| | | 6 | 2 | SS | 40 | | | | OH | W | 80 | 85 | 68 | C, V-lab |
| 10 | | 7 | 18 | SS | 60 | | -5.00 | | OH | W | | | | |
| | | 8 | – | D | 80 | | | Multicolored clayey c-f sand, micaceous (residual soil), compact | SC | W | 26 | 23 | 19 | C,T |
| | | 9 | 20 | SS | 60 | | | | SC | M | | | | |
| 15 | | 10 | 27 | SS | 65 | | -10.30 | Grading to saprolite; multicolored silty c-f sand, micaceous | SP | M | | | | |
| 20 | | 11 | 50/2" | SS | 70 | | -14.80 | Grading to moderately decomp gneiss; light grey; soft to hard closely jointed mod. foliated | WM | | | | | |
| | | 12 | | DS (NX) | 85 | | -18.60 | Gneiss; light grey, very hard, weathering slight, texture: c-m fabric: mod. foliated minerals: qtz, feldspar, hornblende | 60% | | | | | |
| 25 | | | | | | | | | WS | | | | | |
| | | 13 | | DS (NX) | 95 | | | | 75% | | | | | |
| 30 | | | | | | | | End of boring at 37 ft | | | | | | |

Notes:

(1) Boring type: W—wash, UD—undisturbed, C—core (Size NX).

(2) Samplers: SS—split spoon, S—Shelby, P—piston, D—Denison, O—other (Describe).
Core barrels: ST—single tube, DT—double tube, DS—double tube swivel.

(3) N=sum of last 12 in.

(4) Soil moisture: D—dry, M—moist, W—wet, GWL —▼ (date).

(5) Other tests: G—grain size, C—consolidation, T—triaxial, U—unconfined, V—field vane, K—field perm.

**FIG. 2.92 Example of test boring report log for soil and rock boring.**

Happy Engineering Co.

Core boring report log

Boring No. CB-12

Project: Power Plant — Location: Bundocks, N. J. — Sheet 1 of 1

Client: Candle Power Co. — Coordinates: N980,708.81/E603.931.02 — Project No. 78-103

Contractor: Drill Good Inc. — Driller: W. Dowell — Inspector: J. Eyesharp

Date started: 7.30.78 — Date finished: 8.2.78 — Surface elevator: +358.71 ft — Datum: MSL

| Depth, ft | Soil samples: Sample No. | N value | Sampler | Recovery, % | Rock gores: Run No. | Core pieces | % recovery | RQD, % | Hardness (1) | Weathering grade (2) | Soil or rock description | Structure | Rock structure joint spacing, filling, aperture, orientation | Remarks |
|---|---|---|---|---|---|---|---|---|---|---|---|---|---|---|
| 0–4 (▼ GWL) | 1 | 6 | 2 in SS | 50 | | | | | | RS | Brown clayey silt to silty clay - bec. dark grey | | | Moist GWL from augerhole |
| 5 | | | M series DS barrel w/ NX diamond bit | | 1 | 1@4 in | 80 | 30 | V | WH | Shale, black w/pyrite inclusions platy texture thinly bedded, | | Very close w/ clay seams, @ 30° 4 in clay sm @6.4 ft | @6.8 ft core bbl blocked |
| | | | | | 2 | 1@12 in, 1@4 in | 80 | 67 | V | WH | | | | |
| 10 | | | | | 3 | 1@10 in, 1@6 in, 3@5 in | 92 | 53 | V / IV | WH / WM | | | Thinly bedded w/ clay sms. 6 in seam soft clay at 10.7 ft Beds @20-25° - Slickenside at 13.6 ft | |
| 15 | | | | | 4 | 1@12 in, 1@7 in, 3@6 in, 1@5 in | 83 | 70 | IV / III | WM / WS | Limestone, light grey, fine-grained amorphous | | Mod. close, horizontal, w/small (1/4 in) voids, joints stained, but clean and tight | - No water loss |
| 20–25 | | | | | 5 | 2@6 in, 2@5 in, 2@10 in, 1@12 in, 1@24 in, 1@30 in | 95 | 90 | III / II / I | WS / WS / F | Becoming argillaceous, light grey Limestone, fine-grained, lightly bedded, horizontal | | Mod. close, hor. joints occ. vertical clean and tight | - Coring very slow below 20.6 ft |
| 27–30 | | | | | | | | | | | End of boring at 27.0 ft | | | |

NOTES:

(1) Hardness: I—extremely hard, II—very hard, III—hard, IV—soft, V—very soft (Table 5.20).

(2) Weathering Grade: F—fresh, WS—slightly weathered, WM—moderately weathered, HW—highly weathered. WC—completely weathered (saprolite), RS—residual soil (Table 5.21).

**FIG. 2.93 Example of a report log for rock-core boring.**

## REFERENCES


1. Rogers, F. C. (1950) "Engineering Soil Survey of New Jersey, Report No. 1," *Engineering Research Bulletin No. 15*, College of Engineering, Rutgers Univ., Edwards Bros. Inc., Ann Arbor, Mich.
2. USDA (1977) *Soil Survey of Autauga County, Alabama*, U.S. Dept. of Agriculture, Soil Conservation Service, 64 pp. and maps.
3. Godfrey, K. A. Jr. (1979) "What Future for Remote Sensing in Space," *Civil Engineering*, ASCE, July, pp. 61–65.
4. Way, D. S. (1978) *Terrain Analysis*, 2d ed., Dowden, Hutchinson & Ross, Stroudsburg, Pa.
5. Lueder, D. R. (1959) *Aerial Photographic Interpretation: Principles and Applications*, McGraw-Hill Book Co., New York.
6. A.S.P. (1960) *Manual of Photo Interpretation*, American Society of Photogrammetry, Washington, D.C.
7. Belcher, D. J. (1948) "The Engineering Significance of Landforms," *Highway Research Board Pub. No. 13*, Washington, D.C.
8. Griffiths, D. H. and King, R. F. (1969) *Applied Geophysics for Engineers and Geologists*, Pergamon Press, London.
9. Mooney, H. M. (1973) *Handbook of Engineering Seismology*, Bison Instruments Inc., Minneapolis, Minn.
10. Auld, B. (1977) "Cross-hole and Down-hole $V_s$ by Mechanical Impulse," *Proc. ASCE, J. Geotech. Engrg. Div.*, Vol. 103, No. GT12, December, pp. 1381–1398.
11. Ballard, R. F., Jr., (1976) "Method for Crosshole Seismic Testing," *Proc. ASCE, J. Geotech. Engrg. Div.*, Vol. 102, No. GT12, December, pp. 1261–1273.
12. Dobecki, T. L. (1979) "Measurements of Insitu Dynamic Properties in Relation to Geologic Conditions," *Geology in the Siting of Nuclear Power Plants, Reviews in Engineering Geology IV*, The Geological Society of America, Boulder, Colo., pp. 201–225.
13. Ladd, J. W., Buffler, R. T., Watkins, J. S., Worzel, J. L. and Carranza, A. (1976) "Deep Seismic Reflection Results from the Gulf of Mexico," *Geology*, Geological Society of America, Vol. 4, No. 6, pp. 365–368.
14. ASTM (1951) *Symposium on Surface and Subsurface Reconnaissance*, Spec. Pub. No. 122, American Society for Test. and Mat., Philadelphia, June.
15. Greenfield, R. J. (1979) "Review of Geophysical Approaches to the Detection of Karst," *Bull. Assoc. Engrg. Geol.*, Vol. XVI, No. 3, Summer, pp. 393–408.
16. Cook, J. C. (1974) "Status of Ground Probing Radar and Some Recent Experience," *Subsurface Exploration for Underground Excavation and Heavy Construction, Proc. ASCE*, New York, pp. 175–194.
17. Moffatt, B. T. (1974) "Subsurface Video Pulse Radars," *Subsurface Exploration for Underground Excavation and Construction, Proc. ASCE*, New York, pp. 195–212.
18. Morey, R. M. (1974) "Continuous Subsurface Profiling by Impulse Radar," *Subsurface Exploration in Underground Excavation and Heavy Construction, Proc. ASCE*, New York, pp. 213–232.
19. Alperstein, R. and Leifer, S. A. (1976) "Site Investigation with Static Cone Penetrometer," *Proc. ASCE, J. Geotech. Engrg. Div.*, Vol. 102, No. GT5, May, pp. 539–555.
20. Sanglerat, G. (1972) *The Penetrometer and Soil Exploration*, Elsevier Publ. Co., Amsterdam, p. 464.
21. Schmertmann, J. H. (1977) *Guidelines for CPT Performance and Design*, U.S. Dept. of Transportation, Federal Highway Admin., Offices of Research and Development, Washington, D.C.
22. ENR (1977) "Impact Drill Drives through Hard Rock Fast," *Engineering News-Record*, Nov. 3, p. 14.
23. USBR (1974) *Earth Manual*, U.S. Bureau of Reclamation, Denver, Colo.
24. Myung, J. T. and Baltosser, R. W. (1972) "Fracture Evaluation by the Borehole Logging Method," *Stability of Rock Slopes*, ASCE, New York, pp. 31–56.
25. Cotecchia, V. (1978) "Systematic Reconnaissance Mapping and Registration of Slope Movements," *Bull. No. 17*, Intl. Assoc. Engrg. Geol., June, pp. 5–37.
26. Underwood, L. B. (1974) "Exploration and Geologic Prediction for Underground Works," *Subsurface Exploration for Underground Excavation and Heavy Construction*, ASCE, New York, pp. 65–83.
27. Broms, B. B. (1980) "Soil Sampling in Europe: State-of-the Art," *Proc. ASCE, J. Geotech. Engrg. Div.*, Vol. 106, No. GT1, January, pp. 65–98.
28. ENR (1952) "Soil Sampling Techniques," *Engineering News-Record*, April 24.
29. Eide, O. (1974) "Marine Soil Mechanics—Applications to North Sea Offshore Structures," Norwegian Geotechnical Institute, Oslo, Pub. No. 103, p. 19.
30. Deere, D. U. (1963) "Technical Description of Rock Cores for Engineering Purposes," *Rock Mech. Engrg. Geol.*, 1, pp. 18–22.

## BIBLIOGRAPHY

AEG (1978) "A Guide to Core Logging for Rock Engineering," *Bull. Assoc. Engrg. Geol.*, So. African Section, Core Logging Comm. Vol XV, summer, pp. 295–328.

Bartelli, L. J., Klingebiel, A. A., Baird, J. V. and Heddleson, M. R. (1966) "Soil Surveys and Land Use Planning," Soil Science of America and American Society of Agronomy.

Birdsall, L. E. (1973) "Sources of Geologic Data," *Geology Seismicity and Environmental Impact*, Assoc. Engrg. Geol., University Publishers, Los Angeles, pp. 57–64.

Focht, J. A. and Kraft, L. M. Jr. (1977) "Progress in Marine Geotechnical Engineering," *Proc. ASCE, J. Geotech. Engrg. Div.*, Vol. 103, No. GT10, October, pp. 1097–1118.

Grau, G. (1979) "Seabed Reconnaissance by Seismic Methods," *Proc. Offshore Brazil Conf.*, Rio de Janeiro, June, pp. OB-78 01.1–01.8.

Hvorslev, J. J. (1949) *Subsurface Exploration and Sampling of Soils for Civil Engineering Purposes*, Waterways Experimental Station, U.S. Army Engineers, Vicksburg, Miss., November.

Hunt, R. E. (1972) "The Geologic Environment—Definition by Remote Sensing," *Proc. International Conf. on Microzonation*, Seattle, Wash., Vol. II, pp. 577–593.

Hunt, R. E. (1975) "Interpretation of Remote Sensing Imagery: A New Tool for Land Use Planning and Protection," *Consulting Engineer*, June, pp. 54–59.

Hunt, R. E. and Santiago, W. B. (1976) "A Funcão Crítica do Engenheiro Geólogo em Estudos de Implantacão de Ferrovias," *1st Cong. Bras. de Geol. de Eng.*, Associacão Brasileira de Geologia de Engenharia, August, Vol. I, pp. 79–98.

Kempe, W. F. (1967) *Core Orientation*, Christenson Diamond Products Co., Salt Lake City, Utah.

Knill, J. L. (1975) "Suggested Methods for Geophysical Logging of Borehole and Measurements of Insitu Seismic Characteristics," *Intl. Soc. for Rock Mechs.*, Comm. Standardization of Laboratory and Field Tests, 4th Report Draft, Comm. of Field Tests, Dept. of Geology, Imperial College, London.

Lowe, J. III and Zaccheo, P. F. (1975) "Subsurface Explorations and Sampling," *Foundation Engineering Handbook*, Winterkorn and Fang, eds., Van Nostrand Reinhold Co., New York, Chap. 1, pp. 1–66.

Lundstrom, R. and Stanberg, R. (1965) "Soil-Rock Drilling and Rock Locating by Rock Indicator," *Proc. 6th Intl. Conf. Soil Mechs. and Found. Engrg.*, Montreal.

Merritt, A. H. (1974) "Exploration Methods," *Proc. ASCE, Spec. Conf. on Subsurface Exploration for Underground Excavation and Heavy Construction*, Henniker, New Hampshire, August, pp. 56–64.

McEldowney, R. C. and Pascucci, R. F. (1979) "Applications of Remote-sensing Data to Nuclear Power Plant Site Investigations," *Geology in the Siting of Nuclear Power Plants, Reviews in Engineering Geology IV*, The Geological Society of America, Boulder, Colo., pp. 121–139.

Mohr, H. A. (1962) *Exploration of Soil Conditions and Sampling Operations*, Harvard University Press, Cambridge, Mass.

Stimpson, W. E., Brierly, G. S. and Liu, T. K. (1976) "Determining Bedrock Elevation by Acoustic Sounding Device," *Proc. ASCE, Rock Engineering for Foundations and Slopes*, ASCE, New York, Vol. I, pp. 1–12.

USGS (1976) *Earth Science Information in Land-Use Planning—Guidelines for Earth Sciences and Planners*, Geological Survey Circular No. 721, U.S. Geological Survey, Arlington, Va.

Williams, R. S., Jr. and Carter, W. D. (1976) *ERTS-1: A New Window on Our Planet*, Geological Society Prof. Paper 929, U.S. Govt. Printing Office, Washington, D.C., 362 pp.

# CHAPTER THREE

# MEASUREMENTS OF PROPERTIES

## 3.1 INTRODUCTION

### 3.1.1 OBJECTIVES

The properties of the geologic materials are measured to provide the basis for:

1. Identification and classification *(see Chap. 5)*
2. Correlations between properties including measurements made during other investigations in similar materials
3. Engineering analysis and evaluations

### 3.1.2 GEOTECHNICAL PROPERTIES

#### Basic Properties

Basic properties include the fundamental characteristics of the materials and are used for identification and correlations. Some are used in engineering calculations.

#### Index Properties

Index properties define certain physical characteristics used basically for classifications, but also for correlations with engineering properties.

#### Hydraulic Properties

Hydraulic properties, expressed in terms of permeability, are engineering properties. They involve the flow of fluids through geologic media.

#### Mechanical Properties

Rupture strength and deformation characteristics are mechanical properties. They are also engineering properties, and are grouped as static or dynamic.

#### Correlations

Measurements of the hydraulic and mechanical properties, which provide the basis for all engineering analysis, are often costly or difficult to obtain, especially with reliable accuracy. Correlations based on basic or index properties, with data obtained from other investigations in which extensive testing was employed or engineering properties were evaluated by back-analysis of failures, provide data for preliminary engineering studies as well as a check on the reasonableness of data obtained during investigation.

Data on typical basic, index, and engineering properties are given throughout the book for general reference. A summation of the tables and figures providing these data is given in Appendix E.

### 3.1.3 METHODS OF MEASUREMENT

#### Laboratory Testing

Soil samples and rock cores are tested in the laboratory for the most part, and occasionally in the field.

*Rock cores* are tested primarily for basic and index properties, since engineering properties of significance are not usually represented by an intact specimen.

*Soil samples* are tested for basic and index properties, and for engineering properties when

**TABLE 3.1**
**MEASUREMENT OF GEOTECHNICAL PROPERTIES OF ROCK AND SOIL**

| Property | Laboratory test | | In situ test | |
|---|---|---|---|---|
| | Rock | Soil | Rock | Soil |
| *(a)* BASIC PROPERTIES | | | | |
| Specific gravity | x | x | | |
| Porosity | x | x | | |
| Void ratio | | x | | |
| Moisture content | x | x | x | x |
| Density: | x | x | x | x |
| Natural | | x | | x |
| Maximum | | x | | |
| Minimum | | x | | |
| Relative | | x | | x |
| Optimum moisture density | | x | | |
| Hardness | x | | | |
| Durability | x | | | |
| Reactivity | x | x | | |
| Sonic-wave characteristics | x | x | x | x |
| *(b)* INDEX PROPERTIES | | | | |
| Grain-size distribution | | x | | |
| Liquid limit | | x | | |
| Plastic limit | | x | | |
| Plasticity index | | x | | |
| Shrinkage limit | | x | | |
| Organic content | | x | | |
| Uniaxial compression | x | | | |
| Point-load index | x | | | |
| *(c)* ENGINEERING PROPERTIES | | | | |
| Permeability | x | x | x | x |
| Deformation moduli: static or dynamic | x | x | x | x |
| Consolidation | | x | | x |
| Expansion | x | x | x | x |
| Extension strain | x | | x | |
| Strength: | | | | |
| Unconfined | x | x | | |
| Confined: | | | | |
| Static, | x | x | x | x |
| Dynamic | | x | | |
| California bearing ratio (CBR) | | x | | x |

high-quality undisturbed samples are obtained (generally limited to soft to hard intact specimens of cohesive soils lacking gravel-size or larger particles).

### In Situ Testing

Geologic formations are tested in situ within boreholes, on the surface of the ground, or within an excavation.

*Rock masses* are usually tested in situ to measure their engineering properties, as well as their basic properties.

*Soils* are tested in situ to obtain measures of engineering properties to supplement laboratory data, and in conditions where undisturbed sampling is difficult or not practical such as with highly organic materials, cohesionless granular soils, fissured clays, and cohesive soils with

**TABLE 3.2**
**INTACT ROCK SPECIMENS: LABORATORY TESTING**

| Property/test | Applications | Article |
|---|---|---|
| Basic properties | Correlations, analysis | |
| Specific gravity | Mineral identification | 3.2.1 |
| Porosity | Property correlations | 3.2.1 |
| Density | Material and property correlations<br>Engineering analysis | 3.2.1 |
| Hardness | Material correlations | 3.2.1 |
| | Tunneling machine excavation evaluation | |
| Durability<br>LA abrasion<br>British crushing | Evaluation of construction aggregate quality | 3.2.1 |
| Reactivity | Reaction between cement and aggregate | 3.2.1 |
| Sonic velocities | Computations of dynamic properties | 3.5.3 |
| Index properties | Classification/correlations | |
| Uniaxial compression | See rupture strength | 3.4.3 |
| Point-load test | See rupture strength | 3.4.3 |
| Permeability | Not normally performed in the lab | |
| Rupture strength | Measurements of: | |
| Triaxial shear | Peak drained or undrained strength | 3.4.3 |
| Unconfined compression | Unconfined (uniaxial) compressive strength used for correlations | 3.4.3 |
| Point-load test | Tensile strength for correlation with uniaxial compression | 3.4.3 |
| Uniaxial tensile strength | Strength in tension | 3.4.3 |
| Flexural or beam strength | Strength in bending | 3.4.3 |
| Deformation (static) | Measurements of: | |
| Triaxial test | Deformation moduli, $E_i$, $E_s$, $E_t$ | 3.4.3 |
| Unconfined compression | Deformation moduli $E_i$, $E_s$, $E_t$ | 3.4.3 |
| Dynamic properties | Measurements of: | |
| Resonant column | Compression and shear wave velocities $V_p$, $V_s$ | 3.5.3 |
| | Dynamic moduli $E$, $G$, $D$ | |

large granular particles (glacial till, residual soils, etc.).

### 3.1.4 TESTING METHODS SUMMARIZED

#### General

A general summary of the significant basic, index, and engineering properties of soil and rock, and an indication of whether they are performed in the laboratory, in situ, or both, is given on Table 3.1.

#### Intact Rock and Rock Masses

*Laboratory tests* of intact specimens, the property measured, and the application of the test in terms of the data obtained are summarized on Table 3.2.

**TABLE 3.3**
**ROCK MASSES—IN SITU TESTING**

| Category—tool or method | Applications | Limitations | Article |
|---|---|---|---|
| | BASIC PROPERTIES | | |
| Gamma-gamma borehole probe | Continuous measure of density. | Density measurements. | 2.3.6 |
| Neutron borehole probe | Continuous measure of moisture. | Moisture measurements. | 2.3.6 |
| | INDEX PROPERTIES | | |
| Rock coring | Measure the RQD (rock quality designation) used for various empirical correlations. | Values very dependent on drilling equipment and techniques. | 2.4.5 |
| Seismic refraction | Estimate rippability on the basis of *P* wave velocities. | Empirical correlations. Rippability depends on equipment used. | 2.3.2 |
| | PERMEABILITY | | |
| Constant-head test | In boreholes to measure *k* in heavily jointed rock masses. | Free-draining materials. Requires ground saturation. | 3.3.4 |
| Falling-head test | In boreholes to measure *k* in jointed rock masses. Can be performed to measure $k_{mean}$, $k_v$, or $k_h$. | Slower draining materials or below water table. | 3.3.4 |
| Rising-head test | Same as for falling-head test | Same as for falling-head test. | 3.3.4 |
| Pumping tests | In wells to determine $k_{mean}$ in saturated uniform formations. | Not representative for stratified formations. Measures average *k* for entire mass. | 3.3.4 |
| Pressure testing | Measures $k_h$ in vertical borehole. | Requires clean borehole walls. Pressures can cause joints to open or to clog from migration of fines. | 3.3.4 |
| | SHEAR STRENGTH | | |
| Direct shear box | Measure strength parameters along weakness planes of rock block. | Sawing block specimen and test setup costly. A surface test. Several tests required for Mohr's envelope. | 3.4.3 |
| Triaxial or uniaxial compression | Measure triaxial or uniaxial compressive strength of rock block. | Same as for direct shear box. | 3.4.3 |
| Dilatometer or Goodman jack | Measures limiting pressure $P_L$ in borehole. | See Dilatometer under Deformation. Limited by rock-mass strength. | 3.5.3 |

| Category—tool or method | Applications | Limitations | Article |
|---|---|---|---|
| DEFORMATION MODULI (STATIC) | | | |
| Dilatometer or Goodman jack | Measures $E$ in lateral direction. | Modulus values valid for linear portion of load-deformation curve. Results affected by borehole roughness and layering. | 3.5.3 |
| Large-scale foundation-load test | Measure $E$ under footings or bored piles. Measure shaft friction of bored piles. | Costly and time consuming. | 3.5.4 |
| Plate-jack test | Measure $E$; primarily used for tunnels and heavy structures. | Requires excavation and heavy reaction or adit. Stressed zone limited by plate diameter and disturbed by test preparation. | 3.5.3 |
| Flat-jack test | Measure $E$ or residual stresses in a slot cut into the rock. | Stressed zone limited by plate diameter. Test area disturbed in preparation. Requires orientation in same direction as applied construction stresses. | 3.5.3 |
| Radial jacking tests (pressure tunnels) | Measure $E$ for tunnels. Data most representative of in situ rock tests and usually yields the highest values for $E$. | Very costly and time-consuming and data difficult to interpret. Preparation disturbs rock mass. | 3.5.3 |
| Triaxial compression test | Measure $E$ of rock block. | Costly and difficult to set up. Disturbs rock mass during preparation. | 3.5.3 |
| DYNAMIC PROPERTIES | | | |
| Seismic direct methods | Obtain dynamic elastic moduli, $E$, $G$, $K$, and $\nu$ in boreholes. | Very low strain levels yield values higher than static moduli. | 3.5.3 |
| 3-D velocity logger | Measure velocity of shear and compression waves ($V_s$, $V_p$) from which moduli are computed. Borehole test. | Penetrates to shallow depth in borehole. | 3.5.3 |
| Vibration monitor | Measure peak particle velocity, or frequency, acceleration, and displacement for monitoring vibrations from blasting, traffic, etc. | Surface measurements, low energy level. | 4.2.5 |

**TABLE 3.4**
**SOILS: LABORATORY TESTING**

| Property/test | Applications | Article |
|---|---|---|
| Basic properties | Correlations, classification | |
| Specific gravity | Material identification | 3.2.3 |
| | Void ratio computation | |
| Moisture or water content | Material correlations in the natural state | 3.2.3 |
| | Computations of dry density | |
| | Computations of Atterberg limits | |
| Density: natural (unit weight) | Material correlations | 3.2.3 |
| | Engineering analysis | |
| Density: maximum | Relative density computations | 3.2.3 |
| | Moisture-density relationships | |
| Density: minimum | Relative density computations | 3.2.3 |
| Optimum-moisture density | Moisture-density relationships for field compaction control | 3.2.3 |
| Sonic velocities | Computations of dynamic properties | 3.5.3 |
| Index properties | Correlations, classification | |
| Gradation | Material classification | 3.2.3 |
| | Property correlations | |
| Liquid limit | Computation of plasticity index | 3.2.3 |
| | Material classification | |
| | Property correlations | |
| Plastic limit | Computation of plasticity index | 3.2.3 |
| | Field identifications | 5.3.6 |
| Shrinkage limit | Material correlations | 3.2.3 |
| Organic content | Material classification | 3.2.3 |
| Permeability | Measurements of: | |
| Constant head | $k$ in free-draining soil | 3.3.3 |
| Falling head | $k$ in slow-draining soil | 3.3.3 |
| Consolidometer | $k$ in very slow draining soil (clays) | 3.5.4 |

*In situ tests* in rock masses, their applications, and their limitations are summarized on Table 3.3.

### Soils

*Laboratory soil tests,* properties measured, and the application of the tests in terms of the data obtained are summarized on Table 3.4.

*In situ soil tests,* properties measured, applications, and limitations are summarized on Table 3.5.

## 3.2 BASIC AND INDEX PROPERTIES

### 3.2.1 INTACT ROCK

#### General

Testing is normally performed in the laboratory on a specimen of fresh to slightly weathered rock free of defects.

*Basic properties* include volume-weight relationships, hardness (for excavation resistance), and durability and reactivity (for aggregate quality).

| Property/test | Applications | Article |
|---|---|---|
| Rupture strength | Measurements of: | |
| Triaxial shear (compression or extension) | Peak undrained strengh $s_u$, cohesive soils (UU test) | 3.4.4 |
| | Peak drained strength, $\phi$, $c$, $\bar{\phi}$, $\bar{c}$, all soils | |
| Direct shear | Peak drained strength parameters | 3.4.4 |
| | Ultimate drained strength $\bar{\phi}_r$, cohesive soils | |
| Simple shear | Undrained and drained parameters | 3.4.4 |
| Unconfined compression | Unconfined compressive strength for cohesive soils. Approximately equals $2s_u$ | 3.4.4 |
| Vane shear | Undrained strength $s_u$ for clays | 3.4.4 |
| | Ultimate undrained strengths $s_r$ | |
| Torvane | Undrained strengths $s_u$ | 3.4.4 |
| | Ultimate undrained strength $s_r$ (estimate) | |
| Pocket penetrometer | Unconfined compressive strength (estimate) | 3.4.4 |
| California bearing ratio | CBR value for pavement design | 3.4.4 |
| Deformation (static) | Measurements of: | |
| Consolidation test | Compression vs. load and time in clay soil | 3.5.4 |
| Triaxial shear test | Static deformation moduli | 3.5.4 |
| Expansion test | Swell pressures and volume change in the consolidometer | 3.5.4 |
| Dynamic properties | | |
| Cyclic triaxial | Low-frequency measurements of dynamic moduli ($E$, $G$, $D$), stress vs. strain and strength | 3.5.5<br>3.4.4 |
| Cyclic torsion | Low-frequency measurements of dynamic moduli, stress vs. strain | 3.5.5 |
| Cyclic simple shear | Low-frequency measurements of dynamic moduli, stress vs. strain and strength | 3.5.5<br>3.4.4 |
| Ultrasonic device | High-frequency measurements of compression- and shear-wave velocities $V_p$, $V_s$ | 3.5.5 |
| Resonant column device | High-frequency measurements of compression and shear-wave velocities and the dynamic moduli | 3.5.5 |

*Index tests* include the uniaxial compression test *(see Art. 3.4.3)*, the point load index test *(see Art. 3.4.3)*, and sonic velocities which are correlated with field sonic velocities to provide a measure of rock quality *(see Art. 3.5.3)*.

**Volume-Weight Relationships**

Included are specific gravity, density, and porosity as defined and described on Table 3.6.

**Hardness**

*General*

Hardness is the ability of a material to resist scratching or abrasion. Correlations can be made between rock hardness, density, uniaxial compressive strength, and sonic velocities and between hardness and the rate of advance for tunneling machines and other excavation methods. The predominant mineral in the rock specimen and the degree of weathering decomposition are controlling factors.

**TABLE 3.5**
**SOILS—IN SITU TESTING**

| Category test or method | Applications | Limitations | Article |
|---|---|---|---|
| | BASIC PROPERTIES | | |
| Gamma-gamma borehole probe | Continuous measure of density. | Density measurements. | 2.3.6 |
| Neutron borehole probe | Continuous measure of moisture content. | Moisture-content measures. | 2.3.6 |
| Sand-cone density apparatus | Measure surface density. | Density at surface. | 3.2.3 |
| Balloon apparatus | Measure density at surface. | Density at surface. | 3.2.3 |
| Nuclear density moisture meter | Surface measurements of density and moisture. | Moisture and density at surface. | 3.2.3 |
| | PERMEABILITY | | |
| Constant-head test | In boreholes or pits to measure $k$ in free-draining soils. | Free-draining soils. Requires ground saturation. | 3.3.4 |
| Falling-head test | In boreholes in slow-draining materials, or materials below GWL. Can be performed to measure $k_{mean}$, $k_v$, or $k_h$. | Slow-draining materials or below water table. | 3.3.4 |
| Rising-head test | Similar to falling-head test. | Similar to falling-head test. | 3.3.4 |
| Pumping tests | In wells to measure $k_{mean}$ in saturated uniform soils. | Results not representative in stratified formations. | 3.3.4 |
| | SHEAR STRENGTH (DIRECT METHODS) | | |
| Vane shear apparatus | Measure undrained strength $s_u$ and remolded strength $s_r$ in soft to firm cohesive soils in a test boring. | Not performed in sands or strong cohesive soils. Affected by soil anisotropy and construction time-rate differences. | 3.4.4 |
| Pocket penetrometer | Measures approximate $U_c$ in tube samples, test pits in cohesive soils. | Not suitable in granular soils. | 3.4.4 |
| Torvane | Measures $s_u$ in tube samples and pits. | Not suitable in sands and strong cohesive soils. | 3.4.4 |
| | SHEAR STRENGTH (INDIRECT METHODS) | | |
| Static cone penetrometer (CPT) | Cone penetration resistance is correlated with $s_u$ in clays and $\phi$ in sands. | Not suitable in very strong soils. | 3.4.5 |
| Pressuremeter | Undrained strength is found from limiting pressure correlations. | Strongly affected by soil anisotropy. | 3.5.4 |
| Camkometer (self-boring pressuremeter) | Provides data for determination of shear modulus, shear strength, pore pressure, and lateral stress $K_o$. | Affected by soil anisotropy and smear occurring during installation. | 3.5.4 |

| Category test or method | Applications | Limitations | Article |
|---|---|---|---|
| | PENETRATION RESISTANCE | | |
| Standard penetration test (SPT) | Correlations provide measures of granular soil $D_R$, $\phi$, $E$, allowable bearing value, and clay soil consistency. Samples recovered. | Correlations empirical. Not usually reliable in clay soils. Sensitive to sampling procedures. | 3.4.5 |
| Static cone penetrometer test (CPT) | Continuous penetration resistance can provide measures of end bearing and shaft friction. Correlations provide data similar to SPT. | Samples not recovered, material identification requires borings or previous area experience. | 3.4.5 |
| California bearing ratio | CBR value for pavement design. | Correlations are empirical. | 3.4.5 |
| | DEFORMATION MODULI (STATIC) | | |
| Pressuremeter | Measures $E$ in materials difficult to sample undisturbed such as sands, residual soils, glacial till, and soft rock in a test boring. | Modulus values only valid for linear portion of soil behavior; invalid in layered formations; not used in weak soils. | 3.5.4 |
| Camkometer | See shear strength. | | |
| Plate-load test | Measures modulus of subgrade reaction used in beam-on-elastic-subgrade problems. Surface test. | Stressed zone limited to about 2 plate diameters. Performed in sands and overconsolidated clays. | 3.5.4 |
| Lateral pile-load test | Used to determine horizontal modulus of subgrade reaction. | Stressed zone limited to about 2 pile diameters. Time deformation in clays not considered. | Not described |
| Full-scale foundation load tests | Obtain $E$ in sands and design parameters for piles. | Costly and time-consuming. | 3.5.4 |
| | DYNAMIC PROPERTIES | | |
| Seismic direct methods | Borehole measurements of $S$ wave velocity to compute $E_d$, $G_d$, and $K$. | Very low strain levels yield values higher than static moduli. | 2.3.2 |
| Steady-state vibration method | Surface measurement of shear wave velocities to obtain $E_d$, $G_d$, and $K$. | Small oscillators provide data only to about 3 m. Rotating mass oscillator provides greater penetration. | 3.5.5 |
| Vibration monitors | Measure peak particle velocity or frequency, acceleration, and displacement for monitoring vibrations from blasting, traffic, etc. | Surface measurements at low energy level for vibration studies. | 4.2.5 |

**TABLE 3.6**
**VOLUME-WEIGHT RELATIONSHIPS FOR INTACT ROCK SPECIMENS**

| Property | Symbol | Definition | Expression | Units |
|---|---|---|---|---|
| Specific gravity (absolute) | $G_s$ | The ratio of the unit weight of a pure mineral substance to the unit weight of water at 4°C. $\gamma_w = 1$ g/cm$^3$ or 62.4 pcf. | $G_s = \gamma_m/\gamma_w$ | |
| Specific gravity (apparent) | $G_s$ | The specific gravity obtained from a mixture of minerals composing a rock specimen. | $G_s = \gamma_m/\gamma_w$ | |
| Density | $\rho$ or $\gamma$ | Weight $W$ per unit volume $V$ of material. | $\rho = W/V$ | t/m$^3$ |
| Bulk density | $\rho$ | Density of rock specimen from field. (Also g/cm$^3$, pcf.) | $\rho = W/V$ | t/m$^3$ |
| Porosity | $n$ | Ratio of pore or void volume $V_v$ to total volume $V_t$ | $n = V_v/V_t$ | % |
| | | In terms of density and the apparent specific gravity. | $n = 1 - \rho/G_s$ | % |

NOTES:

*Specific Gravities:* Most rock-forming minerals range from 2.65 to 2.8, although heavier minerals such as hornblende, augite, or hematite vary from 3 to 5 and higher *(see Table 5.5).*

*Porosity:* Depends largely on rock origin. Slowly cooling igneous magma results in relatively nonporous rock, whereas rapid cooling associated with escaping gases yields a porous mass. Sedimentary rocks depend on amount of cementing materials present and on size, grading, and packing of particles.

*Density:* Densities of fresh, intact rock do not vary greatly unless they contain significant amounts of the heavier minerals.

*Porosity and density:* Typical value ranges are given on Table 3.12 and Table 3.26, respectively.

*Significance:* Permeability of intact rock often related to porosity, although normally the characteristics of the in situ rock govern rock-mass permeability. There are strong correlations between density, porosity, and strength.

### *Measurement Criteria*

The following criteria are used to establish hardness values:

1. Moh's system of relative hardness for various minerals *(see Table 5.4).*
2. Field tests for engineering classification *(see Table 5.20).*
3. "Total" hardness concept of Deere,[1] based on laboratory tests and developed as an aid in the design of tunnel boring machines (TBMs). Ranges in total hardness of common rock types are given on Fig. 3.1.
4. *Testing Methods for Total Hardness* [*Tarkoy (1975)*[2]]: Total hardness $H_T$ is defined as:

$$H_T = H_R\sqrt{H_A} \qquad g^{-1/2} \tag{3.1}$$

where $H_R$ = Schmidt hardness and $H_A$ = abrasion test hardness.

- *Schmidt rebound hardness test:* An L-type concrete test hammer, with a spring in tension, impels a known mass onto a plunger held against the specimen (energy = 0.54 ft-lb, or 0.075 m-kg). The amount of energy reflected from the rock-hammer interface is measured by the amount that the hammer mass is caused to rebound.

- *Shore (C-2) sclerescope* is also used to measure rebound hardness. The rebound height of a small diamond-tipped weight falling vertically down a glass tube is measured and compared with the manufacturer's calibration.

- *Abrasion hardness test* is performed on a thin disk specimen which is rotated a specific number of times against an abrading wheel and the weight-loss recorded.

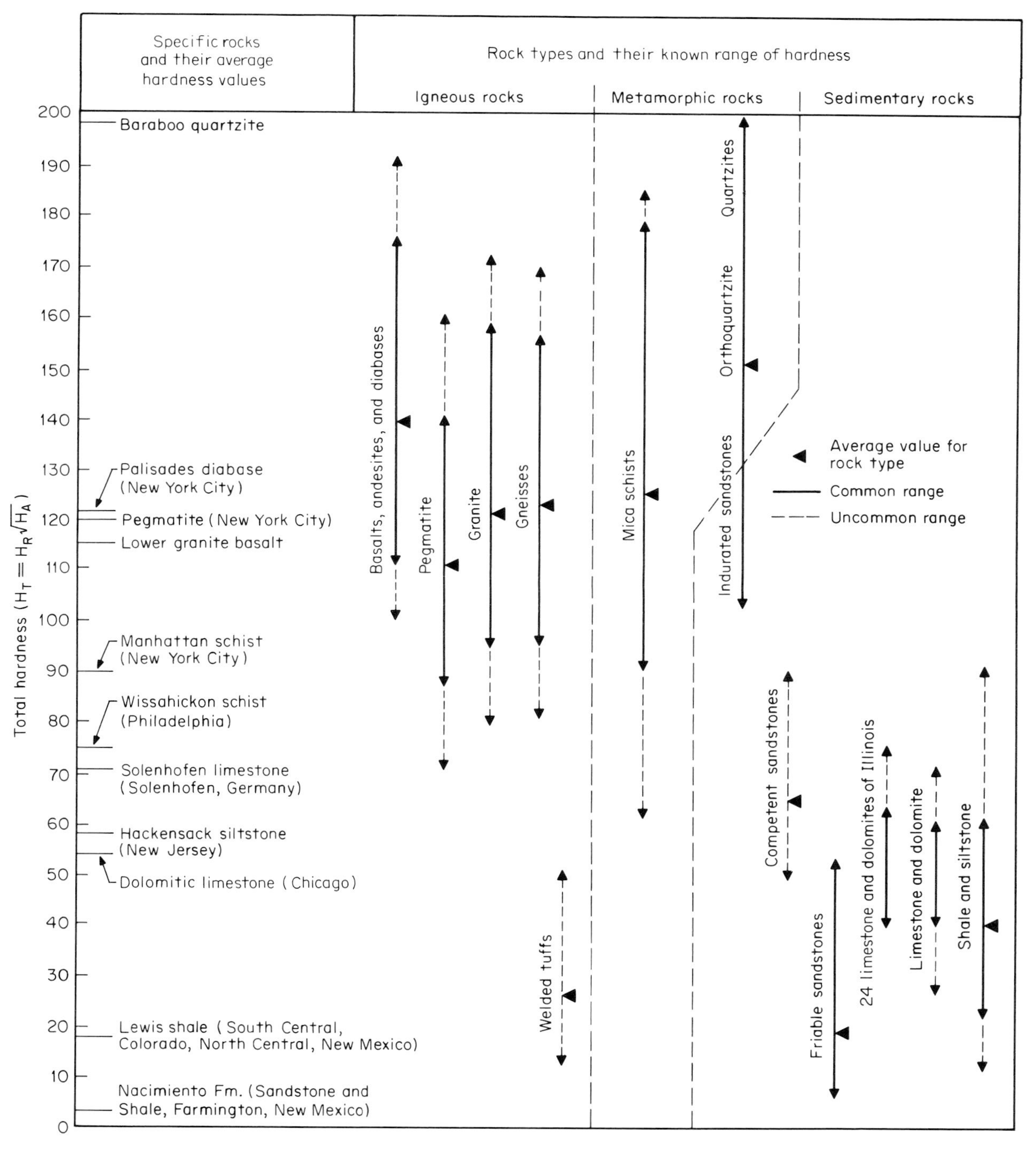


**FIG. 3.1 Range of "total" hardness for common rock types. Data are not all inclusive, but represent the range for rocks tested in the Rock Mechanics Laboratory, University of Illinois, over recent years. $H_R$ = Schmidt hardness; $H_A$ = abrasion test hardness.** [*From Tarkoy (1975)*[2].]

### Durability

*General*

Durability is the ability of a material to resist degradation by mechanical or chemical agents. It is the factor controlling the suitability of rock material used as aggregate for roadway base course, or in asphalt or concrete. The predominant mineral in the specimen, the microfabric (fractures or fissures), and the decomposition degree are controlling factors.

*Test Methods*

LOS ANGELES ABRASION TEST (ASTM C131-47) Specimen particles of a specified size are placed in a rotating steel drum with 12 steel balls (1 ⅞ in diameter). After rotation for a specific period, the aggregate particles are weighed and the weight loss compared with the original weight to arrive at the LA abrasion value. The maximum acceptable weight loss usually is about 40% for bituminous pavements and 50% for concrete.

BRITISH CRUSHING TEST Specimen particles of a specified size are placed in a 4-in-diameter steel mold and subjected to crushing under a specified static force applied hydraulically. The weight loss during test is compared with the original weight to arrive at the British crushing value. Examples of acceptable value ranges, which may vary with rock type and specifying agency are as follows: particle size (maximum weight loss): ¾–1 in (32%), ½–¾ in (30%), ⅜–½ in (28%), and ⅛–³⁄₁₆ in (26%).

## REACTIVITY: CEMENT-AGGREGATE

### Description

Crushed rock is used as aggregate to manufacture concrete. A reaction between soluble silica in the aggregate and the alkali hydroxides derived from portland cement can produce abnormal expansion and cracking of mortar and concrete, often with severely detrimental effects to pavements, foundations, and concrete dams. There is often a time delay of about 2 to 3 years after construction, depending upon the aggregate type used.

### The Reaction

Alkali-aggregate reaction can occur between hardened paste of cements containing more than 0.6% soda equivalent and any aggregate containing reactive silica. The soda equivalent is calculated as the sum of the actual $Na_2O$ content and 0.658 times the $K_2O$ content of the clinker.[3] The alkaline hydroxides in the hardened cement paste attack the silica to form an unlimited-swelling gel which draws in any free water by osmosis and expands, disrupting the concrete matrix. Expanding solid products of the alkali-silica reaction help to burst the concrete, resulting in characteristic map cracking on the surface. In severe cases the cracks reach significant widths.

### Susceptible Rock Silicates

(*See Art. 5.2.2* for descriptions.) Reactive silica occurs as opal or chalcedony in certain cherts and siliceous limestones and as acid and intermediate volcanic glass, cristobolite, and tridymite in volcanic rocks such as rhyolite, dacites, and andesites, including the tuffs. Synthetic glasses and silica gel are also reactive. All of these substances are highly siliceous materials which are thermodynamically metastable at ordinary temperatures and can exist also in sand and gravel deposits. Additional descriptions are given in Krynine (1957).[4]

### Reaction Control

Reaction can be controlled[5] by:

1. Limiting the alkali content of the cement to less than 0.6% soda equivalent. Even if the aggregate is reactive, expansion and cracking should not result.
2. Avoiding reactive aggregate.
3. Replacing part of the cement with a very finely ground reactive material (a pozzolan) so that the first reaction will be between the alkalis and the pozzolan, which will use up the alkalis, spreading the reaction and reaction products throughout the concrete.

### Tests to Determine Reactivity

Tests include:

- The mortar-bar expansion test (ASTM C227) made from the proposed aggregate and cement materials.
- Quick chemical test on the aggregates (*ASTM*

**TABLE 3.7**
**ROCK RIPPABILITY* AS RELATED TO SEISMIC *P*-WAVE VELOCITIES (*COURTESY CATERPILLAR TRACTOR CO.*)**

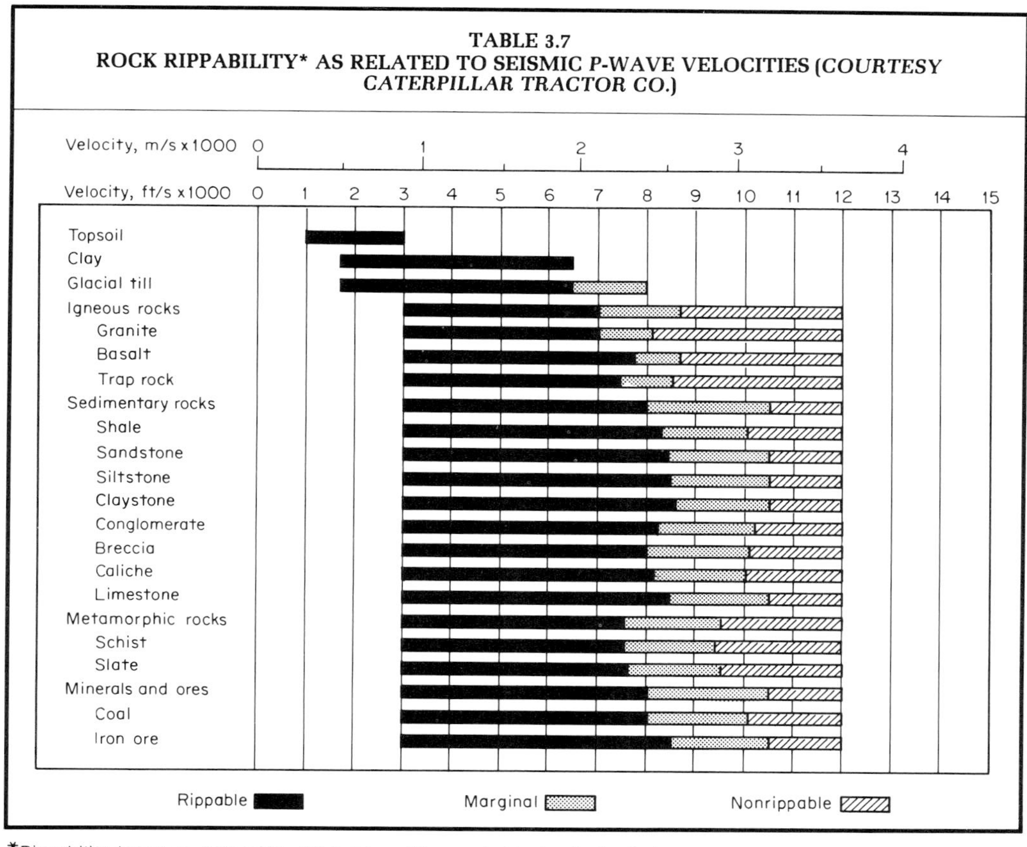


*Rippability based on Caterpillar D9 tractor with mounted hydraulic No. 9 ripper.

*Tentative Method of Test for Potential Reactivity of Aggregates—Chemical Method,* C289).

- Petrographic examination of aggregates to identify the substances (ASTM C295).

## 3.2.2 ROCK MASSES

### General

The rock mass, often referred to as *in situ rock,* may be described as consisting of rock blocks, ranging from fresh to decomposed, and separated by discontinuities *(see Art. 5.2.7).* Mass density is the basic property. Sonic-wave velocities and the rock quality designation (RQD) are used as index properties.

### Mass Density

Mass density is best measured in situ with the gamma-gamma probe *(see Art. 2.3.6),* which generally allows for weathered zones and the openings of fractures and small voids, all serving to reduce the density from fresh rock values.

### Rock Quality Indices

*Sonic wave velocities* from seismic direct surveys *(see Art. 2.3.2)* are used in evaluating rock-mass quality and dynamic properties.

*Rock quality designation* may be considered as an index property *(see Art. 2.4.5).*

### Rippability

Rippability refers to the ease of excavation by construction equipment. Since it is related to rock quality in terms of hardness and fracture density, which may be measured by seismic refraction surveys *(see Art. 2.3.2),* correlations have been made between rippability and seismic *P* wave velocities as given on Table 3.7. If the material is not rippable by a particular piece of equipment, then jackhammering and blasting are required.

**TABLE 3.8**
**VOLUME-WEIGHT RELATIONSHIPS FOR SOILS***

| | Property | Saturated sample ($W_s$, $W_w$, $G_s$ are known) | Unsaturated sample ($W_s$, $W_w$, $G_s$, $V$ are known) |
|---|---|---|---|
| Volume components | Volume of solids $V_s$ | $\frac{W_s}{G_s\gamma_w}$† | |
| | Volume of water $V_w$ | $\frac{W_w}{\gamma_w}$‡ | |
| | Volume of air or gas $V_a$ | zero | $V - (V_s + V_w)$ |
| | Volume of voids $V_v$ | $\frac{W_w}{\gamma_w}$‡ | $V - \frac{W_s}{G_s\gamma_w}$ |
| | Total volume of sample $V$ | $V_s + V_w$ | Measured |
| | Porosity $n$ | $\frac{V_v}{V}$ or $\frac{e}{1+e}$ | |
| | Void ratio $e$ | $\frac{V_v}{V_s} = \frac{G_S\gamma_w}{\gamma_d} - 1$ | |
| Weights for specific sample | Weight of solids $W_s$ | Measured | |
| | Weight of water $W_w$ | Measured | |
| | Total weight of sample $W_t$ | $W_s + W_w$ | |
| Weights for sample of unit volume | Dry-unit weight $\gamma_d$ | $\frac{W_s}{V_s + V_w}$ | $\frac{W_s}{V}$ |
| | Wet-unit weight $\gamma_t$ | $\frac{W_s + W_w}{V_s + V_w}$ | $\frac{W_s + W_w}{V}$ |
| | Saturated-unit weight $\gamma_s$ | $\frac{W_s + W_w}{V_s + V_w}$ | $\frac{W_s + V_v\gamma_w}{V}$ |
| | Submerged (buoyant) unit weight $\gamma_b$ | $\gamma_s - \gamma_w$‡ | |
| Combined relations | Moisture content $w$ | $\frac{W_w}{W_s}$ | |
| | Degree of saturation $S$ | 1.00 | $\frac{V_w}{V_v}$ |
| | Specific gravity $G_s$ | $\frac{W_s}{V_s\gamma_w}$ | |

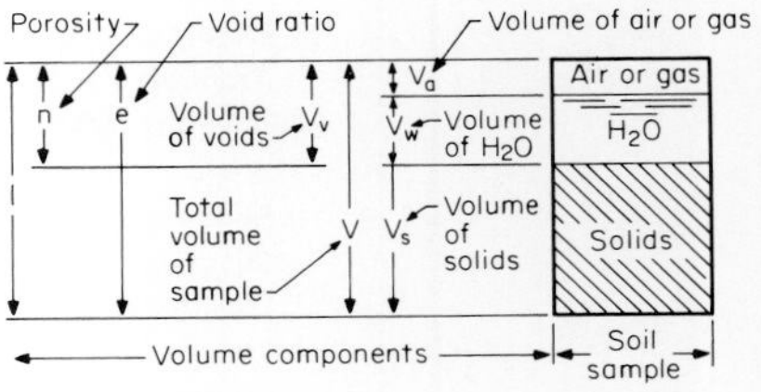

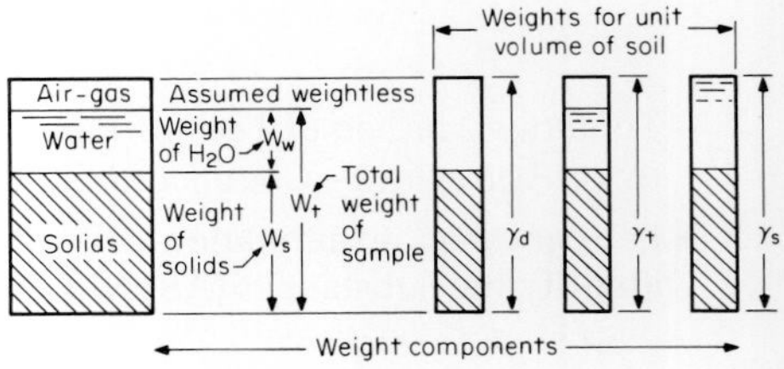

$$\gamma_d = \frac{\gamma_t}{1+w} \qquad \gamma_s = \gamma_d + \gamma_w\left(\frac{e}{1+e}\right)$$

*After NAVFAC (1971).[6]

†$\gamma_w$ is unit weight of water, which equals 62.4 pcf for fresh water and 64 pcf for sea water (1.00 and 1.025 g/cm$^3$). Where noted with ‡ the actual unit weight of water surrounding the soil is used. In other cases use 62.4 pcf. Values of $w$ and $s$ are used as decimal numbers.

| TABLE 3.9 DETERMINATION OF BASIC SOIL PROPERTIES | | |
|---|---|---|
| | Determination | |
| Basic soil property | Laboratory test | Field test |
| Unit weight or density, $\gamma_d$, $\gamma_t$, $\gamma_s$, $\gamma_b$ | Weigh specimens | Cone density device (Fig. 3.2), ASTM D1556<br>Rubber ballon device, ASTM D2167<br>Nuclear moisture-density meter, ASTM D2922 |
| Specific gravity $G_s$ | ASTM D854 | None |
| Moisture content $w$ | ASTM D2216 | Moisture meter<br>Nuclear moisture-density meter |
| Void ratio $e$ | Computed from unit dry weight and specific gravity | |

## 3.2.3 SOILS

### General

The basic and index properties of soils are generally considered to include volume-weight and moisture-density relationships, relative density, gradation, plasticity, and organic content.

### Volume-Weight Relationships

*Definitions* of the various volume-weight relationships for soils are given on Table 3.8.

*Commonly used relationships* are void ratio $e$, soil unit weight (also termed density or mass density and reported as total or wet density $\gamma_t$, dry density $\gamma_d$, and buoyant density $\gamma_b$), moisture (or water) content $w$, saturation degree $S$, and specific gravity of solids $G_s$.

*Determinations* of basic soil properties are summarized on Table 3.9, which includes brief descriptions of the common field tests. A nomograph for determination of basic soil properties is given on Fig. 3.2.

*Cone Density Device (Fig. 3.3)*

A hole 6 in deep and 6 in in diameter is dug and the removed material stored in a sealed container. The hole volume is measured with calibrated sand and the density calculated from the weight of the material removed from the hole.

*Rubber Balloon Device*

A hole is dug and the material stored as described above. The hole volume is measured by a rubber balloon inflated by water contained in a metered tube.

*Nuclear Moisture-Density Meter*

A surface device, the nuclear moisture-density meter measures wet density from either the direct transmission or backscatter of gamma rays and moisture content from the transmission or backscatter of neutron rays in a manner similar to that of the borehole nuclear probes (see *Art. 2.3.6*). In use since 1949, for many years the device had been found to give a large standard error, as much as ± 2.60 pcf. A rapid but at times approximate method, measurement with the meter yields satisfactory results with modern equipment frequently calibrated and is most useful on large projects where soil types used as fills do not vary greatly.

*Borehole Tests*

Borehole tests measure natural density and moisture content. Tests using nuclear devices are described in Art. 2.3.6.

*Moisture content*

The moisture meter is used in the field. Calcium carbide mixed with a soil portion in a closed container generates gas, causing pressure which is read on a gage to indicate moisture content. Results are approximate for some clay soils.

For cohesive soils, moisture content is most reli-

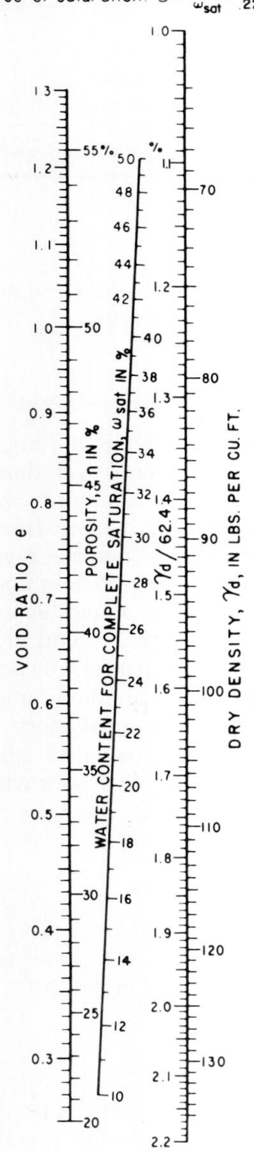


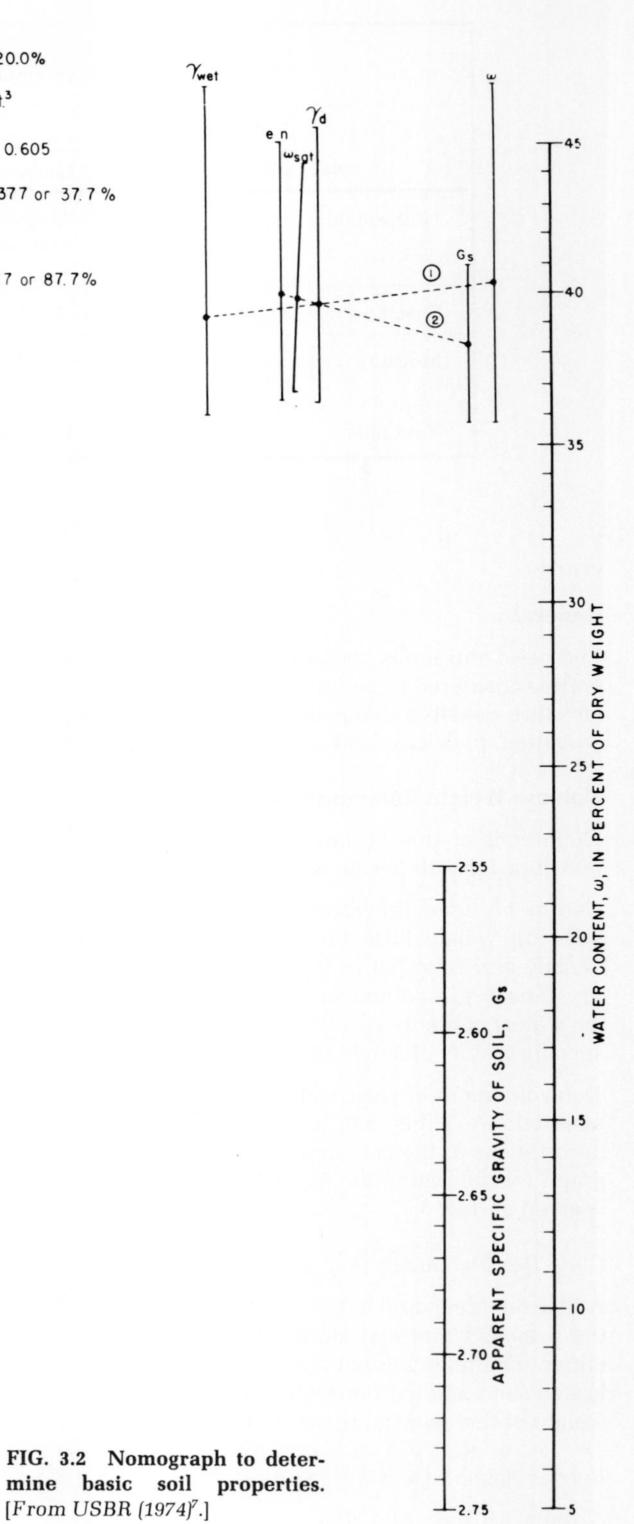


**FIG. 3.2 Nomograph to determine basic soil properties.** [*From USBR (1974)*[7].]

FIG. 3.3 Sand cone density device being used to measure in situ density of a compacted subgrade test section for an airfield pavement.

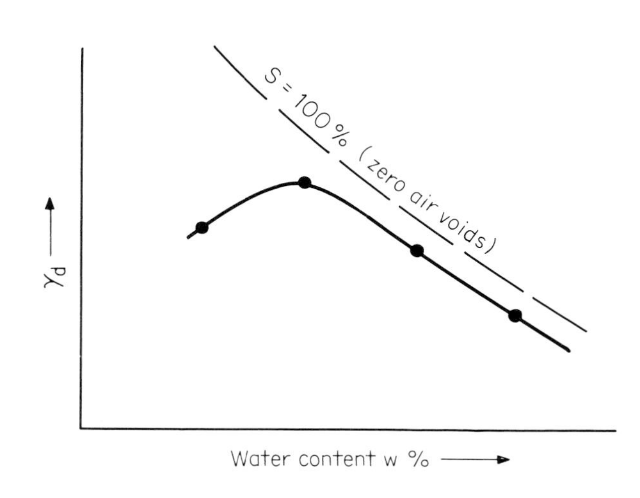


FIG. 3.4 The moisture-density relationship. The soil does not become fully saturated during the compaction test.

ably determined by oven drying for at least 24 h at 104°C.

### Moisture-Density Relationships (Soil Compaction)

*Optimum moisture content* and *maximum dry density relationships* are commonly used to specify a standard degree of compaction to be achieved during the construction of a load-bearing fill, embankment, earth dam, or pavement. Specification is in terms of a percent of maximum dry density, and a range in permissible moisture content is often specified as well.

*Description*

The density of a soil can be increased by compaction with mechanical equipment. If the moisture content is increased in increments, the density will also increase in increments under a given compactive effort, until eventually a peak or maximum density is achieved for some particular moisture content. The density thereafter will decrease as the moisture content is increased. Plotting the values for *w* percent vs. $\gamma_t$ or *w* percent vs. $\gamma_d$ will result in curves similar to that given on Fig. 3.4; 100% saturation is never reached because air remains trapped in the specimen.

*Factors Influencing Results*

The shape of the moisture-density curve varies for different materials. Uniformly graded cohesionless soils may undergo a decrease in dry density at lower moisture as capillary forces cause a resistance to compaction or arrangement of soil grains (*bulking*). As moisture is added, a relatively gentle curve with a poorly defined peak is obtained (Fig. 3.5). Some clays, silts, and clay-sand mixtures usually have well-defined peaks, whereas low-plasticity clays and well-graded sands usually have gently rounded peaks (Fig. 3.6). Optimum moisture and maximum density values also will vary with the compacted energy (Fig. 3.7).

*Test Methods*

STANDARD COMPACTION TEST OR STANDARD PROCTOR TEST (ASTM D698 OR AASHO T99) An energy of 12,400 ft-lb is used to compact 1 $ft^3$ of soil, which is accomplished by compacting three sequential layers with a 5 ½-lb hammer dropped 25 times from a 12-in height, in a 4-in diameter mold with a volume of $\frac{1}{30}$ $ft^3$.

MODIFIED COMPACTION TEST (ASTM D1557 OR AASHO T180) An energy of 56,250 ft-lb is used to compact 1 $ft^3$ of soil, which is accomplished by com-

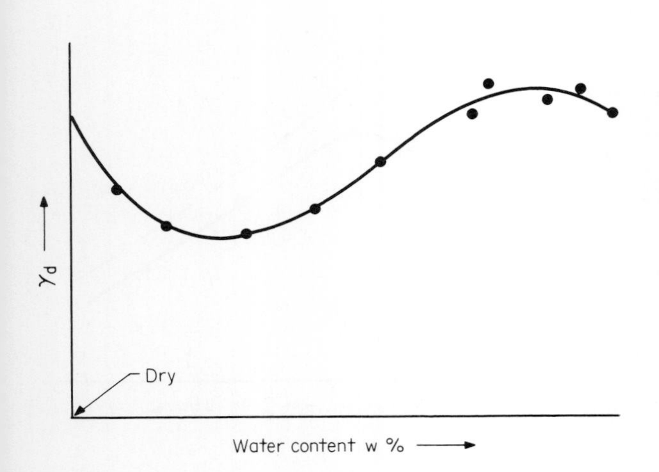


**FIG. 3.5 Typical compaction curve for cohesionless sands and sandy gravels.** [*From Foster (1962).*[8] *Reprinted with permission of McGraw-Hill Book Company.*]

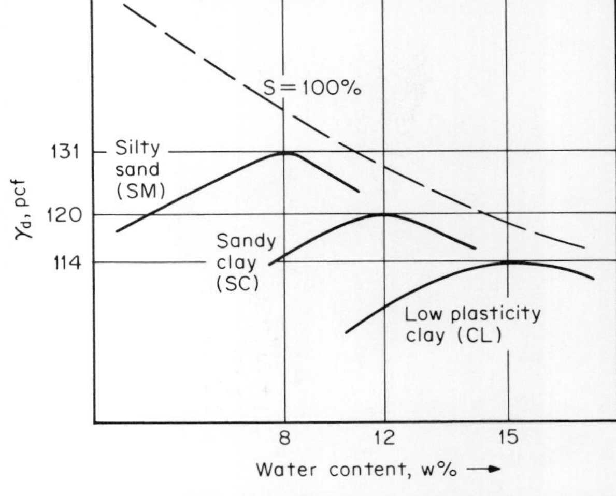


**FIG. 3.6 Typical standard Proctor curves for various materials.**

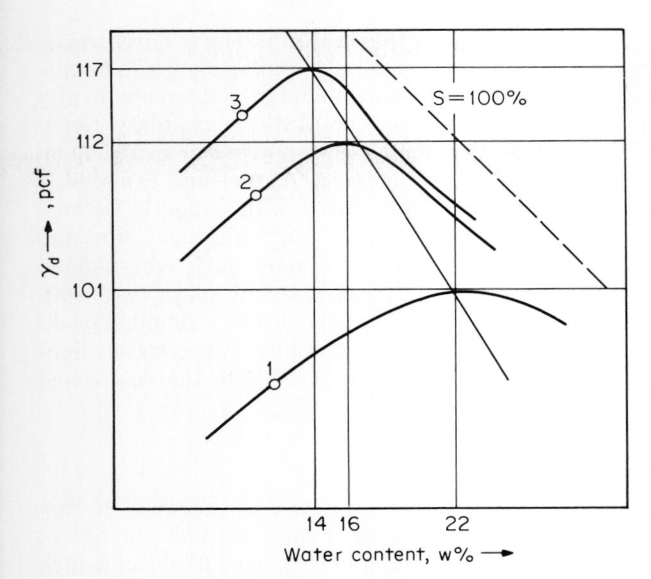


**FIG. 3.7 Effect of different compactive energies on a silty clay.** [*After Turnbull (1950).*]

pacting five sequential layers with a 10-lb hammer dropped 25 times from an 18-in height in a standard mold. Materials containing significant amounts of gravel are compacted in a 6-in-diameter mold (0.075 ft$^3$) by 56 blows on each of five layers. Methods are available for correcting densities for large gravel particles removed from the specimen before testing.

### Relative Density $D_R$

*Relative density* refers to an in situ degree of compaction, relating the *natural density* of a cohesionless granular soil to its *maximum density* (the densest state to which a soil can be compacted, $D_R = 100\%$) and the *minimum density* (the loosest state that dry soil grains can attain, $D_R = 0\%$). The relationship is illustrated on the diagram given as Fig. 3.8, which can be used to find $D_R$ when $\gamma_N$ (natural density), $\gamma_D$ (maximum density), and $\gamma_L$ (loose density) are known. $D_R$ may be expressed as

$$D_R = \frac{1/\gamma_L - 1/\gamma_N}{1/\gamma_L - 1/\gamma_D} \qquad (3.2)$$

*Significance*

$D_R$ is used for classification of the degree of in situ compactness as given on Fig. 3.8 or, more commonly, to classify in situ density as follows: very loose (0–15%), loose (15–35%), medium dense (35–65%), dense (65–85%), and very dense (85–100%). (*See Table 3.23* for correlations with $N$ values of the SPT.)

Void ratio and unit weight are directly related to $D_R$ and gradation characteristics. Permeability, strength, and compressibility are also related directly to $D_R$ and gradation characteristics.

*Measurements of $D_R$*

LABORATORY TESTING See ASTM D2049 and Burmister (1948).[10] *Maximum density* is determined by compaction tests as described in the foregoing section, or by vibratory methods wherein the dry material is placed in a small mold in layers and densified with a hand-held vibrating tool. *Minimum density* is found by very lightly pouring dry sand with a funnel into a mold. $D_R$ measurements are limited to material with less than about 35% nonplastic soil passing the no. 200 sieve because fine-grained soils falsely affect the loose density. A major problem is the determination of the *natural density* of sands which cannot be sampled undisturbed. The shear-pin piston (*see Art. 2.4.2*) has been used to obtain values for $\gamma_N$, or borehole logging with the gamma probe is used to obtain values (*see Art. 2.3.6*).

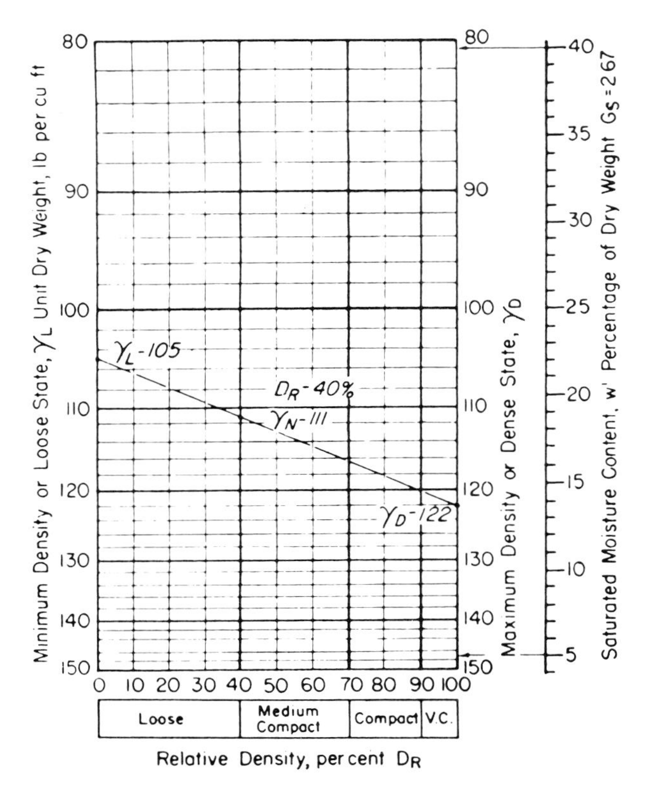


**FIG. 3.8 Relative density diagram** [*From Burmister (1948).*[10] *Reprinted with permission of the American Society for Testing and Materials.*]

FIELD TESTING The SPT and CPT methods are used to obtain estimates of $D_R$.

CORRELATIONS Relations such as those given on Fig. 3.10 for various gradations may be used for estimating values for $\gamma_D$ and $\gamma_L$.

## Gradation (Grain Size Distribution)

*Gradation* refers to the distribution of the various grain sizes in a soil specimen plotted as a function of the percent by weight passing a given sieve size (Fig. 3.9):

- *Well-graded*—a specimen with a wide range of grain sizes
- *Poorly graded*—a specimen with a narrow range of grain sizes
- *Skip-graded*—a specimen lacking a middle range of grain sizes
- *Coefficient of Uniformity* $C_u$—the ratio between the grain diameter at 60% finer to the grain diameter corresponding to the 10% finer line, or:

$$C_u = D60/D10 \tag{3.3}$$

*Significance*

Gradation relationships are used as the basis for soil classification systems. Gradation curves

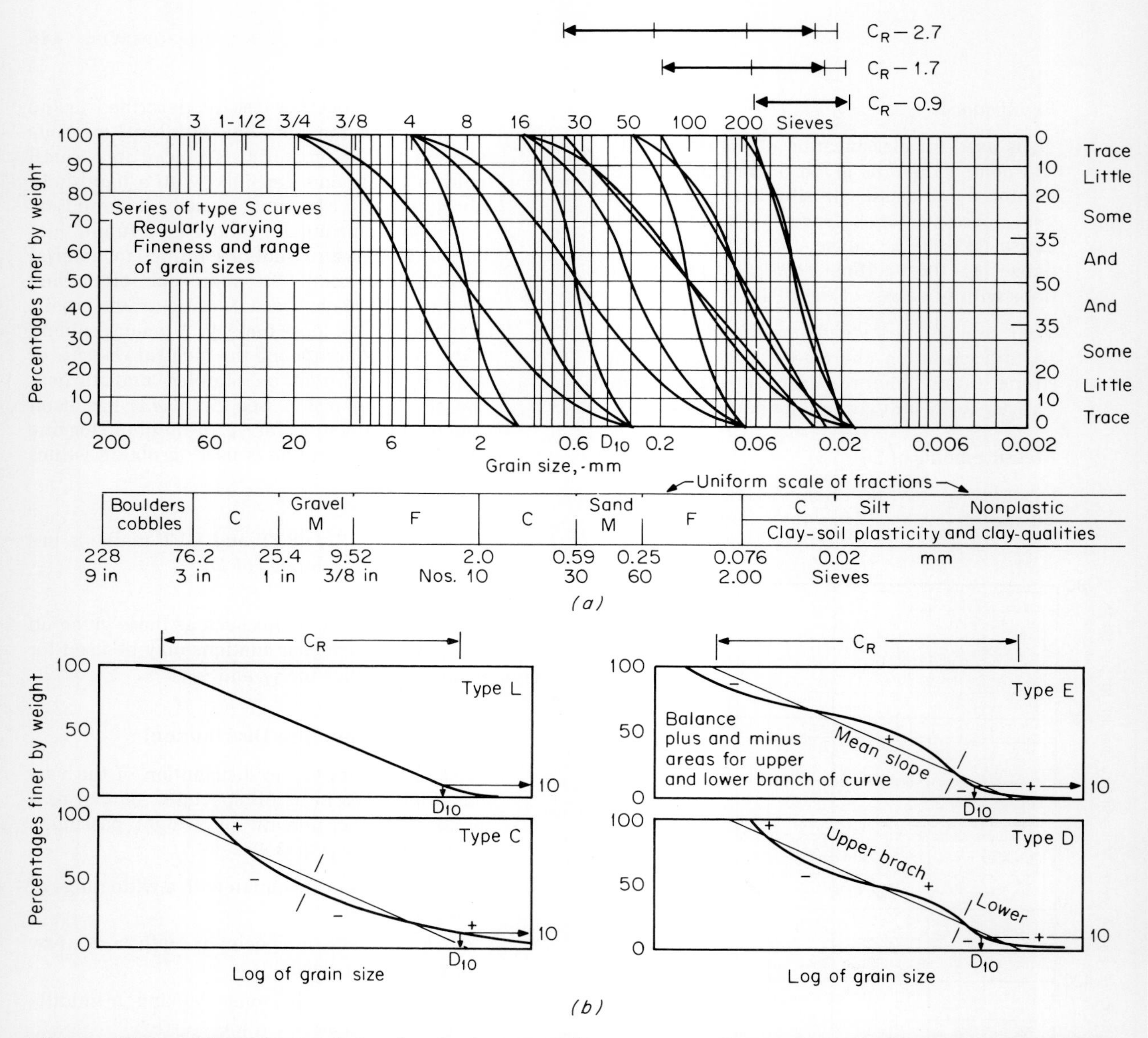


**FIG. 3.9 Distinguishing characteristics of grain size curves: fineness, range of grain sizes, and shape: (a) type S grain size curves and (b) type of grain size curve.** [*From Burmister (1948).*[10] *Reprinted with permission of the American Society for Testing and Materials.*]

from cohesionless granular soils may be used to estimate $\gamma_D$ and $\gamma_L$, and, if $\gamma_N$ or $D_R$ is known, estimates can be made of the void ratio, porosity, internal friction angle, and coefficient of permeability.

*Gradation Curve Characteristics*[10,11,12]

The gradation curves and characteristic shapes, when consideration is given to the range in sizes, can be used for estimating engineering properties. The range of sizes $C_R$ represents fractions of a uniform division of the grain size wherein the divisions 0.02 to 0.06, 0.06 to 0.02, etc. on Fig. 3.9 each represents a $C_R$ = 1. Curve shapes are defined as *L*, *C*, *E*, *D*, or *S* as given on Fig. 3.9 and are characteristic of various types of soil formations as follows:

- *S* shapes are the most common, characteristic of well-sorted (poorly graded) sands deposited by flowing water, wind, or wave action.
- *C* shapes have a high percentage of coarse and fine particles compared with sand particles and are characteristic of some alluvial valley deposits in an arid climate where the native rocks are quartz-poor.
- *E* and *D* shapes include a wide range of particle sizes characteristic of glacial tills and residual soils.

*Relationships*

General relationships among gradation characteristics and maximum compacted densities, minimum densities, and grain angularity are given on Fig. 3.10 (note the significance of grain angularity). Probable depositional values for $D_R$ in terms of gradation characteristics for soils of various geologic origins (see Chap. 7) are given on Fig. 3.11.

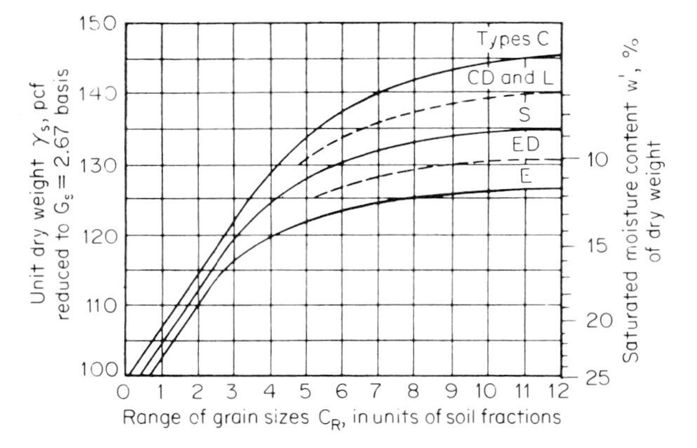


**(a) Maximum compacted densities: $D_R$ obtained by vibrations = 100%.**

**(b) APPROXIMATE MINIMUM DENSITIES, 0% $D_R$**

| Range in grain sizes, $C_R$ | Decrease in density, pcf | | |
|---|---|---|---|
| | Coarser soils | | Finer soils |
| 1 to 3 | 10 | to | 20 |
| 3 to 5 | 20 | to | 25 |
| 5 and greater | 25 | to | 30+ |

**(c) APPROXIMATE INFLUENCE OF GRAIN SHAPE ON DENSITY**

| Grain shape | Change in density, pcf |
|---|---|
| | — |
| Very angular | 10 to — 15 |
| Subangular | 0 to normal |
| Rounded or waterworn | +2 to +5 |
| 0.5% mica | −2 to −5 |

**FIG. 3.10 Maximum compacted densities, approximate minimum densities, and influence of grain shape on density for various gradations.** [*From Burmister (1948).*[10] *Reprinted with permission of the American Society for Testing and Materials.*]

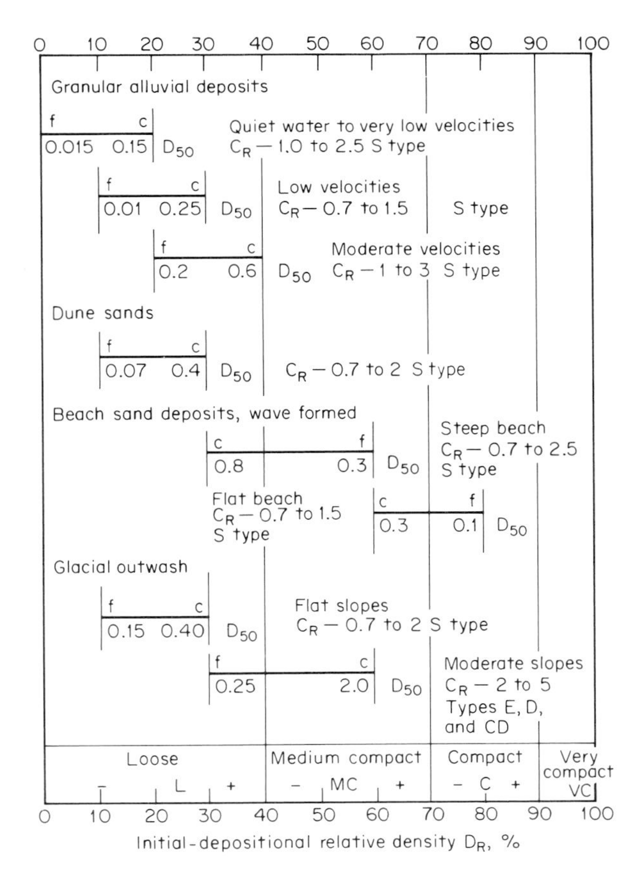


**FIG. 3.11 Probable initial depositional relative densities produced by geologic process of granular soil formation as a tentative guide showing dependence on grain-size parameters, grading-density relations, and geological processes.** [*From Burmister (1962).*[13] *Reprinted with permission of the American Society for Testing and Materials.*]

*Test Methods*

Gradations are determined by *sieve analysis* (ASTM D422) and *hydrometer analysis* (ASTM D422), the latter test being performed on material finer than a no. 200 sieve. For the sieve analysis, a specimen of known weight is passed dry through a sequence of sieves of decreasing size of openings and the portion retained is weighed, or a specimen of known weight is washed through a series of sieves and the retained material dried and weighed. The latter procedure is preferred for materials with cohesive portions because dry sieving is not practical and will yield erroneous results.

## Plasticity

*Definitions and Relationships*

*Atterberg limits,* which include the liquid limit, plastic limit, and the shrinkage limit, are used to define plasticity characteristics of clays and other cohesive materials.

*Liquid limit* (LL) is the moisture content at which a soil passes from the liquid to the plastic state as moisture is removed. At the LL, the undrained shear strength $s_u \approx 0.03\ \text{kg/cm}^2$.

*Plastic limit* (PL) is the moisture content at which a soil passes from the plastic to the semisolid state as moisture is removed.

*Plasticity index* (PI) is defined as PI = LL − PL.

*Shrinkage limit* (SL) is the moisture content at which no more volume change occurs upon drying.

*Activity* is the ratio of the PI to the percent by weight finer than 2 $\mu$.[14] (*See Table 5.28 and Art. 10.6.2* for significance in identifying expansive clays.)

*Liquidity index* (LI) is used for correlations and is defined as:

$$\text{LI} = \frac{w - \text{PL}}{\text{LL} - \text{PL}} = \frac{w - \text{PL}}{\text{PI}} \tag{3.4}$$

*Significance*

A plot of PI vs. LL provides the basis for cohesive soil classification as shown on the plasticity

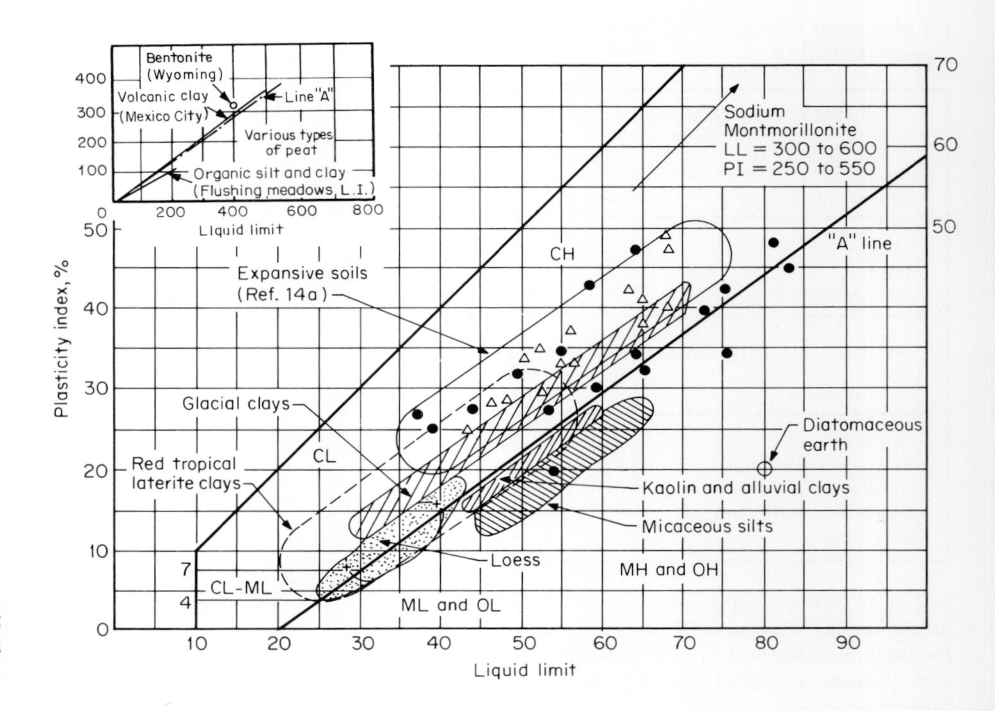


**FIG. 3.12 Plasticity chart for Unified Classification System** (*see Table 5.33*).

chart (Fig. 3.12). Correlations can be made between test samples and characteristic values of natural deposits, for example, predominantly silty soils plot below the *A* line, and predominantly clayey soils plot above. In general, the higher the value for the PI and LL, the greater is the tendency of a soil to shrink upon drying and swell upon wetting. The relationship between the natural moisture content and LL and PI is an indication of the soil's *consistency*, which is related to strength and compressibility *(see Table 3.29)*. The liquidity index expresses this relationship quantitatively. The controlling factors in the values of PI, PL, and LL for a given soil type are the clay mineral present, and the percentages of silt, fine sand, and organic materials.

*Test Methods*

*Liquid limit* (ASTM D423) is performed in a special device containing a cup which is dropped from a controlled height. A pat of soil (only material passing a no. 40 sieve) is mixed thoroughly with water and placed in the cup, and the surface is smoothed and then grooved with a special tool. The LL is the moisture content at which 25 blows of the cup are required to close the groove for a length of 1 cm. There are several test variations [Lambe (1951)[15]].

*Plastic limit* (ASTM D424) is the moisture content at which the soil can just be rolled into a thread ⅛ in in diameter.

*Shrinkage limit* (ASTM D427) is performed infrequently. See Lambe (1951)[15] for discussion.

### Organic Content

*General*

Organic materials are found as pure organic matter or as mixtures with sand, silt, or clay.

*Basic and Index Properties*

Organic content is determined by the *loss by ignition* test which involves specimen combustion at 440°F until constant weight is attained [Arman (1970)[16]]. Gradation is determined after loss by ignition testing. Plasticity testing (PI and LL) provides an indication of organic matter as shown on Fig. 3.12.

## 3.3 HYDRAULIC PROPERTIES (PERMEABILITY)

### 3.3.1 INTRODUCTION

### Flow Through Geologic Materials

*Definitions and Relationships*

*Permeability*, the capacity of a material to transmit water, is described in detail in Art. 8.3.2, along with other aspects of subsurface flow, and is only summarized in this article. Flow through a geologic medium is quantified by a material characteristic termed the *coefficient of permeability k*, expressed in terms of Darcy's law, valid for laminar flow in a saturated, homogeneous material, as

$$k = \frac{q}{iA} \quad \text{cm/s} \tag{3.5}$$

where $q$ = quantity of flow per unit of time, $\text{cm}^3/\text{s}$

$i$ = hydraulic gradient, i.e., the head loss per length of flow $h/L$ (a dimensionless number)

$A$ = $\text{cm}^2$

Values for $k$ are often given in units other than centimeters per second; a conversion chart for various unit systems of units including relative permeabilities is given as Fig. 3.13.

*Secondary Permeability* refers to the rate of flow through rock masses, as contrasted to that through intact rock specimens, and is often given in Lugeon units *(see Art. 3.3.4)*.

*Factors Affecting Flow Characteristics*

SOILS In general, gradation, density, porosity, void ratio, saturation degree, and stratification affect $k$ values in all soils. Additional significant factors are relative density in granular soils and mineralogy and secondary structure in clays.

ROCKS Intact-rock $k$ values relate to porosity and saturation degree. In situ rock $k$ values relate to fracture characteristics (concentration, opening width, nature of filling), degree of saturation, and level and nature of imposed stress form (compressive or tensile). Tensile stresses, for example, beneath a concrete dam can cause

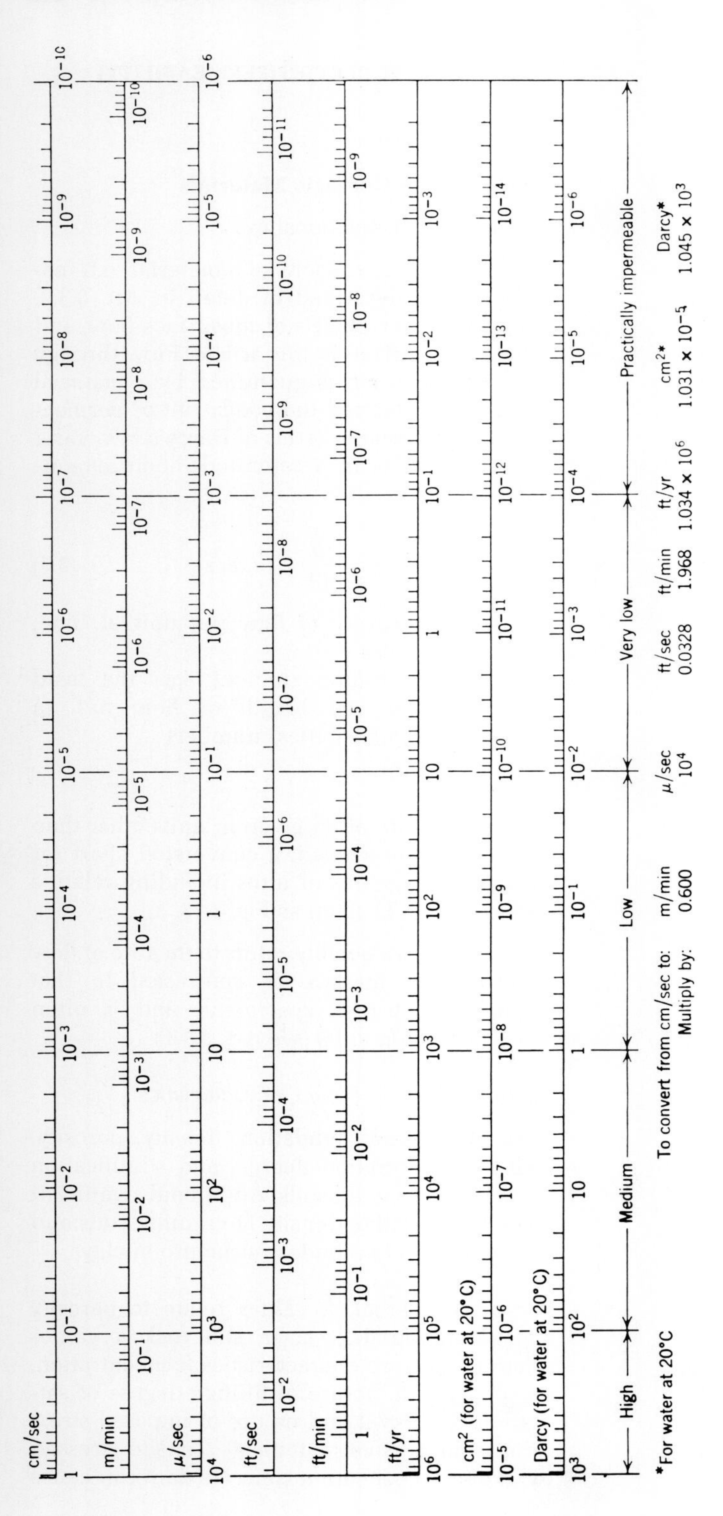


**FIG. 3.13 Permeability conversion chart.** [*From Lambe and Whitman (1969).*[30] *Reprinted by permission of John Wiley & Sons, Inc.*]

opening of joints and foliations, significantly increasing permeability.

### Permeability Considerations

*$k$ Value Determinations*

Values for $k$ are often estimated from charts and tables *(see Art. 3.3.2)* or can be measured in laboratory tests *(see Art. 3.3.3)* or in situ tests *(see Art. 3.3.4)*.

*Applications*

Values of $k$ as estimated or measured in the laboratory are used for:

- Flow net construction and other analytical methods to calculate flow quantities and seepage forces
- Selection of groundwater control methods for surface and underground excavations
- Design of dewatering systems for excavations
- Evaluation of capillary rise and frost susceptibility
- Evaluation of yield of water-supply wells

*In situ measurements of $k$* are made for evaluations of:

- Percolation rates for liquid-waste disposal systems
- Necessity for canal linings (as well as for designing linings)
- Seepage losses beneath and around dam foundations and abutments
- Seepage losses in underground-cavern storage facilities
- Groundwater control during excavation

### Associated Phenomena: Capillarity, Piping, and Liquefaction

*Capillarity* is the tendency of water to rise in "soil tubes," or connected voids, to elevations above the groundwater table. It provides the moisture that results in heaving of foundations and pavements from freezing (frost heave) and swelling of expansive soils. Rating criteria for drainage, capillarity, and frost heave in terms of soil type are given on Table 3.10.

*Piping* refers to two phenomena: (1) water seeping through fine-grained soil, eroding the soil grain by grain and forming tunnels or pipes and (2) water under pressure flowing upward through a granular soil with a head of sufficient magnitude to cause soil grains to lose contact and capability for support. Also termed *boiling* or *liquefaction*, piping is the cause of a "quick" condition (as in quicksand) during which the sand essentially liquefies (*see also Art. 8.3.2* for piping).

*"Cyclic" liquefaction* refers to the complete loss of supporting capacity occurring when dynamic earthquake forces cause a sufficiently large temporary increase in pore pressures in the mass *(see Art. 11.3.3)*.

## 3.3.2 ESTIMATING THE PERMEABILITY COEFFICIENT $k$

### General

*Basis*

Since $k$ values are a function of basic and index properties, various soil types and formations have characteristic ranges of value. Many tables and charts have been published by various investigators relating $k$ values to geologic conditions, which are based on numerous laboratory and field investigations and which may be used for obtaining estimates of $k$ of sufficient accuracy for many applications.

*Partial Saturation Effects*

In using tables and charts, one must realize that the values given are usually for saturated conditions. If partial saturation exists, as often obtains above the groundwater level, the voids will be clogged with air and permeability may be only 40 to 50% of that for saturated conditions.

*Stratification Effects*

In stratified soils, lenses and layers of fine materials will impede vertical drainage, and horizontal drainage will be much greater than that in the vertical direction.

### Relationships

Permeability characteristics of soils and their methods of measurement are given on Table 3.11. Typical permeability coefficients for var-

**TABLE 3.10**
**TENTATIVE CRITERIA FOR RATING SOILS WITH REGARD TO DRAINAGE, CAPILLARITY, AND FROST HEAVING CHARACTERISTICS***

| | | | | | |
|---|---|---|---|---|---|
| Fineness identification† | "Trace fine sand" | "Trace silt" | "Little silt" (coarse and fine) | "Some fine silt" "Little clayey silt" (fissured clay soils) | "Some clayey silt" (clay soils dominating) |
| Approx. effective size, $D_{10}$ mm‡ | 0.4 0.2 | 0.2 0.074 | 0.074 0.02 | 0.02 0.01 | 0.01 |
| Drainage | Free drainage under gravity excellent | Drainage by gravity good | Drainage good to fair | Drains slowly, fair to poor | Poor to impervious |
| Approx. range of $k$, cm/s | 0.5 0.10<br>0.2 | 0.020<br>0.04 | 0.0010<br>0.006 | 0.0002<br>0.0004 | 0.0001 |
| | ←——— Deep wells | ←——— Well points successful | | ———→ | |
| Capillarity | Negligible | Slight | Moderate | Moderate to high | High |
| Approx. rise in feet, $H_c$ | 0.5 | 1.5<br>1.0 | 7.0<br>3.0 | 15.0<br>10.0 | 25.0 |
| Frost heaving susceptibility | Non-frost-heaving | Slight | Moderate to objectionable | Objectionable | Objectionable to moderate |
| Groundwater within 6 ft or $H_c/2$ | | | | | |

*Criteria for soils in a loose to medium-compact state. From Burmister (1951a).[12]

†Fineness classification is in accordance with the ASEE Classification System (*Table 5.34*).

‡Hazen's $D_{10}$: The grain size for which 10% of the material is finer.

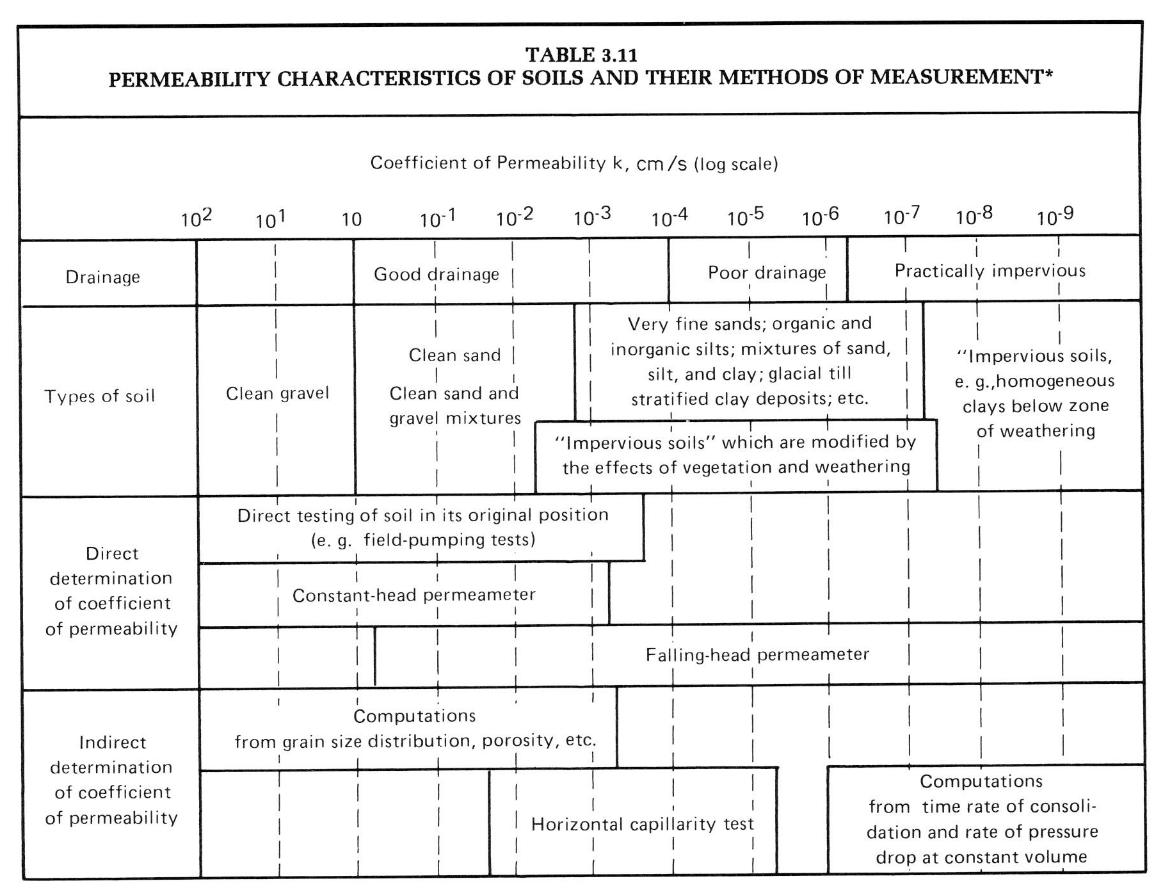

**TABLE 3.11**
**PERMEABILITY CHARACTERISTICS OF SOILS AND THEIR METHODS OF MEASUREMENT***

Coefficient of Permeability k, cm/s (log scale)

$10^2$ $10^1$ $10$ $10^{-1}$ $10^{-2}$ $10^{-3}$ $10^{-4}$ $10^{-5}$ $10^{-6}$ $10^{-7}$ $10^{-8}$ $10^{-9}$

| | |
|---|---|
| Drainage | Good drainage; Poor drainage; Practically impervious |
| Types of soil | Clean gravel; Clean sand; Clean sand and gravel mixtures; Very fine sands; organic and inorganic silts; mixtures of sand, silt, and clay; glacial till stratified clay deposits; etc.; "Impervious soils" which are modified by the effects of vegetation and weathering; "Impervious soils, e. g., homogeneous clays below zone of weathering |
| Direct determination of coefficient of permeability | Direct testing of soil in its original position (e. g. field-pumping tests); Constant-head permeameter; Falling-head permeameter |
| Indirect determination of coefficient of permeability | Computations from grain size distribution, porosity, etc.; Horizontal capillarity test; Computations from time rate of consolidation and rate of pressure drop at constant volume |

*After Casagrande and Fadum (1940)[18]; from Leonards (1962)[19] Reprinted with permission of McGraw-Hill Book Co.

ious conditions are given on the following tables: rock and soil formations, Table 3.12; some natural soil formations, Table 3.13; and various materials for turbulent and laminar flow, Table 3.14. Values of $k$ for granular soils in terms of gradation characteristics ($D_{10}$, $C_R$, curve type) are given on Figs. 3.14 and 3.15, with the latter figure giving values in terms of $D_R$.

### 3.3.3 LABORATORY TESTS

#### Types and Applications

*Constant-head tests* are used for coarse-grained soils with high permeability. *Falling-head tests* are used in fine-grained soils with low permeability. *Consolidometer tests* are used for essentially impervious soils as described in Art. 3.5.4.

#### Constant- and Falling-Head Tests

The two types of laboratory permeameters are illustrated on Fig. 3.16. In both cases, remolded or undisturbed specimens which are completely saturated with gas-free distilled water are used.

*Constant-Head Test*

A quantity of water is supplied to the sample while a constant head is maintained and the discharge quantity $q$ is measured. From Darcy's law, $k$ is found from

$$k = qL/Ah \tag{3.6}$$

*Falling-Head Test*

Flow observations are made on the rate of fall in the standpipe (Fig. 3.16*b*). During time $t_1$ the water level drops from $h_0$ to $h_1$. The value of $k$ is found from the expression:

$$k = \frac{aL}{At_1} \cdot \ln \frac{h_0}{h_1} \tag{3.7}$$

**TABLE 3.12**
**TYPICAL PERMEABILITY COEFFICIENTS FOR ROCK AND SOIL FORMATIONS***

| | k, cm/s | Intact rock | Porosity $n$, % | Fractured rock | Soil |
|---|---|---|---|---|---|
| Practically impermeable | $10^{-10}$<br>$10^{-9}$<br>$10^{-8}$<br>$10^{-7}$ | Massive low-porosity rocks | 0.1–0.5<br>0.5–5.0 | | Homogeneous clay below zone of weathering |
| Low discharge, poor drainage | $10^{-6}$<br>$10^{-5}$<br>$10^{-4}$<br>$10^{-3}$ | Sandstone ($10^{-6}$ to $10^{-3}$)<br>Weathered granite<br>Schist | 5.0–30.0 | Clay-filled joints | Very fine sands, organic and inorganic silts, mixtures of sand and clay, glacial till, stratified clay deposits |
| High discharge, free draining | $10^{-2}$<br>$10^{-1}$<br>1.0<br>$10^{1}$<br>$10^{2}$ | | | Jointed rock<br>Open-jointed rock<br>Heavily fractured rock | Clean sand, clean sand and gravel mixtures<br>Clean gravel |

*After Hoek and Bray (1977).[20]

**TABLE 3.13**
**PERMEABILITY COEFFICIENTS FOR SOME NATURAL SOIL FORMATIONS***

| Formation | Value of $k$, cm/s |
|---|---|
| RIVER DEPOSITS | |
| Rhone at Genissiat | Up to 0.40 |
| Small streams, eastern Alps | 0.02 to 0.16 |
| Missouri | 0.02 to 0.20 |
| Mississippi | 0.02 to 0.12 |
| GLACIAL DEPOSITS | |
| Outwash plains | 0.05 to 2.00 |
| Esker, Westfield, Mass. | 0.01 to 0.13 |
| Delta, Chicopee, Mass. | 0.0001 to 0.015 |
| Till | Less than 0.0001 |
| WIND DEPOSITS | |
| Dune sand | 0.1 to 0.3 |
| Loess | 0.001 ± |
| Loess loam | 0.0001 ± |
| LACUSTRINE AND MARINE OFFSHORE DEPOSITS | |
| Very fine uniform sand, $C_u$ = 5 to 2† | 0.0001 to 0.0064 |
| Bull's liver, Sixth Ave., N.Y., $C_u$ = 5 to 2 | 0.0001 to 0.0050 |
| Bull's Liver, Brooklyn, $C_u$ = 5 | 0.00001 to 0.0001 |
| Clay | Less than 0.0000001 |

*From Terzaghi and Peck (1967).[21] Reprinted with permission of John Wiley & Sons, Inc.
†$C_u$ = uniformity coefficient.

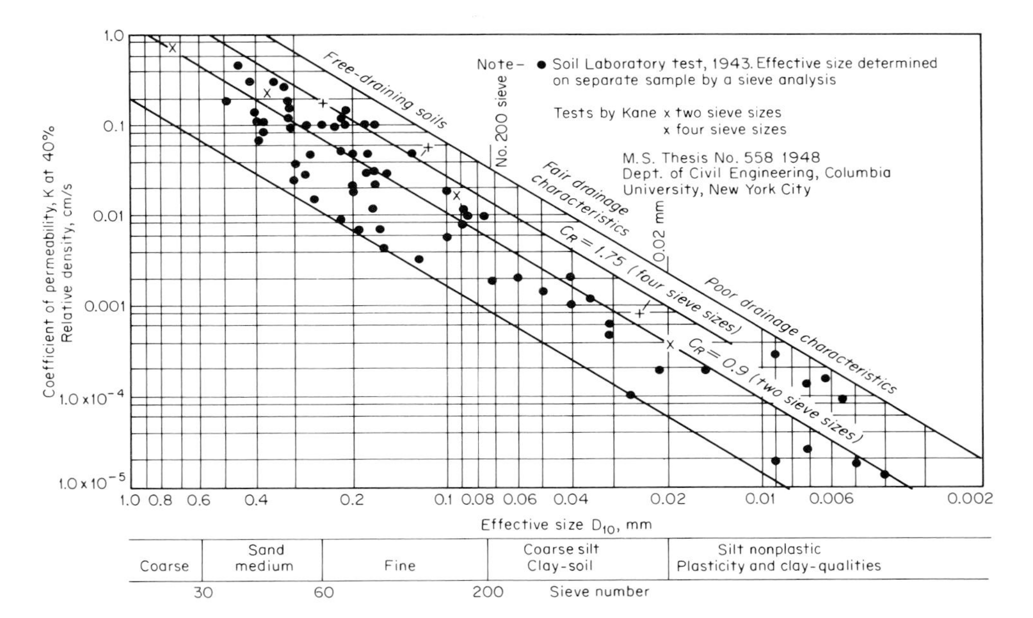


**FIG. 3.14 Relationships between permeability and Hazen's effective size $D_{10}$. Coefficient of permeability reduced to basis of 40% $D_R$ by Fig. 3.15.** *[From Burmister (1948).[10] Reprinted with permission of the American Society for Testing and Materials.]*

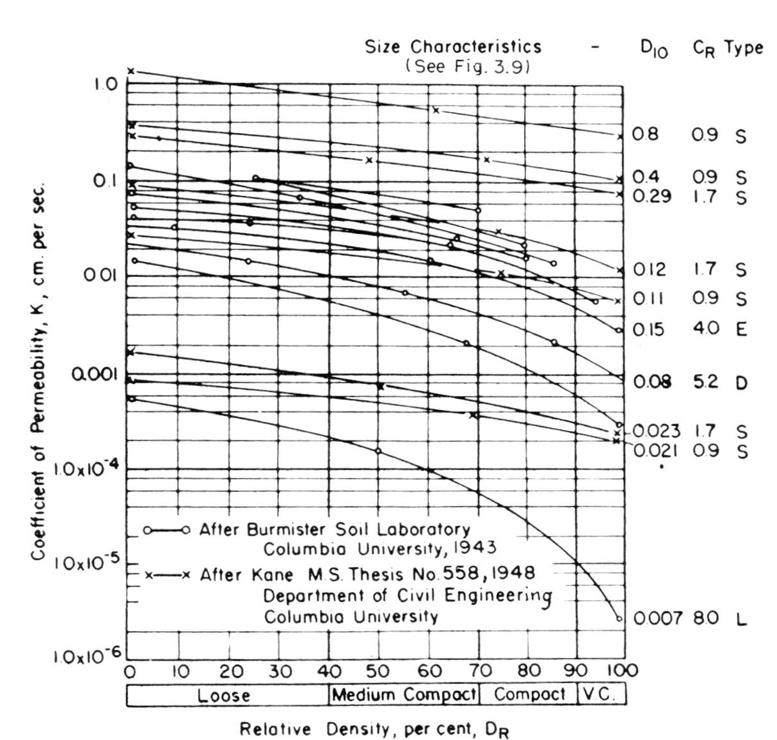


**FIG. 3.15 Permeability–relative density relationships.** *[From Burmister (1948).[10] Reprinted with permission of the American Society for Testing and Materials.]*

**TABLE 3.14**
**TYPICAL PERMEABILITY COEFFICIENTS FOR VARIOUS MATERIALS***

| | Particle-size range | | | | "Effective" size | | Permeability coefficient $k$ | | |
|---|---|---|---|---|---|---|---|---|---|
| | Inches | | Millimeters | | | | | | |
| | $D_{max}$ | $D_{min}$ | $D_{max}$ | $D_{min}$ | $D_{20}$, in | $D_{10}$, mm | ft/year | ft/month | cm/s |
| TURBULENT FLOW | | | | | | | | | |
| Derrick stone | 120 | 36 | | | 48 | | $100 \times 10^6$ | $100 \times 10^5$ | 100 |
| One-man stone | 12 | 4 | | | 6 | | $30 \times 10^6$ | $30 \times 15^5$ | 30 |
| Clean, fine to coarse gravel | 3 | ¼ | 80 | 10 | ½ | | $10 \times 10^6$ | $10 \times 10^5$ | 10 |
| Fine, uniform gravel | ⅜ | 1/16 | 8 | 1.5 | ⅛ | | $5 \times 10^6$ | $5 \times 10^5$ | 5 |
| Very coarse, clean, uniform sand | ⅛ | 1/32 | 3 | 0.8 | 1/16 | | $3 \times 10^6$ | $3 \times 10^5$ | 3 |
| LAMINAR FLOW | | | | | | | | | |
| Uniform, coarse sand | ⅛ | 1/64 | 2 | 0.5 | | 0.6 | $0.4 \times 10^6$ | $0.4 \times 10^5$ | 0.4 |
| Uniform, medium sand | | | 0.5 | 0.25 | | 0.3 | $0.1 \times 10^6$ | $0.1 \times 10^5$ | 0.1 |
| Clean, well-graded sand and gravel | | | 10 | 0.05 | | 0.1 | $0.01 \times 10^6$ | $0.01 \times 10^5$ | 0.01 |
| Uniform, fine sand | | | 0.25 | 0.05 | | 0.06 | 4000 | 400 | $40 \times 10^{-4}$ |
| Well-graded, silty sand and gravel | | | 5 | 0.01 | | 0.02 | 400 | 40 | $4 \times 10^{-4}$ |
| Silty sand | | | 2 | 0.005 | | 0.01 | 100 | 10 | $10^{-4}$ |
| Uniform silt | | | 0.05 | 0.005 | | 0.006 | 50 | 5 | $0.5 \times 10^{-4}$ |
| Sandy clay | | | 1.0 | 0.001 | | 0.002 | 5 | 0.5 | $0.05 \times 10^{-4}$ |
| Silty clay | | | 0.05 | 0.001 | | 0.0015 | 1 | 0.1 | $0.01 \times 10^{-4}$ |
| Clay (30 to 50% clay sizes) | | | 0.05 | 0.0005 | | 0.0008 | 0.1 | 0.01 | $0.001 \times 10^{-4}$ |
| Colloidal clay ($-2\mu \leq$ 50%) | | | 0.01 | 10 Å | | 40 Å | 0.001 | $10^{-4}$ | $10^{-9}$ |

*From Hough (1957).[22] Reprinted with permission of John Wiley & Sons, Inc.

**TABLE 3.15**
**SEEPAGE TESTS IN SOILS**

| Test | Field conditions | Method | Procedure |
|---|---|---|---|
| Constant head | Unsaturated granular soils | (*a*) Shallow-depth small pit, 12 in deep and square (percolation test)<br>(*b*) Moderate depth, hand-auger hole<br>(*c*) Greater depth, install casing (open-end pipe test)* | 1. Uncased holes, backfill with fine gravel or coarse sand.<br>2. Saturate ground around hole.<br>3. Add metered quantities of water to hole until quantity decreases to constant value (saturation).<br>4. Continue adding water to maintain constant level, recording quantity at 5-min intervals.<br>5. Compute $k$ as for laboratory test. |
| Falling or variable head | Below GWL, or in slow-draining soils | Performed in cased hole* | 1. Fill casing with water and measure rate of fall.<br>2. Computations.† |
| Rising-head test | Below GWL in soil of moderate $k$ | Performed in cased hole* | 1. Bail water from hole.<br>2. Record rate of rise in water level until rise becomes negligible.<br>3. After testing, sound hole bottom to check for quick condition as evidenced by rise of soil in casing. Computations.† |

*Tests performed in casing can have a number of bottom-flow conditions. These are designed according to geologic conditions, to provide measurements of $k_{mean}$, $k_v$, or $k_h$.

- $k_{mean}$: Determined with the casing flush with the end of the borehole in uniform material, or with casing flush on the interface between an impermeable layer over a permeable layer.
- $k_v$: Determined with a soil column within the casing, similar to the laboratory test method, in thick, uniform material
- $k_h$: Determined by extending an uncased hole some distance below the casing and installing a well-point filter in the extension.

†References for computations of $k$ with various boundary conditions:

- NAVFAC Desigh Manual DM-7 (1971)[6]
- Hoek and Bray (1974)[20]
- Lowe and Zaccheo (1975)[23]
- Cedergren (1967)[24]

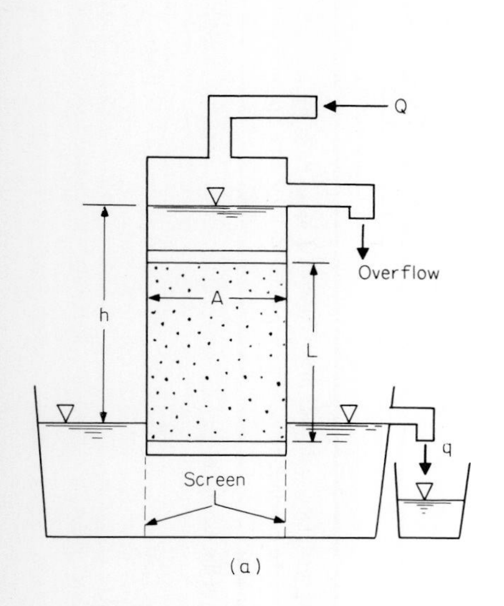


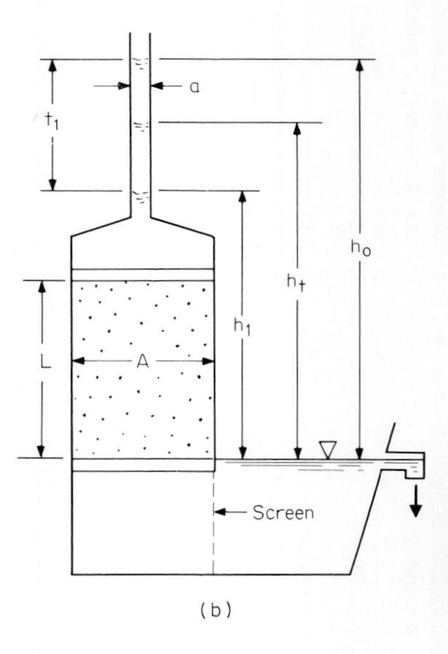


**FIG. 3.16 Two types of laboratory permeameters: (*a*) constant-head test and (*b*) falling-head test.**

**TABLE 3.16**
**PUMPING TESTS***

| Test | Field conditions | Method | Procedure | Disadvantages |
|---|---|---|---|---|
| Gravity well (Fig. 8.33) | Saturated, uniform soil (unconfined aquifer) | Pump installed in screened and filtered well and surrounded by a pattern of observation wells. | Well is pumped at constant rate until cone of drawdown measured in observation wells has stabilized. (Recharge equals pumping rate.) | Provides values for $k_{mean}$. |
| Gravity well | Rock masses | Similar to above. | Similar to above. | Flow from entire hole measured. Provides an average value. |
| Artesian well (Fig. 8.34) | Confined aquifer (Pervious under thick impervious layer) | Similar to above. | Similar to above. | Provides values for $k_{mean}$ in aquifer. |

*For field arrangement and evaluation of data see Art 8.3.3.

## 3.3.4 IN SITU TESTING

### Seepage Tests in Soils

Tests include constant head, falling or variable head, and rising head. They are summarized in terms of applicable field conditions, method, and procedure on Table 3.15.

### Pumping Tests

Tests are made from gravity wells or artesian wells in soils or rock masses as described on

Table 3.16. *See Art. 8.3.3* for additional discussion.

### Pressure Testing In Rock Masses

*General Procedures*

The general arrangement of equipment is illustrated on Fig. 3.17, which shows two packers in a hole. One of two general procedures is used, depending on rock quality.

*The common procedure*, used in poor to moderately poor rock with hole collapse problems, involves drilling the hole to some depth and performing the test with a single packer. Casing is installed if necessary, and the hole is advanced to the next test depth.

*The alternate procedure*, used in good-quality rock where the hole remains open, involves drilling the hole to the final depth, filling it with water, surging it to clean the walls of fines, then bailing it. Testing proceeds in sections from the bottom up with two packers.

*Packer spacing* depends on rock conditions and is normally 1, 2, or 3 m, or at times 5 m. The wider spacings are used in good-quality rock and the closer spacings in poor-quality rock.

*Testing Procedures*

1. Expand the packers with air pressure.
2. Introduce water under pressure into the hole, first between the packers and then below the lower packer.
3. Record elasped time and volume of water pumped.
4. Test at several pressures; usually 15, 30, and 45 psi (1, 2, and 3 kg/cm$^2$) above the natural piezometric level.[23] To avoid rock-mass deformation, the excess pressure above the natural piezometric level should not exceed 1 psi for each foot (0.2 kg/cm$^2$·m) of soil and rock above the upper packer.

*Data Evaluation*

Curves are plotted of flow vs. pressure to permit evaluation of changes in the rock mass during testing:

- Concave upward curves indicate that fractures are opening under pressure
- Convex curves indicate that fractures are being clogged (permeability decreasing with increased pressure)
- Linear variation indicates no change occurring in the fractures

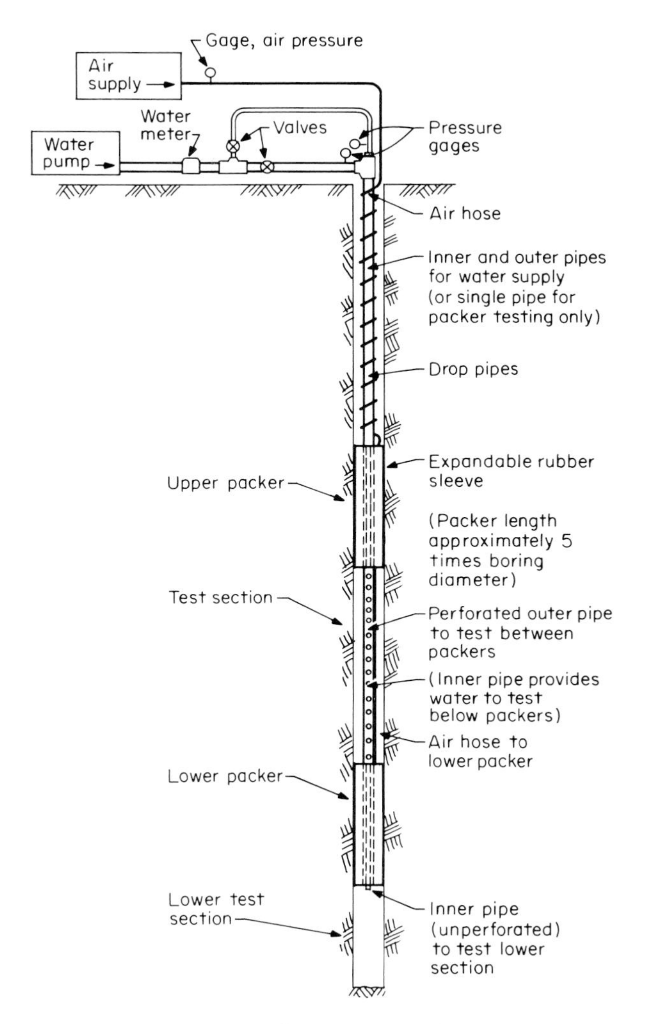


**FIG. 3.17 Apparatus for determining rock permeability in situ using pressure testing between packers.**

Approximate values for $k$ are computed from the expressions[7]:

$$\text{For } L \geq 1\text{: } k = \frac{Q}{2\pi LH} \ln \frac{L}{r} \qquad (3.8)$$

$$\text{For } 10\,r > L \geq r\text{: } k = \frac{Q}{2\pi LH} \sinh^{-1} \frac{L}{2r} \qquad (3.9)$$

where $k$ = coefficient of permeability
$Q$ = constant flow rate into hole
$L$ = length of test section
$H$ = differential head on test section (see explanation below)
$r$ = hole radius
$\sinh^{-1}$ = inverse hyperbolic sine

In determining the value for $H$, *head losses* in the system should be accounted for. Since most head losses occur in the drop pipe, they can be minimized by using as large a diameter pipe as practical. Head loss can be estimated from the relationship[25]

$$h_f = f\frac{L}{d} \cdot \frac{v^2}{2g} \qquad (3.10)$$

where $L$ = pipe length
$d$ = pipe diameter
$v$ = flow velocity
$g$ = gravitational acceleration
$f$ = frictional component (obtained from charts for various pipe diameters, materials, and discharges)

*Lugeon Test*

In the Lugeon test, used commonly in Europe, the hole is drilled to test depth and a packer installed about 5 m from the bottom. Flow is measured after 5 or 10 min of test under pressure, and the test is performed under several pressures. The standard measurement of pressure is 10 kg/cm² and the results are given in Lugeon units.

*Lugeon unit* is defined as a flow of 1 liter of water per minute per meter of borehole length at a pressure of 10 kg/cm². (1 Lugeon unit equals about $10^{-5}$ cm/s.)

*Disadvantages of Pressure Testing*

Values can be misleading because high pressures cause erosion of fines from fractures as well as deformation of the rock mass and closure of fractures.[26]

*Additional References*

Dick (1975)[27] and Hoek and Bray (1977)[20] provide further information.

## 3.4 RUPTURE STRENGTH

### 3.4.1 INTRODUCTION

**Basic Definitions**

*Stress $\sigma$*

Force $P$ per unit of area, expressed as

$$\sigma = P/A \qquad (3.11)$$

| System | Equivalent units for stress |
|---|---|
| English | 13.9 psi = 2000 psf = 2 ksf = 1 tsf = 1 bar |
| Metric | 1 kg/cm² = 10 T/m² (≈ 1 tsf) |
| SI | 100 kN/m² = 100 kPa = 0.1 MPa (≈ 1 tsf) |

*Strain $\epsilon$*

Change in length per unit of length caused by stress. Can occur as compressive or tensile strain. Compressive and tensile strain are expressed as

$$\epsilon = \Delta L/L \qquad (3.12)$$

*Shear*

Shear is the displacement of adjacent elements along a plane or curved surface.

*Shear Strain $\zeta$*

The angle of displacement between elements during displacement.

*Shear Stress $\lambda$*

Shear stress is the stress causing shear.

*Shear Strength $S$ or $s$*

Shear strength is a characteristic value at which a material fails in rupture or shear under an applied force.

*Dilatancy*

Dilatancy is the tendency of the volume to *increase* under increasing shear or stress difference.

## Strength of Geologic Materials

*Components: Friction and Cohesion*

*Friction* is a resisting force between two surfaces as illustrated on Fig. 3.18. It often is the only source of strength in geologic materials and is a direct function of the normal force.

*Cohesion* results from a bonding between surfaces of particles caused by electrochemical forces and is independent of normal forces.

*Influencing Factors*

Strength is not a constant value for a given material but rather is dependent upon many factors, including material properties, magnitude and direction of the applied force and the rate of application, drainage conditions in the mass, and the magnitude of the confining pressure.

## Stress Conditions in Situ

*Importance*

A major factor in strength problems is stress conditions existing in the ground, primarily because normal stresses on potential failure surfaces result from overburden pressures.

*Geostatic Stresses*

*Overburden pressures*, consisting of both vertical and lateral stresses, exist on an element in the ground as a result of the weight of the overlying materials. Stress conditions for level ground are illustrated on Fig. 3.19; sloping ground results in more complex conditions. (Changes in geostatic stresses are invoked by surface foundation loads, surface and subsurface excavations, lowering of the groundwater level, and natural phenomena such as erosion and deposition.) Vertical earth pressures from overburden weight alone are found by summing the weights from the various strata as follows:

$$\sigma_v = \sum_{0}^{z} \gamma_n z_n \quad (3.14)$$

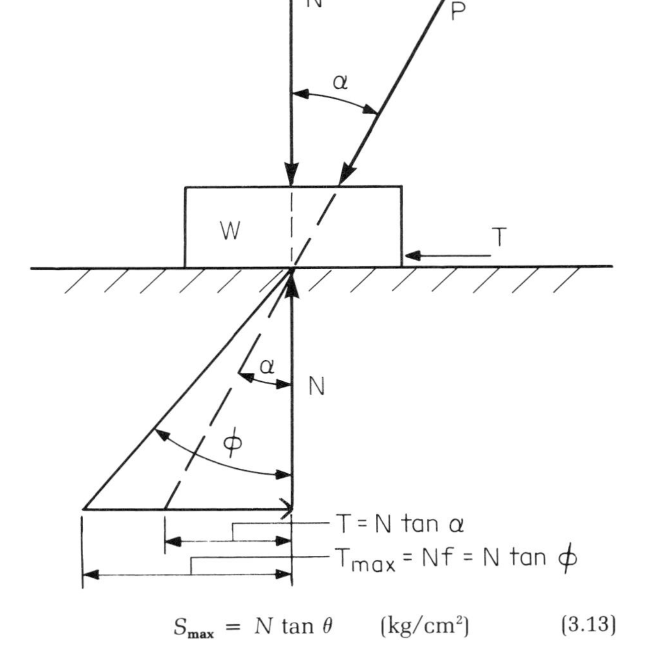


$$S_{max} = N \tan \theta \quad (\text{kg/cm}^2) \quad (3.13)$$

**FIG. 3.18 Frictional force $f$ resisting shearing force $T$. [$P$ = force applied in increments until slip occurs; $N$ = normal force component including block weight $W$; $T$ = shearing stress component; $f$ = frictional resistance; $\alpha$ = angle of obliquity (resultant of $N$ and $T$); $\phi$ = friction angle, or $\alpha_{max}$ at slip; $S_{max}$ = maximum shearing resistance = $T_{max}$.]**

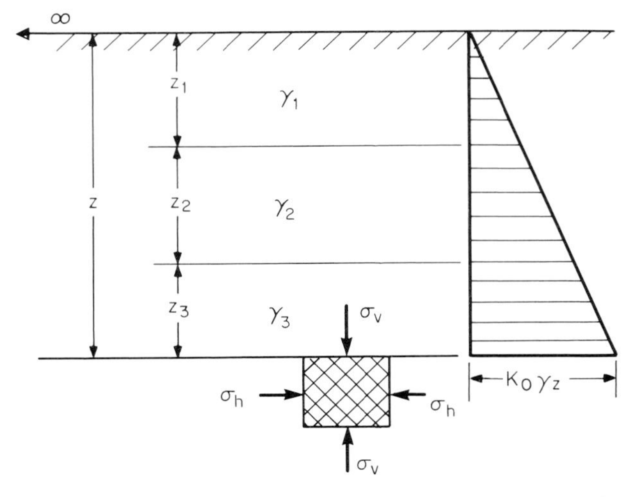


**FIG. 3.19 The geostatic stress condition and "at-rest" earth pressures.**

GLACIAL ICE LOAD

Ice

Ice

Ice Marginal Lake

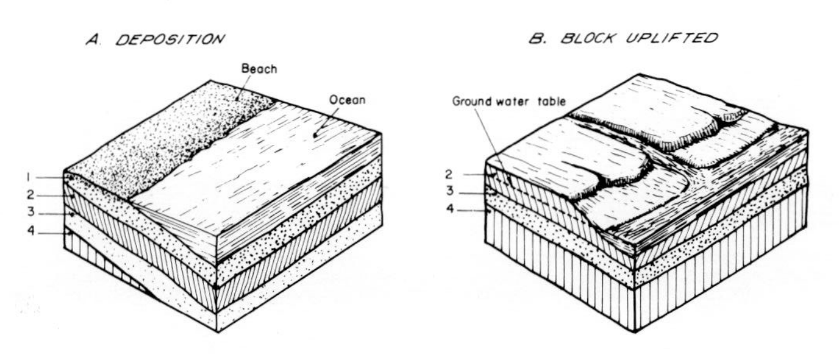


*Block rises in elevation, stratum 1 removed by erosion. Strata 2,3,4 have been prestressed by weight of overburden removed and reduction of bouyant affect of water in stratum 2. (The block rises, or rebounds, as load is removed from the earth's surface; for example, during eons of erosion or Glacial recession).*

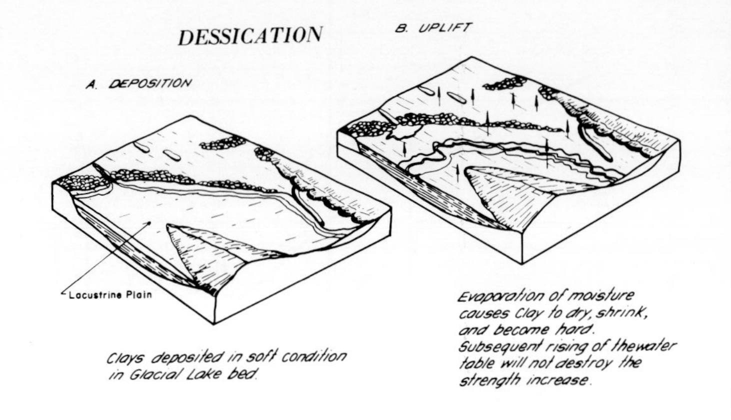


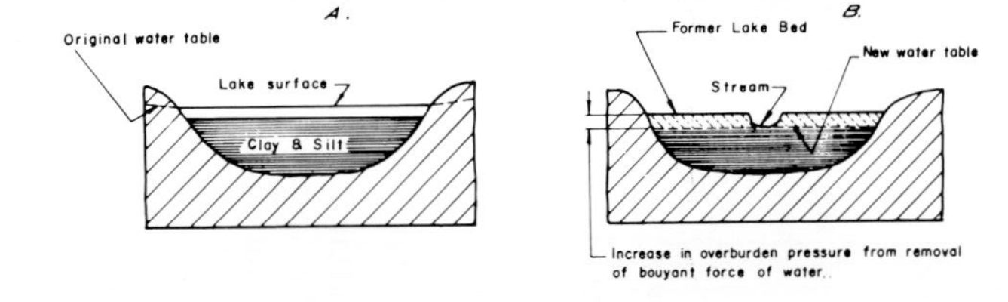


**FIG. 3.20 Soil profile strengthening processes.** [From Ward (1967).[28]]

*Coefficient of lateral earth pressure "at-rest"* $K_0$ is the ratio of the lateral to vertical stress in a natural deposit that has not been subject to lateral strain, the values for which vary substantially with material types and properties *(see Art. 3.4.2)*. It is expressed as

$$K_0 = \sigma_h/\sigma_v \quad \text{or} \quad \sigma_h = K_0\sigma_v \tag{3.15}$$

For an elastic solid:

$$K_0 = \frac{\nu}{1 - \nu} \tag{3.16}$$

In the above expressions, $\gamma$ = material unit weight ($\gamma$ above GWL; $\gamma_b$ below GWL) and $\nu$ = Poisson's ratio *(see Art. 3.5.1)*.

### *Total and Effective Stresses*

The *total stress* on the soil element in Fig. 3.19 at depth Z is

$$\sigma_v = \gamma_t Z \tag{3.17}$$

If the static water table is at the surface, however, and the soil to depth Z is saturated, there is pressure on the water in the pores because of a piezometric head $h_w$ and the unit weight of water $\gamma_w$, which is termed the *neutral stress* (acting equally in all directions), or the *pore-water pressure* $u_w$ or $u$, given as

$$u = \sigma_w h_w \tag{3.18}$$

The *effective stress* $\sigma_v$, or actual intergranular stresses between soil particles, results from a reduction caused by the neutral stress and is equal to the total stress minus the pore-water pressure, or:

$$\bar{\sigma}_v = \sigma_v - u \tag{3.19}$$

or

$$\bar{\sigma}_v = \gamma_b Z \tag{3.20}$$

where $\gamma_b$ = the effective or submerged soil weight ($\gamma_b = \gamma_t - \gamma_w$).

In calculations, therefore, above the groundwater level the effective soil weight is the total weight $\gamma_t$, and below the groundwater level (or any other water surface), the effective soil weight is the submerged soil weight $\gamma_b$.

### *Prestress in Soil Formations*

GENERAL Soils compress naturally under the weight of overlying materials, or some other applied load, resulting in strength increase over values inherent as deposited, or shortly thereafter. Three categories of prestress are defined according to the degree of compression that has occurred.

NORMALLY CONSOLIDATED (NC) The soil element has never been subjected to pressures greater than existing overburden pressures.

OVERCONSOLIDATED (OC) The soil element has at some time in its history been subjected to pressures in excess of existing overburden, such as result from glacial ice loads, removal of material by erosion, desiccation, or lowering of the groundwater level as shown on Fig. 3.20.

UNDERCONSOLIDATED (UC) The soil element exists at a degree of prestress less than existing overburden pressures, such as can result from hydrostatic pressures reducing overburden load as shown in Fig. 3.21. Such soils are normally relatively weak. Weakening of strata can also occur from removal by solution of a cementing agent or other mineral constituents.

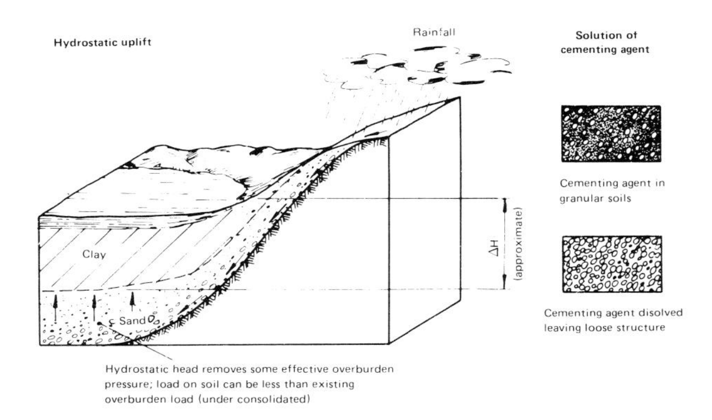


**FIG. 3.21 Soil profile weakening processes.** [*From Ward (1967).*[28]]

## Principal Stresses and the Mohr Diagram

*Importance*

Fundamental to the strength aspects of geologic materials are the concepts of principal stresses and the Mohr diagram on which their relationships may be illustrated.

*Principal Stresses*

Stresses acting on any plane passed through a point consist of a normal stress $\sigma$ (compression or tension) and a shearing stress $\tau$. (Soil mechanics problems are normally concerned with compressive stresses.) On one particular plane the normal stress will be the maximum possible value and the shearing stress will be equal to zero. On one plane perpendicular to this plane the normal stress will be the minimum possible value, with shear stress also equal to zero. On a second plane perpendicular to this plane the normal stress will have an intermediate value and the shearing stress will also be zero. These planes are termed the *principal planes*.

The principal stresses are the stresses acting perpendicular to the principal planes including the maximum (major) principal stress $\sigma_1$, the minimum (minor) principal stress $\sigma_3$ and the intermediate principal stress $\sigma_2$. The relationship between principal stresses and the normal stress and the shear stress acting on a random plane through a point is shown on Fig. 3.22. The intermediate principal stress is the plane of the paper and in soil mechanics problems is normally considered to be equal to $\sigma_3$.

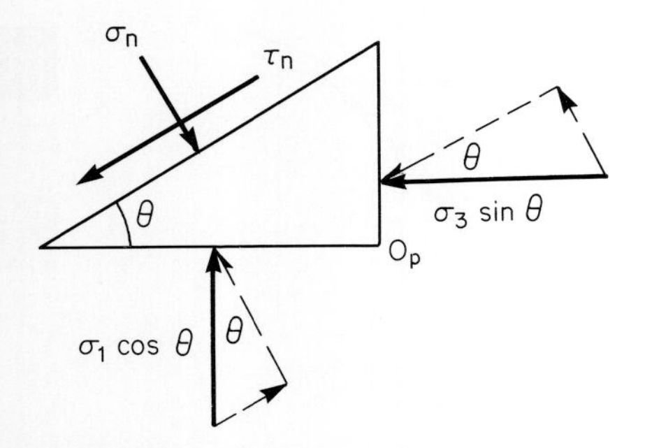


**FIG. 3.22 Stresses on a random plane through a point ($\sigma_2$ is the plane of the paper).**

*The Mohr Diagram*

Equilibrium requires that the sum of the forces given on Fig. 3.22 equal zero. Therefore, $\sigma_n$ and $\tau$ can be expressed in terms of the principal stresses and the angle $\theta$ as

$$\sigma_n = \frac{\sigma_1 + \sigma_3}{2} + \frac{\sigma_1 - \sigma_3}{2} \cos 2\theta \qquad (3.21)$$

$$\tau = \frac{\sigma_1 - \sigma_3}{2} \sin 2\theta \qquad (3.22)$$

If points are plotted to represent coordinates of normal and shearing stresses acting on a particular plane for all values of $\theta$ given in Eqs. 3.21 and 3.22, their loci form a circle which intersects the abscissa at coordinates equal to the major ($\sigma_1$) and minor ($\sigma_3$) principal stresses. The circle is referred to as the *Mohr diagram*, or *Mohr's circle*, given on Fig. 3.23.

## Applications of Strength Values

*Stability Analysis*

Values for strength are used in stability analyses, the discussion of which is beyond the scope of this book, except for evaluations of slopes. In general terms, stability is based on *plastic equilibrium*, or a condition of maximum shear strength with failure by rupture imminent. When the imposed stresses cause the shear strength to be exceeded, rupture occurs in the mass along one or more failure surfaces. Analyses are normally based on the *limit equilibrium approach*, i.e., a limiting value that can be reached when the forces acting to cause failure are in balance with the forces acting to resist failure. Resistance to failure is provided by the shear strength mobilized along the failure surface.

*Typical Problems*

Some field conditions involving failure by rupture are illustrated on Fig. 3.24, showing the relationships between the force acting to cause failure, the strength acting along the failure surface, and the principal and normal stresses.

### 3.4.2 SHEAR STRENGTH RELATIONSHIPS

#### Basic Concepts

Shear strength may be given in several forms, depending on various factors, including the drained strength, the undrained strength, the peak strength, the residual or ultimate strength, and strength under dynamic loadings. In addition, strength is the major factor in determining active and passive earth pressures.

Under an applied force a specimen will strain until rupture occurs at some *peak stress;* in some materials, as strain continues, the resistance reduces until a constant minimum value is reached, termed the *ultimate* or *residual strength.*

*Factors Affecting Strength*

MATERIAL TYPE Some materials exhibit only a frictional component of resistance $\phi$; others exhibit $\phi$ as well as cohesion $c$. In soft clays at the end of construction it is normally the undrained strength $s_u$ that governs.

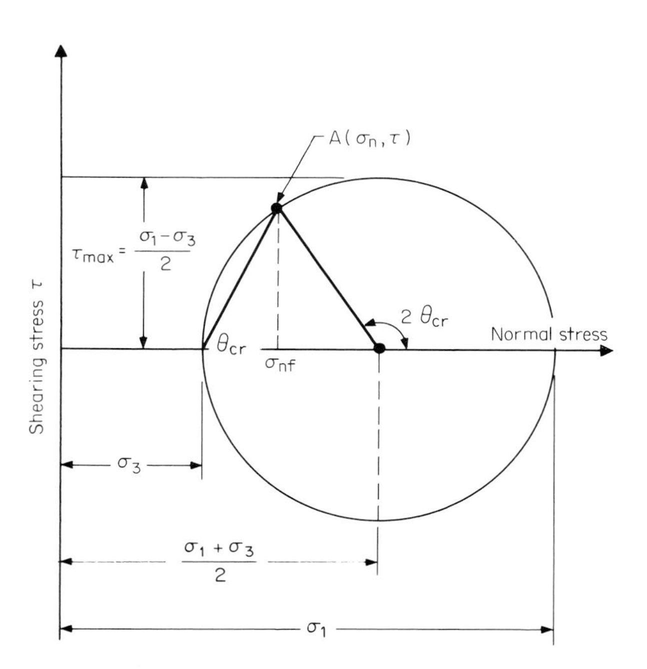


**FIG. 3.23** **The Mohr diagram relating $\tau$, $\sigma_{nf}$, $\sigma_3$, and $\sigma_1$ ($\theta_{cr}$ = $\theta$ at failure; $\sigma_{nf}$ = normal stress at failure).**

CONFINING PRESSURE In materials with $\phi$ acting, the strength increases as the confining pressure increases.

UNDRAINED OR DRAINED CONDITIONS Relates to the ability of a material to drain under applied stress and determines whether total or effective stresses act.

LOADING DIRECTION Forces applied parallel to weakness planes as represented by stratification in soils, or foliation planes or joints in rock masses, will result in lower strengths than if the force is applied perpendicularly. Compressive strengths are much higher than tensile strengths.

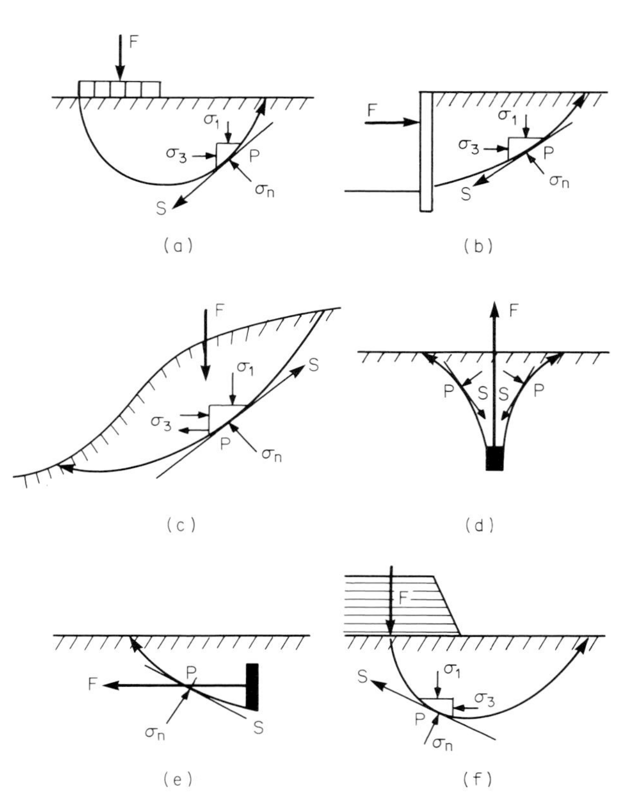


**FIG. 3.24** **Some field conditions involving failure by rupture: (*a*) foundations; (*b*) retaining structures; (*c*) slopes; (*d*) ground anchors; (*e*) wall anchors; (*f*) embankments.**

DISPLACEMENT AND NORMAL STRESS In some cohesive materials, such as overconsolidated fissured clays and clay shales, the strain at which the peak stress occurs depends on the normal stress level and the magnitude of the peak strength varies with the magnitude of normal stress[29] as shown in Fig. 3.25. As strain continues the ultimate strength prevails. These concepts are particularly important in slope stability problems.

### *Angle of Internal Friction $\phi$*

The stresses acting on a confined specimen of *dry cohesionless soil*, either in the ground or in a triaxial testing device (*see Art. 3.4.4*), are illustrated on Fig. 3.26. $p$ is the applied stress, $\sigma_3$ is the confining pressure, and $\sigma_1 = p + \sigma_3$ (in testing $p$ is termed the deviator stress $\sigma_d$). As stress $p$ is gradually increased, it is resisted by the frictional forces acting between grains until some characteristic stress level is reached at which resistance is exceeded and rupture occurs; the stress is termed the *peak strength*.

If $\sigma_3$ and $\sigma_1$ (for the peak strength) are plotted to define a Mohr's circle at failure, a line drawn tangent to the circle passes through the origin at an angle $\phi$ as shown on Figs. 3.27 and 3.28. Other specimens of the same material loaded to failure, but at different confining pressures, will have Mohr's circles tangent to the line defined by $\phi$. The circle tangent line represents the limits of stability and is termed *Mohr's envelope*. It defines shear strength in terms of the friction angle $\phi$, and the normal or total stresses, expressed as

$$s = \tau_{max} = \sigma_n \tan \phi \tag{3.23}$$

(In actuality, the envelope line will not be straight, but will curve downward slightly at higher confining pressures.)

### *Total vs. Effective Stresses ($\phi$ vs. $\bar{\phi}$)*

In a fully saturated cohesionless soil the $\phi$ value will vary with the drainage conditions prevailing during failure. In undrained conditions, total stresses prevail, and in drained conditions, only effective stresses are acting and the friction angle is expressed as $\bar{\phi}$.

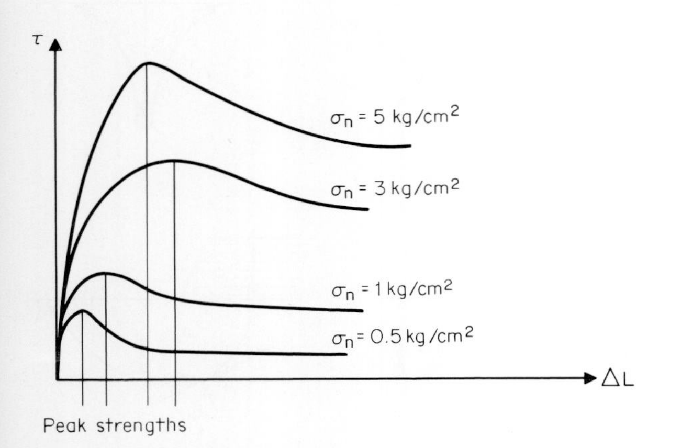


**FIG. 3.25 Peak strength vs. displacement and normal stress.** [*From Peck (1969).*[29]]

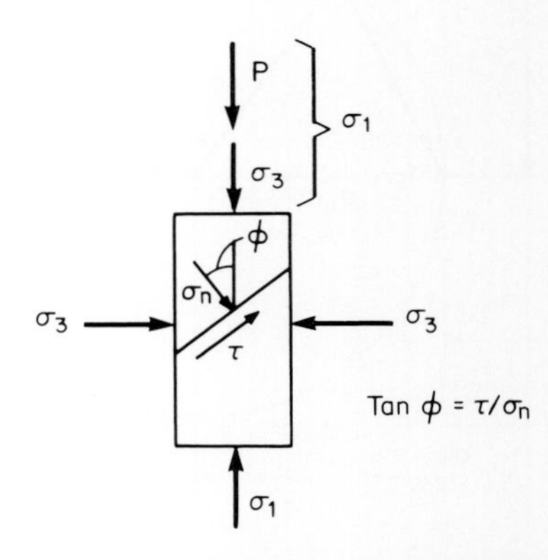


**FIG. 3.26 Stresses on a specimen in the confined state.**

### Total Stresses

*Saturated Soils*

If no drainage is permitted from a fully saturated soil as load is applied, the stress at failure is carried partially by the pore water and partially by the soil particles which thus develop some intergranular stresses. The peak "undrained" strength parameters mobilized depend on the soil type:

- Cohesionless soils—the envelope passes through the origin (Figs. 3.27 and 3.28) and strength is expressed in terms of $\phi$ as in Eq. 3.23.
- Soils with cohesion—the envelope intercepts the shear stress ordinate at a value taken as cohesion $c$, like the case shown on Fig. 3.29, and the strength is expressed by the *Mohr-Coulomb* failure law:

$$s = c + \sigma_n \tan \phi \qquad (3.24)$$

- Normally to slightly overconsolidated clays have no frictional component ($\phi = 0$) and the strength is expressed as $s_u$, the undrained strength, as will be discussed.

*Applicable Field Conditions*

- Initial phases of slope excavations or retaining wall construction in slowly draining soils where time is too short to permit drainage. (Results are conservative if drainage occurs.)
- Sudden drawdown conditions in the upstream slope of a dam embankment, or when flood waters recede rapidly from stream banks and the soils remain saturated.
- Rapid placement of embankments or loading of storage tanks over soft clays, or high live loads applied to slow-draining soils.

### Effective Stresses

*Saturated Soils*

If the soil is permitted to drain as load is applied, the stress is initially carried partly by the pore water, but as drainage occurs the load is transferred to the soil grains, and the effective stresses (intergranular stresses) are mobilized. The effective stress is equal to the total normal stress minus the pore-water pressure, and the peak "drained" shear strength is expressed as

$$s = (\sigma_n - u) \tan \bar{\phi} = \bar{\sigma}_n \tan \bar{\phi} \qquad (3.25)$$

As shown on Figs. 3.27 and 3.28, $\phi$ values based on effective stresses are often higher than those based on total stresses as long as $u$ is positive. In dense sands, $u$ is negative and the undrained strength is higher, i.e., $\phi$ is higher. If pore pressures are measured during *undrained* loadings, effective stresses can be computed.

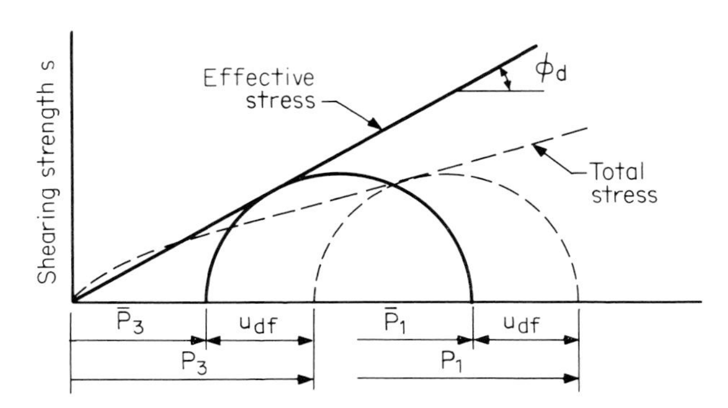


**FIG. 3.27 Mohr's envelopes for total and effective stresses from CU triaxial tests on loose saturated sands.**

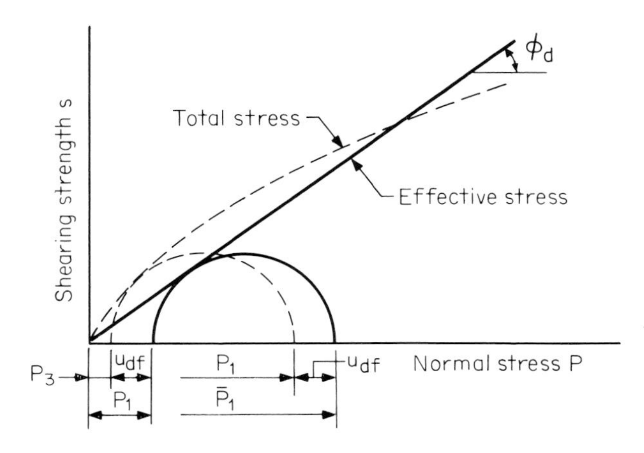


**FIG. 3.28 Mohr's envelopes for total and effective stresses from CU triaxial tests on dense saturated sands.**

For soils with a cohesion intercept (Fig. 3.29), the drained strength is expressed by the *Coulomb-Terzaghi equation* as

$$s = \bar{c} + (\sigma_n - u) \tan \bar{\phi} = \bar{c} + \bar{\sigma}_n \tan \bar{\phi} \quad (3.26)$$

*Applicable Field Conditions*

Drained strength prevails in the field under relatively slow loading conditions during which pore pressures can dissipate. Such conditions generally exist in:

- Most foundations, except for cases involving rapid load application
- Natural slopes, except for the sudden drawdown case
- Cut slopes, embankments, and retaining structures some time after construction completion

*Partially Saturated Soils*

In partially saturated soils, strength is controlled by effective stresses, but the effective stress concept cannot be applied directly because of pressures in the air or gas in the partially saturated voids. Strength should be estimated from tests performed to duplicate in situ conditions as closely as possible in terms of percent saturation, total stress, and pressure on the liquid phase.[30]

*Apparent cohesion* results from capillary forces in partially saturated fine-grained granular soils such as fine sands and silts and provides a temporary strength which is lost upon saturation or drying. The apparent cohesion has been expressed in terms of the depth $D$ to the water table [Lambe and Whitman (1969)[30]] as

$$c_a = D\gamma_w \tan \bar{\phi} \quad (3.27)$$

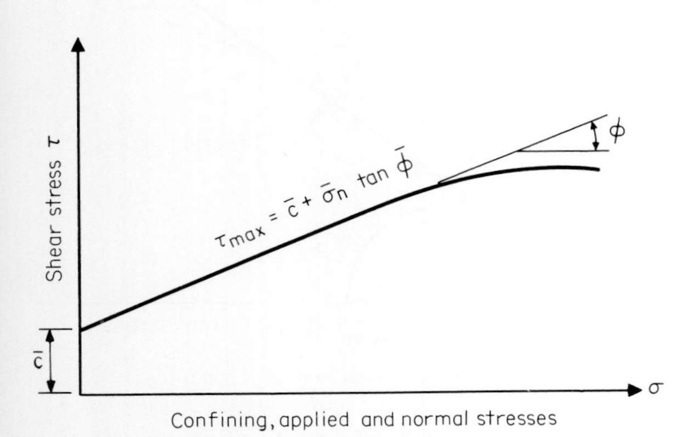


**FIG. 3.29 Drained triaxial test, soil with cohesion.**

**Pore-Pressure Parameters**

*Definition*

Pore-pressure parameters express the portion of a stress increment carried by the pore fluid in terms of the ratio of the pore-pressure increment ($\Delta u$) to the total stress increment ($\Delta\sigma$).

As indicated in Lambe and Whitman (1969)[30] and Bishop and Henkel (1962),[31] the parameters are

$$C = \frac{\Delta u}{\Delta \sigma_1} \quad (3.28)$$

for loading in the odeometer (one-dimensional compression)

$$B = \frac{\Delta u}{\Delta \sigma} \quad (3.29)$$

for isotropic loading (three-dimensional compression)

$$A = \frac{\Delta u - \Delta \sigma_3}{\Delta \sigma_1 - \Delta \sigma_3} \quad (3.30)$$

for triaxial loading

$$A = \frac{\Delta u}{\Delta \sigma_1} \quad (3.31)$$

for the normal undrained test where $\sigma_3 = 0$.

*Pore-pressure parameter A* is the most significant in practice. Values depend on soil type, state of stress, strain magnitude, and time. Typical values are given on Table 3.17 for conditions at failure but important projects always require measurement by test.

High values occur in soft or loose soils. Minus values indicate negative pore pressures, which occur in dense sands and heavily preconsolidated clays as the result of volume increase during shear (dilatancy). Pore pressures are most responsive to applied stress in sensitive clays and loose fine sands which are subject to liquefaction. In soft or loose soils higher shear strain magnitudes result in higher values for parameter $A$.

*Applications*

Pore-pressure parameter $A$ is used to estimate the magnitude of initial excess pore pressure produced at a given point in the subsoil by a change in the total stress system. With piezometers used to measure in situ pore pressures, the validity of the estimates can be verified and the loading rate during surcharging or embankment construction can be controlled to avoid failure caused by exceeding the undrained shear strength.

**TABLE 3.17**
**TYPICAL VALUES OF PORE-PRESSURE PARAMETER *A* AT FAILURE***

| Soil type | Parameter *A* |
|---|---|
| Sensitive clay | 1.5–2.5 |
| Normally consolidated clay | 0.7–1.3 |
| Overconsolidated clay | 0.3–0.7 |
| Heavily overconsolidated clay | −0.5–0.0 |
| Very loose fine sand | 2.0–3.0 |
| Medium fine sand | 0.0 |
| Dense fine sand | −0.3 |
| Loess | −0.2 |

*From Lambe (1962),[32] Wu (1966).[33]

### Undrained Shear Strength $s_u$

*Concepts*

In soft to firm saturated clays during undrained loading the applied stress is carried partly by the soil skeleton and partly by the pore water. Increasing the confining pressure does not increase the diameter of Mohr's circle, since the pore pressure increases as much as the confining pressure. The undrained strength, therefore, is independent of an increase in normal stress ($\phi = 0$), and as shown on Fig. 3.30, is given by the expression:

$$s_u = \frac{1}{2}\sigma_d \tag{3.32}$$

where $\sigma_d$ = applied or deviator stress = $\sigma_1 - \sigma_3$. On the figure, $U_c$ represents the unconfined compressive strength where $\sigma_3 = 0$. As shown, $s_u = \frac{1}{2}U_c$.

For NC clays $s_u$ falls within a limited fraction of the effective overburden pressure ($\bar{\sigma}_v$ or $p'$), usually in the range $s_u/p' = 0.16$ to $0.4$ (based on field vane and $K_0$ triaxial tests). Therefore, if $s_u > 0.5p'$ the clay may be considered as overconsolidated *(see Art. 3.5.2)*.

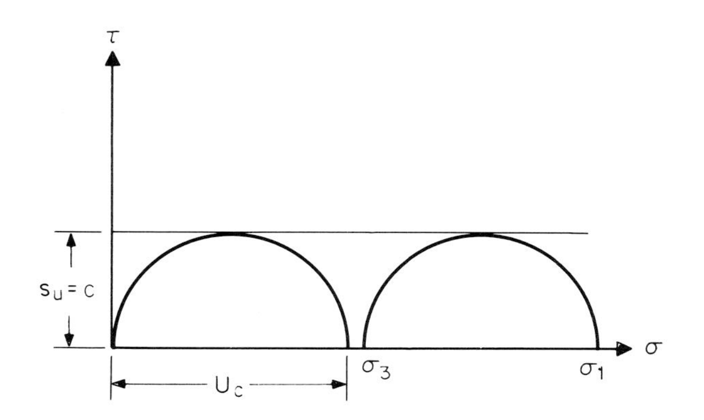


**FIG. 3.30 Undrained tests on soil with cohesion.**

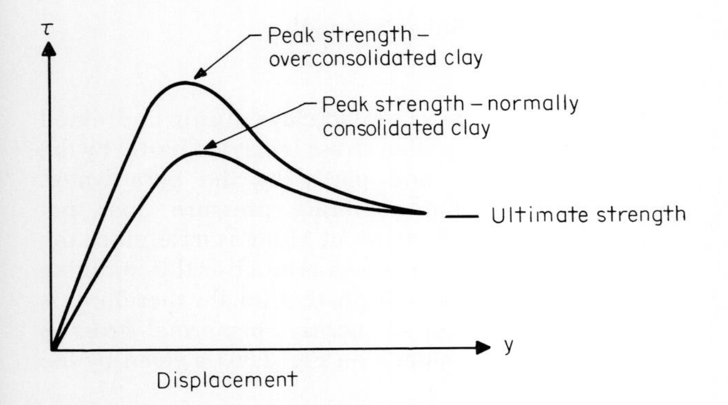


**FIG. 3.31 Peak and ultimate strength vs. displacement.**

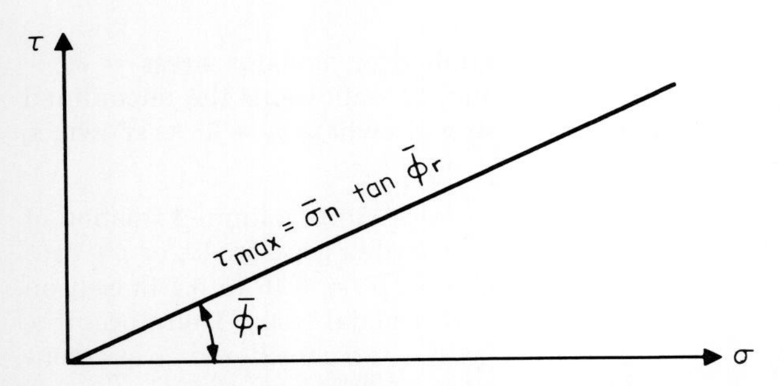


**FIG. 3.32 Mohr's envelope for ultimate drained strength in clay.**

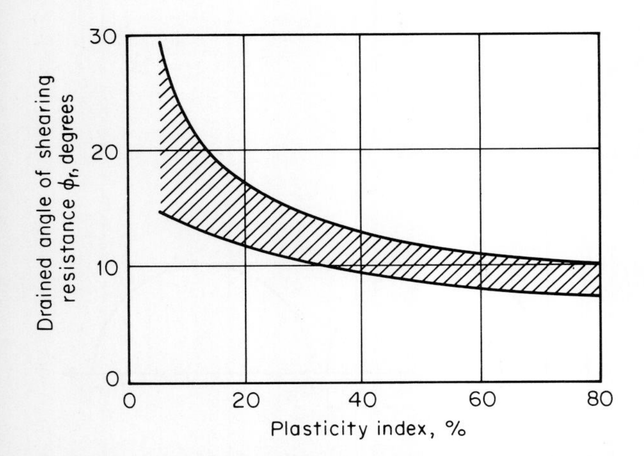


**FIG. 3.33 Approximate relationship between the drained angle of residual shearing resistance and plasticity index for rock gouge material.** [*From Patton and Hendron (1974),*[37] *Kanji (1970).*[38]]

*Factors Affecting Values for* $s_u$

TIME RATE OF LOADING For soft clays, $s_u$ measured by vane shear tests has been found to be greater than the field strength mobilized under an embankment loading, and the difference increases with the plasticity of the clay [Casagrande (1959),[34] Bjerrum (1969),[35] and Bjerrum (1972)[36]]. The difference is attributed to the variation in loading time rates; laboratory tests are performed at much higher strain rates than occur during the placement of an embankment in the field, which may take several months. A correction factor, therefore, is often applied to test values for $s_u$ *(see Art. 3.4.4).*

DIRECTION OF LOADING Soil anisotropy may cause shear strength measured in a horizontal direction to be significantly higher than when measured in a vertical direction (see *In Situ Vane Shear Test, Art. 3.4.4*).

## Residual or Ultimate Shear Strength $s_r$ or $\bar{\phi}_r$

*Concepts*

The lowest strength that a cohesive material can attain in the confined state is termed the ultimate strength. In many cohesive materials, if strain continues past the peak strength under continued stress, the strength will decrease until some minimum value is reached. Thereafter strength remains constant with increasing strain as shown on Fig. 3.31.

*Residual strength* is a strength lower than peak strength remaining after failure has occurred.

In a normally consolidated clay the *remolded undrained* shear strength is considered to be equal to the ultimate strength $s_r$.

The envelope for the ultimate *drained* strength passes through the origin as a straight line on the Mohr diagram and has *no cohesion intercept* (even in cohesive materials) as shown on Fig. 3.32 [Lambe and Whitman (1969)[30]]. The drained ultimate shear strength is expressed as

$$s = \bar{\sigma}_n \tan \bar{\phi}_r \qquad (3.33)$$

### Other Factors

NATURAL SLOPES The residual strength, rather than the peak strength, often applies, as discussed in Art. 9.3.2. An approximate relationship between $\phi_r$ and plasticity index for rock gouge materials is given in Fig. 3.33.

SENSITIVITY $S_t$ The ratio of the natural peak strength to the ultimate undrained strength when a sample is completely remolded at unaltered water content is referred to as the soil sensitivity, expressed as:

$$S_t = s_u/s_r \qquad (3.34)$$

A clay classification based on $S_t$ is given on Table 3.18. Soils termed as "quick" have high sensitivities and are extremely sensitive to vibrations and other forms of disturbance. They can quickly lose their strength, change to a fluid, and flow, even on very flat slopes *(see Art. 7.6.5, Glacial-Marine Clays)*.

*Thixotropy* refers to the regain in strength occurring in remolded soils because of a "rehabilitation of the molecular structure of the adsorbed layers" [Lambe and Whitman (1969)[30]].

## Dynamic Shear

### Concepts

Under dynamic or cyclic loading, a soil specimen subject to shear initially undergoes deformations that are partially irreversible irrespective of the strain amplitude; hence, stress-strain curves in loading and unloading do not coincide. If strain amplitude is small, the difference

**TABLE 3.18**
**CLAY CLASSIFICATION BY SENSITIVITY $S_t$***

| Sensitivity $s_u/s_r$† | Classification |
|---|---|
| 2 | Insensitive |
| 2–4 | Moderately sensitive |
| 4–8 | Sensitive |
| 8–16 | Very sensitive |
| 16–32 | Slightly quick |
| 32–64 | Medium quick |
| 64 | Quick |

*From Skempton (1953),[14] Bjerrum (1954).[39]

†$s_u$ = peak undrained strength, $s_r$ = remolded strength.

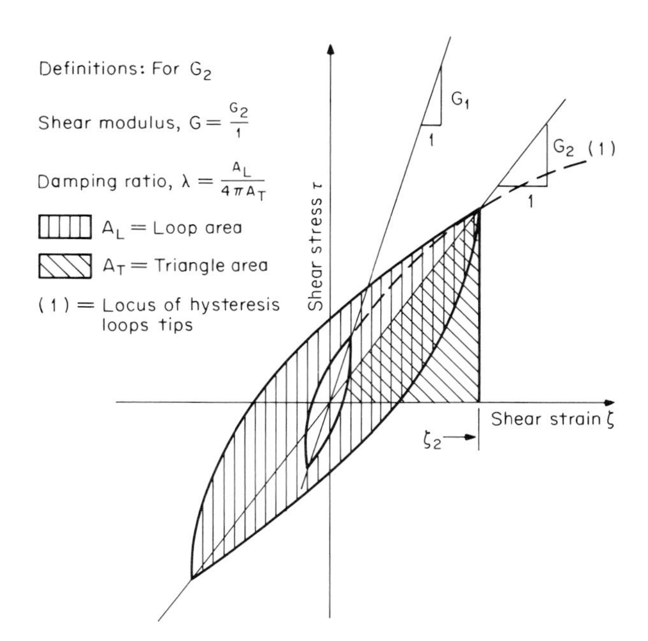


**FIG. 3.34 Hysteretic stress-strain relationship from cyclic shear test at different strain amplitudes.** [*After USAEC (1972).*[40]]

between successive reloading curves tends to disappear after a few loading cycles of a similar amplitude and the stress-strain curve becomes a closed loop that can be defined by two parameters as shown on Fig. 3.34:

- *Shear modulus (Art. 3.5.2)* is defined by the average slope and has been found to decrease markedly with increasing strain amplitude.
- *Internal damping* refers to energy dissipation and is defined by the enclosed areas as shown on the figure (damping ratio λ).

In *clay soils*, if strain amplitude is large there is a significant reduction in the undrained strength.

In *cohesionless soils*, pore pressure increases almost linearly with the number of cycles until failure occurs (Fig. 3.35). The rate of increase of the ratio of the change in pore pressure normalized with respect to the applied stress vs. the number of cycles is shown on Fig. 3.35*b* to be essentially linear. *Liquefaction* occurs when pore pressures totally relieve effective stresses *(see Art. 11.3.3)*.

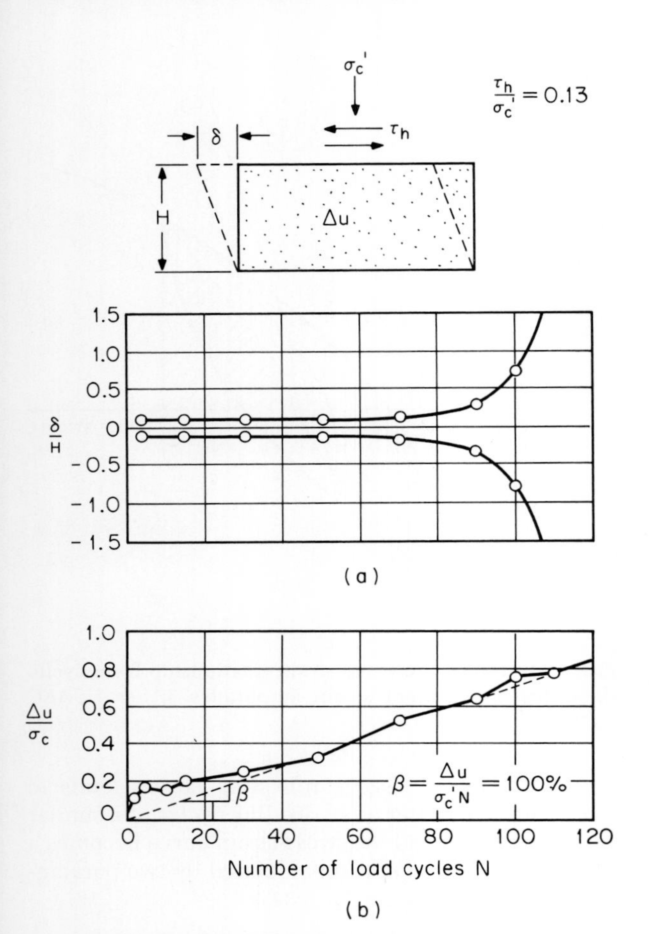


**FIG. 3.35 Results of cyclic shear tests on sands in simple shear apparatus.** [*From Eide (1974).*[41]]

*Field Occurrence*

Wave forces against offshore structures cause low-frequency cyclic loads. Seismic waves from earthquakes cause low- to high-frequency cyclic loads as discussed in Art. 11.3.2.

### At-Rest, Passive, and Active Stress States

*At-Rest Conditions ($K_0$)*

The coefficient of lateral at-rest earth pressure has been defined as the ratio of lateral to vertical stress; $K_0 = \sigma_h/\sigma_v$ (Eq. 3.15).

For *sands* and *NC clays* it is normally in the range of 0.4 to 0.5 and is a function of $\phi$ in accordance with:

$$K_0 = 1.0 - \sin\phi \qquad (3.35)$$

In *clay soils*, $K_0$ has been found to be directly related to the amount of prestress or preconsolidation in the formation, ranging from 0.5 to almost 3. The degree of prestress is given in terms of the overconsolidation ratio (OCR), i.e., the ratio of the maximum past pressure to the existing overburden pressure *(see Art. 3.5.2).*

*Passive State of Stress ($K_p$)*

The passive state exists when a force pushes against a soil mass and the mass exerts its maximum resistance to the force. The principal stresses for the passive state are shown on Fig. 3.36*a*; as the soil element is pushed, the vertical stress remains unchanged but the horizontal stress increases ($\sigma_h > \sigma_v$). As movement continues the shear stress increases from the at-rest condition (1) until $\sigma_h = \sigma_v$ (2), then continues to increase as slip lines (rupture planes) form and finally failure occurs at (4). At this point of plastic equilibrium, the passive state has been reached.

The relationship between $\phi$ and the principal stresses at failure, as shown on the Mohr diagram of Fig. 3.36*a*, may be expressed for a cohesionless granular soil with a horizontal ground surface as

$$\frac{\sigma_1}{\sigma_3} = \frac{\sigma_v}{\sigma_h} = \frac{1 + \sin\phi}{1 - \sin\phi} \qquad (3.36)$$

or

$$\frac{\sigma_v}{\sigma_h} = \tan^2(45° + \phi/2) = K_p \qquad (3.37)$$

where $K_p$ = the *coefficient of passive stress.*

*Active State of Stress*

The active state exists when a soil mass is allowed to stretch, for example, when a retaining wall tilts, and $\sigma_v > \sigma_h$. In Fig. 3.36*b*, as the mass stretches, $\sigma_v$ remains unchanged and $\sigma_h$ decreases (2) until the induced shear stress is sufficient to cause failure (3). At this point of plastic equilibrium the active state has been reached.

*Coefficient of active stress* $K_a$, the ratio $\sigma_h/\sigma_v$, represents a *minimum* force, expressed for a cohesionless soil with a horizontal ground surface as

$$K_a = \tan^2 (45° - \phi/2) \quad (3.38)$$

and

$$K_a = 1/K_p \quad (3.39)$$

*Applications*

The *at-rest coefficient* $K_0$ has a number of practical applications. It is used to compute lateral thrusts against earth retaining structures where lateral movement is anticipated to be too small to mobilize $K_a$. It is fundamental to the reconsolidation of triaxial test specimens according to an anisotropic stress path resembling that which occurred in situ ($CK_0U$ tests). It is basic to the computation of settlements in certain situations [Lambe (1964)[42]]. It has been used for the analysis of progressive failure in clay slopes [Lo and Lee (1973)[43]], the prediction of pore-water pressure in earth dams [Pells (1973)[44]], and the computation of lateral swelling pressures against friction piles in expansive soils [Kassif and Baker (1969)[45]].

*Coefficients of lateral earth pressure* is the term often used to refer to $K_p$ and $K_a$. When rupture occurs along some failure surface, earth pressures are mobilized. The magnitudes of the pressures are a function of the weight of the mass in the failure zone and the strength acting along the failure surface. These pressures ($P_p$ or $P_a$) are often expressed in terms of the product of the weight of the mass times the active or passive coefficients, as follows:

$$P_p = K_p \gamma z \quad (3.40)$$

$$P_a = K_a \gamma z \quad (3.41)$$

Examples of the occurrence of $P_p$ and $P_a$ in practice are given on Fig. 3.37.

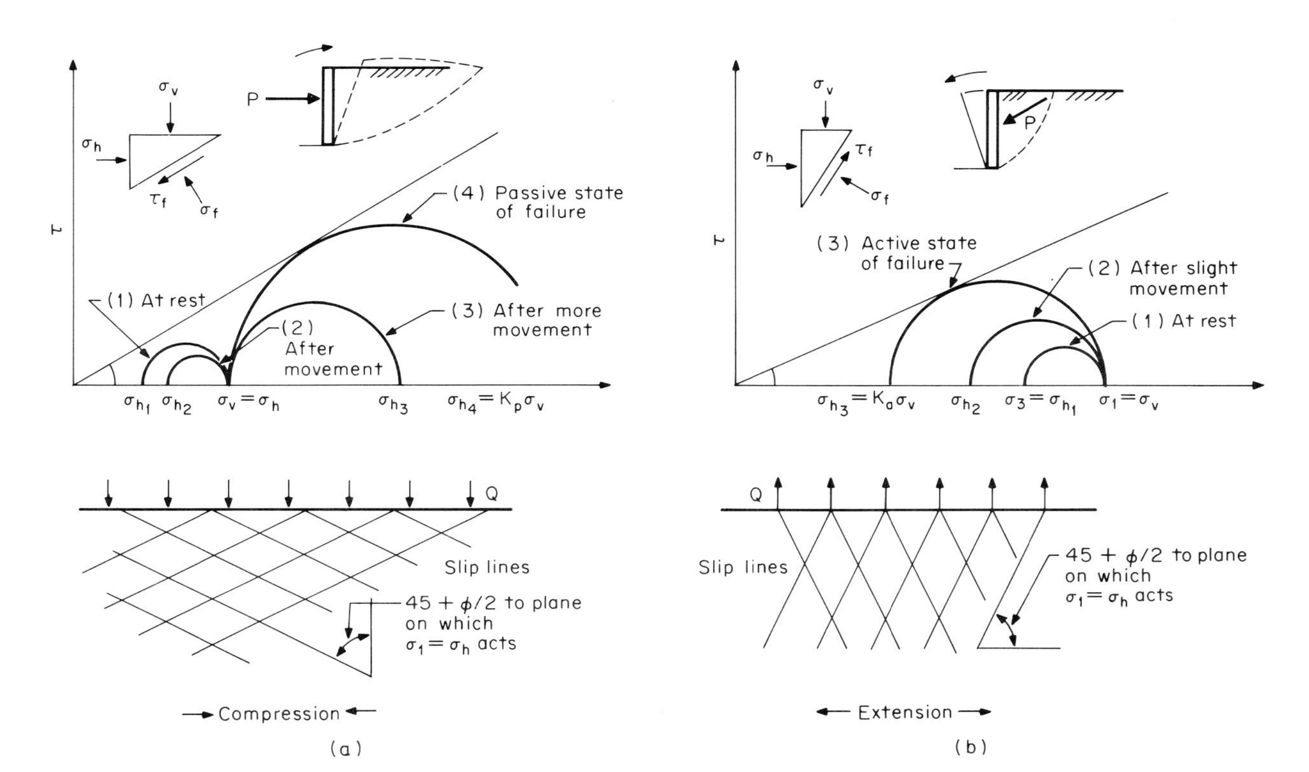


**FIG. 3.36 Mohr diagrams for the (a) passive and (b) active states of stress.**

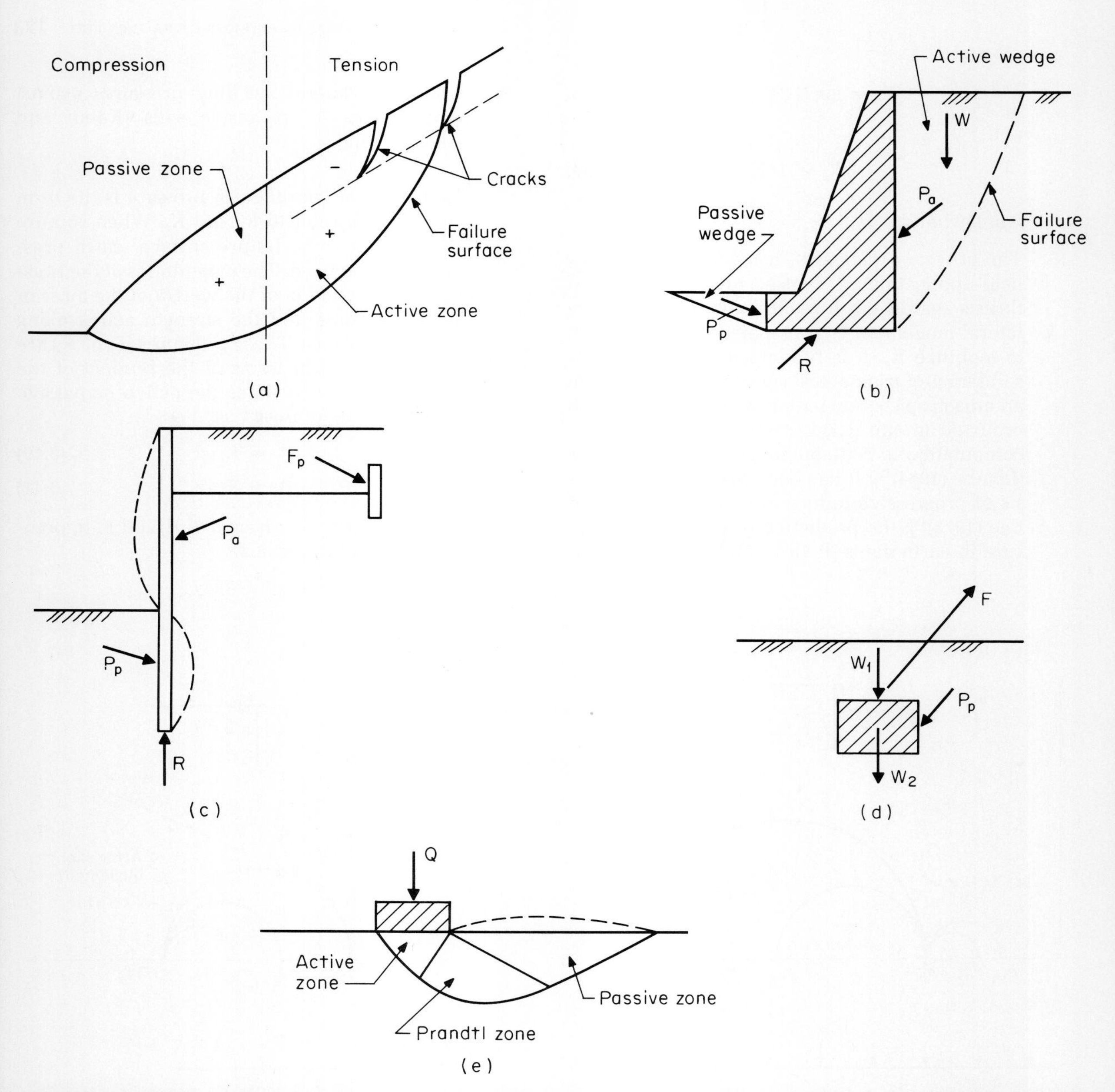


**FIG. 3.37 Examples of the occurrence of active and passive stresses encountered in practice: (*a*) slope; (*b*) retaining wall; (*c*) anchored bulkhead; (*d*) anchor block; (*e*) foundation.**

### 3.4.3 ROCK STRENGTH MEASUREMENTS

#### General

*Intact Rock*

The confined strength of fresh, intact rock is seldom of concern in practice because of the relatively low stress levels imposed.

Brittle shear failure occurs under very high applied loads and moderate to high confining pressures, except for the softer rocks such as halite, foliated and schistose rocks, and lightly cemented sandstones. In softer rocks rupture occurs in a manner similar to that in soils, and the parameters described under Art. 3.4.2 pertain.

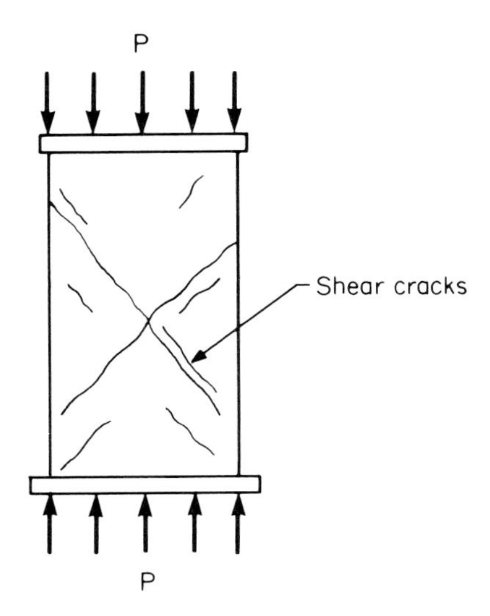


FIG. 3.38 Uniaxial compression test.

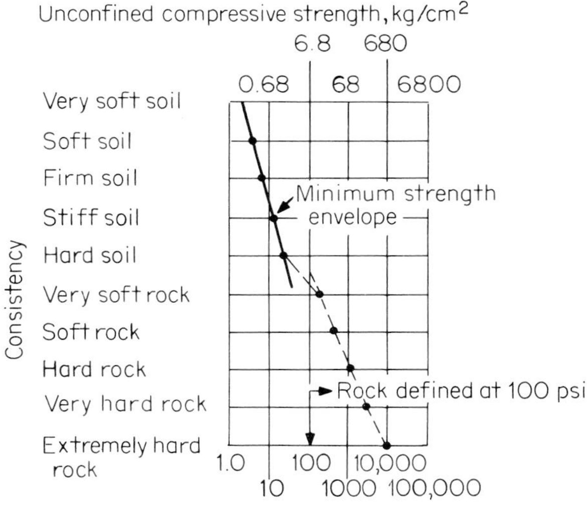


**FIG. 3.39 Relationship between "consistency" and $U_c$ (100 psi = 6.8 kg/cm² = 689.5 kN/m²).** [*After Jennings (1972).*[46a]]

Under very high confining pressures (approximately 45,000 psi or 3000 bars), some competent rocks behave ductilely and failure may be attributed to plastic shear [Murphy (1970)[46]].

Intact specimens are tested in uniaxial compression or tension to provide data for classification and correlations. Other tests include those for flexural strength and triaxial compressive strength.

### *Rock Masses*

Rock-mass strength normally is controlled either by the joints and other discontinuities or by the degree of decomposition, and the strength parameters described under Art. 3.4.2 apply. Strength is measured in situ by direct shear equipment or special triaxial shear equipment.

## Uniaxial Compressive Strength $U_c$

### *Procedure*

An axial compressive force is applied to an unconfined specimen (Fig. 3.38) until failure occurs.

### *Data Obtained*

A stress-strain curve and the unconfined or uniaxial compressive strength (in tsf, kg/cm², etc.) result from the test. Stress-strain curves for various rock types are given on Table 3.32, and $U_c$ typical value ranges for various rock types on Table 3.26.

### *Data Applications*

Primarily used for correlations as follows:

- Material "consistency" vs. $U_c$—Fig. 3.39 [Jennings (1972)[46a]]
- Schmidt hardness vs. $U_c$—Fig. 3.40
- Hardness classification—Table 5.20

## Uniaxial Tensile Strength

### *Cable-Pull Test*

Caps are attached to the ends of a cylindrical specimen with resins. The specimen is then pulled apart by cables exerting tension axially. The method yields the lowest values for tensile strength, which generally ranges from 5 to 10% of the uniaxial compression strength.

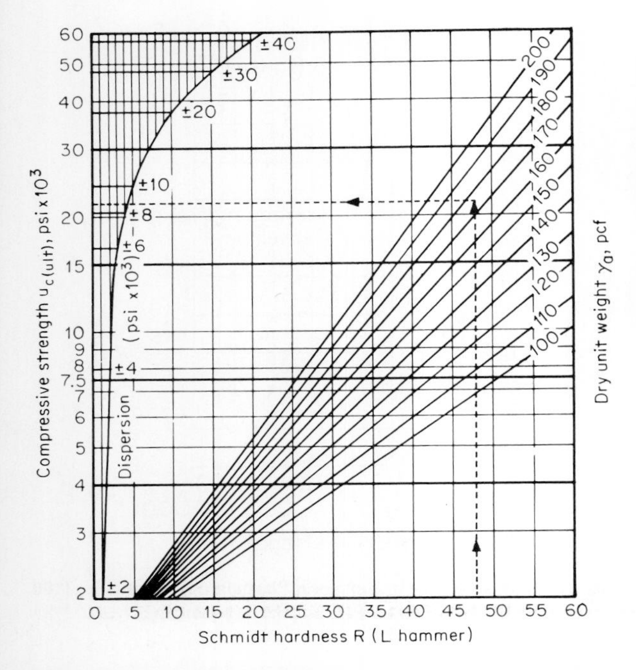


**FIG. 3.40 Correlation between $U_c$ and Schmidt hardness** ***(see Fig. 3.1).*** **Hammer vertical downward; dispersion limits defined for 75% confidence.** [*From Deere and Miller (1966).*[100]]

**FIG. 3.41 Point load strength test apparatus.**

*Point-Load Test* [*Broch and Franklin* (1972)[48a]]

Compressive loads *P* are applied through hardened conical points to diametrically opposite sides of a core specimen of length of at least 1.4*D* until failure occurs. The equipment is light and portable and (Fig. 3.41) is used in the field and the laboratory.

*Point-load index* is the strength factor obtained from the test which is given by the empirical expression [Hoek and Bray (1977)[20]]

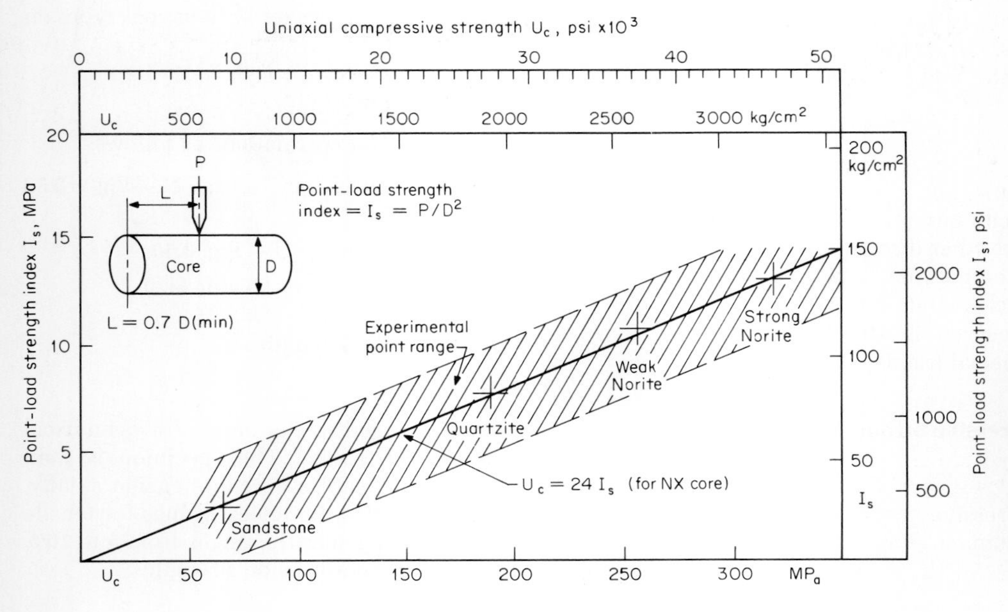


**FIG. 3.42 Relationship between point load strength index $I_s$ and uniaxial compressive strength $U_c$.** [*After Bieniawski (1974).*[47] *Reprinted with permission of the National Academy of Sciences.*]

$$I_s = P/D^2 \tag{3.42}$$

where $D$ is the diameter.

Values for $I_s$ are used to estimate $U_c$ through various correlations as shown on Fig. 3.42, where

$$U_c = 24I_s \tag{3.43}$$

for a core diameter of 50 mm.

### Flexural Strength or Modulus of Rupture

*Procedure*

A rock beam is supported at both ends and loaded at midpoint until failure (Fig. 3.43).

*Data Obtained*

The flexural strength is proportional to the tensile strength but is about 3 times as great [Leet (1960)[48]].

### Triaxial Shear Strength

*Apparatus and Procedures*

General description is given in Art. 3.4.4 and, as applicable to rock testing, in Art. 3.5.3 and on Table 3.36.

*Strength Values*

Studies have been made relating analysis of petrographic thin sections of sandstone to estimates of the triaxial compressive strength [Fahy and Guccione (1979)[49]]. Relationships have been developed for approximating peak strengths for rock masses [Hoek and Brown (1980)[50]].

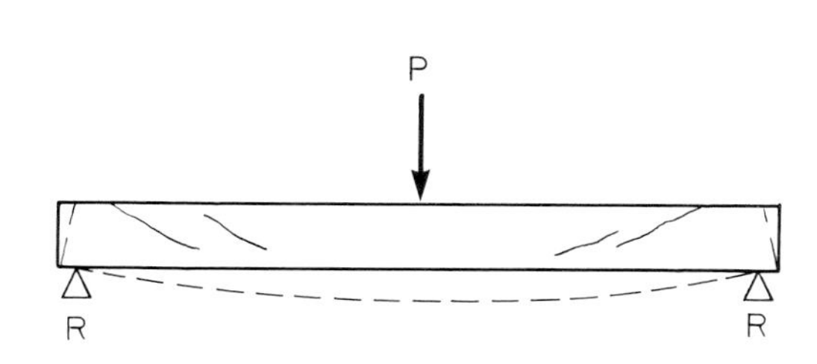


**FIG. 3.43 Flexural strength test.**

### Direct Shear Strength

*Purpose*

The purpose is to obtain measurements in situ of the parameters $\phi$ and $c$.

*Procedure*

A diamond saw is used to trim a rock block from the mass with dimensions 0.7 to 1.0 m square and 0.3 m in height, and a steel box is placed over the block and filled with grout [Haverland and Slebir (1972)[51]].

Vertical load is imposed by a hydraulic jack while a shear force is imposed by another jack (Fig. 3.44) until failure. All jack forces and block movements are measured and recorded.

Deere (1976)[52] suggests at least five tests for each geologic feature to be tested, each test being run at a different level of normal stress to allow construction of Mohr's envelope.

*Laboratory direct shear tests* are described in Art. 3.4.4.

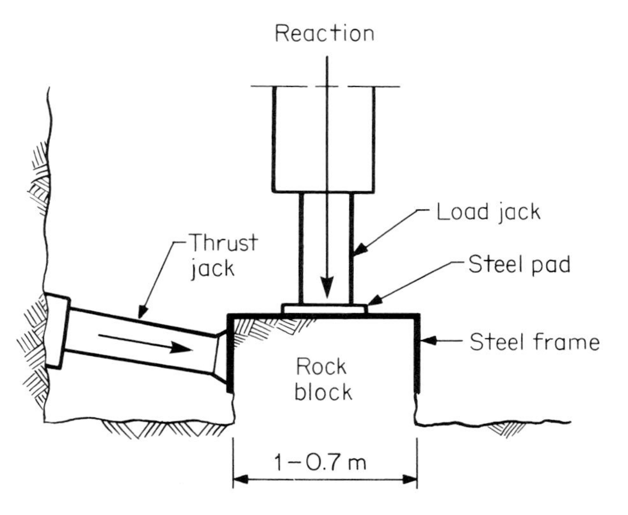


**FIG. 3.44 In situ direct shear test.**

## 3.4.4 SOIL STRENGTH MEASUREMENTS

### General

*Selection of Test Method*

A number of factors require consideration for the selection of the testing method, including the following:

- Loading conditions: static or dynamic
- Loading duration in the field: long-term (drained conditions) or short-term (undrained conditions)
- Parameter desired: peak or ultimate strength
- Material suitability for undisturbed sampling and the necessity or desirability for in situ testing
- Orientation of the field failure surface with that in the test; some cases are shown on Figs. 3.45 and 3.46. Stability analysis is often improperly based on compression tests only, whereas direct shear and extension tests often apply. Their strength values may differ significantly from the compressive parameters.

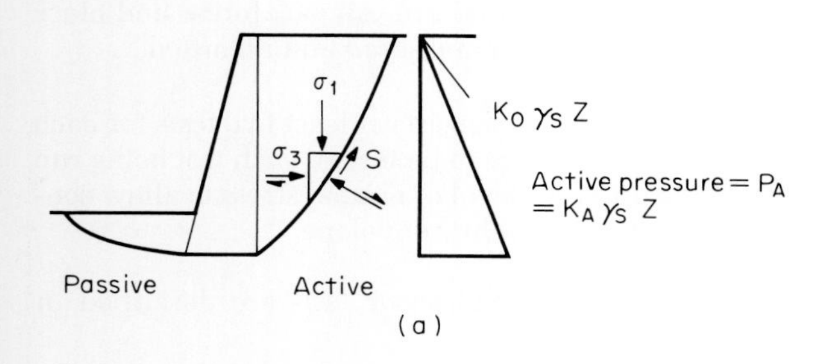


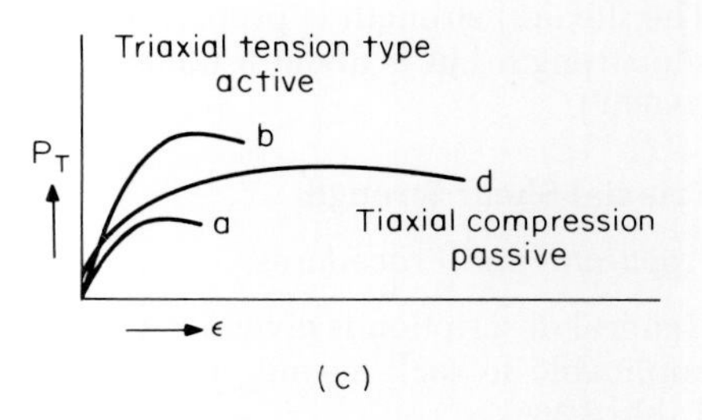


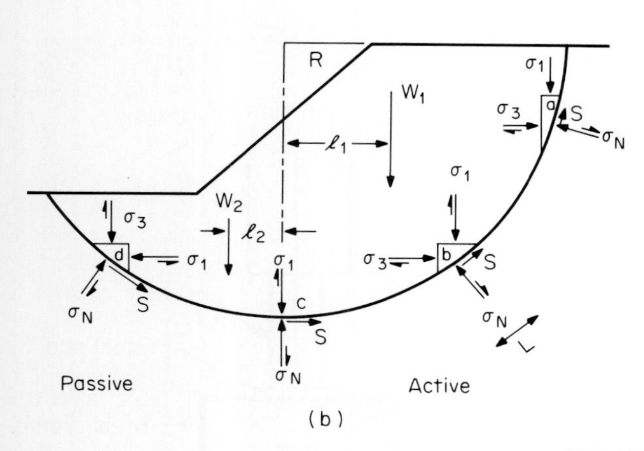


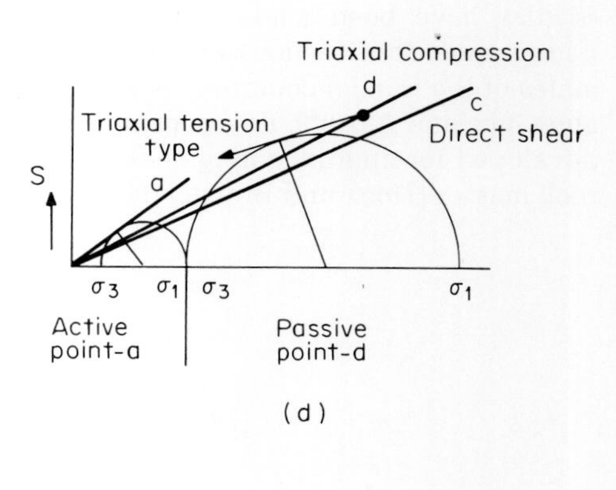


$$\text{Factor of safety} = \frac{\Sigma \text{ resisting forces and masses}}{\Sigma \text{ active driving forces and masses}} = \frac{W_2 l_2 + R\,\Sigma\,S - L}{W_1 l_1}$$

**FIG. 3.45 Probable natural stress and strain restraint conditions: (*a*) Retaining wall influence of lateral yielding on stresses. (*b*) Mass slide of excavated slope. Influence of lateral yielding. (*c*) Stress-strain relations corresponding to lateral yield conditions in (*b*). (*d*) Angle of friction relations corresponding to lateral yield conditions in (*b*).** [*From Burmister (1953).*[53] *Reprinted with permission of the American Society for Testing and Materials.*]

**TABLE 3.19**
**SUMMARY OF SOIL LABORATORY STATIC STRENGTH TESTS**

| Test (F = also field test) | Parameters measured | Reference | Comments |
|---|---|---|---|
| Triaxial compression | | | |
| CD | $\bar{\phi}$, $\bar{c}$ | Fig. 3.29 | Most reliable method for effective stresses |
| CU | $\phi$, $c$, $\bar{\phi}$, $\bar{c}$<br>$s_u$ | Fig. 3.47<br>Fig. 3.48<br>Table 3.21 | Strength values higher than reality because disturbance causes lower $w\%$ upon reconsolidation. (Table 3.21, footnote‡.) |
| UU | $s_u$ | Fig. 3.30 | Most representative laboratory value for undrained shear strength in compression. |
| Triaxial extension | $\bar{\phi}$, $\bar{c}$, $s_u$ | Table 3.21 | Normally consolidated clays yield values approximately one-third those of compression tests because of soil anisotropy [Bjerrum et al. (1972)[55]]. |
| Plain strain compression or extension | $\phi$ | Table 3.21 | Values are a few degrees higher than those of normal triaxial test except for loose sands; more closely approach reality for retaining structure [Lambe and Whitman (1969)[30]]. |
| Direct shear box | $\bar{\phi}$, $\bar{c}$, $\bar{\phi}_r$ | Fig. 3.49 | Values most applicable where test failure surface has same orientation with field failure surface. Values generally lower than triaxial compression values for a given soil, but higher than triaxial extension. Most suitable test for determination of residual strength $\phi_r$ from UD samples. |
| Simple shear | $s_u$, $\bar{\phi}$, $\bar{c}$ | Fig. 3.35 | Horizontal plane becomes plane of maximum shear strain at failure. |
| Unconfined compression | $s_u = ½U_c$ | Table 3.22 | Strength values generally lower than reality. |
| Vane shear (F) | $s_u$, $s_r$ | Table 3.22 | Applies shear stress on vertical planes. |
| Torvane (F) | $s_u$, $s_r$ | Table 3.22 | Shear occurs in a plane perpendicular to the axis of rotation. |
| Pocket penetrometer (F) | $s_u = ½U_c$ | Table 3.22 | Yields approximate values in clays. Used primarily for soil classification by consistency. |
| California bearing ratio (F) | CBR value | Fig. 3.61 | Used for pavement design. Empirical strength correlates roughly with $U_c$. |

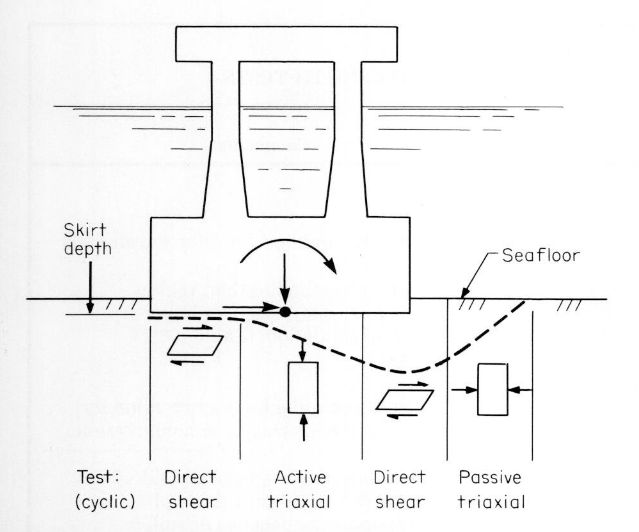


**FIG. 3.46 Relevance of laboratory tests to shear strength along potential slip surface beneath offshore gravity structure.** [*From Kjerstad and Lunne (1979).*[54]]

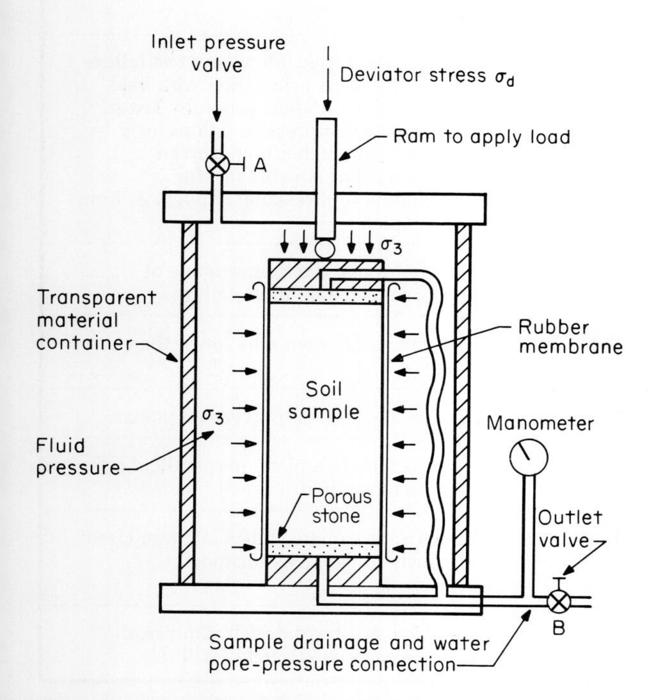


**FIG. 3.47 Triaxial compression chamber arrangement.**

*Testing Methods Summarized*

- Soil laboratory static strength tests—Table 3.19
- In situ static strength tests—Table 3.20
- Laboratory dynamic strength tests—Table 3.43

### Triaxial Shear Test

*Purpose*

Total or effective stress parameters, either in compression or extension, are measured in the triaxial shear apparatus. The test method is generally unsuited for measuring ultimate strength because displacement is limited and testing parallel to critical surfaces is not convenient.

*Apparatus* includes a compression chamber to contain the specimen (Fig. 3.47) and a system to apply load under controlled stress or strain rates, and to measure load, deflection, and pore pressures (Fig. 3.48).

*Specimens* are usually 2.78 in in diameter as extruded from a Shelby tube, or 1.4 in in diameter as trimmed from an undisturbed sample. Specimen height should be between 2 and 2.6 times the diameter.

*General Procedures*

*Rate of loading or strain* is set to approximate field-loading conditions and the test is run to failure [Bishop and Henkel (1962)[31]].

CONFINING PRESSURES Tests are usually made on three different specimens with the same index properties, to permit defining a Mohr diagram. Test method variations are numerous. Specimens can be preconsolidated or tested at their stress conditions as extruded. Tests can be performed as drained or undrained, or in compression or extension. The various methods, parameters measured, and procedures are summarized on Table 3.21.

### Direct Shear Test

*Purpose*

The purpose normally is to measure the drained strength parameters $\bar{\phi}$, $\bar{c}$, and $\bar{\phi}_r$.

**TABLE 3.20**
**SUMMARY OF SOIL IN SITU STATIC STRENGTH TESTS**

| Test | Parameters measured | Reference | Comments |
|---|---|---|---|
| Vane shear | $s_u$, $s_r$ (direct test) | Fig. 3.51 | Measures undrained strength by shearing two circular horizontal surfaces and a cylindrical vertical surface; therefore, affected by soil anisotropy. |
| SPT | $\bar{\phi}$, $s_u$ (indirect test) | Art. 3.4.5 | $D_r$ is estimated from $N$ and correlated with soil gradation to obtain estimates of $\phi$ (Fig. 3.63, Table 3.28). Consistency is determined from $N$ and correlated with plasticity to obtain estimates of $U_c$ (Fig. 3.64, Table 3.29). |
| CPT | $\bar{\phi}$, $s_u$ (indirect test) | Art. 3.4.5 | Various theoretical and empirical relationships have been developed relating $q_c$ to $\bar{\phi}$ [summarized by Mitchell and Lunne (1978)[56]]. See Fig. 3.59. $s_u$ is expressed as $$s_u = \frac{q_c - \Sigma\gamma z}{N_c} \quad (3.48)$$ where $N_c$ is the cone-bearing capacity factor (deep foundation depth correction factor). Pore-water pressure $u$ is measured by some cones (piezocones). |
| Pressuremeters | $s_u$ (indirect test) | Art. 3.5.4 | Affected by material anisotropy, $s_u$ is expressed as[57] $$s_u = \frac{P_L - P_o}{2K_b} \quad (3.83)$$ |
| California bearing ratio | CBR value | Art. 3.4.5 | Field values generally less than lab values because of rigid confinement in the lab mold. |

*Apparatus*

The test apparatus is illustrated on Fig. 3.49.

*Procedure*

Specimen is trimmed to fit into the shear box between two plates, which can be pervious or impervious, depending upon drainage conditions desired, and a normal load applied which remains constant throughout the test. The test is normally run as a consolidated drained (CD) test (sample permitted to consolidate under the normal load), but it has been performed as a consolidated undrained (CU) test when load application rates are high, although some drainage always occurs during shear. The shear force is applied either by adding deadweights (stress-controlled) or by operating a motor acting through gears (controlled strain) and increasing the load gradually until failure occurs.

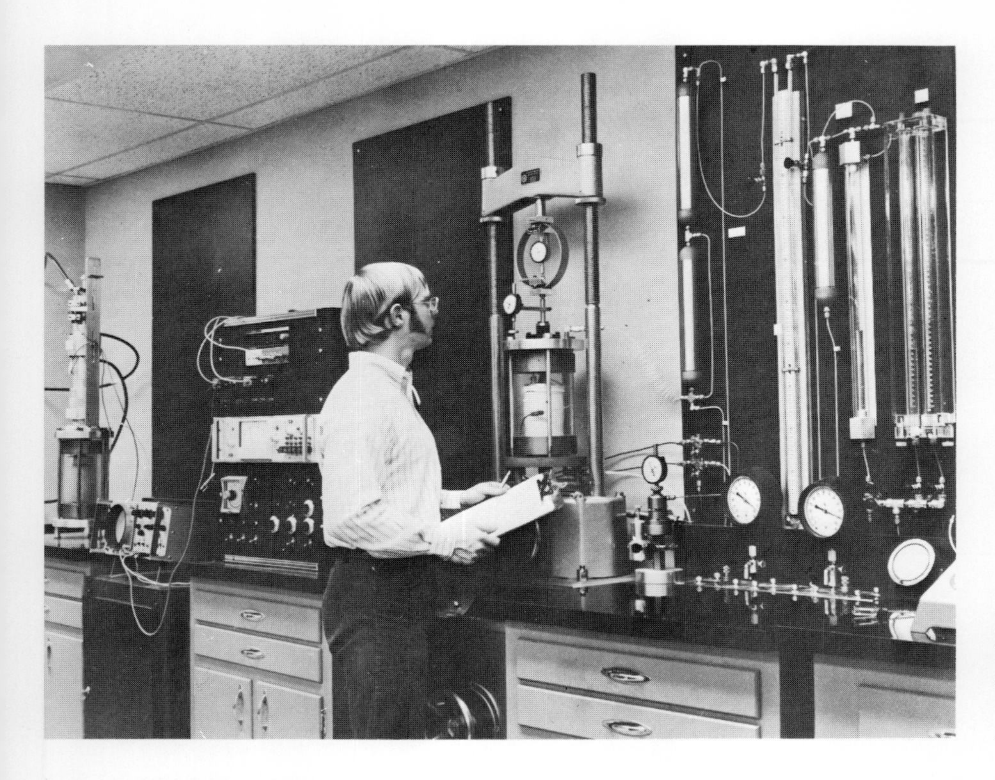

**FIG. 3.48 The triaxial compression chamber, load application, and measurement system.** *(Courtesy of Joseph S. Ward & Associates.)*

*Mohr's Envelope*

Plotting the results for a sequence of tests under different normal stresses as points $(\tau, \sigma)$ results in a Mohr's envelope. Because of the form of the test, only the normal and shear stresses on a single plane are known. If it is assumed that the horizontal plane through the shear box is identical with the theoretical failure plane (usually an inexact assumption), it can be assumed that the stresses at failure are in the ratio of $\tau/\sigma = \tan \phi$ to provide a point on the envelope.

*Residual or Ultimate Shear Strength*

The direct shear test is the best method for measuring residual strength. The test is performed by back-and-forth sequential movement of the shear box after failure of an undisturbed specimen has occurred, or by testing a remolded specimen.

The relationship between peak and ultimate strength for normally consolidated and overconsolidated clays is given on Fig. 3.31, and Mohr's envelope for the ultimate drained strength $\bar{\phi}_r$ in overconsolidated clay is given on Fig. 3.32.

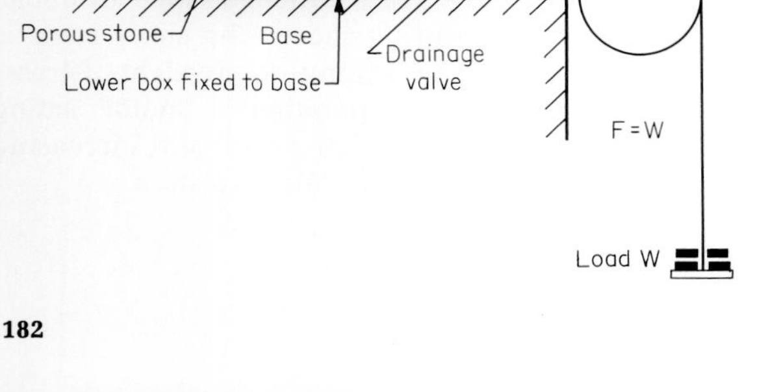

**FIG. 3.49 Direct shear box.**

**TABLE 3.21**
**TRIAXIAL TEST METHODS***

| Test | To measure | Procedure |
|---|---|---|
| Consolidated-drained (CD) or (S) compression test | Effective stress parameters $\bar{\phi}$, $\bar{c}$ (Fig. 3.29) | Specimen permitted to drain and consolidate under confining pressure until $u = 0$. Deviator stress applied slowly to failure while specimen drains during deformation.† |
| Consolidated-undrained (CU) or (R) compression test | Total stress parameters $\phi$, $c$ | Specimen permitted to drain and consolidate under confining pressure until $u = 0$. Deviator stress applied slowly to failure, but specimen drainage not permitted. |
| | Effective stress parameters $\bar{\phi}$, $\bar{c}$ | Pore pressures are measured during test (see Pore-Pressure Parameters). |
| $CK_0U$ test | $s_u$ | See Notes ‡ and § below. |
| Unconsolidated-undrained (UU) or (Q) compression test | Undrained strength $s_u$ *(Fig. 3.30)* | Confining pressure applied but no drainage or consolidation permitted to reduce test time. Deviator stress applied slowly to failure with no drainage permitted. |
| Extension tests as CD, CU, UU | Lateral shear strength | Maintain confining pressure constant and reduce axial stress, or maintain axial stress constant and increase confining pressure until failure. |
| Plane strain compression or extension test | Parameter $\phi$ in cohesionless granular soils | Modified triaxial apparatus in which specimen can strain only in axial direction and one lateral direction while its dimension remains fixed in the other lateral direction. |

*See Table 3.19 for comments on test results and test comparisons.

†Back pressure is applied to the pore water to simulate in situ pore-water pressures, or to saturate partially saturated specimens.

‡$CK_0U$ test: Specimen anisotropically consolidated with lateral pressure at $K_0p_0$ and vertical pressure at $p_0$

§SHANSEP procedure [Ladd and Foott (1974)[58]] attempts to minimize the effects of sample disturbance and assumes normalized behavior of clay. Specimens are consolidated to $p_c$ stresses higher than $p_0$, rebounded to selected values of OCR, and then tested in undrained shear to establish the relationship between $s_u/p_0$ vs. OCR for different modes of failure. Normalized behavior requires this relationship to be independent of $p_c$. The $s_u$ profile is then calculated from the values of $p_0$ and OCR vs. depth [Baligh et al. (1980)[59]].

## Simple Shear Test

*Purpose*

The purpose is to measure drained or undrained strength parameters.

*Apparatus*

The soil specimen is contained either in a cylindrical rubber membrane reinforced by wire, which permits shear deformation to be distributed fairly uniformly throughout the specimen [NGI apparatus—Kjellman (1951)[60]], or in a box which provides a rigid confinement as lateral stress is applied [Roscoe (1953)[61]]. A vertical stress is applied to cause consolidation and a lateral stress is applied to induce shear failure (Fig. 3.35).

*Procedure*

The specimen is consolidated anisotropically under a vertical stress and then sheared by application of lateral stress. In the undrained test, zero volume change during shear can be maintained by adjusting the vertical stress con-

| TABLE 3.22 LABORATORY TESTS FOR UNDRAINED SHEAR STRENGTH | | | |
|---|---|---|---|
| **Test** | **Parameters** | **Procedure** | **Comments** |
| Triaxial compression | $s_u$ | See Table 3.21. | UU or $CK_0U$ test |
| Unconfined compression (Fig. 3.50)* | $U_c = 2s_u$ (Fig. 3.30) | Unconfined specimen ($\sigma_3 = 0$) strained to failure by axially applied load. | Test yields stress-strain curve and $U_c$. Slight drainage may occur during shear. Values usually conservative because of nonrecovered decrease in effective stress occurring from disturbance during extraction. |
| Vane shear (miniature vane) (Fig. 3.51) | $s_u$, $s_r$ | Small two-bladed vane is pushed about 2 in into a UD-tube specimen and then rotated while the torque indicated on a gauge is recorded to measure peak and ultimate strength. | Test limited to clay soils of very soft to firm consistency (Table 3.29). |
| Torvane* (Fig. 3.52) | $s_u$, $s_r$ | One-inch-diameter disk containing eight small vanes is pushed about ¼-in penetration in UD sample and rotated while torque is recorded. | Test limited to clay soils of very soft to firm consistency. Tested zone may be disturbed by sample trimming and cutting. |
| Pocket penetrometer* (Fig. 3.53) | $U_c = 2s_u$ | Calibrated spring-loaded rod (¼ in diameter) is pushed into soil to a penetration of 6 mm and the gauge read for $U_c$. | Penetration is limited to soils with $U_c < 4.5$ tsf. Data are representative for soils with PI > 12. Below this value the $\phi$ of granular particles increases strength to above the measured value of $s_u$. |
| Simple shear | $s_u$ | See Art. 3.4.4. | |

*These tests are considered generally as classification tests to provide an index of soil consistency (Table 3.29).

tinuously during test [Bjerrum and Landva (1966)[62]].

*Principal Axis Orientation*

Initially the principal axes are in the vertical and horizontal directions. At failure, the horizontal plane becomes the plane of maximum shear strain (the lower point on the failure circle in Fig. 3.45*b*).

### Undrained Shear Tests: Laboratory

Laboratory tests used to measure the undrained shear strength, and in some cases the residual or remolded shear strength, are described on Table 3.22. Included are: triaxial compression, unconfined compression (Fig. 3.50), miniature vane (Fig. 3.51), torvane (Fig. 3.52), pocket penetrometer (Fig. 3.53), and the simple shear device.

### In Situ Vane Shear Test

*Purpose*

The purpose is to measure $s_u$ and $s_r$ in situ in normally consolidated to slightly overconsolidated clays.

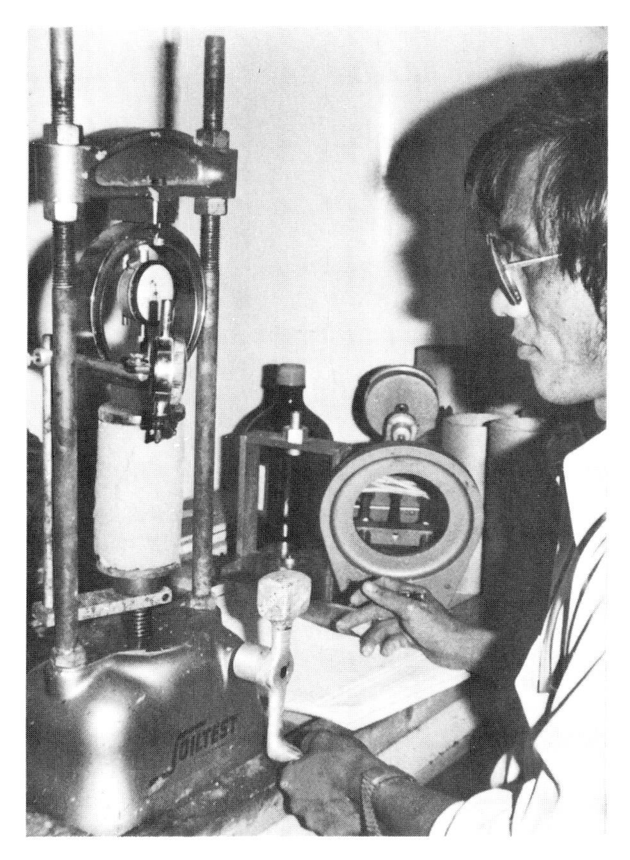


**FIG. 3.50** Unconfined compression apparatus.

**FIG. 3.51** Miniature vane-shear apparatus.

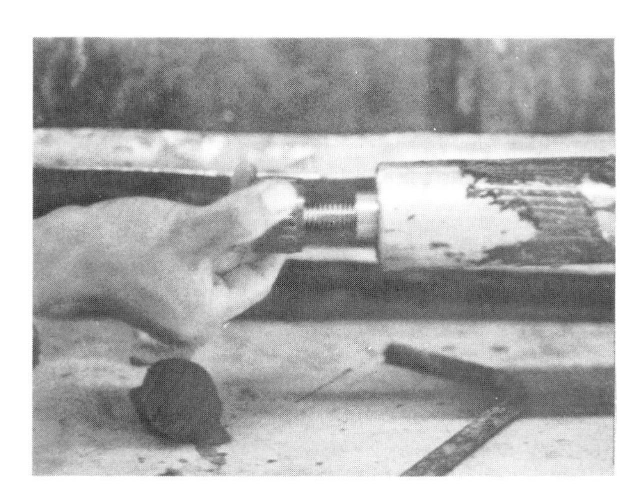

**FIG. 3.52** *(Above)* Torvane applied to the end of a Shelby tube specimen.

**FIG. 3.53** *(Right)* Pocket penetrometer applied to a Shelby tube specimen.

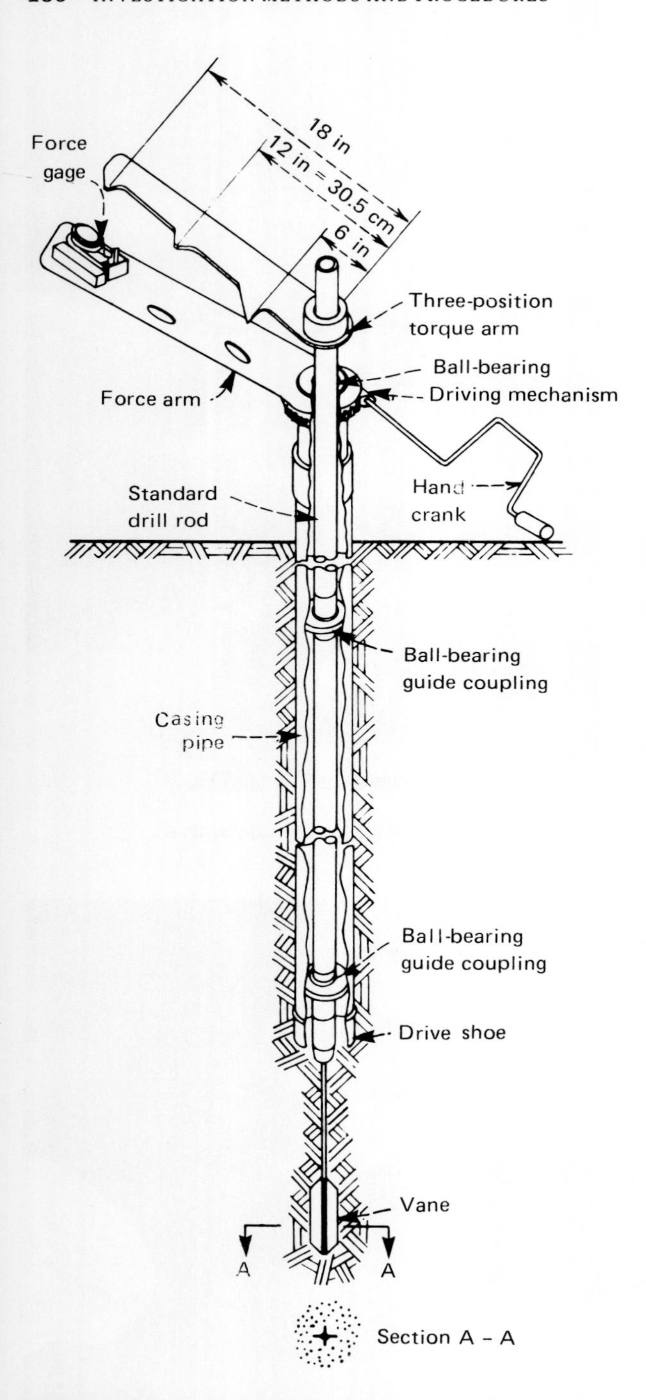


**FIG. 3.54 The in situ vane-shear test arrangement.** *(Courtesy of Acker Drill Co.)*

*Apparatus*

Test apparatus is illustrated on Fig. 3.54.

*Procedure*

A standard 4-in-diameter boring is made and cased to the desired test depth.

A 3-in-wide steel vane on a 1-in-diameter rod is lowered to the test depth (other vane sizes are available for smaller casing). Spacers at 30-ft intervals center the rod in the casing.

The vane is pushed to a depth 2½ ft below the casing and torsion applied at a constant rate through a friction-free mechanism while the torque is measured on a proving ring. Failure is evidenced by a sudden loss of torsion; the test is stopped for a short interval and then rerun to measure $s_r$. Charts give values for $s_u$ and $s_r$ for the measured torque and vane size used.

*Comments*

The test generally is considered to give the most reliable values for $s_u$ and $s_r$, but since the test applies shear directionally, the values are affected by soil anisotropy. Failure actually takes place by shearing of two circular horizontal surfaces and a cylindrical vertical surface. Vanes of different height-to-diameter ratios have been used by the Norwegian Geotechnical Institute[63] to evaluate anisotropy approximately. Relating $s_u$ as measured by the field vane to $s_u$ as measured in triaxial tests, Bjerrum et al. (1972)[55] found that

- $s_u$ (compression) $\approx$ 1.5 $s_u$ (vane)
- $s_u$ (extension) $\approx$ 0.5 $s_u$ (vane) in low- to medium-PI clays
- $s_u$ (extension) $\approx$ 1.0 $s_u$ (vane) in highly plastic clays

Field vane loading time rate is very rapid compared with actual field loadings, and during construction values are generally lower (Art. 3.4.2). Rate correction factor $\mu_r$ is proposed as a function of plasticity (*Fig. 3.55*) by Bjerrum et al. (1972)[55] on the basis of analyses of a number of embankment failures constructed over soft clays.

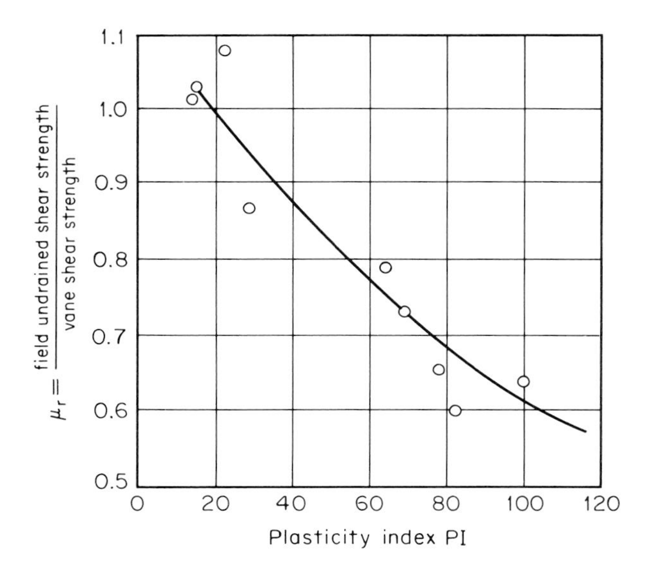


**FIG. 3.55 Loading time-rate correction factor vs. plasticity index for undrained shear strength as measured by the field vane.** [*From Bjerrum et al. (1972).*[55]]

### Pressuremeter Testing

See Art. 3.5.4.

### Dynamic Strength Tests

*Cyclic Triaxial Test*

PURPOSE The purpose is to measure strength parameters under low-frequency loadings (intermediate to large strain amplitudes). Also to measure cyclic compression modulus $E_d$ of cohesive soil [dynamic shear modulus $G_d$ is computed from $E_d$ and Eq. (3.64)] and the liquefaction resistance of undisturbed and reconstituted specimens of cohesionless soils under low-frequency loadings.

APPARATUS The apparatus is similar to standard triaxial equipment except that the deviator stress is applied cyclically. Sophisticated equipment allows variation in the frequency of applied load as well as control of the waveform (sinusoidal or square, ramp or random). In random loadings actual earthquake histories are imposed.

PROCEDURE The specimen is placed in the chamber and subjected to all-around confining pressure. Back-pressure may be applied to the pore water to saturate partially saturated specimens and to simulate in situ pore pressures. Constant-amplitude cyclic axial load is applied while specimen drainage is prevented, and pore-water pressure, axial load, and axial deformations are recorded. For many soils, after a number of stress cycles, the pore pressure increases to a value approximately equal to the cell pressure at which point the soil will begin to deform excessively.

LIMITATIONS The test only approximately duplicates field stress conditions during earthquake loadings [USAEC (1972)[40]].

*Cyclic Simple Shear Test*

PURPOSE The purpose is to measure strength parameters under intermediate to large strain amplitudes and to measure the dynamic shear modulus $G_d$ and the damping ratio of cohesive and cohesionless soils.

APPARATUS The simple shear device, described previously, in which the lateral load is applied cyclically. Loading frequency and waveform are controlled.

PROCEDURE The specimen is placed in the chamber and subjected to confining pressure. Back-pressure may be applied. A vertical stress is applied, and a cyclic shear force is applied laterally while no drainage is permitted.

COMMENT The test most nearly duplicates field conditions for soils subject to strong earthquake motion (large strains), although test boundary conditions will cause values to be somewhat lower than reality [USAEC (1972)[40]].

## 3.4.5 SOIL PENETRATION TESTS

### Standard Penetration Test (SPT) (ASTM D1586)

*Purpose*

SPT values are correlated with the compactness of granular soils, from which $\phi$ values are estimated, and with the consistency of cohesive soils.

**FIG. 3.56 Quick-release drop hammer used in Israel and Europe for SPT sampling.**

*Test Method*

The split-barrel sampler *(see Art. 2.4.2)*, 2-in OD and 1½-in ID, is driven into the ground by a 140-lb hammer dropping in free fall from a height of 30 in (energy = 4200 in · lb). Number of blows to advance the sampler each 6 in of penetration is recorded.

*Standard penetration resistance*, or $N$ value, is taken normally as the penetration of the second plus the third 6 in to provide "blows per foot of penetration." The first 6 in is disregarded because of the possibility of cuttings in the hole or disturbance during washing. The penetration resistance is composed of end resistance plus shaft friction. End resistance is the greater component in granular soils, and shaft resistance is the greater component in clay soils.

*Operational Factors Affecting N Values*

DRIVING ENERGY Standard hammer weight and drop should always be used. The weight is lifted by various methods: rope or cable on a drum, block and tackle, or by a quick-release grab (Fig. 3.56). To ensure that fall is always completely free of any resistance requires care by the operator, especially with the rope-and-drum method.

SAMPLER The sampler is illustrated on Fig. 2.60. Diameters larger than the standard 2-in sampler require correction. One relationship [Burmister (1962)[13]] provides for hammer weight variation $N$, drop height variation $H$, and sampler diameter variations $D_o$ and $D_i$ to determine adjusted blow count $B$ for the measured blow count $B'$ as follows:

$$B = B' \cdot \frac{4200}{WH} \cdot \left[ \frac{D_o^2 - D_i^2}{2.0^2 - 1.375^2} \right] \tag{3.44}$$

Damaged or blunt cutting edge on a sampler will increase resistance. Liner-designed tubes will produce lower $N$ values when used without liners [Schmertmann (1979)[64]].

DRILL RODS AND CASING One study indicates that the difference in weight between $N$ rod and $A$ rod has negligible effect on $N$ values [Brown (1977)[65]], although there are differences of opinion in the literature regarding the effects. Drill rods that are loosely connected or bent, or casing that is out of plumb, will affect penetration resistance because of energy absorption and friction along casing walls.

HOLE BOTTOM CONDITIONS *(See Art. 2.4.2 and Fig. 2.59.)* Soil remaining in casing, even as little as 2 in of sand, will cause a substantial increase in $N$ (plugged casing). Boiling and rising of fine sands in casing when the tools are removed, which causes a decrease in $N$, is common below the water table and is prevented by keeping the casing filled with water unless conditions are artesian. Gravel particles remaining in the mud-cased hole will increase $N$; values of 30 to 50 can result in a loose sand that should only be 10 blows. Overwashing or washing with a bottom-discharge bit with high pressure will loosen the soils to be sampled.

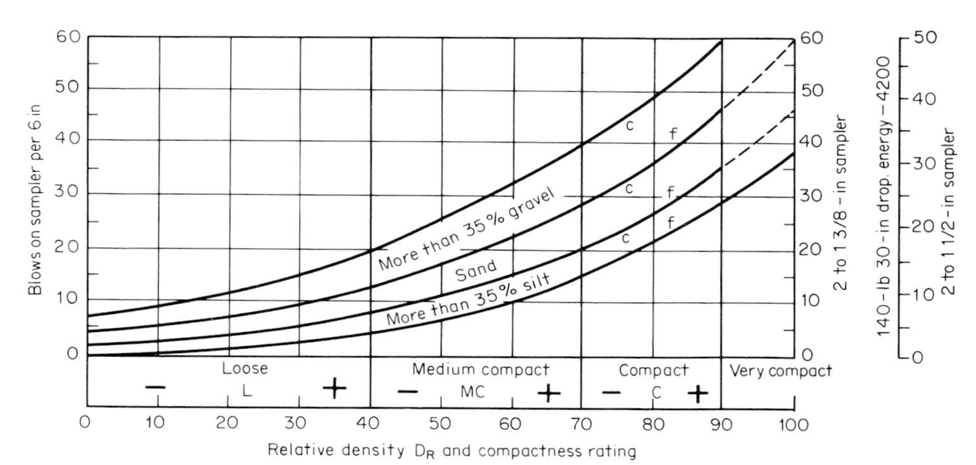


**FIG. 3.57 Relationship between sampler blows per 6 in, sampler diameter, gradation, and compactness.** [*From Burmister (1962).*[13] *Reprinted with permission of the American Society for Testing and Materials.*]

### *Natural Factors Affecting N Values*

SOIL TYPE AND GRADATION Gravel particles in particular have a very significant effect on $N$ as shown on Fig. 3.57.

COMPACTNESS AND CONSISTENCY *Compactness* of granular soils ($D_R$) and *consistency* of clay soils affect $N$ values.

SATURATION In gravels and coarse sands, saturation has no significant effect. In very fine or silty sands, pore-pressure effects during driving are corrected approximately for test values $N' > 15$ by the expression [Terzaghi and Peck (1948)[66] and Sanglerat (1972)[67]]:

$$N = 15 + \frac{1}{2}(N' - 15) \tag{3.45}$$

DEPTH CORRECTION FACTOR $C_N$ Effective overburden pressures cause a fictitious increase in $N$ values with increasing depth as shown by studies by various investigators [Gibbs and Holtz (1957),[68] Bazaraa (1967),[69] Marcuson and Bieganousky (1977*a*),[70] and Marcuson and Bieganousky (1977*b*)[71]]. Peck et al. (1973)[72] and Seed (1976)[73] proposed a *corrected penetration resistance* $N_1$ expressed as

$$N_1 = C_N N, \quad C_N = 0.77 \log_{10} \frac{20}{\bar{\sigma}_o} \text{ [Peck et al.]} \qquad C_N = 1 - 1.25 \log \bar{\sigma}_o/\bar{\sigma}_1 \text{ (Seed)} \tag{3.46}$$

where $C_N$ = a correction factor
$\bar{\sigma}_o$ = effective overburden pressure $(\gamma_t - \gamma_w)z$, tsf
$\bar{\sigma}_1$ = 1 tsf (95 kN/m²) ($N$ value at a depth corresponding to $\sigma_1$ is considered to be a standard)

Seed (1979)[74] presents recommended curves for the determination of $C_N$ based on averages for results obtained in various studies [Marcuson and Bieganousky (1977*a*),[70] (1977*b*)[71]] (Fig. 3.58).

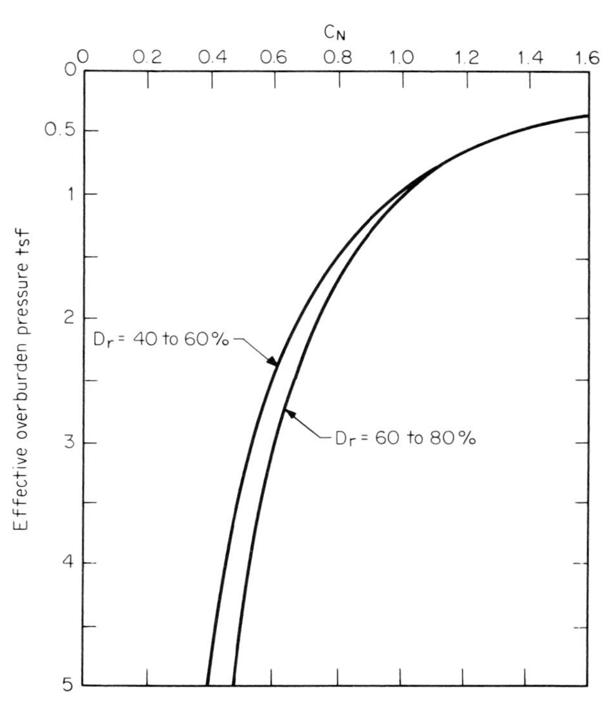


**FIG. 3.58 Recommended curves for determination of $C_N$ based on averages from WES tests.** [*From Seed (1979).*[74]]

*Applications of the N value*

- *Correlations with compactness and $D_R$*—see Table 3.23. Marcuson and Bieganousky[71] give an expression for computing $D_R$ which accounts for $N$, $\bar{\sigma}_o$, and $C_u$ (coefficient of uniformity, Eq. 3.3) as follows:

$$D_R = 11.7 + 0.76[|222N + 1600 - 53\bar{\sigma}_o - 50C_u^2|]^{1/2} \quad (3.47)$$

- From $D_R$ and *gradation* curves estimates can be made for:
  - $E$ (Young's modulus) from Table 3.33
  - $k$ from Fig. 3.15
  - $\phi$ from Table 3.28
- *Correlations with consistency and $U_c$*—Table 3.29, Fig. 3.64.
- *Allowable soil-bearing value for foundations*—correlations with $N$ [see, for example, Terzaghi and Peck (1967),[21] p. 491]

*Limitations*

$N$ value must always be considered as a rough approximation of soil compactness or consistency in view of the various influencing factors. Even in controlled laboratory testing conditions, a single value for $N$ can represent a spread of $\pm 15\%$ in $D_R$ [Marcuson and Bieganousky (1977b)[71]].

**TABLE 3.23**
**CORRELATIONS FOR COHESIONLESS SOILS BETWEEN COMPACTNESS, $D_R$, AND $N$**

| Compactness | Relative density $D_R$* | $N$ (SPT) |
|---|---|---|
| Very loose | <0.15 | <4 |
| Loose | 0.15–0.35 | 4–10 |
| Medium dense | 0.35–0.65 | 10–30 |
| Dense (compact) | 0.65–0.85 | 30–50 |
| Very dense | 0.85–1.0 | >50 |

*From Gibbs and Holtz (1957).[68]

## Cone Penetrometer Test (CPT) (ASTM D3441)

*Purpose*

Correlations are made between the cone tip resistance $q_c$ and $\phi$ and $s_u$.

*Test Method*

See Art. 2.3.4

*Estimating $\bar{\phi}$*

The Trofimenkov method [Trofimenkov (1974)[75]] relates $\bar{\phi}$ directly to $q_c$ and effective overburden pressure (Fig. 3.59), and although the values for $\bar{\phi}$ are generally lower than laboratory values, it is useful for a first estimate [Mitchell and Lunne (1978)[56]].

The Schmertmann method [Schmertmann (1977)[76]] estimates $D_R$ from $q_c$ (Fig. 3.60) and then correlates $D_R$ with $\bar{\phi}$ (Fig. 3.63). Mitchell and Lunne (1978)[56] evaluated a number of methods for determining $\bar{\phi}$ from $q_c$ and found good agreement with laboratory results using the method of Schmertmann.

*Estimating $s_u$*

Sanglerat (1972)[67] gives:

- $s_u \approx q_c/15$ for soft to stiff clays
- $s_u \approx q_c/30$ for stiff fissured clays

Schmertmann (1977)[76] gives:

$$s_u = \frac{q_c - \Sigma\gamma z}{N_c} \quad (3.48)$$

where $\Sigma\gamma z$ = total overburden pressure at depth z and $N_c$ = deep foundation depth correction factor varying from 10 for stiff clays to 16 for soft clays with the Fugro electric cone. $N_c$ should be confirmed by other strength tests for each new location.

For correlations with pressure meter data, see Art. 3.5.4.

*Correlations between $q_c$ and $N$*

Correlations are given on Tables 3.24 and 3.25.

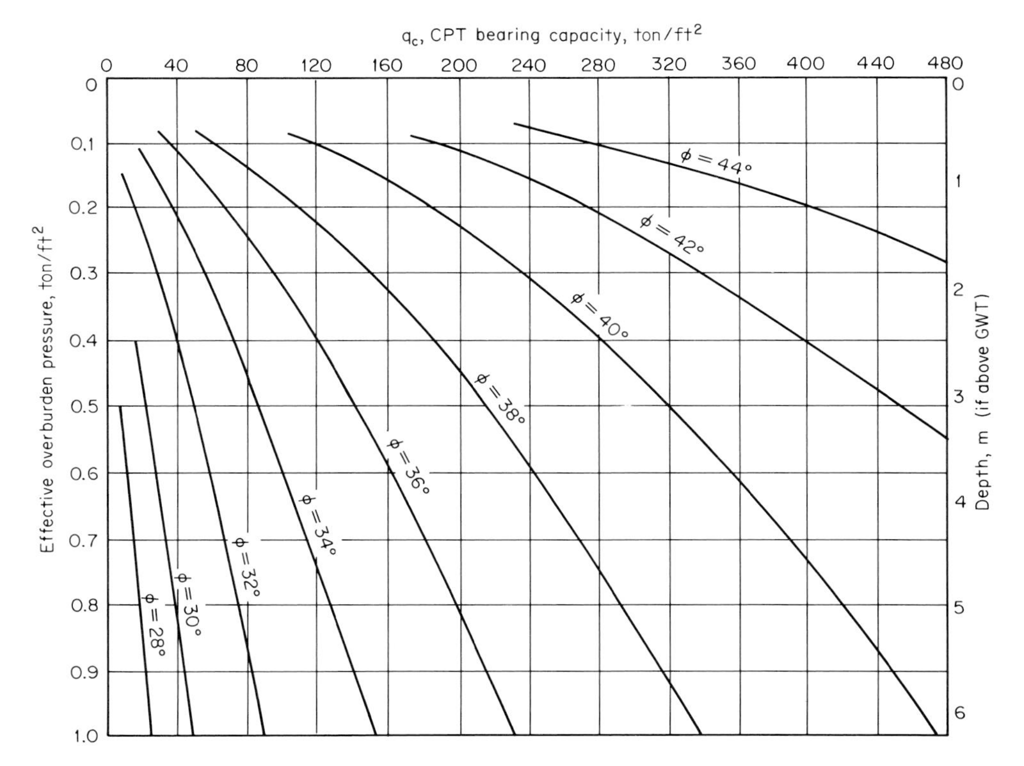


**FIG. 3.59 Correlation between effective overburden pressure, $q_c$ and $\phi$.** [*From Trofimenkov (1974).*[75]]

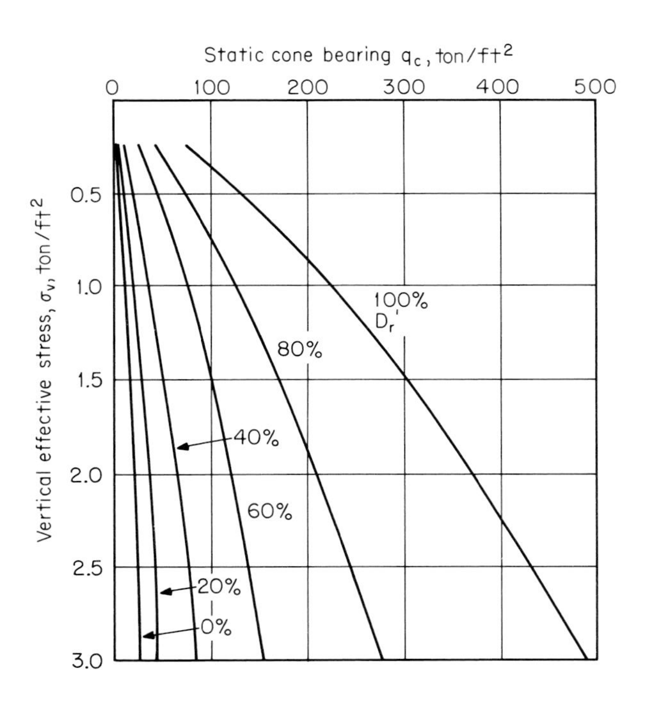


**FIG. 3.60 Static cone resistance $q_c$ vs. $D_R$.** [*From Schmertmann (1977).*[76] *Reprinted with permission of the Federal Highway Administration.*]

**TABLE 3.24**
**TYPICAL $q_c$ (ton/ft$^2$)/$N$(SPT blows/ft) RATIOS***

| Type soil | Fugro tip | Delft mechanical tips |
|---|---|---|
| Sand and gravel mixtures | 8 | 6 |
| Sand | 5 | 4 |
| Sandy silts | 4 | 3 |
| Clay-silt-sand mixtures | 2 | 2 |
| Insensitive clays | 1 | 1½ |
| Sensitive clays | Ratios can get very high because $N \rightarrow 0$. | |

*From Schmertmann (1977).[76] Reprinted with permission of the Federal Highway Administration.

**TABLE 3.25**
**CORRELATIONS BETWEEN DUTCH CONE BEARING VALUE $q_c$ AND THE STANDARD PENETRATION TEST VALUE $N$ (BEGEMANN CONE)***

| | $q_c/N$† | | | |
|---|---|---|---|---|
| **Material type** | **Ref. 77** | **Ref. 78** | **Ref. 79** | **Ref. 80** |
| Organic silty clay (OH) | 2 | | | 1–2 |
| Clayey silt to silty clay (MH-CH) | 3.5 | 2 | 2.5 | 2–4 |
| Micaceous fine sand, trace to some silt (SP-SM) | 5.5 | 3–4 | 4 | 4–6 |
| Clay, trace fine gravel, occasional gravel lenses (CL) | 8 | | | |
| Clay–fine sand, trace to some gravel (SP-SW) | 9 | 5–6 | 8 | 6–8 |
| Micaceous silty clay–fine sand, trace to some clay (SM-SC) | 10 | 2 | 4 | 4–6 |

*From Alperstein and Leifer (1976).[77]
†Units of $q_c$ are kg/cm$^2$; $N$ is in blows per foot.

### California Bearing Ratio Test (CBR)

*Purpose*

A penetration test performed to determine the California bearing ratio (CBR) value for base, subbase, and subgrade materials upon which pavement design thicknesses are based.

*Laboratory Test Procedure (ASTM D1883)*

A series of specimens are compacted in 6-in-diameter molds to bracket either the standard or modified optimum moisture (*see Art. 3.2.3*). A surcharge weight is placed on the soil surface and the specimens immersed in a water tank for 4 days to permit saturation while swelling is measured.

A mold is placed in the testing apparatus (Fig. 3.61) and a plunger 1.91 in in diameter is forced to penetration of 0.1 to 0.2 in (penetrations to 0.5 in are used) while the penetration resistance is recorded.

The CBR value is determined as the ratio of the test resistance at 0.1-in penetration to the standard resistance of crushed stone at the same penetration (1000 psi or 70.3 kg/cm²) [Asphalt Institute (1969)[81]].

*Field Test Procedure*

The plunger is jacked against the reaction of a truck weight. The test should not be performed during the dry season, for resistances will be higher than those to be relied upon for pavement support. Field values are usually lower than laboratory values, probably because of the effect of confinement in the lab.

*Soil Rating System*

A soil rating system for subgrade, subbase, or base for use in the design of light traffic pavements in terms of CBR values is given on Fig. 3.62. Typical CBR values for compacted materials are given on Table 3.31.

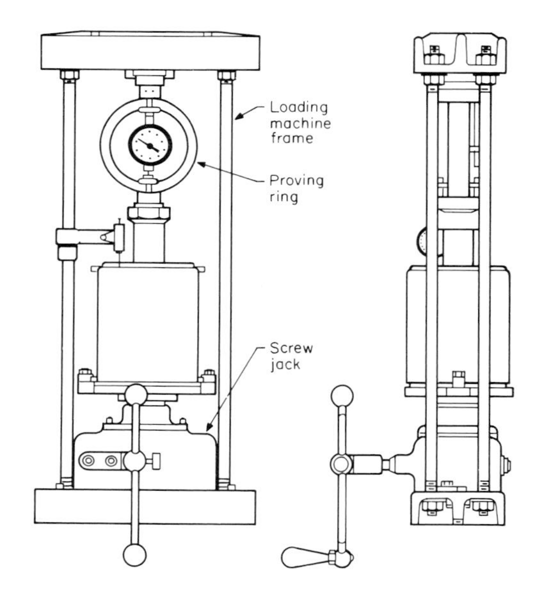


**FIG. 3.61 Laboratory CBR loading device. (Electric motor is often used to drive jack at rate of 0.05 in/min.)**

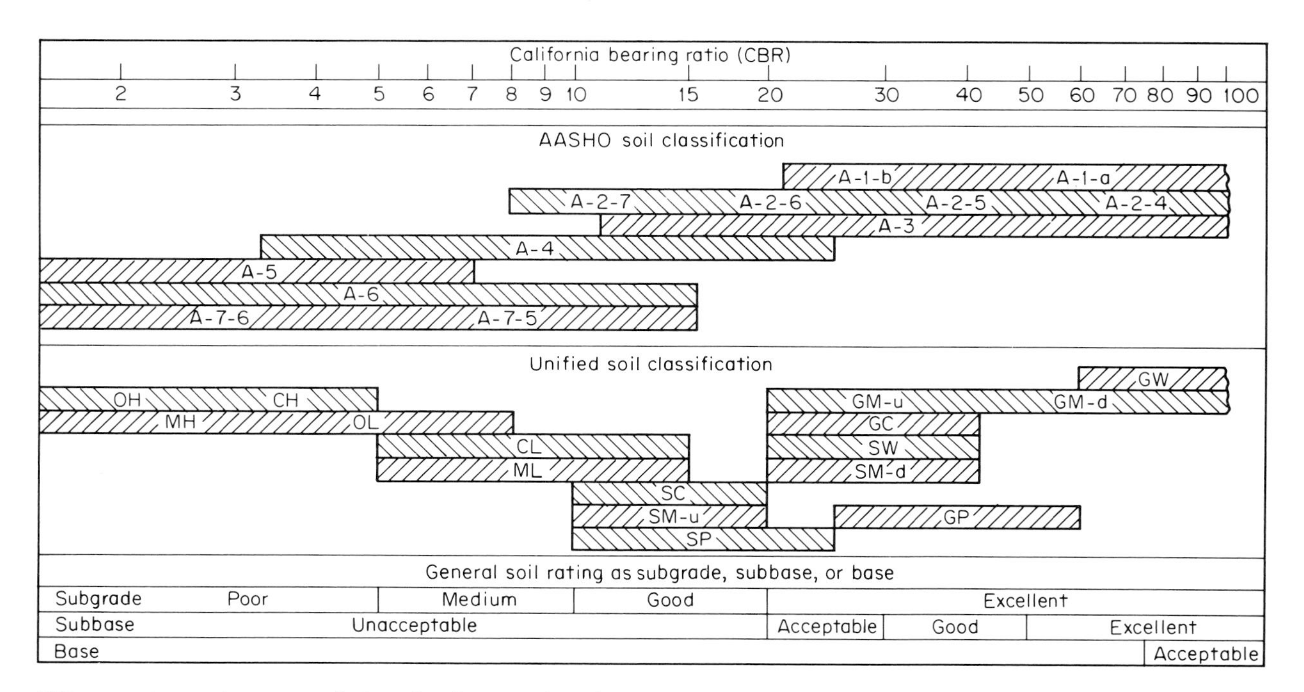


**FIG. 3.62 Approximate correlation of soil ratings based on CBR values for use in design of light-traffic pavements.** [*From The Asphalt Institute (1970).*[82]]

## 3.4.6 TYPICAL VALUES OF BASIC, INDEX, AND STRENGTH PROPERTIES

### Fresh Intact Rock

Common properties including unconfined compressive strength, density, and deformation modulus for common rock types are given on Table 3.26.

### Weathered Rock and Residual Soil

Shear strength parameters for various rock types and in situ conditions are given on Table 3.27.

### Cohesionless Soils

Common properties including relative density, dry density, void ratio, and strength as related to gradation and $N$ are given on Table 3.28. Relationships between $\phi$ and $D_R$ for various gradations are given on Fig. 3.63.

### Clay Soils

Common properties, including relationships between consistency, strength, saturated weight and $N$, are given on Table 3.29.

Correlation of $N$ with $U_c$ for cohesive soils of various plasticities are given on Fig. 3.64.

Typical properties of cohesive materials classified by geologic origin, including density, natural moisture contents, plasticity indices, and strength parameters, are given on Table 3.30.

### Compacted Materials

Typical properties, including maximum dry weight, optimum moisture content, compression values, strength characteristics, coefficient or permeability, CBR value ranges, and range of subgrade modulus are given on Table 3.31.

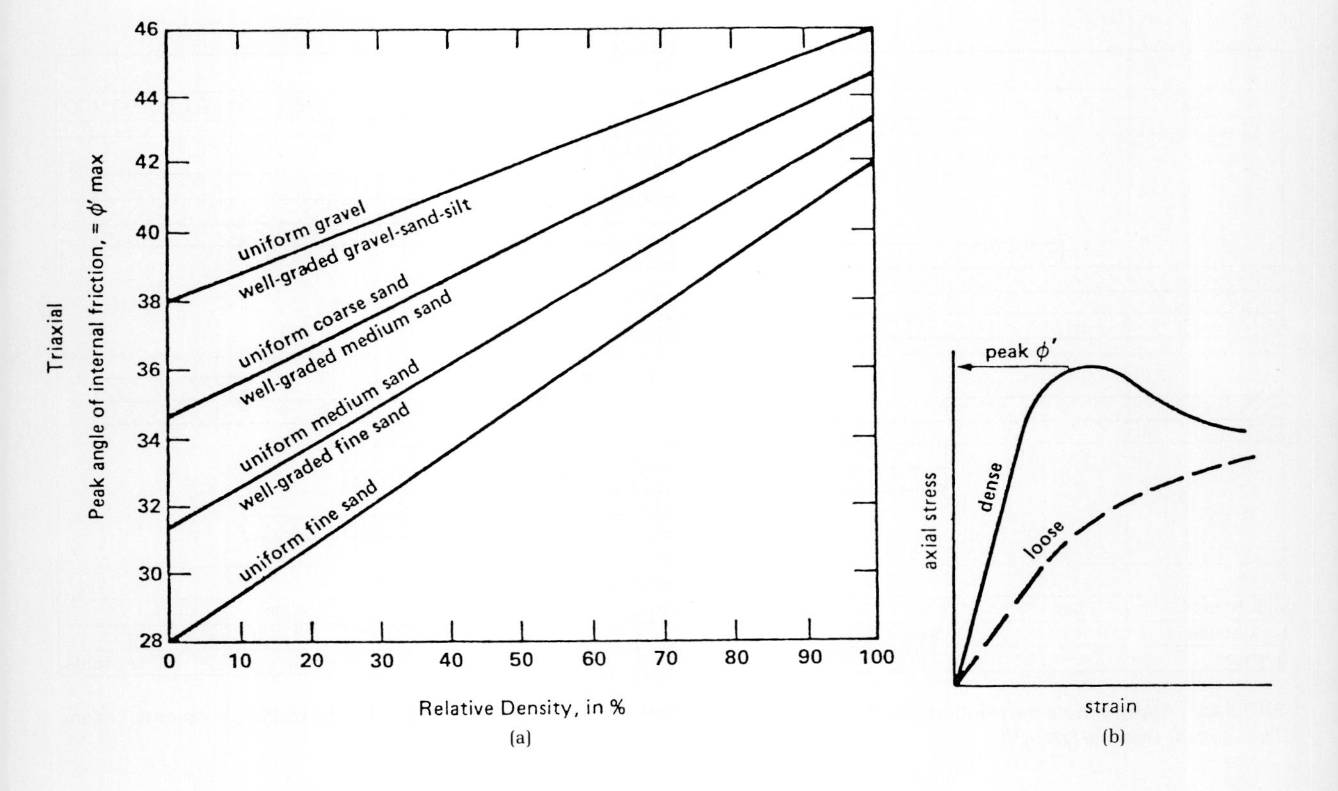

**TABLE 3.26**
**COMMON ROCK TYPES AND TYPICAL ENGINEERING PROPERTIES***

| Rock type† | Texture | Fabric structure | $\gamma_d$, g/cm³ | $U_c$, kg/cm² | $E_r$, $10^4$ kg/cm² |
|---|---|---|---|---|---|
| | | IGNEOUS | | | |
| Granite | Coarse to medium | Massive, relatively tight, and widely spaced joints | 2.69 | 700–1750 | 28–49 |
| Diorite | Coarse to medium | | 2.82 | 700–1750 | 35–56 |
| Gabbro | Coarse to medium | | 2.88 | 1050–2100 | 49–84 |
| Rhyolite | Fine | Massive, extensive jointing, often vesicular | 2.59 | 700–1750 | 35–56 |
| Andesite | Fine | | 2.66 | 700–1750 | 42–63 |
| Basalt | Fine | | 2.85 | 1750–2800 | 49–90 |
| Obsidian | Glassy | Massive, continuous | 2.20 | 140– 560 | 7–28 |
| Tuff | Coarse | Cemented ash, porous | 1.60 | 14– 70 | 1– 7 |
| | | METAMORPHIC | | | |
| Gneiss | Coarse to medium | Banded to foliated | 2.70 | 700–1400 | 28–56 |
| Schist | Fine | Foliated | 2.67 | 350–1050 | 14–35 |
| Slate | Fine | Platy | 2.69 | 700–1400 | 35–56 |
| Quartzite | Fine | Massive, fine and widely spaced joints | 2.66 | 1050–2450 | 42–56 |
| Marble | Fine to very fine | | 2.69 | 840–2100 | 49–70 |
| Serpentine | Various | Massive, often soft | 2.53 | 70– 700 | 7–35 |
| | | SEDIMENTARY | | | |
| Conglomerate | Coarse, rounded | Layered, cemented | 2.48 | 350–1050 | 7–35 |
| Breccia | Course, angular | Layered, cemented | 2.53 | 350–1050 | 7–35 |
| Sandstone | Medium | Layered, cemented | 2.35 | 280– 840 | 7–21 |
| Siltstone | Fine | Layered, cemented | 1.8–2.4 | 7– 350 | 3–14 |
| Shale‡ | Very fine | Laminated, compaction shales unstable, cemented shales stable | 1.6–2.2 | 7– 350 | 3–14 |
| Limestone | Fine | Massive, stratified, soluble, cavities form | 2.64 | 350–1050 | 14–42 |
| Dolomite | Fine | Massive, some recrystallization | 2.67 | 490–1400 | 28–56 |

*After NAVFACS (1971).[6] Properties are for sound, unweathered specimens without voids or fractures, tested dry in the laboratory. Elasticity and strength depend on porosity, cementation, and in foliated, platy or laminated rocks, on loading direction. Saturated values for $U_c$ and $E_r$ are usually 80 to 90% of the dry values given.

†For detailed descriptions of rock types, composition, textures, fabrics and structure, see Art. 5.2.

‡See also Table 3.30.

◀ **FIG. 3.63 Friction angle and relative density relationships for granular soils: (*a*) chart for the approximate evaluation of the peak angle of internal friction from relative density—Schmertmann modification of Burmister (1948)[10]; (*b*) in problems where the sand may strain past the peak strength value before a general failure occurs, then a reduced value of $\phi$ must be used, particularly in the denser cohensionless soils. [NOTES: (1) For quartz sands. (2) Angular grains can increase $\phi$ by about 15% in the loose state and 30% in the dense state over rounded grains.]** *(Reprinted with permission of the Federal Highway Administration.)*

**TABLE 3.27**
**SHEAR STRENGTH PARAMETERS OF RESIDUAL SOIL, WEATHERED ROCKS, AND RELATED MATERIALS***

| Rock type | Weathering degree (see *Table 6.15*) | Strength parameters: $c, \bar{c}$, kg/cm$^2$ | Strength parameters: $\phi, \bar{\phi}, \phi_r$, degrees | Remarks |
|---|---|---|---|---|
| IGNEOUS ROCKS | | | | |
| Granite | Decomposed | $c = 0$ | $\phi = 27–31$<br>$\phi_{avg} = 29$ | 500 tests, Cherry Hill Dam |
| | Quality index $i$ | $c$ | $\phi$ | In situ direct shear tests, Alto Rabagão |
| | 15 | 1 | 41 | |
| | 10 | 2 | 45–46 | |
| | 7 | 3 | 49–52 | |
| | 5 | 5 | 57 | |
| | 3 | 6–13 | 62–63 | |
| | Weathered, zone IIB<br>Partly weathered, zone IIB<br>Relatively sound, zone III | | $\phi_r = 26–33$<br>$\phi_r = 27–31$<br>$\phi_r = 29–32$ | Lab direct shear tests, Alto Lindosa |
| | Red earth, zone IB<br>Decomposed, zone IC | | $\bar{\phi} = 28$<br>$\bar{\phi}_{avg} = 35$ | |
| | Decomposed, fine-grained<br>Decomposed, coarse-grained<br>Decomposed, remolded | $c = 0$ if saturated | $\phi = 25–34$<br>$\phi = 36–38$<br>$\phi = 22–40$ | |
| Quartz diorite | Decomposed; sandy, silty | $c = 0.1$ | $\phi = 30+$ | Lab tests, UD samples |
| Diorite | Weathered | $c = 0.3$ | $\phi = 22$ | CU triaxial tests |
| Rhyolite | Decomposed | | $\phi = 30$ | |
| METAMORPHIC ROCKS | | | | |
| Gneiss (micaceous) | Zone IB<br>Decomposed | $c = 0.6$<br>$c = 0.3$ | $\phi = 23$<br>$\phi = 37$ | Direct shear tests |
| Gneiss | Decomposed, zone IC<br>Decomposed (fault zone)<br>Much decomposed<br>Medium decomposed<br>Unweathered | ·<br>$c = 1.5$<br>$c = 4.0$<br>$c = 8.5$<br>$c = 12.5$ | $\phi = 18.5$<br>$\phi = 27$<br>$\phi = 29$<br>$\phi = 35$<br>$\phi = 60$ | CU triaxial tests<br>Direct shear tests on concrete-rock surfaces |
| Schist | Weathered (mica schist soil) | | $\phi = 24½$ | From analysis of slides |

| | | | | |
|---|---|---|---|---|
| | Partly weathered mica schists and phyllites (highly fractured) | $c = 0.7$ | $\phi = 35$ | Perpendicular to schistosity |
| | Weathered, intermediate zone IC | $\bar{c} = 0.5$<br>$\bar{c} = 0.7$ | $\phi, \bar{\phi} = 15$<br>$\phi = 18$<br>$\bar{\phi} = 21$ | CU tests, $S = 50\%$<br>CU tests, $S = 100\%$ |
| | Weathered | | $\phi = 26\text{–}30$ | Compacted rock fill, field direct shear tests |
| Phyllite | Residual soil, zone IC | $c = 0$ | $\phi = 24$ | Perpendicular to schistosity |
| | | $c = 0$ | $\phi = 18$ | Parallel to schistosity (both from analysis of slides) |
| | SEDMENTARY ROCKS | | | |
| Keuper marl | Highly weathered | $\bar{c} \leq 0.1$ | $\bar{\phi} = 25\text{–}32$;<br>$\phi_r = 18\text{–}24$ | 2% carbonates |
| | Intermediately weathered | $\bar{c} \leq 0.1$ | $\bar{\phi} = 32\text{–}42$<br>$\phi_r = 22\text{–}29$ | 14% carbonates |
| | Unweathered | $\bar{c} \leq 0.3$ | $\bar{\phi} = 40$<br>$\phi_r = 23\text{–}32$ | 20% carbonates (all triaxial tests, $D$ and CU and cut planes) |
| London clay | Weathered (brown) | $\bar{c} =$ 0.1–0.2 | $\bar{\phi} = 19\text{–}22$<br>$\phi_r = 14$ | |
| | Unweathered | $\bar{c} =$ 0.9–1.8 | $\bar{\phi} = 23\text{–}30$<br>$\phi_r = 15$ | |
| | JOINT FILLING | | | |
| "Black seams" | In zone IC | | $\phi_r = 10.5$<br>$\phi_r = 14.5$ | Seam with slickensides<br>Seam without slickensides (both CU tests) |
| | SHEAR ZONES† | | | |
| Metamorphic rocks<br>Shales<br>Fault gouge, general | | | $\phi_r = 15\text{–}25$<br>$\phi = 10\text{–}20$<br>$\phi_r = 15\text{–}30$ | Foliation shear<br>Mylonite seam |

*From Deere and Patton (1971).[83]

†From Deere (1974).[84]

**TABLE 3.28**
**COMMON PROPERTIES OF COHESIONLESS SOILS**

| Material | Compactness | $D_R$, % | $N$* | $\gamma$ dry,† g/cm$^3$ | Void ratio e | Strength‡ $\phi$ |
|---|---|---|---|---|---|---|
| GW: well-graded gravels, gravel-sand mixtures | Dense | 75 | 90 | 2.21 | 0.22 | 40 |
| | Medium dense | 50 | 55 | 2.08 | 0.28 | 36 |
| | Loose | 25 | <28 | 1.97 | 0.36 | 32 |
| GP: poorly graded gravels, gravel-sand mixtures | Dense | 75 | 70 | 2.04 | 0.33 | 38 |
| | Medium dense | 50 | 50 | 1.92 | 0.39 | 35 |
| | Loose | 25 | <20 | 1.83 | 0.47 | 32 |
| SW: well-graded sands, gravelly sands | Dense | 75 | 65 | 1.89 | 0.43 | 37 |
| | Medium dense | 50 | 35 | 1.79 | 0.49 | 34 |
| | Loose | 25 | <15 | 1.70 | 0.57 | 30 |
| SP: poorly graded sands, gravelly sands | Dense | 75 | 50 | 1.76 | 0.52 | 36 |
| | Medium dense | 50 | 30 | 1.67 | 0.60 | 33 |
| | Loose | 25 | <10 | 1.59 | 0.65 | 29 |
| SM: silty sands | Dense | 75 | 45 | 1.65 | 0.62 | 35 |
| | Medium dense | 50 | 25 | 1.55 | 0.74 | 32 |
| | Loose | 25 | <8 | 1.49 | 0.80 | 29 |
| ML: inorganic silts, very fine sands | Dense | 75 | 35 | 1.49 | 0.80 | 33 |
| | Medium dense | 50 | 20 | 1.41 | 0.90 | 31 |
| | Loose | 25 | <4 | 1.35 | 1.0 | 27 |

*$N$ is blows per foot of penetration in the SPT. Adjustments for gradation are after Burmister (1962).[13] See Table 3.23 for general relationships of $D_R$ vs. $N$.

†Density given is for $G_s = 2.65$ (quartz grains).

‡Friction angle $\phi$ depends on mineral type, normal stress, and grain angularity as well as $D_R$ and gradation (see Fig. 3.63).

**TABLE 3.29**
**COMMON PROPERTIES OF CLAY SOILS**

| Consistency | N | Hand test | $\gamma_{sat}$,* g/cm³ | Strength† $U_c$, kg/cm² |
|---|---|---|---|---|
| Hard | >30 | Difficult to indent | >2.0 | >4.0 |
| Very stiff | 15–30 | Indented by thumbnail | 2.08–2.24 | 2.0–4.0 |
| Stiff | 8–15 | Indented by thumb | 1.92–2.08 | 1.0–2.0 |
| Medium (firm) | 4–8 | Molded by strong pressure | 1.76–1.92 | 0.5–1.0 |
| Soft | 2–4 | Molded by slight pressure | 1.60–1.76 | 0.25–0.5 |
| Very soft | <2 | Extrudes between fingers | 1.44–1.60 | 0–0.25 |

*$\gamma_{sat} = \gamma_{dry} + \gamma_w\left(\frac{e}{1+e}\right)$

†Unconfined compressive strength $U_c$ is usually taken as equal to twice the cohesion $c$ or the undrained shear strength $s_u$. For the drained strength condition, most clays also have the additional strength parameter $\phi$, although for most normally consolidated clays $c = 0$ [Lambe and Whitman (1969)[30]]. Typical values for $s_u$ and drained strength parameters are given on Table 3.30.

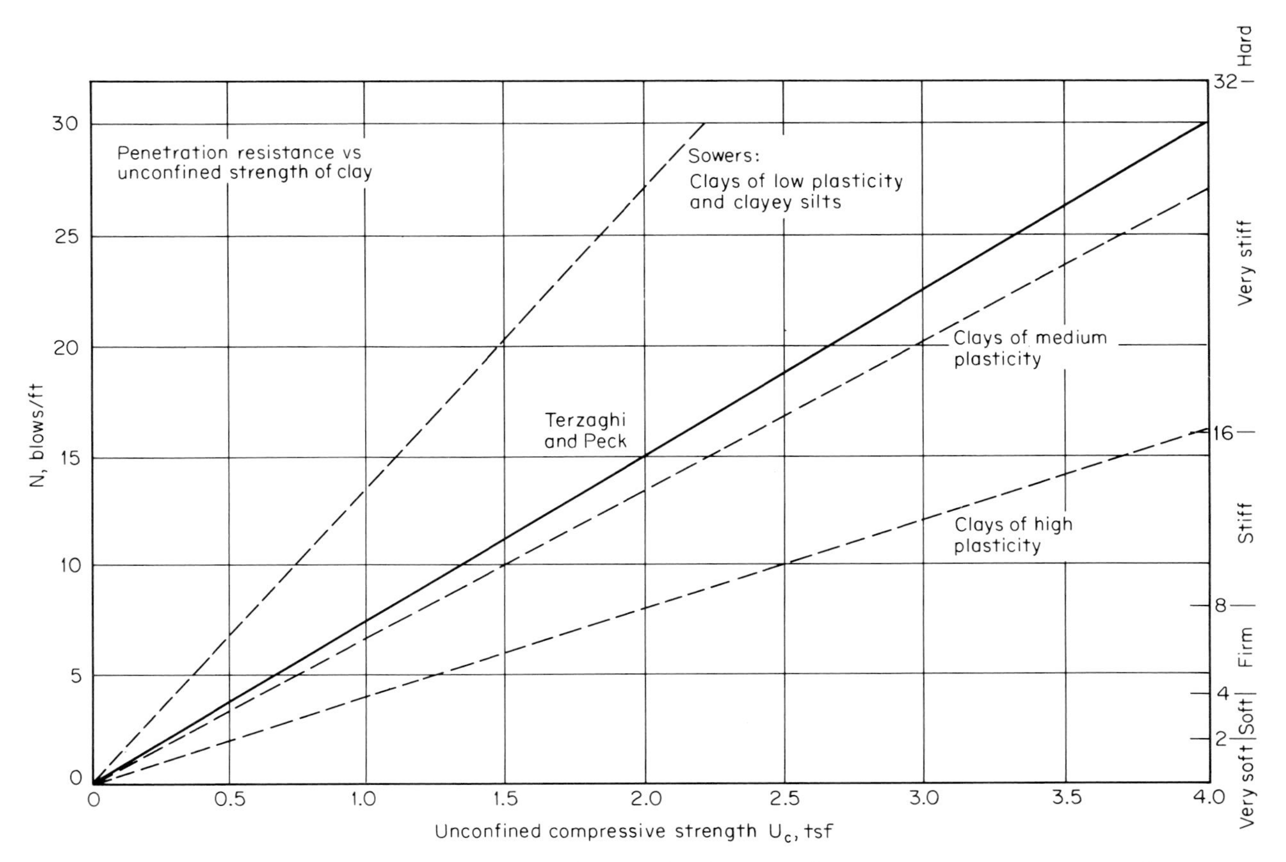


**FIG. 3.64 Correlations of SPT N values with $U_c$ for cohesive soils of varying plasticities.** [*From NAVFAC (1971).*[6]]

**TABLE 3.30**
**TYPICAL PROPERTIES OF FORMATIONS OF COHESIVE MATERIALS**

| Material | Type* | Location | $\gamma_d$, g/cm³ | $w$, % | LL, % | PI, % | $s_u$, kg/cm² | $\bar{c}$, kg/cm² | $\bar{\phi}$ | Remarks |
|---|---|---|---|---|---|---|---|---|---|---|
| CLAY SHALES (WEATHERED) | | | | | | | | | | |
| Carlisle (Cret.) | CH | Nebraska | 1.48 | 18 | | | | 0.5 | 45 | $\phi$ extremely |
| Bearpaw (Cret.) | CH | Montana | 1.44 | 32 | 130 | 90 | | 0.35 | 15 | variable |
| Pierre (Cret.) | | South Dakota | 1.47 | 28 | | | | 0.9 | 12 | |
| Cucaracha (Cret.) | CH | Panama Canal | | 12 | 80 | 45 | | | | $\phi_r = 10°$ |
| Pepper (Cret.) | CH | Waco, Texas | | 17 | 80 | 58 | | 0.4 | 17 | $\phi_r = 7°$ |
| Bear Paw (Cret.) | CH | Saskatchewan | | 32 | 115 | 92 | | 0.4 | 20 | $\phi_r = 8°$ |
| Modelo (Tert.) | CH | Los Angeles | 1.44 | 29 | 66 | 31 | | 1.6 | 22 | Intact specimen |
| Modelo (Tert.) | CH | Los Angeles | 1.44 | 29 | 66 | 31 | | 0.32 | 27 | Shear zone |
| Martinez (Tert.) | CH | Los Angeles | 1.66 | 22 | 62 | 38 | | 0.25 | 26 | Shear zone |
| (Eocene) | CH | Menlo Park, Calif. | 1.65 | 30 | 60 | 50 | | Free swell 100%; $P$ = 10 kg/cm² | | |
| RESIDUAL SOILS | | | | | | | | | | |
| Gneiss | CL | Brazil; buried | 1.29 | 38 | 40 | 16 | | 0 | 40 | $e_0 = 1.23$ |
| Gneiss | ML | Brazil; slopes | 1.34 | 22 | 40 | 8 | | 0.39 | 19 | $c$, $\phi$— |
| Gneiss | ML | Brazil; slopes | 1.34 | | 40 | 8 | | 0.28 | 21 | unsoaked |
| COLLUVIUM | | | | | | | | | | |
| From shales | CL | West Virginia | | 28 | 48 | 25 | | 0.28 | 28 | $\phi_r = 16°$ |
| From gneiss | CL | Brazil | 1.10 | 26 | 40 | 16 | | 0.2 | 31 | $\phi_r = 12°$ |
| ALLUVIUM | | | | | | | | | | |
| Back swamp | OH | Louisiana | 0.57 | 140 | 120 | 85 | 0.15 | | | |
| Back swamp | OH | Louisiana | 1.0 | 60 | 85 | 50 | 0.1 | | | |
| Back swamp | MH | Georgia | 0.96 | 54 | 61 | 22 | 0.3 | | | $e_0 = 1.7$ |
| Lacustrine | CL | Great Salt Lake | 0.78 | 50 | 45 | 20 | 0.34 | | | |
| Lacustrine | CL | Canada | 1.11 | 62 | 33 | 15 | 0.25 | | | |
| Lacustrine (volcanic) | CH | Mexico City | 0.29 | 300 | 410 | 260 | 0.4 | | | $e_0 = 7$, $S_t = 13$ |
| Estuarine | CH | Thames River | 0.78 | 90 | 115 | 85 | 0.15 | | | |
| Estuarine | CH | Lake Maricaibo | | 65 | 73 | 50 | 0.25 | | | |
| Estuarine | CH | Bangkok | | 130 | 118 | 75 | 0.05 | | | |
| Estuarine | MH | Maine | | 80 | 60 | 30 | 0.2 | | | |

| | | | | | | | | | | |
|---|---|---|---|---|---|---|---|---|---|---|
| MARINE SOILS (OTHER THAN ESTUARINE) | | | | | | | | | | |
| Offshore | MH | Santa Barbara, Calif. | 0.83 | 80 | 83 | 44 | 0.15 | | | $e_0$ = 2.28 |
| Offshore | CH | New Jersey | | 65 | 95 | 60 | 0.65 | | | |
| Offshore | CH | San Diego | 0.58 | 125 | 111 | 64 | 0.1 | | | Depth = 2 m |
| Offshore | CH | Gulf of Maine | 0.58 | 163 | 124 | 78 | 0.05 | | | |
| Coastal Plain | CH | Texas (Beaumont) | 1.39 | 29 | 81 | 55 | 1.0 | 0.2 | 16 | $\phi_r$ = 14, $e_0$ = 0.8 |
| Coastal Plain | CH | London | 1.60 | 25 | 80 | 55 | 2.0 | | | |
| LOESS | | | | | | | | | | |
| Silty | ML | Nebraska-Kansas | 1.23 | 9 | 30 | 8 | | 0.6 | 32 | Natural $w$% |
| Silty | ML | Nebraska Kansas | 1.23 | (35 ) | 30 | 8 | | 0 | 23 | Prewetted |
| Clayey | CL | Nebraska-Kansas | 1.25 | 9 | 37 | 17 | | 2.0 | 30 | Natural $w$% |
| GLACIAL SOILS | | | | | | | | | | |
| Till | CL | Chicago | 2.12 | 23 | 37 | 21 | 3.5 | | | |
| Lacustrine (varved) | CL | Chicago | 1.69 | 22 | 30 | 15 | 1.0 | | | $e_0$ = 0.6 (OC) |
| Lacustrine (varved) | CL | Chicago | | 24 | 30 | 13 | 0.1 | | | $e_0$ = 1.2 (NC) |
| Lacustrine (varved) | CH | Chicago | 1.18 | 50 | 54 | 30 | 0.1 | | | |
| Lacustrine (varved) | CH | Ohio | 0.96 | 46 | 58 | 31 | 0.6 | | | $S_t$ = 4 |
| Lacustrine (varved) | CH | Detroit | 1.20 | 46 | 55 | 30 | 0.8 | | | $e_0$ = 1.3 (clay) |
| Lacustrine (varved) | CH | New York City | | 46 | 62 | 34 | 1.0 | | | $e_0$ = 1.25 (clay) |
| Lacustrine (varved) | CL | Boston | 1.35 | 38 | 50 | 26 | 0.8 | | | $S_t$ = 3 |
| Lacustrine (varved) | CH | Seattle | | 30 | 55 | 22 | | | 30 | $\phi_r$ = 13° |
| Marine† | CH | Canada–Leda clay | 0.89 | 80 | 60 | 32 | 0.5 | | | $S_t$ = 128 |
| Marine† | CL | Norway | 1.34 | 40 | 38 | 15 | 0.13 | | | $S_t$ = 7 |
| Marine† | CL | Norway | 1.29 | 43 | 28 | 15 | 0.05 | | | $S_t$ = 75 |

*See Figure 3.12.

†Marine clays strongly leached.

**TABLE 3.31**
**TYPICAL PROPERTIES OF COMPACTED SOILS***

| | | | | Typical value of compression | | Typical strength characteristics | | | |
|---|---|---|---|---|---|---|---|---|---|
| | | | | Percent of original height | | | | | |
| Group symbol | Soil type | Range of maximum dry unit weight, pcf | Range of optimum moisture, % | At 1.4 tsf (20 psi) | At 3.6 tsf (50 psi) | Cohesion (as compacted), psf | Cohesion (saturated), psf | Effective stress envelope $\phi$, degrees | tan $\phi$ |
| GW | Well-graded clean gravels, gravel-sand mixtures | 125–135 | 11–8 | 0.3 | 0.6 | 0 | 0 | >38 | >0.79 |
| GP | Poorly graded clean gravels, gravel-sand mix | 115–125 | 14–11 | 0.4 | 0.9 | 0 | 0 | >37 | >0.74 |
| GM | Silty gravels, poorly graded gravel-sand silt | 120–135 | 12–8 | 0.5 | 1.1 | . . . | . . . | >34 | >0.67 |
| GC | Clayey gravels, poorly graded gravel-sand-clay | 115–130 | 14–9 | 0.7 | 1.6 | . . . | . . . | >31 | >0.60 |
| SW | Well-graded clean sands, gravelly sands | 110–130 | 16–9 | 0.6 | 1.2 | 0 | 0 | 38 | 0.79 |
| SP | Poorly-graded clean sands, sand-gravel mix | 100–120 | 21–12 | 0.8 | 1.4 | 0 | 0 | 37 | 0.74 |
| SM | Silty sands, poorly graded sand-silt mix | 110–125 | 16–11 | 0.8 | 1.6 | 1050 | 420 | 34 | 0.67 |
| SM-SC | Sand-silt clay mix with slightly plastic fines | 110–130 | 15–11 | 0.8 | 1.4 | 1050 | 300 | 33 | 0.66 |
| SC | Clayey sands, poorly graded sand-clay mix | 105–125 | 19–11 | 1.1 | 2.2 | 1550 | 230 | 31 | 0.60 |
| ML | Inorganic silts and clayey silts | 95–120 | 24–12 | 0.9 | 1.7 | 1400 | 190 | 32 | 0.62 |
| ML-CL | Mixture of inorganic silt and clay | 100–120 | 22–12 | 1.0 | 2.2 | 1350 | 460 | 32 | 0.62 |
| CL | Inorganic clays of low to medium plasticity | 95–120 | 24–12 | 1.3 | 2.5 | 1800 | 270 | 28 | 0.54 |
| OL | Organic silts and silt-clays, low plasticity | 80–100 | 33–21 | . . . | . . . | . . . | . . . | . . . | . . . |
| MH | Inorganic clayey silts, elastic silts | 70–95 | 40–24 | 2.0 | 3.8 | 1500 | 420 | 25 | 0.47 |
| CH | Inorganic clays of high plasticity | 75–105 | 36–19 | 2.6 | 3.9 | 2150 | 230 | 19 | 0.35 |
| OH | Organic clays and silty clays | 65–100 | 45–21 | . . . | . . . | . . . | . . . | . . . | . . . |

*From NAVFAC Manual DM 7 (1971).[6] All properties are for condition of "standard Proctor" maximum density, except values of $k$ and CBR which are for "modified Proctor" maximum density. Typical strength characteristics are for effective strength envelopes and are obtained from USBR data. Compression values are for vertical loading with complete lateral confinement. ( . . .) Indicates insufficient data available for an estimate.

| Typical coefficient of permeability, ft/min | Range of CBR values | Range of subgrade modulus $k_s$, lb/in³ |
|---|---|---|
| $5 \times 10^{-2}$ | 40–80 | 300–500 |
| $10^{-1}$ | 30–60 | 250–400 |
| $>10^{-6}$ | 20–60 | 100–400 |
| $>10^{-7}$ | 20–40 | 100–300 |
| $>10^{-3}$ | 20–40 | 200–300 |
| $>10^{-3}$ | 10–40 | 200–300 |
| $5 \times 10^{-5}$ | 10–40 | 100–300 |
| $2 \times 10^{-6}$ | . . . | |
| $5 \times 10^{-7}$ | 5–20 | 100–300 |
| $10^{-5}$ | 15 or less | 100–200 |
| $5 \times 10^{-7}$ | . . . | |
| $10^{-7}$ | 15 or less | 50–200 |
| . . . | 5 or less | 50–100 |
| $5 \times 10^{-7}$ | 10 or less | 50–100 |
| $10^{-7}$ | 15 or less | 50–150 |
| . . . | 5 or less | 25–100 |

## 3.5 DEFORMATION WITHOUT RUPTURE

### 3.5.1 INTRODUCTION

#### Forms of Deformation

*Ideal Materials*

ELASTIC DEFORMATION Stress is directly proportional to strain; the material recovers all deformation upon removal of stress (Fig. 3.65*a*).

PLASTIC DEFORMATION Permanent and continuous deformation occurs when the applied stress reaches a characteristic stress level (Fig. 3.65*b*). Geologic materials often combine deformation modes under stress (Fig. 3.65*c*).

VISCOUS DEFORMATION Rate of deformation is roughly proportional to applied stress.

*Geologic Materials*

COMPRESSION Varies from essentially ideal to less than ideal, i.e., elastic compression under relatively low stress levels followed by plastic compression resulting from closure of soil voids or rock fractures. Some materials may deform only plastically, others may exhibit plastic-elastic-plastic deformation in stages of stress levels.

CREEP A time-dependent deformation at constant stress level below failure level.

EXPANSION An increase in volume from swelling, or elastic or plastic extension strain.

*Geotechnical Parameters Used to Define Deformations*

*Deformation moduli* are quantities expressing the measure of change in the form of dimensions of a body occurring in response to changing stress conditions. The quantities include the elastic moduli (static and dynamic), moduli from the stress-strain curve, compression moduli from pressure-meter testing, and the modulus of subgrade reaction.

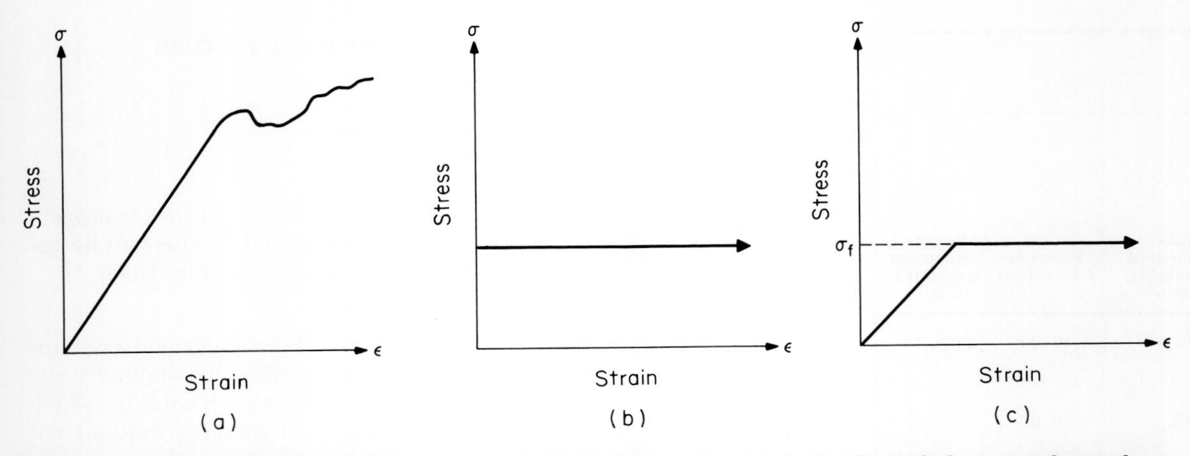


**FIG. 3.65 Stress-strain relationships of elastic and plastic deformation: (*a*) elastic; (*b*) rigid-plastic; (*c*) elastic-plastic.**

*Consolidation parameters* define various aspects of the slow process of compression under applied stress that occurs as water is extruded from the voids of clay soils.

### *Influencing Factors*

Magnitude and form of deformation in geologic materials are influenced by material properties; in situ stress conditions; level, direction, and rate of application of applied stress; temperature of environment; and time interval under stress.

### *Deformation in Practice*

*Foundations* undergo settlement or heave; bending occurs in mats, continuous footings, and piles from differential deflection.

*Flexible retaining structures* undergo bending; deflections may result in significant backslope subsidence.

*Tunnels,* particularly in rock masses, result in arching, which affects loads on support systems; in soils, deformations may result in ground surface subsidence.

*Pavements* undergo settlement or heave, and bending.

## Elastic Body Characteristics

### *Elastic Media*

In an elastic medium, strain is instantly and totally recoverable. Stress is directly proportional to strain as related by Hooke's law, expressed by *Young's modulus* E (or *modulus of elasticity*) as

$$E = \sigma/\epsilon \qquad \text{tsf, etc.} \qquad (3.49)$$

*Poisson's ratio* $\nu$ is the inverse ratio between strain in the direction of applied stress and the induced strain $\epsilon_L$ in a perpendicular direction:

$$\nu = \frac{\epsilon_L}{\epsilon} = \frac{\epsilon_L}{E\sigma} \qquad (3.50)$$

Young's modulus and Poisson's ratio are referred to as the *elastic constants.*

### *Dependent Factors for Elasticity*

For a body to exhibit elastic characteristics defined by only one value of E and of $\nu$ it must be isotropic, homogeneous, and continuous.

ISOTROPY Particles are oriented so that the ratio of stress to strain is the same regardless of the direction of applied stress; i.e., the elastic properties in every direction through any point are identical.

HOMOGENEITY The body has identical properties at every point in *identical* directions.

CONTINUITY Refers to structure; if a mass is continuous it is free of planes of weakness or breaks.

*Validity for Geologic Materials*

*Sound, intact, massive rock* approaches an elastic material under most stress levels prior to rupture.

*Most rocks* generally are to some extent anisotropic, nonhomogeneous, and discontinuous, and are termed as quasi-elastic, semi-elastic, or nonelastic. In *intact rock* specimens, deformation varies with the rock type as shown on Table 3.32 in regard to mineral hardness, grain bonding, and fabric *(see Art. 5.2)*. *Nonintact rock* or rock masses are basically discontinuous, usually undergoing plastic deformation as fractures

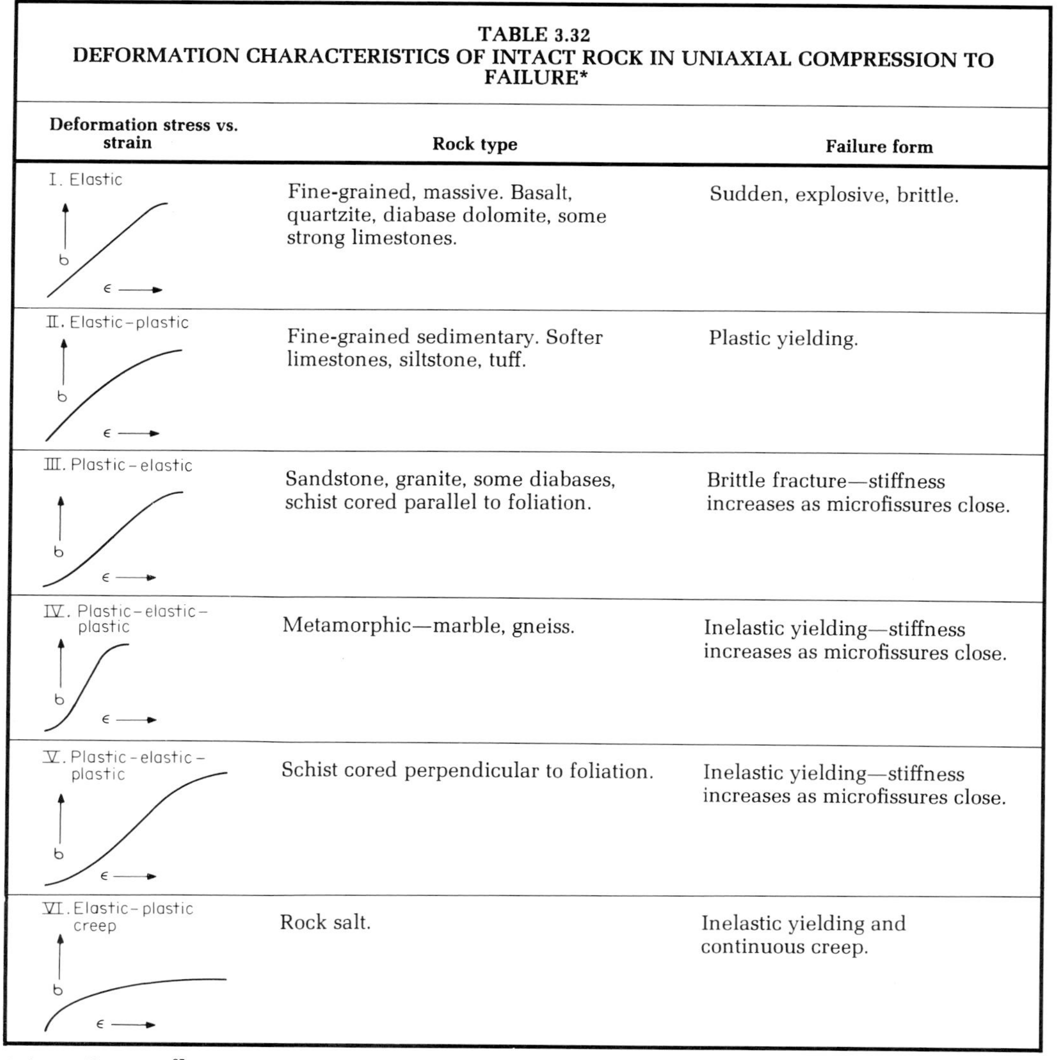

**TABLE 3.32**
**DEFORMATION CHARACTERISTICS OF INTACT ROCK IN UNIAXIAL COMPRESSION TO FAILURE***

| Deformation stress vs. strain | Rock type | Failure form |
|---|---|---|
| I. Elastic | Fine-grained, massive. Basalt, quartzite, diabase dolomite, some strong limestones. | Sudden, explosive, brittle. |
| II. Elastic–plastic | Fine-grained sedimentary. Softer limestones, siltstone, tuff. | Plastic yielding. |
| III. Plastic–elastic | Sandstone, granite, some diabases, schist cored parallel to foliation. | Brittle fracture—stiffness increases as microfissures close. |
| IV. Plastic–elastic–plastic | Metamorphic—marble, gneiss. | Inelastic yielding—stiffness increases as microfissures close. |
| V. Plastic–elastic–plastic | Schist cored perpendicular to foliation. | Inelastic yielding—stiffness increases as microfissures close. |
| VI. Elastic–plastic creep | Rock salt. | Inelastic yielding and continuous creep. |

*After Miller (1965).[85]

**TABLE 3.33**
**TYPICAL RANGES FOR ELASTIC CONSTANTS OF VARIOUS MATERIALS***

| Material | Young's modulus $E_s$,† kg/cm² | Poisson's ratio $\nu$‡ |
|---|---|---|
| SOILS | | |
| Clay: | | |
| Soft sensitive | 20–40 ($500s_u$) | |
| Firm to stiff | 40–80 ($1000s_u$) | 0.4–0.5 |
| Very stiff | 80–200 ($1500s_u$) | (undrained) |
| Loess | 150–600 | 0.1–0.3 |
| Silt | 20–200 | 0.3–0.35 |
| Fine sand: | | |
| Loose | 80–120 | |
| Medium dense | 120–200 | 0.25 |
| Dense | 200–300 | |
| Sand: | | |
| Loose | 100–300 | 0.2–0.35 |
| Medium dense | 300–500 | |
| Dense | 500–800 | 0.3–0.4 |
| Gravel: | | |
| Loose | 300–800 | |
| Medium dense | 800–1000 | |
| Dense | 1000–2000 | |
| ROCKS | | |
| Sound, intact igneous and metamorphics | $6\text{–}10 \times 10^5$ | 0.25–0.33 |
| Sound, intact sandstone and limestone | $4\text{–}8 \times 10^5$ | 0.25–0.33 |
| Sound, intact shale | $1\text{–}4 \times 10^5$ | 0.25–0.30 |
| Coal | $1\text{–}2 \times 10^5$ | |
| OTHER MATERIALS | | |
| Wood | $1.2\text{–}1.5 \times 10^5$ | |
| Concrete | $2\text{–}3 \times 10^5$ | 0.15–0.25 |
| Ice | $7 \times 10^5$ | 0.36 |
| Steel | $21 \times 10^5$ | 0.28–0.29 |

*After CGS (1978)[116] and Lambe and Whitman (1969).[30]

†$E_s$ (soil) usually taken as secant modulus between a deviator stress of 0 and ½ to ⅓ peak deviator stress in the triaxial test (Fig. 3.69) [Lambe and Whitman (1969)[30]]. $E_r$ (rock) usually taken as the initial tangent modulus [Farmer (1968)[86a]] *(Art. 3.5.2)*. $E_s$ (clays) is the slope of the consolidation curve when plotted on a linear $\Delta h/h$-vs.-$p$ plot [CGS (1978)[116]]. Values for $E_r$ given also on Table 3.26.

‡Poisson's ratio for soils is evaluated from the ratio of lateral strain to axial strain during a triaxial compression test with axial loading. Its value varies with the strain level and becomes constant only at large strains in the failure range [Lambe and Whitman (1969)[30]]. It is generally more constant under cyclic loading; cohesionless soils range from 0.25–0.35 and cohesive soils from 0.4–0.5.

close, then elastic deformation, often followed by plastic deformation.

*Creep* deformation occurs in some rocks over long time intervals at stress levels substantially less than those required to cause short-term deformation or failure. In rock masses the problem has most practical significance in soft rocks such as halite or in overstressed rocks (high residual stresses), where relaxation occurs along joints [Hendron (1969)[86]].

*Soils* are essentially nonelastic, usually plastic, and occasionally viscous. They demonstrate pseudo-elastic properties under low stress levels, as evidenced by initial stress-strain linearity. "Elastic" deformation, however, is immediate deformation, and in many soil types does not account for the total deformation occurring over long time intervals because of consolidation.

*Elastic Constant Values*

Typical values for the elastic constants for a variety of materials are given on Table 3.33. Typical value ranges for Young's modulus for various rock types are also given on Table 3.26.

*Poisson's ratio* for soils is evaluated from the ratio of lateral strain to axial strain during a triaxial compression test with axial loading. Its value varies with the strain level and becomes constant only at large strains in the failure range [Lambe and Whitman (1969)[30]]. It is generally more constant under cyclic loading: in cohesionless soils it ranges from 0.25 to 0.35 and in cohesive soils from 0.4 to 0.5. Values above 0.5 can indicate dilatant material; i.e., the lateral strain under applied vertical stress can exceed one-half the vertical strain.

### Deformation Analyses

*Induced Stresses*

Evaluation of deformation under imposed loads requires the determination of the magnitude of stress increase (or decrease) within some depth below a loaded (or unloaded) area with respect to the existing in situ stress conditions. The determination of these stresses is based on elastic theory.

THEORY Equations were presented by Boussinesq in 1885 expressing stress components, caused by a perpendicular point surface load, at points within an elastic, isotropic, homogeneous mass extending infinitely in all directions below a horizontal surface, as shown on Fig. 3.66. The stresses are directly proportional to the applied load and inversely proportional to the depth squared (a marked decrease with depth).

*Distributions* of stresses beneath a uniformly loaded circular area are given on Fig. 3.67 and have been determined for other shapes, including square and rectangular areas, and uniform and triangular loadings, for example [Lambe and Whitman (1969)[30]]. Distributions vary from the Boussinesq with thinly layered soil formations [Westergaard distribution, Taylor (1948),[87] for example], layered soil formations [strong

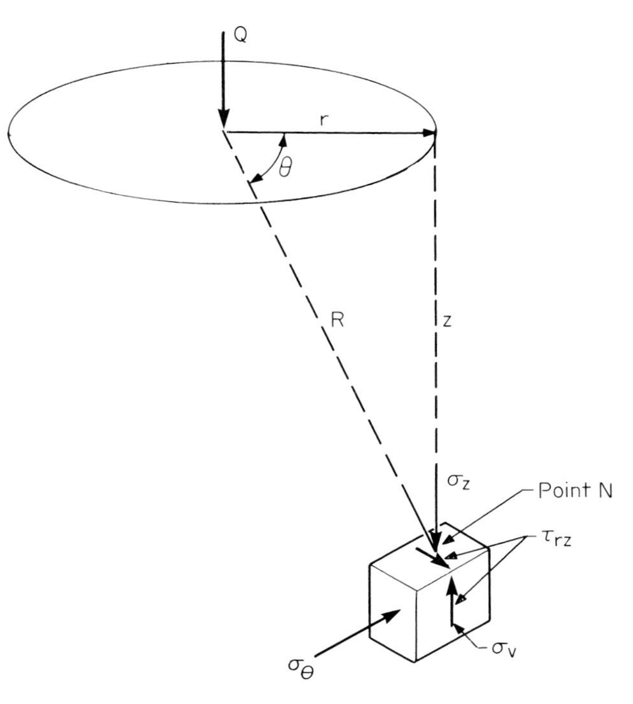


$$\sigma_z = \frac{Q}{z^2}\frac{3/2\pi}{[1 + (r/z)^2]^{5/2}}\,;\; N_B = \frac{3/2\pi}{[1 + (r/z)^2]^{5/2}}\,;\; \sigma_z = \frac{Q}{z^2}N_B$$

**FIG. 3.66 Stresses in an elastic medium caused by a surface, vertical point load.**

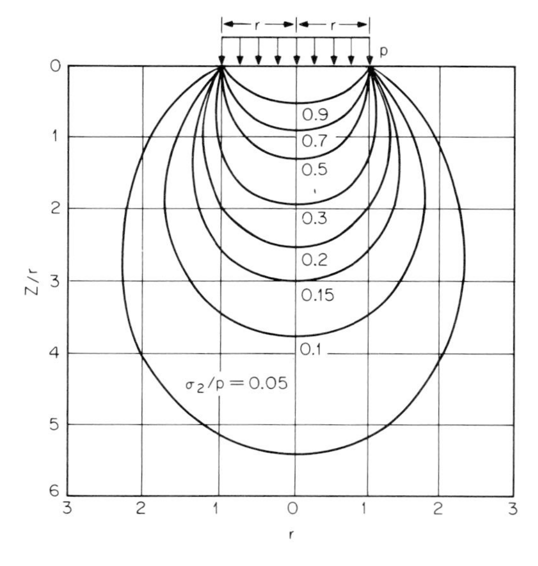


**FIG. 3.67 The "pressure bulb" of stress distribution: contours of vertical normal stress beneath a uniformly loaded circular area on a linear elastic half space.**

over weak layer, Burmister (1962*b*),[88] or rigid base under weak layer, Burmister (1956)[89]], and jointed rock masses [Gaziev and Erlikhman (1971)[90]].

VALIDITY Although geologic materials are essentially nonelastic, the theory is considered applicable for relatively low stress levels, although magnitude and distribution of stresses are very much affected by soil layering and fracture patterns in rock masses.

*Computing Deformations*

FROM MODULI In general, settlement may be expressed crudely in terms of elastic moduli as

$$S = \Delta H = \Delta p \frac{H}{E} \quad \text{in, cm} \qquad (3.51)$$

where $\Delta p$ = pressure increase over effective overburden pressure at the midpoint of the "pressure bulb" (zone of influence, or stress distribution zone beneath the loaded area as shown on Fig. 3.67)

$H$ = depth of pressure bulb to some arbitrary boundary such as $\sigma_z/p$ = 0.05 (insignificant stress), or the stratum thickness, if less

$E$ = $\sigma/\epsilon$ as determined by in situ or laboratory testing $\epsilon = \Delta L/L$; in in situ testing $\Delta L$ is measured but $L$ is not known; it is estimated, on the basis of Boussinesq stresses, as the depth to which the significant stress occurs

NOTE: If $\Delta p$ is applied over a large area compared to the stratum thickness, the constrained modulus ($D$) is used to account for the effect of confining pressures. (Fig. 3.68).

FROM CONSOLIDATION TEST DATA (*See Art. 3.5.4*) Settlement ($S$ or $\rho$) is expressed as:

$$S = \Delta H = \frac{\Delta e}{1 + e_0} H \quad \text{in, cm} \qquad (3.52)$$

where $H$ = stratum thickness

$\Delta e$ = the change in void ratio resulting from an increase in pressure from effective overburden pressure $\bar{p}_o$, to the pressure imposed on the stratum at middepth by the foundation load $p_f$.

$e_0$ = the initial void ratio corrected for the slight expansion occurring during sampling and extrusion in the laboratory.

NOTE: Other than this simplified introduction, analysis of deformations is beyond the scope of this book.

## 3.5.2 DEFORMATION RELATIONSHIPS

### Deformation Moduli

*General*

Deformation moduli are expressed in a number of ways which relate to either the form of deformation or the stress-strain relationships obtained from a particular test method as summarized on Table 3.34.

*Elastic Moduli*

Young's modulus, shear modulus, bulk modulus, and constrained modulus are defined on Fig. 3.68.

*Dynamic Elastic Moduli*

Compression and shear wave velocities of sonic waves (*see Art. 2.3.2*) are functions of the elastic properties and mass density of the transmitting medium as shown on Table 3.35. Measurements of $V_p$ and $V_s$, therefore, provide the basis for computing the dynamic properties of Poisson's ratio, Young's modulus, shear modulus, and bulk modulus as given on Table 3.35.

*Moduli from the Stress-Strain Curve*

Derived from the triaxial compression or uniaxial compression test, these moduli include the initial tangent modulus, tangent modulus, and secant modulus defined on Fig. 3.69. The initial tangent modulus $E_i$ is usually reported as the elastic modulus for rock (given as $E_r$), whereas the secant modulus $E_{se}$ is usually reported as the elastic modulus for soils, $E_s$, to be used for calculating deformations. The secant modulus $E_{se}$ is used also for defining a stress or limiting deformation.

**TABLE 3.34**
**PARAMETERS OF DEFORMATION**

| Parameter* | Reference | Definition | Expression | Normal application |
|---|---|---|---|---|
| | | ELASTIC MODULI (STATIC) | | |
| Young's modulus | Fig. 3.68 | Relates stress to strain. | $E = \sigma/\epsilon$ (3.49) | Rock masses, sands, strong granular cohesive soils. |
| Shear modulus | Fig. 3.68 | Relates shear strain to shear force (modulus of rigidity). | $G = \frac{E}{2(1+\nu)}$ (3.53) | Not commonly used statically. |
| Bulk modulus | Fig. 3.68 | Ratio of all-around pressure to change in volume per unit volume (modulus describing incompressibility). | $B = \frac{E}{3(1-2\nu)}$ (3.54) | Not commonly used statically. |
| Constrained modulus | Fig. 3.68 | Deformation occurring in confined compression. | $D = \frac{E(1-\nu)}{(1+\nu)(1-2\nu)}$ (3.55) | Structure of large areal extent underlain by relatively thin compressible soil deposit. |
| | | DYNAMIC ELASTIC PARAMETERS—SEE TABLE 3.35 | | |
| | | MODULI FROM THE STRESS-STRAIN CURVE | | |
| Initial tangent modulus | Fig. 3.69 | Initial portion of curve. | $E_i = \sigma/\epsilon$ (3.56) | Usually taken as $E$ for geologic materials. |
| Secant modulus | Fig. 3.69 | $\sigma$ and $\epsilon$ taken between two particular points. | $E_{se} = \frac{\Delta\sigma}{\Delta\epsilon}$ (3.57) | Define $E$ for a particular stress limit. |
| Tangent modulus | Fig. 3.69 | Modulus at specific point. Often taken at point of maximum curvature before rupture. | $E_t = \frac{d\sigma}{d\epsilon}$ (3.58) | The lowest value for $E$ usually reported. |
| | | OTHER MODULI | | |
| Compression modulus | Fig. 3.93 | Lateral deformation caused by applied stress from pressuremeter. | $E_c = E\alpha$ (3.59) | In situ measures of $E$ for materials difficult to sample. |
| Modulus of subgrade reaction | Fig. 3.96 | A unit of pressure to produce a unit of deflection. | $k_s = p/y$ kg/cm$^3$ (3.60) | Beam or plate on an elastic subgrade problems. |

*Unless noted, units are tsf, kg/cm$^2$, kN/m$^2$.

### *Other Moduli*

These include the *compression modulus*, as determined by pressuremeter testing, which varies in its relationship to Young's modulus with soil type and the *modulus of subgrade reaction*, measured by plate load test.

## Plastic Deformation

### *Consolidation in Clay Soils*

Compression in clay soils is essentially plastic deformation defined in terms of consolidation, i.e., the decrease in volume occurring under applied stress caused primarily by expulsion of

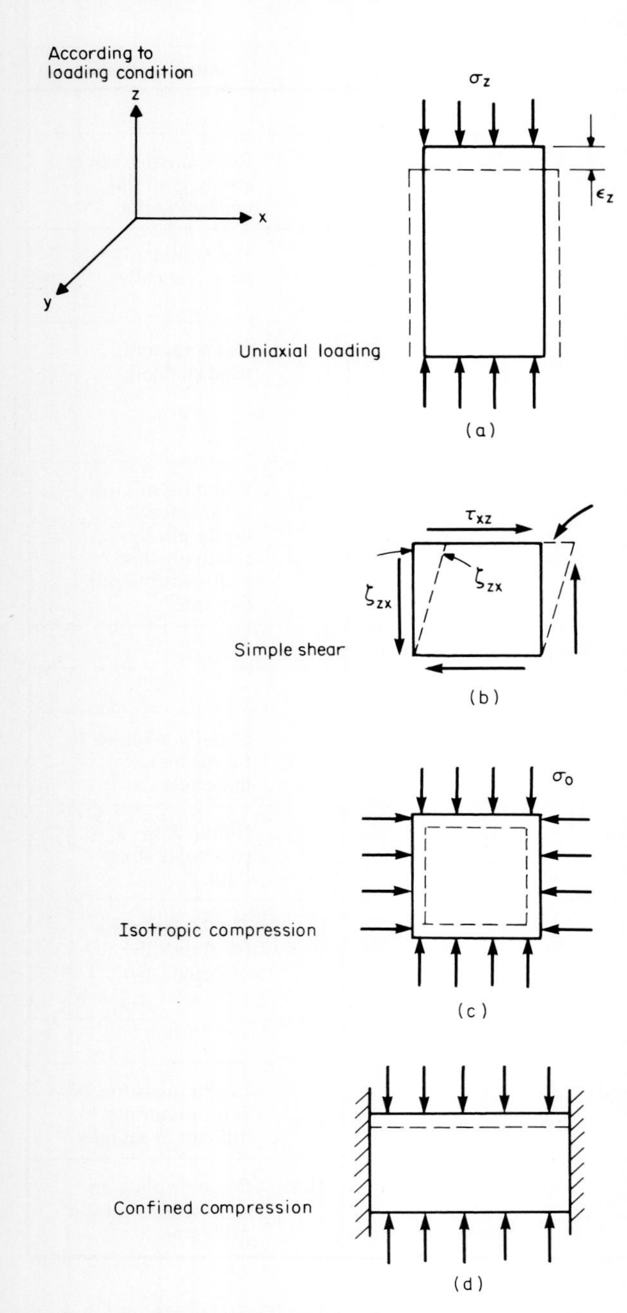


**FIG. 3.68 Various types of elastic moduli: (*a*) Young's modulus—$E = \sigma_z/\epsilon_z = \sigma/\epsilon$; $\epsilon_x = \epsilon_y = -\nu\epsilon_z$; (*b*) shear modulus—$G = \tau_{zx}/\zeta_{zx} = E/2(1 + \nu)$; (*c*) bulk modulus—$B = \sigma_o/3\epsilon_z = E/3(1 - 2\nu)$; (*d*) constrained modulus—$D = \sigma_z/\epsilon_z = E(1 - \nu)/(1 + \nu)(1 - 2\nu)$.** [*From Lambe and Whitman (1969).*[30] *Reprinted by permission of John Wiley & Sons, Inc.*]

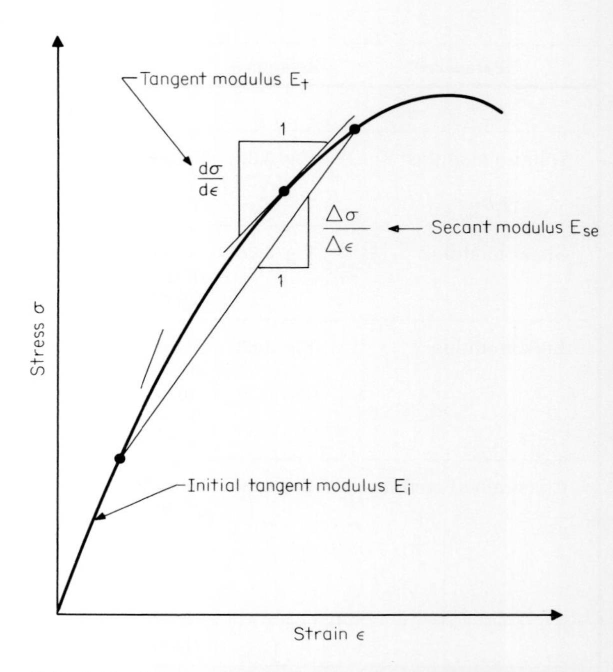


**FIG. 3.69 Forms of moduli from the stress-strain curve.**

water from interstices *(see Art. 3.5.4)*. Most soils exhibit primary and secondary consolidation, and in some soils the latter phenomenon, which is not well understood, can be of very significant magnitude. A substantial time delay in compression occurs in clay soils under a given applied stress which increases generally as the plasticity of the clay increases.

STRESS-STRAIN RELATIONSHIPS When a series of increasing loads is applied to a clay soil, the amount of compression occurring up to the magnitude of the maximum past pressure or preconsolidation pressure $p_c$ is relatively small to negligible. Stresses exceeding $p_c$ enter the "virgin" portion of a stress-strain curve, and substantially greater compression occurs. It may be expressed by the *compression index* $C_c$, the slope of the virgin compression or void ratio vs. $\log_{10}$ time curve expressed as

$$C_c = \frac{e_1 - e_2}{\log (p_1/p_2)} \tag{3.66}$$

$C_c$ is useful for correlating data for normally consolidated clays of low to moderate sensitivity, or for computing settlements when data from

| TABLE 3.35 DYNAMIC ELASTIC PARAMETERS | |
|---|---|
| **Parameter** | **Expression** |
| Compression-wave velocity | $V_p = \{[K + (4/3)G]/\rho\}^{1/2}$ m/s *(Art. 2.3.2)* (3.61) |
| Shear-wave velocity | $V_s = (G/\rho)^{1/2}$ m/s *(Art. 2.3.2)* (3.62) |
| Mass density of materials | $\rho = \gamma/g$ kg/m$^3$ (Determined by gamma probe, *Art. 2.3.6*) |
| Dynamic Poisson's ratio | $\nu = (V_p^2/2V_s^2 - 1)/(V_p^2/V_s^2 - 1)$ *Appropriate Values* Igneous rocks—0.25 Sedimentary rocks—0.33 Soils—see *Art. 11.3.2* |
| Dynamic Young's modulus* | $E_d = \rho(3V_p^2 - 4V_s^2)/(V_p^2/V_s^2 - 1)$ or $E_d = 2\rho V_s^2(1 + \mu)$ (3.63) |
| Dynamic shear modulus* | $G_d = \rho V_s^2 = E_d/2(1 + \nu)$ (3.64) |
| Dynamic bulk modulus* | $K = \rho(V_p^2 - 4V_s^2/3) = E_d/3(1 - 2\nu)$ (3.65) |

*Units are pascals or tsf or kg/cm$^2$.

a number of consolidation tests are available. It can be estimated from the expression [Terzaghi and Peck (1967)[21]]:

$$C_c \approx 0.009(\text{LL} - 10\%) \qquad (3.67)$$

*Overconsolidation ratio* is defined as the ratio of the maximum past pressure $p_c$ to the existing effective overburden pressure $p_o'$. In general, for an NC clay, $p_c/p_o' = 0.8$ to 1.5 and for an OC clay, $p_c/p_o'$ is greater than 1.5 [Clemence and Finbarr (1980)[91]]. OCR is used to estimate consolidation in clays [Schmertmann (1977)[76]], for correlation of strength properties [Terzaghi and Peck (1967)[21]], and for estimating $K_0$ in terms of PI [Brooker and Ireland (1965)[92]]. $K_0$ as a function of OCR and PI is given on Fig. 3.70.

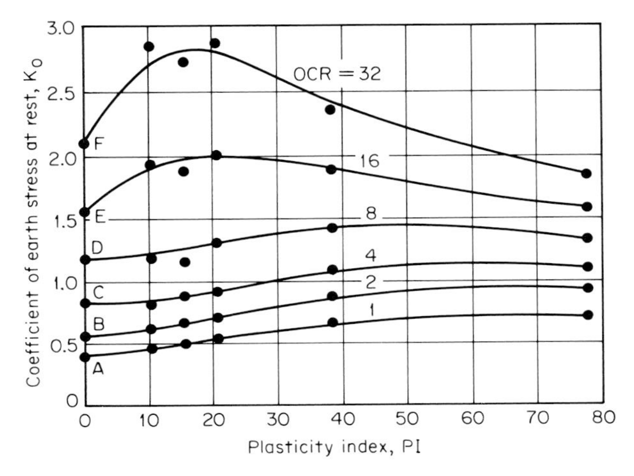


**FIG. 3.70 Relationship between $K_0$, OCR, and PI.** [*From Brooker and Ireland (1965).[92] Reprinted with permission of the National Research Council of Canada.*]

The undrained shear strength $s_u$ of NC clay normally falls within a limited fraction of the effective overburden stress, usually ranging from $s_u/p' = 0.16$ to 0.4 (Art. 3.4.2), and for estimating can be taken as an average value of

$$(s_u/p')_{NC} = 0.33 \qquad (3.68)$$

or in terms of PI as[21]

$$s_u/p' = 0.11 + 0.0037\text{PI} \qquad (3.69)$$

Schmertmann (1977)[76] has presented a normalized ratio of $s_u/p'$ as measured by the cone penetrometer test [Eq. (3.48)] to $(s_u/p')_{NC}$ in terms of the overconsolidation ratio, given on Fig. 3.71. It provides an approximation of OCR, but variations with clay types will occur. For example, OCR for fresh water montmorillonite clays will be overestimated and OCR for "quick" clay will be underestimated.

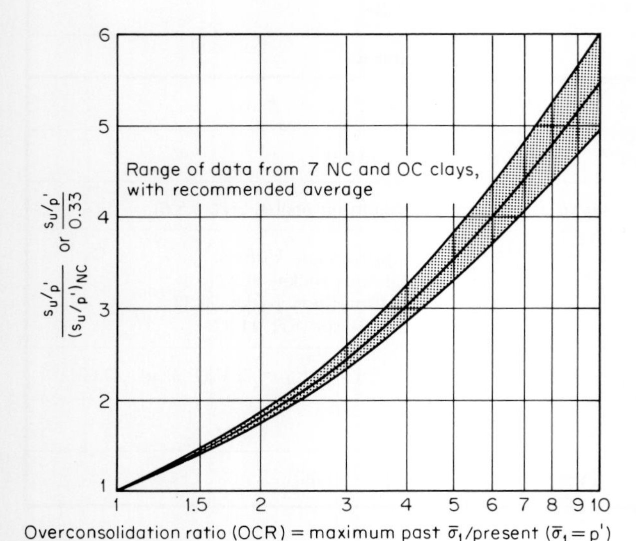


**FIG. 3.71 Normalized $s_u/p'$ ratio vs. OCR for use in estimating OCR from $q_c$ in clays.** [*From Schmertmann (1977).*[76] *Reprinted with permission of the Federal Highway Administration.*]

### *Compression in Sands*

Sands and other cohesionless granular materials undergo a decrease in void volume under applied stress, caused primarily by rearrangement of grains (Art. 3.5.4). Small elastic compression of quartz grains may occur. In most cases the greater portion of compression is essentially immediate upon application of load.

### *Expansion*

An increase in volume occurs as a result of reduction in applied stress, increase in moisture content, or mineralogical changes in certain soil and rock materials *(see Arts. 3.5.4 and 10.6).*

## 3.5.3 ROCK DEFORMATION MEASUREMENTS

### **Methods Summarized**

#### *Laboratory Testing*

Intact specimens are statically tested in the laboratory in the triaxial and unconfined compression apparatus; dynamic properties are measured with the resonant column device or by ultrasonic testing (ASTM 2845). A summary of parameters measured, apparatus description, and test performance is given on Table 3.36.

The data are normally used for correlations with in situ test data.

#### *In Situ Testing*

IMPORTANCE In situ testing provides the most reliable data on the deformation characteristics of rock masses because of the usual necessity to account for the effects of mass defects from discontinuities and decomposition.

REQUIREMENTS Determination of moduli in situ requires that the deformation and the stress producing it are measurable and that an analytical method of describing the geometry of the stress deformation relationship is available.

*Analytical methods* are governed by the testing method. Modulus is the ratio of stress to strain, and since strain is the change in length per total length, the deflection that is measured during in situ testing must be related to the depth of the stressed zone to determine strain. The depth of the stressed zone may be determined by instrumentation *(see Chap. 4)*, or the Boussinesq equations may be used to determine stress distributions. The values for the modulus $E$ are given in terms of the test geometry, the applied pressure, the deflection, and Poisson's ratio.

STATIC MODULI Determined from plate-jack tests, radial jacking and pressure tunnel tests, flat-jack tests, borehole tests (dilatometer and Goodman jack), and triaxial compression tests.

DYNAMIC MODULI Determined from seismic direct velocity tests *(see Art. 2.3.2)* and the 3-D velocity probe (sonic logger) *(see Art. 2.3.6).* Relationships between seismic velocities and dynamic moduli are given on Table 3.35. In moduli computations the shear-wave velocity $V_s$ is used rather than the compression-wave velocity $V_p$ because water in rock fractures does not affect $V_s$, whereas it couples the seismic energy across joint openings, allowing much shorter travel times for $P$ waves than if an air gap existed. Dynamic moduli are always higher than static moduli because the seismic pulse is of short duration and very low stress level,

**TABLE 3.36**
**LABORATORY TESTS TO DETERMINE DEFORMATION MODULI OF INTACT ROCK**

| Test method | Parameters measured | Apparatus | Performance | Comments |
|---|---|---|---|---|
| | | STATIC MODULI | | |
| Triaxial compression apparatus | $E_i$, $E_{se}$, $E_t$ | See Fig. 3.47 and Art. 3.4.4. Rock testing equipment similar but larger. USBR machine can test cores of $D = 15$ cm and $L = 30$ cm under 800 tons axial load and 6000 kg/cm$^2$ lateral pressure. | Specimen subjected to confining pressure and loaded axially to failure to obtain a stress-strain curve. Test repeated at various confining pressures to bracket in situ lateral pressure. Strain rate can be varied to suit field loading conditions. | Modulus values controlled by orientation of applied load in rocks with bedding planes, foliations, schistosity, as well as major joints. |
| Unconfined compression | $E_i$, $E_{se}$, $E_t$ | See Art. 3.4.4. Can use concrete testing apparatus. | Specimen loaded to failure as stress-strain curve is recorded. | Yields very conservative values for most practical uses. Often directly applicable for mine pillars. |
| | | DYNAMIC MODULI | | |
| Resonant column device (sonic column) | $E_d$, $G_d$ | Core specimen held fixed at the ends and subjected to vibrations while axial load applied equivalent to overburden. (See also ASTM D2845.) | Specimen subjected first to vibrations in the torsional mode, then in the longitudinal mode. End displacement is monitored for various frequencies. | From specimen length and end displacement as a function of frequency, $V_s$ is computed from the torsional test, and $V_p$ ("rod" compression-wave velocity) from the longitudinal test. |

although the ratio of $E_{static}$ to $E_d$ will normally approach unity as rock-mass quality approaches sound, intact rock. $E_d$ as determined from field testing is often referred to as $E_{seis}$ and is correlated with other field and laboratory data to obtain a design modulus, as will be discussed.

### Plate-Jack Test (In Situ Compression Test)

*Performance*

A load is applied with hydraulic jacks to a plate in contact with the rock mass using the roof of an adit as a reaction, as illustrated in Fig. 3.72.

When the tests are set up, areas representative of rock-mass conditions are selected, and the size of the loaded area is scaled to the structural elements of the rock. (Depending on mass-defect spacings, the larger the loaded area the more representative are the results.) Borehole extensometers *(see Art. 4.3.4)* are grouted into the rock mass for measurements of strain vs. depth beyond the upper and lower plates, and concrete facing is poured over the rock mass to provide flat bearing surfaces. As load is applied, the extensometers sense compression strains which are recorded on electronic readouts.

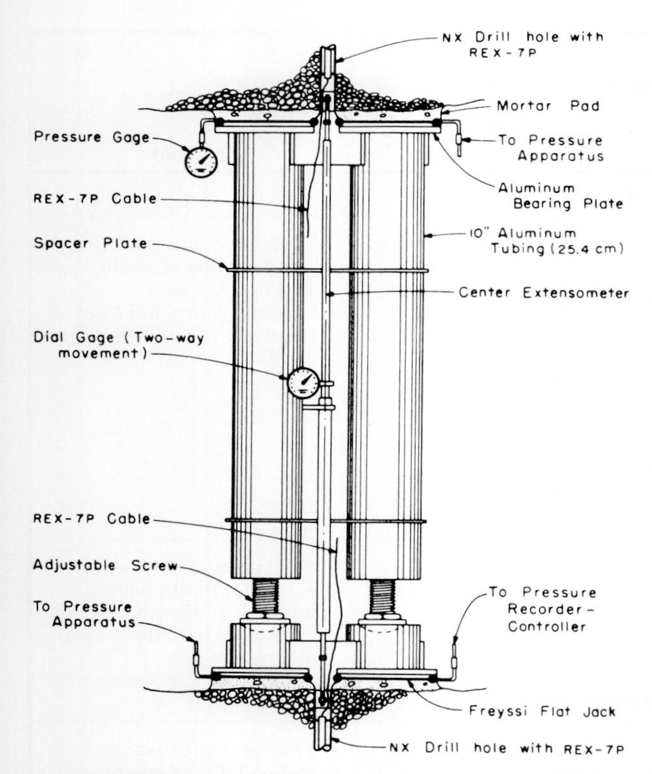


**FIG. 3.72 Arrangement for uniaxial jacking test in an adit.** [*From Wallace et al. (1970).[93] Reprinted with permission of the American Society for Testing and Materials.*]

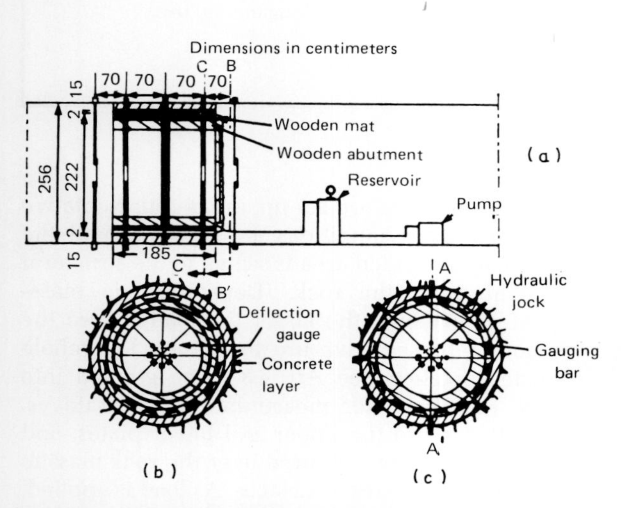


**FIG. 3.73 The radial jacking test arrangement.** [*From Stagg (1969).[96] Reprinted by permission of John Wiley & Sons, Inc.*]

*Computing Moduli*

Theoretical basis is the Boussinesq solution. For Young's modulus $E_r$, surface displacements y are related to the applied load P [Jaeger (1972)[94]]:

$$y = \frac{P(1 - \nu)^2}{\pi E_r r} \tag{3.70}$$

where r = plate radius.

Moduli $E_i$, $E_{se}$, and $E_t$ are determined from the stress-strain curve. A *recovery modulus* $E_{sr}$ can be taken from the unload portion of the curve to provide a measure of rock elasticity. If $E_t = E_{sr}$ the material is perfectly elastic. A value of $E_{se}$ greatly lower than $E_t$ usually indicates the closure of fractures and plastic deformation.

*Test Limitations*

Rock-mass response is affected by disturbance during excavation and surface preparation for testing and by the relatively small stressed zone, often to the degree that results are not representative of response under construction loading, although the results will be usually conservative [Rocha (1970)[95]].

## Radial Jacking and Pressure Tunnel Tests

*Performance*

RADIAL JACKING TEST Pressure to cause deformation of the wall of an adit excavated into the rock mass is applied mechanically as illustrated on Fig. 3.73. Displacements are measured with extensometers or some other strain-measuring devices.

PRESSURE TUNNEL TEST A portion of the adit is sealed with concrete and water pumped in under high pressure to cause deformation (Fig. 3.74). Displacements are measured with extensometers or some other strain-measuring device. Flow into the rock mass may cause large errors from pressure drop.

*Computing Moduli*

Analysis is based on the Boussinesq stress distribution and theories of stresses and deflections about a circular hole, or a hole cut in a plate [Wallace et al. (1970)[93] and Misterek (1970)[97]].

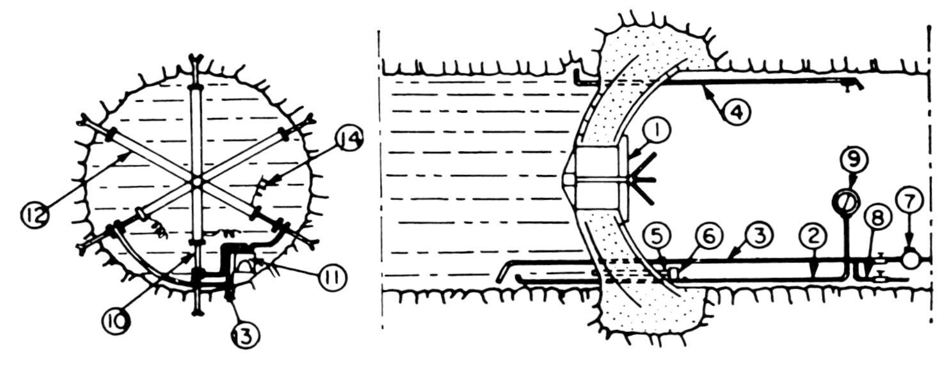


| | |
|---|---|
| 1 Manhole | 8 Water outlet |
| 2 Pressure gauge tube | 9 Pressure gauge |
| 3 Water inlet | 10 Vibrating wire meter |
| 4 Air outlet | 11 Air-pressure equalizing chamber |
| 5 Cable admission tube | 12 Invar rods |
| 6 Cable tube seal | 13 Air-pressure equalizing tube |
| 7 Water meter | 14 Cable |

**FIG. 3.74 Pressure-tunnel test in a rock gallery.** [*From Stagg (1969).*[96] *Reprinted by permission of John Wiley & Sons, Inc.*]

*Comments*

Although both tests are relatively costly to perform, larger areas are stressed than is possible with plate-jack tests, which allows a better assessment of rock-mass deformation. Measurements of differential wall movements permit assessment of the anisotropic properties of the rock mass. Excavation, however, results in straining of the rock mass, which affects its properties.

### Flat-Jack Tests

*Performance*

A slot is cut into rock with a circular diamond saw, or by a series of line-drilled holes. A steel flat jack, or Freyssinet jack (Fig. 3.75) is grouted into the slot to ensure uniform contact with the rock face. Jacks measuring 1 m in width by 1.25 m in depth and capable of applying pressures to 100 kg/cm$^2$ have been used.[95]

Strain meters measure the increase in distance between points on opposite sides of the slot as pressure is applied. *(See also Art. 4.4.4.)*

Jacks at different orientations provide a measure of rock anisotropy.

*Computing Moduli*

The increase in distance between points in the rock arranged symmetrically on opposite sides of the slot can be related to the pressure in the jack by elastic theory. $E_r$ is determined from

$$E_r = c\frac{P}{\delta} \tag{3.71}$$

where $P$ = pressure in flat jack
$c$ = a laboratory constant for the jack and geometry of the test
$\delta$ = the measured increase in distance between points

The constant $c$ depends on the position, shape, and size of the area under pressure as well as the location of the measured points, and has the dimension of length. It may be determined by model tests in large plaster blocks for a given jack and strain gauge arrangement.

### Borehole Tests

*Apparatus*

The *dilatometer* (Fig. 3.76) and the *Goodman jack* [Goodman et al. (1968)[98]], devices similar to the pressuremeter *(see Art. 3.5.4)*, are lowered into boreholes to measure moduli in situ.

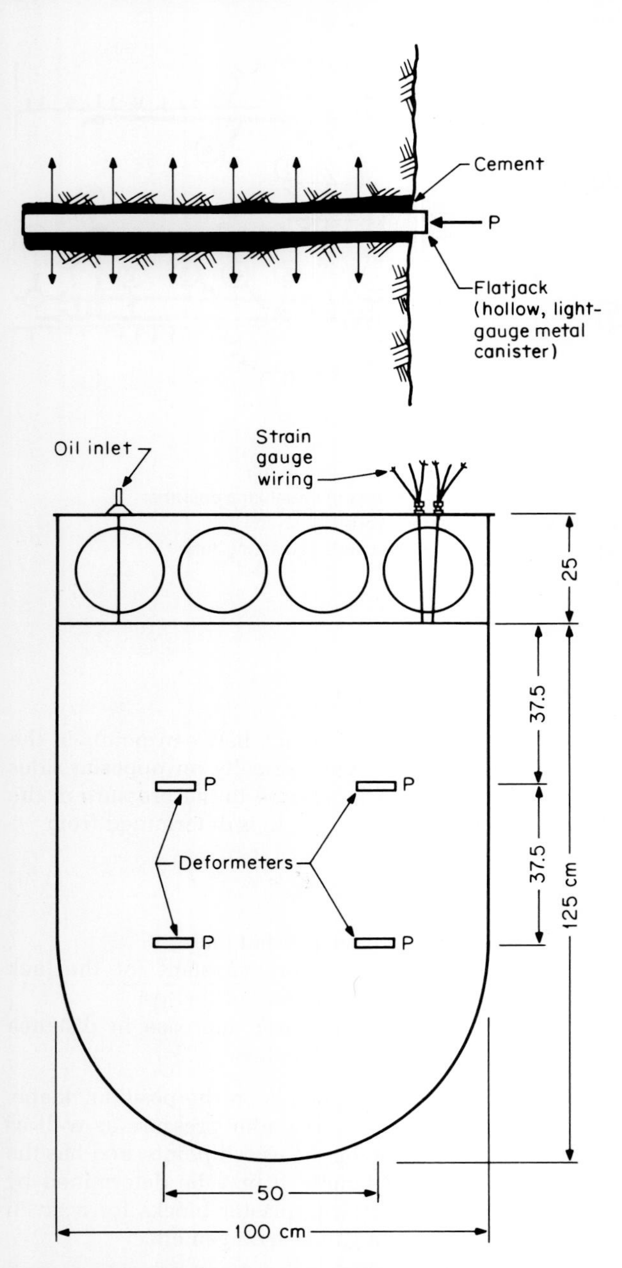


**FIG. 3.75 The flat jack or Freyssinet jack used in slots sawed into rock.** [*From Rocha (1970).*[95] *Reprinted with permission of the American Society for Testing and Materials.*]

Hydraulic pressure is applied between a metal jacket and a deformable rubber or metallic jacket in the apparatus which presses against the borehole walls. Linear variable differential transducers (LDVTs) in the instrument allow measurement of four diameters 45° apart to account for lateral rock anisotropy.

Designed to operate underwater in NX-size boreholes, the dilatometer has been used to depths of 200 m, exerting radial pressures[95] to 150 kg/cm$^2$, and the 12-piston model of the Goodman jack, manufactured by Sinco, can apply and maintain pressures up to approximately 10,000 psi (680 kg/cm$^2$).

*Computing Moduli*

Rocha (1970)[95] gives the following expression for computing $E_r$:

$$E_r = 2r \frac{1 - \nu}{\Delta} P \qquad (3.72)$$

where $2r$ = hole diameter
$\nu$ = Poisson's ratio
$\Delta$ = deformation when pressure $P$ is applied

**Methods Compared**

A series of tests was performed in jointed granite for the Tokyo Electric Power Company using a number of methods [Hibino et al. (1977)[99]] including in situ triaxial testing, and the results were compared.

The *in situ triaxial compression test* was performed on a 1-m$^3$ block of rock carved out of the rock mass but with its base remaining intact with the mass. The block was surrounded with 1500-ton-capacity flat jacks to provide confinement, the space between the rock and the jacks was backfilled with concrete, and loads were applied to the rock by the jacks.

*Modulus values* measured by the various tests are summarized on Fig. 3.77. A wide range in values is apparent. The largest values were obtained from laboratory sonic column and uniaxial compression tests, which would not be expected to be representative of mass conditions. The larger plate jack test resulted in less deflection (higher modulus) than did the smaller plate and would be considered as more repre-

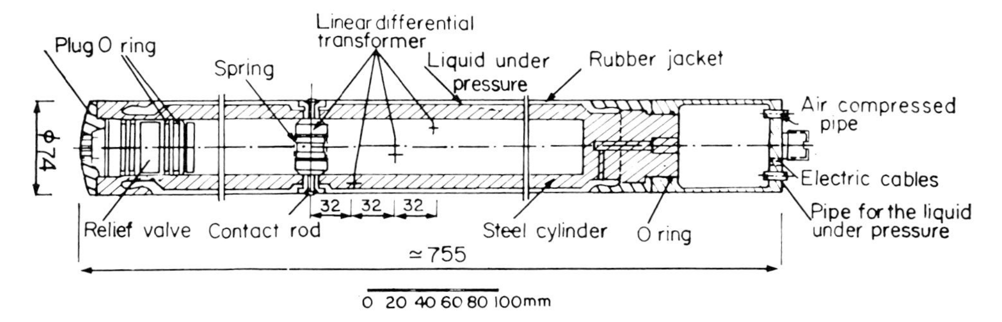


**FIG. 3.76 The dilatometer developed by Laboratorio Nacional de Engenharia Civil (LENC) of Lisbon.** [*From Stagg (1969).*[96] *Reprinted by permission of John Wiley & Sons, Inc.*]

sentative of deformation characteristics, as would the water chamber test.

## Modulus and Rock-Quality Relationships

*Selection of Design Moduli*

Deformation is related to rock-mass quality, which because of fabric, discontinuities, and decomposition can be extremely variable. Most practical problems involve poor-quality rock, which is the most difficult to assess.

The various test methods, in general, individually often do not yield representative values for $E_r$. All static methods disturb the rock surface during test setup, and most stress a relatively small area. Dynamic tests apply short-duration pulses at very low stress levels. Laboratory tests are performed on relatively intact specimens, seldom providing a representative model except for high-quality rock approaching an elastic material.

Various investigators have studied relationships

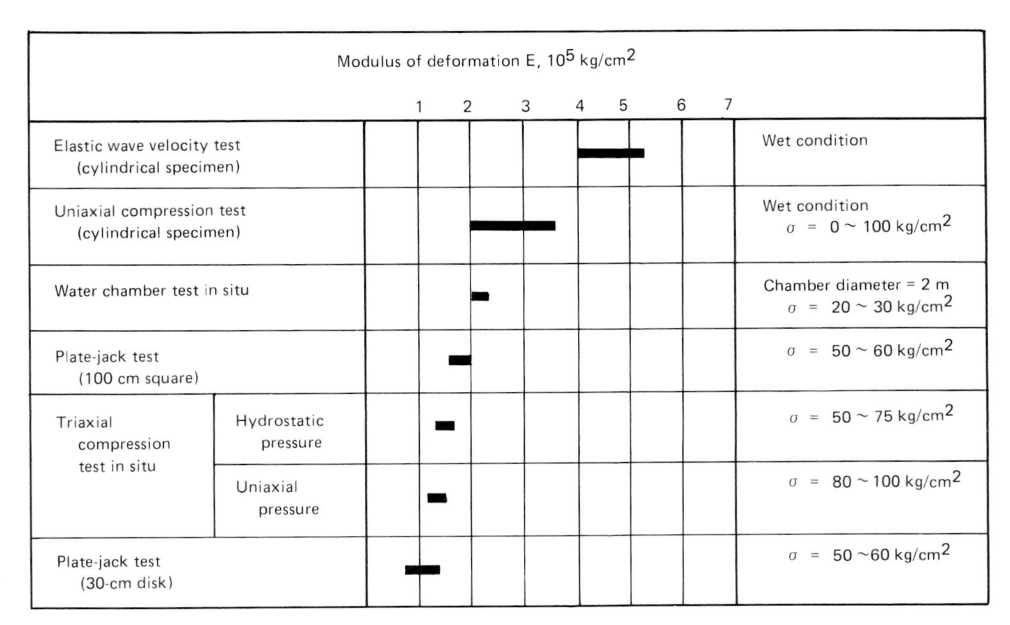


**FIG. 3.77 Deformation moduli obtained by various types of tests in a jointed granite.** [*From Hibino et al. (1977).*[99]]

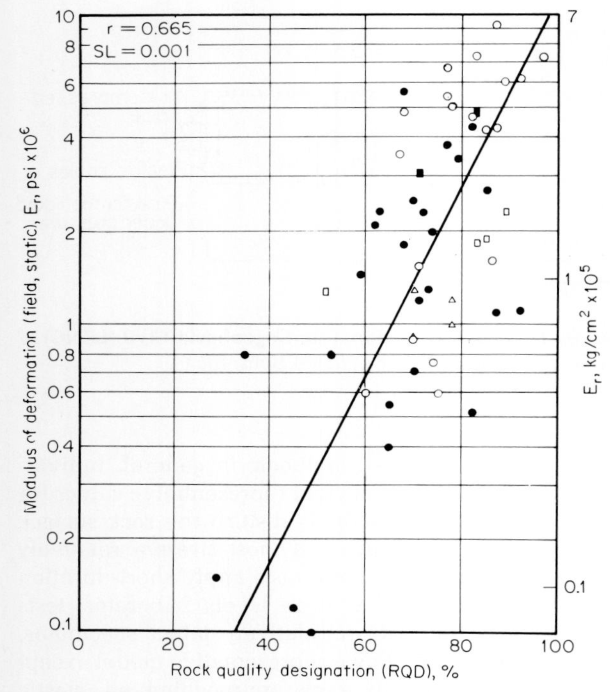


**FIG. 3.78 Comparison of the RDQ and the in situ static modulus of deformation.** [*From Coon and Merritt (1970).*[101] *Reprinted with permission of the American Society for Testing and Materials.*]

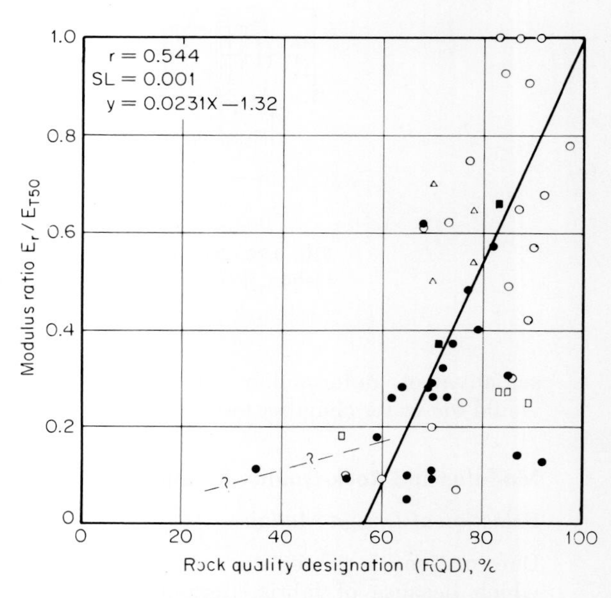


**FIG. 3.79 Comparison of the RDQ and the modulus ratio $E_r/E_{t50}$.** [*From Coon and Merritt (1970).*[101] *Reprinted with permission of the American Society for Testing and Materials.*]

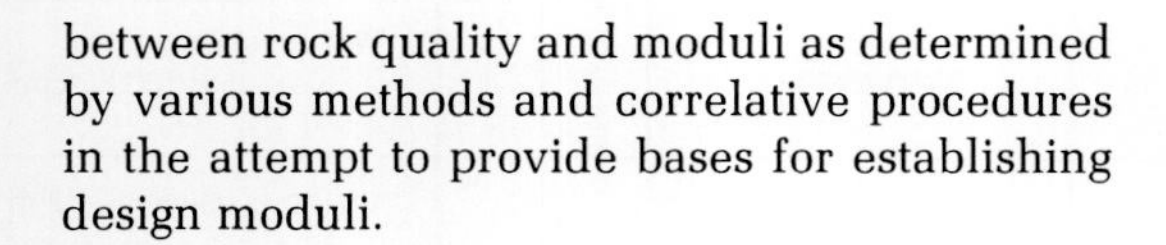

between rock quality and moduli as determined by various methods and correlative procedures in the attempt to provide bases for establishing design moduli.

*Measuring Rock Quality*

*Rock-quality designation* (RQD) as determined from core examination *(see Art. 2.4.5)* provides an index to rock quality.

*Modulus ratio* $E_{t50}/U_{ult}$ is defined [Deere and Miller (1966)[100]] as the ratio of the tangent modulus $E_{t50}$, taken at a stress level of one-half the ultimate strength, to the uniaxial compressive strength $U_{ult}$ as obtained by laboratory test. It was selected as a basis for engineering classification of intact rock *(see Art. 5.2.7)* because it is related to deformation and strength and provides a measure of material anisotropy. Low modulus values result from large deformations caused by closure of foliation and bedding planes; high modulus values are representative of material with interlocking fabric and little or no anisotropy.

*Modulus ratio* $E_r/E_{t50}$ is defined [Coon and Mer-

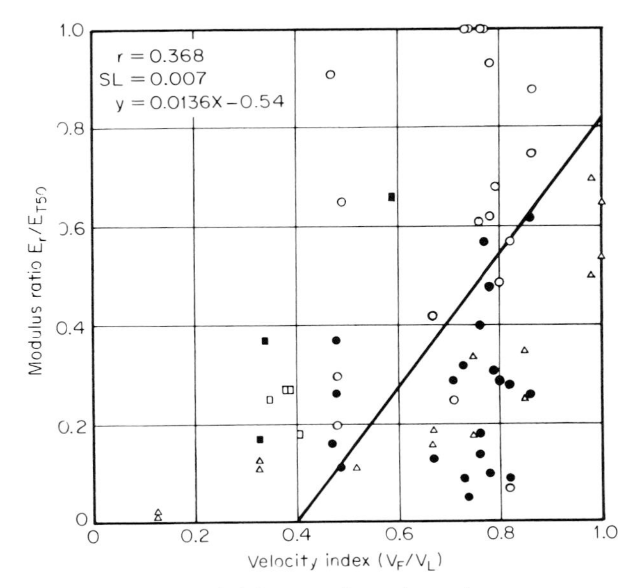


**FIG. 3.80 Relationship between modulus ratio and the velocity index.** [*From Coon and Merritt (1970).*[101] *Reprinted with permission of the American Society for Testing and Materials.*]

ritt (1970)[101]] as the ratio of the in situ static modulus $E_r$ to the intact static modulus $E_{t50}$.

*Velocity ratio* $V_F/V_L$ is defined [Coon and Merrit (1970)[101]] as the ratio of the field *P*-wave velocity, as determined by uphole or crosshole seismic methods or the 3-D velocity logger, to the laboratory *P*-wave velocity as determined by the resonant column device or ultrasonic testing (ASTM 2845). The velocity ratio approaches unity in high-quality massive rocks with few joints.

*Velocity index* $(V_F/V_L)^2$ is defined [Coon and Merritt (1970)[101]] as the square of the ratio of the field seismic velocity to the laboratory seismic velocity, or the square of the ratio of the in situ velocity to the intact velocity. The velocity ratio

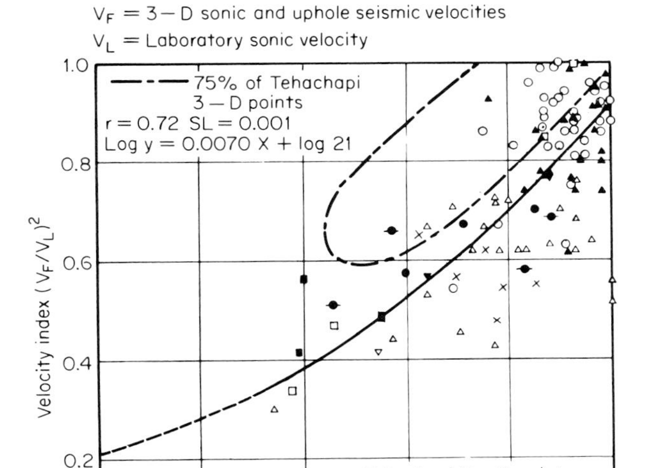


**FIG. 3.81 Relationship between RDQ and velocity index.** [*From Coon and Merritt (1970).*[101] *Reprinted with permission of the American Society for Testing and Materials.*]

is squared to make the velocity index equivalent to the ratio of the dynamic moduli.

*Rock Quality and Modulus Relationships*

- RQD vs. the in situ static modulus is given on Fig. 3.78.
- RQD vs. the modulus ratio is given on Fig. 3.79.
- RQD vs. the velocity index is given on Fig. 3.80.
- Modulus ratio vs. the velocity index is given on Fig. 3.81.

*Modulus Reduction Factor β*

Proposed by Deere et al. (1967),[102] the modulus reduction factor $\beta$ expresses the extent to which $E_r$ is always lower than $E_{seis}$ because of the short pulse duration and low stress level of the in situ seismic test. It is given as

$$\beta = E_r/E_{seis} \tag{3.73}$$

$\beta$ approaches unity as the rock quality approaches an elastic material. A relationship among $\beta$, RQD, and the velocity index is given on Fig. 3.82. The reduction factor is applied to $E_{seis}$ to arrive at a design value for $E_r$.

The application of such relationships requires substantial judgment and experience and should be developed for a particular site on important projects.

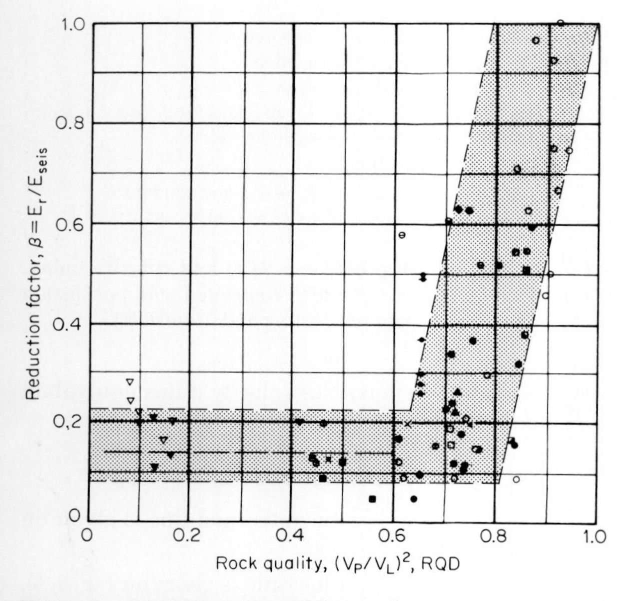


**FIG. 3.82 Variation in modulus reduction factor $\beta$ with rock quality. [$E_r$ = static rock modulus (from load tests); $E_{seis}$ = modulus from seismic velocity; $V_p$ = field seismic velocity (compressional); $V_L$ = laboratory sonic velocity of sound cores].** [*From Deere et al. (1967).*[102]]

3.5.4 SOIL DEFORMATION MEASUREMENTS (STATIC)

### Methods Summarized

*Laboratory Testing*

Undisturbed specimens are statically tested in the laboratory in the triaxial compression apparatus to obtain measures of moduli ($E_s$ or $E_t$, $E_i$, and $E_{se}$), and in the consolidometer to obtain measures of consolidation and expansive properties. The dynamic deformation moduli ($E_d$, $G_d$, and $K$) are usually measured in the cyclic triaxial, cyclic simple shear, or resonant column devices. Soils suitable for testing undisturbed samples include intact clays and cohesive granular soils containing a minimum of fine gravel and coarser particles.

Specimens of sand may be reconstituted at various values of $D_R$ and tested in the consolidometer for static compression characteristics, or in the shaking table device or other devices for dynamic properties.

*In Situ Testing*

Materials in which undisturbed sampling is difficult or impossible, such as cohesionless sands and gravels, or materials with coarse particles such as some residual soils, decomposed rock, or glacial till, and fissured clays, are tested in situ to measure deformation characteristics.

Tests include:

- Pressuremeter tests for compression modulus
- Penetration tests for correlations for estimating moduli
- Full-scale load tests for the compression of sands, organic soils, and other materials
- Plate-load test to measure the vertical modulus of subgrade reaction
- Lateral pile-load tests to measure the horizontal modulus of subgrade reaction
- Dynamic tests to measure shear modulus or peak particle velocity

### Consolidation Test for Clays

*General*

PURPOSES The purposes of the consolidation test are to obtain relationships for compression vs. load and compression vs. time for a given load.

VALIDITY The one-dimensional theory of consolidation of Terzaghi, (1943)[103] is valid for the assumptions that strains are small, the soil is saturated, and flow is laminar.

APPARATUS The consolidometer (oedometer) for a one-dimensional test includes a ring to contain the specimen, a small tank to permit sample submergence, a device for applying loads in stages, and a dial gage for deflection measurements as illustrated on Figs. 3.83 and 3.84. In the apparatus shown, loads are applied by adding weights to a loading arm. Some machines apply loads hydraulically through a bellows arrangement; these apparatus often are designed to permit application of back-pressure to the specimen to achieve saturation, or to permit measurement of pore pressures.

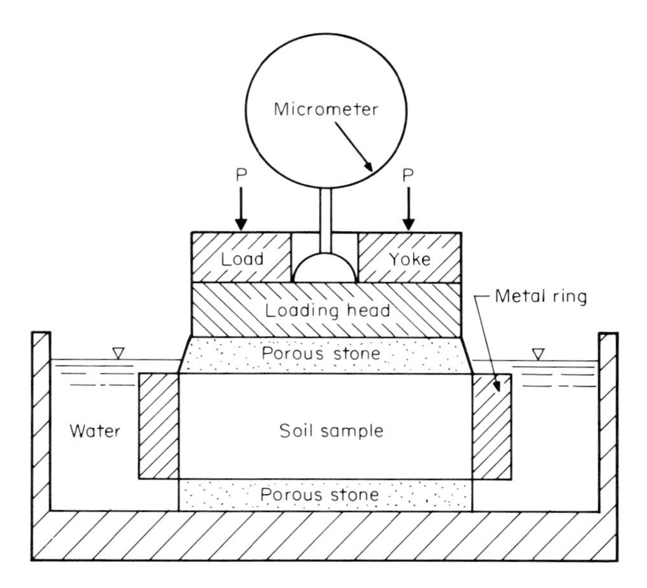


**FIG. 3.83 The floating-head-type consolidometer.**

*Procedures*

A specimen from an undisturbed clay sample is trimmed carefully into a rigid ring (brass for most soils; Teflon for organic soils because of corrosion), and porous stones are placed on top and bottom to permit vertical drainage. (A variation to permit horizontal drainage uses a ring of porous stone and solid end platens.)

The assembly is placed in a loading frame and subjected to a sequence of loads starting initially with very small loads. Then (normally) the load is increased by doubling until the test load significantly exceeds the anticipated field load, and the "virgin" compression curve and preconsolidation stress $p_c$ have been defined (Fig. 3.85). Usually three loads beyond the range of $p_c$ are required to define the virgin compression.

After the initial seating loading, the specimen is immersed in water to maintain saturation (unless the clay is expansive, as will be discussed). Each load remains until pore pressures are essentially dissipated and consolidation terminated; for most clays, 24 hr is adequate for each load cycle. When first applied, the load is carried by the pore water. As water drains from

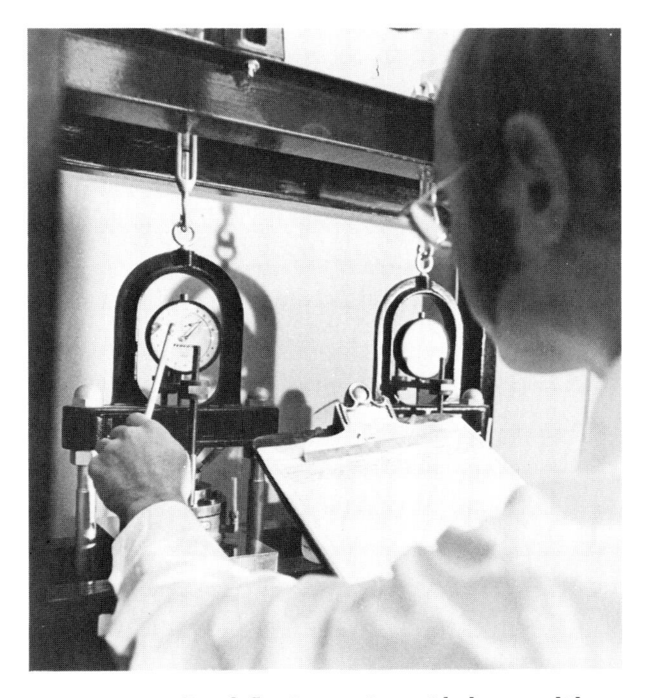

**FIG. 3.84 Reading deflection vs. time with the consolidometer.** *(Courtesy of Joseph S. Ward & Associates.)*

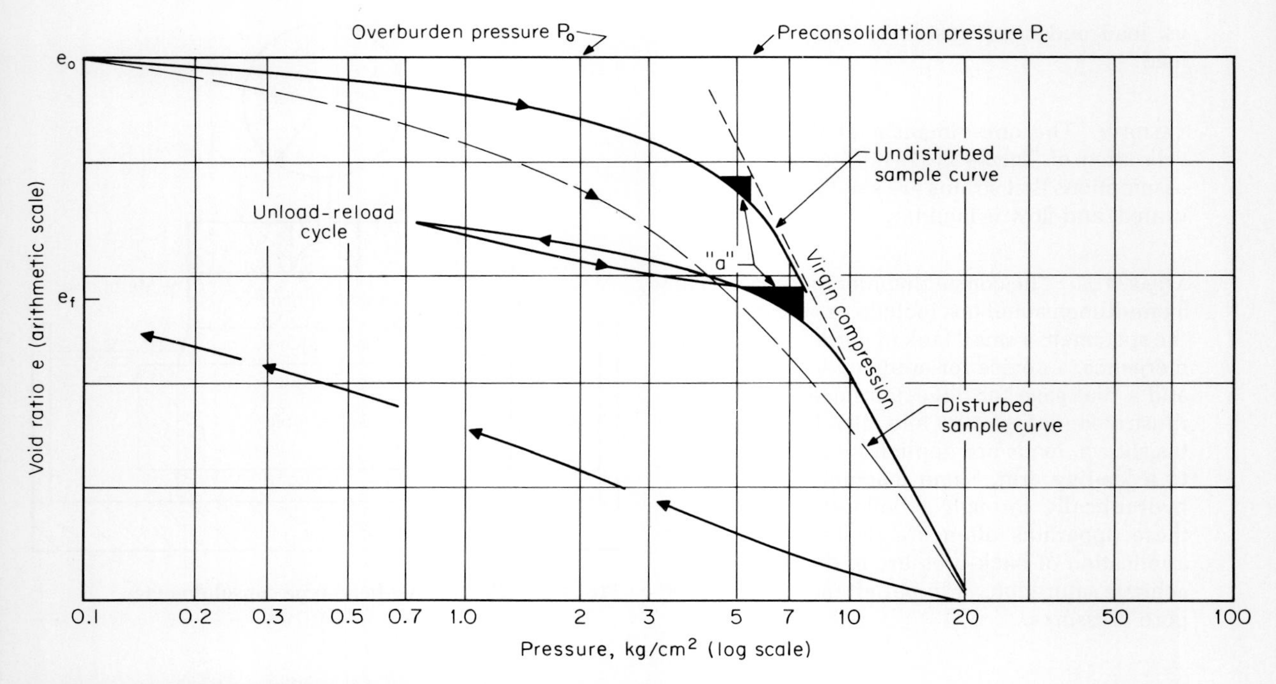


**FIG. 3.85 Consolidation test: pressure-void ratio curve (*e*-log-*p*).**

the specimen, the voids close, the soil compresses, and the strength increases until it is sufficient to support the load and the extrusion of water ceases. At this point "primary compression" has terminated and the next load increment is applied (see discussion of *secondary compression* below). Measurements are made and recorded during the test of deflection vs. time for each load increment. (Plotting the data during the test shows the experienced technician when full consolidation under a given load has occurred, at which point the next load is applied. This practice eliminates the need to wait 24 hr between loadings. If pore-water pressures are measured, when they reach zero the next load increment is applied.)

### *Pressure vs. Void Ratio Relationships*

The total compression that has occurred after application of each load is plotted to yield a curve of either pressure vs. strain or pressure vs. void ratio (*e*-log-*p* curve), where pressure is plotted on a log scale and strain or void ratio on an arithmetic scale (Fig. 3.85).

CURVE SHAPE SIGNIFICANCE A truly undisturbed specimen will yield a curve with a relatively flat initial portion to the range of prestress, then a distinct change in slope near the maximum $p_c$, and thereafter a steep drop under loads causing virgin compression. Disturbed samples or samples with a large granular component yield a curve with a gradual change in slope and a poorly defined $p_c$ as shown on Fig. 3.85.

METHODS TO DETERMINE $p_c$ The Casagrande construction [Casagrande (1936)[104]] is given on Fig. 3.86 and the Burmister construction [Burmister (1951*b*)[17]] on Fig. 3.85.

In the Burmister construction, triangle *a* from the unload-reload cycle is moved upward along the *e*-log-*p* curve until the best fit is found, where the pressure is taken to be $p_c$. The method compensates for slight disturbance during sampling and laboratory handling, which cause the initial curve to be steeper than in situ.

SETTLEMENT ANALYSIS CURVE CORRECTION Unload-reload cycles, usually made soon after $p_c$

has been identified, provide data for correction of the settlement curve to reduce conservatism caused by the slight sample expansion occurring during sampling and handling. The recompression portion of the unload-reload curve is extended back from the initial curve at $p_c$ to form a new curve with a flatter initial slope.

*Soil characteristics* determined from *e*-log-*p* curves, including the compression index $C_c$ and the overconsolidation ratio, are discussed in Art. 3.5.2. Typical *e*-log-*p* curves for various soil types are given on Fig. 3.87. They serve to illustrate that soils of various geologic origins *(see Chap. 7)* have characteristic properties.

### *Compression vs. Time Relationships*

SIGNIFICANCE The *e*-*p* curve provides an estimate of the compression occurring at 100% consolidation. In practice, it is important to estimate the amount of settlement that will occur under a given stress increment in some interval of time (end of construction, 15 years, etc.). The *time rate of consolidation* is analyzed from the compression-vs.-log time curve for a particular load increment (Fig. 3.88).

CURVE CHARACTERISTICS The curve is divided into two portions for analysis: *primary consolidation* occurs while the excess pore-water pressures dissipate and consolidation proceeds in accordance with theory; *secondary consolidation* is a slow, continuing process of compression beyond primary consolidation after the excess pore pressures have been dissipated. The phenomenon is not clearly understood.

PRIMARY CONSOLIDATION Time *t* to reach a given percent consolidation *U* is expressed as

$$t = \frac{T_V}{c_v} H^2 \qquad (3.74)$$

where $T_v$ = the *theoretical time factor*
$H$ = one-half the stratum thickness if there is drainage at both interfaces, or the stratum thickness if there is a drainable layer at only one interface
$c_v$ = *the coefficient of consolidation*

*Theoretical time factor* $T_v$ is a pure number that has been determined for all conditions of importance, given in terms of percent consolidation *U*. The values given on Fig. 3.89 apply to the common cases of (1) a consolidating stratum free to drain through both its upper and lower boundaries, regardless of the distribution of the consolidation pressure or (2) a uniform distribution of the consolidation pressure throughout a layer free to drain through only one surface. The cases of consolidation pressure increasing or decreasing through a consolidating stratum with an impervious boundary can be found in Terzaghi and Peck (1967),[21] p. 181.

*Coefficient of consolidation* $c_v$ is found from the compression-log time curve. Tangents are drawn to the primary section of the curve at its point of inflection and to the secondary sections of the curve to locate $e_{100}$ of primary consolidation. The initial void ratio $e_0$ is found by taking the amount of compression between 0.25 and 1.0 min and adding this value to the void ratio for

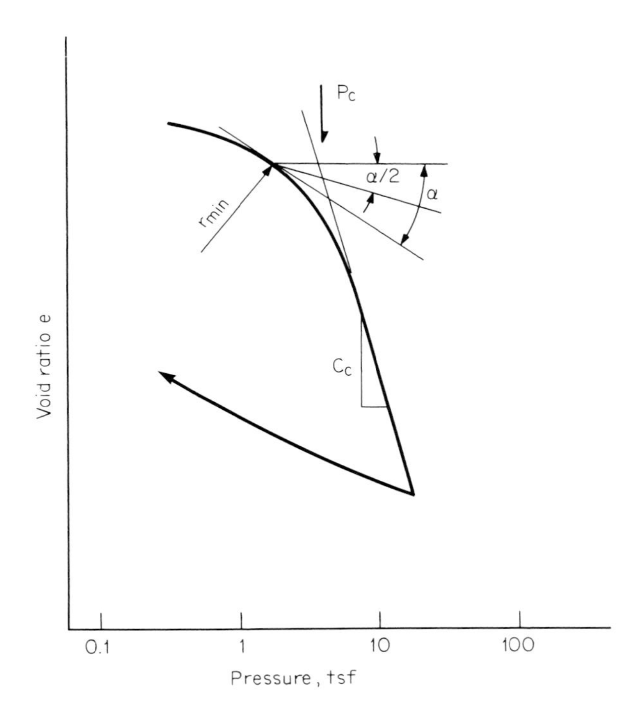


**FIG. 3.86 Casagrande construction for determining $p_c$.** [*From Casagrande (1936).*[104]]

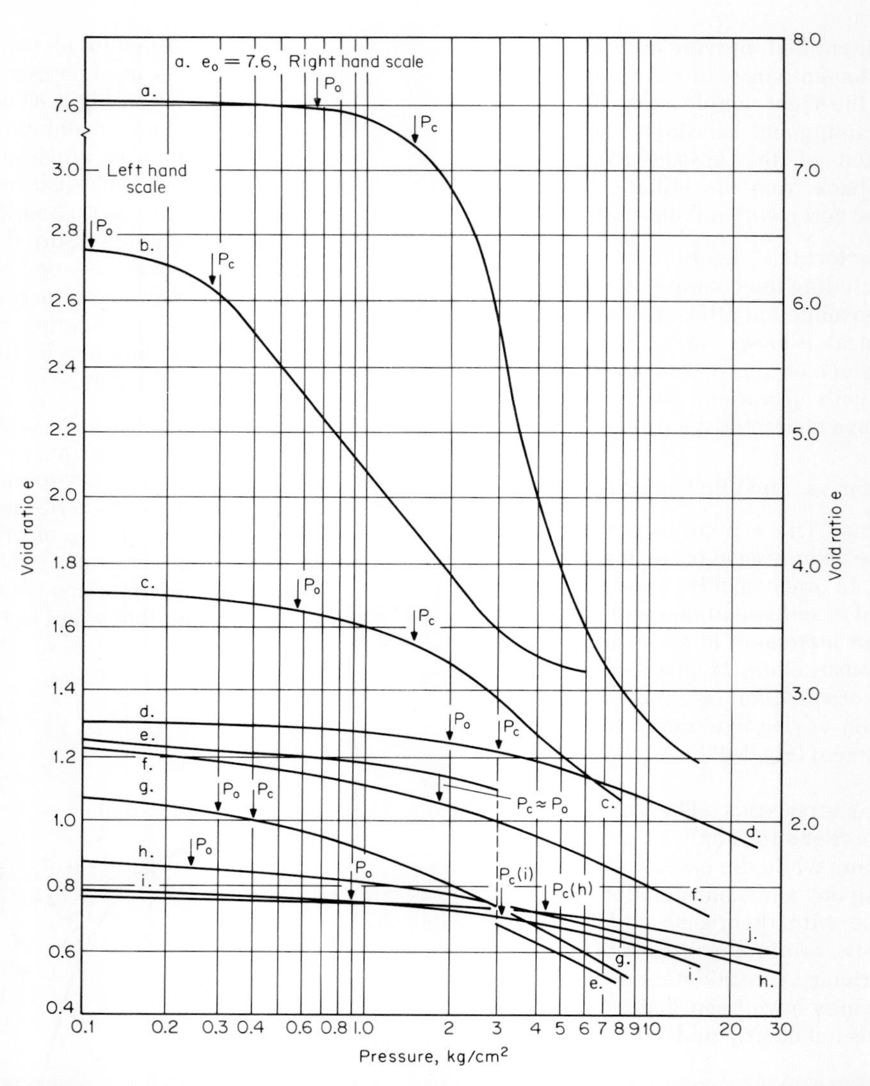


| Material | $\gamma_d$ | w | LL | PI | Material | $\gamma_d$ | w | LL | PI |
|---|---|---|---|---|---|---|---|---|---|
| ***a.*** **Soft silty clay (lacustrine, Mexico)** | **0.29** | **300** | **410** | **260** | ***f.*** **Medium sandy, silty clay (residual gneiss, Brazil)** | **1.29** | **38** | **40** | **16** |
| ***b.*** **Soft organic silty clay (alluvium, Brazil)** | **0.70** | **92** | | | ***g.*** **Soft silty clay (alluvium, Texas)** | | **32** | **48** | **33** |
| ***c.*** **Soft organic silty clay (backswamp, Georgia)** | **0.96** | **65** | **76** | **31** | ***h.*** **Clayey silt, some fine sand (shallow alluvium, Georgia)** | **1.46** | **29** | **53** | **24** |
| ***d.*** **Stiff clay varve (glaciolacustrine, New York)** | | **46** | **62** | **34** | ***i.*** **Stiff clay (Beaumont clay, Texas)** | **1.39** | **29** | **81** | **55** |
| ***e.*** **"Porous" clay (residual, Brazil)** | **1.05** | **32** | **43** | **16** | ***j.*** **Silt varve (glaciolacustrine, New York)** | | | | |

**FIG. 3.87 Typical pressure-void ratio curves for various clay soils.**

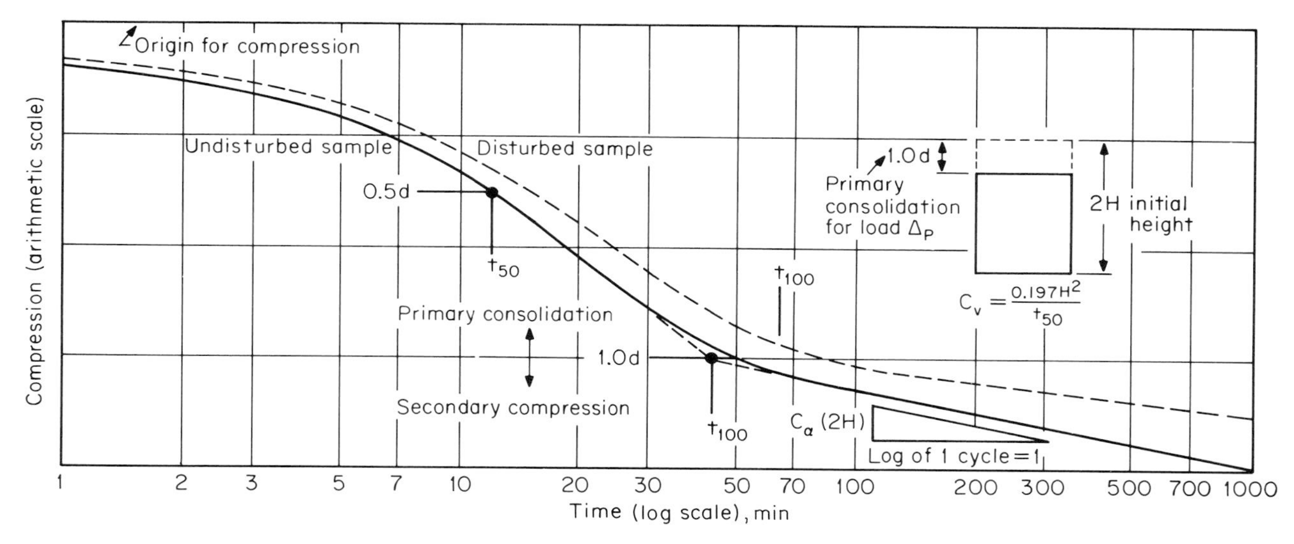


**FIG. 3.88 Compression vs. time for one load cycle.**

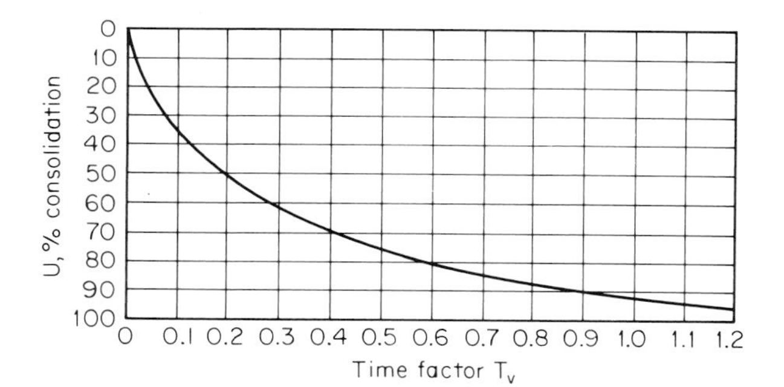


| $U$ | $T_v$ | $U$ | $T_v$ |
|---|---|---|---|
| 10 | 0.0077 | 60 | 0.286 |
| 20 | 0.0314 | 70 | 0.403 |
| 30 | 0.0707 | 80 | 0.567 |
| 40 | 0.126 | 90 | 0.848 |
| 50 | 0.197 | 100 | Infinity |

**FIG. 3.89 Theoretical time factor $T_v$ vs. percent consolidation $U$.**

0.25 min. The void ratio corresponding to $U = 50\%$ is midway between $e_0$ and $e_{100}$, and the corresponding time $t_{50}$ represents the time for 50% consolidation. The theoretical time factor for $U = 50\%$ is 0.197 (Fig. 3.89), and $c_v$ is found from:

$$c_v = \frac{T_v h^2}{t} = \frac{0.197\ h^2}{t_{50}} \qquad \text{cm}^2/\text{s, in}^2/\text{min, ft}^2/\text{day} \qquad (3.75)$$

where $h$ = one-half the thickness of sample, since it has double drainage. The *square root of time method* is an alternative procedure [Lambe and Whitman (1969)[30]].

ANALYSIS OF TIME RATES OF CONSOLIDATION Consolidation is essentially complete at $t_{90}$ (time factor = 0.848) since $T_v$ for $U_{100}$ is infinity. Field time rates of consolidation can be quickly determined from $c_v$ with the chart given as Fig. 3.90, or computed from Eq. 3.74. The values are usually only approximations because field time rates are normally much higher than laboratory rates primarily because stratification permits lateral drainage. Studies where time rates are critical employ piezometers to measure in situ pore pressures (*see Art. 4.4.2*).

COMPUTING COEFFICIENT OF PERMEABILITY $k$ The relationship between $k$ and $c_v$ may be expressed as

$$c_v = \frac{k(1 + e)}{\gamma_w a_v} = \frac{k}{\gamma_w m_v} \qquad \text{cm}^2/\text{s} \qquad (3.76)$$

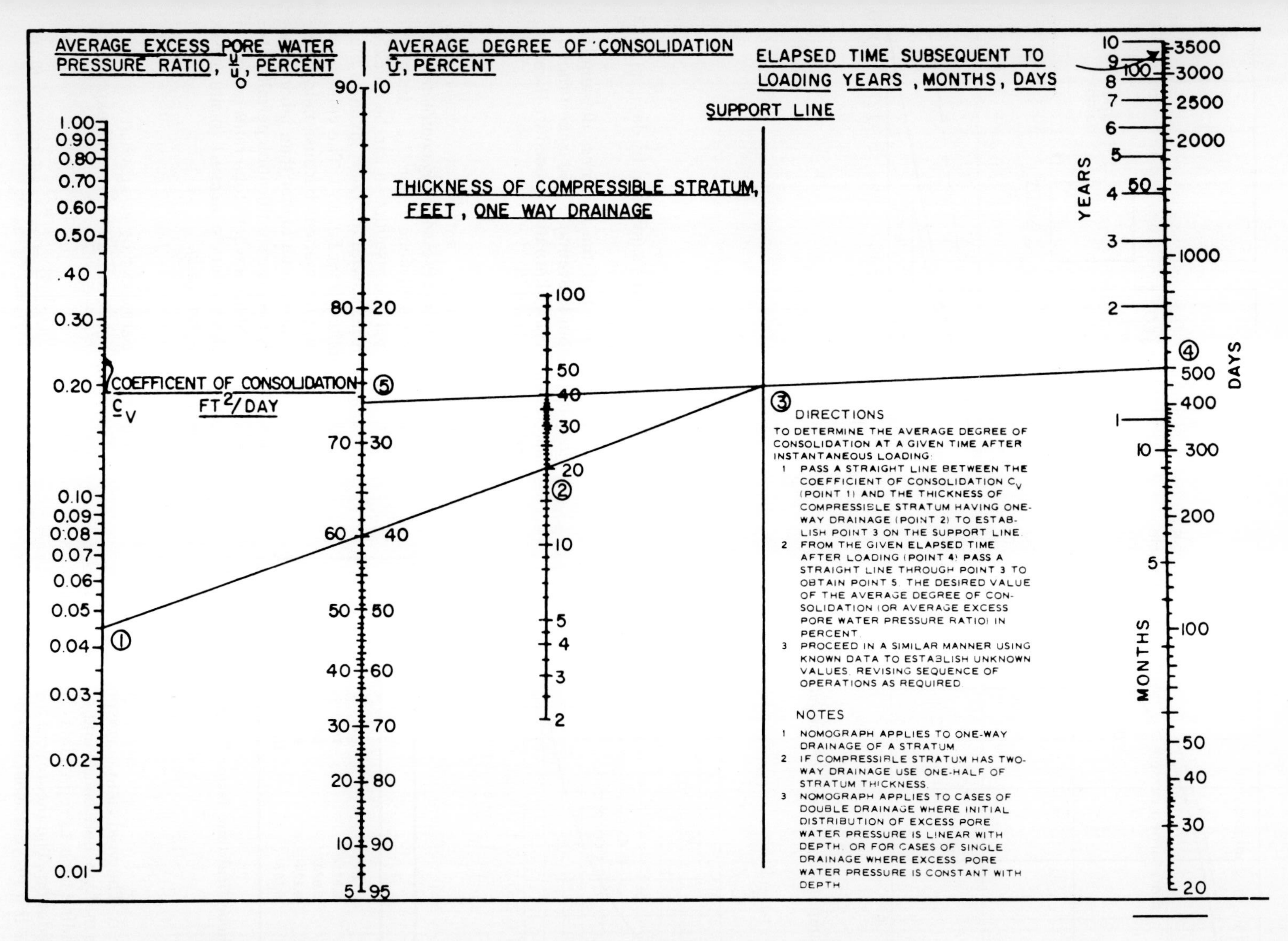


**FIG. 3.90 Nomograph for consolidation with vertical drainage (1 cm²/s = 93 ft²/day).** [*From NAVFAC (1971).*[6]]

where $a_v$ = the *compressibility coefficient,* or the ratio between the change in void ratio and the change in vertical effective stress for the given increment, expressed as

$$a_v = -\frac{e_0 - e_1}{p_1 - p_0} \quad \text{cm}^2/\text{g} \qquad (3.77)$$

$m_v$ = the *coefficient of volume change,* or the ratio of the change in vertical strain to the change in vertical stress, expressed as

$$m_v = \frac{\Delta\epsilon_v}{\Delta\sigma_v} = \frac{a_v}{1 - e_0} \quad \text{cm}^2/\text{g} \qquad (3.78)$$

*Settlement Analysis*

Settlement from *primary consolidation* ($S$ or $\rho$) is determined from the $e$-log-$p$ curve (Fig. 3.85) from the expression

$$\rho = \frac{\Delta e}{1 + e_0} H \qquad (3.52)$$

or, if representative values for $C_c$ have been obtained from a number of tests on similar materials, from the expression

$$\rho = H \frac{C_c}{1 + e_0} \log \frac{p_o - \Delta p}{p_o} \qquad (3.79)$$

where $\Delta p$ = the average change in pressure resulting from the imposed stress *(see Art. 3.5.1).*

*Secondary compression* can result in substantial compression in addition to primary consolidation in very soft clays and organic soils. It can be estimated from the compression vs. log time curve for a desired load. Since the relationship between compression and time on the semilog plot is essentially a straight line, it can be expressed[105] as

$$\Delta e = -C_\alpha \log (t_1/t_2) \qquad (3.80)$$

where $C_\alpha$ = the *coefficient of secondary compression* represented by the slope of the time curve

$t_2$ = the time at which secondary compression is desired

$t_1$ = the time at the start of secondary consolidation.

Secondary compression is a form of *creep.*

*Lateral Strains*

OCCURRENCE Lateral strains, prevented in the one-dimensional consolidation test, can be significant under conditions of relatively rapid loadings of soft clays.

ANALYSIS Based on triaxial test data [Lambe and Whitman (1969),[30] Lambe (1964),[42] Skempton and Bjerrum (1957),[106] Hansen (1967),[107] Davis and Poulos (1968),[108] Lambe (1973),[109] and Navazanjian and Mitchell (1980)[110]].

## Expansion of Clay Soils

*General*

Expansive clay characteristics are described in terms of activity in Art. 5.3.3 and in general in Art. 10.6.2, which includes tests for their identification.

Clays with expansion potential usually are less than 100% saturated in the field; therefore, the consolidation theory does not apply. Tests are performed in the consolidometer or the California bearing ratio mold to obtain measures of percent swell or volume change under a given load, or the maximum swell pressure that may be anticipated.

*Consolidometer Testing of Undisturbed Samples*

PERCENT SWELL MEASUREMENTS

- Place specimen in consolidometer at natural moisture content and provide protection against changes in $w$.
- Add loads in the same manner as the consolidation test, although initial loads may be higher, measuring and recording compression until the final design foundation load is attained.
- Immerse specimen in water, permitting saturation, and measure the volume increase as a function of time, until movement ceases.

MAXIMUM SWELL PRESSURE

- Place specimen in consolidometer under an initial seating load.
- Immerse in water and add loads as necessary to prevent specimen from swelling to determine the maximum swell pressure.

*CBR Mold Testing of Compacted Samples*

Procedures are similar to the percent swell test in the consolidometer.

## Compression in Cohesionless Sands

*Measurements*

Cohesionless sands are seldom tested in the laboratory for compression characteristics since undisturbed samples cannot be obtained; they are occasionally tested in the consolidometer as reconstituted specimens placed at varying values of $D_R$.

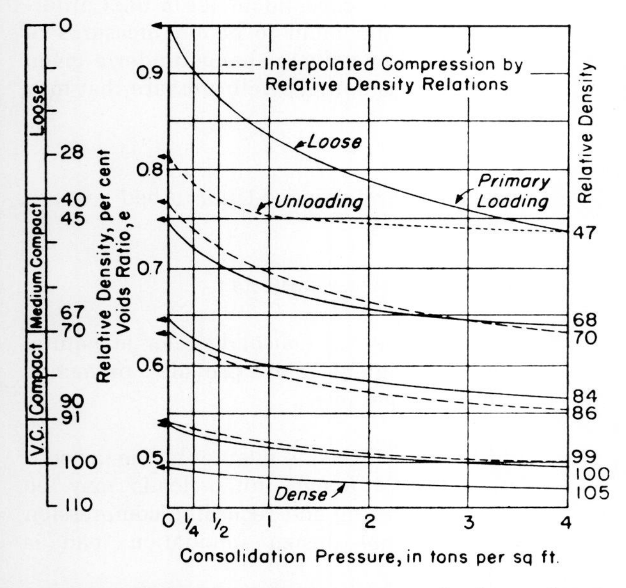


**FIG. 3.91 Pressure-void ratio curves and relative density relationships for a "coarse to fine sand, little silt."** [*From Burmister (1948).*[10] *Reprinted with permission of the American Society for Testing and Materials.*]

Evaluations of compression are normally based on in situ test data such as obtained with the pressure meter, SPT or CPT, or load test.

*Load vs. Compression Characteristics*

Under normal loading conditions, compression in quartz sands is essentially plastic and results from void closure. Compression of individual grains is insignificant except for sands composed of soft materials such as shell fragments, gypsum, or lightly cemented calcareous sands.

Compression magnitude in quartz sands is related to $D_R$, gradation characteristics, and magnitude of applied static load or characteristics of dynamic loadings. A family of curves representing pressure vs. void ratio for various values of $D_R$, obtained by testing reconstituted samples in a consolidometer, is given on Fig. 3.91.

*Compression vs. Time*

Compression under applied load results essentially in immediate closure of the voids as the grains compact, although in saturated silty soils some time delay will occur as pore pressures dissipate. The normal process of consolidation, however, does not occur.

Swiger (1974)[111] reports a case where primary settlements occurred within 1 hr of load application during large-scale field load tests, but secondary compression, ultimately as large as primary compression, continued over a period of several years *(see Art. 3.5.5)*.

## Pressuremeter Testing

*General*

PURPOSES Pressuremeters are used for in situ measurements of deformation moduli and strength.

TYPES

- Menard pressuremeter for soils and soft rock
- Camkometer for soils
- Dilatometer and Goodman jack for rock (*see Art. 3.5.3*)

*Menard Pressuremeter*

See Menard (1963),[112] Menard (1965),[113] Menard (1975),[114] Dixon (1970),[115] and CGS (1978).[116]

APPLICABILITY Although used in all soil types and soft rocks, the Menard pressuremeter is most useful in materials for which undisturbed sampling is difficult or not possible, such as sands, residual soils, glacial till, and soft rock (in a smooth borehole).

PROCEDURE The apparatus is illustrated on Fig. 3.92. A cylindrical flexible probe is lowered into an NX-size borehole to the test depth, and increments of pressure are applied to the probe by gas while radial expansion of the borehole is measured in terms of volume changes. The test can be carried to failure of the surrounding materials; the limit of the radial pressure is in the range of 25 to 50 kg/cm$^2$.

TEST DATA The volume of the expanded probe is plotted vs. the "corrected" pressure which is equal to the gage pressure at the surface minus the probe inflation pressure plus the piezometric head between the probe and the gage, as shown on Fig. 3.93. The quantities obtained from the test include:

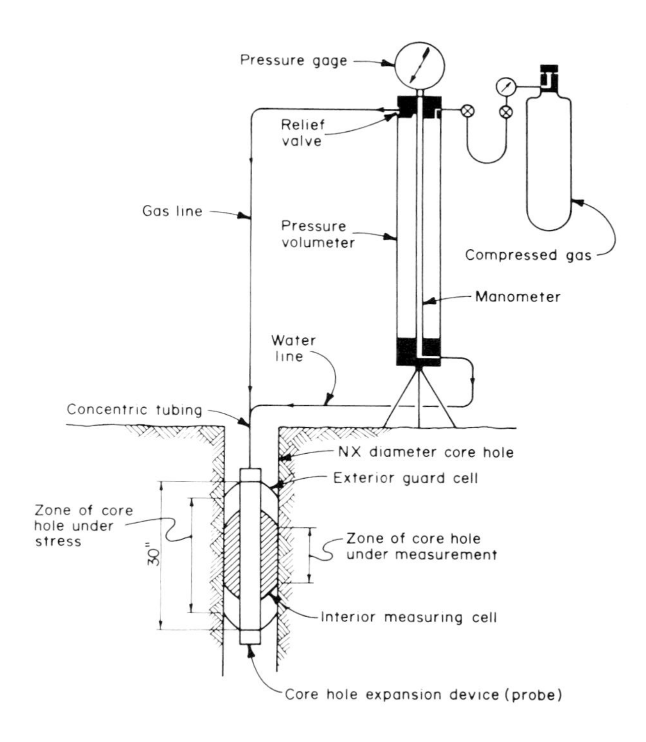


**FIG. 3.92 Schematic of pressuremeter equipment.** [*From Dixon (1970).*[115] *Reprinted with permission of the American Society for Testing and Materials.*]

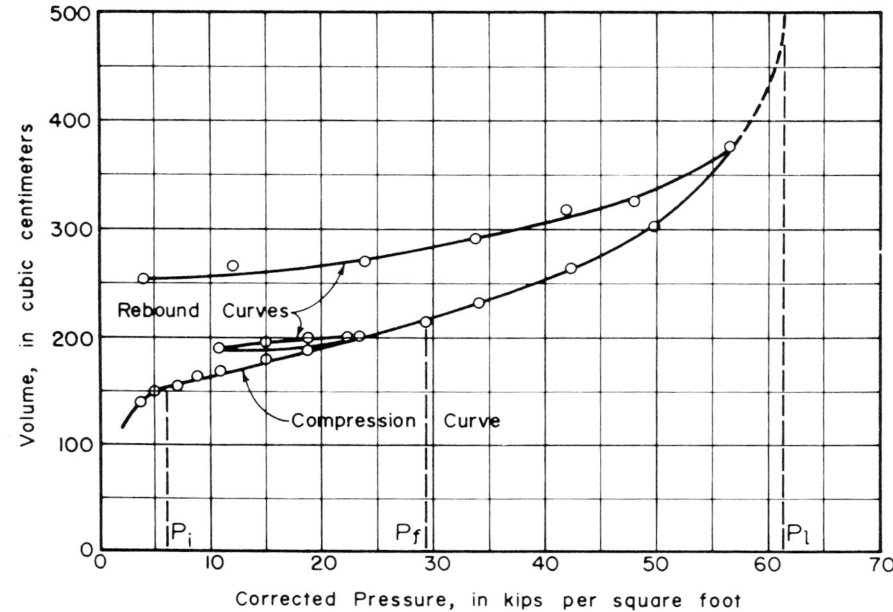


**FIG. 3.93 Typical pressuremeter test results.** [*From Dixon (1970).*[115] *Reprinted with permission of the American Society for Testing and Materials.*]

$P_i$ = initial or seating pressure at the beginning of the elastic stress stage, generally considered equal to $P_o$, the at-rest horizontal stress
$P_f$ = creep pressure at the end of the elastic stress stage
$P_L$ = limiting pressure, or ultimate pressure; the failure pressure
$E_c$ = the *compression modulus;* obtained from the slope of the compression curve between $P_i$ and $P_L$ (Eq. 3.81)
$E_f$ = the rebound modulus; obtained from the slope of the rebound curve

COMPUTING THE COMPRESSION MODULUS $E_c$

$$E_c = K \frac{dp}{dv} \quad \text{psi, kg/cm}^2 \qquad (3.81)$$

where $K$ = a constant of the pressuremeter accounting for borehole diameter, probe size, and Poisson's ratio (usually taken as 0.3 to 0.4)
$K = 2(1 + \nu)(V_0 + V_m)$
$V_0$ = initial hole volume over length of cell (790 cm³ for NX hole)
$V_m$ = fluid volume injected into cell for the average applied pressure
$K$ = 2700 cm³ typically, for an NX hole and $\nu = 0.33$.

*Moduli and Strength Parameters*

Typical values for $E_c$ and $P_L$ for various materials are given on Table 3.37.

ESTIMATING YOUNG'S MODULUS $E$ FROM $E_c$ Menard (1965)[113] suggests the use of a "rheological" factor $\alpha$ to convert the compression modulus to Young's modulus, expressed in:

$$E = E_c/\alpha \qquad (3.82)$$

Menard's values for $\alpha$ are given in Table 3.38 in terms of the ratio $E_c/P_L$ which have been found to be related to the amount of precompression in the material. There is some controversy over the selection of the $\alpha$ factor [Silver et al. (1976)[119]] but most practitioners appear to use Menard's values.

DATA APPLICATIONS

- Estimating values for deformation modulus $E_s$
- Estimating the undrained shear strength from the expression [Menard (1975)[113]]:

$$s_u = \frac{P_L - P_o}{2K_b} \qquad (3.83)$$

**TABLE 3.37**
**VALUES OF COMPRESSION MODULUS $E_c$ AND ULTIMATE PRESSURE $P_L$ FROM MENARD PRESSUREMETER TESTING***

| Soil type | $E_c$ kg/cm² | $P_L$ kg/cm² |
|---|---|---|
| Peat and very soft clays | 2–15 | 0.2–1.5 |
| Soft clays | 5–30 | 0.5–3.0 |
| Firm clays | 30–80 | 3.0–8.0 |
| Stiff clays | 80–400 | 6–25 |
| Loose silty sands | 5–20 | 1– 5 |
| Silts | 20–100 | 2–15 |
| Sands and gravels | 80–400 | 12–50 |
| Till | 75–400 | 10–50 |
| Recent fill | 5–50 | 0.5–3 |
| Ancient fill | 40–150 | 4.0–10 |
| Glacial till (Ohio) | 600–1500 | Ward (1972)[117] |
| Residual soil (schist) | 20–160 | Ward (1972)[117] |
| Decomposed schist | 200–500 | Ward (1972)[117] |
| Residual soils (schist and gneiss; SM, ML, saprolite) | 50–2000 | Martin (1977)[118] |

*After *Canadian Foundation Engineering Manual,* Part I (1978),[116] unless otherwise noted. Reprinted with permission of the National Research Council of Canada.

where $P_L$ = limiting pressure
$P_o$ = at-rest horizontal stress $\bar{\sigma}_v K_0$
$K_b$ = a coefficient varying with $E_c/P_L$; typically 5.5 [Lukas and deBussy (1976)[57]]

- Estimating allowable bearing value for foundations [CGS (1978)[116]]
- Settlement analysis of shallow foundations [CGS (1978),[116] Menard (1972)[120]]
- Analysis of deep foundations [CGS (1978),[116] Menard (1972)[120]]
- Estimating horizontal subgrade reaction modulus $k_h$ [Poulos (1971)[121]]:

$$k_h = 0.8E_c/d \quad (3.84)$$

where $d$ = pile diameter

*Limitations of Pressuremeter Test Data*

The modulus is valid only for the linear portion of soil behavior ($P_f$ on Fig. 3.93).

Stratified or otherwise anisotropic materials may have modulus values much lower in the vertical direction than in the horizontal direction. Since foundation stresses are applied vertically in most cases, values for $E$ may be overestimated from pressuremeter tests.

Stiffness of the device may be significant compared with the compressibility of the tested material, as in very loose sands, soft clay, or organic soils.

Borehole wall disturbance and irregularities greatly affect test results.

Modulus values should be correlated with data obtained by other test methods.

*Camkometer*

DESCRIPTION The Camkometer is a *self-boring* pressuremeter developed in the early 1970s by Cambridge University [Clough and Denby (1980)[122]]. The 80-mm-diameter device is covered by a rubber membrane that incorporates two very small cells for pore-pressure measurements. It is drilled into position and the membranes expanded; transducers permit the pressure response to be converted to electrical impulses. An effective stress-strain curve is plotted from the data.

APPLICABILITY It is used primarily in soft clays and sands for measurements in situ of shear modulus, shear strength, pore-water pressure, and lateral stress $K_0$. Formerly, values for $K_0$ were obtained only by empirical methods or laboratory tests.

ADVANTAGES OVER CONVENTIONAL PRESSUREMETER Lateral stress is measured directly because the instrument is self-boring and stress relief is not permitted. Records are obtained from precise electrical impulses, whereas conventional pressuremeters record total volume changes of hydraulic fluid within the flexible membrane and furnish only average values.

**TABLE 3.38**
**RHEOLOGICAL FACTOR $\alpha$ FOR VARIOUS SOIL CONDITIONS***

| | Peat | | Clay | | Silt | | Sand | | Sand and gravel | |
|---|---|---|---|---|---|---|---|---|---|---|
| Material | $E_c/P_L$ | $\alpha$ | $E_c/P_L$ | $\alpha$ | $E_c/P_L$ | $\alpha$ | $E_c/P_L$ | $\alpha$ | $E_c/P_L$ | $\alpha$ |
| Preconsolidated | | | >16 | 1 | >14 | 0.67 | >12 | 0.5 | >10 | 0.33 |
| Normally consolidated | | 1 | 9–16 | 0.67 | 8–14 | 0.5 | 7–12 | 0.33 | 6–10 | 0.25 |
| Underconsolidated | | | 7–9 | 0.5 | 5–8 | 0.5 | 5–7 | 0.50 | | 0.25 |

*Menard (1965).[113]

**TABLE 3.39**
**CORRELATION OF $q_c$, $C_c$, and $w$***

| Point resistance $q_c$, bar | Water content $w$, % | Compressibility index $C_c$ |
|---|---|---|
| $q_c > 12$ | >30 | <0.2 |
| $q_c < 12$ | <25 | <0.2 |
| | 25–40 | 0.2–0.3 |
| | 40–100 | 0.3–0.7 |
| $q_c < 7$ | 100–300 | 0.7–1.0 |
| | >130 | >1.0 |

*From Sanglerat (1972).[67] Reprinted with permission of Elsevier Scientific Publishing Company.

**TABLE 3.40**
**GUIDE FOR ESTIMATING PRESSUREMETER $E_c$ AND $P_L$ FROM $q_c$ USING DELFT MECHANICAL TIPS***

| Soil type | $E/q_c$† | $q_c/P_L$† |
|---|---|---|
| Sand, dense | 1 | 10 |
| Sand, loose | 1.5 | 5 |
| Silt | 2 | 6 |
| Clay, insensitive | 3 | 3 |
| Clay, very sensitive | 20 | 1.5 |

*From Schmertmann (1977).[76] Reprinted with permission of the Federal Highway Administration.

†Typical values only—they depend on soil stress-strain curve. Error easily ±25%, maybe ±100%.

DISADVANTAGES As with all pressuremeters, soil anisotropy is not accounted for. Smear caused by drilling in soft clays reduces the true permeability and affects the pore-pressure measurements. To reduce drainage effects, undrained tests are performed at high strain rates.

**Penetration Tests (SPT, CPT)**

*General*

The standard penetration test and the cone penetrometer test and intercorrelations between test values are described in Art. 3.4.5. Data from the CPT, in particular point resistance $q_c$, have been correlated with various soil properties.

*CPT Test Correlations*

- Estimating $E_s$:
- Design values for granular soils [Schmertmann (1970)[78] and Meyerhof (1956)[123]:

$$E_s \approx 2q_c \tag{3.85}$$

- To adjust for soil compressibility, a higher $E_s/q_c$ ratio to include $D_R$ is suggested by Vesić (1967)[124]:

$$E_s = 2(1 + D_R)^2 q_c \tag{3.86}$$

- Correlations between $q_c$, $C_c$, and $w$—see Table 3.39
- Estimating OCR for clays—Fig. 3.71
- Correlations between Delft mechanical cone and pressuremeter $E$ and $P_L$—Table 3.40

**Plate-Load Test (ASTM D1194)**

*Purposes*

Vertical modulus of subgrade reaction $k_{sv}$ is obtained from plate-load tests.

Results are used occasionally for estimating settlements in sands. In direct application, however, it must be considered that the depth of the stressed zone is usually much smaller than the zone that will be stressed by a larger footing, and the results will not necessarily be represen-

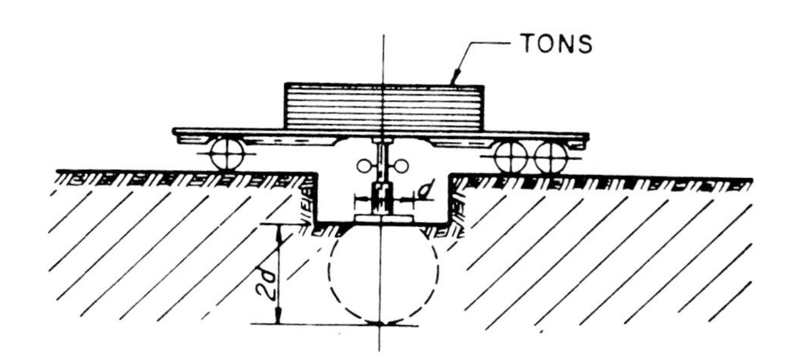


**FIG. 3.94 Reaction for plate-load test and the stressed zone.**

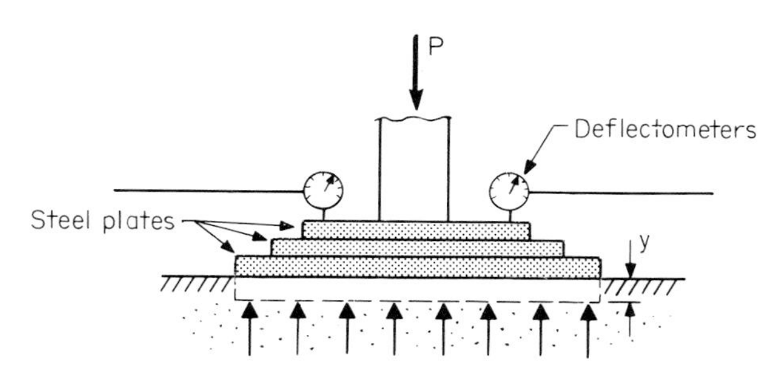


**FIG. 3.95 Load applied to plate in increments while deflection is measured.**

tative. This constraint is overcome by performing tests at various depths to stress the entire zone to be stressed by the footing or by performing full-scale load tests. Plate-load tests are not performed on clay soils for settlement measurement because of long-term consolidation effects.

*Procedure*

A 12- or 30-in-diameter plate is jack-loaded against a reaction to twice the design load, and the plate deflection measured under each load increment. The test setup is shown on Figs. 3.94 and 3.95, and a plot of the test results in terms of load vs. deflection is given on Fig. 3.96, from which the test modulus $k_t$ is determined.

*Modulus of Vertical Subgrade Reaction, $k_{sv}$ and $k_t$*

DETERMINATION $k_{sv}$ is defined as the unit pressure required to produce a unit of deflection expressed as:

$$k_{sv} = p/y \qquad \text{(force/length}^3 = \text{tons/ft}^3) \qquad (3.87)$$

where $p$ = pressure per unit area between the contact surface of the loaded area and the supporting subgrade

$y$ = the deflection produced by the load; length cubed results from area times the deflection in the denominator

Values for $p$ and $y$ are taken at one-half the yield point as estimated on a log plot of the field data (Fig. 3.96) to determine subgrade modulus test value $k_t$. Values for $k_{sv}$ depend on the dimensions of the loaded area as well as elastic properties of the subgrade; therefore the test value $k_t$ always require adjustment for design. Various expressions[21] have been presented for converting $k_t$ to $k_{sv}$.

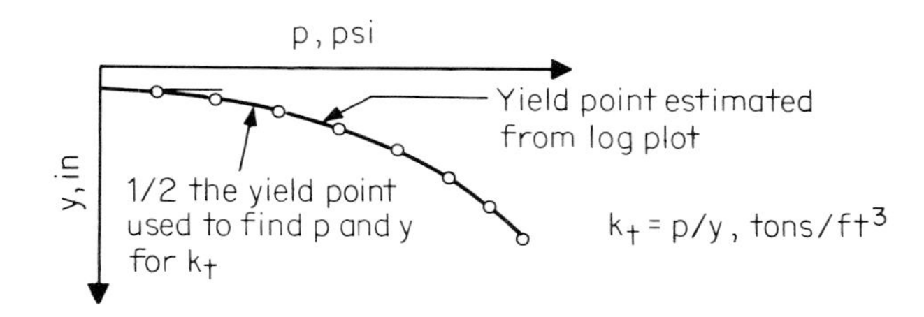


**FIG. 3.96 Load-deflection curve for load test on 1-ft-square plate to find the subgrade modulus $k_t$.**

VALUE RANGES FOR $k_t$ For a given soil the value for $k_t$ decreases with increasing plate diameter, and the modulus for sands is reduced substantially below the water table. Typical value ranges for plates 1 ft square or beams 1 ft wide have been given by Terzaghi (1955)[125] for sands as shown on Table 3.41 and for clays as shown on Table 3.42. In sands the modulus increases with depth. In overconsolidated clays the modulus remains more or less constant as long as the consistency remains constant. The test is not applicable to weaker clays because of the effect of consolidation.

*Applications*

The modulus of vertical subgrade reaction is used to determine shears and bending moments resulting from deflections of pavements, mat

**TABLE 3.41**
**VALUES OF VERTICAL MODULUS OF SUBGRADE REACTION ($k_t$) FOR PLATES ON SAND***

| Relative density of sand | Loose† | Medium† | Dense† |
|---|---|---|---|
| Dry or moist sand, limiting values for $k_t$ | 20–60 | 60–300 | 300–1000 |
| Dry or moist sand, proposed values for $k_t$ | 40 | 130 | 500 |
| Submerged sand, proposed values for $k_t$ | 25 | 80 | 300 |

*From Terzaghi (1955).[125]
†$k_t$ in ton/ft³ for square plates, 1 × 1 ft, or beams 1 ft wide (1 ton/ft³ = 0.032 kg/cm³ = 32 tonne/m³).

**TABLE 3.42**
**VALUES OF VERTICAL MODULUS OF SUBGRADE REACTION ($k_t$) FOR PLATES ON PRECONSOLIDATED CLAY***

| Consistency of clay | Stiff† | Very stiff† | Hard† |
|---|---|---|---|
| Values of $u_c$, ton/ft² | 1–2 | 2–4 | >4 |
| Ranges for $k_t$, square plates | 50–100 | 100–200 | >200 |
| Proposed values, square plates | 75 | 150 | 300 |

*From Terzaghi (1955).[125]
†$k_t$ in ton/ft³ for square plates 1 × 1 ft, or long strips 1 ft wide (1 ton/ft³ = 1.15 lb/in³).

foundations, and continuous footings by employing the concept of a beam or plate on an elastic subgrade.

The modulus of horizontal subgrade reaction is determined by a pile-load test and is used for the evaluation of shears and bending moments in flexible retaining structures and laterally loaded piles employing the concept of a beam or plate on an elastic subgrade.

### Full-Scale Load Tests

*Footings*

The test is performed in a similar manner to the plate-load test except that a poured concrete footing is used rather than a plate.

Tests performed on various sizes and at various depths provide much useful information on the compressibility of sand subgrades and are the most reliable procedure for measuring settlements in these materials if representative sites are tested.

*Pile-Load Tests*

Vertical pile-load tests are performed to obtain measures of $E$ and shaft friction in large-diameter piles, or end-bearing capacity and shaft friction in slender piles. A load test setup is illustrated on Fig. 4.35.

Lateral load tests are performed on piles to obtain data on the horizontal modulus of subgrade reaction $K_{sh}$.

*Embankment Tests*

APPLICATIONS Embankments are constructed to obtain information on the compressibility of loose fills, or thick deposits of soft clays and weak organic soils, and are often designed to preload weak soils.

PROCEDURE An embankment with height and width adequate to stress the weak soils to a substantial depth is constructed with fill. Instrumentation is installed to monitor deflections and pore pressures as functions of time *(see Art. 4.5.2)*. Interpretation of the field data and correlations with laboratory test data provide information on the magnitude and time rate of settlements to be anticipated, as well as the height limitations during actual construction required to avoid shear failures. This is the most reliable procedure for evaluating characteristics of weak organic deposits.

### 3.5.5 DYNAMIC DEFORMATION MODULI (SOILS)

#### Methods Summarized

*Laboratory Methods*[40]

Dynamic deformation moduli are measured in the laboratory in the cyclic triaxial *(see Art. 3.4.4)*, cyclic simple shear *(see Art. 3.4.4)*, cyclic torsion, ultrasonic, and resonant column devices as summarized on Table 3.43.

The shaking table test has been used to date primarily in university research studies and can be found described in Novacs et al. (1971)[127] and De Alba et al. (1976).[128] Dynamic testing procedures are described in detail in USAEC (1972).[40]

*In Situ Methods*

Dynamic moduli are measured in the field by seismic direct methods and steady-state vibration methods. Vibration monitors obtain data on ground motion *(see Art. 4.2.5)*.

#### Resonant Column Devices

*Apparatus*

Several types are in use as described in USAEC (1972).[40]

*Procedure*

Specimen is placed in a chamber, subjected to a confining pressure stimulating overburden pressure, vibrated first in the torsional mode and then in the longitudinal mode while end displacements are monitored.

Shear-wave velocity is computed from the torsional test results and the compression-wave velocity from the longitudinal test results as functions of specimen and end displacements. $E_d$ and $G_d$ are computed from equations given on Table 3.35. When applicable, factors are applied to include the effects of damping and end conditions during test.

#### Seismic Direct Methods

*Purpose*

As described in Art. 2.3.2, seismic direct methods are used to obtain values of $E_d$ and $G_d$, and have been used for estimating values of $E_s$ in medium-dense to dense sands for settlement computations where small strains are critical.

*Estimating Values for $E_s$*

Moduli for strain levels in the order of $10^{-6}$ can be estimated from shear-wave velocities from crosshole or uphole seismic surveys [Swiger (1974)[111]]. Shear-wave velocities are used because they can be measured above and below the groundwater level, whereas the compression-wave velocity can be measured only above groundwater level, since it is obscured by the compression-wave velocity for water.

Under loads in the order of 2 to 3 kg/cm$^2$, strains in dense sands are small, approximately $10^{-3}$, but higher than the strains occurring during seismic testing which require adjustment for analysis. In granular soils $E_d$ and $G_d$ have been found to decrease with increasing strain levels. [Hardin and Drnevich (1972)].[130] A relationship between shear strain and axial strain as a function of strain level is given on Fig. 3.97. The ratios given on the abscissa are used to reduce the field shear and compression moduli for use in analysis.

*Case Study*

In a study reported by Swiger (1974)[111] good agreement was found between settlements computed from seismic direct surveys and large-scale in situ load tests and the actual settlements measured on the structure for which the study was made. For a Poisson's ratio of 0.3, the values for $E_s$ were in the order of $4 \times 10^6$ psf (2000 kg/cm$^2$).

The primary settlements occurred within about 1 hr of load application, but the magnitude of the secondary settlement appeared ultimately to approximate that of the primary and to continue over a period of some years. Approximately 25%

**TABLE 3.43**
**LABORATORY METHODS FOR DETERMINING DYNAMIC SOIL PROPERTIES***

| Test conditions | Test and reference | Properties measured† | Stress or strain conditions: Initial | Stress or strain conditions: Dynamic | Strain amplitude |
|---|---|---|---|---|---|
| Low frequency | Cyclic triaxial (Seed and Chan, 1966; Castro, 1969) | $E$, $D$; stress vs. strain; strength | Axisymmetric consolidation | Pulsating axial or confining stress; constant-amplitude stress | $10^{-4}$ to $10^{-1}$ |
| | Cyclic torsion (Zeevaert, 1967; Hardin and Drnevich, 1972a and b) | $G$, $D$; stress vs. strain | Axisymmetric consolidation | Pulsating shear stress; constant-amplitude stress or free vibration | $10^{-4}$ to $10^{-2}$ |
| | Cyclic simple shear (Seed and Wilson, 1967) | $G$, $D$; stress vs. strain; strength | $K_0$ consolidation | Pulsating shear stress; constant-amplitude stress (strain) or free vibration | $10^{-4}$ to $3 \times 10^{-2}$ |
| High frequency | Ultrasonic (Lawrence, 1965; Nacci and Taylor, 1967) | $c_p$ or $c_s$ | Axisymmetric consolidation | Dilation or shear; single pulse wave | $10^{-6}$ |
| | Resonant column (Afifi, 1970) | $c_p$ or $c_s$; $E$ or $G$; $D$ | Uniform or axisymmetric consolidation | Pulsating axial or shear (torsional) stress; constant-amplitude strain | $10^{-6}$ to $10^{-2}$ |

*From Faccioli and Reséndiz (1976).[126] Reprinted with permission of Elsevier Scientific Publishing Company.

†$E$ = dynamic Young's modulus, $G$ = dynamic shear modulus, $D$ = damping ratio, $c_p$ = compression-wave velocity, $c_s$ = shear-wave velocity.

occurred in the first year after load application (about 4 to 8 kg/cm² foundation pressure).

### Steady-State Vibration Methods

*Purpose*

Steady-state vibration methods are performed to obtain in situ values of $E_d$ and $G_d$.

*Principles*

Ground oscillations are induced from the surface causing Rayleigh waves. The Rayleigh wave velocity $V_r$ is used directly as the shear-wave velocity because, for Poisson's ratios of 0.35 to 0.45, $V_r$ is 0.935 to 0.95 of $V_s$, a difference which is of little engineering significance [Richart (1975)[132]]. $E_d$ and $G_d$ are then computed from $V_r$ (for $V_s$) with equations given on Table 3.35.

*Procedure*

A source of harmonic vibration is applied to the surface to generate the Rayleigh waves in a strain range of $10^{-3}$ to $10^{-5}$. The wavelengths of these surface waves are determined by measuring the distance between two points vibrating in phase with the source. Velocities are computed from the measured wavelength and the vibrating frequency.

The source is either an electromagnetic oscillator modified to produce vibrations or a rotating-mass mechanical oscillator. *Electromagnetic oscillators* are used for high-frequency vibrations of about 30 to 1000 Hz, but wavelengths produced in this frequency range are generally less than about 6 m, and therefore produce penetrations only to depths of about 3 m or less. *Rotating mass oscillators* produce longer wavelengths with input forces adequate to cause ground motions greater than the ambient level.

The generated vibrations are measured with a velocity or acceleration transducer and the wavelengths are measured by comparing the phase relationship of vibrations at various radii from the source with the vibrations of the source.

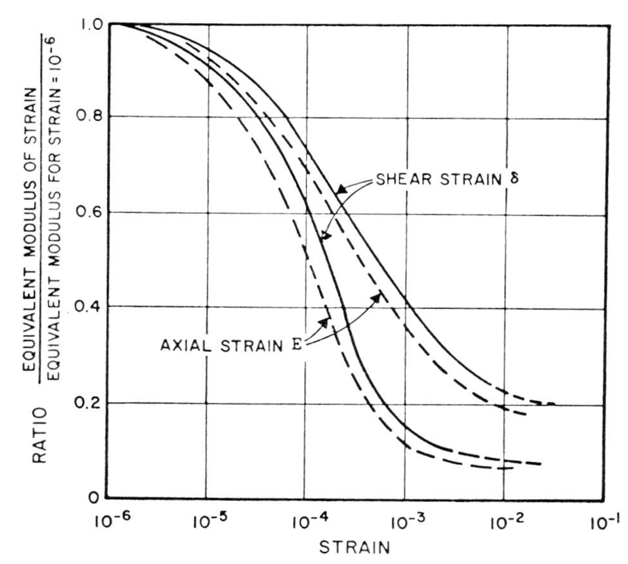


**FIG. 3.97 Strain modulus relations for sands.** [*After Seed (1969),*[131] *from Swiger (1974).*[111]]

## REFERENCES


1. Deere, D. U. (1970) "Indexing Rock for Machine Tunneling," *Proc. Tunnel and Shaft Conf.*, Soc. Mining Engrs. of Amer., Minneapolis, May 1968, pp. 32–38.
2. Tarkoy, P. J. (1975) "A Study of Rock Properties and Tunnel Boring Machine Advance Rates in Two Mica Schist Formations," *Applications of Rock Mechanics, Proc. 15th Symp. on Rock Mechs.*, Custer State Park, South Dakota, September 1973, ASCE, New York, pp. 415–447.
3. NCE (1980) "No Known Cure for Jersey Dam," *NCE International*, Inst. Civil Engrs. London, July, pp. 38–39
4. Krynine, D. and Judd, W. R. (1957) *Principles of Engineering Geology and Geotechnics*, McGraw-Hill Book Co., New York.
5. Mather, K. (1956) "Cement-Aggregate Reaction: What is the Problem?", paper prepared for presentation, *Panel Discussion on Cement Aggre-*

*gate Reaction*, Annual Mtg., Amer. Inst. Mining and Metallurgical Engs., New York, Feb. 21, pp. 83–85.

6. NAVFAC (1971) *Design Manual DM-7, Soil Mechanics, Foundations and Earth Structures*, Naval Facilities Engineering Command, Alexandria, Va.
7. USBR (1974) *Earth Manual*, U.S. Bureau of Reclamation, Denver, Colo.
8. Foster, C. R. (1962) "Field Problems: Compaction," *Foundation Engineering*, G. A. Leonards, ed., McGraw-Hill Book Co., New York, pp. 1000–1024.
9. Turnbull, W. J. (1950) "Compaction and Strength Tests on Soil," paper presented at annual ASCE meeting, January.
10. Burmister, D. M. (1948) "The Importance and Practical Use of Relative Density in Soil Mechanics," *ASTM, Vol. 48*, Philadelphia, Pa.
11. Burmister, D. M. (1949) "Principles and Techniques of Soil Identification," *Proc. 29th Annual Mtg.*, Highway Research Board, December.
12. Burmister, D. M. (1951*a*) "Identification and Classification of Soils—An Appraisal and Statement of Principles," *ASTM Spec. Pub. 113*, Amer. Soc. for Testing and Materials, Philadelphia, Pa., pp. 3–24, 85–91.
13. Burmister, D. M. (1962) "Physical, Stress-Strain, and Strength Responses of Granular Soils," *ASTM Spec. Tech. Pub. No. 322*, pp. 67–97.
14. Skempton, A. W. (1953) "The Colloidal Activity of Clays," *Proc. 3rd Intl. Conf. Soil Mechs. and Found. Engrg.*, Switzerland, Vol. I, pp. 57–61.

14*a*. Vijavergiya, V. N. and Ghazzaly, O. I. (1973) "Prediction of Swelling Potential for Natural Clays," *Proc. 3rd Intl. Conf. on Expansive Soils*, Haifa, Vol. 1.

15. Lambe, T. W. (1951) *Soil Testing for Engineers*, John Wiley & Sons, New York.
16. Arman, A. (1970) "Engineering Classification of Organic Soils, *HRB No. 30*, Highway Research Board, Washington, D.C., pp. 75–89.
17. Burmister, D. M. (1951*b*) "The Application of Controlled Test Methods in Consolidation Testing," *ASTM Spec. Pub. No. 126*, pp. 83–97.
18. Casagrande, A. and Fadum, R. E. (1940) "Notes on Soil Testing for Engineering Purposes," *Harvard Univ. Soil Mechs. Series No. 8*, Bull. 268, Cambridge, Mass.
19. Leonards, G. A. (1962) *Foundation Engineering*, McGraw-Hill Book Co., New York, Chap. 2.
20. Hoek, E. and Bray, J. W. (1977) *Rock Slope Engineering*, Inst. of Mining and Metallurgy, London.
21. Terzaghi, K. and Peck, R. B. (1967) *Soil Mechanics in Engineering Practice*, 2nd ed., John Wiley & Sons, New York.
22. Hough, K. B. (1957) *Basic Soils Engineering*, The Ronald Press, New York.
23. Lowe, J., III and Zaccheo, P. F. (1975) "Subsurface Explorations and Sampling," *Foundation Engineering Handbook*, Winterkorn and Fang, eds., Van Nostrand Reinhold, New York, Chap. 1.
24. Cedergren, H. R. (1967) *Seepage, Drainage and Flow Nets*, John Wiley & Sons, New York.
25. Davis, C. V. and Sorensen, K. E. (1970) *Handbook of Applied Hydraulics*, McGraw-Hill Book Co., New York.
26. Serafim, J. L. (1969) "Influence of Interstitial Water on the Behavior of Rock Masses," *Rock Mechanics in Engineering Practice*, Stagg and Zienkiewicz, eds., John Wiley & Sons, New York, Chap. 3.
27. Dick, R. C. (1975) "In Situ Measurement of Rock Permeability: Influence of Calibration Errors on Test Results," *Bull. Assoc. Engrg. Geol.*, Vol. XII, No. 3.
28. Ward, J. S. (1967) "The Significance of Geology in Determining Soil Profiles and Stress Histories," *Soils*, Chap. 1, J. S. Ward and Assocs., Caldwell, New Jersey.
29. Peck, R. B. (1969) "Stability of Natural Slopes," *Proc. ASCE, Stability and Performance of Slopes and Embankments*, Berkeley, Calif., August, pp. 437–451.
30. Lambe, T. W. and Whitman, R. V. (1969) *Soil Mechanics*, John Wiley & Sons, New York.
31. Bishop, A. W. and Henkel, D. J. (1962) *The Triaxial Test*, Edward Arnold Ltd., London.
32. Lambe, T. W. (1962) "Pore Pressures in a Foundation Clay," *Proc. ASCE, J. Soil Mechs. and Found. Engrg. Div.*, Vol. 88, No. SM2, pp. 19–47.
33. Wu, T. H. (1966) *Soil Mechanics*, Allyn and Bacon, Inc., Boston.
34. Casagrande, A. (1959) "An Unsolved Problem of Embankment Stability on Soft Ground," *Proc. 1st Panamerican Conf. Soil Mech. and Found. Engrg.*, Mexico City, Vol. II, pp. 721–746.
35. Bjerrum, L. (1969) Contribution to panel discussion, Session 5, "Stability of Slopes and Embankments," *Proc. 7th Intl. Conf. Soil Mech. and Found. Engrg.*, Mexico City, Vol. 3, pp. 412–413.

36. Bjerrum, L. (1972) "Embankments on Soft Ground," State-of-the-Art Paper, *Proc. ASCE Spec. Conf. on Performance of Earth and Earth-Supported Structures*, Purdue Univ., Lafayette, Ind.

37. Patton, F. D. and Hendron, A. J., Jr. (1974) "General Report on Mass Movements," *Proc. 2d Intl. Cong., Intl. Assoc. Engrg. Geol.*, São Paulo, p. V-GR 1.

38. Kanji, M. A. (1970) "Shear Strength of Soil Rock Interfaces," M.S. Thesis, Dept. of Geology, Univ. of Illinois, Urbana.

39. Bjerrum, L. (1954) "Geotechnical Properties of Norwegian Marine Clays," *Geotechnique*, Vol. 4, No. 2, p. 49.

40. USAEC (1972) *Soil Behavior under Earthquake Loading Conditions*, National Technical Information Service Pub. TID—25953, U.S. Dept. of Commerce, Oak Ridge National Laboratory, Oak Ridge, Tenn., January.

41. Eide, O. (1974) "Marine Soil Mechanics—Applications to North Sea Offshore Structures," *Norwegian Geotechnical Institute Pub. 103*, Oslo, p. 19.

42. Lambe, T. W. (1964) "Methods of Estimating Settlements," *Proc. ASCE, J. Soil Mechs. and Found. Engrg. Div.*, Vol. 90, No. SM5, September, pp. 47-71.

43. Lo, K. Y. and Lee, C. F. (1973) "Analysis of Progressive Failure in Clay Slopes," *Proc. 8th Intl. Conf. Soil Mechs. and Found. Engrg.*, Moscow, Vol. 1, pp. 251-258.

44. Pells, P. J. N. (1973) "Stress Ratio Effects on Construction Pore Pressures," *Proc. 8th Intl. Conf. Soil Mechs. and Found. Engrg.*, Moscow, Vol. 1, pp. 327-332.

45. Kassif, G. and Baker, R. (1969) "Swell Pressure Measured by Uni- and Triaxial Techniques, *Proc. 7th Intl. Conf. Soil Mechs. and Found. Engrg.*, Mexico City, Vol. 1, pp. 215-218.

46. Murphy, D. J. (1970) "Soils and Rocks: Composition, Confining Level and Strength," doctoral thesis, Dept. of Civil Engrg., Duke Univ., Durham, N.C.

46*a* Jennings, J. E. (1972) "An Approach to the Stability of Rock Slopes Based on the Theory of Limiting Equilibrium with a Material Exhibiting Anisotropic Shear Strength," *Stability of Rock Slopes, Proc. ASCE, 13th Symp. on Rock Mechs.*, Univ. of Illinois (1971), Urbana, pp. 269-302.

47. Bieniawski, Z. T. (1974) "Geomechanics Classification of Rock Masses and Its Application to Tunneling," *Proc. 3rd Intl. Cong. for Rock Mechs.*, Intl. Soc. for Rock Mechs., Vol. IIA, Denver, pp. 27-32.

48. Leet, L. D. (1960) *Vibrations from Blasting Rock*, Harvard Univ. Press, Cambridge, Mass.

48*a*. Broch, E. and Franklin, J. A. (1972) "The Point Load Strength Test," *Intl. J. Rock Mechs.*, Min. Sci., Vol. 8.

49. Fahy, M. P. and Guccione, M. J. (1979) "Estimating Strength of Sandstone Using Petrographic Thin-section Data," *Bull. Assoc. Engrg. Geol.*, Vol. XVI, No. 4, Fall, pp. 467-486.

50. Hoek, E. and Brown, E. T. (1980) "Empirical Strength Criterion for Rock Masses," *Proc. ASCE, J. Geotech. Engrg. Civ.*, Vol. 106, No. GT9, September, pp. 1013-1035.

51. Haverland, M. L. and Slebir, E. J. (1972) "Methods of Performing and Interpreting In Situ Shear Tests," *Stability of Rock Slopes, ASCE*, New York, pp. 107-137.

52. Deere, D. U. (1976) "Dams on Rock Foundations—Some Design Questions," *Proc. ASCE, Rock Engineering for Foundations and Slopes*, Speciality Conf., Boulder, Colo., Vol. II, pp. 55-85.

53. Burmister, D. M. (1953) "The Place of the Direct Shear Test in Soil Mechanics," *ASTM Spec. Tech. Pub. No. 131*.

54. Kjerstad and Lunne (1979), paper in *Intl. Conf. on Behavior of Offshore Structures*, London (from *Construction Industry Intl.*, December 1979).

55. Bjerrum, L., Clausen, C. J. F. and Duncan, J. M. (1972) "Earth Pressures on Flexible Structures: A State-of-the-Art Report," *Proc. 5th European Conf. Soil Mech. and Found. Engrg.*, Madrid, Vol. II, pp. 169-196.

56. Mitchell, J. K. and Lunne, T. A. (1978) "Cone Resistance as a Measure of Sand Strength," *Proc. ASCE, J. Geotech. Engrg. Div.*, Vol. 104, No. GT7, July, pp. 995-1012.

57. Lukas, R. G. and deBussy, B. L. (1976) "Pressuremeter and Laboratory Test Correlations for Clays," *Proc. ASCE, J. Geotech. Engrg. Div.*, Vol. 102, No. GT9, September, pp. 945-962.

58. Ladd, C. C. and Foott, R. (1974) "New Design Procedure for Stability of Soft Clays," *Proc. ASCE J. Geotech. Engrg. Div.*, Vol. 100, No. GT7, July, pp. 753-786.

59. Baligh, M. M., Vivatrat, V. and Ladd, C. C. (1980) "Cone Penetration in Soil Profiling," *Proc. ASCE, J. Geotech. Engrg. Div.*, Vol. 106, No. GT4, April, pp. 447-461.

60. Kjellman, W. (1951) "Testing the Shear Strength of Clay in Sweden," *Geotechnique*, Vol. 2, No. 3, June, pp. 225–235.

61. Roscoe, K. H. (1953) "An Apparatus for the Application of Simple Shear to Soil Samples," *Proc. 3rd Intl. Conf. Soil Mechs. and Found. Engrg.*, Switzerland, Vol. 1, pp. 186–191.

62. Bjerrum, L. and Landva, A. (1966) "Direct Simple Shear Tests on a Norwegian Quick Clay," *Geotechnique*, Vol. 16, No. 1, pp. 1–20.

63. Aas, G. (1965) "A Study of the Effect of Vane Shape and Rate of Strain on the Measured Values of In-Situ Shear Strength of Clays," *Proc. 6th Intl. Conf. Soil Mechs. and Found. Engrg.*, Montreal, Vol. 1, pp. 141–145.

64. Schmertmann, J. H. (1979) "Statics of SPT," *Proc. ASCE, J. Geotech. Engrg. Div.*, Vol. 105, No. GT5, May, pp. 655–670.

65. Brown, R. E. (1977) "Drill Rod Influence on Standard Penetration Test," *Proc. ASCE, J. Geotech. Engrg Div.*, Vol. 103, No. GT11, November, pp. 1332–1336.

66. Terzaghi, K. and Peck, R. B. (1948) *Soil Mechanics in Engineering Practice*, John Wiley & Sons, New York.

67. Sanglerat, G. (1972) *The Penetrometer and Soil Exploration*, Elsevier Pub. Co., Amsterdam.

68. Gibbs, H. J. and Holtz, W. G. (1957) "Research on Determining the Density of Sands by Spoon Penetration Testing," *Proc. 4th Intl. Conf. Soil Mechs. and Found. Engrg.*, London, Vol. I, pp. 35–39.

69. Bazaraa, A. R. S. (1967) "Use of SPT for Estimating Settlements of Shallow Foundations on Sand," Doctoral thesis, Univ. of Illinois, Urbana, Ill.

70. Marcuson, W. F. and Bieganousky, W. A. (1977*a*) "Laboratory Penetration Tests on Fine Sands," *Proc. ASCE, J. Geotech. Engrg. Div.*, Vol. 103, No. GT6, June, pp. 565–588.

71. Marcuson, W. F. and Bieganousky, W. A. (1977b) "SPT and Relative Density in Coarse Sands," *Proc. ASCE J. Geotech. Engrg. Div.*, Vol. 102, No. GT11, November, pp. 1295–1309.

72. Peck, R. B., Hanson, W. E. and Thornburn, T. H. (1973) *Foundation Engineering*, 2d ed., John Wiley & Sons, New York.

73. Seed, H. B. (1976) "Evaluation of Soil Liquefaction Effects on Level Ground During Earthquakes," *Liquefaction Problems in Geotechnical Engineering*, Preprint 2752, ASCE National Convention, Philadelphia, Pa., pp. 1–104.

74. Seed, H. B. (1979) "Soil Liquefaction and Cyclic Mobility Evaluation for Level Ground During Earthquakes," *Proc. ASCE, J. Geotech. Engrg. Div.*, Vol. 105, No. GT2, February, pp. 201–255.

75. Trofimenkov, J. G. (1974) "Penetration Testing in USSR," *State-of-the-Art Report, European Symp. Penetration Testing*, Stockholm, Vol. I.

76. Schmertmann, J. H. (1977) *Guidelines for CPT Performance and Design*, Pub. No. FHWA-TS-78-209, Federal Highway Administration, Washington, D.C.

77. Alperstein, R. and Leifer, S. A. (1976) "Site Investigation with Static Cone Penetrometer," *Proc. ASCE, J. Geotech. Engrg. Div.*, Vol. 102, No. GT5, May, pp. 539–555.

78. Schmertmann, J. H. (1970) "Static Cone to Compute Static Settlement Over Sand," *Proc. ASCE, J. Soil Mechs. and Found. Engrg. Div.*, Vol. 96, No. SM3, May, pp. 1011–1043.

79. Simons, N. E. (1972) "Prediction of Settlements of Structures on Granular Soils," *Ground Engineering*, Vol. 5, No. 1.

80. Lacroix, Y. and Horn, H. M. (1973) "Direct Determination and Indirect Evaluation of Relative Density and Its Use on Earthwork Construction Projects," *Evaluation of Relative Density and Its Role in Geotechnical Projects Involving Cohesionless Soils*, ASTM STP 523, Amer. Soc. for Test. and Mat., pp. 251–280.

81. Asphalt Institute (1969) *Soils Manual for Design of Asphalt Pavement Structures*, The Asphalt Inst., Manual Series No. 10, College Park, Md.

82. Asphalt Institute (1970) *Thickness Design*, The Asphalt Inst., Manual Series No. 1, College Park, Md.

83. Deere, D. U. and Patton, F. D. (1971) "Slope Stability in Residual Soils," *Proc. 4th Panamerican Conf. Soil Mechs. and Found. Engrg.*, San Juan, Vol. I, pp. 87–100.

84. Deere, D. U. (1974) "Engineering Geologist's Responsibilities in Dam Foundation Studies," *Foundations for Dams*, ASCE, New York, pp. 417–424.

85. Miller, R. P. (1965) "Engineering Classification and Index Properties for Intact Rock," Ph.D thesis, Univ. of Illinois, Urbana.

86. Hendron, A. J. Jr. (1969) "Mechanical Properties of Rocks, *Rock Mechanics in Engineering Practice*, Stagg and Zienkiewicz, eds., John Wiley & Sons, New York, Chap. 2.

86*a*. Farmer, I. W. (1968) *Engineering Properties of Rocks*, E & F. N. Spon, London.

87. Taylor, D. W. (1948) *Fundamentals of Soil Mechanics*, John Wiley & Sons, New York.

88. Burmister, D. M. (1962*b*) "Prototype Load-Bearing Tests for Foundations of Structures and Pavements," *Symposium on Field Testing of Soils*, STP No. 322, Amer. Soc. for Test. and Mat., Philadelphia, Pa., pp. 98–119.

89. Burmister, D. M. (1956) "Stress and Displacement Characteristics of a Two-Layer Rigid Base Soil System: Influence Diagrams and Practical Applications," *Proc. HRB*, Vol. 35, pp. 773–814, Highway Research Board, Washington, D.C.

90. Gaziev, E. G. and Erlikhman, S. A. (1971) "Stresses and Strains in Anisotropic Rock Foundation (Model Studies)," *Symp. Intl. Soc. for Rock Mechs.*, Nancy.

91. Clemence, S. P. and Finbarr, A. O. (1980) "Design Considerations and Evaluation Methods for Collapsible Soils," *Proc. ASCE, Preprint 80-116*, April.

92. Brooker, E. W. and Ireland, H. O. (1965) "Earth Pressures at Rest Related to Stress History," *Canadian Geotechnical J.*, Vol. II, No. 1, February.

93. Wallace, B., Slebir, E. J. and Anderson, F. A. (1970) "In Situ Methods for Determining Deformation Modulus Used by the Bureau of Reclamation," *Determining the In-Situ Modulus of Deformation of Rock*, STP 477, Amer. Soc. for Test. and Mat., Philadelphia, Pa., pp. 3–26.

94. Jaeger, C. (1972) *Rock Mechanics and Engineering*, Cambridge Univ. Press, England.

95. Rocha, M. (1970) "New Techniques in Deformability Testing of in Situ Rock Masses," *Determination of the In-Situ Modulus of Deformation of Rock*, ASTM STP 477, Amer. Soc. for Test. and Mat., Philadelphia, Pa.

96. Stagg, K. G. (1969) "In-Situ Tests in the Rock Mass," *Rock Mechanics in Engineering Practice*, Stagg and Zienkiewicz, eds., John Wiley & Sons, New York, Chap. 5.

97. Misterek, D. L. (1970) "Analysis of Data from Radial Jacking Tests," *Determination of the In-Situ Modulus of Deformation for Rock*, ASTM STP 477, Amer. Soc. for Test. and Mat., Philadelphia, Pa.

98. Goodman, R. E., Van Tran, D. and Heuze, F. E. (1968) "The Measurement of Rock Deformability in Boreholes," *Symp. on Rock Mechs.*, Univ. of Texas, Austin.

99. Hibino, S., Hayashi, M., Kanagawa, T., and Motojima, M. (1977) "Forecast and Measurement of the Behavior of Rock Masses During Underground Excavation Works," *Proc. Intl. Symp. on Field Measurements in Rock Mechs.*, Zurich, Vol. 2, pp. 935–948.

100. Deere, D. U. and Miller, R. P. (1966) "Classification and Index Properties for Intact Rock," *Tech. Report AFWL-TR-65-116*, AF Special Weapons Center, Kirtland Air Force Base, New Mexico.

101. Coon, H. H. and Merritt, A. H. (1970) "Predicting In Situ Modulus of Deformation Using Rock Quality Indexes," *Determination of the In-Situ Modulus of Deformation of Rock, ASTM Spec. Tech. Pub. 477*, Amer. Soc. for Test. and Mat., Philadelphia, Pa., 154–173.

102. Deere, D. U., Hendron, A. J., Patton, F. D. and Cording, E. J. (1967) "Design of Surface and Nearsurface Construction in Rock," *Proc. 8th Symp. on Rock Mechs.*, Univ. of Minnesota, Amer. Min. and Metal. Engrs., Chap. 11.

103. Terzaghi, K. (1943) *Theoretical Soil Mechanics*, John Wiley & Sons, New York.

104. Casagrande, A. (1936) "The Determination of the Preconsolidation Load and Its Practical Significance," *Proc. 1st Intl. Conf. Soil Mechs. and Found. Engrg.*, Cambridge, Mass., Vol. 3, p. 60.

105. Perloff, W. H. (1975) "Pressure Distribution and Settlement," *Foundation Engineering Handbook*, Winterkorn and Fang, eds., Van Nostrand Reinhold, New York, Chap. 4.

106. Skempton, A. W. and Bjerrum, L. (1957) "A Contribution to the Settlement Analysis of Foundation on Clay, *Geotechnique*, Vol. VII, No. 4, London.

107. Hansen, B. J. (1967) "Refined Calculations of Foundation Movements," *Proc. 3rd Intl. Conf. Soil Mechs. and Found. Engrg.*, Caracas, Vol. III.

108. Davis, E. H. and Poulos, H. G. (1968) "The Use of Elastic Theory for Settlement Prediction under Three Dimensional Conditions," *Geotechnique*, Vol. XVII, No. 1, London.

109. Lambe, T. W. (1973) "Predictions in Soil Engineering," *Geotechnique*, Vol. 23, No. 2, London.

110. Kavazanjian, E., Jr. and Mitchell, J. K. (1980) "Time-Dependent Deformation of Clays," *Proc. ASCE, J. Geotech. Engrg. Div.*, Vol. 106, No. GT6, June, pp. 593–610.

111. Swiger, W. F. (1974) "Evaluation of Soil Moduli," *Proc. ASCE, Conf. on Analysis and Design in Geotechnical Engineering*, Univ. of Texas, Austin, Vol. II, pp. 79–92.

112. Menard, L. F. (1963) "Calculations of the Bearing Capacity of Foundations Based on Pressuremeter Results" *Sols-Soils*, Paris, Vol. 2, No. 5, June, and Vol. 2, No. 6, September.

113. Menard, L. F. (1965) "Rules for the Computation of Bearing Capacity and Foundation Settlement Based on Pressuremeter Results," *Proc. 6th Intl. Conf. Soil Mechs. and Found. Engrg.*, Montreal, Vol. 2, pp. 295–299.

114. Menard, L. F. (1975) "Interpretation and Application of Pressuremeter Test Results," *Sols-Soils*, Paris, Vol. 26, pp. 1–43.

115. Dixon, S. J. (1970) "Pressuremeter Testing of Soft Bedrock," *Determination of the In Situ Modulus of Deformation of Rock, ASTM Spec. Tech. Pub. 477*, Amer. Soc. for Test. and Mat., Philadelphia, Pa.

116. CGS (1978) *Canadian Foundation Engineering Manual*, Canadian Geotech. Soc., Montreal.

117. Ward, G. S. (1972) "In Situ Pressuremeter Testing on Two Recent Projects," *Soils*, J. S. Ward and Assocs., Caldwell, New Jersey, November, pp. 5–7.

118. Martin, R. E. (1977) "Estimating Settlements in Residual Soils," *Proc. ASCE, J. Geotech. Engrg. Div.*, Vol. 103, No. 6T3, March, pp. 197–212.

119. Silver, V. A., Clemence, S. P. and Stephenson, R. W. (1976) "Predicting Deformations in the Fort Union Formation," *Rock Engineering for Foundations and Slopes, Proc. ASCE*, Vol. I, pp. 13–33.

120. Menard, L. F. (1972) "Rules for the Calculation of Bearing Capacity and Foundation Settlement Based on Pressuremeter Tests," Draft Translation 159, U.S. Army Corps of Engineers, Cold Regions Research and Engineering Lab.

121. Poulos, H. G. (1971) "Behavior of Laterally Loaded Piles: I—Single Piles," *Proc. ASCE, J. Soil Mechs. and Found. Engrg. Div.*, Vol. 97, No. SM5, May, p. 722.

122. Clough, G. W. and Denby, G. M. (1980) "Self-Boring Pressuremeter Study of San Francisco Bay Mud," *Proc. ASCE, J. Geotech. Engrg. Div.*, Vol. 106, No. GT1, January, pp. 45–64.

123. Meyerhof, G. G. (1956) "Penetration Tests and Bearing Capacity of Cohesionless Soils," *Proc. ASCE, J. Soil Mechs. and Found. Engrg. Div.*, Vol. 82, No. SM1, p. 866.

124. Vesić, A. S. (1967) "A Study of Bearing Capacity of Deep Foundations," *Final Report, Project B-189*, Georgia Inst. of Technology, March, p. 170.

125. Terzaghi, K. (1955) "Evaluation of the Modulus of Subgrade Reaction," *Geotechnique*, Vol. 4, London.

126. Faccioli, E. and Resendiz, D. (1976) "Soil Dynamics," *Seismic Risk and Engineering Decisions*, Lomnitz and Rosenblueth, eds., Elsevier Pub. Co., Amsterdam, Chap. 4.

127. Novacs, W. D., Seed, H. B. and Chan, C. K. (1971) "Dynamic Moduli and Damping Ratios for a Soft Clay," *Proc. ASCE, J. Soil Mechs. and Found. Engrg. Div.*, Vol. 97, No. SM1, January pp. 59–75.

128. De Alba, P., Seed, H. B. and Chan, C. K. (1976) "Sand Liquefaction in Large Scale Simple Shear Tests," *Proc. ASCE, J. Geotech. Engrg. Div.*, Vol. 102, No. GT9, September, pp. 909–927.

129. Skogland, G. R., Marcuson, W. F. III and Cunny, R. W. (1976) "Evaluation of Resonant Column Test Device," *Proc. ASCE, J. Geotech. Engrg. Div.*, Vol. 102, No. GT11, November, pp. 1147–1158.

130. Hardin, B. M. and Drnevich, V. P. (1972) "Shear Modulus and Damping in Soils: Measurement and Parameter Effects," *Proc. ASCE, J. Soil Mechs. and Found. Engrg. Div.*, Vol. 98, No. SM6, June.

131. Seed, H. B. (1969) "Influence of Local Soil Conditions in Earthquake Damage," *Speciality Session 2, Soil Dynamics, Proc. 7th Intl. Conf. Soil Mechs. and Found. Engrg.*, Mexico City.

132. Richart, F. E. Jr. (1975) "Foundation Vibrations," *Foundation Engineering Handbook*, Winterkorn and Fang, eds., Van Nostrand Reinhold, New York, Chap. 24.

## BIBLIOGRAPHY


ASTM (1965) "Testing Techniques for Rock Mechanics," *ASTM Spec. Tech. Pub. No. 402*, Amer. Soc. for Test. and Mat., Philadelphia, Pa.

Barton, N. (1974) "Estimating Shear Strength on Rock Joints," *Proc. 3rd Intl. Cong. on Rock Mechs.*, Intl. Soc. for Rock Mechs., Denver, Vol. 11A, pp. 219–221.

Deklotz, E. J. and Boisen, B. (1970) "Development of Equipment for Determination of Deformation Modulus and In-Situ Stress by Large Flatjacks," *Determination of In-Situ Modulus of Deformation of Rock, ASTM Spec. Tech. Pub. 477*, Amer. Soc. For Test. and Mat., Philadelphia, Pa.

DeMello, V. F. B. (1971) "The Standard Penetration Test," State-of-the-Art Session 1, *Proc. 4th Panamerican Conf. Soil Mechs. and Found. Engrg.*, San Juan, Puerto Rico, Vol. 1, pp. 1–86.

Goodman, R. E. (1974) "The Mechanical Properties of Joints," *Proc. 3rd Intl. Cong. on Rock Mechs.*, Intl Soc. for Rock Mechs., Denver, Vol. 1A, pp. 127–140.

Grubbs, B. R. and Nottingham, L. C. (1978) "Applications of Cone Penetrometer Tests to Gulf Coast Foundation Studies," presented at meeting of Texas Section, ASCE, Corpus Christi, April.

Heuze, F. E. (1971) "Sources of Errors in Rock Mechanics Field Measurements and Related Solutions," *Intl. J. Rock Mechs.*, Min. Sci., Vol. 8.

Hilf, J. W. (1975) "Compacted Fill," *Foundation Engineering Handbook*, Winterkorn and Fang, eds., Van Nostrand Reinhold Co., New York, Chap. 7

Mitchell, J. K. (1976) *Fundamentals of Soil Behavior*, John Wiley & Sons, New York.

NCE (1978) "Instruments—Cambridge Camkometer Goes Commercial," *NCE International (The New Civil Engineer)*, Inst. Civil Engrs., London, October, p. 53.

Nelson, J. D. and Thompson, E. G. (1977) "A Theory of Creep Failure in Overconsolidated Clay," *Proc. ASCE, J. Geotech. Engrg. Div.*, Vol. 102, No. GT11, November, pp. 1281–1294.

Obert, L. and Duval, W. F. (1967) *Rock Mechanics and the Design of Structures in Rock*, John Wiley & Sons, New York.

Patton, F. D. (1966) "Multiple Modes of Shear Failure in Rock," *Proc. 1st Intl. Cong. for Rock Mechs.*, Intl. Soc. for Rock Mechs., Lisbon, Vol. 1, pp. 509–513.

Poulos, H. G., deAmbrosis, L. P. and Davis, E. H. (1976) "Method of Calculating Long-Term Creep Settlements," *Proc. ASCE, J. Geotech. Engrg. Div.*, Vol. 102, No. GT7, July, pp. 787–804.

Rapheal, J. M. and Goodman, R. E. (1979) "Strength and Deformability of Highly Fractured Rock," *Proc. ASCE, J. Geotech. Engrg. Div.*, Vol. 105, No. GT11, November, pp. 1285–1300.

Richart, F. E. Jr. (1960) "Foundation Vibrations," *Proc. ASCE, J. Soil Mechs. and Found. Engrg. Div.*, Vol. 88, August.

Richart, F. E., Jr., Hall, J. R., Jr. and Woods, R. D. (1970) *Vibrations of Soils and Foundations*, Prentice-Hall Inc., Englewood Cliffs, New Jersey.

Sharp, J. C., Maini, Y. N. and Brekke, T. L. (1973) "Evaluation of the Hydraulic Properties of Rock Masses," *New Horizons in Rock Mechanics, Proc. 14th Symp. on Rock Mechs.*, ASCE, University Park, Pa., pp. 481–500.

Woodward, R. J., Gardner, W. S. and Greer, D. M. (1972) *Drilled Pier Foundations*, McGraw-Hill Book Co., New York.

# CHAPTER FOUR

# FIELD INSTRUMENTATION

## 4.1 INTRODUCTION

### 4.1.1 OBJECTIVES

Instrumentation is installed to measure and monitor field conditions subject to change including:

- Surface movements
- Subsurface deformations
- In situ earth and pore pressures
- Stresses on structural members

### 4.1.2 APPLICATIONS

**Foundation Design Studies**

Measurements are made of deflections and stresses during load tests and preloading operations.

**Construction Operations**

During construction of buildings, embankments, retaining structures, open excavations, tunnels, caverns, and large dams, instrumentation is installed to monitor loads, stresses, and deformations to confirm design assumptions and to determine the need for changes or remedial measures.

**Postconstruction**

After construction has been completed, stresses and deformations may be monitored to provide an early warning against possible failure of slopes, retaining structures, earth dams, or concrete dams with rock foundations and abutments.

**Existing Structures**

Structures undergoing settlement are monitored to determine if and when remedial measures will be required and how stability will be achieved.

Instrumentation is installed to monitor the effects on existing structures of: (1) dynamic loadings from earthquake forces, vibrating machinery, or nearby blasting or other construction operations and (2) of deformations resulting from nearby tunneling or open excavations.

**Mineral Extraction**

The extraction of oil, gas, water, coal, or other minerals from the subsurface can result in surface deformations which require monitoring, as discussed in Arts. 10.2 and 10.3.

**Tectonic Movements**

In areas of crustal activity, movements and stresses related to surface warping and faulting are monitored as discussed in Art. 11.3.1.

### 4.1.3 PROGRAM ELEMENTS

**General**

In planning an instrumentation program, the initial steps are to determine what quantities should be measured and to select the instrument or instruments. The choice of instruments requires consideration of a number of factors, followed by a design layout of the instrumentation arrays for field installation.

Execution requires installation and calibration, data collection, recording, processing, and interpretation.

### Instrumentation Selection

#### *Condition to Be Monitored*

Instrumentation types and methods are generally grouped by the condition to be monitored, i.e., surface movements, subsurface deformation, and in situ pressures and stresses.

#### *Application*

Instruments vary in precision, sensitivity, reliability, and durability. The relative importance of these factors varies with the application or purpose of the program, such as collecting design data or monitoring construction and postconstruction works, existing structures, or the effects of mineral extraction or tectonic activity.

#### *Instrumentation Characteristics*

PRECISION Refers to the degree of measurement agreement required in relation to true or accepted reference values of the quantity measured.

SENSITIVITY Refers to the smallest unit detectable on the instrument scale, which generally decreases with the instrument measurement range.

REPEATABILITY Refers to obtaining sequential readings with similar precision, and can be the most significant feature since it indicates the trend in change of the measurement quantity and is often more important than sensitivity or precision.

DRIFT Refers to the instrument error progression with time and affects precision.

RELIABILITY Refers to the resistance to large deformations beyond the desired range, high pressures, corrosive elements, temperature extremes, construction activities, dirty environment, humidity and water, erratic power supply, accessibility for maintenance, etc. For a particular application, the less complicated the instrument the greater is its reliability. Thus, mechanical instruments are preferred to electrical; no moving parts are preferred to moving parts. Electrical instruments should be battery-powered for greater reliability.

SERVICE LIFE REQUIRED Can vary from a few weeks, to months, to a year, or even several years and is influenced by the reliability factors.

### Calibration

#### *Instrument Precision*

Precision is verified upon receipt of an instrument from the manufacturer by checking it against a standard, and corrections are considered for temperature and drift. After installation, precision is verified, then checked periodically during monitoring, and upon project completion.

#### *Bench Marks or Other Reference Data*

Reference points such as bench marks are installed or established and protected to ensure reliability against all possible changes that may occur during the project life. "Temporary" reference points which require transfer to permanent references should be avoided.

### Installation and Maintenance

Some systems are relatively simple to install and maintain, since they are portable and require only surface reference points; others require boreholes or excavations and are difficult to install and maintain. These factors substantially affect costs.

### Operations

#### *Data Collection and Recording*

Data collection should be on a planned basis. Data may be collected and recorded manually or automatically, directly or remotely, and on a periodic or continuous basis. Automatic collection and remote recording can be connected to alarm systems and are useful in dangerous situations, or in situations where inactivity may continue for long time intervals followed by some significant occurrence such as an earthquake or heavy rainfall.

Readout systems measure and display the measured quantity.

*Frequency of Observations*

The construction progress, data trends, and interpretation requirements influence the frequency of the observations. Periodic monitoring is always required and the frequency of readings increases when conditions are critical.

*Data Processing and Interpretation*

Data are recorded, plotted, reviewed, and interpreted on a programmed basis, either periodic or continuous. Failures have occurred during monitoring programs when data were recorded, collected, and filed, but not plotted and interpreted.

## 4.1.4 TRANSDUCERS

### General

Transducers, important elements of complex instruments, are devices actuated by energy from one system to supply energy, in the same or some other form, to a second system. They function generally on the membrane principle, the electrical resistance gage principle, or the vibrating-wire principle. Occasionally they function on the linear-displacement principle.

### Membrane Principle

Pressure against a membrane is measured either hydraulically (for example, by hydraulic piezometers) or pneumatically (for example, by the Gloetzl pressure cell, the pneumatic piezometer, or geophones).

### Resistance Strain Gages

*Applications*

Resistance strain gages are cemented to structural members or used as sensors in load cells, piezometers, extensometers, inclinometers, etc.

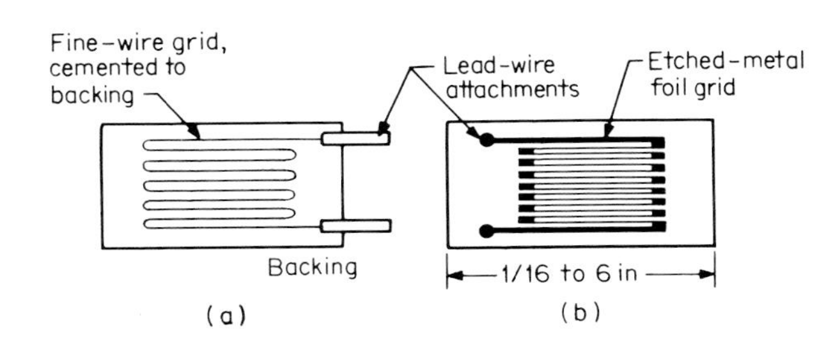


**FIG. 4.1 Bonded resistance strain gages: (*a*) wire grid and (*b*) foil grid.**

*Principle*

The straining of a wire changes its cross-sectional area and consequently its electrical resistance. When a strain-gage wire is attached to a structural member, either externally or as the component of an instrument, measurements of changes in electrical resistance are used to determine strains in the structural element. There are several types.

*Bonded Gages*

The most common form is the bonded gage consisting of a thin wire filament or metal foil formed into a pattern and bonded to a backing of paper, thin plastic, or epoxy (Fig. 4.1). The backing is then cemented to the surface where the measurements are to be made.

Gage lengths range from 1/16 to 6 in, and strain sensitivity is usually in the order of 2 to 4 microstrains with ranges up to 20,000 to 50,000 microstrains (2 to 5% strain) for normal gages. High elongation gages ranging up to strains of 10 to 20% are available.

*Encapsulated Gages*

Bonded gages mounted and sealed in the factory into a stainless steel or brass envelope are termed encapsulated gages. They are welded to the measurement surface, or embedded in concrete. Encapsulated gages provide better protection against moisture than do bonded gages.

*Unbonded, Encapsulated Gages*

A fine wire is strung under tension over ceramic insulators mounted on a flexible metal frame to

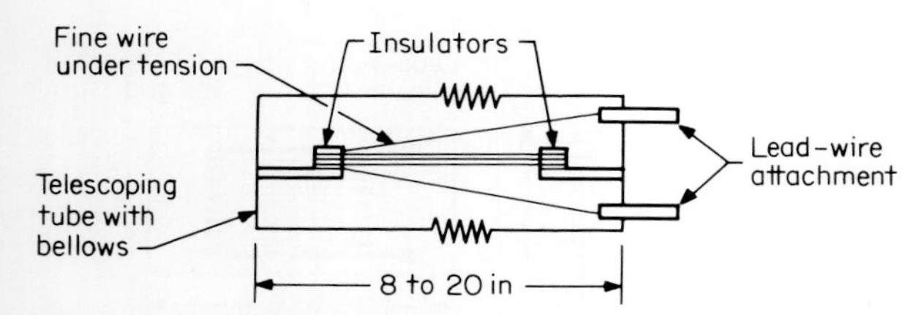


**FIG. 4.2 Unbonded resistance strain gage.** [*From Cording et al. (1975).*[1]]

form a resistance coil in unbonded, encapsulated gages (Fig. 4.2). Usually two coils are used, arranged so that one contracts while the other expands when the frame is strained. The coils and the frame are factory-sealed into a tubular metal cover to provide moisture protection. Mounting is achieved by bolting the gage to saddle brackets previously tack-welded or bolted to the measuring surface. The *Carlson strain meter* is one version of the unbonded, encapsulated resistance gage.

*Readouts*

Strain indicators measure the resistance changes of the gages. They consist of a power supply, various fixed and variable resistors, a galvanometer, and a bridge circuit that can be switched to connect the strain gages, resistors, and galvanometers in various configurations.

*Temperature Compensation*

Temperature compensation is always required and can be achieved by a dummy gage mounted so that it responds only to temperature-induced resistance changes or by the use of self-compensating gages.

### Vibrating-Wire Devices

*Applications*

Vibrating-wire devices are used for measurements of:

- Strain with the transducer mounted on steel or embedded in concrete
- Displacement and deformation with extensometers and joint meters
- Tensile and compressive forces
- Pore and joint water pressure
- Changes in rock stresses
- Changes in surface or subsurface inclinations
- Temperature changes

*Principle*

A vibrating steel wire is attached to the object on which a measurement is to be made. A change in the quantity to be measured causes a change in the stress in the measuring wire and consequently a change in its frequency.

The measuring wire oscillates within a magnetic field, inducing in a coil an electrical oscillation of the same frequency, which is transmitted by cable to a receiving instrument.

*Types*

*Intermittent vibrating-wire* systems, used for static and semistatic measuring systems, consist of a transducer, containing the measuring wire and an electromagnet, and a receiver.

*Continuous vibrating-wire* systems contain the measuring wire, an oscillator, and two electromagnets in the transducer, and are used for static and dynamic measuring devices and for alarm and control systems [Dreyer (1977)[2]].

*Characteristics*

Gage lengths generally range from 4 to 14 in, and strain sensitivity has typical maximum ranges from 600 to 7000 microstrains.

### Linear-Displacement Transducers

*Applications*

Linear-displacement transducers are used generally in extensometers.

*Linear Potentiometer*

Resistance devices, linear potentiometers consist of a mandrel wound with fine wire or conductive film. A wiper attached to a shaft rides along the mandrel and divides the mandrel resistance into two parts. The resistance ratio of these parts is measured with a Wheatstone bridge circuit to determine the displacement of the wiper and shaft.

Movement ranges are generally about 0.5 to 24 in with an average sensitivity of 0.01 to 0.001.

Linear potentiometers are extremely sensitive to moisture and require absolute sealing.

*Linear Variable Differential Transformer (LVDT)*

The LVDT converts a displacement into a voltage change by varying the reluctance path between a primary coil and two or more secondary coils when an excitation voltage is applied to the primary coil. Variations in the output signal are calibrated to displacements of the LVDT core.

Movement ranges are generally about 0.1 ft to several feet; sensitivity is about 10 microstrains.

LVDTs are much less sensitive to moisture and less affected by temperature than are linear potentiometers.

#### 4.1.5 METHODS AND INSTRUMENTS SUMMARIZED

A comprehensive reference on instrumentation has been prepared by Cording et al. (1975).[1] Methods and instruments are summarized for general reference on the following tables:

- Table 4.1—method or instrument vs. condition to be monitored
- Table 4.2—applications of simple vs. complex instruments
- Table 4.3—methods or instrument summarized on basis of category of application

## 4.2 SURFACE MOVEMENTS

### 4.2.1 FORMS AND SIGNIFICANCE

#### Vertical Displacements

Settlement or heave is vertical surface displacement which results in detrimental distortion of structures, especially if differential, when significant magnitudes are reached.

#### Tilt

Tilt can indicate differential settlement of a structure, impending failure of walls and slopes, or major tectonic movements.

#### Lateral Displacement

Lateral surface displacements can signify slope failure or fault activity. The surfaces of tunnels and other underground openings undergo *convergence* from stress relief, which may lead to failure.

#### Vibrations

Construction activity or earthquakes can induce vibrations in structures resulting in overstress and failure.

#### Summary

Surface movement forms, occurrence, and applicable monitoring methods are summarized on Table 4.4.

### 4.2.2 OPTICAL SYSTEMS

#### Survey Nets

*Applications*

Survey nets making use of optical systems are used to monitor deflections of slopes, walls, buildings, and other structures, as well as ground subsidence and heave.

*Methods*

Ranges, accuracy, advantages, limitations, and reliability of various surveying methods are summarized on Table 4.5.

*Procedures*

*Reference points* are installed to provide a fixed framework. They include ground monuments, pins in structures, and immovable bench marks.

*Close-distance surveys* are usually performed with a first-order theodolite, and critical distances are measured by chaining with Invar or steel tapes, held under a standard tension.

*Long-distance surveys*, such as for slope movements, are more commonly performed with the *geodimeter* (electronic distance-measuring unit or EDM) because of a significant reduction in measurement time and the increased accuracy for long ranges. Distances are determined by measuring the phase difference between trans-

**TABLE 4.1**
**FIELD INSTRUMENTATION: METHODS AND DEVICES**

| Condition to be monitored | Surface movements: Survey nets | Water level device | Settlement plates | Vertical extensometer | Tiltmeters | Inverse pendulum | Convergence meters | Strain meter | Terrestrial photography | Vibration monitors |
|---|---|---|---|---|---|---|---|---|---|---|
| **Settlements:** | | | | | | | | | | |
| Steel and masonry structures | x | x | x | x | x | x | | | x | |
| Fill embankments | x | | x | x | | | | | | |
| Dam embankments | x | | x | x | | | | x | | |
| Ground subsidence | x | | x | x | x | | | x | | |
| **Stability:** | | | | | | | | | | |
| Soil slopes | x | | | | | | | x | | |
| Rock slopes | x | | | | x | | x | x | x | |
| Embankment and dam foundation | x | | | | | | | | | |
| Retaining structures | x | | | | x | | | x | x | |
| Underground openings in rock | x | | | | | | x | | | |
| Tunnel linings | x | | | | | | x | | | |
| **Deformations:** | | | | | | | | | | |
| Subsurface: | | | | | | | | | | |
| Soils, vertical | x | | x | x | | | | x | | |
| Soils, lateral | x | | | | | | | | | |
| Rock | x | | | | | | x | | | |
| Faults | x | | | | | | | x | | x |
| Piles during load test | x | | | | | | | | | |
| Structural elements | | | | | | | | | | |
| **Pressures:** | | | | | | | | | | |
| Pore water | | | | | | | | | | |
| Embankments | | | | | | | | | | |
| Against walls, beneath foundations | | | | | | | | | | |
| **Loads:** | | | | | | | | | | |
| On structural elements | | | | | | | | | | |
| **Residual stresses:** | | | | | | | | | | |
| Rock masses | | | | | | | | | | |
| **Vibrations:** | | | | | | | | | | |
| Seismic, man-induced | | | | | | | | | | x |

TABLE 4.1
FIELD INSTRUMENTATION: METHODS AND DEVICES (*CONTINUED*)

| | Subsurface deformation | | | | | | | Loads and stresses | | | | | | | | |
|---|---|---|---|---|---|---|---|---|---|---|---|---|---|---|---|---|
| | **Settlement points** | **Inclinometer** | **Deflectometer** | **Shear-strip indicator** | **Borehole extensometer** | **Electric strain meter** | **Acoustic emissions** | **Piezometers** | **Pressure cells** | **Load cells** | **Tell tales** | **Strain gages** | **Strain meters** | **Stress meters** | **Shallow-rock stress** | **Deep-rock stress** |
| | x | | | | | | | x | | | | | | | | |
| | x | | | | | | | x | | | | | | | | |
| | x | | | | | x | | x | | | | | | | | |
| | x | | | | | x | | x | | | | | | | | |
| | | x | | x | | | x | x | | | | | | | | |
| | | x | x | x | x | | x | x | | | | | x | | | |
| | | x | | | | x | x | x | x | | | | | | | |
| | | | x | | | | | x | x | x | | | | | | |
| | | | x | | x | | x | x | | | | | | x | x | x |
| | | | x | | x | | | | x | x | | | | | | |
| | x | | | | | | | x | | | | | | | | |
| | | x | | x | | | | x | | | | | | | | |
| | | | x | x | x | | | x | | | | | | x | x | x |
| | | x | x | x | | x | | x | | | | | | | | |
| | | x | | | | | | | x | x | x | x | x | | | |
| | | | | | | | | | x | x | | x | x | | | |
| | | | | | | | | x | | | | | | | | |
| | | | | | | | | | x | | | | | | | |
| | | | | | | | | | x | | | | | | | |
| | | | | | | | | | | x | | x | x | | | |
| | | | | | | | | | | | | x | x | | x | x |
| | | | | | | | | | | | | | | | | |

**TABLE 4.2**
**INSTRUMENTATION: SIMPLE VS. COMPLEX METHODS AND DEVICES**

| Applications | Simple* | Complex† |
|---|---|---|
| Surface movements | Optical surveys of monuments, settlement plates<br>Water-level device<br>Simple strain meter<br>Wire extensometer | Electrical tiltmeter<br>Electrical strain meter<br>Vertical extensometer |
| Subsurface deformations | Settlement points<br>Borros points<br>Rock bolt and rod-type MPBX extensometers<br>Shear-strip indicator | Inclinometer<br>Deflectometer<br>Wire-type MPBX (extensometer)<br>Acoustical emissions |
| In situ pressures and stresses | Tell tales (pile load tests)<br>Open-system piezometers | Pneumatic and electric piezometers<br>Strain gages<br>Pressure cells<br>Vibrating-wire stress meter |
| Residual rock stresses | Strain meter on rock surface<br>Flat jacks | Borehole devices |

**Simple types* are read optically or with dial gages. They are less costly to install and less subject to malfunction.

†*Complex devices* are attached to remote readouts, and many can be set up to monitor and record changes continuously. The device and installation are more costly, and reliability will usually be less than that of the simpler types. In many cases, however, there is no alternative but to use the more complex type.

mitted and reflected light beams, using the laser principle. Calculation and data display can be automated by using an electronic geodimeter with a punched-tape recorder. Accuracy for slope monitoring is about 30 mm (as compared with 300 mm for the theodolite) [Blackwell et al. (1975)[3]], depending on the sight distance. An example of a scheme for geodetic control of a landslide is given on Fig. 4.3. Sequential plotting of the data reveals the direction and velocity of the movements.

### Water-Level Device

*Application*

Floor, wall, or column deflections inside structures undergoing settlements or heave may be monitored with the water-level device. Accuracy ranges are $\pm 0.001$ to 0.5 in.

*Procedure*

Monument pins are set at a number of locations on walls, columns, and floors. Elevations relative to some fixed point are measured periodically with the device, which consists of two water-level gages connected by a long hose. In operation, one gage is set up at a bench mark elevation and the other is moved about from pin to pin and the elevation read at the control gage. The principle is similar to that of the theodolite except that the water-level device is quicker and simpler for working around walls and columns inside buildings. Temperature differential, as

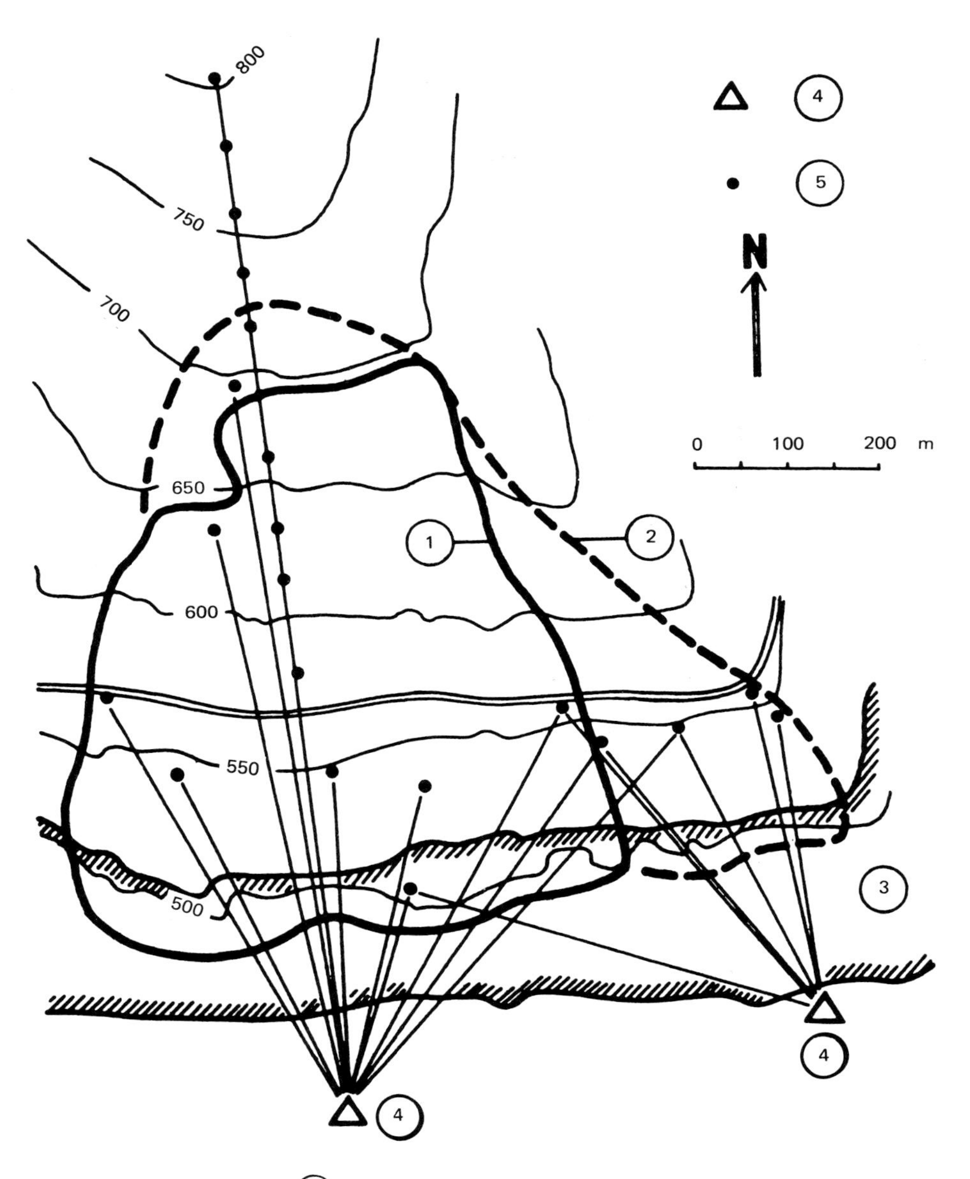


(1) Boundary of old landslide
(2) Demarcation of potential landslide
(3) Sance Reservoir on the Ostravice River
(4) Fixed reference points on the opposite stable slope
(5) Concrete monuments set in monitored slope

**FIG. 4.3 Scheme of the geodetic control of the Recice landslide. Note that a laser geodimeter was used in the survey because of significant reduction in measurement time and increased accuracy.** [*From Novosad (1978)*[4].]

**TABLE 4.3**
**INSTRUMENTATION OF MOVEMENTS, DEFORMATIONS, AND STRESSES**

| Method/instrument | Applications |
|---|---|
| **SURFACE MOVEMENTS** | |
| Survey nets | Vertical and horizontal movements of slopes, walls; settlements by precise leveling, theodolite, or laser geodimeter. Requires stable bench mark. Slow except with laser. |
| Water-level device | To monitor building settlements optically. |
| Settlement plates | Installed at base of fills and read optically for settlement monitoring. |
| Remote settlement monitor (settlement extensometer) | Installed at base of fills, over tunnels in soft ground, adjacent to excavations to monitor vertical deflections. Reference point installed below in rock or strong soils. Instrument connected electrically to readout or recorder. |
| Tiltmeters | Measure rotational component of deflection electronically. Used in buildings, on walls, and rock-cut benches. |
| Pendulums | Monitor tilt in buildings adjacent to excavations. |
| Convergence meters | Measure convergence in tunnels, between excavation walls, and down cut slopes in rock. |
| Surface extensometers or strain meters | Measure linear strains downslope or across faults. Small strain meters for joints. |
| Terrestrial stereophotography | Monitor movements of slopes, retaining structures, and buildings. Less accurate than optical systems. |
| Vibration monitoring | Monitor vibrations caused by blasting, pile driving, traffic, etc. |
| **SUBSURFACE DEFORMATIONS** | |
| Vertical-rod extensometer | Settlement points installed at various depths allow measurement of increments of vertical deflection. Borros points, cross-arm devices, and the remote settlement monitor also are used. |
| Inclinometer | Measure lateral deflections. Used behind walls, in lateral pile-load tests, for measuring deflections beneath loaded areas over soft soils, and to locate the failure surface in a slope and monitor slope movements. |
| Deflectometers | Used in rock as permanent installation to monitor movements perpendicular to the borehole in rock slopes, open-pit mines, and fault zones. |
| Shear-strip indicators | Used to locate failure surface in earth mass and to give alarm when failure occurs. |
| Borehole extensometers | Installed singly or in series (MPBX) in boreholes to monitor deflections occurring parallel to hole. Used to monitor slopes in rock, tunnels, and caverns. Installed in any orientation. |

**TABLE 4.3**
**INSTRUMENTATION OF MOVEMENTS, DEFORMATIONS, AND STRESSES (continued)**

| Method/instrument | Applications |
|---|---|
| **SUBSURFACE DEFORMATIONS** | |
| Electrical strain meters | Installed below the surface in earth dams to monitor longitudinal strains between embankment and abutment and to locate transverse cracks. |
| Acoustical emissions device | Detect and monitor subaudible noise in soil and rock resulting from distress caused by slope movements and mine collapse, and along faults. Also used to locate leakage paths in dams. |
| **IN SITU PRESSURES AND STRESSES** | |
| Piezometers | Monitor pore-water pressures in slopes, dewatered excavations, beneath embankments, in dams, beneath buildings, and during preloading. Various systems available. Application depends on soil/rock conditions, response time required, and necessity for remote readout and recording. |
| Stress or pressure cells | Measure stresses behind walls, in tunnel linings, beneath foundations during load test, and in embankments. |
| Load cells | Measure loads in anchors, wall braces, and tunnel lining. |
| Tell tales | Measure deflections at various depths in a pile during load test. Used to compute side friction and end bearing. |
| Strain gages | Measure strains in piles during load test, bracing for retaining structures, earth and rock anchors, and steel storage tank walls during hydrostatic testing. |
| Strain meters | Purposes similar to tiltmeters above, but meters are encased so as not to be susceptible to short circuits, and are welded to the structure so as to not be subject to long-term creep of a cementing agent. |
| Stress meters | Installed in boreholes to measure stress changes during tunneling and mining operations. |
| **RESIDUAL ROCK STRESSES** | |
| *Shallow-depth methods* | |
| Strain meters or rosettes | Stresses a short distance behind the wall remain unknown and excavation for test relieves some residual stress. |
| Flat jacks | Relatively low costs. Used in good quality rock. |
| *Deep methods* | |
| Borehole devices<br>Deformation gage<br>Inclusion stress meter<br>Strain gage | Permits deep measurement of residual stresses by borehole overcoring techniques. Any borehole orientation is possible, but installation and overcoring are difficult operations. Practical depth limit is about 10 to 15 m. |
| Hydraulic fracturing | Allows very deep measurement, about 300 to 1500 m. Boreholes are limited to vertical or near-vertical. Technique is in development stages. |

**TABLE 4.4**
**SURFACE MOVEMENTS: FORMS, OCCURRENCE, AND MONITORS**

| Movement | Occurrence | Monitoring method | Reference |
|---|---|---|---|
| Settlement or subsidence | Fills, surcharge; buildings | Optical survey nets with monuments or settlement plates | Art. 4.2.2 |
| | Over tunnels, behind walls | Settlement extensometers | Art. 4.2.4 |
| Heave or rebound | Fills, surcharge; buildings | Optical survey nets | Art. 4.2.2 |
| | | Settlement extensometers | Art. 4.2.4 |
| Differential settlement | Within buildings | Optical survey nets | Art. 4.2.2 |
| | | Water level device | Art. 4.2.2 |
| Tilt | Retaining structures, buildings, rock-slope benches | Portable electric tiltmeter | Art. 4.2.3 |
| | | Pendulums | Art. 4.2.3 |
| | | Terrestrial stereophotography | Art. 4.2.2 |
| | Ground surface | Mercury pools set in piers | Art. 4.2.3 |
| Lateral displacement | Slopes | Survey nets | Art. 4.2.2 |
| | | Terrestrial photography | Art. 4.2.2 |
| | | Wire extensometer | Art. 4.2.4 |
| | | Precision electric strain meter | Art. 4.2.4 |
| | | Sliding wire contact | Art. 4.2.4 |
| | Tension cracks, joints | Simple strain meter with pins | Art. 4.2.4 |
| | Faults | Survey nets | Art. 4.2.2 |
| | | Precision electric strain meter | Art. 4.2.4 |
| Convergence | Tunnels | Wire extensometer | Art. 4.2.4 |
| Vibrations | Blasting, pile driving, etc. | Vibration monitors | Art. 4.2.5 |

might be caused by radiators or sunshine affecting the hose, may introduce considerable error.

*Data Reduction*

Plots of movements measured periodically reveal the rate and magnitude of differential settlement or heave within the structure. Determination of total movements requires an immovable reference point such as a bench mark set into rock.

**Settlement Plates**

*Application*

Ground surface settlements under fills and surcharges are monitored with settlement plates.

*Procedure*

Plates, at least 3 ft (90 cm) square, made of heavy plywood or steel, are placed carefully on prepared ground, and a 1-in rod is attached to the plate and its elevation determined by survey.

TABLE 4.5
SURVEYING METHODS SUMMARIZED*

| Method | Range | Accuracy | Advantages | Limitations and precautions | Reliability |
|---|---|---|---|---|---|
| Chaining | Variable | $\pm\frac{1}{5000}$ to $\pm\frac{1}{10,000}$ Ordinary (3d order) survey. $\pm\frac{1}{20,000}$ to $\pm\frac{1}{200,000}$ Precise (first-order) survey. | Simple, direct observation. Inexpensive. | Requires clear, relatively flat surface between points and stable reference monuments. Corrections for temperature and slope should be applied and a standard tension used. | Excellent |
| Electronic distance measuring | 50 to 10,000 ft | $\pm\frac{1}{50,000}$ to $\pm\frac{1}{300,000}$ of distance | Precise, long range, fast. Usable over rough terrain. | Accuracy influenced by atmospheric conditions. Accuracy at short ranges (under 100–300 ft) is limited. | Good |
| Optical leveling | | | | | |
| Ordinary—2d and 3d order | | ±0.01 to ±0.02 ft | Simple, fast (particularly with self-leveling instruments). | Limited precision. Requires good bench mark nearby. | Excellent |
| Precise parallel-plate micrometer attachment, special rod, first-order techniques. | | ±0.002 to ±0.004 ft | More precise. | Requires good bench mark and reference points, and careful adherence to standard procedures. | Excellent |
| Offsets from a baseline | | | | | |
| Theodolite and scale | 0 to 5 ft off baseline | ±0.002 to ±0.005 ft | Simple direct observation. | Requires baseline unaffected by movements and good monuments. Accuracy can be improved by using a target with a vernier and by repeating the sight from the opposite end of the baseline. | Excellent |
| Laser and photocell detector | 0 to 5 ft off baseline | ±0.005 ft | Faster than transit. | Seriously affected by atmospheric conditions. | Good |
| Triangulation | | ±0.002 to 0.04 ft | Usable when direct measurements are not possible. Good for tying into points outside of construction area. | Requires precise measurement of base distance and angles. Good reference monuments are required. | Good |
| Photogrammetric methods | | $\pm\frac{1}{5000}$ to $\pm\frac{1}{50,000}$ | Can record hundreds of potential movements at one time for determination of overall displacement pattern. | Weather conditions can limit use. | Good |

*From Cording et al. (1975).[1]

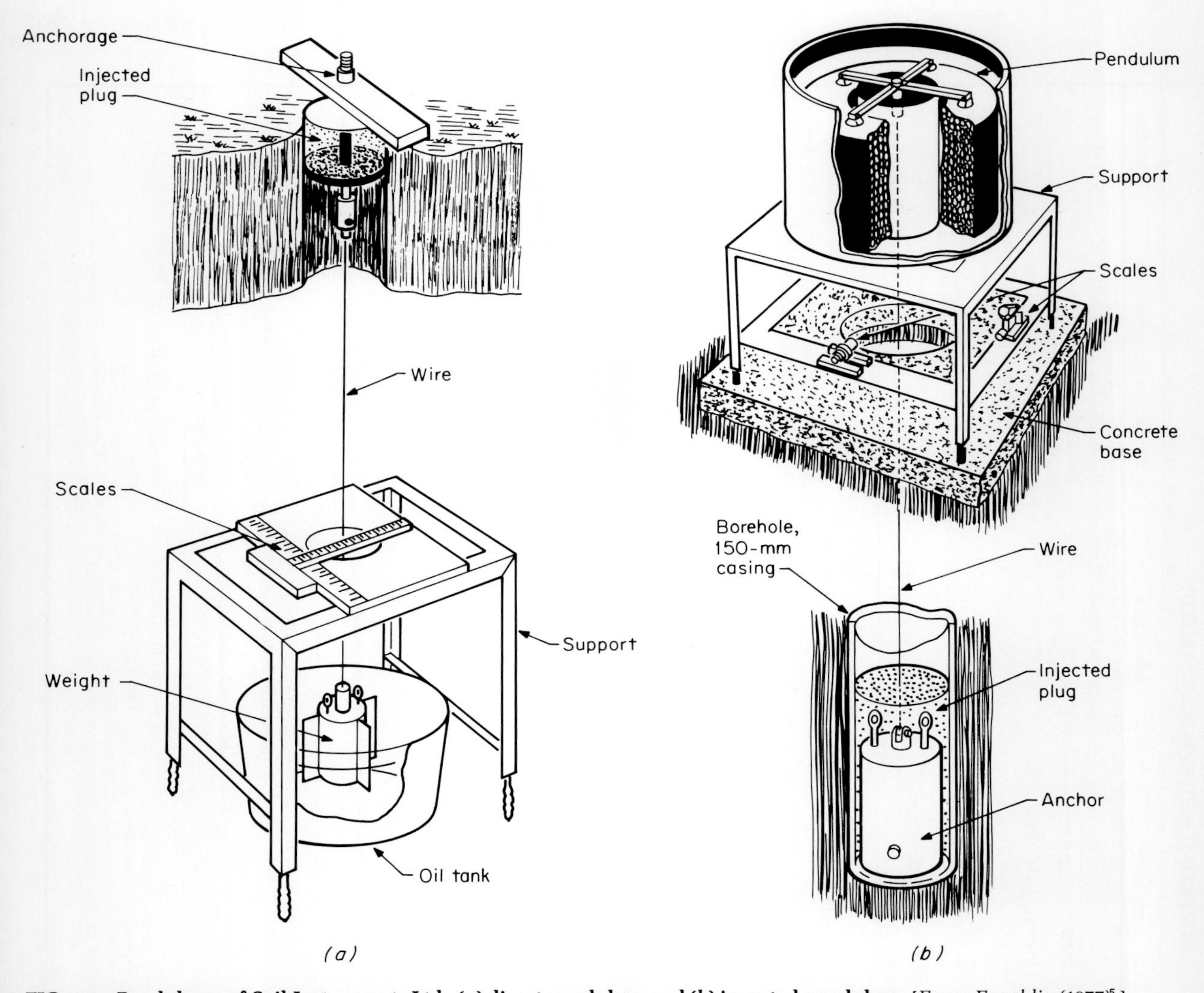


**FIG. 4.4 Pendulums of Soil Instruments Ltd.: (*a*) direct pendulum and (*b*) inverted pendulum.** [*From Franklin (1977)*[5].]

As fill is placed, readings are taken of attachments, and when necessary, additional rod sections are attached. A relatively large number of plates should be installed on a given site, not only to provide data on differential movements, but also because invariably some will be disturbed or destroyed during filling operations.

*Data Reduction*

Plots of movements provide data on the rate and magnitude of settlement when referenced to an immovable bench mark.

**Terrestrial Photography**

*Applications*

Slopes, retaining structures, buildings, and other structures suffering movements may be monitored on a broad basis with terrestrial photography.

*Procedure*

A bench mark reference point is established and photographs are taken periodically from the same location, elevation, and orientation with a high-resolution camera and film. Moving the camera a short distance laterally (about 1 m) and taking a subsequent photo provides photo pairs suitable for stereoscopic viewing when enlarged.

*Data Obtained*

The direction and relative magnitude of movements are detected by comparing sequential photos. The precision is not equal to that of standard survey techniques, but it may be adequate for detecting changes in locations of survey reference points and providing for approximate measurements if a reference scale is included in the photo.

Stereo terrestrial photographs are extremely useful for mapping joints and other discontinuities on rock slopes.

### 4.2.3 TILTMETERS

**Electrical Tiltmeter**

*Application*

Tilting or rotational movements of retaining walls, structures, and rock benches are monitored with electrical tiltmeters.

*Instrument*

The portable tiltmeter manufactured by Sinco utilizes a closed-loop-force-balanced servoaccelerator to measure tilt with an accuracy reported to be equivalent to a surface displacement of 0.06 in of a rock mass rotating about an axis 100 ft beneath the surface, or a displacement of 200 $\mu$in over the 4-in length of the instrument.

*Procedure*

Ceramic plates are cemented to the surface to be monitored. The plates contain pegs which serve as reference points for aligning the tiltmeter and measuring the angular deflection.

**Pendulums**

*Application*

Pendulums are used to measure the horizontal distance between two points at different elevations. They are installed in the highest elevations of structures adjacent to areas of open excavations or over tunneling operations where ground subsidence may cause differential settlement and tilting.

*Instrument*

The two general pendulum types, direct and inverted, are illustrated on Fig. 4.4.

**Mercury Pools Set in Piers**

*Application*

Mercury pools set in piers were installed by National Oceanic and Atmospheric Administration's (NOAA) Earthquake Mechanism Laboratory, Stone Canyon, California, adjacent to the San Andreas fault to monitor ground surface movements over a large area [Bufe (1972)[6]].

*Instrument*

Similar in principle to the water-level device, mercury pools 20 m apart in an interconnected array were set in piers attached to Invar rods driven to resistance in the ground.

### 4.2.4 EXTENSOMETERS

**Simple Strain Meter**

*Application*

Joint or tension crack deformations in rock masses may be monitored with simple strain meters; they are particularly useful on slopes.

*Installation*

Pins are installed in a triangular array adjacent to the crack and initial reference measurements are made. Periodic measurements monitor vertical, lateral, and shear displacements along the crack.

**Simple Sliding Wire Contacts**

*Application*

Slope movements are monitored with simple sliding wire contacts which give warning of large deformations [Kennedy (1972)[7]], a particularly useful feature for open-pit-mine slopes.

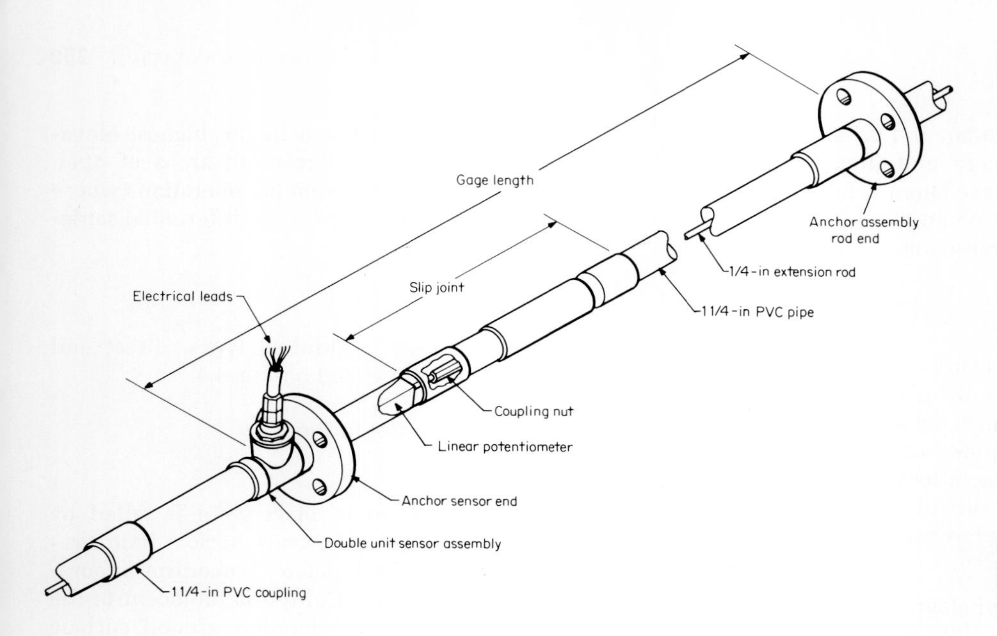


**FIG. 4.5 In-line assembly of multiple electrical strain meters.** *(Courtesy of Sinco.)*

*Installation*

A stake is driven into the ground and attached by wire to a stake driven into an unstable area. The wire contains two separated copper contacts and a spring system. The spacing between the contacts is set to some desired distance, and, when contact is made during slope movement, a relay is tripped and a warning light or siren activated.

### Electrical Strain Meter

*Application*

Linear strains occurring during slope movements or fault displacements are monitored with electrical strain meters which can be used in nets and set as an early warning system. (They are also used to monitor differential vertical displacements, Art. 4.3.4.)

*Instrument*

Linear potentiometers are mounted in slip-jointed polyvinyl chloride (PVC) pipe, as shown on Fig. 4.5.

*Performance*

Installed in shallow trenches, they are read and continuously recorded at remote receiving stations, and can be set in series to extend over long distances.

Changes of several inches can be measured over gage lengths of several feet with a sensitivity of $10^{-3}$ times the movement range.

### Vertical Extensometer

*Application*

Settlement measurements can be made automatically and recorded continuously with the remote settlement extensometer.

*Instrument*

Telescoping tubing is installed through a compressible stratum and founded in a rigid material. A weight and pulley system is connected to a detector which is in turn connected to a remote readout station, as shown on Fig. 4.6. Surface settlement beneath the fill or surcharge load lowers the weight attached to a stainless steel cable which causes the pulley in the detector to rotate. Deflection is presented on a digital readout located at the remote station, with a range in the order of 0.001 ft.

### Wire Extensometer (Convergence Meter)

*Applications*

Wire extensometers are used to measure movements of soil- or rock-cut benches in the downslope direction as indicators of potential failure

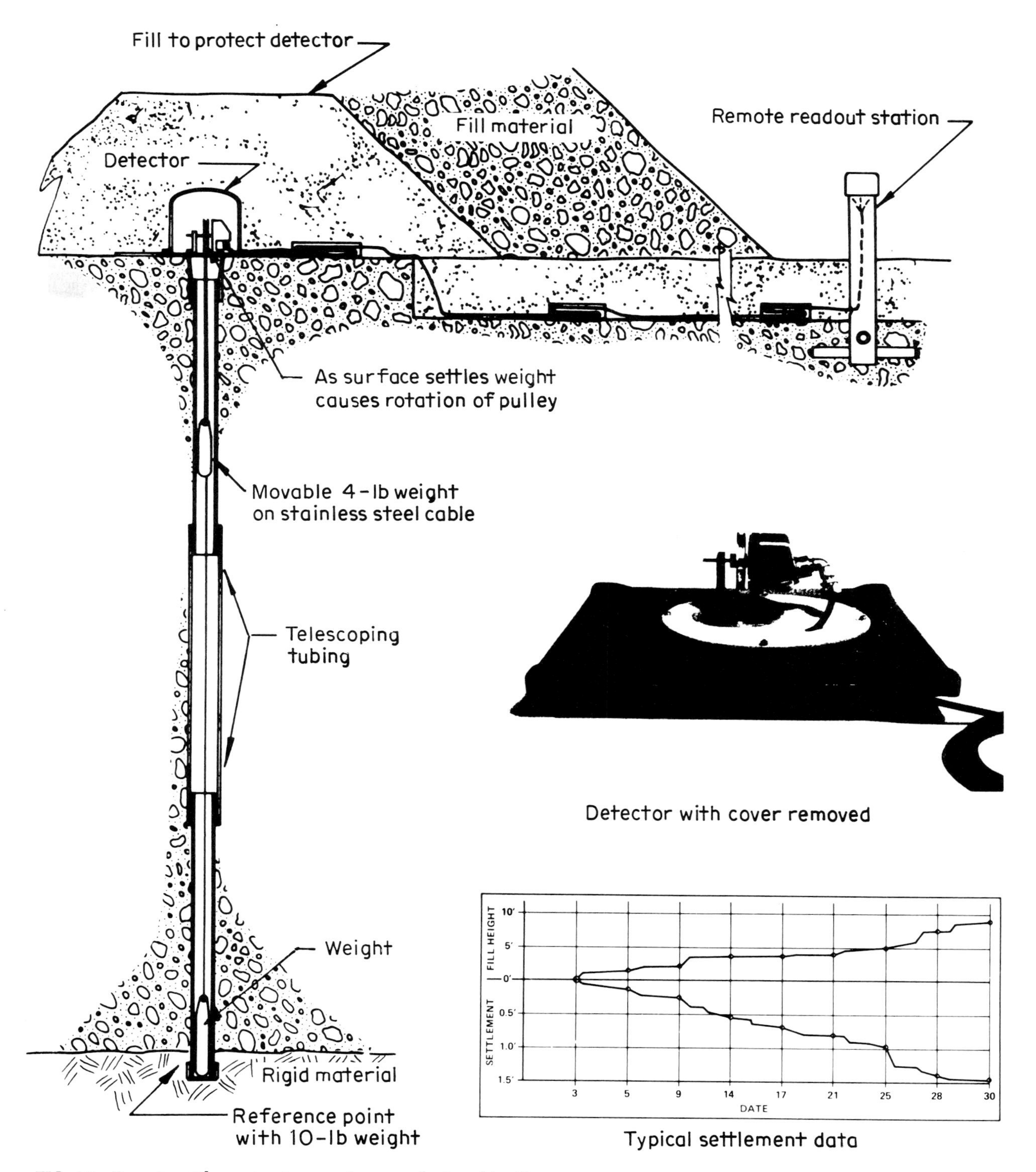


**FIG. 4.6 Remote settlement extensometer manufactured by Sinco.**

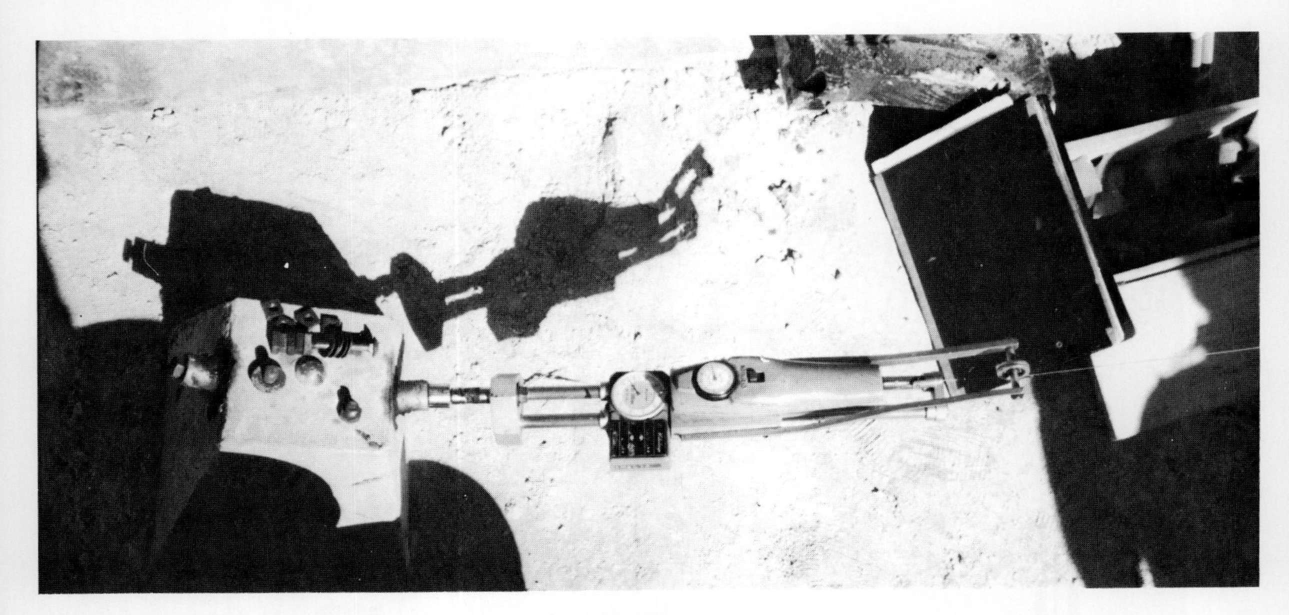

**FIG. 4.7 Tensioning device and dial indicator used to measure convergence.**

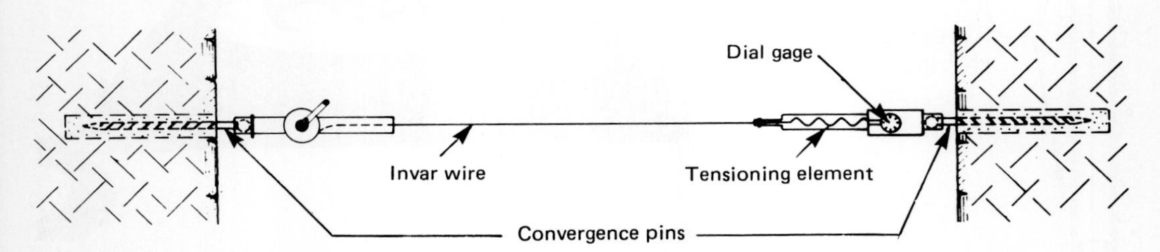


**FIG. 4.8 Convergence meter.** [*From Silveira (1976).*[9]]

or to monitor convergence in tunnels, particularly along shear zones where movements are likely to be relatively large, to provide data on the need for additional supports [Hartmann (1966)[8]].

*Instrument*

An Invar steel wire is attached to a tensioning device and dial indicator as shown on Fig. 4.7. Reference pins are grouted into tunnel walls or mounted in monuments on slopes.

One end of the wire is connected to a pin (Fig. 4.7) and the instrument is attached to the opposite pin (Fig. 4.8). The wire is adjusted to a calibrated tension and the distance between pins is measured with the dial gage.

Well-made instruments, such as those of Interfels, used properly, have an accuracy of $1 \times 10^{-5}$ times the measurement length [Silveira (1976)[9]].

*Slope Monitoring*

As shown on Fig. 4.9, pins are set in monuments on slope benches in a series of parallel lines running up- and down-slope, with each line starting well behind the slope crest and any potential failure surface, and extending to the bottom as shown on Fig. 4.40.

Either positive or negative movements will be measured between any two adjacent pins. Interpretation of where movements are occurring along the slope requires plotting the readings between each two pins on a section showing all of the pins.

*Tunnel Convergence Monitoring*

There are two basic layouts for positioning the pins in the tunnel walls: one provides for measurements across the diameter but has the often serious disadvantage of interfering with operations, and the other provides for measurements around the perimeter with less interference in tunneling operations.

For measurements across the diameter of tunnels in rock, the pins are installed in arrays positioned on the basis of the geologic structure. In Fig. 4.10, an array for horizontally bedded rock provides a concentration around the roof arch, since roof deflections are likely to be the most significant movements. The base pins at points I and II are reference pins. In Fig. 4.11, the pin array is positioned for measurements when the rock structure is dipping. The pins are concentrated where the structure dips out of the tunnel wall, since at this location failure is most likely. These arrays provide relatively precise measures of tunnel closure.

Stringing the wire through a series of rollers attached to pins located around the tunnel wall, as in Fig. 4.12, leaves the tunnel unobstructed for construction activity. Measurements are made of the net change in tunnel diameter, but the locations of maximum deflection are not known. If creep occurs in the wire, accuracy

**FIG. 4.9 Installation of convergence monuments on a cut slope.**

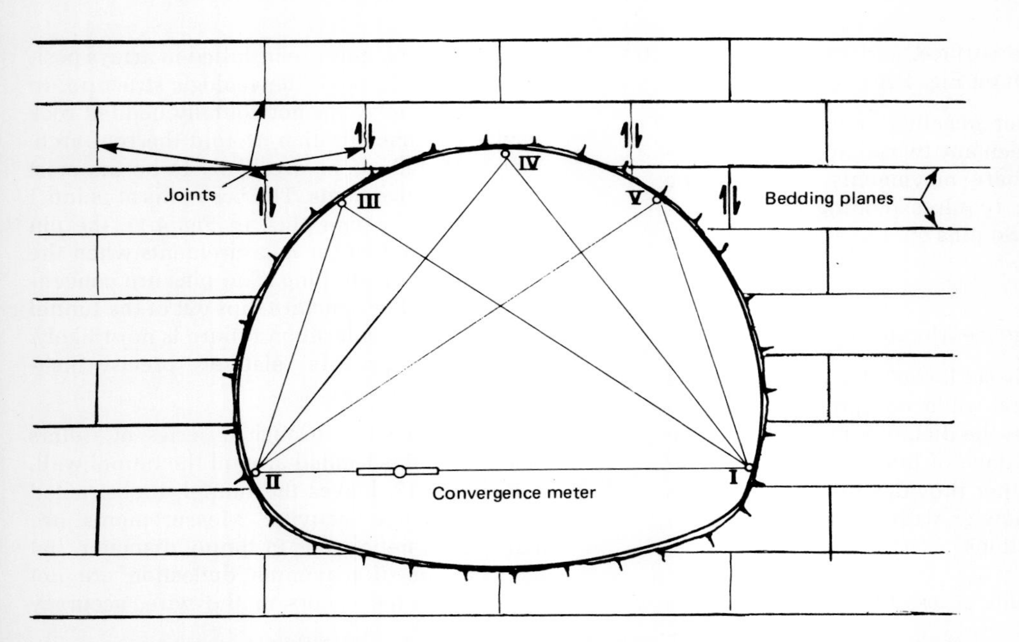


**FIG. 4.10 Convergence meter pin array for horizontally bedded rock.** [*From Silveira (1976).*[9]]

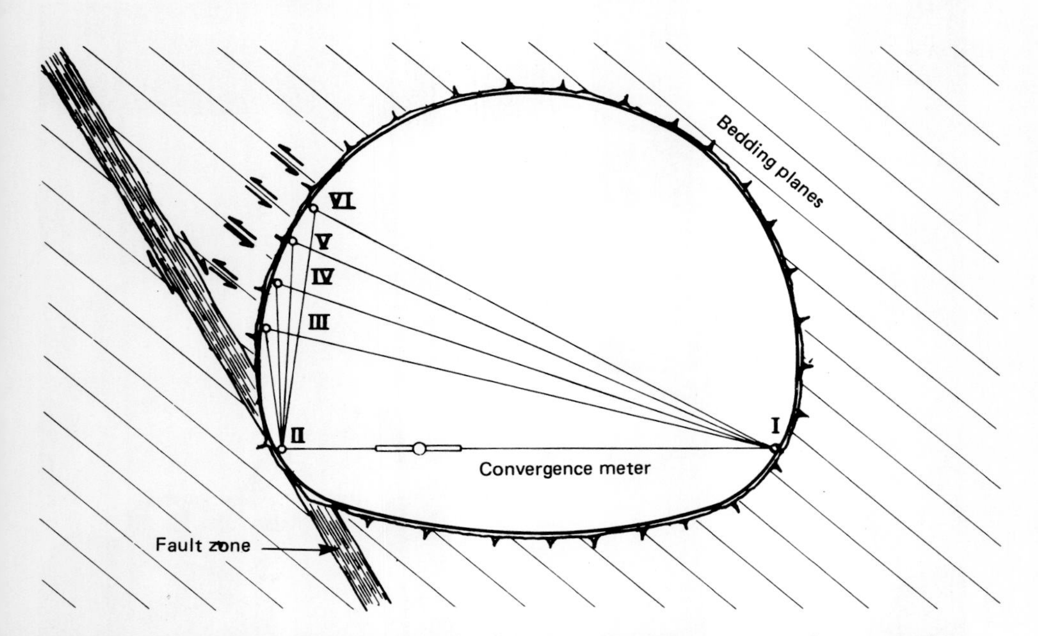


**FIG. 4.11 Convergence meter pin array for dipping rock structure.** [*From Silveira (1976).*[9]]

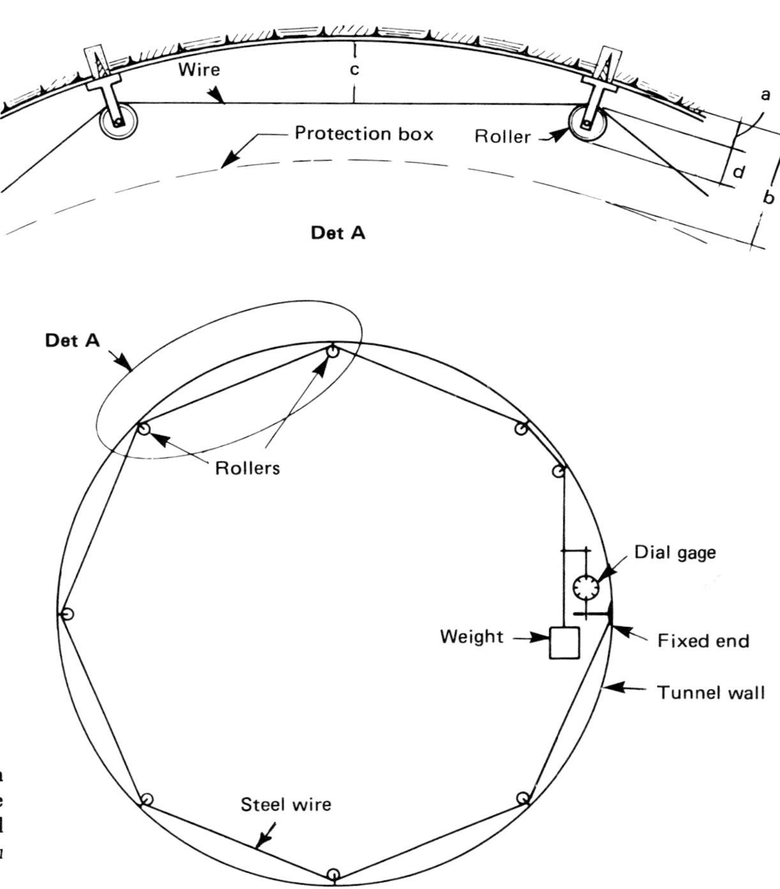


**FIG. 4.12 Stringing the wire through a series of rollers attached to pins around the tunnel wall leaves the tunnel unobstructed for construction activity.** [*From Silveira (1976).*[9]]

reduces to values of about 0.2 to 0.8 in [Silveira (1976)[9]].

### 4.2.5 VIBRATION MONITORING

#### Purpose

Ground motion is monitored to obtain data for evaluation of vibration problems associated with blasting, road and rail traffic, pile driving, the operation of heavy equipment, and rotating or reciprocating machinery.

#### Equipment

Small portable seismographs (particle-velocity seismographs) are used to determine frequency, acceleration, and displacement of surface particles in three orthogonal directions of vibration motion *(see Art. 11.2.3)*.

#### Principles

Frequency and amplitude are the basic elements of waveform of harmonic motion from which can be determined acceleration, force, particle velocity, and kinetic energy. Acceleration or displacement, when combined with the measured frequency, can be used to control or measure vibration damage by providing a measure of the energy transmitted by the vibration source.

Some instruments measure directly the peak particle velocity. The energy transmitted is directly proportional to the square of the peak velocity *(see Art. 11.4.2)*.

*Ground motion* is the result of the induced vibrations. A scale of vibration damage to structures and the limits of human perception in

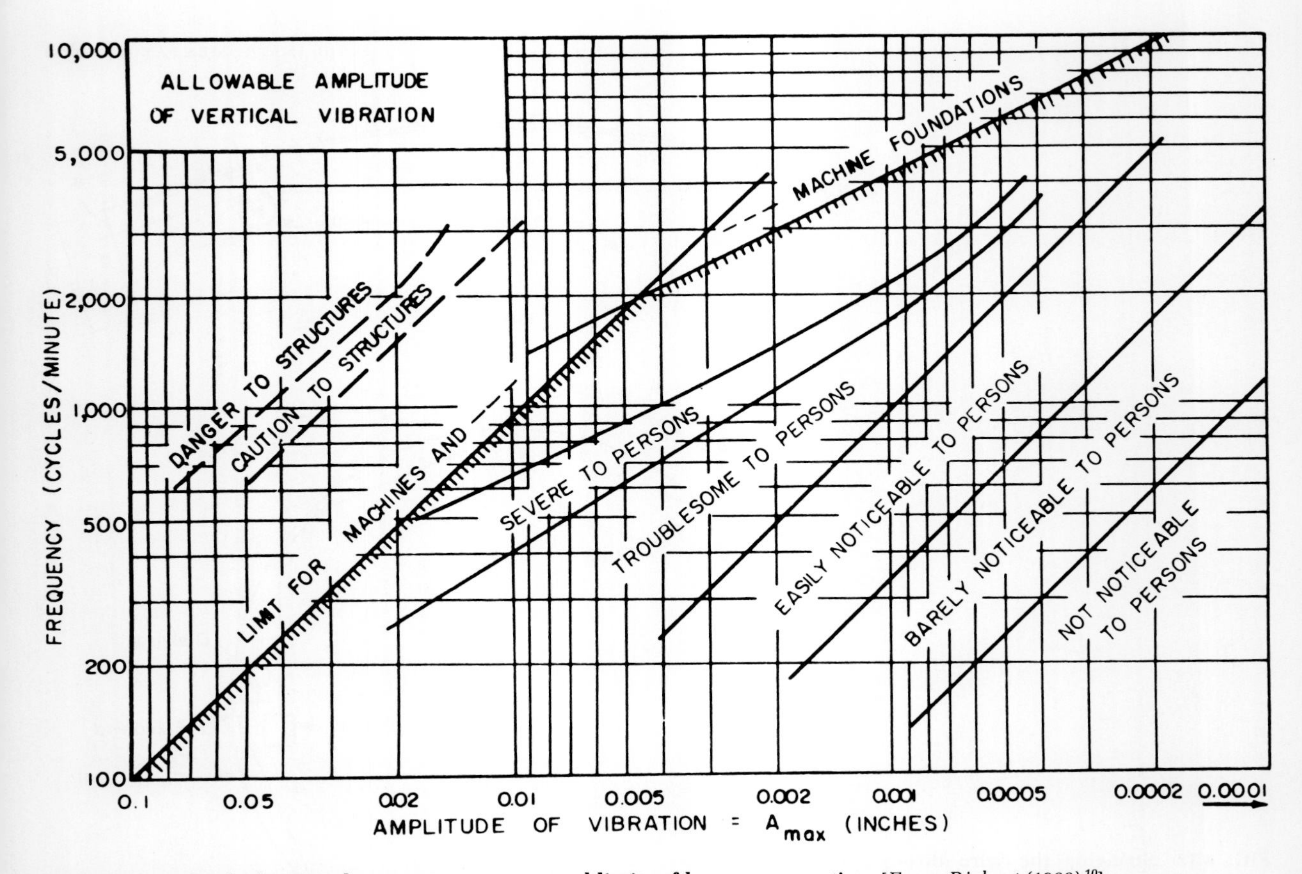


**FIG. 4.13 Scale of vibration damage to structures and limits of human perception.** [*From Richart (1960).*[10]]

terms of amplitude and frequency is given on Fig. 4.13.

## 4.3 SUBSURFACE DEFORMATIONS

### 4.3.1 FORMS AND SIGNIFICANCE

#### Vertical Displacements

Compression or consolidation of strata under applied stress results in surface settlements as described in Art. 4.2.4. In some cases it is important to determine which strata are contributing most significantly to the vertical displacements.

#### Lateral Displacements

Slopes, retaining structures, pile foundations, soft ground beneath embankment loads, and faults all undergo lateral deformations. Earth pressures against retaining walls cause tilting or other wall deflections. Slope failures often occur along a zone or surface, resulting in lateral displacements. Surface loads can cause lateral displacement of buried, weak soil strata. Foundation piles deflect laterally during load tests and while in service.

#### Strain Gradients

Differential deflection of earth dam embankments and reinforced earth walls, the closure of excavations in rock masses, and the imposition of high loads in rock masses result in strain gradients.

#### Internal Erosion

Seepage forces through, around, or beneath earth embankments or beneath or around concrete dams or cofferdams cause internal erosion.

**TABLE 4.6**
**SUBSURFACE DEFORMATIONS: FORMS, OCCURRENCE, AND MONITORS**

| Movement | Occurrence | Monitoring method | Reference |
|---|---|---|---|
| Vertical displacement | Ground compression at various depths | Vertical-rod extensometer<br>Borros points<br>Cross-arm device | Art. 4.3.2 |
| | Mine roof deflection | Acoustical emissions device | Art. 4.3.5 |
| Lateral displacement | Slope failure zone (*a*, *b*, *c*, *d*)<br>Wall or pile deflection (*a*)<br>Weak soils under stress (*a*)<br>Fault movements (*a*, *b*, *d*) | (*a*) Inclinometer for soil and rock masses<br>(*b*) Deflectometers for rock<br>(*c*) Shear strip indicators<br>(*d*) Slope failure sensor<br>(*e*) Acoustical emission device | Art. 4.3.3 |
| Linear strain gradients | Dam foundations or abutments<br>Open or closed rock excavations | Rock bolt extensometers<br>Multiple position borehole extensometers (rod or wire) | Art. 4.3.4 |
| | Differential settlement of dam bodies and foundations | Electrical strain meters | Art. 4.2.4 |
| Piping erosion | Through, around, or beneath dams<br>Broken underground pipes | Acoustical emission device<br>Dye or radioactive tracers | Art. 4.3.5<br>Art. 8.4.1 |

**Summary**

The monitoring methods are summarized in terms of the type of ground movement and its occurrence on Table 4.6.

4.3.2 VERTICAL DISPLACEMENT

**Vertical Rod Extensometers (Settlement Reference Points)**

*Applications*

The vertical rod extensometer is a simple device for measuring deformations between the surface and some depth in most soil types, except for very soft materials.

*Instrument*

An inflatable bag is set into an HX-size borehole at the depth above which the deformations are to be monitored. The bag is connected to the surface with a 1-in galvanized pipe installed inside a 3-in-diameter pipe which acts as a sleeve permitting freedom of movement of the inner pipe. A dial gage is mounted to read relative displacement of the 1-in-diameter pipe and a reference point on the surface.

*Installation*

An HX-size hole is drilled to test depth and 3-in-pipe installed as shown on Fig. 4.14*a*. The 1-

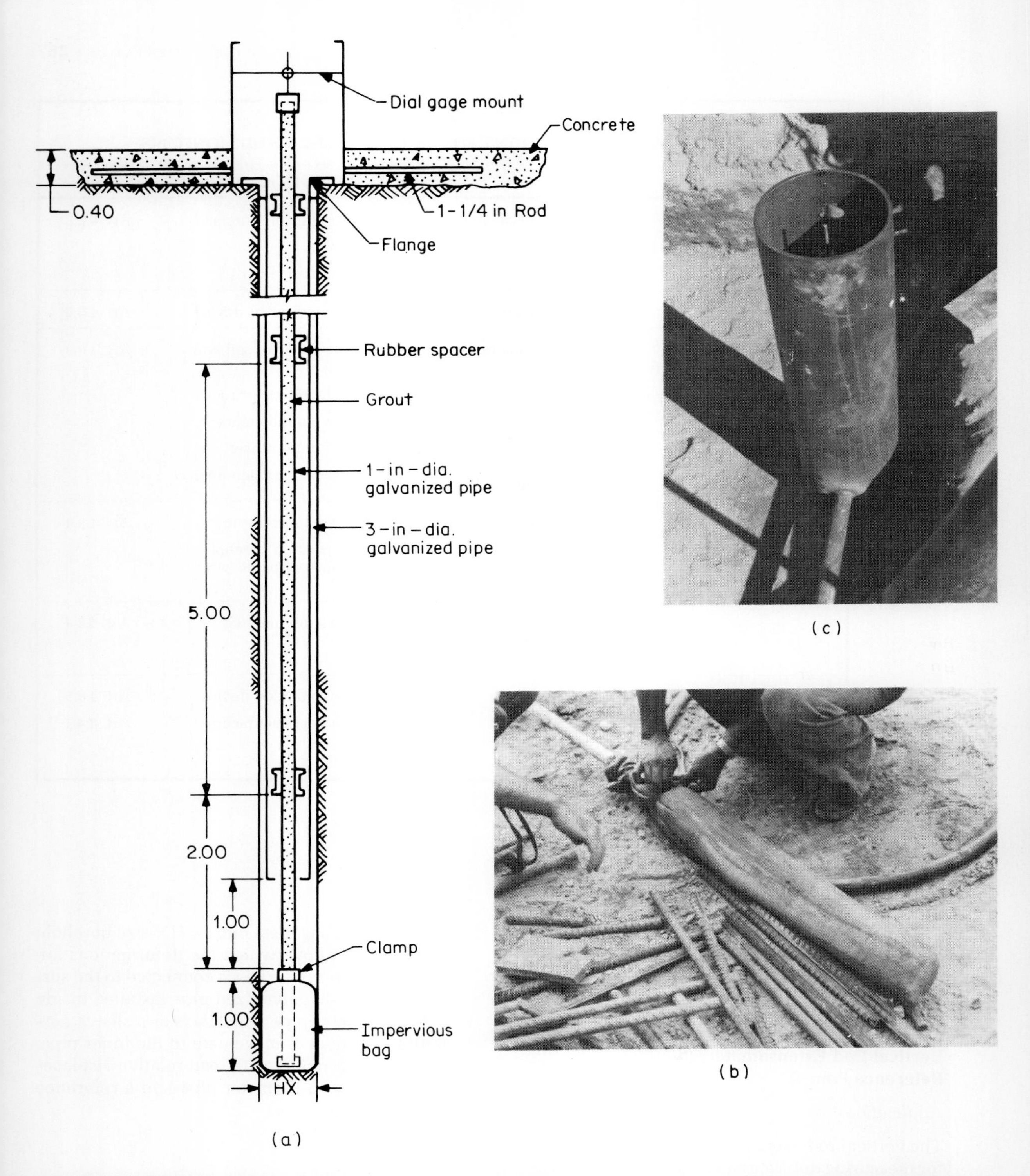


**FIG. 4.14 Settlement point installation: (*a*) schematic of the settlement point (depths are in meters); (*b*) clamp and bag being tested on the surface; (*c*) dial gage mount and protection.**

in pipe and bag are lowered to test depth with the pipe centered with rubber spacers free to move in the 3-in pipe.

Cement grout is injected into the inflatable bag through the 1-in pipe under pressures of about 50 psi to expand the bag and secure it against the sides of the hole, below the casing. An expanded bag under surface test is illustrated in Fig. 4.14*b*. When the bag and pipe are filled, the grout returns to the surface. The return can be signaled by the bursting of a short length of rubber tubing installed in the top of the grout pipe.

A 10-in-diameter steel pipe set in a concrete pad at the ground surface, in which is mounted a dial gage, serves as the reference measuring component (Fig. 4.14*c*).

Piezometers are also installed *(see Art. 4.4.2)* to monitor pore-water pressures during the settlement observations.

**Borros Points**

*Application*

Borros points are used to monitor settlements from compression of soft soils.

*Instrument*

A three-pronged anchor which is expanded at test depth in the soft soils.

*Installation*

The anchor, with prongs folded to the sides of a driving point, is attached to a ¼-in rod contained within a 1-in pipe which is pushed or driven through a 2½-in casing to the observation depth. The 1-in pipe is held fixed while the ¼-in pipe is advanced about 16 mm, extending the prongs into the soil.

The 2½-in casing is removed and the 1-in pipe withdrawn a short distance. The ¼-in rod, attached to the points, is free to move within the 1-in pipe.

**Cross-Arm Device**

*Application*

Cross-arm devices are installed at various elevations in earth dam embankments to provide for measurements of embankment compression [USBR (1974)[11]].

*Instrument*

Cross arms, fabricated of 3-in metal channels, are installed at 10-ft intervals (or some other spacing) and connected by metal pipe as the embankment is raised. A final assembly is illustrated on Fig. 4.15. The bottom section is grouted into the dam foundation.

*Installation and Measurements*

The channels, 6 ft long, are placed one by one in excavated trenches as the embankment is constructed. At each channel a 1½-in standard pipe is attached, over which is placed a 2-in pipe free to slide, forming a telescoping member. The lower tip of the 1½-in pipe provides a measuring point.

The vertical settlement of each segment is measured with a torpedo device lowered into the pipe on a steel surveyor's tape. At each measurement point the torpedo is raised until collapsible fins catch on the end of the 1½-in pipe. The fins are collapsed for final retrieval. Probes are also lowered for water-level readings.

The pipe must be protected from disturbance during construction.

Plant growth and corrosion within the pipe may in time inhibit torpedo movement.

4.3.3 LATERAL DISPLACEMENT

**Inclinometer**

*Application*

Continuous measurements of lateral deflections are made from the surface with the inclinometer. Commonly used in slope studies, the device may be used to measure deformations in soft soils beneath fills, or installed behind retaining structures or in pile foundations.

*Instruments*

There are two general types of inclinometers. One type contains a pendulum-actuated Wheatstone bridge circuit with a sensitivity of 1 in per 100 ft, which can readily detect movements of ¹⁄₁₆ in (2 mm).

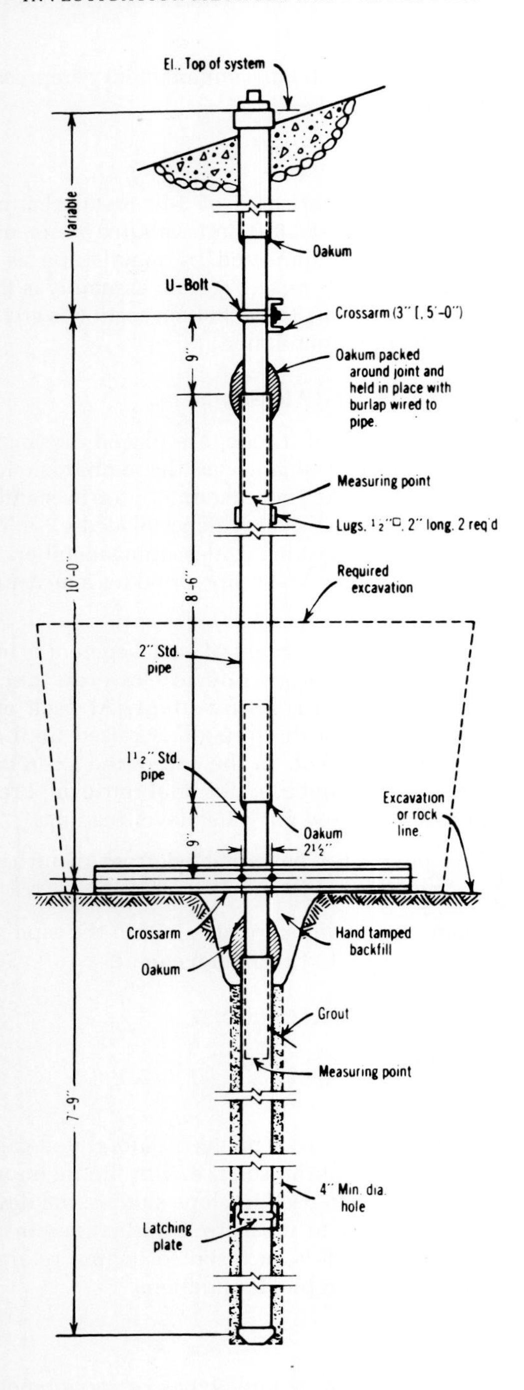


**FIG. 4.15 Cross-arm device for measuring embankment compression.** [*From Sherard et al. (1963),*[12] *after USBR (1974).*[11] *Reprinted with permission of John Wiley & Sons, Inc.*]

The more sensitive instrument contains servoaccelerometers which can detect lateral movements in the order of ±0.0001 ft per 2 ft of casing (the normal depth increment at which readings are taken). Since the voltage output is proportional to the sine of the angle of inclination of the long axis of the sensor from the vertical, it can be used to measure true deviations from verticality.

*Installation and Operation*

The inclinometer is lowered and raised in specially grooved casing (3.38 or 2.79 in OD) installed in a borehole and extended to a depth below the anticipated movement zone. The casing, the servo-type inclinometer, cable, and digital readout are shown on Fig. 4.16.

In soil formations it is advisable to install 6-in casing to allow for packing sand between the casing and the grooved casing as the 6-in casing is withdrawn, to provide a sure contact between the inclinometer casing and the borehole wall. The casing bottom is often grouted into place to assure fixity.

The casing is grooved at 90° intervals to guide the inclinometer wheels and to allow measurements along two axes. Since the casing is relatively flexible, it deflects freely during ground movements. From readings taken at regular depth intervals, a profile of the casing is constructed. Repeating measurements periodically provides data on the location, magnitude, direction, and rate of movement.

For monitoring pile and wall movements, the casing is attached directly to the structural member.

## Deflectometer

*Application*

Deflectometers are used as permanent installations in boreholes to measure movements normal to the hole axis such as might occur along a fault zone, or some other weakness plane in a rock mass.

*Operation*

The elements of the deflectometer are illustrated on Fig. 4.17. Any movement in the mass perpendicular to the axis of the hole causes an angular distortion between consecutive rods or wires which is monitored by transducers.

Chains of deflectometers can be made with eight transducers to depths of about 60 m. The system is connected to a remote readout and can be attached to an early warning system. Precision is about 0.025 mm for a 5-m distance between transducers.

### Shear-Strip Indicator

*Application*

Shear-strip indicators are used to locate the failure surface in a moving earth mass.

*Instrument*

A row of electrical transducers is wired in parallel and mounted on a bakelite strip, spaced about 15 cm apart.

*Installation and Operation*

The strip is mounted in a borehole. When shear displacement at any point exceeds 2 to 3 mm, the electrical circuit is broken. The failure surface is determined from the resistance of the resistors remaining in the circuit [Broms (1975)[13]].

### Slope Failure Plane Sensor

*Application and Instrument*

A simple, low-cost instrument, a 2-m-long rod on a steel fishing line, is used to locate the failure surface in a moving mass [Brawner (1975)[14]].

*Installation and Operation*

A borehole is made through the sliding mass and a 2-m-long rod on a steel fishing line is lowered into the borehole. The rod is raised each day until eventually an elevation is reached where the rod stops during lifting, indicating the bottom of the failure surface. Another rod is lowered from the ground surface until it stops, indicating the location of the top of the failure zone.

## 4.3.4 LINEAR STRAIN GRADIENTS

### Occurrence and Measurements

*Rock Masses*

Linear strains occur around the openings for tunnels, caverns, open-pit mines, or other rock slopes as a result of excavation or in the abutments or foundations of concrete dams as a result of excavations and imposed loads.

In tunnels and caverns *borehole extensometers* are installed to measure the tension arch or zone of relaxation which determines the depth to which support is required, the type of support, and the excavation methods. Deformations are also monitored during and after construction in critical zones characterized by faulting, intense fracturing, or swelling ground (in which horizontal stresses can be many times vertical stresses).

**FIG. 4.16 Sinco inclinometer, grooved casing, cable, and digital readout.**

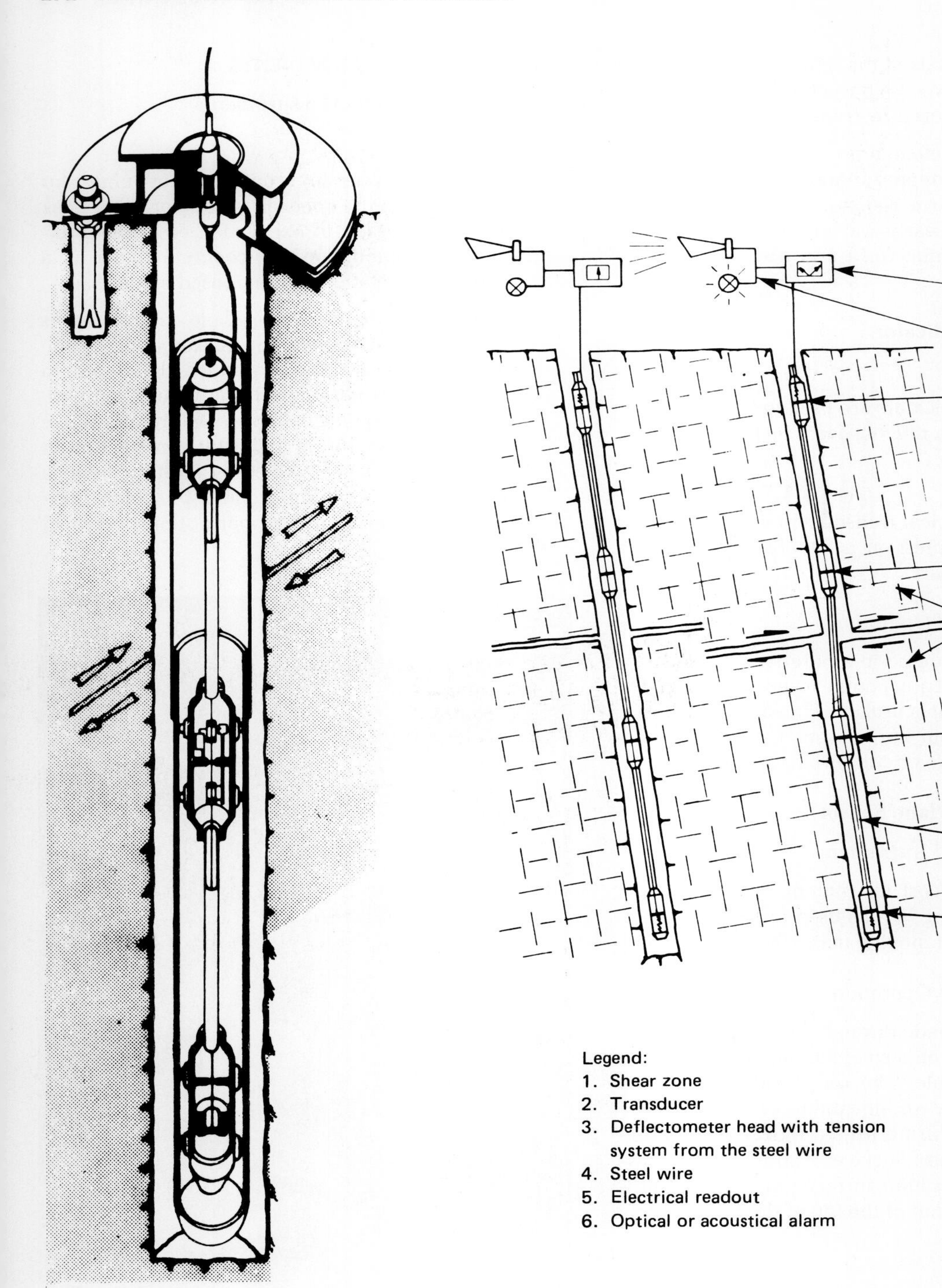


**FIG. 4.17 Schematic of the function of the Interfels deflectometer.** [*From Silveira (1976).*[9]]

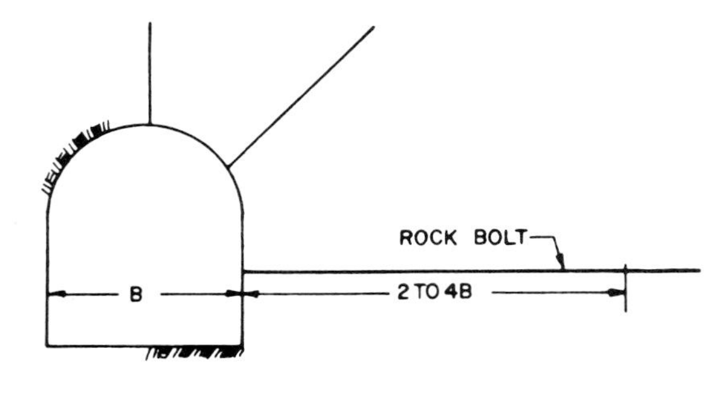


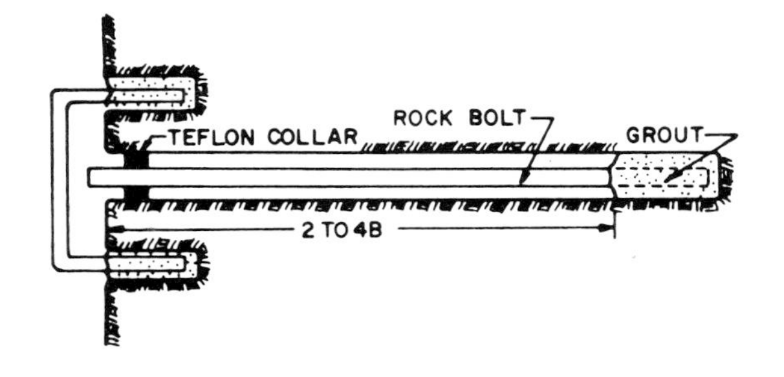


**FIG. 4.18** **Simple rock-bolt extensometer.**

*Earth Dams and Reinforced Earth Walls*

Linear strains occurring in earth dams and reinforced earth walls are measured with *electrical strain meters* (*see Art. 4.2.4*).

In earth dams the meters are installed in trenches at various depths to monitor strains and detect and locate cracks resulting from differential movement of the dam foundation or the embankment itself and between the embankment and the abutment (*see Fig. 4.37*).

In reinforced earth walls the meters are installed to monitor horizontal movements in the body of the backfill.

### Rock-Bolt Borehole Extensometer

*Application*

Rock bolts (*see Fig. 9.114*) are installed to monitor deflections occurring parallel to a borehole in a rock mass. They can be installed in any orientation and are rugged and low in cost because they can be installed rapidly and are relatively simple. Precision is on the order of 0.001 in.

*Installation*

A small-diameter borehole is drilled to some depth and a 1-in-diameter steel rod is grouted into a fixed position as shown on Fig. 4.18. A Teflon collar is fitted on the rod as a spacer near the tunnel face and a head is fixed to the wall for mounting a strain gage for deflection readings.

The bolts are installed to various depths to measure differential strains and to determine the extent of the zone in which significant strains occur, which is a function of the tunnel diameter.

### Multiple-Position Borehole Extensometer (MPBX): Rod Type

*Application*

The rod-type MPBX is installed to monitor deflections occurring parallel to a borehole at a number of positions in a single hole. Vertical overhead installation is difficult and lengths are limited to about 50 ft. Sensitivity is in the order of 0.001 in.

*Installation*

The ends of up to six metal rods are cemented into a borehole at various distances from the excavation face and the rods attached to a measuring platform fixed to the excavation wall as shown on Fig. 4.19*a*.

### Multiple-Position Borehole Extensometer (MPBX): Wire Type

*Application*

The wire-type MPBX is used like the rod type but can be installed to far greater depths.

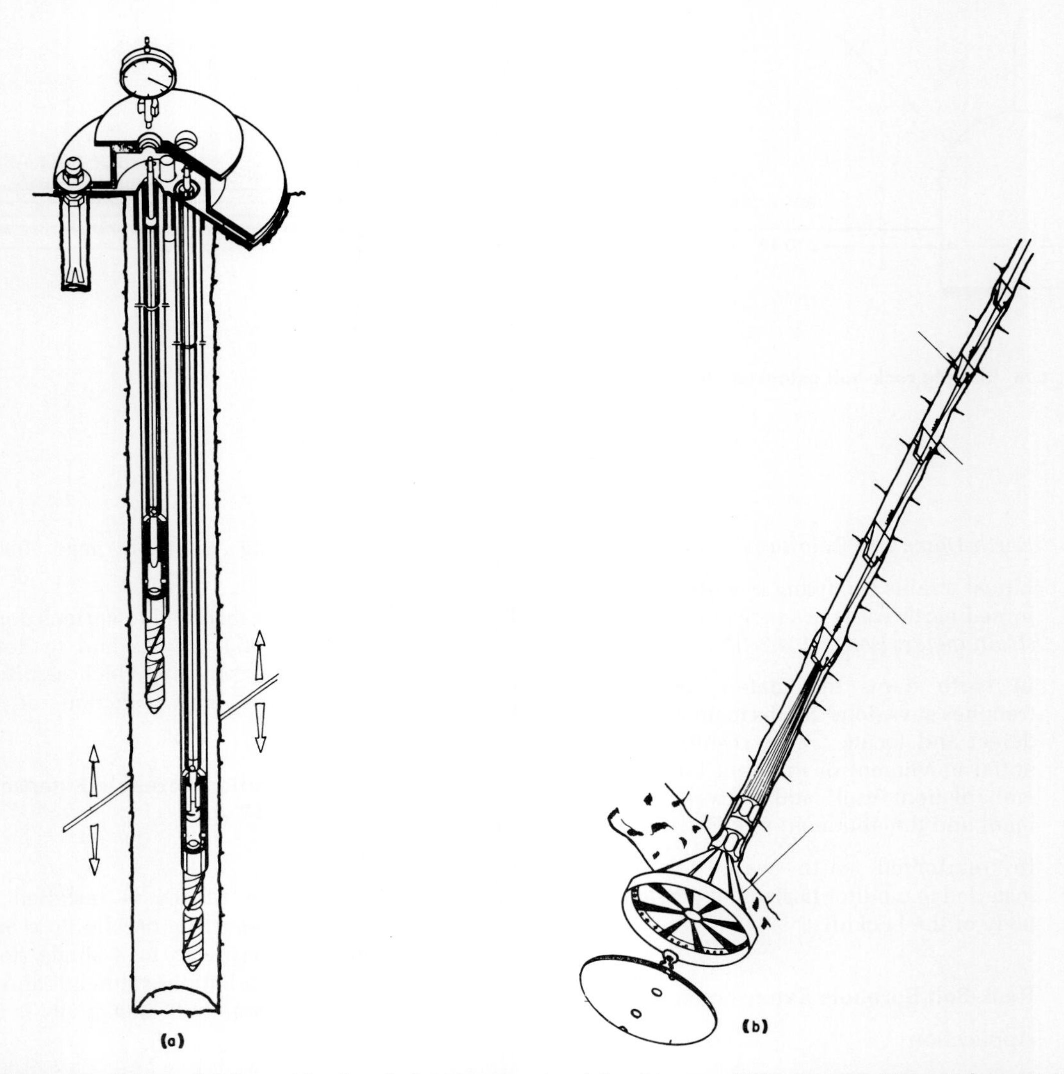


**FIG. 4.19 Schematics of the multiple-position borehole extensometer: (*a*) rod type and (*b*) wire type.** [*From Silveira (1976).*[9]]

*Installation*

Up to eight stainless steel tension wires are installed in a borehole and anchored by grouting, or some other means, to the side of the hole at various positions along the hole as shown on Fig. 4.19*b*. A measuring head is threaded over the wires and installed in the collar of the hole, and each wire is connected to a measuring element in the head. The measuring element can be a steel rod running in a track, pulling against a tension spring or pushing against a compression spring, which is read by dial gages; or it can be an electric sensor, such as an LVDT, linear potentiometer, or strain gage connected to an electrical readout.

Calibration is difficult and allowances are made for wire stretch, friction between the wires and the borehole, and temperature. Wire-type MPBXs can be read to the nearest 0.001 in, but because of calibration difficulties, the deeper the installation, the greater should be the changes in readings before they are judged significant.

## 4.3.5 ACOUSTICAL EMISSIONS

### Description

Acoustical emissions refer to subaudible noise resulting from distress in soil and rock masses and are also termed microseismic activity, microseisms, seismic-acoustic activity, stress-wave activity, and rock noise.

### Applications

Monitoring acoustical emissions provides an aid to the anticipation of failure by rupture or internal erosion. The method was first used in the underground mining industry to detect instability of the mine roof, face, or pillar rock.

Recent applications include:

- Monitoring natural slopes, open-pit mines, and other cut slopes; excavations for tunnels, caverns and underground storage facilities and pressure chambers; and the stability of dams and embankments
- Locating leakage paths in dams, reservoirs, pressure pipelines, and caverns
- Monitoring fault zones
- Inspecting steel and concrete structures under test loading

### The Phenomenon

*General*

A material subject to stress emits elastic waves as it deforms, termed acoustic emissions.

*Rock Masses*

Testing has shown that both the amplitude and number of emissions increase continuously as macroscopic cracks initiate and propagate first in a stable manner, then in an unstable manner. Near rupture, friction along crack surfaces, as well as crack propagation and coalescence, contribute to the acoustical emission activity. Mineralogical and lithological differences, moisture content, and stress conditions affect the emissions [Scholz (1968)[15]].

*Soils*

Frictional contact in well-graded soils produces the greatest amount of "noise" during stress, and activity increases with the confining pressure. Clays exhibit a different form of response than sands (Fig. 4.20), and the emission amplitude for

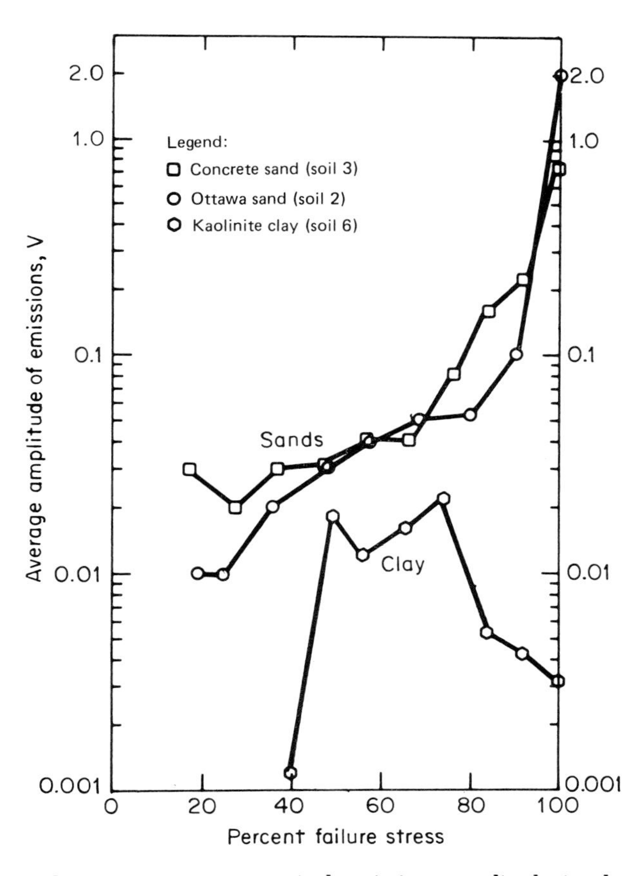


**FIG. 4.20 Average acoustical emissions amplitude (peak signal voltage output) vs. percentage of failure stress in triaxial creep (5-psi confining pressure). [From Koerner et al. (1977).[16]]**

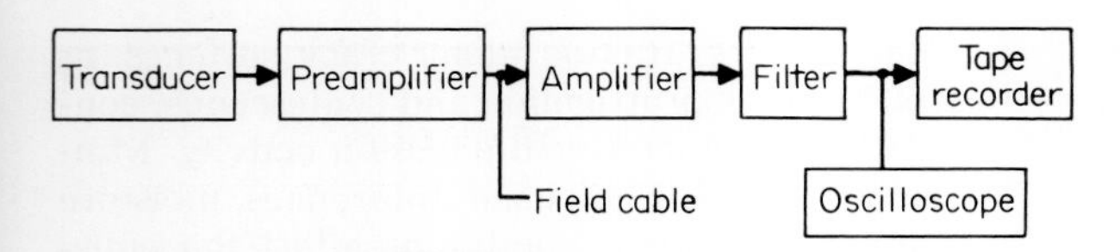


**FIG. 4.21 Simplified diagram of typical microseismic field-monitoring system.** [*From Hardy and Mowrey (1977).*[17]]

sands can be 400 times that for clay [Koerner et al. (1977)[16]].

### Detection

*General*

Sensors detect microseismic activity at a specific location in an earth mass by monitoring displacements, velocities, or accelerations generated by the associated stress waves. At times the sounds are audible, but usually they are subaudible because of either low magnitude or high frequency, or both.

In application, if after all extraneous environmental noise is filtered and there are no emissions, the mass can usually be considered as stable. If, however, emissions are observed, a nonequilibrium condition exists which may eventually lead to failure.

*Instrumentation*

Piezometric sensors (accelerometer or transducer) are used to detect acoustic emissions by converting mechanical energy associated with the microseisms into an electrical signal proportional to the amplitude of sound or vibration being detected. The detected signal is then passed through a preamplifier, amplifier, and filter, and finally to a display on a cathode-ray oscilloscope or into a recorder as shown on Fig. 4.21.

The system components are selected for a specific study to provide suitable frequency response, signal-to-noise ratio, amplification, and data recording capacity. The design, construction, and calibration of equipment is discussed in Hardy and Leighton (1977).[18]

*Sensors*

Geophones can detect signals in the frequency range of 1 Hz to 1000 Hz, the normal seismic energy range for large geologic structures. They are generally used as sensors to monitor mines and tunnels because of greater sensitivity in the lower frequency ranges [Hardy and Mowrey (1977)[17]].

Accelerometers are employed for high-frequency components ($f > 1000$ Hz), and displacement gages are employed for very low frequency ranges ($f < 1$ Hz).

*Installation*

Geophones can be installed at shallow depths below the surface, cemented into boreholes, or mounted on a rock face in an excavation as shown on Fig. 4.22. The best results for rock studies are obtained when the transducer is located in the rock mass.

### Case Studies

*Cut-Slope Failure*[19]

OPERATIONS The toe excavation of a large fill was monitored by an acoustic emissions device. Five cuts were required to cause failure over a 21-day period. After each cut the "acoustic counts" were high, then tapered off. During one operation rainfall caused an increase in acoustic counts. During the fifth cut, the emission rate followed the trend for the previous cuts, but 30 min after the cut was made the rate increased rapidly and reached its maximum when a large portion of the fill slid downslope.

EQUIPMENT USED Included in the equipment were a Columbia 476-R accelerometer, a Columbia VM-103 amplifier, and a Hewlett-Packard 5300A counter. The accelerometer was a piezometric transducer with a relatively flat frequency response from 500 to 5000 Hz, a resonance at 5700 Hz, and a voltage sensitivity of approximately 100 mV per peak g.

ACOUSTIC COUNT For the above equipment, acoustic count was defined as being registered each time the amplitude of the electric signal exceeded the threshold level of 0.025 V as recorded on the electronic counter. From several studies [Koerner et al. (1978)[19]] it was concluded that:

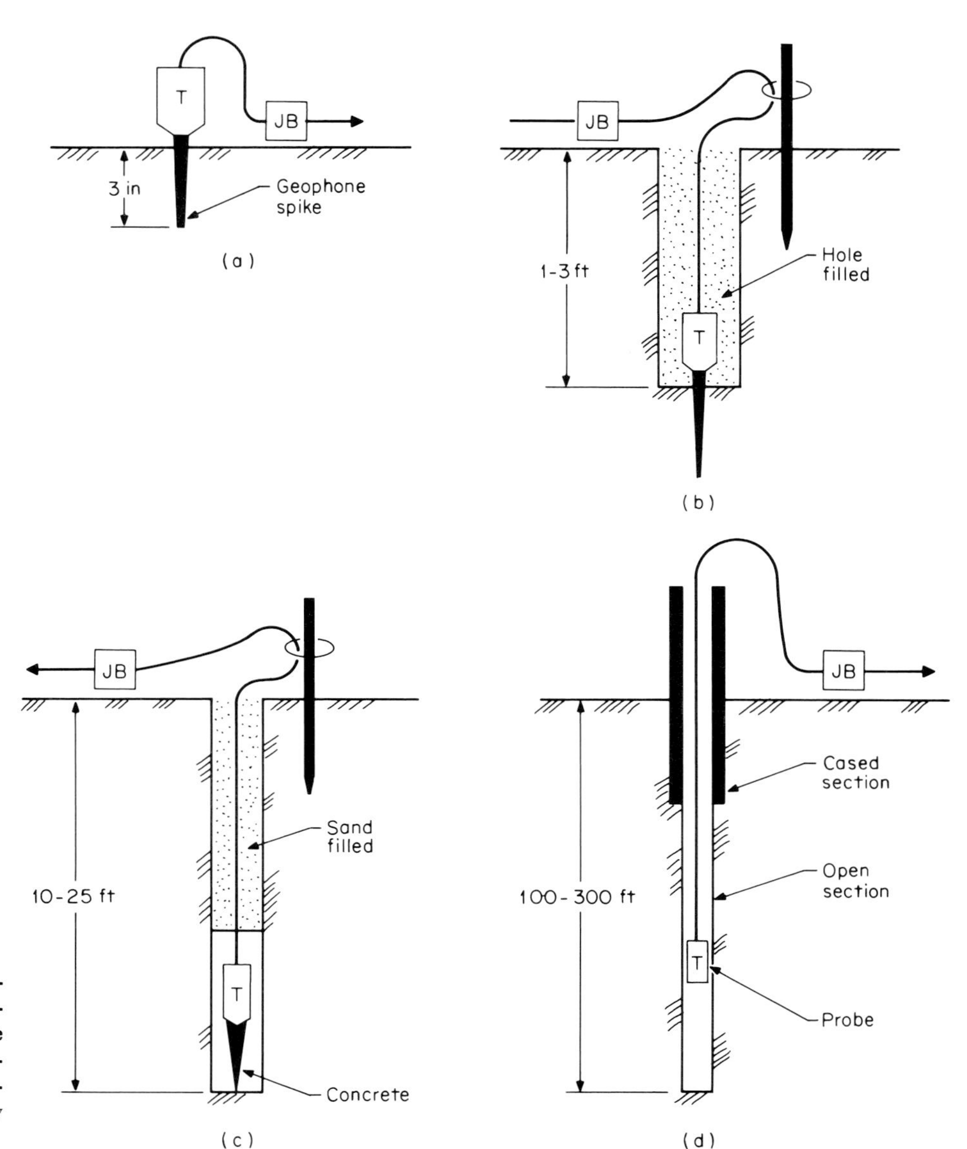


**FIG. 4.22 Types of microseismic transducer mounting techniques: (*a*) surface mounting; (*b*) shallow burial; (*c*) deep burial; (*d*) borehole probe.** [*From Hardy and Mowrey (1977).*[17]]

1. Soil masses with no emissions are stable
2. Granular soil masses generating moderate levels of emissions (from 10 to 100 counts per minute) are considered as marginally stable
3. High emission levels (100 to 500 counts per minute) indicate that the soil mass is deforming substantially and immediate remedial measures are required
4. Above 500 counts per minute the mass is in the failure state.

*Active Slide*[4]

The inclinometer and the acoustic emissions device were compared during the study of an active landslide. The emissions detected and located the shear zone after only three measurements, even though movements were very slow, in the order of 5 to 10 mm/year.

Novosad concluded that the emissions method had the advantages of a simple, common bore-

**TABLE 4.7**
**IN SITU PRESSURES AND STRESSES: OCCURRENCE AND MEASUREMENTS**

| Phenomenon | Occurrence | Measurement | Reference |
|---|---|---|---|
| Pore-water pressures | Natural static GWL | Standpipe piezometer | Art. 4.4.2 |
| | Artesian conditions | Casagrande piezometer | |
| | Excess pore pressures under static or dynamic loads | Hydraulic piezometer | |
| | | Pneumatic piezometer | |
| | | Electrical piezometer | |
| Loads and stresses | Beneath foundations | Pressure cells | Art. 4.4.3 |
| | In embankments | Load cells | Art. 4.4.3 |
| | In structural members | Tell tales | Art. 4.4.3 |
| | Against retaining walls | Strain gages | Art. 4.4.3 |
| | Tunnel linings and supports | Acoustical-emissions devices | Art. 4.3.5 |
| | | Vibrating-wire stress meter | Art. 4.4.3 |
| Residual rock stresses | In closed and open excavations in rock masses | Flat-jack test | Art. 3.4.4 |
| | | Strain meters and overcoring | Art. 4.4.4 |
| | | Borehole methods<br>Deformation meter<br>Inclusion stress plug<br>Strain gage (Leeman doorstopper)<br>Hydraulic fracturing | Art. 4.4.4 |

hole casing, smaller casing diameter (40 cm ID is sufficient), and a deeper reach. The disadvantage was the qualitative character of the measurements.

*Open-Pit-Mine Slopes*[7]

Rock noises emanating from an unstable zone in an open-pit-mine slope are identifiable by their characteristic envelope shape, a frequency between 6 and 9 Hz, and a characteristic irregular amplitude. Experience is required to distinguish extraneous noise from drills, trucks, shovels, locomotives, etc. operating in the pit. Earthquakes are easily identifiable because they have a very definite waveform envelope.

In application, records are kept of the number of rock noises for a given period (counts) and of the frequency, duration, and equivalent earth motion of each seismic disturbance. As the instability increases, so does the number of rock noises released during a given period. Plotting the cumulative number of rock noises against time may give a useful prediction chart similar in shape to those obtained from plotting displacement vs. time.

Advantages of the method include its extreme sensitivity and continuous recording capabilities. Early warning of impending failure allows the removal of workers and equipment from the threatened area.

## 4.4 IN SITU PRESSURES AND STRESSES

### 4.4.1 GENERAL

#### Significance

Deformations occur as the result of response to changes in stress conditions in situ, which can be measured under certain conditions. The data obtained by instrumentation are used as a design basis, as a basis for checking design assumptions and determining the need for additional support, or as a warning of impending failure.

### Categories

Pressure and stress conditions measured in situ include:

1. Groundwater levels and pore pressures
2. Stresses in embankments, beneath loaded areas, or in structural members
3. Earth pressures against retaining structures or tunnel linings and supports
4. Residual stresses in rock masses

### Summary

In situ pressures and stresses, their occurrence, and methods of measurement are summarized on Table 4.7.

## 4.4.2 PORE-WATER PRESSURES (PIEZOMETERS)

### Applications and Summary

Piezometers are used to monitor water levels (static, perched, artesian) for excavations, slopes, and dam embankments, to measure excess hydrostatic pressures beneath dams and embankments, and to aid in control of preloading operations and placement of fill over soft ground.

The types, applications, advantages, and disadvantages are summarized on Table 4.8. Piezometers are divided into open systems, in which measurements are made from the surface and the water level is generally below the surface, and closed systems, in which measurements are made remotely and the water level may be at any location.

### Single-Tube Open Piezometer

The simplest piezometer consists of a tube or pipe which connects a tip or sensor (porous stone, well point, or slotted pipe) to the surface and which is installed in a borehole as the casing is withdrawn.

Readings are made by plumbing with a chalked tape or with an electric probe contacting the water in the standpipe. The static head measured is the average head existing over the depth of the inflow part of the borehole below the water table. The head may be higher or lower than the free water table, and in moderately impervious to impervious soils is subject to time lag.

### Casagrande Open-Tube Piezometer

Casagrande-type piezometers are similar to the open-tube system, except that the tip is surrounded by a bulb of clean sand and a clay seal is placed above the tip as shown on Fig. 4.23 to confine response to a particular depth interval.

Before the clay seal is placed, the annular space around the tip is filled with gravel or sand to prevent soil migration and system clogging. The hole is backfilled above the clay seal. The casing is withdrawn slowly as the clean sand, clay seal, and backfill are placed.

In granular soils, large-diameter (1-in) tubes adequately reflect changes in pore pressure, but

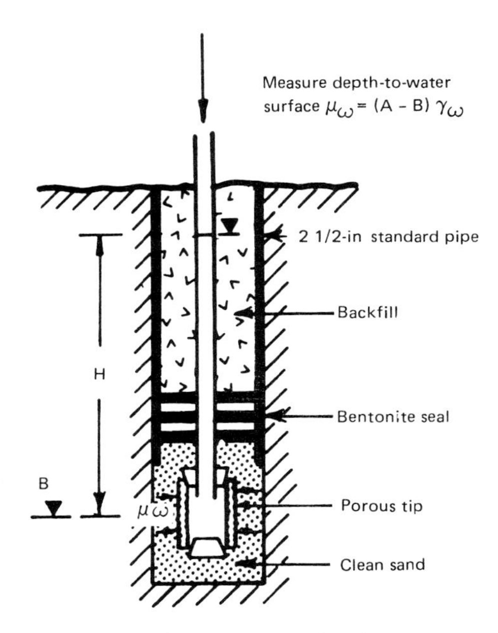


**FIG. 4.23 Casagrande-type open-tube piezometer.**

**TABLE 4.8**
**PIEZOMETER TYPES AND APPLICATIONS**

| Type | Application | Advantages | Disadvantages | Figure |
|---|---|---|---|---|
| **OPEN SYSTEMS (VERTICAL READOUT; WATER LEVEL BELOW GROUND SURFACE)** | | | | |
| Single-tube open system (Standpipe) | Coarse-grained granular soils, free-draining rock masses | Simple, rugged, inexpensive. | Indicates average head, relatively insensitive, time lag in impervious soils. | None |
| Casagrande type (single tube or double tube) | Coarse-grained to silty soils and rock masses | Measures response at a particular depth.<br>Decreased time lag.<br>Double tube allows flushing entrapped air or gas from lines. | Time lag in relatively impervious soils.<br>Low sensitivity.<br>Single-tube system subject to clogging from entrapped air or gas. | 4.23 |
| **CLOSED SYSTEMS (REMOTE READOUT; WATER LEVEL ANY LOCATION)** | | | | |
| Pneumatic (diaphragm principle) | Fine-grained soils, slow-draining rock masses | Negligible time lag.<br>Increased sensitivity.<br>Durable and reliable.<br>Lines readily purged and extended to avoid construction. | Relatively costly device and installation.<br>Requires protection against pinching of lines. | 4.24 |
| Hydraulic | Earth dams with soils of low to medium permeability | System can be flushed to remove air and gas.<br>Remote sensing of pore pressure to leave construction area free. | Slow response time.<br>Long tubing lines require careful flushing.<br>Fittings may be subject to leaks.<br>Requires protection against pinching. | 4.25 |
| Electric (diaphragm principle) | Fine-grained soils and slow-draining rock masses | Extreme sensitivity.<br>Fast response.<br>Continuous recording possible (only type with this capability). | Relatively costly.<br>Decreased durability and reliability over other closed systems because of electrical circuitry. | 4.26 |

small tubes are required for soils of low permeabilities. Water levels are read in the same manner as for simple systems, except that in small tubes readings are made with electric probes with ⅛-in OD.

*Double-tube* systems are a variation which permits flushing to remove entrapped air or gas which clogs lines, a common occurrence in single-tube systems of small diameter. In cold weather, tubes are filled with alcohol or antifreeze.

### Pneumatic-Type Piezometers

In pneumatic piezometers the sensor is a sealed porous tip which contains a gas- or fluid-operated diaphragm and valve, which are connected by two lines to a pressure supply and outlet system on the surface as shown on Fig. 4.24. When air pressure applied to the connecting line is equal to the pore-water pressure acting on the diaphragm, the valve closes. This pressure is assumed equal to the pore-water pressure in soil, or the cleft-water pressure in rock masses.

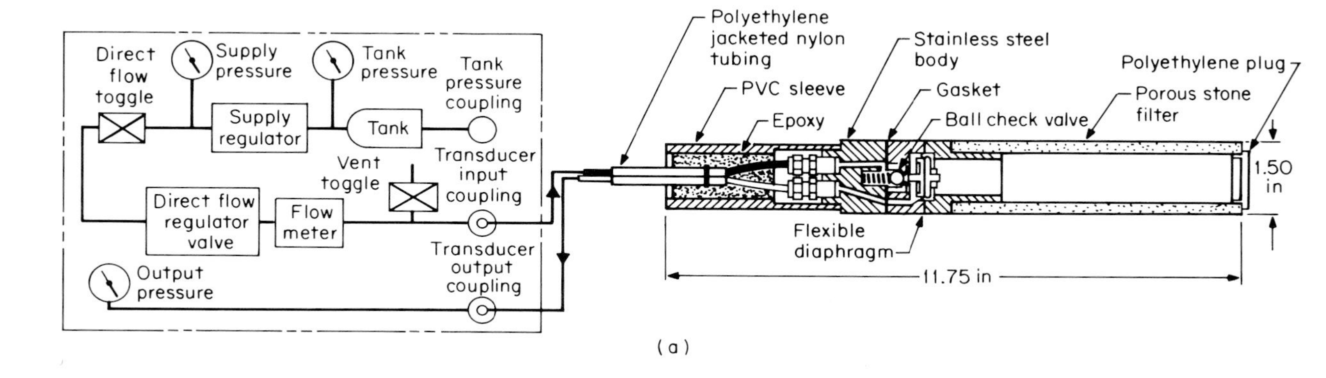


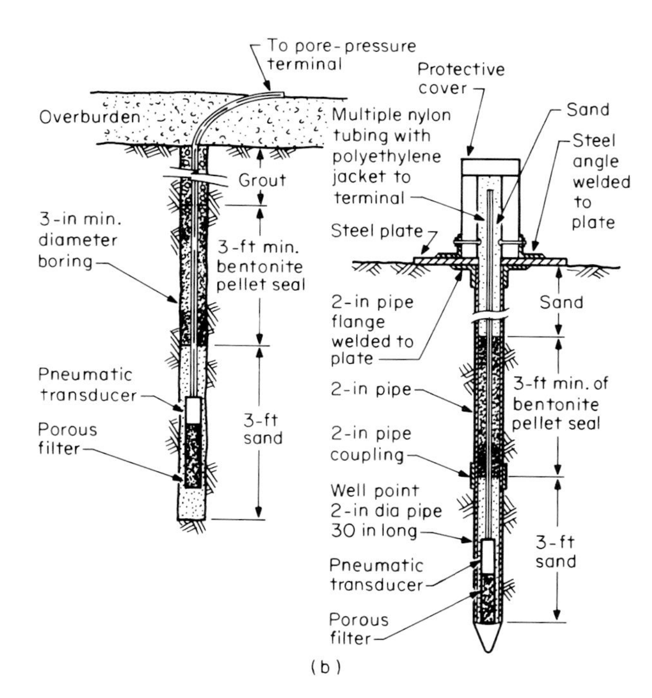


**FIG. 4.24 Pneumatic-type piezometer: (*a*) *(above)* pore-pressure measurement system and (*b*) *(left)* installation of system.** (*Courtesy of Sinco.*)

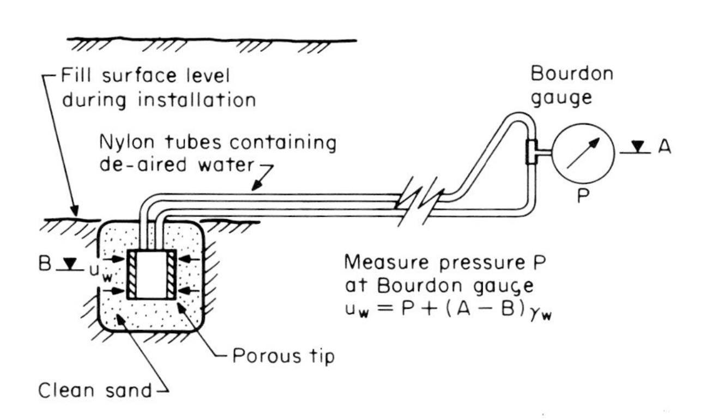


**FIG. 4.25 Hydraulic-type piezometer.**

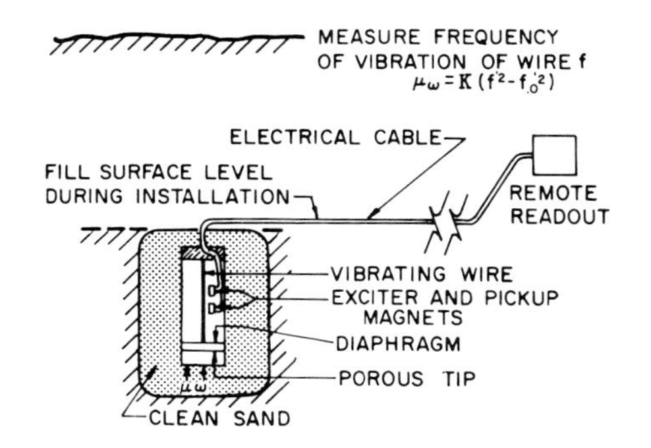


**FIG. 4.26 Electrical-transducer-type piezometer.**

Other types of pneumatic transducers include the Warlam piezometer, which operates with air, and the Glötzl piezometer, which employs a hydraulic fluid.

In all cases the lines can be extended through an embankment to a readout so that they do not interfere with construction operations. The lines must be protected against pinching.

### Hydraulic Piezometers

Commonly installed in earth dams, the hydraulic piezometer used by the U.S. Bureau of Reclamation [USBR (1974)[11]] uses a porous ceramic disk as a sensor. The disk is connected by two plastic tubes to a pressure gage near the downstream face of the embankment (Fig. 4.25). The pressure gage and its housing are located at an elevation slightly higher than the piezometer tip.

The system requires long tubing lines. Deairing by water circulation must be done carefully, and all fittings must be well made to avoid leaks. The lines require protection against pinching.

For operation, fluid is pumped into the system until a desired balancing pressure is obtained as read on the gage; then an inlet valve to the piezometer tip is opened slightly and the response pressure observed.

### Electrical Piezometers

The tip of an electrical piezometer contains a diaphragm that is deflected by pore pressure against one face. The deflection is proportional to the pressure and is measured by means of various types of electrical transducers. The system components are illustrated on Fig. 4.26.

Extremely sensitive and having negligible time lag, electrical piezometers are used in impervious materials where measurements are critical and continuous monitoring and recording are required.

## 4.4.3 LOADS AND STRESSES

### Applications and Summary

Loads and stresses are measured to check design assumptions, provide early warning against failure, and to permit the installation of remedial measures to provide additional support where necessary. Measurements are made primarily during construction. The various devices and their applications are summarized on Table 4.9.

### Strain Gages

*Purpose*

Strain gages are used to measure deformations of structural members to provide the basis for estimating stresses and loads. There are two general types: electrical *(see Art. 4.1.4)* and mechanical.

*Resistance Strain Gages (see Figs. 4.1 and 4.2)*

*Bonded gages* are used primarily for short-term surface measurements. Installation requires meticulous care and skill in the preparation of the receiving surface and cementing the gage in place. Moisture sensitivity is high and adequate field waterproofing is difficult to achieve. Reliability generally is low but may be partially compensated for by the installation of a relatively large number of gages on a given member.

*Encapsulated gages* are factory waterproofed and should have a service life of about a year or more if installed properly. Installation is relatively simple, but as with normal bonded gages, lead wires require protection.

*Unbonded, encapsulated gages* are relatively simple to install and have a reliability much higher than other types. Carlson strain meters installed in concrete dams have performed successfully for periods of over 20 years. They are also left inserted in boreholes to measure rock stresses. Deflection ranges, however, are somewhat less than for the other types.

*Vibrating-Wire Strain Gage*

Installation is relatively simple and the gages are recoverable, remote-reading with excellent long-term stability, strong, and waterproof. The gage, however, is a relatively large instrument and its measurement range is more limited than that of the resistance-gage types. Cost is generally higher than that of most resistance gages.

**TABLE 4.9**
**LOAD OR STRESS VS. DEVICE SUITABLE FOR MEASUREMENTS**

| Condition | Load or stress to be measured | Device suitable |
|---|---|---|
| Soil formations | Stress distribution beneath foundations or in embankments | Pressure cell |
| | Earth pressure distributions behind retaining structures | Pressure cell |
| Rock formations | Stress distribution beneath foundations or in abutment walls, during or after construction | Carlson strain meter<br>Borehole stress meter |
| Structural members | Stress distribution in piles to differentiate between end bearing and shaft friction during load test | Tell tale<br>Strain gage |
| | Stresses on retaining structure bracing, earth and rock anchors, steel storage-tank walls | Strain gage |
| | Stresses on tunnel linings | Pressure cell |
| | Loads on tunnel lining elements, earth and rock anchors, pile tips | Load cell |

*Mechanical Strain Gages*

Used to measure deformations of accessible metal surfaces, mechanical strain gages have two pointed arms that fit into conical holes (gage points) drilled or punched into the surface to be measured, or into studs set into or on the surface. The change in distance between gage holes is measured by determining the distance between arms when inserted into the gage points. Measurement is with a dial gage. Whittemore gages are a popular type.

Gage lengths of 2 to 80 in are available. The 10-in length is used for most engineering studies; sensitivity is in the order of 10 microstrains, and gages are read usually to the nearest 0.0001 in.

Mechanical gages are relatively simple and reliable. Temperature corrections are easily made and gage calibration is easily checked. However, they are suited only for measuring surface strains, cannot be read remotely, and have a lower sensitivity than the electrical types. Repeatability depends on the skill and experience of the operator.

### Pressure Cells

*Purpose*

Pressure cells are used to measure pressures against retaining walls or tunnel linings, or stresses beneath foundations or in embankments.

*Device*

The cell consists of a circular (or rectangular) double-wall metal pad, extremely thin relative to its diameter (ratio of about 1.1 cm to 23 cm).

A common type is the Glötzl cell, which is filled with oil or antifreeze liquid and functions hydraulically as shown on Fig. 4.27. Pressurized fluid is delivered to the small pressure diaphragm by pump or compressor until the pressure equals the resisting pressure on the outside of the diaphragm. The diaphragm deflects a slight amount and opens a bypass orifice which permits the fluid in the system to return to the reservoir. The pressure creating a balance is read on gages which provide a measure of the external pressures acting on the cell.

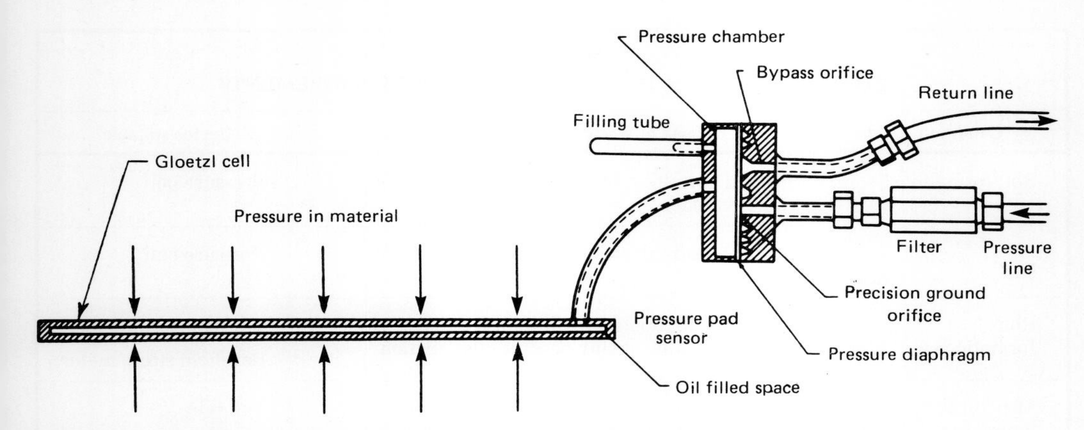


**FIG. 4.27 Schematic of the hydraulic pressure cell.** [*From Hartmann (1966).*[8]]

*Installations*

EMBANKMENTS Cells can be oriented to measure stresses in the three principal planes, but, when vertical or near vertical, may result in erroneous measurements of lateral stress because compression of the soil adjacent to the cell face causes arching and redistribution of stresses. The major problem is to compact the material surrounding the cell to a density that will provide a deformation modulus representative of the embankment material. During installation (Fig. 4.28), cell orientation should be accurately recorded.

TUNNELS Installations made to measure lining stresses have not been particularly successful, primarily because of uneven stress distributions over short distances [Cording et al. (1975)[1]].

BOREHOLES Installations to measure stresses behind walls have met with little success [Sauer and Sharma (1977)[20]].

### Load Cells

*Purpose*

Load cells are used to measure loads on earth or rock anchors, in the various elements of tunnel linings, in struts in braced cuts, in the bottom of piles (total pile load minus tip load equals shaft friction), and at the top of piles in pile load tests.

*Device*

Cell function is based on the electric transducer principle. The cells are mounted within variously shaped containers. Various capacities are available.

### Tell Tales

*Purpose*

Tell tales are installed in piles at various depths for strain measurements during load testing, from which are calculated shaft friction and end bearing.

*Device and Installation*

A small-diameter rod or pipe is fixed at its bottom in the concrete or to a steel pipe, and extended up through the pile top, encased in a slightly larger pipe to permit free movement. A dial gage at the top reads deflections.

A group of rods is usually installed at the same depth in large-diameter piles, and often at various depths in long piles. They are distributed as a function of the soil profile to measure shaft friction for the various soil types.

A tell-tale set in the pile bottom provides a measure of both base movement and shaft shortening when compared with the total pile deflection measured at the top by separate instruments.

*Analysis*

The average load $P_s$ from shaft friction carried by a segment of the pile $\Delta L$ from the measurement of two tell tales is expressed as

$$P_s = a_s E_p \frac{R_1 - R_2}{\Delta L} \tag{4.1}$$

where $a_s$ = section area
$E_p$ = modulus of concrete (or steel) (or steel and concrete) composing the pile section
$R_1$ and $R_2$ = the deflection readings of tell tales at depths 1 and 2

Subtracting the values for shaft friction from the total applied load provides a measure of the load reaching the pile tip.

### Vibrating-Wire Stress Meter

*Purpose*

Changes in in situ stresses occurring in rock masses during mining and tunneling operations have been measured with vibrating-wire stress meters [Sellers (1977)[21]].

*Device*

A transducer consisting of a hollow steel cylinder (proving ring), across which a steel wire is tensioned, is wedged into a 1½-in-diameter borehole. Rock stress release flexes the proving ring, changing the wire's tension and natural frequency or period of vibration. An electromagnet excites the vibrations in the wire and their frequency is measured as excavation proceeds in the rock. The frequency provides a measure of the proving-ring strain and, therefore, the stress change.

## 4.4.4 RESIDUAL ROCK STRESSES

### General

*Occurrence*

The geologic conditions causing overstress or residual stresses in rock masses in excess of overburden stresses are described in Art. 6.6. In practice they are significant in open excavations and tunnels where their effects are mitigated with rock bolts or bracing systems.

*Measurements*

Measurements are made at the rock surface where disturbance and stress relief from excavation may affect values obtained with strain meters, strain rosettes, or flat jacks. Alternatively, measurements are made in boreholes. Although they are difficult to perform, borehole

**FIG. 4.28 Installation of a pressure cell in an embankment.** *(Courtesy of Joseph S. Ward & Associates.)*

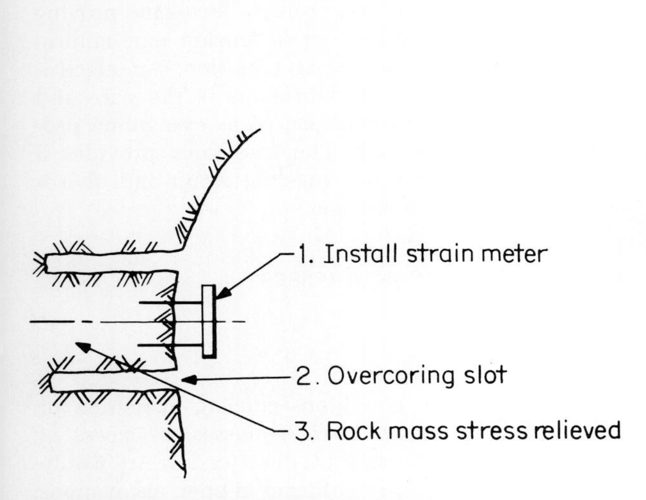


**FIG. 4.29 Strain meter for residual rock stresses.**

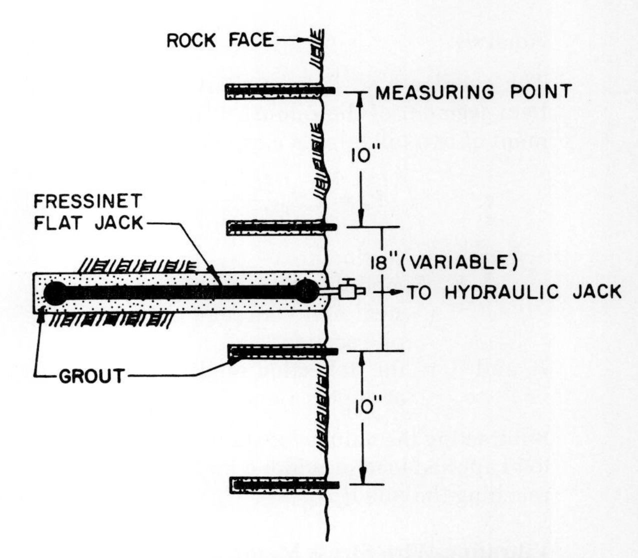


**FIG. 4.30 Flat-jack test to measure in situ rock stresses.**

measurements provide more representative data. Borehole devices require overcoring. Hydraulic fracturing is also performed in boreholes.

**Strain Meters or Strain Rosettes**

SR-4 strain gages (bonded gages) are attached to the rock wall, a zero reading is taken, and slots are cored around the gage to relieve rock stresses as shown on Fig. 4.29, and the gages are read again. By a suitable array of gages the principal stresses can be calculated.

Results are difficult to interpret unless rock modulus values have been measured.

**Flat Jacks**

Flat-jack tests have been described in Art. 3.5.3. Reference pins or strain gages are installed on the rock face and distance measurements made accurately (or strains are measured by gages mounted on the jack).

A slot is drilled, and the jack inserted and grouted (Fig. 4.30), then expanded under pressure to restore the original strain or reference-point measurements. The restoration pressure is taken as equal to the residual rock stress.

Flat jacks as well as strain meters are effective only in tightly jointed rock of high quality.

**Borehole Devices**

*Devices*

Residual stresses are measured in boreholes with the borehole deformation meter (Fig. 4.31*a*), the high-modulus stress plug or inclusion stress meter (Fig. 4.31*b*), or the Leeman "door-stopper" strain gage (Fig. 4.31*c*).

*Procedure*

The device is inserted into a small-diameter borehole (NX) and the stresses relieved by overcoring. Stresses are read directly with the inclusion stress meter, or computed from strains measured with the deformation meter or strain gages. To compute stresses when only strains are measured requires either a measurement or an assumption for the rock modulus. Installation and interpretation are described by Roberts (1969).[22] Maximum applicable depths are in the order of 10 to 15 m because of the difficulties of making accurate overcores, especially in holes not vertical.

*Retractable Cable Method*

Has been used to measure stresses to depths of 300 to 1500 m but its application is not a routine procedure [de la Cruz (1972)[23]].

**Hydraulic Fracturing**

*Application*

Hydraulic fracturing has been used to measure stresses in deep boreholes [Haimson (1977)[24]], and is still in the development stages.

*Technique*

A section of borehole is sealed off at the depth to be tested by means of two inflatable packers. The section is pressurized hydraulically with drilling fluid until the surrounding rock mass ruptures in tension.

The pressure required to initiate failure and the subsequent pressure required to maintain an open hole after failure are used to calculate the magnitude of the in situ stress.

## 4.5 INSTRUMENTATION ARRAYS FOR TYPICAL PROBLEMS

### 4.5.1 IMPORTANCE

**Soil Formations**

*Design Data Procurement*

Design data are usually based on properties measured in the laboratory or in situ. Load tests on piles or other types of foundations, and preloading with embankments provide additional design information.

Instrumentation as related to the soil-structure interaction problem is illustrated on Fig. 4.32. Practical problems in soil-structure interaction include laterally loaded piles, flexible retaining structures, continuous footings or mat foundations, and pavements.

*Construction Control*

Instrumentation is required where analytical limitations result in marginal safety factors, if excessive deformations are anticipated, or if remedial measures must be invoked. Such conditions exist for retaining structures constructed in soft ground, structures on shallow foundations, embankments on soft ground, and earth dams and cut slopes.

**Rock Masses**

*Design Data Procurement*

Design is usually based on assumed properties because of the cost and limitations of the validity of in situ tests; therefore, careful construction monitoring is required on important projects. In situ load tests and in situ stress measurements require instrumentation.

*Construction Control*

Instrumentation is used to monitor the changes in in situ conditions caused by construction, to

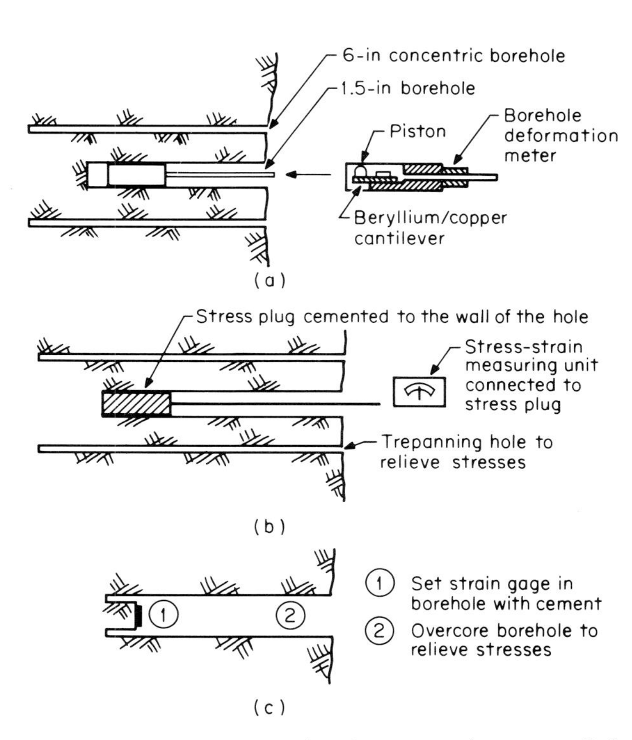


**FIG. 4.31 Measurements of in situ stresses by stress relief in boreholes: (*a*) borehole deformation meter; (*b*) high-modulus stress plug or inclusion stress meter (a rigid or near-rigid device calibrated directly in terms of stress); (*c*) Leeman "doorstopper" strain gage.**

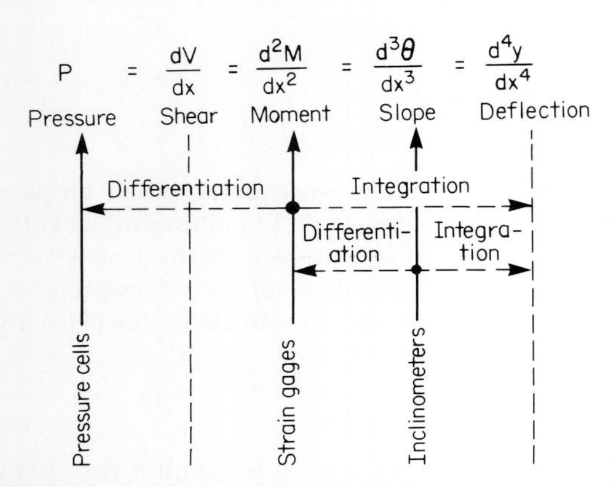


**FIG. 4.32 Instrumentation and the soil-structure interaction problem.** [*From Tschebotarioff (1973).*[25] *Reprinted with permission of McGraw-Hill Book Company.*]

provide an early warning of impending failures, and to assess the effectiveness of the installation of supports or other changes to improve stability of tunnels, caverns, mines, cut slopes, and other excavations.

### 4.5.2 SETTLEMENT OF STRUCTURES

#### Case 1: Structure Undergoing Differential Settlement

*Problem*

The structure in Fig. 4.33*a* is undergoing excessive differential settlements resulting in wall cracks, distortions of doors and windows, misaligned machinery, and floor warping, as a result of consolidation of a weak soil stratum.

*Objectives*

It is desired to determine the magnitude and anticipated rate of the remaining settlements of the floors and structural frame. This information will provide the basis for determining the remedial measures.

*Instrumentation*

The amounts and time rates of settlement at various locations in the structure and pore pressures at various depths are monitored with the devices shown on the figure. The inclinometer is installed to measure possible lateral deflections. The data obtained are analyzed with laboratory test data to determine the magnitude and time rates of the remaining settlement.

#### Case 2: Structure Undergoing Tilt

*Problem*

The relatively rigid, mat-supported structure in Fig. 4.33*b* is undergoing tilt from differential settlement which is affecting the balance of turbines.

*Objectives*

It is desired to determine which strata are contributing to the settlements, to estimate the magnitude of the remaining settlement, and to judge the time required for its essential completion in order to arrive at remedial treatments.

*Instrumentation*

As shown on the figure, building deflections are monitored, as are the compression occurring in each stratum and the pore pressures.

#### Case 3: Construction over Soft Ground and Preloading

*Problem*

Embankment (Fig. 4.34*a*) or steel storage tank (Fig. 4.34*b*) constructed over soft ground. Preloading is achieved by adding fill or loading the tank with water.

*Objectives*

Loading rates must be controlled to prevent foundation failure. Data on the magnitudes and rate of settlement and on the pore pressures provide the basis for determining when the surcharge may be removed. (After the storage tank is surcharged, it is raised, releveled, and used for storing products lighter than water.)

*Instrumentation*

Precise optical surveys monitor settlements as a function of time, as well as heave occurring beyond loaded areas. Settlement plates and rod extensometers are monitored at the fill surface. Remotely read settlement extensometers may be installed in the fill and beneath the tank. (With

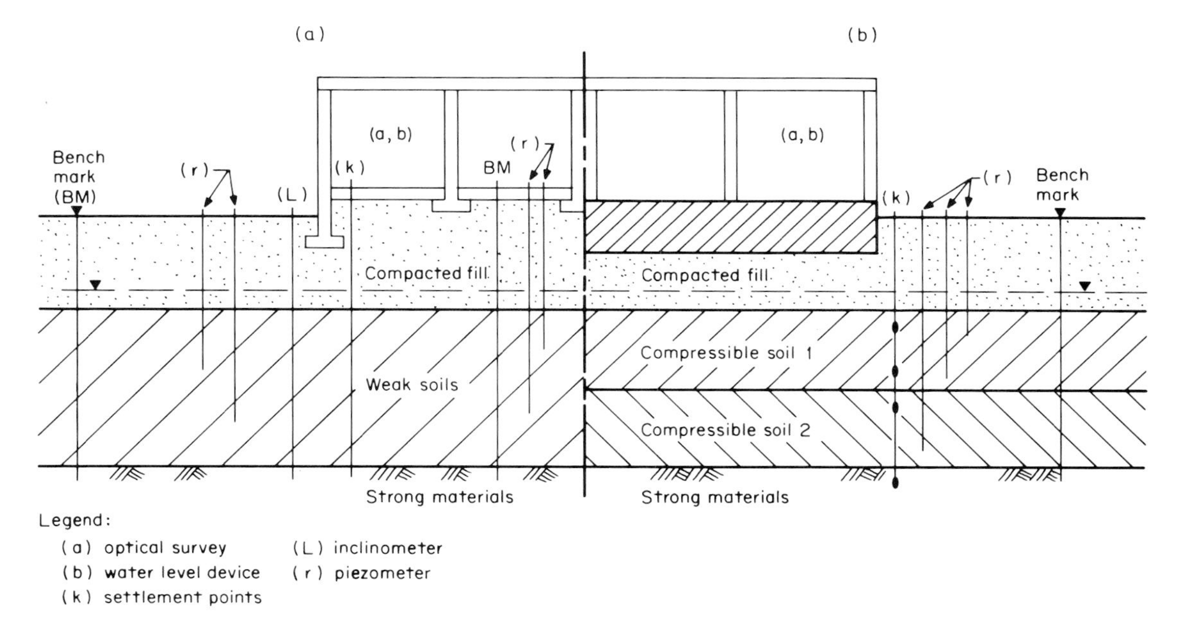


**FIG. 4.33 Instrumentation for building settlements: (*a*) Spread footing-supported light structure and (*b*) mat-supported heavy structure.**

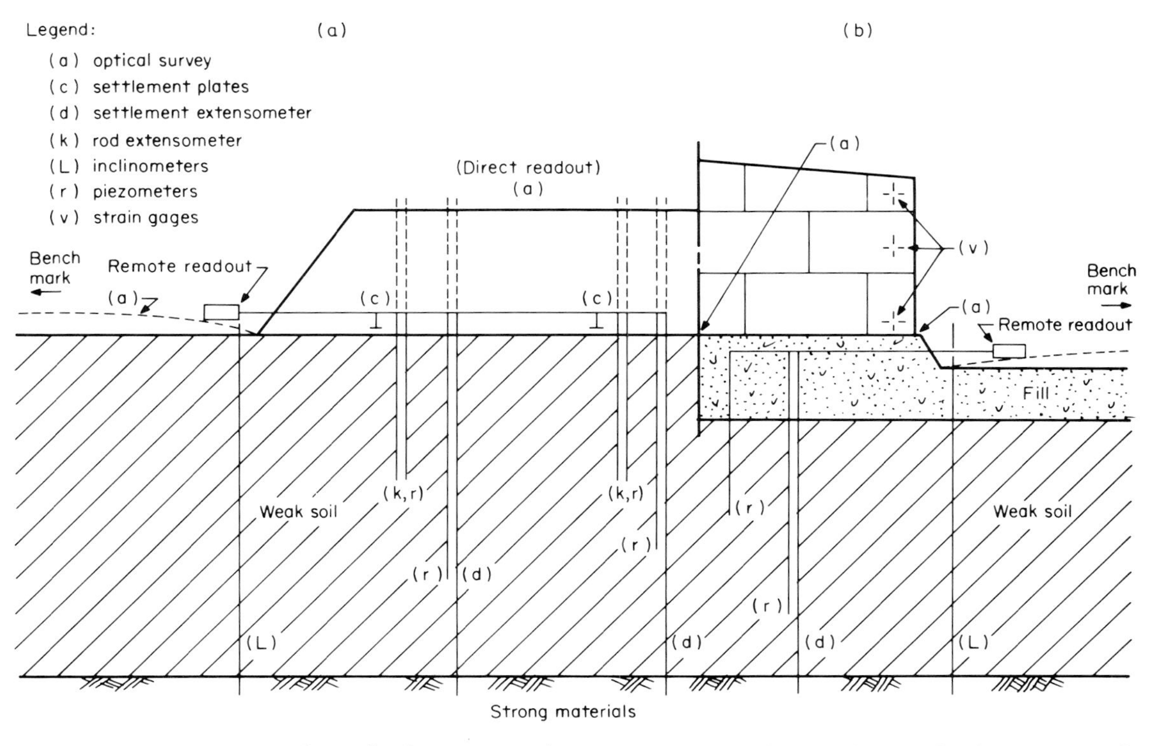


**FIG. 4.34 Instrumentation for embankment or steel storage tank over soft ground: (*a*) Embankment over soft ground and (*b*) steel storage tank over soft ground.**

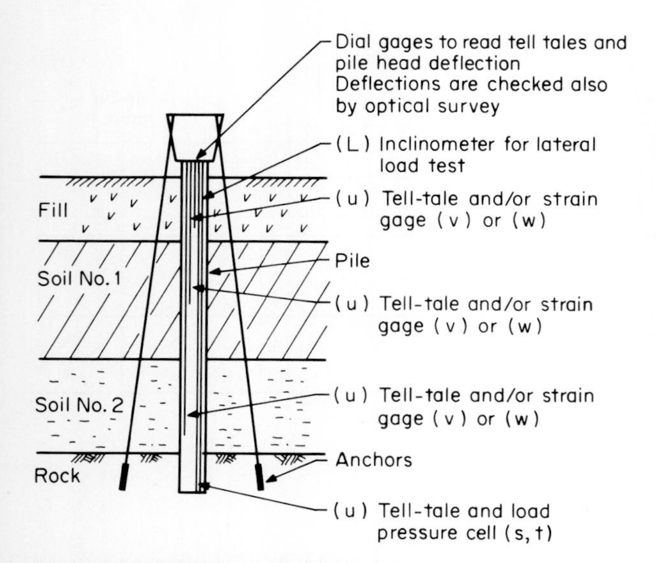


**FIG. 4.35 Instrumentation for pile load test.**

respect to tank rupture, the critical differential settlements are between the center and the bottom and along the perimeter.)

Stresses in the tank wall are monitored with strain gages; mechanical types are preferred.

The piezometers, inclinometer, and heave reference points are necessary to provide control against failure by foundation rupture. Pore-pressure data are needed to determine when the preload may be removed.

Bench marks must be installed beyond any possible ground movements.

### Case 4: Pile Load Testing

*Objectives*

A pile load test may be performed (1) to determine the capacity of a pile as required to satisfy a building code, (2) to determine the proportion of the total load carried in end bearing and that carried in shaft friction during vertical load testing, and (3) to measure horizontal deflections during lateral load testing for the determination of the horizontal modulus of subgrade reaction.

The information is particularly important for the design of long, high-capacity piles penetrating several formations where design analysis based on laboratory test data often is not reliable, especially in formations difficult to sample undisturbed.

*Instrumentation*

Deflections of the pile top shown on Fig. 4.35 are monitored by dial gages and optical survey. Deflections at the pile tip and at various locations along the pile's length are monitored with tell tales and strain gages.

A load or pressure cell installed at the base provides data on the end bearing pressure.

During horizontal testing, dial gages and optical surveys monitor deflections at the top, and the inclinometer monitors deflection along the length.

*Dynamic Load Testing*

The Case-Goble method of dynamic pile testing [Goble and Rausche (1970),[26] Rausche et al. (1971),[27] La Fond (1977)[28]] utilizes transducers attached either on the pile top or along the shaft to measure the force and acceleration caused by hammer impact. Instrumentation removes the need to evaluate the unknown factors of energy loss in the driving system and allows indirect elimination of the dynamic pile characteristics. It thus provides more accurate measurement of driving resistance. The data are input to the wave equation method of dynamic pile analysis.

## 4.5.3 EXCAVATION RETENTION

### Objectives

A retaining structure in an open excavation must be designed to provide adequate support for the excavation sides and to restrict wall deflection to a minimum in urban areas. As the wall deflects, the ground surface behind the wall subsides, and significant movements can result in detrimental settlements of adjacent structures. The problem is particularly serious in soft to firm clays and in clays underlain by sands or silts under water pressure.

### Wall Design Criteria

The design of a braced excavation in soft ground is based on calculations of earth pressure magnitudes and distributions which can be determined only by empirical or semiempirical methods. Calculations of wall deflections and backslope subsidence cannot be made with the classical theories, although the finite element method is being used in current analysis for such predictions. As a rule, however, the engineer cannot be certain of the real pressures and deflections that will be encountered, especially in soft ground. The engineer also may have little control over construction techniques which will influence wall performance.

A wall may become unnecessarily costly if design provides for all contingencies. Because the nature of wall construction provides the opportunity for strengthening by placing additional braces, anchors, or other supports, initial design can be based on relatively low safety factors but with provisions for contingency measures. This approach to design and construction is feasible only if construction is monitored with early warning systems.

### Construction Monitoring

Elements to be monitored during construction include:

- Movements of the backslope area and adjacent structures, heave of the excavation bottom, and movements in the support system
- Groundwater levels and the quality and quantity of water obtained from the dewatering system
- Strains and loads in the support system
- Vibrations from blasting, pile driving, and traffic

### Instrumentation

As illustrated on Fig. 4.36, vertical deflections of the wall, excavation bottom, backslope, and adjacent buildings are monitored by precise leveling (*a*). In important structures, measurements

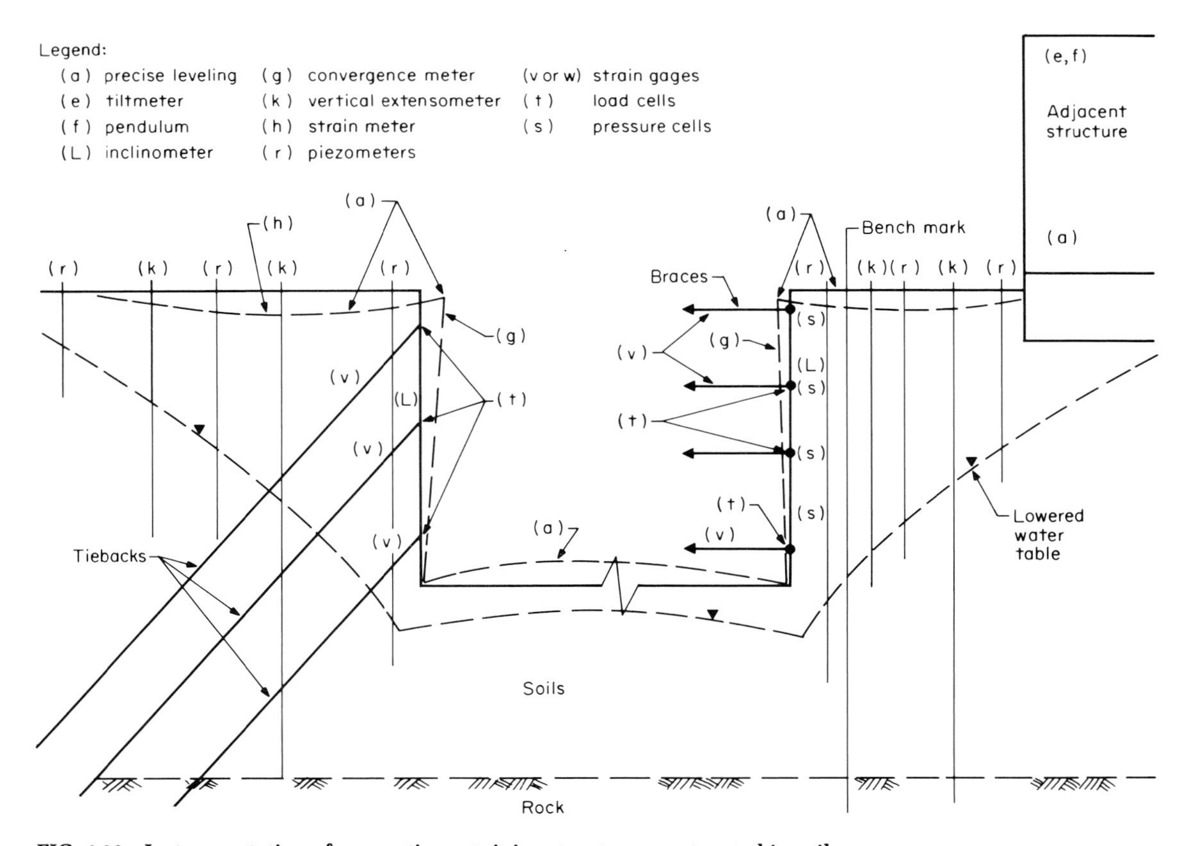


**FIG. 4.36 Instrumentation of excavation retaining structure constructed in soils.**

should be made with tiltmeters (*e*) or pendulums (*f*). Inclinometers (*l*) are installed immediately behind the wall to measure its lateral deflections. Excavation closure can be measured with convergence meters (g), and backslope subsidence with vertical extensometers (*k*) or shallow buried strain meters (*h*).

Loads in anchors and braces are monitored with strain gages (*v* or *w*) or load cells (*t*), or checked periodically by jacks. Strain gages must be protected against temperature changes, or the temperatures must be monitored, to provide useful data. Pressure cells (*s*) installed behind the wall have the advantage that they can be installed between supports. Groundwater variations are measured with piezometers (*r*).

An immovable bench mark is necessary for reliable measurements of deflections of adjacent ground, structures, and the excavation bottom.

4.5.4 EARTH DAMS

### Objectives

An earth dam as actually constructed may differ from that designed because of the necessity of using earth materials for fabrication which may

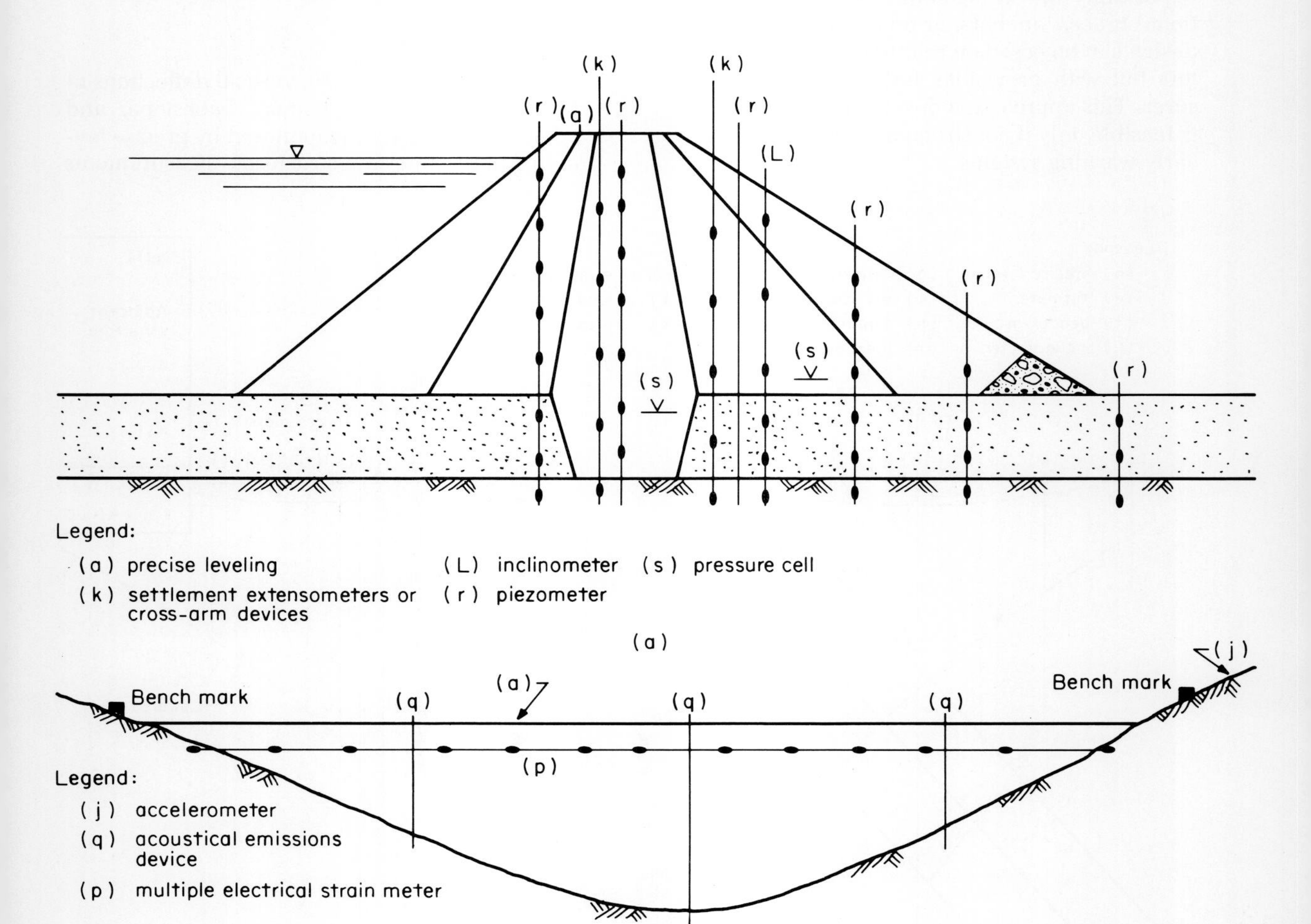


**FIG. 4.37 Instrumentation for an earth dam: (*a*) cross section and (*b*) longitudinal section.**

vary in properties and distributions from those assumed originally.

Most important dam structures, therefore, are instrumented and their performance monitored during construction, during impoundment, and while in service. Deformations in the foundation and embankment and seepage forces are monitored.

#### Instrumentation

*Deformations*

Most large dams undergo compression under their own weight, and often are subjected to large foundation settlements. Cracks can develop across the core near the crest soon after construction and during reservoir filling. The cracks generally appear near the abutments as a result of differential settlement along the valley walls, or over irregularities in an underlying rock surface. As illustrated on Fig. 4.37, external evidence of deformations is measured by optical survey (*a*). Internal evidence is monitored by settlement extensometers or cross-arm devices (*k*) and strain meters (*p*). Longitudinal strain meters (*p*) permit the detection and location of the internal cracking when it first occurs.

*Pore Pressures and Seepage*

Monitoring of pore-water pressures and seepage is necessary in the embankment, abutments, and foundation materials since these phenomena normally are the most critical factors of dam stability. Critical zones are at the toe and in front of and below seepage cutoffs such as core trenches and grout curtains. Piezometers (*r*) measure pore pressures and the acoustic emissions device (*g*) may locate seepage paths and piping zones. Seepage may often be collected in drainage ditches and measured by weirs.

*Slope Stability*

Inclinometers (*l*) monitor deflections in the slopes and pressure cells (*s*) monitor stresses.

*Seismic Areas*

Accelerographs (*j*) are installed on the dam and abutments to monitor earthquake loadings (*see Art. 11.2.3*). At times seismoscopes have been used. Automatic-recording piezometers provide useful data because of the pore-pressure buildup occurring during seismic excitation.

### 4.5.5 TUNNELS, CAVERNS, AND MINES

#### Case 1: Closed Excavations in Rock

*Objectives*

Closed excavations in rock masses may or may not require support depending on rock quality, seepage forces, and the size of the opening. Support systems may be classed as temporary to permit construction to proceed safely and permanent to provide the final structure. In large openings there is the need to evaluate progressive deformation and the possibility of full-scale failure. In small openings the most frequent problems involve partial or confined failures such as falling blocks and running ground. It is not enough to provide adequate support to the opening itself; consideration must be given also to the possibility of surface subsidence and detrimental deflections of adjacent structures.

*Design Criteria*

Support requirements may be selected on the basis of experience and empirical relationships with rock quality or on the basis of finite element methods applied to assess stress conditions and deformations. Analytical methods require information on rock-mass stress conditions and deformation moduli, which are measured in situ (*see Art. 3.5.3*). In situ test results may not be representative of real conditions for various reasons depending on the type of test, and are generally most applicable to rock of good to excellent quality in large openings.

A tunnel may become unnecessarily costly if all contingencies are provided for in design, especially in regard to the problem of identifying all significant aspects of underground excavation by exploration methods. The nature of tunnel construction allows design on a contingency basis in which the support systems are modified to suit the conditions encountered. This approach requires monitoring in situ conditions as excavation proceeds because of the relative

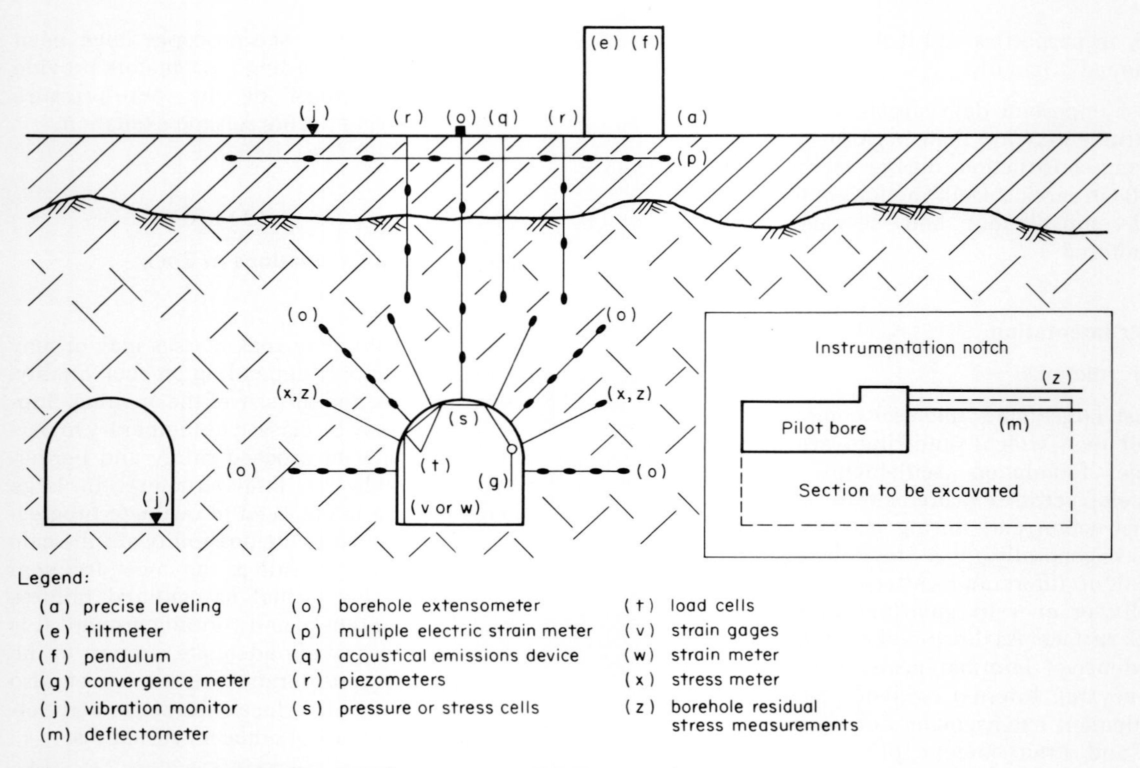


**FIG. 4.38 Instrumentation for tunnels, caverns, and mines in rock.**

nonpredictability of rock-mass response to an opening.

*Construction Monitoring*

Elements to be monitored during construction may include:

1. Closure of the opening and strain gradients in the rock mass
2. Support system strains and loads
3. Ground surface subsidence in urban areas over mines, caverns, and tunnels (surface subsidence may result from closure of the opening, from running ground into the opening, or from groundwater lowering causing compression in overlying soil formations)
4. Vibrations from blasting affecting surface structures or adjacent underground openings

*Instrumentation (Fig. 4.38)*

Borehole extensometers (o) are installed to measure strain gradients and closure, and serve a number of purposes. Rock-mass loads can be deduced for support design, and the modulus can be determined in good-quality rock when rock stresses are known from borehole stress relief or flat-jack tests.

Rock moduli computed from radial deflections may be lower than those computed from flat-jack tests. During tunnel driving, radial stresses diminish to zero at the edge of the opening while

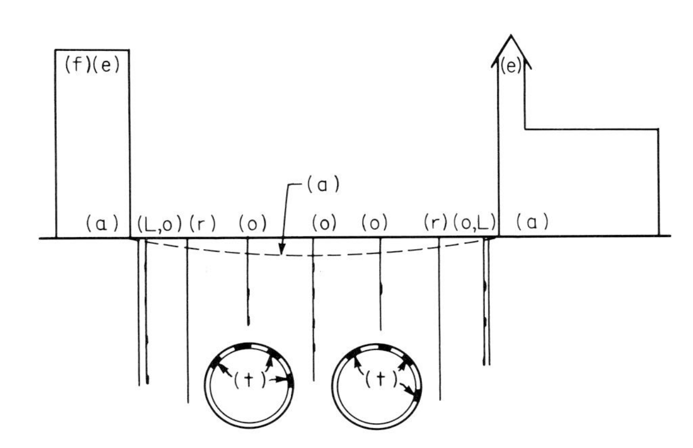


Legend:
(a) precise leveling
(e) tiltmeter or
(f) pendulum
(L) inclinometer
(o) borehole extensometer
(r) piezometer
(t) load cells

**FIG. 4.39 Instrumentation for tunnels in soil in urban areas.**

tangential stresses become concentrated [Kruse (1970)[29]]. Joints which have nearly tangential orientations have the greatest tendency to open. This condition influences borehole extensometer measurement as well as the compression generated by plate bearing loads. Borehole extensometer data are also used to estimate rock-bolt lengths.

Convergence meters (g) monitor closure. Deflectometers (*m*) installed in advance of the tunnel from a small pilot bore provide warning of mass deformations occurring in major shear zones; groundwater conditions are revealed during the drilling process and pilot holes serve to provide predrainage.

In situ stress meters (x,z) provide data for estimating roof and wall pressures to be retained by the support system. Pressure cells (s) installed between the tunnel lining and the roof monitor stress changes occurring with time and forewarn of the necessity for additional supports. Strain gages (*v* or *w*) are installed on metal supports. Load cells (*t*) installed in the tunnel lining or on rock anchors monitor load changes. Load reductions in anchors indicate movements of the rock and a relaxation of the anchor load and the anchor's retention capacity. In this case, additional anchors, probably of greater length, are required.

Vertical extensometers (*o*), surface strain meters (*p*) and precise leveling (*a*) are used to monitor ground subsidence, and tiltmeters (*e*) and pendulums (*f*) monitor deflections in overlying structures.

Acoustical emission devices (*q*) provide warning of impending collapse of mines or large caverns.

Vibration monitoring (*j*) is required when blasting is being done near an adjacent existing tunnel or near overlying structures.

### Case 2: Tunnels in Soil in Urban Areas

*Objectives*

The most important objective in instrumenting tunnels in urban areas is the monitoring of surface deflections and the prevention of damage to overlying structures.

*Instrumentation*

As illustrated on Fig. 4.39, surface movements are monitored by precise leveling (*a*), but important data are obtained from tiltmeters (*e*) or pendulums (*f*) installed in structures. Borehole extensometers provide data on subsurface deformations in the vertical mode and inclinometers (*l*) in the lateral mode, serving as early warnings of excessive deformations. Pore pressures are monitored with piezometers (*r*), and load cells (*t*) installed in tunnel linings provide data on earth pressures.

## 4.5.6 NATURAL AND CUT SLOPES

### Case 1: Rock Cuts

*Objectives*

The basic objective is to maintain slope stability, which is related to steepness, height, the orientation of weakness planes with respect to the slope angle, and cleft-water pressures. Failure

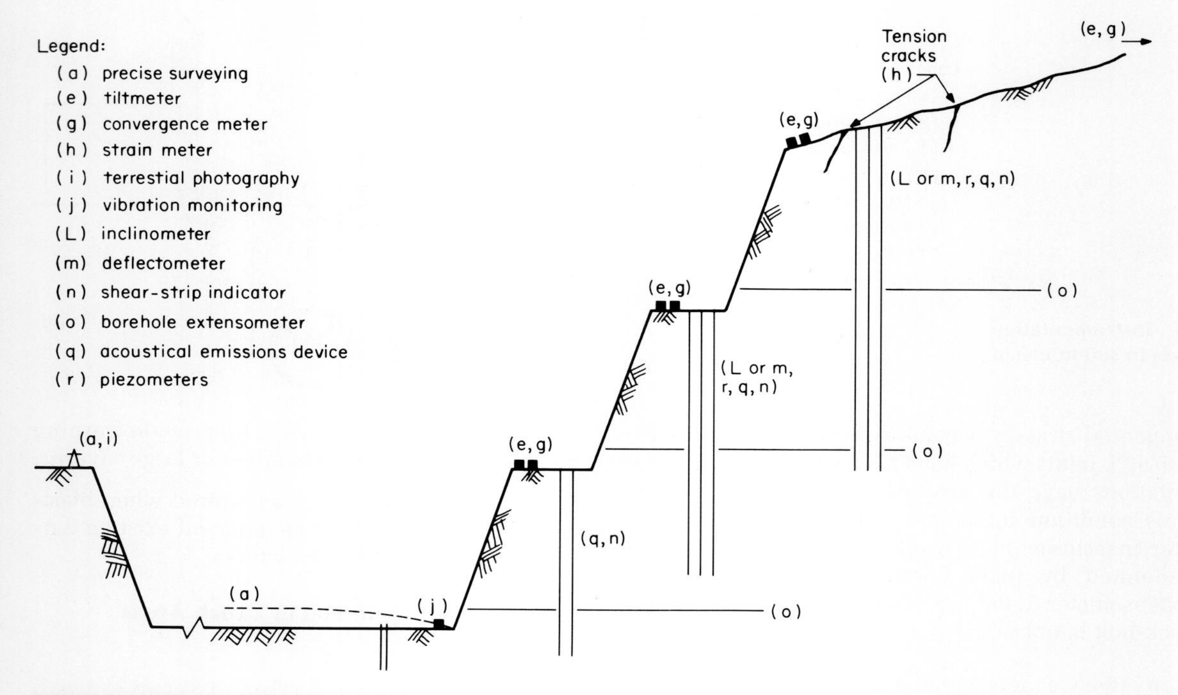


**FIG. 4.40 Instrumentation for a rock cut.**

in rock slopes usually occurs suddenly, with rapid movement.

*Slope Stability Problems*

Natural slopes may become unstable from weathering effects, erosion, frost wedging, or the development of high cleft-water pressures.

Open-pit mines are excavated with the steepest possible stable slope, and low safety factors are accepted for economic reasons. Instrumentation monitoring generally is accepted as standard procedure for deep pits.

Hillside cuts for roadways and other construction are usually less steep than open-pit mine slopes in the same geologic conditions, and require a higher safety factor against failure. Evaluations are usually based on experience and empirical relationships, and instrumentation is used normally only in critical situations.

*Instrumentation (Fig. 4.40)*

Internal lateral movements are monitored in terms of displacement vs. time with inclinometers (*l*) and deflectometers (*m*) which can be attached to alarm systems, or with borehole extensometers (*o*). Shear-strip indicators (*n*) and the acoustical-emissions device (*q*) provide indications of mass movements.

External lateral movements are monitored with the convergence meter (*g*) or tiltmeter (*e*) on benches, strain meters (*h*) on tension cracks, optical surveys (*a*) with the laser geodimeter which provide rapid measurements of the entire slope, and terrestrial stereophotography (*i*) which provides a periodic record of the entire slope face.

Bottom heave is monitored by optical survey of monuments or "settlement points"; it often precedes a major slope failure.

Vibrations from blasting and traffic, which may affect stability, are monitored (*j*), and cleft-water pressures are monitored with piezometers (*r*).

*Installations*

Inclinometers, deflectometers, piezometers, and extensometers often are limited to early excavation stages to obtain data for the determinations of stable slope inclinations. They are expensive instruments to install and monitor, and not only will they be lost if failure occurs, but in mining operations many will be lost as excavation proceeds. In addition, they monitor only limited areas by section.

Tiltmeters, convergence meters, and optical surveys are both economical and "retrievable," and provide for observations of the entire slope rather than a few sections; therefore, they provide the basic monitoring systems. In critical areas, where structures or workers are endangered if a collapse occurs during mining operations, MPBX extensometers and the acoustical-emissions device are used to provide early warning systems.

### Case 2: Soil Slopes

*Objectives*

In soil slopes the objectives are to detect movements when they first occur since in many instances slope failures develop gradually; and, when movement occurs, to monitor the rate of movement and accelerations, locate the failure surface, and monitor pore pressures. These data provide the basis for anticipation of total and perhaps sudden failure.

*Slope Stability Problems*

In potentially unstable natural or cut slopes, failure is usually preceded by the development of high pore-water pressures, an increase in rate of slope movement, and the occurrence of tension cracks. Slope movement is not necessarily indicative of total failure, however, since movements are often progressive, continuing for many years.

Stability evaluations require information on the failure surface location, pore pressures, and rates of movement as well as the geologic and climatic factors *(Chap. 9)*.

*Instrumentation*

Surface movements of the natural slope shown on Fig. 4.41 are monitored by precise leveling with the laser geodimeter (*a*), convergence meters (g), and strain meters (*h*). The meters may be attached to alarm systems.

Subsurface deformations are monitored with the inclinometer (*l*), shear-strip indicator (*n*), or steel-wire sensor to locate the failure surface. Nuclear probes *(see Art. 2.3.6)* have also been used. Acoustical emissions (*q*) may indicate approaching failure.

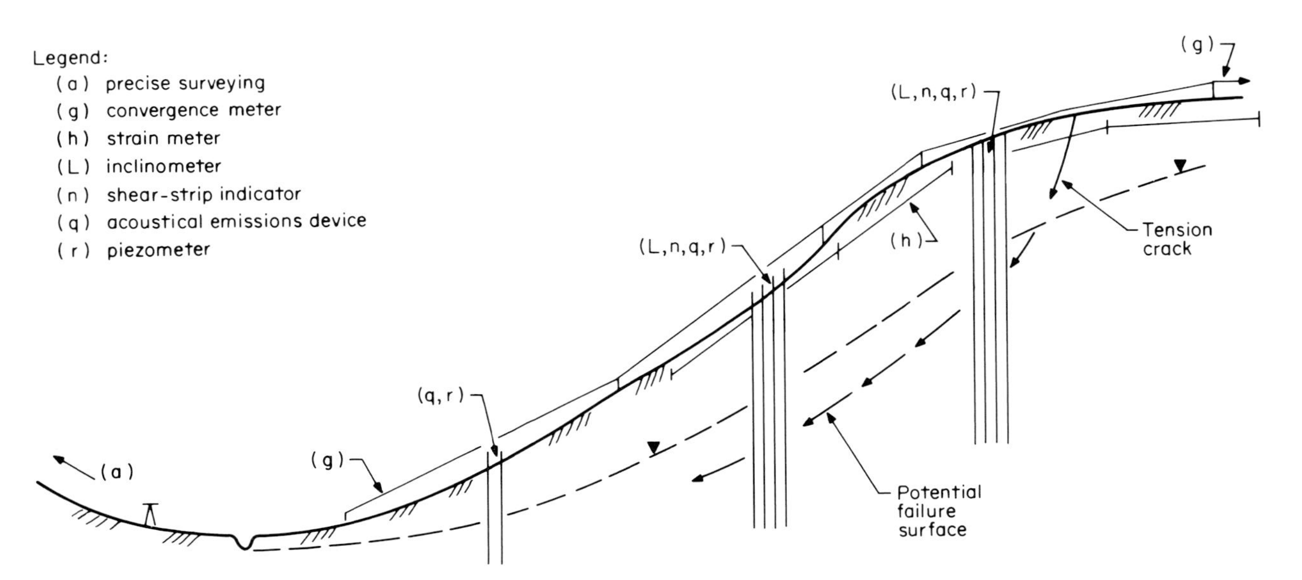


**FIG. 4.41 Instrumentation for a potentially unstable soil slope.**

Legend:
- (a) precise leveling
- (e) tiltmeter
- (h) strain meter
- (j) accelerometer
- (m) deflectometer
- (n) shear-strip indicator
- (q) acoustical emissions device
- (r) piezometer
- (x) stress meter

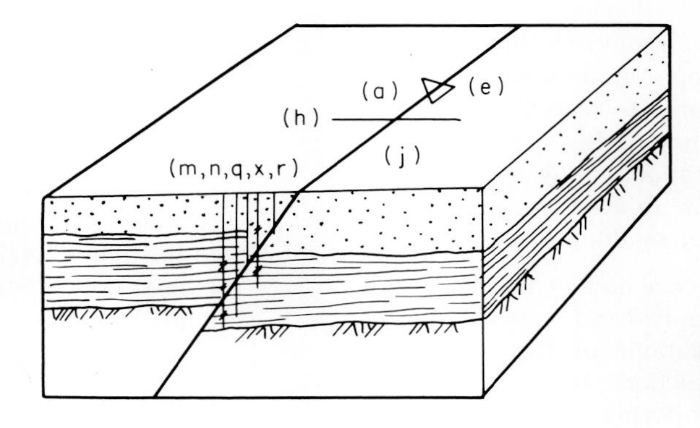


**FIG. 4.42 Instrumentation to monitor fault movements.**

Pore-water pressures are monitored with piezometers (*r*). Knowing the failure surface location and soil shear strength, one can estimate the pore pressures required for total failure.

*Installations*

Piezometers, inclinometers, extensometers, and other devices used in boreholes may not be monitoring the most critical areas, especially in slopes in a stable condition. Therefore, optical surveys, which provide information over the entire study area, are always important, although, because of terrain conditions, perhaps not always possible.

### 4.5.7 FAULT MOVEMENTS

#### Objectives

Earthquakes have been associated with fault rupture (*see Art. 11.3.1*), which may be preceded by ground warping, slippage along the fault, and increase in ground stresses and pore-water pressures. Fault monitoring is still in the experimental stages; eventually some basis for predicting or anticipating rupture may be developed.

#### Instrumentation

Ground warping and fault movements on a regional scale are being monitored by the ARIES system (Astronomical Radio Interferometric Earth Survey), a system of portable and fixed radio telescopes using differing reception times from a distant quasar to triangulate the exact position of three ground antennas on a periodic basis.

Surface indications of displacement along the San Andreas fault are being monitored by an early warning system set up by NOAA's Earthquake Mechanism Laboratory at Stone Canyon, California (Fig. 4.42). The system consists of an interconnected 20-m triangular array of mercury pools set in piers attached to Invar rods to monitor tilting (*e*) and strain meters consisting of three 30-m long extensometers (*h*) to measure creep [Bufe (1972)[6]]. Accelerometers (*j*) set on the surface monitor ground motion accompanying fault activity.

Subsurface deformations may be monitored with deflectometers (*m*) and shear strips (*n*) installed in boreholes across the fault, set to sound alarms if desired. Stress meters (*x*) monitor stress increase as does the acoustic-emissions device (*q*) on a qualitative basis, when set in or near the fault zone. Piezometers (*r*) set in the fault zone monitor water pressures that may indicate stress changes in the rock mass.

## REFERENCES


1. Cording, E. J., Hendron, A. J., Jr., Hansmire, W. H., Mahar, J. W., MacPherson, H. H., Jones, R. A., and O'Rourke, T. D. (1975) *Methods for Geotechnical Observations and Instrumentation in Tunneling*, Dept. of Civil Engrg., Univ. of Illinois, Urbana, Vols. 1 and 2.
2. Dreyer, H. (1977) "Long Term Measurements in Rock Mechanics by Means of Maihak Vibrating Wire Instrumentation," *Field Measurements in Rock Mechanics*, Vol. I, *Proc. 4th Intl. Symp.*, Zurich, April, pp. 109–135 (publ. A. A. Balkema, Rotterdam).
3. Blackwell, G., Pow, D. and Klast, L. (1975) "Slope Monitoring at Brenda Mine," *Proc. 10th Canadian Rock Mechs. Symp.*, Kingston, Ontario, September, pp. 45–79.
4. Novosad, S. (1978) "The Use of Modern Methods in Investigating Slope Deformations," *Bull. Intl. Assoc. of Engrg. Geol.*, No. 17, June, pp. 71–73.
5. Franklin, J. A. (1977) "Monitoring Rock Structures," *Intl. J. Rock Mech., Min Sci. and Geomech.*, abstr., Vol. 14, July, pp. 163–192.
6. Bufe, C. G. (1972) "Strain and Tilt Measurements Near an Active Fault," *Proc. ASCE, Stability of Rock Slopes, 13th Symp. on Rock Mechs.*, Urbana, Ill. (1971) pp. 691–716.
7. Kennedy, B. A. (1972) "Methods of Monitoring Open Pit Slopes," *Stability of Rock Slopes, Proc. ASCE, 13th Symp. Rock Mechs.*, Urbana, Illinois (1971) pp. 537–572.
8. Hartmann, B. E. (1966) *Rock Mechanics Instrumentation for Tunnel Construction*, Terrametrics, Wheatridge, Colo.
9. Silveira, J. F. A. (1976) "A Instrumentação de Mecânica das Rochas em Tuneis, Métodos de Observação e Históricos de Casos," *Proc. 1st Cong. Brasileira de Geologia de Engenharia*, Rio de Janeiro, August, Vol. 1, pp. 131–154.
10. Richart, F. E., Jr. (1960) "Foundation Vibrations," *Proc. ASCE, J. Soil Mechs. and Found. Engrg. Div.*, Vol. 86, August.
11. USBR (1960, 1st ed.) (1974, 2d ed.) *Earth Manual*, U.S. Bureau of Reclamation, Denver, Colo.
12. Sherard, J. L., Woodward, R. J., Gizienski, S. F. and Clevenger, W. A. (1963) *Earth and Earth-Rock Dams*, John Wiley & Sons, New York.
13. Broms, B. B. (1975) "Landslides," *Foundation Engineering Handbook*, Winterkorn and Fang, eds., Van Nostrand Reinhold Pub., New York, Chap. 11.
14. Brawner, C. O. (1975) "Case Examples of Instability of Rock Slopes," *J. Assoc. Prof. Engrs*, British Columbia, February, Vol. 26, No. 2.
15. Scholz, C. H. (1968) "Mechanism of Creep in Brittle Rock," *J. Geophysical Research*, Vol. 73, pp. 3295–3302.
16. Koerner, R. M., Lord, A. E., Jr., and McCabe, W. M. (1977) "Acoustic Emission Behavior of Cohesive Soils," *Proc. ASCE, J. Geotech. Engrg. Div.*, Vol. 103, No. GT8, August, pp. 837–850.
17. Hardy, H. R., Jr. and Mowrey, G. L. (1977) "Study of Underground Structural Stability Using Near-surface and Downhole Microseismic Techniques," *Field Instrumentation in Rock Mechanics, Proc. 4th Intl. Conf.*, Zurich, April, pp. 75–92 (publ. A. A. Balkema, Rotterdam).
18. Hardy, H. R., Jr. and Leighton, F. (1977) "Design, Calibration and Construction of Acoustical Emissions Equipment," *Proc. 1st Conf. on Acoustical Emission/Microseismic Activity in Geologic Structures and Materials*, Penn. State Univ., June (1975), Trans. Tech. Pub. Co., Clausthal, Germany.
19. Koerner, R. M., Lord, A. E., Jr., and McCabe, W. M. (1978) "Acoustic Monitoring of Soil Stability," *Proc. ASCE, J. Geotech. Engrg. Div.*, Vol. 104, No.GT5, pp. 571–582.
20. Sauer, G. and Sharma, B. (1977) "A System for Stress Measurements in Constructions in Rock," *Field Instrumentation in Rock Mechanics, Proc. 4th Intl. Symp.*, Zurich, Vol. I, pp. 317–329 (Publ. A. A. Balkema, Rotterdam).
21. Sellers, J. B. (1977) "The Measurement of Stress Changes in Rock Using the Vibrating Wire Stress Meter," *Field Instrumentation in Rock, Proc. 4th Intl. Symp.*, Zurich, Vol. I, pp. 317–329 (publ. A. A. Balkema, Rotterdam).
22. Roberts, A. (1969) "The Measurement of Stress and Strain in Rock Masses," *Rock Mechanics in Engineering Practice*, Stagg and Zienkiewicz, eds., John Wiley & Sons, New York, Chap. 6.
23. de la Cruz, R. V. (1972) "Stress Measurements by the Retractable Cable Method," *Proc. ASCE, Stability of Rock Slopes*, 13th Symp. on Rock Mechs., Univ. of Illinois, Urbana (1971), pp. 856–882.
24. Haimson, B. C. (1977) "Stress Measurements Using the Hydrofracturing Technique," *Field Measurements in Rock Mechanics, Proc. 4th Intl. Symp.*, Zurich, April, Vol. I, pp. 223–242 (publ. A. A. Balkema, Rotterdam).
25. Tschebotarioff, G. P. (1973) *Foundations, Retaining and Earth Structures*, McGraw-Hill Book Co., New York.

26. Goble, G. G. and Rausche, F. (1970) "Pile Load Test by Impact Driving," paper presented to *Highway Research Board Annual Mtg.*, Washington, D.C., January.

27. Rausche, F., Goble, G. G. and Moses, F. (1971) "A New Testing Procedure for Axial Pile Strength," OTC paper 1481, *Offshore Technology Conf.*, Houston, Texas (preprint).

28. La Fond, K. J. (1977) "Applications of Dynamic Pile Analysis," from *Piletips*, Assoc. Pile and Fitting Corp., Clifton, N.J., May–June.

29. Kruse, G. H. (1970) "Deformability of Rock Structures, California State Water Project," *Determination of the in-Situ Modulus of Rock*, ASTM STP 477, Amer. Soc. for Test. and Mat., Philadelphia, Pa.

## BIBLIOGRAPHY

ASTM (1966) *Determination of Stress in Rock: A State of the Art Report*, SPT 429, Amer. Soc. for Test. and Mat., Philadelphia, Pa.

Brawner, C. O., Stacey, P. F. and Stark, R. (1975) "Monitoring of the Hogarth Pit Highwall, Steep Rock Mine, Atikokan, Ontario," *Proc. 10th Canadian Rock Mechs. Symp.*, Kingston, Ontario, September.

Cotecchia, V. (1978) "Systematic Reconnaissance Mapping and Registration of Slope Movements," *Intl. Assoc. Engrg. Geol.*, Bul. No. 17, June, pp. 5–37.

Gartung, I. and Bauernfiend, P. (1977) "Subway Tunnel at Nurnberg—Predicted and Measured Deformations," *Field Measurements in Rock Mechanics, Proc. 4th Intl. Symp.*, Zurich, April, pp. 473–483 (publ. A. A. Balkema, Rotterdam).

Koerner, R. M., Lord A. E., Jr., McCabe, W. M. and Curran, J. W. (1976) "Acoustic Behavior of Granular Soils," *Proc. ASCE, J. Geotech. Engrg. Div.*, Vol. 102, No. GT7, July.

Koerner, R. M., Reif, J. S. and Burlingame, M. J. (1979) "Detection Methods for Locating Subsurface Water and Seepage," *Proc. ASCE, J. Geotech. Engrg. Div.*, Vol. 105, No. GT11, November, pp. 1301–1316.

Sinco (The Slope Indicator Co.), Catalog, Seattle, Washington.

Terrametrics, Catalog, Golden, Colo.

Wilson, S. D. (1967) "Investigation of Embankment Performance," *Proc. ASCE, J. Soil Mechs. and Found. Engrg. Div.*, Vol. 93, No. SM4, July.

# PART II

# CHARACTERISTICS OF GEOLOGIC MATERIALS AND FORMATIONS

## CHAPTERS



## PURPOSE AND SCOPE

Part II provides the basis for the recognition, identification, and classification of the various soil and rock types; describes them in terms of their origin, mode of occurrence, and structural features in situ, and presents their typical characteristics of engineering significance.

Part II also provides the basis for recognizing the elements of engineering significance of surface and subsurface water; for the selection of methods of controlling flooding, erosion, subsurface flow, and seepage forces; and for environmental conservation as related to water.

## SIGNIFICANCE

Proper identification and classification of rock masses and soil formations permit estimations of their characteristic properties from correlations.

Formations of rocks and soils have characteristic features which provide the basis for interpretation and determination of their constituents, and therefore the basis for estimating their engineering properties. In rock masses, properties are related to the characteristics of intact blocks, the mass discontinuities such as fractures and cavities, and the degree of weathering. In soil formations, properties are strongly related to their origin and mode of occurrence.

The characteristics of groundwater relate to its mode of occurrence and the nature of the materials in which it is contained, since these factors control flow quantities, rates, and seepage forces.

## CORRELATIONS

Correlations among the various types of geologic materials and their characteristic engineering properties are given in Chaps. 3 and 5 for intact rock, Chaps. 3, 5, and 6 for rock masses in general, and Chaps. 3 and 7 for soil formations. A general summary is given in Appendix D.

## GEOLOGIC HISTORY AND TECTONICS

These aspects of geologic material are described in Appendix A.

# CHAPTER FIVE

# ROCK AND SOIL: IDENTIFICATION AND CLASSIFICATION

## 5.1 INTRODUCTION

### 5.1.1 THE GEOLOGIC MATERIALS

**Definitions**

Precise definitions of the two general constituents, rock and soil, applicable to all cases are difficult to establish because of the very significant transition zone in which rock is changing to soil or in which a soil formation has acquired rock-like properties, or various other conditions. In general terms, the constituents may be defined as follows.

*Rock*

Material of the earth's crust, composed of one or more minerals strongly bonded together, that are so little altered by weathering that the fabric and the majority of the parent minerals are still present.

*Soil*

A naturally occurring mass of discrete particles or grains, at most lightly bonded together, occurring as a product of rock weathering either in situ or transported, with or without admixtures of organic constituents, in formations with no or only slight lithification.

**Comments**

The definitions given are geologic and not adequate for application to engineering problems in which the solution relates to hydraulic and mechanical properties as well as to certain other physical properties, such as hardness. For most practical engineering problems, it is more important to describe and classify the materials in terms of their physical conditions and properties than to attempt in every case to define the material as a soil or a rock.

The most important practical distinction between soil and rock in engineering works arises in excavations, since soil normally is much less costly to remove than is rock, which may require blasting. For pay quantities, soils are usually defined as materials that can be removed by machine excavation, and other materials are defined as rock. This can be an ambiguous definition since the success of machine excavation depends on the size, strength, and condition of the equipment, as well as the effort applied by the operator.

In rock masses, the material factors of major significance in excavations, as well as in other engineering problems, are mineral hardness, the frequency and orientations of fractures, and the degree of decomposition. Some geologic materials defined as rock, such as halite, many shales, and closely jointed masses, may be removable by machine, or may be more deformable under applied stress than are some materials defined as soils, such as glacial tills and cemented soils. Compounding the problem of definitions is rock decomposition which at a

TABLE 5.1
A BROAD CLASSIFICATION OF GEOLOGIC MATERIALS

| Category | Material | | Description |
|---|---|---|---|
| Rock | Fresh rock | Intact | Unweathered rock free of fractures and other defects. |
| | | Nonintact | Unweathered, but divided into blocks by fractures. |
| (Transition zone) | Decomposed rock | Intact | Weathered, the rock structure and fabric remain but mineral constituents are altered and the mass softened. |
| | | Nonintact | Rock softened and altered by weathering and containing discontinuities (fractures). |
| Soil | Residual soils | | Most minerals changed by advanced decomposition; fabric remains or is not apparent; material friable. |
| | Transported soils | | The residual soils are removed and graded by the agents of transportation: wind, water, gravity, and glaciers. Degree of grading varies with transport mode. |
| | Sedentary soils | | Organic deposits develop in situ from decomposition of vegetation. |
| Either | Duricrusts | | Deposits from evaporation of groundwater: caliche, laterite, ferrocrete, silcrete. |

given location can vary in short lateral and vertical distances resulting in materials ranging from "hard" to "soft."

In the literature, rock can be found defined for engineering purposes as "intact specimens with a uniaxial compressive strength in the order of 100 psi (6.8 kg/cm$^2$) or greater." For the foregoing reasons, this definition is not applicable to rock masses in many instances.

### General Classification

A broad classification which generally includes all geologic materials, together with their brief description, is given on Table 5.1.

## 5.1.2 ROCK GROUPS AND CLASSES

### Geologic Bases

From geologic aspects, rocks are grouped by origin as igneous, sedimentary, or metamorphic, and classed according to petrographic characteristics which include their mineral content, texture, and fabric.

### Engineering Bases

On an engineering basis, rock is often referred to as either intact or in situ.

Intact rock refers to a block or fragment of rock free of defects, in which the hydraulic and mechanical properties are controlled by the petrographic characteristics of the material, whether in the fresh or decomposed state. Classification is by uniaxial compressive strength and hardness.

In situ rock refers to the rock mass which normally contains defects, such as fractures or cavities, which separate the mass into blocks of intact rock and control the hydraulic and mechanical properties. Classification is by rock quality with the mass termed generally as competent or incompetent. Various practitioners have presented systems for describing incompetent rock, which can contain a wide range of characteristics, but a universally accepted nomenclature has not been established.

### 5.1.3 SOIL GROUPS AND CLASSES

#### Geologic Bases

Geologically, soils are grouped or classed on a number of bases, as follows:

- Origin: residual, colluvial, alluvial, aeolian, glacial, sedentary
- Mode of occurrence: floodplain, estuarine, marine, moraine, etc.
- Texture: particle size and gradation
- Pedology: climate and morphology

#### Engineering Bases

*Classes*

Soils are classed on an engineering basis by gradation, plasticity, and organic content, and described generally as cohesionless or cohesive, granular or nongranular.

*Groups*

Soils are grouped by their engineering characteristics as strong or weak, sensitive or insensitive, compressible or incompressible, swelling (expansive) or nonswelling, pervious or impervious; or grouped by physical phenomena as erodible, frost-susceptible, or metastable (collapsible or liquefiable, with the structure becoming unstable under certain environmental changes).

Soils are also grouped generally as gravel, sand, silt, clay, organics, and mixtures.

## 5.2 ROCK

### 5.2.1 THE THREE GROUPS

#### Igneous

Igneous rocks are formed by crystallization of masses of molten rock originating from below the earth's surface.

#### Sedimentary

Sedimentary rocks are formed from sediments which have been transported and deposited, sometimes as chemical precipitates, or from the remains of plants and animals which have been lithified under the tremendous heat and pressure of overlying sediments or by chemical reactions.

#### Metamorphic

Metamorphic rocks are formed from other rocks by the enormous shearing stresses of orogenic processes which cause plastic flow, in combination with heat and water, or by the heat of molten rock injected into adjoining rock, which causes chemical changes and produces new minerals.

### 5.2.2 PETROGRAPHIC IDENTIFICATION

#### Significance

Rocks are described and classified by their petrographic characteristics of mineral content, texture, and fabric.

A knowledge of the mineral constituents of a given rock type is useful also in predicting the engineering characteristics of the residue from chemical decomposition in a particular climatic environment. Residual soils are commonly clayey materials, and the "activity" of the formation is very much related to the original rock minerals.

#### Rock Composition

*Minerals*

Rock minerals are commonly formed of two or more elements, although some rocks consist of only one element such as carbon, sulfur, or a metal.

*Elements*

Oxygen, silicon, aluminum, iron, calcium, sodium, potassium, and magnesium comprise 98% of the earth's crust. Of these, oxygen and silicon represent 75% of the elements. These elements combine to form the basic rock minerals.

*Groups*

The mineral groups are silicates, oxides, hydrous silicates, carbonates, and sulfates. Sili-

**TABLE 5.2**
**COMMON ROCK-FORMING MINERALS AND THEIR CHEMICAL COMPOSITION**

| Group | Mineral | Variety | Chemical composition | Comments |
|---|---|---|---|---|
| Silicates | Feldspars | Orthoclase | $KAlSi_3O_8$ | Very abundant |
| | | Plagioclase | $NaAlSi_3O_8$ / $CaAl_2Si_2O_8$ } variable | |
| | Micas | Muscovite (white mica) | $KAl_2(Si_3Al)O_{10}(OH)_2$ | Very abundant |
| | | Biotite (dark mica) | $K_2(MgFe)_6(SiAl)_8O_{20}(OH)_4$ | Ferromagnesian minerals* |
| | Amphiboles | Hornblende | Na, Ca, Mg, Fe, Al silicate | Ferromagnesian mineral* |
| | Pyroxenes | Augite | Ca, Mg, Fe, Al silicate | Ferromagnesian mineral* |
| | Olivine | | $(MgFe)_2SiO_4$ | Ferromagnesian mineral* |
| Hydrous silicates | Kaolinite | | $Al_2Si_2O_5(OH)_4$ | Secondary origin† |
| Oxides | Quartz | | $SiO_2$ | Very abundant |
| | Aluminum | | $Al_2O_3$ | |
| | Iron oxides | Hematite | $Fe_2O_3$ | * |
| | | Limonite | $2Fe_2O_33H_2O$ | |
| | | Magnetite | $Fe_3O_4$ | ‡ |
| Carbonates | Calcite | | $CaCO_3$ | Very abundant |
| | Dolomite | | $CaMg(CO_3)_2$ | |
| Sulfates | Gypsum | | $CaSO_42(H_2O)$ | |
| | Anhydrite | | $CaSO_4$ | |

*Ferromagnesian minerals and the iron oxides stain rock and soils orange and red colors upon chemical decomposition.

†Hydrous silicates form from previously existing minerals by chemical weathering and include the clay minerals (such as kaolinite and the chlorites), serpentine, talc, and zeolites.

‡Magnetite occurs in many igneous rocks and provides them with magnetic properties even though often distributed as very small grains.

cates and oxides are the most important. The groups, mineral constituents, and chemical compositions are summarized on Table 5.2.

### Texture

Texture refers to the size of grains or discrete particles in a specimen and is generally classed as given on Table 5.3.

### Fabric

Fabric refers to grain orientation, which can be described in geologic or in engineering terminology.

*Geologic Terminology*

- Equigranular: grains essentially of equal size

**TABLE 5.3**
**TEXTURAL CLASSIFICATION OF MINERAL GRAINS**

| Class | Size range | Recognition |
|---|---|---|
| Very coarse-grained | >2.0 mm | Grains measurable |
| Coarse-grained | 0.6–2.0 | Clearly visible to eye |
| Medium grained | 0.2–0.6 | Clearly visible with hand lens |
| Fine-grained | 0.06–0.2 | Just visible with hand lens |
| Very fine-grained | <0.06 | Not distinguishable with hand lens |

- Porphyritic: mixed coarse and fine grains
- Amorphous: without definite crystalline form
- Platy: schistose or foliate

*Engineering Terminology*

- Isotropic: the mineral grains have a random orientation and the mechanical properties are the same in all directions.
- Anisotropic: the fabric has planar or linear elements from mineral cleavage, foliations, or schistosity, and the properties vary with the fabric orientation.

### Mineral Identification Factors

*Crystal Form*

Distinct crystal form is encountered only occasionally in rock masses. Some common minerals with distinct forms are garnet, quartz, calcite, and magnetite.

*Color*

Clear or light-colored minerals include quartz, calcite, and feldspar.

Dark green, dark brown, or black minerals contain iron as the predominant component.

*Streak*

Streak refers to the color produced by scratching a sharp point of the mineral across a plate of unglazed porcelain. It is most significant for dark-colored minerals, since streak does not always agree with the apparent color of the mineral. Some feldspars, for example, appear black but yield a white streak.

*Luster*

Luster refers to the appearance of light reflected from a mineral, which ranges from metallic to nonmetallic to no luster. Pyrite and galena have metallic luster on unweathered surfaces. Nonmetallic luster is described as vitreous (quartz), pearly (feldspar), silky (gypsum), and greasy (graphite). Minerals with no luster are described as earthy or dull (limonite and kaolinite).

*Cleavage*

Cleavage is used to describe both minerals and rock masses. In minerals it refers to a particular plane or planes along which the mineral will split when subjected to the force of a hammer striking a knife blade. Cleavage represents a weakness plane in the mineral, whereas crystal faces represent the geometry of the mineral structure, although appearances may be similar. For example, quartz exhibits strong crystal faces but has no cleavage. Types of mineral cleavage are given on Fig. 5.1; rock-mass cleavage is described in Art. 6.3.2.

*Fracture*

The appearance of the surface obtained by breaking the mineral in a direction other than that of the cleavage, or by breaking a mineral that has no cleavage, provides fracture characteristics. Fracture can be fibrous, hackly (rough and uneven), or conchoidal. The last form is

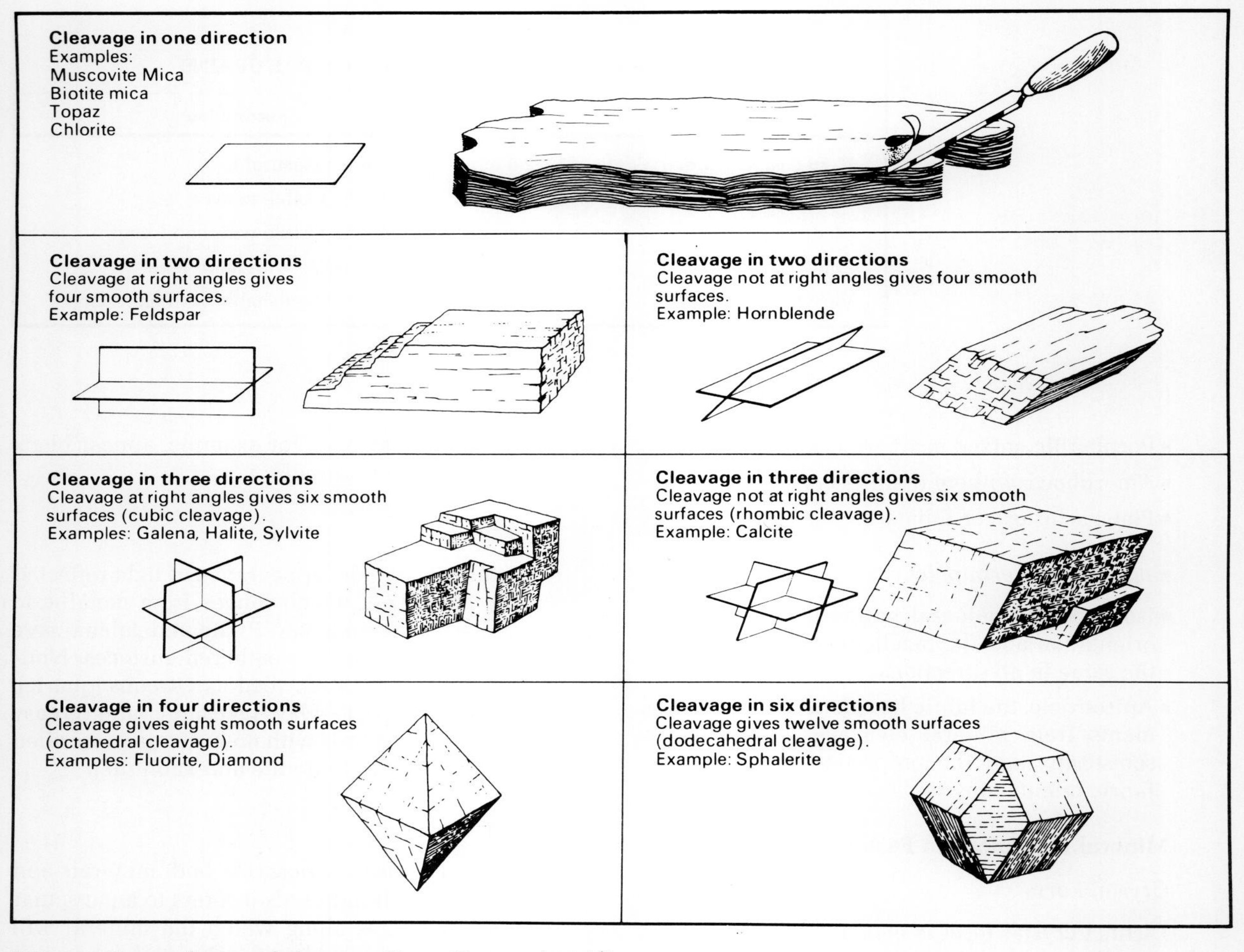


**FIG. 5.1 Types of mineral cleavage.** [*From Simpson (1974).*[1]]

common in fine-grained and homogeneous minerals such as quartz and volcanic glass.

*Specific Gravity (SG or $G_s$)*

The ratio between the mass of a mineral and the mass of an equal volume of water defines specific gravity, expressed as

$$G_s = W_a/(W_a - W_w) \tag{5.1}$$

where $W_a$ is the weight of the test specimen and $W_w$ is the weight of the specimen submerged in water. Quartz and calcite, for example, have $G_s$ = 2.65, and any variation from that amount is caused by impurities.

*Hardness*

Hardness refers to the ability of a mineral to resist scratching relative to another mineral. The hardness scale of minerals assigned by Friedrich Mohs is given on Table 5.4. It signifies that each mineral, if used in the form of a sharply pointed fragment, will scratch smooth surfaces of all minerals preceding it on the table.

Some useful hand tests are:

- Window glass has a hardness of about 5.5.
- Pocket knife blade has a hardness of about 5.
- Brass pinpoint has a hardness a little over 3 (can scratch calcite).

- Fingernail is a little over 2 (can scratch gypsum).

*Summary*

The identification characteristics of some common minerals including streak, luster, color, hardness, specific gravity, cleavage, and fracture are given on Table 5.5.

### Laboratory Methods

*Chemical Tests*

A simple test is reaction to hydrochloric acid. Calcite is differentiated from most other minerals by its vigorous effervescence when treated with cold hydrochloric acid. Dolomite will react to hydrochloric acid only if the specimen is powdered.

*Petrographic Microscope*

Polarized light is used in the petrographic microscope for studying *thin sections* of mineral or rock specimens. To prepare a thin section, a sample about 25 mm in diameter is ground down to a uniform thickness of about 0.03 mm by a sequence of abrasives. At this thickness it is usually translucent. The specimen is enveloped in balsam and examined in polarized light. The minerals are identified by their optical properties.

*Other Methods*

Minerals are identified also by a table microscope, the electron microscope, blow-pipe analysis, or X-ray diffraction.

## 5.2.3 IGNEOUS ROCKS

### Origin and Occurrence

Molten rock charged with gases *(magma)* rises from deep within the earth. Near the surface a volcanic vent is formed, the pressures decrease, the gases are liberated, and the magma cools and solidifies. Igneous rocks occur in two general forms *(see also Art. 6.2.2)*.

*Intrusive*

The magma is cooled and solidified beneath the surface, forming large bodies (plutons) which generally consist of coarser-grained rocks, or small bodies such as dikes and sills, and volcanic necks, which generally consist of finer-grained rocks because of more rapid cooling.

*Extrusive*

Associated with volcanic activity, extrusive rocks originate either as *lava*, quiet outwellings of fluid magma flowing onto the earth's surface and solidifying into an *extrusive sheet*, or as *pyroclastic rocks*, magma ejected into the air by the violent eruption of gases which then falls as numerous fragments.

### Classification

Igneous rocks are classified primarily according to mineral content and texture as presented on Table 5.6.

*Mineral Composition and the Major Groups*

The important minerals are quartz, feldspar, and the ferromagnesians, as given on Fig. 5.2. Modern classification is based primarily on silica content ($SiO_2$) [Turner and Verhoogan (1960)[4]].

- *Sialic rocks* (acid rocks) are light-colored, composed chiefly of quartz (silica) and feldspar (silica and alumina, $Al_2O_3$), with silica $> 66\%$.
- *Intermediate group* rocks have a silica content between 52 and 66%.
- *Mafic rocks* (basic rocks) are the ferromagnesian group containing the dark-colored minerals (biotite mica, pyroxine, hornblende, olivine, and the iron ores), with a silica content between 45 and 52%.
- *Ultramafic rocks* have a silica content $< 45\%$.

**TABLE 5.4**
**THE MOHS SCALE OF MINERAL HARDNESS**

| | |
|---|---|
| 1. Talc | 6. Feldspar |
| 2. Gypsum | 7. Quartz |
| 3. Calcite | 8. Topaz |
| 4. Fluorite | 9. Corundum |
| 5. Apatite | 10. Diamond |

**TABLE 5.5**
**CHARACTERISTICS OF SOME COMMON MINERALS***

| Mineral | Streak | Luster | Color | | Hardness | Specific gravity | Characteristics |
|---|---|---|---|---|---|---|---|
| Galena $PbS$ | Gray | Metallic | Silver-gray | | 2.5 | 7.6 | Perfect cubic cleavage |
| Magnetite $Fe_3O_4$ | Black | Metallic | Black to dark gray | | 6 | 5.2 | Magnetic |
| Graphite C | Black | Metallic | Steel gray | | 1 | 2 | Greasy feel |
| Chalcopyrite $CuFeS_2$ | Greenish-black | Metallic | Golden yellow | | 4 | 4.3 | May tarnish purple |
| Pyrite $FeS_2$ | Greenish-black | Metallic | Brass yellow | | 6–6.5 | 5 | Lacks cleavage |
| Hematite $Fe_2O_3$ | Reddish-brown | Metallic | Black–dark brown | | 5–6.5 | 5 | Lacks cleavage |
| Limonite $Fe_2O_3 \cdot H_2O$ | Yellow brown | Metallic | Yellow, brown, black | | 5–5.5 | 3.5–4 | Hard structureless or radial fibrous |
| Pyroxene group (see Table 5.2) Augite | | Nonmetallic dark color | Dark green–black | | 6 | 3.5 | Cleavage, 2 at 90°. Prismatic 8-sided crystals. |
| Amphibole group (see Table 5.2) Hornblende | | Nonmetallic dark color | Dark green, black, brown | | 6 | 3–3.5 | Cleavage, 2 at 60° and 90°. Long 6-sided crystals. |
| Olivine $(MgFe)_2SiO_4$ | | Nonmetallic dark color (glassy) | Olive green | Hard | 6.5–7 | 3.5–4.5 | Conchoidal fracture, transparent to translucent. |
| Garnet group (Fe, Mg, Ca, Al silicates) | | Nonmetallic dark color (glassy) | Red, brown, yellow | | 7–7.5 | 3.5–4.5 | Conchoidal fracture, 12-sided crystals. |
| Biotite (see Table 5.2) | | Nonmetallic dark color | Brown to black | | 2.5–3 | 3–3.5 | 1 perfect cleavage, thin sheets. |
| Chlorite (hydrous Mg, Fe, Al silicate) | | Nonmetallic dark color | Green to very dark green | | 2–2.5 | 2.5–3.5 | 1 cleavage direction foliated or scaly masses. |
| Sphalerite ZnS | Yellow brown to white | Nonmetallic dark color (resinous) | Yellowish-brown | Soft | 3.5–4 | 4 | Cleavage 6 directions. |
| Hematite $Fe_2O_3$ earth variety | Red | Nonmetallic dark color | Red | | 1.5 | | Earthy appearance. No cleavage. |
| Limonite $Fe_2O_3 \cdot H_2O$ | Yellow brown | Nonmetallic dark color | Yellow brown to dark brown | | 1.5 | | Compact earth masses. No cleavage. |

| | | | | | | |
|---|---|---|---|---|---|---|
| Feldspar group<br>Potassium feldspar<br>$KAlSi_3O_8$ | Nonmetallic light color (pearly to vitreous) | Pink, white, green | | 6–6.5 | 2.5 | Cleavage 2 directions at 90°. |
| Plagioclase feldspar<br>$NaAlSi_3O_8$ to $CaAl_2Si_2O_8$ | Nonmetallic light color | White, blue gray | | 6–6.5 | 2.5 | Cleavage 2 directions. Striations on some cleavage planes. |
| Quartz<br>$SiO_2$ (silica) | Nonmetallic light color (vitreous) | Various | Hard | 7 | 2.65 | Conchoidal fracture, 6-sided crystals, transparent to translucent. |
| Cryptocrystalline<br>Quartz $SiO_2$<br>(Agate)<br>(Flint)<br>(Chert)<br>(Jasper)<br>(Opal) | Nonmetallic (dull or clouded)<br>(waxy) | Various<br>(banded)<br>(dark)<br>(light)<br>(red)<br>(light) | | 6–6.5 | | Translucent to opaque. No cleavage. |
| Halite<br>NaCl | Nonmetallic | Colorless to white | | 2–2.5 | 2 | Perfect cubic cleavage. Soluble in water. Salty. |
| Gypsum<br>$CaSO_4 \cdot 2H_2O$<br>Many varieties | Nonmetallic | White | | 2 | 2.3 | Perfect cleavage in one direction. Transparent. |
| Calcite<br>$CaCO_3$ | Nonmetallic | White or pale yellow or colorless | | 3 | 2.7 | Perfect cleavage 3 directions ≈75°. Transparent to opaque. Effervesces in HCl. |
| Dolomite<br>$CaMg(CO_3)_2$ | Nonmetallic | Variable, commonly white, pink | | 3.5–4 | 2.8 | Cleavage as in calcite. Effervesces in HCl only when powdered. |
| Fluorite<br>$CaF_2$ | Nonmetallic | Colorless, blue, green, yellow, violet | Soft | 4 | 3 | Good cleavage 4 directions. Cubic crystals. Transparent to translucent. |
| Muscovite<br>(see Table 5.2) | Nonmetallic | Colorless in thin sheets | | 2–3 | 2.8 | Perfect cleavage one direction, producing thin elastic sheets. |
| Talc<br>$KAl_2(AlSi_3O_{10})(OH)_2$ | Nonmetallic (pearly) | Green to white | | 1 | 2.8 | Soapy feel, foliated or compact masses. |
| Kaolinite<br>$Al_4Si_4O_{10}(OH)_8$ | Nonmetallic (earthy) | White to red | | 1.2 | | No cleavage visible. Plastic when wet. |
| Nephelite<br>$Na_8K_2Al_8Si_9O_{34}$ | Vitreous to greasy | Various | | 5–6 | 2.6 | Conchoidal fracture (also nepheline). |

*After Hamblin and Howard (1971).[2]

**TABLE 5.6**
**CLASSIFICATION OF IGNEOUS ROCKS[3]**

| Major classes | Subdivisions of major classes based on mineral composition | | | | | | | |
|---|---|---|---|---|---|---|---|---|
| Minerals† | Light-colored rocks | | | | Medium-colored | | Dark-colored rocks | |
| | Orthoclase feldspar | | Ortho- or Plagio- clase feldspar | | Plagioclase | | Plagio- clase | No feld- spar |
| | BHP | | BHP | | BHP | HBP | PHOA | OPHBA |
| Grain size | With Q | Without Q | With Q | Without Q | With Q | Without Q | Without Q | No Q |
| Coarse > 1 mm | Pegmatite Granite | | | | | | | |
| Phanerites Equigranular > 1 mm | Granite | Syenite | Grano- diorite | Monzonite | Tonalite (quartz diorite) | Diorite | Gabbro | Peridotite Pyroxenite Dunite (O) |
| Micro- phanerites Equigranular < 1 mm | Aplite | Micro- syenite | Micro- grano- diorite | Micro- monzomite | Micro- tonalite | Micro- diorite | Dolerite (diabase) | |
| Porphyries | All phanerites are found with phenocrysts (granite porphyry, etc.) | | | | | | | |
| Aphanites and aphanite porphyries | Rhyolite | Trachyte | Quartz, latite | Latite | Dacite | Andesite | Basalt | |
| | Felsite (and felsophyre) | | | | | | | |
| Classes | Obsidian and pitchstone | | | | | | | |
| Porous | Pumice | | | | Scoria | | Vesicular basalt | |

Legend: Plutonic rocks — [plain box] Volcanic rocks — [hatched box] Border rocks — [bold box]

† Minerals: A = augite, B = biotite, H = hornblende, P = pyroxene, O = olivine, Q = quartz.

*After Pirsson and Knopf (1955).[3] (Excludes pyroclastic rocks.) Adapted with permission of John Wiley & Sons, Inc.

- *Alkaline rocks* contain a high percentage of $K_2O$ and $Na_2O$ compared with the content of $SiO_2$ or $Al_2O_3$.

*Texture*

*Intrusives and lavas* are grouped as follows:

- *Phanerocrystalline (phanerites)* have individual grains large enough to be distinguished by the unaided eye and are classed by grain size:
  - Coarse-grained—>5 mm diameter (pea size)
  - Medium-grained—1 to 5 mm diameter
  - Fine-grained—<1 mm diameter
- *Microcrystalline (microphanerites)* have grains that can be perceived but are too small to be distinguished.
- *Porphyries* are phanerites with large conspicuous crystals *(phenocrysts)*.
- *Aphantic (aphanites)* contain grains too small to be perceived with the unaided eye.
- *Glassy* rocks have no grain form that can be distinguished.

*Pyroclastic rocks* are grouped as follows:

- *Volcanic breccia* are the larger fragments which fall around the volcanic vent and build a cone including:
  - Blocks—large angular fragments
  - Bombs—rounded fragments the size of an apple or larger
  - Cinders—which are the size of nuts
- *Tuff* is the finer material carried by air currents to be deposited at some distance from the vent, including:
  - Ash—the size of peas
  - Dust—the finest materials

*Fabric*

Igneous rocks generally fall into only two groups:

- *Equigranular,* in which all of the grains are more or less the same size
- *Porphyritic,* in which phenocrysts are embedded in the ground mass or finer material (the term refers to grain size, not shape)

*Grain Shape*

Grains are described as rounded, subrounded, or angular.

*Structure Nomenclature*

- *Continuous* structure is the common form, a dense, compact mass.
- *Vesicular* structure contains numerous pockets or voids resulting from gas bubbles.
- *Miarolitic cavities* are large voids formed during crystallization.
- *Amygdaloidal* refers to dissolved materials carried by hot waters permeating the mass and deposited to fill small cavities or line large ones, forming *geodes.*
- *Jointed structure* is described in Art. 6.4.

### Characteristics

Photos of some of the more common igneous rocks are given as Plates 5.1 to 5.13.

The characteristics of igneous rocks are summarized on Table. 5.7.

## 5.2.4 SEDIMENTARY ROCKS

### Origin

Soil particles resulting from the decay of rock masses or from chemical precipitates, deposited in sedimentary basins in increasing thicknesses, eventually lithify into rock strata from heat, pressure, cementation, and recrystallization.

### Rock Decay or Weathering

*Processes (See also Art. 6.7)*

In mechanical weathering, the rock mass is broken into fragments as the joints react to freeze-thaw cycles in cold climates, expansion-contraction, and the expansive power of tree roots.

In chemical weathering, the rock mass is acted upon chemically by substances dissolved in water, such as oxygen, carbon dioxide, and weak acids, causing the conversion of silicates, oxides, and sulfides into new compounds such as carbonates, hydroxides, and sulfates, some of which are soluble.

*Materials Resulting*

The residue can include rock fragments of various sizes, consisting essentially of unaltered rock; particles of various sizes, consisting of materials resistant to chemical decomposition, such as quartz; and clays or colloidal particles, which are insoluble products of chemical decomposition of less-resistant rocks such as feldspar and mica.

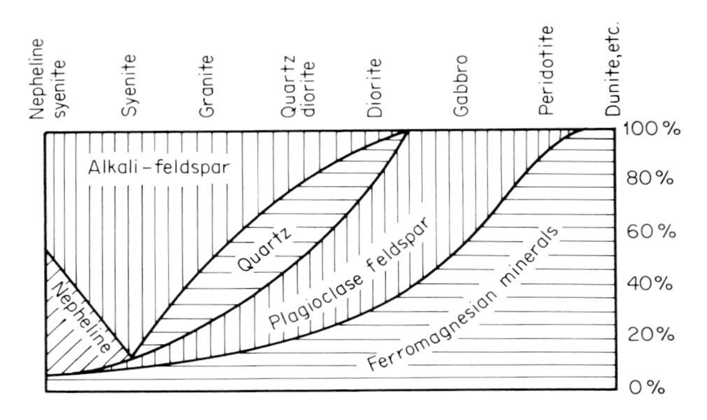


**FIG. 5.2 Minerals composing the important igneous rocks.** [*From Pirsson and Knopf (1955).*[3] *Reprinted with permission of John Wiley & Sons, Inc.*]

**TABLE 5.7**
**CHARACTERISTICS OF SOME IGNEOUS ROCKS**

| Rock | Characteristics | Plate |
|---|---|---|
| | COARSE TO MEDIUM GRAINED—VERY SLOW TO SLOW COOLING | |
| Pegmatite | Abundant as dikes in granite masses and other large bodies. Chiefly quartz and feldspar appearing separately as large grains ranging from a centimeter to as large as a meter in diameter. | 5.1<br>5.2 |
| Granite | The most common and widely occurring igneous rock. Fabric roughly equigranular normally. Light colors contain chiefly quartz and feldspar; gray shades contain biotite mica or hornblende. | 5.3<br>5.4 |
| Syenite | Light-colored rock differing from granite in that it contains no quartz, consisting almost entirely of feldspar but often containing some hornblende, biotite, and pyroxine. | 5.5 |
| Diorite | Gray to dark gray or greenish, composed of plagioclase feldspar and one or more of the ferromagnesian minerals. Equigranular fabric. | 5.6 |
| Gabbro | Dark-colored rock composed chiefly of ferromagnesian minerals and plagioclase feldspar. | 5.7 |
| Peridotite | Dark-colored rocks composed almost solely of ferromagnesian minerals. Olivine predominant; negligible feldspar. Hornblende or pyroxenes associated. Readily altered. | 5.8 |
| Pyroxenite | As above but pyroxene alone or predominant. | |
| Hornblendite | As above but hornblende alone or predominant. | |
| Dunite | Major constituent is olivine, which alters readily to serpentine. | |
| Dolerite (or diabase) | Dark-colored rock intermediate in grain size between gabbro and basalt. Abundant as thick lava flows that have cooled slowly. | |

NOTE: The more common types are italicized.

Soluble products of decomposition go into solution.

### Transport and Deposition

*Clastic Sediments (Detritus)*

The particle products of weathering are transported primarily by flowing water to be deposited eventually in large water bodies or basins. The products are generally segregated by size as defined on Table 5.8 into boulders, cobbles, pebbles, granules, sand, silt, and clay. Wind currents provide transport for finer sand grains and silt sizes.

*Chemical Precipitates (Nondetrital)*

Materials are carried in solution in flowing water to the sea or other large water bodies where they precipitate from solution. Chemical precipitates include the immense thicknesses of marine carbonates (limestones and dolomites) and the less abundant evaporites (gypsum, anhydrite and halite).

In addition to being formed from physical-chemical processes, many nondetrital rocks are formed from the dissolved matter precipitated into the seas by the physiological activities of living organisms.

**TABLE 5.7**
**CHARACTERISTICS OF SOME IGNEOUS ROCKS (*CONTINUED*)**

| Rock | Characteristics | Plate |
|---|---|---|
| | FINE-GRAINED—RAPID COOLING | |
| *Andesite* | Generally dark gray, green, or red. Pure andesite is relatively rare, and it is usually found with phenocrysts. Porphyritic andesite and basalt compose about 95% of all volcanic materials. | |
| *Basalt* | Most abundant extrusive rock; found in all parts of the world and beneath the oceans. Colors range from grayish to greenish black to black. Fine-grained with a dense compact structure. Often contains numerous voids (vesicular basalt). | 5.9<br>5.10 |
| *Rhyolite* | The microcrystalline equivalent of granite formed at or near the surface. Characteristically white, gray, or pink and nearly always contains a few phenocrysts of quartz or feldspar. | 5.11 |
| *Felsite* | Occurs as dikes, sills, and lava flows. The term felsite is used to define the finely crystalline varieties of quartz-porphyries or other light-colored porphyries that have few or no phenocrysts and give but slight indications to the unaided eye of their actual mineral composition. | |
| | GLASSY ROCKS—VERY RAPID CHILLING | |
| Obsidian | Solid natural glass devoid of all crystalline grains, generally black with a brilliant luster and a remarkable conchoidal fracture. | 5.12 |
| Pitchstone | A variety of obsidian with a resinous luster. | |
| Pumice | Extremely vesicular glass; a glass froth. | 5.13 |
| Scoriae | Formations that have as much void space as solids. | |

*Organics*

Beds of decayed vegetation remain in place to eventually form coal.

### Depositional Characteristics

*Horizontal Bedding*

Under relatively uniform conditions, the initial deposition is often in horizontal beds.

*Cross-bedding*

Wave and current action produces cross-bedded stratification as shown on Fig. 5.3.

*Ripple Marks*

Wave and current action can also leave ripple marks on the top of some beds.

*Unconformity*

When a stratum is partially removed by erosion and a new stratum subsequently is deposited, providing an abrupt change in material, an unconformity exists.

*Disconformity*

A lack of parallelism between beds, or the depo-

**TABLE 5.8**
**BROAD CLASSIFICATION OF SEDIMENTARY ROCKS**

| Rock type | Material* | Diameter, mm | Composition | Depositional environment |
|---|---|---|---|---|
| | | | DETRITAL | |
| Conglomerate | Boulders | >256 | Same as source rock. | Along stream bottoms. Seldom found in rock masses. |
| | Cobbles | 256–64 | Same as source rock. | Along stream bottoms. Deposited as alluvial fans and in river channels. |
| | Pebbles | 64–4 | As for cobbles or sand. | As for cobbles; also deposited in beaches. |
| | Granule | 4–2 | As for cobbles or sand. | As for pebbles and sand. |
| Sandstone | Sand | 2–0.02 | Primarily quartz; also feldspar, garnet, magnetite. Some locales: hornblende, pyroxene, shell fragments. | All alluvial deposits: stream channels, fans, floodplains, beaches, deltas. Occasionally aeolian. |
| Siltstone | Silt | 0.02–0.002 | As for sand; often some clay particles. | Deltas and floodplains. |
| Shales | Clay | <0.002 | Colloidal sizes of the end result of decomposition of unstable minerals yielding complex hydrous silicates *(see Art. 5.3.3)*. | Quiet water. Salt water: Clay particles curdle into lumps and settle quickly to the bottom. Show no graded beds. Freshwater: Settle slowly; are laminated and well-stratified, showing graded bedding. |
| | | | NONDETRITAL | |
| Limestone | Calcareous precipitate | | Massive calcite ($CaCO_3$). | Deep, quiet water. |
| Coquina | Calcareous precipitates | | Cemented shells. | Along beaches, warm water. |
| Chalk | Calcareous precipitates | | Microscopic remains of organisms. | Clear, warm, shallow seas. |
| Dolomite | Calcareous precipitates | | Dolomite—$CaMg(CO_3)_2$. | Seawater precipitation or alteration of limestone. |
| Gypsum† | Calcareous precipitates | | Gypsum—$CaSO_4 \cdot 2H_2O$. | Saline water. |

**TABLE 5.8**
**BROAD CLASSIFICATION OF SEDIMENTARY ROCKS (*CONTINUED*)**

| Rock type | Material* | Diameter, mm | Composition | Depositional environment |
|---|---|---|---|---|
| | | DETRITAL | | |
| Anhydrite† | Calcareous precipitate | | Anhydrite—$CaSO_4$. | Saline water. |
| Halite† | Saline precipitates | | Sodium chloride. | Saline water. |
| Coal | Organic | | Carbonaceous matter. | Swamps and marshes. |
| Chert | Silicate | | Silica, opal. | Precipitation. |

*The Wentworth scale.
†Evaporites.

**FIG. 5.3 Cross-bedding in sandstone (Santa Amara, Bahia, Brazil).**

**TABLE 5.9**
**CHARACTERISTICS OF DETRITAL SEDIMENTARY ROCKS**

| Rock type | Characteristics | Plate |
|---|---|---|
| | COARSE-GRAINED (RUDITES) (>2 mm) | |
| Conglomerates | | |
| General | Rounded fragments of any rock type, but quartz predominates. Cementing agent chiefly silica, but iron oxide, clay, and calcareous material also common. | 5.14 |
| Pudding stone | Gap-graded mixtures of large particles in a fine matrix. | |
| Basal | First member of a series; deposited unconformably. | |
| Breccia | Angular fragments of any rock type. Resulting from glaciation, rock falls, cave collapse, fault movements. | |
| | MEDIUM-GRAINED (ARENITES) (>50% SIZES BETWEEN 0.02 AND 2.0 mm) | |
| Sandstones | Predominantly quartz grains cemented by silica, iron oxide, clay, or a carbonate such as calcite. Color depends on cementing agent: yellow, brown, or red—iron oxides predominate; lighter sandstones—silica or calcareous material predominates. Porous and pervious with porosity ranging from 5 to 30% or greater. Material hard, and thick beds are common. | 5.15 |
| Arkose | Similar to sandstone but with at least 25% feldspar. | 5.16 |
| Graywacke | Angular particles of a variety of minerals in addition to quartz and feldspar, in a clay matrix. Gray in color; a strongly indurated, impure sandstone. | |
| | FINE-GRAINED (LUTITES) | |
| Siltstone | Composition similar to sandstone but at least 50% of grains are between 0.002 and 0.02 mm. Seldom forms thick beds, but is often hard. | |

sition of a new stratum without the erosion of the underlying stratum after a time gap, results in a disconformity.

### Lithification

Rock forms by lithification, which occurs as the thickness of the overlying material increases. The detritus or precipitate becomes converted into rock by compaction, the deposition of cementing agents into pore spaces, and physical and chemical changes in the constituents. At the greater depths "consolidation" by cementation is a common process, caused by the increase in the chemical activity of interstitial water that occurs with the increase in temperature associated with depth.

### Classification

Sedimentary rocks have been divided into two broad groups: detrital and nondetrital. A general classification is given on Table 5.8 and more detailed classifications and descriptions on Tables 5.9 and 5.10. A special classification system for carbonate rocks formed in the middle latitudes is given on Table 7.10.

*Detrital Group (Clastic Sediments)*

Classified by particle size as conglomerate,

TABLE 5.9
CHARACTERISTICS OF DETRITAL SEDIMENTARY ROCKS (*CONTINUED*)

| Rock type | Characteristics | Plate |
|---|---|---|
| Shale | | |
| General | Predominant particle size <0.002 mm (colloidal); a well-defined fissile fabric. Red shales are colored by iron oxides and gray to black shales are often colored by carbonaceous material. Commonly interbedded with sandstones and relatively soft. Many varieties exist. | 5.17 |
| Argillites | Hard, indurated shales devoid of fissility; similar to slates but without slaty cleavage. | |
| Calcareous shales | Contain carbonates, especially calcite. With increase in calcareous content becomes shaly limestone. | |
| Carbonaceous shales | Black shales containing much organic matter, primarily carbon, often grading to coal formations. | |
| Oil shales | Contain carbonaceous matter that yields oil upon destructive distillation. | |
| Marine shales | Commonly contain montmorillonite clays that are subject to very large volume changes upon wetting or drying (*see Art. 6.7.3 and Fig. 6.91*). | |
| Clay shales | Moderately indurated shales. | 5.18 |
| Claystones and mudstones | Clay-sized particles compacted into rock without taking a fissile structure (the geologist's term for clay soils in the stiff to hard consistency). | |

NOTE: 1. Sandstones and siltstones are frequently interbedded and grade into one another unless an unconformity exists.
2. Flysch: A term used in Europe referring to a very thick series of sandstone, shales, and marls (impure limestones) well-developed in the western Alps.

sandstone, siltstone, and shale. *Arenaceous rocks* are predominantly sandy. *Argillaceous rocks* are predominantly clayey.

*Nondetrital Group*

Includes chemical precipitates and organics. Chemical precipitates are classed by texture, fabric, and composition. Organics include only the various forms of coal.

**Characteristics**

Photos of some of the more common sedimentary rocks are given as Plates 5.14 to 5.22.

The characteristics of the detrital rocks are summarized on Table 5.9 and the nondetrital rocks on Table 5.10.

### 5.2.5 METAMORPHIC ROCKS

**Origin**

The constituents of igneous and sedimentary rocks are changed by metamorphism.

**Metamorphism**

*Effects*

Tremendous heat and pressure in combination with the activity of water and gases promote recrystallization of rock masses, including the

**TABLE 5.10**
**CHARACTERISTICS OF NONDETRITAL SEDIMENTARY ROCKS**

| Rock type | Characteristics | Plate |
|---|---|---|
| | CALCAREOUS PRECIPITATES | |
| Limestone* General | Contains more than 50% calcium carbonate (calcite); the remaining percentages consist of impurities such as clay, quartz, iron oxide, and other minerals. The calcite can be precipitated chemically or organically, or it may be detrital in origin. There are many varieties; all effervesce in HCl. | |
| Crystalline limestone* | Relatively pure, coarse to medium texture, hard. | 5.19 |
| Micrite | Microcrystalline form, conchoidal fracture, pure, hard. | |
| Oolitic limestone | Composed of pea-size spheres (oolites), usually containing a sand grain as a nucleus around which coats of carbonate are deposited. | |
| Fossiliferous limestone | Parts of invertebrate organisms such as mollusks, crinoids, and corals cemented with calcium carbonate. On Barbados, dead coral reefs reach 30 m thickness, and although very porous, are often so hard as to require drilling and blasting for excavation. | 5.20 5.21 |
| Coquina | Weak porous rock consisting of lightly cemented shells and shell fragments. Currently forming along the U.S. south Atlantic coast and in the Bahama Islands. | |
| Chalk | Soft, porous, and fine-textured; composed of shells of microscopic organisms; normally white color. Best known are of Cretaceous Age. | |
| Dolomite | Harder and heavier than limestone (bulk density about 2.87 g/cm$^3$ compared with 2.71 g/cm$^3$ for limestone). Forms either from direct precipitation from seawater or from the alteration of limestone by "dolomitization." Effervesces in HCl only when powdered. Hardness $>5$. | |
| Gypsum* | An evaporite, commonly massive in form, white-colored and soft. | |
| Anhydrite* | An evaporite, soft but harder than gypsum, composed of grains of anhydrite. Ranges from microcrystalline to phanerocrystalline. Normally a splintery fracture, pearly luster, and white color. | |
| Halite* | An evaporite; a crystalline aggregate of salt grains, commonly called *rock salt*. Soft, tends to flow under | |

*Rocks readily soluble in groundwater.

**TABLE 5.10**
**CHARACTERISTICS OF NONDETRITAL SEDIMENTARY ROCKS (*CONTINUED*)**

| Rock type | Characteristics | Plate |
|---|---|---|
| | CALCAREOUS PRECIPITATES | |
| | relatively low pressures and temperatures, and forms *salt domes*. Since the salt is of substantially lower specific gravity than the surrounding rocks, it rises toward the surface as the overlying rocks are eroded away, causing a dome-shaped crustal warping of the land surface. The surrounding beds are warped and fractured by the upward thrust of the salt plug, forming traps in which oil pools are found. | |
| | ORGANIC ORIGIN (COMPOSED OF CARBONACEOUS MATTER) | |
| Coal | Composed of highly altered plant remains and varying amounts of clay, varying in color from brown to black. Coalification results from the burial of peat and is classified according to the degree of change that occurs under heat and pressure. *Lignite* (brown coal) changes to *bituminous coal* (soft coal) which changes to *anthracite* (hard coal). | |
| | BIOGENIC AND CHEMICAL ORIGIN (SILICEOUS ROCKS) | |
| Chert | Formed of silica deposited from solution in water both by evaporation and the activity of living organisms, and possibly by chemical reactions. Can occur as small nodules or as relatively thick beds of wide extent and is common to many limestone and chalk formations. Hardness is 7 and as the limestone is removed by weathering, the chert beds remain prominent and unchanged, often covering the surface with numerous rock fragments. *Flint* is a variety of chert; *jasper* is a red or reddish-brown chert. | 5.22 |
| Diatomite | Soft, white, chalklike, very light rock composed of microscopic shells of diatoms (one-celled aquatic organisms which secrete a siliceous shell); porous. | |
| | OTHER MATERIALS OFTEN INCLUDED BY GEOLOGISTS | |
| Duricrusts<br>Caliche<br>Laterite<br>Ferrocrete<br>Silcrete<br>Loess<br>Marl | Discussed in Chap. 7. | |

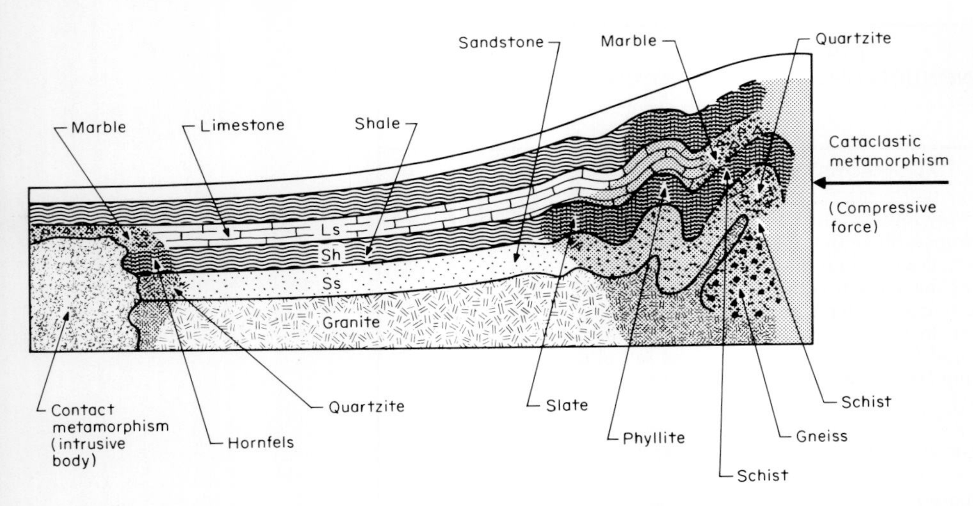


**FIG. 5.4 Formation of metamorphic rocks from heat and pressure.**

**FIG. 5.5 Banding in the highly foliated Fordham gneiss, New York City (photo near natural scale).**

formation of minerals into larger grains, the deformation and rotation of the constituent grains, and chemical recombination and growth of new minerals, at times with the addition of new elements from the circulating waters and gases.

*Metamorphic Forms*

*Contact or thermal metamorphism* (Fig. 5.4) has a local effect, since it is caused by the heat from an intrusive body of magma recrystallizing the enveloping rock into a hard, massive body. The effect diminishes rapidly away from the intrusive body.

*Cataclastic metamorphism* (Fig. 5.4) involves mountain building processes (orogenic processes), which are manifestations of huge compressive forces in the earth's crust producing tremendous shearing forces. These forces cause plastic flow, intense warping and crushing of the rock mass, and in combination with heat and water bring about chemical changes and produce new minerals.

*Regional metamorphism* combines high temperatures with high stresses, during which the rocks are substantially distorted and changed.

### Classification

Classification is based primarily on fabric and texture as given on Table 5.11.

*Massive fabric* is homogeneous, often with equigranular texture.

*Foliated fabric* is a banded or platy structure providing weakness planes that result from high directional stresses, and include three forms:

- Banded or lenticular as shown on Fig. 5.5.
- *Schistose* as shown on Fig. 5.6.
- *Slaty cleavage* as shown on Fig. 6.35.

*Metamorphic Derivatives of Igneous and Sedimentary Rocks*

The general derivatives are summarized on Table 5.12. It is seen that some metamorphic rocks can be derived from a large number of other rock types, whereas a few are characteristic of a single other rock type.

### Characteristics

Photos of some of the more common metamorphic rocks are given as Plates 5.23 to 5.33.

Characteristics of metamorphic rocks with foliate fabric are summarized on Table 5.13 and those with massive or other fabric are summarized on Table 5.14.

## 5.2.6 ENGINEERING CHARACTERISTICS OF ROCK MASSES

### General

Engineering characteristics of rock masses are examined from the aspects of three general conditions:

- Fresh intact rock (competent rock)—*Arts. 5.2.3 through 5.2.5*
- Decomposed rock—*Art. 6.7*
- Nonintact rock—*Arts. 6.3 through 6.6*

Engineering properties in general for the three rock-mass conditions are summarized on Table 5.15.

**FIG. 5.6 Phyllite schist showing strong platy fabric that frequently results in unstable slopes—note small fault (Ouro Preto, M.G., Brazil).**

**TABLE 5.11**
**CLASSIFICATION OF COMMON METAMORPHIC ROCKS**

| Texture | Fabric: Foliated | Fabric: Massive |
|---|---|---|
| Coarse | Gneiss<br>Amphibolite | Metaconglomerate<br>Granite gneiss (imperfect foliations) |
| Medium | Schist (mica, chlorite, etc.) | Quartzite<br>Marble<br>Serpentinite<br>Soapstone |
| Fine to microscopic | Phyllite<br>Slate | Hornfels |
| Other forms | Migmatite: Complex composite rocks; intermixtures of metamorphic and igneous rocks. | |
| | Mylonites: Formed by intense mechanical metamorphism; show strong laminations, but original mineral constituents and fabric crushed and pulverized. Formed by differential shearing movement between beds. | |

**TABLE 5.12**
**METAMORPHIC DERIVATIVES OF IGNEOUS AND SEDIMENTARY ROCKS***

| Parent rock | Metamorphic derivative |
|---|---|
| SEDIMENTARY ROCKS | |
| Conglomerate | Gneiss, various schists, metaconglomerate |
| Sandstone | Quartzite, various schists† |
| Shale | Slate, phyllite, various schists |
| Limestone | Marble† |
| IGNEOUS ROCKS | |
| Coarse-grained feldspathic, such as granite | Gneiss, schists, phyllites |
| Fine-grained feldspathic, such as felsite and tuff | Schists and phyllites |
| Ferromagnesian, such as dolerite and basalt | Hornblende schists, amphibolite |
| Ultramafic, such as peridotite and pyroxene | Serpentine and talc schist |

*After Pirsson and Knopf (1955).[3] Reprinted with permission of John Wiley & Sons, Inc.

†Depends on impurities.

**TABLE 5.13**
**CHARACTERISTICS OF SOME METAMORPHIC ROCKS WITH FOLIATE FABRIC**

| Rock | Characteristics | Plate |
|---|---|---|
| Gneiss | Coarse-grained; imperfect foliation resulting from banding of different minerals *(see Fig. 5.5)*. The foliation causes lenticular planes of weakness resulting in slabbing in excavations *(see Fig. 6.47)*. Chief minerals are quartz and feldspar, but various percentages of other minerals (mica, amphibole, and other ferromagnesians) are common. The identification of gneiss includes its dominant *accessory* mineral such as hornblende gneiss *(see Plate 5.23)*, biotite gneiss *(see Plate 5.24)*; or general composition, i.e., granite gneiss *(see Fig. 2.85)*. | 5.23<br>5.24 |
| Paragneiss | Derived from sedimentary rocks | 5.2 |
| Orthogneiss | Derived from feldspathic igneous rocks | |
| Schist | Fine-grained, well-developed foliation, resulting from the parallel arrangement of platy minerals (termed *schistosity*). The important platy minerals are muscovite, chlorite, and talc. Schist is identified by the primary mineral as mica schist *(see Plate 5.25)*, chlorite schist, etc. Garnet is a common accessory mineral to mica schist and represents intense metamorphism. Schists and gneisses commonly grade into each other and a clear distinction between them is often not possible. | 5.25 |
| Amphibolite | Consist largely of amphibole and show more or less schistose form of foliation. Composed of darker minerals and, in addition to hornblende, can contain quartz, plagioclase feldspar, and mica. They are hard and have densities ranging from 3.0 to 3.4. Association with gneisses and schists is common in which they form layers and masses that are often more resistant to erosion than the surrounding rocks. | 5.26 |
| Phyllite | Soft, with a satinlike luster and extremely fine schistosity. Composed chiefly of chlorite. Very unstable in cut slopes. Grades to schists as the coarseness increases *(see Fig. 5.6)*. | 5.27 |
| Slate | Extremely fine-grained, exhibiting remarkable planar cleavage *(see Fig. 6.35)*. Generally hard plates split from formations; once used for roofing materials. | 5.28 |

**TABLE 5.14**
**CHARACTERISTICS OF METAMORPHIC ROCKS WITH MASSIVE FABRIC AND OTHER FORMS**

| Rock | Characteristics | Plate |
|---|---|---|
| Metaconglomerate | Heat and pressure cause the pebbles in a conglomerate to stretch, deform, and fuse. | |
| Quartzite | Results from sandstone so firmly fused that fracture occurs across the grains, which are often imperceptible. | 5.29 |
| Marble | Results from metamorphism of limestone or dolomite and is found with large and small crystals, and in many colors including white, black, green, and red. Metamorphosed limestone does not normally develop cavities. Very hard. | 5.30 |
| Serpentinite | Derived from serpentine. Generally compact, dull to waxy luster, smooth to splintery fracture, generally green in color and often soft unless it contains significant amounts of quartz. Can have foliate fabric. | 5.31 |
| Soapstone | Derived from talc; generally gray to green color, very soft and easily trimmed into shapes with a knife, without cleavage or grain, and resists well the action of heat or acids. | 5.32 |
| Hornfels | Rocks baked by contact metamorphism into hard aphanitic material, with conchoidal fracture, dark gray to black color, often resembling a basalt. | |
| | OTHER FORMS | |
| Migmatite | Signifies a rock that is a complex intermixture of metamorphic and granular igneous rocks such as formed by the injection of granite magma into foliated rocks. | 5.33 |
| Mylonites | Produced by intense mechanical metamorphism; can show strong lamination but the original mineral constituents and fabric have been crushed and pulverized by the physical processes rather than altered chemically. Common along the base of overthrust sheets and can range from very thin, to a meter or so, to several hundreds of meters thick. Shale mylonites form very unstable conditions when encountered in cut slopes or tunneling. They are formed by differential movement between beds. | |

### Competent Rock

Intact rock that is fresh, unweathered, and free of discontinuities and reacts to applied stress as a solid mass is termed competent or sound rock in engineering nomenclature. Permeability, strength, and deformability are related directly to hardness and density *(see Art. 3.2.1)* as well as to fabric and cementing. The general engineering properties of common rocks are summarized on Table 5.16.

### Decomposed Rock

Decomposition from weathering causes rock to become more permeable, more compressible, and weaker. As the degree of decomposition advances, affecting the intact blocks and the discontinuities, the properties approach those of soils. The final product and its thickness are related closely to the mineral composition of the parent rock and to the climate and other environmental factors *(see Art. 6.7)*.

**TABLE 5.15**
**ROCK-MASS PROPERTIES SUMMARIZED**

| Property | Fresh, intact rock | Decomposed rock | Nonintact rock |
|---|---|---|---|
| Permeability *(see Art. 3.3)* | Essentially impermeable except for porous sandstones, vesicular and porous rocks. Significant as aquifers for water supply, or seepage beneath dams. *Table 3.12.* | Increases with degree of decomposition. | Water moves with relative freedom along fractures, and flow quantity increases as joint openings, continuity, and pattern intensities increase. Significant in cut slopes and other excavations, seepage pressures beneath dam foundations, or seepage loss, and for water supply. |
| Rupture strength *(see Art. 3.4)* | Most rocks essentially nonrupturable in the confined state, although under conditions of high tensile stresses and high pore pressures under a foundation not confined totally, rupture can occur, especially in foliated rocks. | Decreases with degree of decomposition. | Seldom exceeded in the confined state, but in an unconfined condition, such as slopes or tunnels, strength can be very low along weakness planes. Normally controls mass strength. |
| Deformability (from stress increase) *(see Art. 3.5)* | Compression under foundation loads essentially elastic, although some rocks such as halite deform plasticly and undergo creep. Plastic deformation can also occur along foliations in a partially confined situation such as an excavation, or from high loads applied normal to the foliations. | Increases with degree of decomposition. | Occurs from the closure of fractures and displacement along the weakness plane. When confined, displacements are usually negligible. In open faces, such as tunnels and slopes, movements can be substantial and normally control mass deformation. |
| Expansion (stress decrease) *(see Art 3.5)* | Occurs in shales with montmorillonite clays or pyrite, causing excavation and foundation heave, slope collapse and tunnel closing, when in contact with free water or moisture in the air. | Increases with degree of decomposition. | *Residual stresses* are locked into the rock mass during formation or tectonic activity and can far exceed overburden stresses, causing deflections of walls and floors in excavations and underground openings, and even violent rock bursts in deep mines (intact or nonintact). |

**TABLE 5.16**
**GENERAL ENGINEERING PROPERTIES OF COMMON ROCKS***

| Rock type | Characteristics | Permeability | Deformability | Strength |
|---|---|---|---|---|
| | | IGNEOUS | | |
| Phanerites | Welded interlocking grains, very little pore space. | Essentially impermeable. | Very low. | Very high. |
| Aphanites | Similar to above, or can contain voids. | With voids can be highly permeable. | Very low to low. | Very high to high. |
| Porous | Very high void ratio. | Very high. | Relatively low. | Relatively low. |
| | | SEDIMENTARY | | |
| Sandstones | Voids cement-filled. | Low. | Low. | High. |
| | Partial filling of voids by cement coatings. | Very high. | Moderate to high. | Moderate to low. |
| Shales* | Depend on degree of lithification. | Impermeable. | High to low. Can be highly expansive. | Low to high. |
| Limestone | Pure varieties normally develop caverns. | High through caverns. | Low except for cavern arch. | High except for cavern arch. |
| | Impure varieties. | Impermeable. | Generally low. | Generally high. |
| Dolomite | Seldom develops cavities. | Impermeable. | Lower than limestone. | Higher than limestone. |
| | | METAMORPHIC | | |
| Gneiss† | Weakly foliated. | Essentially impermeable. | Low. | High. |
| | Strongly foliated. | Very low. | Moderate normal to foliations. Low parallel to foliations. | High normal to foliations. Low parallel to foliations. |
| Schist† | Strongly foliated. | Low. | As for gneiss. | As for gneiss. |
| Phyllite† | Highly foliated. | Low. | Weaker than gneiss. | Weaker than gneiss. |
| Quartzite | Strongly welded grains. | Impermeable. | Very low. | Very high. |
| Marble | Strongly welded. | Impermeable. | Very low. | Very high. |

*Fresh intact condition.
†Anisotropic fabric.

### Nonintact Rock

*Discontinuities* or defects, representing weakness planes in the mass, control the engineering properties by dividing the mass into *blocks* separated by fractures such as faults, joints, foliations, cleavage, bedding, and slickensides, as described on Table 5.17. Joints are the most common defect in rock masses. They have the physical properties of spacing, width of opening, configuration, and surface roughness. They can be tight, open, or filled with some material, and can display the strength parameters of cohesion and friction along their surfaces (*see Art. 6.4.4*).

*Blocks* will have the characteristics of intact rock. As the degree of decomposition increases, the significance of the discontinuities decreases, but even in highly decomposed rock and residual soils, relict fractures can represent potential failure surfaces.

In general, experience shows that the response of a rock mass during tunneling operations will be governed by intact properties, and the rock may be considered as competent if joints are tight, their spacing is about 1 m or more, and the rock is fresh [Hartmann (1966)[5]]. In slopes and excavations, however, any size block of fresh rock within an exposed wall can fail if the bounding planes of the block incline downward and out of the slope.

### Engineering Properties Given in Chap. 3

*Hardness*

Range in "total hardness" for common rock types is in Fig. 3.1. Rock rippability as related to seismic velocities is in Table 3.7.

*Permeability*

Typical permeability coefficients for rock and soil formations are in Table 3.12.

*Rupture Strength*

Relationship between "consistency" and uniaxial compressive strength is in Fig. 3.39. Correlation between unconfined compressive strength, density, and Schmidt hardness is in Fig. 3.40. Common properties of fresh intact rock (dry density, strength, and moduli) are given in Table 3.26 (*see also Table 5.20*). Shear-strength parameters of residual soil and weathered rock are in Table 3.27.

*Deformation*

Typical values for the elastic constants are in Tables 3.26 and 3.33. Deformation characteristics of intact rock in uniaxial compression to failure are in Table 3.32.

Field modulus of deformation vs. RQD is given in Fig. 3.78.

## 5.2.7 ROCK-MASS DESCRIPTION AND CLASSIFICATION

### Importance

Systems that provide an accurate description and classification of the rock mass are necessary as a basis for the formulation of judgments regarding the response to engineering works including:

- Excavation difficulties
- Stability of slopes and open and closed excavations
- Capacity to sustain loads
- Capacity to transmit water

### Rock-Mass Description

*General*

The degree of complexity of description can be based on the nature of the problem under study and the relative importance of the rock-mass response. For routine problems, such as the average building foundation on good-quality rock, simple descriptions suffice, whereas for nonroutine problems, the rock mass is described in terms of intact rock characteristics, discontinuities, and groundwater conditions.

Description is made from the examination of outcrops, exploration pits and adits, and boring cores.

*Intact Rock Characteristics*

Descriptions should include the hardness, weathering grade, rock type, coloring, texture, and fabric.

**TABLE 5.17**
**ROCK-MASS DISCONTINUITIES**

| Discontinuity | Definition | Characteristics |
|---|---|---|
| Fracture | A separation in the rock mass, a break. | Signifies joints, faults, slickensides, foliations, and cleavage. |
| Joint | A fracture along which essentially no displacement has occurred. | Most common defect encountered. Present in most formations in some geometric pattern related to rock type and stress field. Open joints allow free movement of water, increasing decomposition rate of mass. Tight joints resist weathering and the mass decomposes uniformly. |
| Faults | A fracture along which displacement has occurred due to tectonic activity. | Fault zone usually consists of crushed and sheared rock through which water can move relatively freely, increasing weathering. Waterlogged zones of crushed rock are a cause of *running ground* in tunnels. |
| Slickensides | Preexisting failure surface; from faulting, landslides, expansion. | Shiny, polished surfaces with striations. Often the weakest elements in a mass, since strength is often near residual. |
| Foliation planes | Continuous foliation surface results from orientation of mineral grains during metamorphism. | Can be present as open joints or merely orientations without openings. Strength and deformation relate to the orientation of applied stress to the foliations. |
| Foliation shear | Shear zone resulting from folding or stress relief. | Thin zones of gouge and crushed rock occur along the weaker layers in metamorphic rocks. |
| Cleavage | Stress fractures from folding. | Found primarily in shales and slates; usually very closely spaced. |
| Bedding planes | Contacts between sedimentary rocks. | Often are zones containing weak materials such as lignite or montmorillonite clays. |
| Mylonite | Intensely sheared zone. | Strong laminations; original mineral constituents and fabric crushed and pulverized. |
| Cavities | Openings in soluble rocks resulting from groundwater movement, or in igneous rocks from gas pockets. | In limestone range from caverns to tubes. In rhyolite and other igneous rocks range from voids of various sizes to tubes. |

*Discontinuities*

Joint spacing and joint characteristics are described, and details of joint orientations and spacing should be illustrated with photographs and sketches to allow preparation of two- or three-dimensional joint diagrams (*see Art. 6.1.3*).

Other mass characteristics, including faults, slickensides, foliation shear zones, bedding, and cavities are provided in an overall mass description.

*Groundwater Conditions*

Observations of groundwater conditions made in cuts and other exposures must be related to recent weather conditions, season, and regional climate to permit judgments as to whether seepage is normal, high, or low, since such conditions are transient.

*Rock-Quality Indices (see Art. 2.4.5)*

Indices to rock quality are determined from a number of relationships as follows:

1. Bulk density of intact specimens
2. Percent recovery from core borings
3. Rock-quality designation RQD from core borings
4. Point-load index $I_s$ from testing core specimens in the field
5. Field shear-wave velocity $V_{Fs}$ for dynamic Young's modulus
6. Field compression-wave velocity $V_F$ for rock type and quality
7. Laboratory compression-wave velocity $V_L$ on intact specimens to combine with $V_F$ to obtain the velocity index
8. Laboratory shear-wave velocity $V_{Ls}$ to compare with $V_{Fs}$ for rock quality

*Summary*

A suggested guide to the field description of rock masses is given on Table 5.18.

## Intact Rock Classification

*General*

Intact rock is classified by most current workers on the basis of hardness, degree of weathering, and uniaxial compressive strength, although a universally accepted system has yet to be adopted. A strength classification system must consider the rock type and degree of decomposition because the softer sedimentary rocks such as halite can have strengths in the range of decomposed igneous rocks.

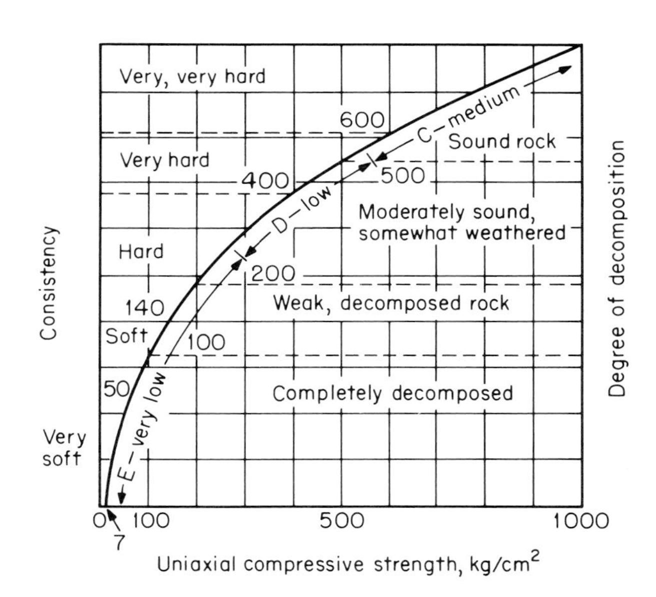


**FIG. 5.7 Rock classification based on uniaxial compressive strength as given by various investigators: consistency [Jennings (1972)[6]], degree of composition [Jaeger (1972)[7]], and strength [Deere (1969)[8]]. (E, D, and C from Table 5.19.)**

*Some Approaches*

Jennings (1972)[6] relates "consistency" to strength without definition of rock types or degree of weathering, as given on Fig. 5.7.

Jaeger (1972)[7] relates strength to the degree of decomposition, also as given on Fig. 5.7. "Completely decomposed" is given as below 100 kg/cm$^2$, but the boundary between very soft rock and hard soil usually is accepted as 7 kg/cm$^2$ (*see Fig. 3.39*).

Deere (1969)[8] proposed an engineering classification system based on strength, which he related to rock type (Table 5.19*a*), and the modulus ratio, which he related to rock fabric (Table 5.19*b*), providing for anisotropic conditions caused by foliations, schistosity, and bedding. It is to be noted that very low strength rock is given

**TABLE 5.18**
**FIELD DESCRIPTION OF ROCK MASSES**

| Characteristic | Description of grade or class | Reference |
|---|---|---|
| | INTACT ROCK | |
| Hardness | Class I–V, extremely hard to very soft. | Table 5.20 |
| Weathering grade | F, WS, WM, WC, RS, fresh to residual soil. | Table 5.21 |
| Rock type | Identify type, minerals, and cementing agent. | Arts. 5.2.3 to 5.2.5 |
| Coloring | Red, gray, variegated, etc. | |
| Texture (gradation) | Coarse, medium, fine, very fine. | Table 5.3 |
| Fabric: Form | Equigranular, porphyritic, amorphous, platy, (schistose or foliate), isotropic, anisotropic. | Arts. 5.2.3 to 5.2.5 |
| Orientation | Horizontal, vertical, dipping (give degrees). | |
| | DISCONTINUTIES | |
| Joint spacing | Very wide, wide, moderately close, close, very close. Give orientations of major joint sets. Solid, massive, blocky, fractured, crushed mass. | Table 5.23 |
| Joint conditions Form | Stepped, smooth, undulating, planar. | Art. 6.4 |
| Surface | Rough, smooth, slickensided. | |
| Openings | Closed, open (give width). | |
| Fillings | None, sand, clay, breccia, other minerals. | |
| Other discontinuities and mass characteristics | Faults, slickensides, foliation shear zones, cleavage, bedding, cavities, and groundwater conditions. Included in overall mass description. | |

NOTE: Example as applied to a geologic unit: "Hard, moderately weathered GNEISS, light gray, medium grained (quartz, feldspar, and mica); strongly foliated (anisotropic), joints moderately spaced (blocky), planar, rough, open (to 1 cm), clay filled."

as less than 275 kg/cm$^2$, which covers the category of rocks composed of soft minerals as well as those in the decomposed state.

*Suggested Classification Systems*

Hardness classifications based on simple field tests and related to uniaxial compressive strength ranges is given on Table 5.20.

Weathering grade, class, and diagnostic features are given on Table 5.21.

## Rock-Mass Classification

*General*

Historically, rock-mass classification has been based on percent core recovery, which is severely limited in value *(see Art. 2.4.5)*. Core recovery depends on many factors including equipment used, operational techniques, and rock quality, and provides no direct information on hardness, weathering, and defects. Even good core recovery cannot provide information

**TABLE 5.19**
**ENGINEERING CLASSIFICATION OF INTACT ROCK***

(A) ON THE BASIS OF UNIAXIAL COMPRESSIVE STRENGTH

| Class | Strength | Uniaxial compression, $kg/cm^2$ | Point-load index† | Rock type |
|---|---|---|---|---|
| A | Very high | >2200 | >95 | Quartzite, diabase, dense basalts |
| B | High | 1100–2200 | 50–95 | Majority of igneous rocks, stronger metamorphics, well-cemented sandstone, hard shales, limestones, dolomites |
| C | Medium | 550–1100 | 25–50 | Shales, porous sandstones, limestones, schistose metamorphic rocks |
| D | Low | 275–550 | 13–25 | Porous and low-density rocks, friable sandstones, clay-shales, chalk, halite, and all altered rocks |
| E | Very low | <275 | <13 | As for class D |

(B) ON THE BASIS OF MODULUS RATIO‡

| Class | Modulus ratio | Value | Rock fabric |
|---|---|---|---|
| H | High | >500 | Steeply dipping schistosity or foliation |
| M | Medium (average) | 200–500 | Interlocking fabric, little or no schistosity |
| L | Low | <200 | Closure of foliations or bedding planes affects deformation |

*After Deere (1969).[8] Reprinted with permission of John Wiley & Sons, Inc.

†Point-load index values from Hoek and Bray (1974).[9]

‡Modulus ratio: defined as the ratio of the tangent modulus at 50% ultimate strength to the uniaxial compressive strength *(see Fig. 3.69)*.

equivalent to that obtained by field examination of large exposures, although ideal situations combine core recovery with exposure examinations.

*Building Codes*

Many codes classify rock in terms of hardness using nomenclature such as sound, hard, medium hard, and soft, but without defining the significance of the terms. The New York City Building Code (Table 5.22) provides a relatively comprehensive nomenclature, but its applicability is geographically somewhat limited to the local rock types, which are not deeply weathered.

*Simple Classification Systems*

Early workers in rock mechanics developed systems to classify joints according to spacing, as given in Table 5.23, and rock quality based on RQD and the velocity index, as given in Table 5.24.

**TABLE 5.20**
**SUGGESTED HARDNESS CLASSIFICATION FOR INTACT ROCK***

| Class | Hardness | Field test | Strength,† kg/cm$^2$ |
|---|---|---|---|
| I | Extremely hard | Many blows with geologic hammer required to break intact specimen | >2000 |
| II | Very hard to hard | Hand-held specimen breaks with hammer end of pick under more than one blow | 2000–700<br>700–250 |
| III | Moderate | Cannot be scraped or peeled with knife, hand-held specimen can be broken with single moderate blow with pick. | 250–100 |
| IV | Soft | Can just be scraped or peeled with knife. Indentations 1 mm to 3 mm deep show in specimen with moderate blow of pick. | 100–30 |
| V | Very soft | Material crumbles under moderate blow with sharp end of pick and can be peeled with a knife, but is too hard to hand-trim for triaxial test specimen. | 30–10 |

*After ISRM Working Party (1975).[10]

†Uniaxial compressive strength [Core Logging Comm. (1978)[11]].

**TABLE 5.21**
**SUGGESTED CLASSIFICATION FOR WEATHERED ROCK***

| Grade | Symbol | Diagnostic features |
|---|---|---|
| Fresh | F | No visible sign of decomposition or discoloration. Rings under hammer impact. |
| Slightly weathered | WS | Slight discoloration inward from open fractures, otherwise similar to F. |
| Moderately weathered | WM | Discoloration throughout. Weaker minerals such as feldspar decomposed. Strength somewhat less than fresh rock, but cores cannot be broken by hand or scraped by knife. Texture preserved. |
| Highly weathered | WH | Most minerals somewhat decomposed. Specimens can be broken by hand with effort or shaved with knife. Core stones present in rock mass. Texture becoming indistinct but fabric preserved. |
| Completely weathered | WC | Minerals decomposed to soil but fabric and structure preserved (saprolite). Specimens easily crumbled or penetrated. |
| Residual soil | RS | Advanced state of decomposition resulting in plastic soils. Rock fabric and structure completely destroyed. Large volume change. |

*After ISRM Working Party (1975).[10]

**TABLE 5.22**
**ALLOWABLE BEARING PRESSURES FOR ROCK***

| Characteristic | Rock material class: Hard, sound | Medium hard | Intermediate | Soft |
|---|---|---|---|---|
| $Q_{all}$, tsf†‡ | 60 | 40 | 20 | 8 |
| Rock type | Crystalline: gneiss, diabase, schist, marble, serpentinite | Same as hard rock | Same as hard to medium-hard rock, and cemented sandstones and shales | All rocks and uncemented sandstones |
| Struck with pick or bar | Rings | Rings | Dull sound | Penetrates |
| Exposure to air or water | Does not disintegrate | Does not disintegrate | Does not disintegrate | May soften |
| Fractures: Appearance | Sharp, fresh breaks | Cracks slightly weathered | Show weathered surfaces | Contain weathered zones |
| Width | <3 mm | <6 mm | Weathered zone to 25 mm width | Weathered zone to 30 cm width |
| Spacing | <1 m | <60 cm | Weathered zone spaced as close as 30 cm | Weathered zone filled with stiff soil |
| Core recovery§ | >85% | >50% | >35% | >35% |
| SPT | | | | >50 blows/ft |

*After New York City Building Code (1968).[12]

†Allowable bearing value applies only to massive crystalline rocks or to sedimentary or foliated rocks where strata are level or nearly level, and if area has ample lateral support. Tilted strata and their relation to adjacent slopes require special consideration.

‡Allowable bearing for hard to intermediate rock applies to foundations bearing on sound rock. Values can be increased by 10% for each 30 cm of penetration into rock of equal or better quality, but shall not exceed twice the basic values.

§Rock cored with double-tube, diamond core barrel, in 5-ft (1.5-m) run.

**TABLE 5.23**
**CLASSIFICATION OF JOINTS BASED ON SPACING**

| Description | Joint spacing | Rock-mass designation |
|---|---|---|
| Very wide | ≥3 m | Solid |
| Wide | 1 to 3 m | Massive |
| Moderately close | 30 cm to 1 m | Blocky/seamy |
| Close | 5 cm to 30 cm | Fractured |
| Very close | <5 cm | Crushed |

*From Deere (1963)[13] and Bieniawski (1974).[14]

**TABLE 5.24**
**ENGINEERING CLASSIFICATION FOR IN SITU ROCK QUALITY***

| RQD, % | Velocity index | Rock-mass quality |
|---|---|---|
| 90–100 | 0.80–1.00 | Excellent |
| 75–90 | 0.60–0.80 | Good |
| 50–75 | 0.40–0.60 | Fair |
| 50–25 | 0.20–0.40 | Poor |
| 25–0 | 0–0.20 | Very poor |

*After Coon and Merritt (1970).[15] Reprinted with permission of the American Society for Testing and Materials.

*Complex Classification Systems*

Recently developed systems provide information on rock quality in detail that includes joint factors such as orientation, opening width, irregularity, water conditions, and filling materials, as well as other factors. They are most applicable to tunnel engineering.

The *Geomechanics Classification System for Jointed Rock Masses* proposed by Bieniawski (1974, 1976)[14,16] is given on Table 5.25. It is based on grading six parameters: uniaxial compressive strength (from the point-load test), RQD, joint spacing, joint conditions, joint orientation, and groundwater conditions. Each parameter is given a rating, the ratings are totaled, and the rock is classed from "very good" to "very poor." In application to tunnel engineering the classes are related to stand-up time and unsupported tunnel span, and to ranges in rock-strength parameters of friction and cohesion. The system defines "poor rock" as having an RQD between 25 and 50% and $U_c$ between 250 and 500 kg/cm$^2$, which places poor rock in the range of "moderately sound, somewhat weathered" of Jaeger (1972)[7] (Fig. 5.7) and in the "hard" range given on Table 5.20.

*Engineering Classification of Rock Masses for Tunnel Support Design* proposed by Barton, Lien, and Lunde (1974, 1977)[17,18] is given as Table 5.26. It is based on a very detailed grading of six basic parameters including RQD, description of joint sets, joint roughness, joint alteration, joint water conditions, and a stress reduction factor which provides for rating major zones of weakness in the mass, residual stresses, squeezing rock, and swelling rock. The system does not relate RQD to strength parameters, but instead factors in the table are used to estimate roof pressures and tunnel support requirements directly.

## 5.3 SOILS

### 5.3.1 COMPONENTS

**Basic**

Defined by grain size, soil components are generally considered to include boulders, cobbles, gravel, sand, silt, and clay. Several of the more common and current classification systems are given on Table 5.27.

**Major Groupings**

On the basis of grain size, physical characteristics, and composition, soils may be placed in a number of major groups:

1. Boulders and cobbles, which are individual units.
2. Granular soils including gravel, sand, and silt are cohesionless materials (except for apparent cohesion evidenced by partially saturated silt).
3. Clay soils are cohesive materials.
4. Organic soils are composed of, or include, organic matter.

**Other Groupings**

Soils are also placed in general groups as:

1. Coarse-grained soils including gravel and sand
2. Fine-grained soils including silt and clay
3. Cohesive soils, which are clays mixed with granular soils or pure clays

### 5.3.2 GRANULAR OR COHESIONLESS SOILS

**Characteristics**

*General*

Boulders and cobbles normally respond to stress as individual units. Gravel, sand, and silt

**PLATE 5.1 Pegmatite: quartz and feldspar** (Portland, Connecticut).

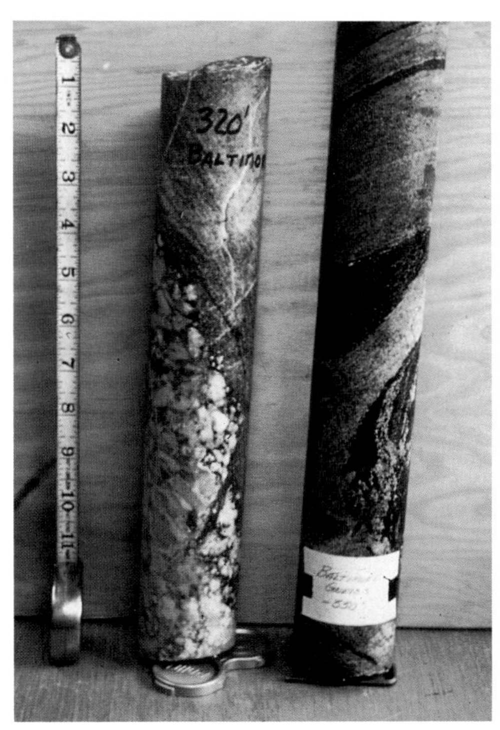


**PLATE 5.2 Pegmatite intruded into gneiss** (Baltimore, Maryland).

**PLATE 5.3 Porphyritic biotite granite with phenocrysts of feldspar** (St. Cloud, Minnesota).

**PLATE 5.4 Muscovite-biotite granite** (Concord, New Hampshire).

**PLATE 5.5 Syenite** (Victor, Colorado).

**PLATE 5.6 Diorite** (Salem, Massachusetts).

**PLATE 5.7 Hornblende gabbro** (Salem, Massachusetts).

**PLATE 5.8 Mica-augite peridotite** (Pike County, Arkansas).

**PLATE 5.9 Basalt** (Chimney Rock, New Jersey).

**PLATE 5.10 Vesicular basalt** (Salida, Colorado).

**PLATE 5.11** **Rhyolite** (Castle Rock, Colorado).

**PLATE 5.12** **Obsidian** (Lake County, Oregon).

**PLATE 5.13 Pumice** (Millard County, Utah).

**PLATE 5.14 Triassic conglomerate** (Rockland County, New York).

**PLACE 5.15** **Sandstone** (Potsdam, New York).

**PLATE 5.16** **Arkose** (Mt. Tom, Massachusetts).

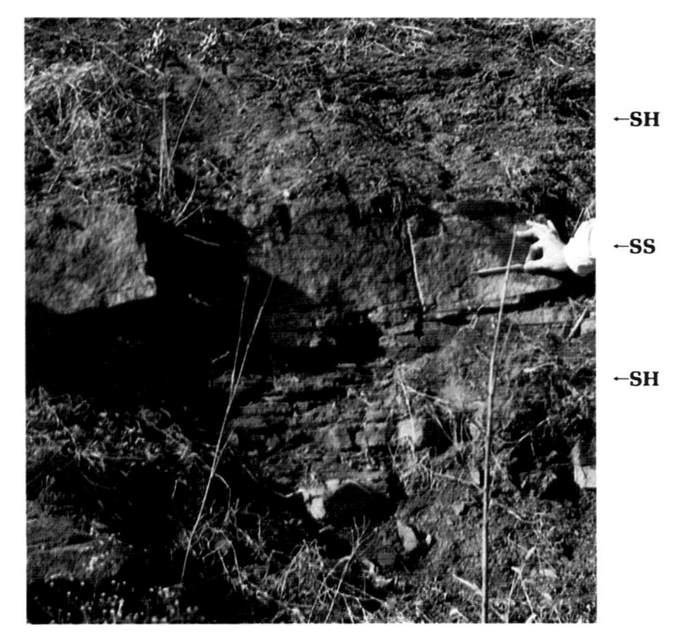


**PLATE 5.17 Interbedded Triassic sandstones and shales** (Rockland County, New York).

**PLATE 5.18 Clay shale** (Rio do Sul, Santa Catarina, Brazil).

**PLATE 5.19 Edwards limestone** (Cretaceous) (Round Rock, Texas).

**PLATE 5.20 Fossiliferous limestone** (Rochester, New York).

**PLATE 5.21 Dead coral reef exposed in cut** (Bridgetown, Barbados, West Indies).

**PLATE 5.22 Chert interbedded in the Edwards limestone** (Round Rock, Texas).

**PLATE 5.23 Hornblende gneiss with phenocrysts of feldspar** (Rio de Janeiro, Brazil).

**PLATE 5.24** **Biotite gneiss** (Oxbridge, Massachusetts).

**PLATE 5.25** **Mica schist** (Manhattan, New York City).

**PLATE 5.26** **Amphibolite** (Tres Ranchos, Goias, Brazil).

**PLATE 5.27** **Phyllite** (Minas Gerais, Brazil).

**PLATE 5.28 Slate** (Rio Itajai, Santa Catarina, Brazil).

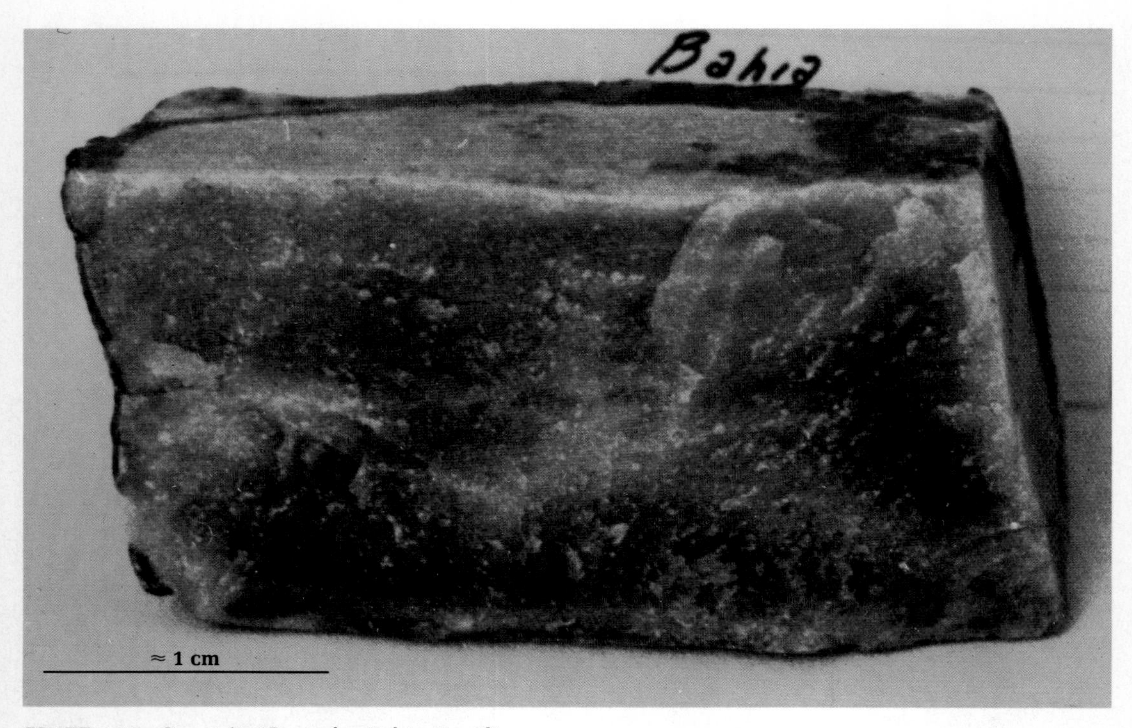


**PLATE 5.29 Quartzite** (Cocorobo, Bahia, Brazil).

**PLATE 5.30** **Marble** (Minas Gerais, Brazil).

**PLATE 5.31** **Serpentinite** (Cardiff, Maryland).

**PLATE 5.32 Soapstone** (Minas Gerais, Brazil).

**PLATE 5.33 Migmatite** (Rio Santos Highway, Rio de Janeiro, Brazil).

respond to stress as a mass and are the most significant granular soils.

*Particles*

Shape is bulky and usually equidimensional, varying from rounded to very angular. The shape results from abrasion and in some cases solution, and is related to the mode and distance of transport. Subangular sand grains are illustrated on Fig. 5.8.

Behavior is mass-derived because of pore spaces between individual grains which are in contact.

*Properties*

Cohesionless, nonplastic.

### Grain Minerals

*Types*

The predominant granular soil mineral is quartz which is essentially stable, inert, and nondeformable. On occasion, sands and silts will include garnet, magnetite, and hornblende.

In climates where mechanical disintegration is rapid and chemical decomposition is minor, mica, feldspar or gypsum may be present, depending on the source rock.

Shell fragments are common in many beach deposits, especially in areas where quartz-rich rocks are lacking, and in the middle latitudes offshore calcareous or carbonate sands are common (*see Art. 7.4.6*). The weaker minerals such as shells, mica, and gypsum have low crushing strengths, and calcareous sands can have deleterious effects on concrete.

*Identification*

Simple tests to identify grain minerals include application of hydrochloric acid to test for calcareous materials and determination of specific gravity (Table 5.5).

### Silt

*General*

Although it consists of bulky particles, silt is often grouped with clays as a fine-grained soil since its particle size is defined as smaller than 0.074 mm. Nonplastic silt consists of more or less

**FIG. 5.8 Subangular grains of coarse to medium quartz sand (≈14×).**

equidimensional quartz grains and is at times referred to as "rock flour." Plastic silt contains appreciable quantities of flake-shaped particles.

Silts are classed as *inorganic*, ranging from nonplastic to plastic (*see Fig. 3.12*), or *organic*, containing appreciable quantities of organic matter.

The smooth texture of silt when wet gives it the appearance of clay.

*Properties*

DILATANCY Silts undergo changes in volume with changes in shape, whereas clays retain their volume with changes in shape (plasticity). Grains are fine, but compared with clays, pore spaces are relatively large, resulting in a high sensitivity to pore-pressure changes, particularly from increases due to vibrations. Because of their physical appearance and tendency to quake under construction equipment, silts are often referred to as "bull's liver."

**TABLE 5.25**
**THE GEOMECHANICS CLASSIFICATION SYSTEM FOR JOINTED ROCK MASSES***

A. CLASSIFICATION PARAMETERS AND THEIR RATINGS

| | Parameter | | | | | | | | |
|---|---|---|---|---|---|---|---|---|---|
| 1 | Strength of intact rock material | Point-load strength index, MPa | >8 | 4–8 | 2–4 | 1–2 | Use of aniaxial compressive test preferred | | |
| | | Uniaxial compressive strength, MPa | >200 | 100–200 | 50–100 | 25–50 | 10–25 | 3–10 | 1–3 |
| | Rating | | 15 | 12 | 7 | 4 | 2 | 1 | 0 |
| 2 | Drill core quality RQD | | 90–100% | 75–90% | 50–75% | 25–50% | <25% | | |
| | Rating | | 20 | 17 | 13 | 8 | 3 | | |
| 3 | Spacing of joints | | >3 m | 1–3 m | 0.3–1 m | 50–300 mm | <50 mm | | |
| | Rating | | 30 | 25 | 20 | 10 | 5 | | |
| 4 | Condition of joints | | Very rough surfaces. Not continuous. No separation. Hard joint wall rock. | Slightly rough surfaces. Separation <1 mm. Hard joint wall rock. | Slightly rough surfaces. Separation <1 mm. Soft joint wall rock. | Slickensided surfaces or gouge < 5 mm thick or joints open 1–5 mm. Continuous joints | Soft gouge > 5 mm thick or joints open > 5 mm. Continuous joints | | |
| | Rating | | 25 | 20 | 12 | 6 | 0 | | |
| 5 | Groundwater | Inflow per 10 m tunnel length | None | | <25 liters/min | 25–125 liters/min | >125 liters/min | | |
| | | Ratio, joint water pressure / major principal stress | or 0 | | or 0.0–0.2 | or | or >0.5 | | |
| | | General conditions | or Completely dry | | or moist only (interstitial water) | or water under moderate pressure | or severe water problems | | |
| | Rating | | 10 | | 7 | 4 | 0 | | |

B. ADJUSTMENT FOR JOINT ORIENTATIONS

| Strike and dip orientations of joints | | Very favorable | Favorable | Fair | Unfavorable | Very unfavorable |
|---|---|---|---|---|---|---|
| | Tunnels | 0 | −2 | −5 | −10 | −12 |
| Ratings | Foundations | 0 | −2 | −7 | −15 | −25 |
| | Slopes | 0 | −5 | −25 | −50 | −60 |

C. ROCK MASS CLASSES AND THEIR RATINGS

| Class no. | I | II | III | IV | V |
|---|---|---|---|---|---|
| Description | Very good rock | Good rock | Fair rock | Poor rock | Very poor rock |
| Rating | 100 ← 90 | 90 ← 70 | 70 ← 50 | 50 ← 25 | <25 |

D. MEANING OF ROCK MASS CLASSES

| Class no. | I | II | III | IV | V |
|---|---|---|---|---|---|
| Average stand-up time | 10 years for 5-m span | 6 months for 4-m span | 1 week for 2-m span | 5 hr for 1.5-m span | 10 min for 0.5-m span |
| Cohesion of the rock mass, kPa | >300 | 200–300 | 150 | 100–150 | <100 |
| Friction angle of the rock mass | >45° | 40–45° | 35–40° | 30–35° | <30° |
| Caveability of ore | Very poor | Will not cave readily. Large fragments. | Fair | Will cave readily. Good fragmentation. | Very good |

*From Bieniawski (1974).[14] Reprinted with permission of the National Academy of Sciences. Units are International System: 1 MPa = 1 megapascal = 10 kg/cm$^2$; 100 kPa = 100 kilopascals = 0.96 kg/cm$^2$.
Geomechanics classification of jointed rock masses.

## TABLE 5.26
## ENGINEERING CLASSIFICATION OF ROCK MASSES FOR TUNNEL SUPPORT DESIGN (From Barton et al. (1977).[18]

| (a) Rock quality designation | RQD[a] |
|---|---|
| Very poor | 0–25 |
| Poor | 25–50 |
| Fair | 50–75 |
| Good | 75–90 |
| Excellent | 90–100 |

| (b) Joint set number[b] | $J_n$ |
|---|---|
| Massive, no, or few joints | 0.5–1.0 |
| One joint set | 2 |
| One joint set plus random | 3 |
| Two joint sets | 4 |
| Two joint sets plus random | 6 |
| Three joint sets | 9 |
| Three joint sets plus random | 12 |
| Four or more joint sets, random, heavily jointed, "sugar cube," etc. | 15 |
| Crushed rock, earthlike | 20 |

| (c) Joint roughness number | $J_r$ |
|---|---|
| Rock wall contact and rock wall contact before 10-cm shear[c] | |
| Discontinuous joints | 4 |
| Rough or irregular, undulating | 3 |
| Smooth, undulating | 2 |
| Slickensided, undulating | 1.5 |
| Rough or irregular, planar | 1.5 |
| Smooth, planar | 1.0 |
| Slickensided, planar | 0.5 |
| No rock wall contact when sheared[d] | |
| Zone containing clay minerals thick enough to prevent rock wall contact | 1.0 |
| Sandy, gravelly or crushed zone thick enough to prevent rock wall contact | 1.0 |

| (d) Joint alteration number | $J_a$ | $\phi_r$ (approx.) |
|---|---|---|
| Rock wall contact | | |
| Tightly healed, hard, nonsoftening, impermeable filling, e.g., quartz or epidote | 0.75 | |
| Softening or low-friction clay mineral coatings, e.g., kaolinite or mica. Also chlorite, talc, gypsum, graphite, etc., and small quantities of swelling clays | 4.0 | (8–16) |
| Rock wall contact before 10-cm shear | | |
| Sandy particles, clay-free disintegrated rock, etc. | 4.0 | (25–30) |
| Strongly overconsolidated, nonsoftening clay mineral fillings (continuous, but <5-mm thickness) | 6.0 | (16–24) |
| Medium or low overconsolidation, softening, clay mineral fillings (continuous but <5-mm thickness) | 8.0 | (12–16) |
| Swelling-clay fillings, e.g., montmorillonite (continuous, but <5-mm thickness). Value of $J_a$ depends on percent of swelling clay-size particles, and access to water, etc. | 8–12 | (6–12) |
| No rock wall contact when sheared | | |
| Zones or bands of disintegrated or crushed rock and clay (see above for description of clay condition) | 6, 8, or 8–12 | (6–24) |
| Zones or bands of silty- or sandy-clay, small clay fraction (nonsoftening) | 5.0 | |
| Thick, continuous zones or bands of clay (see above for description of clay condition) | 10, 13, or 13–20 | (6–24) |

| (e) Joint-water reduction factor[e] | $J_w$ | Approx. water pressure, kg/cm² |
|---|---|---|
| Dry excavations or minor inflow, i.e. <5 liter/min locally | 1.0 | <1 |
| Exceptionally high inflow or water pressure at blasting, decaying with time | 0.2–0.1 | >10 |
| Exceptionally high inflow or water pressure continuing without noticeable decay | 0.1–0.05 | >10 |

| (f) Stress reduction factor | SRF |
|---|---|
| Weakness zones intersecting excavation, which may cause loosening of rock mass when tunnel is excavated[f] | |
| Multiple occurrences of weakness zones containing clay or chemically disintegrated rock, very loose surrounding rock (any depth) | 10 |
| Single weakness zones containing clay or chemically disintegrated rock (depth of excavation ≤50 m) | 5 |
| Single weakness zones containing clay or chemically disintegrated rock (depth of excavation >50 m) | 2.5 |
| Multiple shear zones in competent rock (clay-free), loose surrounding rock (any depth) | 7.5 |
| Single shear zones in competent rock (clay-free) (depth of excavation ≤50 m) | 5.0 |
| Single shear zones in competent rock (clay-free) (depth of excavation >50 m) | 2.5 |
| Loose open joints, heavily jointed or "sugar cube," etc. (any depth) | 5.0 |

| | $\sigma_c/\sigma_1$ | $\sigma_t/\sigma_1$ | SRF |
|---|---|---|---|
| Competent rock, rock stress problems[g] | | | |
| Low stress, near surface | >200 | >13 | 2.5 |
| Medium stress | 200–10 | 13–0.66 | 1.0 |
| High stress, very tight structure (usually favorable to stability, may be unfavorable for wall stability) | 10–5 | 0.66–0.33 | 0.5–2 |
| Mild rock burst | 5–2.5 | 0.33–0.16 | 5–10 |

| | | |
|---|---|---|
| Unaltered joint walls, surface staining only | 1.0 | (25–35) |
| Slightly altered joint walls. Nonsoftening mineral coatings, sandy particles, clay-free disintegrated rock, etc. | 2.0 | (25–30) |
| Silty-, or sandy-clay coatings, small clay fraction (nonsoft.) | 3.0 | (20–25) |

| | | |
|---|---|---|
| Medium inflow or pressure, occasional outwash of joint fillings | 0.66 | 1–2.5 |
| Large inflow or high pressure in competent rock with unfilled joints | 0.5 | 2.5–10 |
| Large inflow or high pressure, considerable outwash of joint fillings | 0.33 | 2.5–10 |

| | | | |
|---|---|---|---|
| Heavy rock burst (massive rock) | <2.5 | <0.16 | 10–20 |
| | | | **SRF** |
| Squeezing rock: plastic flow of incompetent rock under the influence of high rock pressure | | | |
| Mild squeezing rock pressure | | | 5–10 |
| Heavy squeezing rock pressure | | | 10–20 |
| Swelling rock: chemical swelling activity depending on presence of water | | | |
| Mild swelling rock pressure | | | 5–10 |
| Heavy swelling rock pressure | | | 10–15 |

[a]Where RQD is reported or measured as ≦10 (including 0), a nominal value of 10 is used to evaluate Q in Eq. (1). RQD intervals of 5, i.e., 100, 95, 90, etc., are sufficiently accurate:

Rock mass quality Q = $(RDQ/J_n)(J_r/J_a)(J_w/SRF)$ (1)

[b]For intersections use $3.0 \times J_n$. For portals use $2.0 \times J_n$.

[c]Descriptions refer to small-scale features and intermediate-scale features, in that order.

[d]Add 1.0 if the mean spacing of the relevant joint set is greater than 3 m. $J_r = 0.5$ can be used for planar slickensided joints having lineations, provided the lineations are orientated for minimum strength.

[e]Last four factors above are crude estimates. Increase $J_w$ if drainage measures are installed. Special problems caused by ice formation are not considered.

[f]Reduce these values of SRF by 25–50% if the relevant shear zones only influence but do not intersect the excavation.

[g]For strongly anisotropic virgin stress field (if measured): when $5 \leq \sigma_1/\sigma_3 \leq 10$, reduce $\sigma_c$ and $\sigma_t$ to $0.8\sigma_c$ and $0.8\sigma_t$. When $\sigma_1/\sigma_3 > 10$, reduce $\sigma_c$ and $\sigma_t$ to $0.6\sigma_c$ and $0.6\sigma_t$, where $\sigma_c$ = unconfined compression strength, $\sigma_t$ = tensile strength (point load), and $\sigma_1$ and $\sigma_3$ are the major and minor principal stresses. Few case records available where depth of crown below surface is less than span width. SRF increase from 2.5 to 5 is suggested for such cases (see low stress, bear surface).

ADDITIONAL NOTES: When rock-mass quality Q is estimated, the following guidelines should be followed, in addition to the notes listed in parts a to f:

1. When borecore is unavailable, RQD can be estimated from the number of joints per unit volume, in which the number of joints per meter for each joint set are added. A simple relation can be used to convert this number to RQD for the case of clay-free rock masses:

$$RQD = 115 - 3.3J_v \quad \text{(approx.)}$$

where $J_v$ = total number of joints per cubic meter (RQD = 100 for $J_v < 4.5$).

2. The parameter $J_n$ representing the number of joint sets will often be affected by foliation, schistocity, slatey cleavage, or bedding, etc. If strongly developed these parallel "joints" should obviously be counted as a complete joint set. However, if there are few "joints" visible, or only occasional breaks in bore core because of these features, then it will be more appropriate to count them as "random joints" when evaluating $J_n$ in part b.

3. The parameters $J_r$ and $J_a$ (representing shear strength) should be relevant to the *weakest significant joint set or clay filled discontinuity* in the given zone. However, if the joint set or discontinuity with the minimum value of $(J_r/J_a)$ is favorably oriented for stability, then a second, less favorably oriented joint set or discontinuity may sometimes be of more significance, and its higher value of $J_r/J_a$ should be used when evaluating Q from the equation above. *The value of $J_r/J_a$ should in fact relate to the surface most likely to allow failure to initiate.*

4. When a rock mass contains clay, the factor SRF appropriate to *loosening loads* should be evaluated (part f). In such cases the strength of the intact rock is of little interest. However, when jointing is minimal and clay is completely absent, the strength of the intact rock may become the weakest link, and the stability will then depend on the ratio rock-stress/rock-strength (see part f). A strongly anisotropic stress field is unfavorable for stability and is roughly accounted for as in Note g, part f.

5. The compressive and tensile strengths $\sigma_c$ and $\sigma_t$ of the intact rock should be evaluated in the saturated condition if this is appropriate to present or future in situ conditions. A very conservative estimate of strength should be made for those rocks that deteriorate when exposed to moist or saturated conditions.

**TABLE 5.27**
**SOIL CLASSIFICATION SYSTEMS BASED ON GRAIN SIZE**

| System | Grain diameter, mm |
|---|---|
| | 0.0006 \| 0.002 \| 0.006 \| 0.02 \| 0.06 \| 0.2 \| 0.6 \| 2.0 \| 4.76 \| 19 \| 76 |
| M.I.T. and British Standards Institute | f \| m \| c \| f \| m \| c<br>Clay \| Silt \| Sand \| Gravel |
| American Association of State Highway Officials (AASHO) | 0.001 \| 0.005 \| 0.074 \| 0.25 \| 2.0 \| 9 \| 24 \| 76<br>f \| c \| f \| m \| c<br>Colloids \| Clay \| Silt \| Sand \| Gravel \| Boulders |
| U.S. Dept. of Agriculture (USDA) | 0.002 \| 0.05 \| 0.25 \| 0.5 \| 2.0 \| 76<br>vf \| f \| m \| c \| vc \| f \| m<br>Clay \| Silt \| Sand \| Gravel \| Cobbles |
| Unified Soil Classification system (USBR, USAEC) | f \| m \| c \| f \| c<br>Clay and silt \| Sand \| Gravel \| Cobbles<br>No. 200 \| 40 \| 10 \| 4 \| ¾ in \| 3 in |
| | **Grain diameter in U.S. standard sieve sizes** |
| American Society for Engineering Education (ASEE) (Burmister) | No. 200 \| 60 \| 30 \| 10 \| 3/16 \| 3/8 in \| 1.0 in \| 3 in<br>Silt f \| c \| f \| m \| c \| f \| m \| c<br>Clay or silt \| Sand \| Gravel |
| Field identification | Not discernible \| Hand lens \| Visible to eye \| Measurable |

STABILITY When saturated and unconfined, silts have a tendency to become "quick" and flow as a viscous fluid (liquefy).

"APPARENT COHESION" Results from capillary forces providing a temporary bond between particles which is destroyed by saturation or drying.

## 5.3.3 CLAYS

### Characteristics

*General*

Clays are composed of elongate mineral particles of colloidal dimensions, commonly taken as less than 2 $\mu$ in size [Gillott (1968)[19]]. Behavior is controlled by surface-derived forces rather than by mass-derived forces. A spoon sample of lacustrine clay is illustrated in Fig. 5.9.

*Mass Structures*

Clay particles form two general forms of structures: flocculated or dispersed, as shown on Fig. 5.10.

*Flocculated* structure consists of an edge-to-face orientation of particles which results from electrical charges on their surfaces during sedimen-

**FIG. 5.9 "Spoon" sample of lacustrine clay with root fibers showing the characteristic smooth surfaces made by a knife blade in plastic material.**

tation. In salt water, flocculation is much more pronounced than in fresh since clay particles curdle into lumps and settle quickly to the bottom without stratification. In fresh water the particles settle out slowly, forming laminated and well-stratified layers with graded bedding.

*Dispersed* structure consists of face-to-face orientation or parallel arrangement which occurs during consolidation (compaction.)

*Properties*

COHESION Results from a bond developing at the contact surfaces of clay particles, caused by electrochemical attraction forces. The more closely packed the particles, the greater is the bond and the stronger is the cohesion. It is caused by two factors: the high specific surface of the particles (surface area per unit weight), and the electrical charge on the basic silicate structure resulting from ionic substitutions in the crystal structure (Table 5.30).

ADHESION Refers to the affinity of a clay to adhere to a foreign material, i.e., stickiness.

PLASTICITY Material undergoes a change in shape without undergoing a change in volume, with its moisture content held constant.

CONSISTENCY With decreasing moisture content, clays pass from the fluid state (very soft) through a plastic state, then a semisolid state, to finally a hard brick-like state (Table 3.29). The moisture contents at the transitions between these various states are defined by the Atterberg limits (*see Art. 3.2.3*), which vary with the clay type and its purity. Clay soils are commonly identified by the relationship between the plasticity index and the liquid limit as given on Fig. 3.12.

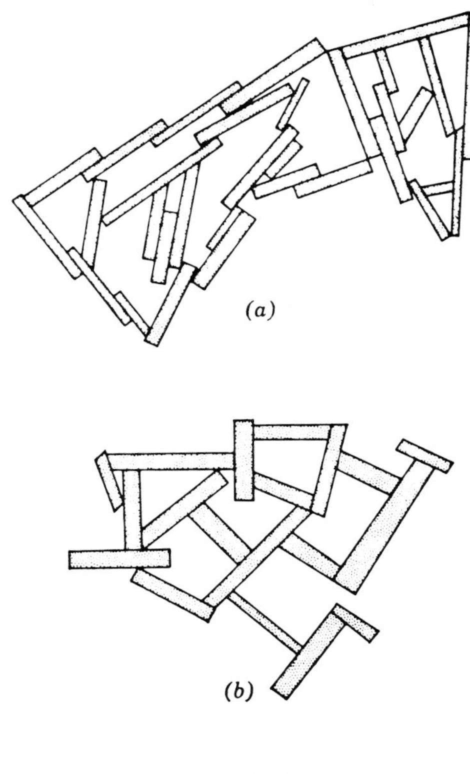


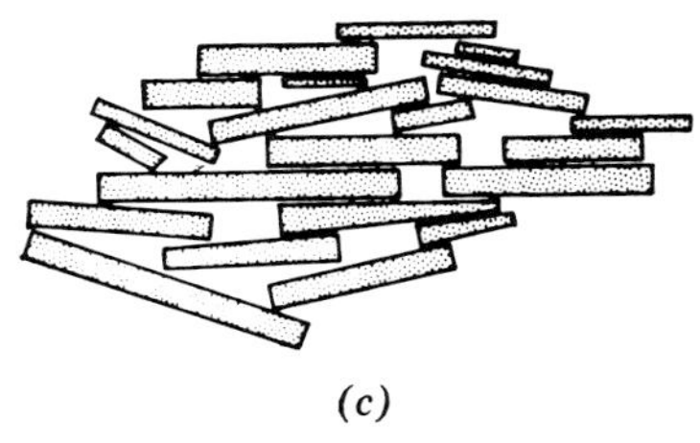


**Fig. 5.10 Clay structures from sedimentation: (*a*) flocculated structure—saltwater environment; (*b*) flocculated structure—freshwater environment; (*c*) dispersed particles.**

**FIG. 5.11 Shrinkage pattern development in clayey alluvium with silt lenses. Shrinkage is caused by capillary tension developing in pore water during drying (Fronteira, Piaui, Brazil).**

ACTIVITY Refers to an affinity for moisture resulting in large volume changes with increase in moisture content (swelling) or decrease in moisture content (shrinking) (Fig. 5.11), which is due to the crystal structure and chemistry.

The degree of activity is related to the percent of the clay fraction in the specimen and the type of clay mineral, and has been defined as the ratio of the plasticity index to the percent by weight finer than 2 $\mu$. A clay classification based on activity is given on Table 5.28. Expansive materials and their characteristics are described in Art. 10.6.2.

## Clay Mineralogy and Chemistry[19,23–25]

### *Clay Minerals*

Clays are hydrous aluminum silicates which are classed in a number of groups based on their crystal structure and chemisty. Common groups include kaolinite, halloysite, illite, and montmorillonite. Less common groups include vermiculite and chlorite, which although common in decomposing rock masses, alter readily to the other types.

Characteristics of the common clay minerals are summarized on Table 5.29.

**TABLE 5.28**
**CLAY ACTIVITY***

| Activity† | Classification |
|---|---|
| <0.75 | Inactive clays (kaolinite) |
| 0.75–1.25 | Normal clays (illite) |
| >1.25 | Active clays (montmorillonite) |

*After Skempton (1953).[20]

†Ratio of PI to percent by weight finer than 2 $\mu$.

**TABLE 5.29**
**CHARACTERISTICS OF COMMON CLAY MINERALS**

| Mineral | Origin (see Art. 6.7.3) | Activity | Particles |
|---|---|---|---|
| Kaolinite | Chemical weathering of feldspars.<br>Final decomposition of micas and pyroxenes in humid climates or well-drained conditions.<br>Main constituents of clay soils in humid-temperate and humid-tropical regions. | Low. Relatively stable material in the presence of water. | Platy but lumpy. |
| Halloysite | Similar to kaolinite, but from feldspars and mica. (Primarily sialic rocks.) | Low, except properties are radically altered by intense drying. Process not reversible.* | Elongated rodlike units, or hollow cylinders. |
| Illite | Main constituent of many clay shales, often with montmorillonite. | Intermediate between kaolinite and montmorillonite. | Thin plates. |
| Montmorillonite (smectite) | Chemical decomposition of olivine (mafic rocks).<br>Partial decomposition of micas and pyroxene in low rainfall or poor drainage environment.<br>Constituent of marine and clay shales.<br>Alteration of rock during shearing by faulting.<br>Volcanic dust. | Highly expansive and the most troublesome of the clay minerals in slopes and beneath foundations.<br>Used as an impermeabilizing agent. | In electron microscope appears as a mass of finely chopped lettuce leaves. |

*In compaction tests on halloysites, it was found that higher densities were obtained on material air-dried and then brought back to the desired moisture content, than with material at natural moisture content that was either wet or dried to the desired moisture [Gibbs et al. (1960)[21]]. Therefore, when halloysites are used as embankment material, testing procedures should duplicate field placement procedures.

Classification of the clay minerals based on chemistry and crystal structure is given on Table 5.30.

### *Clay Chemistry*

CLASSES Clays are classed also on the basis of the cations adsorbed on the particle surfaces of the mineral (H, Ca, K, Mg, or Na). *Sodium clays* may be the product of the deposition of clay in seawater, or of their saturation by saltwater flooding or capillary action. *Calcium clays* are formed essentially in freshwater deposition. *Hydrogen clays* are the result of prolonged leaching by pure or acid water, with the resulting removal of all other exhangeable bases.

*Base exchange* refers to the capacity of colloidal particles to change the cations adsorbed on the surfaces. Thus a hydrogen clay can be changed to a sodium clay by constant percolation of water containing dissolved sodium salts. The permeability of a clay can be decreased by such changes and the sensitivity increased. Base exchange may explain the susceptibility of some soils to the phenomenon termed "dispersion" or erosion by piping. Soils with a high percentage of sodium cation relative to calcium and mag-

**TABLE 5.30**
**CLASSIFICATION OF CLAY MINERALS***

| LAYERS | EXPANSION | GROUP | SPECIES | CRYSTALLOCHEMICAL FORMULA | STRUCTURE (SCHEMATIC) |
|---|---|---|---|---|---|
| TWO-SHEET (1:1) | NON-SWELLING | KAOLINITE | KAOLINITE<br>DICKITE<br>NACRITE | $Al_4(OH)_8[Si_4O_{10}]$ | KAOLINITE 7,1A°<br>· Si<br>• Al, Mg, Fe<br>exchangeable cations<br>⊙ K<br>○ O<br>● OH<br>$H_2O$ |
| | NON-SWELLING and SWELLING | HALLOYSITE | HALLOYSITE<br>METAHALLOYSITE | $Al_4(OH)_8[Si_4O_{10}]\cdot(H_2O)_4$<br>$Al_4(OH)_8[Si_4O_{10}]\cdot(H_2O)_2$ | HALLOYSITE 10,0A° |
| Three-Sheet (2:1) | SWELLING | MONTMORILLONITE (SMECTITE) | MONTMORILLONITE<br>BEIDELLITE<br>NONTRONITE | $\{(Al_{2-x}Mg_x)(OH)_2[Si_4O_{10}]\}^{-x}Na_x\cdot nH_2O$<br>$\{Al_2(OH)_2[(Al,Si)_4O_{10}]\}^{-x}Na_x\cdot nH_2O$<br>$\{(Fe_{2-x}Mg)(OH)_2[Si_4O_{10}]\}^{-x}Na_x\cdot nH_2O$ | MONTMORILLONITE 14,2A° |
| | NON-SWELLING | ILLITE (HYDROMICA) | ILLITE-VARIETIES | $(K,H_3O)Al_2(H_2O,OH)_2[AlSi_3O_{10}]$ | ILLITE 10,0A° |
| | SWELLING | VERMICULITE | VERMICULITE | $(Mg,Fe)_3(OH)_2[AlSi_3O_{10}]Mg\cdot(H_2O)_4$ | VERMICULITE 14,2A° |
| Three-Sheet + One-Sheet (2:2) | NON-SWELLING | 14A-CHLORITE (NORMAL CHLORITE) | CHLORITE-VARIETIES | $(Al,Mg,Fe)_3(OH)_2[(Al,Si)_4O_{10}]Mg_3(OH)_6$ | 14A° CHLORITE 14,2A° |

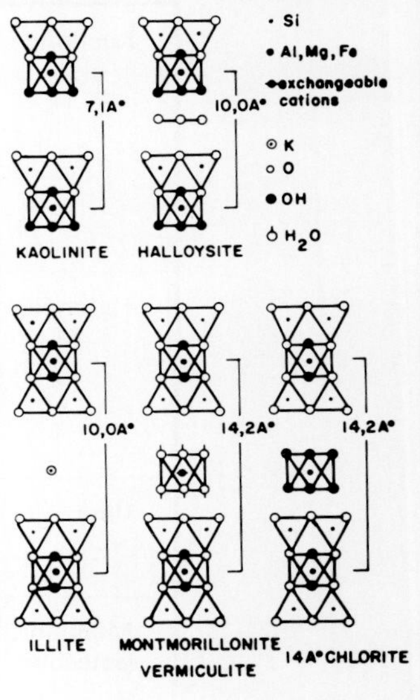

*From Morin and Tudor (1975).[22]

nesium cations appear to have a high susceptibility (*see Art. 10.5.5*).

*Exchange capacity* refers to the quantity of exchangeable cations in a soil; not all cations are exchangeable. It increases with the acidity of the soil crystals.

*Acidity* of a clay is expressed by lower values of pH, or higher values of the *silica-sequioxide ratio* $SiO_2/R_2O_3$, where:

$$R_2O_3 = Fe_2O_3 + Al_2O_3 \tag{5.2}$$

For soils the reference is mainly to the acidity of the soluble particles. Corrosion of iron or steel embedded in soil, in the presence of moisture, increases with soil acidity.

### *Identification*

SUSPENSION If a specimen of clay is mixed with pure water to form a paste and then dispersed in pure water, particles generally smaller than about 1 $\mu$ ($10^{-3}$ mm) will remain in suspension almost indefinitely and are considered as colloidal.

ELECTRON MICROSCOPE Can view particles down to about 1 $\mu$m ($10^{-6}$ mm). The crystal shape is used to identify the clay type [for examples see Osipor and Sokolov (1978)[26]].

X-RAY DIFFRACTION Used to identify particles to about $10^{-8}$ mm (1 angstrom, Å = $10^{-7}$ mm). In the diffractometer a powdered mineral sample mounted on a glass slide is rotated at a fixed angular rate in an x-ray beam. A pick-up device, such as a Geiger tube, rotates about the same axis, detecting the diffracted beams. The impulse is transmitted and recorded on a strip chart. Expressions are available relating wavelength of radiation and the angle of rotation $\theta$ to $d$, where $d$ is the spacing of a particular set of crystal planes, from which the mineral is identified [Walhstrom (1973)[27]].

DIFFERENTIAL THERMAL ANALYZER Measures the temperatures and magnitudes of exothermic and endothermic changes occurring in a sample as it is heated at a uniform rate. Measurements made by thermocouples embedded in the specimen

**FIG. 5.12 Rootmat and silty clay exposed in excavation (New Jersey Meadowlands).**

and changes occurring in the specimen during heating are recorded on a strip chart. The curve forms are characteristic for various clay types.

5.3.4 ORGANIC MATERIALS

## Origin and Formation

*Origin*

Organic matter is derived primarily from decayed plant life and occasionally from animal organisms.

*Formation*

*Topsoil* is formed as plant life dies and becomes fixed with the surficial soils. The layer thickness and characteristics are a function of climate and drainage; the latter is related to slope and soil type. Well-drained granular soils above the water table, poor in minerals other than quartz, develop very thin topsoil layers, even in humid climates. Thick topsoil layers develop in mineral-rich soils and humid climates, particularly where the soil is cool. Soil temperatures above 30°C destroy humus because of bacterial activity, whereas humus accumulates below 20°C [Mohr et al. (1972)[28]]. This phenomenon is evident in tropical countries where topsoil is usually thin, except where drainage is poor.

*Rootmat* forms in marshy regions *(Art. 7.4.5)* and is a thick accumulation of living and dead marsh growth as illustrated on Fig. 5.12.

*Peat* is fibrous material with a spongelike structure, composed almost entirely of dead organic matter which can form to extensive thicknesses.

*Organic silts and clays* form in lakes and estuarine environments, where they can attain thicknesses of 25 m or more.

*Occurrence*

Although surface deposits during formation, organic layers can be found deeply buried by alluvium as shown on Fig. 5.13, by beach deposits *(see Fig. 7.46)*, colluvium *(see Fig. 7.11)*, glacial till *(see Fig. 7.81)*, and aeolian soils, thereby representing a zone of weakness in otherwise strong formations.

## Characteristics

Organic deposits are characterized by very low natural densities, very high natural water contents, a loss in mass upon ignition, and substantial shrinkage upon drying.

**FIG. 5.13 Undisturbed sample of clayey sand overlying organic silty sand at a 25-ft depth (Leesburg, New Jersey).**

## 5.3.5 RELATED ENGINEERING PROPERTIES

### General

The major soil groupings have distinguishing characteristics which relate directly to their engineering properties.

*Characteristics*

- Granular soils: gradation, relative density, grain shape, and mineral composition
- Clay soils: mineral type, chemistry, plasticity, and stress history
- Organic materials: percentage of organic matter vs. soil particles, and stress history
- Mixtures: Combine the characteristics, but relative density quickly becomes insignificant

A general summary of engineering properties of the various soil groups is given on Table 5.31.

### Gravels and Sands

*Hydraulic Properties*

PERMEABILITY Gravels and sands are free-draining materials with large storage capacity, acting as aquifers or natural reservoirs, providing the sources of water flowing into excavations, or through, around, and beneath dams.

CAPILLARITY Negligible.

FROST HEAVING Essentialy nonsusceptible.

LIQUEFACTION AND PIPING Potential increases with increasing fineness. Loose fine sands are most susceptible; gravel is nonsusceptible.

*Rupture Strength*

Strength is derived from intergranular friction.

FAILURE CRITERIA General shear failure of shallow foundations does not occur because compression occurs simultaneously with load application, and a deep failure surface cannot develop. Failure occurs by local shear, the displacement around the edge of a flexible foundation, or punching shear, failure by rupture of a deep foundation. In slopes, failure is relatively shallow in accordance with the infinite-slope criteria. Collapse of soil structure occurs in lightly cemented loose sands.

*Deformability*

Response to load is immediate as the voids close and the grains compact by rearrangement. Deformation is essentially plastic, with some elastic compression occurring within the grains. The amount of compression is related to gradation, relative density, and the magnitude of the applied stress.

Susceptibility to densification by vibrations is high and the materials are readily compactable.

Crushing can occur in grains of shell fragments, gypsum, or other soft materials, even under relatively low applied stresses.

### Silts (Inorganic)

*Hydraulic Properties*

- Permeability: slow draining

**TABLE 5.31**
**ENGINEERING PROPERTIES OF SOILS SUMMARIZED**

| Property | Gravel and sand | Silt | Clay | Organics |
|---|---|---|---|---|
| HYDRAULIC PROPERTIES | | | | |
| Permeability | Very high to high | Low | Very low to impermeable | Very high to very low |
| Capillarity | Negligible | High | Very high | Low to high |
| Frost-heaving susceptibility | Nil to low | High | High | Low to high |
| Liquefaction susceptibility | Nil to high in fine sands | High | None | High in organic silts |
| RUPTURE STRENGTH | | | | |
| Derivation | Intergranular friction $\phi$ | Friction $\phi$, apparent cohesion | Drained: $\bar{\phi}$ and $\bar{c}$; undrained: $s_u$ | Organic silts and clay, $\phi$ and $c$ |
| Relative strength | High to moderate | Moderate to low | High to very low | Very low |
| Sensitivity | None | None | Low to very high | As for clay |
| Collapsing formations | Lightly cemented sands | Loess | Porous clays | Not applicable |
| DEFORMABILITY | | | | |
| Magnitude (moderate loads) | Low to moderate | Moderate | Moderate to high | Very high |
| Time delay | None | Slight | Long | None to long |
| Compactability | Excellent | Very difficult | Moderate difficulty; requires careful moisture control | Not applicable |
| Expansion by wetting | None | None | Moderate to very high | Slight |
| Shrinkage upon drying | None | Slight | Moderate to very high | High to very high |
| CORROSIVITY | | | | |
| | Occasional; calcareous sands troublesome to concrete | Occasional | Low to high | High to very high |

- Capillarity: high
- Frost heaving susceptibility: high
- Liquefaction and piping susceptibility: high

*Rupture Strength*

Strength is derived from intergranular friction and apparent cohesion when silt is partially saturated. Strength is destroyed by saturation or drying.

Collapse may occur in lightly cemented formations, such as loess, upon saturation.

*Deformability*

Slow draining characteristics result in some time delay in compression under applied load. Compaction in fills, either wet or dry, is relatively difficult.

**Clays**

*Hydraulic Properties*

PERMEABILITY Clays are relatively impervious, but permeability varies with the mineral composition. Sodium montmorillonite with void ratios from 2 to as high as 15 can have $k = 10^{-8}$ cm/s$^2$ and is used as an impermeabilizing agent in drilling fluid for test borings or in a slurry trench cutoff wall around an excavation. Kaolinite, with void ratios of about 1.5, can have $k$ values 100 times higher than montmorillonite [Cornell University (1951)[29]].

CAPILLARITY High, but in excavations evaporation normally exceeds flow.

FROST SUSCEPTIBILITY Many thin ice layers can form in cold climates, resulting in ground heave.

LIQUEFACTION SUSCEPTIBILITY Nonsusceptible.

PIPING Occurs in dispersive clays.

*Rupture Strength*

Consistency provides a general description of strength identified by the relationship between the natural moisture content and the liquid and plastic limits and by the unconfined compressive strength.

Parameters include the peak drained strength ($\bar{c}$ and $\bar{\phi}$), the peak undrained strength ($s_u$, $\phi = 0$), the residual drained shear strength ($\phi_r$), and the residual undrained strength ($s_r$).

Sensitivity is defined as $s_u/s_r$, and is a measure of the loss in strength upon remolding.

Failure occurs by general shear, local shear, or punching shear. Collapse upon saturation or under a particular stress level occurs in certain clays (for porous clays, *see Art. 10.5.2*) from which minerals have been leached, leaving an open, porous structure.

*Deformability*

Compression, by plastic deformation, occurs in clays during the process of consolidation. Clay soils retain their "stress history" as overconsolidated, normally consolidated, or underconsolidated *(see Art. 3.4.1)*. During consolidation there is substantial time delay caused by low permeabilities slowing the neutralization of porewater pressures. Overconsolidated, fissured clays, however, deform in a manner similar to in situ rock; i.e., displacements occur at the fissures, possibly combined with consolidation.

Expansion is a characteristic of partially saturated clays in the presence of moisture. The amount varies with mineral type, and swelling pressures and volume changes can reach substantial magnitudes. Not all clays or clay-mixtures are susceptible.

**Organic Soils**

*Hydraulic Properties*

Permeability of peat and rootmat, primarily fibrous matter, is usually very high and, for organic silts and clays, is usually low. In the latter cases, systems of root tunnels can result in $k$ values substantially higher than for inorganic clays.

*Rupture Strength*

Peat and rootmat tend to crush under applied load, but shallow cuts will stand open indefinitely because of the low unit weight, as long as surcharges are not imposed.

Organic silts and clays have very low strengths, and generally the parameters for clay soils pertain. Embankments less than 2 m in height placed over these soils often undergo failure.

*Deformability*

Organic materials are highly compressible, even under relatively low loads. Fibers and gas pockets cause laboratory testing to be unreliable in the measurement of compressibility, which is best determined by full-scale instrumented load tests. Compression in peat and rootmat tends to be extremely rapid, whereas in organic silts and clays there is a substantial time delay, although significantly less than for inorganic clays.

Rootmat undergoes substantial shrinkage upon drying. The shrinkage can reach 50% or more within a few weeks when excavations are open and dewatered by pumping.

*Corrosivity*

Because of high acidity, organic materials are usually highly corrosive to steel and concrete.

**Mixtures**

Sand and silt mixed with clay commonly assume the properties of clay soils to a degree increasing with the increasing percentage of clay included in the mixture.

The plasticity chart *(see Fig. 3.12)* relates PI and LL to the behavior of remolded clays, mixtures of clays with sand and silt, and organic materials.

**Engineering Properties Given in Chap. 3**

*Hydraulic Properties*

DRAINAGE, CAPILLARITY, AND FROST-HEAVING CHARACTERISTICS Table 3.10 provides general rating criteria.

PERMEABILITY Various relationships based on the coefficient of permeability $k$ are given on Tables 3.11, 3.12, 3.13, and 3.14, and Figs. 3.14 and 3.15.

*Rupture Strength and Other Properties*

COHESIONLESS SOILS Relative density, dry density, void ratio, and strength as related to gradation and $N$ are given on Table 3.28; relationships between $\phi$ and $D_R$ are given on Fig. 3.63.

CLAY SOILS Relationships among consistency, strength, saturated weight, and $N$ are given on Table 3.29. Correlations between $N$ and $U_c$ for cohesive soils of various plasticities are given on Fig. 3.64. Cohesive soils classed by geologic origin with properties including density, natural moisture contents, plasticity indices, and strength parameters are given on Table 3.30.

COMPACTED MATERIALS Maximum dry weight, optimum moisture content, compression values, strength characteristics, permeability coefficient, CBR value range, and subgrade modulus range for various compacted materials are given on Table 3.31.

*Deformability*

Typical ranges for the elastic properties of various materials are given on Table 3.33.

Typical pressure-void ratio curves for various clay soils are given on Fig. 3.87, and pressure-void ratio curves and relative density relationships for a "coarse to fine sand, little silt" are given on Fig. 3.91.

## 5.3.6 CLASSIFICATION AND DESCRIPTION OF SOILS

**General**

Current classification systems provide the nomenclature to describe a soil sample or formation in terms of gradation, plasticity, and organic content as determined visually or as based on laboratory index tests. They do not provide the nomenclature to describe mineral type, grain shape, stratification, or fabric.

A complete description of each soil stratum is necessary to provide the basis for anticipating engineering properties and for the selection of representative samples for laboratory testing or for representative conditions for in situ testing, as well as for the correlation of test results with data from previous studies.

**TABLE 5.32**
**AMERICAN ASSOCIATION OF STATE HIGHWAY OFFICIALS CLASSIFICATION OF SOILS AND SOIL-AGGREGATE MIXTURES AASHO DESIGNATION M-145**

| General classification* | Granular materials (35% or less passing no. 200) | | | | | | | Silt-clay materials (more than 35% passing no. 200) | | | |
|---|---|---|---|---|---|---|---|---|---|---|---|
| Group classification | A-1 | | A-3 | A-2 | | | | A-4 | A-5 | A-6 | A-7 |
| | A-1-a | A-1-b | | A-2-4 | A-2-5 | A-2-6 | A-2-7 | | | | A-7-5, A-7-6 |
| Sieve analysis, percent passing: | | | | | | | | | | | |
| No. 10 | 50 max | | | | | | | | | | |
| No. 40 | 30 max | 50 max | 51 min | | | | | | | | |
| No. 200 | 15 max | 25 max | 10 max | 35 max | 35 max | 35 max | 35 max | 36 min | 36 min | 36 min | 36 min |
| Characteristics of fraction passing no. 40: | | | | | | | | | | | |
| Liquid limit | | | | 40 max | 41 min | 40 max | 41 min | 40 max | 41 min | 40 max | 41 min |
| Plasticity index | 6 max | | NP† | 10 max | 10 max | 11 min | 11 min | 10 max | 10 max | 11 min | 11 min |
| Usual types of significant constituent materials | Stone fragments—gravel and sand | | Fine sand | Silty or clayey gravel and sand | | | | Silty soils | | Clayey soils | |
| General rating as subgrade | Excellent to good | | | | | | | Fair to poor | | | |

*Classification procedure: With required test data in mind, proceed from left to right in chart; correct group will be found by process of elimination. The first group from the left consistent with the test data is the correct classification. The A-7 group is subdivided into A-7-5 or A-7-6 depending on the plastic limit. For $w_p < 30$, the classification is A-7-6; for $w_p > 30$, A-7-5.

†NP denotes nonplastic.

### Current Classification Systems

A general summary defining grain size components is given on Table 5.27.

#### *American Association of State Highway Officials (AASHO M-145)*

Given on Table 5.32, the system is a modification of the U.S. Bureau of Public Roads system dating from 1929, which is used commonly for highway and airfield investigations.

#### *Unified Classification System (ASTM D2487)*

The unified system (Table 5.33) appears to be the most common in current use. It was developed by A. Casagrande in 1953 from the Airfield Classification system [AC or Casagrande system (1948)[30]] for the U.S. Army Corps of Engineers and has been adopted by the U.S. Bureau of Reclamation, and many other federal and state agencies.

#### *American Society for Engineering Education System*

The ASEE or Burmister system, presented in 1940, is given on Table 5.34. It is not universally used, but is applied in the northeastern United States, particularly for the field description of granular soils, for which it is very useful in defining component percentages.

#### *M.I.T. Classification System*

Presented by Gilboy in 1931, the M.I.T. system was the basic system used by engineering firms for many years, and still is used by some engineering firms in the United States and other

## TABLE 5.33
## UNIFIED SOIL CLASSIFICATION SYSTEM (ASTM D-2487) [After USAWES (1967)[30a]]

| *Major Divisions* | | | *Group Symbols* | | *Typical Names* | *Laboratory Classification Criteria* | | |
|---|---|---|---|---|---|---|---|---|
| Coarse-grained soils (More than half of material is larger than No. 200 sieve size) | Gravels (More than half of coarse fraction is larger than No. 4 sieve size) | Clean gravels (Little or no fines) | GW | | Well-graded gravels, gravel-sand mixtures, little or no fines | Determine percentages of sand and gravel from grain-size curve. Depending on percentage of fines (fraction smaller than No. 200 sieve size), coarse-grained soils are classified as follows: Less than 5 per cent: GW, GP, SW, SP; More than 12 per cent: GM, GC, SM, SC; 5 to 12 per cent: *Borderline* cases requiring dual symbols[b] | $C_u = \dfrac{D_{60}}{D_{10}}$ greater than 4; $C_c = \dfrac{(D_{30})^2}{D_{10} \times D_{60}}$ between 1 and 3 | |
| | | | GP | | Poorly graded gravels, gravel-sand mixtures, little or no fines | | Not meeting all gradation requirements for GW | |
| | | Gravels with fines (Appreciable amount of fines) | GM[a] | d | Silty gravels, gravel-sand-silt mixtures | | Atterberg limits below "A" line or P.I. less than 4 | Above "A" line with P.I. between 4 and 7 are *borderline* cases requiring use of dual symbols |
| | | | | u | | | | |
| | | | GC | | Clayey gravels, gravel-sand-clay mixtures | | Atterberg limits below "A" line with P.I. greater than 7 | |
| | Sands (More than half of coarse fraction is smaller than No. 4 sieve size) | Clean sands (Little or no fines) | SW | | Well-graded sands, gravelly sands, little or no fines | | $C_u = \dfrac{D_{60}}{D_{10}}$ greater than 6; $C_c = \dfrac{(D_{30})^2}{D_{10} \times D_{60}}$ between 1 and 3 | |
| | | | SP | | Poorly graded sands, gravelly sands, little or no fines | | Not meeting all gradation requirements for SW | |
| | | Sands with fines (Appreciable amount of fines) | SM[a] | d | Silty sands, sand-silt mixtures | | Atterberg limits above "A" line or P.I. less than 4 | Limits plotting in hatched zone with P.I. between 4 and 7 are *borderline* cases requiring use of dual symbols |
| | | | | u | | | | |
| | | | SC | | Clayey sands, sand-clay mixtures | | Atterberg limits above "A" line with P.I. greater than 7 | |
| Fine-grained soils (More than half material is smaller than No. 200 sieve) | Silts and clays (Liquid limit less than 50) | | ML | | Inorganic silts and very fine sands, rock flour, silty or clayey fine sands, or clayey silts with slight plasticity | Plasticity Chart | | |
| | | | CL | | Inorganic clays of low to medium plasticity, gravelly clays, sandy clays, silty clays, lean clays | | | |
| | | | OL | | Organic silts and organic silty clays of low plasticity | | | |
| | Silts and clays (Liquid limit greater than 50) | | MH | | Inorganic silts, micaceous or diatomaceous fine sandy or silty soils, elastic silts | | | |
| | | | CH | | Inorganic clays of high plasticity, fat clays | | | |
| | | | OH | | Organic clays of medium to high plasticity, organic silts | | | |
| | Highly organic soils | | Pt | | Peat and other highly organic soils | | | |

Plasticity Chart

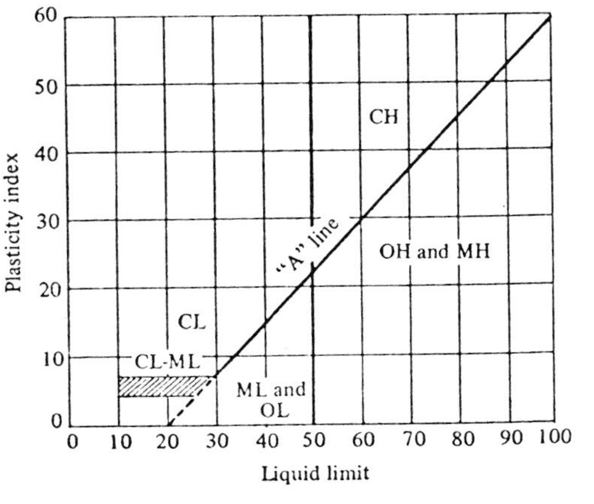

[a]Division of GM and SM groups into subdivisions of d and u are for roads and airfields only. Subdivision is based on Atterberg limits; suffix d used when L.L. is 28 or less and the P.I. is 6 or less; the suffix u used when L.L. is greater than 28.

[b]Borderline classifications, used for soils possessing characteristics of two groups, are designated by combinations of group symbols. For example: GW-GC, well-graded gravel-sand mixture with clay binder.

countries. Summarized on Table 5.27, it is similar to the British Standards Institution system.

### Field Identification and Description

*Important Elements*

Field description of soils exposed in cuts, pits, or test boring samples should include: gradation, plasticity, organic content, color, mineral constituents, grain shape, compactness or consistency, field moisture, homogeneity (layering or other variations in structure or fabric), and cementation.

*Significance*

Precise identification and description permit preliminary assessment in the field of engineering characteristics without the need for the delay involved with laboratory testing. Such an assessment is necessary in many instances to provide data of the accuracy required for thorough site evaluation.

GRANULAR SOILS Undisturbed sampling is often very difficult, and disturbed sample handling, storage, and preparation for gradation testing usually destroy all fabric. Test results, therefore, may be misleading and nonrepresentative, especially in highly stratified soils. Precise description provides the basis for estimating permeability, frost susceptibility, height of capillary rise, use of materials as compacted fill, and general supporting capabilities.

CLAY SOILS Unless a formation contains large particles such as are found in residual soils and glacial tills, precise description is less important than for granular formations because undisturbed samples are readily obtained.

*Gradation*

M.I.T. SYSTEM: Materials are described from visual examination in terms of the major component with minor components and sizes as modifiers, such as "silty fine to medium sand," "clayey silt," or "clayey fine sand."

*Unified Classification System Nomenclature*

- Soil particles: G—gravel, S—sand, M—silt, C—clay, O—organic
- Granular soil gradations: W—well graded, P—poorly graded
- Cohesive soils: L—low plasticity, H—high plasticity
- Major divisions:
  - Coarse-grained soils: more than one-half retained on no. 200 sieve
    - Gravels: more than one-half retained on no. 4 sieve ($\frac{3}{16}$ in)
    - Sands: more than one-half passing no. 4 sieve
  - Fine-grained soils: more than one-half passing no. 200 sieve
    - Low plasticity: LL $< 50$ (includes organic clays and silts)
    - High plasticity: LL $> 50$ (includes organic clays and silts)
  - Highly organic soils

Subdivisions are based on laboratory test results.

*Burmister System (ASEE)*

The system provides a definitive shorthand nomenclature. Percentage ranges in weight for various granular components are given as: AND, $>50\%$; and, 35–50%; some, 20–35%; little, 10–20%; trace, 1–10%. The percentages are estimated from experience, or by the use of the "ball moisture test" [see Burmister (1949)[32] and Table 5.36].

Silts and clays can be identified by the smallest-diameter thread that can be rolled with a saturated specimen as given on Table 5.35.

An example sample description is "Coarse to fine SAND, some fine gravel, little silt," or in shorthand nomenclature: "c-f S, s.f G, l. S̸." From field descriptions it is possible to construct reasonably accurate gradation curves, such as the example given on Fig. 5.14, which have many applications. (*see Art. 3.2.3*)

*Field Determinations*

A guide to determining the various soil components on the basis of characteristics and diagnostic procedures is given on Table 5.36, and a

**TABLE 5.34**
**ASEE SYSTEM OF DEFINITION FOR VISUAL IDENTIFICATION OF SOILS***

**DEFINITION OF SOIL COMPONENTS AND FRACTIONS**

| Granular material | Symbol | Fraction | Sieve size and definition |
|---|---|---|---|
| Boulders | Bldr | | 9 in + |
| Cobbles | Cbl | | 3 to 9 in |
| Gravel | G | Coarse (c) | 1 to 3 in |
| | | Medium (m) | ⅜ to 1 in |
| | | Fine (f) | No. 10 to ⅜ in |
| Sand | S | Coarse (c) | No. 30 to no. 10 |
| | | Medium (m) | No. 60 to no. 30 |
| | | Fine (f) | No. 200 to no. 60 |
| Silt | S | | Passing no. (0.074 mm). (Material nonplastic and exhibits little or no strength when air-dried.) |
| Organic silt | OS | | Material passing no. 200, exhibiting: (1) plastic properties within a certain range of moisture content and (2) fine granular and organic characteristics. |
| Clay | See below | | Material passing no. 200 which can be made to exhibit plasticity and clay qualities within a certain range of moisture content, and which exhibits considerable strength when air-dried. |

| Clay material | Symbol | Plasticity | Plasticity index |
|---|---|---|---|
| Clayey SILT | CyS | Slight (SL) | 1 to 5 |
| SILT and CLAY | S&C | Low (L) | 5 to 10 |
| CLAY and SILT | C&S | Medium (M) | 10 to 20 |
| Silty CLAY | SyC | High (H) | 20 to 40 |
| CLAY | C | Very high (VH) | 40+ |

**DEFINITION OF COMPONENT PROPORTIONS**

| Component | Written | Portions | Symbol | Percentage range by weight† |
|---|---|---|---|---|
| Principal | CAPITALS | | | 50 or more |
| Minor | Lower case | And | a | 35 to 50 |
| | | some | s | 20 to 35 |
| | | little | l | 10 to 20 |
| | | trace | t | 1 to 10 |

*After Burmister (1948).[31]

†Minus sign (−) signifies lower limit, plus sign (+) upper limit, no sign middle range.

guide to the identification of the fine-grained fractions is given on Table 5.37.

*Field Descriptions*

The elements of field descriptions, including the significance of color, and nomenclature for structure and fabric are given on Table 5.38. The importance of complete field descriptions can not be overstressed, since they provide the basic information for evaluations.

**TABLE 5.35**
**IDENTIFICATION OF COMPOSITE CLAY SOILS ON AN OVERALL PLASTICITY BASIS***

| Degree of overall plasticity | PI | Identification (Burmister system) | Smallest diameter of rolled threads, mm |
|---|---|---|---|
| Nonplastic | 0 | SILT | None |
| Slight | 1–5 | Clayey Silt | 6 |
| Low | 5–10 | SILT and CLAY | 3 |
| Medium | 10–20 | CLAY and SILT | 1.5 |
| High | 20–40 | Silty CLAY | 0.8 |
| Very high | >40 | CLAY | 0.4 |

*After Burmister (1951*a*).[33] Reprinted with permission from the *Annual Book of ASTM Standards*, Part 19, copyright, American Society for Testing and Materials.

**TABLE 5.36**
**FIELD DETERMINATION OF SOIL COMPONENTS***

| Component | Characteristic | Determination |
|---|---|---|
| Gravel | Dia. 5–76 mm | Measurable. |
| Sand | | |
| Coarse | Dia. 2–5 mm | Visible to eye, measurable. |
| Medium | Dia. 0.4–2.0 mm | Visible to eye. |
| Fine | Dia. 0.074–0.4 mm | Barely discernible to unaided eye. |
| Silt: coarse | Dia. 0.02–0.074 mm | Distinguishable with hand lens. |
| Sand-silt mixtures | Apparent cohesion | Measured by ball test [Burmister (1949)[32]]. Form ball in hand by compacting moist soil to diameter 1½ in (37 mm). |
| | | Medium to fine sand forms weak ball with difficulty; cannot be picked up between thumb and forefinger without crushing. |
| | | Ball can be picked up with difficulty: 20% silt |
| | | Ball readily picked up: 35 to 50% silt. |
| Silt vs. clay | Dia. <0.074 mm | See also Table 5.37. |
| Silt | Strength | Low when air-dried, crumbles easily. |
| | Dilatancy test | Mixed with water to thick paste consistency. Appears wet and shiny when shaken in palm of hand, but when palm is cupped and sample squeezed, surface immediately dulls and dries. |

*See also ASTM D2488.

**TABLE 5.37**
**IDENTIFICATION OF FINE-GRAINED SOIL FRACTIONS FROM MANUAL TESTS***

| Material | Dry strength | Dilatency reaction | Toughness of plastic thread | Plasticity description |
|---|---|---|---|---|
| Sandy silt | None–very low | Rapid | Weak, soft | None–low |
| Silt | Very low–low | Rapid | Weak, soft | None–low |
| Clayey silt | Low–medium | Rapid–slow | Medium stiff | Slight–medium |
| Sandy clay | Low–high | Slow–none | Medium stiff | Slight–medium |
| Silty clay | Medium–high | Slow–none | Medium stiff | Slight–medium |
| Clay | High–very high | None | Very stiff | High |
| Organic silt | Low–medium | Slow | Weak, soft | Slight |
| Organic clay | Medium–very high | None | Medium stiff | Medium–high |

*From ASTM D2488. Reprinted with permission of the American Society for Testing and Materials.

**TABLE 5.36**
**FIELD DETERMINATION OF SOIL COMPONENTS* (CONTINUED)**

| Component | Characteristic | Determination |
|---|---|---|
| | Dispersion test | Mixed with water in container; particles settle out in ¼ to 1 hour ($L = >10$ cm). |
| | Thread test | Rolls into thin threads in wet state but threads break when picked up by one end. |
| Clay | Strength | High when air-dried, breaks with difficulty. |
| | Plasticity | When mixed with water to form paste and squeezed in hand, specimen merely deforms and surface does not change in appearance. |
| | Dispersion test | Remains in suspension from several hours to several days in container. |
| | Thread test | Can be rolled into fine threads that remain intact. Fineness depends on clay content and mineralogy.<br>Thread diameter when saturated vs. PI and identification given on Table 5.35. |
| | Adhesion | Sticky and greasy feel when smeared between fingers. |
| Organic Soils | Strength | Relatively high when air-dried. |
| | Odor | Decayed organic matter; gases. |
| | Organic matter | Root fibers, etc. |
| | Shrinkage | Very high. |

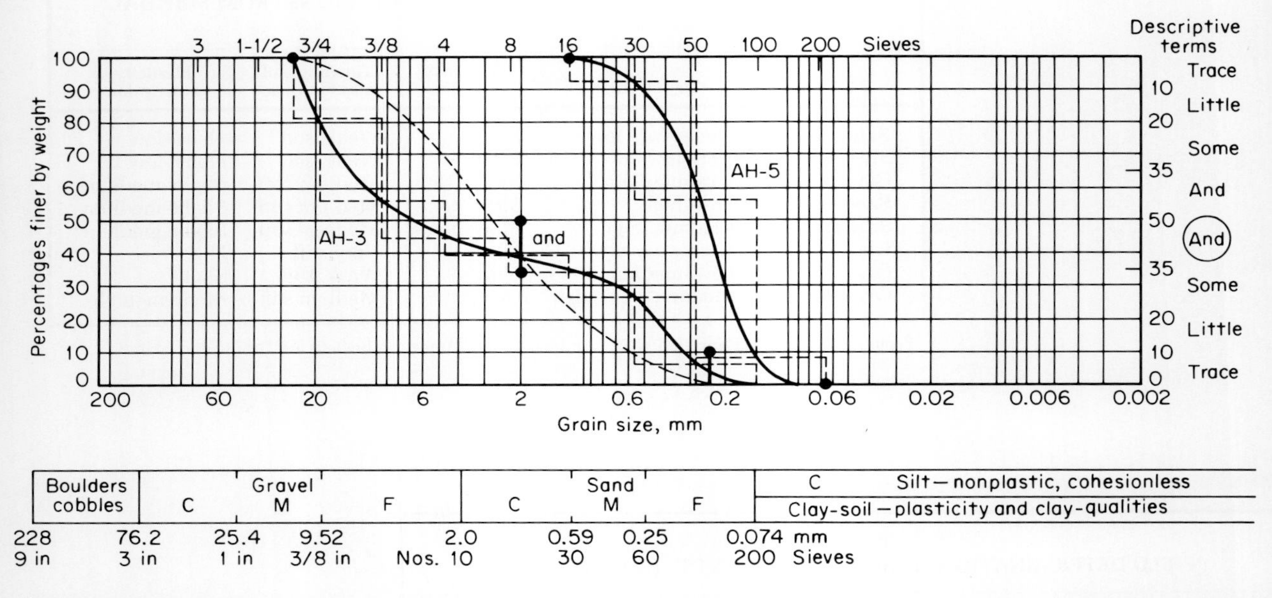


| Sample AH-3 | Long Island, N.Y. | Depth 1 to 3 ft |
|---|---|---|
| Identification: Light gray medium to fine gravel, and coarse to medium sand (Symbol form – lg m + fG, a·cmS) | | |
| Gravel—water-worn, sand—subangular | | |
| Sample AH-5 | Long Island, N.Y. | Depth 0 to 2 ft |
| Identification: Gray medium to fine sand—beach sand (Symbol form – g mfS) | | |
| Few shell fragments | | |

**FIG. 5.14 Gradation curves constructed from field identifications. Curve characteristics may be used for estimating maximum and minimum densities of k and $\phi$ if $D_R$ is known (see *Art. 3.2.3*, discussion of soil gradation).** [*From Burmister (1951).*[33] *Reprinted with permission of the American Society for Testing and Materials.*]

**TABLE 5.38**
**SOIL IDENTIFICATION: ELEMENTS OF FIELD DESCRIPTIONS**

| Elements | Importance | Description |
|---|---|---|
| Gradation | Components | See Table 5.36. |
| Grain shape | Strength | Rounded, subrounded, subangular, angular. |
| Mineral constituents | Strength | From Table 5.5. |
| Color | Provides information on soil minerals and environment | Tone: Function of soil moisture; the wetter the deeper the color.<br>Red, yellow, brown: Good drainage and aeration.<br>Deep reds: Indicate iron oxides.<br>Pale yellow, yellow browns: Hydrated iron oxides.<br>Bluish gray: Reduced bivalent iron compound, poor drainage and aerobic conditions.<br>Light grays: Due to leaching.<br>Mottled colors: Restricted permeability, or poor drainage and aeration.<br>Black, dark brown, or gray: Organic soils; or caused by dark minerals (manganese, titanium, magnetite).<br>Green: Glauconite (hydrous silicate, K and Fe).<br>White: Silica, lime, gypsum, kaolin clay. |
| Compactness in situ | Compressibility of granular soils | From SPT (*see Table 3.28*) or visual estimate. |
| Consistency in situ | Strength of clay soils | From hand test or SPT (*see Table 3.29*). |
| Field moisture | Estimate GWL depth | From sample appearance: Dry, moist, wet (saturated). |
| Homogeneity | Permeability estimates ($k_h$ vs. $k_v$) | Fabric or structure: Terms not universally defined.<br>Homogeneous: Without stratification; uniform fabric.<br>Stratified: Partings—very fine, barely visible, form weakness planes<br>Lenses—from very fine to 5 mm<br>Seams—5 mm to 2 cm<br>Layers—>2 cm<br>Varves—interbedded seams<br>Pockets: Foreign irregularly shaped mass in matrix.<br>Heterogeneous: very irregular, without definite form. |
| Cementation | Strength | Reaction with dilute HCl: None, weak, strong. |

NOTE: Example: "Medium compact, tan, silty coarse to fine sand (subrounded, quartz, with some shell fragments) with lenses and seams of dark gray silt; moist."

## REFERENCES


1. Simpson, B. (1974) *Minerals & Rocks,* Octopus Books Ltd., London.
2. Hamlin, W. K. and Howard, J. L. (1971) *Physical Geology Laboratory Manual,* Burgess Pub. Co., Minneapolis.
3. Pirsson, L. V. and Knopf, A. (1955) *Rocks and Rock Minerals,* John Wiley & Sons, New York.
4. Turner, F. J. and Verhoogan, J. (1960) *Igneous and Metamorphic Petrology,* McGraw-Hill Book Co., New York.
5. Hartmann, B. E. (1966) *Rock Mechanics Instrumentation for Tunnel Construction,* Terrametrics Inc., Golden, Colo.
6. Jennings, J. E. (1972) "An Approach to the Stability of Rock Slopes Based on the Theory of Limiting Equilibrium with a Material Exhibiting Anisotropic Shear Strength," *Stability of Rock Slopes, Proc. ASCE, 13th Symp. on Rock Mechs.,* Univ. of Illinois, Urbana (1971), pp. 269–302.
7. Jaeger, C. (1972) *Rock Mechanics and Engineering,* Cambridge Univ. Press, England.
8. Deere, D. U. (1969) "Geologic Considerations," *Rock Mechanics in Engineering Practice,* Stagg and Zienkiewicz, eds., John Wiley & Sons, New York, Chap. 1.
9. Hoek, E. and Bray, J. W. (1974) *Rock Slope Engineering,* Inst. of Mining and Metallurgy, London.
10. ISRM Working Party (1975) "Suggested Methods for the Description of Rock Masses, Joints and Discontinuities," *Intl. Soc. of Rock Mechs.,* 2d Draft, August, Lisbon.
11. Core Logging Comm. (1978) "A Guide to Core Logging for Rock Engineering," *Bull. Assoc. Engrg. Geol.,* Summer, Vol. XV, No. 3, pp. 295–328.
12. Building Code, City of New York (1968) "Building Code—Local Law No. 76 of the City of New York," effective December 6, 1968, amended to August 22, 1969, *The City Record,* New York, N.Y.
13. Deere, D. U. (1963) "Technical Description of Rock Cores for Engineering Purposes," *Rock Mechs. Engrg. Geol. 1,* pp. 18–22.
14. Bieniawski, Z. T. (1974) "Geomechanics Classification of Rock Masses and Its Application to Tunneling," *Proc. 3d Intl. Cong. Rock Mechs.,* Denver, Colo., Vol. IIA, pp. 27–32.
15. Coon, J. H. and Merritt, A. H. (1970) "Predicting Insitu Modulus of Deformation Using Rock Quality Indexes," *Determination of the In-Situ Modulus of Deformation of Rock,* ASTM STP 477, Amer. Soc. for Test. and Mat., Philadelphia, Pa.
16. Bieniawski, Z. T. (1976) "Classification System is Used to Predict Rock Mass Behavior," *World Construction,* May.
17. Barton, N., Lien, R. and Lunde, J. (1974) "Engineering Classification of Rock Masses for Tunnel Support," *Rock Mechanics,* Vol. 6, No. 4, *J. Intl. Soc. of Rock Mechs.,* December.
18. Barton, N., Lien, R. and Lunde, J. (1977) "Estimation of Support Requirements for Underground Excavation," *Design Methods in Rock Mechanics, Proc. ASCE 16th Symp. on Rock Mechs.,* Univ. of Minnesota, September 1975, pp. 163–178.
19. Gillott, J. E. (1968) *Clay in Engineering Geology,* Elsevier Pub. Co., Amsterdam.
20. Skempton, A. W. (1953) "The Colloidal Activity of Clays," *Proc. 3d. Intl. Conf. Soil Mechs. and Found. Engrg.,* Switzerland, Vol. I, pp. 57–61.
21. Gibbs, H. J., Hilf, J. W., Holtz, W. G. and Walker, F. C. (1960) "Shear Strength of Cohesive Soils," *Research Conf. Shear Strength of Cohesive Soils, Proc. ASCE,* Boulder, Colo., June, pp. 33–162.
22. Morin, W. J. and Tudor, P. C. (1976) *Laterite and Lateritic Soils and Other Problem Soils of the Tropics,* AID/csd 3682, U.S. Agency for International Development, Washington, D.C.
23. Grim, R. E. (1962) *Clay Mineralogy,* 2d ed., McGraw-Hill Book Co., New York.
24. Leonards, G. A. (1962) *Foundation Engineering,* McGraw-Hill Book Co., New York, Chap. 2.
25. Lambe, T. W. and Whitman, R. V. (1969) *Soil Mechanics,* John Wiley & Sons, New York.
26. Osipov, V. I. and Sokolov, V. N. (1978) "Structure Formation in Clay Sediments," *Bull. Intl. Assoc. Engrg. Geol.,* No. 18, December, pp. 83–90.
27. Wahlstrom, E. C. (1973) *Tunneling in Rock,* Elsevier Scientific Pub. Co., New York.
28. Mohr, E. C. J., van Baren, F. A. and van Schyenborgh, J. (1972) *Tropical Soils,* 3d ed., Mouon-Ichtiar-Van Hoeve, The Hague.
29. Cornell Univ. (1951) "Final Report on Soil Solidification Research," Cornell Univ., Ithaca, New York.
30. Casagrande, A. (1948) "Classification and Identification of Soils," *Trans. ASCE,* Vol. 113, p. 901–992.

30*a*. USAWES (1967) *The Unified Soil Classification System,* Tech. Memo No. 3-357, U. S.Army Engineer Waterways Experiment Station, Corps of Engineers, Vicksburg, Miss.

31. Burmister, D. M. (1948) "The Importance and Practical Use of Relative Density in Soil Mechanics," ASTM Vol. 48, Amer. Soc. for Test. and Mat., Philadelphia, Pa.

32. Burmister, D. M. (1949) "Principles and Techniques of Soil Identification," *Proc. 29th Annual Mtg.*, Highway Research Board, Washington, D.C.

33. Burmister, D. M. (1951) "Identification and Classification of Soils—An Appraisal and Statement of Principles," ASTM STP 113, Amer. Soc. for Test. and Mat., Philadelphia, Pa., pp. 3–24, 85–91.

## BIBLIOGRAPHY

Dennen, W. H. (1960) *Principles of Mineralogy*, The Ronald Press, New York.

Kraus, E. H., Hunt, W. F. and Ramsdell, L. S. (1936) *Mineralogy—An Introduction to the Study of Minerals and Crystals*, McGraw-Hill Book Co., New York.

Rice, C. M. (1954) *Dictionary of Geologic Terms*, Edwards Bros., Ann Arbor, Mich.

Skempton, A. W. and Northey, R. D. (1952) "The Sensitivity of Clays," *Geotechnique*, 3, No. 1, pp. 30–53, March.

Travis, R. B. (1955) "Classification of Rock," *Quarterly of the Colorado School of Mines*, Boulder.

Vanders, I. and Keer, P. F. (1967) *Mineral Recognition*, John Wiley & Sons, New York.

Zumberge, J. H. and Nelson, C. A. (1972) *Elements of Geology*, John Wiley & Sons, New York.

# CHAPTER SIX

# ROCK-MASS CHARACTERISTICS

## 6.1 INTRODUCTION

### 6.1.1 CHARACTERISTICS SUMMARIZED

#### General

Rock-mass characteristics can be discussed under four stages of development:

1. The original mode of formation with a characteristic structure
2. Deformation with characteristic discontinuities
3. Development of residual stresses that may be several times greater than overburden stresses
4. Alteration by weathering processes in varying degrees from slightly modified to totally decomposed

Rock-mass response to human-induced stress changes are controlled normally by the degree of alteration and the discontinuities. The latter, considered as mass defects, range from weakness planes (faults, joints, foliations, cleavage, and slickensides) to cavities and caverns. In the rock mechanics literature, all fractures often are referred to as joints.

Terrain analysis is an important method for identifying rock-mass characteristics.

Structural features are mapped and presented on diagrams for analysis.

#### Original Mode of Formation

*Igneous Rocks*

Intrusive masses form large bodies (batholiths and stocks), smaller irregular-shaped bodies (lapoliths and laccoliths), and sheetlike bodies (dikes and sills). Extrusive bodies form flow sheets.

*Sedimentary Rocks*

Deposited generally as horizontal beds, sedimentary rocks can be deformed in gentle modes by consolidation warping or local causes such as currents.

Cavities form in the purer forms of soluble rocks, often presenting unstable surface conditions.

*Metamorphic Rocks*

Their forms relate to the type of metamorphism (*see Art. 5.2.5*). Except for contact metamorphism, the result is a change in the original form of the enveloping rocks.

#### Deformation and Fracturing

*Tectonic Forces (See Appendix A)*

Natural stress changes result from tectonic forces, causing the earth's crust to undergo elastic and plastic deformation and rupture. Plastic deformation is caused by steady long-term stresses resulting in folding translation of beds and some forms of cleavage, and occurs when stresses are in the range of the elastic limit and temperatures are high. Creep, slow continuous strain under constant stress, deforms rock and can occur even when loads are substantially below the elastic limit, indicating that time is an element of deformation. Short-term stresses within the elastic limit at normal temperatures

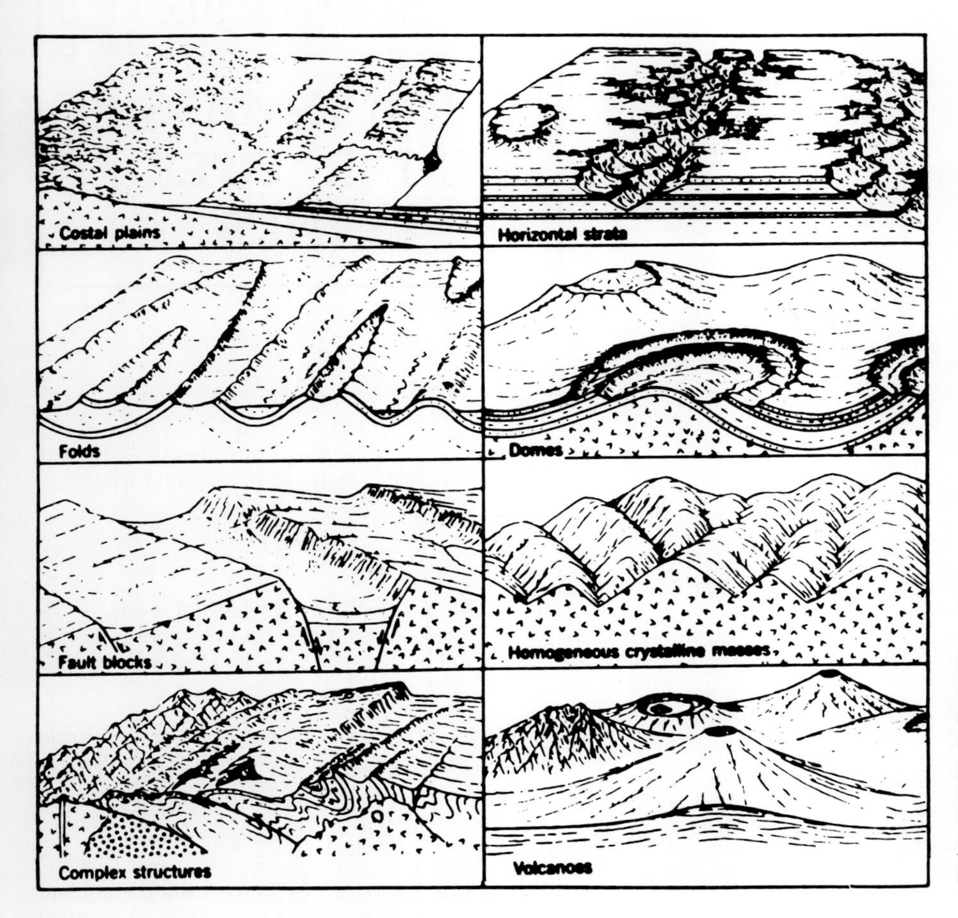


**FIG. 6.1 Diagrams showing effect of varying geologic structure and lithology on morphology of landscapes.** [*After A. N. Strahler, Physical Geology, John Wiley & Sons, Inc., from Thornbury (1969).*[1] *Reprinted with permission of John Wiley & Sons, Inc.*]

leave no permanent effect on masses of competent rock.

Rupture and fracture occur at conditions of lower temperatures and more rapid strain, resulting in faults, joints, and some forms of cleavage.

*Other Forces and Causes*

Tensile forces occur during cooling and contraction and cause jointing of igneous rocks. They are created by uplift (rebound) following erosion and cause jointing in sedimentary rocks. Slope movements result in failure surfaces evidenced by slickensides.

*Significance*

Faults are associated with earthquake activity and surface displacements, and produce slickensides; small faults are associated with folding and produce foliation shear and mylonite shear zones.

Faults, joints, bedding planes, etc. divide the mass into blocks. All discontinuities represent weakness planes in the mass, controlling deformation and strength as well as providing openings for the movement of water.

## Residual Stresses

Residual stresses are locked into the mass during folding, metamorphism, and slow cooling at great depths. They can vary from a few kilograms per square centimeter to many times overburden stresses, and can result in large deflections in excavations and "rock bursts," violent explosive ejections of rock fragments and blocks occurring in deep mines and tunnels.

### Alteration

New minerals are formed underground by chemical reactions, especially when heated, and by heat and pressure associated with faulting and metamorphism. New minerals are formed from the surface by chemical weathering processes.

The result of chemical weathering is decomposition of minerals, the final product being residual soil. Weathering products are primarily functions of the parent rock type and the climate.

Mechanical weathering also results in the deterioration of the rock mass.

Altered masses generally have higher permeability and deformability, and lower strengths than the mass as formed originally or deformed tectonically.

## 6.1.2 TERRAIN ANALYSIS

### Significance

Regional and local rock types and structural features are identified through terrain analysis *(see also Art. 2.2.3).*

### Interpretative Factors

Rock masses have characteristic features, as deposited, which result in characteristic landforms which become modified by differential weathering. Deformation and rupture change the rock mass, which subsequently develops new landforms. Some relationships between morphological expression, structure, and lithology are given on Fig. 6.1.

Interpretation of these characteristic geomorphological expressions provides the basis for identifying rock type and major structural features. The more significant factors providing the interpretative basis are landform (surface shape and configuration), drainage patterns, and lineations.

*Lineations* are weakness planes intersecting the ground surface and providing strong rectilinear features of significant extent when emphasized by differential weathering. The causes include faults, joints, foliations, and bedding planes of tilted or folded structures.

### Stream Forms and Patterns

*Significance*

Rainfall runoff causes erosion which attacks rock masses most intensely along weakness planes, resulting in stream forms and patterns which are strongly related to rock conditions.

*Stream Forms*

Stream channels are classified with respect to shape as either straight, crooked, braided, or meandering (also classed as young, mature, old age, or rejuvenated). Although influenced by rock-mass conditions, stream forms are described under Art. 7.4.1 because of their characteristic channel and valley soil deposits.

*Drainage Patterns*

Formed by erosion channels on the land surface, the drainage pattern is controlled by, or related to, the geologic conditions. Patterns are readily apparent on topographic maps and remote-sensing imagery. When traced onto an overlay, a clear picture is provided for interpretation. Pattern classes (dendritic, rectangular, etc.) are described on Table 6.1, and typical patterns for various geologic conditions are illustrated on Table 6.2. (For soil formations *see Table 7.3.*)

*Texture* refers to pattern intensity and is given by Way (1978)[2] as fine, medium, and coarse, as apparent on vertical aerial photos at a scale of 1:20,000 (1667 ft/in) and defined in terms of the distance between streams measured at this scale (*see Table 6.2*). Texture is controlled primarily by the amount of runoff, which is related to the imperviousness of the materials as follows:

- Fine-textured patterns indicate high levels of runoff and intense erosion such as occurs with impervious soils or rocks at the surface.
- Coarse-textured patterns indicate pervious materials and low runoff levels.
- No stream channels or drainage patterns develop where subsurface materials are highly pervious and free-draining. Vegetation is thin, even in wet climates. These conditions occur in clean sands and in coral and limestone formations on a regional basis. On a local level they can occur in porous clays and in saprolite resulting from the decomposition of schist,

**TABLE 6.1**
**CLASSES OF DRAINAGE PATTERNS IN ROCK FORMATIONS**

| Class | Associated formations | Characteristics |
|---|---|---|
| Dendritic | Sedimentary rocks (except limestone) and uniform, homogeneous soil formations. Other patterns are modifications of the basic dendritic pattern and are characteristic of other rock conditions. | Tributaries join the gently curving mainstream at acute angles and rock structure control is weak. The more impervious the material, the finer the texture. Intrusive granite domes cause curvilinear alignments. |
| Rectangular | Controlled by rock structure; primarily joints and foliations. | A strongly modified dendritic pattern with tributaries connecting in a regular pattern at right angles. The stronger the pattern, the thinner the soil cover. |
| Angulate | Controlled by rock structure that includes major faults and joints. Also gneiss and impure limestone. | A strongly modified dendritic pattern that is highly irregular. |
| Trellis | Tilted, interbedded sedimentary rocks. | A regular, parallel pattern. The main tributaries follow the strike of the beds and the branches follow the dip. The mainstream cuts across ridges to form gaps during rejuvenation. |
| Barbed | Regional uplift and warping changes flow direction. | Reverse dendritic pattern. |
| Parallel | Gentle, uniform slopes of basalt flows; also mature coastal plains. | Modified dendritic with parallel branches entering the mainstream. |
| Radial | Domes, volcanoes | Patterns radiate out from central high ground to connect with mainstream. |
| Annular | Domes with some joint control. | Radial pattern with cross tributaries. |
| Centripedal | Basins, or ends of anticlines or synclines | Radial drainage toward a central connecting stream. |
| Deranged | Young landforms (floodplains and thick till plains) and karst. | Lack of pattern development. Area contains lakes, ponds, and marshes. In karst, channels end on surface where runoff enters limestone through cavities and joints. |

NOTE: For forms in soil formations see Table 7.3. Include pinnate, meandering, radial braided, parallel braided, and thermokarst as well as dendritic, parallel, and deranged.

although on a large-area basis they fall into the coarse-textured category.

Pattern shape is controlled by rock-mass form (horizontally bedded, dipping beds, folds, domes, flows, batholiths, etc.), rock structure (faults, joints, foliations), and soil type where the soil is adequately thick. In residual soils, even where thick, the pattern shape is usually controlled by the rock conditions.

## 6.1.3 MAPPING AND PRESENTING STRUCTURAL FEATURES

### Purpose

The elements of the geologic structure are defined in terms of spatial orientation and presented on maps and geometric diagrams.

### Map Presentation

*Dip and Strike*

The position of a planar surface is defined by its dip and strike.

*Strike* is the bearing or compass direction of a horizontal plane through the plane of the geologic bed, as shown on Fig. 6.2*a*. It is determined in the field with a leveling compass (such as the Brunton compass, Fig. 6.2*b*), by laying the level on an outcrop of the planar surface and noting its compass direction from either magnetic or true north. The compass direction can also be plotted directly from remote-sensing imagery if the north direction is known.

*True dip* is the angle of inclination from the horizontal measured in a vertical plane at right angles to the strike, as shown on Fig. 6.2*a*.

*Apparent dip* is the angle measured between the geologic plane and the horizontal plane in a vertical direction *not* at right angles to the strike, Fig. 6.2*b* (strike orientation is not defined).

*Map Symbols*

Various map symbols are used to identify structural features such as folded strata, joints, faults, igneous rocks, cleavage, schistosity, etc. as given on Fig. 6.3.

### Geometric Presentation

*General*

The attitude of lineations and planes may be plotted statistically for a given location in a particular area to illustrate the concentrations of their directions. Planar orientation is used for strike, and stereographic orientation is used for strike and dip.

*Planar Orientation*

*Star diagram* or *joint rose* is constructed to illustrate the planar orientation of concentrations of lineaments.

The strike of the lineations on a large scale is determined from interpretation of remote-sensing imagery (such as the SLAR image, Fig. 6.4), or on a small scale from surface reconnaissance and patterns are drawn of the major "sets" as shown on Fig. 6.5.

All joints with directions occurring in a given sector of a compass circle (usually 5 to 10°) are counted and a radial line representing their strike is drawn in the median direction on a polar equal-area stereonet (Fig. 6.6). The length of the line represents the number of joints and is drawn to a scale given by the concentric circles. The end of the line is identified by a dot. The star diagram or joint rose is constructed by closing the ends of the lines represented by the dots as shown on Fig. 6.7.

*Spherical Orientations*

Planes can be represented in normal, perspective, isometric, or orthographic projections. Geologic structures are mapped on the spherical surface of the earth; therefore, it is frequently desirable to study structures in their true spherical relationship, which can be represented on stereonets.

Stereonets are used to determine:

- Strike and dip from apparent dips measured in the field
- The attitude of two intersecting plane surfaces
- The attitude of planes from oriented cores from borings
- The solution of other problems involving planar orientations below the surface

## TABLE 6.2
## GEOLOGIC CONDITIONS AND TYPICAL DRAINAGE PATTERNS*

| Geologic condition | Predominant drainage pattern | Geologic condition | Predominant drainage pattern |
|---|---|---|---|
| **I. UNIFORM ROCK COMPOSITION WITH RESIDUAL SOIL BUT WEAK OR NO ROCK STRUCTURE CONTROL** | | | |
| Sandstone (coarse texture†) | Coarse dendritic | Interbedded thick beds of sandstones and shales | Coarse to medium dendritic |
| Interbedded thin beds of sandstone and shale (medium texture‡) | Medium dendritic | Shale | Medium to fine dendritic |
| Clay shale, chalk, volcanic tuff (fine texture§) | Fine dendritic | Intrusive igneous (granite) | Domes<br>Medium to fine dendritic with curvilinear alignments |
| **II. CONTROLLED BY ROCK STRUCTURE** | | | |
| Joint and foliation systems in regular pattern | 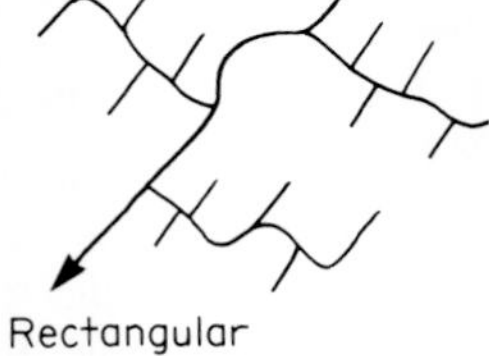<br>Rectangular | Joint and fault systems in irregular pattern | 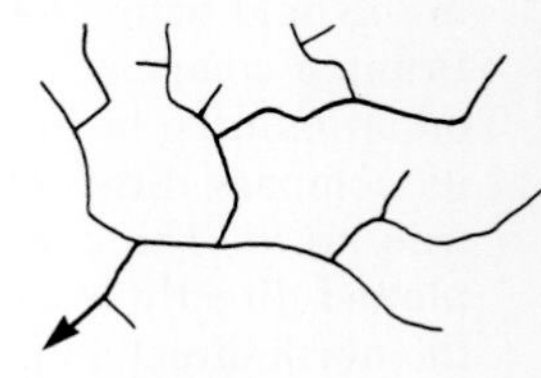<br>Angulate |
| Gneiss | 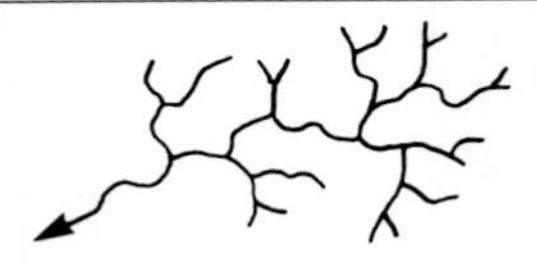<br>Fine to medium angular dendritic | Slate | 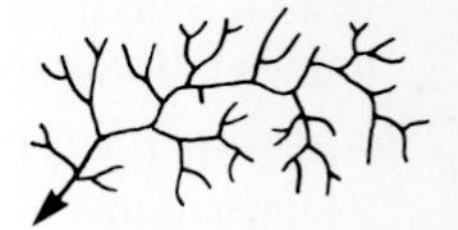<br>Fine rectangular dendritic |

*Drainage patterns given are for humid climate conditions (categories I and II). Arid climates generally produce a pattern one level coarser.

†Coarse: First-order streams over 2 in (5 cm) apart (on map scale of 1:20,000) and carry relatively little runoff.

‡Medium: First-order stream ¼ to 2 in (5 mm to 5 cm) apart.

§Fine: Spacing between tributaries and first-order stream less than ¼ in (5 mm).

**TABLE 6.2**
**GEOLOGIC CONDITIONS AND TYPICAL DRAINAGE PATTERNS* (*Continued*)**

| Geologic condition | Predominant drainage pattern | Geologic condition | Predominant drainage pattern |
|---|---|---|---|
| | II. CONTROLLED BY ROCK STRUCTURE | | |
| Schist | 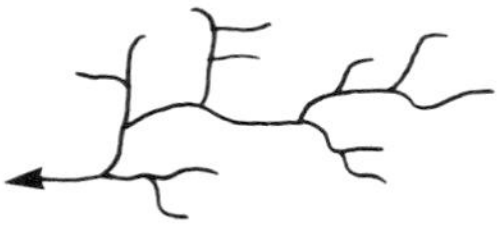 Medium to fine rectangular dendritic | Impure limestone | 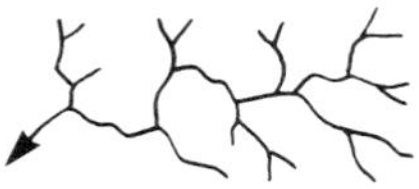 Medium angular dendritic |
| | III. CONTROLLED BY REGIONAL ROCK FORM | | |
| Tilted interbedded sedimentary | 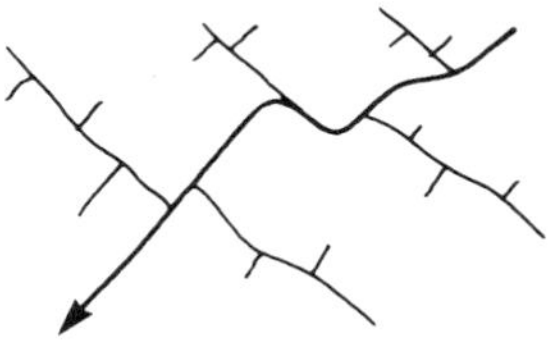 Trellis | Basalt flows over large areas |  Parallel, coarse |
| Regional warping and uplift | 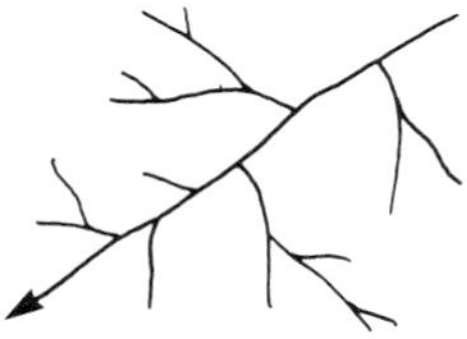 Barbed | Domes, volcanoes | 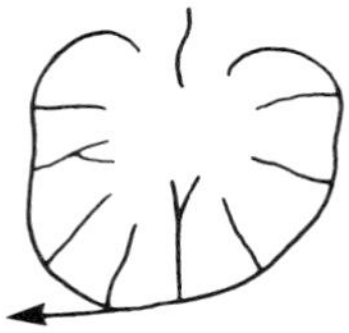 Radial |
| Domes with some joint control |  Annuler | Soluble limestone; glaciated crystalline shields |  Deranged with intermittent drainage |
| Basin | 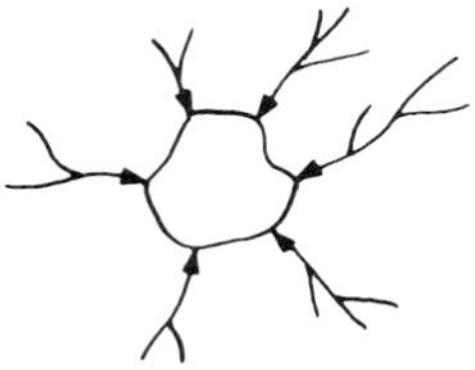 Centripedal | Coral | No surface drainage |

**FIG. 6.2 Determination of strike and dip in the field: (*a*) dip and strike defined, with strike here being N45°E and (*b*) measuring apparent dip of bedding plane with Brunton compass. Finger points to cleavage. Care is required to establish if planar surface being measured is bedding, a joint, or cleavage. (The rock being measured is a shale.)**

If the structural plane in orthographic projection on Fig. 6.8 is drawn through the center of a sphere (Fig. 6.9), it intersects the surface as a great circle which constitutes *spherical projection* of the plane. Spherical projections on a plane surface can be represented by *stereographic projections* which are geometric projections of spherical coordinates onto a horizontal plane. If various points on the great circle are projected to the zenithal point of the sphere (Fig. 6.10), a stereographic projection of the plane (stereogram) appears on the lower half-hemisphere. The projection of a series of planes striking north-south and dipping east-west results in a net of meridional (great circle) curves as shown on Fig. 6.10.

The hemisphere is divided by lines representing meridians and by a series of small circles of increasing diameter to form a *stereonet*, or *stereographic net*, of which there are two types:

1. Non-true-area stereonet or Wulff net (Fig. 6.11)
2. Lamberts equal-area plot or Schmidt net (Fig. 6.12), which is used for statistical analysis of a large number of smaller planar features such as joint sets

(1) Symbols used on maps of igneous rocks—*a*, flow layers, strike as plotted (N18°E), dip 25° eastward; *b*, flow layers, strike N45°W, dip 60° NE (dips below 30° shown as open triangles; over 30°, as solid triangles); *c*, flow lines, trend plotted, pitch 30° nearly north; *d*, horizontal flow lines, tend as plotted; *e*, vertical flow lines; *f*, combination of flow layers and flow lines. *(After U.S. Geological Survey with some additions by Balk.)*

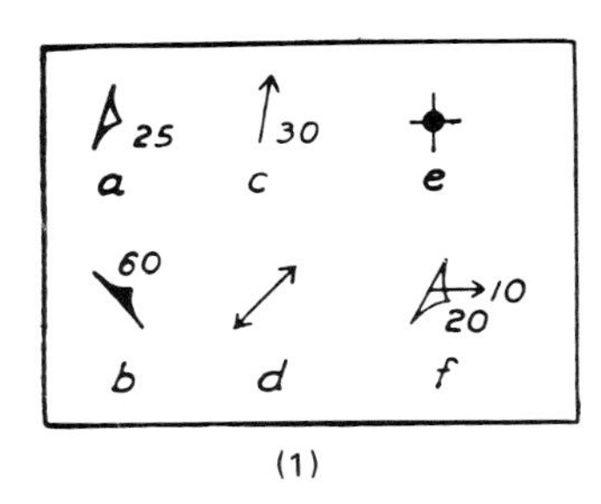


(1)

(2) Symbols for strata—*a*, strike plotted (N10°E), dip 30°; *b*, strike plotted (N10°E), dip overturned 80°. (These symbols may be used with or without the arrowhead. In *b*, the strata have been turned through an angle of 100°, i.e., up to 90° and then 10° beyond the vertical.) *c*, strike east-west, dip vertical; *d*, beds horizontal. *(After U.S. Geological Survey.)*

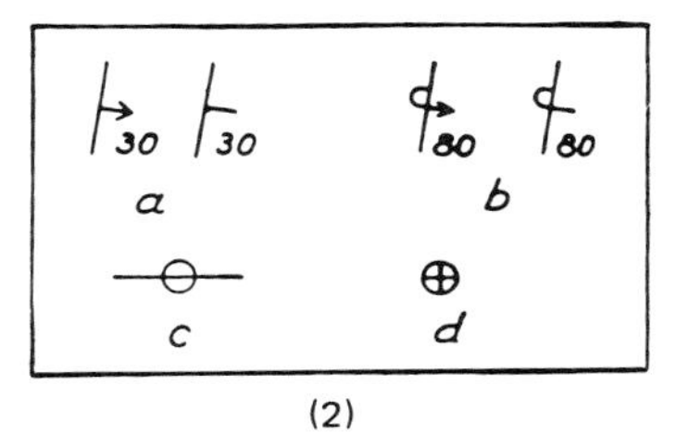


(2)

(3) Symbols for rock type combined with symbol for dip and strike—*a*, shale or slate; *b*, limestone, *c*, sandstone; *d*, conglomerate.

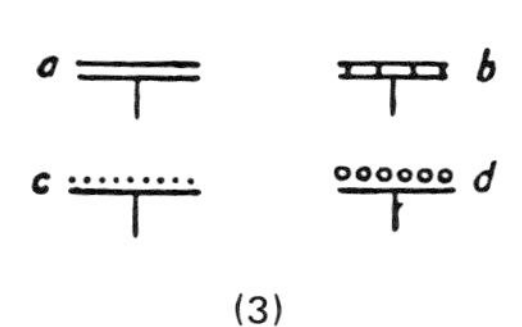


(3)

(4) Symbols used for a folded strata—*a*, general strike and dip of minutely folded beds; *b*, direction of pitch of minor anticline; *c*, same for minor syncline; *d*, axis of anticline; *e*, axis of syncline; *f*, pitch of axis of major anticline; *g*, same for major syncline; *h*, axis of overturned or recumbent anticline, showing direction of inclination of axial surface; *i*, same for overturned or recumbent syncline. *(After U.S. Geological Survey.)*

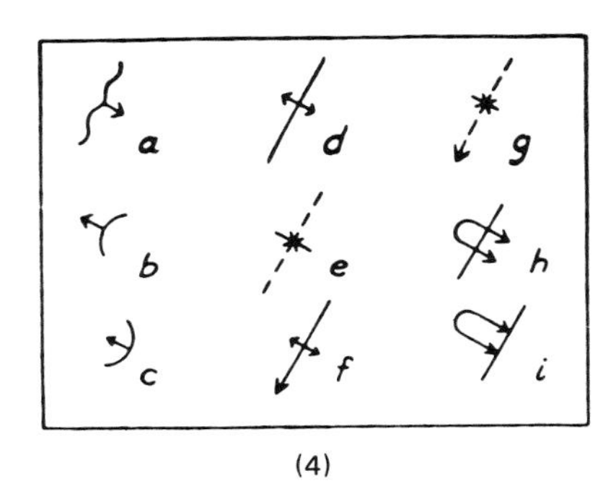


(4)

(5) Symbols used for joints on maps—*a*, strike and dip of joint; *b*, strike of vertical joint; *c*, horizontal joint; *d*, direction of linear elements (striations, grooves, or slickensides) on joint surfaces and amount of pitch of these linear elements on a vertical joint surface. Linear elements here shown in horizontal projection. *(After U.S. Geological Survey.)*

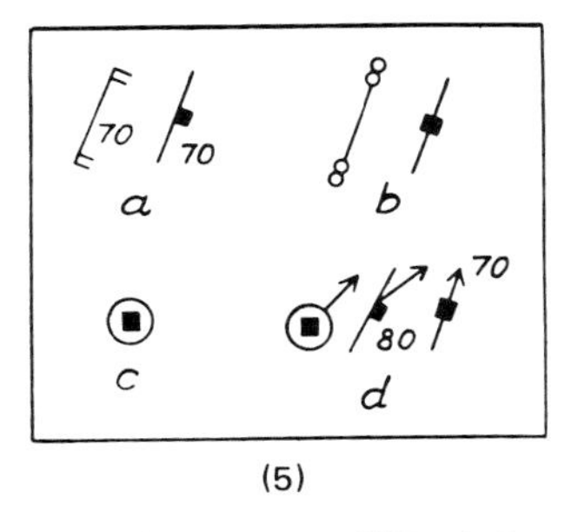


(5)

**FIG. 6.3 Symbols used on geologic maps.** *[From Lahee (1941).[3] Reprinted with permission of McGraw-Hill Book Company.]*

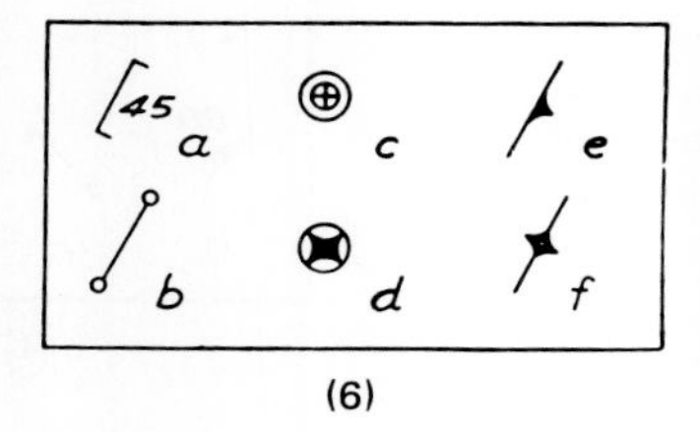


(6)

**(6) Symbols used for cleavage and schistosity on maps—*a*, strike (long line) and dip (45° in direction of short lines) of cleavage of slate; *b*, strike of vertical cleavage of slate; *c*, horizontal cleavage of slate; *d*, horizontal schistosity or foliation; *e*, strike and dip of schistosity or foliation; *f*, strike of vertical schistosity or foliation.** *(After U.S. Geological Survey.)*

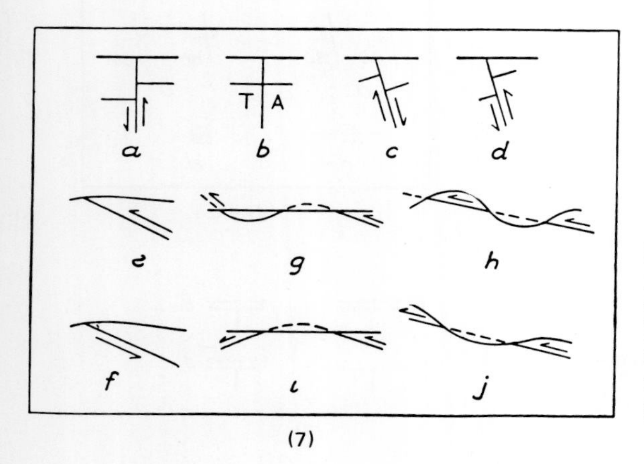


(7)

**(7) Symbols for faults in sections—*a* to *d*, high-angle faults; *e* to *j*, low-angle faults. *a*, vertical fault, with principal component of movement vertical; *b*, vertical fault with horizontal movement, block *A* moving *away* from the observer and block *T* moving toward the observer; *c*, normal fault; *d*, reverse fault; *e*, overthrust; *f*, underthrust; *g* and *k*, klippen or fault outliers; *i* and *j*, fenster, windows or fault inliers.** *(After U.S. Geological Survey.)*

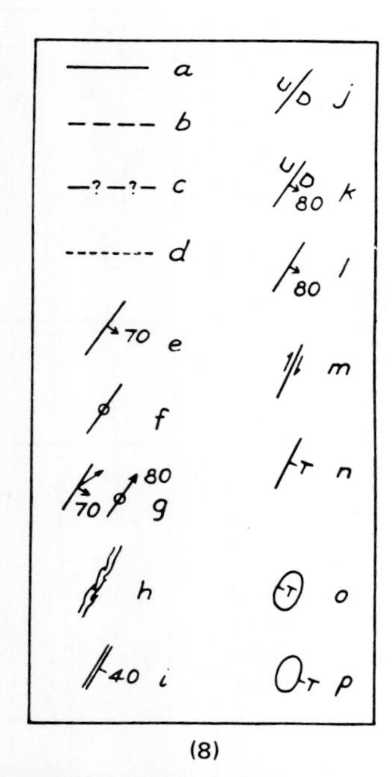


(8)

**(8) Symbols for faults on maps—*a*, known fault; *b*, known fault, not accurately located; *c*, hypothetical or doubtful fault; *d*, concealed fault (known or hypothetical) covered by later deposits; *e*, dip and strike of fault surface; *f*, strike of vertical fault; *g*, direction of linear elements (striation, grooves, slickensides, shown by longer arrow) caused by fault movement, and amount of pitch of striations on vertical surface; *h*, shear zone; *i*, strike and dip of shear zone; *j*, high-angle fault, normal or reverse, with upthrow *U* and downthrow *D* shown; *k*, normal fault; *l*, reverse fault; *m*, relative direction of horizontal movement in shear or tear fault, or flaw; *n*, overthrust low-angle fault, *T* being the overthrust (overhanging) side; *o*, klippe, or outlier remnant of low-angle fault plate (*T*, overthrust side); *p*, window, fenster, or hole in overthrust plate (*T*, overthrust side).** *(After U.S. Geological Survey.)*

**FIG. 6.3 Symbols used on geologic maps *(continued)*.** [*From Lahee (1941).*[3] *Reprinted with permission of McGraw-Hill Book Company.*]

**FIG. 6.4 Joints and faults evident on side-looking airborne image, Esmeralda quadrangle, Venezuela.** *(Courtesy of International Aero Service Corp., imagery obtained by Goodyear Aerospace Corp.)*

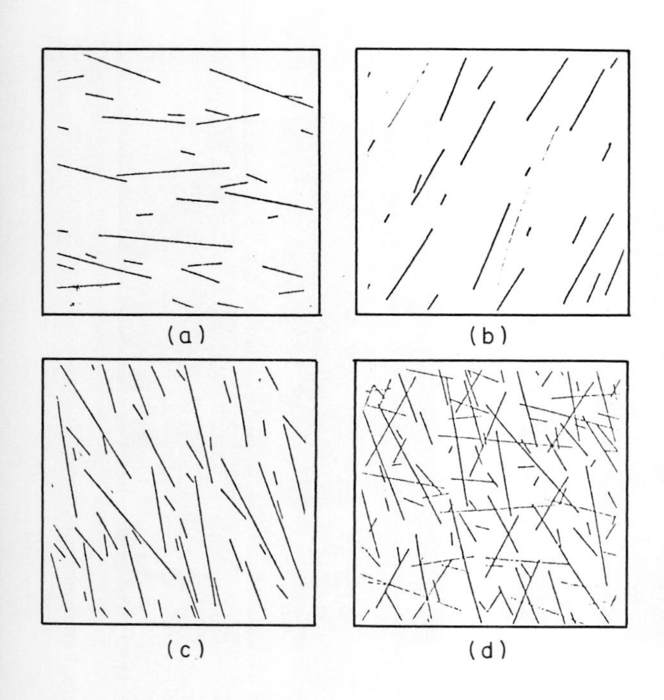


**FIG. 6.5 Singular joint sets and their combination: (*a*) joint set *A* alone; (*b*) joint set *B* alone; (*c*) joint set *C* alone; (*d*) joint sets *A*, *B*, and *C* superimposed.**

**FIG. 6.6** ***(Opposite)*** **Polar equal-area stereonet in 2° intervals.** [*Computer-drawn by Dr. C. M. St. John of the Royal School of Mines, Imperial College, London. From Hoek and Bray (1977).*[4]]

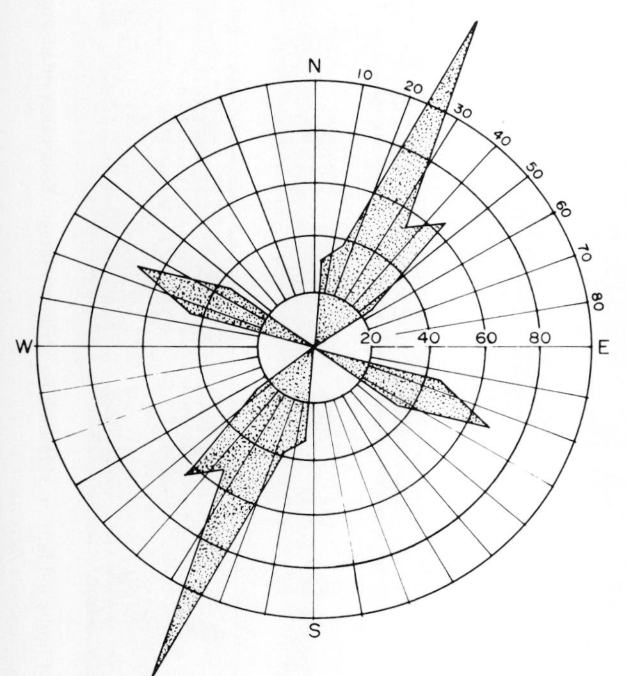


**FIG. 6.7 Joint rose or star diagram showing the number of joints counted in each 10° sector of a polar equal-area stereonet. The plot shows two sets of joints with average strike about N25°E and N65°W. Plot is made only when strike is determined.** [*From Wahlstrom (1973).*[5] *Reprinted with permission of Elsevier Scientific Publishing Company.*]

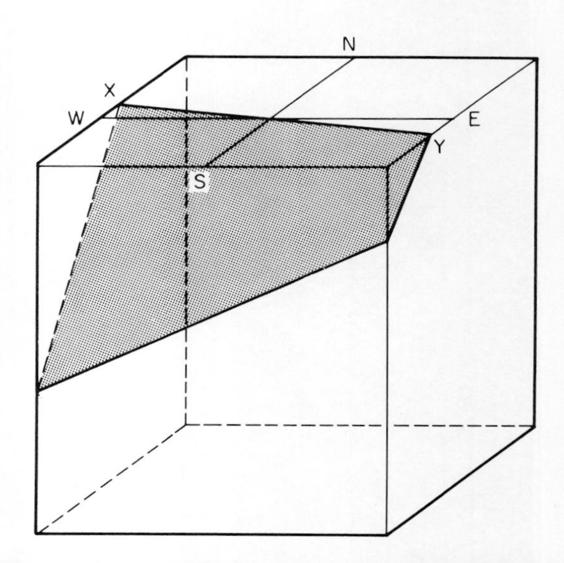


**FIG. 6.8 A structural plane as visualized in orthographic projection.** [*From Badgley (1959).*[6]]

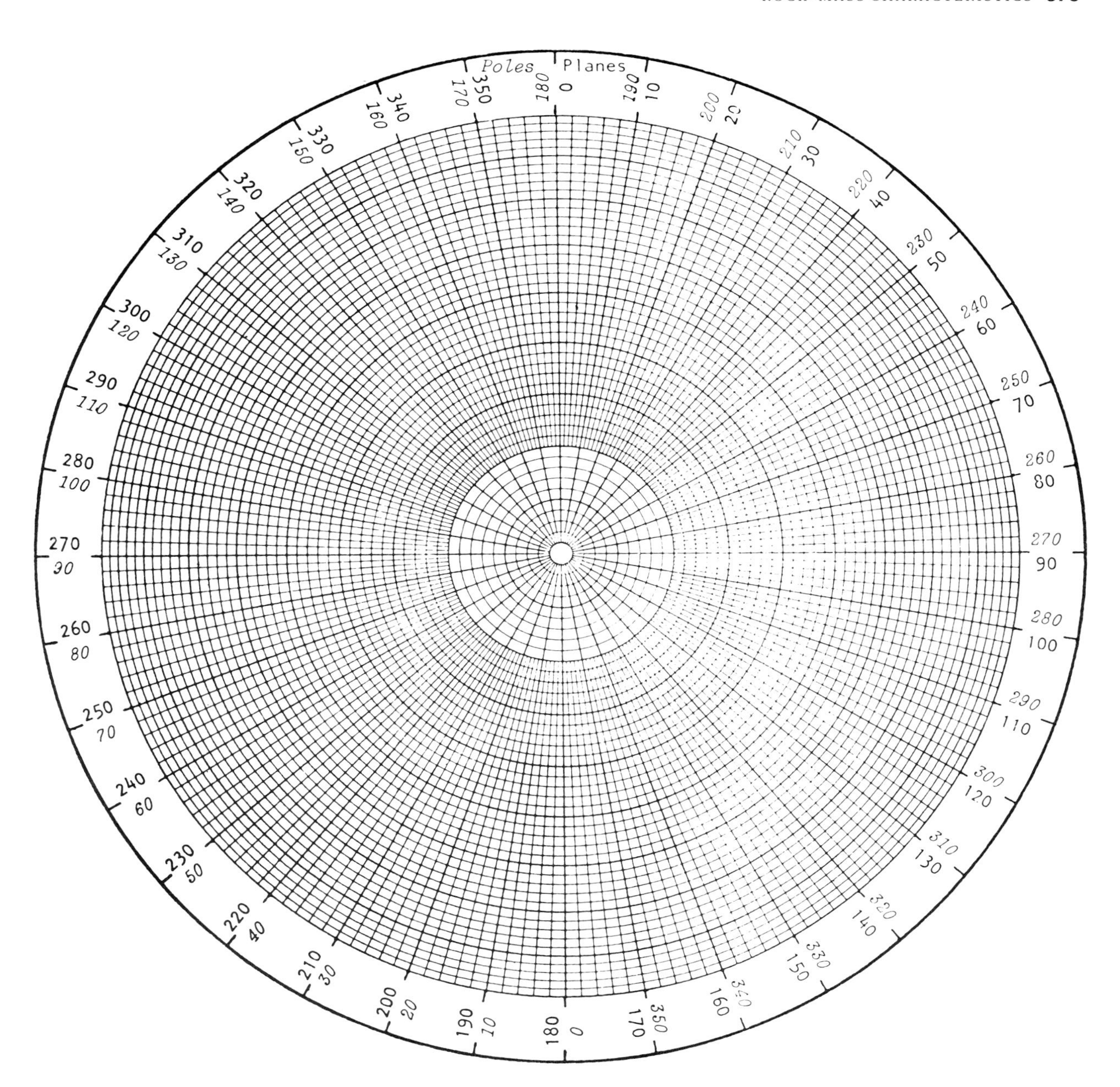
Poles
Planes

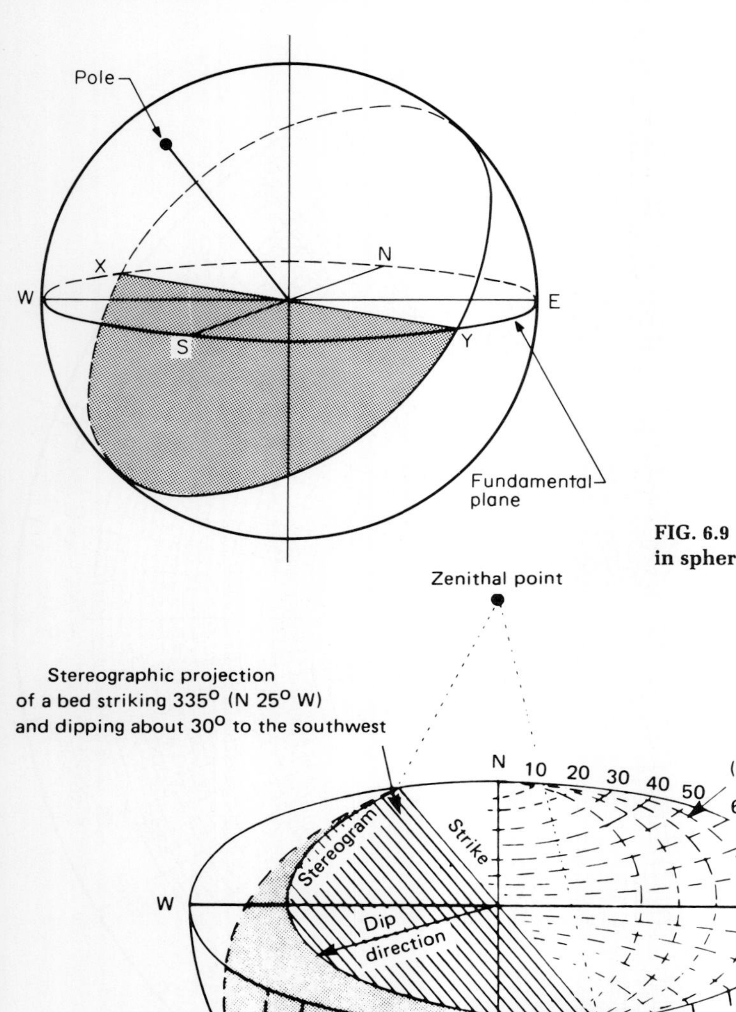


FIG. 6.9 The same structural plane of Fig. 6.8 as visualized in spherical projection. [*From Badgley (1959).*[6]]

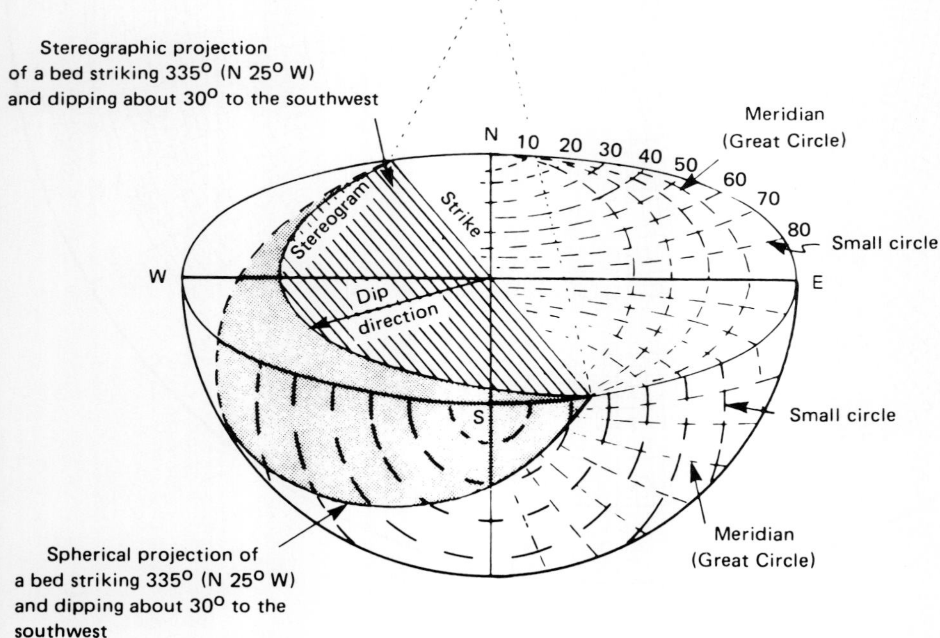


**FIG. 6.10 Stereographic projection on a planar surface. [Three-dimensional view of a bed striking 335° (N25°W) and dipping about 30° to the southwest. This illustration shows the relationship between spherical and stereographic projections. Study of this diagram shows why all dip angles on stereographic projection should be measured only when the dip direction (plotted on overlay tracing paper) is oriented east-west. Similarly the stereographic curve can only be drawn in accurately when the strike direction on overlay paper is oriented parallel to the north-south line on the underlying stereonet.]** [*From Badgley (1959).*[6]]

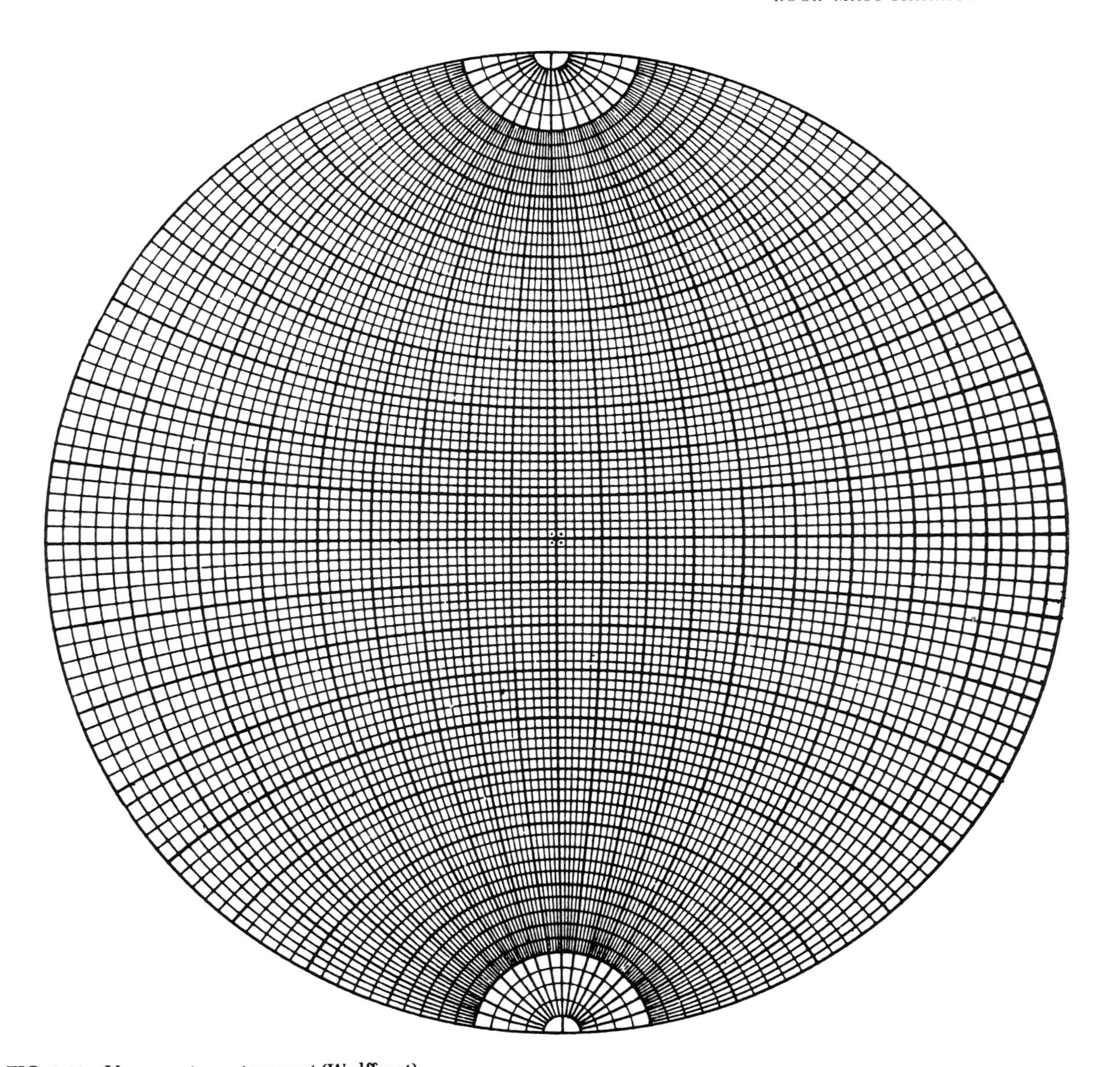

**FIG. 6.11 Nonarea true stereonet (Wulff net).**

The spherical projection of the plane in Fig. 6.9 representing strike and dip can be fully represented as a *pole* on the surface of the sphere, the location of which is a line drawn through the center of the sphere and intersecting it as shown. The pole is found by geometric construction and plotted on the net. (The meridians are marked off from the center to represent dip degrees and the circles represent strike.) The procedure is followed for other joints and the joint concentrations are contoured as shown on Fig. 6.13. An example of a Wulff net showing the plot of two joint sets in relationship to the orientation of a proposed cut slope for an open-pit mine is given as Fig. 6.14.

The geometrical construction of stereonets and their applications are described in texts on structural geology such as Badgley (1959),[6] Hills

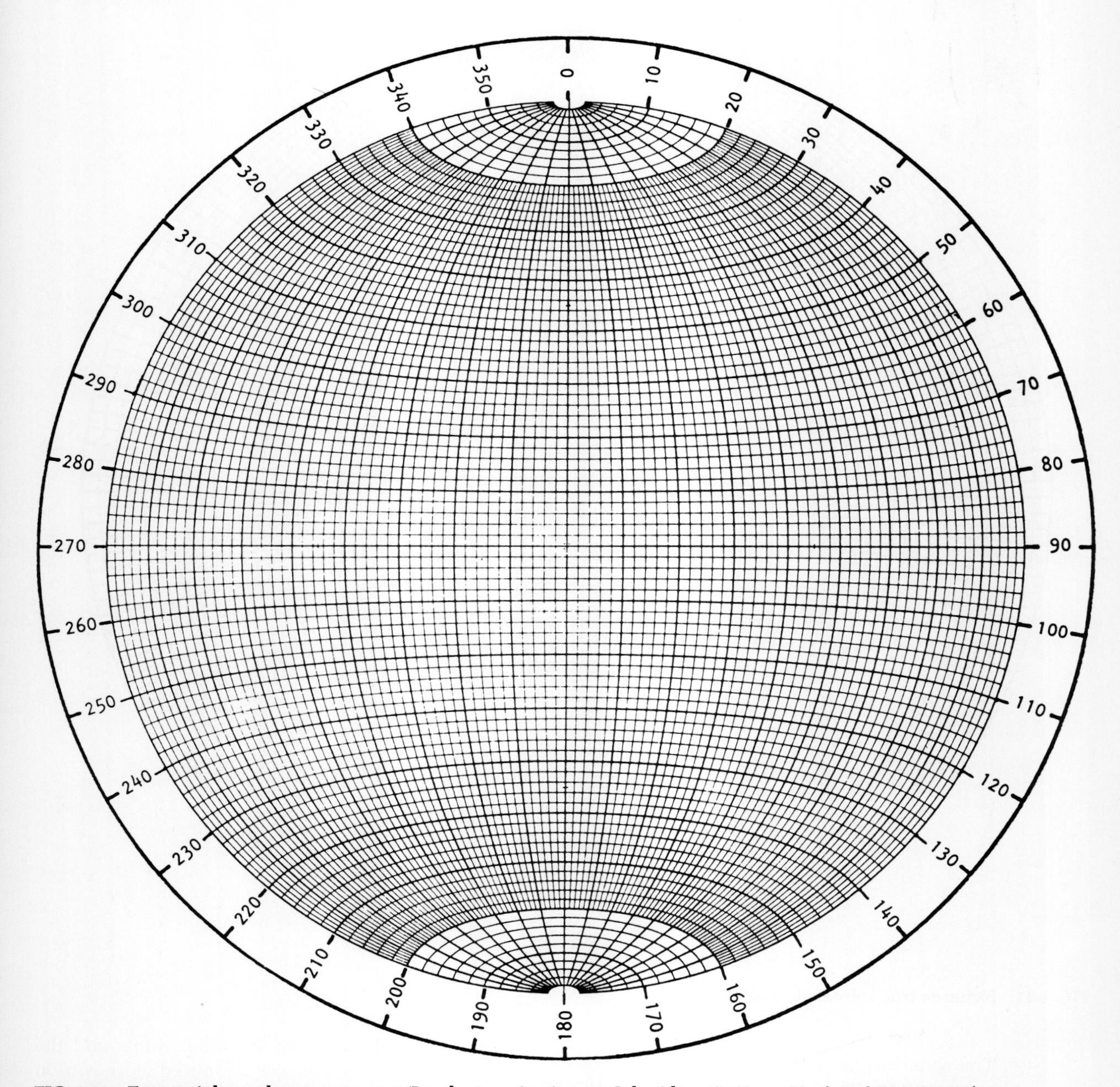


**FIG. 6.12 Equatorial equal-area stereonet (Lambert projection, or Schmidt net).** [*From Hoek and Bray (1977).*[4]]

(1972),[7] Wahlstrom (1973),[5] and Hoek and Bray (1977).[4]

## 6.2 ORIGINAL ROCK-MASS FORMS

### 6.2.1 SIGNIFICANCE

#### General

Igneous and sedimentary rocks are formed with bodies of characteristic shapes, and some igneous bodies change the land surface during formation. Metamorphic rocks modify the igneous and sedimentary formations.

#### Landforms

Weathering processes attack the rock masses, eroding them differentially, resulting in characteristic landforms *(see also Art. 6.7.3)*. The landforms reflect the rock type and its resistance to weathering agents, but climate exerts a strong influence. For a given rock type, humid climates result in more rounded surface features because of greater chemical weathering, and arid climates result in more angular features because of the predominance of mechanical weathering processes.

### 6.2.2 IGNEOUS ROCKS

#### Formation Types

Igneous rocks are grouped on the basis of origin as intrusive or extrusive, and their formational types are defined according to shape and size as batholiths, stocks, laccoliths, etc. The characteristics and surface expression of the various types are described on Table 6.3 and illustrated on Fig. 6.15. Examples of dikes and sills are illustrated on Figs. 6.16 and 6.17 respectively. Associated rock types for the various igneous bodies are given on Table 6.4.

#### Landforms

The general surface expression of the various igneous formations is described on Table 6.3. An example of a batholith is illustrated on the topographic map given as Fig. 6.18 and sills are illustrated on the topographic map given as Fig. 6.19. Differential erosion of weaker formations around a stock may leave a monadnock as shown on the topographic map given as Fig. 6.76.

### 6.2.3 SEDIMENTARY ROCKS

#### Sandstones and Shales

*Formations*

Sandstones and shales are commonly found interbedded in horizontal beds covering large areas, often forming the remnants of peneplains which are dissected by valleys from stream erosion. The drainage patterns are dendritic unless structure-controlled.

*Humid Climate*

Chemical weathering results in rounded landforms and gentle slopes as shown on the topographic map (*Fig. 6.20*). The more resistant sandstone strata form the steeper slopes and the

**FIG. 6.13 Contoured equal-area plot of a joint system containing two sets. The concentration of poles in the figure indicates that the intersection of the sets plunges at a small angle from the horizontal. Plot is on the lower hemisphere.** *[From Wahlstrom (1973).[5] Reprinted with permission of Elsevier Scientific Publishing Company.]*

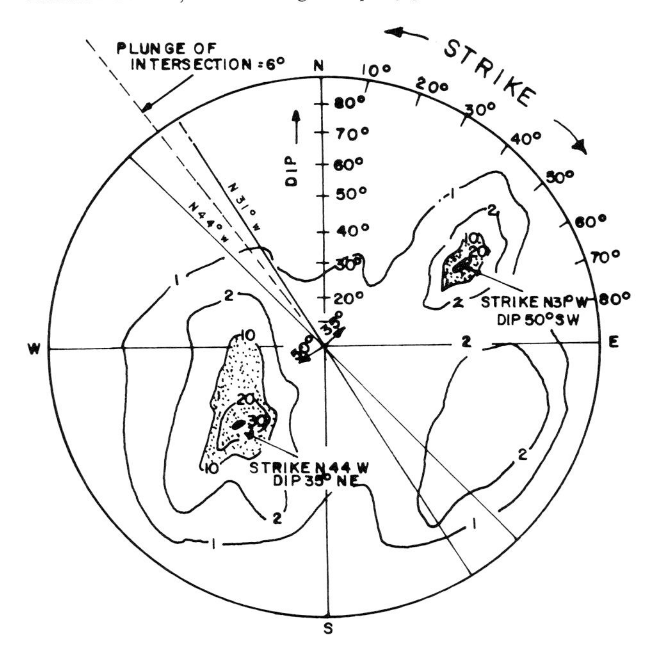

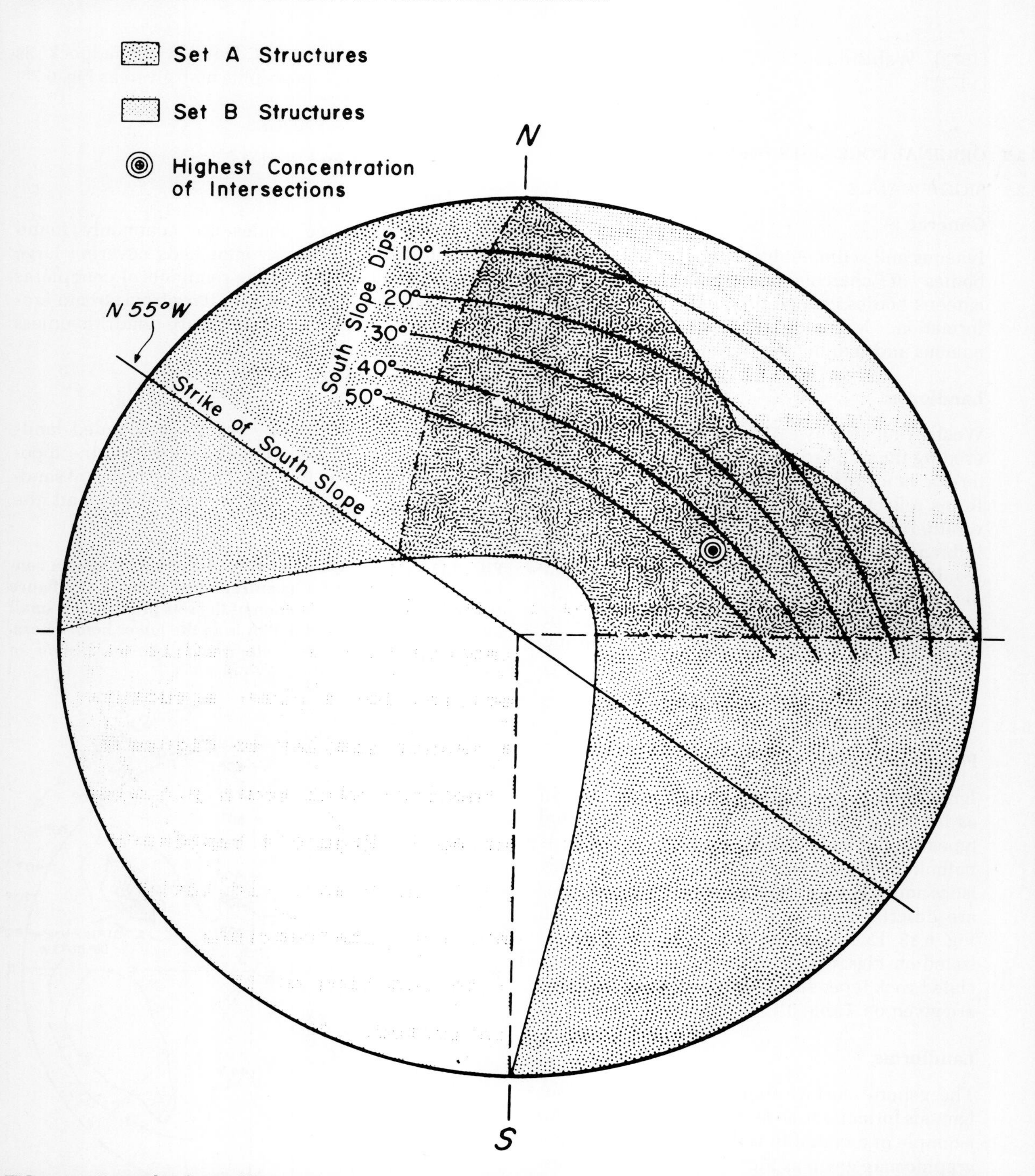


**FIG. 6.14 Example of a Wulff stereonet showing relationship of two joint sets on the south slope of an open-pit mine. The greatest concentration of joints in set *A* dips at about 45°E and strike N55°E; therefore, a cut slope made at this orientation or greater would be essentially unstable since the cut slope would intersect the dip.** [*From Seegmiller (1971).*[8]]

**TABLE 6.3**
**IGNEOUS ROCKS: FORMATIONS AND CHARACTERISTICS**

| Formation | Characteristics | Surface expressions |
|---|---|---|
| | INTRUSIVE | |
| Batholith | Huge body, generally accepted as having an exposed surface area greater than 90 km$^2$ and no known floor. | Irregular; drainage develops along discontinuities, in medium to fine dendritic with curvilinear alignments, or rectangular or angulate along joints (*see Fig. 6.18*). |
| Stock | Small batholith. | Same as for batholiths. |
| Laccolith | Deep-seated, lenticular shape, intruded into layered rocks. More or less circular in plan with flat floor. | Dome-shaped hill often formed by lifting and arching of overlying rocks. Drainage: annular. |
| Lappolith | Basin-shaped body probably caused by subsidence of the crust following a magma intrusion. Deep-seated. | Basins; drainage centripedal. |
| Pluton | Body of any shape or size which cannot be identified as any of the other forms. | Various; generally similar to batholiths. |
| Dike | Formed by magma filling a fissure cutting through existing strata. Often thin, sheetlike (*see Fig. 6.16*). | Younger than enveloping rocks, they often weather differentially, leaving strong linear ridges. See sill below. |
| Sill | Formed by magma following a stratum. Rocks are baked and altered on both sides (flows bake only on the underside). When formed close to the surface, sills develop the characteristic columnar jointing of flows (*see Fig. 6.17*). | Weather differentially in tilted strata forming ridges (*see Fig. 6.19*). In humid climates rocks of dikes and sills often decompose more rapidly than surrounding rocks, resulting in depression lineaments, often associated with faults. |
| Volcanic neck | The supply pipe of an extinct volcano formed of solidified magma. | Distinctly shaped steep-sided cone remains after erosion removes the pyroclastic rocks. |
| | EXTRUSIVE | |
| Flows | Outwellings of fluid magma solidifying on the surface, forming sheets often of great extent and thickness. May have characteristic columnar jointing near surface. Only underlying rocks are baked. | Young age: varies from broad flat plains to very irregular surfaces.<br>Maturity develops a dissected plateau, eventually forming mesas and buttes in arid climates.<br>Drainage: parallel, coarse on young, flat plains. |

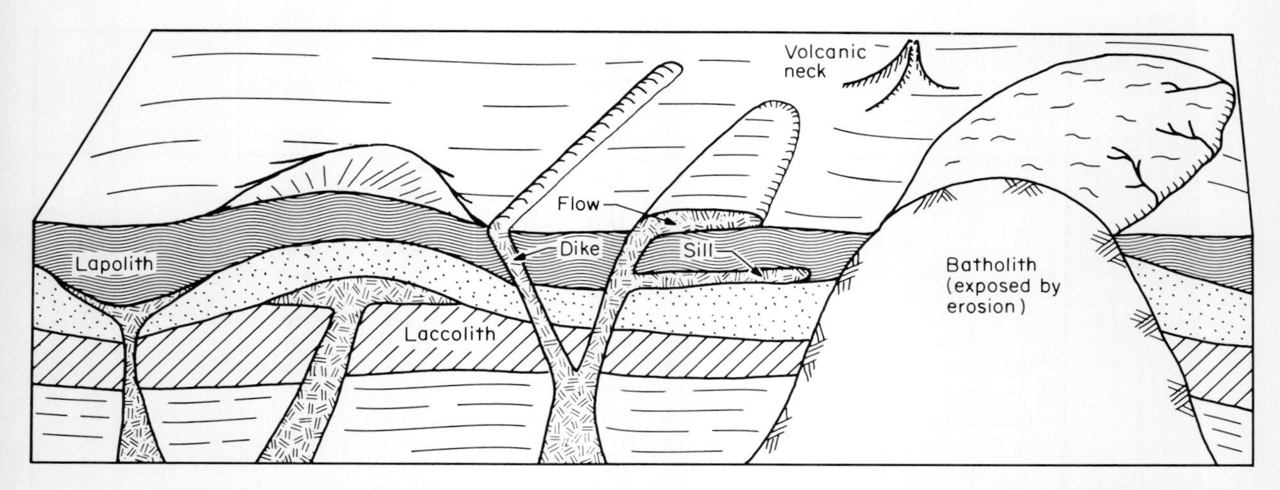


**FIG. 6.15 Forms of intrusive and extrusive igneous rock types (batholith, dike, and volcanic neck are exposed by erosion).**

**TABLE 6.4**
**ROCK TYPES CHARACTERISTIC OF INTRUSIVE AND EXTRUSIVE IGNEOUS BODIES**

| | Rock type | Batholith | Stock | Lapolith | Laccolith | Dike | Sill | Flow | Neck |
|---|---|---|---|---|---|---|---|---|---|
| | Pegmatite | x | x | | | x | x | x | |
| Medium- to coarse-grained | Granite | x | x | | | | | | |
| | Syenite | x | x | | | | | | |
| | Diorite | x | x | | | | | | |
| | Gabbro | x | x | | | | | | |
| | Peridotite | x | x | x | x | x | x | | |
| | Pyroxenite | x | x | x | x | x | x | | |
| | Hornblendite | x | x | x | x | x | x | | |
| | Dunite | x | x | x | x | x | x | | |
| | Dolerite | x | x | | | x | x | x | |
| Fine-grained | Rhyolite | | | | | x | x | x | |
| | Andesite | | | | | x | x | x | |
| | Basalt | | | | | x | x | x | |
| | Felsite | | | | | x | x | x | |
| Glass | Obsidian | | | | | x | x | x | x |
| | Pitchstone | | | | | x | x | x | x |
| | Pumice | | | | | x | x | x | x |
| | Scoria | | | | | x | x | x | x |
| | Porphyries | x* | x* | | | x | x | x* | x |

*Contact zones.

**FIG. 6.16 Diabase dike intruded into granite, Rio-Santos Highway, Rio de Janeiro, Brazil.** *(Photo by Geraldo Lauria.)*

**FIG. 6.17 Sills (Yellowstone National Park, Wyoming).**

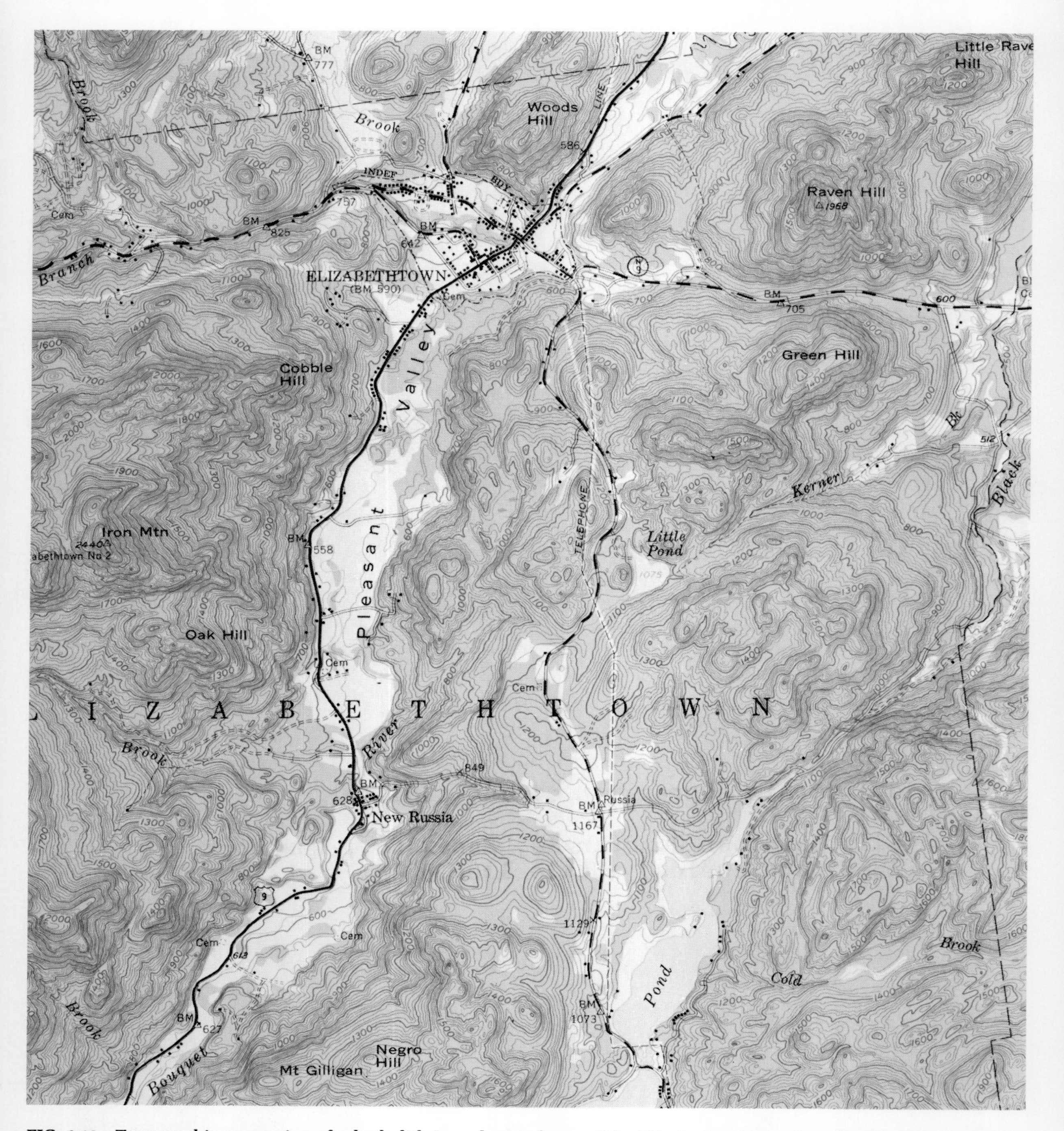


**FIG. 6.18** Topographic expression of a batholith in a glaciated area. (The Elizabethtown quadrangle, Essex County, New York, 15-min series. Adirondack Mountains, New York.)

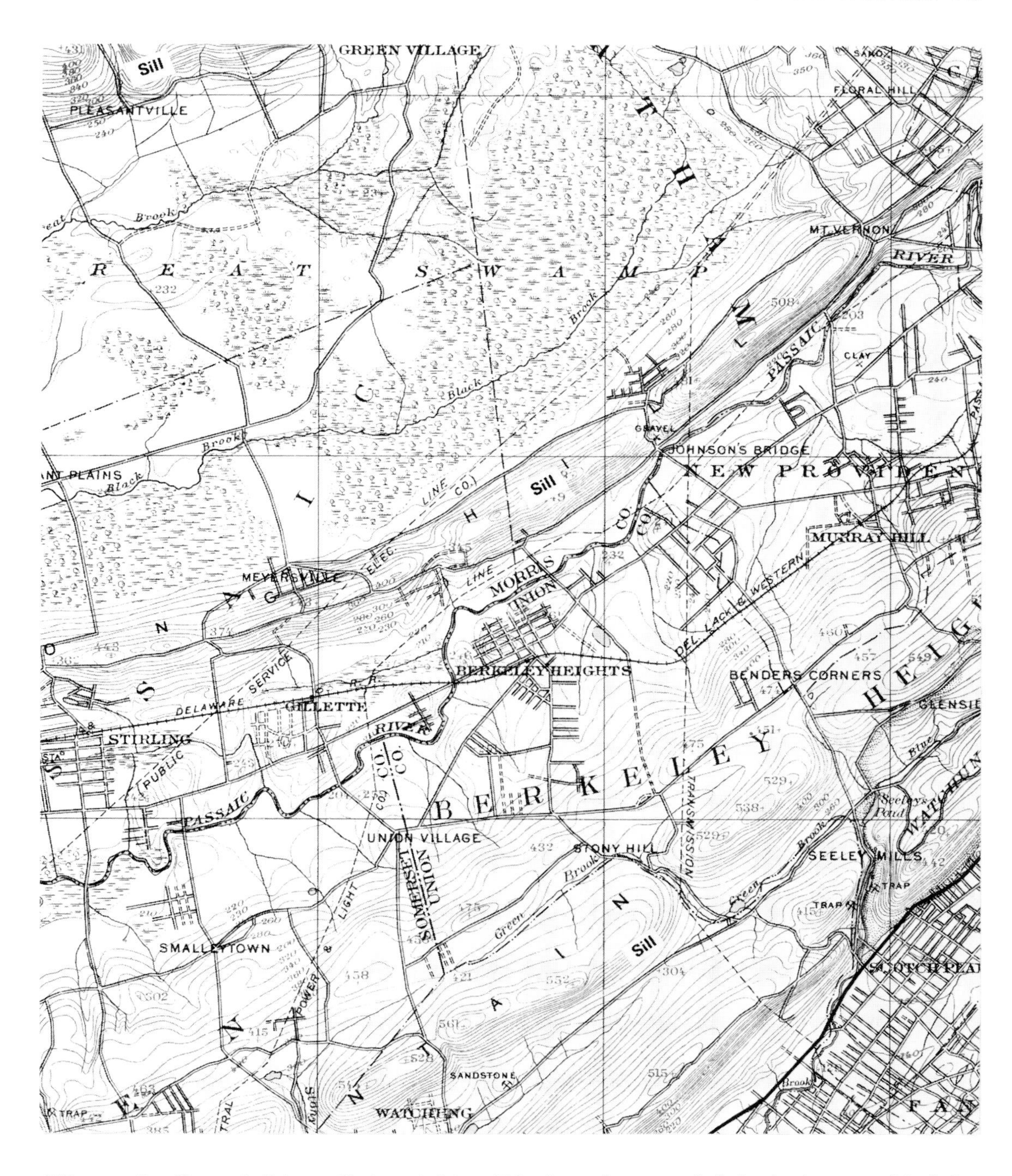


FIG. 6.19 Landform of diabase sills intruded into Triassic sandstones and shales in the area of Berkeley Heights, New Jersey. The area shown as the "great swamp" is former glacial Lake Passaic *(see Art. 7.6.4)*. (Sheet 25 of the New Jersey Topographic Map Series; Scale 1 in = 1 mile.)

**FIG. 6.20** Humid climate landform; horizontally bedded sandstones and shales. The ridge crests all have similar elevations. (Warren County, Pennsylvania. Scale 1:62,500.)

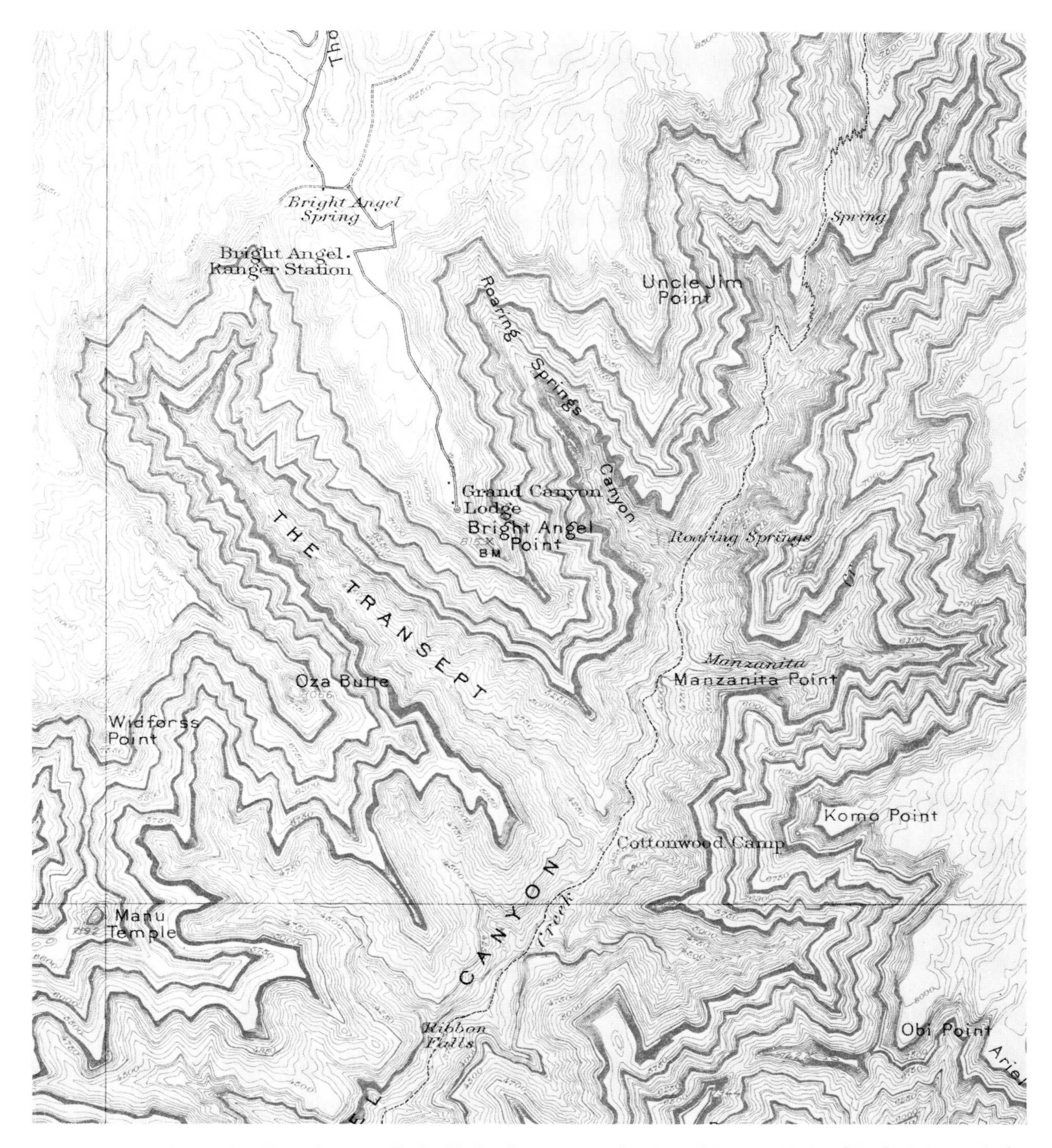


**FIG. 6.21 Arid climate landform: horizontally bedded sedimentary rocks. (Grand Canyon National Park, Arizona. Scale 1:48,000.)**

**FIG. 6.22 Cleaning soil from solution cavity in limestone prior to pouring lean concrete (Versailles, Kentucky).** *(Photo by R. S. Woolworth.)*

shales the flatter slopes. Thick formations of marine shales develop very gentle slopes and are subjected to slumping and landsliding *(see Art. 6.7.3)*.

*Arid Climate*

Weathering is primarily mechanical in arid climates, and erosion proceeds along the stream valleys, leaving very steep slopes with irregular shapes as shown on the topographic map (Fig. 6.21). The resistant sandstones form vertical slopes and the softer shales form slopes with relatively gentler inclinations.

**Limestone**

*Formations*

The most significant features of pure limestone result from its solubility in water. Impure limestones usually take the forms of other sedimentary rocks during decomposition.

Solution results in cavities, providing the mass with an irregular rock surface, large and small openings within the mass, and distinct surface expressions *(see also Art. 10.4.2)*. These features are extremely significant to engineering works, since they often present hazardous conditions as discussed in Art. 10.4.

**FIG. 6.23 Stalactites (hanging) and stalagmites in a limestone cavern. (Cave-of-the-Winds, Colorado Springs, Colorado.)**

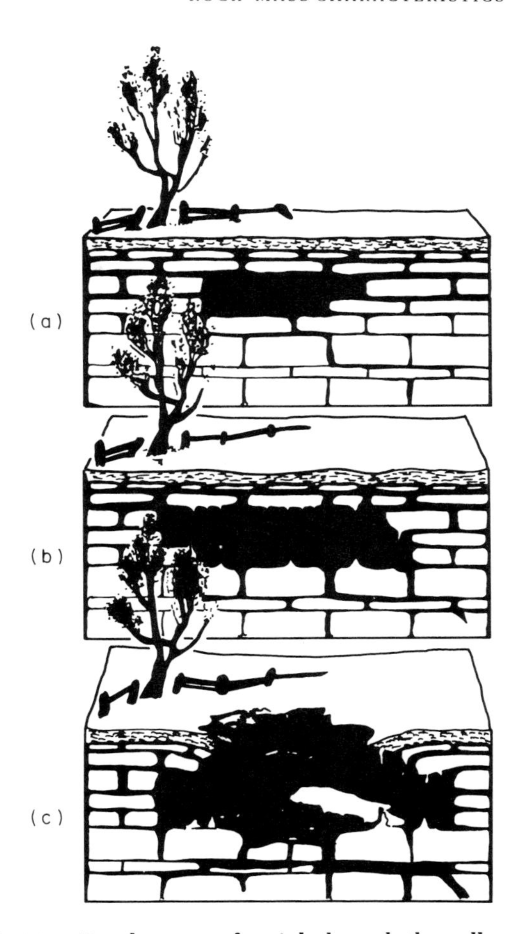


**FIG. 6.24 Development of a sink through the collapse of the cavern roof.**

*Solution Characteristics*

Solution source is rainfall infiltration, especially the slightly acid water of humid climates penetrating the surface vegetation. This water is effective as a solvent as it moves through the rock fractures. Its effectiveness increases with temperature and flow velocity.

Cavity growth develops as the rainwater attacks the joints from the surface, causing erosion to proceed downward, creating cavities as shown on Fig. 6.22.

Attack proceeds horizontally at depth to create caverns, as shown on Fig. 6.23, which can be many meters in diameter and extend many meters in length. The precipitation of calcium carbonate in the caverns results in the formations of stalactites and stalagmites.

Solution proceeds outward from the cavity, causing it to grow. Eventually the roof arch cannot support the overburden load, and it begins to sag to form a dish-shaped area *(see Fig. 10.20)* which can be quite large in extent. Finally collapse occurs and a sinkhole is formed as illustrated on Fig. 6.24 *(see also Figs. 10.15 and 10.18)*. The surface expression of sinks, as apparent on aerial photos, is given on Fig. 10.19.

*Landforms*

Landforms developing in limestone are referred to as *karst* topography. Characteristic features

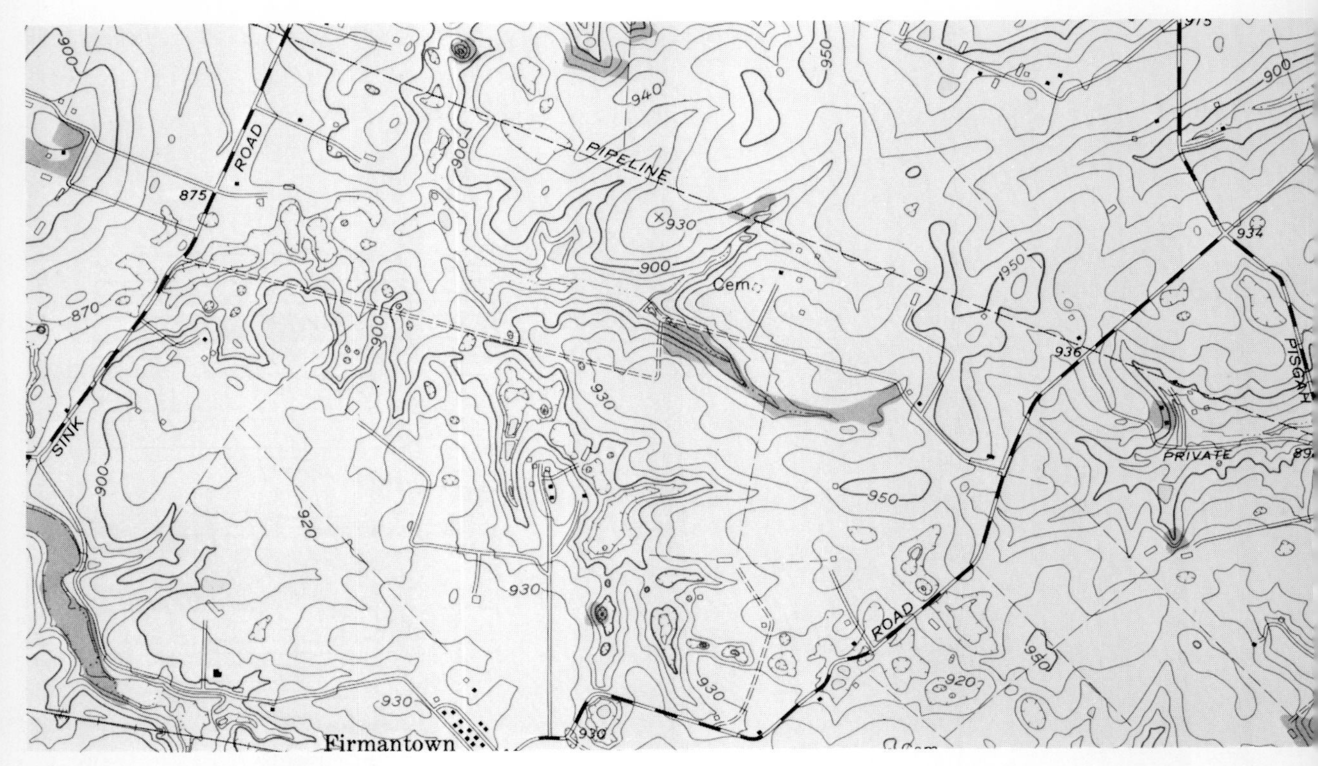
ROAD
PIPELINE
SINK
Cem
ROAD
PRIVATE
PISGAH
Firmantown

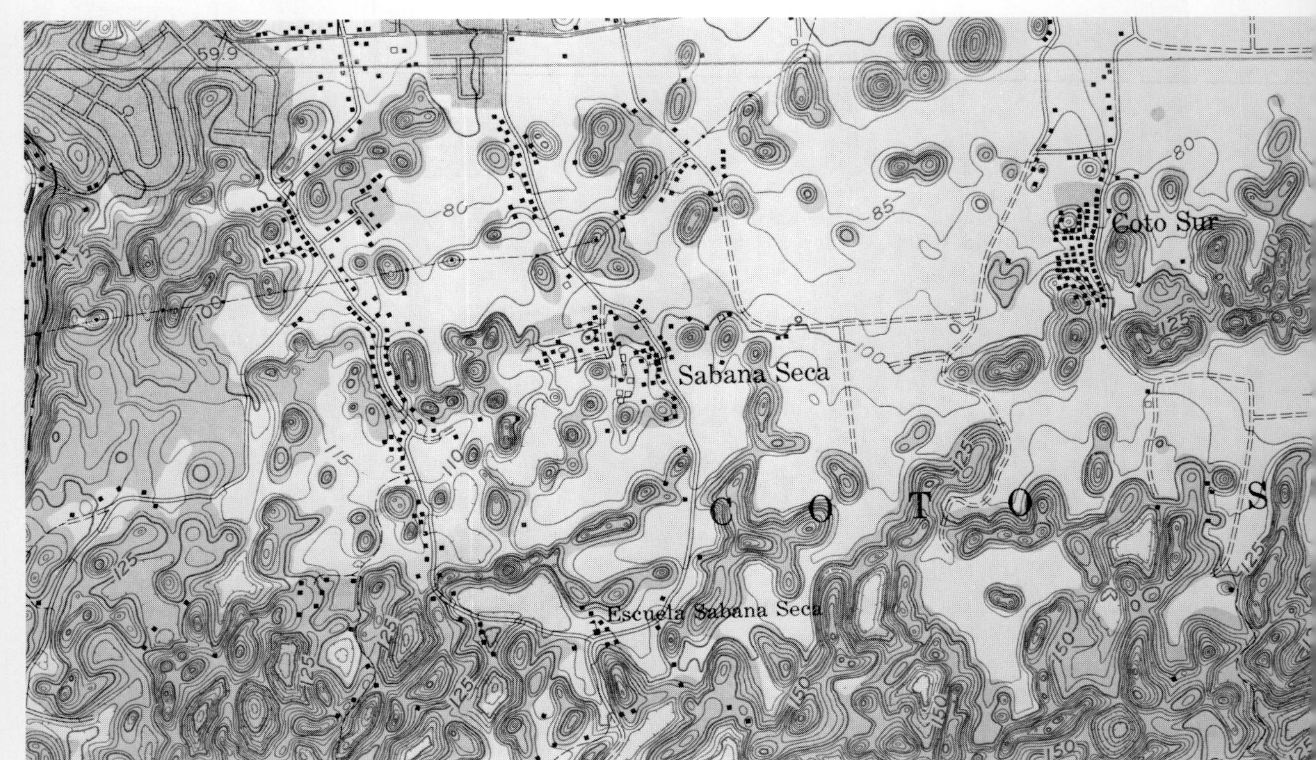
Coto Sur
Sabana Seca
C O T O S
Escuela Sabana Seca

**FIG. 6.25** *(Above left)* **Limestone landform in a relatively cool, moist climate. Note sinks and intermittent, disappearing streams. (Versailles, Kentucky, quadrangle. Scale 1:24,000. The photograph in Fig. 6.22 was taken in this general area.)**

**FIG. 6.26** *(Below left)* **Limestone landform developing in a tropical climate. Note numerous "haystack" hills. (Manati, Puerto Rico, quadrangle. Scale 1:20,000. Advanced tropical karst.)**

**FIG. 6.27 Color composite ERTS 1 image mosaic of Florida showing central lake district, a region of extensive sinkhole development. In the lower regions the lakes are often filled with 30 m of soils; in the higher regions sinks are still developing, especially in areas subjected to groundwater withdrawal.** *(Mosaic prepared by General Electric Company, Beltsville, Maryland, in cooperation with USGS, NASA, and the Southern Florida Flood Control District.)*

**FIG. 6.28 Landsat image of the state of Rio de Janeiro, Brazil, from the Baia de Guanabara on the east to Parati on the west (Scale 1:1,000,000); an area characterized by metamorphic and igneous rocks in a tropical climate. Major fault systems trend southwest-northeast. (Image No. 17612113328, dated July 9, 1976, by NASA, reproduced by USGS, EROS Data Center.)**

vary with climate, the duration of exposure, and the purity of the rock, but in general, in moist climates the terrain varies from flat to gently rounded to numerous dome-shaped hills. Rounded depressions and sinkholes are common and there are few or no permanent streams. Intermittent streams disappear suddenly into the ground. Drainage is termed deranged or intermittent.

Terrain features developing in the relatively cool, moist climate of Kentucky are shown on Fig. 6.25. Numerous depressions, sinkholes, and suddenly terminating intermittent streams are apparent. (This is the area in which the photo of Fig. 6.22 was taken.)

Terrain features developing in the warm, moist climate of Puerto Rico are shown on Fig. 6.26. Sinks, depressions, and the lack of surface drainage are evident, but the major features are the numerous dome-shaped hills ("haystacks"). The landform is referred to as tropical karst.

The numerous lakes of central Florida, illustrated on Fig. 6.27, an ERTS image mosaic, are formed in sinks that developed in an emerged land mass.

*Conditions for Karst Development*

Thinly bedded limestone is usually unfavorable for karst development because the thin beds are often associated with insoluble ferruginous constituents in the parting planes, and by shale and

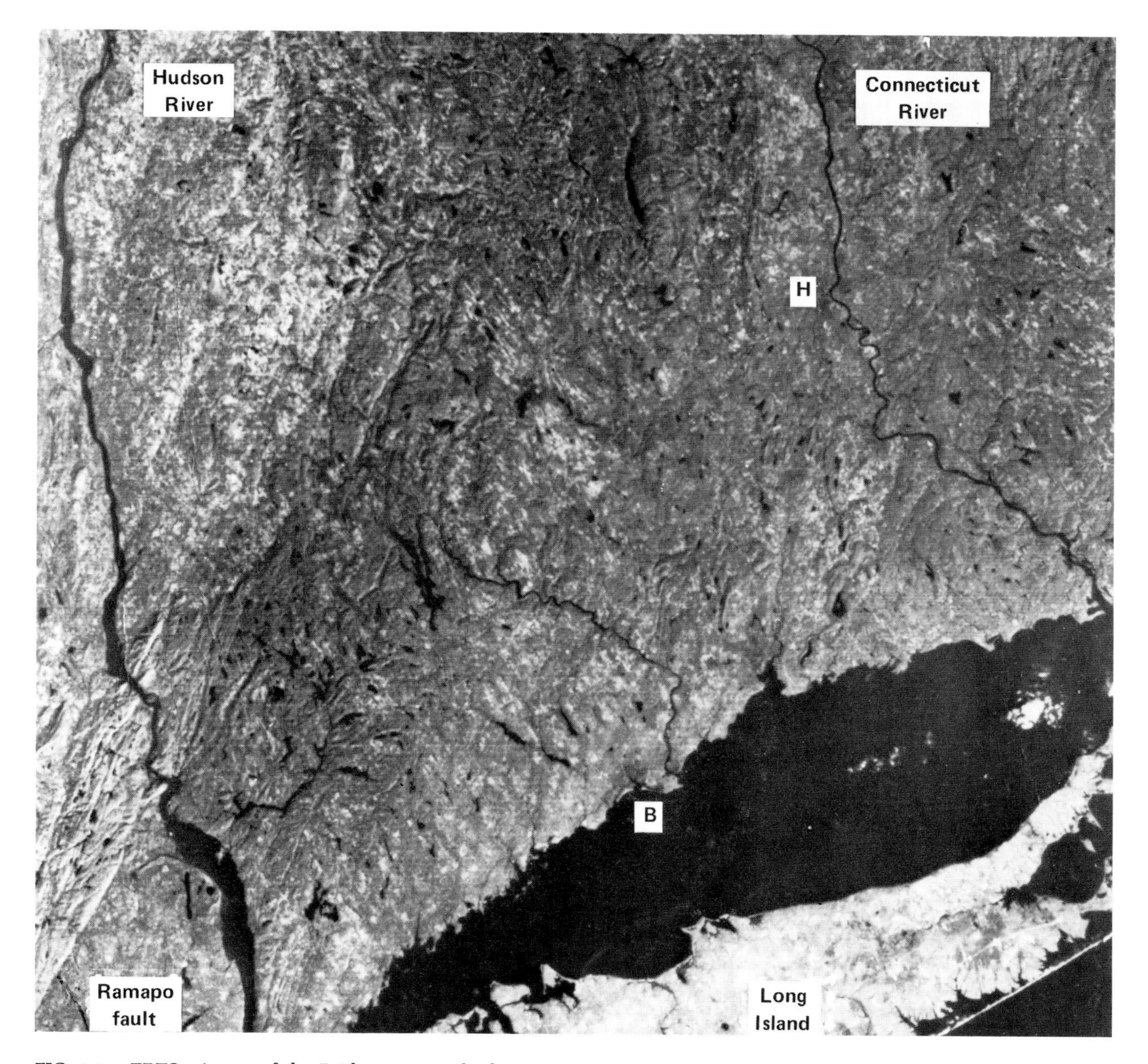


**FIG. 6.29 ERTS 1 image of the Bridgeport-Hartford, Connecticut, area showing the landform developing in a cool, moist climate in igneous and metamorphic rocks denuded by glaciation. The texture is extremely variegated. The strong lineament in the lower left is the Ramapo fault. The northern portions of the Hudson and Connecticut Rivers flow through valleys of sedimentary rocks (Scale 1:1,000,000).** *(Original image by NASA, reproduced by USGS, EROS Data Center.)*

clay interbeds which block the internal drainage necessary for solution. These formations develop medium angular, dendritic drainage patterns.

Thick uniform beds are conducive to solution and cavity and cavern development as long as water is present. At least 60% of the rock must be carbonate materials for karst development, and a purity of 90% or greater is necessary for full development [Corbel (1959)[9]].

Limestones, even when pure, located above the water table in arid climates do not undergo cavern development, although relict caverns from previous moist climates may be present.

The most favorable factor for cavity develop-

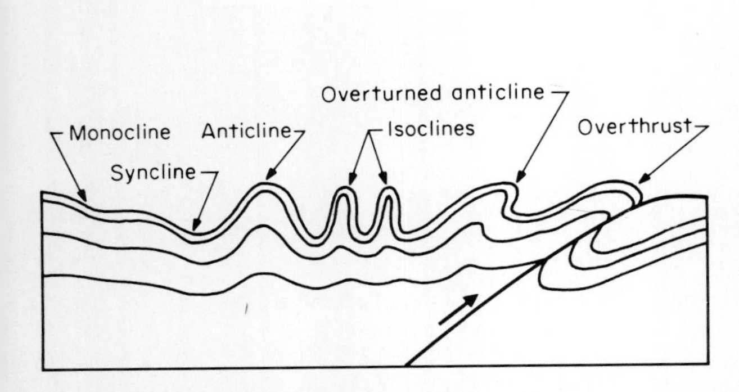


**FIG. 6.30 Common geologic fold structures.**

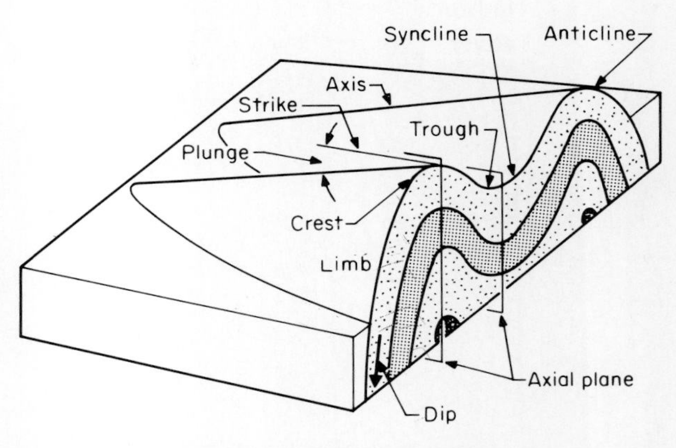


**FIG. 6.31 Nomenclature of plunging folds.**

ment and collapse in limestones is the withdrawal of groundwater by pumping as described in Art. 10.4.2.

6.2.4 METAMORPHIC ROCK

### Formations

Metamorphic rocks often extend over very large areas, normally intermixed with igneous rocks and associated with mountainous and rugged terrain, the irregularity of which relates to climate.

The landform results from differential weathering between rocks of varying resistance and develops most intensely along weakness planes. Drainage patterns depend primarily on joint and foliation geometric forms: rectangular patterns result from regular forms, and angulate patterns result from irregular forms.

### Gneiss

*Tropical Climate*

The Landsat image in Fig. 6.28 covers a portion of the Brazilian crystalline complex in the state of Rio de Janeiro. Differential weathering has produced numerous parallel ridges which generally trend southwest-northeast. The ridges are steep-sided with sharp crests, and together with the lineations, follow the foliations of a great thickness of biotite gneiss and occasionally schists, into which granite has intruded, as well as basalt dikes. The texture pattern created by foliation and jointing is intense at the scale shown. A number of the longer lineations represent major faults, evident on the left-hand portion of the image.

*Cool, Moist Climate: Glaciated Area*

Landsat image of the Bridgeport–Hartford, Conn., area showing the landform developing in a cool, moist climate in metamorphic and igneous rocks denuded by glaciation is given as Fig. 6.29. The texture is extremely variegated and the landform is much more subdued than that illustrated on Fig. 6.28, which is at the same scale. The strong lineament in the lower left-hand corner is the Ramapo fault *(see Art. 6.5.3).* The area identified as Long Island is composed of Cretaceous, Pleistocene, and Recent soil formations.

### Schist

Landforms in schist are characterized by rounded crests which follow the schistosity, shallow side slopes in humid climates, and more rugged slopes in dry climates. Drainage is medium to fine rectangular dendritic.

### Slate

Slate weathers quickly in a moist climate, developing rugged topography intensely patterned with many sharp ridges and steep hillsides. Drainage is fine rectangular dendritic.

**FIG. 6.32** Crest of anticline in High Falls shale (Delaware River, Pennsylvania). Tension cracks are calcite-filled.

## 6.3 DEFORMATION BY FOLDING

### 6.3.1 GENERAL

**Significance**

*Mass Deformation*

Deformation from compressive forces in the crust can result in gentle warping of horizontal strata, or in intense and irregular folding of beds (*see also discussion of cataclastic metamorphism, Art. 5.2.5*). It can proceed to overturn beds completely, or it can result in faulting and overthrusting (*see Art. 6.5*).

*Rock Fracturing*

Distortions during folding result in many types of rock fractures, including fracture cleavage (*see Art. 6.3.2*) and other forms of joints (*see Art. 6.4*), and foliation and mylonite shear (*see Table 6.10*).

**Nomenclature**

Common fold structures include monoclines, synclines, anticlines, isoclines, overturned anticlines, and overthrusts. There are also recumbent folds, drag folds, and plunging folds. Their characteristics are summarized on Table 6.5 and most are illustrated on Fig. 6.30.

The nomenclature of plunging folds is given on Fig. 6.31. The crest of an anticline is illustrated on Fig. 6.32, and a recumbent fold is illustrated on Fig. 6.33.

### 6.3.2 FRACTURE CLEAVAGE

**Cause**

Fracture cleavage results from the development of shear forces in the folding of weak beds with schistosity such as shales and slates.

**FIG. 6.33 Intense folding in the Tertiary clay-shales of Santa Monica, California.**

**TABLE 6.5**
**GEOLOGIC FOLD STRUCTURES AND CHARACTERISTICS**

| Structure | Characteristics |
|---|---|
| Monocline | Gentle warping, fold has only one limb |
| Syncline | Warping concave upward |
| Anticline | Warping concave downward (*see Fig. 6.32*) |
| Isocline | Tight folds with beds on opposite limbs having same dip |
| Overturned anticline | Warping stronger in one direction, causing overturning |
| Overthrust | Overturning continues until rupture occurs (*see Fig. 6.52*) |
| Recumbent folds | Strata overturned to cause axial plane to be nearly or actually horizontal (*see Fig. 6.33*) |
| Drag folds | Form in weak strata when strong strata slide past; reveal direction of movement of stronger strata |
| Plunging folds | Folding in dipping beds (*see Fig. 6.31*) |

### Formation

Considered as a form of jointing, fracture cleavage is independent of the arrangement of mineral constituents in the rock; therefore, it is often referred to as *false cleavage* to differentiate it from mineral cleavage.

Since fracture cleavage results from a force couple (Fig. 6.34), its lineation is often parallel to the fold axis, and therefore when it is exposed in outcrops it is useful in determining the orientation of the overall structure. Fracture cleavage in a slate is illustrated on Fig. 6.35.

### Flow Cleavage

The recrystallization of minerals that occurs during folding is referred to as flow cleavage.

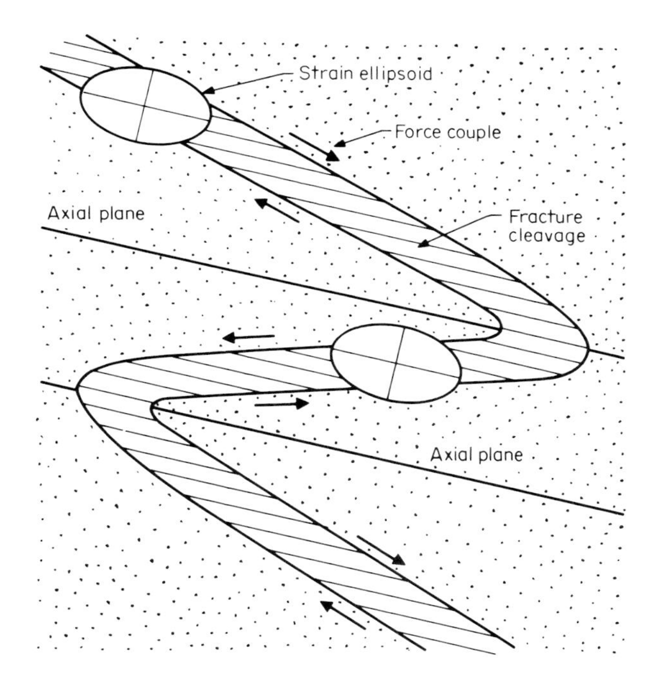


**FIG. 6.34 Fracture cleavage forms during folding.**

## 6.3.3 LANDFORMS

### Origin

Erosional agents attack the beds differentially in sedimentary rocks in the general sequence from resistant sandstone to less-resistant shale and limestone. The resulting landforms relate to the warped bedding configuration and can be parallel ridges, plunging folds, domes, and basins at macro scales.

**FIG. 6.35 Fracture cleavage in overturned fold in Martinsburg slate; sheared by dragging (Warren County, New Jersey).**

### Parallel Ridges

Horizontal beds folded into anticlines and synclines produce parallel ridges (Fig. 6.36), typical of the Appalachian Mountains and a moist climate. Starting from a peneplain, the soluble limestones weather the quickest and form the valleys. Shales form the side slopes, and the more resistant sandstones form the crests and ridges.

Horizontal beds folded into a monocline produce parallel ridges (hogbacks) and mesas in an arid climate where weathering is primarily mechanical, as shown on Fig. 6.37.

### Ridges in Concentric Rings

Beds deformed into a *dome* will erode to form ridges in concentric rings, with the shallower side slopes *dipping away* from the center of the former dome (see *Fig. 6.1*).

Beds deformed into a *basin* will erode to form ridges in concentric rings with the shallow side slopes *dipping into* the center of the former basin.

### Plunging Folds

The landform of plunging folds is illustrated on the topographic map (Fig. 6.38); also included are several sections showing the folded beds. The ERTS 1 image (Fig. 6.39) shows prominent parallel ridges of folded sedimentary rocks, plunging folds, and two major wrench faults which indicate very large dislocation of strata.

## 6.4 JOINTING

### 6.4.1 GENERAL

#### Significance

Joints are rock fractures along which essentially no displacement has occurred. They represent planes of weakness in the rock mass, substantially influencing its competency to support loads or to remain stable when partially unconfined as in slopes.

#### Causes

They are caused by tensile, compressive, or shearing forces, the origins of which can be crustal warping and folding, cooling and contraction of igneous rocks, displacement from injection of igneous masses into adjacent bodies, stress relief by erosion or deglaciation, or stress increase by dehydration.

### 6.4.2 FORMS AND CHARACTERISTICS

#### Nomenclature

The various joint types and joint systems are summarized on Table 6.6.

#### Engineering Classifications

*By Origin*

- *Shear joints* are formed by shearing forces such as occur during faulting and represent displacement in one direction that is parallel to the joint surface.
- *Tension joints* result from stress relief, cooling of the rock mass, etc. and represent displacement in one direction that is normal to the joint surface.
- *Displacement joints* represent displacement in two directions, one normal to the fracture surface and the other parallel, indicating that displacement joints begin as tension joints.

*By Spacing*

The rock mass is designated according to spacing between joints as solid (> 3 m), massive (1 to 3 m), blocky or seamy (30 cm to 1 m), fractured (5 to 30 cm), and crushed (< 5 cm), as given on Table 5.23.

#### Characteristics

*Physical*

Joints are characterized by spacing as defined above, by width of opening, by continuity or length, and by surface roughness or configuration. They can be clean or they can contain fill-

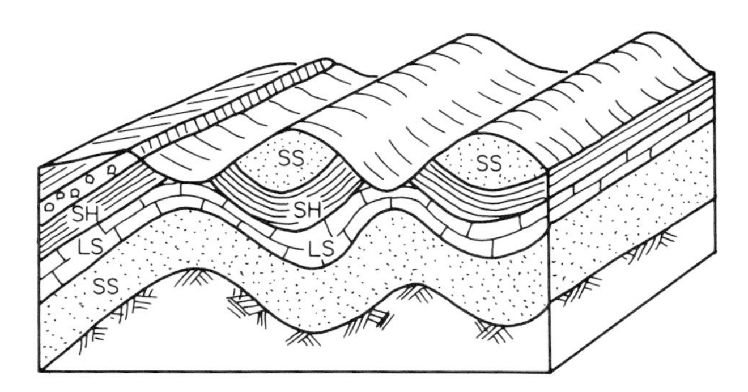


**FIG. 6.36 Parallel ridges in sedimentary rocks folded into synclines and anticlines.**

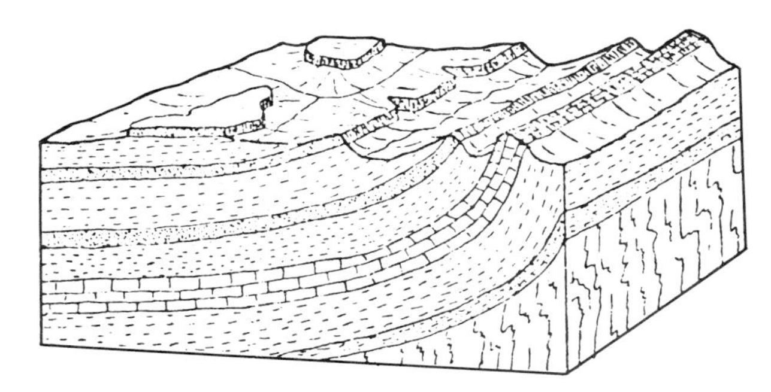

**FIG. 6.37 Parallel ridges in sedimentary rocks folded into a monocline in an arid climate (hogbacks). Mesas are formed where beds are horizontal.**

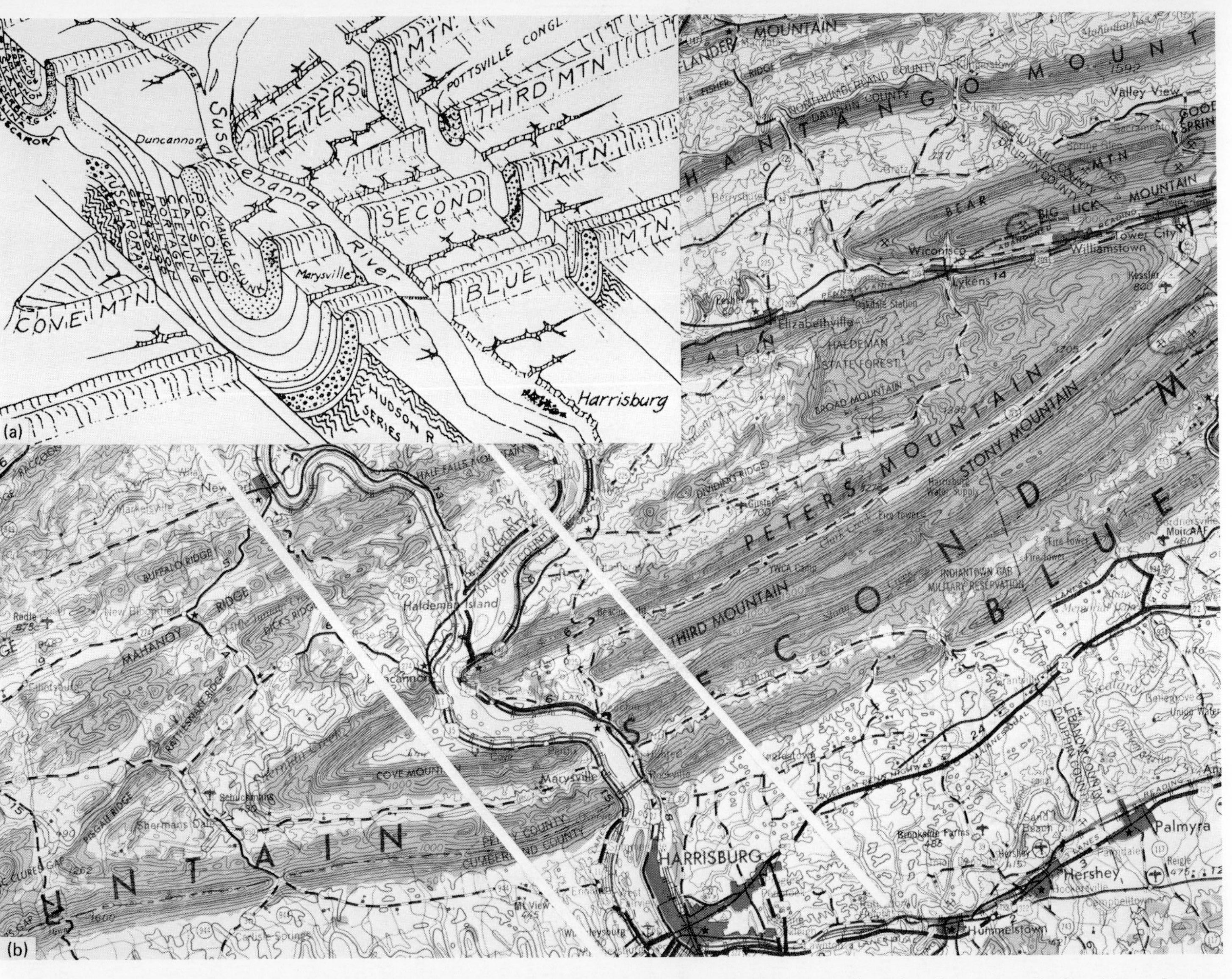


**FIG. 6.38 Landform expression of plunging folds. (Harrisburg, Pennsylvania. Scale 1:300,000.)** *(From the Atlas of American Geology.)* **Inset diagram shows structure of the folded Appalachians at Harrisburg.** *(Inset © Hammond, Inc. Reprinted with permission.)*

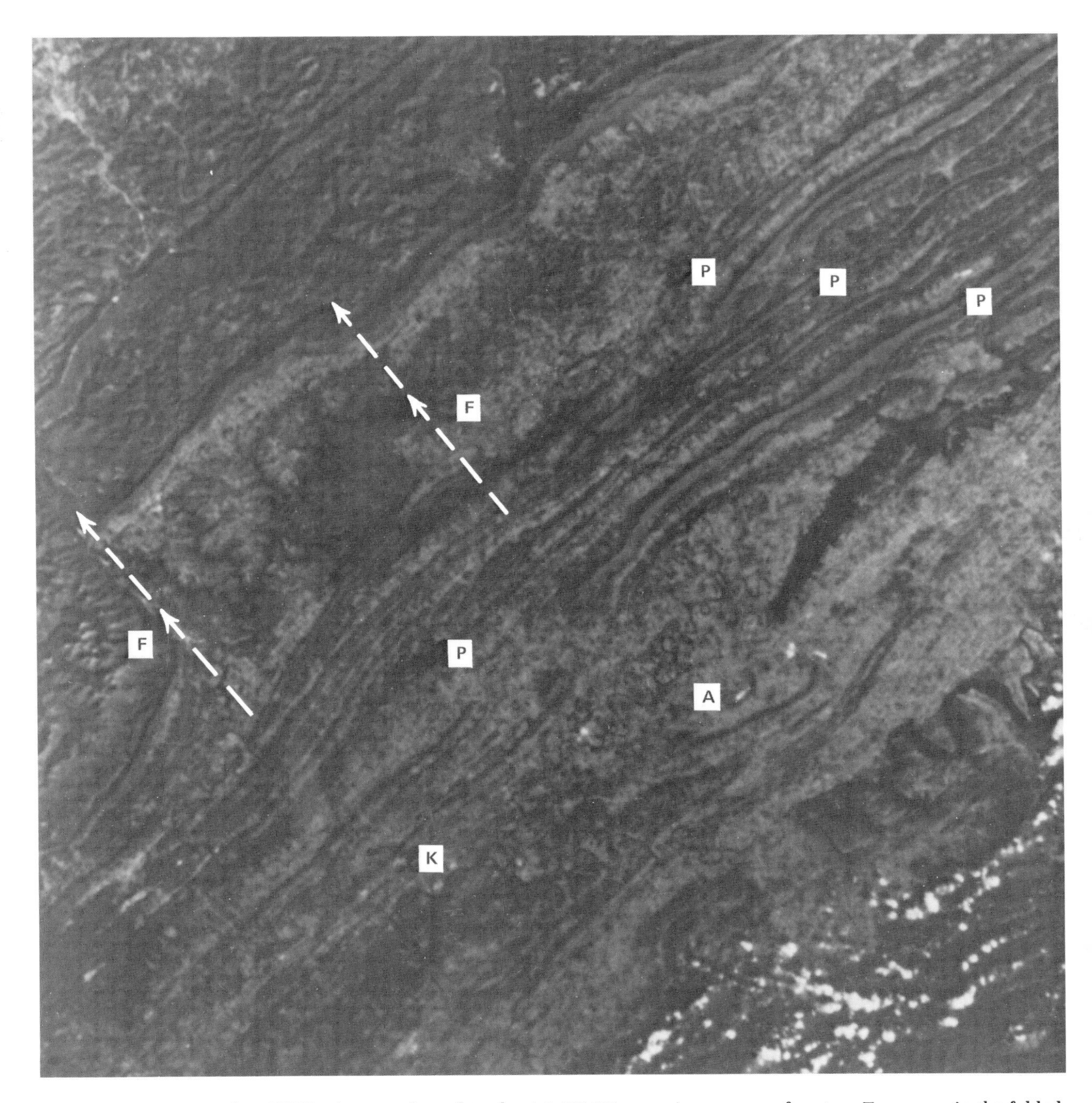


**FIG. 6.39** A portion of an ERTS 1 image enlarged to about 1:670,000, covering an area of eastern Tennessee in the folded sedimentary rocks of the Appalachian Mountains (for location and geology, *see Fig. 2.2*). Knoxville is at *K*. The trend of the folded mountains is clearly shown; *P* indicates plunging folds. Numerous lineaments are evident and two wrench faults are identified at *F*. The stream patterns are very much controlled by rock structure, except at *A* where a mature stream flows in a limestone valley. The larger water bodies are man-made. *(Original image by NASA, reproduction by U.S. Geological Survey, EROS Data Center.)*

**TABLE 6.6**
**GEOLOGIC NOMENCLATURE FOR JOINTS**

| Joints | Characteristics | Figure |
|---|---|---|
| Joint type | | |
| Longitudinal | Parallels bedding or foliation planes | 6.48 |
| Normal or cross | Intersects bedding or foliations at right angles | 6.48 |
| Diagonal or oblique | Intersects bedding or foliations obliquely | 6.48 |
| Foliation | Parallels foliations | 6.47 |
| Curvilinear | Forms parallel sheets or slabs; often curved | 6.43 |
| Joint systems | | |
| Joint set | Group of parallel joints | |
| Joint system | Two or more sets or a group of joints with a characteristic pattern | 6.46 |
| Conjugate system | Two intersecting sets of continuous joints | |
| Orthogonal system | Three sets intersecting at right angles | 6.41 |
| Cubic system | Forming cubes | 6.41 |
| Rhombic | Three sets parallel but with unequal adjacent sides and oblique angles | |
| Pyramidal | Sets intersecting at acute angles to form wedges | 9.14 |
| Columnar | Divide mass into columns; three- to eight-sided, ideally hexagonal | 6.42 |
| Intense | Badly crushed and broken rock without system; various shapes and sizes of blocks | 6.45 |

ing material. All of these features affect their engineering characteristics. Silt, clay, and mylonite are common fillings. A classification of surface configuration or roughness is given on Table 6.7.

*Engineering*

Some characteristics of various joint sets in hard rocks are given on Table 6.8. In general, deformation occurs by closure of joints and by displacement of the bounding blocks. Strength is derived from joint surfaces in contact, or from the joint filling materials *(see also Art. 6.4.4)*.

### 6.4.3 JOINTING IN VARIOUS ROCK TYPES

#### Igneous Rocks

*Solid Masses*

In solid masses joints are widely spaced and tight as shown on Fig. 6.40.

*Systematic Orientation*

Systematic orientation of joints results from cooling and contraction of the mass during formation from magma. Normally three joint sets form more or less according to a definite system (orthogonal, rhombic, etc.). In granite and similar massive formations, joints often form cubic blocks as shown on Fig. 6.41.

*Columnar Jointing*

Rapid cooling of magma in dikes, sills, and flows results in columnar jointing as shown on Figs. 6.17 and 6.42.

*Curvilinear Joints*

Curvilinear joints form on the surface of granite masses from stress relief during uplift and expansion of minerals (feldspar decomposing to kaolin) during weathering, resulting in loose

**TABLE 6.7**
**CLASSIFICATION OF JOINT SURFACE ROUGHNESS**

| Surface | Roughness | Origin |
|---|---|---|
| Undulating | Rough or irregular | Tension joints, sheeting, bedding |
| | Smooth | Sheeting, nonplanar foliation, bedding |
| | Slickensided | Faulting or landsliding |
| Planar | Rough or irregular | Tension joints, sheeting, bedding |
| | Smooth | Shear joints, foliation, bedding |
| | Slickensided | Faulting and landsliding |

**TABLE 6.8**
**CHARACTERISTICS OF DIFFERENT JOINT SETS IN HARD ROCKS***

| Characteristics | Shear joints | Displacement joints | Tension joints |
|---|---|---|---|
| Evenness | Very even | Even to uneven | Rarely even |
| Roughness | Smooth to mirror smooth | Smooth to rough | Rough |
| Degree of joint continuity | High | Medium | Low |
| Length of joint | Faults to very large joints | Very large joints to small joints | Large joints to small joints |
| Joint opening | Close | Variable | Wide |
| Fill material | Mylonite | Product of abrasion | |
| Traces of movement | Slickensides | | |
| Fractures on molecular scale | Shearing of mesh | Fracture of crystallites | Fracture of crystallites |
| Orientation with respect to microstructure | Diagonal joints | Diagonal joints | Transverse joints = ac joints<br>Longitudinal joints = bc joints<br>Sedimentation joints = ab joints |
| Fracture angle $2\alpha$ | $\sigma_1$; $\sigma_1 > \sigma_3$; $\sigma_3$ | $\sigma_1$; 70-80°; $\sigma_3$ | $\sigma_1$; 0°; $\sigma_3$ |
| Friction angle (in unrestrained dilatation) | | | |
| Initial (peak) | | 30 to 55° | 40 to 55° |
| Residual | About 15 to 30° | 20 to 40° | 30 to 45° |

*From Fecker (1978).[10]

FIG. 6.40 Dike in solid, tightly jointed granite, displaced by small fault (Cabo Frio, Rio de Janeiro, Brazil).

FIG. 6.41 Cubic blocks formed by jointing. Frost wedging has increased intensity (Mt. Desert Island, Maine).

slabs as shown on Fig. 6.43. The phenomenon, referred to as *exfoliation* or *sheeting*, results in the formation of distinctive domes, termed *inselbergs* such as Sugar Loaf Mountain in Rio de Janeiro (Fig. 6.44).

*Intense Jointing*

Crustal warping, faulting, or the injection of magma into adjacent bodies can result in breaking the mass into numerous fragments of various shapes and sizes as shown on Fig. 6.45. Uncontrolled blasting can have the same result.

**Metamorphic Rocks**

*Systematic Orientation*

In massive bodies the system more or less follows the joint orientation of the original igneous rock, but new joints usually form along mineral orientations to form a blocky mass as illustrated in Fig. 6.46.

*Intense Jointing*

Forces as described for igneous rocks break a massive formation into numerous blocks of various shapes and sizes.

**FIG. 6.42 Columnar joints in a basalt sill injected over Triassic sandstones and shales (West Orange, New Jersey).**

FIG. 6.43 Exfoliation jointing in granite in a tropical climate (Rio de Janeiro, Brazil).

FIG. 6.44 Exfoliation of granite resulted in Sugar Loaf, Rio de Janeiro, Brazil.

*Foliation Jointing*

Jointing develops along the foliations in gneiss and schist creating a "foliate" structure as shown on Fig. 6.47. In the photo, the joints opened from the release of high residual stresses which produced large strains in the excavation walls.

*Platy Jointing*

Platy jointing follows the laminations in slate.

**Sedimentary Rocks**

*Systematic Jointing*

Normal, diagonal, and longitudinal joints form in the more brittle sandstones and limestones because of erosion of overlying materials, increased stresses from expanding materials in adjacent layers, and increased stresses from desiccation and dehydration.

Joints in brittle rocks usually are the result of the release of strain energy, stored during compression and lithification, that occurs during subsequent uplift, erosion, and unloading. Because of this dominant origin, most joints are normal to the bedding planes. Major joints, cutting across several beds, usually occur in parallel sets and frequently two sets intersect at about 60° in a conjugate system. Jointing in a siltstone is illustrated on Fig. 6.48, but the joint intensity has been substantially increased by the intrusion of an underlying sill.

**FIG. 6.45 Intense jointing in migmatite (Rio Santos Highway, Brazil). (Hard hat gives scale.)**

**FIG. 6.46 Block jointing in granite gneiss (Connecticut Turnpike, Middlesex County, Connecticut).**

**FIG. 6.47 Foliation jointing in the Fordham gneiss (Welfare Island, New York). (At this location the 30-m-deep excavation encountered high residual stresses.)**

**FIG. 6.48 Normal, diagonal, and longitudinal jointing in siltstone with thin shale beds. The intense jointing was caused by the intrusion of a sill beneath the Triassic sedimentary rocks (Montclair, New Jersey). Excavation was made for a high-rise building.**

In interbedded sandstones and shales, expansion of the shales causes normal jointing in the sandstones as shown on Fig. 6.95, especially when exposed in cut.

*Folding*

Folding develops joints normal to the bedding and along the bedding contacts in the brittle rocks. Cleavage joints form in shales.

### 6.4.4 BLOCK BEHAVIOR

**General**

The planes of weakness divide the mass into blocks, and it is the mechanical interaction of the blocks and joints and other discontinuities that normally govern rock-mass behavior under applied stress. Under stress, however, the behavior of weak, decomposed rock may be governed by the properties of the intact blocks, rather than fractures.

Blocks of rock, intact or decomposed, have the parameters of dimension, unit weight, Young's modulus, Poisson's ratio, cohesion, friction angle, and total strength at failure. These properties normally will account for the smaller-scale defects of foliation, schistosity, and cleavage.

The orientation of the discontinuities with respect to the direction of the applied stress has a most significant effect on the strength parameters and deformation.

### Deformation

As stresses are imposed, either by foundation load or the removal of confining material, the fractures may begin to close, or to close and undergo lateral displacement. Closure is common in tunneling operations.

Under certain conditions, such as on slopes or other open excavations, and in tunnels, removal of confining material may result in strains and the opening of fractures.

### Rupture Strength

*Shearing Resistance*

As the joints close under applied load, resistance to deformation is provided as the strength along the joint is mobilized. If the joints are filled with clay, mylonite, or some other material, the strength will depend on the properties of the fillings. If the joints are clean, the strength depends on the joint surface irregularities (asperities).

*Strength of Asperities*

Small asperities of very rough surfaces may easily shear under high stresses. On a smooth undulating surface, however, displacement occurs without shearing of the asperities.

Under low normal stresses, such as in slopes, the shearing of asperities is considered uncommon.

*Peak Shear Strength*

CASE 1 Expressions to provide for both asperity strength and the joint friction angle in terms of normal stress are presented by Ladanyi and Archambault (1970)[11] and Hoek and Bray (1977).[4]

CASE 2 An expression to provide for joint shear strength for low normal stresses is given by Barton (1973)[12] as follows:

$$\tau = \sigma_n \tan\left(\phi + \text{JRC} \cdot \log_{10}\frac{\sigma_j}{\sigma_n}\right) \qquad (6.1)$$

where $\tau$ = joint shear strength
$\phi$ = joint friction angle
$\sigma_n$ = effective normal stress
$\sigma_j$ = joint uniaxial compressive strength
JRC = joint roughness coefficient (JRC = 20 for rough undulating joint surfaces, 10 for smooth undulating joint surfaces, 5 for smooth, nearly planar joint surfaces)

The JRC results from tests and observations and accounts for the average value of the angle $i$ of the asperities, which affects the shear and normal stresses acting on the failure surface. Hoek and Bray (1977)[4] suggest that Barton's equation is probably valid in the range for $\sigma_n/\sigma_j$ = 0.01 to 0.03. Ranges for friction angles for various joint types are given on Table 6.8. If movement has occurred along the joint in the past, the residual strength may govern.

CASE 3 Strength parameters for joints are also given as *joint stiffness* $k_s$ and *joint stiffness ratio* $k_s/k_n$. Joint stiffness is the ratio of shear stress to shear displacement, or the unit stiffness *along* the joint. Joint normal stiffness $k_n$ is the ratio of the normal stress to normal displacement, or the unit stiffness *across* the joint [Goodman et al. (1968)[13]].

### Applications

Block behavior is a most important consideration in the foundation design of concrete gravity or arch dams, and in rock-slope stability analysis. Rigorous analysis requires accurate modeling of the rock-mass structure and the definition of the strength parameters, which is a very difficult undertaking from the practical viewpoint.

## 6.5 FAULTS

### 6.5.1 GENERAL

Faults are fractures in the earth's crust along which displacement has occurred. Their description and identification are discussed in this article. Fault activity, displacement, and their significance to earthquake engineering is presented in Art. 11.3.1.

Natural causes are tectonic activity and the compressive forces in the earth's crust. Strains increase with time until the crust ruptures and adjacent blocks slip and translate vertically, horizontally, and diagonally.

Unnatural causes are the extraction of fluids

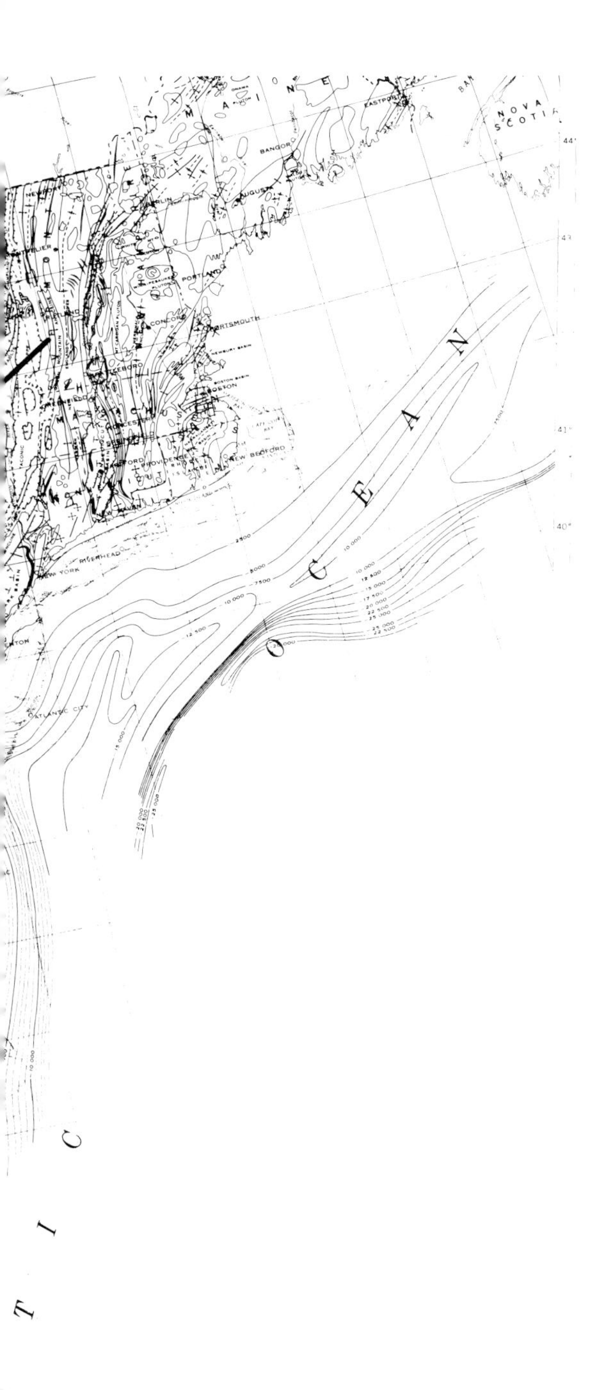


FIG. 6.49 **A portion of the tectonic map of the eastern United States showing the location of the New York-Alabama lineament in relation to Appalachian deformation patterns.** [*From U.S. Geological Survey and American Association of Petroleum Geologists (1962) and King and Zeitz (1978).*[13a]]

from beneath the surface as described in Art. 10.2.2.

Faults are shown on geologic and tectonic maps as lineaments such as are illustrated on Fig. 6.49.

## 6.5.2 TERMINOLOGY

### Fault Systems

*Master faults* refer to major faults, generally active in recent geologic time, that extend for tens to many hundreds of kilometers in length. They are usually strike-slip in type, and are the origin of much of the earth's earthquake activity. Examples are the San Andreas fault in California, the Anatolian system in Turkey, the Philippine fault, the Alpine and Wellington faults of New Zealand, and the Atacama fault of northern Chile.

*Minor faults* are fractures connected to or adjacent to the major fault, referred to as *secondary, branch,* or *subsidiary* faults, forming a fault system.

*Major faults* refer to faults, other than master faults, that are significant in extent.

### Fault Nomenclature

Fault nomenclature is illustrated on Fig. 6.50.

- *Net slip* is the relative displacement along the fault surface of adjacent blocks.
- *Strike slip* is the component of net slip parallel to the strike of the fault in horizontally or diagonally displaced blocks.
- *Dip slip* is the component of the net slip parallel to the dip of the fault in vertically or diagonally displaced blocks.
- *Hanging wall* is the block above the fault.
- *Footwall* is the block below the fault.
- *Heave* is the horizontal displacement.
- *Throw* is the vertical displacement.
- *Fault scarp* is the cliff or escarpment exposed by vertical displacement.

### Fault Types

Fault types are defined by their direction of movement as illustrated on Fig. 6.51.

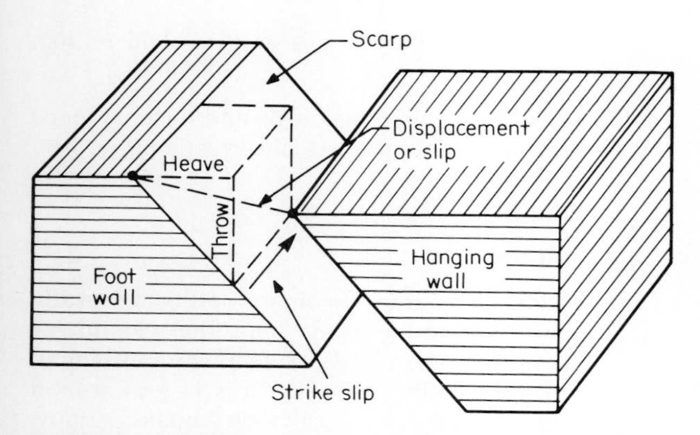


**FIG. 6.50 Nomenclature of obliquely displaced fault blocks.**

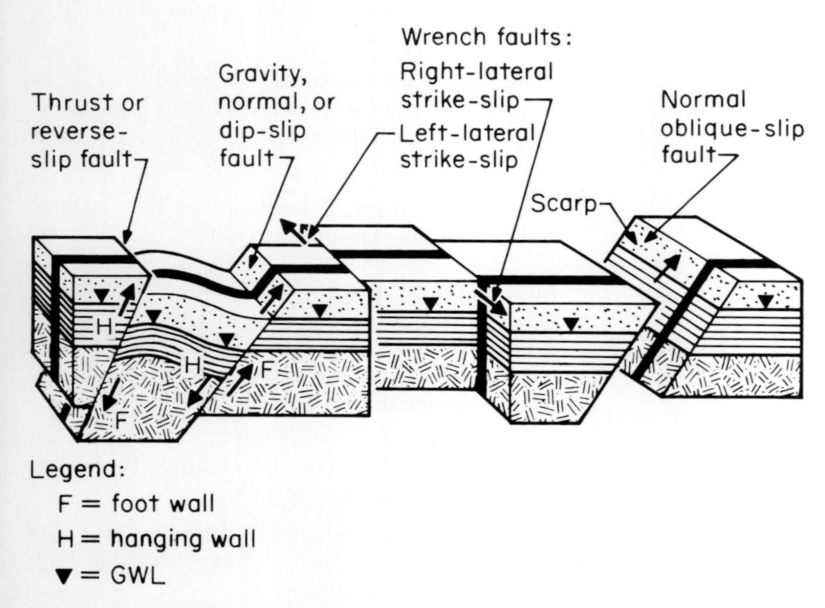


**FIG. 6.51 Various types of faults.**

- *Normal or gravity fault:* Displacement is in the direction of dip.
- *Reverse or thrust fault:* Hanging wall rides up over foot wall.
- *Wrench or strike fault:* Displacement is lateral along the strike.
- *Oblique fault:* Movement is diagonal with both strike and dip components.
  - Normal oblique has diagonal displacement in the dip direction.
  - Reverse oblique has diagonal displacement

with the hanging wall riding over the foot wall.

- *Overthrusts:* Result in the overturning of beds such as occurs with the rupture of an anticline as in Fig. 6.52. Large overthrusts result in unconformities when removal of beds by erosion exposes older beds overlying younger beds.

### 6.5.3 CHARACTERISTICS AND IDENTIFICATION

#### Surface Evidence

Surface evidence of faulting is given by lineations, landforms, drainage features, secondary features, and seismological data as summarized and described on Table 6.9. Some characteristic landforms are illustrated on Figs. 6.53, 6.54, and 6.55.

#### Internal Evidence

Internal evidence as disclosed by borings and excavations is given by stratum discontinuity, slickensides, fault zone materials (breccia, gouge, mineral alteration), groundwater levels, and foliation and mylonite shear zones as summarized and described on Table 6.10. Examples of slickensides are given on Figs. 6.56 and 6.57, various kinds of fault fillings on Fig. 6.58, and some effects of the circulation of hydrothermal solutions along faults on Fig. 6.59.

#### Normal and Thrust Faults

The Ramapo fault of northern New Jersey is an example of normal faulting and a tilted fault block. The Ramapo River flows along the contact between the Precambrian crystalline rocks in the western highlands and the Triassic lowlands to the east as shown on Figs. 6.60 and 6.61. The steep slopes on the western side of the river are the fault scarp; blocked drainage and sag ponds are apparent on the easterly lowlands as shown on the topographic map. The area is shown also on the ERTS image given as Fig. 6.29.

The characteristics of thrust and normal faults during initial displacement are illustrated on Fig. 6.62. Block diagrams illustrating the surface features of a normal-slip and a reverse-slip fault after substantial displacement are given on Fig. 6.63. Thrust faults are illustrated also on Figs. 6.64 and 6.65.

#### Strike Slip or Wrench Faults

The San Andreas fault is the best known example of a strike-slip fault. Located parallel to the coastline in southern California *(see Fig. 11.11)*, it is almost 1000 km in length and extends vertically to a depth of at least 30 km beneath the surface. Along with branch faults and other major faults, it is clearly evident as strong lineations on the ERTS image given as Fig. 6.66 and the SKYLAB photo given as Fig. 2.9.

The fault is represented by a complex zone of crushed and broken rock ranging generally from about 100 to 1500 m in width at the surface. It provides many evidences of faulting including linear valleys in hilly terrain (Fig. 6.67), offset stream drainage, and offset geologic formations.

### 6.5.4 ENGINEERING SIGNIFICANCE

#### Earthquake Engineering

*Relationship*

As discussed in Art. 11.3.1, it is generally accepted that large earthquakes are caused by rupture of one or more faults. The dominant fault is termed the *causative fault*.

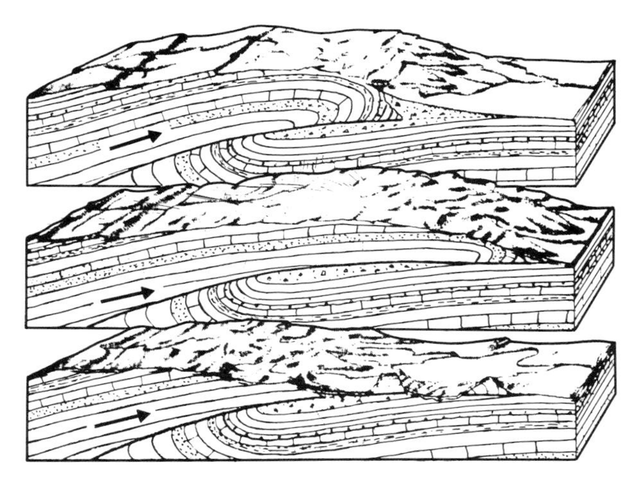

**FIG. 6.52 The stages of development of a thrust fault, or overthrust.**

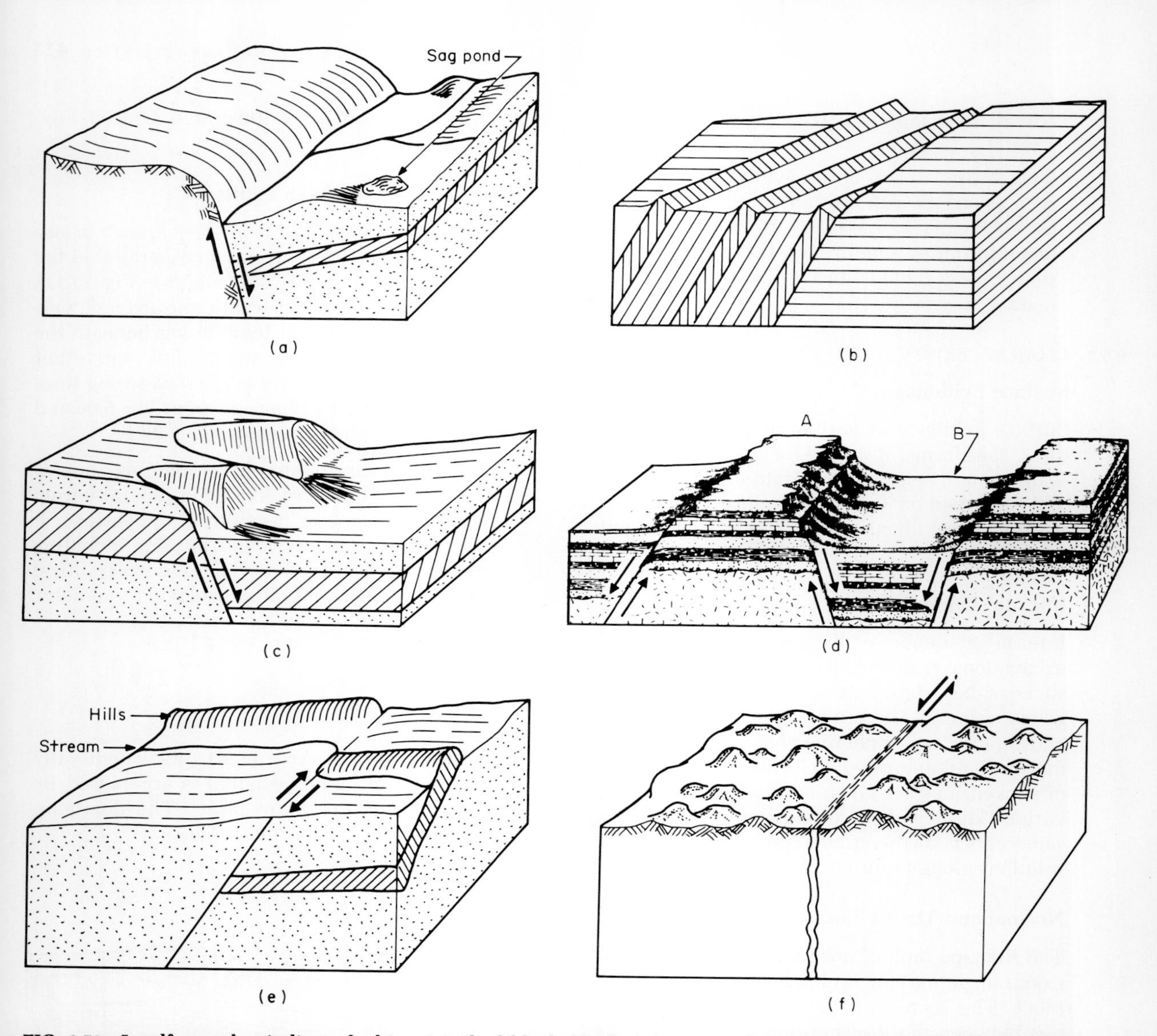


**FIG. 6.53** **Landforms that indicate faulting: (*a*) tilted block, blocked drainage of dip-slip fault; (*b*) truncated landform of strike-dip fault; (*c*) triangular facets along scarp of dip-slip fault; (*d*) horst (*A*) and graben (*B*) block faulting forming rift valleys; (*e*) offset drainage shows recent horizontal movement of strike-slip fault; (*f*) lineation in landform caused by differential erosion along fault zone of strike-slip fault.**

**FIG. 6.54** ***(Opposite)*** **Evolution of faceted spurs along the Wasatch fault (Salt Lake City, Utah). Diagrams show evolution of faceted spurs produced by periods of movement separated by periods of stability. (*a*) Undissected fault scarp; (*b*) development of faceted spurs by streams cutting across scarp; (*c*) period of stability with slope retreat and development of a narrow pediment; (*d*) recurrent movement; (*e*) dissection of new segment of scarp by major streams and by those formed in *b*; (*f*) new period of stability with slope retreat and development of another narrow pediment at base of mountain front upthrown block; (*g*) recurrent movement; (*h*) dissection of scarp formed in g resulting in a line of small faceted spurs at base of mountain front. Remnants of narrow pediments are preserved at apices of each set of faceted spurs; (*i*) progressive slope retreat is accompanied with age by a decrease in slope angle of faceted spurs.** [*From Hamblin (1964).*[14]]

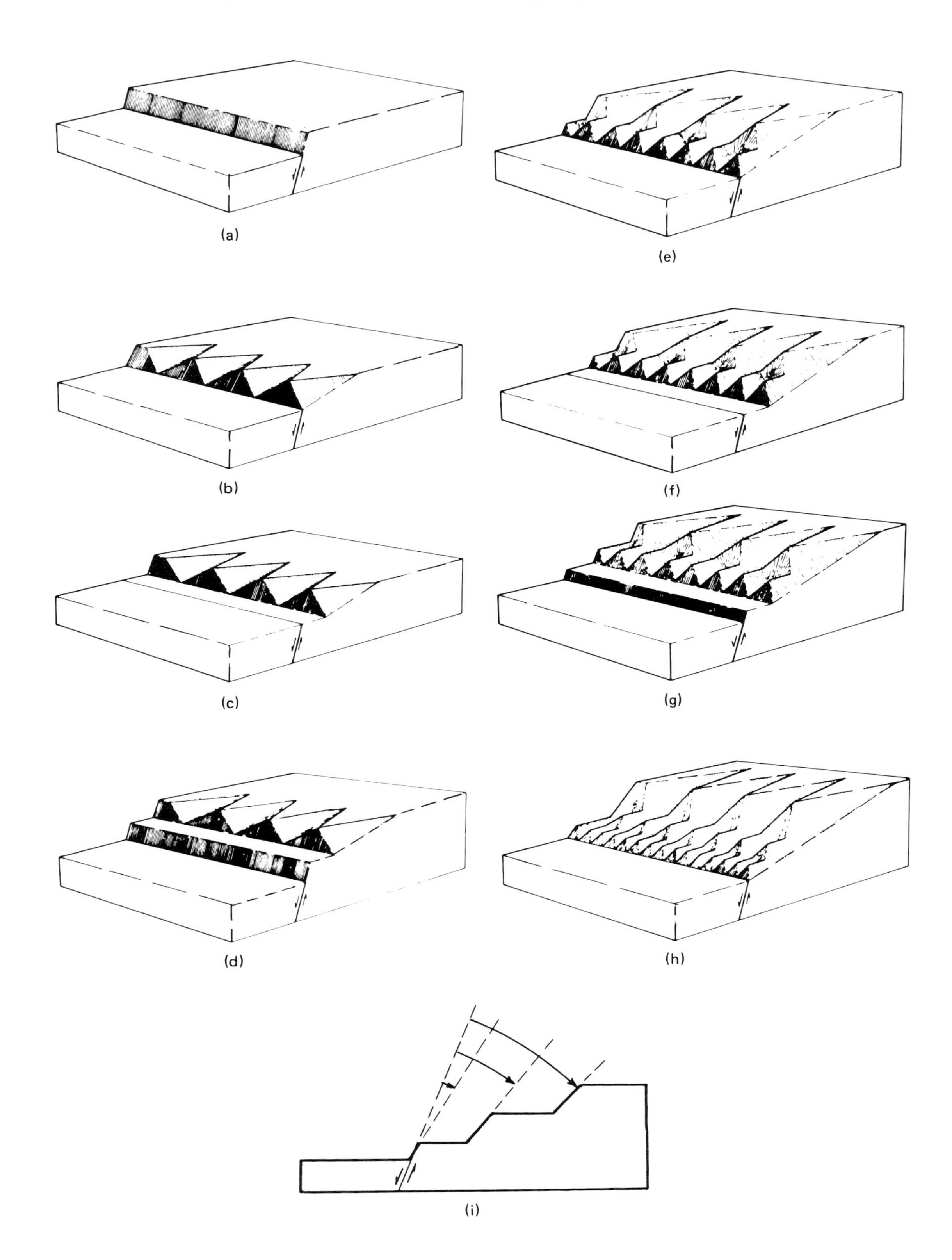
(a)
(e)
(b)
(f)
(c)
(g)
(d)
(h)
(i)

**FIG. 6.55** *(Left)* **Faceted spurs along the base of a mountain range (western United States).**

**FIG. 6.56** *(Below)* **Slickenside in shale. Thirty-six-inch calyx core taken during explorations for the Tocks Island Dam site (Delaware River, New Jersey-Pennsylvania).**

**FIG. 6.57 Slickensides in dike of decomposed basalt (Highway BR 277, Curitiba, Parana, Brazil).**

*Seismic Design Requirements*

Seismic design studies require the identification of all faults that may represent an earthquake source of significance to construction, the determination of fault characteristics, and finally the selection of the causative fault.

*Significant Characteristics*

The fault length is determined and the amount and form of displacement that might occur are judged. Worldwide evidence indicates that the maximum displacements along faults during earthquakes is generally less than 5 to 7 m, and the average displacement is generally less than 1 m [Sherard et al. (1974)[15]] *(see Art. 11.3.1)*.

A most important element is the determination of fault activity, i.e., active, potentially active, or inactive *(see Tables 11.7 and 11.9)*. The U.S. Nuclear Regulatory Commission (NRC) defines an active fault basically as one having movement within the past 35,000 years, or essentially during the Holocene epoch. The definition requires dating the last movement and determining whether creep is currently occurring. An active or potentially active fault is termed a *capable fault* and its characteristics become a basis for seismic design criteria.

*Conclusion*

From these factors the earthquake magnitude and duration that might be generated by fault rupture are estimated. A relationship between earthquake magnitude and fault displacement is given on Fig. 11.23, and a relationship between length of surface rupture and magnitude is given on Fig. 11.24.

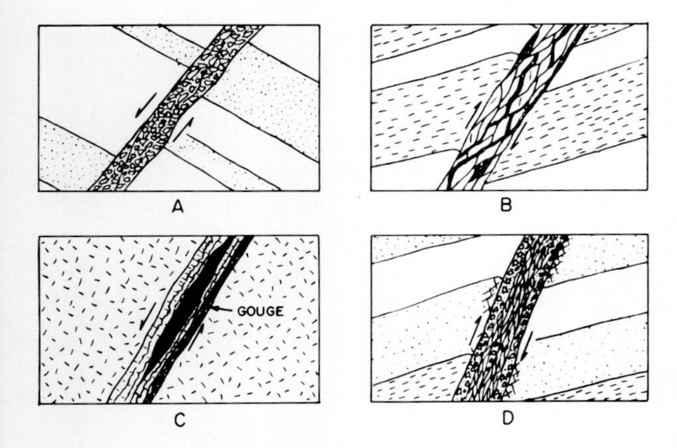


**FIG. 6.58 Various kinds of fault fillings: (*a*) fault breccia; (*b*) braided fault with intersecting seams of gouge; (*c*) vein in an igneous rock with a gouge layer adjacent to foot wall; (*d*) gouge which has rotated and forced larger angular fragments towards walls of fault because of movement concentrated in the center of the fault.** [*From Wahlstrom (1973).*[5] *Reprinted with permission of Elsevier Scientific Publishing Company.*]

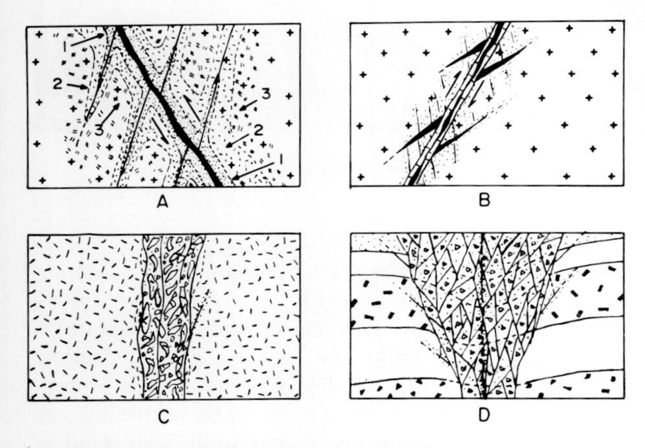


**FIG. 6.59 Some effects of circulation of hydrothermal solutions along faults. (*a*) A metalliferous vein (black) is accompanied by zoned alteration in igneous wall rocks: A zone of silicification and sericitization, near the vein and subsidiary fractures (1) grades outward into a zone of kaolinized feldspars (2), and finally into a zone of alteration containing mixed-layer illite—montmorillonite (3). (*b*) Fault and adjacent gash fractures are filled with minerals deposited from solution. Argillic alteration permeates wall rocks (stippling). (*c*) A corrosive hydrothermal fluid enlarges opening along a fault by solution of the wall rocks. Walls collapse into opening to form a *slab breccia*. (*d*) Corrosive hydrothermal solution dissolves wall rock along a fault and its subsidiary fractures to produce a *collapse breccia*. This phenomenon is sometimes described as *solution stoping*.** [*From Wahlstrom (1973).*[5] *Reprinted with permission of Elsevier Scientific Publishing Company.*]

### General Construction Impact

Fault zones provide sources for large seepage flows into tunnels and open excavations, or seepage losses beneath and around concrete dams. Particularly hazardous to tunneling operations, they are the source of "running ground." Concrete deterioration can be caused by the attack from sulfuric acid which results from reaction of oxidizing surface waters with sulfides such as pyrite and marcasite deposited in the fault zone.

They present weakness zones in tunnels and potential failure zones in slopes or concrete dam foundations or abutments.

## 6.5.5 INVESTIGATION METHODOLOGY SUMMARIZED

*See also Art. 11.5.3.*

### Terrain Analysis

Remote-sensing imagery and topographic maps are interpreted to identify lineaments and other characteristic landforms and drainage features evident on small scales. Interpretation of low-level imagery, such as aerial photos at scales of 1:10,000 and 1:25,000, is made to identify fault scarps, offset drainage, sag ponds, truncated geologic formations, etc. Stereo and nonstereo aerial photos taken when the sun is at a low angle clearly delineate low fault scarps and other small surface features by shadow effects [Sherard et al. (1974)[15]]. Field reconnaissance is performed to confirm desk-top interpretations.

### Explorations

Geophysical methods are used to investigate shallow and deep-seated anomalies. The methods include seismic refraction, seismic reflection (marine surveys), gravimeter measurements, and magnetometer measurements *(see Art. 2.3.2).*

Test borings, vertical and inclined, are drilled to explore and sample the suspected fault zone *(see Art. 2.3.5).*

Trenches excavated across the postulated fault zone enable close examination of the overburden soils and the procurement of samples. Displacement of Holocene strata is the most reliable indicator of "recent" fault activity.

**TABLE 6.9**
**SURFACE EVIDENCE OF FAULTING**

| Feature | Characteristics | Fault type | Figure |
|---|---|---|---|
| Lineations | Strong rectilinear features of significant extent are indicative but not proof. Can also represent dikes, joints, foliations, bedding planes, etc. (Fig. 2.8). | All types | 6.28<br>6.29<br>6.39<br>6.66 |
| Landforms | | | |
| Scarps | Long, relatively smooth-faced, steep-sided cliffs. | Normal | 6.53*a* |
| Truncated ridges | Lateral displacement of ridges and other geomorphic features. | Wrench | 6.53*b*,*e*<br>6.39 |
| Faceted spurs | Erosion-dissected slopes form a series of triangular-shaped faces on the foot wall. | Normal | 6.53*c*<br>6.54<br>6.55 |
| Horst and graben | Block faulting. A sunken block caused by downfaulting or uplifting of adjacent areas forms a rift valley (graben). An uplifted block between two faults forms a horst. Soils forming in the valley are more recent than those on the uplands. Examples: Lake George, New York; Dead Sea; Gulf of Suez; Rhine Valley, West Germany; Great Rift Valley of Kenya; parts of Paraiba River, Sao Paulo, Brazil. | Normal | 6.53*d* |
| Drainage | | | |
| Rejuvenated streams | Direction of flow reversed by tilting. | Normal | 6.53*a* |
| Blocked or truncated | Flow path blocked by scarp and takes new direction. | Normal | 6.53*a* |
| Offset | Flow path offset laterally. | Wrench | 6.53*e* |
| Sag pond | Lakes formed by blocked drainage. | Normal | 6.53*a* |
| Secondary features | Practically disappear within less than 10 years in moist climates, but may last longer in dry [Oakeshott (1973)[13b]]. | | |
| Mole tracks | En-echelon mounds of heaved ground near base of thrust fault or along wrench fault. | Thrust, wrench | 6.63*a* |
| Step-scarps | En-echelon fractures form behind the scarp crest in a reverse fault (tension cracks). | Normal | 6.63*b* |
| Seismological | Alignment of epicenters. | All types | 11.29 |

### Laboratory Testing

Radiometric dating of materials recovered from core borings and trenches provides the basis for estimating the most recent movements. Dating techniques are summarized on Tables 11.8 and A.3.

### Instrumentation

Monitoring of fault movements is performed with acoustical emission devices, shear-strip indicators, tiltmeters, extensometers, seismographs, etc., as described in Art. 4.5.7.

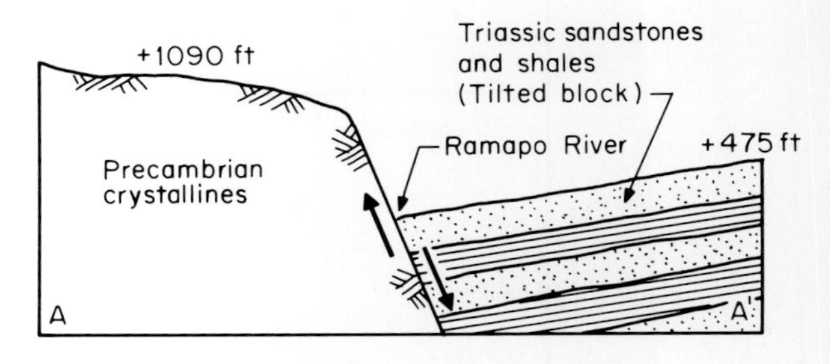


**FIG. 6.60 A section across the Ramapo fault.**

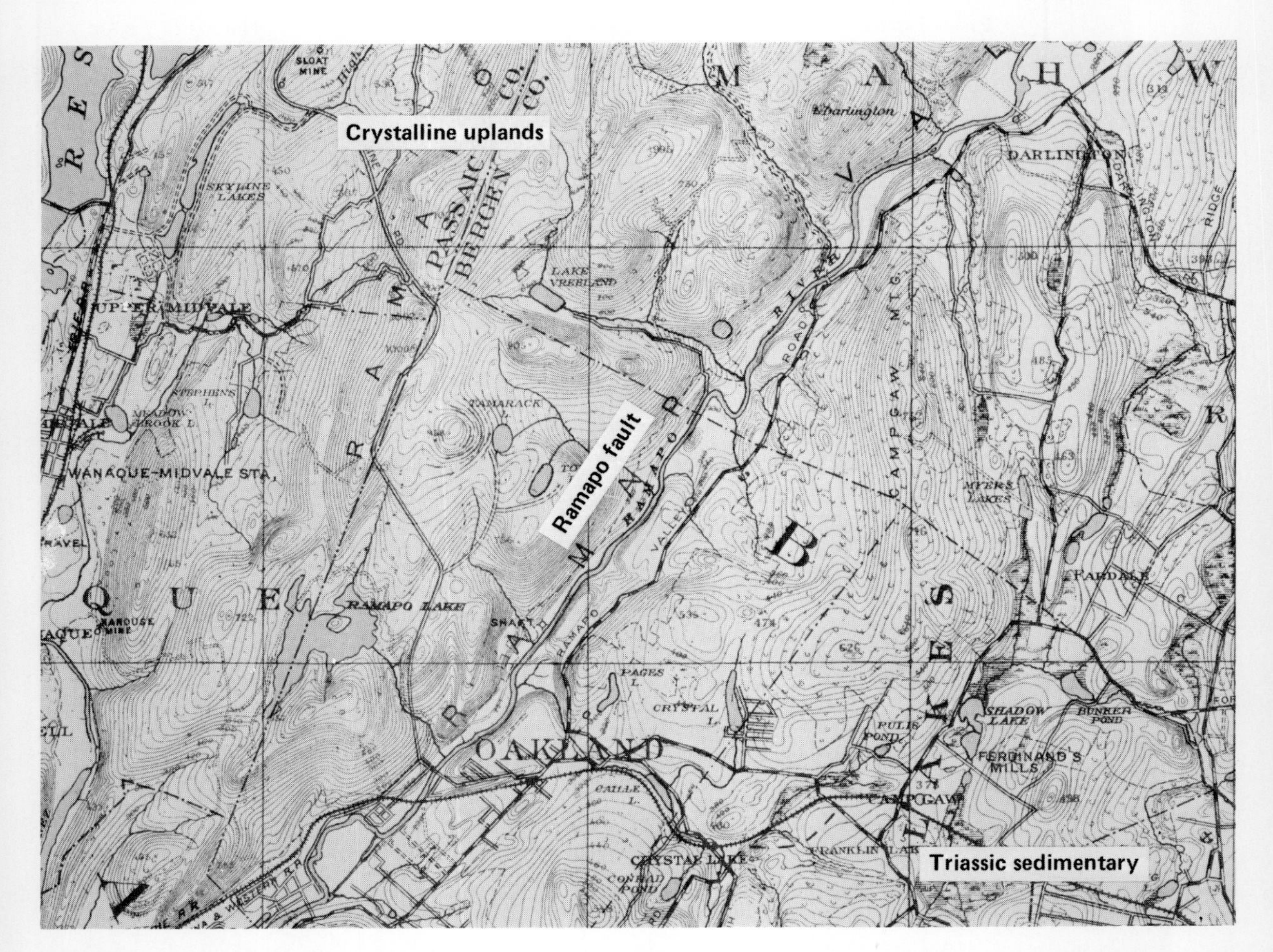


**FIG. 6.61 Topographic expression along the Ramapo fault. (Passaic County, New Jersey. Scale 1 in = 1 mile. *See also Figs. 2.4 and 6.29.*)**

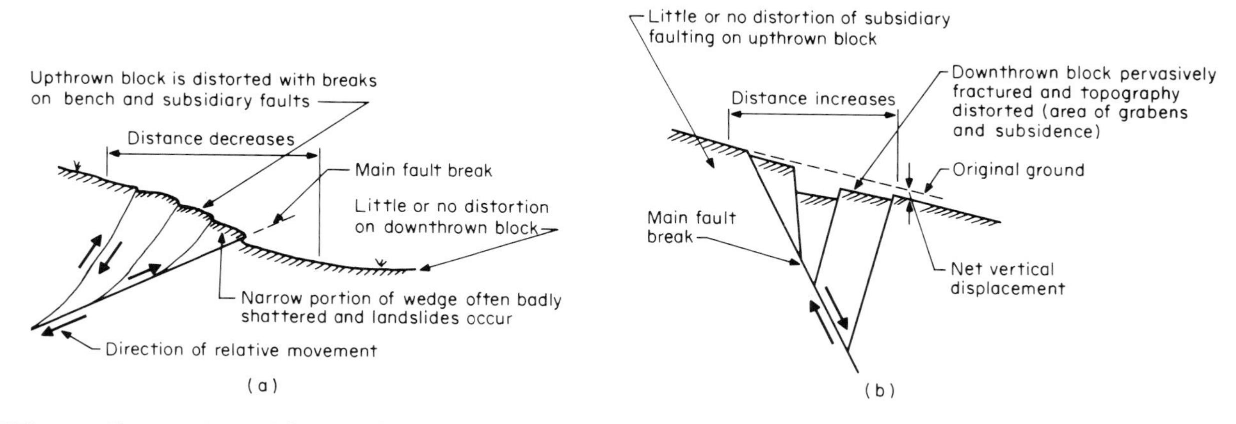


**FIG. 6.62 Characteristics of thrust and normal faults during initial displacement: (*a*) thrust fault and (*b*) normal fault.** [*From Sherard et al. (1974).*[15]]

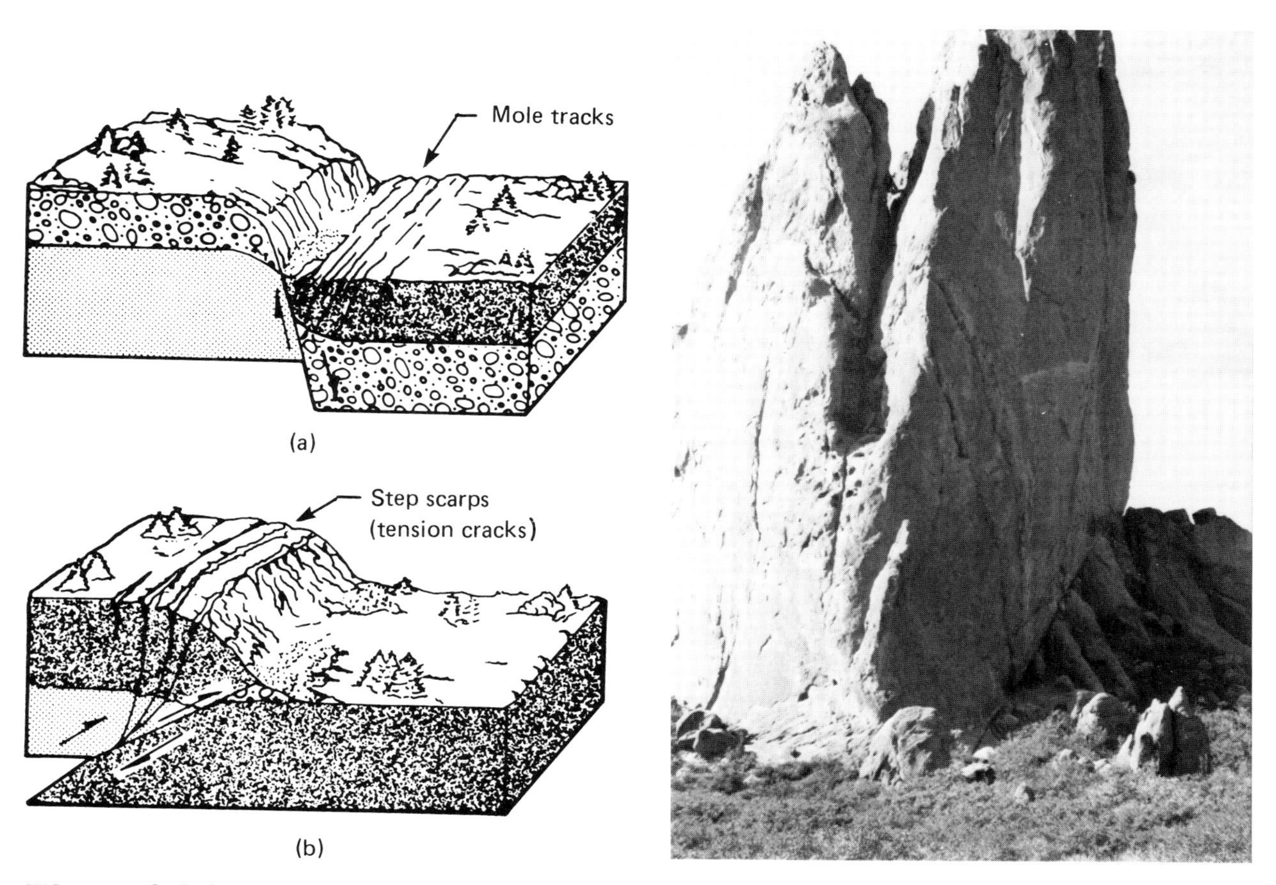


**FIG. 6.63 Block diagrams showing the effects of surface displacement along: (*a*) normal-slip and (*b*) reverse-slip fault. A strike-slip component in either direction results in either a normal-oblique-slip or a reverse-oblique-slip fault.** [*From Taylor and Cluff (1977).*[16]]

**FIG. 6.64 Thrust fault in sandstone (Garden of the Gods, Colorado Springs, Colorado).**

**TABLE 6.10**
**INTERNAL EVIDENCE OF FAULTING**

| Feature | Characteristics | Figure |
|---|---|---|
| Stratum discontinuity | Abrupt change in strata; discontinuous, omitted, or repeated | 6.64 |
| Slickensides | Polished and striated surfaces resulting from shearing forces; characteristic of weaker rocks. | 6.56<br>6.57 |
| Breccia | Angular to subangular fragments in a finely crushed matrix in the fault zone in strong rocks. | 6.58*a* |
| Gouge | Pulverized material along the fault zone; typically clayey; characteristic of stronger rocks. | 6.58*b,c,d* |
| Mineral alteration | Groundwater deposits minerals in the pervious fault zone, often substantially different from the local rock.<br>Circulating waters can also remove materials.<br>Radiometric dating of the altered minerals aids in dating the fault movement. | 6.59 |
| Groundwater levels | Clayey gouge causes a groundwater barrier and results in a water table of varying depths on each side of the fault. The difference in water levels can result in a marked difference in vegetation on either side of the fault, especially in an arid climate. Tree lines in arid climates often follow faults. | 6.51 |
| Foliation shear | Short faults caused by folding result in foliation shear in weaker layers in metamorphic rocks (typically mica, chlorite, talc, or graphite schist in a sequence of harder massive rocks) [Deere (1974)[14a]].<br>Shear zone thickness typically a few centimeters including the gouge and crushed rock. Adjacent rock is often heavily jointed, altered, and slightly sheared for a meter or so on each side. The zones can be continuous for several hundred meters and can be spaced through the rock mass. | |
| Shale mylonite seam | A bedding shear zone caused by differential movement between beds of sedimentary rock during folding or during relief of lateral stress by valley cutting. Concentrated in the weaker beds such as shale, or along a thin seam of montmorillonite or lignite, and bounded by stronger beds such as sandstone or limestone. Sheared and crushed shale gouge is usually only a few centimeters thick but it can be continuous for many tens of meters [Deere (1974)[14a]]. Both foliation shear and mylonite, when present in slopes, represent potential failure surfaces. | |

**FIG. 6.65 Thrust fault in Cambro-Ordovician shales exposed in cut (Rio do Sul, Santa Catarina, Brazil).**

## 6.6 RESIDUAL STRESSES

### 6.6.1 GENERAL

Residual stresses are high generally horizontal compressive stresses stored in a rock mass, like force stored in a spring, in excess of overburden stresses. They have their origin in folding deformation, metamorphism, and slow cooling of magma at great depths, and the erosion of overburden causing stress relief.

Relatively common, they can be in the range of 3 to 10 times, or more, greater than overburden stresses, causing heaving of excavation bottoms, slabbing of rock walls in river gorges and rock cuts, and rock bursts in deep mines and tunnels. During investigation they are measured in situ in shallow-depth excavations with strain meters, strain rosettes, and flat jacks, and, at greater depths in boreholes, with deformation gages, inclusion stress meters, and strain-gage devices as described in Art. 4.4.4.

### 6.6.2 TENSILE STRAINING

#### Condition

Tensile straining occurs at low residual stresses, insufficient to cause bursting. Significant deflections can occur in river valley and excavation walls in some relatively flat-lying, massive sedimentary rocks, and intensely folded and massive rock bodies.

**FIG. 6.66 ERTS image of the San Francisco Bay area showing the lineaments of (1) the San Andreas fault; (2) the Calaveras fault; and (3) the Hayward fault. Compare with the SKYLAB image in Fig. 2.9.** *(Original image by NASA; reproduction by USGS, EROS Data Center.)*

**FIG. 6.67 Aerial oblique of San Andreas Lake, formed in the trough of the San Andreas fault, a strike-slip fault (San Pedro, California).**

### Examples

A quarry floor in limestone in Ontario suddenly experienced a heave of 2.5 m [Coates (1964)[17]].

The walls of a 30-m-deep excavation in the Fordham gneiss in New York City (see *Fig. 6.47*) expanded to cause rock bolt failures [Ward (1972)[18]]. Residual stresses were determined to have been as much as 10 times overburden stresses.

In a Canadian National Railways Tunnel [Mason (1968)[19]] pressures substantially over overburden pressures, in the order of 80 kg/cm$^2$, were measured in the tunnel lining. Displacements due to stress relief essentially reached equilibrium within 8 to 30 days, but creep displacements continued over a period of several years.

At the Snowy Mountains project in Australia [Moye (1964)[20]] flat-jack tests indicated that lateral stresses in granite were about 2.6 times vertical. Measured lateral stresses ranged from 40 kg/cm$^2$ to a high over 200 kg/cm$^2$.

## 6.6.3 ROCK BURSTS

### Significant Factors

Rock bursts are sudden explosive separations of slabs, often weighing as much as several hundred kilos or more, from the walls or ceilings of underground openings. They are more common in deeper mines and tunnels, especially those about 600 to 1000 m in depth, although they have been noted in quarries and shallow mines.

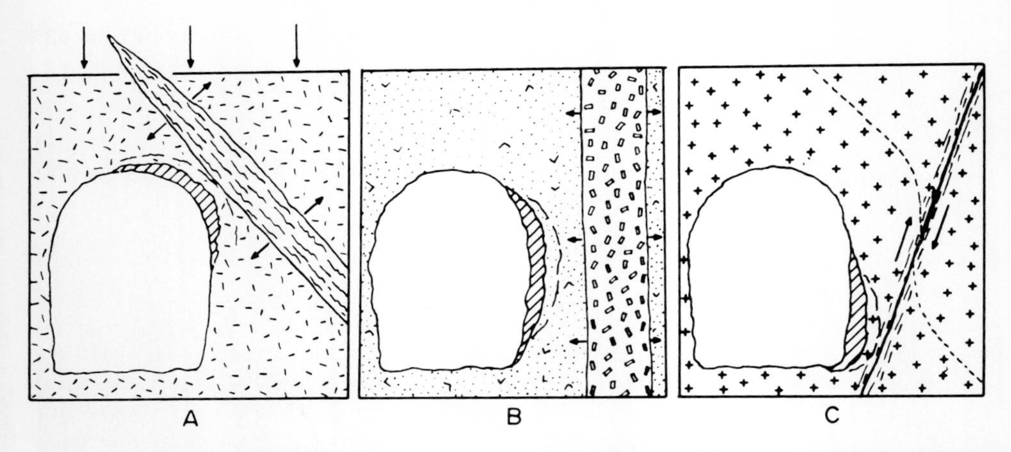


**FIG. 6.68 Some geologic conditions creating forces resulting in rock bursts. (a) Load of overlying rock in deep tunnel in brittle granite causes semiplastic movement in a weak lens of mica-rich schist. Rock burst (hatched) occurs in tunnel at point closest to lens. (b) An igneous dike which was forcefully injected into a hard quartzite is accompanied by residual stresses promoting a rock burst in the wall of the tunnel. (c) Forces that have caused elastic strain in strong igneous rock are directed toward the tunnel opening to cause a rock burst.** [*From Wahlstrom (1973).*[5] *Reprinted with permission of Elsevier Scientific Publishing Company.*]

They can occur immediately after the opening is made, or at any future time. Their activity has been correlated with earthquakes [Rainer (1974)[21]].

### Occurrence

There are a number of conditions particularly favorable for the occurrence of high in situ stresses and rock bursts including: massive rock such as granite or gneiss, with few joints, located at substantial depths; geologically complex conditions characterized by rock fracture and anisotropy, such as exists where highly competent, brittle rocks are interbedded with less competent, somewhat plastic rocks; and concentrations of large stresses accumulated near fractures.

Several conditions which create forces to cause rock bursts in tunnels and mines are illustrated on Fig. 6.68.

## 6.6.4 ANTICIPATING UNSTABLE CONDITIONS

### Instability Factors

The extent of rock instability in excavation from loosening and overall displacement is a function of the nature of the jointing, foliation, and schistosity [Paulmann (1966)[22]], the magnitude of the residual stress [Denkhaus (1966)[23]], and the significance of the in situ stress in terms of rock strength [Hawkes (1966)[24]].

### Stress Ratio Criteria

Deere (1966)[25] stated that tensile stress will not prevail in the vicinity of an opening unless the ratio of lateral to vertical stress falls outside the range of ⅓ to 3 for circular or near-circular openings in rock. In prismatic openings stress concentrations occur around corners and the stress fields are complex.

### Stability Rating Criteria

Hawkes (1966)[24] evaluated the potential for instability and related the ratio of the intact rock stress ($\sigma_1$, the residual stress) to the uniaxial compressive strength (termed the rock stress index), to the major principal stresses. His stability rating system is given on Fig. 6.69.

## 6.7 ALTERATION OF ROCK

## 6.7.1 GENERAL

### Significance

Various rock types decompose to characteristic soil types. The type of residual soil and the approximate depth to fresh rock at a given location are generally predictable if the climate, topography, and basic rock type are known and the processes of rock alteration are understood.

### Definitions

*Alteration* refers to any physical or chemical change in a rock or mineral subsequent to its formation [Rice (1954)[26]].

*Weathered rock* is surficial rock that has undergone physical and/or chemical changes due to atmospheric agents.

*Disintegration* refers to the breaking of rock into smaller fragments, which still retain the identity of the parent rock, through the action of physical agents (wind, water, ice, etc.).

*Decomposition* refers to the process of destroying the identity of mineral particles and changing them into new compounds through the activity of chemical agents.

*Hydrothermal alteration* refers to changes in rock minerals occurring deep beneath the surface, caused by percolating waters and high temperatures.

### Weathering Agents and Processes

*Mechanical Fragmentation*

Mechanical fragmentation is a product of rock joints forced open and fractured under the influence of freezing water, growing tree roots, and expanding minerals; slabs freed by exfoliation and stress relief; and blowing sand causing erosion and abrasion.

*Talus* is the accumulation along a slope of large fragments that have broken free and migrated downward. They can accumulate in large masses as shown on Fig. 6.70.

*Chemical Decomposition*

Chemical decomposition occurs through the processes of oxidation, leaching, hydrolysis, and reduction as described on Table 6.11, and from the engineering viewpoint is the most important aspect of rock alteration since the result is residual soils (see Art. 7.2).

*Hydrothermal Alteration*

Occurring deep beneath the surface at temperatures of 100 to 500°C, hydrothermal alteration changes rock minerals and fabrics, producing weak conditions in otherwise sound rock. It is particularly significant in deep mining and tunneling operations.

*Argillization* is the most significant of many forms of hydrothermal alteration from the aspect of construction, since it represents the conversion of sound rock to clay. When encountered in tunnels, the clay squeezes into the excavation, often under very high swelling pressures, and control can be very difficult.

## 6.7.2 FACTORS OF DECOMPOSITION

### The Factors

Climate, parent rock, time, rock structure, topography, and the depth of groundwater table all affect the rate of decomposition, the depth of penetration, and the mineral products. Urbanization may accelerate decomposition in some rock types through percolation of polluted waters.

### Climate

Precipitation and temperature are major factors in decomposition. General relationships among climate, type of weathering (mechanical vs. chemical), and activity resulting in rock decay are given on Table 6.12. Climatic regional

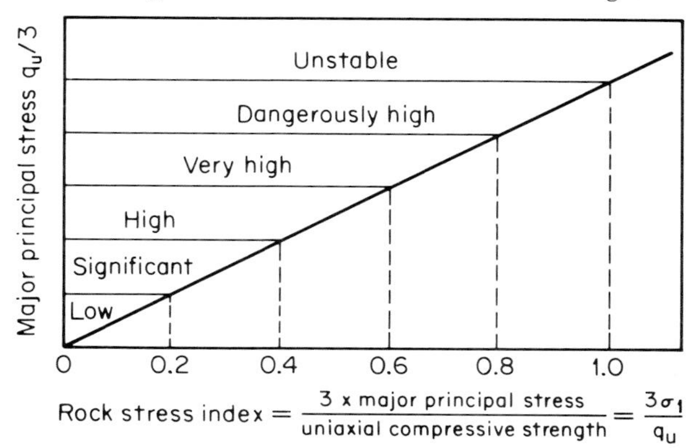


**FIG. 6.69 Significance of residual stress and excavation wall stability in terms of uniaxial compressive strength.** [*From Hawkes (1966).*[24]]

| TABLE 6.11 CAUSES OF CHEMICAL DECOMPOSITION OF MINERALS | |
|---|---|
| **Cause** | **Process** |
| Oxidation | Oxygen from the atmosphere replaces the sulfur element in rock minerals such as pyrite, containing iron and sulfur, to form a new compound—limonite. The sulfur combines with water to form a weak solution of sulfuric acid that attacks the other minerals in the rock. |
| Leaching | Chemical components are removed by solution. Calcite, the chief component of limestone, dissolves readily in carbonic acid, formed when carbon dioxide from the air dissolves in water or when rainwater permeates organic soils. The greater the amount of dissolved oxygen, the greater is the leaching activity. |
| Hydrolysis | The chemical change that occurs in some minerals, such as the feldspars, when some of the mineral constituents react with the water molecule itself. Orthoclase feldspar is changed to kaolin by hydrolysis, and appears in the rock mass as a very white, friable material. The formation of kaolinite causes further disintegration of the rock mechanically by expansion because the kaolinite has a greater volume than the feldspar. It is considered, therefore, a cause of exfoliation. |
| Reduction | Humic acids, produced from decaying vegetation in combination with water, attack the rock mass. |

| TABLE 6.12 CLIMATE VS. ROCK DECAY | | |
|---|---|---|
| **Climate** | **Weathering** | **Activity** |
| Cold, dry | Strong mechanical weathering | Freezing temperatures cause rock breakup. |
| Cool, wet | Moderate chemical weathering, some mechanical | Organic material decays to produce active humic acid to react with the parent rock and cause decay. |
| Hot, wet | Strong chemical weathering | High moisture and high temperatures cause rapid decay of organic material and an abundance of humic acid to cause rapid rock decay. |
| Continuously wet | Some chemical weathering | Water movement is downward, causing leaching by removal of soluble salts and other minerals. |
| Alternating wet and dry | Retarded | Prevailing water movement may be upward during the dry period, concentrating and fixing oxides and hydroxides of iron (laterization), which eventually results in a barrier preventing downward movement of water and retarding decay below the laterized zone *(see Art. 7.7.2)*. In predominantly dry climates evaporites such as caliche form *(see Duricrusts, Art. 7.7.2)*. |
| Hot, dry | Very slight | Chemical and mechanical activity is very low. |

**FIG. 6.70 Talus formation at the base of a cliff (Mt. Desert Island, Maine). The figures give scale.**

boundaries in terms of mean annual rainfall and temperature vs. the intensity of types of weathering are given on Fig. 6.71, and climate vs. weathering profiles in tectonically stable areas are given on Fig. 6.72. In general, the higher the temperature and precipitation, the higher is the degree of decomposition activity.

### Parent Rock Type

*Significance*

The parent rock type influences the rate of decomposition, which depends on the stability of the component minerals for a given climatic condition and other environmental factors.

*Mineral Stability*

The relative stability of the common rock-forming minerals is given on Table 6.13. Clay minerals are usually the end result, unless the parent mineral is stable or soluble.

Quartz is the most stable mineral; minerals with intermediate stability, given in decreasing order of stability, are muscovite, orthoclase feldspar, amphibolite, pyroxene, and plagioclase. Olivine is the least stable, rapidly undergoing decomposition.

*Igneous and Metamorphic Rocks*

*Sialic crystalline rocks* (granite, granite gneiss, rhyolite, etc.), composed of quartz, potash and soda feldspars, and accessory micas and amphiboles, represent the more resistant rocks.

*Mafic crystalline rocks* (basalt, gabbro, dolerite, etc.) and metamorphic rocks of low silica content contain the minerals least resistant to decomposition, i.e., the ferromagnesians. The relative susceptibility to decomposition of rocks composed of silicate minerals is given on Table 6.14. Clays are the dominant products of decomposition of igneous rocks and are summarized in terms of rock type and original mineral constituents on Fig. 6.73.

*Sedimentary Rocks*

*Argillaceous and arenaceous types* are already composed of altered minerals (clay and sand particles) and normally do not undergo further

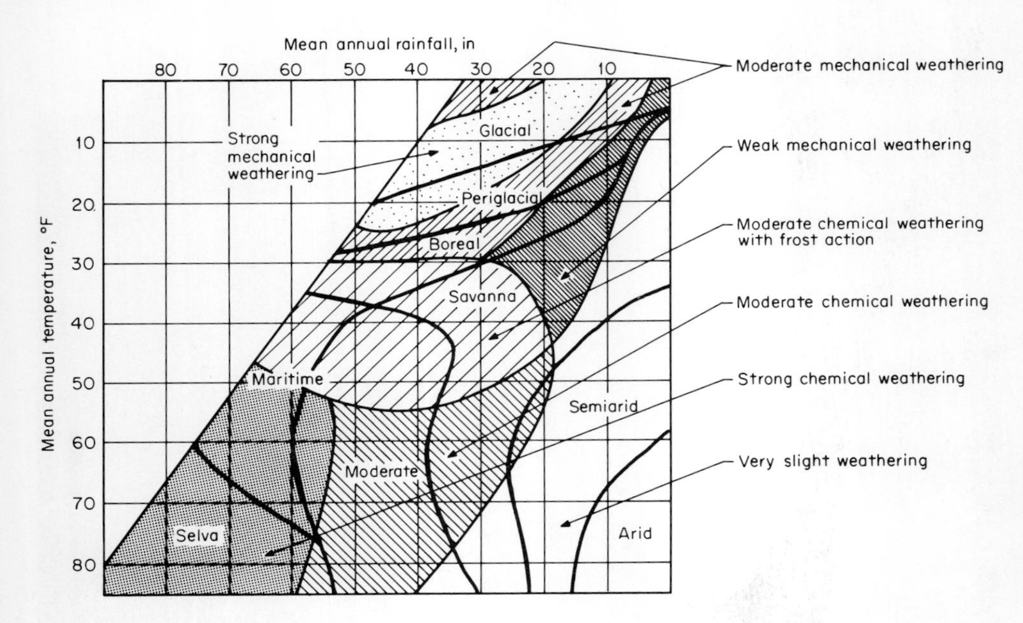


**FIG. 6.71 Diagram of climatic boundaries of regions and intensity of various types of weathering.** *[After Fookes et al. (1971).*[27]*]*

decomposition except for impurities such as feldspar and other weatherable minerals.

*Carbonates and sulfates* go into solution faster than they decompose, except for mineral impurities, which decompose to form a residue in the normal manner.

*Final Product and Thickness*

The final product is usually mixtures of quartz particles (which are relatively stable), iron oxides, and clay minerals (kaolinite, halloysite, montmorillonite, illite, vermiculite, and chlorite—*see Art. 5.3.3*). Vermiculite and chlorite are uncommon since they alter readily to montmorillonite, illite, and kaolinite. The clay minerals result from the most common groups of silicates (feldspars and ferromagnesians). The ferromagnesians usually contain iron, which decomposes to iron oxides that impart the reddish color typical of many residual soils.

The thickness of the decomposed zone is related directly to rock type in a given climate, as well as to topography (see "Other Factors" below). Froelich (1973)[30] describes the depth of decomposition in the Baltimore-Washington, D.C., area of the eastern United States. Formerly a peneplain, the typical topographic expression in the area is given on Fig. 6.74. The quartz veins and dikes and indurated quartzites have little or no overburden, and the massive ultramafic bodies (serpentinite) generally have less than 5 ft of overburden in either valley bottoms or ridgetops. Foliated mafic rocks (greenstones) commonly have 5 to 20 ft of cover; gneisses and granitic rocks are mantled by as much as 60 ft of saprolite but may contain fresh corestones (*see Art. 6.7.3*); and schists and phyllites commonly have from 80 to 120 ft of overburden. Saprolite is thickest beneath interstream ridges and thins towards valley bottoms. The larger streams are incised in, and may flow directly on, hard fresh rock.

## Other Factors

*Time*

The length of geologic time during which the rock is exposed to weathering is a significant factor. Decomposition proceeds very slowly, except for the case of water flowing through soluble rocks. Millions of years are required to decompose rocks to depths of 30 to 50 m. Erosion and other forms of mass wasting, however, are continually reducing the profile during formation.

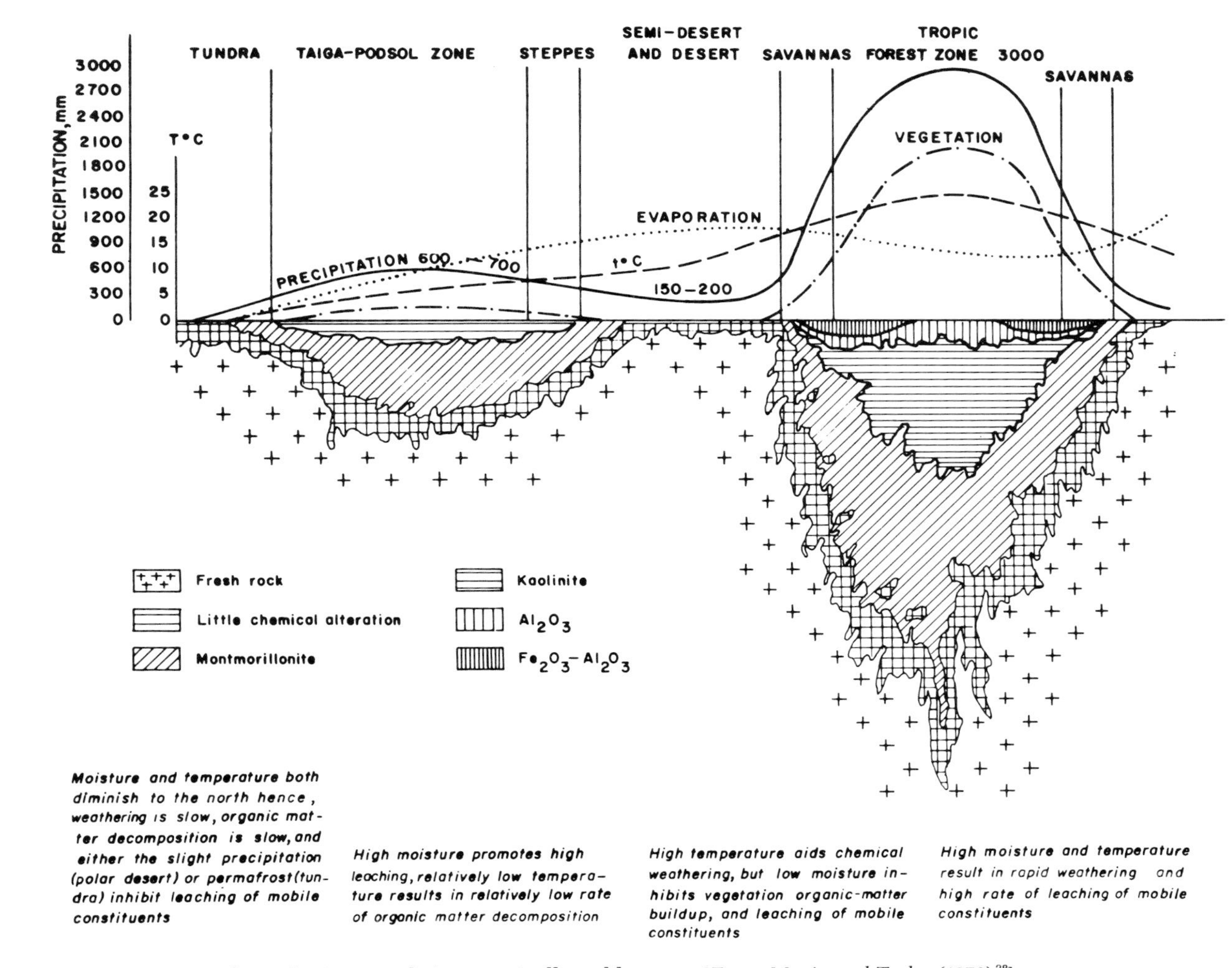


**FIG. 6.72 Formation of weathering mantle in tectonically stable areas.** [*From Morin and Tudor (1976).*[28]]

The greatest depths of decomposition occur in tectonically stable areas. In unstable tectonic areas the weathered zone is thinner because topographic changes increase the erosion activity. Glaciation removes decomposed materials, often leaving a fresh rock surface. In glaciated areas the geologic time span for decomposition has been relatively short, and the depth of weathering is shallow. An example is the basalt sills and dikes of Triassic age in the glaciated northeastern United States which form prominent ridges with very little soil cover (*see Fig. 6.19*). To the south the dikes, which have been exposed for millions of years longer, have weathered to clayey soils more rapidly than the adjacent crystalline rocks and form depressions in the general surface as a result of differential erosion. Although the climate to the south is somewhat warmer, the precipitation is similar to that of the northeast.

### *Rock Structure*

Differential weathering reflects not only rock type, but also structure. Decomposition proceeds much more rapidly in strongly foliated or fractured rocks, and along fault zones, than in sounder masses. The depth of decomposition

**TABLE 6.13**
**RELATIVE STABILITY OF COMMON ROCK-FORMING MINERALS***

| Group | Minerals | Relative stability |
|---|---|---|
| Silicates | Feldspars | |
| | Orthoclase (potash) | Most persistent of feldspars. |
| | Plagioclase (soda lime) | Weathers to kaolinite. |
| | Micas | |
| | Muscovite (white mica) | Persistent, weathers to illite. |
| | Biotite (dark mica) | Easily altered to vermiculite, iron element causes staining. |
| | Amphiboles (hornblende) | Persistent. |
| | Pyroxines (augite) | Less persistent than hornblende. Decomposes to montmorillonite. |
| | Olivine | Readily decomposes to montmorillonite. |
| Oxides | Quartz | Most stable; slightly soluble. |
| | Iron oxides | |
| | Hematite | Relatively unstable. |
| | Limonite | Stable; product of alteration of other oxides of iron. |
| Carbonates | Calcite | Readily soluble. |
| | Dolomite | Less soluble than calcite. |
| Sulfates | Gypsum | More soluble than calcite. |
| | Anhydrite | Like gypsum. |
| Hydrous aluminum silicates (clay minerals) | Kaolinite | The most stable clay mineral. |
| | Illite | Alters to kaolinite or montmorillonite. |
| | Vermiculite | Alteration product of chlorite and biotite. Alters to kaolinite or montmorillonite. |
| | Montmorillonite | Alters to kaolinite. |
| | Chlorite | Least stable of clay minerals. Alters readily to any or all of the others. |

*After Hunt (1972).[29] Adapted with permission of W. H. Freeman and Company.

**TABLE 6.14**
**SUSCEPTIBILITY TO WEATHERING OF COMMON ROCK-FORMING SILICATE MINERALS AND THEIR OCCURRENCE IN VARIOUS IGNEOUS ROCKS***

| Susceptibility to weathering | Dark minerals | Light minerals | Rock types: Volcanic | Rock types: Intrusive | Sequence of crystal-lization |
|---|---|---|---|---|---|
| Least resistant | Olivine $(Mg,Fe)_2SiO_4$ | Calcic plagioclase (anorthite) $CaAl_2Si_2O_8$ | | | Early |
| | Augite $(Ca,Mg,Fe,Al)_2(Al,Si)_2O_6(OH)_2$ | Calcic plagioclase (labradorite) with sodium | Basalt | Gabbro | |
| | Hornblende $(Ca,Na,Fe,Mg,Al)_7(Al,Si)_8O_{22}(OH)_2$ | Sodium plagioclase with calcium (andesine, oligoclase) | Andesite | Diorite | |
| | | Sodium plagioclase (albite) $NaAlSi_3O_8$ | Latite | Monzonite | |
| | Biotite (dark mica) $K(Mg,Fe)_3(Al,Si_3)O_{10}(OH,F)_2$ | Potash feldspar (orthoclase, microcline) $KAl_2Si_3O_8$ | Rhyolite | Granite | |
| | | Muscovite (white mica) $KAl_2(Al,Si_3)O_{10}(OH,F)_2$ | | | |
| Most resistant | | Quartz, $SiO_2$ | | | Late |

*After Hunt (1972).[29] © W. H. Freeman and Company. Reprinted with permission. Adapted from Goldrich (1938).

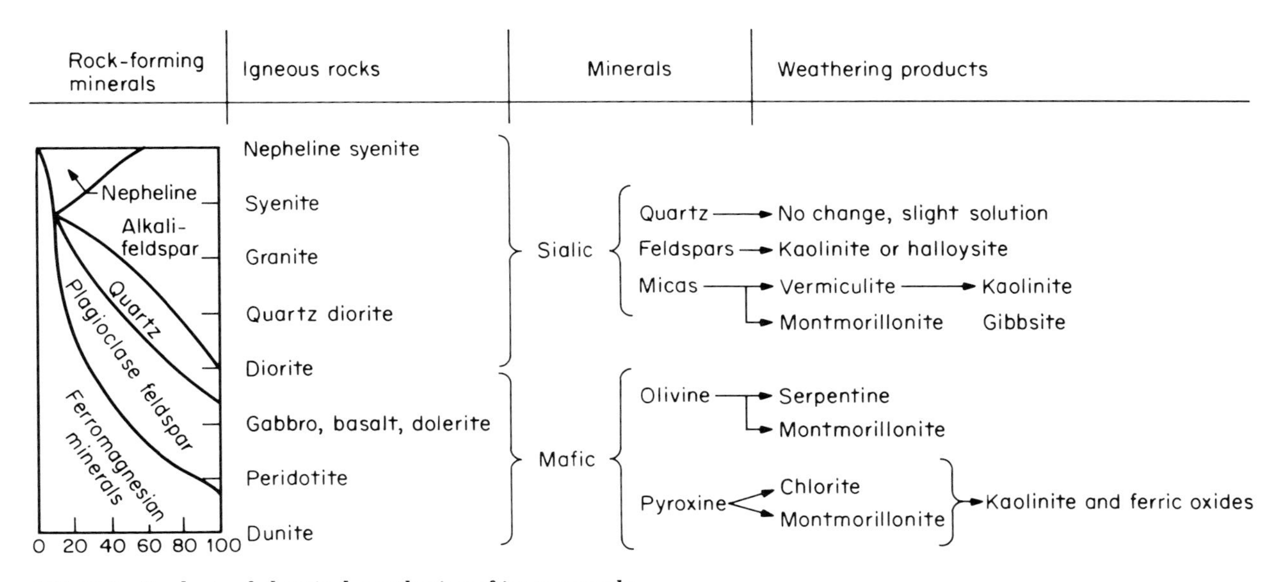


**FIG. 6.73 Products of chemical weathering of igneous rocks.**

FIG. 6.74 Landform expressions near Kensington, Maryland, developing from decomposition of metamorphic rocks in a warm, humid climate. (*See test boring log in Fig. 7.5.* Kensington, Maryland, quadrangle sheet. Scale 1:24,000.)

can be extremely irregular in areas of variable rock type and structural features as shown on Fig. 6.75. Where granite masses have been intruded into foliated metamorphic rocks, differential weathering produces resistant domes, termed *monadnocks*, as illustrated on the topographic map in Fig. 6.76.

*Topography and Groundwater Depth*

Topography influences the movement of water through materials and the rate of erosion removing the products of decomposition. On steep slopes rainfall runs off instead of infiltrating, and oxidation and reduction activity are much less severe than on moderately to slightly inclined slopes. On flat slopes and in depressions that are almost continuously saturated, oxidation, reduction, and leaching are only feebly active.

It is the partially saturated zones where vertical water movement can occur that provide the optimum conditions for oxidation, reduction, and leaching, and where decomposition is most active. A relationship between topography and depth of decomposition is illustrated on Fig. 6.77, a location with a warm, moist climate and hilly to mountainous terrain. Average rainfall is more than 2000 mm annually, and the average temperature is 22°C (mean range is 15 to 40°C). Decomposition depth, often to depths in excess of 30 m, is greatest beneath the crests of the hills composed of foliated crystalline rocks. Along the sideslopes, where erosion occurs, the depth is about 10 m. In the narrow valleys, where rock is permanently saturated, the decomposed depth is usually only a few meters at most, and streams often flow on fresh rock surfaces. The overburden cover is thicker in the wider valleys.

There is little decomposition activity below the permanent water table. Limestone cavities do not increase substantially in size unless the water is caused to flow, for example by well pumping, in which case solution increases rapidly. Laterite and caliche (*see Art. 7.7.2*) are usually found above the permanent water table in well-drained, nonsaturated zones.

*Urbanization and Construction Excavations*

Foliated metamorphic rocks and mafic igneous rocks are the least resistant to chemical decomposition, but in an urban environment where polluted rain water, or sewage from broken sewer lines may seep downward into the rock mass, deterioration may be substantially accelerated. Feld (1966)[32] reported a number of instances in New York City where the foundation rock (Manhattan mica-schist) was "hard, ringing" material when foundations were installed, and in a matter of 10 to 20 years was found during adjacent excavation to be altered and disintegrated and easily removed by pick and shovel. Depths of alteration were as great as 8 m. Bearing pressures as high as 25 kg/cm² were reported, but no indication of distress in the buildings was given. In one case, a broken steam line 13 m distant from a tunnel excavation was attributed to the rock softening.

Marine shales and clay shales may undergo very rapid disintegration and softening when exposed to humidity during excavation, as discussed in Art. 6.7.3.

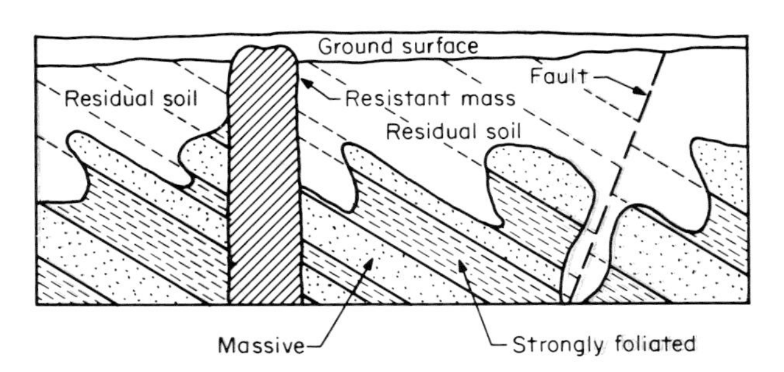


**FIG. 6.75 Irregular depth of weathering as affected by rock type and structure in the southern Piedmont of the United States. The rocks are predominantly gneisses and schists into which granite masses have been injected, as well as basalt dikes. Decomposition often reaches 30 m.** [*From Sowers (1954).*[31]]

### 6.7.3 WEATHERING PROFILE IN VARIOUS ROCK TYPES (*See also Art. 7.2,* "Residual Soils")

**Igneous Rocks**

*Quartz-Rich Sialic Rocks*

Quartz-rich sialic rocks go through four principal stages of profile development as illustrated on Fig. 6.78.

STAGE 1 Weathering proceeds first along the joints of the fresh rock surface, and decomposition is most rapid where the joints are closely

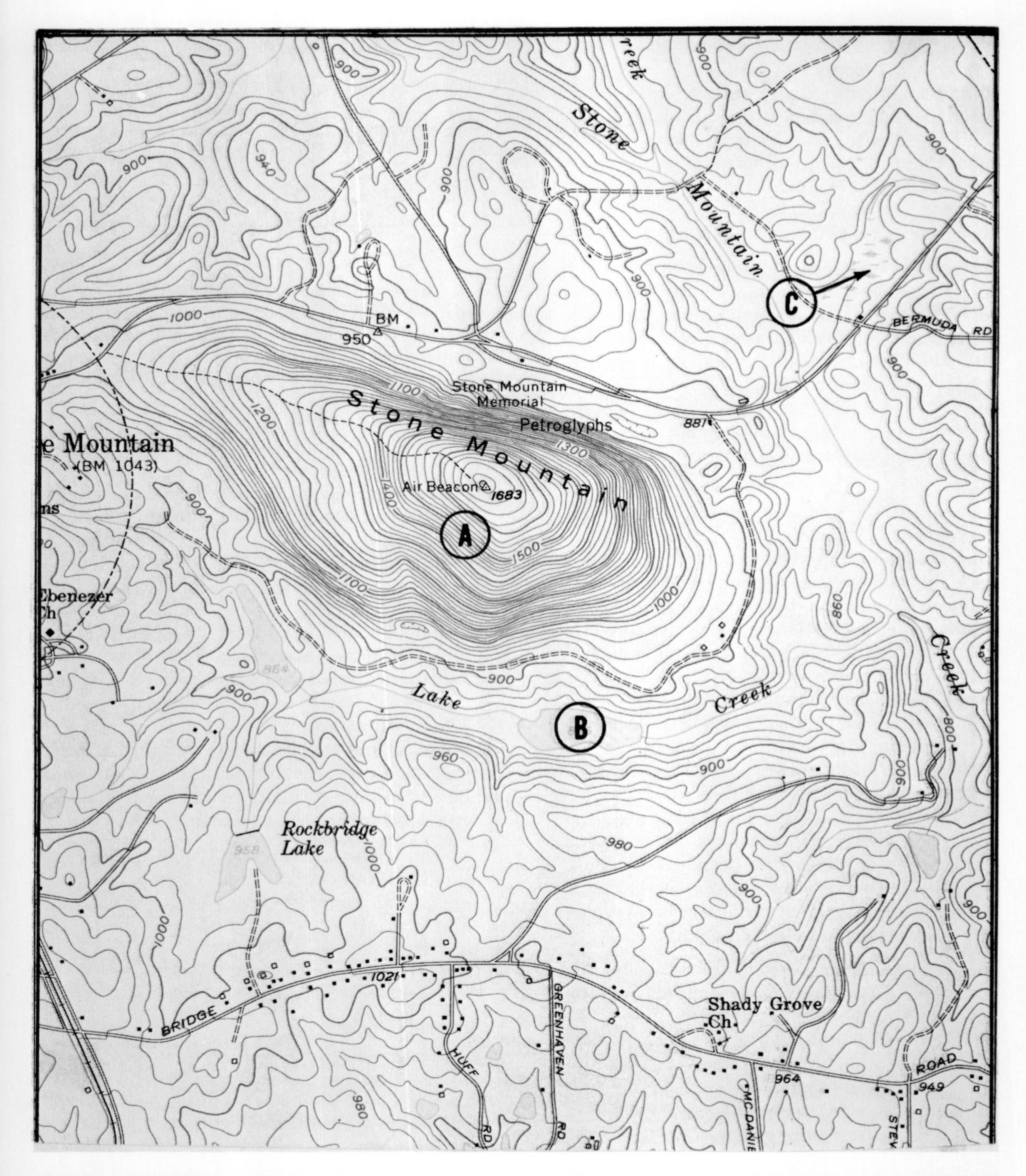


**FIG. 6.76 Differential erosion between a granite mass and surrounding foliated metamorphic rocks has resulted in the formation of Stone Mountain, Georgia, a monadnock.** *(From USGS topographic quadrangle. Scale 1:24,000.)*

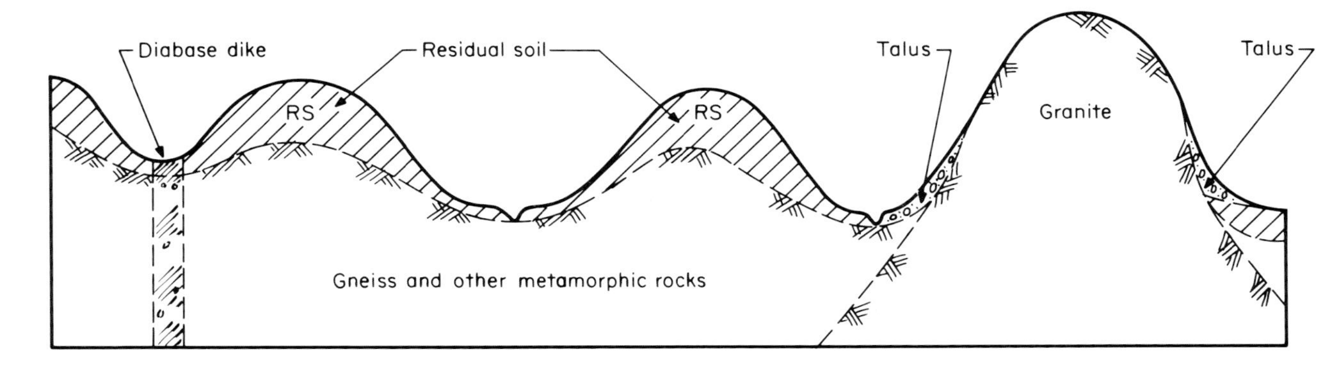


**FIG. 6.77** *(Above)* **Typical section of rock decomposition in the warm, moist climate of the coastal mountains of Brazil.**

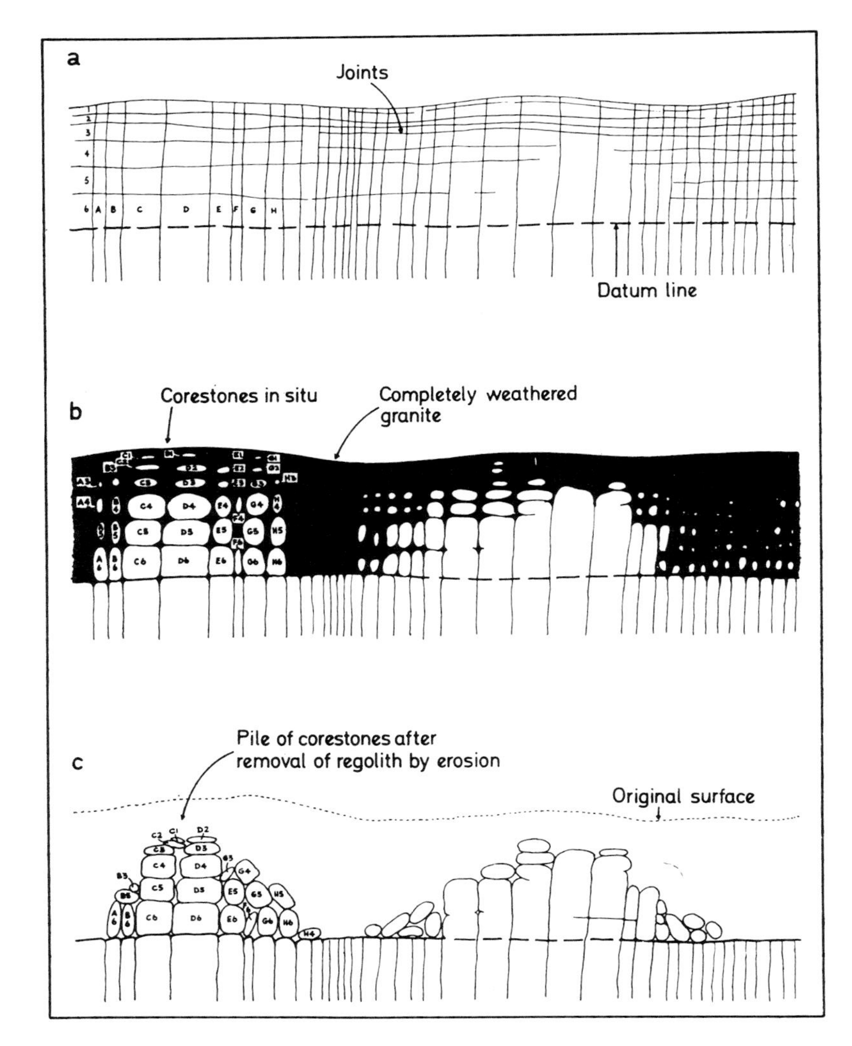


**FIG. 6.78 Successive stages in the progressive weathering and erosion of Dartmoor granite from original ground surface to the development of tors: (*a*) stage 1; (*b*) stages 2 and 3; (*c*) stage 4. Soil development is most advanced in the heavily jointed zones.** [*After Fookes et al. (1971).*[27]]

**FIG. 6.79 Fresh granite and slightly weathered granite. Mica in the specimen on the left has begun to decompose and stain the rock with iron oxide (Paranagua, Parana, Brazil).**

spaced. In temperate regions the initial change is largely mechanical disintegration and the resulting material does not differ significantly from the original rock. The granite begins to alter in appearance; the biotite tends to bleach and lighten in color, and the ferrous compounds of biotite tend to become ferric hydrate and to migrate, staining the rock yellowish-red to reddish-brown. Fresh and slightly weathered granite specimens are illustrated on Fig. 6.79. During stage 1 the joints fill with sand or clay.

STAGE 2 During intermediate decomposition the granite loses its coherence and becomes crumbly. In humid climates a sandy matrix forms around spherical boulders (corestones) as shown in Fig. 6.80, especially in partially saturated but continuously moist zones. The corestone size reflects the fracture spacing. In well-drained zones, stage 2 soil cover is often relatively thin, and on steep slopes in granite the soil is removed quickly, and slabbing by exfoliation occurs.

STAGE 3 During final decomposition, a sandy soil is formed, composed chiefly of angular particles of quartz and feldspar. Further decomposition yields clayey soils [Grim (1962)[33]]. In poorly drained zones, rocks such as diorite and syenite, containing potassium and magnesium, yield illite and montmorillonite, depending on which mineral predominates. In well-drained zones the potash and magnesium are removed quickly, and kaolinite is formed. The transition to slightly decomposed rock often is abrupt and the rock surface highly irregular as shown on Fig. 6.81.

STAGE 4 As the soils are removed by erosion the virginal corestones rise to the surface as boulders, as shown on Fig. 6.82. In mountainous or hilly terrain they eventually move downslope to accumulate in groups or in a matrix of colluvium at the base of the hill. They can be of very large diameters as shown in Fig. 6.83. On level terrain the boulders form "tors" as illustrated on Fig. 6.78.

### *Quartz-Poor Mafic Rocks*

In general, the quartz-poor mafic rocks develop a weathering profile, also typical of many metamorphic rocks, characterized by four zones:

1. An upper zone of residual soils that are predominantly clays with small amounts of organic matter (equivalent to the A and B horizons of pedological soils, *see Art. 7.8.1*).

**FIG. 6.80 Granite corestones and spheroidal weathering in a sandy soil matrix (Itaorna, Rio de Janeiro, Brazil).**

2. An intermediate zone of residual soil, predominantly clayey, but with decomposition less advanced than in the upper zone.
3. A saprolite zone in which relict rock structure is evident and the materials are only partially decomposed, which grades to a weathered rock zone.
4. The weathered rock zone where rock has only begun alteration.

Clay soils are the decompositional product. The clay type is related strongly to the rainfall and drainage environment [Grim (1962)[33]] as follows:

- Low rainfall or poor drainage; montmorillonite forms as magnesium remains.
- High rainfall and good drainage; kaolinite forms as magnesium is removed.
- Hot climates, primarily wet but with dry periods; humic acids are lacking, silica is dissolved and carried away, and iron and aluminum are concentrated near the surface (laterization).
- Cold, wet climates; potent humic acids remove aluminum and iron and concentrate silica near the surface (silcrete).

A decomposition profile in basaltic rocks in a warm, moist climate is given on Fig. 6.84. The area is characterized by rolling hills, mean annual temperature of 25°C, and annual rainfall of about 1300 mm, most of which falls between September and April. These climatic conditions favor the maximum development of the porous clays *(see Art. 10.5.2)*, and produce some laterites. Laterite development is stronger to the north of the area where rainfall exceeds 1500 mm annually and temperatures are higher on the average than 26°C *(Art. 7.7.2)*.

VSL

**FIG. 6.83** Granite boulder 10 m in diameter at base of slope (Rio Santos Highway, Brazil).

**FIG. 6.81** *(Above left)* Differential weathering and the development of boulders in granite (Rio-São Paulo Highway, São Paulo, Brazil).

**FIG. 6.82** *(Below left)* Granite boulders on the surface after soil erosion (Frade, Rio de Janeiro, Brazil).

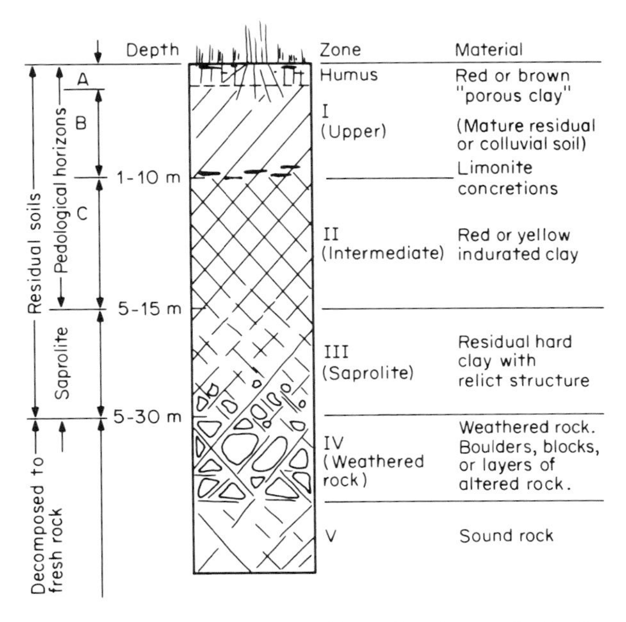


**FIG. 6.84** Decomposition profile in basaltic rocks in the Interland Plateau of south-central Brazil. (NOTE: Saprolite is usually considered to be in the C horizon.) [*From Vargas (1974).*[34]]

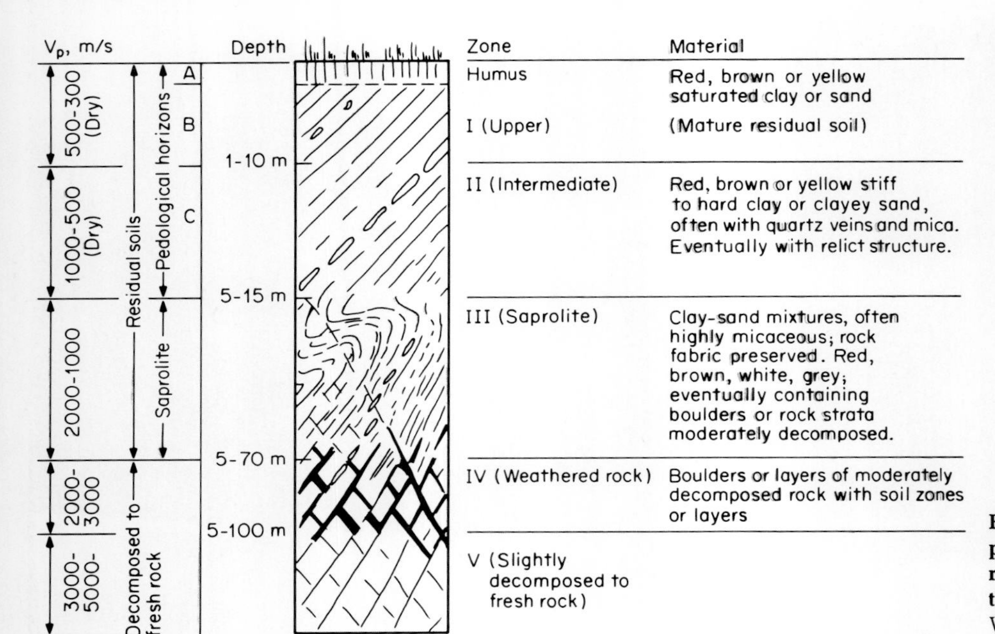


**FIG. 6.85 Decomposition profile in metamorphic rocks of the coastal mountain range of Brazil.** [*From Vargas (1974).*[34]]

## Metamorphic Rocks

### *General*

Foliated metamorphic rocks such as gneiss and schist decompose with comparative ease because water enters and moves with relative freedom through the foliations. They tend to decompose to substantial depths, often greater than 30 m in moist climates. The typical decomposition profile in the metamorphic rocks of the coastal mountain range of Brazil is given on Fig. 6.85, which includes ranges in depths and typical seismic refraction velocities for each zone. Seismic velocities are strong indicators of the profile development and depths.

Slate, amphibolite, and massive metamorphic rocks are relatively resistant to chemical decomposition.

### *Foliated Rocks: Typical Profile Development*

STAGE 1 *Fresh rock grades to moderately decomposed rock* (Fig. 6.86). Feldspars and micas have just begun to decompose and the clayey alteration of minerals can be seen with a microscope but the minerals are still firm. Brown, reddish-brown, or yellow brown discoloration with iron oxides occurs, especially along partings, and the density reduces to about 5 to 10% less than that of fresh rock. The material is tough and requires a hammer to break it, but gives a dull thud, rather than ringing, when struck. In dense rocks the layer is thin, but in the more porous types it can be many meters thick.

STAGE 2 *Saprolite* develops; the structure of the parent rock is preserved, but the mass is altered largely to clay stained with iron oxide. Leaching removes sodium, potassium, calcium, and magnesium; alumina, silica, and iron remain. In Fig. 6.87*a*, the feldspar has formed a white kaolinite and the biotite mica is only partially decomposed; the remaining minerals are primarily quartz grains. In Fig. 6.87*b*, a sample from a shallower depth, the minerals remaining are kaolin, iron oxides, and quartz. Saprolites are generally grouped as soils [Vargas (1974)[34] and Deere and Patton (1971)[35]] because of their engineering properties. In the Piedmont of the southeastern United States void ratios typically range from 0.7 to 1.3 and as high as 2.0, indicating the relative compressibility of the material [Sowers (1954)[31]]. Excavation by pick and shovel is relatively easy.

| TABLE 6.15<br>DESCRIPTION OF A WEATHERING PROFILE IN IGNEOUS AND METAMORPHIC ROCKS* | | | | | |
|---|---|---|---|---|---|
| **Zone (class)** | **Description** | **RQD, %** | **Core recovery, %** | **Relative permeability** | **Relative strength** |
| | | RESIDUAL SOIL | | | |
| A Horizon | Top soil, roots. Zone of leaching and eluviation may be porous | — | 0 | Medium to high | Low to medium |
| B Horizon | Usually clay-enriched with accumulations of Fe, Al, and Si. Hence may be cemented, no relict structure present. | — | 0 | Low | Commonly low, high if cemented |
| C Horizon (saprolite) | Relict rock structure retained. Silty grading to sandy material. Less than 10% corestones. Often micaceous. | 0 | 0–10 | Medium | Low to medium (relict structure very significant) |
| | | WEATHERED ROCK | | | |
| Transition | Highly variable, soil to rocklike. Commonly c-f sand, 10 to 95% corestones. Spheroidal weathering common. | 0–50 | 10–90 | High (water losses common) | Medium to low where weak or relict structures present |
| Partly weathered rock | Soft to hard rock joints stained to altered. Some alteration of feldspars and micas. | 50–75 | >90 | Medium to high | Medium to high for intact specimens |
| | | UNWEATHERED ROCK | | | |
| | No iron stains to trace along joints. No weathering of feldspars and micas | >75 (generally >90) | Generally 100 | Low to medium | Very high for intact specimens |

NOTE: The specimens provide the only reliable means of distinguishing the zones. *From Deere and Patton (1971).[35]

STAGE 3 *Residual soil without relict structure* is the final stage of decomposition *(see Fig. 7.3)*. Clay soils (kaolin) predominate; some parts are white, but most are stained with iron oxides and range from brightly colored red and purple to browns and yellows. In places the iron oxides are concentrated in nodules, fissure veins, or blanket veins. Desiccation and some cementation following leaching can cause the upper zone of 1 to 3 m to form a stiff crust where located above the water table. The soils are classed usually as ML or CL-ML.

*Engineering Properties*

A general description of the weathering profile in igneous and metamorphic rocks together with some relative engineering properties is given on Table 6.15. Strength parameters of weathered igneous and metamorphic rocks are given on Table 3.27.

**FIG. 6.86 Moderately decomposed gneiss (João Monlevade, Minas Gerais, Brazil).**

**FIG. 6.87 Saprolite taken with split spoons in decomposed gneiss. White material, kaolinite; black material, biotite mica; gray material, stained with iron oxides; (*above*) lower sample (15 m) and (*right*) upper sample (12 m) (Jacarepagua, Rio de Janeiro, Brazil).**

## Sedimentary Rocks (Excluding Marine Shales)

### *General*

Sedimentary rocks are composed of stable minerals (quartz and clays) or soluble materials, with minor amounts of unstable materials present either as cementing agents or components of the mass. The depth of decomposition is normally relatively thin in comparison with foliated crystalline rocks, but increases with the amount of impurities in the mass.

### *Sandstones*

Composition minerals of sandstones are chiefly quartz grains cemented by silica, calcite, or iron oxide, which are all stable except for calcite, which is soluble. Decomposition occurs in feldspars and other impurities to form a clayey sandy overburden, generally not more than a few meters thick.

### *Shales: Freshwater*

Freshwater shales are composed of clay minerals and silt grains and decomposition is generally limited to impurities. Weathering is primarily mechanical, especially in temperate or cooler zones. The characteristic weathering product is small shale fragments in a clay matrix usually only a few meters in thickness at most.

Triassic shales, considered to be freshwater deposits in shallow inland seas, predominantly

**FIG. 6.88 Quarry wall in slightly metamorphosed high-purity limestone (Cantagalo, Rio de Janeiro, Brazil). Cavities are clay-filled and about 6 m deep and 1 to 2 m wide. Man at bottom left gives scale.**

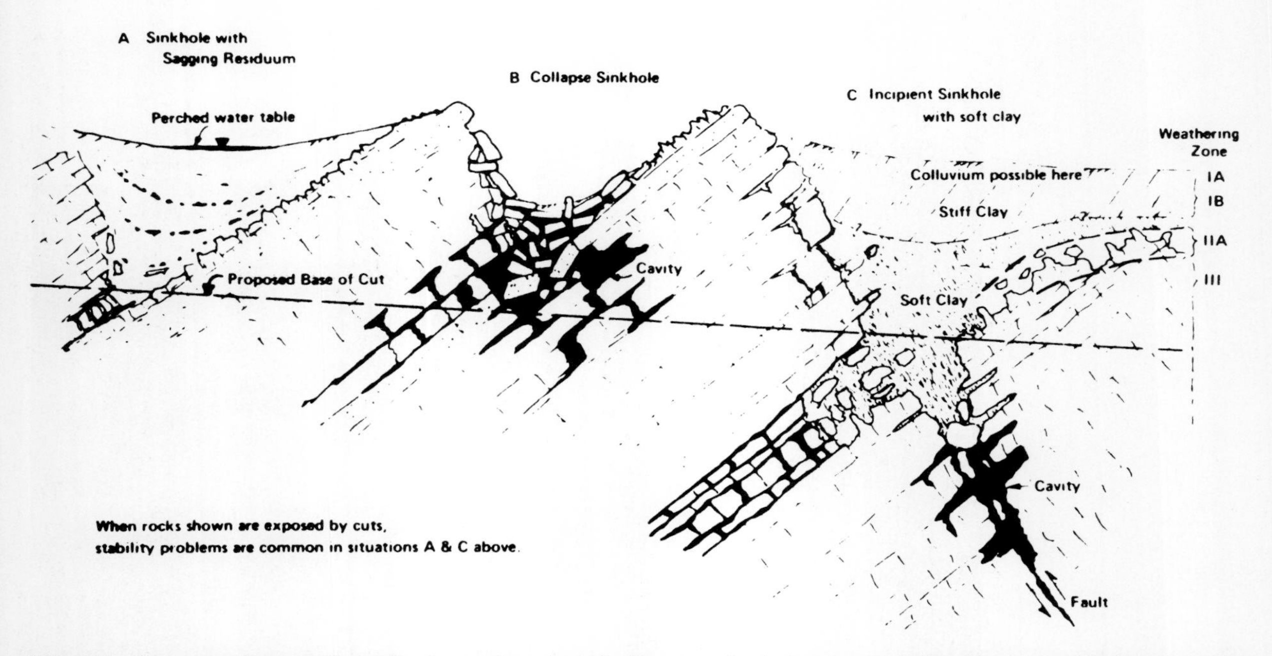


**FIG. 6.89 Decomposition profile in dipping carbonate rocks. Decay and solution proceed along the joints. The impure strata form soil deposits which may fill a cavity when collapse occurs.** [*From Deere and Patton (1971)*[35]]

contain inactive clays, and normally develop a thin reddish clayey overburden.

*Limestones and Other Carbonates*

Composed chiefly of calcite, the pure limestones are readily soluble and do not decompose to soil. It is the impurities that decompose, but normally there is no transition zone between the soil and the rock surface (*see Fig. 6.22*), as is normal for other rock types, and the rock surface can be very irregular as shown on Figs. 6.88 and 6.89. The residual soils in warm, wet climates are typically clayey and colored red ("terra rossa") to yellow to reddish brown and, in cooler, less moist climates, grayish brown.

### Marine Shales

*Significance*

Marine shales, particularly of the Tertiary, Cretaceous, and Permian periods, normally contain montmorillonite. They are the most troublesome shales from an engineering viewpoint because of their tendency to form unstable slopes and to heave in excavations (*see Arts. 9.2.6 and 10.6.3*). Well-known troublesome formations include the Cucaracha shales (Tertiary) encountered during the construction of the Panama Canal, the Cretaceous shales covering large areas of the northwestern United States and adjoining Canada (Fig. 6.90), the Permian shales of central and western Brazil, and the Cretaceous shales of Bahia, Brazil.

*Weathering Processes*

Marine shale formations have characteristically been prestressed by high overburden pressures, often as great as 100 kg/cm$^2$ or more. When uplifted and subjected to erosion, stress release and the resulting strains cause intense fracturing in the mass. Water enters the fractures and the montmorillonite clay minerals expand to break the mass into numerous small fragments as shown on Fig. 6.91, thus making it susceptible to further weathering and reduction to a soil.

Weathering, however, is primarily mechanical, although some chemical decomposition occurs.

*Landforms*

Landforms are characteristically gently rolling topography with shallow slopes, often in the order of 8 to 15°, with the shallow-depth materials subjected to movements as shown on Fig. 6.92.

*Characteristics*

PROFILES Colors are predominantly gray to black. A log of a deep test pit excavated in the Bearpaw shale (Upper Cretaceous) shows a weathering profile consisting of four distinct zones (Fig. 6.93). It is the amount of montmorillonite in a given zone that has the greatest effect on the intensity of disintegration because of nonuniform swelling. The Bearpaw shale is found in Montana and south Saskatchewan (*see Fig. 6.90*). Farther south its correlative is the Pierre shale.

BASIC AND INDEX PROPERTIES A profile of the natural water content in the Bearpaw shale is given on Fig. 6.93; in the unweathered zones it is about equal to the plastic limit. In the upper, soft-shale zone, LL = 115%, PI = 92%, activity is 1.8, and the content of particles smaller than 2 $\mu$ is 50%.

STRENGTH Undrained strength (unconfined compression) of the Bearpaw ranges from 5 to 15 kg/cm$^2$ [Bjerrum (1967)[37]]. Disintegration occurs rapidly when fresh rock is exposed to moisture in the air during excavation, accompanied by a rapid reduction in strength. Typical properties of some marine and clay shales are given on Table 3.30. Residual shear strengths $\phi_r$, as determined by direct shear tests, vs. the liquid limit for various marine shales from the northwestern United States are given on Fig. 6.94.

**Interbedded Shales**

*Marine Shales*

Sandstone interbeds retard deep weathering which proceeds along exposed shale surfaces. In the northwestern United States, Banks (1972)[36] observed that slopes capped with or underlain by resistant sandstone strata had inclinations ranging from 20 to 45°, whereas slopes containing only marine shales ranged from 8 to 15°.

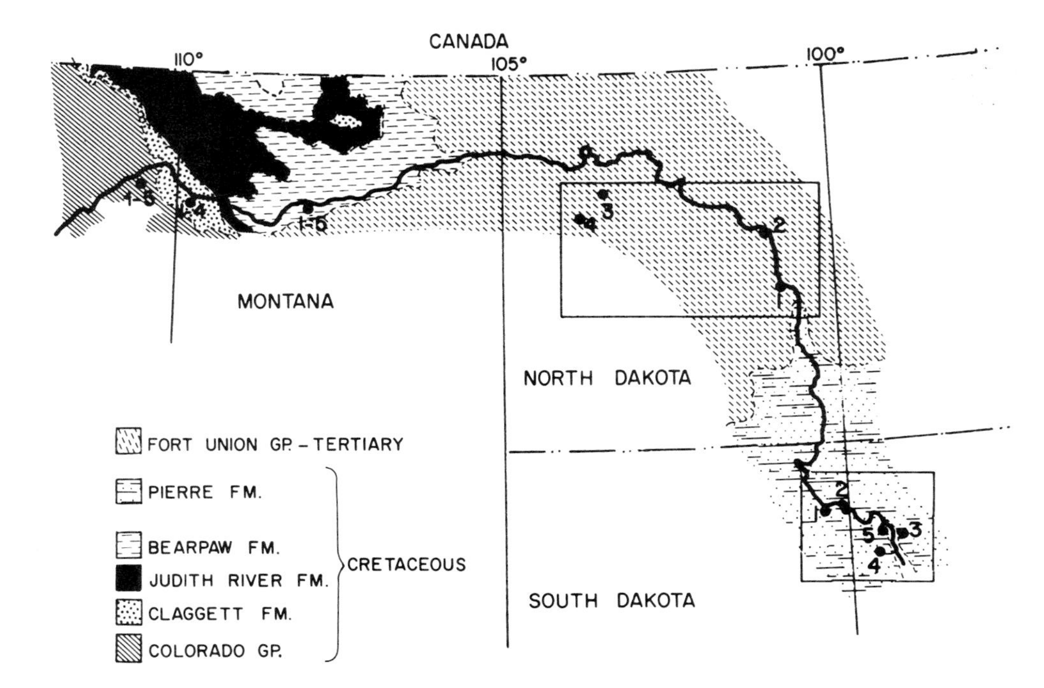


**FIG. 6.90 Distribution of Tertiary and Cretaceous marine shales in the northwestern United States.** [*From Banks (1972).*[36]]

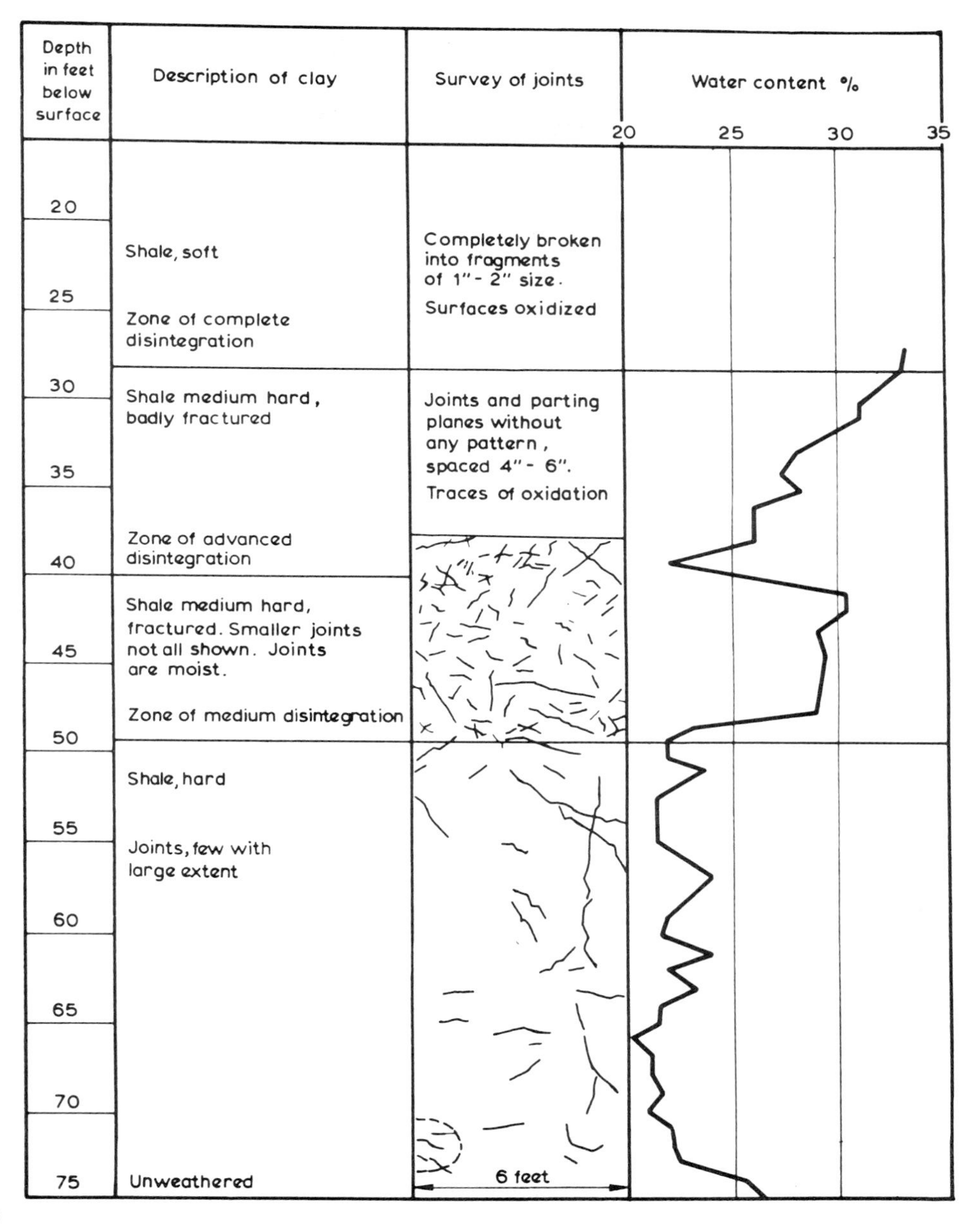


FIG. 6.93 **Profile in a test pit in Bearpaw clay shale at South Saskatchewan Dam with joint survey.** [*From Bjerrum (1967),*[37] *published with permission of the Prairie Farm Rehabilitation Administration, Canadian Department of Agriculture.*]

FIG. 6.91 *(Above left)* **Highly disintegrated marine shales of the Permian, Mineiros, Goias, Brazil.**

FIG. 6.92 *(Below left)* **Gentle topography and sloughing slopes of the Permian marine shales (Mineiros, Goias, Brazil). At this location an approach fill placed on the shales failed and caused the sequential failure of all of the columns supporting a bridge.**

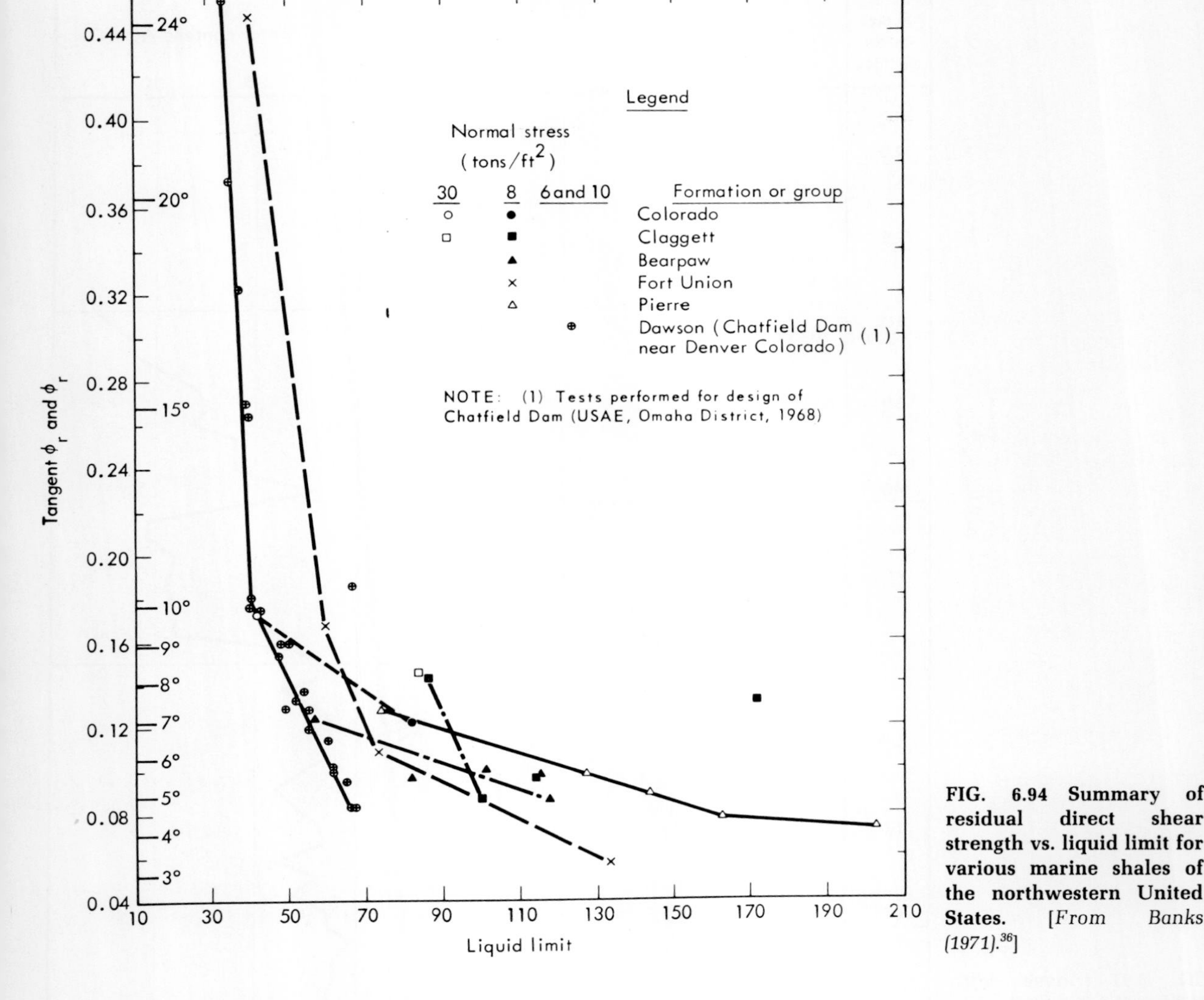


**FIG. 6.94 Summary of residual direct shear strength vs. liquid limit for various marine shales of the northwestern United States.** [*From* Banks (1971).[36]]

Dipping interbeds of sandstone and shale have caused severe problems in a development in Menlo Park, California [Meehan et al. (1975)[38]]. The shales have decomposed to a black, expansive clay. Differential movement of foundations and pavements founded over alternating and dipping beds of sandstone and shale have resulted as discussed in Art. 10.6.3.

*Sandstones Containing Thin Shale Beds*

Relatively steep overall stable slopes are characteristic of sandstones interbedded with thin shale layers, but such slopes are subject to falling blocks of sandstone when the shales contain expansive materials. Thin seams of montmorillonite in the shale expand, fracturing and wedging blocks of sandstone loose. Differential weathering of the shale causes it to recede beneath the sandstone as shown on Fig. 6.95, resulting in loss of support and blocks falling to the roadway.

**FIG. 6.95 Interbedded sandstones and shales in a 60-m-high cut in the Morro Pelado formation (late Permian) (Highway BR 116, km 212, Santa Cecilia, Santa Catarina, Brazil). The montmorillonite in the shale expands, fracturing the sandstone; as the shale decomposes, support of the sandstone is lost and blocks fall to the roadway. Shotcreting the slope retards moisture changes in the shale.**

## REFERENCES


1. Thornbury, W. D. (1969) *Principles of Geomorphology*, John Wiley & Sons, New York.
2. Way, D. S. (1978) *Terrain Analysis*, 2d ed., Dowden, Hutchinson and Ross, Stroudsburg, Pa.
3. Lahee, F. H. (1941) *Field Geology*, McGraw-Hill Book Co., New York.
4. Hoek, E. and Bray, J. W. (1977) *Rock Slope Engineering*, 2d ed., Inst. of Mining and Metallurgy, London.
5. Walhstrom, E. C. (1973) *Tunneling in Rock*, Elsevier Scientific Pub. Co., New York.
6. Badgley, P. C. (1959) *Structural Methods for the Exploration Geologist*, Harper & Bros., New York.
7. Hills, E. S. (1972) *Elements of Structural Geology*, John Wiley & Sons, New York.
8. Seegmiller, B. L. (1972) "Rock Slope Stability Analysis at Twin Buttes," *Stability of Rock Slopes, Proc. ASCE, 13th Symp. on Rock Mechs.*, Univ. of Illinois, Urbana (1971), pp. 511–536.
9. Corbel, J. (1959) "Erosion em terrain calcaire," *Ann. Geog.*, 68, pp. 97–120.
10. Fecker, E. (1978) "Geotechnical Description and Classification of Joint Surfaces," *Bul. Intl. Assoc. Engrg. Geol.*, No. 18, December, pp. 111–120.
11. Ladanyi, B. and Archambault, G. (1970) "Simulation of Shear Behavior of a Jointed Rock Mass," *Proc. 11th Symp. on Rock Mechs.*, AIME, New York, pp. 105–125.
12. Barton, N. R. (1973) "Review of a New Shear Strength Criterion for Rock Joints," *Engineering Geology*, Vol. 7, Elsevier Pub. Co., Amsterdam, pp. 287–332.
13. Goodman, R.E., Taylor, R. L. and Brekke, T. L. (1968) "A Model for the Mechanics of Jointed Rock," *Proc. ASCE, J. Soil Mechs. and Found. Engrg. Div.*, Vol. 94, No. SM3, May.

13*a*. King, E. R. and Zeitz, L. (1978) "The New York-Alabama Lineament: Geophysical Evidence for a Major Crustal Break in the Basement Beneath the Appalachian Basin," *Geology*, GSA Bull., Vol. 6, No. 5, May, pp. 312–318.

13*b*. Oakeshott, G. B. (1973) "Patterns of Ground Rupture in Fault Zones Coincident with Earthquakes:

Some Case Histories," *Geology, Seismicity and Environmental Impact*, Spec. Pub. Assoc. Engrg. Geol., University Publishers, Los Angeles, pp. 287–312.

14. Hamblin, W. K. (1976) "Patterns of Displacement Along the Wasatch Fault," *Geology*, Vol. 4, No. 10, October, pp. 619–622.

14*a*. Deere, D. U. (1974) "Engineering Geologist's Responsibilities in Dam Foundation Studies," *Foundation for Dams*, ASCE, New York, pp. 417–424.

15. Sherard, J. L., Cluff, L. S. and Allen, C. R. (1974) "Potentially Active Faults in Dam Foundations," *Geotechnique*, Vol. XXIV, No. 4, September, pp. 367–428.

16. Taylor, C.L. and Cluff, L. S. (1977) "Fault Displacement and Ground Deformation Associated with Surface Faulting," *Proc. ASCE, The Current State of Knowledge of Lifeline Earthquake Engineering, Spec. Conf.*, Univ. of California, Los Angeles, pp. 338–353.

17. Coates, D. F. (1964) "Some Cases of Residual Stress Effects in Engineering Work," *State of Stress in the Earth's Crust*, W. R. Judd, ed., American Elsevier Pub. Co., New York, p. 679.

18. Ward, J. S. and Assoc. (1972) "Bedrock Can Be a Hazard—Wall Movements of a Deep Rock Excavation Analyzed by the Finite Element Method," *Soils*, Vol. 1, February, Caldwell, New Jersey.

19. Mason, R. E. (1968) "Instrumentation of the Shotcrete Lining in the Canadian National Railways Tunnel," M.S. Thesis, Univ. of British Columbia.

20. Moye, D. G. (1964) "Rock Mechanics in the Investigation of T 1 Underground Power Station, Snowy Mountains, Australia," *Engineering Geology Case Histories 1–5*, Geol. Soc. of Amer., New York, pp. 123–154.

21. Rainer, H. (1970) "Are There Connections Between Earthquakes and the Frequency of Rock Bursts in the Mine at Blieburg?," *J. Intl. Soc. on Rock Mechs.*, Vol. 6, No. 2, August.

22. Paulmann, H. G. (1966) "Measurements of Strength Anisotropy of Tectonic Origin on Rock Mechanics," *Proc. 1st Cong. Intl. Soc. Rock Mechs.*, Vol. 1, Lisbon.

23. Denkhaus, H. (1966) "Residual Stresses in Rock Masses," *Proc. 1st Intl. Cong. Intl. Soc. of Rock Mechs.*, Vol. 3, Lisbon.

24. Hawkes, I. (1966) "Significance of In-Situ Stress Levels," *Proc. 1st Intl. Cong. Intl. Soc. of Rock Mechs.*, Vol. 3, Lisbon.

25. Deere, D. U. (1966) "Residual Stresses in Rock Masses," *Proc. 1st Intl. Cong. Intl. Soc. Rock Mechs.*, Vol. 3, Lisbon.

26. Rice, C. M. (1954) *Dictionary of Geological Terms*, Edwards Brothers, Inc., Ann Arbor, Michigan.

27. Fookes, P. G., Dearman, W. R. and Franklin, J. A. (1971) "Some Engineering Aspects of Rock Weathering with Field Examples from Dartmoor and Elsewhere," *Quarterly of Engrg. Geol.*, Vol. 4, No. 3, July, The Geologic Soc. of London.

28. Morin, W. J. and Tudor, P. C. (1976) "*Laterite and Lateritic Soils and Other Problem Soils of the Tropics*," AID/csd 3682, U.S. Agency for International Development, Washington, D.C.

29. Hunt, C. B. (1972) *Geology of Soils*, W. H. Freeman and Co., San Francisco.

30. Froelich, A.J. (1973) *Geol. Survey Prof. Paper 850*, p. 211, U.S. Govt. Printing Office, Washington, D.C.

31. Sowers, G. F. (1954) "Soil Problems in the Southern Piedmont Region," *Proc. ASCE, J. Soil Mechs. and Found. Engrg. Div.*, Separate No. 416, Vol. 80, March.

32. Feld, J. (1966) "Age Change in the Bearing Capacity of Mica Schist," *Proc. 1st Intl. Cong. Intl. Soc. for Rock Mechs.*, Vol. II, pp. 523–524.

33. Grim, R. E. (1962) *Clay Mineralogy*, 2d ed., McGraw-Hill Book Co., New York.

34. Vargas, M. (1974) "Engineering Properties of Residual Soils from South-Central Brazil," *Proc. 2nd Intl. Cong. Intl. Assoc. Engrg. Geol.*, São Paulo, Vol. 1.

35. Deere, D. U. and Patton, F. D. (1971) "Slope Stability in Residual Soils," *Proc. 4th Panamerican Conf. Soil Mechs. and Found. Engrg.*, San Juan, Vol. 1, pp. 87–100.

36. Banks, D. C. (1972) "Study of Clay Shale Slopes," *Stability of Rock Slopes, Proc. ASCE, 13th Symp. on Rock Mechs.*, Univ. of Illinois (1971), pp. 303–328.

37. Bjerrum, L. (1967) "Progressive Failure in Slopes of Overconsolidated Clays and Clay Shales," *Terzaghi Lectures*, ASCE, New York (1974), pp. 139–187.

38. Meehan, R. L., Dukes, M. T. and Shires, B. O. (1975) "A Case History of Expansive Clay Stone Damage," *Proc. ASCE, J. Geotech. Engrg. Div.*, Vol. 101, No. GT9, September.

## BIBLIOGRAPHY

*Atlas of American Geology* (1932) The Geographic Press.

Billings, M. P. (1959) *Structural Geology*, 2d ed., Prentice-Hall, Englewood Cliffs, N.J.

Blake, W. (1972) "Rock Burst Mechanics," *Quarterly, Colo. School of Mines*, Boulder, Vol. 67, No. 1, January.

Dearman, W. H. (1976) "Weathering Classification in the Characterization of Rock: A Revision," *Intl. Assoc. Engrg. Geol.*, Bull. No. 13, June, pp. 123–128.

Holmes, A. (1964) *Principles of Physical Geology*, The Ronald Press, New York.

Jennings, J. N. (1971) *Karst*, The MIT Press, Cambridge, Mass.

Twidale, C. R. (1971) *Structural Landforms*, The MIT Press, Cambridge, Mass.

# CHAPTER SEVEN

# SOIL FORMATIONS: GEOLOGIC CLASSES AND CHARACTER-ISTICS

## 7.1 INTRODUCTION

### 7.1.1 GEOLOGIC CLASSIFICATION OF SOIL FORMATIONS

#### General

Soils are classed geologically by their origin as residual, colluvial, alluvial, eolian, glacial, or secondary soils. They are subclassed on the basis of their mode of occurrence, which refers to the landform or surface expression of a deposit or its location relative to the regional physiography. The various geologic classes have characteristic modes of occurrence.

Classifying soils by geologic origin, describing the formation in terms of its mode of occurrence, and considering both as related to climate provide information on the characteristics of gradation, structure, and stress history for a given deposit.

Knowledge of gradation, structure, and stress history provides the basis for formulating preliminary judgments on the engineering properties of permeability, strength, and deformability; for intelligent planning of exploration programs, especially in locations where the investigator has little or no prior experience; and for extending the data obtained during exploration from a relatively few points over the entire study area.

#### Classes and Mode of Occurrence

*Residual Soils*

Developed in situ from the decomposition of rock as discussed in Art. 6.7, residual soils have geomorphic characteristics very much related to the parent rock.

*Colluvial Soils*

Colluvium refers to soils transported by gravitational forces. Their modes of occurrence relate to forms of landsliding and other slope movements such as falls, avalanches, and flows.

*Alluvial Soils*

Alluvium is transported by water. The mode of occurrence can take many forms generally divided into four groups and further subdivided. In this chapter the term is used broadly to include marine deposits.

Fluvial or river deposits include streambed, alluvial fan, and floodplain deposits (point bar, clay plugs, natural levees, back swamp), those laid down under rejuvenated stream conditions (buried valleys, terraces), and those deposited in the estuarine zone (deltas, estuary soils).

Lacustrine deposits include those laid down in lakes and playas.

Coastal deposits include spits, barrier beaches, tidal marshes, and beach ridges.

Marine deposits include offshore soils and coastal-plain deposits.

*Eolian Soils*

Eolian deposits are transported by wind and occur as dunes, sand sheets, loess, and volcanic dust.

*Glacial Soils*

Soils deposited by glaciers or glacial waters can take many forms, subdivided into two groups:

- Moraines are deposited directly from the glacier as ground moraine (basal till, ablation till, drumlins) or as end, terminal, and interlobate moraines.
- Stratified drift is deposited by the meltwaters as fluvial formations (kames, kame terraces, eskers, outwash, kettles), or lacustrine (freshwater or saltwater deposition).

*Secondary Deposits*

Original deposits modified in situ by climatic factors to produce duricrusts, permafrost, and pedological soils are referred to as secondary deposits. The duricrusts include laterite, ironstone (ferrocrete), caliche, and silcrete.

**Some Engineering Relationships**

*General*

The various classes and subclasses have characteristic properties which allow predictions of the impact of a particular formation on construction. A classification of soils by origin and mode of occurrence is given in Table 7.1. Included are the depositional environment, the occurrence either as deposited or as subsequently modified, and the typical material associated with the formation. A general distribution of soils in the United States, classed by origin, is illustrated on Fig. 7.1. The nomenclature for Fig. 7.1 is given on Table 7.2.

*Foundation Conditions*

Generally favorable foundation conditions are associated with (1) medium dense or denser soils characteristic of some stream channel deposits, coastal deposits, and glacial moraines and stratified drift; (2) overconsolidated inactive clays of some coastal plains; and (3) clay-granular mixtures characteristic of residual soils formed from sialic rocks.

Marginal foundation conditions may be associated with glacial lacustrine clays and soils with a potential for collapse such as playa deposits, loess, and porous clays.

Poor foundation conditions may be associated with colluvium, which is often unstable on slopes; granular soils deposited in a loose condition in floodplains, deltas, estuaries, lakes, swamps, and marshes; active clays resulting from the decomposition of mafic rocks and marine shales, or deposited as marine clays and uplifted to a coastal plain, or deposited by ancient volcanic activity; and all organic deposits.

### 7.1.2 TERRAIN ANALYSIS

**General**

Terrain analysis *(see Arts. 2.2.3 and 6.1.2)* provides a basis for identifying the mode of occurrence of soil formations and classifying them by origin.

**Interpretative Factors**

Landform, stream forms and patterns, gully characteristics, vegetation, tone and color (of remote-sensing imagery), and land use are the basic interpretative factors for terrain analysis. Classes of drainage patterns in soil formations are described on Table 7.3, and the typical drainage patterns for various soil types and formations are given on Table 7.4. Gully characteristics are illustrated on Fig. 7.2.

## 7.2 RESIDUAL SOILS

### 7.2.1 INTRODUCTION

**General**

Residual soils develop in situ from the disintegration and decomposition of rock *(see Art. 6.7)*. The distinction between rock and soil is difficult when the transition is gradual, as is the

**TABLE 7.1**
**CLASSIFICATION OF SOILS BY ORIGIN AND MODE OF OCCURRENCE**

| Origin | Depositional environment | Occurrence | | Typical material‡ |
|---|---|---|---|---|
| | | Primary* | Secondary† | |
| Residual | In situ | Syenite<br>Granite<br>Diorite | Saprolite | Low activity clays and granular soils |
| | | Gabbro<br>Basalt<br>Dolerite | Saprolite | High activity clays |
| | | Gneiss<br>Schist | Saprolite | Low activity clays and granular soils |
| | | Phyllite | | Very soft rock |
| | | Sandstone | | Thin cover depends on impurities |
| | | Shales | Red | Thin clayey cover |
| | | | Black, marine | Friable and weak mass, high activity clays |
| | | Carbonates | Pure | No soil, rock dissolves |
| | | | Impure | Low to high activity clay |
| Colluvial | Slopes | Falls | Talus | Boulders to cobbles |
| | | Slides | Structure preserved | Parent material |
| | | Flows | Structure destroyed | Parent or mixed material |
| Alluvial | Fluvial | Streambed | Youthful stage | Very coarse granular |
| | | | Mature, braided | Coarse granular |
| | | | Old age, meandering | Coarse to fine, loose |
| | | Alluvial fan | | Coarse to fine, loose |
| | | Floodplain | Point bar | Medium-fine sand, loose |
| | | | Clay plugs | Soft clay |
| | | | Natural levees | Coarse-fine sand, loose |
| | | | Backswamp | Organic silt and clay |
| | | | Lateral accretion | Medium-granular, loose |
| | | Rejuvenated | Buried valleys | Coarse granular, dense |
| | | | Terraces | Variable, medium dense |
| | Estuarine | Delta | Parent delta | Variable, soft/loose |
| | | | Subdelta | Chiefly sands, loose |
| | | | Prodelta | Soft clays |

(*Continued*)

**TABLE 7.1**
**CLASSIFICATION OF SOILS BY ORIGIN AND MODE OF OCCURRENCE (*Continued*)**

| Origin | Depositional environment | Occurrence: Primary* | Occurrence: Secondary† | Typical material‡ |
|---|---|---|---|---|
| | | Estuary | | Primarily fine-grained grading to coarse |
| | Lacustrine | Lakes | Various forms | Primarily fine-grained |
| | | | Swamps and marshes | Organic soils |
| | | Playas | Evaporites | Fine-grained with salts |
| | Coastal | Marine depositional coast | Spits | Coarse-fine sand, medium dense |
| | | | Barrier beach | Coarse-fine sand, dense |
| | | | Tidal marsh | Very soft organics |
| | | | Beach ridges | Coarse-fine sand, dense |
| | Marine | Offshore | Varies with water depth and currents | Marine clays<br>Silica sands<br>Carbonate sands |
| | | Coastal plain | | Various, preconsolidated |
| Aeolian | Ground moisture deficient | Dunes | | Medium-fine sand, loose |
| | | Sand sheets | | Medium-fine sand, loose |
| | | Loess | | Silts, clays, lightly cemented |
| | | Volcanic clay | | Expansive clays |
| Glacial | In situ | Ground moraine | Basal till | Extremely hard mixture |
| | | | Ablation till | Relatively loose mixtures |
| | | | Drumlins | Various, with rock core |
| | | End moraine | | Coarse mixtures, medium dense |
| | | Terminal moraine | | Coarse mixtures, medium dense |
| | | Interlobate moraine | | Coarse mixtures, medium dense |
| | Fluvial | Ice-contact stratified drift | Kanes | Poorly sorted, granular |

**TABLE 7.1**
**CLASSIFICATION OF SOILS BY ORIGIN AND MODE OF OCCURRENCE (*Continued*)**

| Origin | Depositional environment | Occurrence: Primary* | Occurrence: Secondary† | Typical material‡ |
|---|---|---|---|---|
| | | | Kane terraces | Poorly sorted, loose |
| | | | Eskers | Sand and gravel |
| | | Pro-glacial stratified drift | Outwash | Gravels to silts, loose to medium dense |
| | | | Kettles | Organics |
| | Lacustrine | Lakes | Seasonal affects | Varved clays |
| | Marine | Offshore | Quiet waters | Sensitive clays |
| Secondary | In situ | Duricrusts | Laterite | Iron, aluminum rich |
| | | | Ironstone | Iron rich |
| | | | Caliche | Carbonate rich |
| | | | Silcrete | Silica rich |
| | | Permafrost | | Ice and soil |
| Pedological (modern soils) | In situ | Soil profile | A Horizon } Solum | Leached, elluviated, organic |
| | | | B Horizon } Solum | Illuviated zone, clays, organic |
| | | | C Horizon | Partly altered parent material |
| | | | D Horizon | Unaltered parent material |
| | | Soil groups | Tundra | Arctic soils |
| | | | Podzol | Grayish forest soils, organic rich |
| | | | Laterites | Reddish tropical soils, low organic iron rich |
| | | | Chernozems | Black prairie soils, organic rich |
| | | | Chestnut soils | Brownish grassland soils, organic rich |
| | | | Brown aridic soils | Low organic, calcareous |
| | | | Gray desert soils | Very low organic |
| | | | Noncalcic brown | Former forest areas, organic |

*Denotes the original form of the deposited material: i.e., rock type for residual soils, type of slope failure for colluvial soils, mode of occurrence for soils of other origins (alluvial, aeolian, glacial).

†Refers to the general characteristics of residual and colluvial soils; for alluvial, aeolian and glacial, refers to a distinctive landform, or geographical or physical position representing a modification of the primary mode.

‡The soil type generally characteristic of the deposit.

TABLE 7.2
DISTRIBUTION OF PRINCIPAL SOIL DEPOSITS IN THE UNITED STATES

| Origin of principal soil deposits | Symbol for area in Fig. 7.1 | Physiographic province | Physiographic features | Characteristic soil deposits |
|---|---|---|---|---|
| Alluvial | Al | Coastal plain | Terraced or belted coastal plain with submerged border on Atlantic. Marine plain with sinks, swamps, and sand hills in Florida. | Marine and continental alluvium thickening seaward. Organic soils on coast. Broad clay belts west of Mississippi. Calcareous sediments on soft and cavitated limestone in Florida. |
| Do | A2 | Mississippi alluvial plain | River floodplain and delta. | Recent alluvium, fine grained and organic in low areas, overlying clays of coastal plain. |
| Do | A3 | High Plains section of Great Plains province | Broad intervalley remnants of smooth fluvial plains. | Outwash mantle of silt, sand, silty clay, lesser gravels, underlain by soft shale, sandstone, and marls. |
| Do | A4 | Basin and range province | Isolated ranges of dissected block mountains separated by desert plains. | Desert plains formed principally of alluvial fans of coarse-grained soils merging to playa lake deposits. Numerous nonsoil areas. |
| Do | A5 | Major lakes of basin and range province | Intermontane Pleistocene lakes in Utah and Nevada, Salton Basin in California. | Lacustrine silts and clays with beach sands on periphery. Widespread sand areas in Salton basin. |
| Do | A6 | Valleys and basins of Pacific border province | Intermontane lowlands, Central Valley, Los Angeles Basin, Willamette Valley. | Valley fills of various gradations, fine grained and sometimes organic in lowest areas near drainage system. |
| Residual | R1 | Piedmont province | Dissected peneplain with moderate relief. Ridges on stronger rocks. | Soils weathered in place from metamorphic and intrusive rocks (except red shale and sandstone in New Jersey). Generally more clayey at surface. |
| Do | R2 | Valley and ridge province | Folded strong and weak strata forming successive ridges and valleys. | Soils in valleys weathered from shale, sandstone, and limestone. Soil thin or absent on ridges. |
| Do | R3 | Interior low plateaus and Appalachian plateaus | Mature, dissected plateaus of moderate relief. | Soils weathered in place from shale, sandstone, and limestone. |
| Do | R4 | Ozark plateau, Ouachita province, portions of Great Plains and central lowland, Wisconsin driftless section | Plateaus and plains of moderate relief, folded strong and weak strata in Arkansas. | Soils weathered in place from sandstone and limestone predominantly, and shales secondarily. Numerous nonsoil areas in Arkansas. |

| | | | | |
|---|---|---|---|---|
| Do | R5 | Northern and western sections of Great Plains province | Old plateau, terrace lands, and Rocky Mountain piedmont. | Soils weathered in place from shale, sandstone, and limestone including areas of clay-shales in Montana, South Dakota, Colorado. |
| Do | R6 | Wyoming basin | Elevated plains. | Soils weathered in place from shale, sandstone, and limestone. |
| Do | R7 | Colorado plateaus | Dissected plateau of strong relief. | Soils weathered in place from sandstone primarily, shale and limestone secondarily. |
| Do | R8 | Columbia plateaus and Pacific border province | High plateaus and piedmont. | Soils weathered from extrusive rocks in Columbia plateaus and from shale and sandstone on Pacific border. Includes area of volcanic ash and pumice in central Oregon. |
| Loessial | L1 | Portion of coastal plain | Steep bluffs on west limit with incised drainage. | 30 to 100 ft of loessial silt and sand overlying coastal plain alluvium. Loess cover thins eastward. |
| Do | L2 | Southwest section of central lowland; portions of Great Plains | Broad intervalley remnants of smooth plains. | Loessial silty clay, silt, silty fine sand with clayey binder in western areas, calcareous binder in eastern areas. |
| Do | L3 | Snake River plain of Columbia plateaus | Young lava plateau. | Relatively thin cover of loessial silty fine sand overlying fresh lava flows. |
| Do | L4 | Walla Walla plateau of Columbia plateaus | Rolling plateau with young incised valleys. | Loessial silt as thick as 75 ft overlying basalt. Incised valleys floored with coarse-grained alluvium. |
| Glacial | G1 | New England Province | Low peneplain maturely eroded and glaciated. | Generally glacial till overlying metamorphic and intrusive rocks, frequent and irregular outcrops. Coarse, stratified drift in upper drainage systems. Varved silt and clay deposits at Portland, Boston, New York, Connecticut River Valley, Hackensack area. |
| Do | G2 | Northern section of Appalachian plateau, Northern section of Central lowland | Mature glaciated plateau in northeast, young till plains in western areas. | Generally glacial till overlying sedimentary rocks. Coarse stratified drift in drainage system. Numerous swamps and marshes in north central section. Varved silt and clay deposits at Cleveland, Toledo, Detroit, Chicago, northwestern Minnesota. |
| Do | G3 | Areas in southern central lowland | Dissected old till plains. | Old glacial drift, sorted and unsorted, deeply weathered, overlying sedimentary rocks. |

*(Continued)*

**TABLE 7.2**
**DISTRIBUTION OF PRINCIPAL SOIL DEPOSITS IN THE UNITED STATES (*Continued*)**

| Origin of principal soil deposits | Symbol for area in Fig. 7.1 | Physiographic province | Physiographic features | Characteristic soil deposits |
|---|---|---|---|---|
| Do | G4 | Western area of northern Rocky Mountains | Deeply dissected mountain uplands with intermontane basins extensively glaciated. | Varved clay, silt, and sand in intermontane basins, overlain in part by coarse-grained glacial outwash. |
| Do | G5 | Puget trough of Pacific border province | River valley system, drowned and glaciated. | Variety of glacial deposits, generally stratified, ranging from clayey silt to very coarse outwash. |
| Do | G6 | Alaska peninsula | Folded mountain chains of great relief with intermontane basins extensively glaciated. | In valleys and coastal areas widespread deposits of stratified outwash, moraines, and till. Numerous nonsoil areas. |
| | | Hawaiian Island group | Coral islands on the west, volcanic islands on the east. | Coral islands generally have sand cover. Volcanic ash, pumice, and tuff overlie lava flows and cones on volcanic islands. In some areas volcanic deposits are deeply weathered. |
| Nonsoil areas | | Principal mountain masses | Mountains, canyons, scablands, badlands. | Locations in which soil cover is very thin or has little engineering significance because of rough topography or exposed rock. |

SOURCE: *Design Manual DM-7, Soil Mechanics, Foundations, and Earth Structures,* Naval Facilities Engineering Command, Alexandria, Va., 1971.

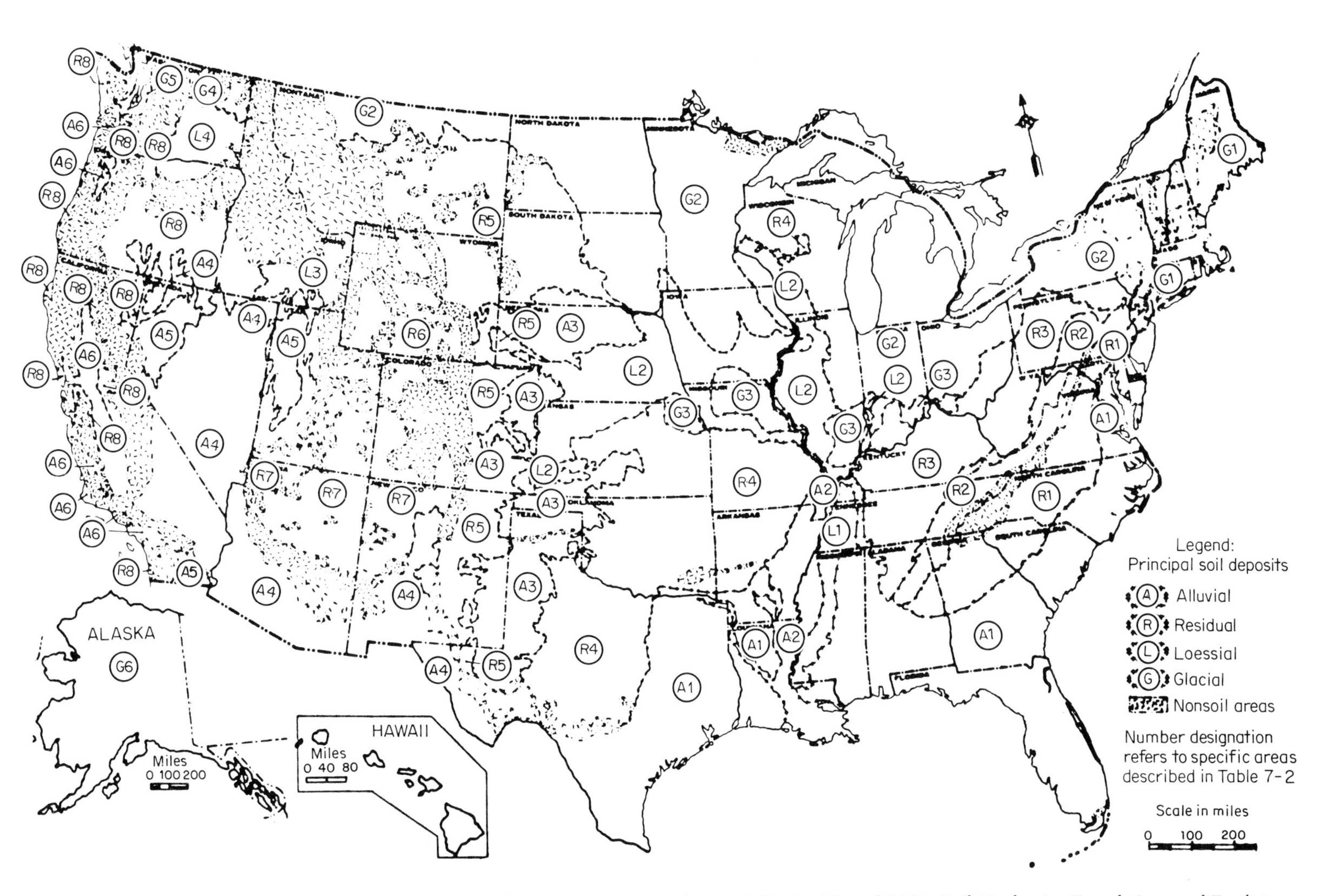


**FIG. 7.1 Distribution of soils in the United States classed by origin.** [*From Navfac (1971), Design Manual DM-7, Soil Mechanics, Foundations, and Earthstructures, Naval Facilities Engineering Command, Alexandria, Va.*]

**TABLE 7.3**
**CLASSES OF DRAINAGE PATTERNS IN SOIL FORMATIONS**

| Class | Associated formations | Characteristics |
|---|---|---|
| Dendritic | Uniform, homogeneous formations | Tributaries join the gently curving mainstream at acute angles. The more impervious the materials, the finer the texture. |
| Pinnate | Loess and other easily eroded materials | Intense pattern of branches enters the tributaries almost at right angles, or slightly upstream. |
| Parallel | Mature coastal plains | Modified dendritic with parallel branches entering the mainstream. |
| Deranged | Young landforms: floodplains and thick till plains | Lack of pattern development. Area contains lakes, ponds, and marshes in which channels terminate. |
| Meandering | Floodplains, lake beds, swamps | Sinuous, curving mainstream with cutoffs and oxbow lakes. |
| Radial braided | Alluvial fans | Mainstream terminating in numerous off-parallel branches radiating outward. |
| Parallel braided | Sheet wash, coalescing alluvial fans | Parallel and subparallel streams in fine-textured pattern. |
| Thermokarst | Poorly drained soils in thermofrost regions | Ground freezing and heaving causes hexagonal patterns; subsequent thawing results in a sequence of lakes, giving a beaded appearance along the stream. |

NOTE: Patterns in rock formations *(see Table 6.1)* include rectangular, angulate, trellis, barbed, radial, annular, and centripedal as well as dendritic, parallel, and deranged.

case with most rocks. In some rocks, such as limestone, the transition is abrupt.

Chemical decomposition produces the most significant residual soil deposits. Mechanical weathering generally produces primarily granular particles of limited thickness, except in marine shales.

The depth and type of soil cover that develops are often erratic, since they are a function of the mineral constituents of the parent rock, climate, the time span of weathering exposure, orientation of weakness planes (permitting the entry of water), and topography *(see Art. 6.7.2)*.

### Rock-Type Relationships

Igneous and metamorphic rocks composed of silicates and oxides produce thick, predominantly clay soils.

Sandstones and shales are composed chiefly of stable minerals (quartz and clay) which undergo very little additional alteration. It is the impurities (unweathered particles and cementing agents) that decompose to form the relatively thin soil cover.

Carbonates and sulfates generally go into solution before they decompose. The relatively thin soil cover that develops results from impurities.

Marine and clay shales generally undergo mechanical weathering from swelling with some additional chemical decomposition.

### Climate

Soil profile development is related primarily to rainfall, but temperature is an important factor. Very little soil cover develops in either a cool-dry or hot-dry climate. Cool-wet zones produce relatively thick soil cover, but tropical climates with combinations of high temperatures and high rainfall produce the greatest thicknesses.

**Mode of Occurrence**

As an integral part of the rock mass, the surface expression of residual soils relates to the rate of rock decay and the effects of differential erosion, and therefore generally reflects the parent rock type and the original landform (peneplain, mountainous, etc.) as described in Art. 6.7.3.

## 7.2.2 IGNEOUS AND METAMORPHIC ROCKS

**General**

*Products of Decomposition*

Igneous and metamorphic rocks are composed of silicates and oxides as summarized on Tables 6.13 and 6.14. Silicates decompose to yield primarily clay soils. Feldspars produce kaolin clays generally of light colors ranging from white to cream. Ferromagnesians *(see Table 5.2)* produce clays and iron oxides which in humid climates give residual soils a strong red color, or when concentrations are weaker, yellow or brown colors. Black, highly expansive clays occur under certain conditions. Muscovite (white mica) is relatively stable and is often found unaltered, since it is one of the last minerals to decompose (biotite, black mica, decomposes relatively rapidly).

Quartz, the predominant oxide, essentially undergoes no change, except for some slight solution, and produces silt and sand particles.

*Effects of Climate (See Also Fig. 6.72)*

Temperate and semitropical climates tend to produce predominantly kaolinite clays, and occasionally halloysite and hydrated and anhydrous oxides of iron and aluminum. The end product usually is sand-silt-clay mixtures with mica, with low activity.

In tropical climates the decomposition is much more intense and rapid, in some locations reaching depths of 30 m or more in foliated rocks. The predominant end products are silty clay mixtures with minor amounts of sand. The common red tropical clays contain hydrated oxides of iron and aluminum in the clay fraction. The type of clay is a function of the parent rock minerals and time and can vary from inactive to active as given in Fig. 6.73 and Table 5.29.

In hot climates with alternating wet and dry seasons, as decomposition advances, the clay minerals are essentially destroyed and the silica leached out. The material remaining consists of aluminum hydroxide (bauxite, in pure form), or hydrous iron oxide, such as limonite. This is the process of *laterization (see Art. 7.7.2).*

In hot climates with low to moderate rainfall, mafic rocks in an environment of poor drainage produce black, highly expansive clays ("black cotton soils").

*Prediction of Soil Type*

Known factors required for prediction of the soil type are the basic rock type (from a geologic map), climate, and topography.

The procedure is as follows. Determine the mineral composition from Table 5.7 for igneous rocks or Table 5.13 for metamorphic rocks. Determine the possible physical extent of the formation on the basis of typical forms vs. the rock type from Table 6.4, and evaluate the influence of topography from terrain analysis. Evaluate the relative stability of the minerals in terms of the end product from Table 6.13. Estimate the soil composition and activity from Fig. 6.73, and consider the climate factor as described above and given on Fig. 6.72 in terms of the topographic expression as it relates to formation drainage *(see Art. 6.7.2).*

**Typical Residual Soil Profiles and Characteristics**

*General*

Soil profile and engineering characteristics at a given site can be extremely variable, even for a given rock type, because of variations in mineral content, rock structure, and topography.

*Sialic Rocks*

Granite, syenite, etc. decompose to mixtures of quartz sands and kaolin clays with muscovite mica fragments. Biotite mica decomposes to produce iron oxides that impart reddish or brown-

**TABLE 7.4**
**TYPICAL DRAINAGE PATTERNS FOR VARIOUS SOIL TYPES AND FORMATIONS***

| Geologic condition | Predominant drainage pattern | Geologic condition | Predominant drainage pattern |
|---|---|---|---|
| Sand sheets, terraces | No surface drainage | | |
| Clayey soils† | 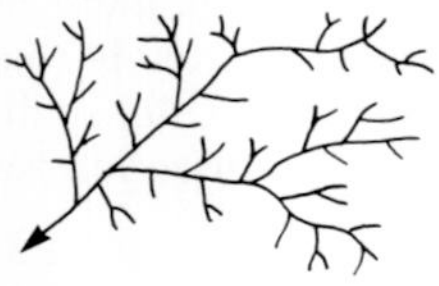 Fine dendritic | Young till plains, uplifted peneplains with impervious soils | 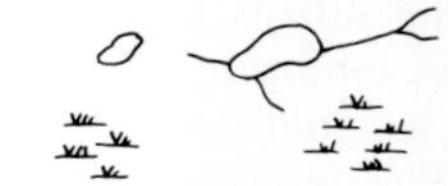 Deranged with ponds and swamps |
| Clay-sand mixtures‡ |  Medium dendritic | Old till plains, thick | 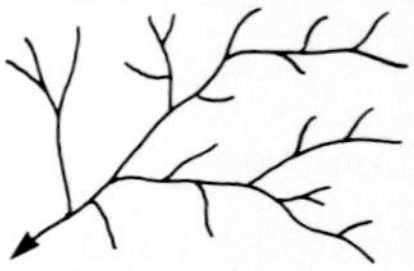 Medium dendritic |
| Sandy soils, some cohesion, porous clays§ |  Coarse dendritic | Loess | 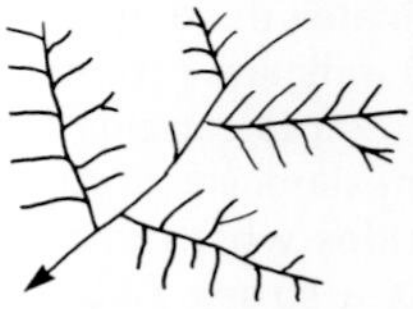 Pinnate and dendritic |
| Coastal plains, mature with questas | 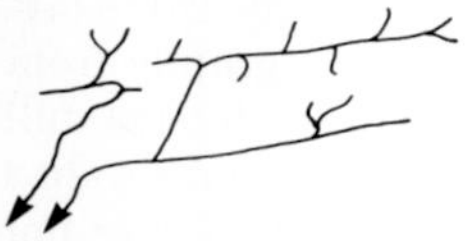 Coarse parallel | Lake beds, floodplains |  Meandering with oxbow lakes |
| Alluvial fans | 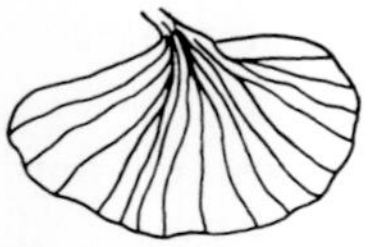 Radial braided | Permafrost regions | 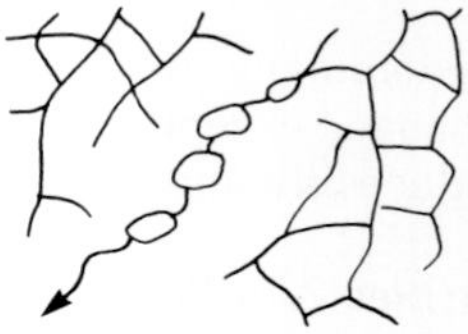 Thermokarst |
| Sheet wash, coalescing alluvial fans |  Parallel braided | | |

*Drainage patterns given in terms of coarse, medium, and fine textures are for humid climate conditions. Arid climates generally produce a pattern one level coarser.

†Coarse: First-order streams over 2 in (5 cm) apart (on map scale of 1:20,000) and carrying relatively little runoff.

‡Medium: First-order streams ¼ to 2 in (5 mm to 5 cm) apart.

§Fine: Spacing between tributaries and first-order stream less than ¼ in (5 mm).

ish color. Clays are of low activity. Boulders form in massive to blocky formations as described in Art. 6.7.3.

*Mafic Rocks*

Gabbro, basalt, periodotite, etc. decompose primarily to plastic red clays in tropical climates, and black expansive clays in hot climates with low to moderate rainfall and poor drainage. The black clays are often associated with basalt flows.

A typical decomposition profile of basalt in south central Brazil is given on Fig. 6.84. Some typical engineering properties including index and strength values are given on Table 7.5. "Porous clays," a type of collapsible soil, are common in the residual soils of basalt in Brazil (as well as sandstones and Tertiary clays, *see Art. 10.5.2*).

*Foliated Rocks*

Decomposition in gneiss varies with the original rock (prior to metamorphism, Table 5.12), and can range from red micaceous sandy clay (low activity) to micaceous sandy silt depending upon the amount of feldspar and biotite in the original rock. Foliations facilitate the entry of water, and decomposition is relatively deep compared with other rocks. Unweathered quartz veins are typical of gneiss formations.

As shown on Fig. 6.85, decomposition of gneiss results in two general zones, residual soils without relict structure (at times referred to as massive saprolite) and *saprolite*, residual soils with relict rock structures (*see Fig. 6.87*).

Typical soil profile zones and engineering characteristics of gneiss are given on Table 7.6 and on the boring logs (Fig. 7.5*c* and *d*). A photo of an exposure is given as Fig. 7.3. Gradation, index test results, and virtual preconsolidation pressure in a boring to a depth of 15 m in Belo Horizonte, Brazil, are given on Fig. 7.4. Typical engineering properties including index and strength values are given on Tables 3.27 and 7.5.

*Various Rock Types*

Boring logs from four different locations are given on Fig. 7.5. Included are SPT values and some index test results for decomposed schist, volcanics, and gneisses.

### Tropical Residual Soils

*Red Tropical or Lateritic Soils*

Red tropical or lateritic soils cover large parts of the world in the middle latitudes including much of Brazil, the southern third of Africa, southeast Asia, and parts of India. They are associated with numerous rock types including basalt, diabase, gneiss, seritic schist, and phyllite.

Decomposition forms a reddish soil rich in iron, or iron and aluminum, and tropical climate conditions are required. The degree of *laterization* is estimated by the silica-sesquioxide ratio $[SiO_2/(Fe_2O_3 + Al_2O_3)]$ (*see Art. 5.3.3*) as follows:

- Less than 1.33—true laterite, a hard rocklike material

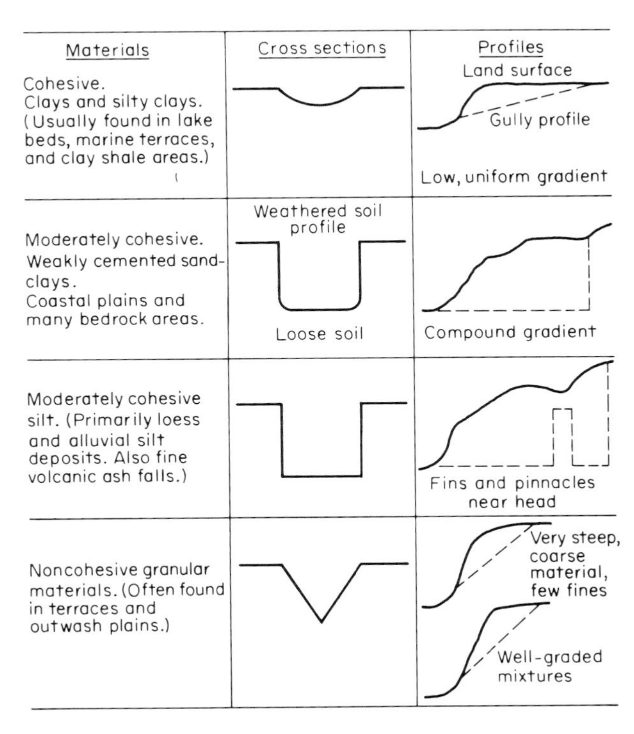


**FIG. 7.2 Gully characteristics: cross sections, profiles, and associated soils.** [*From Way (1978)*[2].]

**TABLE 7.5**
**TYPICAL ENGINEERING PROPERTIES OF RESIDUAL SOILS OF BASALT AND GNEISS**

| Parent rock | Zone | Location | *N* value (SPT) | LL | PI | *e* | $\phi$ | c, kg/cm² | Ref.‡ |
|---|---|---|---|---|---|---|---|---|---|
| Granite gneiss | Upper | Georgia | 10–25 | 30–50 | 9–25 | 0.7–0.8 | | | (*a*) |
| Granite gneiss | Intermediate | Georgia | 5–10 | 20–40 | 0–5 | 0.8–1.2 | | | |
| Granite gneiss | Saprolite | Georgia | 17–70 | | | 0.8–0.4 | | | |
| Gneiss | | Brazil (coastal mountains) | | 20–70 | 0–35 | 1.4–1.0 | 25–31 | 0.4–0.6 | (*b*) |
| Basalt (porous clays) | Upper | Brazil | | 35–75 | 15–40 | 1.1–1.0 | 27–31 | 0.1–0.2 | |
| Gneiss*† | Upper | Rio de Janeiro | 5–10 | 20–55 | 5–25 | 2.3–0.8 | 23–43 | 0–0.4 | (*c*) |
| Gneiss*† | Intermediate | Rio de Janeiro | 10–30 | | | | | | |
| Gneiss*† | Saprolite | Rio de Janeiro | 30–50 | | | | 23–38 | 0.2–0.5 | |
| Gneiss | Upper | São Paulo | 8–28 | 50–70 | 30–35 | | | | (*d*) |
| Gneiss | Intermediate | São Paulo | 7–10 | 40–50 | 30–20 | | | | |
| Gneiss | Saprolite | São Paulo | 10–30 | 48–50 | 20–25 | | | | |
| Basalt | Upper | Parana, Brazil | | 53–60 | 26–18 | | | | |
| Basalt | Intermediate | Parana, Brazil | | 65–45 | 17–10 | | 24–31 | 0.1–0.7 | |

*Strength tests performed at natural moisture; not necessarily saturated.

†Natural density range: 1.6–1.8 g/cm³; saturated, 1.8–2.0 g/cm³.

‡References: (*a*) Sowers (1954),[3] (*b*) Vargas (1974),[4] (*c*) data courtesy of Tecnosolo S.A., (*d*) Medina (1970).[5]

**TABLE 7.6**
**TYPICAL RESIDUAL SOIL PROFILE AND ENGINEERING CHARACTERISTICS OF GNEISS IN HUMID CLIMATE***

| Zone | Description | Engineering characteristics† |
|---|---|---|
| Upper (A and B horizons) | Surficial humus zone (A horizon) grades to mature residual soil, usually reddish- to yellowish-brown, heterogeneous mixtures of silt, sand, and clay (ML or CL-ML groups) with kaolin clays (Fig. 7.3). Desiccation or cementation following leaching can result in a stiff crust from 1 to 2 m, and occasionally 3 m thick. Laterite gravel can be found on the surface and at the contact with the intermediate zone in alternating wet and dry climates. | A horizon is usually porous since it is the zone of maximum leaching. Strengths are low to medium. B horizon: If desiccated or cemented, strengths will be high, compressibility low (see *Fig. 7.4b*). Relatively impervious. If uncemented and not desiccated, as often exists when deposit is permanently below GWL: low permeability, but can be compressible with moderate strengths.‡ SPT values often range from 5 to 10; void ratios, 1.0 to 1.5 and higher. |
| Intermediate | Red, brown, yellow stiff to hard clays or heterogeneous sand-clay mixtures, often micaceous with quartz veins. | Can be strong and relatively incompressible with allowable bearing of about 10 kg/cm²§ with SPT ranging from 30 to 50; or it can be relatively weak and compressible (SPT 5 to 9) as shown on the boring log from Atlanta (Fig. 7.5*d*). |
| Saprolite | Relict rock: has the fabric and structure of parent rock but the consistency of soil; with quartz and mica sand particles and kaolin clays (*see Fig. 6.87*). Colors vary from white to gray to brown or reddish. Grades erratically to moderately fresh rock. | Permeability can be relatively high. SPT values are high and void ratios low. Partially saturated specimens can have strengths 20% higher than saturated specimens. Strength is higher perpendicular to the foliations than parallel; compressibility is higher perpendicular to foliations. Compaction characteristics: In situ particles are angular, but soft and readily crushable by compaction equipment in the field, particularly the micas. |

*For general profile see Fig. 6.85.

†At a given site characteristics can be extremely variable and differential settlements require careful consideration. See Table 7.5 for typical engineering properties.

‡In one case, a large structure imposing contact pressures of about 3 kg/cm³ suffered differential settlements of 10 cm within the first 3 years, and the rate thereafter continued at 2.5 cm/yr.

§From a full-scale load test on a bored-pile bearing at 22 m depth, with shaft friction eliminated by bentonite.

**FIG. 7.3 (*Left*) Reddish-brown residual soils of zones I and II (silty clay) grading to saprolite (São Paulo, Brazil).**

**FIG. 7.4 *(Above right)* Characteristics of residual clay from gneiss (Belo Horizonte, Brazil): (*a*) variation of consistency, grain-size distribution, and porosity with depth and (*b*) virtual preconsolidation vs. depth.** [*From Vargas (1953).*[6]]

**FIG. 7.5 *(Below right)* Typical test boring logs from residual soils from igneous and metamorphic rocks: (*a*) schist (Kensington, Maryland); (*b*) igneous (Humacao, Puerto Rico); (*c*) gneiss (Jacarepaquá, Rio de Janeiro, Brazil); (*d*) gneiss (Atlanta, Georgia).** [*Parts a, b, c from Joseph S. Ward & Associates. Part d from Sowers (1954).*[3]]

- From 1.33 to 2.0—lateritic soil
- Greater than 2.0—nonlateritic, tropical soil

True laterites are hard, forming crusts of gravel and cobble-size fragments on or close to the surface (*see Fig. 7.106*). Most of the clay minerals and quartz have been removed, and iron minerals predominate. Lateritic soils commonly have a surface layer of laterite gravel as illustrated in Fig. 7.6, and another layer frequently is found near the limit of saturation.

From the engineering aspect, lateritic soils are not particularly troublesome, since they consist mainly of kaolin clays and are relatively inactive and nonswelling (CL clays). In areas where granular quartz materials are lacking, laterite gravels have been used for aggregate, and large fragments have been used for building-facing stone, since they are very resistant to weathering.

*Tropical Black Clays*

Tropical black clays are common in large areas of Africa, India, Australia, and Southeast Asia. They are associated with mafic igneous rocks such as basalt, in an environment where rainfall is generally under 1250 mm annually, temperatures are high, and drainage is poor, resulting in

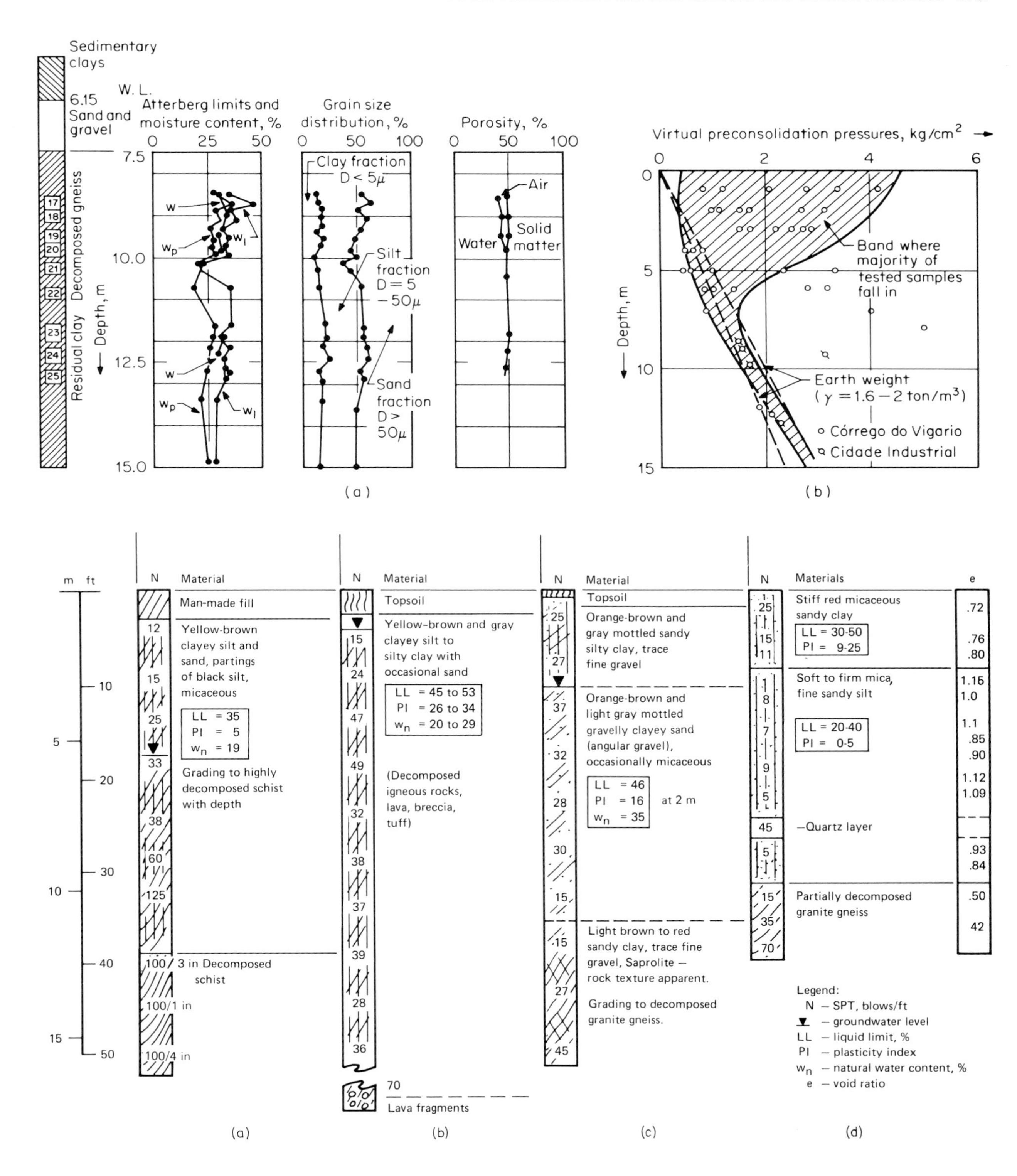

Sedimentary clays
W. L.
6.15
Sand and gravel
Atterberg limits and moisture content, %
Grain size distribution, %
Porosity, %
Residual clay Decomposed gneiss
Depth, m
Clay fraction D < 5μ
Silt fraction D = 5 − 50μ
Sand fraction D > 50μ
Air
Water
Solid matter
(a)
Virtual preconsolidation pressures, kg/cm²
Band where majority of tested samples fall in
Earth weight (γ = 1.6 − 2 ton/m³)
Córrego do Vigario
Cidade Industrial
(b)
N
Material
Man-made fill
Yellow-brown clayey silt and sand, partings of black silt, micaceous
LL = 35
PI = 5
wn = 19
Grading to highly decomposed schist with depth
3 in Decomposed schist
100/1 in
100/4 in
Topsoil
Yellow-brown and gray clayey silt to silty clay with occasional sand
LL = 45 to 53
PI = 26 to 34
wn = 20 to 29
(Decomposed igneous rocks, lava, breccia, tuff)
Lava fragments
Orange-brown and gray mottled sandy silty clay, trace fine gravel
Orange-brown and light gray mottled gravelly clayey sand (angular gravel), occasionally micaceous
LL = 46
PI = 16
wn = 35
at 2 m
Light brown to red sandy clay, trace fine gravel, Saprolite — rock texture apparent.
Grading to decomposed granite gneiss.
Materials
Stiff red micaceous sandy clay
LL = 30-50
PI = 9-25
Soft to firm mica, fine sandy silt
LL = 20-40
PI = 0-5
—Quartz layer
Partially decomposed granite gneiss
Legend:
N — SPT, blows/ft
▼ — groundwater level
LL — liquid limit, %
PI — plasticity index
wn — natural water content, %
e — void ratio
(a)
(b)
(c)
(d)

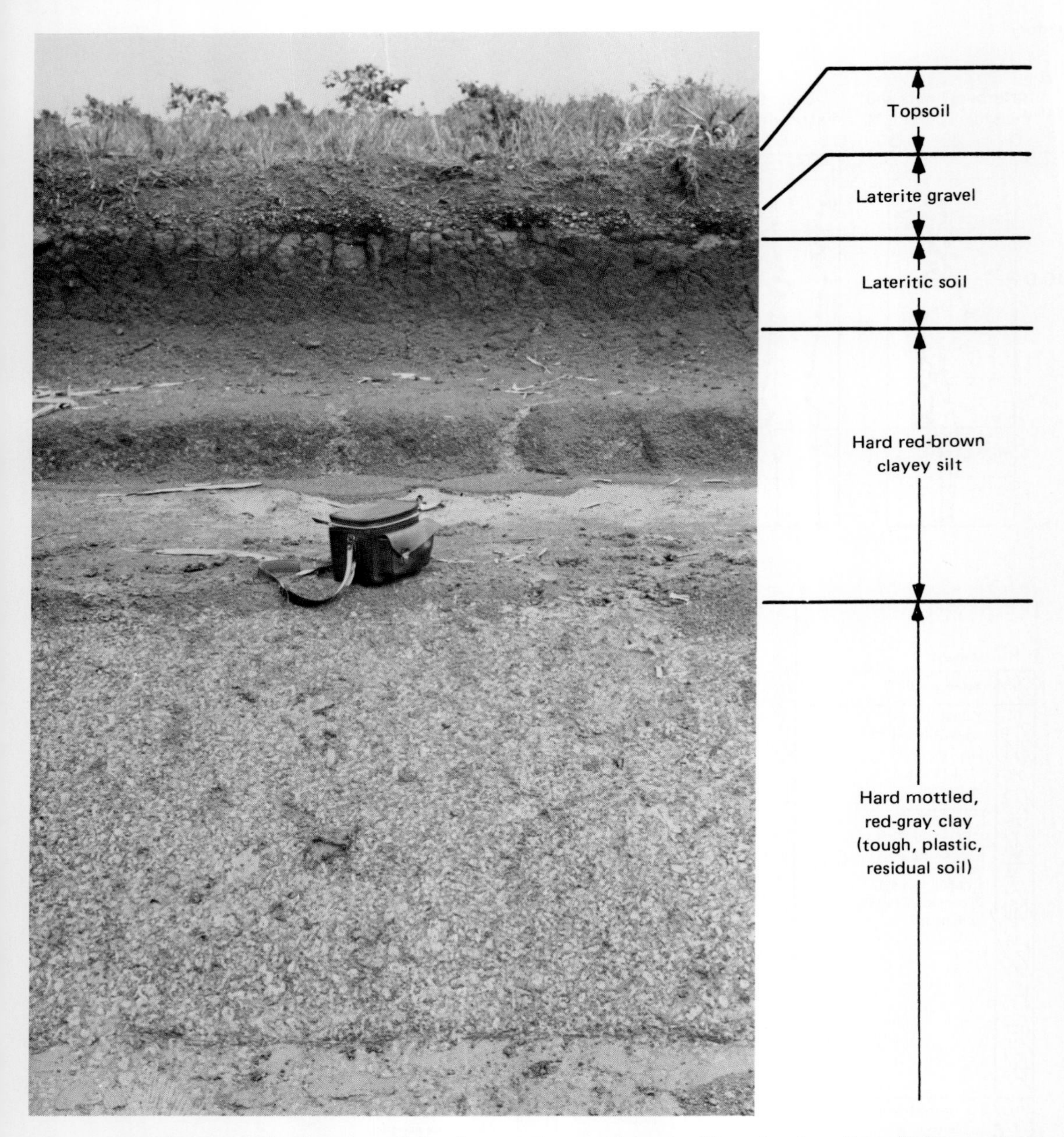


**FIG. 7.6** Cut showing typical laterite soil development in granitic gneiss. New airstrip at Porto Velho, territory of Rondonia, Brazil.

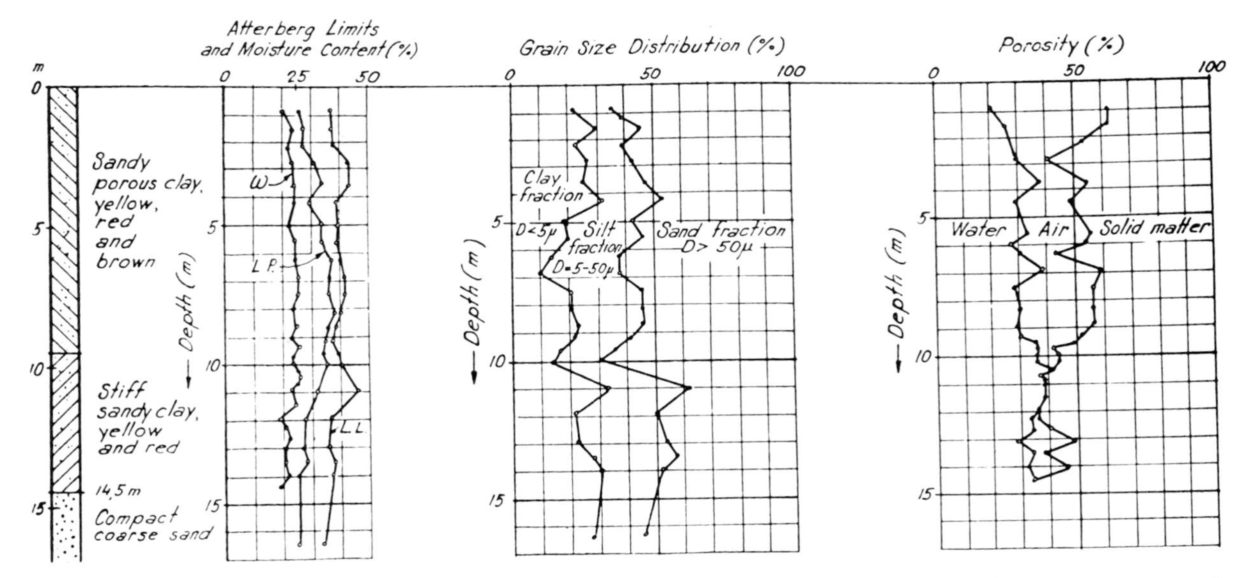


**FIG. 7.7 Characteristics of residual clay from decomposition of clayey sandstone (Campinas, Brazil).** [*From Vargas (1953).*[6]]

alkaline conditions [Morin and Tudor (1976)[7]]. They are not associated with sialic rocks.

Characteristically black in color, they are composed chiefly of montmorillonite clay and are highly expansive. Road construction and maintenance are difficult, not only because of their volume change characteristics, but also because they cover large areas (lava flows) in which there is no granular material for construction aggregate.

7.2.3 SEDIMENTARY ROCKS

### Sandstones

Composed chiefly of quartz grains, most sandstones develop a profile of limited thickness which results from decomposition of impurities, or unweathered materials, such as feldspar. The resulting material is a low-plasticity clayey sand.

Where the percentage of impurities is high, a substantial thickness of clayey soil can develop, as shown on Fig. 7.7, a soil profile in the "porous clays" of Brazil *(see Art. 10.5.2)*.

### Shales (Other than Marine Shales)

Composed primarily of clay minerals, already totally reduced, most shales develop a soil profile of limited depth from the decomposition of the impurities. A typical formation of clayey residual soils from the decomposition of Triassic shales, and the weathered rock zone, is illustrated on Fig. 7.8. Approximately 5 m of overburden has been removed from the site, which is located in a warm, moist climate. The clays are relatively inactive; a log of a test boring from the site is given on Fig. 7.9*a*.

### Marine Shales

As described in Art. 6.7.3, marine shales undergo disintegration and microfissuring from the stress release and expansion of clay minerals (montmorillonite), and in some cases develop a clay soil profile from decomposition of clay minerals. The depth of disintegration can reach 15 m or more (*see Fig. 6.93*). The result normally is a mass of hard fragments in a clay matrix (*see Fig. 6.91*).

In some cases the shales decompose to leave a residue of highly expansive clay without shale fragments. In Texas the parent formation is characteristically a hard gray shale or claystone of Cretaceous age or younger, containing montmorillonite and some calcium and iron pyrite. Decomposition transforms the shale into a tan

**FIG. 7.8 Residual soils from Triassic shales (Leesburg, Virginia).**

jointed clay which can extend to depths of 10 to 20 m [Meyer and Lytton (1966)[8]]. The deposits frequently contain bentonite layers or limey layers due to changes in the marine environment during deposition. The expansive "hard dark gray clay" shown on the boring log (Fig. 7.9*c*) is believed to be decomposed Del Rio shale. The distribution of shales and expansive clays in Texas is given on Fig. 7.54 and discussed in Art. 7.4.4.

In Menlo Park, California, the marine shales are interbedded with sandstones *(see Art. 10.6.3)*. The shales have weathered to form highly expansive black clays which cause differential heaving of foundations and pavements because of the tilted sandstone interbeds *(see Art. 10.6.3)*.

### Limestone

Carbonates, such as limestone and sulfates, pass into solution more quickly than they decompose, and characteristically thin clayey residual soil results from decomposition and impurities. The greater the percentage of the impurities, the greater is the thickness of the soil cover. Transition between the soil and the limestone is normally very abrupt when the limestone is relatively pure, as shown on Fig. 6.22 and 6.88. The reddish-brown to red clays, typical of many limestones in tropical climates, are called "terra rossa." A log of a test boring from Versailles, Kentucky, is given on Fig. 7.9*b* *(see also Fig. 6.22)*.

## 7.3 COLLUVIUM

### 7.3.1 INTRODUCTION

#### General

Colluvial soils are deposits displaced from their original location of formation or deposition by gravity forces during slope failures.

Their engineering characteristics vary with the nature of the original formation and the form of slope failures, as described in Chap. 9. They typically represent an unstable mass, often of relatively weak material, and frequently are found burying very weak alluvial soils. Their recognition, therefore, is important.

### Mode of Occurrence

Colluvium is found on slopes, at the toe of slope, or far beyond the toe of slope. Displacements from the origin can vary from a few centimeters to meters for creep movements, to tens of meters for rotational slides, to many meters or even kilometers for avalanches and flows.

## 7.3.2 RECOGNITION

### Terrain Features (See Also Chap. 9)

*General Criteria*

Terrain features vary with the type of slope failure, its stage of development, and the time elapsed since failure occurred. At the beginning of failure, tension cracks appear along the upper slope. During long-term gradual movements, tree trunks tilt or bend.

After failure, rotational slides result in distinctive spoon-shaped depressions readily identifiable on aerial photographs *(see Figs. 2.13, 9.28, and 9.29)*. Slides, avalanches, and flows can result in hillside scars often devoid of vegetation *(see Fig. 9.51)*, also readily identifiable on aerial photos.

The failure mass (colluvium) of slides, avalanches, and flows results in an irregular, hummocky topography that is distinctly different from the adjacent topography of stable areas *(see Figs. 9.28 and 9.29)*.

Vegetation is an indicator of colluvium in some climates: banana trees, for example, favor colluvial soils over residual.

*Slope Failure Development Stages*

Terrain features will vary with the stage of the slope failure:

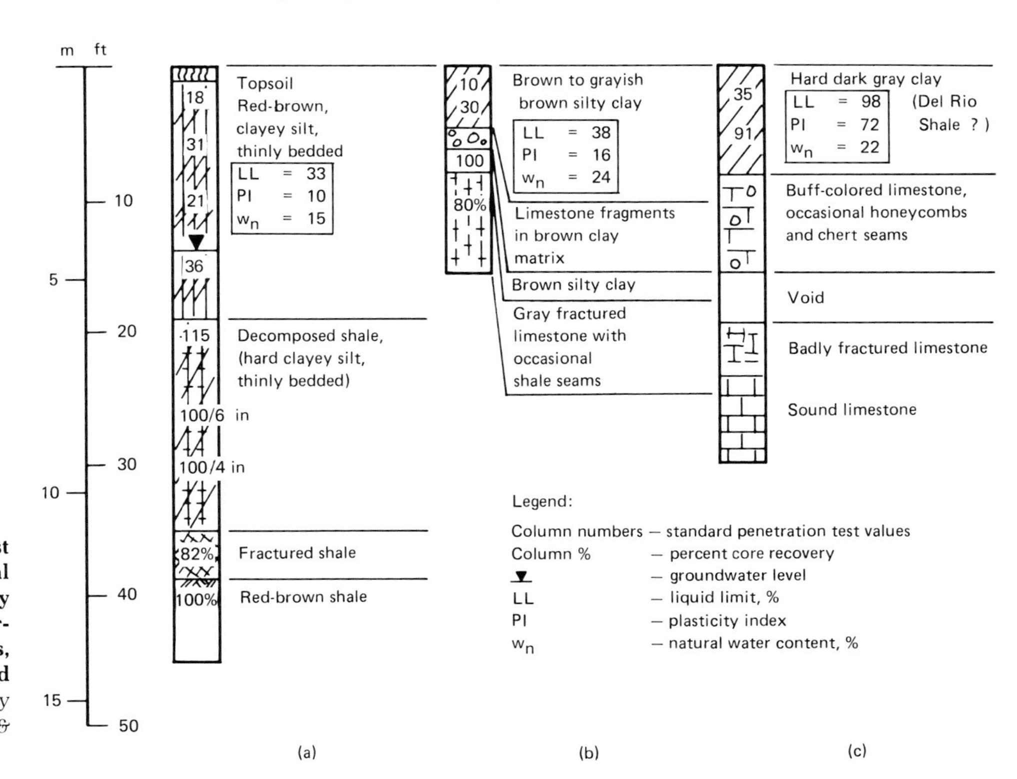


**FIG. 7.9 Typical test boring logs, residual soils from sedimentary rocks: (a) Leesburg, Virginia; (b) Versailles, Kentucky; (c) Round Rock, Texas.** *(Courtesy of Joseph S. Ward & Associates.)*

**TABLE 7.7**
**CHARACTERISTICS OF COLLUVIAL DEPOSITS**

| Origin* | Depositional extent† | Movement rate‡ | Material characteristics§ |
|---|---|---|---|
| Falls | Along slope to beyond | Very rapid | Rock blocks and fragments heterogeneously assembled (talus, *see Fig. 6.70*). |
| Creep | Few centimeters to meters | Very slow | Original structure distorted but preserved. |
| Rotational slides | Few meters to tens of meters | Slow to rapid | Original structure preserved in blocks but planar orientation altered. Debris mass at slide toe. |
| Translational slides in rock | Few to hundreds of meters | Slow to very rapid | Original structure preserved in blocks which are dislocated; or, in major movements, a mass of mixed debris at and beyond the toe of slope. |
| Lateral spreading | Few to many tens of meters | Slow to very rapid | Same as for translational slides. |
| Avalanche | Many tens of meters to kilometers | Very rapid | Completely heterogeneous mixture of soil and rock debris; all fabric destroyed (*see Fig. 7.10*). |
| Flows | Many tens of meters to kilometers** | Very rapid | Completely heterogeneous mixture of soil and rock, or of only soil. All fabric destroyed. |

*Origin refers to type of slope failure; see Chap. 9.

†Significant to the size of the deposit at final failure.

‡Ranges refer to failure stage (initial to total). *See Art. 9.1.2,* Slope Activity.

§Relate to the materials of the parent formation prior to failure.

**The Achocallo mudflow near La Paz, Bolivia, extends for 25 km (*see Art. 9.2.11*).

1. Initial stage: The early stage is characterized by slow, often discontinuous movements, tension cracks, and tilted and bent trees. Recent movements are characterized by tilted but straight trunks; older, continuing movements result in bent trunks.
2. Intermediate stage: Significant movement occurs, often still discontinuous, accompanied by large displacements. This stage is characteristic of slides, especially in the progressive mode. It is during the intermediate stage that the soils may be considered as colluvium, although the displaced blocks are often intact.
3. Final stage: Total displacement has occurred, leaving a prominent failure scar and resulting in a mature colluvial deposit. Additional movement can still occur. Normally only falls, avalanches, and flows constitute the final stage because failure is usually sudden.

*Time Factor*

Erosion modifies and generally smooths and reduces the irregular surface of a fresh failure mass. In time it may make detection difficult.

### Depositional Characteristics

*Materials*

The composition of the failure mass varies with the type of slope failure and the extent of mass displacement as summarized on Table 7.7, as well as with the original materials.

Materials can be placed in four general categories based on structure and fabric. *Talus* is the heterogeneous assembly of rock blocks and fragments at the toe of steep slopes (*see Fig. 6.70*). Distorted surficial beds result from creep. Displaced and reoriented intact blocks with some mixed debris at the toe are common to rotational and translational slides. Heterogeneous mixtures in which all fabric is destroyed (Fig. 7.10) is common to avalanches and flows. These latter mixtures can be difficult to detect when they originate from residual soils with a heterogeneous fabric.

*Unconformities*

In slides the colluvium will be separated from the underlying material by a failure surface which may demark a material change or a change in fabric or structural orientation.

In avalanches and flows, the underlying material often will be completely different in texture and fabric from the colluvium, especially where large displacements occur.

**FIG. 7.10 Colluvial deposit of boulders in a clayey matrix, originating from residual soils failing in an avalanche, exposed in a road cut along the Rio Santos Highway, Brazil. Material is called talus in Brazil because of the large number of boulders.**

*Properties*

Distinguishing colluvium from residual soils is often difficult, but, in general, the colluvium will be less dense with higher moisture contents and lower strengths and will be completely without structure (if of avalanche origin), whereas the residual soil may have some relict structure. A few typical properties are given on Table 3.30.

In some cases seismic refraction surveys will detect the contact between colluvium and residual soils, especially if the contact is above the saturated zone. Typical velocity ranges for soils derived from gneiss are: colluvium, $V_p$ = 300 to 600 m/s; residuum, $V_p$ = 600 to 900 m/s.

### 7.3.3 ENGINEERING SIGNIFICANCE

**Unstable Masses**

Where colluvium rests on a slope it normally represents an unstable condition, and further slope movements are likely. In slides the mass is bounded by a failure surface along which residual (or slightly higher) strengths prevail, representing a weakness surface in the mass and often evidenced by slickensided surfaces. The unconformable mass on the slope blocks the normal slope seepage and evaporation because of its relative impermeability, resulting in pore-pressure buildup during the rainy season *(see Fig. 9.84)* and further decrease in stability.

Slope movements of colluvium are common, and before total failure range from the barely perceptible movements of creep to the more discernible movements of several centimeters per week. Movements normally are periodic, accelerating, decelerating, and stopping completely for some period of time (the slip-stick phenomenon). The natural phenomena causing movements are rainfall, snow and ice melt, earthquake-induced vibrations, and changing levels of adjacent water bodies resulting from floods and tides. Cuts made in colluvial soil slopes can be expected to become much less stable with time and usually lead to failure, unless retained.

**Buried Weak Alluvium**

Colluvial deposits overlying soft alluvium in valleys are a little-recognized but common phenomenon in hilly or mountainous terrain. An example is illustrated on Figs. 7.11 through 7.14.

In Fig. 7.11, the colluvium from the debris avalanche has moved out onto the valley floor and formed a relatively level blanket extending over an area of several hectares, a large part of which overlies alluvium. If test borings were not deep enough (Fig. 7.12), the colluvium (Fig. 7.13) could easily be mistaken for residual soils, especially where it has been strengthened by drying, as indicated at this location by the high SPT values. In this case the underlying soft gray organic silt with shell fragments (Fig. 7.14) (alluvium) is somewhat underconsolidated because of artesian pressures in the underlying sand stratum. Foundations supported in the upper strong soils could be expected to undergo severe settlements, and possibly even failure by rupture of the soft soils. The hills at this particular location are over 500 m distant, making surface detection very difficult. These conditions are common in Brazil along the coastal mountain ranges. It is also usual to find large boulders buried in the alluvium along the coast adjacent to granite mountains.

Another case in different regional conditions is described by D'Appolonia et al. (1969).[9] The colluvium originated as a residual soil derived from horizontally bedded shales and claystones in Weirton, West Virginia. Carbon dating of the underlying soils yielded an age of 40,000 years. The ancient slide was determined to be 1.6 km in length, rising 60 m in elevation, and extending laterally for about 300 m. Stabilization was required to permit excavations at the toe of slope to heights as great as 18 m *(see Fig. 9.106)*.

## 7.4 ALLUVIAL DEPOSITS

### 7.4.1 FLUVIAL ENVIRONMENT

**General**

*Fluvial* refers to river or stream activity. *Alluvia* are the materials carried and deposited by streams. The *stream channel* is the normal extent of the flow confined within banks, and the *floodplain* is the area adjacent to the channel which is covered by overflow during periods of high runoff. It is often defined by a second level of stream banks. Intermittent streams flow periodically, and a wash, wadi, or arroyo is a normally dry stream channel in an arid climate.

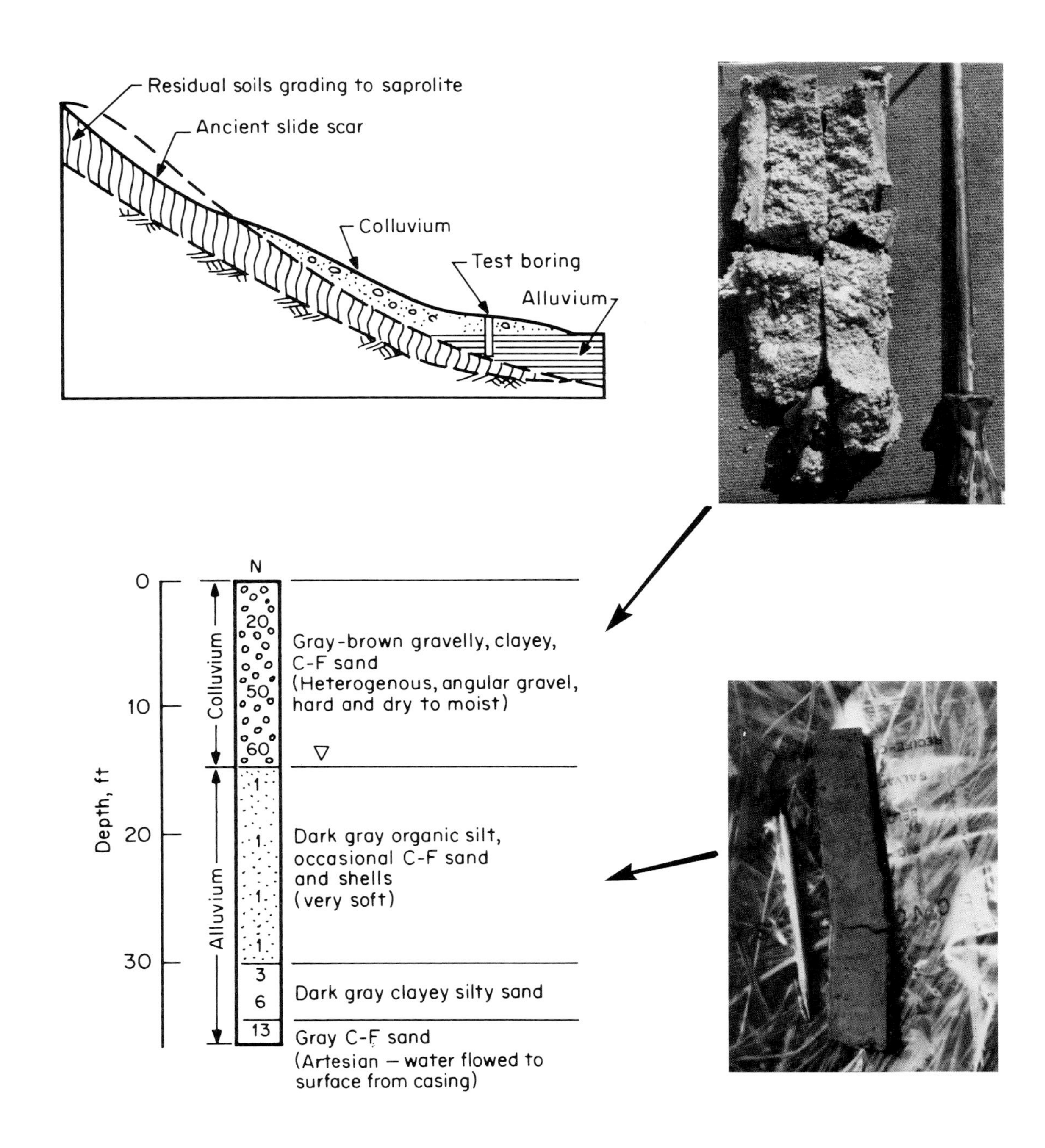


**FIG. 7.11** *(Above left)* **Slope and valley condition of colluvium overlying alluvium (Jacarepaquá, Rio de Janeiro, Brazil).**

**FIG. 7.12** *(Below left)* **Log of test boring in deposit of colluvium overlying alluvium as shown on geologic section in Fig. 7.11.**

**FIG. 7.13** *(Above right)* **Strong colluvium in Fig. 7.12.**

**FIG. 7.14** *(Below right)* **Soft clayey alluvium in Fig. 7.12.**

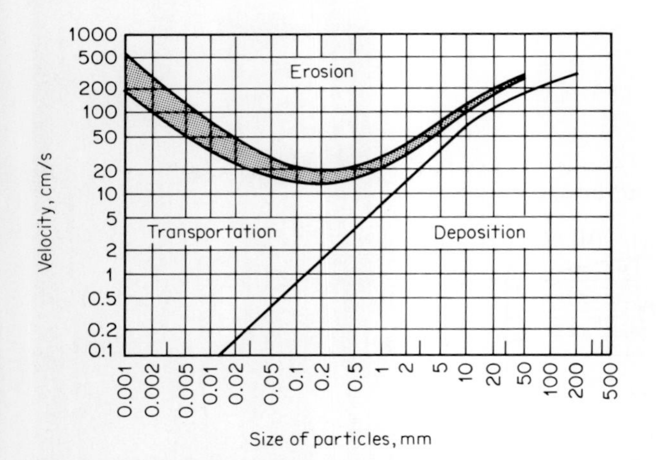


**FIG. 7.15 Relationship between water velocity and particle size, and transportation, erosion, and deposition of a stream.** [*From Hjulström (1935).*[10]]

## Stream Activity

The elements of stream activity are erosion, transport, and deposition.

### *Erosion*

Erosion is caused by hydraulic action, abrasion, solution, and transport. It occurs most significantly during flood stages; banks are widened and the channel is deepened by scouring.

### *Transport*

The greater the stream velocity, the greater is its capacity to move materials. Under most conditions the coarser particles (cobbles, gravel, and sand) are moved along the bottom where velocities are highest, forming the *bed load materials.* The finer materials (clay, silt, and sand, depending on velocities) are carried in the stream body and are referred to as *suspended load materials.*

### *Deposition*

As velocities subside, the coarser particles come to rest on the bottom or settle out of suspension. The finer particles continue to be carried until a quiet water condition is reached where they settle slowly out of suspension. Relationships between erosion, transport, and deposition in terms of flow velocity and particle size are given on Fig. 7.15.

The materials in the stream depend on the source materials. Silts and clays, causing muddy waters, result from shales and chemically altered rocks in the drainage basin. Clear waters, under normal flow conditions, are common to mountainous regions composed of predominantly sialic crystalline rocks.

## Stream Classification and Deposition

### *General*

Streams are classed by geomorphic development and shape and are associated with characteristic soil deposits, modified by stream velocities, flood conditions, source materials, and climate.

SHAPE Streams have been classed by shape [Tanner (1968)[11]] as either straight, crooked, braided, or meandering. These classes relate in a general manner to those based on geomorphic development.

GEOMORPHIC CLASSES Several classifications can be found in the literature:

- Boulder zone (also headwater tract, young or early stage)
- Floodway zone (also valley tract, mature or middle stage)
- Pastoral zone (also plain tract, old age or late stage)
- Estuarine zone (at the river's mouth, Art. 7.4.2)

Rejuvenated streams have been uplifted by tectonic movements which change their characteristics.

The terms boulder zone, floodway zone, etc. describe river morphology based on the criteria of Bauer [Palmer (1976)[12]] for tectonically stable areas. All are illustrated on Fig. 7.16, an aerial mosaic of a 5-km-long river flowing over crystalline rocks in a tropical climate.

### *Boulder Zone*

The boulder zone (Fig. 7.17*a*) represents the early stage of the fluvial cycle. In the drainage basin the stream density per unit area is low. Youthful streams form the headwater tract and have V-shaped valleys that can be shallow or

**FIG. 7.16 Aerial photo mosaic of the life cycle of a river—early, middle, and late development stages. Length: about 5 km; environment: crystalline rocks in a tropical climate. See Fig. 7.17 for stereo pairs of each stage.**

deep, and are narrow, separated by steep-sided divides.

In the stream, gradients exceed 5 m/km, rapids are common, and erosion is the dominant feature. The stream fills the valley from side to side and is degrading or deepening its channel. Stream shape, or alignment, is controlled by geologic structure and can vary from straight to crooked. Sediments are coarse and bouldery.

*Floodway Zone*

The floodway zone (Fig. 7.17*b*) represents the middle stage of fluvial development. The drainage basin contains a well-integrated drainage system with many streams per unit area. The valleys are deeper and wider with interstream divides narrow and rounded; local relief is at a maximum, leaving most of the surface as eroded slopes.

Stream gradients are moderate, ranging from about 1 to 5 m/km. Braided or shifting channels form an alluvial valley floor, with or without a floodplain, but the stream no longer fills the valley floor. Pools, ripples, eddies, point bars, sand and gravel beaches, and channel islands are common. The stream continues to cut into the valley sides, extending the floodplain, which is formed of relatively coarse soils. Erosion and deposition are more or less in balance; sediment load is coarse (sand and gravel) and is equal to or greater than the carrying capacity of the river which therefore is aggrading. Sediments are stratified.

*Pastoral Zone*

The pastoral zone (Fig. 7.17*c*) represents the late stage of development. The drainage basin has developed into a peneplain, and basin drainage is poor. Interstream divides are areas of low, broad, rolling hills, often with a few erosional remnants as isolated hills, or monadnocks.

Stream gradients range from 0 to 1 m/km. The river is still side-cutting its channel and building up the valley floodplain, but the channel does not extend by meandering to the valley sides. Sediment loads are usually balanced with the transporting capacity of the stream. Fine bed-load materials (silt and sand) form banks. The

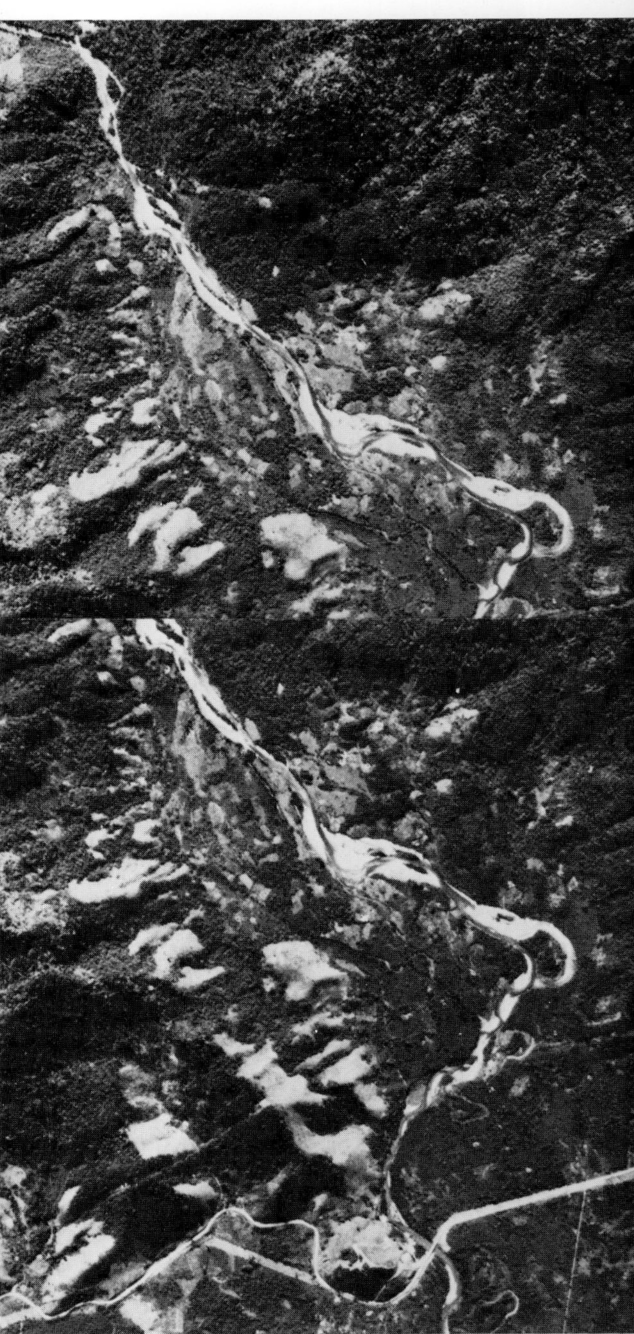

(a)

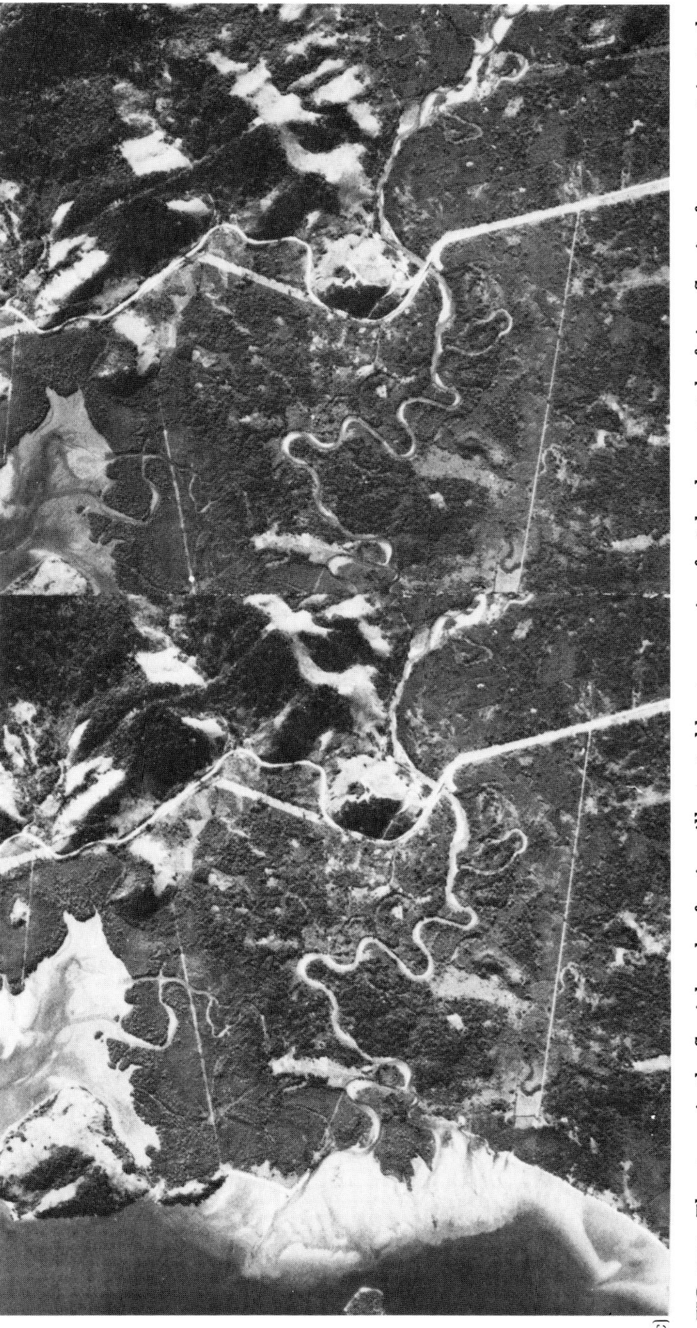

(c)

**FIG. 7.17** The stages in the fluvial cycle of a river illustrated by stereo pairs of a 5-km-long stretch of river flowing from mountains to the sea: (*a*) early stage or boulder zone; (*b*) middle stage or floodway zone; (*c*) late stage or pastoral zone, and estuarine zone. (See Fig. 7.16. Scale 1:20,000.)

**FIG. 7.18 Meandering of the Colorado River in a rejuvenated plateau of sedimentary rocks.**

channel meanders, changing course frequently, building point bars and isolating oxbow lakes. The valley floor is broad and subject periodically to floods during which the oxbow lakes are filled, and natural levees and backswamps are formed. Stratification and interbedding are common.

*Estuarine Zone*

See Art. 7.4.2.

*Rejuvenation*

Regional uplift causes the base level of the stream to rise (in reference to sea level) and the river begins again to incise its channel and dissect the surrounding floodplain. Entrenched meanders can develop deep canyons (Fig. 7.18), such as the Grand Canyon of the Colorado River, and relatively steep gradients out of harmony with the normal meander gradients of the later stages of erosion.

*Water gaps* cut through resistant strata also result from rejuvenation. Examples are the Susquehanna River where it has eroded the folded Appalachian Mountains at Harrisburg, Pa. *(see Fig. 6.38)*, and the Delaware Water Gap between New Jersey and Pennsylvania *(see Figs. 2.5 and 2.8)*.

*Terrace deposits* also result from changing base levels. Uplift, or lowering of sea level, can cause a river to incise a new channel in an old floodplain. The limits of the previous channels remain at higher elevations as terraces, as shown on Fig. 7.19.

*Buried Channels*

In broad bedrock valleys, the river may cut a deep gorge at some location in the valley. Subsequent deposition during periods of *base level lowering* may fill the gorge with very coarse-grained material and the river may change its course to cut a new, shallower channel. The old channel remains buried (Fig. 7.20) and leaves no surface expression indicating its presence. Because such a channel is often narrow, locating it during exploration is difficult. The irregular bedrock surface presents problems for the founding of piers or other foundations, and the boulder and gravel fill in the buried channel permits high seepage losses beneath dams.

### Stream Shape, Channel Characteristics, and Deposition

*General*

Stream shape often reflects the geologic conditions in the area through which it flows, as well as providing indications of the types of materials deposited in its channel and floodplain.

Currents, even in straight, confined channels, exhibit meandering, or swinging from side to side in the channel. If the channel sides are erodible by the current, the channel extends itself from side to side, and is said to *meander*. Meander development depends on the erodibility of the stream banks and the water velocity.

*Straight Channels*

Straight channels develop along weakness zones in the rock, or are constructed.

*Crooked Channels*

Crooked channels are characteristic of rock or hard clay beds resisting meandering. In rock masses, forms are controlled by weakness planes or intrusions of hard rock into softer rock (Table 7.4). The crooked shape of Rock Creek, Washington, D.C., which eroded Precambrian crystalline rocks having varying degrees of resistance, is shown on Fig. 7.21. In such youthful, steep-sided valleys, the stream flows either on the bedrock surface, or over deposited coarse-grained materials including boulders, cobbles, and gravels.

*Braided Streams*

In braided streams, channel widths can range up to several kilometers, and are characterized by an interlocking network of channels, bars, shoals, and islands, found generally in mature streams. The larger braided streams are found downstream of glaciers or in the lower tracts of major rivers where seasonal high runoff carries large quantities of coarse materials downslope until a flat portion or low gradient is encountered. The coarse sediments are dropped and then continue to move downstream by periodically increased flow.

Deposits are typically sand and gravel mixtures in the channel. A moderately crooked channel flowing through formations of hard granite and gneiss is shown on Fig. 7.22. Braided conditions exist within the channel, and the limits of the floodplain (underlain primarily by clayey silts) are shown clearly by the tree line.

An intensely braided channel is illustrated on Fig. 7.23. The materials forming the high ground above the valley are loessial deposits showing typical fine-textured pinnate and dendritic drainage pattern.

*Meandering Streams*

Meandering streams develop in easily erodible, weaker valley sediments composed of the finer

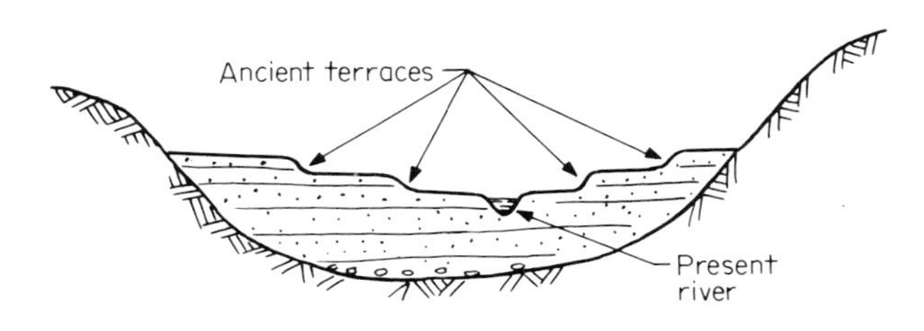


**FIG. 7.19 Uplift starts new erosion cycle, leaving old floodplain limits as terrace formations.**

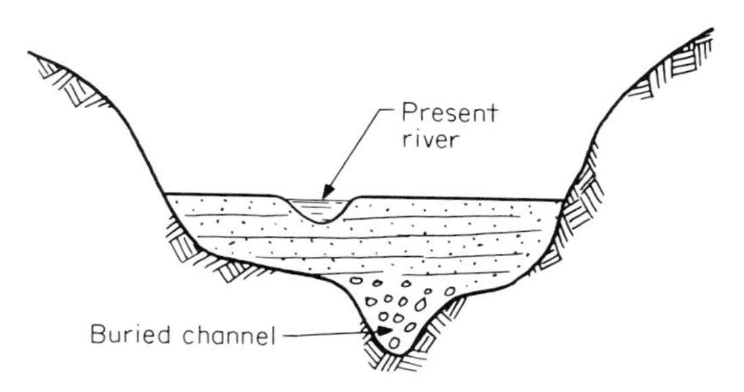


**FIG. 7.20 Ancient buried valley in rock gorge filled with cobbles and boulders.**

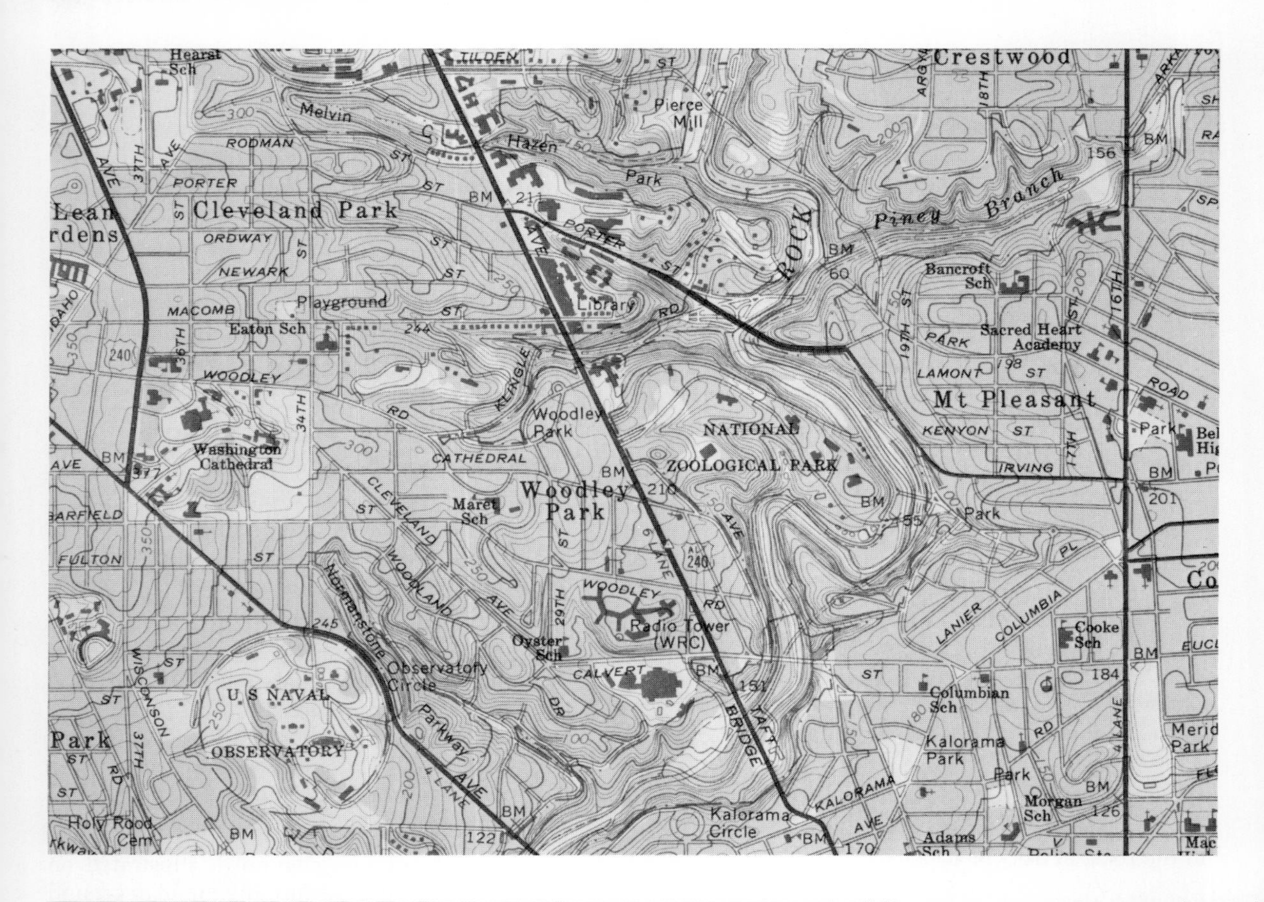

**FIG. 7.21** *(Above)* **Rock Creek, District of Columbia—an example of a young stream eroding rock.** *(From USGS quadrangle sheet, District of Columbia. Scale 1:24,000.)*

**FIG. 7.22 Moderately crooked stream channel flowing through hard granite and gneiss; direction fault controlled. Extent of floodplain apparent as tree line (Rio Itapicuru, Bahia, Brazil).**

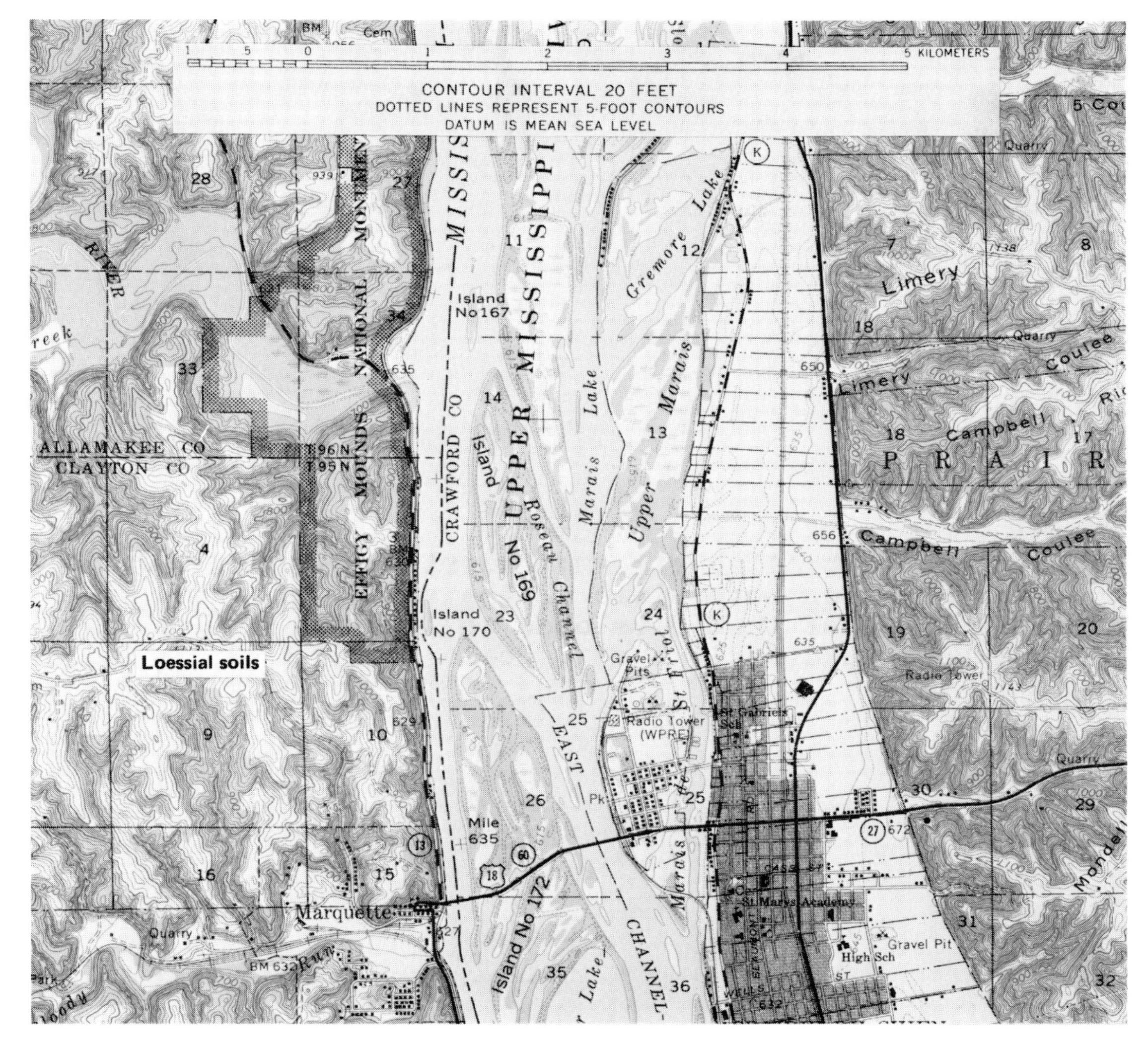


**FIG. 7.23 Example of a braided stream (Prairie du Chien, Iowa). Upland areas on both sides of the valley show the fine-textured relief that develops in loessial soils.** *(From USGS topographic quadrangle sheet. Scale 1:62,500.)*

materials (silt, sand, and clay mixtures). They can occur in any valley with a floodplain, but are more characteristic of broad, mature valleys. (Rejuvenated plateaus are a special case, as mentioned.) Broad, mature valleys are likely to have contained several streams and to have gone through several cycles of erosion and deposition during their life, thereby developing a variable and complex stratigraphy. The geomorphic expression of a meandering stream is given on Fig. 7.24, a stereo pair of aerial photos showing former river channels, cutoff channels (oxbow lakes), and point bars. The modern stream deposits soils over the valley in forms that can be recognized and classified.

POINT BAR DEPOSITS As the stream meanders, the current erodes the outer downstream portions of the channel where velocities are highest, and deposits granular materials, primarily sands, in the quiet waters of the inner, upstream portions. In the early stages of meandering these point bar deposits remain as sand islands. As the channel migrates, a sequence of formations begins.

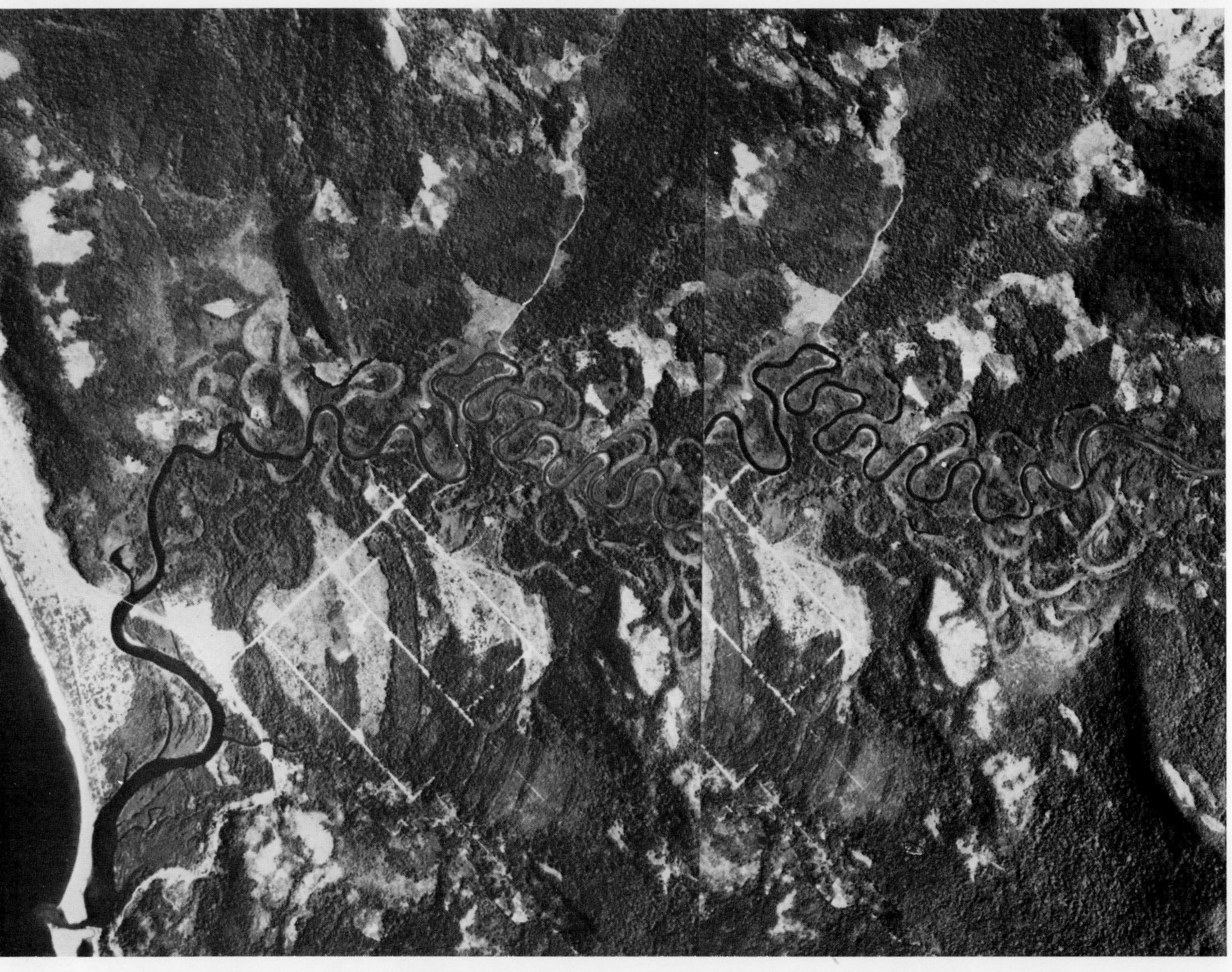

FIG. 7.24 Stereo pair of aerial photos of a meandering river. Apparent are abandoned channels, oxbow lakes, point bars, and back swamps. The river enters the ocean on the left and ancient beach ridges are apparent in the lower portion of the photos. (Scale ≈ 1:40,000.)

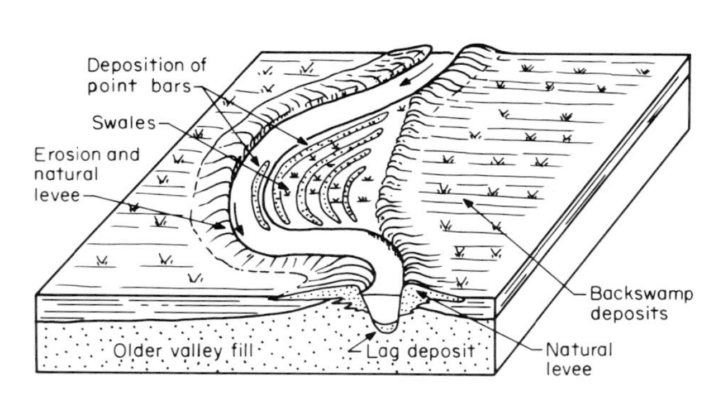


**FIG. 7.25 Meander development and deposition of point bars, swales, and natural levees.**

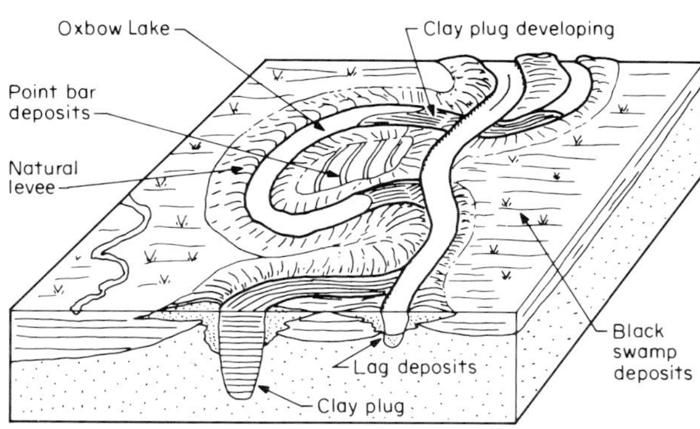


**FIG. 7.26 Meander cutoff and formation of oxbow lake followed by deposition of clay plug.**

**FIG. 7.27 Log of test boring made in a backswamp deposit (New Orleans, Louisiana).** *(From the files of Joseph S. Ward & Associates.)*

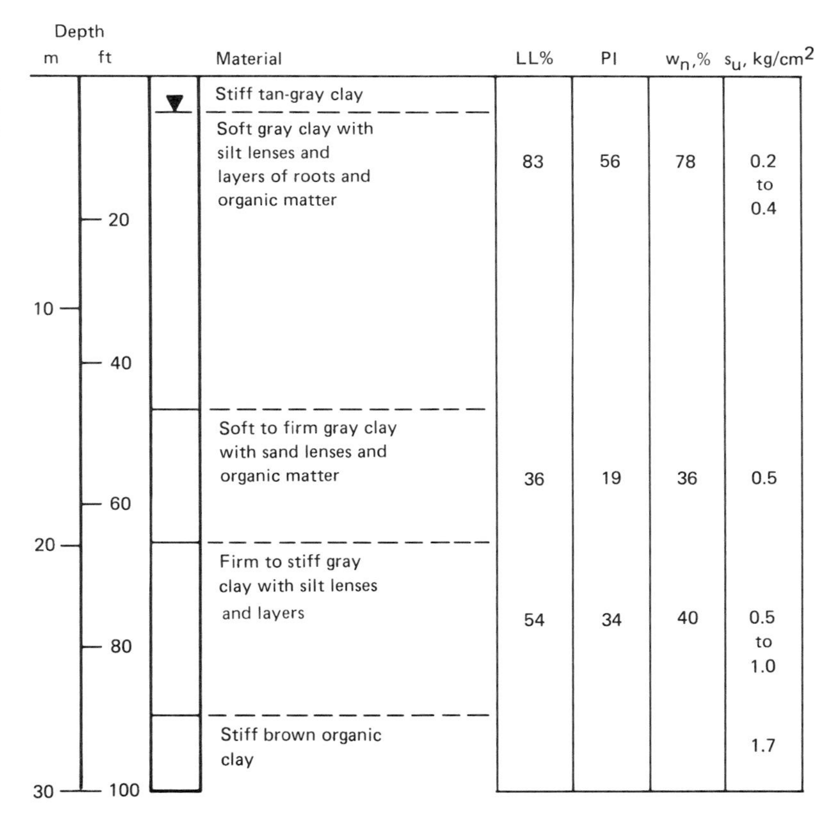


Sandy bars are deposited during high water, and swales filled with fine-grained soils remain between the bar and the bank. As the river continues to migrate, a succession of sandy bars and clay-filled swales remain as illustrated on Fig. 7.25 *(see also Fig. 2.3)*. In the lower Mississippi River valley, in the late stage of development, swales reach depths of 10 to 20 m [Kolb and Shockley (1957)[13]].

OXBOW LAKES AND CLAY PLUGS As time passes, the meander curve grows until a narrow neck separates two portions of the river. Finally the river cuts through the neck, leaving an island and abandoning its former channel, which is cut off with sand fills and forms an *oxbow lake* encircling the point bar deposits as shown on Fig. 7.26. Since there is no drainage out of the lake, it eventually fills with organic soils and other

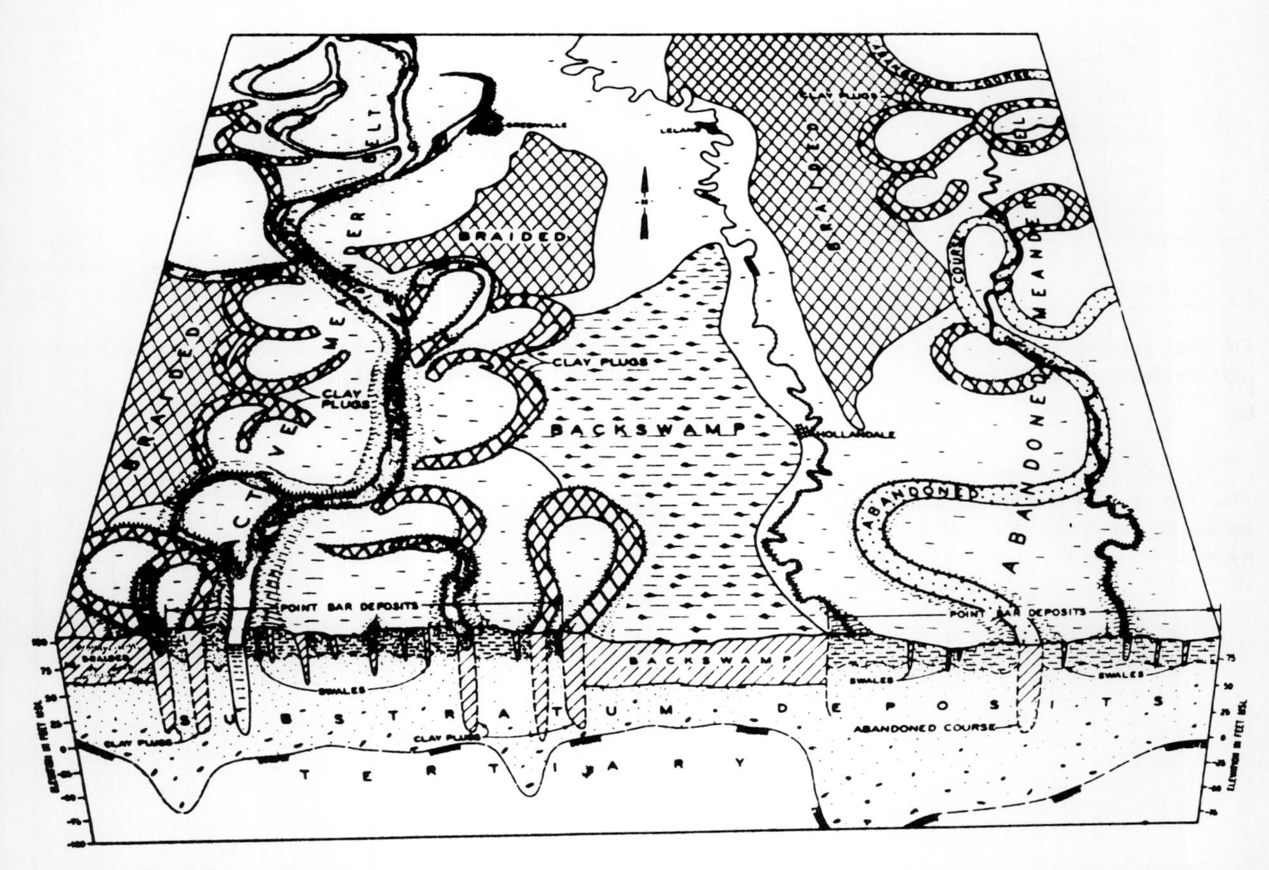


FIG. 7.28 **Deposition in a river valley in the late stage of development (pastoral zone), vicinity of Greenville, Mississippi.** [*From Kolb and Shockley (1957).*[13]]

soft sediments washed in during flood stages. In the Mississippi Valley oxbow lakes are referred to as *clay plugs* because of the nature of the filling, which tends to remain soft and saturated.

*Lag deposits* are the generally granular soils remaining in the stream channel (*see Fig. 7.26*).

*Lateral accretion* occurs during flood stages when the stream overflows its banks and covers the valley floor with sediments. *Natural levees* of sandy soils build up along the banks (*Fig. 7.26*). In extreme floods with higher velocities, coarse-grained sandy soils will be deposited to blanket the valley floor. Normal floods will deposit fine-grained soils filling depressions with clays and organic soils referred to as *backswamp* deposits (*Fig. 7.26*). During low-water periods these deposits are exposed to drying and become preconsolidated to shallow depths. The log of a test boring in a backswamp deposit near New Orleans is given as Fig. 7.27. It shows the weakness of the soils to depths of over 20 m. Fig. 7.28 illustrates the complexity of soil conditions that can develop in a river valley in the pastoral stage.

## Arid-Climate Stream Activity

### *Stream Characteristics*

In arid regions streams are termed washes, wadis, or arroyos; flow is intermittent and stream shapes tend to be crooked or braided because the significant flows are of high velocity, occurring during flash floods.

*Deposition*

In semiarid mountainous regions, such as the southwestern United States, runoff from the mountains has filled the valleys with sediments often hundreds of meters thick. The significant source of the sediments is the young, narrow tributary streams flowing into the valley from higher elevations during storms. Their high velocities provide substantial load-carrying capacity, and when they reach lower gradients as they exit from the mountains, velocity diminishes and they debouch their sediments to build *alluvial fans* such as illustrated on Fig. 7.29.

Where streams are closely spaced, the fans coalesce to form broad "bajadas" which can cover many tens of square kilometers. These areas are intensely dissected with drainage networks as shown on Figs. 2.11 and 7.30, termed *sheet erosion* which results from *sheet wash.* Boulders, cobbles, and gravels are characteristic of the higher elevations of the bajadas; fine soils and soluble salts are carried to the valley floor to form temporary lakes. The water evaporates rapidly to leave a loosely structured, lightly cemented deposit, usually a collapsible soil (*see Art. 10.5.2*) susceptible also to piping.

*Terraces* result from subsequent erosion of the bajadas as illustrated on the topographic map (Fig. 7.31). The stratification and boulders characteristic of the higher elevations of the bajada are shown in the photo of a terrace given as Fig. 7.32.

## Engineering Characteristics of Fluvial Soils

*General*

Fluvial deposits are typically stratified and extremely variable, with frequent interbedding. Permeability in the horizontal direction is significantly greater than in the vertical. Unless subjected to removal of overburden by erosion or desiccation, the deposits are normally consolidated. Clays are soft and sands are in the loose-to medium-dense state.

*Boulders, Cobbles, and Gravel*

The coarser sizes occur in the beds of youthful streams, in buried channels, and in the upper portions of alluvial fans in arid climates. Permeability in these zones is very high.

*Sands and Silts*

Sands and silts are the most common fluvial deposit, occurring in all mature- and late-stage stream valleys as valley fills, terraces, channel deposits, lag deposits, point bars, and natural levees. They are normally consolidated and compressible unless prestressed by overburden subsequently removed by erosion, or by water table lowering during uplift.

Permeability can be high, particularly in the

**FIG. 7.29 Alluvial fan (Andes Mountains, Peru).**

FIG. 7.30 Stereo pair of aerial photos at a scale of 1:20,000 covering a portion of the Cañada del Oro near Tucson, Arizona. *(See small-scale stereo pair in Fig. 2.11.)* The features of erosion and deposition at the base of mountain slopes in an arid climate are apparent. Upland topography is angular and rugged in the metamorphic rocks. The myriad of dry-drainage channels results from sheet erosion during flash floods. At this location, at the apex of an alluvial fan, the soils are predominantly coarse-grained and bouldery.

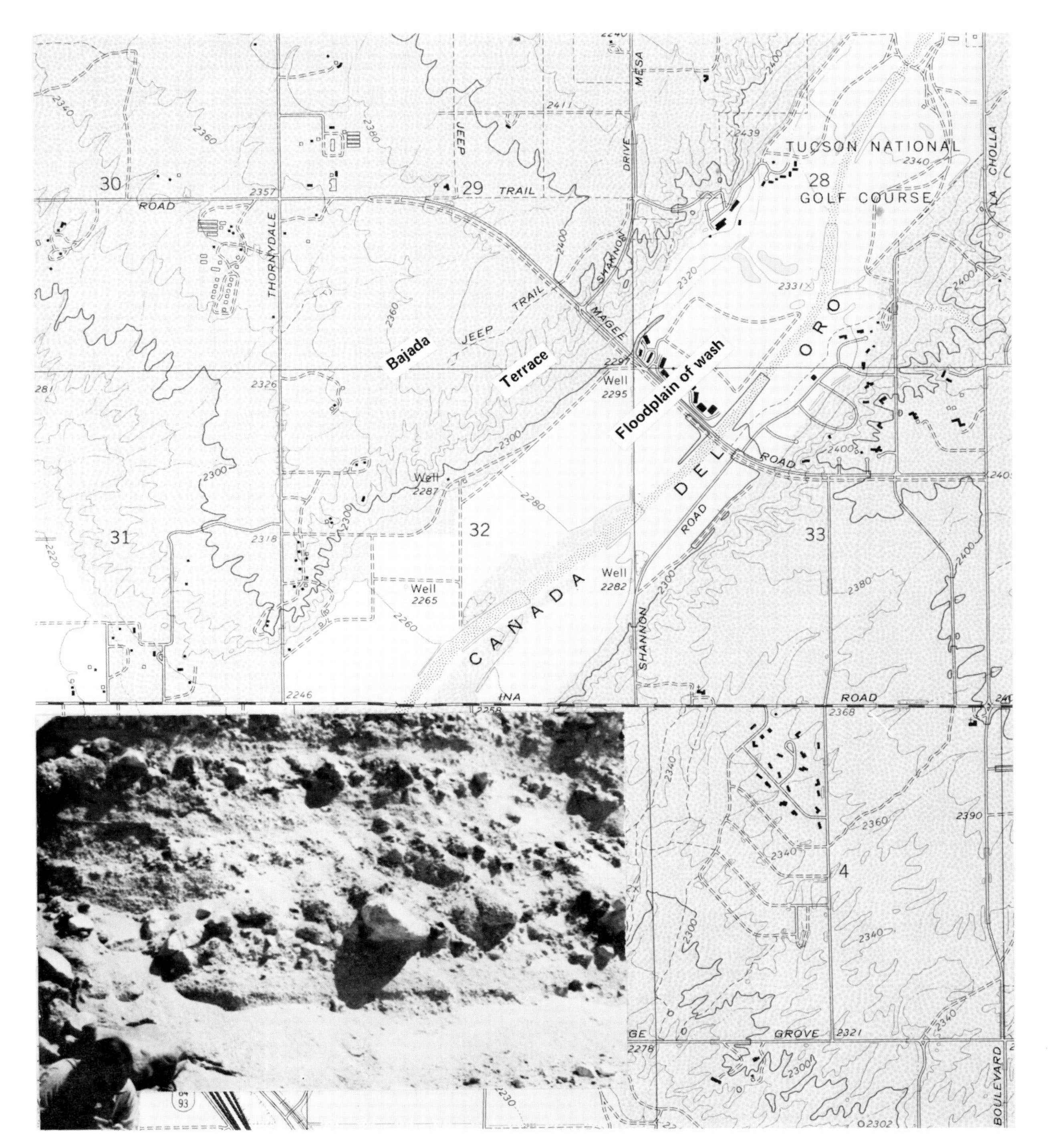


FIG. 7.31 Topographic expression of a bajada dissected by the floodplain of a wash in an arid climate. (Cañada del Oro, Tucson, Arizona. Scale 1:24,000.)

FIG. 7.32 *(Inset)* Terrace deposit of boulders, cobbles, gravel, and sand. (Cañada del Oro, Tucson, Arizona. Photo taken in the area of Fig. 7.31 by R. S. Woolworth.)

**FIG. 7.33 LANDSAT image of portion of the Mississippi River delta, including New Orleans and Lake Pontchartrain. The upland soils are sands and clays of the coastal plain.** (*Scale 1:1,000,000. Original image by NASA, reproduction by USGS, EROS Data Center.*)

horizontal direction. In the pastoral zone, where silty sands are common, high pore-water pressures build up during flood stages, often resulting in stability failure of flood-control levees.

Silty sands deposited as valley fill in an arid environment are frequently lightly cemented with salts and other agents, and prone to collapse upon saturation following the initial drying.

*Clays*

Clay soils are not encountered in the boulder zone and are relatively uncommon in the floodway zone except for deposits along the river banks and on the valley floor during flood stages. In these deposits the soils are usually clayey silts.

In the pastoral zone clays are deposited as clay plugs in oxbow lakes, where they remain soft, or as backswamp deposits where they are prestressed by desiccation. Even so, they remain predominantly soft materials. In the backswamp areas the clays are generally interbedded with silts and sands and are often organic, containing root fibers. The largest extent of fluvial clays in the United States is in the lower portions of the Mississippi alluvial plain.

### 7.4.2 ESTUARINE ENVIRONMENT

#### Estuarine Zone

Located at the river's mouth, the estuarine zone can be affected by periodic reversal of the river gradient from tidal activity, as well as by flood stages. Sediments are primarily fine-grained when located at the terminus of the pastoral zone, and channel branches, islands, and marshes are common.

#### Deltas

*Occurrence*

Deltas form where a river enters a large water body and deposits its load in the relatively quiet, deep waters. As flow velocity decreases, the gravel, sand, and silt particles are segregated and deposited in beds dipping toward the bottom of the water body. The colloidal particles remain in suspension until a condition of very quiet water occurs to permit sedimentation. The delta form that occurs is related to the sediment load and the energy conditions of currents and waves. *No delta* forms when high-energy conditions carry away all but the coarser particles, which remain as spits and bars along the shoreline *(see Art. 7.4.3).*

*Deltaic Deposits*

Deltaic deposits are characterized by well-developed cross-bedding of mixtures of sands, silts, and even clays and organic soils. Delta formation requires that the river provide materials in such quantities that they are not removed by tides, waves, or currents. This requires either a low-energy environment with little water movement, or a river carrying tremendous quantities of material. The Mississippi delta (Figs. 7.33 and 7.34) and the Nile delta are built in low-energy environments. The locus of active deposition occurs outward from the distributary mouths forming the delta front complex. The front advances into the water body, resulting in a sheet of relatively coarse detritus which thickens locally in the vicinity of channels. Seaward of the delta front is an area of fine clay accumulation, termed *prodelta* deposits, illustrated on Fig. 9.57. Over long periods of time, deltas will shift their locations as shown on Fig. 7.34.

*Subdeltas*

Subdeltas are extended into the water body by rivers crossing the parent delta. As the subdeltas extend seaward, the rivers build *bar fingers* along their routes. The bar fingers are composed chiefly of sands. Between them are thick deposits of silts and clays often interbedded or covered with thick organic formations. Considerable quantities of *marsh gas* can be generated in the buried marsh or organic deposits *(see Fig. 2.32).*

*New Orleans, an Example*

The city of New Orleans is built over deltaic deposits. The complexity of the formations in the area is illustrated on Fig. 7.35.

#### Estuaries

*Occurrence*

Streams whose cycle of erosion has been interrupted by a rise in sea level, resulting in drowning of the river valley, form *estuaries.* During the

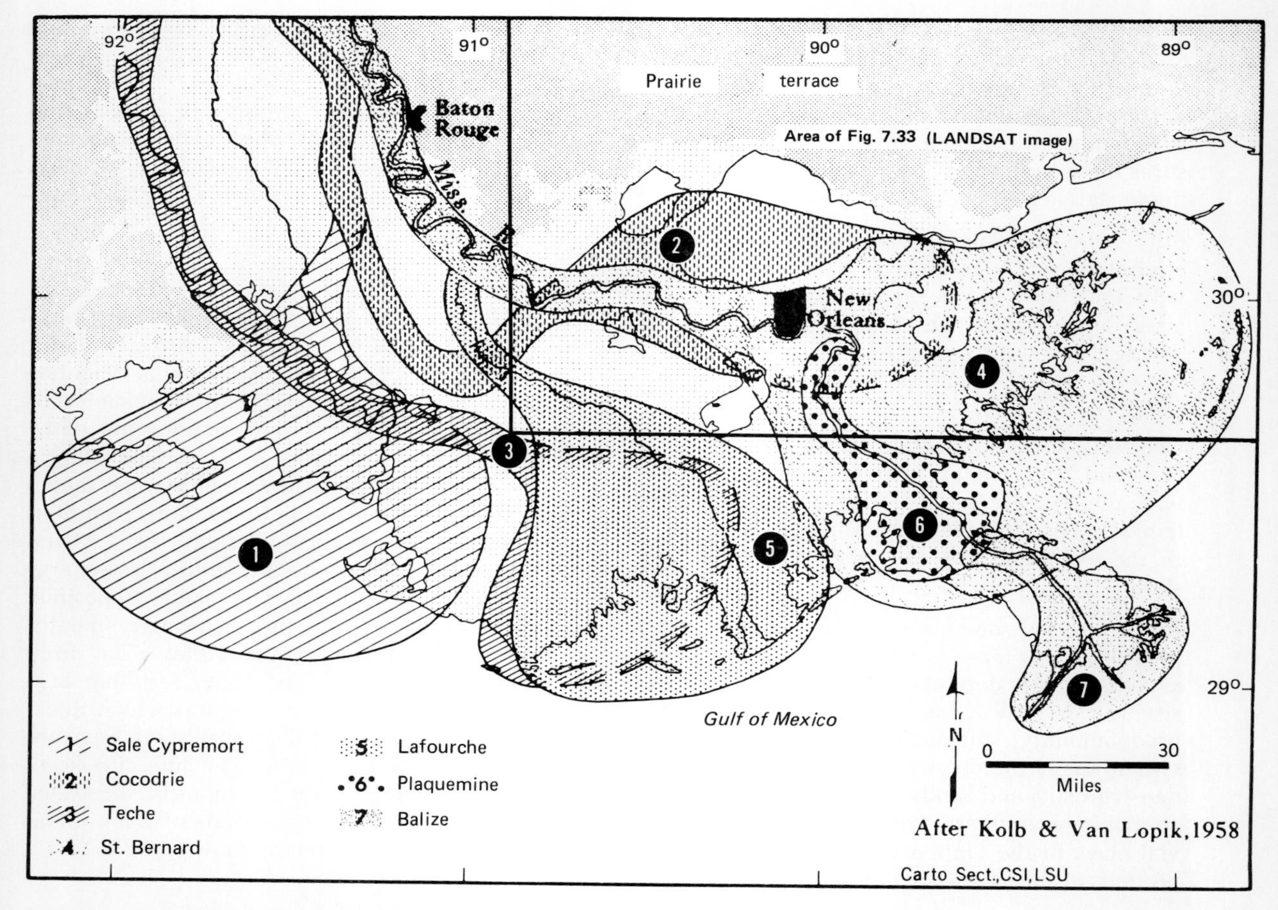


**FIG. 7.34 Recent delta lobes of the Mississippi River.** [*From Coleman (1968).*[14] *Reprinted with permission of Dowden, Hutchinson & Ross, Inc.*]

last glacial age of the Pleistocene, mean sea level was lowered by about 100 m or more in the northeastern United States, and river valleys, eroding downward, made new profiles. When the glaciers began to melt, sea level rose to form *drowned valleys* which are found in many seacoast regions of the world. Along the east coast of the United States, for example, the Hudson, Delaware, Susquehanna, and Potomac Rivers have drowned valleys at their mouths. Where the rivers enter the sea to form bays, estuarine conditions exist. The landforms of the drowned valleys in the northeastern United States are shown on the LANDSAT image mosaic (Fig. 7.36).

*Deposition*

Common to these buried valleys are strata of recent alluvia deposited on the eroded surface of much older formations. The lower alluvia are characteristically coarse sand and gravel mixtures interbedded with sands, becoming finer-grained with decreasing depth. Overlying the granular soils are soft clays grading upward to organic soils usually interbedded with thin layers of silty sands. The organic soils are very soft, often ranging from 10 to 30 m thick, and frequently containing shells. Typical sections from two locations are given as Figs. 7.37 and 7.38.

*Engineering Properties*

Test boring logs from four East Coast locations are given on Fig. 7.39, which also includes some index property data. Test boring logs and engineering property data from Portland, Maine, and the Thames estuary clay in England are given in Figs. 7.40 and 7.41. Additional data are given in Table 3.30.

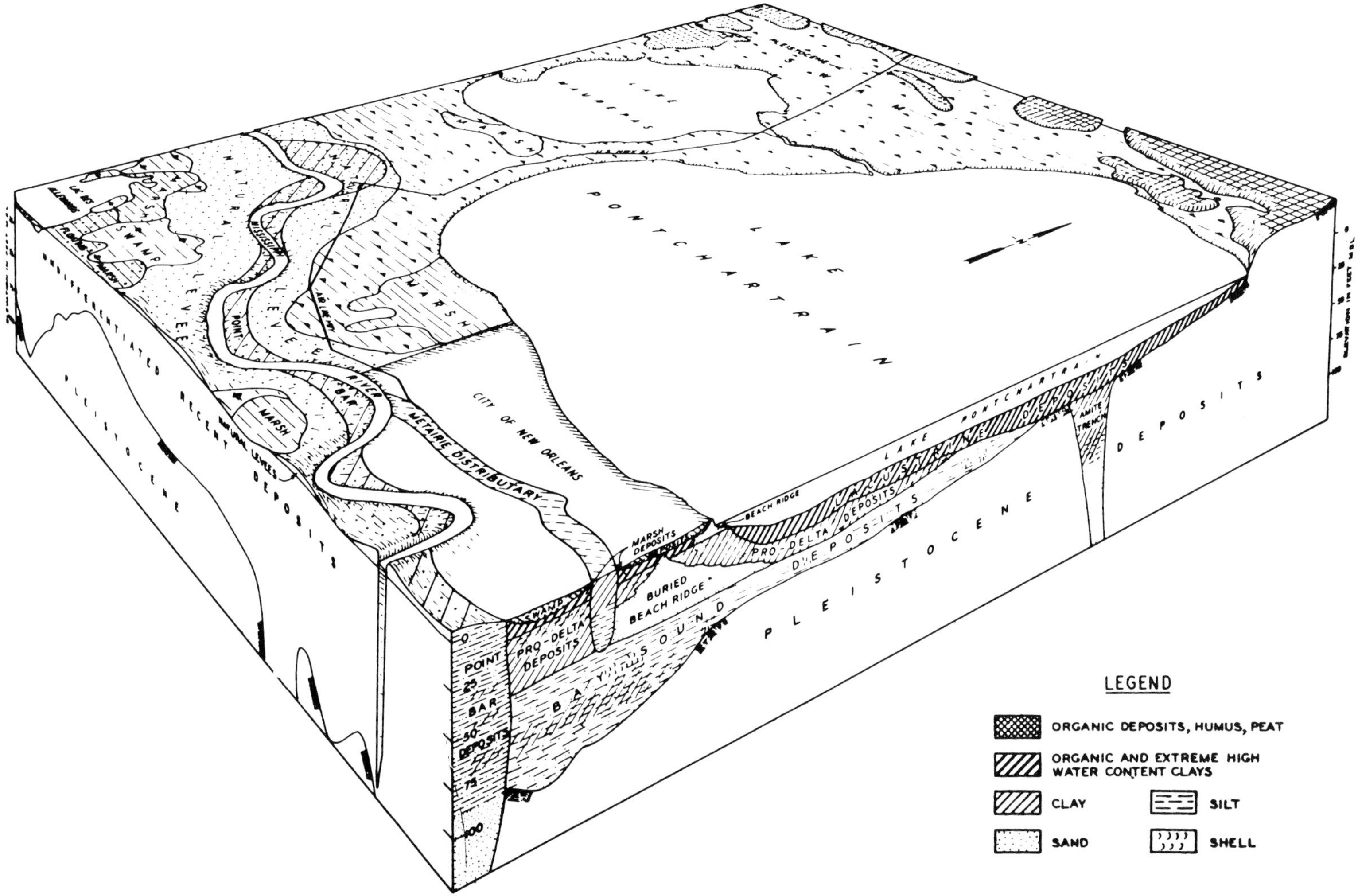

**FIG. 7.35 Major environments of deposition and associated soil types in the vicinity of New Orleans, Louisiana.** [*From Kolb and Schockley (1957).*[13]]

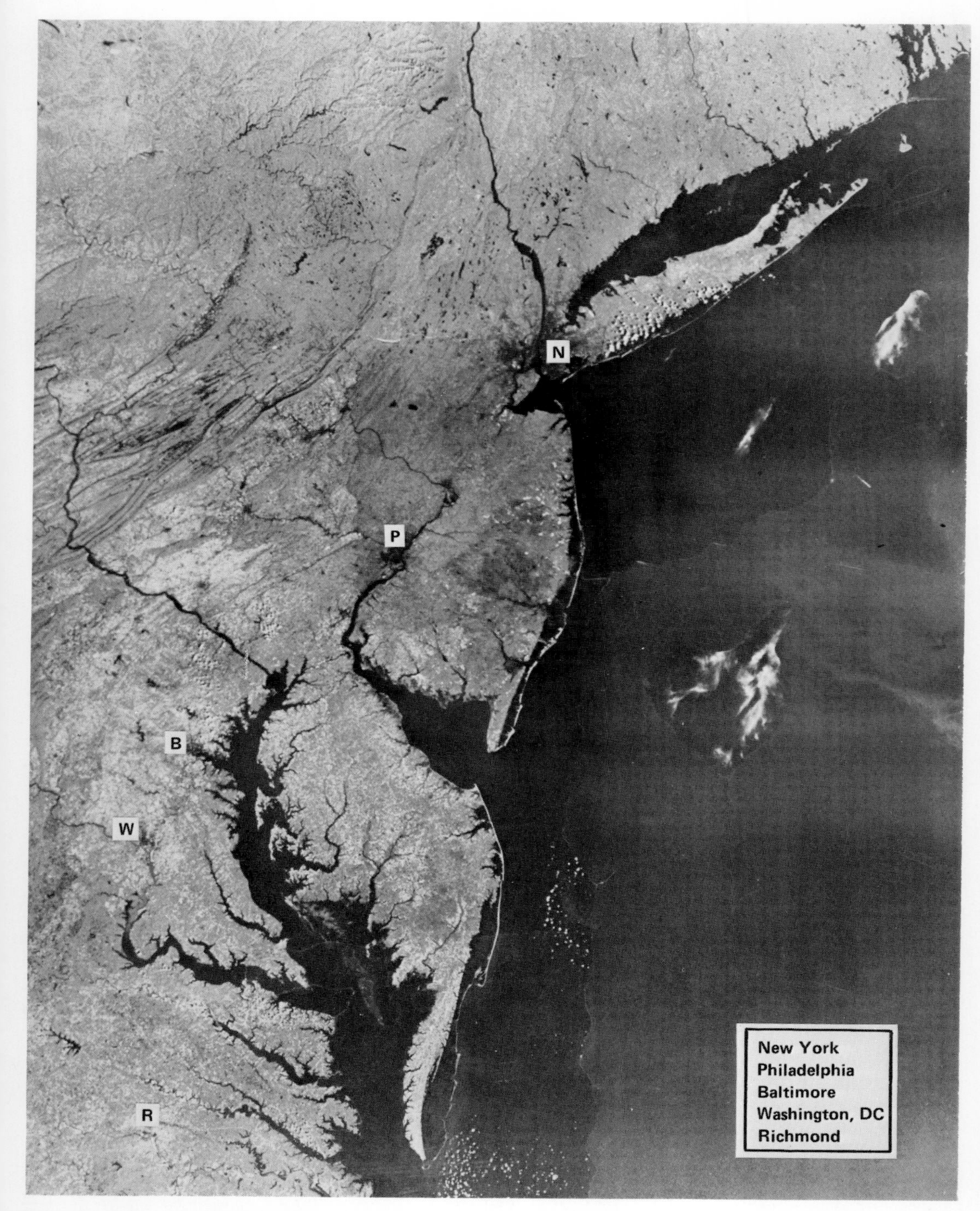


**FIG. 7.36 Mosaic of LANDSAT images of the U.S. east coast from New York City to Richmond, Virginia, illustrating the drowned river valleys of the region.** *(Original image by NASA, reproduction by U.S. Geological Survey, EROS Data Center.)*

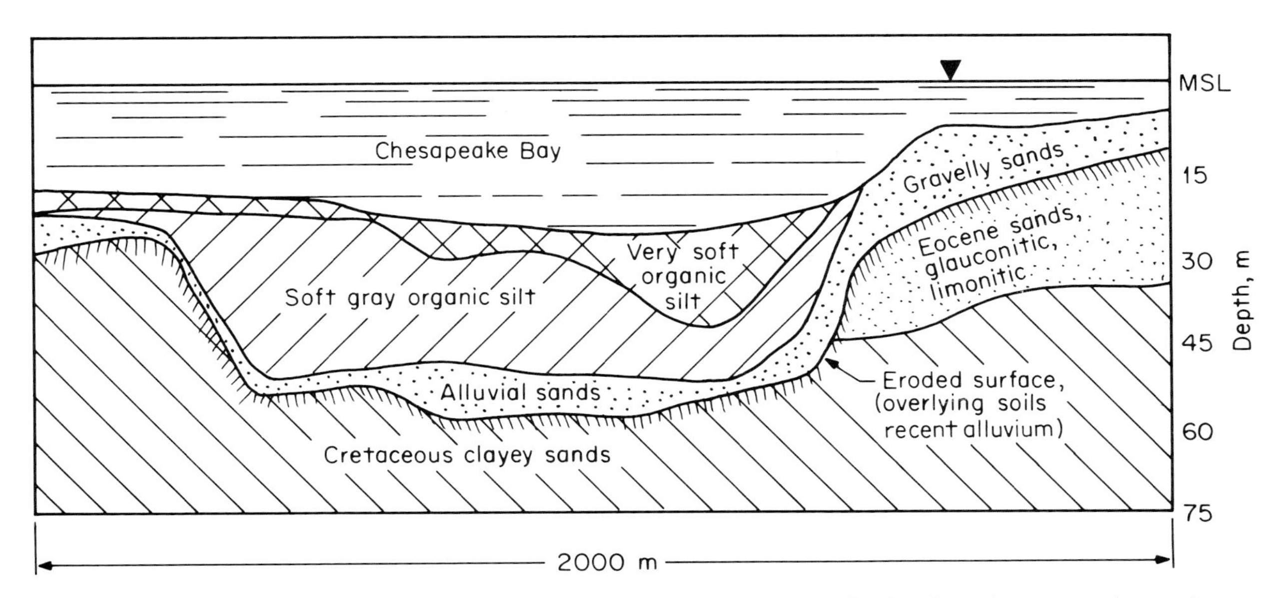


**FIG. 7.37 Geologic section in channel area of Chesapeake Bay between Kent Island and Sandy Point. A drowned river valley.** [*After Supp (1964).*[15]]

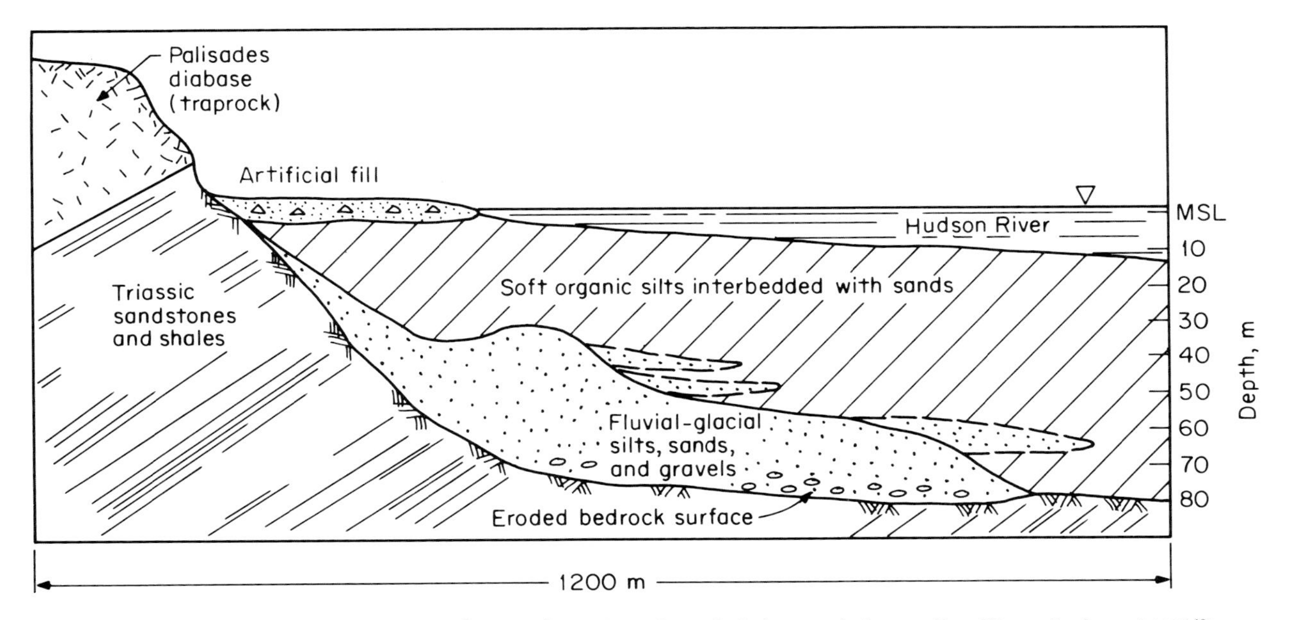


**FIG. 7.38 Geologic section, Lincoln Tunnel, west shore (New Jersey). A drowned river valley.** [*From Sanborn (1950).*[16]]

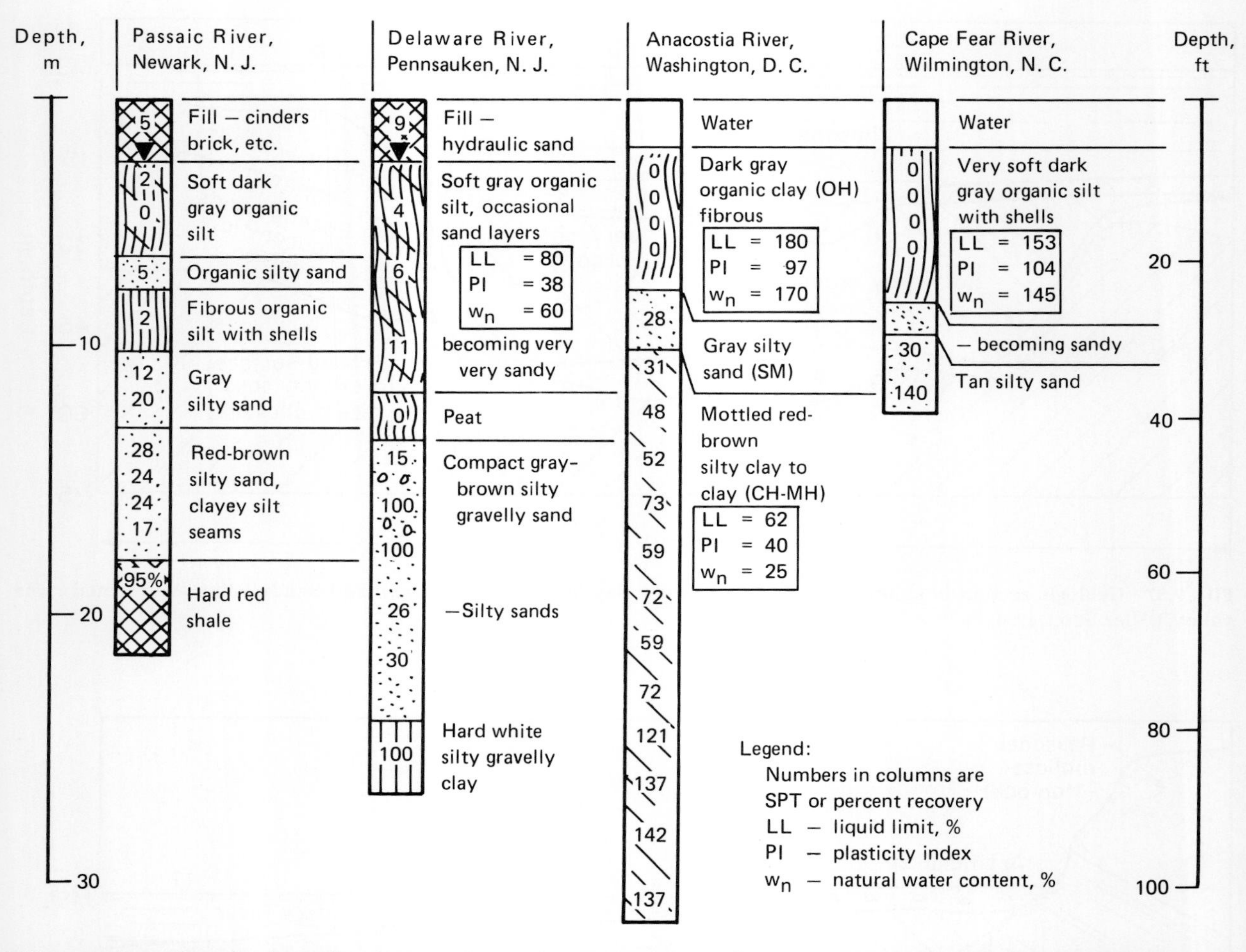


**FIG. 7.39 Logs of test borings taken in estuarine environments at several U.S. east coast locations. All borings terminated in the eroded surface that preceded submergence.** *(From the files of Joseph S. Ward & Associates.)*

## 7.4.3 COASTLINE ENVIRONMENT

### Classification of Coastlines

*Emergent and Submergent Coasts*

Since the Pleistocene the elevations of many coastlines have been changing, some even fluctuating. In general terms, they may be classed relative to sea level as rising (emergent), subsiding (submergent), or stable. The general conditions of the coastlines of the world are given on Fig. 7.42. Erosion dominates emergent coastline and deposition dominates submergent coastlines. The landform expressions indicate soil conditions to be anticipated, since they reflect whether erosional or depositional processes are occurring.

Modern coastlines are primarily the result of the activity of the Pleistocene glaciers. The earth's crust deflected beneath their huge masses and then, as they melted, crustal rebound occurred and is still active today in the extreme northern and southern hemispheres. Simultaneously with rebound, the glacial ice melted, causing sea level to rise as much as 100 m, and in many areas around the middle latitudes sea levels are still rising and the land submerging. Exceptions are

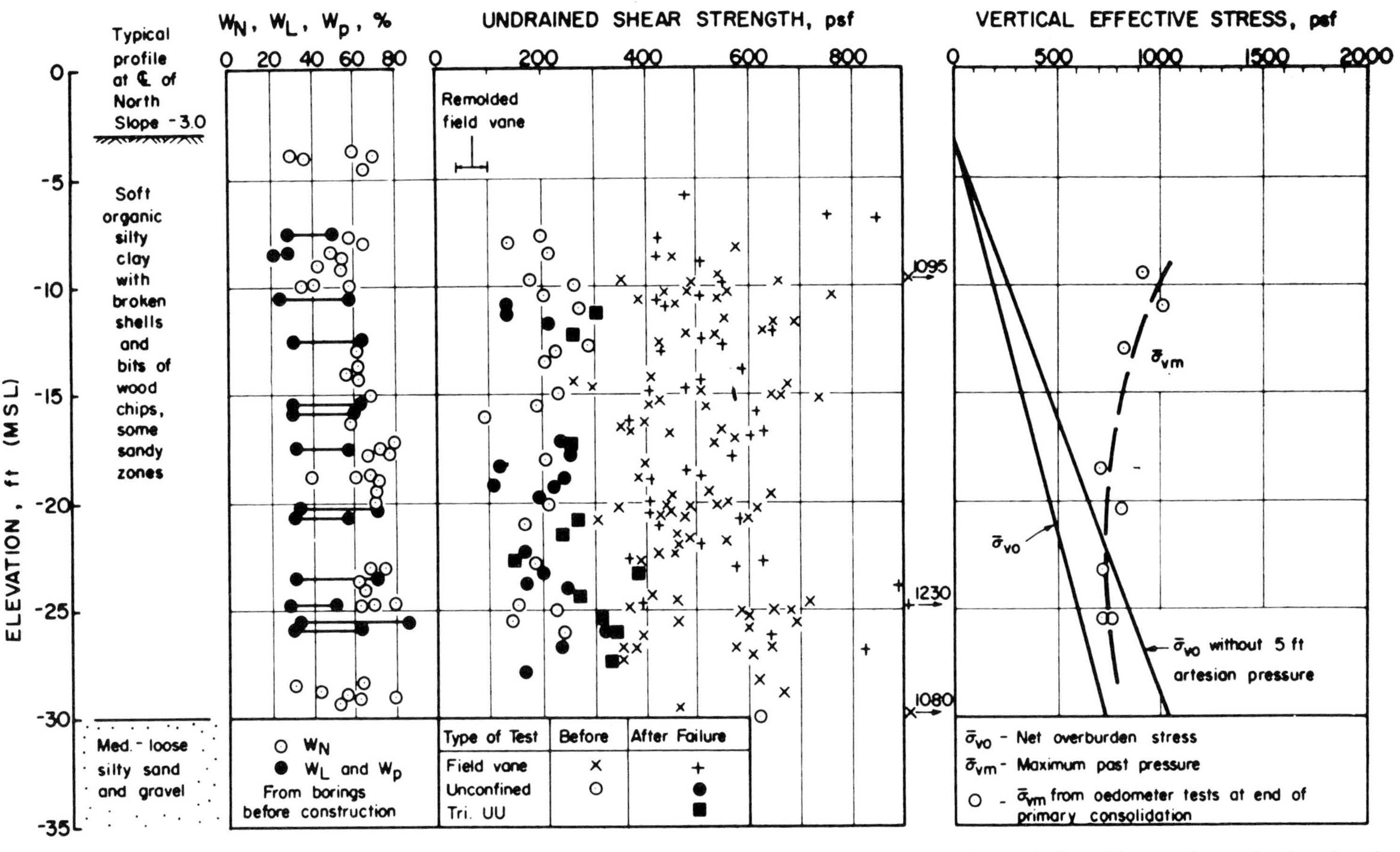


**FIG. 7.40** Properties of estuarine soils (Fore River, Portland, Maine). Natural moisture contents at or above the liquid limit indicate the deposit to be normally or underconsolidated. [*From Simon et al. (1974).*[17]]

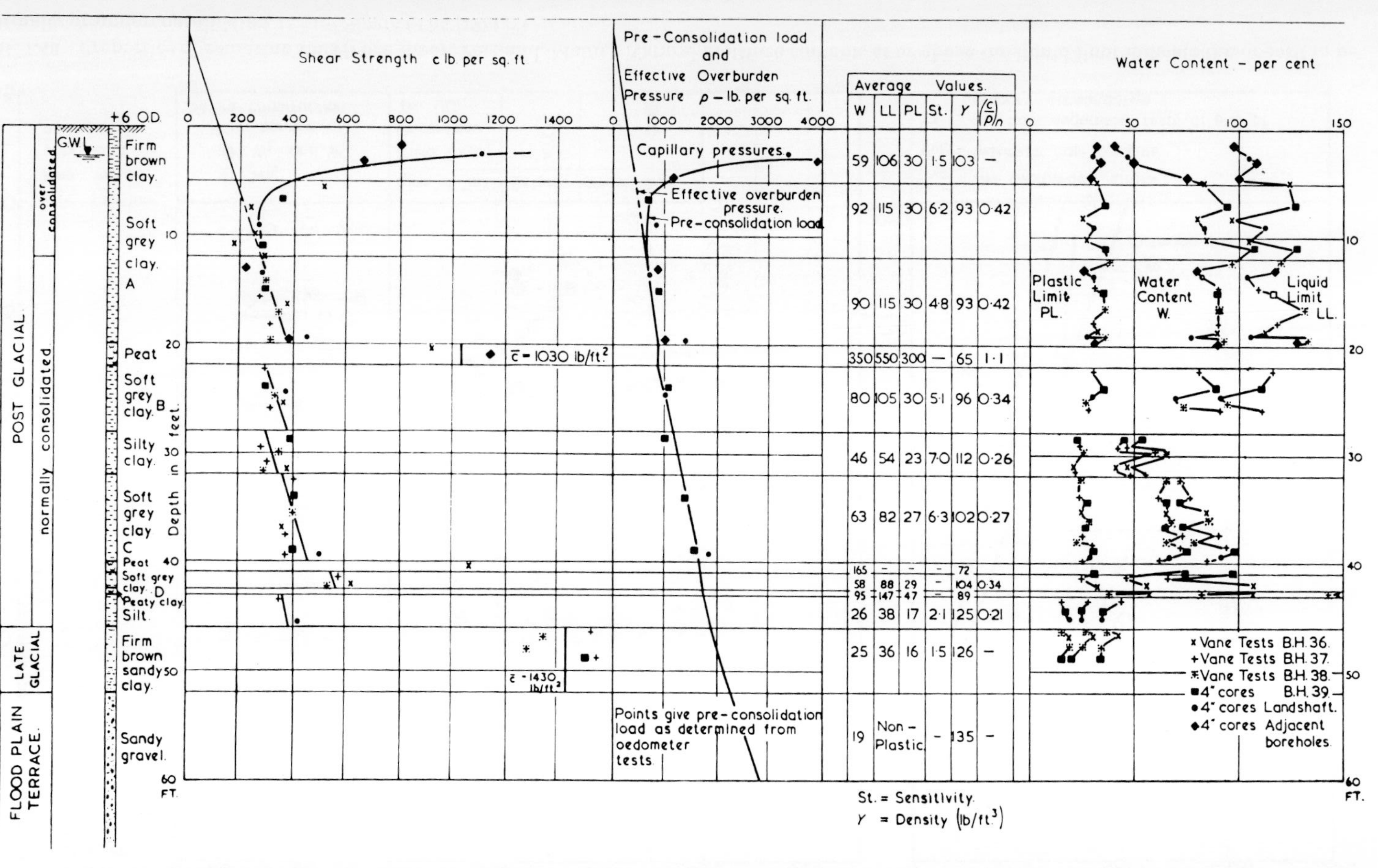


**FIG. 7.41 Characteristics of the Thames estuary clay at Shellhaven. During postglacial times the formation has been uplifted and allowed to develop a shallow crust from desiccation.** [*From Skempton and Henkel (1953).*[18]]

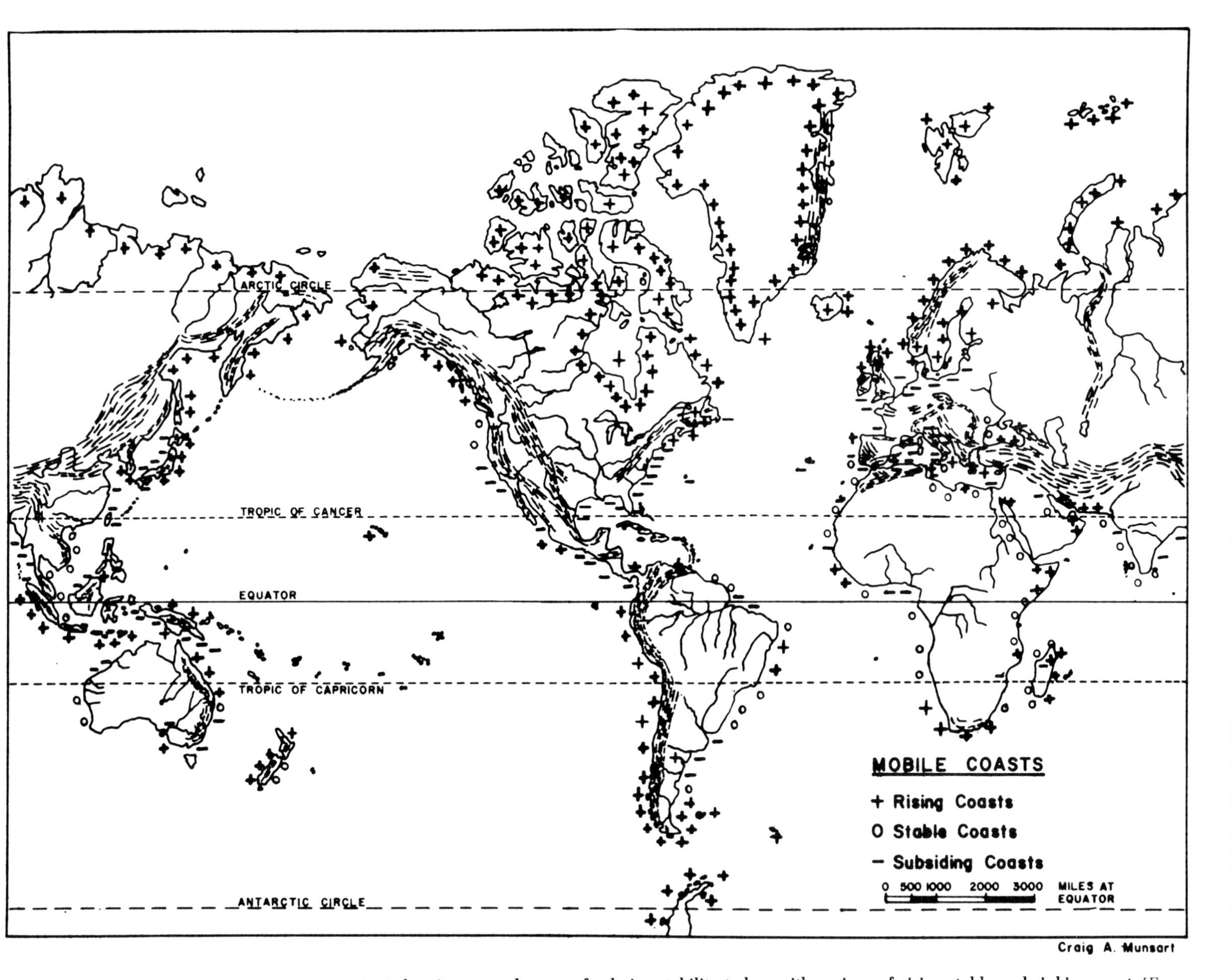


**FIG. 7.42** **World map (Mercator projection) showing coastal areas of relative stability today, with regions of rising stable and sinking coast.** [*From Newman (1968).*[19] *Reprinted with permission of Dowden, Hutchinson & Ross, Inc.*]

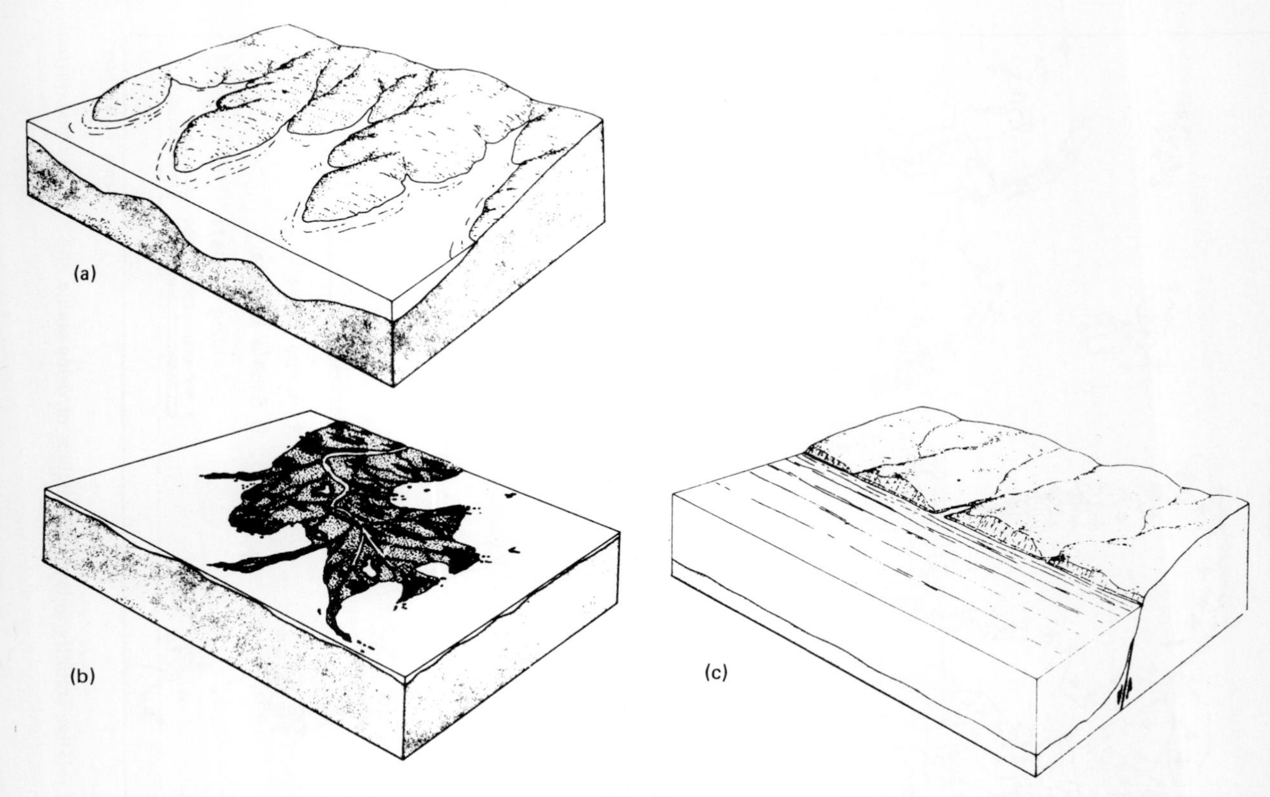


**FIG. 7.43 Primary shorelines and coasts. (*a*) Subaerial erosion coast. The landform along the shoreline was developed by erosion during a period of emergence above sea level; then sea level rose or the land subsided leaving drowned valleys and offshore islands. The coast of Maine is a classic example. (*b*) Subaerial depositional coast. Features resulting from deposition of sediments from rivers, glaciers, wind, or landslides. The most common and significant are deltas *(see Art. 7.4.2)*. (*c*) Structural or Diastrophic coast. The most common are formed by faulting along the coastline. Wave attack usually results in very straight cliffs with steep slopes and deep offshore water, such as common to California coasts *(see Fig. 7.49)*. (*d*) Volcanic coastlines (not shown). Result from volcanic activity and include volcanoes or lava flows.** [*From Hamblin and Howard (1971).*[21]]

found in areas with high tectonic activity such as the west coast of South America, where emergence is the dominant factor.

Evidence of sea level transgression is given by Pleistocene terrace formations, which along the Serra do Mar coast of Brazil are 20 m above sea level and along the east coast of the United States are high as 80 m above sea level (the Brandywine terrace). The eroded surface of the Cretaceous clays in Chesapeake Bay (see *Fig. 7.37*) indicates that sea level was once 60 m lower in that area, during the immediate postglacial period.

*Primary Shoreline*

Landforms resulting from some terrestrial agency of erosion, deposition, or tectonic activity are classed as *primary shorelines* [Shepard (1968)[20]]. Illustrated on Fig. 7.43 are the subaerial erosional coast, subaerial depositional coast, and structural or diastrophic coast. Volcanic coastlines are not shown.

*Secondary Coastlines*

Secondary coastlines result from marine processes and include wave-erosion coasts, marine

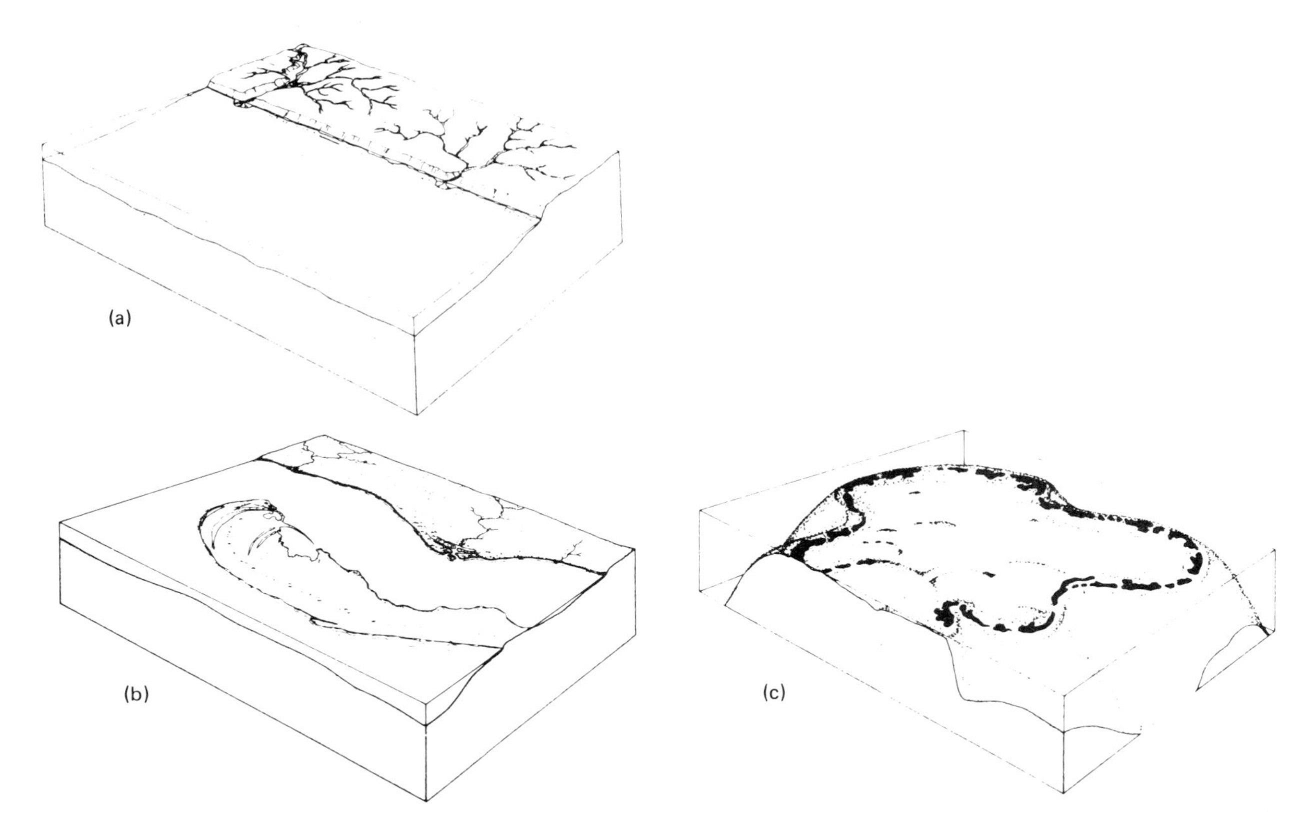


**FIG. 7.44 Secondary coastlines. (*a*) Wave erosion coast. Wave attack usually results in a straight coastline, unless the rocks vary from soft to resistant, in which case an irregular coastline develops. Characteristic of the straight coastline are inshore cliffs and a shallow, gently inclined seafloor. (*See ERTS image in Fig. 2.8*, lower right corner, inside the offshore bar.) (*b*) Marine depositional coast. Waves and currents deposit barrier islands and spits along the shoreline, fluvial activity fills in the lagoons and tidal marshes develop. From the aspect of soil deposition and engineering problems, it is the most important coastline class because of the variation of material types and properties and its common occurrence along thousands of kilometers of shoreline of the U.S. east and Gulf coasts. Offshore waters are characteristically shallow. The ERTS image in Fig. 2.8 illustrated a typical marine depositional coastline. (*c*) Organically built coastline. Includes coral reefs and mangrove growths, common in the tropics. The mangrove trees grow out into the water, particularly in shallow bays, resulting in the deposition of mud around the roots.** [*From Hamblin and Howard (1971).*[21]]

depositional coasts, and organically built coastlines as illustrated on Fig. 7.44 and classed by Shepard (1968).[20]

*Significant Classes*

From the aspect of soil formations, the most significant classes are the marine depositional coast and the emerged depositional coast.

### Marine Depositional Coastline Deposits

*Barrier Beach or Offshore Bar*

As a result of abundant stream deposition from the land and shallow offshore conditions, waves and currents pick up sand particles and return them landward. Eventually a *barrier beach* or *offshore bar* is built parallel to the shore, creating a lagoon between the bar and the mainland. Strong currents extend the bar along the shore-

**FIG. 7.45 Migrating barrier beach forming a spit and dislocating the mouth of the Rio Itapicuru, Bahia, Brazil. Note large dune deposits in foreground and tidal marsh inland.**

line which conflicts with river flow, causing the channel to migrate around a *spit* as illustrated on Fig. 7.45 (*see also Fig. 7.24*). Wave and wind forces, particularly during storms, push the bar inland.

Because the bars and spits are deposited by current and wave action, they are composed of well-sorted materials, usually in a medium-compact to compact state, free of fine-grained particles. During periods of high current and wave activity, gravel beds are deposited and become interbedded with the sands.

*Tidal Marsh*

The tidal lagoon fills with sediments from stream activity, and when the water becomes sufficiently shallow to provide protection against wind and currents, biotic activity begins and a *tidal marsh* with organic soils is formed. If sea level is rising, as is the case along the east coast of the United States, these organic deposits thicken and enlarge to fill depressions in the lagoon bottom. Along the Atlantic coastline, these organic deposits can range from 2 to 20 m thick and cover large areas.

*Dunes*

Wind action moves the finer sands from the beach area to form *dunes*, which eventually migrate to cover the organic soils.

*Engineering Characteristics*

A geologic section through the barrier beach at Jones Beach, Long Island, New York, is given as Fig. 7.46.

The logs of the test borings from which the section was made show the materials of an older barrier bar to be gravelly and very compact on the shoreward side. The very high SPT values are caused by the gravel particles. These older deposits have interbedded landward with thin strata of soft organic silt, and strata of interbedded sand and organic silt. Overlying the older beach soils are more recent beach and dune soils; the latter are migrating inland to cover the tidal marsh soils. Fine-grained soils in tidal lagoons are usually extremely soft, since the salinity of lagoon waters causes the clayey particles to flocculate as they settle out.

Logs of test borings from three other locations of barrier beaches and tidal marshes are given on Fig. 7.47. It is apparent that soil conditions attendant to marine depositional coastlines can be extremely variable and include weak soils interbedded with strong. The buried soft organic soils can be expected to be very irregular in thickness and lateral distribution, and careful exploration is required to identify their presence. If not accounted for in design, significant differential settlements of foundations results.

### Emerged Depositional Coast

The significant features of emerged depositional coastlines are a series of wave-cut terraces, or beach ridges, as illustrated on Fig. 7.48, on which tidal lagoons are also apparent. The oldest beach ridge on the figure, now 30 m above sea level, migrated inland and its tidal lagoon filled in; the land uplifted and a second beach ridge formed. Eventually its lagoon filled and a third beach ridge formed as a barrier beach. The offshore waters at this location are very shallow, indicative, along with the beach ridges, of an emerging shoreline (*see also Fig. 2.10*).

### Combined Shorelines

The features of a structurally shaped coast along the ocean and a subaerial erosion and marine depositional coast on a large bay are shown on Fig. 7.49. The geologic formations inland are primarily rock, such as quartz diorite. The area is shown also on the SKYLAB photo (*see Fig. 2.9*).

On the ocean side, strong wave forces have cut a linear cliff in the rock; a narrow beach has been formed and fine sands from the beach blow inland to form dunes.

In Drakes Bay, longshore currents are depositing sands to form a *bay bar* and spits. The drowned valley (Schooners Bay) has been filled with sediments to form shallow waters. Current in the bay is minimal and deposition is limited to fine-grained soils forming mud flats. The relatively steep slopes and high elevations of the adjacent hills indicate that the bay deposits may be as

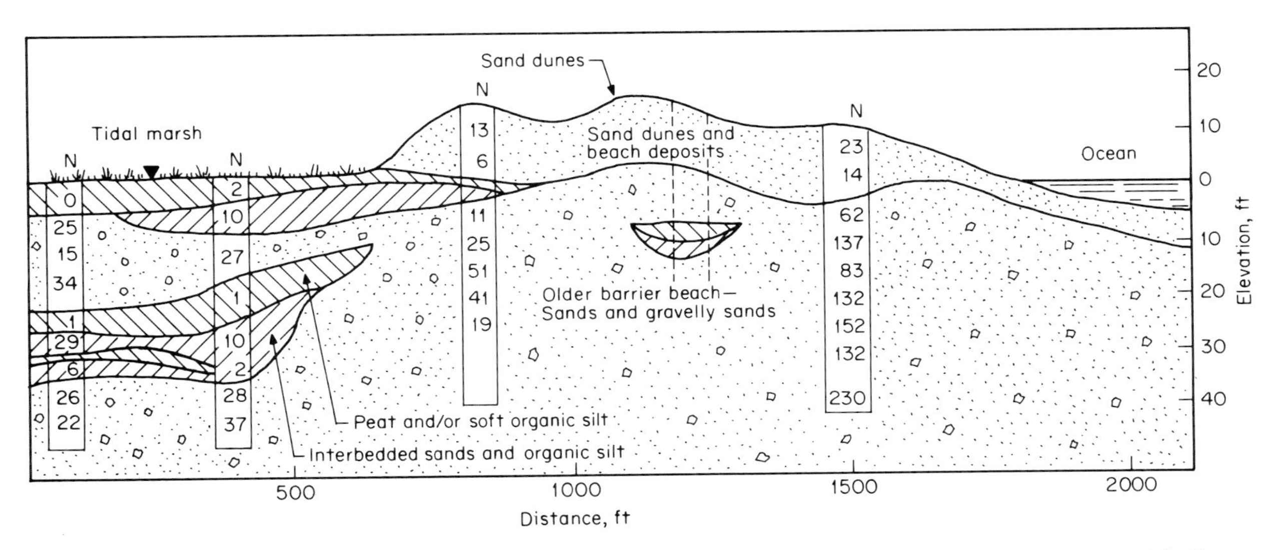


**FIG. 7.46 Geologic section illustrating migration of dune and beach deposits over tidal marsh organic materials (Jones Beach, Long Island, New York).** *(Courtesy of Joseph S. Ward & Associates.)*

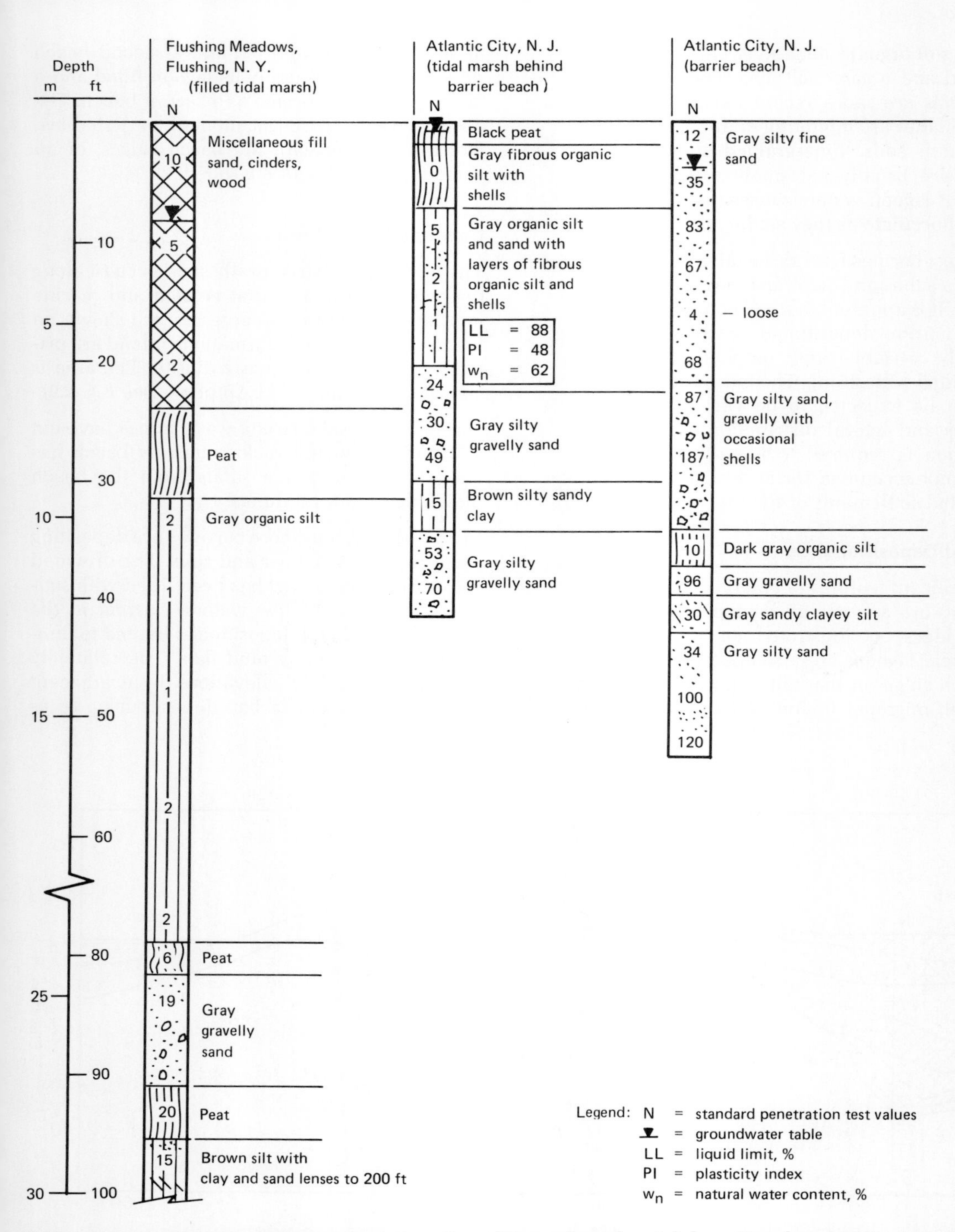


**FIG. 7.47 Logs of test borings representative of conditions of a subaerial depositional coast.** *(Courtesy of Joseph S. Ward & Associates.)*

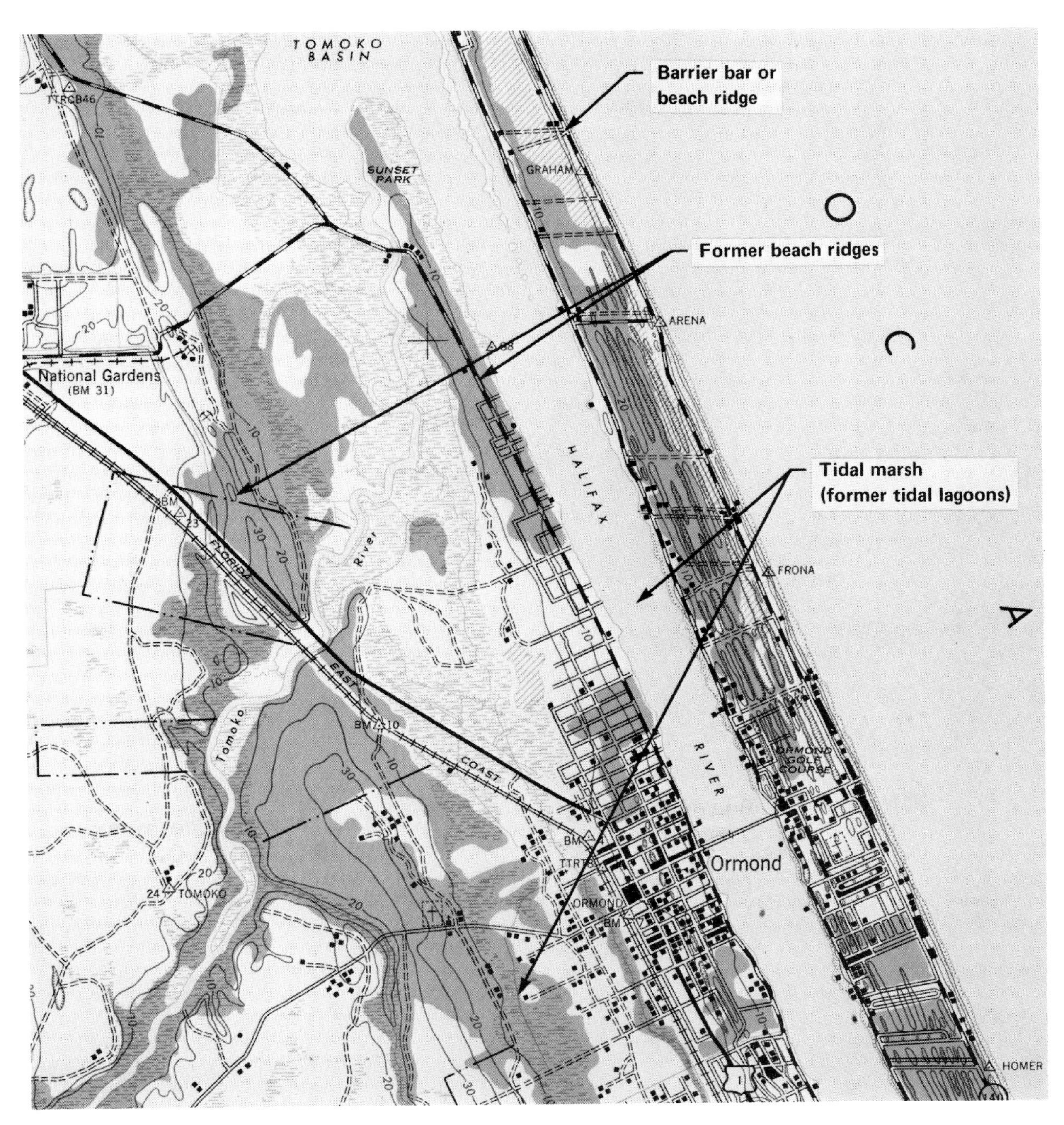


FIG. 7.48 Beach ridges of an emerging shoreline near Ormond Beach, Florida. (Scale 1:62,500.)

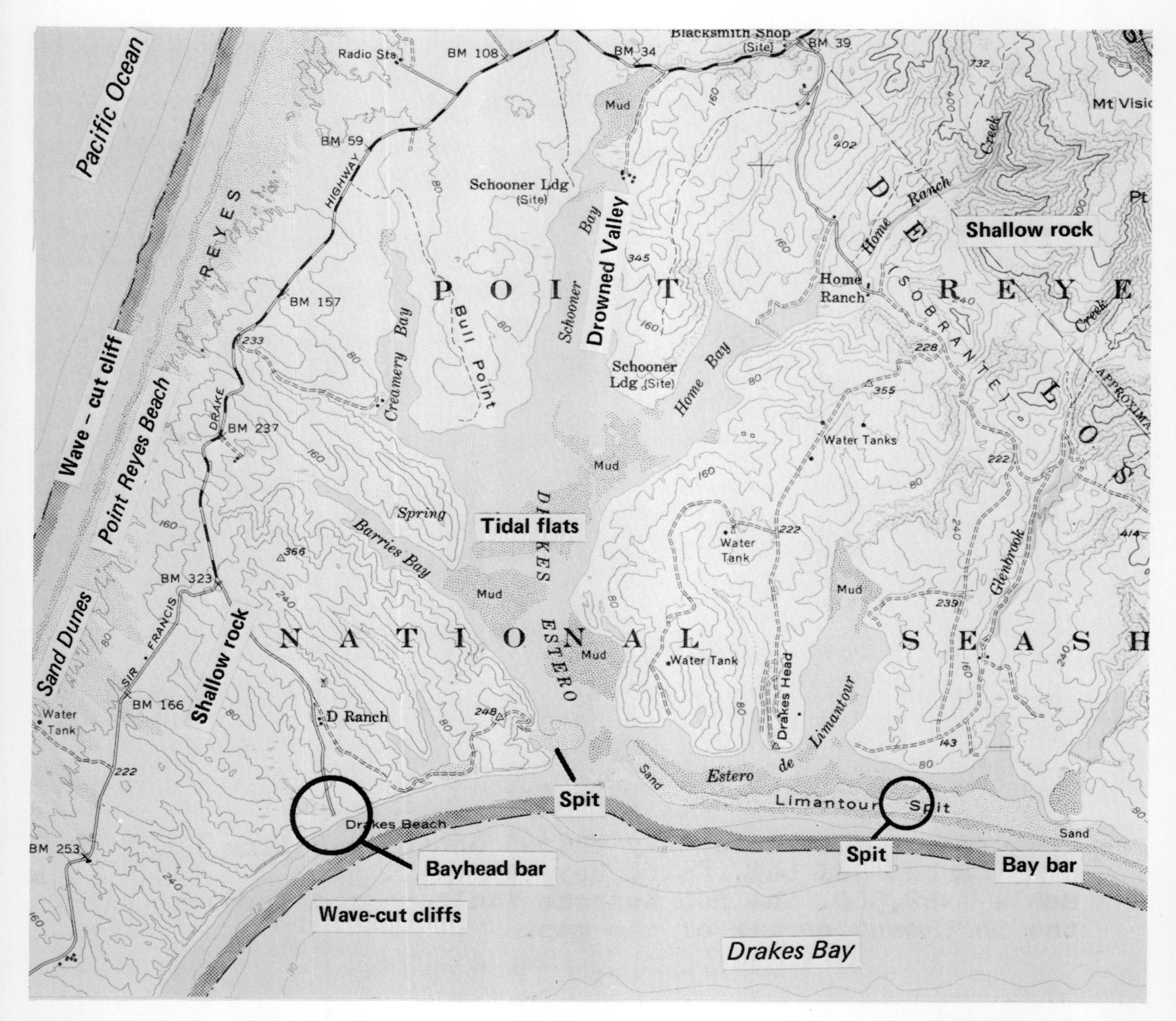


**FIG. 7.49 Features of a structurally shaped coastline and a submerged bay (Point Reyes, California). (Scale 1:62,500. Contour interval: 80 ft.** ***See also the SKYLAB image in Fig. 2.9.*****)**

much as 30 m thick. At Drakes Beach, a *bayhead bar* has formed, blocking the flow of a small stream into the bay. The lagoon behind the bay bar is filling with fine sediments from landward.

### 7.4.4 COASTAL PLAIN DEPOSITS

#### General

Coastal plains result from regional uplift or a lowering of sea level, or both, causing the seafloor to emerge and become land. Geographically they are defined as regional features of relatively low relief bounded seaward by the shore and landward by highlands [Freeman and Morris (1958)[22]]. They are of engineering significance as a geologic class because of their low relief, characteristic engineering properties, and worldwide distribution (Table 7.8). There are two major coastal plains in the United States: the Gulf and Atlantic coastal plain, which covers the largest land area of any in the world, and the

**TABLE 7.8**
**AREA OF MAJOR AND SOME MINOR COASTAL PLAINS OF THE WORLD***

| Geographic coastal plain name | Area ($km^2$; exclusive of continental shelf and landward portions of major drainage basins) |
|---|---|
| Africa | |
| Egyptian-North African (Egypt, Libya, Tunisia) | 370,000 |
| Niger | 90,000 |
| Mauritania, Spanish Sahara | 300,000 |
| Mozambique | 130,000 |
| Somali | 110,000 |
| Asia | |
| Bengal, Pakistan-India | 220,000 |
| Coromandel-Colconda, India | 40,000 |
| Irrawaddy, Burma | 40,000 |
| Kanto plain, Japan | 5,000 |
| Karachi, Pakistan-India | 370,000 |
| Malabar-Konkan, India | 25,000 |
| Mekong, Vietnam-Cambodia | 100,000 |
| Ob-Khatanga-Lena, U.S.S.R. | 800,000 |
| Persia, Saudi Arabia, Iraq | 325,000 |
| Sumatra, Indonesia | 160,000 |
| Yellow-Yangtze plains, China | 125,000 |
| Australia | |
| Nullarbor | 120,000 |
| Europe | |
| Aquitaine, France | 25,000 |
| Baltic, Poland | 6,000 |
| Flandrian and Netherlands (Belgium, Holland, Germany) | 150,000 |
| Po, Italy | 25,000 |
| North America | |
| Arctic, U.S.-Canada | 130,000 |
| Atlantic and Gulf Coastal Plain | 940,000 |
| Costa de Mosquitas, Nicaragua-Honduras | 28,000 |
| Los Angeles, U.S. | 21,000 |
| Yucatan-Tabasco-Tampeco, Mexico | 125,000 |
| South America | |
| Amazon, Brazil | 245,000 |
| Buenos Aires, Uruguay | 270,000 |
| Orinoco-Guianan (Venezuela, Guyana, Surinam, French Guiana) | 120,000 |

*From Colquhoun (1968).[23] Reprinted with permission of Dowden, Hutchinson & Ross, Inc.

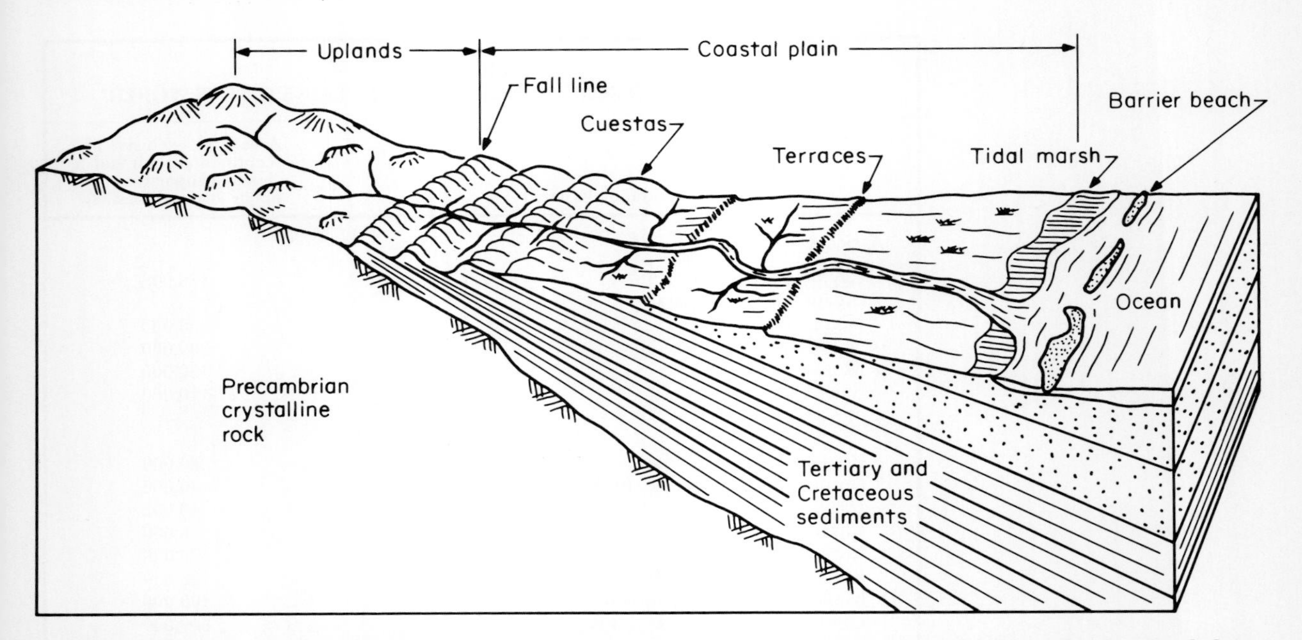


**FIG. 7.50 Schematic of coastal plain in mature stage of development, generally representative of the Atlantic coastal plain.**

Los Angeles coastal plain. Their distribution is illustrated on Fig. 7.1.

Nearly all coastal plains contain Quaternary sediments, and most contain Tertiary, Jurassic and Cretaceous strata as well. These strata typically have been preconsolidated under the load of hundreds of meters of material removed by erosion during emergence. L. Casagrande (1966)[24] reported preconsolidation pressures as high as 10 kg/cm$^2$ from a depth of 17 m in Richmond, Virginia, where overburden pressure was about 2 kg/cm$^2$.

## Atlantic Coastal Plain

### *Landform Characteristics*

The Atlantic coastal plain sediments lie unconformably on the Precambrian rocks dipping seaward and extending out beneath the ocean as shown on Fig. 7.50. The contact on the surface between the older crystalline rocks and the coastal plain soils is called the *fall line*, and is a characteristic of the Atlantic coastal plain.

As the land emerged from the sea, stream erosion proceeded along the strike of the exposed beds, eventually forming a series of low, parallel ridges of low relief, or *cuestas*. Emergence continued at a slow, barely perceptible rate for some time, then accelerated for a relatively short time during which wave action cut *terraces* along the plain (*see Art. 7.4.3*). In modern times the characteristics of a marine depositional coast have developed. The coastal plain in New Jersey is shown on the LANDSAT image (*see Fig. 2.8*); the similarity of the tone pattern and texture with that of the Gulf coastal plain given in Fig. 7.33 is apparent.

### *Stratigraphy*

The sediments of the New Jersey coastal plain reach thicknesses greater than 2000 m beneath the shoreline. The beds dip seaward, and uplift and erosion have exposed the older beds of Cretaceous and Tertiary periods to surface as long narrow bands trending in a northeast-southwest direction. These older beds include alternating thick strata of clays, marls, and sands, with clayey soils and marls the dominant material near the surface.

The term *marls* is used by geologists to refer to loose earthy materials that engineers call soils. They consist chiefly of calcium carbonate mixed with clay. The carbonate content is readily detectable by its effervescence in hydrochloric acid. Sandy marl contains grains of quartz sand or other minerals; shell marl is a whitish material containing shells of organisms mixed with clay; greensand marl contains many grains of glauconite (a hydrous silicate of iron and potassium colored bright green on freshly exposed nodule surfaces).

Tertiary soils overlie the Cretaceous clays and often consist of well-defined interbedded layers of brightly colored sands and clays (Figs. 7.51 and 7.52).

Late Tertiary and Quaternary sands and gravels cover much of the coastal plain region. They are the most recent deposits (except for recent alluvium and swamp soils). Composed chiefly of gravelly soils, more resistant to erosion than sands and clayey sands, they cap the hills that are characteristic of the seaward portions of the region. There are remnants of wave-cut terraces in New Jersey, but terrace features are much more evident in the coastal region of Maryland, Virginia, and North Carolina.

**FIG. 7.51 Tertiary clays from a depth of 8.0 m. (Leesburg, New Jersey. A low-plasticity silty clay.)**

**FIG. 7.52 Tertiary clays interbedded with sands from a depth of 6 m (Jesup, Georgia).**

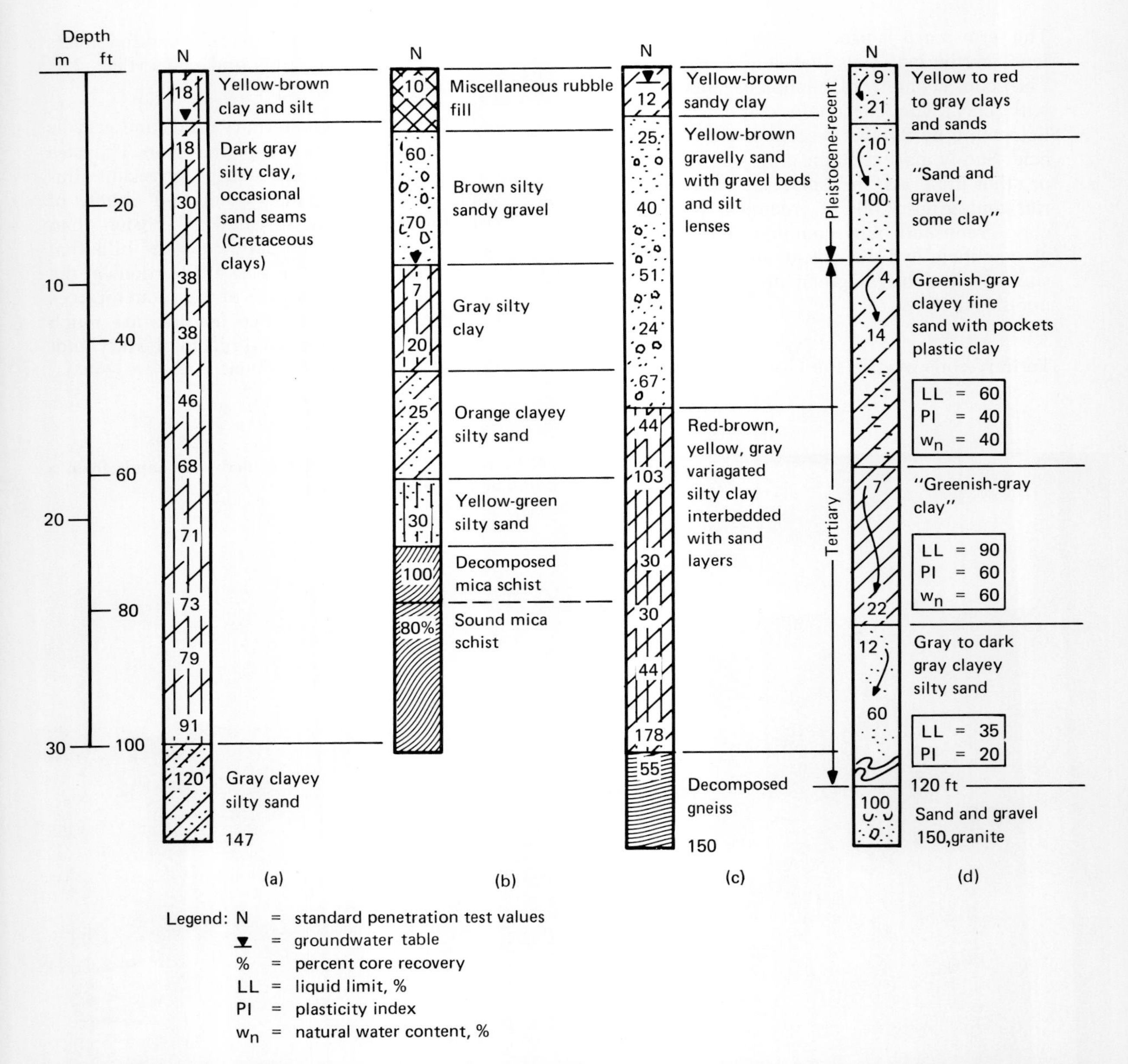


**FIG. 7.53 Logs of test borings made in Atlantic coastal plain deposits: (*a*) Keyport, New Jersey; (*b*) Philadelphia, Pennsylvania; (*c*) Wilmington, Delaware; (*d*) Richmond, Virginia.** [*Parts a, b, c courtesy of Joseph S. Ward & Associates. Part d after Casagrande (1966).*[24]]

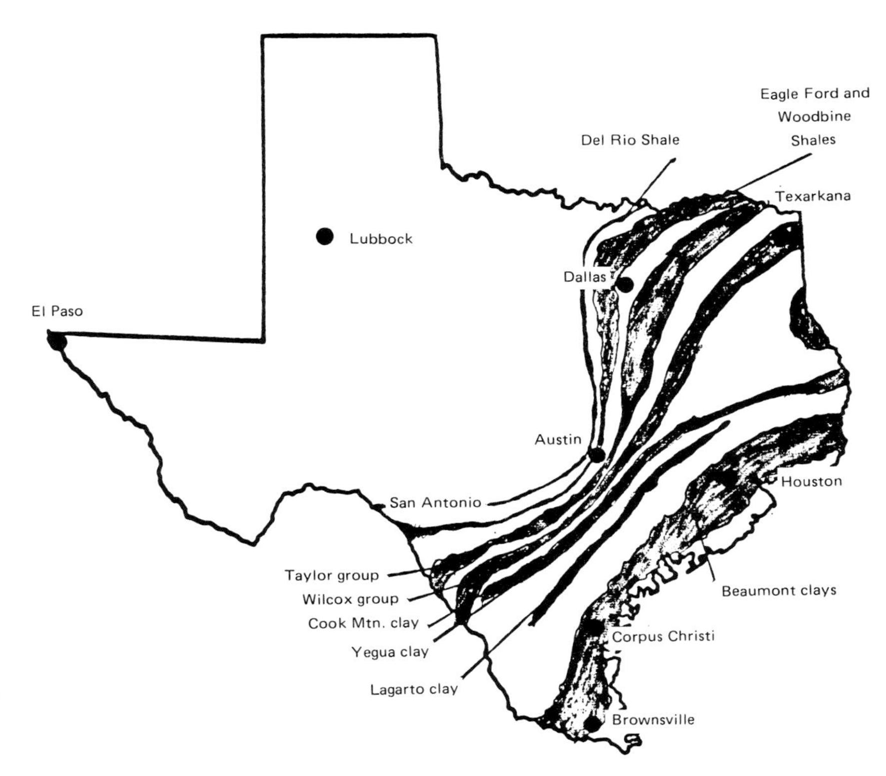


**FIG. 7.54 Location and extent of the most active clay soils in Texas.** [*From Meyer and Lytton (1966).*[8]]

*Engineering Characteristics*

Selected boring logs from several locations are given on Fig. 7.53. In general, the soils of the Atlantic coastal plain do not present difficult foundation conditions. The clays are overconsolidated and of low activity, and the Quaternary formations provide sources of sand and gravel borrow.

### Gulf Coastal Plain

Soil conditions can be troublesome to construction in the East Gulf and West Gulf sections. Cretaceous formations, including the Selma chalk of Alabama and the Austin chalk and Taylor marl of Texas, have weathered to produce black plastic clays. The marine clays of the outer coastal plain in Texas and Louisiana are of the Beaumont formation.

The most important characteristics of these clays are their high activity and tendency to swell. Particularly in the western portions of the region, where seasonal dry periods are common, these clays are active, undergoing large volume changes by shrinking and swelling (*see Art. 10.6.2*). The location and extent of the most active clay soils in Texas are given on Fig. 7.54. Typical properties are as follows [from Meyer and Lytton (1966)[8]]:

- Description: Tan and gray clay (CH); very stiff, jointed, and slickensided. Black discoloration along the joints. Clay type, montmorillonite.
- Index properties: $w$ = 13 to 30%; LL = 40 to 90%; PL = 17 to 25%; PI = 25 to 70%; bar linear shrinkage from LL = 12 to 25%; percent passing no. 200 seive, 70 to 100%.
- Strength: $U_c$ = 2.0 to 8.0 kg/cm$^2$; failure usually occurs along joints.
- Swell pressure at zero volume change: 2.0 to 11.0 kg/cm$^2$.
- Volume change at 1 psi (0.7 kg/cm$^2$) confining pressure: 5 to 20%.

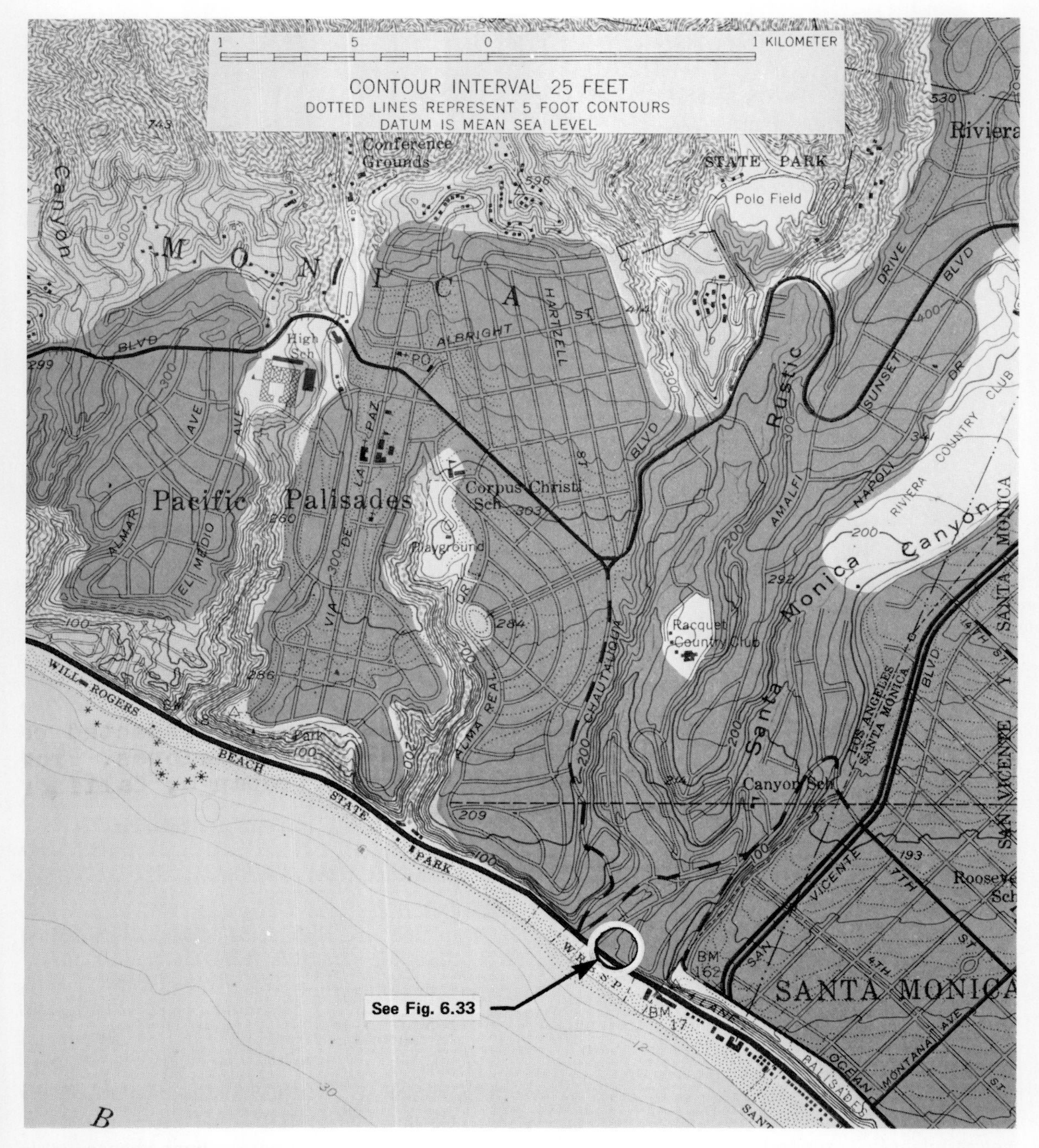


**FIG. 7.55 Topographic expression of a dissected coastal plain with unstable natural slopes.** *(From USGS quadrangle sheet, Topango, California. Scale 1:24,000. For a structural coastline, see Fig.7.43.)*

### Los Angeles Coastal Plain

*General*

The heavily preconsolidated marine clays of Tertiary age evident along the coast north and south of Los Angeles are of particular interest because of their instability in slopes and their tendency toward frequent landslides. The Miocene formations and the large slide at Palos Verdes Hills are described in Art. 9.2.6. The topographic relief of the Pacific Palisades area near Los Angeles where these formations are exposed is shown on Fig. 7.55 as a dissected coastal plain and a structural coastline. Figure 6.33 shows an exposure.

*Stratigraphy*

See Gould (1960).[25] Recent *alluvium* caps the mesas and ancient slide masses at lower levels; it grades from a very compact sand and gravel at the formation base to a hard, brown silty clay at the surface.

MIOCENE FORMATION Preconsolidation from removal of about 1000 m of overburden has caused the natural water content to be 5 to 10% *below* the plastic limit. The lower portion, the *Modelo*, is referred to as a shale in geologic reports, but in engineering terms it is described as "hard, dark gray silty clay" (montmorillonite) thinly bedded with laminations and partings of fine sand. It can be found massive and highly bituminous, or in sandy phases. Gradations and plasticity limits vary considerably. Intense tectonics has resulted in intricately distorted bedding *(see Fig. 6.33)*, numerous small fractures, randomly oriented, and occasional slickensides. The latter are revealed when a specimen is broken open or sheared. The *weathered Modelo*, the upper portion, is lighter colored and iron stained, containing growths of gypsum crystals, with lower SPT values and higher water contents than the parent material. The average identification properties for these materials are given on Table 7.9 and shear strength values on Table 3.30.

PALEOCENE FORMATION The Martinez is described as a hard, dark green clay with pockets of medium to fine sand and occasional calcareous nodules and cemented fragments; heavily preconsolidated; and highly fractured, slickensided, and distorted by tectonics. Two phases have been identified: a predominantly clayey phase, and a hard and partially cemented sandy phase. The clayey stratum, of major importance in slope stability, lacks the thin-bedded character of the Modelo and actually has been so badly distorted that it is marbled with sworls and pockets of sand. The parent materials weather to lighter and brighter-colored soils of lower SPT values and high water content; identification properties are given on Table 7.9 and strength values on Table 3.30.

### London Clays

The famous clays underlying the city of London were deposited under marine conditions during the Eocene [Skempton and Henkel (1957)[26]]. Uplift and erosion removed from one-third to two-thirds of the original thickness, and overconsolidation by maximum past pressure has been about 20 $kg/cm^2$. A geologic map of the central part of the city is given as Fig. 7.56*a*, and a typical section as Fig. 7.56*b*. The clay thickness is as great as 200 ft. Boring logs, including index properties, and strength and compressibility characteristics from two locations, are given on Fig. 7.57. It is interesting to compare these clays with those from the Thames estuary (*Fig. 7.41*).

## 7.4.5 LACUSTRINE ENVIRONMENT (NONGLACIAL)

### General

Materials deposited in a lake environment are termed *lacustrine*, and include granular soils along the shoreline, in the forms of beaches, dunes, and deltas, and silts and clays filling the lake basin.

In geologic time many lakes are relatively short-lived features terminating as large flat areas, swamps, and marshes. Land areas that were former lake beds can be expected to present difficult foundation conditions.

### Origin and Occurrence of Lakes

*Significance*

Because lakes occur in depressions, which normally do not result from erosion activity, the

**TABLE 7.9**
**IDENTIFICATION PROPERTIES OF THE TERTIARY FORMATIONS IN THE LOS ANGELES AREA***

| Material | Description | Dry unit weight, pcf | Natural water content, % | Liquid limit | Plastic index | Plastic limit | % smaller than 0.002 mm | % smaller than 200 sieve | Activity |
|---|---|---|---|---|---|---|---|---|---|
| Weathered clayey modelo | Stiff brown and gray silty clay thin-bedded with some light gray fine sand | 89.8 | 28.7 | 66 | 31 | 35 | 34 | 85 | 0.91 |
| Unweathered clayey modelo | Hard, dark gray silty clay thin-bedded with some light gray fine sand | 98.2 | 23.3 | 69 | 37 | 32 | 31 | 84 | 1.19 |
| Sandy modelo | Compact, light gray and tan, fine to medium sand, some silt, occasional clay seams | 110.4 | 16.5 | NP | — | NP | 6 | 26 | — |
| Weathered clayey Martinez | Soft to medium stiff, brown clayey silt and silty clay | 103.6 | 22.1 | 62 | 38 | 24 | 30 | 73 | 1.26 |
| Weathered sandy Martinez | Medium compact, rust brown and yellow silty fine to medium sand | 108.2 | 17.6 | NP | — | NP | 14 | 42 | — |
| Unweathered sandy Martinez | Very compact, gray-green clayey fine to medium sand with calcareous concretions | 120.7 | 12.8 | NP | — | NP | 16 | 38 | — |
| Unweathered clayey Martinez | Hard, dark gray-green clay with some pockets of gray medium to fine sand, occasional shale fragments | 105.6 | 19.2 | 68 | 40 | 28 | 30 | 80 | 1.33 |

*From Gould (1960).[25]

cause of their existence aids in understanding regional geology. Numerous lakes are made by beavers and humans; the natural causes are described below.

*Tectonic Basins*

Many of the largest lakes are formed from movements of the earth's crust, which cause faulting, folding, and gentle uplifting of the surface. Uplifting can cut off a portion of the sea leaving relict seas or large lakes such as the Caspian Sea, or raise the seafloor to become land surfaces with lakes such as Okeechobee in Florida forming in relict depressions, or raise margins of areas to form basins such as Lake Victoria in Africa. Tilting of the land surface causes drainage reversal such as in Lake Kioga in East Africa and, adjacent to faults, small lakes or sag ponds. Block faulting creates grabens, resulting in some of the largest and deepest lakes such as Tanganyika and Nyasa in Africa, and Lake Baikal in central Asia. Faulting and tilting combined created lakes of the Great Basin and Sierra Nevada portions of North America including Lake Bonneville and its relict, the Great Salt Lake.

*Glacial Lakes*

Glaciers form lakes by scouring the surface (the Great Lakes of North America, Lake Agassiz in northwestern United States), by deposition of damming rivers (Fig. 7.85), by melting blocks of

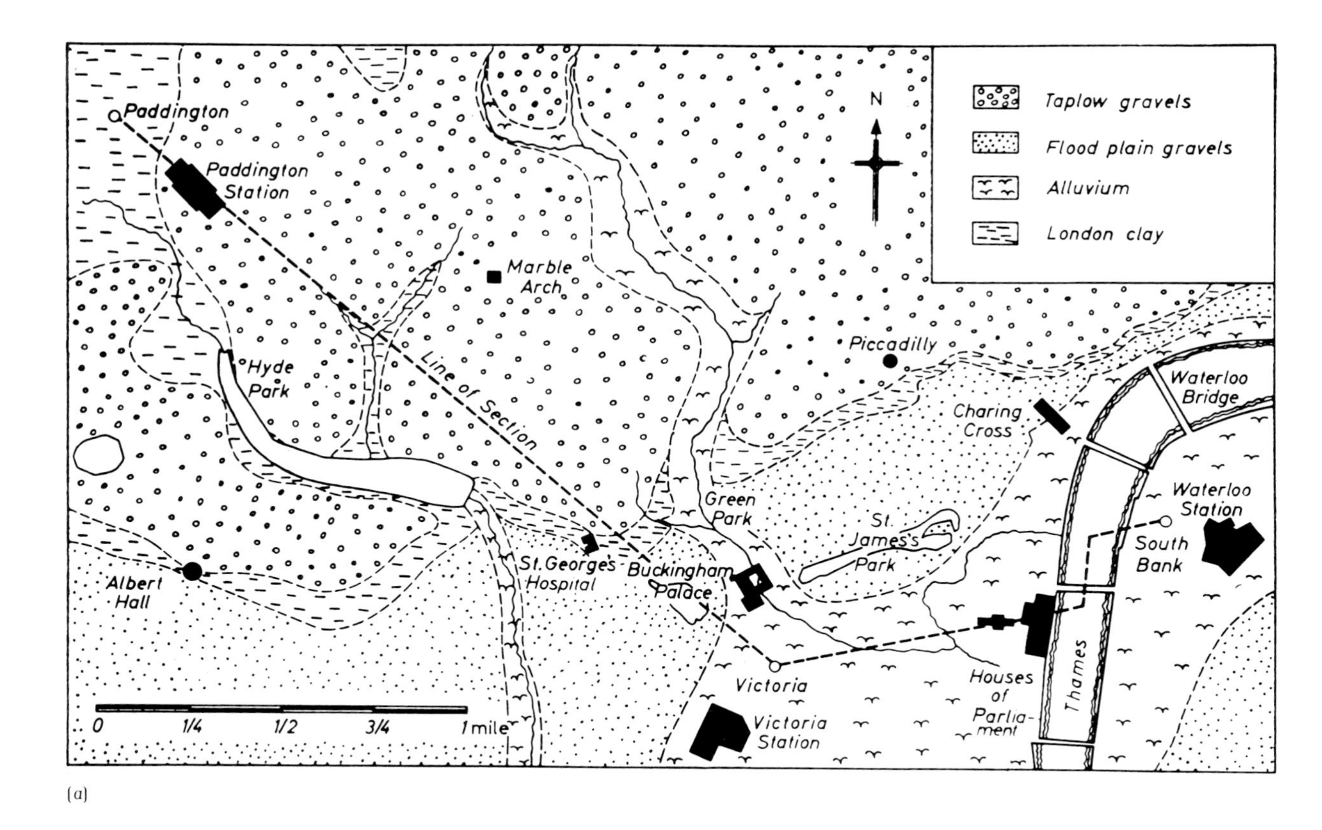


(a)

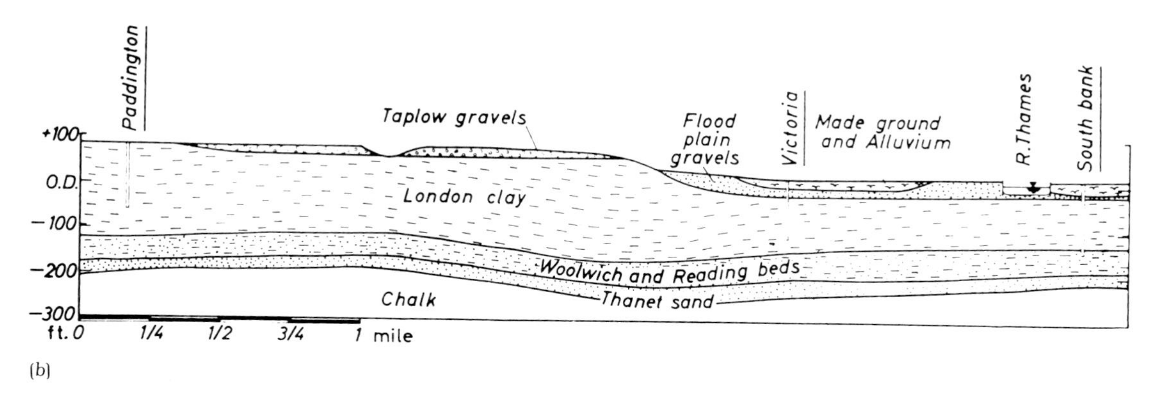


(b)

**FIG. 7.56 Geologic map and section from Paddington to the south bank of the Thames, London: (a) geologic map showing the position of the three sites of Fig. 7.57 and (b) geologic section.** [*From Skempton and Henkel (1957).*[26]]

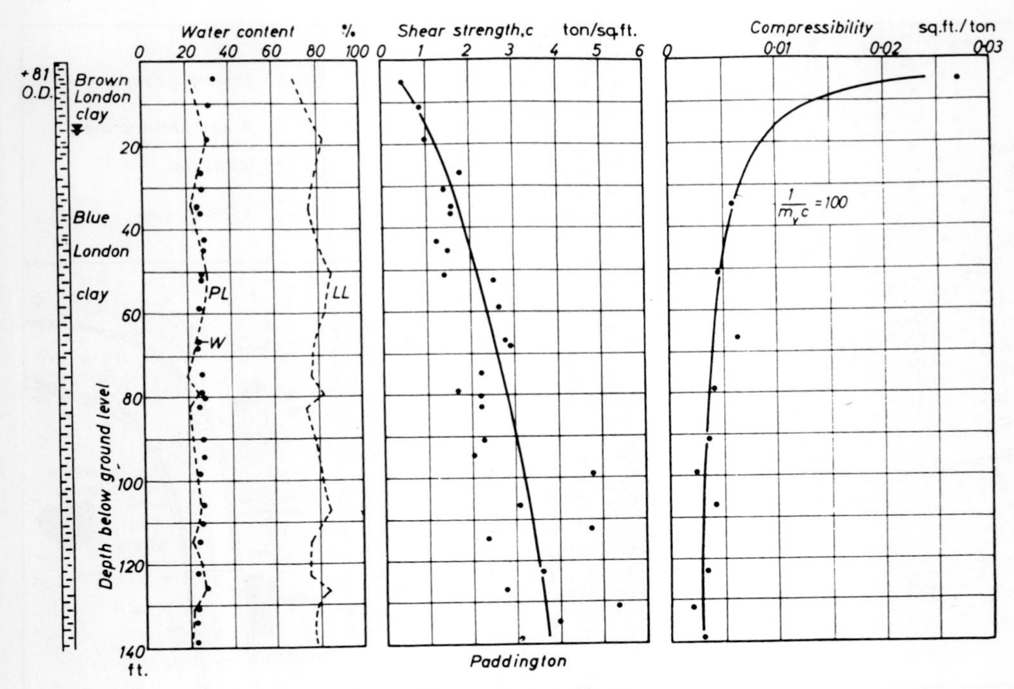


Test results at Paddington

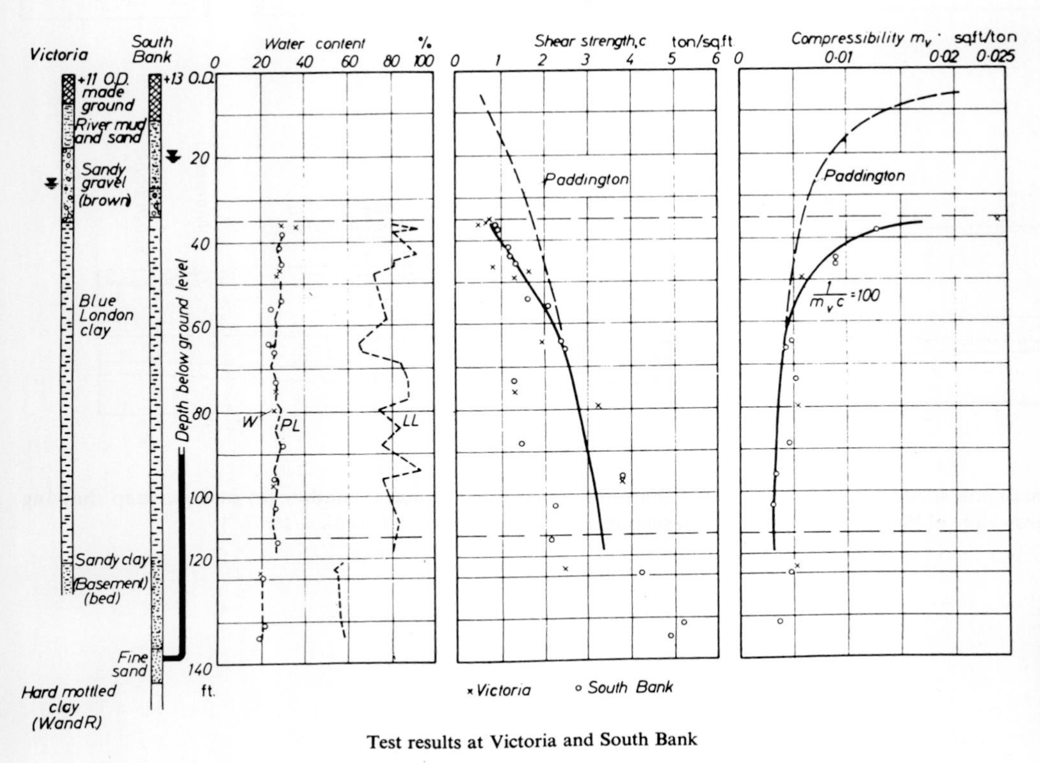


Test results at Victoria and South Bank

**FIG. 7.57 Characteristics of the London clay from the locations given in Fig. 7.56*a*.** [From *Skempton and Henkel (1957).*[26]]

ice leaving depressions (kettles), and by the very irregular surface of ground moraine (*Fig. 7.82*).

*Volcanic Activity*

Lakes form in extinct volcanic craters and calderas, and in basins formed by tectonic and volcanic activity combined (Mexico City basin).

*Landslides*

Natural dams are formed by landslides in valleys to create lakes.

*Solution Lakes*

Depressions resulting from the collapse of caverns in limestone form solution lakes.

*Floodplain Lakes*

Floodplain lakes form in cut off meanders (oxbows) or in floodplain depressions created by natural levees (the backswamp zone).

*Deflation Basins*

Formed by wind erosion, deflation basins are found in arid or formerly arid regions common to the Great Plains of the United States, northern Texas, Australia, and South Africa.

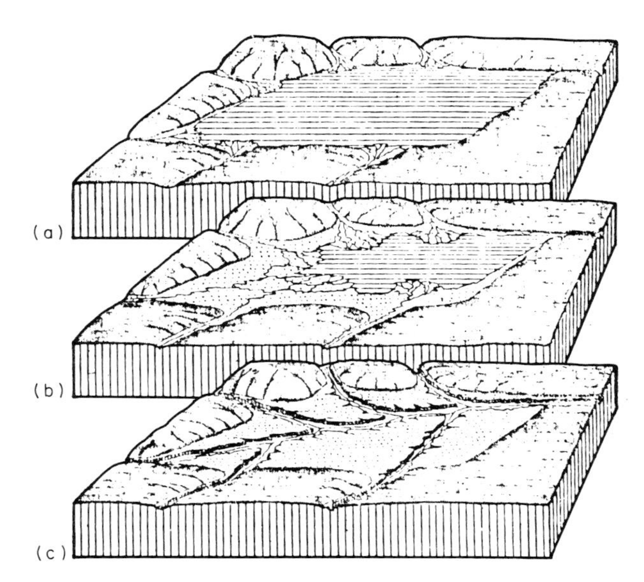


**FIG. 7.58 Life cycle of a typical lake: (*a*) stream system formed by gentle upwarp forming a shallow lake; (*b*) the streams entering the lake build deltas which enlarge and coalesce as the lake body is filled with fines; (*c*) the outfall channel slowly degrades its outlet, lowering the lake level until eventually only the lake bottom remains. Entrenching by the stream leaves terraces.** [*From Longwell et al. (1948).*[26a] *Reprinted with permission of John Wiley & Sons, Inc.*]

### Depositional Characteristics

*Sedimentation*

The life cycle of a typical lake is illustrated on Fig. 7.58.

Deposition includes a wide range of materials, the nature of which depends on the source materials, the velocity of the streams entering the lake, and the movement of the lake water. Streams enter the lake to form deltas; in large lakes currents carry materials away to form beaches and other shoreline deposits.

The finer sediments are carried out into the lake where they settle out in quiet deep water, accumulating as thick deposits of silts and clays. Since freshwater clays settle out slowly they tend to be laminated and well-stratified, and can be extremely weak. In relatively small lakes with an active outlet, the fines may be carried from the lake by the exiting stream.

As the lake reaches maturity and filling creates shallow areas with weak current and wave action, plants grow and the accumulation of organic material begins.

*Modern Lakes in Dry Climates*

In arid climates water enters the basin, but flow is inadequate to replenish the loss to evaporation, and the lake quickly dries forming a *playa*. As the water evaporates, salts are precipitated. The salt type reflects the rock type in the drainage area, as well as other factors, and can vary from sodium chloride (common salt) to sodium sulfate to sodium and potassium carbonates (alkali lakes) to borax [Hunt (1972)[27]]. The salts impart a light cementation to the soil particles which dissolves upon saturation, resulting in ground collapse and subsidence.

*Ancient Lakes in Dry Climates*

The remains of once-enormous Pleistocene lakes, shrunken by evaporation, are evident as

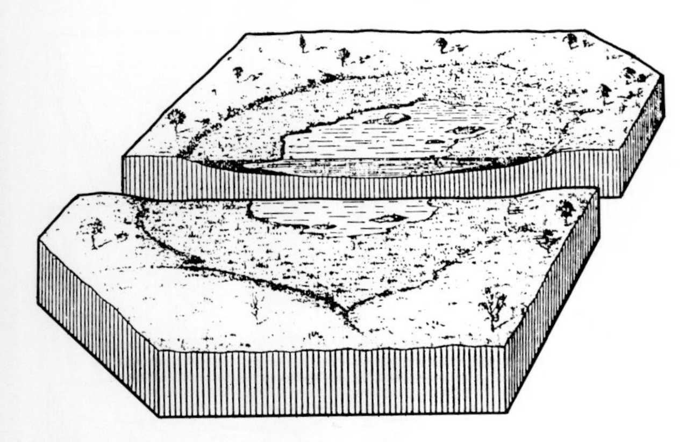

**FIG. 7.59 Gradual destruction of a lake by filling with marsh and plant growth.** [*From Longwell et al. (1948).*[26a] *Reprinted with permission of John Wiley & Sons, Inc.*]

salt flats extending over many hundreds of square kilometers of the Bolivian plateau, western Utah, and western Nevada, as well as other arid regions of the world.

The Great Salt Lake is all that remains of Lake Bonneville, which once covered 50,000 $km^2$ (20,000 $mi^2$) and was 300 m deep. Sediments from melting glaciers filled in the depression formed by crustal warping and faulting. Many depositional forms are evident in the former lake basin, including deltas and wave-cut terraces along former shore lines at several different elevations, gravel and sand bars, spits, and fine-grained lake bottom sediments. These latter materials, soft clays and salt strata, were the cause of numerous failures and the subsequent high costs during the construction of the 20-mile-long embankment for the Southern Pacific Railroad, which was only 4 m in height [A. Casagrande (1959)[28]]. C. B. Hunt (1972)[27] notes that shorelines about islands in the central part of the basin are about 50 m higher than along the eastern and western shores, which he attributes to crustal warping during rebound from unloading by evaporation of hundreds of meters of water.

### Swamps and Marshes

*Occurrence*

Swamps and marshes develop in humid regions over permanently saturated ground. A *swamp* has shrub or tree vegetation, and a *marsh* has grassy vegetation.

They are commonly associated with lakes and poorly drained terrain in any location and with coastal environments such as estuaries and tidal marshes. In upland areas they are common to coastal plains, peneplains, and glacial terrain, which all have characteristically irregular surfaces and numerous depressions.

*Formation*

The growth of a swamp from the gradual destruction of a lake is illustrated on Fig. 7.59. Vegetation growing from the shores toward the center is gradually filling in the lake.

Various types of plants are responsible for the accumulation of the organic material. As each type is adapted to a certain water depth, one succeeds another from the shore outward. Some plants float on the surface, others are rooted to the bottom. Floating plants form mats that live above and die below. These catch sediments and in the course of time may form a thick cover. Beneath the mat is water or a thick black sludge. In time the mat may support large plants and bushes, and eventually the basin becomes filled completely with semidecayed organic matter.

The lush growth, by accelerated transpiration, can dry what remains of the lake water, and swamp trees move onto the firmer ground. As the vegetable matter decomposes, peat deposits are formed and continue to build in thickness with time. Some that have been forming since the end of the Wisconsinin glaciation have reached thicknesses of 6 m.

*Diatomaceous Earth or Diatomite*

Diatomite, also referred to as diatomaceous earth, is an accumulation of diatoms, microscopic plants that secrete siliceous material in lakes or swamps. It has a very low unit weight, high porosity, and an absence of plasticity accounted for by the round shape of its hollow silica shells.

### Engineering Characteristics

*Granular Deposits*

Granular materials are deposited around the lake perimeter in the forms of beaches, deltas, and dunes. The coarser materials are borrow sources. Beach deposits may be moderately compact, but most deposits are in a loose state.

*Lake Body Soils*

Silts, clays, and organic materials generally compose the lake body soils. Freshwater clays evidence stratification; saltwater clays do not. In an existing lake the materials are soft, weak, and compressible.

Former lake beds are large flat areas which can be found covered with swamp or marsh vegetation, or tilled as farmland because of rich modern soils, or existing in remnants as level benches or terraces high above some river valley. Downcutting by river erosion as uplift occurs permits the lake bed soils to strengthen as internal drainage results in consolidation; as uplift continues and the permanent groundwater table lowers, evaporation causes prestress by desiccation. The prestress remains even if the lake bed is resubmerged.

MEXICO CITY CLAYS The famous clays of Mexico City, which extend to depths of 70 m, are volcanic materials washed down from the nearby mountains and deposited in ancient Lake Texcoco. The general geology, the ground subsidence phenomenon, and foundation problems associated with these clays are described in Art. 10.2.4. A log of a typical test boring is given as Fig. 7.60; included also are values for water contents, unconfined compressive strengths, effective pressures, and specific gravity. As shown on Fig. 7.60, these clays have very high void ratios, often ranging from 10 to 13, indicating their high compressibility. Time-rate consolidation curves show a high amount of secondary compression.

*Marsh and Swamp Soils*

Characteristics of marsh and swamp soils are variable; rootmats and peats can stand in vertical cuts although they are highly compressible, whereas organic silts flow as a fluid when they are saturated and unconfined. Their low strengths and high compressibility make them the poorest of foundation soils, and they tend to become weaker with time as decomposition continues. Usually highly acidic, they are very corrosive to construction materials.

### 7.4.6 MARINE ENVIRONMENT

#### General

*Origin*

Marine deposits originate from two general sources: (1) terrestrial sediments from rivers, glaciers, wind action, and slope failures along the shoreline and (2) marine deposition from organic and inorganic remains of dead marine life and by precipitation from oversaturated solutions.

*Deposition*

Sediments of terrestrial origin normally decrease in particle size and proportion of the total sediment with increasing distance from the land, whereas the marine contribution increases with distance from the land. The selective effect of currents normally produces well-sorted (uniformly graded) formations.

The typical distribution off the east coast of the United States is as follows:

- To depths of about 200 m, in which sea currents are active, deposits include strata of sand, silt, and clay; depositional characteristics depend on geologic source, coastal configuration, and the proximity to rivers.
- Between 200- and 1000-m depths, silts and clays predominate.
- Beyond 1000-m depths are found brown clays of terrestrial origin, calcareous *ooze* and siliceous *ooze* (sediments with more than 30% material of biotic origin).

#### Marine Sands

Of major significance to offshore engineering projects is the composition of marine sands. These are normally considered to be composed of quartz grains which are hard and virtually indestructible, although the deposit may be compressible. Strength is a function of intergranular friction.

In the warm seas of the middle latitudes, however, sands are often composed of calcium or other carbonates with soft grains that are weak and readily crushable. These sands can include ooliths (rounded and highly polished particles of calcium carbonate in the medium to fine sizes) formed by chemical precipitation in highly agi-

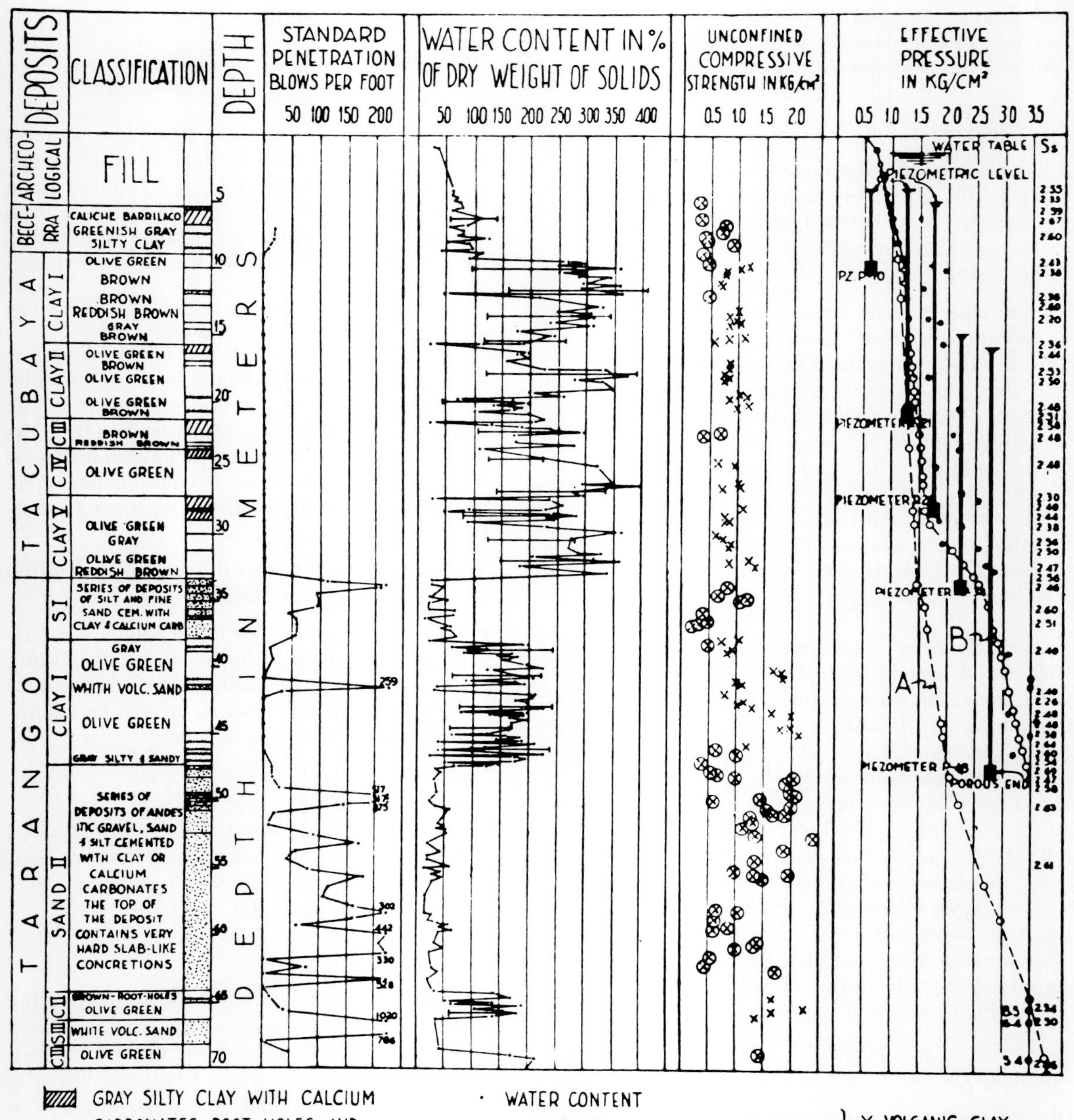


**FIG. 7.60 Soil section and index properties at the site of the Tower Latino Americana, Mexico City.** [*From Zeevaert (1957).*[31]]

tated waters, oblong lumps of clay-size particles of calcite (probably originating as fecal pellets), and sands composed almost entirely of fossil fragments (coral, shells, etc.) [McClelland (1972)[30]]. These "sands" are often found in layers with various degrees of cementation. The general distribution of these materials on continental shelves worldwide is given on Fig. 7.61.

For the design of foundations for offshore structures in calcareous sands, the frictional resistance common for a quartz sand of a particular gradation is reduced by empirical factors which are based on the coefficient of lateral earth pressure [McClelland (1972),[30] Agarwal et al. (1977)[32]].

### Marine Clays

Recent deposits of marine clays are normally consolidated, but at depths below the seafloor of about 100 m or more they are often found in a stiff or greater state of consistency. Typical properties from several locations are given on Table 3.30. Stiff to hard clays can also exist in areas formerly above sea level which subsequently submerged. Heavily overconsolidated clays (stiff to hard consistency), for example, thought to be possibly preglacial in origin, have been encountered in the North Sea [Eide (1974)[33]].

Because clays deposited in seawater tend to flocculate and settle quickly to the bottom, they are generally devoid of laminations and stratifications common to freshwater clays. Marine clays of glacial origin are discussed in Art. 7.6.6.

### Classification of Carbonate Sediments

A system of classification of Middle Eastern sedimentary deposits has been proposed by Clark

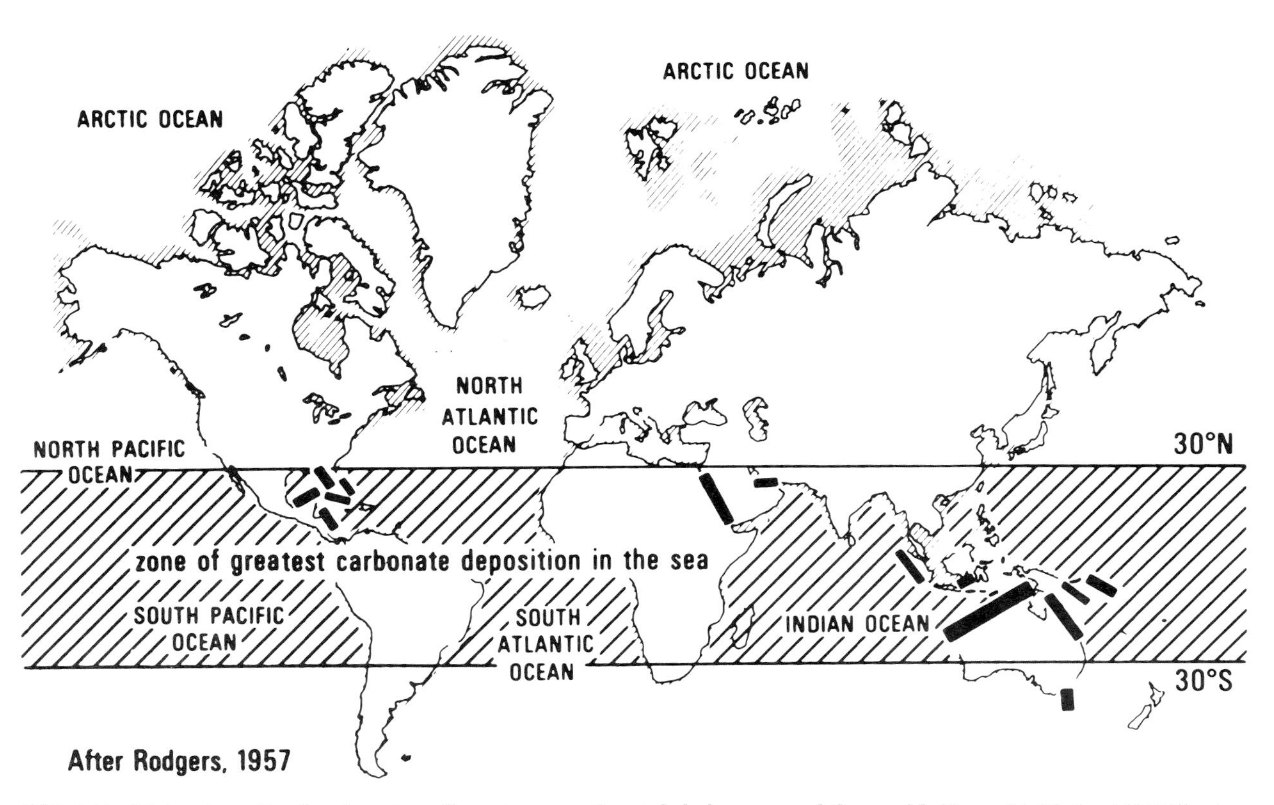


**FIG. 7.61 Major deposits of carbonate sediments on continental shelves around the world.** [*From McClelland (1972)*[30]*; after Rodgers (1957)*[31]*.*]

**TABLE 7.10**
**CLASSIFICATION SYSTEM FOR MIDDLE EASTERN OFFSHORE SEDIMENTARY DEPOSITS***

| Degree of induration | Increasing grain size of particulate deposits: < 0.002 mm | 0.002 mm – 0.06 mm | 0.06 mm – 2 mm | 2 mm – 60 mm | Total carbonate content, % |
|---|---|---|---|---|---|
| Non-indurated | CARBONATE MUD | CARBONATE SILT | CARBONATE SAND | CARBONATE GRAVEL | 90% |
| | Clayey CARBONATE MUD | Siliceous CARBONATE SILT | Siliceous CARBONATE SAND | Mixed carbonate and noncarbonate gravel | 50% |
| | Calcareous CLAY | Calcareous SILT | Calcareous silica SAND | | 10% |
| | CLAY | SILT | Silica SAND | GRAVEL | |
| Slightly indurated | CALCILUTITE (carbonate claystone) | CALCISILTITE (carbonate siltstone) | CALCARENITE (carbonate sandstone) | CALCIRUDITE (carbonate conglomerate or breccia) | 90% |
| | Clayey CALCILUTITE | Siliceous CALCISILTITE | Siliceous CALCARENITE | Conglomeratic CALCIRUDITE | 50% |
| | Calcareous CLAYSTONE | Calcareous SILTSTONE | Calcareous SANDSTONE | Calcareous CONGLOMERATE | 10% |
| | CLAYSTONE | SILTSTONE | SANDSTONE | CONGLOMERATE OR BRECCIA | |
| Moderately indurated | Fine-grained LIMESTONE | | Detrital LIMESTONE | CONGLOMERATE LIMESTONE | 90% |
| | Fine-grained agrillaceous LIMESTONE | Fine-grained siliceous LIMESTONE | Siliceous detrital LIMESTONE | Conglomeratic LIMESTONE | 50% |
| | Calcareous CLAYSTONE | Calcareous SILTSTONE | Calcareous SANDSTONE | Calcareous CONGLOMERATE | 10% |
| | CLAYSTONE | SILTSTONE | SANDSTONE | CONGLOMERATE OR BRECCIA | |
| Highly indurated | CRYSTALLINE LIMESTONE OR MARBLE | | | | 50% |
| | Conventional metamorphic nomenclature applies in this section | | | | |

*From Clark and Walker (1977).[34]

and Walker (1977).[34] Given on Table 7.10, it has been applied by practitioners to the worldwide carbonate belt.

## 7.5 EOLIAN DEPOSITS

### 7.5.1 EOLIAN PROCESSES

#### General

The geologic processes of wind include transportation, erosion (deflation and abrasion), and deposition (sand dunes and sheets, loess, and volcanic clays).

#### Transportation

Winds with velocities of 40 to 50 km/hr can cause medium sand grains to go into suspension; velocities of only 15 km/hr can cause fine sand to move. Grains of fine sand or silt move in suspension, by saltation (hopping along the surface), or by rolling, depending upon wind velocity.

#### Erosion

*Deflation*

The removal of loose particles from an area leaving a denuded surface covered with coarse gravel and cobbles (*lag gravels*) is referred to as *deflation*. These deposits are characteristic of many true deserts and can cover many hundreds of square kilometers as they do in the Sahara and the Middle East. Deflation also excavates large basinlike depressions, some of which are immense, such as the Qattara in Egypt. The bottoms of these depressions are at or near the groundwater table [Holmes (1965)[35]]; the wind cannot excavate lower (deflate) because of the binding action of soil-moisture capillarity forces on the soil particles.

*Abrasion*

Abrasion is a significant cause of erosion in an arid climate. Most of the abrasive particles of quartz sand are carried along at a height of 30 to 60 cm above the surface; therefore, erosion is frequently in the form of undercutting. The results are often evident as balancing rocks (Fig. 7.62). *Ventifacts* are pebbles or cobbles that have had facets eroded into their sides by wind abrasion while lying on the desert floor.

**FIG. 7.62 Wind erosion formed the balancing rock of sandstone (Garden of the Gods, Colorado Springs, Colorado).**

### 7.5.2 DUNES AND SAND SHEETS

#### Occurrence

Dunes and sand sheets occur in arid regions, or along the shores of oceans or large lakes.

#### Dunes

Dunes are depositional features of windblown sand and include any mound or ridge of sand with a crest or definite summit. Deposition begins when windblown sand encounters a surface irregularity or vegetation. As the sand

**FIG. 7.63** Dunes over 100 m in height (southwest coast of France).

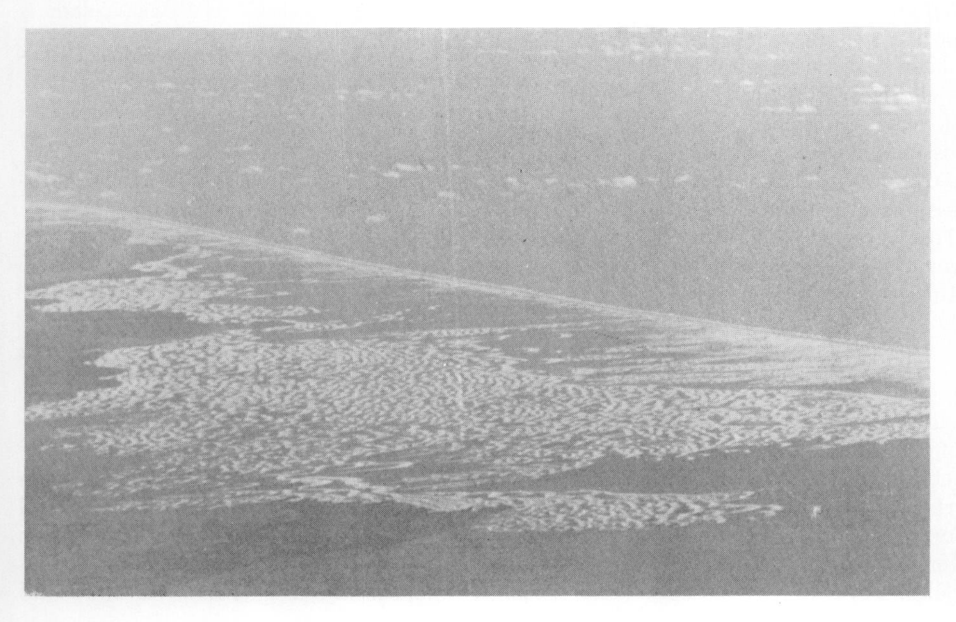

**FIG. 7.64** Transverse dune migration inland for about 25 km (northeast coast of Brazil).

builds a mound, a long windward slope is formed; the sand blows over the crest to come to rest on the leeward side at its angle of repose (30 to 35°). As the base continues to spread, the dune can be mounded to heights of over 100 m, creating sand hills such as shown in Fig. 7.63.

Unless dunes become stabilized by vegetation, they continue to migrate inland, moving very significantly during storms. Dune migration covering a distance of about 25 km is shown on Fig. 7.64.

*Classes of Dunes*

Several classes of dunes are illustrated on Fig. 7.65.

*Transverse dunes* extend at right angles to the wind direction; they are common to the leeward side of beaches (*see Fig. 7.64*).

*Longitudinal dunes* form ridges elongated parallel to the wind direction, and are thought to be the result of crosswinds in a desert environment [Thornbury (1969)[37]].

*Barchans*, or crescent-shaped dunes, are common to desert environments (Fig. 7.66).

*U-shaped dunes* or parabolic dunes are stabilized or partially stabilized. They are typical of the eastern shore of Lake Michigan (Fig. 7.67).

### Sand Sheets

Sand sheets are typical of desert environments and can cover enormous areas. They vary in landform from flat to undulating to complex and can include various forms of barchan, logitudinal, or transverse dunes, as shown on Fig. 7.66, which can extend for hundreds of square kilometers. In the United States, much of Nebraska is covered with ancient sand sheets of postglacial origin (Figs. 7.68 and 7.69).

### Dune Activity

*Active dunes* are typical of desert regions. If winds blow at velocities of 40 to 50 km/hr for long enough intervals, small barchans (to 6 m height) can move at the rate of 20 m per year, and the larger barchans (12 m in height) can move about 10 m per year.

*Inactive dunes* are typical of moist climates and are stabilized by natural vegetation. In the United States, very large areas of stabilized dunes are common to Nebraska. Figure 7.68 shows the very irregularly eroded features of these relict dunes.

*Semistabilized dunes* are common along coastlines in moist climates.

### Dune Stabilization

Stabilization may be achieved by planting tough binding grasses such as bent or marram in climates with adequate moisture. The harsh tufts prevent the movement of the underlying sand, trap wind-blown particles, and continue to grow upward through the accumulating sand.

Migration may also be prevented by light open fences, or by treating the windward side of dunes with a retardant or agent such as calcium chloride, lignin sulfite, or petroleum derivatives.

### Engineering Characteristics

Wind deposited sands are almost always in a loose state of density which results from their uniform gradation (uniformity coefficient about 1.0 to 2.0), grain diameters (0.15 to 0.30 mm), and relatively gentle modes of deposition.

## 7.5.3 LOESS

### Origin and Distribution

Loess originates as eolian deposits in the silt sizes, transported in suspension by air currents. Ancient deposits were derived from silt beds of glacial outwash, or shallow streams when the

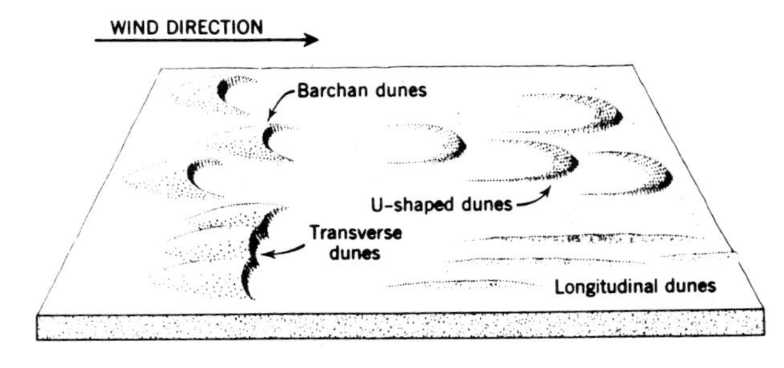


**FIG. 7.65 Various classes of dunes.** [*From Zumberge and Nelson (1972).*[36] *Reprinted with permission of John Wiley & Sons, Inc.*]

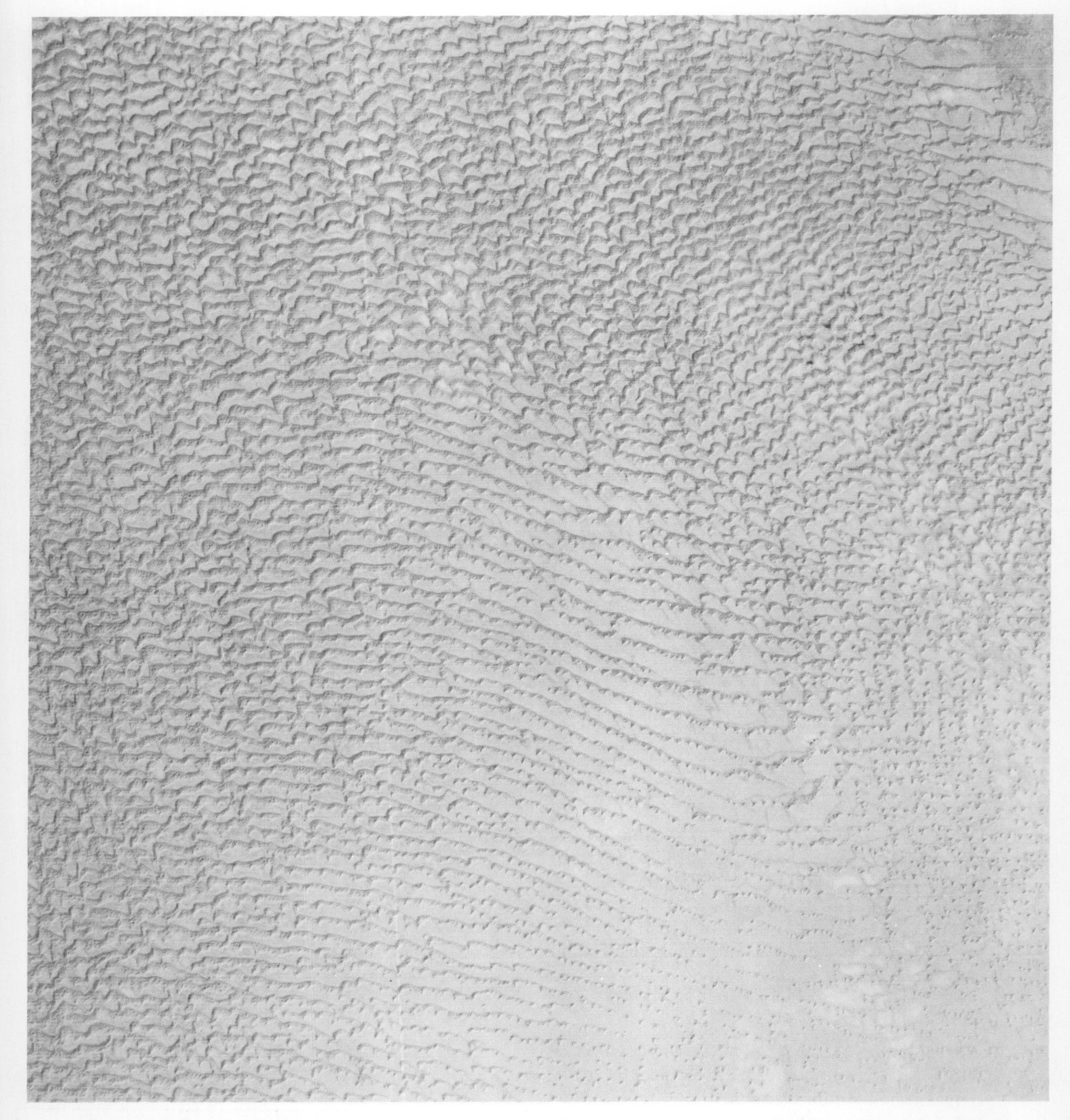

**FIG. 7.66 Portion of ERTS 1 image showing crescentric dunes of a megabarchan desert in Saudi Arabia over a distance of 183 km.** *(Original image by NASA, reproduction by USGS, EROS Data Center.)*

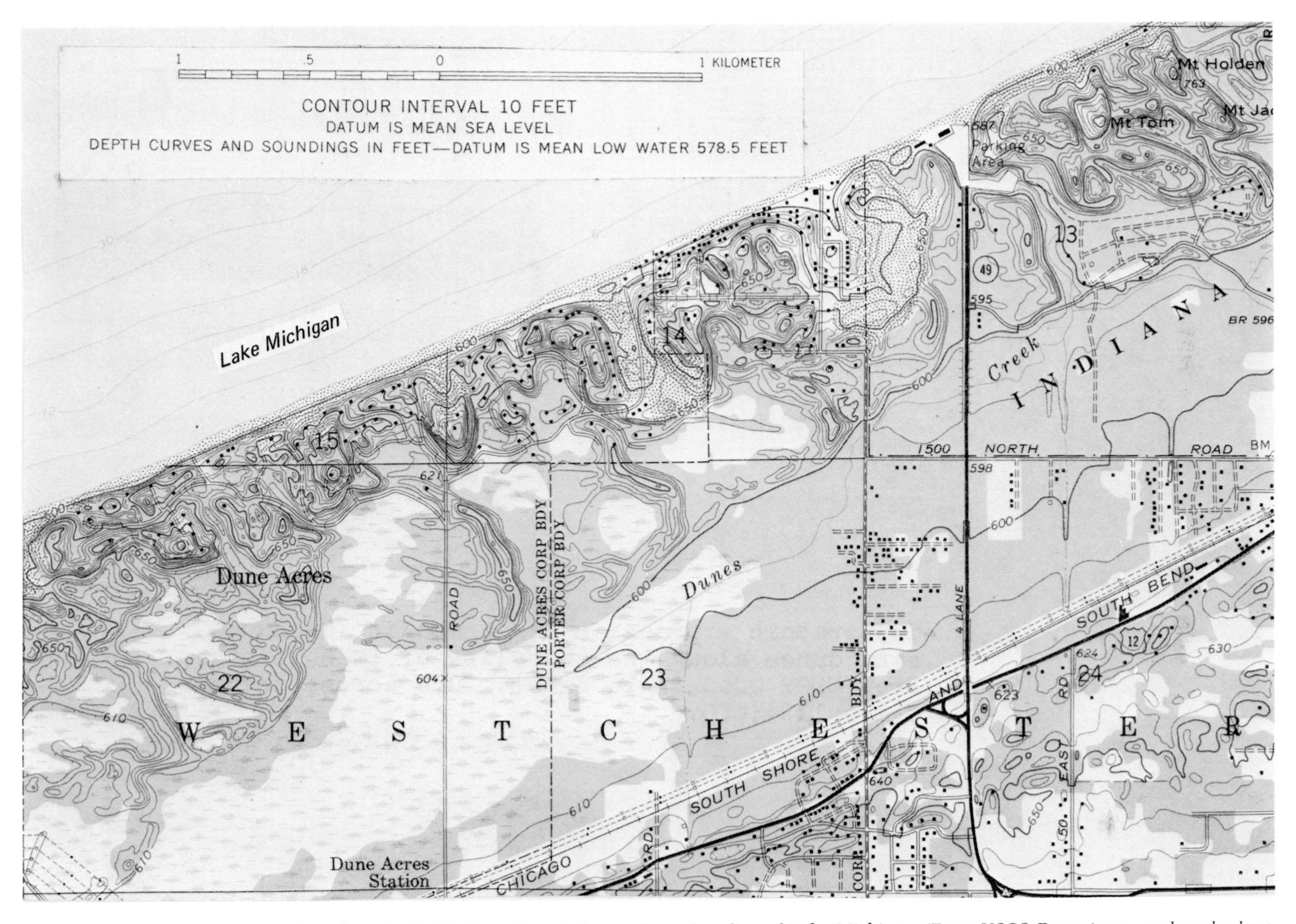


**FIG. 7.67 Topographic expression of parabolic (U-shaped) sand dunes along shoreline of Lake Michigan.** *(From USGS Dune Acres quadrangle sheet, Indiana. Scale 1:24,000.)*

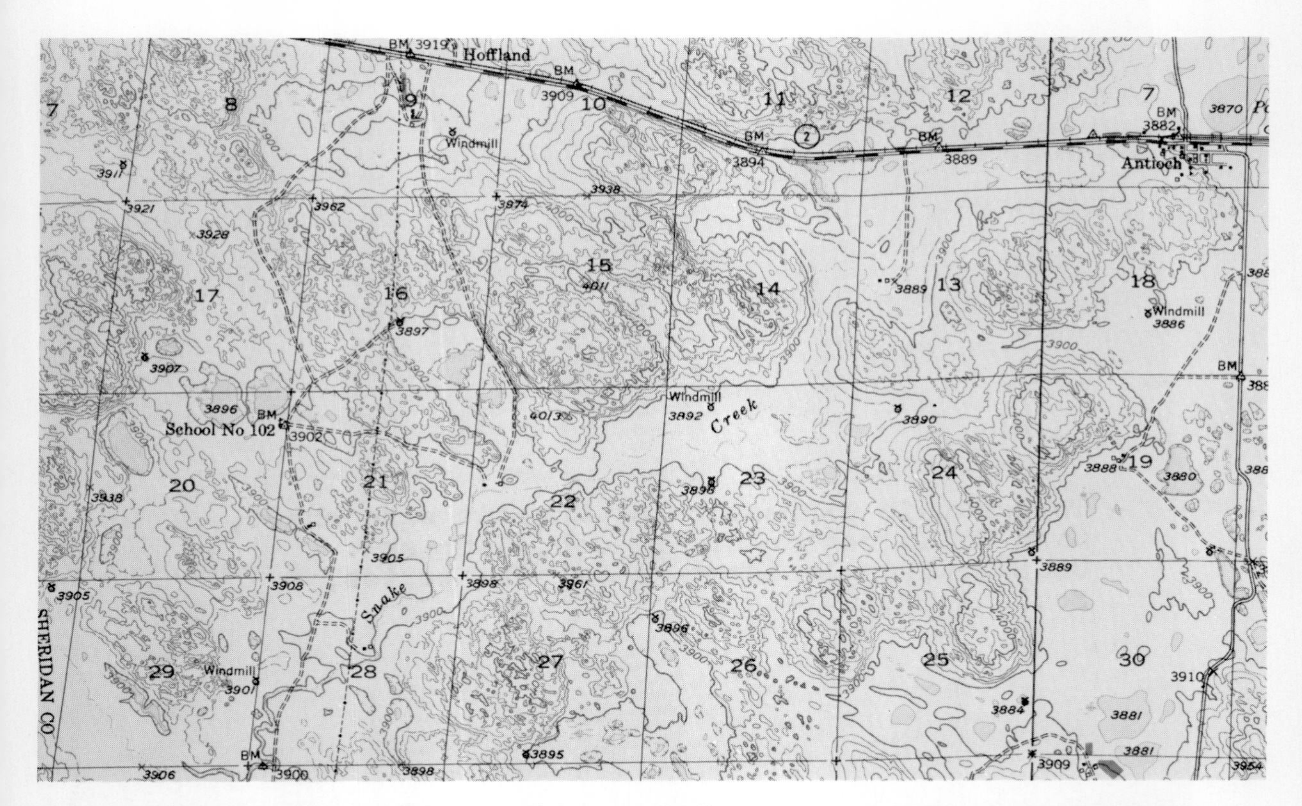


**FIG. 7.68 Topographic expression of ancient sand sheet. (Sand Hills region of Antioch, Nebraska. Scale 1:62,500.)**

last glaciers were beginning their retreat. Modern deposits originate in desert environments, or are associated with glacial rivers.

The most significant occurrences are those associated with the retreat of the Wisconsin glaciers. In the United States, loess covers large areas of Nebraska, Kansas, Iowa, Missouri, Illinois, and Indiana, as well as other states in the Mississippi and Missouri River valleys, Idaho, and the Columbia River plateau of Washington, as shown on Fig. 7.69. In Europe, loess deposits occur along the Rhine, Rhone, and Danube valleys and throughout the Ukraine in Russia as shown on Fig. 7.70.

Substantial deposits also cover the plains regions of Argentina, Uruguay, and central China. The effects of the disastrous earthquake of 1556 in the loess region of the Shensi province of China are described in Art. 11.3.4. Along the Delta River in Alaska today, winds carry silt grains from the floodplain during dry spells, depositing them as loess. Silts of similar origin occur around Fairbanks as a blanket ranging from 3 to 30 m thick on hilltops to more than 100 m thick in valleys [Zumberge and Nelson (1972)[36]].

### Characteristics

*General*

Regardless of location, loess is typically buff- to yellow-colored, lightly cemented (calcareous or clay cements), very fine-grained, permeable (particularly in the vertical direction), and devoid of stratification.

*Formation*

The windborne dust and silt are dropped down from the air and retained by the protective grip of the grasses of the steppe. Each spring the grass grows a little higher on any material collected during the previous year, leaving behind a ramifying system of withered roots. Over immense areas, many hundreds of feet accumulate, burying entire landscapes except for higher peaks which project above the loess blanket. Although loess is friable and porous, the

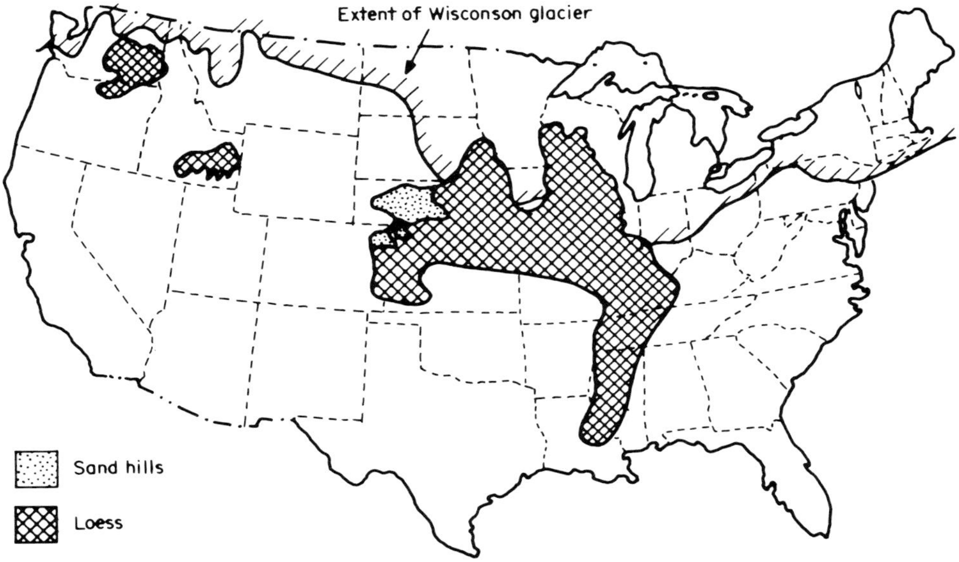

Extent of Wisconson glacier

Sand hills

Loess

**FIG. 7.69 Approximate distribution of loess in the United States. Deposits are generally thicker and coarser near their sources in the floodplains of the major Pleistocene rivers (Missouri, North Platte, and Mississippi), becoming thinner, finer, and discontinuous with distance from the source.**

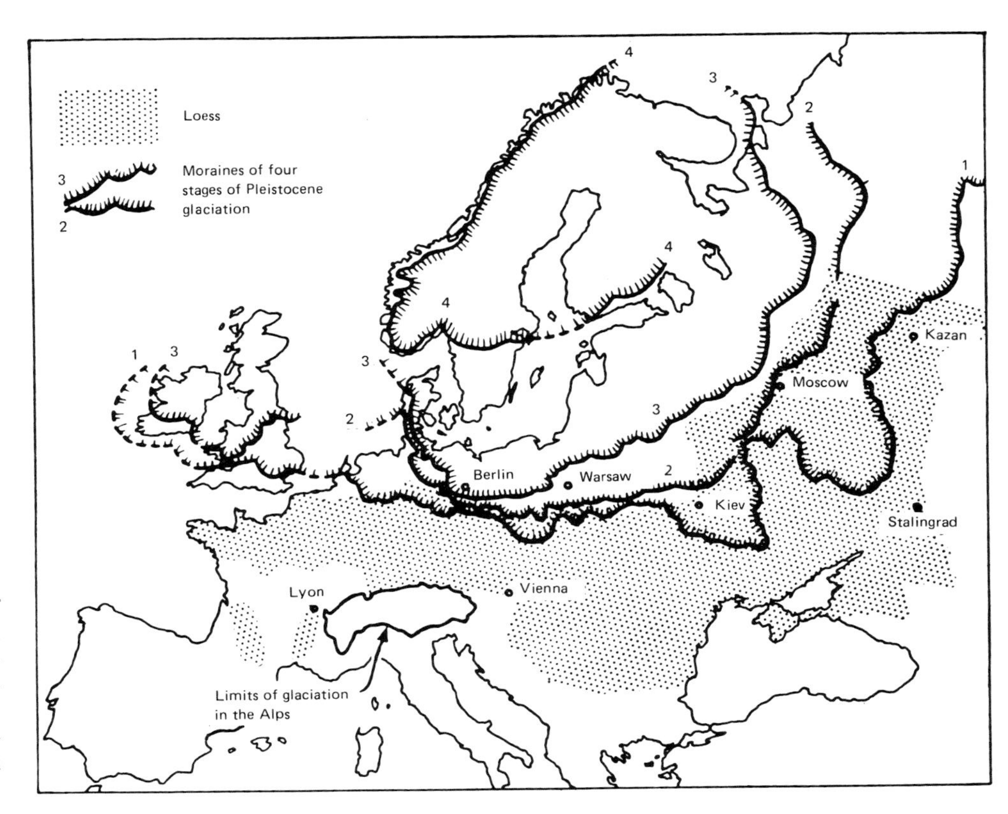


**FIG. 7.70 Extent of the terminal moraines of four glacial advances in Europe and the deposition of loess.** [*From Gilluly et al. (1959).*[38] © *W. H. Freeman and Company. Reprinted with permission.*]

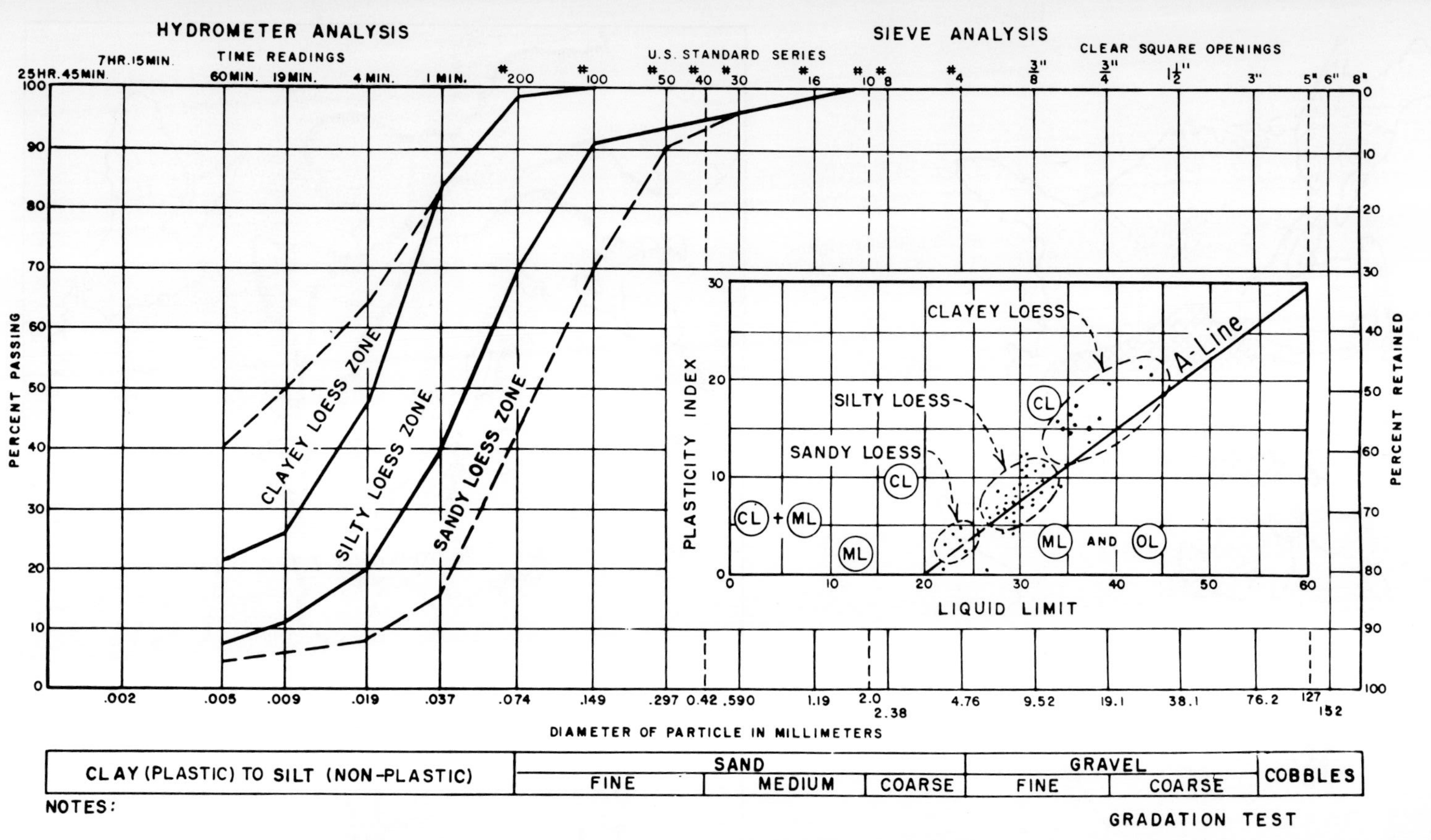


**FIG. 7.71 Trends of gradation and plasticity for loess from Kansas-Nebraska.** [*From Gibbs et al. (1960).*[39]]

successive generations of roots, represented by narrow tubes partly occupied by calcium carbonate, make it sufficiently coherent to stand in vertical walls. [The above description is after Holmes (1964).[35]]

*Structure*

The loose arrangement of the silt particles with numerous voids and rootlike channels can easily be seen by photomicrograph observation of undisturbed loess specimens [Gibbs et al. (1960)[39]]. Petrographic methods determine that a majority of the grains of loess from Kansas-Nebraska are coated with very fine films of montmorillonite clay, which is responsible for the binder in the structure. Upon wetting, the clay bond is readily loosened, causing great loss of strength.

*Collapse Phenomenon*

Loess is one of the deposits referred to as collapsing soils (*see Art. 10.5*). When submerged in water the physicochemical structure may be destroyed and the soil mass may immediately densify causing ground subsidence, often in the order of 30 to 60 cm. The phenomenon does not appear to occur naturally, since the peculiar structure provides loess with a high rate of vertical percolation adequate for rainfall, and a blanket of topsoil and vegetation affords protection against wetting.

*Terrain Features*

The typical landform that develops in thick loess deposits is illustrated on Fig. 7.23. Distinguishing characteristics are the crests of the hills at a uniform elevation (remnants of the old loess plain); the drainage pattern, which is pinnate dendritic (Table 7.4); and the eroded slopes on both sides of the ridges, which are uniform. Concentrated runoff on flat slopes does not cause severe erosion, but when erosion does begin, it proceeds rapidly, and because of the light cementation, vertical slopes develop in streams and gullies. This is particularly true of moist climates. Where loess is thin, erosion proceeds, creating the typical pinnate dendritic pattern, until the underlying formation, usually more resistant, is exposed. Thereafter, erosion patterns reflect the characteristics of the underlying materials.

**Engineering Properties**

*Index Properties*

Loess deposits have distinct physical features that are strikingly similar from location to location. Three types are generally identified: clayey, silty, and sandy. The trends in gradation and plasticity characteristics of loess from the Kansas-Nebraska area are given on Fig. 7.71. Clayey loess is usually in the ML-CL range and porosity of all loess is high: from 50 to 60%.

Loesses are generally uniform in texture, consisting of 50 to 90% silt-size particles exhibiting plasticity, but the fineness increases and the thickness decreases in the downwind direction from the source. In the United States the deposits are thicker and more clayey along the Missouri and Mississippi Valleys, and thinner and more sandy to the west in the high plains area to where they grade into the sand hills of Nebraska.

*Compressibility*

Compressibility is most significantly related to collapse upon saturation, the potential for which is a function of the natural density and moisture content of the deposit as shown on Fig. 7.72.

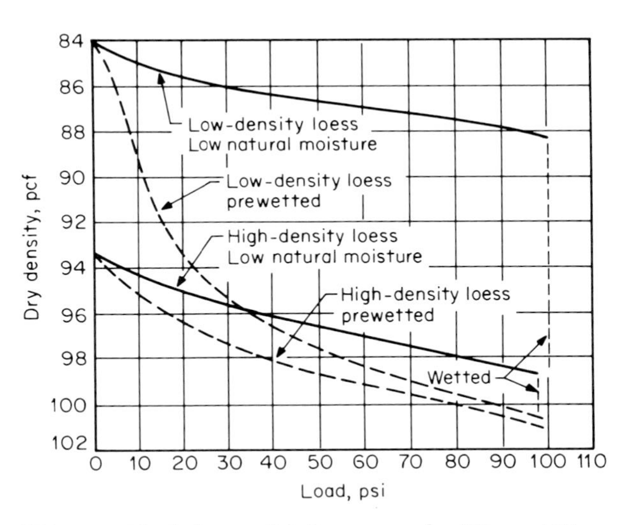


**FIG. 7.72 Typical consolidation curves for Missouri River basin loess.** [*From Clevenger (1958).*[40]]

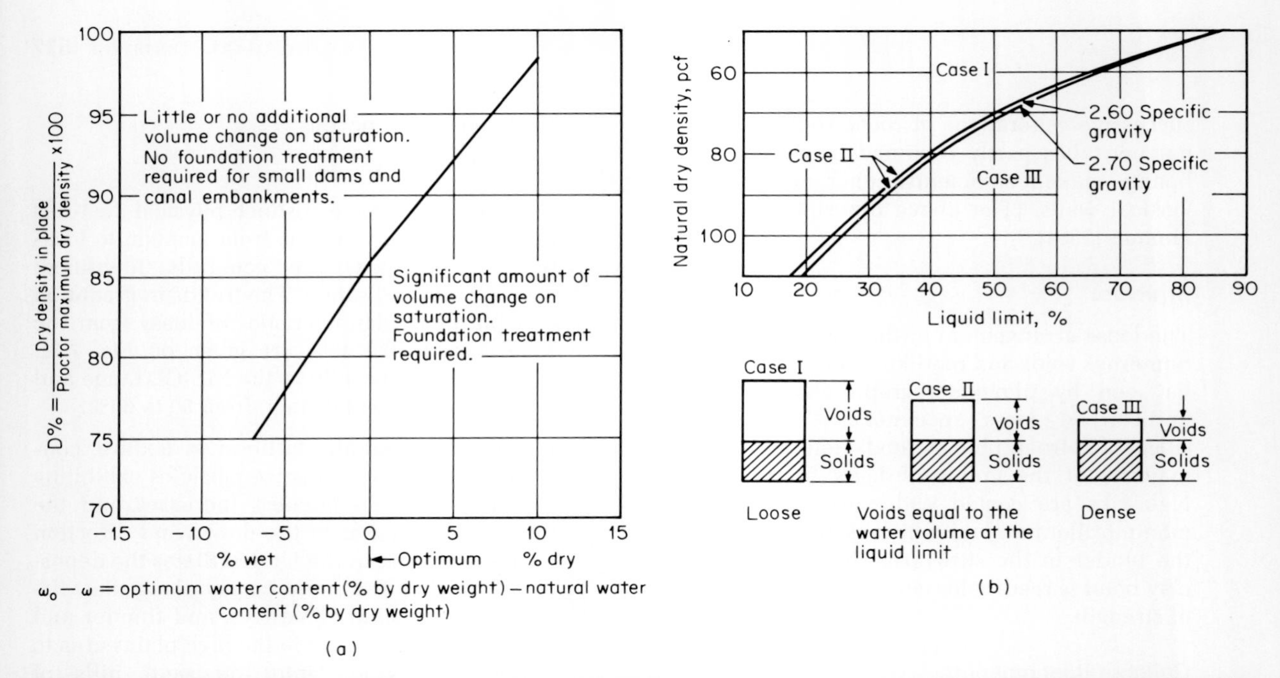


**FIG. 7.73 Criteria for treatment of relatively dry fine-grained foundations: (*a*) "D ratio," the ratio of natural (in place) dry density to Proctor maximum dry density, and $\omega_o - \omega$, optimum water content minus natural water content and (*b*) natural dry density and liquid limit.** [*From USBR (1974).*[41]]

The potential for settlement has been related to the natural dry density and moisture content in terms of the Proctor density and moisture and to the natural dry density in relation to the liquid limit [USBR (1974)[41]], as given on Fig. 7.73. Case I (Fig. 7.73*b*) indicates that a low natural density is associated with void ratios larger than required to contain the liquid limit moisture. Thus, the soil if wetted to saturation can exist at a consistency wet enough to permit settlement. Case III indicates that the natural densities are high enough that the void spaces are too small to contain the liquid limit moisture content and the soil will not collapse upon saturation, but will reach a plastic state in which there will be always particle-to-particle strength. Because of the uniformity of the loessial soils in Kansas and Nebraska (LL = 30 to 40%), criteria for settlement upon saturation vs. natural densities and surface loadings have been developed as given on Table 7.11.

*Shear Strength*

At natural moisture contents loess has relatively high strength, as well as low compressibility, because of its slight cementation. Some typical strength envelopes are given on Fig. 7.74; it is seen that wetting has a severe effect.

Unconfined compressive strength in the dry condition may be several kilograms per square centimeter. At natural moisture contents (usually less than 10%) loess has an apparent cohesion which may be as high as 15 psi (1 kg/cm$^2$) and generally ranges from 5 to 10 psi (0.3 to 0.6 kg/cm$^2$) for the Kansas-Nebraska loess; tan $\phi$ ranges from about 0.60 to 0.65. Effective stress parameters of these magnitudes provide the strength which permits loess to stand vertically in slopes 50 to 80 ft (16 to 24 m) high, even with its characteristic low densities [Gibbs et al. (1960)[39]]. When loess is wetted, cohesion is reduced to less than 1 psi (0.07 kg/cm$^2$) and even for initially dense loess becomes less than 4 psi (0.28 kg/cm$^2$). The breakdown can occur at 20 to 25% moisture content which is in the order of 50 to 60% saturation.

*Compacted Fills*

When loess is dry (typical natural conditions)

**TABLE 7.11**
**SETTLEMENT UPON SATURATION VS. NATURAL DENSITY: LOESSIAL SOILS FROM KANSAS AND NEBRASKA***

| | Density | | | |
|---|---|---|---|---|
| $D_R$ | pcf | g/cm³ | Settlement potential | Surface loading |
| Loose | <80 | <1.28 | Highly susceptible | Little or none |
| Medium dense | 80–90 | 1.28–1.44 | Moderately susceptible | Loaded |
| Dense | >90 | >1.44 | Slight, provides capable support | Ordinary structures |

*From USBR (1974).[41]

NOTES: 1. For earth dams and high canal embankments, $\gamma$ = 85 pcf (1.36 g/cm³) has been used as the division between high-density loess requiring no foundation treatment, and low-density loess requiring treatment.
2. Moisture contents above 20% will generally result in full settlement under load.

compaction is virtually impossible. If placed in an embankment in an excessively wet condition, it can become "quick," suddenly losing strength and flowing.

At proper moisture content loess makes suitable compacted embankment fill, but it must be protected against piping erosion, considered to be one of the causes of the Teton Dam failure (*see Art. 8.3.4*). A large shrinkage factor must be used in estimating earthwork and a thick vegetative cover must be provided for erosion protection.

*Site Preparation*

Stripping the natural vegetation leaves loess vulnerable to rainfall saturation and possible ground collapse. Site grading and drainage require careful planning to avoid the ponding of water, and utilities must be constructed so as to prevent leaks.

## 7.5.4 VOLCANIC CLAYS

### Origin

Volcanic ash and dust are thrown into the atmosphere during volcanic eruptions and can be carried hundreds of kilometers to the leeward of the volcano. The eruptions of recent history are of too short a duration to expel substantial quantities of dust into the air and only very thin deposits settle to the earth, except close to the source. The ashfall from Mount St. Helens during the eruption of May 18, 1980 was reported in *National Geographic*, January 1981, to range from 70 mm near the volcano to 2 to 10 mm about 600 km distant.

In older geologic times, however, eruptions of long duration threw into the atmosphere vast amounts of ash and dust which came to rest as blankets of substantial thickness. These deposits

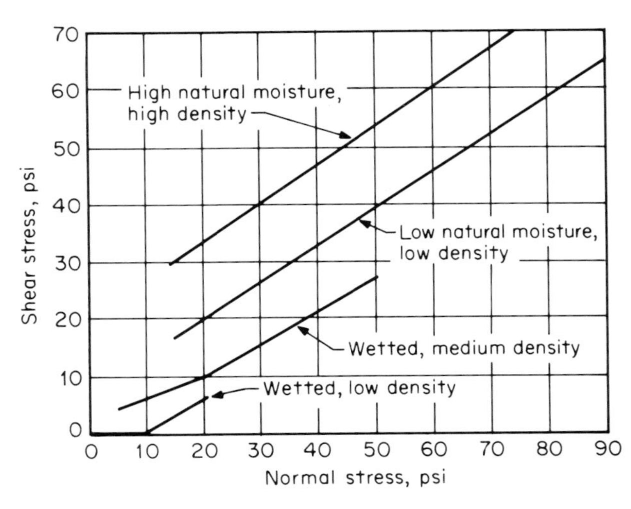


**FIG. 7.74 Typical shear envelopes for loess from Missouri River basin.** [*From Clevenger (1958).*[40]]

were often altered by weathering processes into montmorillonite clay, one form of which is *bentonite*.

### Distribution

Bentonite is found in most states west of the Mississippi, as well as Tennessee, Kentucky, and Alabama. On the island of Barbados, clays thought to be of volcanic origin cover the surface to depths of a meter or more. The extensive volcanic clays washed from the mountains into the basin of Mexico City are discussed in Art. 7.4.5.

## 7.6 GLACIAL DEPOSITS

### 7.6.1 GLACIAL ACTIVITY

#### General

Glaciers are masses of ice, often containing rock debris, flowing under the force of gravity. During long, cold moist periods, vast quantities of snow accumulate and change to ice. Gravity acting on the mass causes it to undergo plastic flow. The tremendous force of the moving glacier causes changes in the landscape over which it passes, and leaves many unique forms of deposition.

#### Classes of Glaciers

*Mountain or Valley Glaciers*, such as the one shown on Fig. 7.75, are common today in most high-mountain regions of the world. The landforms remaining from the erosive action of mountain glaciers are illustrated on Fig. 7.76. The glacier gouges out a *U-shaped valley* which has tributaries, less deeply eroded than the main valley, termed *hanging valleys*. The headward reaches of the valleys end in *cirques* which often contain lakes. *Arêtes* are steep-sided divides separating valleys. As the glacier melts and recedes, a series of moraine-dammed lakes is left behind. In the photo a block of ice has melted to form a *kettle*. The fjords of Norway are drowned valleys eroded by mountain glaciers when the land was above water.

*Piedmont glaciers* result from several valley glaciers coalescing into a single broad mass.

*Continental glaciers*, also termed *ice sheets* or *ice caps*, overspread enormous land masses. It is their deposits that are most significant from an engineering viewpoint. Only two examples exist today: Greenland and Antarctica.

#### Pleistocene Glaciation

*General*

The continental glaciation of the Pleistocene epoch is the most important from an engineering point of view. It sculptured the land to its present form and left significant soil deposits over large areas of the northern United States and Europe. At least four times during the past three million years of the Quaternary period, glaciers expanded to cover about 30% of the earth's land area. About 10,000 to 15,000 years ago they began their latest retreat to their present extent covering less than 10%. The extent of Pleistocene glaciation in the United States is given on Fig. 7.77 and in Europe is given on Fig. 7.70.

The effect on the land was great. The Great Lakes, as well as the Finger Lakes of New York State, are thought to be primarily the result of glacial excavation. The mountain ranges of the Northeast have been reduced and rounded by abrasion and many peaks show striations on their rocky summits. The Ohio and Missouri Rivers flow near the southernmost extent of glaciation and are called ice-marginal rivers. The grasslands of the Great Plains are the result of the rich deposits of loess, picked up by the winds from the outwash plains and redeposited to blanket and level out the landform.

*Erosion*

During long, cold, moist periods of the Pleistocene, the snow and ice accumulated to tremendous thicknesses, estimated to be over 1500 m in the New England states. As the huge mass grew, it flowed as a rheological solid, acquiring a load of soil and rock debris. Loose material beneath the advancing ice was engulfed and picked up, and blocks of protruding jointed rock were plucked from the downstream side of hills to be carried as boulders, often deposited hundreds of kilometers from their source over entirely different rock types (hence the term *glacial erratics*).

As the debris accumulated in the lower portions of the glacier it became a giant rasp, eroding and smoothing the surface. The abrasive action

FIG. 7.75 Receding mountain glacier (Andes Mountains, Peru).

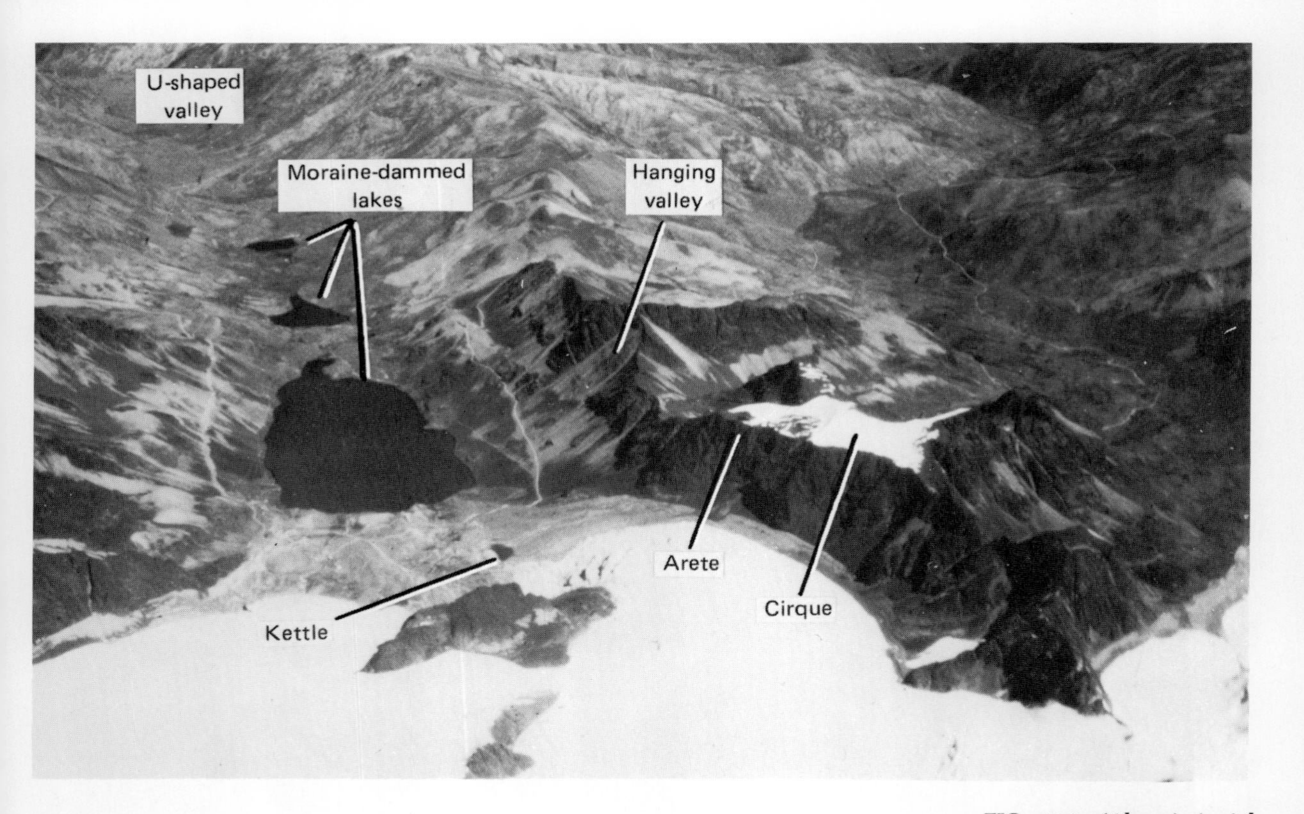


**FIG. 7.76** *(Above)* Aerial oblique showing the many landforms resulting from the activities of a mountain glacier in the Andes Mountains of Bolivia.

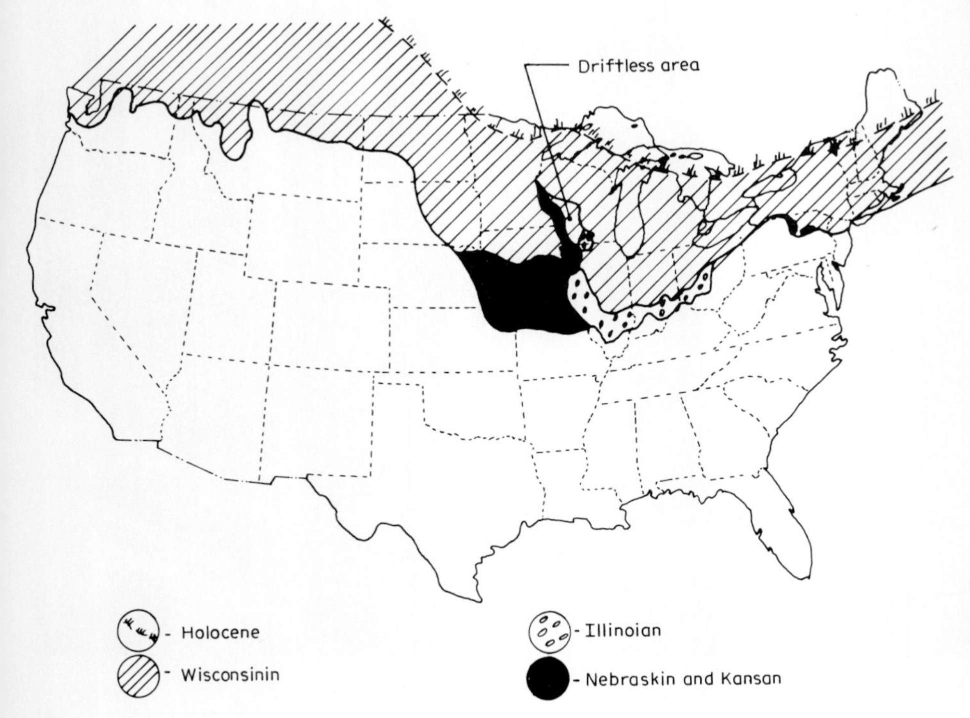


**FIG. 7.77** Drift borders at the time of the maximum advance of the various Pleistocene glaciers and at the beginning of the Holocene in North America.

denuded the rock surface of soil and weathered rock, and often etched the remaining rock surface with striations. As the weather warmed, the glacier began to melt and recede, and streams and rivers formed to flow from the frontal lobes of the ice mass.

*Deposition*

*Glacial drift* is a general term for all deposits having their origin in glacial activity, divided into broad groups: till and stratified drift.

- *Till* is unstratified drift deposited directly from the ice mass, and
- *Stratified drift* is deposited by flowing water associated with melting ice.

Deposits are often classified by landform. The modes of occurrence and depositional features of the more common glacial deposits are given on Fig. 7.78.

### Moraines

The term moraine is often used synonymously with till or drift, but when used with modifiers it more correctly denotes a particular landform which can consist of only till, of mixtures of till and stratified drift, or of only stratified drift.

*Ground moraine* denotes drift (till) deposited beneath the advancing ice, forming sheets over the landscape. The surface is characteristically gently rolling and lacks ridgelike forms.

*Terminal moraine* is a ridgelike feature built along the forward margin of the glacier during a halt in the advance and prior to recession. Composed of various mixtures of drift, a terminal moraine marks the farthest advance of the glacier. The topographic expression as evident on the south shore of Long Island, New York, is given on Fig. 7.87. At this location it rises 30 m above the surrounding outwash plain.

*Moraine plain* (outwash plain) is deposited by the meltwaters of the glacier as its destruction begins during the occurrence of warmer temperatures.

*Kettle moraine* refers to a terminal moraine with a surface marked with numerous depressions (kettle holes) which result from the melting of large blocks of ice remaining on the surface or buried at shallow depths.

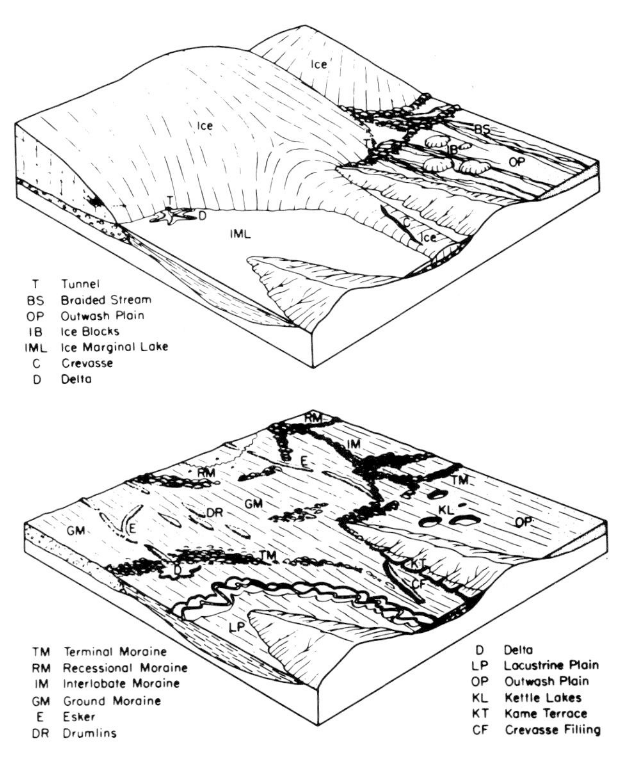


**FIG. 7.78 Schematic diagrams to suggest the modes of origin of some of the more common glacial landforms.** [*Drawing by W. C. Heisterkamp, from Thornbury (1969).*[37] *Reprinted with permission of John Wiley & Sons, Inc.*]

*Recessional moraines* are ridgelike features built of drift along the margins of the glacier as they recede from their location of farthest advance. They represent a temporary stand from retreat allowing an increase in deposition.

*Interlobate or intermediate moraine* is a ridgelike feature formed between two glacier lobes pushing their margins together to form a common moraine between them. They usually trend parallel to the ice movement.

*Frontal moraine* is an accumulation of drift at the terminus of a valley or alpine glacier, but the term is used also to denote a terminal continental moraine.

## 7.6.2 TILL

### Origin

Referred to as glacial till or ground moraine, *till* is material dropped by the ice mass as slow pressure-melting of the flowing mass frees particles

**FIG. 7.79 Exposure of bouldery till: a matrix of sand, silt, and clay (Staten Island, New York).**

and allows them to be plastered to the ground surface.

### Lithology

*General*

Till is a compact, nonsorted mixture of particles which can range from clay to boulders, and has little or no evidence of stratification, as shown on Fig. 7.79. Its lithology is normally related to the bedrock type of the locale, since continental glaciers apparently do not as a rule carry their load great distances, although boulders have been found hundreds of kilometers from their source.

*Materials*

Tills are described as clayey, sandy, gravelly, or bouldery, and in the clayey phases are referred to as "hardpan," "boulder clay," or "gumbotil." In areas of granite rocks, the till is typically gravelly or bouldery; in areas of sandstones, sandy; and, in areas of gneiss, shales, or limestones, clayey. A clayey till derived from shale is shown on Fig. 7.80. In the till formations of the northeast United States, lenses of sand are found indicating occasional fluvial activity within the glacier.

*Boulders*

Throughout the northeast United States, boulders are found strewn over the surface as well as distributed throughout the till, or as concentrations or "nests" at the bottom of the deposit at the bedrock contact. Boulders ranging up to 6 m across, or larger, are common, either on the surface or buried.

### Classes by Mode of Deposition

*Lodgement or basal tills* have been plastered down beneath the actively moving glacier to form an extremely hard and compact mass. Seismic velocities from geophysical surveys can be as high as for some rock types, making identification difficult. Basal till is at times referred to as *tillite*, but the term is more correctly applied to a *lithified till*.

*Ablation or superglacial till* has been dropped in place by stagnant, melting ice, and is a relatively loose deposit. Where flowing water has removed finer particles the material is coarse-granular.

### Tills of North America

Four stages of glaciation and their associated till deposits have been identified in North America. The exposed limits of the various stages are shown approximately on Fig. 7.77. A block diagram of the stratigraphy of the last three stages representative of locations in the midwest United States is given on Fig. 7.81.

*Nebraskan*, the oldest till, has not been clearly found exposed, but underlies most of the area mapped as Kansan. It is a thick sheet of drift spread over an irregularly eroded rock surface and averaging more than 30 m in thickness. *Gumbotil*, a dark, sticky clayey soil, averaging about 2.4 m in thickness, resulted from weathering of the till during the interglacial age. Scat-

**FIG. 7.80 Driven tube sample in clayey till from a depth of 2.1 m (Linden, New Jersey).**

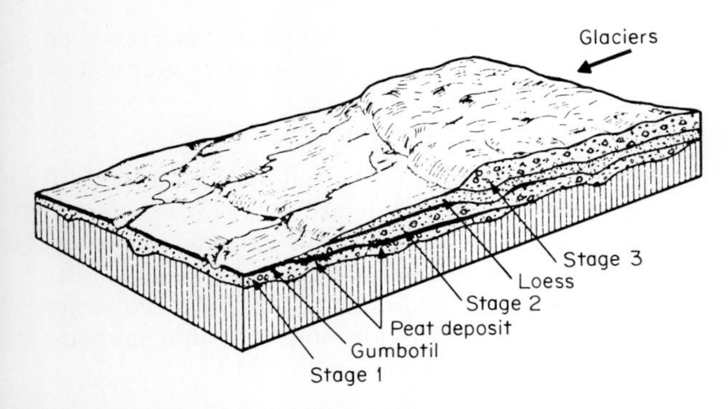


**FIG. 7.81 A stratigraphic section of the last three stages of glaciation representative of locations in the midwestern United States.**

tered deposits of peat have been found sandwiched between the Nebraskan and Kansan tills.

*Kansan* till is exposed over a large area of northern Missouri, northeastern Kansas, Nebraska, and Iowa, averaging about 15 m in thickness. It has been encountered east of the Mississippi beneath the younger drift sheets. Its surface also weathered to form gumbotil averaging about 4 m in thickness, and peat deposits were formed on the surface during the following interglacial age. The peat and gumbotil were subsequently covered with a moderately thick deposit of loess over much of the area.

*Illinoian* till, exposed mainly in Illinois, southern Indiana, and central Ohio, and small areas of Wisconsin, Pennsylvania, and New Jersey, is composed chiefly of silts and clays. Its weathered zone of gumbotil averages 1 m thick, and its surface contains numerous deposits of peat and areas of stratified sand and gravel. In the midwest it is covered by a loess sheet.

*Wisconsinin* till, composed mainly of sands, cobbles and boulders, represents the last stage of deposition that left most of the landforms typical of continental glaciation.

### Landforms

*Till plains* are found in Ohio, Indiana, and Illinois, where the topography is characterized by low relief and an often quite flat surface. The till thickness often exceeds 30 m and the underlying bedrock surface is irregular. A test boring log from Columbus, Ohio (*see Fig. 7.84a*) gives the general stratigraphy.

*Pitted till plains* are common features in the north central states. The surface is extremely irregular, containing numerous lakes and swamps that fill depressions left by melting blocks of ice as shown on Fig. 7.82.

*Drumlins* are hills in the shape of inverted spoons found on till plains. They can be composed entirely of till, or mixtures of till and stratified drift, or can have shallow cores of bedrock which are usually located on their up-glacier end. Drumlin landforms are illustrated on Fig. 7.83.

### Engineering Characteristics

*General*

Basal tills can have densities as high as 150 pcf (2.4 g/cm$^3$) and void ratios as low as 0.21 [Cleaves (1964)[42]]. Typical logs of test borings in Ohio, New Jersey, and Massachusetts are given on Fig. 7.84. The high SPT values are common in till and indicative of its high strength and relative incompressibility. In New York and New Jersey, allowable foundation bearing values in the order of 8 to 12 kg/cm$^2$ are common.

*Midwest Tills*

Care is required in the midwest that heavily loaded foundations are not placed over the relatively compressible peat and gumbotil layers, but in general the till itself is a strong material. An investigation for a high-rise in Toledo, Ohio, encountered "hardpan" at a depth of 22 m described as a "hard, silty clay, mixed with varying percentages of sand and gravel." SPT values were erratic, partially because of the gravel particles, ranging from 12 to 100. On the basis of these data an allowable bearing value of 6 kg/cm$^2$ for caissons bearing in the till was selected. Subsequent tests with the Menard pressure meter (*see Art. 3.5.4*) yielded compression modulus values ranging from 570 to 880 kg/cm$^2$, which permitted the assignment of an allowable bearing value of 12 kg/cm$^2$ at a penetration depth into the till of 2.5 m [Ward (1972)[43]].

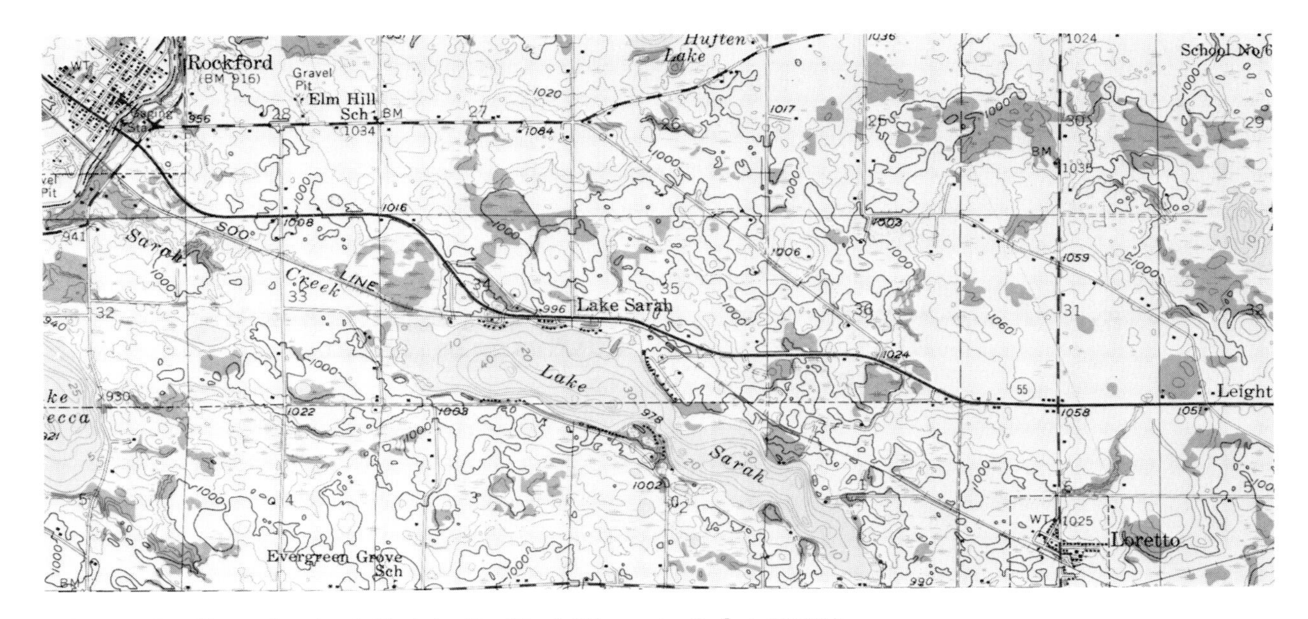


FIG. 7.82 Landform of a pitted till plain. (Rockford, Minnesota. Scale 1:62,500.)

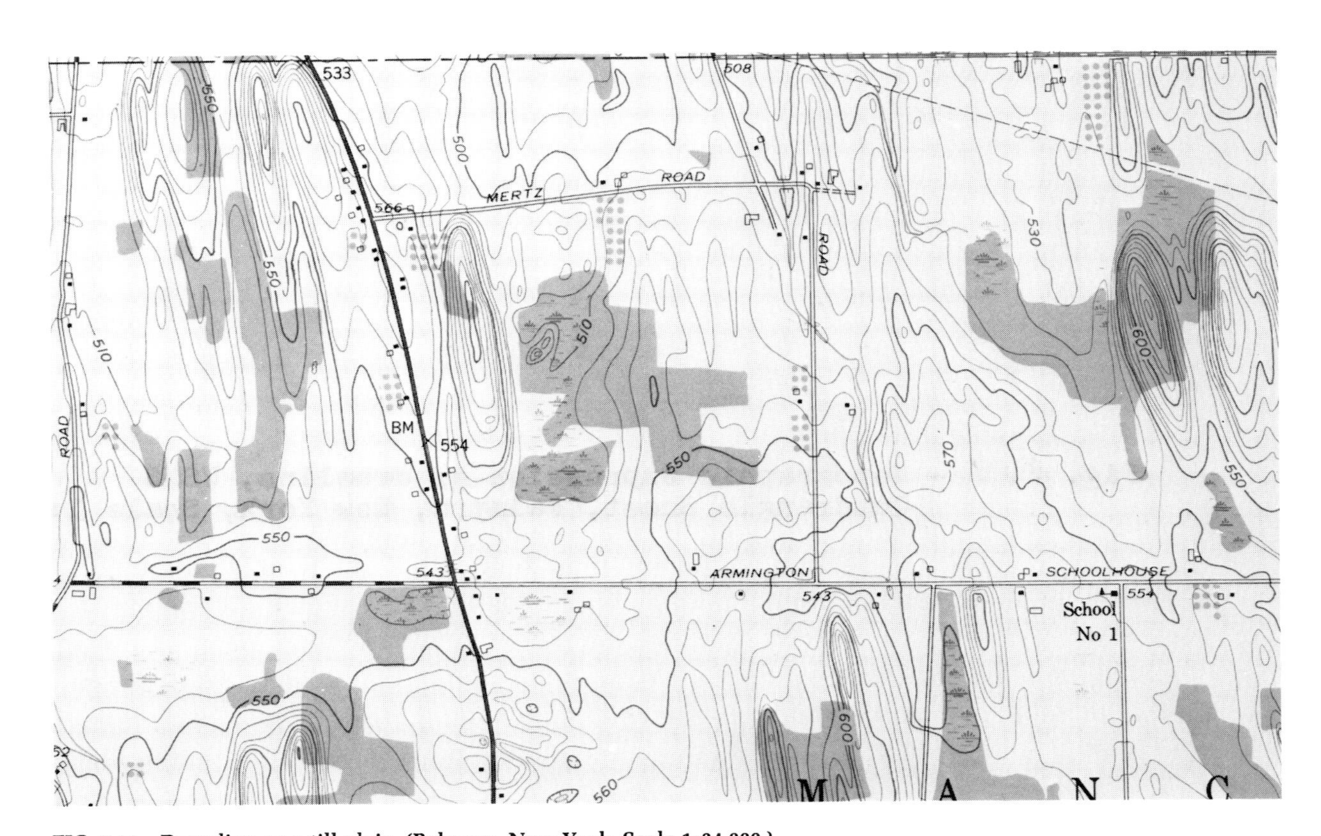


FIG. 7.83 Drumlins on a till plain. (Palmyra, New York. Scale 1:24,000.)

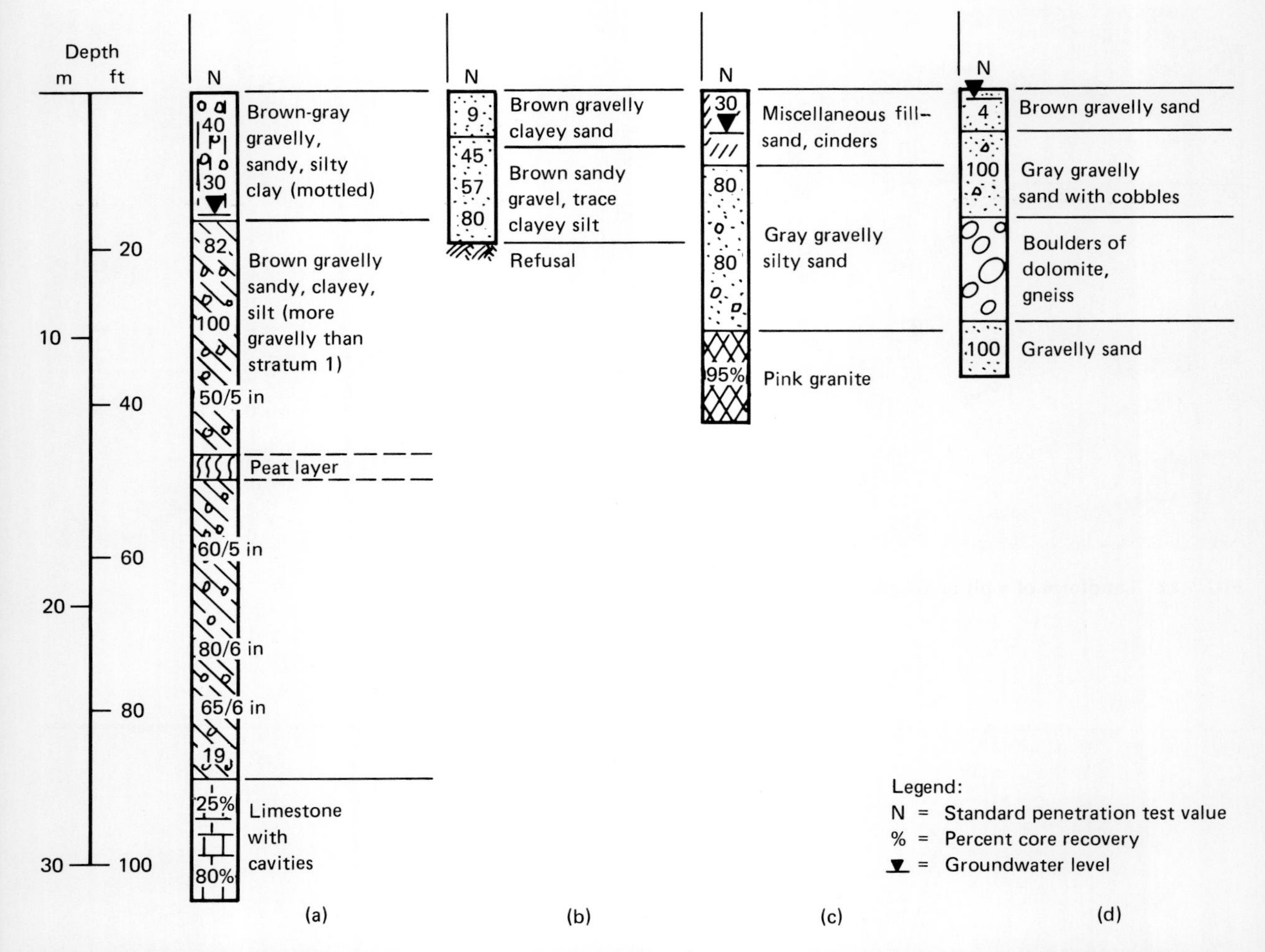


**FIG. 7.84 Logs of typical test borings in till: (*a*) Columbus, Ohio; (*b*) Demarest, New Jersey; (*c*) Weymouth, Massachusetts; (*d*) Edison Township, New Jersey.** *(Courtesy of Joseph S. Ward & Associates.)*

## 7.6.3 GLACIAL-FLUVIAL STRATIFIED DRIFT

### Origin

During warm periods numerous streams flow from the glacier which are literally choked with sediments. The streams are usually braided and shallow, and because of the exceptionally heavy loads being carried, large thicknesses of soils can be deposited in a relatively short time. Some of the streams terminate in moraine-formed lakes where they deposit lacustrine soils; other streams flow to the sea. A general section illustrating the relationship between a recessional moraine and fluvial and lacustrine deposits is given as Fig. 7.85.

### Classes of Stratified Drift

*Proglacial deposits* form beyond the limits of the glacier and include stream, lake, and marine deposits.

*Ice-contact stratified drift* is deposits built in immediate contact with the glacier and includes only fluvial formations.

### Modes of Glacial-Fluvial Deposition (Fig. 7.78)

*Outwash,* the streambed-load materials, consisting of sands and silts, is highly stratified (Fig. 7.86).

*Outwash plain* is formed from the bed load of

several coalescing streams and can blanket large areas. A portion of the terminal moraine and the outwash plain along the south shore of Long Island, New York, is shown on the topographic map (Fig. 7.87).

*Kettles* occur where the outwash is deposited over blocks of ice which subsequently melt and the surface subsides (*Fig. 7.76*). Formerly lakes, these depressions now are commonly filled with deposits of recent soft organic soils, and are particularly troublesome to construction. They can be quite large, extending to 20 m in width or more.

*Pitted outwash plains* result from outwash deposition over numerous ice blocks.

*Valley trains*, often extending for hundreds of kilometers, represent deposition down major drainage ways. Their evidence remains as terrace deposits along the Mississippi, Missouri, Ohio and many other rivers of the north central as well as northeastern United States. They are common also to the northern European plain.

*Ice-contact depositional forms* include kames, kame terraces, and eskers.

- *Kames* are mounds or hummocks composed usually of poorly sorted water-lain materials which represent the filling of crevasses or other depressions in the ice, or between the glacier and the sides of its trough.
- *Kame terraces* are kames with an obvious flat surface of linear extent. They can be expected to be of loose density, since they have slumped subsequent to the melting of the underlying ice.
- *Eskers* are sinuous ridges of assorted and somewhat stratified sand and gravel that represent the fillings of ice channels within the glacier.

### Engineering Characteristics

*Gradations*

The grinding action of the glacier and the relative short transport conditions before deposition cause the outwash soils to be more angular than normal fluvial materials, and the gradation reflects the source rock. In the northeastern United States, the outwash varies from clean sands and gravel south of the Long Island terminal moraine where the rock source was Precambrian crystallines, to fine sands and silts in New Jersey where much of the rock source was Triassic shales and fine-grained sandstones.

*Properties*

Fluvial stratified drift tends to vary from loose to medium compact but normally provides suitable support for moderately loaded foundations in the sandy phases. A log of a test boring from the Long Island, New York, outwash plain is given as Fig. 7.88. The formation thickness at this location was about 11 m.

"*Bull's liver*" refers to the pure reddish silt characteristic of the outwash in many areas of New Jersey. When encountered in foundation excavations below the water table it often "quakes" when disturbed by construction equipment and loses its supporting capacity. When confined and undisturbed, however, it appears firm and can provide suitable support for light to moderately loaded foundations. Several methods of treatment have been used: removal by excavation where thicknesses are limited; changing foundation level to stay above the deposit and avoid disturbance; or, "tightening" and strengthening the silt by dewatering with vacuum well points set into the silt stratum to decrease pore pressures and increase apparent cohesion. This procedure often is followed when preparing foundations for buried pipe in trench excavations. There have been many cases of sewer pipe failures where they were supported on improperly prepared foundations bearing in the silt.

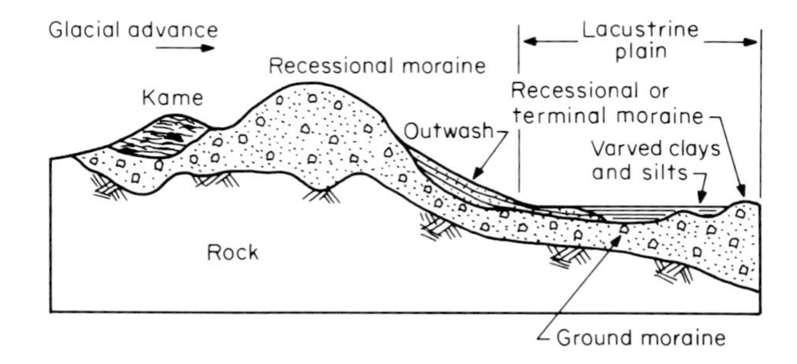


**FIG. 7.85 Geologic section showing relationship between recessional moraine and fluvial and lacustrine deposits.**

**FIG. 7.86** Stratified drift exposed in borrow pit (Livingston, New Jersey).

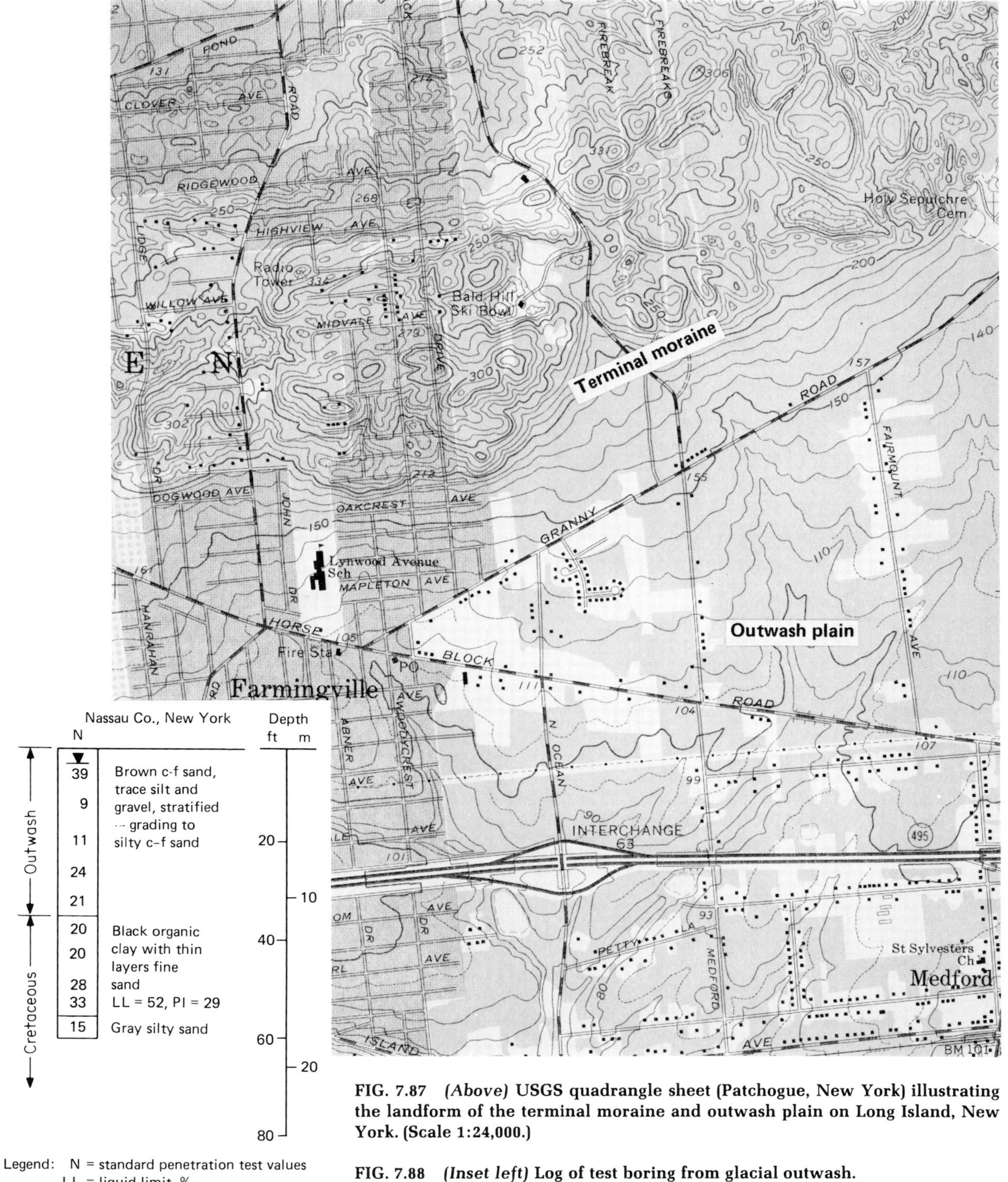


**FIG. 7.87** *(Above)* **USGS quadrangle sheet (Patchogue, New York) illustrating the landform of the terminal moraine and outwash plain on Long Island, New York. (Scale 1:24,000.)**

**FIG. 7.88** *(Inset left)* **Log of test boring from glacial outwash.**

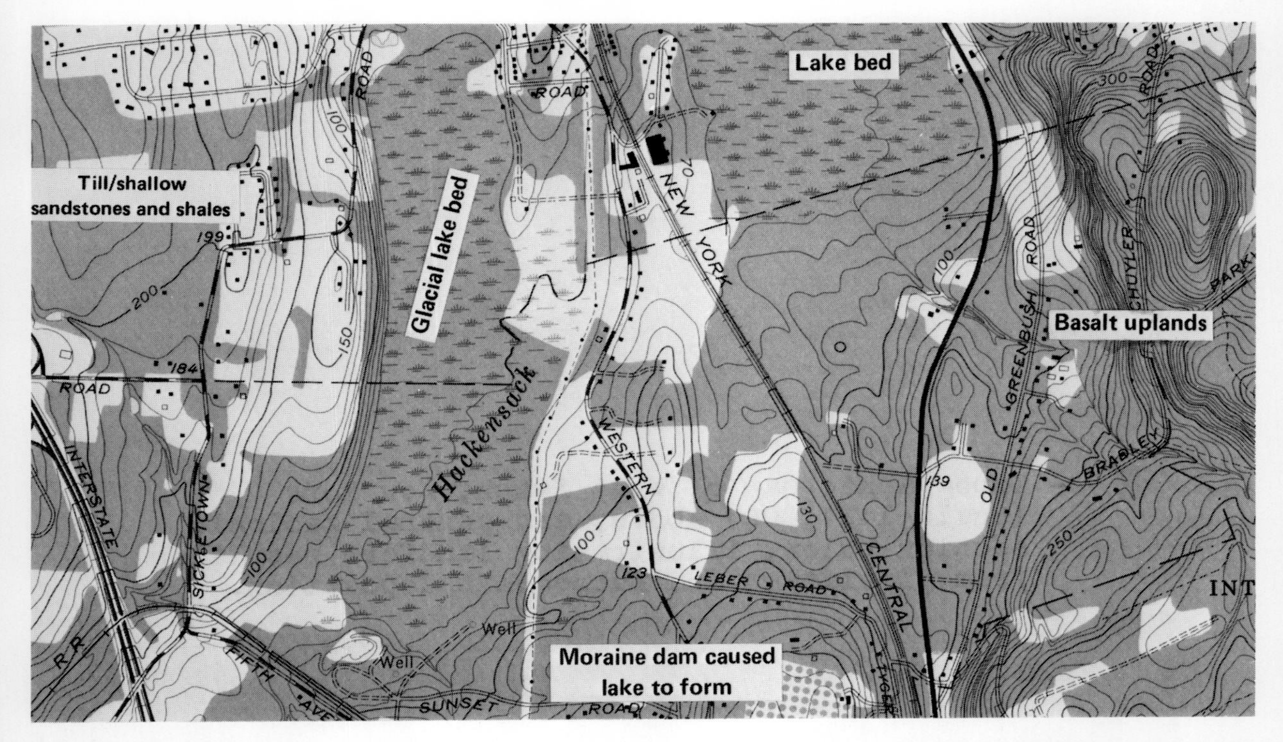


FIG. 7.89 Topographic expression of a freshwater swamp over a glacial lakebed, and adjacent glaciated high ground. (West Nyack, New York. Scale 1:24,000.)

7.6.4 GLACIAL-LACUSTRINE SOILS

### Origin

During the last period of glaciation numerous lakes originated when end moraines formed natural dams across glacial valleys. The lakes were filled subsequently with lacustrine soils from the outwash.

### Mode of Occurrence

Overflowing from the larger lakes formed outlets that permitted them to drain. Subsequent regional uplift as the glacier receded exposed many of the lake bottoms, and during dry spells the upper portion of the lake bed soils became strengthened by desiccation. These lakes remain in several forms as follows.

- *Lacustrine plains* cover vast areas such as Lake Agassiz in North Dakota, Minnesota, and adjacent Canada.
- *Margins of existing lakes* such as underlie Chicago, Cleveland, and Toledo.
- *Terrace deposits* high above present river valleys such as exist along the Hudson River, at Albany, New York, and other rivers of the northeast *(see Fig. 2.17)* and in the Seattle, Washington, area.
- *Saltwater tidal marshes* such as Glacial Lake Hackensack, New Jersey *(see Fig. 2.4)*.
- *Freshwater swamps* such as Glacial Lake Passaic, New Jersey *(see Figs. 2.4 and 6.19)*, and many other lakes in the northeastern United States. The topographic expression and a geologic section through a glacial lake bed in West Nyack, New York, is given on Figs. 7.89 and 7.90, respectively.

### Geographic Distribution

A list of the glacial lakes of North America is given on Table 7.12. The general locations of the larger lakes are within the area identified as G2 on Fig. 7.1. The significance of the characteristics of lake bed soils lies in the fact that many of the larger cities of the United States are located over former lake beds, including Chicago, Cleveland, Toledo, Detroit, Albany, and parts of

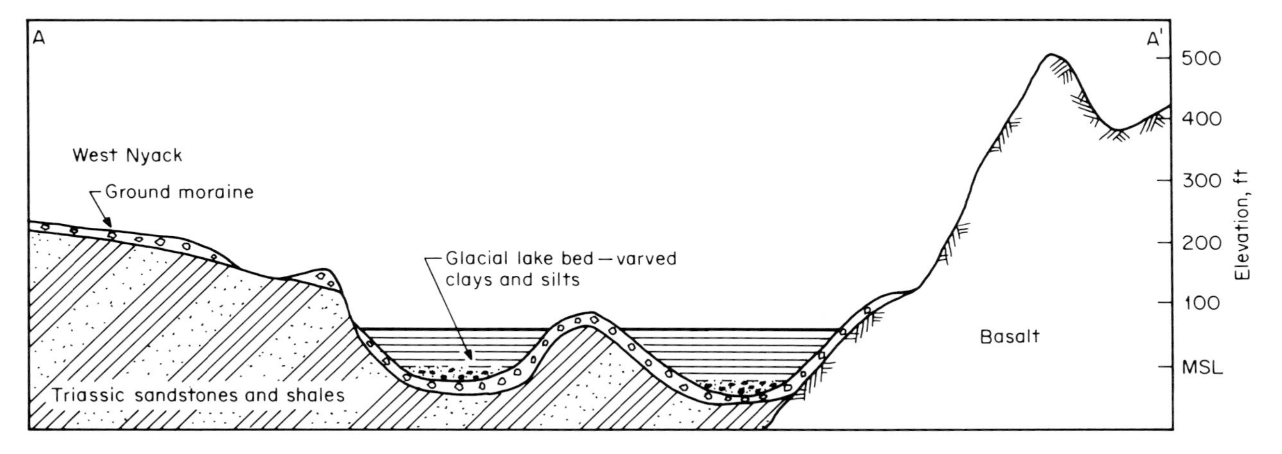


**FIG. 7.90 Geologic section across Fig. 7.89 (West Nyack, New York).**

New York City. Boston is located over glacial-marine soils *(see Art. 7.6.5)*.

### Depositional Sequence

The probable sequence of deposition in the Connecticut River Valley, which can be considered as typical of the formation of many glacial lakes, is given on Fig. 7.91. The present-day landform is illustrated on the topographic map (Fig. 7.92). The ice mass fills the valley and probably was responsible for its excavation (Fig. 7.91*a*). Recession begins, the ice begins to melt, and outwash deposits kame terraces and fills crevasses (Fig. 7.91*b*). The lake grows in size, and fluvial activity fills the lake margins with stratified granular soils and the deeper waters with fine-grained soils (varved silts and clays) (Fig. 7.91*c*). The lake begins to drain from a lower outlet; lake margins are exposed and kettles form (Fig. 7.91*d*). In Fig. 7.91*e*, the lake has drained and the coarser-grained particles are exposed as terraces.

### Varved Clays

*Deposition*

The typical infilling of glacial lakes is varved clays, or alternating thin layers of clay and silt

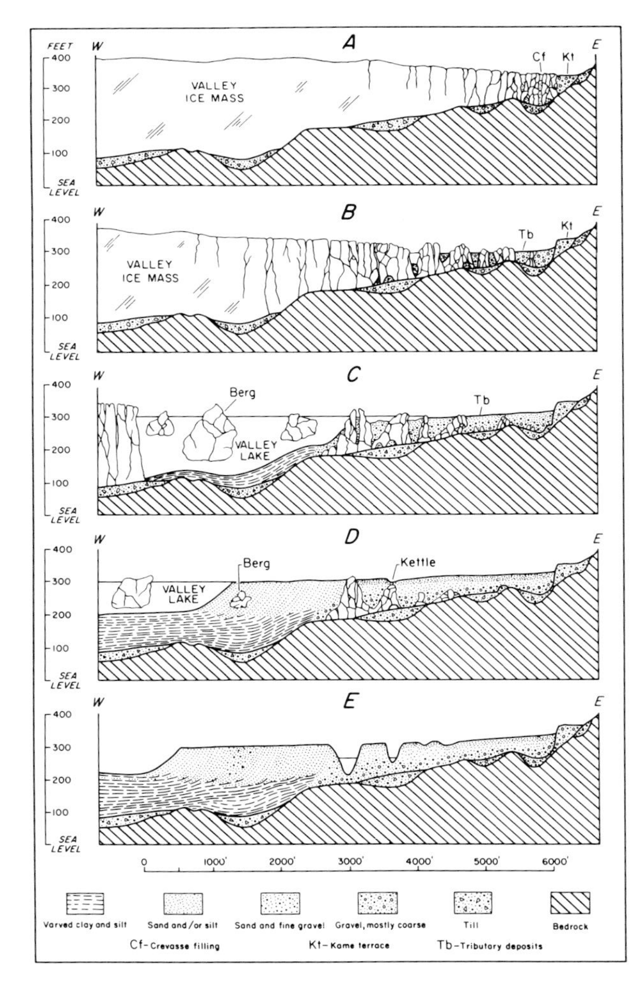


**FIG. 7.91 Suggested depositional sequence of outwash in the Connecticut River Valley.** [*From Thornbury (1967).*[45] *Reprinted with permission of John Wiley & Sons, Inc.*]

**TABLE 7.12**
**GLACIAL LAKES OF NORTH AMERICA***

| |
|---|
| Agassiz (greater Lake Winnipeg, much of Manitoba, western Ontario, North Dakota, and Minnesota) |
| Albany (middle New York State) |
| Algonquin (greater Lake Michigan and Huron) |
| Amsterdam (Mohawk Valley) |
| Arikaree (North and South Dakota) |
| Arkona (a low-level stage in Erie basin and south of Lake Huron) |
| Barlow (see Ojibway) |
| Bascom (New York, Vermont, New Hampshire) |
| Calumet (Lake Michigan) |
| Calvin (Iowa River and Cedar River valleys) |
| Chicago (southern Lake Michigan) |
| Chippewa (Lake Michigan, discharge to Lake Stanley) |
| Coeur d'Alene (Idaho) |
| Columbia (Washington) |
| Dakota (in James River Valley) |
| Dana (see Lundy) |
| Dawson (see Lundy) |
| Duluth (western Lake Superior) |
| Early Lake Erie |
| Glenwood (Lake Michigan) |
| Grassmere |
| Hackensack (New Jersey) |
| Hall (greater Finger Lakes, outflow west to Lake Warren) |
| Herkimer (Mohawk Valley) |
| Houghton (Lake Superior) |
| Iroquois (greater Lake Ontario) |
| Jean Nicolet (Green Bay) |
| Keweenaw (Lake Superior) |
| Lundy, or Dana and Dawson in New York (southern Lake Huron and Erie) |
| Madawaska (St. John River, New Brunswick) |
| Maumee (Lake Erie) |
| McConnell (northern Alberta) |
| Memphremagog (Vermont, province of Quebec) |
| Mignon (Lake Superior) |

*After Fairbridge (1968).[44] Adapted with permission of Dowden, Hutchinson & Ross, Inc.

with occasional sand seams or partings as shown on Fig. 7.93. The varves are climatic in origin; during the summer months meltwaters entered the lake, carrying fine sediments which were distributed throughout the lake in suspension. The silt settled to the bottom and occasional periods of high stream flow resulted in the deposition of sand partings and lenses. When winter arrived, the lake froze over, meltwaters ceased to flow, and the clay particles settled out of the quiet waters. When summer returned each year, a new layer of silt was deposited, and the sequence continued year after year.

In any given location, the varves can range from several millimeters to several centimeters in thickness, or can be so thin as to be discernible only when the specimen is air-dried and broken open. In Connecticut the varves range generally from 6 to 64 mm (¼ to 2½ inches) whereas in New Jersey in Glacial Lake Hackensack, they are much thinner, especially in deeper zones where they range from 3 to 13 mm (⅛ to ½ inches). The New York materials seem to fall between these extremes. Clay and silt varves and sand partings from a tube sample from a depth of about 7 m taken from a site in East Rutherford, New Jersey, are shown as Fig. 7.94.

*Postdepositional Environments*

The engineering characteristics of a given glacial lake deposit will depend directly on the postdepositional environment which is associ-

Minnesota (Driftless area of Minnesota)
Missoula (Washington, Idaho, Montana)
Newberry (united Finger Lakes, outflow to Susquehanna)
Nipissing Great Lakes (postglacial higher Great Lakes, draining through Ottawa Valley; later Port Huron)
Ojibway-Barlow (north central Ontario)
Ontario (Early)
Ontonagon (northern Michigan)
Passaic (eastern New Jersey)

Peace (Alberta)

Rycroft (to Peace River, Alberta)
Saginaw (southwest of Lake Huron)
St. Louis (valley of St. Louis River)
Saskatchewan (midcourse of Saskatchewan River)

Schoharie (middle New York State)

Souris (western Manitoba and North Dakota, draining to ? James River)
Stanley (Lake Huron, draining to Ottawa Valley)
Toleston (Lake Michigan)
Tyrrell (west of Lake Athabasca)

Vanuxem (greater Finger Lakes, outflow east to the Mohawk)
Vermont (Coveville and Fort Ann phases, to Lake Champlain)

Warren (southern Lake Huron, Erie, and Finger Lakes)
Wayne (low-level stage at Erie basin)
Whittlesey (greater Lake Erie, southern end of Lake Huron)
Wisconsin (Wisconsin)
Wollaston (discharged west to Athabasca)

Modern Great Lakes
- Superior (outflow at Sault Ste. Marie rapids)
- Michigan (continuous at Mackinac Straits with Lake Huron)
- Huron (outflow at St. Clair River)
- Erie (Niagara Falls)
- Ontario (Thousand Islands, St. Lawrence)

ated with its mode of occurrence, and is subdivided into three general conditions as follows.

*Normally consolidated* conditions prevail in swamp and marsh areas where the lake bottom has never been exposed as a surface and subjected to drying, desiccation, and the resulting prestress. This condition can be found in small valleys in the northeastern United States, but is relatively uncommon. SPT values are about 0 to 2 blows per foot and the deposit is weak and compressible.

*Overconsolidated* conditions prevail where lake bottoms have been exposed to drying and prestress from desiccation. At present they are either dry land with a shallow water table, or are or were in recent times covered with swamp or marsh deposits. Typically a crust of stiff soil has formed, usually ranging from about a meter to several meters in thickness, which can vary over a given area. The effect of the prestress, however, can extend to substantial depths below the crust. In many locations oxidation of the soils in the upper zone changes their color from the characteristic gray to yellow or brown. This is the dominant condition existing beneath the cities of the northeast and north central United States. These formations can support light to moderately heavy structures on shallow foundations, but some slight settlements normally occur.

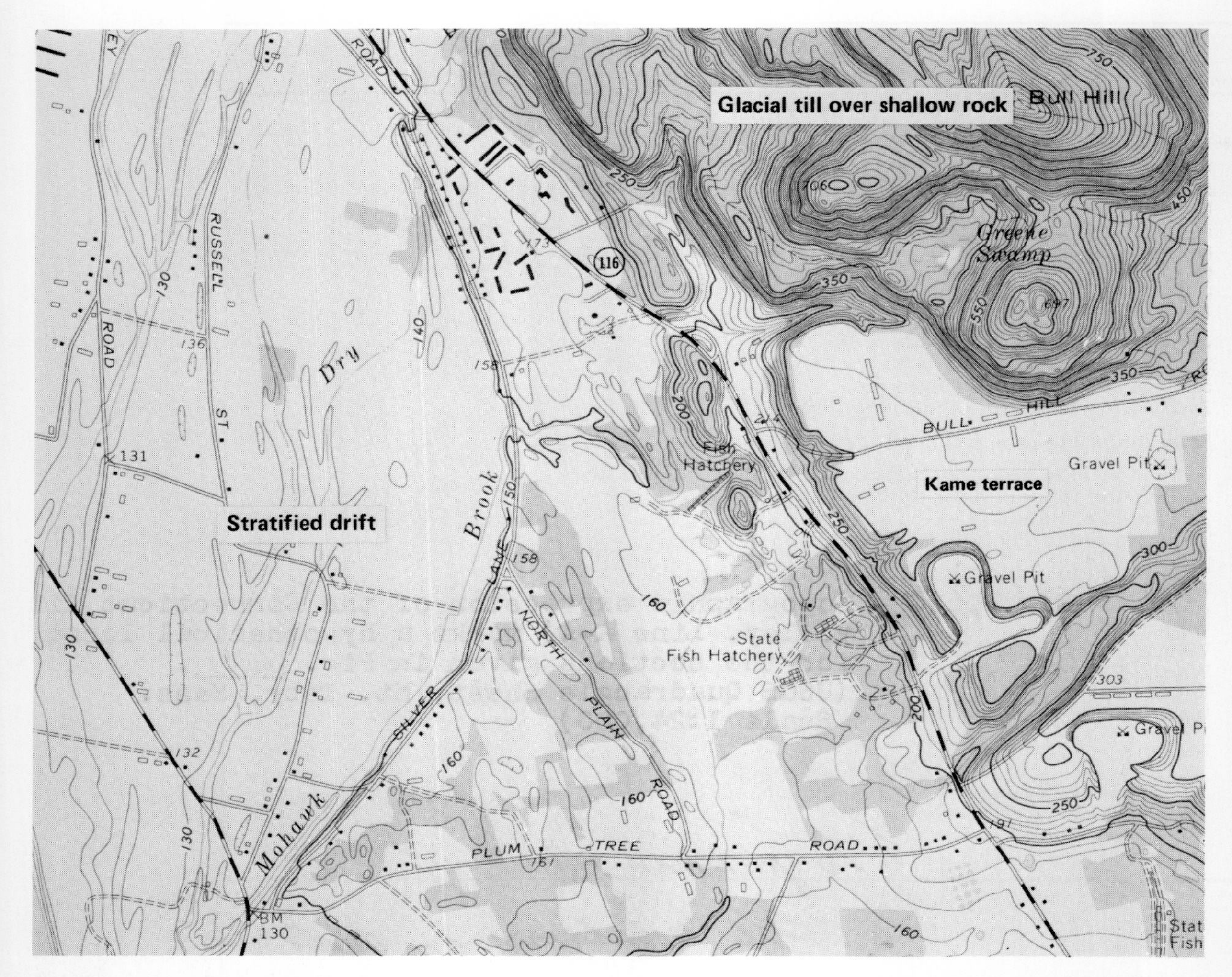


**FIG. 7.92 Portion of USGS quadrangle sheet (Mt. Toby, Massachusetts) giving approximate topographic expression for Fig. 7.91e. (Scale 1:24,000.)**

*Heavily overconsolidated* deposits are characteristic of lake bottoms that have been raised, often many meters, above present river levels and where the varved clays now form the slopes, or are found in depressions above the river valley. This condition is common along the Hudson River in New York and many rivers of northern New England, such as the Barton River in Vermont *(see Fig. 2.17)*. In the last case, desiccation plus the natural tendency for drainage to occur along silt and sand varves have resulted in a formation of hard consistency 100 m or more above river level yielding SPT values of 50 blows per foot or more.

The varved clays along the side slopes of the Hudson Valley (Fig. 7.95) suffer many slope failures. At the campus of Rensselaer Polytechnic Institute in Troy, N.Y., 70 m above sea level, soils exposed on slopes have been prestressed to form stiff fissured clays yielding SPT values from 20 to 50 blows per foot. They are subject to instability and sliding, particularly during spring thaws and rains. Formations in depressions above river level, such as exist in Albany and Troy, both high above the Hudson River, or along the lower valley slopes where the materials tend to remain saturated, have much less prestress, as shown in Fig. 7.97b.

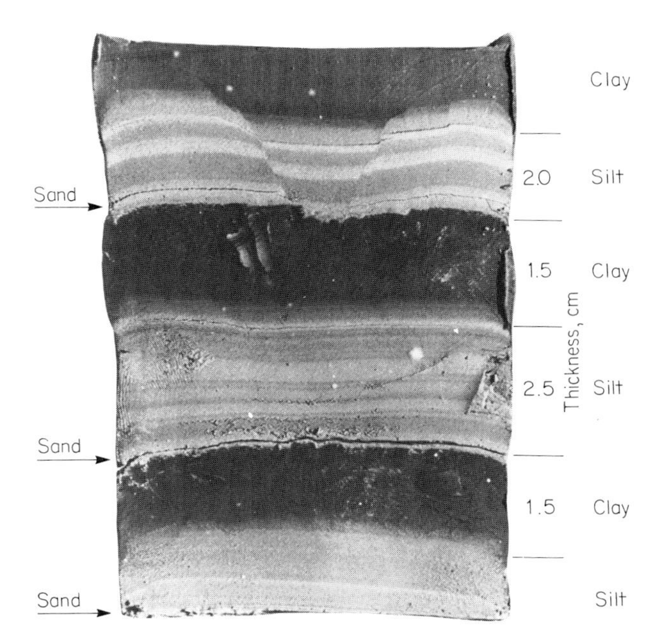


**FIG. 7.93 Failed triaxial test specimen of varved clay (Hartford, Connecticut). Three-inch undisturbed tube specimen from a depth of 30 ft.**

**FIG. 7.94 Shelby tube sample of varved clay (Glacial Lake Hackensack, East Rutherford, New Jersey). Depth about 7 m.**

**FIG. 7.95 Excavated face in varved clays exposed in slope (Roseton, New York). Location is about 10 m above level of the Hudson River.**

*Engineering Characteristics*

GENERAL Because varved clays underlie many large urban areas, their unusual engineering characteristics are very important. Two factors are most significant: the amount of prestress that has occurred in a given formation, and the effect that the sand lenses and silt varves have on internal drainage. The sand and silt permit consolidation of the clay varves to occur much more rapidly than in a normal clay deposit. Both factors may vary substantially from location to location.

Where desiccation has occurred, a stiff crust is formed and the deposit varies from heavily preconsolidated near the surface, decreasing to moderately preconsolidated with depth, and eventually becoming normally consolidated. The crust is evident by the high unconfined compressive strengths of Chicago clays as shown on Fig. 7.96, and by the high SPT values shown on the boring logs from several other cities given on Fig. 7.97. SPT values, however, give only a rough indication of consistency and the materials usually are substantially less compressible than the values would indicate. In engineering studies the properties of the silt and clay varves should be considered separately, when feasible.

NEW JERSEY MEADOWLANDS, GLACIAL LAKE HACKENSACK The area adjacent to New York City, known as the New Jersey Meadowlands, covers several thousand hectares and is a tidal marsh that is considered to have been above sea level at one time and exposed to desiccation. The former lake bottom is covered by a layer of rootmat (see Fig. 5.12) generally about 2 m thick which overlies a stratum of sand that can vary from a meter or so to as much as 8 m or more in thickness. Beneath the sand are varved clays which can extend to depths of over 30 m; the thickness is extremely variable, reflecting the differential erosion of the underlying irregular rock surface. A boring log from a deep trench along the western edge of the Meadowlands is given as Fig. 7.98. The varved clays have been preconsolidated by the desiccation.

Traditional foundation construction for the numerous light industrial buildings constructed in the area has been to excavate the rootmat and replace it with engineered-controlled compacted fill, and then to support the structure on footings bearing in the fill and designed for an allowable bearing value of 2 kg/cm². If a building with a large floor area is supported on relatively thick fill (usually about 3 m), and has floor loads in the order of 0.2 to 0.3 kg/cm², some minor differential settlement may be anticipated from consolidation of the deeper, more compressible varved soils. These settlements usually are in the order of a few centimeters at most, and are generally tolerable [Lobdell (1970)[47]]. Most of the settlement is complete within 2 to 3 years of placing the building in service. The alternate solutions are very long piles or surcharging. Because of the rootmat and other organic soils, proper floor support is always a major concern in the area when deep foundations are used.

NEW YORK CITY (GLACIAL LAKE FLUSHING) A geologic section across upper mid-Manhattan at 113th Street is given as Fig. 7.99. It shows the irregular bedrock surface excavated by the glacier, the variable thickness of the varved clays,

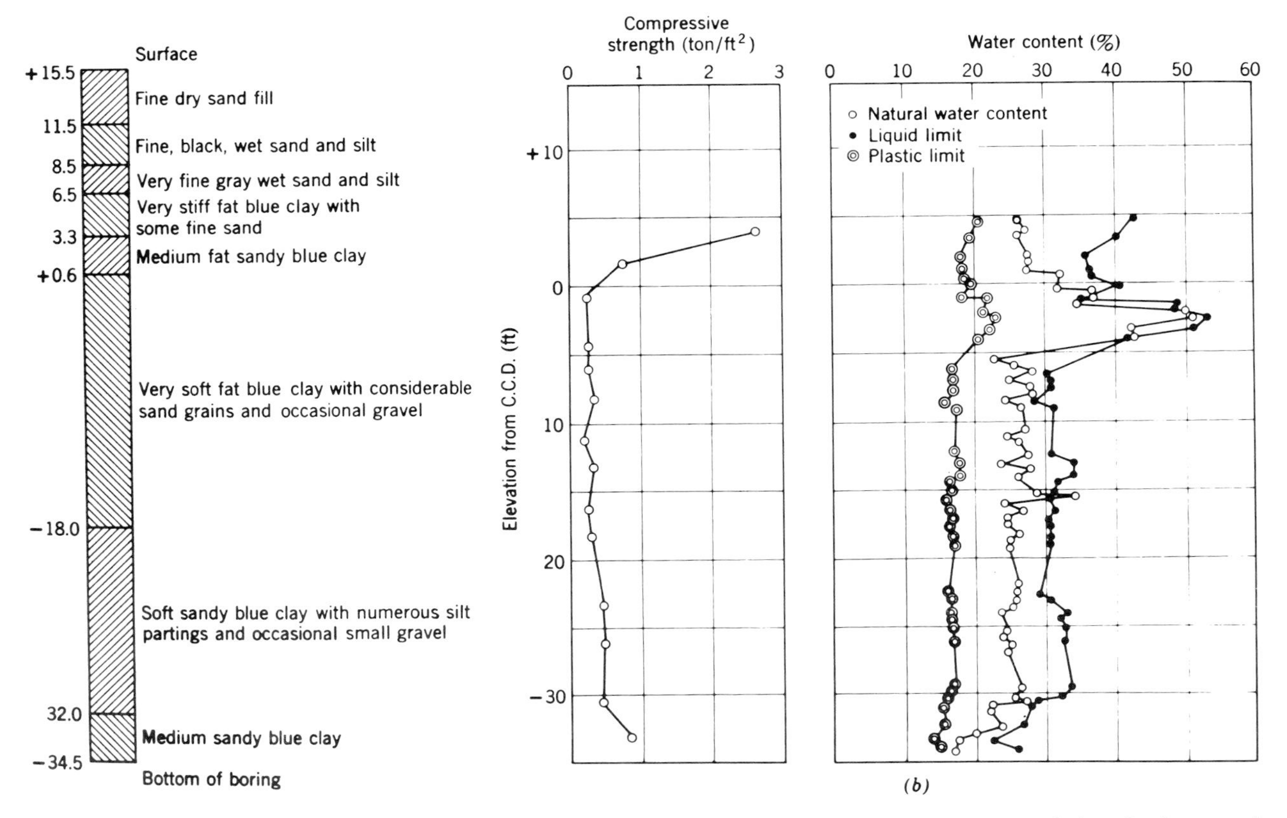


**FIG. 7.96 Characteristics of Chicago lake bed clays showing the general uniformity of their properties below the desiccated zone. Boring at Congress Street and Racine Avenue.** [*From Peck and Reed (1954).*[46]]

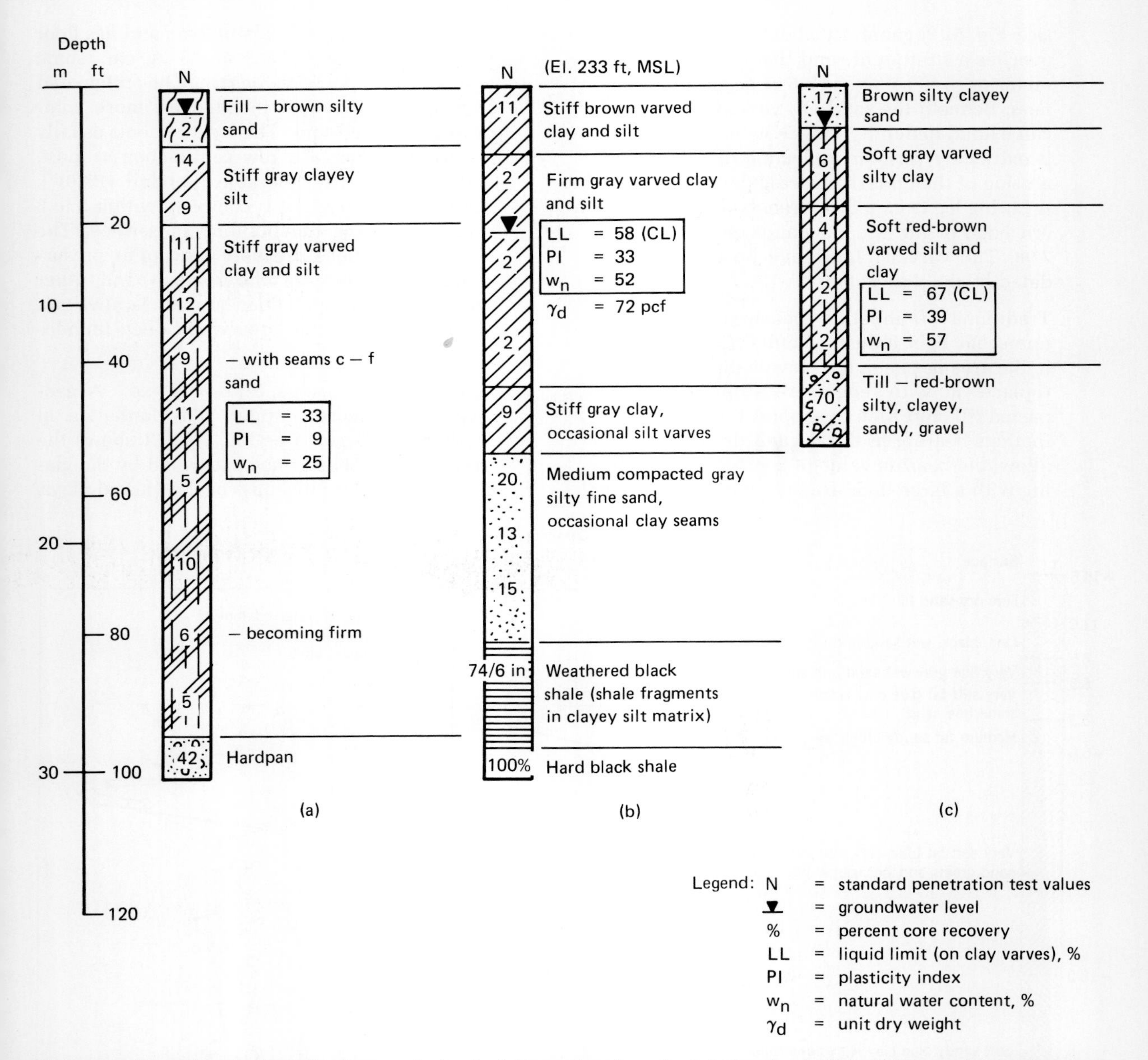


**FIG. 7.97 Logs of typical test borings made in glaciolacustrine deposits: (*a*) Cleveland, Ohio; (*b*) Rensselaer Polytechnic Institute, Troy, New York; (*c*) Hartford, Connecticut.** *(Courtesy of Joseph S. Ward & Associates.)*

and overlying strata of sand, organic silt, and miscellaneous fill. The sand is stratified and probably represents late glacial outwash, whereas the organic silt, common to the New York City area, is a recent estuarine deposit.

In studies of varved clays where an accurate knowledge of their properties is required, it is necessary to count the various varves of sand, silt, and clay; to measure their cumulative thickness; and if possible to perform laboratory tests on representative samples of the silt and clay varves, since their characteristics are distinctly different. Testing a mixture of varves will produce nonrepresentative results. Where varves are very thin, discriminating testing is not feasible. Parsons (1976)[48] has summarized the principal properties of the New York City varved clays as given on Table 7.13. A plasticity chart for silt and clay varves is given on Fig. 7.100.

Buildings ranging from 7 to 21 floors in height have been supported on a number of foundation types including spread footings, mats, and shallow tapered piles with all types apparently bearing in the upper sand stratum. Settlement observations have been carried out on 70 representative buildings through the construction period and generally from 1½ to 3 years thereafter. In several cases readings continued for 17 years. Center settlements of buildings from 20 to 21 stories high totaled from 0.8 to 3.8 in (20 to 96 mm) and maximum differential settlement between center and corner was about 1.5 in (40 mm). Generally between 75 and 85% of primary consolidation occurred during construction, attesting to the rapid drainage characteristics of the varved clays. The foregoing information is cited from Parsons (1976).[48]

STRENGTH VS. OVERCONSOLIDATION RATIO (OCR) In general, glacial lakebed clays have low activity and low sensitivity. Some data on the strength and sensitivity of varved clays from a number of U.S. cities are given on Table 3.30. The shear strength characteristics of varved clays from the New Jersey–Hudson Valley areas are presented

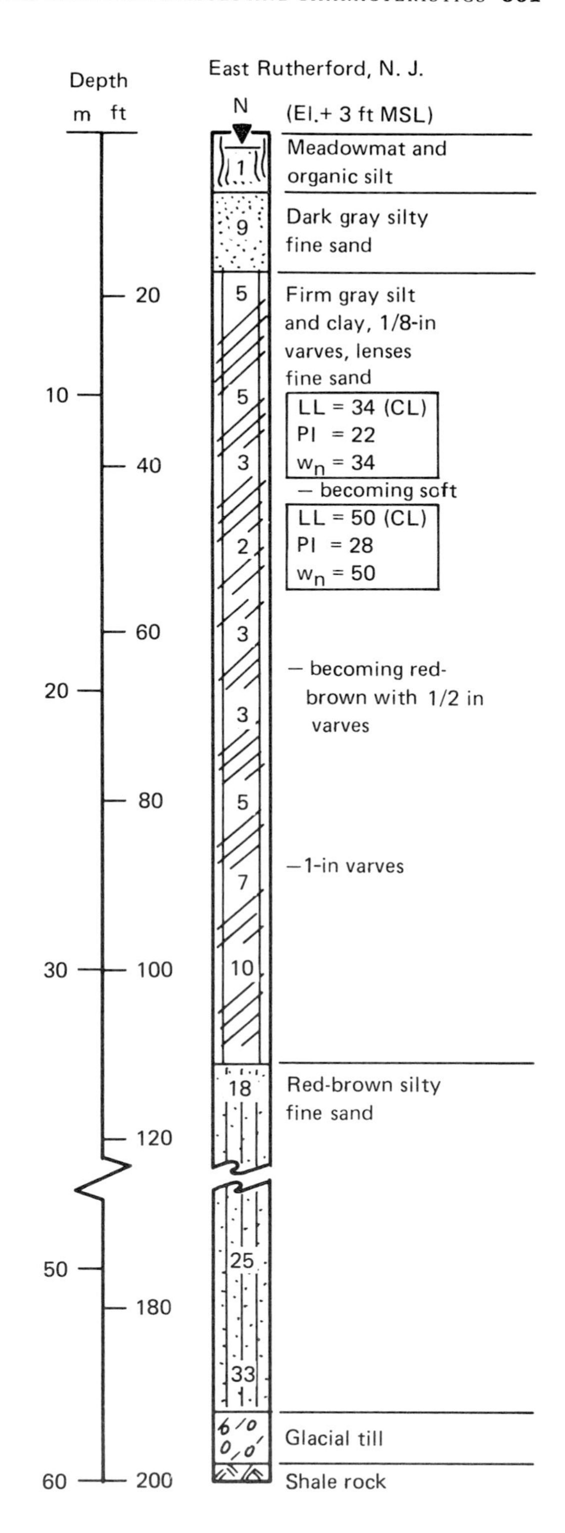


**FIG. 7.98 Log of test boring. (Glacial Lake Hackensack, East Rutherford, New Jersey.)** *(See Fig. 7.94. Courtesy of Joseph S. Ward & Associates.)*

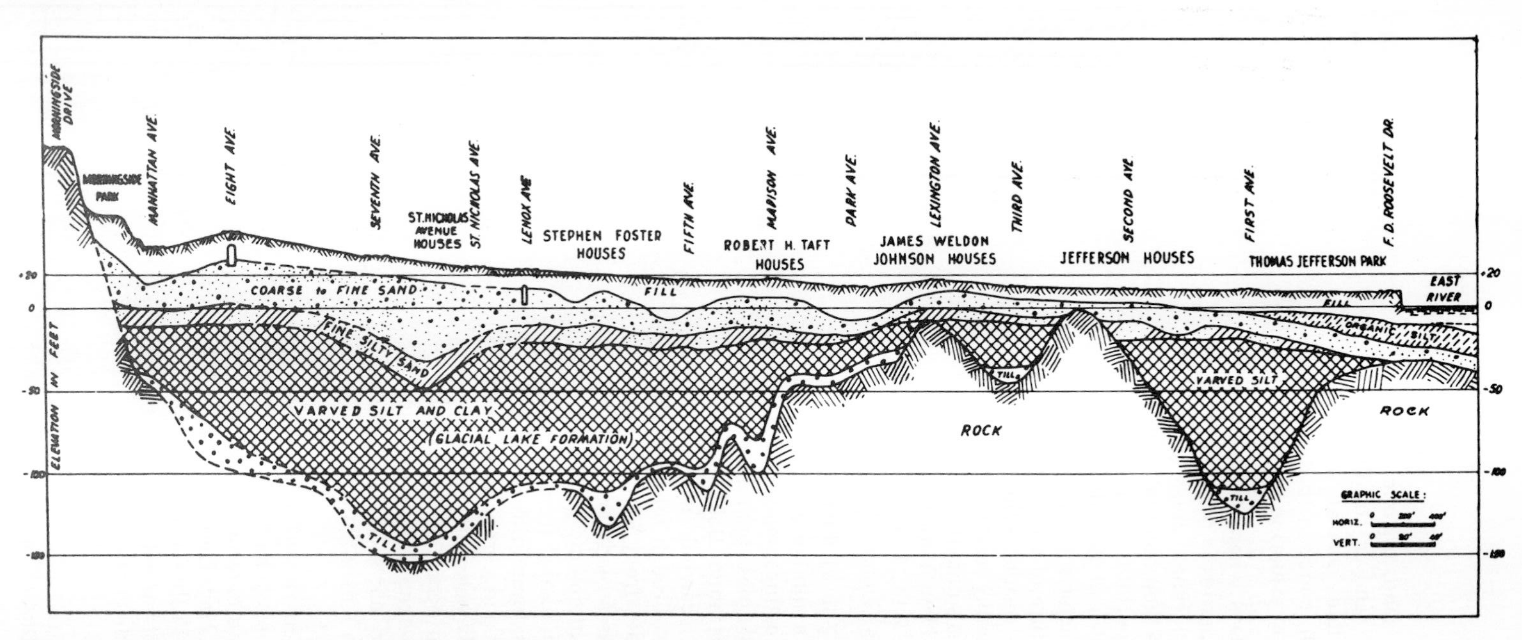


FIG. 7.99 **Geologic section across mid-Manhattan at 113th Street.** [*From Parsons (1976).*[48]]

**TABLE 7.13**
**TYPICAL PROPERTIES OF NEW YORK CITY VARVED CLAYS***

| Characteristic | Clay varves | Silt varves |
|---|---|---|
| Constituents | Gray clay: 10 to 22% of deposit<br>LL = 62%, PL = 28%, W = 46% | Red-brown silt: 40 to 80% of deposit<br>$D_{10}$ = 0.017 mm, $W$ = 28%<br>Gray fine sand: 50 to 40% of deposit<br>$D_{10}$ = 0.03 mm, $D_{60}$ = 0.15 mm |
| Consolidation<br>Preconsolidation | 8–13 kg/cm$^2$ | |
| Overconsolidation ratio | 3–6 | |
| Void ratios | 1.2–1.3 | 0.7–0.85 |
| Void ratio vs. pressure | *See Fig. 3.87d* | *See Fig. 3.87j* |
| Recompression indices | 0.1–0.2 | 0.15–0.04 |
| Coefficient of consolidation $C_v$ (vertical, in recompression) | 0.05 ft$^2$/day (0.005 m$^2$/day) | 1.0 ft$^2$/day (0.09 m$^2$/day) |
| Coefficient of secondary compression | 0.3% strain/time cycle | 0.1% strain/time cycle |
| Shear strength | From undrained triaxial tests: 1–1.5 kg/cm$^2$<br>From field vane shear test: 1.25–1.75 kg/cm$^2$ | |

*After Parsons (1976).[48]

by Murphy et al. (1975)[49] in terms of the undrained strength-effective-stress ratio ($s_u/\bar{\sigma}_v$) vs. the overconsolidation ratio (OCR) and are given on Fig. 7.101. The high degree of overconsolidation characteristic of these soils is apparent. The limits of the varved clays from the Connecticut River Valley as presented by Ladd and Wissa (1970)[50] are also shown. Although these formations are 200 km apart, their postglacial depositional histories are similar as are their general engineering properties. The primary differences are the thicknesses of the silt and clay varves from location to location.

## 7.6.5 GLACIAL-MARINE CLAYS

### Origin

Clays from glacial runoff were deposited in marine estuaries along coastlines and subsequently uplifted to become land areas by isostasy (rebound from removal of the ice load).

### Geographic Distribution

Sensitive marine clays are found in the St. Lawrence and Champlain lowlands of Canada, along the southern Alaskan coastline (*see Art. 9.2.11*), and throughout Scandinavia. The "blue clay" underlying Boston, Mass., has relatively low sensitivity compared with other glacio-marine clays.

### Depositional Characteristics

*Deposition*

Clays of colloidal dimensions tend to remain in suspension for long periods in freshwater, but when mixed with salty water from the sea the

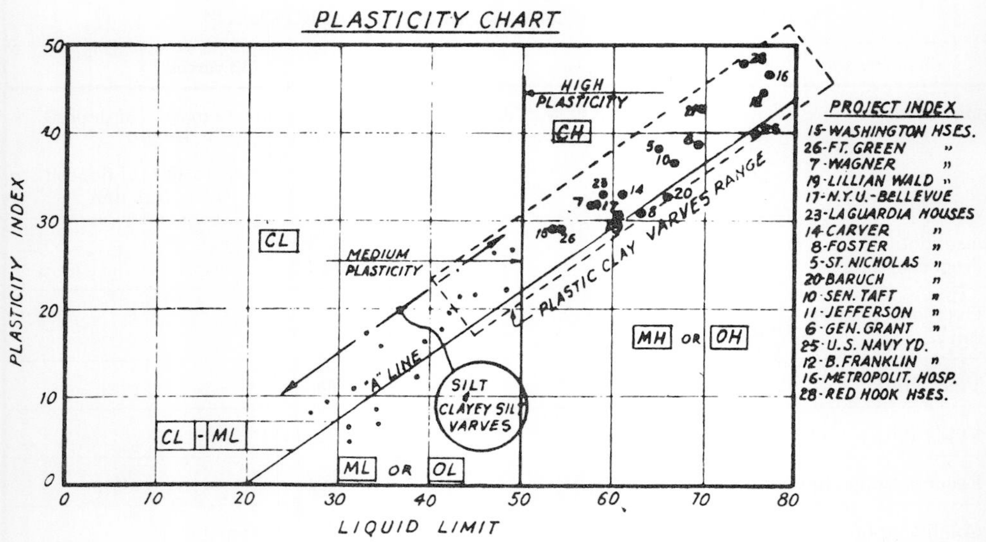


**FIG. 7.100** **Plasticity chart for silt and clay varves; varved clays from New York City.** [From Parsons (1976).[48]]

**FIG. 7.101** **Overconsolidation ratio vs. strength characteristics of varved clays from New York, New Jersey, and Connecticut.** [From Murphy et al. (1975).[49]]

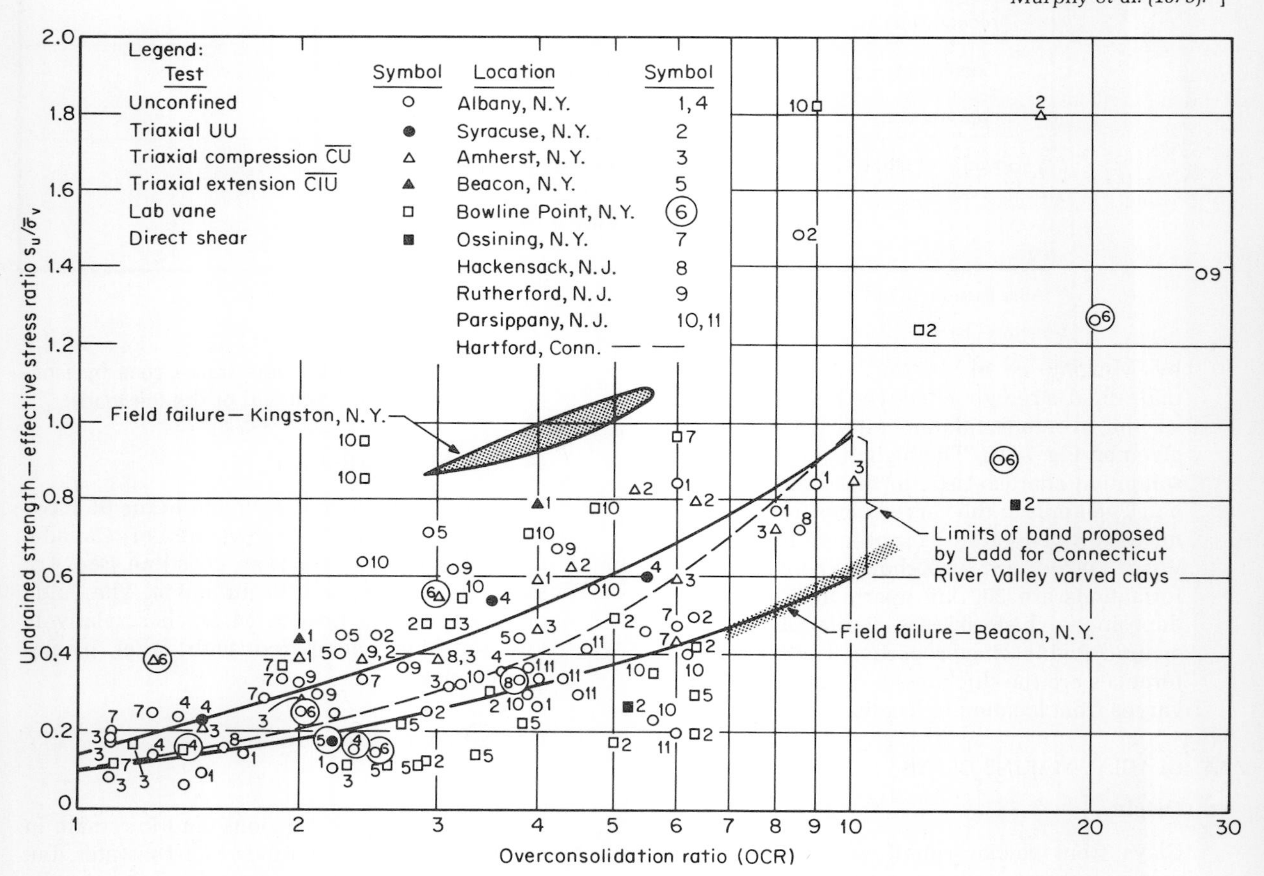

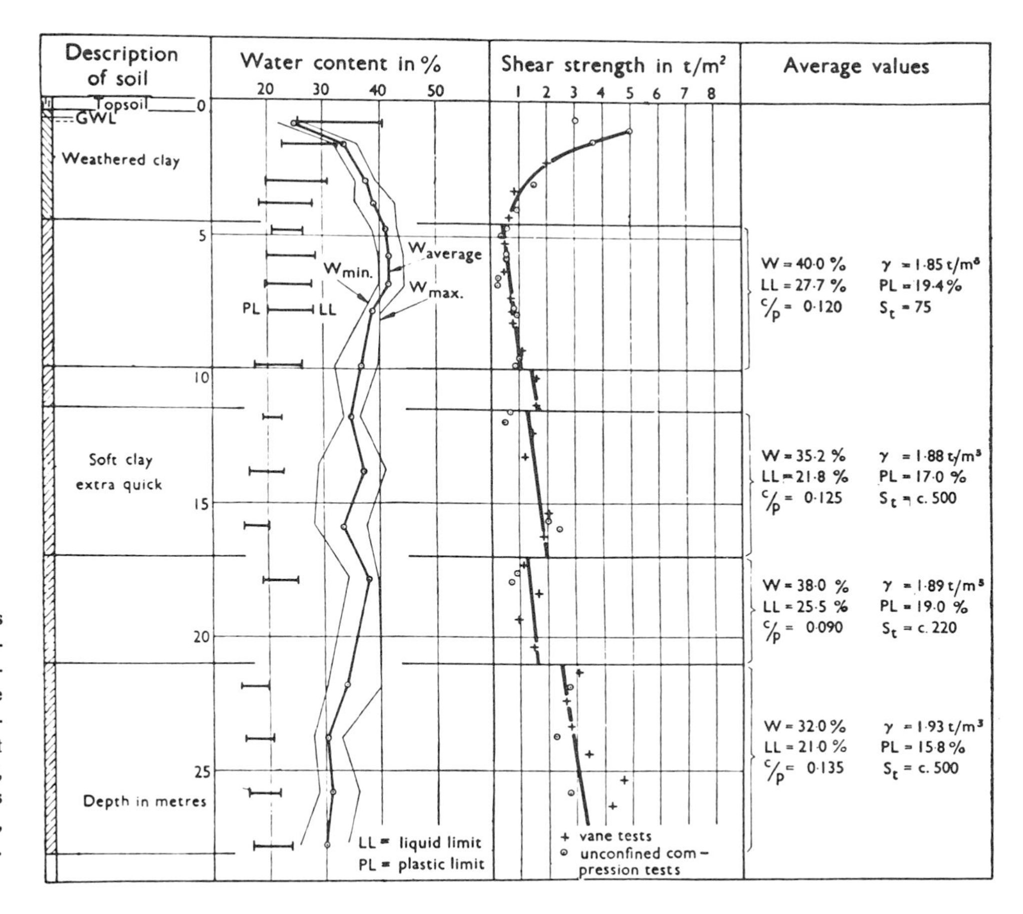


**FIG. 7.102 Properties of Norwegian glacio-marine clays from Manglerud in Oslo. Note the high sensitivity as compared with the clays at Drammen (Fig. 7.103), and the larger thickness of the weathered, preconsolidated zone.** [*From Bjerrum (1954).*[52]]

clay curdles into lumps and flocculates. The flocs settle out quickly, leaving the liquid clear, to form a very weak and compressible structure.

*Postdeposition*

When the formation is lifted above sea level by isostasy, freshwater may leach salt from the deposit, producing a clay of high sensitivity *(see Table 3.18, quick clays)*. The sensitivity increases with time as groundwater continues to leach the salt, slowly weakening the deposit until it can no longer retain its natural slope and a failure results, often in the form of a flow. Leaching occurs both from the downward percolation of rainfall and the lateral movement of groundwater, and the upward percolation caused by artesian pressures in fractured rock underlying the marine clays, a common condition. Bjerrum et al. (1969)[51] found that the sensitivity of Norwegian marine clays, directly related to the amount of leaching and the salt content, is greater where rock is relatively shallow, about 15 to 35 m than where it is deeper.

**Engineering Characteristics**

As described above, marine clays often become extremely sensitive *(see Art. 5.3.3)*, and are known as "quick clays." Land areas can be highly susceptible to vibrations and other disturbances, and even on shallow slopes they readily become fluid and flow *(see Art. 9.2.11)*.

*Norwegian Marine Glacial Clays*

Logs of borings and laboratory test results from two locations are given as Figs. 7.102 and 7.103. At Manglerud, a crust has formed and sensitivities are about $S_t$ = 500; the natural water con-

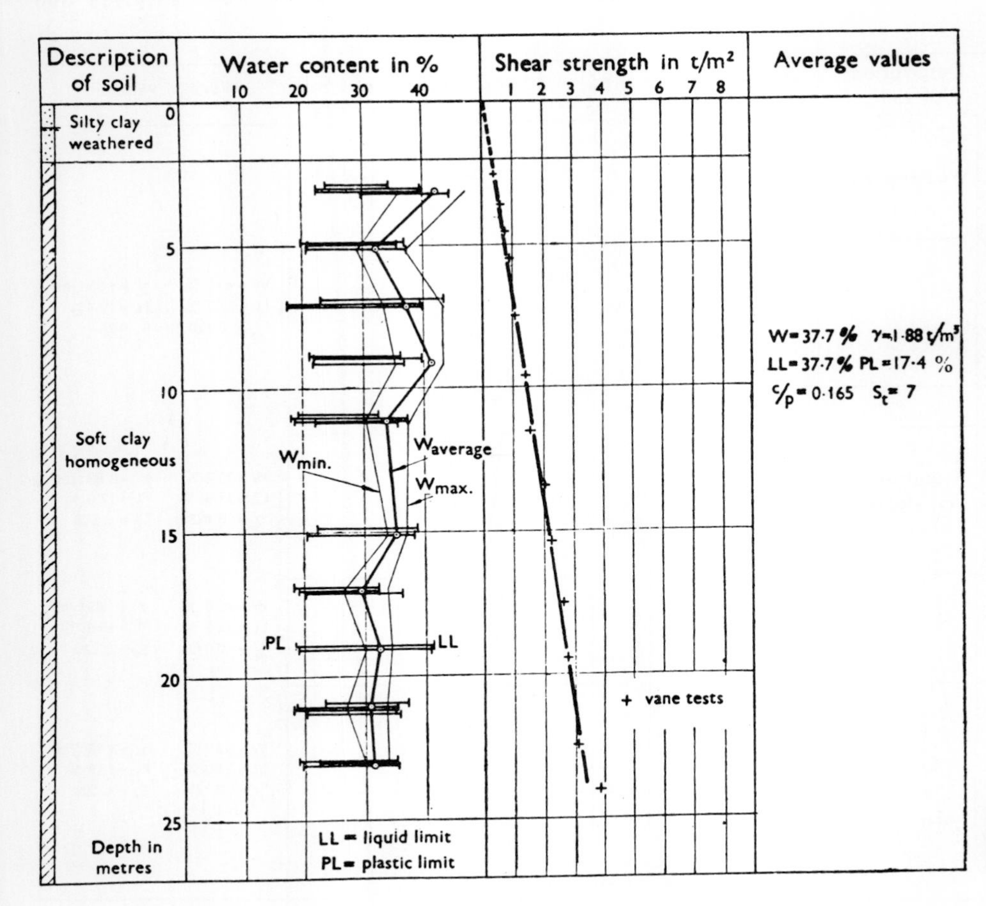


**FIG. 7.103 Properties of Norwegian glaciomarine clays at Drammen.** [*From Bjerrum (1954).*[52]]

tent is high above the liquid limit. At Drammen, where there is no crust, sensitivities are much lower ($S_t = 7$) and the natural water contents are in the range of the liquid limit. The difference is caused by the uplifting and leaching of the Manglerud clays [Bjerrum (1954)[52]].

*Canadian Leda Clays*

The leda clays of the St. Lawrence and Champlain lowlands are essentially nonswelling, but below a typical crust can have void ratios as high as 2.0 [Crawford and Eden (1969)[53]]. Some index and strength properties are given on Fig. 7.104. As with the Norwegian clays, the natural water content is typically above the liquid limit although the values for LL, PL, and $w$ are higher for the Canadian clays. Sensitivity is high, ranging from $S_t = 34$ to 150 below the crust, but lower than the clays at Manglerud. It is seen from Fig. 7.104 that the OCR is about 2.0 which is less than for most varved clay formations.

*Boston Blue Clays*

The Boston blue clays, so named because of their characteristic color, are a glacio-marine deposit that has undergone uplift, submergence, and reuplift [Lambe and Horn (1964)[55]]. As shown on Fig. 7.105, from the shear strength data, the sensitivity is relatively low compared with other glacio-marine clays. They are substantially overconsolidated and the zone of prestress extends far below the surface of the clay. Moisture contents are generally below the liquid limit. Many large structures are supported on mat foundations bearing almost directly on the clay [DiSimone and Gould (1972)[56]]. In one case cited, total settlements were about 0.14 ft

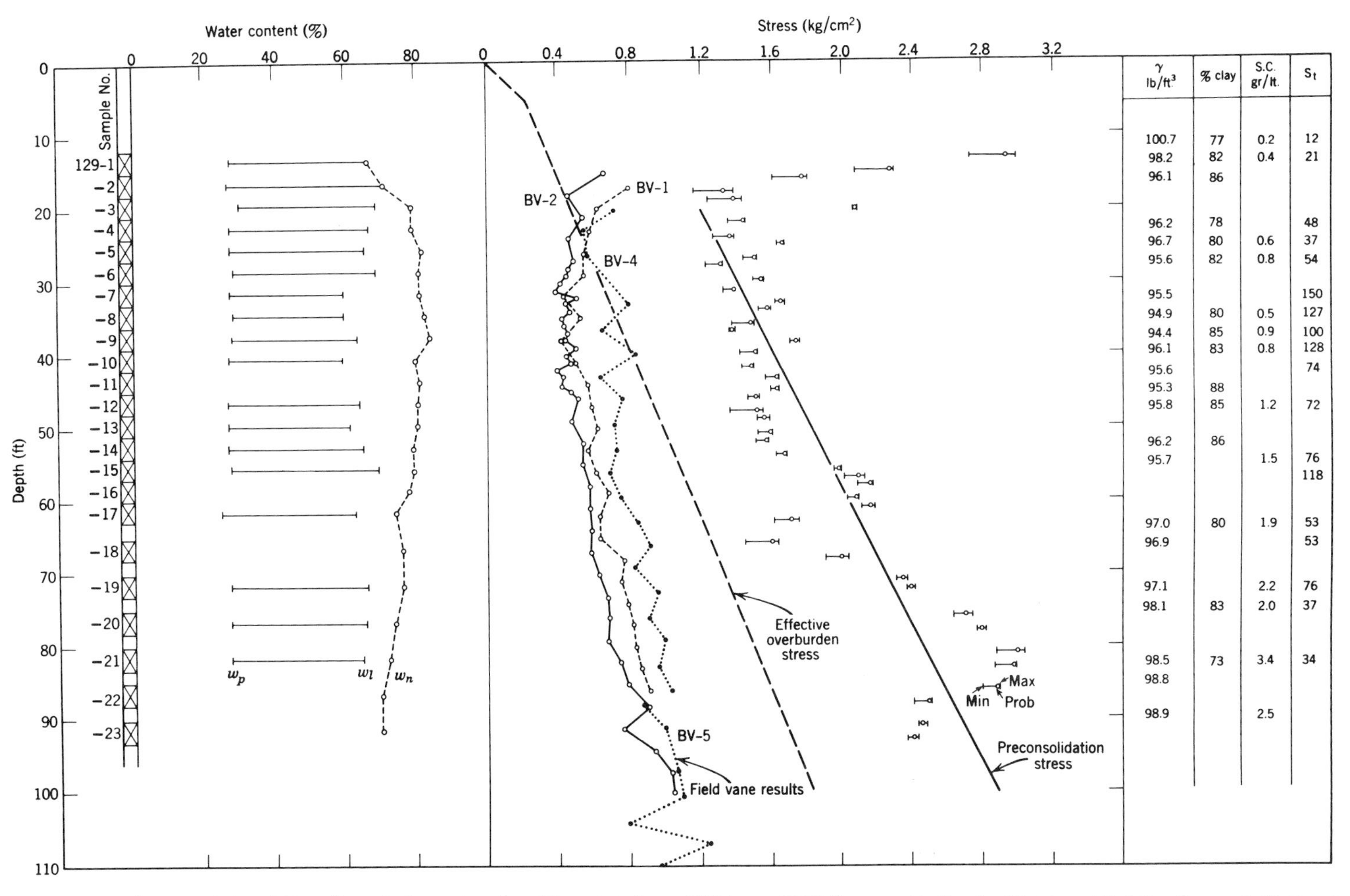


**FIG. 7.104 Characteristics of Canadian glaciomarine clay.** [*From Lambe and Whitman (1969),*[54] *as provided by the Division of Building Research, National Research Council of Canada. Reprinted with permission of John Wiley & Sons, Inc.*]

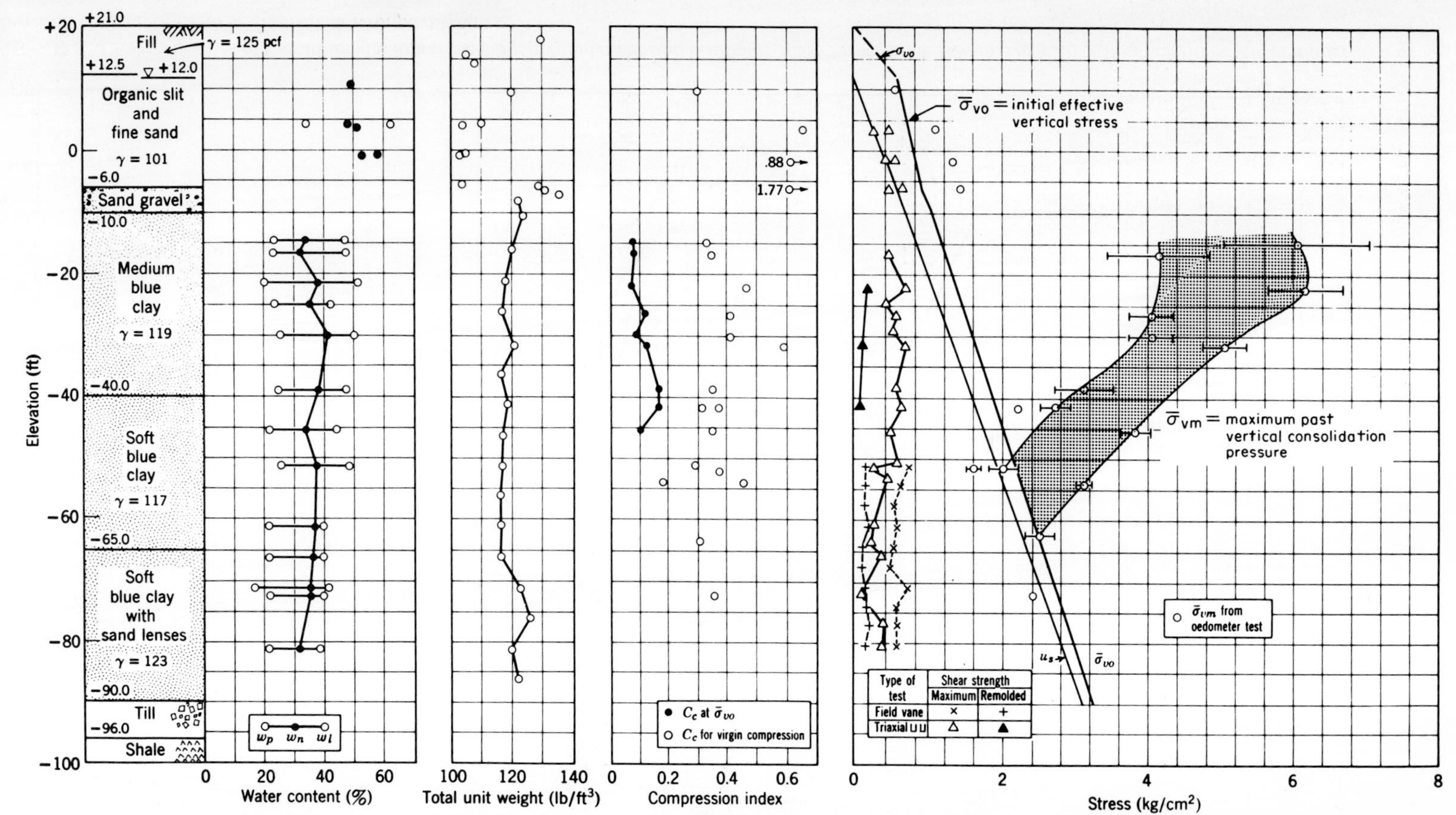


**FIG. 7.105 Section and laboratory test results of glaciomarine clays from Boston, Massachusetts.** [*From Lambe and Whitman (1969).*[54] *Reprinted with permission of John Wiley & Sons, Inc.*]

(43 mm) and were essentially complete within 7 years.

## 7.7 SECONDARY DEPOSITS

### 7.7.1 INTRODUCTION

This book considers *secondary deposits* as a soil classification by origin to include those formations resulting from the deposition of new minerals within a primary soil formation which result in its hardening. Two broad groups are considered:

- *Duricrusts:* the primary formation is hardened by the inclusion of iron, aluminum, carbonate, or silica.
- *Permafrost:* the formation is hardened by ice. Also included is seasonal frost.

### 7.7.2 DURICRUSTS

#### General

Duricrusts are highly indurated zones within a soil formation, often of rocklike consistency, forming normally in the B horizon (*Art. 7.8.1*), and can include either laterite, ironstone, or ferrocrete (iron-rich); bauxite (alumina-rich); calcrete or caliche (lime-rich); or silcrete (silica-rich).

#### Laterites

*Distribution*

Laterites extend over very large areas in tropical regions and are found in Brazil, Thailand, India, and central Africa.

*Description*

In the advanced state of formation, laterite, a residue of hydrous iron and aluminum oxide, is an indurated reddish-brown rocklike deposit which can develop to several meters or more in thickness, as shown on Fig. 7.106. The true rocklike laterites have been defined as having a silica-sesquioxide ratio of less than 1.33 (*Art. 7.2.2*). In some countries it is referred to as ferrocrete because of its hardness.

*Formation*

Laterites occur only in residual soils rich in iron and aluminum. Several basic conditions are required for their formation: basic ferromagnesian rocks; a hot, moist climate with alternating wet and dry periods; and conditions that permit the removal of silicates (*Art. 7.2.2*). If favorable climatic and groundwater conditions remain, the deposit continues to grow in thickness and induration. An important requirement appears to be that the soil remain dry for a substantial period, since induration does not proceed in permanently saturated ground, or in forested areas. The formation is characteristic of gentle slopes on higher ground above the water table. Laterization without induration can occur in rain-forest soils where the vegetation protects the deposit; when the forest is cleared, however, induration can occur dramatically within a few years, severely curtailing agricultural efforts.

**FIG. 7.106 Laterite deposit (Porto Velho, territory of Rondonia, Brazil).**

*Significance*

In the advanced state of induration, laterites are extremely stable and durable, resist chemical change, and will not soften when wet. Often located in areas lacking in quartz sands and gravels, they are an important source of aggregate for road construction, and even have been used for building facing stone.

**FIG. 7.107 Ironstone deposit in Tertiary coastal plain soils (Atlantic Highlands, Monmouth County, New Jersey).**

*Bauxite* is the alumina-rich variety of laterite, with a smaller area-wide distribution than the iron forms, and is of much less engineering significance.

**Ironstone**

Ironstone is a form of limonite found in some coastal plain formations as discontinuous beds usually from 1 to 2 m thick. It is hard, red-brown to reddish-purple, and often mistaken for bedrock in excavations or test borings. In the Atlantic Highlands of New Jersey it caps the hills, making them very resistant to erosion (Fig. 7.107).

In Brazil the deposit is called "canga" and is common to the Tertiary coastal plain sediments. North of Vitoria, in the state of Espirito Santo, the formation extends offshore for about a kilometer, forming a reef that discontinuously follows the shoreline. Its extremely irregular form is shown on Fig. 7.108. Exposed as a capping deposit over the inland hills of the Barrieras formation (Tertiary) from Espirito Santo to Bahia, 1000 km to the north, it has the appearance of laterite and is typically found in cobble-size fragments.

**Caliche (Calcrete)**

*Distribution*

Caliche is common to hot, semiarid regions such as Texas, Arizona, and New Mexico in the United States and Morocco in Africa.

*Description*

Because many factors can affect its deposition, caliche is variable in form. In many areas it may be a hard, rocklike material, and in others it may be quite soft. Characteristically white or buff in color, it can be discontinuous or have large voids throughout, or be layered, or massive. A massive formation is illustrated on Fig. 7.109. Typically it is not more than a meter or so in thickness.

*Formation*

An evaporite, caliche is formed when either surface water or groundwater containing calcium or magnesium carbonate in solution encounters conditions causing the carbonate to precipitate. As long as the general environment (topography, rainfall, and temperature) remains more or less constant, precipitation will continue and a mass of carbonate will form, cementing the soil into a nodular calcareous rock. The lower portion of caliche crust is likely to consist largely of angular rock fragments held together by the carbonate. Because it resembles concrete aggregate, it is commonly termed *calcrete*.

*Occurrence*

Its distribution, although erratic where it does occur, appears to have general boundaries beyond which it does not extend. In particular, waterways and major channels (washes) are free

**FIG. 7.108 Reef formation of "canga" in Tertiary coastal plain soils (Jacariape, Espirito Santos, Brazil).**

**FIG. 7.109 Caliche deposited in a terrace formation (Tucson, Arizona). Upper ½-m-thick layer at the top of the cut is hard, whereas the lower ½-m-thick stratum at the bottom is relatively soft. Both required cutting with a jackhammer.** *(Photo by R. S. Woolworth.)*

of caliche, as are areas where soil is aggrading at a relatively rapid rate, such as some portions of alluvial slopes extending out from the base of a mountain.

*Significance*

It is often necessary to resort to blasting for excavation, and caliche can be highly questionable material for foundation support, since it varies from hard to soft with erratic thickness, and is usually underlain by weaker materials.

**Silcrete**

A siliceous duricrust, silcrete covers up to 300,000 km$^2$ of semidesert in Australia in addition to areas of Africa. In Australia, the formation begins at a depth of about 50 cm and extends to depths of 10 m or more. The parent rock is commonly a granitic gneiss.

## 7.7.3 PERMAFROST AND SEASONAL FROST

### General

Frozen ground can be divided into two general classes:

- *Permafrost*, ground that remains permanently frozen, and
- *Seasonal frost*, ground that thaws periodically.

### Permafrost

#### *Distribution*

Black (1964)[57] has graded permafrost in the Northern Hemisphere into continuous, discontinuous, and sporadic (distributed in patches), corresponding accordingly with mean annual ground surface temperatures and permafrost depths as shown on Fig. 7.110. The southern boundary corresponds roughly with the 0°C mean annual temperature isotherm (about latitude 55°N). Along the southern boundary it is common to find relict permafrost at a depth of 10 to 15 m below the surface and about 10 m thick.

#### *Occurrence*

Where the mean annual temperature drops below 0°C, the depth of ground freezing in winter will exceed the depth of ground thawing in summer and a layer of permanently frozen ground will grow downward from the base of the seasonal frost (the *active zone*). The position of the *top* of the permafrost layer is the depth where the maximum annual temperature is 0°C, whereas the position of the *bottom* of the permafrost is determined by the mean surface temperature acting over long time periods. Heat flow from the earth's interior normally results in a temperature increase in the order of 1°C for every 30 to 60 m of depth; therefore, it could be anticipated that the depth to the bottom of the permafrost layer would be about 30 to 60 m for each degree Celsius below mean ground surface temperature of 0°C. This has, in fact, been found to be a good rule of thumb in places remote from water bodies [Lachenbruch (1968)[58]]. On the Arctic slope of Alaska, where the surface temperatures range from −6 to −9°C, permafrost has been found to extend to depths of 200 to 400 m.

The ground beneath large rivers and deep water bodies such as large lakes remains unfrozen for quite some depth. The depth of the active zone also is influenced by soil conditions; it is much deeper in free-draining soils than in clayey soils, and very thin under swamps and peat beds, which act as insulators.

#### *Terrain Features*

POLYGONAL PATTERNS Thermal contraction of the permafrost in winter generates tensile stresses that often result in tension cracks that divide the surface into polygonal forms. Ranging from 10 to 100 m across, they are strikingly obvious from the air. Summer meltwater draining into the cracks freezes to form veins of ice, which when repeated over long periods of time results in *ice wedges* that can be several meters wide at the top and many meters deep.

PINGOS These are dome-shaped hills resulting from the uplifting pressure of water freezing to form large ice lenses in the ground. Pingos can rise more than 50 m above the surrounding terrain and can measure more than several hundred meters in circumference.

#### *Engineering Characteristics*

ACTIVE ZONE The zone of seasonal freeze-thaw cycles is the most significant factor. Construction of structures, roadways, and fills causes the depth of the active zone to increase, often resulting in a saturated weak material providing poor support for structures. The result is differential settlement in the summer months when the ground thaws, and differential heave in the winter when the ground freezes. Heated structures placed near the ground surface are particularly troublesome. Structures are commonly supported on pile foundations steam-jetted into the permafrost to a depth equal to twice the thickness of the active zone. The pile must be protected from uplift caused by active zone freezing, in the same manner that piles are protected from uplift from swelling clays *(see Art. 10.6.4)*. Insulation between the ground surface and the underside of heated buildings is provided by an airspace or a gravel blanket.

STRENGTH CHARACTERISTICS OF ICE Terzaghi (1952)[59] noted that the unconfined compressive

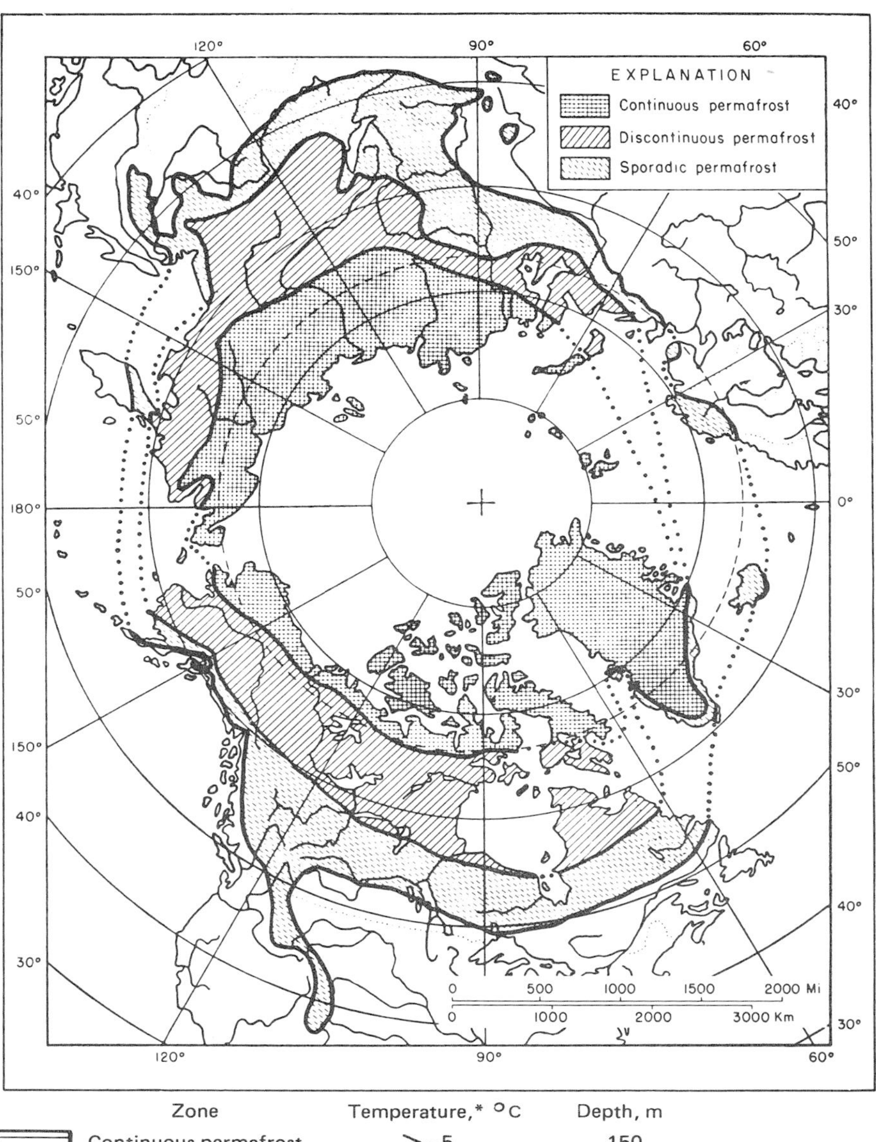


**FIG. 7.110 Contemporary extent of permafrost in the Northern Hemisphere.** [*From Black (1954).*[57]]

| Zone | Temperature,* °C | Depth, m |
|---|---|---|
| Continuous permafrost | > −5 | 150 |
| Discontinuous permafrost | −1 to −5 | 0 to 150 |
| Sporadic permafrost | < 1 | 0 to 30 |

*Mean annual ground surface temperature.

strength of ice depended on ice temperature, structure, and loading rate, and ranged from 21 to 76 kg/cm$^2$. Ice, however, has the capacity to *creep* under constant load. At a load less than about 2 kg/cm$^2$ and a temperature of −5°C, creep was found to be imperceptible, but under greater loads, the creep increased rapidly as the load increased. The tendency for creep to occur under relatively low deviator stress is responsible for the movement of glaciers.

SOLIFLUCTION The downslope movements resulting from the freezing and thawing of silty soils is called solifluction. The phenomenon is most common between the southern boundary of seasonal frost (the 5°C mean annual temperature isotherm) and the southern boundary of the permafrost region. At the foot of slopes sub-

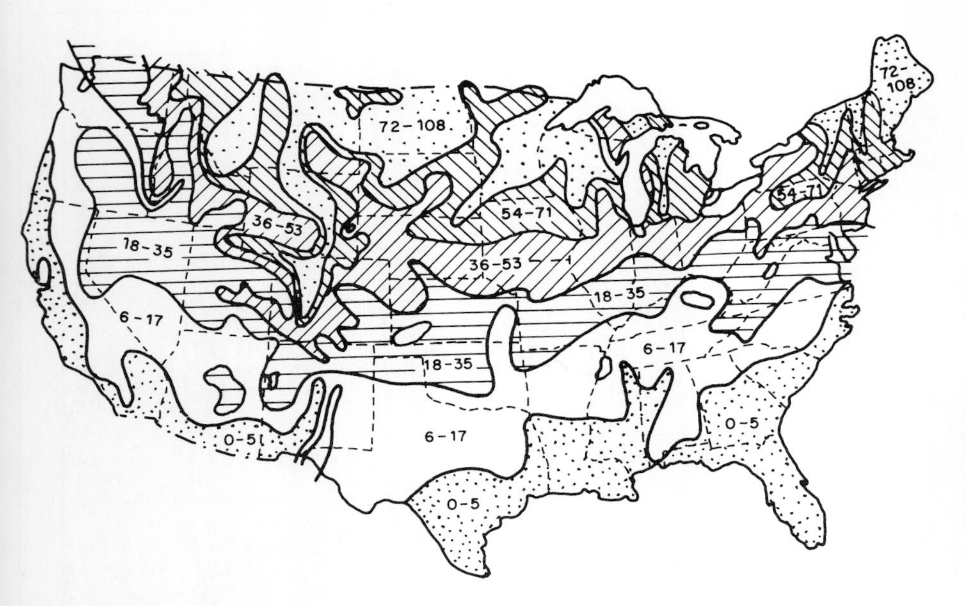


**FIG. 7.111 Approximate maximum depth of frost penetration in the United States given in inches.** *[From HRB Pub. 211 (1952).[60] Adapted with permission of the Transportation Research Board.]*

ject to solifluction, the soil strata may be intricately folded to a depth of more than 3 m.

### Seasonal Frost

*General*

In areas of seasonal frost the depth of frost penetration influences the design of pavements and foundations which is usually based on the maximum depth of frost penetration, given for the United States on Fig. 7.111.

Whether or not frost actually develops to the maximum depths depends on factors other than climate, mainly soil type and the depth to the static water table. Free-draining soils above the water table will develop very little frost. The frost susceptibility of soils increases with increasing fineness, which influences both internal drainage and capillarity as given on Table 3.10. Soil susceptibility to frost is discussed in Art. 5.3.5.

*Pavements*

As the depth of frost penetration increases, pavement thickness is increased accordingly. Full protection against frost heaving is considered to be achieved with a pavement thickness equal to the maximum depth of frost penetration; a thickness of one-half the depth of frost penetration is generally considered as the minimum required protection, as long as some risk of pavement deflection can be tolerated, such as in parking lots, for example.

*Foundations*

Exterior foundations and other foundations in soils subject to freezing are placed normally below the depth of maximum frost penetration. During construction, interior foundations and floors in unheated buildings must be protected from freezing during the winter months, otherwise substantial deflections will result from subsequent thawing.

## 7.8 PEDOLOGICAL SOILS AND PLANT INDICATORS

### 7.8.1 INTRODUCTION

#### Pedology

Although defined as a pure soil science [Rice (1954),[61] Hunt (1972)[27]], pedology is commonly recognized as the science that studies soils primarily from agricultural aspects. *Soils,* to the pedologist or soil scientist, are "a collection of natural bodies on the earth's surface, containing living matter, and capable of supporting plants" [SCS (1960)[62]].

Pedological information provides much useful data to the geologist and engineer, although in general the information pertains basically to depths within a few meters of the surface. It is available in the form of detailed maps *(see Fig.*

2.3) and reports such as published by the Soil Conservation Service of the U.S. Department of Agriculture.

### Modern Soils: The Soil Profile

The soil scientist has divided the modern soil profile into three major morphological units, referred to as *horizons* as shown on Fig. 7.112. The A (or O), B, and C horizons represent weathering zones and the D or R horizon represents unweathered parent materials. As water filters through the upper zone of plant debris and decayed organic matter, weak organic acids are formed. The weak acids and percolating water remove material from the zone beneath the organic layer and redeposit it at some depth below.

### Plant Indicators

For a given climate various plant species favor particular conditions of soil type and ground moisture, and therefore provide useful information about geologic conditions. Much more information on the subject should be provided in engineering publications than is readily available.

## 7.8.2 PEDOLOGICAL CLASSIFICATIONS

### Descriptive Nomenclature [SCS (1960)[62]]

Grain size definitions for soil components are given on Table 5.27. Sand, silt, and clay are less than 2 mm in diameter; gravel, 2 mm to 3 in; cobbles, 3 to 10 in; and boulders, over 10 in in diameter.

*Soil texture* refers to the gradation of particles below 2 mm in diameter, and depending upon relative percentages, soils are termed sand, sandy loam, loam, clay, etc. as given on Fig. 7.113. Loam refers to a detrital material containing nearly equal percentages of sand, silt, and clay.

Organic soils are described as *muck* (well-decomposed material) or *peat* (raw, undecomposed material).

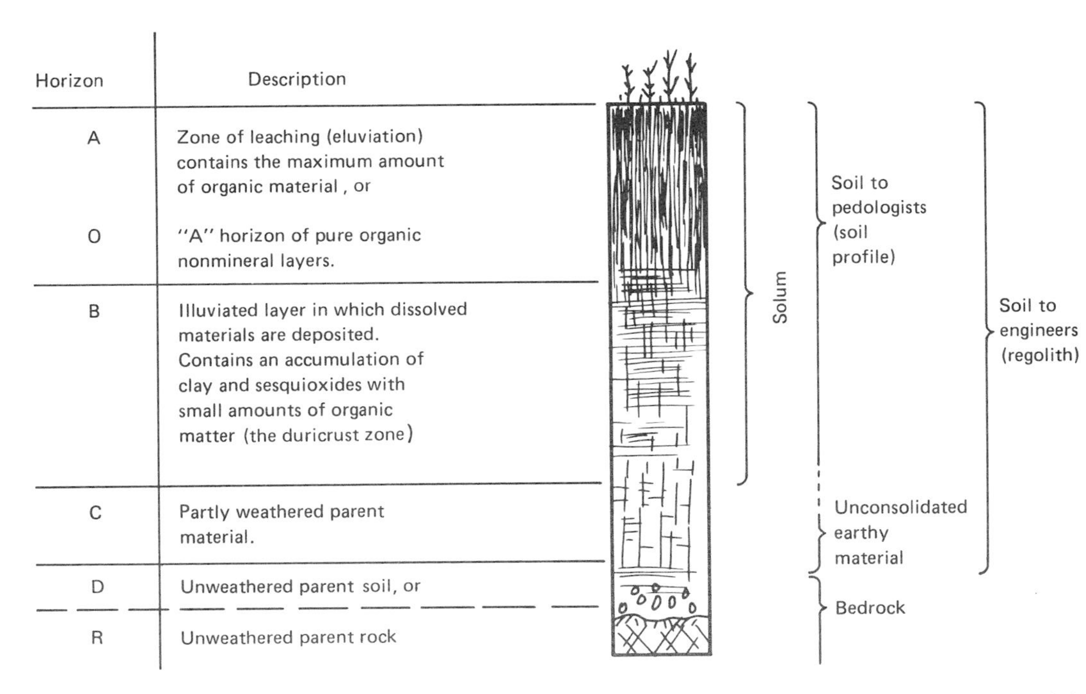


**FIG. 7.112 Generalized soil profile showing the morphological units and soil terminology used by pedologists and engineers. The *Modern Soil Profile* includes three major zones—A, B, and C horizons.**

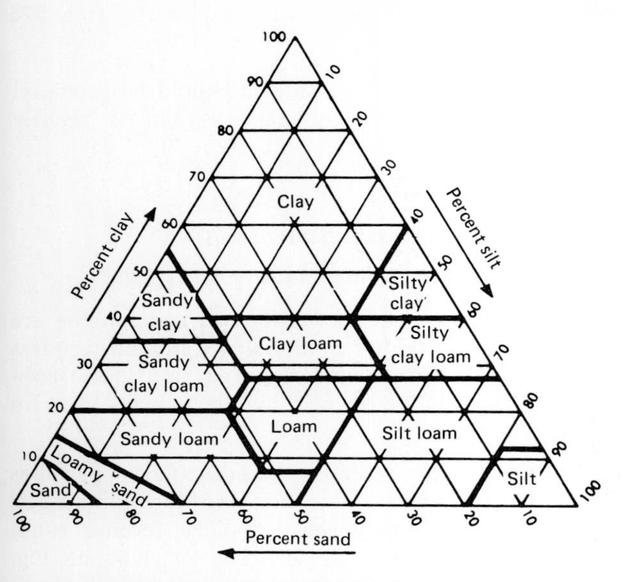


**FIG. 7.113 Soil texture classification of the USDA. Chart shows the percentages of clay (below 0.002 mm), silt (0.002 to 0.005 mm), and sand (0.05 to 2.0 mm) in the basic soil texture classes.** [*From SCS (1960).*[62]]

*Stoniness* describes surface conditions in terms of boulders, ranging in scale from 0 (no stones) to 5 (land essentially paved with stones).

*Rockiness* refers to the relative proportion of bedrock exposures, ranging in scale from 0 (no bedrock exposed) to 5 (land with 90% exposed rock).

*Soil structure* describes the shape of individual particles as prismatic, columnar, blocky, platy, or granular.

*Consistence* is caused by adhesion or cohesion and provides the soil with strength. Descriptive terminology is given for three conditions of soil moisture: dry, moist, and wet.

- Wet soil is described in terms of *stickiness* (adhesion to other objects), classed as 0 (none) to 3 (very sticky), and *plasticity* (the ability to form a thread upon rolling with the finger), classed from 0 (nonplastic) to 3 (very plastic).
- Moist soil is graded from 0 (loose or noncoherent) to 2 (friable) to 5 (very firm, can be crushed only under strong pressure).
- Dry soil is graded from 0 (loose or noncoherent) to 4 (very hard).

*Cemented soils* are described as:

- Weakly cemented—brittle and hard but can be broken in the hands.
- Strongly cemented—can be broken by hammer, but not in hands.
- Indurated—does not soften upon wetting, extremely hard, hammer rings with blow.

*Soil reaction* refers to acidity or alkalinity, given in 10 grades in terms of pH:

- Extremely acid—pH below 4.5
- Neutral—pH from 6.6 to 7.3
- Very strongly alkaline—pH 9.1 and above

*Symbols* used on pedological maps and elsewhere to designate soil horizons and special properties are given on Table 7.14.

## General Group Classifications

*Soil Profile Development*

As described in Art. 6.7.2, soil profile development depends on five factors: parent material, climate, topography, organisms, and time. Holding four variables constant and changing only one can result in a different profile development. For a given climate or locale, and the same parent rock type and time factor, the common variable is topography. Nomenclature has been developed to divide soils into broad groups as follows.

*Soil series* represent a group of soils having similar origin, color, structure, drainage, and arrangement in the soil profile, all derived from a common parent material. For example, a common series derived from a gravelly clay (glacial till) in Ohio, Illinois, and Indiana is the Miami series. Texture is variable and the series ranges from a fine sandy loam to a silty clay loam.

*Soil catena* represents a soil profile derived from the same parent material, but varying with different topographic expressions, i.e., slope and drainage. Its use in mapping is more definitive to the geomorphologist than is the soil series concept. The soil catena of the glacial till of Ohio, Illinois, and Indiana, which includes the Miami series, includes the soils listed on Table 7.15, as a function of slope and drainage.

*Associations* are combinations of series used

**TABLE 7.14**
**DESCRIPTIVE NOMENCLATURE FOR SOIL HORIZONS***

| HORIZON DESCRIPTION | | | |
|---|---|---|---|
| O | Organic layer | $B_1$ | B layer gradational with the A |
| $A_1$ | Organic rich A layer | $B_2$ | Layer of maximum deposition |
| $A_2$ | Layer of maximum leaching | $B_3$ | B layer gradational with the C |
| $A_3$ | A layer gradational with B | C | Weathered parent material |
| | | D | Parent material (R—rock) |

| DESIGNATION OF SPECIAL PROPERTIES | |
|---|---|
| b | Soil layer buried by surface deposit. A leached layer buried under a sand dune would be indicated $A_b$. |
| ca | An accumulation of calcium carbonate. |
| cn | An accumulation of concretions, usually of iron, manganese and iron, or phosphate and iron. |
| cs | An accumulation of calcium sulfate (gypsum). |
| f | Frozen ground; permafrost. |
| g | A waterlogged layer (gleyed). |
| h | An unusual accumulation of organic matter. |
| ir | An accumulation of iron. |
| m | An indurated layer, or hardpan, due to silification or calcification. |
| p | A layer disturbed by plowing; a plowed leached layer would be designated $A_p$. |
| sa | An accumulation of soluble salts. |
| t | An accumulation of clay. |

*After SCS (1960).[62]

when the map scale is such that detailed delineation is not possible or warranted.

*Mapping*

The soil scientist gives the soils names relating to the locale in which the soils occur or were first encountered. Maps can represent *series*, *catenas*, or *associations*. The soil map given as Fig. 2.3 presents a soil series.

*Major Group Classifications*

The Great Soil Groups of the World are classed on the basis of climate.

The Seventh Approximation, or New Soil Taxonomy, divides the soils of the world into 10 main categories, or orders, based on distinguishing characteristics.

### The Great Soil Groups of the World

*General*

Soil scientists considered soils to be of three main types, before the development of the New Soil Taxonomy, as follows:

- *Zonal soils* constitute the *Great Soil Groups of the World* in which climate is the major factor in development. Described below, they are subdivided into two groups by climate. In humid regions, soils are acidic, termed *pedalfers* to emphasize the removal of aluminum (al) and iron (fer) from the leached A horizon. In arid regions there is little leaching and all dissolved matter is precipitated as the water evaporates and layers of carbonates are formed. Soils containing carbonate layers are termed *pedocals*.
- *Interzonal soils* reflect some local conditions that cause a variation in the zonal soils, such as muck.
- *Azonal* soils are without profile development.

*Tundra soils* develop under Arctic type of vegetation at high altitudes and latitudes. Drainage conditions are usually poor and boggy. Underlain by a permanently frozen substratum, the profile is shallow and much decomposed matter is found at the surface.

*Podzol soils* possess well-developed A, B, and C

TABLE 7.15
SOIL CATENA OF GLACIAL TILL OF OHIO, INDIANA, AND ILLINOIS*

| Slope | Drainage | Soil |
|---|---|---|
| Steep (20–55%) | Good | Hennipin |
| Moderate ( 4–15%) | Good | Miami |
| Slight ( 1– 2%) | Good | Crosby |
| Flat ( 0– 1%) | Fair | Bethel |
| Flat ( 0– 1%) | Slight depression | Brookston |
| Flat ( 0– 1%) | Deep depression | Kohoms |
| Flat ( 0– 1%) | Deepest depression | Carlisle (muck) |

*After Thornbury (1969).[37] Reprinted with permission of John Wiley & Sons, Inc.

horizons. The surface material is organic matter under which a whitish or grayish layer develops. The name derives from two Russian words meaning under and ash. Below the gray layer is a zone in which iron and aluminum minerals accumulate. The A and B horizons are strongly acid. They develop under coniferous and mixed hardwood forests.

*Laterites* are soils formed under hot, humid conditions and under forest vegetation. They have a thin organic cover over a reddish leached layer, which in turn is underlain by a still deeper red layer. Hydrolysis and oxidation have been intense. They are rather granular soils, and are confined mainly to tropical and subtropical regions, although some soils in the middle latitudes have been described as lateritic. *(See also Art. 7.7.2.)*

*Chernozems* originate under tall-grass prairie vegetation. The name is the Russian word for black earth and suggests the color and high organic content of the A horizon. The B layer exhibits an accumulation of calcium carbonate rather than leaching. Columnar structure is common, and they are the most fertile soils, typically developing from loess deposits.

*Chestnut soils* are brown or grayish brown soils that develop under short-grass vegetation in areas slightly drier than those that produce chernozems. Secondary lime is found near the surface and the profile is weakly developed.

*Brown aridic soils* are found around the margins of deserts and semiarid regions. They have a low organic content and are highly calcareous.

*Gray desert soils* (sicrozems) and *reddish desert soils* develop under desert or short-grass vegetation. Calcium carbonate accumulates near the surface. The sicrozems are found in the continental deserts and the reddish soils are found in what are termed the subtropical deserts.

*Noncalcic brown soils* form in areas which originally had forest or brush vegetation. Weak podzolification makes the surface layer slightly acidic.

### The Seventh Approximation (New Soil Taxonomy)

#### *General*

In 1960, the USDA published a new classification system referred to as *The Seventh Approximation* [SCS (1960)[62]], now called the *New Soil Taxonomy*. All of the soils of the earth are divided into 10 major categories called *orders*, which are based on distinguishing characteristics rather than climatic factors. The detailed descriptive nomenclature is extremely complex.

#### *Orders*

The 10 soil orders are described in summary form on Table 7.16. Their worldwide distribution is given on Fig. 7.114, and a diagram illustrating the general relationship between The Great Soil Groups and the soils of The Seventh Approximation is given on Fig. 7.115.

**TABLE 7.16**
**THE 10 SOIL ORDERS OF THE SEVENTH APPROXIMATION***

| Soil order | Climatic range | Natural vegetation | Parent materials | Horizon development | Drainage | Colors |
|---|---|---|---|---|---|---|
| Entisol | All climates, arid to humid, tropical to polar. | Highly variable. Forests, grass, desert, tidal marsh. | Primarily free-draining alluvium and aeolian, and low-activity clays. | None, except perhaps for a thin A horizon. | Good; slopes not significant. | Not significant. |
| Vertisols | Subhumid to arid with wet and dry seasons, or arid areas subject to flooding. | Grasses and woody shrubs. | High-activity clays, shrinking when dry, swelling when wet, with montmorillonite common. | B horizon commonly absent. | Poor; slopes flat to gentle. | Black, gray, brown. |
| Inceptisols | Humid, from arctic to tropical. | Mostly forests, occasionally grasslands. | Young soils from residuum, loess, glacial till. Soils moist. | One or more formed without significant illuviation or eluviation. A horizon very organic. | Slopes flat to moderately steep. | Light to dark. |
| Aridsols | Arid to semiarid. | Sparse grasses and other desert vegetation. | Generally alluvium. | Poor horizon development and very little organic matter. Soil rich in lime, gypsum, or sodium chloride. Caliche common. | Slopes flat to gentle. | Light to red. |
| Mollisols | Alpine to tropical; with cool, dry seasons, and hot, moist seasons. | Grasses, sedges, hardwood forests. | Variable: loess, alluvium, till, residuum. | Distinct horizons. Highly organic A layer. B and C horizons may have secondary lime accumulations. | Flat to slight slopes and poor drainage. | Medium to dark colors. |
| Spodosol | Humid regions, alpine to tropical. | Coniferous forest, savannah, or rain forest. | Usually siliceous, granular, and not very clayey. | B horizon illuviated with organic matter, iron and aluminum oxides. | Good. | Black, brown, reddish. |
| Alfisols | Cool humid to subhumid with seasonal rainfall. | Deciduous forest, some tall grasslands. | Variable, but generally young; alluvium, till, loess, or coastal plain. Calcareous in cool, humid climates. | Thin, highly organic A horizon. B horizon illuviated with clay, organic matter, and iron oxide. | Poor. Flat to gentle slopes. | Black, brown, reddish. |
| Ultisols | Humid. | Forest, savannah, marsh, or swamp. | Old and strongly weathered residuum or coastal plain. | Thin A horizon with some humus over illuviated B horizon. Approaching a lateritic soil. | Poor to fair. Slopes moderate to steep. | Variable and often mottled with gray, yellow, and red. |
| Oxisols | Tropics and subtropics with wet and dry seasons. | Forest to savannah. | Lateritic soils primarily from basic rock residuum. | Little organic soil development. Clay content high but formation porous. Concentrations of iron and aluminum at various depths. | Good. | Reds, browns. |
| Histosols | Moist to wet. | Swamp, marsh, and bog. | Organic materials. | None visible. | Very poor. | Gray to black. |

*After SCS (1960).[62]

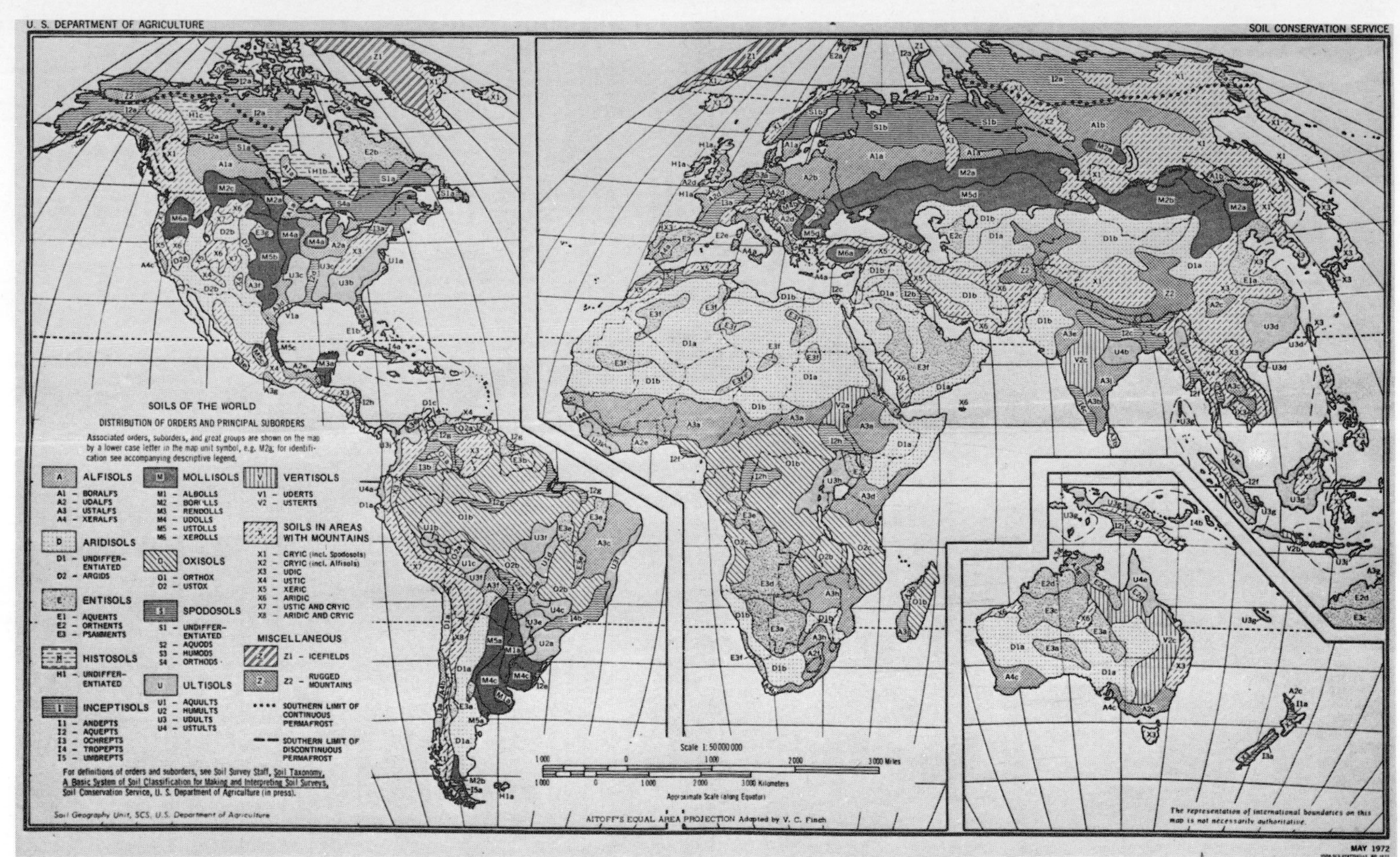


**FIG. 7.114 World distribution of soils as classed by the U.S. Department of Agriculture.** [*From SCS (1960).*[62]]

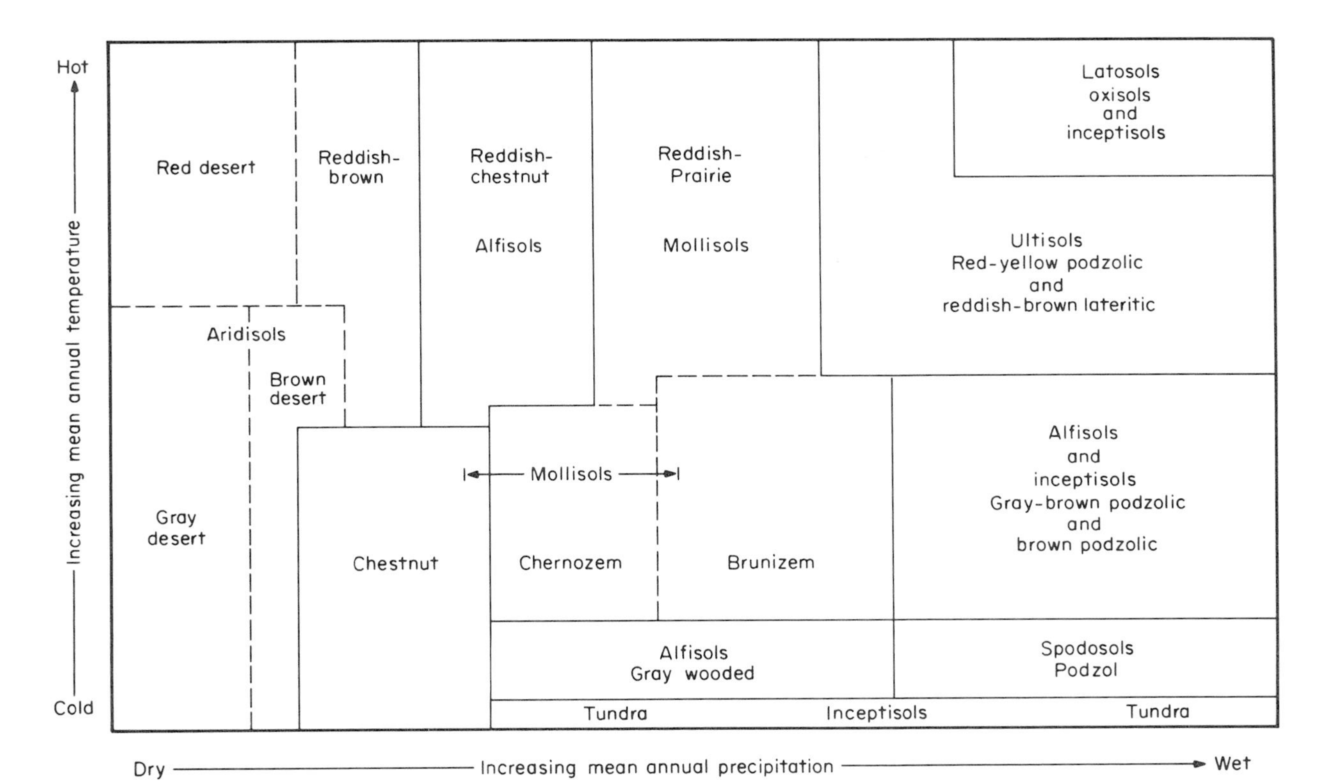


**FIG. 7.115 Diagram showing the relationship of the Great Soil Groups of the World and the classification terminology of the Seventh Approximation.** [*From Zumberge and Nelson (1972).*[36] *Reprinted with permission of John Wiley & Sons, Inc.*]

## 7.8.3 PLANT INDICATORS

### Significance

There is a strong relationship between subsurface conditions and vegetation because the various species of plants that inhabit a particular location differ in their requirements for water and nutrients. A general relationship between climate, vegetation, and shallow subsurface conditions is given on Fig. 7.116.

### Some Indicators

- Trees growing in a line may be indicative of seepage along a terrace edge, or a fault zone in an arid or semiarid climate.
- Orchards are typically found in well-drained areas.
- Willows and hemlocks require substantial amounts of moisture.
- Poplars and scrub oaks are found in areas of low moisture (sandy soils above the water table in a moist climate).
- Banana trees prefer colluvial soils over residual soils on slopes in tropical climates.
- Thin vegetative cover in moist climates indicates either a free-draining soil or a rock formation with little or no soil cover, or a deep water table which can occur in sands, "porous clays," or weathered foliated rocks.

### Some Geographic Relationships

- In the Rocky Mountains, stands of aspen seem to favor damp ground underlain by colluvium with much organic matter.
- In New Jersey good farms and forests grow in the coastal plain region where the Cretaceous clays containing the potassium-rich glaconite

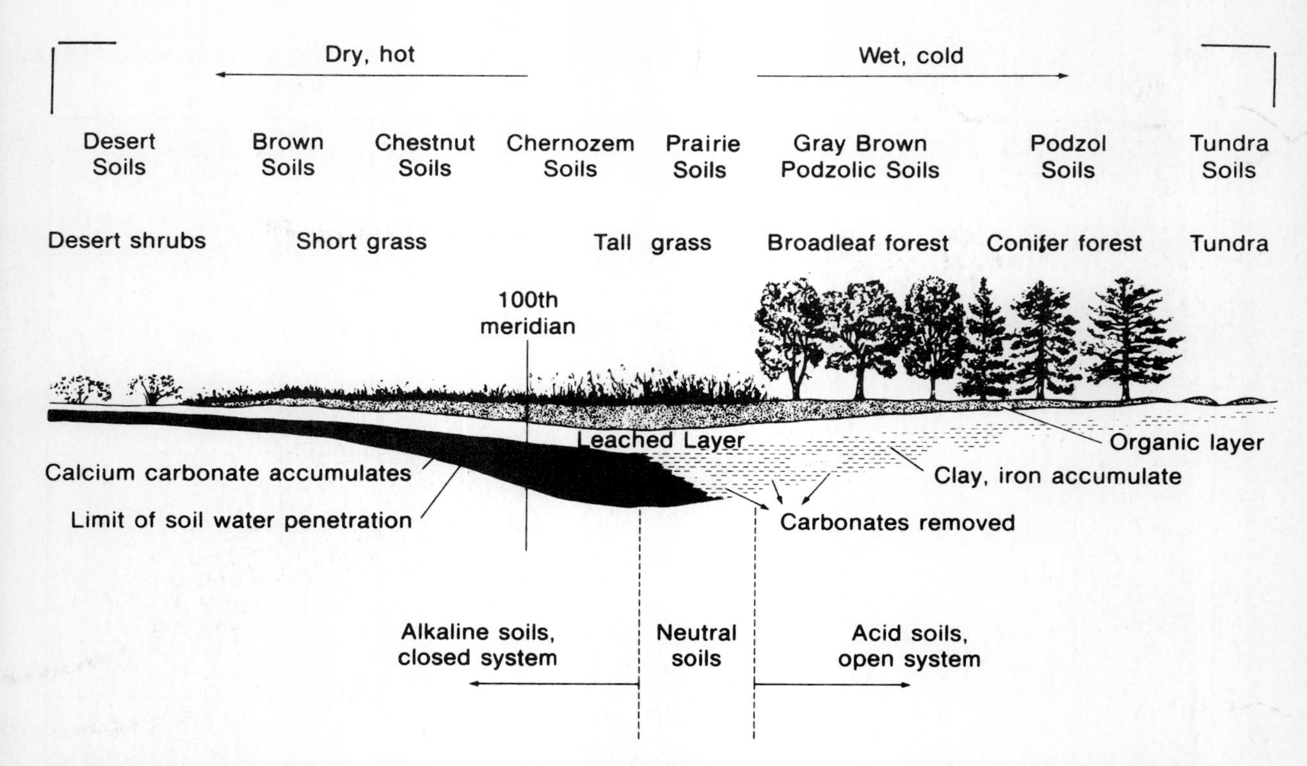


**FIG. 7.116 Transect illustrating changes in soil profiles that accompany changes in vegetation and climate between the tundra in northern Canada and the deserts in the southwestern United States. At the 100th meridian, the annual precipitation averages about 20 in; there and to the west, the soils are alkaline. The easternmost grassland soils are about neutral; farther east, the soils are acid.** [*From Hunt (1972).*[27] *© W. H. Freeman and Company. Reprinted with permission.*]

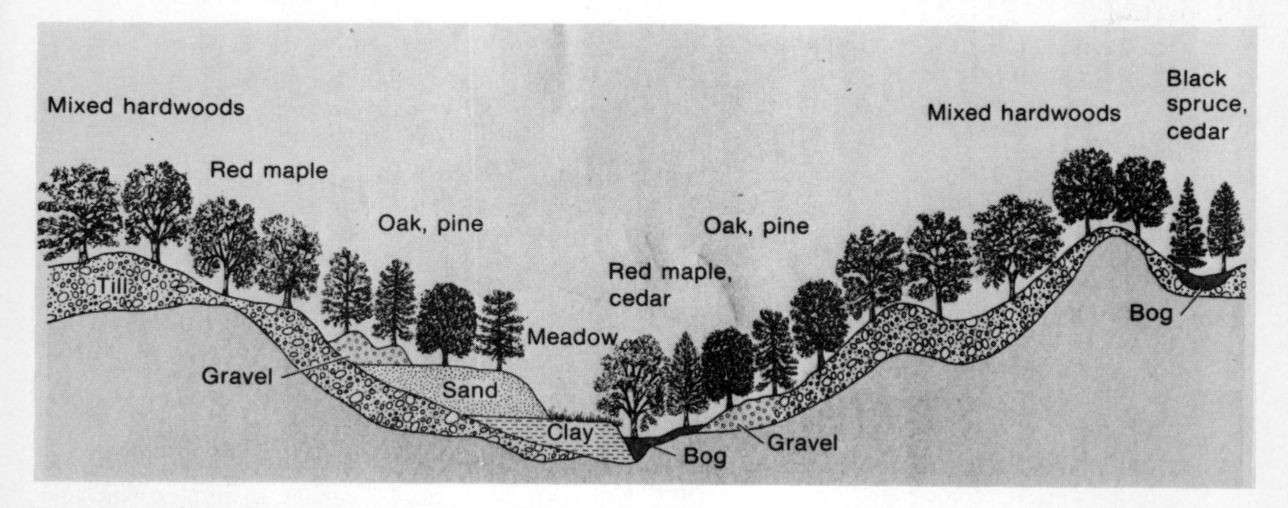


**FIG. 7.117 Relationship between vegetation and kind of ground in a glaciated Connecticut valley. Well-drained uplands have mixed hardwoods; excessively drained gravel and sand have oak and pine. Poorly drained ground, on clay, has meadow; upland bogs have black spruce and cedar; bogs in the alluvial valley have red maple and cedar. Rocky promontories have scarlet, chestnut, and black oak.** [*From Hunt (1972).*[27] *© W.H. Freeman and Company. Reprinted with permission. See also topographic map in Fig. 7.92.*]

Alluvial floodplain

| Species | % |
|---|---|
| Sycamore | 30 |
| Sweet gum | 30 |
| Red maple | 30 |
| Others | 10 |

Gravelly valley side

| Species | % |
|---|---|
| Red maple | 23 |
| Black oak | 20 |
| Hornbeam | 18 |
| Sweet gum | 14 |
| Hickory | 11 |
| Others | 14 |

Colluvium over deep saprolite

| Species | % |
|---|---|
| Sweet gum | 32 |
| Tulip tree | 29 |
| Red maple | 6 |
| Hickory | 6 |
| Black oak | 6 |
| Others | 21 |

Shallow saprolite over gneissic bedrock

| Species | % |
|---|---|
| Beech | 26 |
| Black oak | 16 |
| Red maple | 13 |
| White oak | 13 |
| Hickory | 13 |
| Dogwood | 10 |
| Others | 9 |

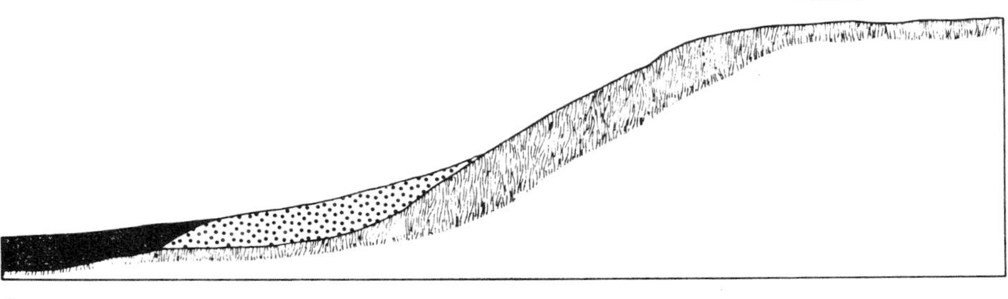

A

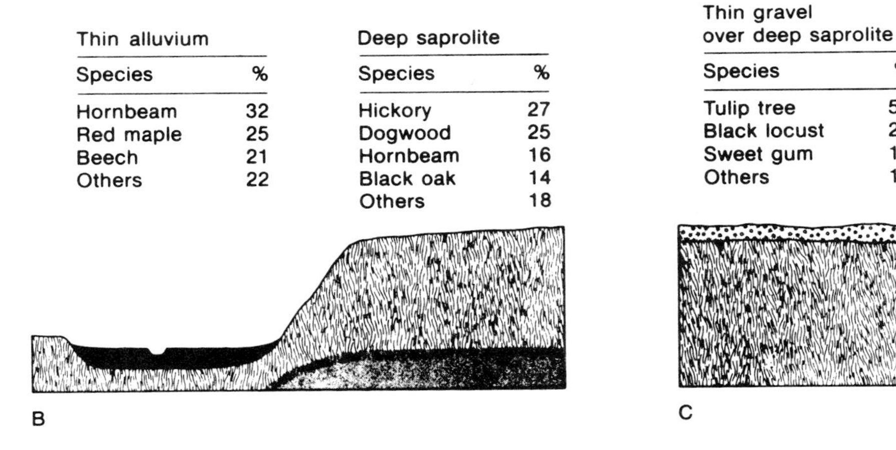

Thin alluvium

| Species | % |
|---|---|
| Hornbeam | 32 |
| Red maple | 25 |
| Beech | 21 |
| Others | 22 |

Deep saprolite

| Species | % |
|---|---|
| Hickory | 27 |
| Dogwood | 25 |
| Hornbeam | 16 |
| Black oak | 14 |
| Others | 18 |

Thin gravel over deep saprolite

| Species | % |
|---|---|
| Tulip tree | 50 |
| Black locust | 22 |
| Sweet gum | 11 |
| Others | 17 |



**FIG. 7.118 Differences in percentages of various plant species, evidently reflecting differences in ground conditions in the Piedmont province, along Kennedy Expressway in Maryland. Geology and plant distribution as revealed by construction in 1962.** [*From Hunt (1972).*[27] © *W. H. Freeman and Company. Reprinted with permission.*]

outcrop, whereas only scrub oak and pines grow on neighboring sandy soils.

- In Maryland, weathered serpentine, rich in magnesium but deficient in other minerals, contains dwarfed vegetation.
- In the Gulf states, from Texas to Alabama, growth on limy formations is marked by belts of tall grass, whereas adjoining sandy formations support pine forests.
- Sections relating ground conditions to vegetation from two locations are shown, including a glaciated Connecticut valley (Fig. 7.117; *see also Fig. 7.92*, a topographic map), and the Piedmont province in Maryland (Fig. 7.118).

## REFERENCES


1. NAVFAC (1971) *Design Manual DM7, Soil Mechanics, Foundations and Earth Structures,* Naval Facilities Engineering Command, Alexandria, Va.
2. Way, D. (1978) *Terrain Analysis,* 2d ed, Dowden, Hutchinson & Ross Publ., Stroudsburg, Pa.
3. Sowers, G. F. (1954) "Soil Problems in the Southern Piedmont Region," *Proc. ASCE, J. Soil Mechs. Found. Engrg. Div.,* March, Separate No. 416.
4. Vargas, M. (1974) "Engineering Properties of Residual Soils from South-Central Brazil," *Proc. 2d. Intl. Cong. Intl. Assoc. Engrg. Geol.,* São Paulo, Vol. I.
5. Medina, J. (1970) "Propriedades Mecânicas dos Solos Residuais," *Pub. No. 2/70,* COPPE, Universidade Federal do Rio de Janeiro, p. 37.
6. Vargas, M. (1953) "Some Properties of Residual Clays Occurring in Southern Brazil," *Proc. 3d Intl. Conf. Soil Mechs. Found. Engrg.,* Zurich, Vol. I, p. 67.
7. Morin, W. J. and Tudor, P. C. (1976) *Laterite and Lateritic Soils and Other Problem Soils of the Tropics,* AID/csd 3682, U.S. Agency for International Development, Washington, D.C.
8. Meyer, K. T. and Lytton, A. M. (1966) "Foundation Design in Swelling Clays," paper presented to Texas Section ASCE, October.
9. D'Appolonia, E., Alperstein, R. A., and D'Appolonia, D. J. (1969) "Behavior of a Colluvial Soil Slope," *Proc. ASCE, Stability and Performance of Slopes and Embankments,* ASCE, New York, pp. 489–518.
10. Hjulstrom, F. (1935) "Studies of the morphological activity of rivers as illustrated by the River Fyris," Uppsala Geological Inst., Bull. 25.
11. Tanner, W. F. (1968) "Rivers—Meandering and Braiding," *Encyclopedia of Geomorphology,* R. W. Fairbridge, ed., Dowden, Hutchinson & Ross Publ., Stroudsburg, Pa., pp. 954–963.
12. Palmer, L. (1976) "River Management Criteria for Oregon and Washington," *Geomorphology and Engineering,* D. R. Coates, ed., Dowden, Hutchinson & Ross Publ., Stroudsburg, Pa., Chap. 16.
13. Kolb, C. R. and Shockley, W. G. (1957) "Mississippi Valley Geology—Its Engineering Significance," *Proc. ASCE, J. Soil Mechs. Found. Engrg. Div.,* July, pp. 1289–1298.
14. Coleman, J. M. (1968) "Deltaic Evolution," *Encyclopedia of Geomorphology,* R. W. Fairbridge, ed., Dowden, Hutchinson & Ross Publ., Stroudsburg, Pa., pp. 255–260.
15. Supp, C. W. A. (1964) "Engineering Geology of the Chesapeake Bay Bridge," *Engineering Geology Case Histories 1-5,* Trask and Kiersch, eds., Geologic Soc. of Amer., New York, pp. 49–56.
16. Sanborn, J. (1950) "Engineering Geology in the Design and Construction of Tunnels," *Engineering Geology* (Berkey Volume), Geologic Soc. of Amer., p. 49.
17. Simon, R. M., Christian, J. T., and Ladd, C. C. (1974) "Analysis of Undrained Behavior of Loads on Clays," *Proc. ASCE, Conf. Analysis and Design in Geotech. Engrg.,* Austin, Texas, June, Vol. I, pp. 51–84.
18. Skempton, A. W. and Henkel, D. J. (1953) "The Post-Glacial Clays of the Thames Estuary at Tilbury and Shellhaven," *Proc. 3rd Intl. Conf. Soil Mechs. Found. Engrg.,* Zurich, Vol. I, p. 302.
19. Newman, W. S. (1968) "Coastal Stability," *Encyclopedia of Geomorphology,* R. W. Fairbridge, Dowden, Hutchinson & Ross Publ., Stroudsburg, Pa., pp. 150–155.
20. Shepard, F. P. (1968) "Coastal Classification," *Encyclopedia of Geomorphology,* R. W. Fairbridge, ed., Dowden, Hutchinson & Ross Publ., Stroudsburg, Pa., pp. 131–133.
21. Hamblin, W. K. and Howard J. L. (1971) *Physical Geology Laboratory Manual,* Burgess Publ. Co., Minneapolis.
22. Freeman, D. W. and Morris, J. W. (1958) *World Geology,* McGraw-Hill Book Co., New York.
23. Colquhoun, D. J. (1968) "Coastal Plains," *Encyclopedia of Geomorphology,* R. W. Fairbridge, ed., Dowden, Hutchinson & Ross Publ., Stroudsburg, Pa., pp. 144–149.
24. Casagrande, L. (1966) "Subsoils and Foundation Design in Richmond, Virginia," *Proc. ASCE, J. Soil Mechs. Found. Engrg. Div.,* September, pp. 106–126.
25. Gould, J. P. (1960) "A Study of Shear Failure in Certain Tertiary Marine Sediments," *Proc. ASCE, Research Conf. on Shear Strength of Cohesive Soils,* Boulder, Colo., June.
26. Skempton, A. W. and Henkel, D. J. (1957) "Tests on London Clay from Deep Borings at Paddington, Victoria and the South Bank," *Proc. 4th Intl. Conf. Soil Mechs. Found. Engrg.,* London, p. 100.

26a. Longwell, C. R., Knopf, A., and Flint, R. F. (1948) *Physical Geology,* 3d ed., John Wiley & Sons, Inc., New York.

27. Hunt, C. B. (1972) *Geology of Soils,* W. H. Freeman and Co., San Francisco.

28. Casagrande, A. (1959) "An Unsolved Problem of Embankment Stability on Soft Ground," *Proc. 1st Panamerican Conf. on Soil Mechs. and Found. Engrg.*, Mexico City, Vol. II, pp. 721–746.

29. Zeevaert, L. (1957) "Foundation Design and Behavior of Tower Latino Americana in Mexico City," *Geotechnique*, Vol. VII, No. 3, September.

30. McClelland, B. (1972) "Design of Deep Penetration Piles for Ocean Structures," *Terzaghi Lectures: 1963–1972*, ASCE, New York (1974), pp. 383–421.

31. Rodgers, J. (1957) "Distribution of Marine Carbonate Sediments," *Regional Aspects of Carbonate Deposition*, Soc. Economic Paleontologists and Mineralogists, Spec. Pub. No. 5, pp. 2–14.

32. Agarwal, S. L., Malhotra, A. K., and Banerjee, R. (1977) "Engineering Properties of Calcareous Soils Affecting the Design of Deep Penetration Piles for Offshore Structures," *Offshore Technology Conf.*, Houston, Texas, paper OTC 2792, pp. 503–512.

33. Eide, O. (1974) "Marine Soil Mechanics," *NGI Pub. 103, Offshore Tech. Conf., Stavanger*, September, Norwegian Geotechnical Inst., Oslo.

34. Clark, A. R. and Walker, F. (1977) "A Proposed Scheme for the Classification and Nomenclature for Use in Engineering Description of Middle Eastern Sedimentary Rocks," *Geotechnique*, Vol. 27, No. 1.

35. Holmes, A. (1964) *Principles of Physical Geology*, The Ronald Press, New York.

36. Zumberge, J. H. and Nelson, C. A. (1972) *Elements of Geology*, John Wiley & Sons, New York.

37. Thornbury, W. D. (1969) *Principles of Geomorphology*, 2d ed., John Wiley & Sons, New York.

38. Gilluly, J., Waters, A. C., and Woodford, A. (1959) *Principles of Geology*, W. H. Freeman and Co., San Francisco.

39. Gibbs, H. J., Hilf, J. W., Holt, W. G., and Walker, F. C. (1960) "Shear Strength of Cohesive Soils," *Proc. ASCE, Research Conf. on Shear Strength of Cohesive Soils*, Boulder, Colo., June, pp. 33–162.

40. Clevenger, W. A. (1958) "Experience with Loess as a Foundation Material," *Proc. ASCE, J. Soil Mechs. Found. Engrg. Div.*, Vol. 82, No. SM3.

41. USBR (1974) *Earth Manual*, 2d ed., U.S. Bureau of Reclamation, Denver, Colo.

42. Cleaves, A. B. (1964) "Engineering Geology Characteristics of Basal Till," *Engineering Geology Case Histories 1–5*, Geol. Soc. of Amer., New York, pp. 235–241.

43. Ward (1972) "Insitu Pressuremeter Testing on Two Recent Project," *Soils*, J. S. Ward & Assoc., Caldwell, New Jersey, November, p. 5.

44. Fairbridge, R. W. (1968) "Glacial Lakes," *Encyclopedia of Geomorphology*, R. W. Fairbridge, ed., Dowden, Hutchinson & Ross Publ., Stroudsburg, Pa., pp. 444–453.

45. Thornbury, W. D. (1967) *Regional Geology of the Eastern U.S.*, John Wiley & Sons, New York.

46. Peck, R. B. and Reed, W. C. (1954) "Engineering Properties of Chicago Subsoils," Univ. of Illinois Experiment Station Bull. No. 423.

47. Lobdell, H. L. (1970) "Settlement of Buildings Constructed in Hackensack Meadows," *Proc. ASCE, J. Soil Mechs. Found. Engrg. Div.*, Vol. 96, No. SM4, July, pp. 1235.

48. Parsons, J. D. (1976) "New York's Glacial Lake Formation of Varved Silt and Clay," *Proc ASCE, J. Geotech. Engrg. Div.*, Vol. 102, No. GT5, June, pp. 605–638.

49. Murphy, D. J., Clough, G. W., and Woolworth, R. S. (1975) "Temporary Excavation in Varved Clay," *Proc. ASCE, J. Geotech. Engrg. Div.*, Vol. 101, No. GT3, March, pp. 279–295.

50. Ladd, C. C. and Wissa, A. E. Z. (1970) "Geology and Engineering Properties of the Connecticut Valley Varved Clays with Special Reference to Embankment Construction," M.I.T., Cambridge, Mass.

51. Bjerrum, L., Loken, T., Heiberg, S., and Foster, H. (1969) "A Field Study of Factors Responsible for Quick Clay Slides," *Proc. 7th Intl. Conf. Soil Mechs. and Found. Engrg.*, Mexico City, Vol. 2, pp. 531–540.

52. Bjerrum, L. (1954) "Geotechnical Properties of Norwegian Marine Clays," *Geotechnique*, Vol. 4, p. 49.

53. Crawford, C. B. and Eden, W. J. (1969) "Stability of Natural Slopes in Sensitive Clay," *Proc. ASCE, Stability and Performance of Slopes and Embankments*, ASCE, New York, pp. 453–475.

54. Lambe, T. W. and Whitman, R. V. (1969) *Soil Mechanics*, John Wiley & Sons, New York.

55. Lambe, T. W. and Horn, H. M. (1965) "The Influence on an Adjacent Building on Pile Driving for the M.I.T. Materials Center," Proc. 6th Intl. Conf. Soil Mech. Found. Engrg., Montreal, Vol. II, p. 280.

56. DeSimone, S. V. and Gould, J. P. (1972) "Performance of Two Mat Foundations on Boston Blue Clay," *Proc. ASCE, Performance of Earth and Earth-Supported Structures*, Vol. I, Part 2, pp. 953–980.

57. Black, R. F. (1954) "Permafrost—A Review," *Geol. Soc. of Amer.*, Bull. 65, pp. 839–856.

58. Lachenbruch, A. H. (1968) "Permafrost," *Encyclopedia of Geomorphology*, R. W. Fairbridge, ed., Dowden, Hutchinson & Ross Publ., Stroudsburg, Pa., pp. 833–839.

59. Terzaghi, K. (1952) "Permafrost," *J. Boston Soc. Civ. Engrs.*, January.

60. HRB (1952) "Frost Action in Roads and Airfields," *HRB Spec. Report No. 1*, Pub. 211, National Academy of Sciences–National Research Council, Washington, D.C.

61. Rice, C. M. (1954) *Dictionary of Geological Terms*, Edwards Brothers, Inc., Ann Arbor, Michigan.

62. SCS (1960) *Soil Classification: A Comprehensive System (7th Approximation)*, Soil Conservation Service, USDA, U.S. Govt. Printing Office, Washington, D.C., and Supplements 1967, 1968, and 1970.

## BIBLIOGRAPHY

Birkeland, P. W. (1974) *Pedology, Weathering and Geomorphic Research*, Oxford University Press, New York.

Building Research Advisory Board (1963) *Proc. Intl. Conf. on Permafrost*, Purdue Univ., Lafayette, Ind., National Academy of Sciences Pub. 1287.

Coates, D. R., ed. (1976) *Geomorphology and Engineering*, Dowden, Hutchinson & Ross Publ., Stroudsburg, Pa.

Flint, R. F. (1957) *Glacial and Pleistocene Geology*, John Wiley & Sons, New York.

Gidigasu, M. D. (1976) *Laterite Soil Engineering*, Elsevier Scientific Publ., Co., New York.

Jumikis, A. R. (1958) "Geology and Soils of the Newark Metropolitan Area," *Proc. ASCE, J. Soil Mechs. Found. Engrg. Div.*, May, Paper 1646.

Koutsoftas, D. and Fischer, J. A. (1976) "In-situ Undrained Shear Strength of Two Marine Clays," *Proc. ASCE, J. Geotech. Engrg. Div.*, Vol. 102, No. GT9, September, pp. 989–1005.

Linell, K. A. and Shea, H. F. (1960) "Strength and Deformation Characteristics of Various Glacial Tills in New England," *Proc. ASCE, Research Conf. Shear Strength of Cohesive Soils*, Boulder, Colo., pp. 275–314.

Lobeck, A. K. (1939) *Geomorphology*, McGraw-Hill Book Co., New York.

Marsal, R. J. (1959) "Unconfined Compression and Vane Shear Tests in Volcanic Lacustrine Clays," *Proc. ASTM Conf. Soils for Engrg. Purposes*, Mexico City.

Noorany, I. and Gizienski, S. F. (1970) "Engineering Properties of Submarine Soils: A State-of-the-Art Review," *Proc. ASCE, J. Soil Mechs. Found. Engrg. Div.*, Vol. 96, No. SM5, September, pp. 1735–1672.

Skempton, A. W. (1948) "A Study of the Geotechnical Properties of Some Post-Glacial Clays," *Geotechnique*, Vol. I, p. 7.

USDA (1951) *Soil Survey Manual*, Handbook No. 18, U.S. Dept. of Agriculture, Washington, D.C.

Winterkorn, H. F. and Fang, H. Y. (1975) "Soil Technology and Engineering Properties of Soils," *Foundation Engineering Handbook*, Winterkorn and Fang, eds., Van Nostrand Reinhold Co., New York, Chap. 2.

# CHAPTER EIGHT

# WATER: SURFACE AND SUBSURFACE

## 8.1 INTRODUCTION

### 8.1.1 GENERAL

**Hydrology and Geohydrology**

*Hydrology* is the science that deals with continental water, its properties, and its distribution on and beneath the earth's surface and in the atmosphere, from the moment of its precipitation until it is returned to the atmosphere through evapotranspiration or is discharged into the oceans.

*Geohydrology* or *hydrogeology* is the science that is concerned with subsurface waters and their related geologic aspects.

**Chapter Scope**

This chapter describes the conditions of engineering significance pertaining to surface and subsurface water (groundwater), analytical procedures, groundwater and seepage control, and environmental planning.

Exploration methods are described in Chap. 2 and the measurement of water tables and pore-water pressures in Chap. 4. Permeability, its measurement, and typical values for the coefficient $k$ are presented in Art. 3.3.

### 8.1.2 ENGINEERING ASPECTS

**Surface Water**

*Flooding* is a geologic hazard that occurs naturally; however, its incidence is increased by human activity.

*Erosion* of the land is also a natural occurrence detrimental to society; its incidence is also increased by human activities.

*Water supply* for human consumption is stored in surface reservoirs created by the construction of dams.

**Subsurface Water**

*Water supply* for human consumption is obtained from underground aquifers that must be protected from pollution, especially since the water often is used without treatment. As groundwater is a depletable natural resource, its extraction, conservation, and recharge require careful planning.

*Land subsidence* results from excessive groundwater extraction for water supply on a regional basis, and from dewatering for excavations on a local basis, as described in Art. 10.2.

*Groundwater and seepage control* is required for a large number of situations including:

- Excavations, to enable construction to proceed in the "dry" and to reduce excessive pressures on the walls and bottom.
- Structures, to provide for dry basements and to prevent hydrostatic uplift on slabs.
- Pavements, to provide protection against "pumping" and frost heave.
- Slopes, to provide for stabilization in either natural or cut conditions.
- Dams, to protect against excessive seepage

through, beneath, or around an embankment, which reduces stability and permits excessive storage loss. Dam construction also can have a significant effect on the regional groundwater regime, sometimes resulting in instability of slopes or surface subsidence.

*Water quality* is of concern in the various consumptive uses as well as for its possible deleterious effect on construction materials, primarily concrete.

## 8.2 SURFACE WATER

### 8.2.1 SURFACE HYDROLOGY

#### The Hydrological Cycle

Precipitation, in the form of rainfall or snowmelt, in part enters the ground by infiltration to become groundwater, in part remains on the surface as runoff, and in part enters the atmosphere by evaporation and transpiration to become a source of precipitation again. The *hydrological equation* relating these factors can be written as:

$$\text{Infiltration} = \text{precipitation} - (\text{runoff} + \text{transpiration} + \text{evaporation})$$

The conditions influencing the factors in the hydrological equation include climate, topography, and geology.

- *Climate* affects all of the factors. In moist climates precipitation is high but when the ground is saturated runoff will also be high, and when vegetation is heavy, transpiration will be high. In arid climates evaporation loss of standing water exceeds precipitation.
- *Topography* impacts most significantly on runoff and evaporation. Steep slopes encourage runoff and preclude significant infiltration. Gentle to flat slopes impede runoff or result in standing water, permitting evaporation to occur.
- *Geologic conditions* impact significantly on runoff and infiltration. Surficial or shallow impervious materials result in high runoff and relatively little infiltration. Pervious surficial materials result in low runoff and high infiltration.

#### Precipitation

In some engineering applications, precipitation data are applied in runoff analyses to evaluate groundwater recharge and flood-prone zones and to design drainage improvements, such as culverts and channels, and spillways for dams.

Recording gages measure rainfall in inches or millimeters on an hourly basis. In many locations rain gages may be read only on a daily basis, or only during storm activity. Snow accumulation is measured visually on a periodic basis and snowmelt estimated.

Daily rainfall records are kept by most countries although locations may be widely dispersed and specific area coverage may be poor. Monthly rainfall charts providing total accumulation are the normal form of presentation in many countries. Mean annual precipitation for countries or other large areas is provided on maps such as that for the United States (Fig. 8.1). Figure 8.2 presents the mean annual pan evaporation for the United States.

Storm data, or rainfall *intensity* and *duration* measured during periods of maximum downfall, are also important data. Storm data are a significant element of rainfall data because maximum runoff and flood flows are likely to occur during storms, with maximum impact on drainage systems and spillways; and the occurrence of erosion, mudflows, avalanches, and slides increases enormously with intensity and duration.

Data are procured from local weather stations and state and federal agencies. In the United States the federal agency is the U.S. Department of Commerce, National Oceanic and Atmospheric Administration (NOAA), National Weather Service, Washington, D.C. The publication *Climatological Data* provides daily precipitation data for each month for each state. The publication *Hourly Precipitation Data* provides hourly data for each month for each state. Locations of the precipitation gaging stations are shown on a series of maps titled "River Basin Maps Showing Hydrological Stations."

#### Runoff and Infiltration

*Land erosion* from runoff results in gullies which grow to streams and finally to rivers,

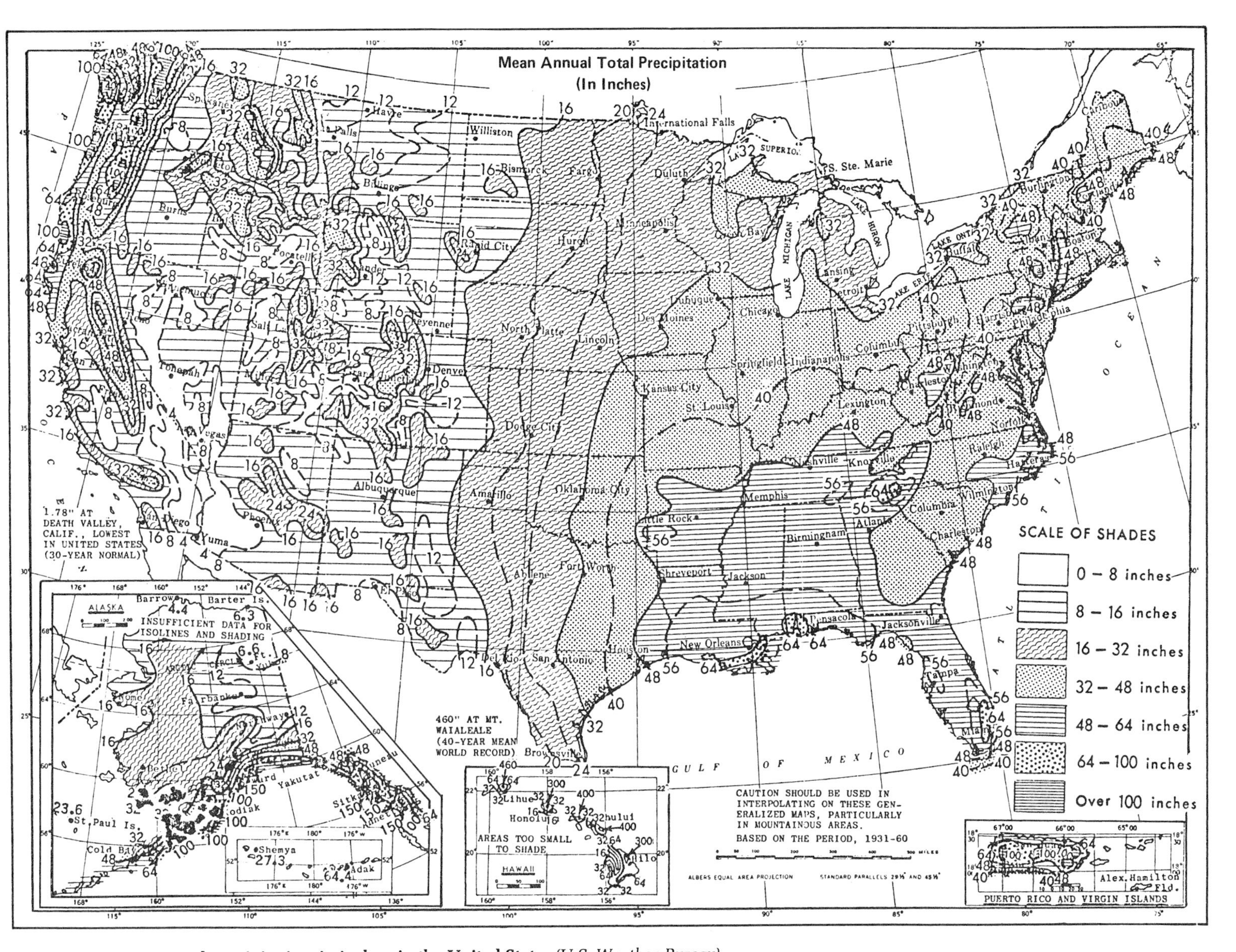


FIG. 8.1 **Average annual precipitation, in inches, in the United States** *(U.S. Weather Bureau).*

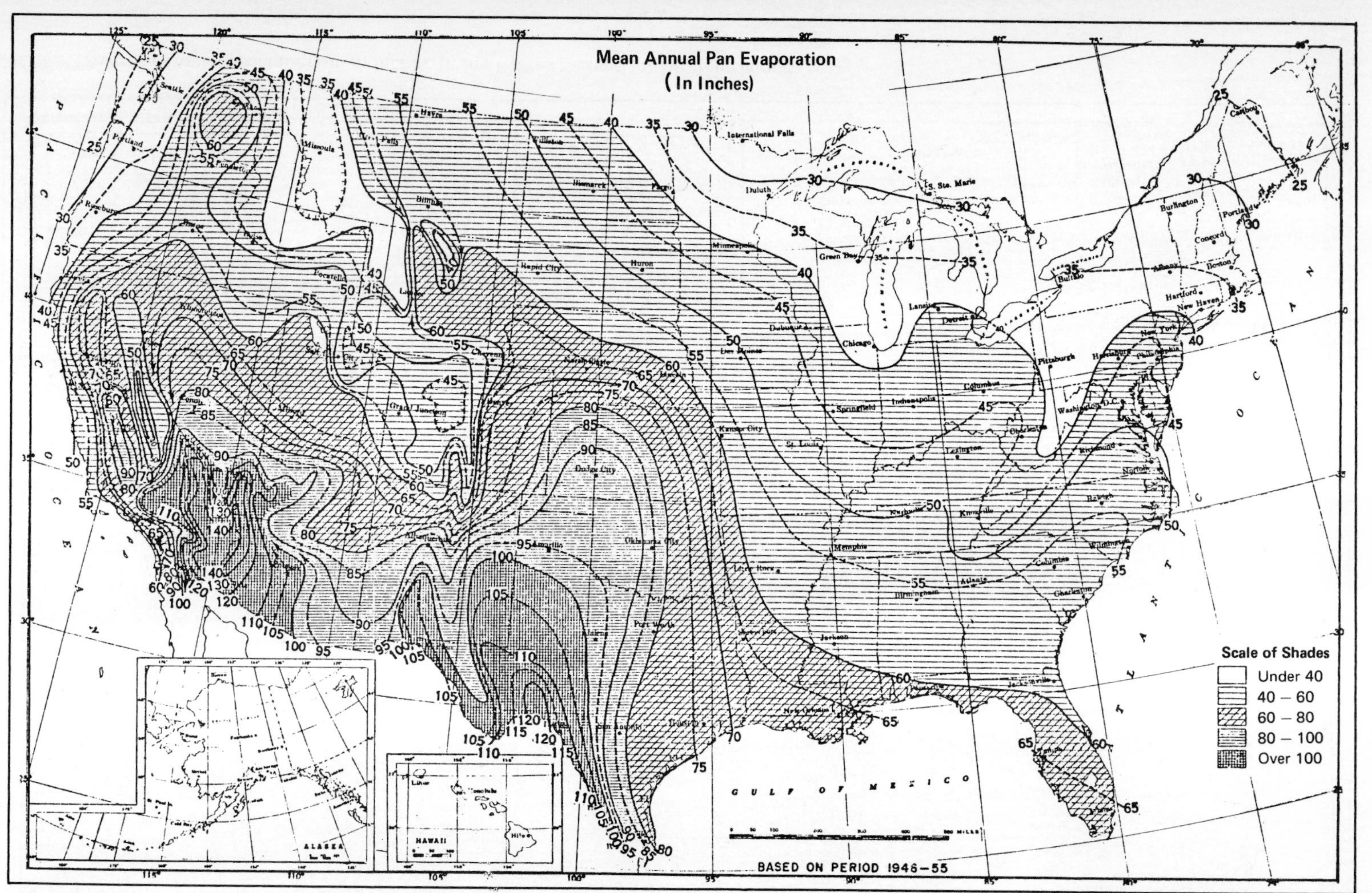


FIG. 8.2 Average annual lake evaporation, in inches, in the United States (*U.S. Weather Bureau*).

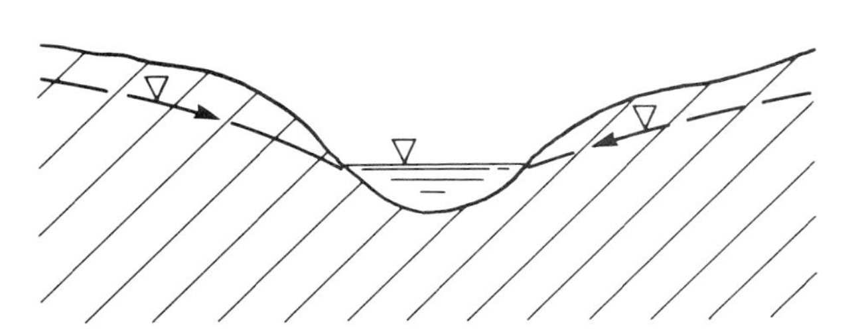

**FIG. 8.3 Effluent stream.**

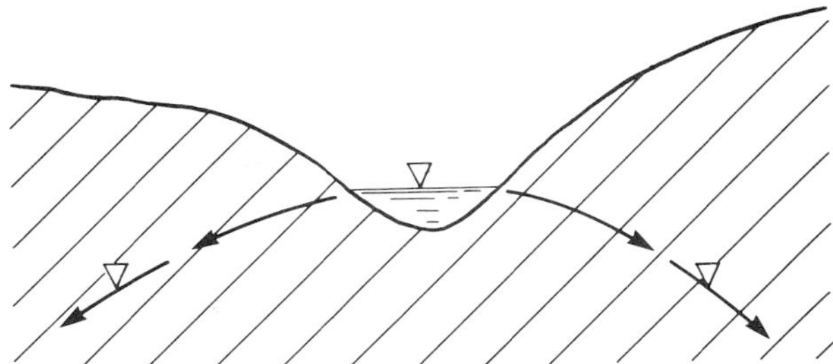

**FIG. 8.4 Influent stream.**

developing into a regional drainage system carrying the runoff into lakes and seas.

*Drainage basins* consist of the rivers and their systems of branches and tributaries. The boundaries of a drainage basin are *divides*, ridge lines, or other strong topographic features separating the basin from adjacent basins.

*Channel* represents the volume within the riverbanks with the capacity to carry flow.

*Floodplain* is that portion of a river valley with a reasonable probability of being inundated during periods of high flow exceeding channel capacity.

*River stage* is the elevation of the water surface at a specific gaging station above some arbitrary zero datum.

*Discharge*, the runoff within the stream channel, is equal to the cross-sectional area times the average velocity ($ft^3/s$ or $m^3/s$, etc.). It is an important element in the determination of the time required to fill a reservoir, in the evaluation of surface erosion, in the evaluation of flood potential, and in the design of flood-protection and drainage-control structures. When discharge quantities are computed, the cross-sectional area is usually based on preflood data, which do not account for the channel deepening that occurs during flooding. Discharge quantities, therefore, are often underestimated.

*Infiltration* occurs as runoff entering the subsurface through pore spaces in soils and openings in rocks. It occurs most readily in porous sands and gravels, through cavities in soluble rocks, and through heavily fractured zones in all rock types. Water moves downward and through the subsurface under gravitational forces.

*Groundwater* results primarily from infiltration.

*Effluent streams* are characterized by the flow of groundwater to the stream, and the stream represents the interception of the surface drainage with groundwater flow as shown on Fig. 8.3. They are characteristic of moist climates.

*Influent streams* supply water to the ground (Fig. 8.4) and are characteristic of intermittent streams in any climate, and most or all streams in arid climates.

## 8.2.2 EROSION

### Causes

*Natural agents* causing erosion include running water, groundwater, waves and currents, wind (*see Art. 7.5.1*), glaciers (*see Art. 7.6.1*), and gravity acting on slopes. Erosion from running water and gravity are the most significant with respect to construction and land development, since such activities often result in an increase in erosional processes.

*Human causes* result from any activity that permits an increase in the velocity of water, thereby increasing its erosional capacity, especially on unprotected slopes. Removal of trees and other vegetation from slopes to clear land for construction, farming, or ranching is probably the greatest cause of unnatural erosion. Severe erosion as a result of tree removal is illustrated on Fig. 8.5. Tree roots and other vegetation hold the soil and also assist in providing slope stability.

FIG. 8.5 Severe land erosion caused by removal of vegetation (state of Rio de Janeiro, Brazil).

FIG. 8.6 Erosion of slope cut in silty residual soils results in slope failures (Sidikalang, Sumatra). Unlined ditch is eroding and undercutting slope and roadway. Ditch in area behind vehicle is filled from slope failures causing runoff to traverse the roadway; the traversal in time will result in the roadway's loss. Such conditions are also a great source of sediment carried to nearby water courses.

### Effects

*Uncontrolled erosion* along riverbanks results in loss of foundation support for structures, pavements, fills, and other works. In hilly or mountainous terrain it increases the incidence of slope failure (Fig. 8.6), and can result in the loss of roadways, such as almost occurred in Fig. 8.7. It removes large land areas from cultivation in all types of terrain.

*Sedimentation* or *siltation* is an important effect of erosion. The construction of highway embankments and cuts and mine-waste fills can create unprotected slopes that erode easily and are sources for sediments to flow into and pollute water bodies. The result is clogging of streams which increases the flood hazard and bank erosion, increased turbidity which can harm aquatic life and spoil water supplies, reduction in the capacity of reservoirs and a shortening of their life, blockage of navigation channels, and even infilling of harbors and estuaries.

### Protection and Prevention

*Riverbanks and channels* are provided with protection by retaining structures, concrete linings, or riprap. Foundations for bridges must be placed at adequate depths to provide protection against *scour*, the most frequent cause of bridge failures. A rule of thumb is to place the foundations at a depth equal to 4 times the distance between the flood and dry-weather levels unless hard rock is at a shallower depth [Smith (1977)[1]].

*Slopes* receive protection by planting fast-growing vegetation and installing surface drainage control as described below and in Chap. 9.

**FIG. 8.7 Deep erosion of unlined ditch almost caused loss of mountain road (Serra do Mar, Brazil).**

**FIG. 8.8 Grain bags or potato sacks soaked in cement provide economical ditch lining for runoff. Rapid planting of grass on graded slope also provides erosion protection (Serra do Mar, Brazil).**

*Construction sites* may be treated in a series of procedures as follows:

1. Identify on-site areas where erosion is likely to occur, and off-site areas where sedimentation and erosion will have detrimental effects.
2. Divert runoff originating from upgrade with ditches to prevent its flow over work areas. Line large ditches with nonerodible material that will not settle and crack, permitting ditch erosion to occur. Unlined ditches are suitable in strong materials where flow velocities will be low. Potato sacks soaked in a cement paste, and then brushed with cement when in place, provide an economical ditch lining that conforms to an irregular surface. They are especially suited to small ditches as shown in Fig. 8.8. Stepped linings of concrete are used on steep slopes to decrease water velocities.
3. Limit the area being graded at any one time and limit the time that the area is exposed to erosion by planting grass or some other fast-growing native vegetation as soon as the slope area is prepared. For example in excavation of benches for a highway cut slope, when the first bench is prepared it should be seeded immediately, even before work proceeds to the next levels. On steep slopes a fabric mat or mesh staked to the slope will prevent the seed from washing away. Cutting numerous shallow benches along the slope also retards runoff and washing.
4. Install bundles of *wattling* composed of live brush along slope contours to provide protection and improve slope stability. The bundles are staked into shallow trenches as shown on Fig. 8.9.
5. Retain heavy runoff in large ditches and diminish water velocity with low dikes of stone or sand bags.
6. Trap sediment-laden runoff in basins, or filter

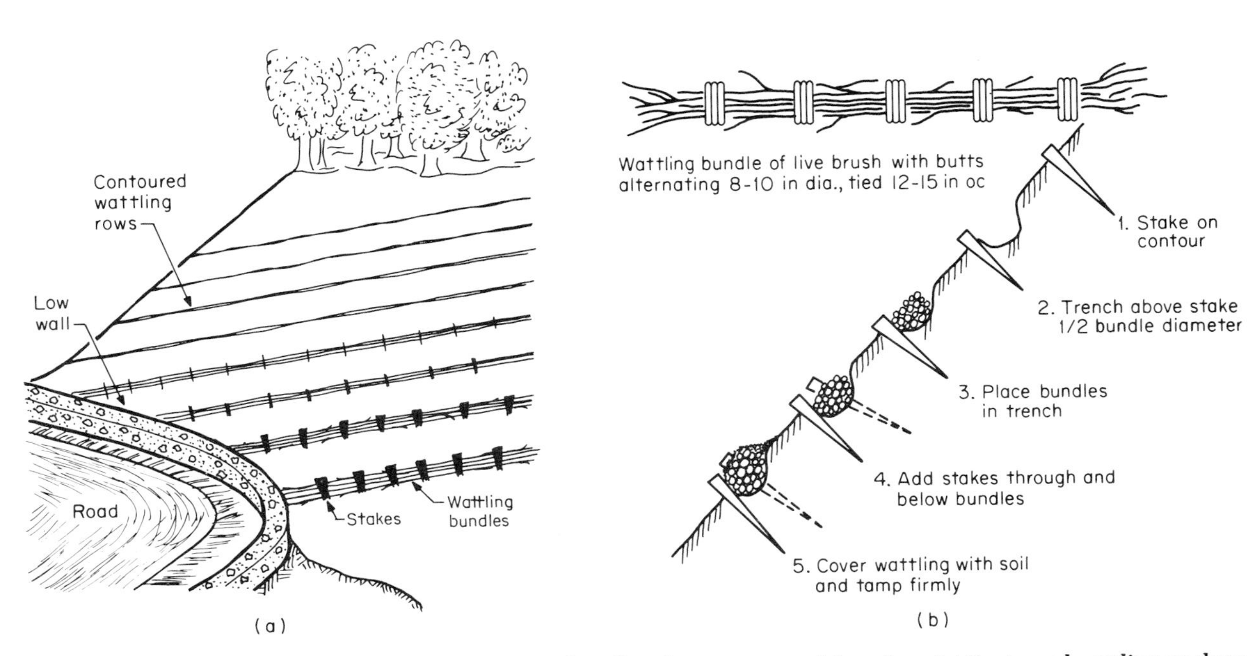


**FIG. 8.9 Erosion protection by installation of wattling bundles along contours of slope face. (*a*) Contoured wattling on slope face. (*b*) Sequence of operations for installing wattling on slope face. Work starts at bottom of cut or fill with each contour line proceeding from step 1 through 5. Cigar-shaped bundles of live brush of species which root are buried and staked along slope. They eventually root and become part of the permanent slope cover.** [*After Gray et al. (1980).*[2] *Adapted with the permission of the American Society of Civil Engineers.*]

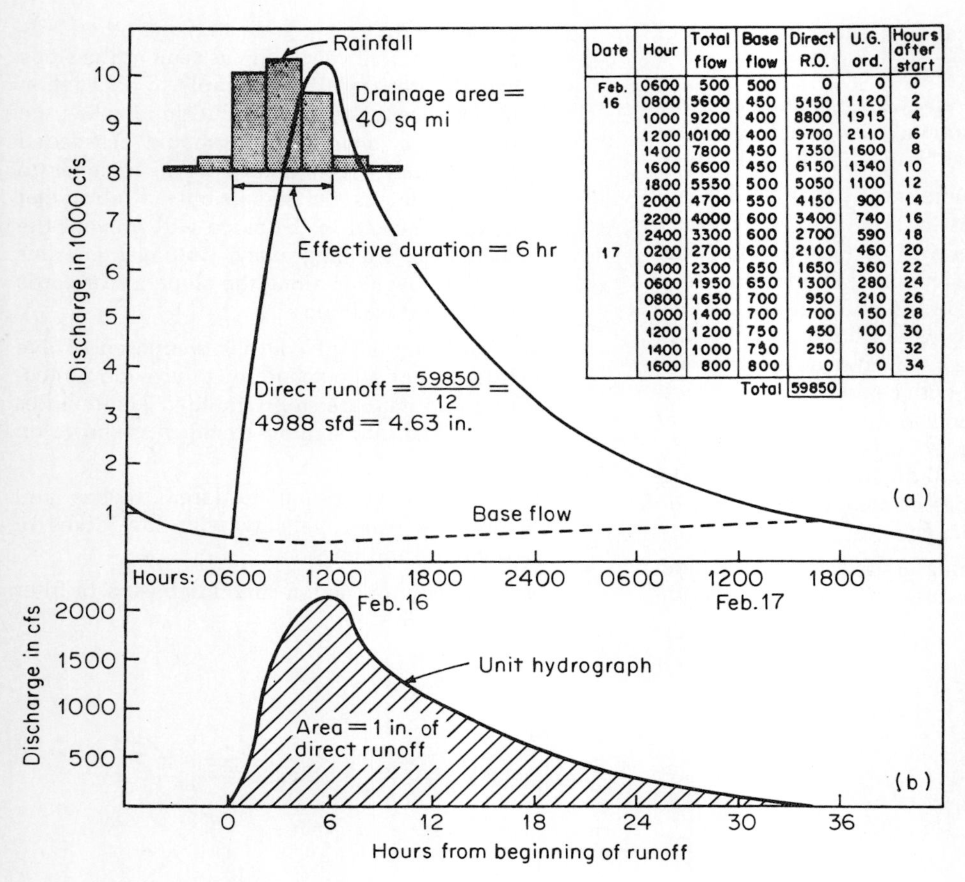


| Date | Hour | Total flow | Base flow | Direct R.O. | U.G. ord. | Hours after start |
|---|---|---|---|---|---|---|
| Feb. 16 | 0600 | 500 | 500 | 0 | 0 | 0 |
| | 0800 | 5600 | 450 | 5450 | 1120 | 2 |
| | 1000 | 9200 | 400 | 8800 | 1915 | 4 |
| | 1200 | 10100 | 400 | 9700 | 2110 | 6 |
| | 1400 | 7800 | 450 | 7350 | 1600 | 8 |
| | 1600 | 6600 | 450 | 6150 | 1340 | 10 |
| | 1800 | 5550 | 500 | 5050 | 1100 | 12 |
| | 2000 | 4700 | 550 | 4150 | 900 | 14 |
| | 2200 | 4000 | 600 | 3400 | 740 | 16 |
| | 2400 | 3300 | 600 | 2700 | 590 | 18 |
| 17 | 0200 | 2700 | 600 | 2100 | 460 | 20 |
| | 0400 | 2300 | 650 | 1650 | 360 | 22 |
| | 0600 | 1950 | 650 | 1300 | 280 | 24 |
| | 0800 | 1650 | 700 | 950 | 210 | 26 |
| | 1000 | 1400 | 700 | 700 | 150 | 28 |
| | 1200 | 1200 | 750 | 450 | 100 | 30 |
| | 1400 | 1000 | 750 | 250 | 50 | 32 |
| | 1600 | 800 | 800 | 0 | 0 | 34 |
| | | | Total | 59850 | | |

**FIG. 8.10 Development of a unit hydrograph.** *[From Linsley et al. (1958).[6] Reprinted with permission of McGraw-Hill Book Company.]*

runoff through brush barriers or silt fences. The latter are made of a filter fabric [Dallaire (1977)[3]].

7. Carry out postconstruction maintenance to clean ditches, replant bare areas, restake loose wattling bundles, etc.

## 8.2.3 FLOODING

### Causes

*Natural floods,* occurring during or after heavy rainfall and snowmelt, cause runoff to exceed the carrying capacity of the normal river channel which consequently overflows its banks and floods the adjacent valley. As natural events, floods can be predicted with some degree of accuracy, but it is very often human activities that cause them to occur in locations where previously they did not occur.

*Human activities* of several kinds increase the incidence of flooding. Construction in the river floodplain involving filling decreases the natural storage capacity, consequently increasing the extent of the floodplain. Removal of vegetation from valley slopes increases runoff volumes, and sedimentation from erosion reduces stream storage capacity. Ground subsidence over large areas can result from the extraction of oil, gas, or water, causing a general lowering of drainage basins (*see Art. 10.2*). Sudden floods, with disastrous potential for destruction, result from the failure of dams.

### Forecasting Flood Levels

*General*

The forecasting of flood levels is necessary for floodplain zoning which imposes restrictions on construction and for the design of flood-control systems such as dikes, upstream holding reservoirs, channel straightening and lining, etc., and the design of culverts and other drainage works. Emergency spillways for earth dams must be designed to prevent overtopping, the most common cause of catastrophic failure of earthen embankments [Sherard et al. (1963)[4]].

The objective of flood-level forecasting is to predict the quantity of flow and the level of flooding that have a probability of occurring with a given frequency such as once every 25, 50, or 100 years. The prediction provides the basis for the selection of protective measures. The design flood is selected with regard to cost of control and to the degree of danger to the public from failure of the proposed flood-control system. If failure will result in loss of life and substantial property damage, design is based on floods of lower probabilities, such as a 1000-year flood, because this flood level is *higher* than those of floods of higher probabilities.

*Analytical Forecasting*

*Factors to consider* in analytical flood level forecasting are:

1. Topography of the total basin contributing runoff to the study area
2. Ground cover including soil and rock type and vegetation (to evaluate runoff vs. infiltration and evaporation)
3. Maximum probable storm in terms of intensity and duration (based on records)
4. Season of the year (affects conditions such as frozen ground, snow cover, and ground saturation, all of which influence runoff)
5. Storage capacity of the river channel and floodplain (possible future downstream changes must be considered)

*Maximum probable flood* computational procedures as described in USBR (1973)[5] require estimates of storm potential and the amount and distribution of runoff within the drainage basin. The general procedure is as follows:

1. A 6-hr point rainfall is selected from an appropriate chart for the geographic location. From graphs, the point rainfall value is adjusted to represent a 6-hr average precipitation over the drainage basin, and is also adjusted to give the accumulated rainfall for longer durations, such as 48 hr, for example.
2. Runoff is determined from an evaluation of the soil and vegetation conditions, and the runoff volume for the drainage area is computed for various time increments. From the data and simple mathematical relationships, runoff hydrographs are prepared.

*Hydrographs* are graphic plots of changes in the flow of water (discharge) or of the water-level elevations (stages) against time as shown on Fig. 8.10*a*. They are often presented in the form of a *unit hydrograph* which is a hydrograph for 1 in of direct runoff from a storm of a specific duration as shown on Fig. 8.10*b*.

*Discharge volumes* for a given period of time are computed from the unit hydrographs. For flood-forecasting purposes the computed flows are converted to stages (water levels) by the application of *stage-discharge* relationships for a given location as illustrated on Fig. 8.11. The curves are prepared from data obtained from field measurements.

*Computer programs* include the HEC series which allows rapid computations for the various elements of a hydrological study once the basic data are collected. Included are the following:

- HEC-1—To compute the flood hydrograph, including routing through channels and reservoirs.
- HEC-2—To determine the water-surface profile under subcritical and supercritical flow conditions.
- HEC-3—Reservoir systems analysis for determining multipurpose routing based on varying storm requirements at reservoirs, diversions, and downstream control points.
- STORM—A program for estimating runoff from small, primarily urban, watersheds and for computing land-surface erosion and water-quality parameters of suspended and settleable solids.

*Site planning* for a location where development is anticipated involves computing peak flows for various storm frequencies by alternative methods, to arrive at unit hydrographs for several

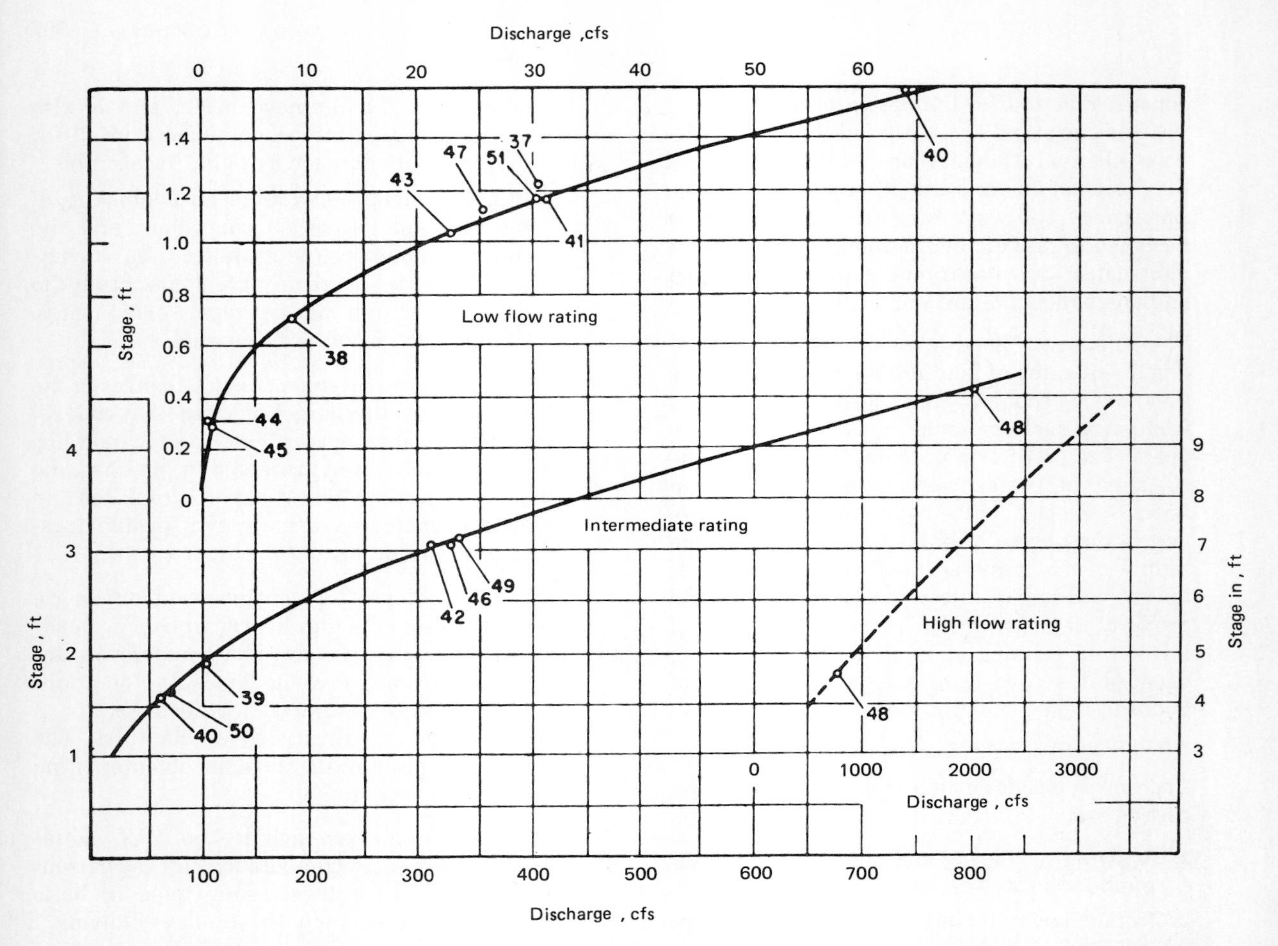


**FIG. 8.11 A simple stage-discharge relationship.** [*From Linsley et al. (1958).*[6] *Reprinted with permission of McGraw-Hill Book Company.*]

conditions including those before, during, and after development as shown on Fig. 8.12. The computed flood levels are checked against the flood levels estimated by geologic techniques (see the following section). Channel capacities are computed and estimated flood levels are derived and plotted in cross section. If the "after development" unit hydrograph results in dangerously high flood levels, then flood-control measures are required, and a storm-water management program ensues to control runoff as also shown on Fig. 8.12.

### *Geologic Forecasting*

The *basis of geologic forecasting* is the delineation of the floodplain boundaries from terrain analysis (*see Arts. 2.2.3 and 7.4.1*) to identify the distribution of recent alluvium or Quaternary soils in the valley, and to identify erosional features in the valley that are flood-related. The result can be more accurate than analytical procedures, especially where long-term rainfall data are lacking, and in any case should always be performed as backup to analysis. A time cannot be placed on flood recurrence except for the conclusion that flooding has occurred in recent geologic history and is likely to return.

EXAMPLE: RAPID CITY, SOUTH DAKOTA [Rahn (1975)[7]] On June 9, 1972, a storm dropped 15 in (352 mm) of rain on some locations on the slopes of the Black Hills in less than 6 hr. The most intensive rain fell in a 133-km$^2$ area of the basin of Rapid Creek, between the city and an upstream dam built for flood control. Streams discharged several times the expected peak discharge and, in the western part of the city where the creek flows into a valley with a 900-m wide floodplain, residents reported local surges of water 6 m high. There were at least 238 deaths. The limits of the flood waters in the city are shown on Fig. 8.13, as is the area mapped previously by the USGS on 7½-min quadrangle maps as Quaternary alluvium. The coincidence is clearly evident. The floodplain area has subsequently been zoned by the city as nonresidential. Most of the area has been converted into parkland but some commercial establishments have been permitted to remain.

EXAMPLE: FLASH FLOODS IN AN ARID CLIMATE In the near-desert environment, stream flow is intermittent and channels are normally dry but easily recognized. Storm runoff from mountain areas flows across the lower "bajadas," forming multiple coalescing alluvial fans along a great number of closely spaced channels (*see Figs. 2.11 and 7.30*). The flow, termed "sheet wash," enters larger channels or washes, connecting with the river channels in the valley, where it fills the floodplains and causes severe bank erosion. A shallow swale, representing a dry wash, is shown on Fig. 8.14.

Stereoscopic interpretation of aerial photographs permits delineation of floodplain limits because of the distinctive boundaries evidenced by low escarpments, in reality intermittent riverbanks. On the aerial photo presented as Fig. 8.15, three flood zones are shown:

1. The channel of the Pantano Wash carries water intermittently, but usually several times a year.
2. The area referred to as the "recent floodplain area" floods only occasionally, usually on at least a yearly basis.
3. The "geologic floodplain" delineates the boundary in which flooding has occurred during recent past geologic history. The mobile home park was placed there because of the flat terrain of the terrace. Its position must be considered as precarious.

### Flood Protection

*Floodplain zoning* affords the best protection against flooding from the aspect of community development, since construction can be prohibited. Floodplains are useful as a natural storage area for floods, and can provide inner-city open space and parkland. They usually also represent the best farmland.

*Construction at adequately high elevations* can be accomplished simply by selecting a site on high natural ground, or by raising grade by filling. In either case consideration should be given to the effect of possible future development on flood levels. Extensive filling will increase upstream flooding.

*Diking to contain water* is a necessary solution for many rapidly growing cities, but it is expensive and not necessarily riskfree because flood levels are not only difficult to predict with certainty, but are affected by natural and development changes.

*Channel straightening and lining* is a solution often applied to small rivers and streams to increase flow velocity and reduce the flood hazard and bank erosion. In addition to aesthetic

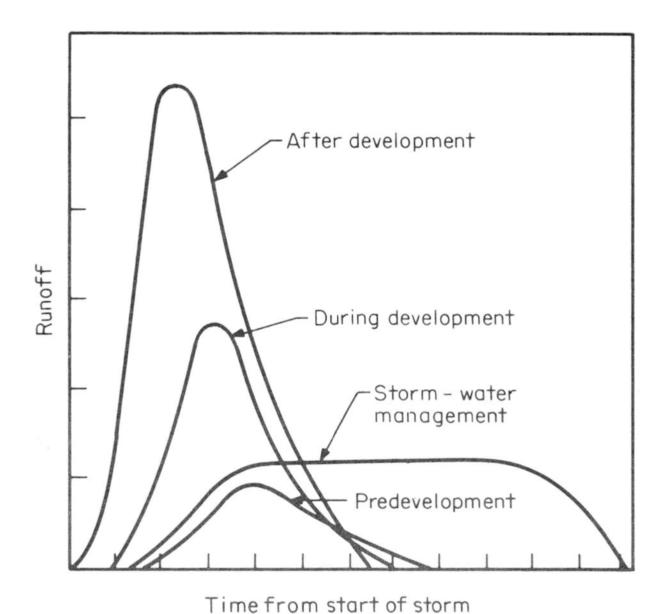


**FIG. 8.12 Unit hydrographs for various conditions at a given site.**

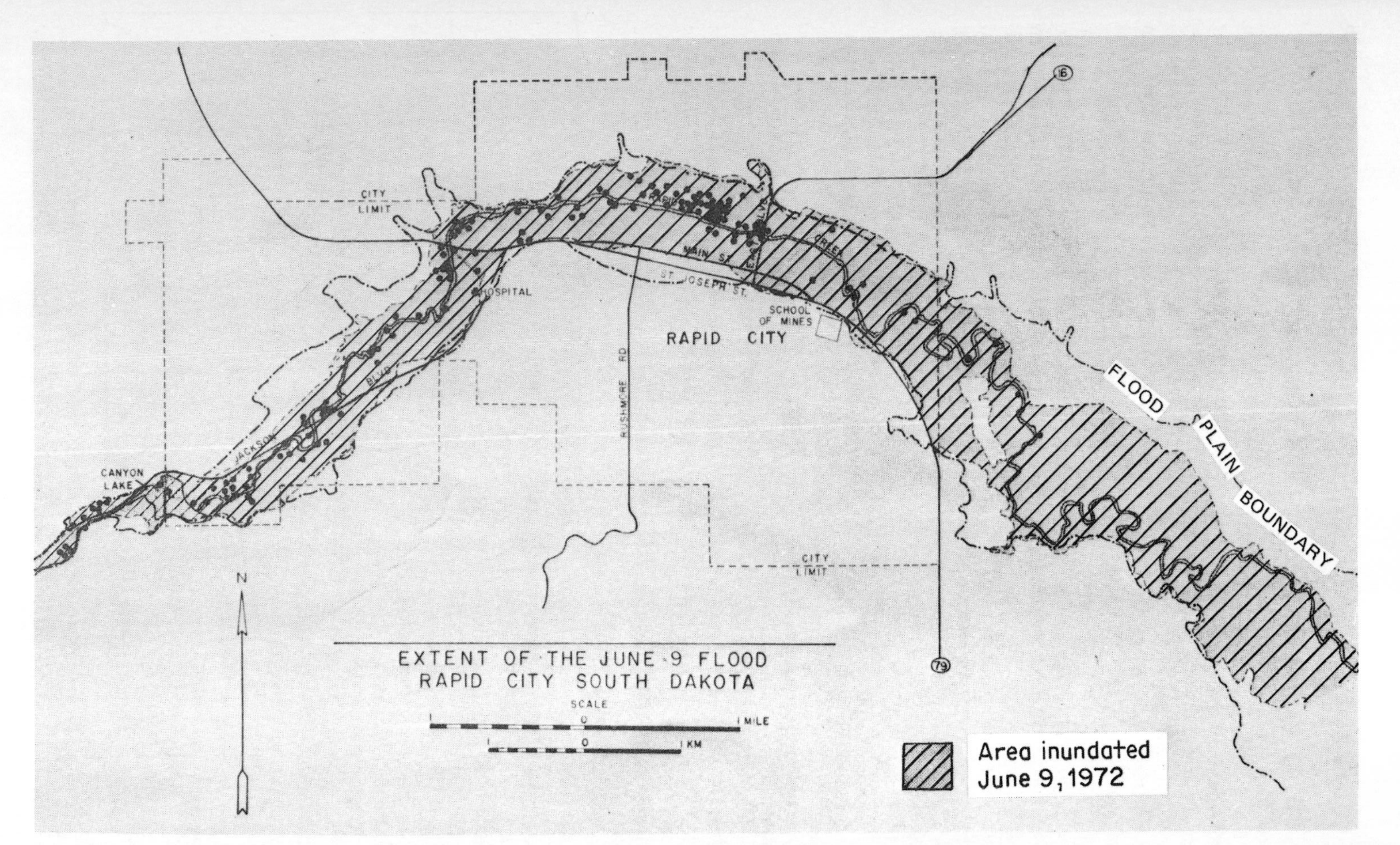


FIG. 8.13 Map of Rapid City showing the area inundated by the June 9, 1972, flood, and the area mapped as Quaternary alluvium on the U.S. Geological Survey 7½-min quadrangle maps. Black dots show locations of bodies recovered after the flood. [*From Rahn (1975).*[7]]

**FIG. 8.14 Roadway crossing a wash, or arroyo (Cañada del Oro area, Tucson, Arizona). Such locations present a severe danger to motorists during high runoff.**

objections, it has the disadvantage of increasing downstream flows and therefore the flood hazard at some other locations.

*Flood-control dams* constructed upstream serve as storage reservoirs and are the most effective construction solution to flood protection. They are often designed for multipurpose uses including hydropower, water supply, and recreation. They are costly to construct, however, and their number and location require careful study to avoid such catastrophes as that at Rapid City, South Dakota, described previously.

## 8.3 SUBSURFACE WATER (GROUNDWATER)

### 8.3.1 OCCURRENCE

#### General Relationships

A portion of the precipitation runoff enters the ground by infiltration, percolating downward under the force of gravity through fractures and pore spaces which below some depth attain saturation or near saturation.

*Porosity* relates the percentage of pore space to the total volume (see *Table 3.6*) and represents the capacity of material to hold water when saturated. Values for various materials are given on Table 8.1.

*Void ratio* is the ratio of the volume of voids to the volume of solids (see *Table 3.8*) and is the term normally used by engineers to describe porosity characteristics of soils. (To distinguish between porosity and void ratio consider the example of a 5-liter can filled with dry, coarse sand to which is added 1 liter of water, which just saturates the sand. The porosity is 20% and the void ratio is 0.25.)

*Seepage* refers, in general, to the movement of water into, out of, or within the ground. *Influent seepage* is movement of water into the ground from the surface. *Effluent seepage* is discharge of groundwater to the surface.

#### Zones and Water Tables

*Static Water Table*

The level within a body of subsurface water at which groundwater pressures are equal to atmo-

**FIG. 8.15 Aerial photo of a portion of Tucson, Arizona, on which three flood zones are delineated by escarpments and terraces.** *(Courtesy of Joseph S. Ward & Associates.)*

spheric pressure is referred to as the *static water table*. Although not truly static, it is so termed to differentiate it from perched water tables. Above the static water table the soil may be saturated by capillarity or it may contain air. Above the saturated zone is the *zone of aeration*. Various modes of groundwater occurrence are illustrated on Fig. 8.16; in general, there are two zones, the *upper zone* and the *saturated zone*.

*Upper Zone*

The pores, fractures, and voids contain both air and moisture, in several forms or conditions.

- *Gravity or vadose water* is "suspended" in the zone of aeration; it moves downward slowly under the force of gravity.
- *Hygroscopic moisture* adheres as a film to soil grains and does not move by gravity.
- *Pellicular water* is moisture adhering to rock surfaces throughout the zone of aeration. Either pellicular or hygroscopic moisture can be extracted by evaporation or transpiration.
- *Perched water* is water in a saturated zone located within the zone of aeration, or unsaturated zone. It is underlain by impervious strata which do not permit infiltration by the force of gravity. Compared to the saturated zone, its water supply is limited and will be rapidly depleted by pumping.
- *Capillary fringe* is the zone immediately above the water table containing capillary water. Capillary activity is produced by the surface tension of water like that which causes water in a tube with its lower end submerged in a reservoir to rise above the reservoir level. The height of the rise varies inversely with the radius of the tube, which may be likened to the interstices of soil masses. In coarse gravels the rise is insignificant and in clean sands it is in the order of a few centimeters to a meter or so. The rise increases substantially as the percentage of fines increases. In predominantly clayey soils, capillary rise occurs very slowly but can be as high as 8 m or more. Capillary rise for various gradations is given on Table 3.10. Capillary rise provides the moisture that results in heaving of buildings and pavements from the volume increase of expansive soils or from freezing, or the destruction of pavements

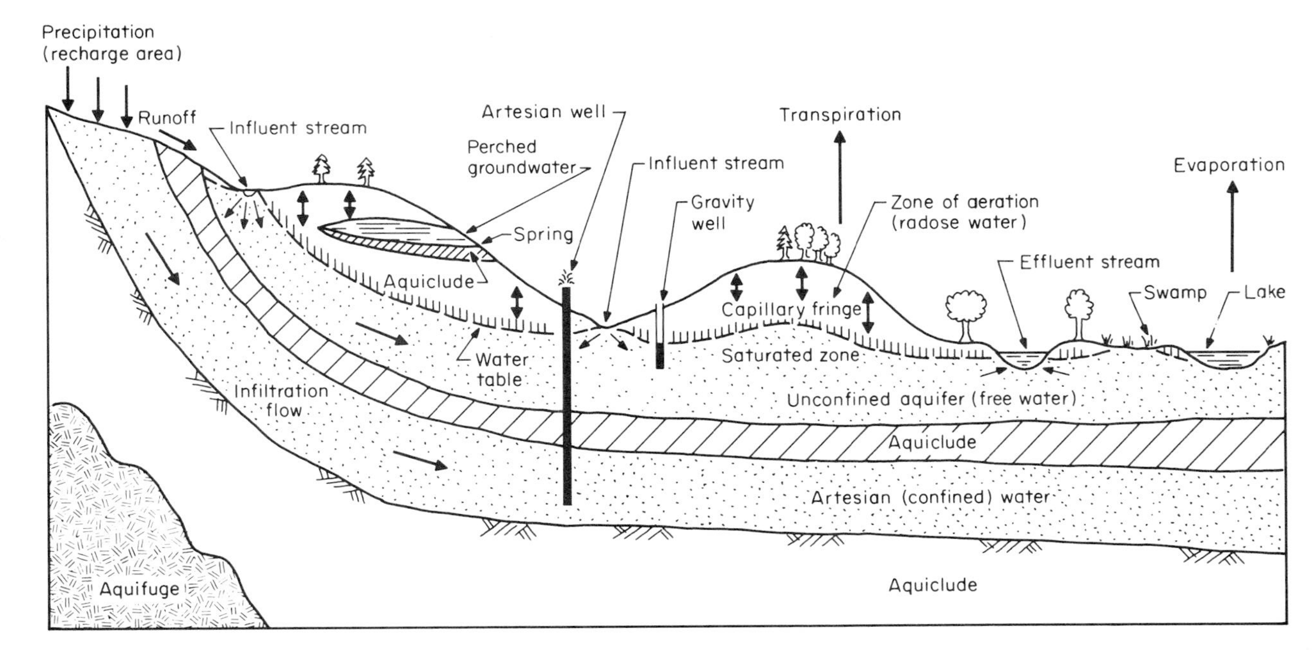


**FIG. 8.16 Cross section illustrating the occurrence of groundwater.**

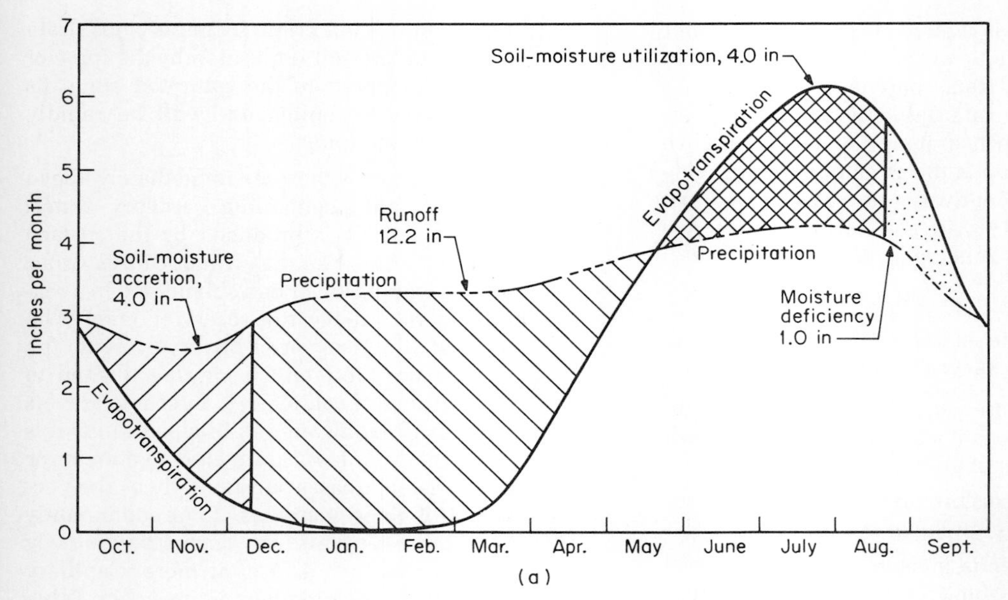


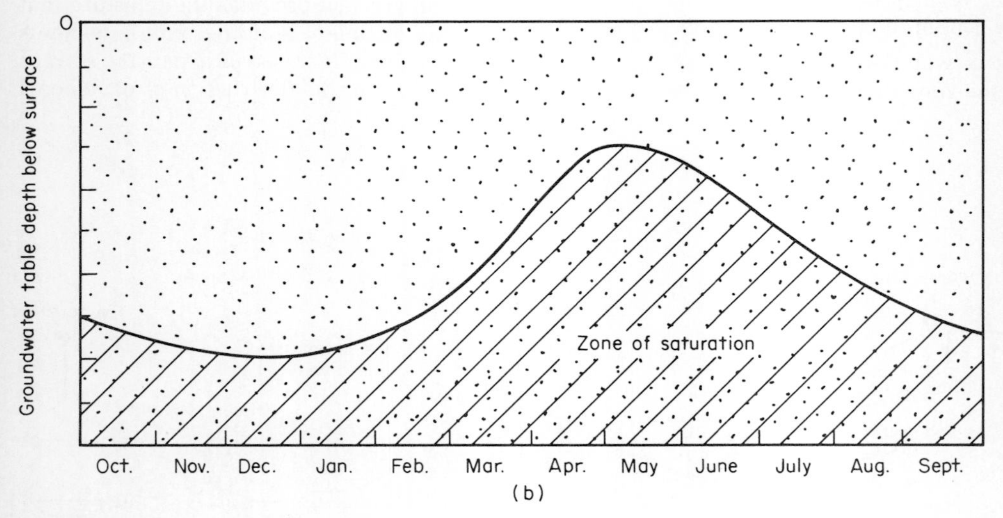


**FIG. 8.17 Relationships between seasonal precipitation and the groundwater table in a cool, moist climate: (*a*) March of normal precipitation and potential evapotranspiration at College Park, Maryland and (*b*) the variation of water-table depth that normally may be anticipated in a cool, moist climate. Even where annual rainfall exceeds evaporation, the infiltration of runoff and groundwater recharge is cyclic.** [*Part (a) from Linsley et al. (1958).*[6] *Reprinted with permission of McGraw-Hill Book Company.*]

from "pumping" under wheel loads where pavement support is provided by fine-grained soils. The potential for the detrimental effects of capillary rise is a function of soil type and the depth to the static groundwater table, or a perched water table.

*Saturated Zone (Free-Water Zone)*

- *Phreatic surface* is the static water table, at which the neutral stress $u_w$ in the soil equals zero. In coarse-grained soils it is approximately the interface between the saturated and unsaturated zones.
- *Confined water* occurs in the free-water zone, but is bounded by impervious or confining strata.
- *Aquifer* refers to a formation which contains water and transmits it from one point to another in quantities sufficient to permit eco-

nomic development. Although a geologic definition, the term is used by engineers to designate a water-bearing stratum. *Specific yield* is the amount of water that can be obtained from an aquifer; it is defined as the ratio of water that drains freely from the material to the total volume of water. *Specific retention* refers to the hygroscopic moisture or pellicular water. Porosity equals the specific yield (effective porosity) plus the specific retention.

- *Aquitard* is a saturated formation, such as a silt stratum, that yields inappreciable quantities of water in comparison to an aquifer, although substantial leakage is possible.
- *Aquiclude* is a formation, such as a clay stratum, which contains water, but cannot transmit it rapidly enough to furnish a significant supply to a well or a spring.
- *Aquifuge* has no connected openings and cannot hold or transmit water; massive granite is an example.
- *Connate water* is the water trapped in rocks or soils at the time of their formation or deposition.

*Transient water-table conditions* are a result of groundwater withdrawal for water supply or construction dewatering, long-term climatic changes such as a series of dry or wet years, and seasonal variations in precipitation. In Fig. 8.17, it is shown that for the period of October through May precipitation at the site substantially exceeds the loss by evaporation, and recharge occurs. During the period of June through September, evapotranspiration exceeds precipitation and the water level drops. The highest water table occurs in the spring after the ground thaws, snow melts, and the spring rains arrive.

### Artesian Conditions

Artesian conditions result from confined groundwater under hydrostatic pressure. If a confined pervious stratum below the water table is connected to free groundwater at a higher elevation, the confined water will have a pressure head (see *Art. 8.3.2*) acting on it equal to the elevation of the free-water surface beyond the confined stratum (less the friction loss during flow). When a well is drilled to penetrate the confined stratum (Fig. 8.16), water will rise above the stratum. The rise is referred to as an *artesian condition*, and the stratum is referred to as an *artesian aquifer*.

An example is the great Dakota sandstone artesian aquifer, the largest and most important source of water in the United States [Gilluly et al. (1959)[8]], which extends under much of North and South Dakota, Nebraska, and parts of adjacent states. This Cretaceous sandstone is generally less than 30 m thick and is overlain by hundreds of meters of other sedimentary rocks, mostly impermeable shales. The principal intake zones are in the west where the formation is upturned and exposed along the edges of the Black Hills as shown on Fig. 8.18. As of 1959, over 15,000 wells had been drilled into the formation.

### Fresh Water Over a Saltwater Body

Fresh water overlying a saltwater body occurs along coastlines or on islands that are underlain to considerable depth by pervious soils, or rocks with large interstices, such as corals. The fresh water floats on the salt water because of density differences. On islands, the elevation of the water table is built up by influent seepage from rainwater; it decreases from its high point at the

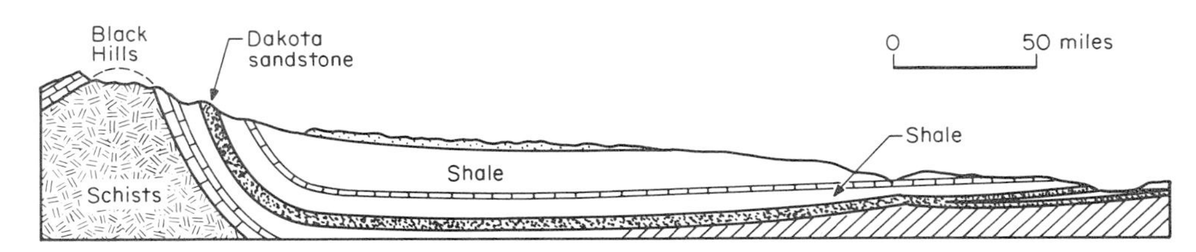


**FIG. 8.18 Section through the Dakota artesian aquifer, from the intake area in the Black Hills of western South Dakota to northern Iowa. Vertical scale is tremendously exaggerated.**

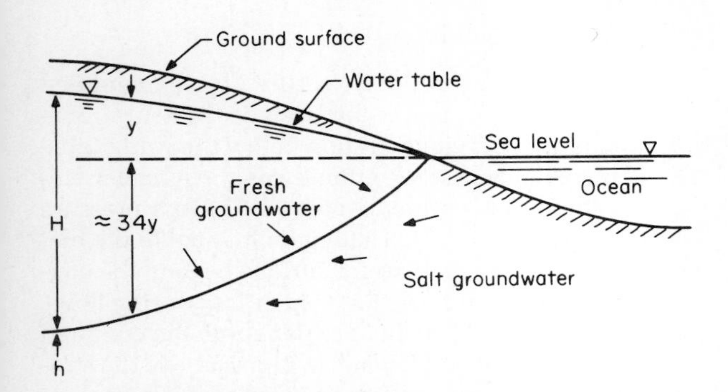


FIG. 8.19 Natural equilibrium between fresh groundwater and salt groundwater along a coastline or beneath an ocean island.

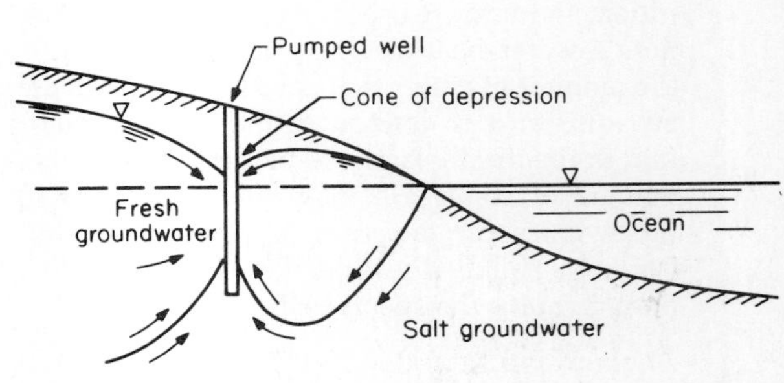


FIG. 8.20 Pumping causing saltwater intrusion into a freshwater well.

island center (if topography is uniformly relatively flat) until it meets the sea. In Fig. 8.19 a column of fresh water $H$ is balanced by a column of salt water $h$, and conditions of equilibrium require that the ratio of $H/h$ should be equal to the ratio of specific gravity of fresh water to that of salt water, or about 1.0 to 1.03. Therefore, if the height of fresh water above sea level is 1 m, then the depth to the saltwater zone will be about 34 m below sea level, if the island is of sufficient size.

*Saltwater intrusion* results from overpumping of freshwater wells. The salt water migrates inland along the coastline, resulting in the pollution of freshwater wells with salt water as shown on Fig. 8.20.

## Springs and Underground Streams

*Springs* represent a concentrated flow of groundwater, or effluent seepage, emerging from the outcrop of an aquifer at the ground surface. The source may be free water moving under control of the water-table slope (a water-table spring), confined water rising under hydraulic pressure (an artesian spring), or water forced up from moderate or great depths by forces other than hydraulic pressures, such as geysers, volcanic, or thermal springs. Springs provide important information about groundwater conditions when observed in the field.

*Underground streams* truly exist naturally only in limestone or other cavernous rocks where large openings are continuous and water can flow freely. Excavations into gravel beds or other free-draining soil or fractured rock below the water table will encounter large quantities of water which may pour into an excavation and give the appearance of a flowing stream. Actually, when confined beneath the surface, movement will be relatively slow even in free-draining materials.

## Significance

### *Geological*

To the geologist, the principal significance of groundwater is as a source of water supply. The primary concern is with the quantities and quality of water available from strata that are relatively free-draining (aquifers), and particularly artesian aquifers since pumping costs are reduced.

### *Engineering*

The primary interests of engineers lie with aquifers and aquitards as sources of water. The water flowing into excavations must be controlled to maintain dry excavations and to reduce pressures on retaining structures. Water flowing through, around, or beneath dams or other retaining structures requires control to prevent excessive losses, seepage pressures, and piping. Aquitards, represented by saturated silts, can allow seepage into excavations and may constitute weakness zones in excavation walls and zones susceptible to "quick" conditions in the bottom of excavations or in slopes. Furthermore, the engineer must be aware that conditions are transient and must realize that groundwater levels measured during investigations are not necessarily representative of those that will exist during construction.

Artesian conditions can result in boiling or piping (*see Art. 8.3.2*) and uplift in excavation bottoms, often with disastrous results if pressures are excessive and not controlled with proper procedures. Underconsolidation results if artesian pressures prevent a soil stratum from draining and consolidating, causing it to remain in a very loose or very soft state. The condition is found in river valleys and other lowlands where buried aquifers extend continuously into nearby hills (Fig. 8.26).

## 8.3.2 SUBSURFACE FLOW

### The Hydraulic Gradient

*General Conditions*

*Hydrostatic conditions* refers to pressures in fluids when there is no flow. The pressure $P$ at depth $h$ in water equals the unit weight of water $\gamma_w$ times the depth, plus atmospheric pressure $P_a$ expressed as:

$$P = \gamma_w h + P_a \tag{8.1}$$

Hydrostatic pressure is equal in all directions; i.e., $P_v = P_h$.

*Groundwater flow* occurs when there is an imbalance of pressure from gravitational forces acting on the water and the water seeks to balance the pressure. Water movement in the ground occurs very slowly in most materials, creating a time lag in the leveling-out process. Typical velocities range from 2 m/day to 2 m/year; therefore, the water table usually follows the ground surface, but at a subdued contour. In dry climates or free-draining materials, however, the groundwater level is approximately horizontal.

*Hydraulic gradient* and *permeability* are the two factors upon which groundwater movement is dependent. The *hydraulic gradient* between two points on the water table is the ratio between the difference in elevation of the two points and the distance between them. It reflects the friction loss as the water flows between the two points.

*Flow-Condition Nomenclature*

Flow-condition nomenclature is illustrated on Fig. 8.21.

- *Static condition* refers to *no flow*, and in Fig. 8.21*a*, water will rise to the same piezometric level in any tube extending from the inclined sand-filled glass tube.
- *Pressure surface* is created when flow is allowed to occur. The levels in the tubes drop as shown on Fig. 8.21*b*.
- *Hydraulic head h* is the difference in water-level elevation between the two tubes, or the head lost during flow.
- *Hydraulic gradient i* is the ratio of the

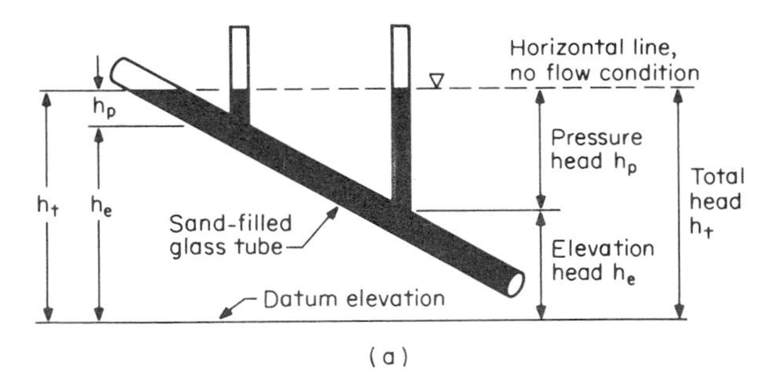


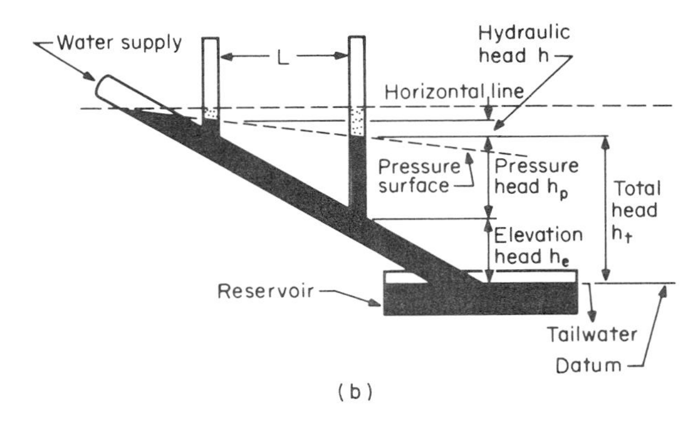


**FIG. 8.21 Diagrams illustrating the hydraulic gradient of groundwater: (*a*) no-flow condition and (*b*) flow creates an hydraulic gradient, $i = h/L$.**

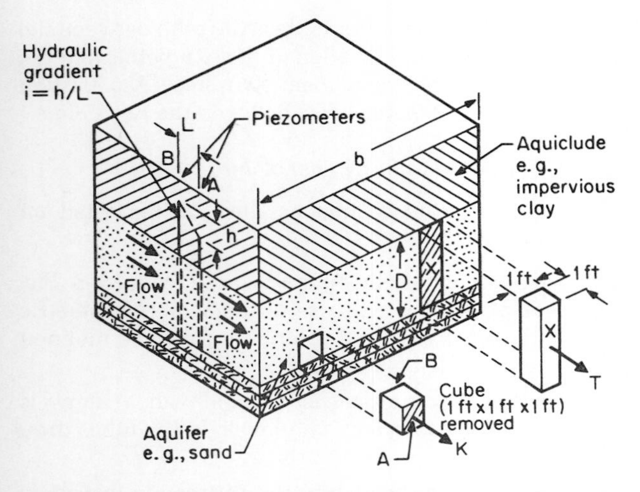


**FIG. 8.22 Coefficient of permeability and transmissibility as used by hydrogeologists.** [*After Krynine and Judd (1957).*[10] *Adapted with permission of McGraw-Hill Book Company.*]

hydraulic head $h$ to the length of flow path $L$, expressed as

$$i = h/L \tag{8.2}$$

*Pressure head* $h_p$ is the height to which water will rise in the vertical tube from the point of interest or reference (also referred to as *piezometric head*).

*Elevation head* $h_e$ is the height of the point of interest or reference with respect to some arbitrary datum.

*Tailwater elevation,* such as a lake or pool where the elevation is constant, is the reference datum selected for most seepage problems.

*Total head* $h_t$ equals the pressure head plus the elevation head.

*Steady-state condition* usually means a state of constant flow, no acceleration or deceleration or changes in piezometric levels.

**Permeability** *(See Also Art. 3.3)*

*Permeability* is the capacity of a material to transmit water.

*Darcy's law* expresses the relationship governing the flow of water through a subsurface medium but is valid only for the conditions of laminar flow through a saturated, incompressible material:

$$q = kiA \tag{8.3}$$

where $q$ = the rate of flow or quantity per unit of time $(Q/t)$ given as liters/minute, $cm^2/s$, etc.
$k$ = the coefficient of permeability in cm/s
$i$ = the hydraulic gradient, or total head loss per flow length, $h/L$
$A$ = cross section of the material through which the flow occurs, in $cm^2$

Darcy's initial expression was $v_d = ki$, where $v_d$ equals the *discharge velocity;* or total volume flow rate per unit of cross section perpendicular to the flow direction $(V = Q/A)$.

Values for $k$ as used by engineers are given in units of velocity (centimeters per second) at a temperature of 20°C, since temperature affects the viscosity of water. Typical values for various materials are given in Art. 3.3. Engineers for convenience refer to permeability as the *superficial or discharge velocity* per unit of gradient, as if the flow occurs through the total volume of the medium, not only the void area.

Geologists use the symbol $K$ to signify permeability. It is expressed as a discharge and defined as "the rate of flow in gallons per day through an area of one square foot under a hydraulic gradient of unity (one foot per foot)" by Krynine and Judd (1957),[9] as illustrated on Fig. 8.22.

*Transmissibility* $T$ is used by geologists to represent the flow in gallons per day through a section of aquifer 1 ft in width and extending the full length of the stratum under a unit head (slope of 1 ft) as shown on Fig. 8.22. Transmissibility equals the coefficient of permeability of the aquifer times its thickness.

*Specific yield* has been defined (*see Art. 8.3.1*) as the amount of water that can be obtained from an aquifer, or the ratio of water that drains freely from the formation to the total volume of water. A comparison of specific yield, porosity, and permeability for various materials is given on Table 8.1.

**TABLE 8.1**
**APPROXIMATE AVERAGE POROSITY, SPECIFIC YIELD, AND PERMEABILITY OF VARIOUS GEOLOGIC MATERIALS**

| Material | Porosity, % | Specific yield, % | Permeability, gal/day·ft$^2$ |
|---|---|---|---|
| Clay | 45 | 3 | 1 |
| Sand | 35 | 25 | 800 |
| Gravel | 25 | 22 | 15,000 |
| Gravel and sand | 20 | 16 | 2,000 |
| Sandstone | 15 | 8 | 700 |
| Limestone, shale | 5 | 2 | 1 |
| Quartzite, granite | 1 | 0.5 | 0.1 |

*From Linsley and Franzini (1964).[10] Reprinted with permission of McGraw-Hill Book Co.

*Determination of k values* is discussed in Art. 3.3 as follows:

- Estimation from charts and tables—*see Art. 3.3.2*
- Laboratory tests—*see Art. 3.3.3*
- In situ tests—*see Art. 3.3.4*
- *Stratification* affects permeability significantly. Strata of granular soils are rarely homogeneous throughout and layering is more often the rule. Water will flow much more freely horizontally in a clean sand or gravel bed than through silty or clayey soils, a fact which causes horizontal permeabilities to be often greater than vertical permeabilities by factors of 10 or 100 or more. An evaluation of the effect of stratification is given on Fig. 8.23.
- *Rock-mass* permeability was referred to by Terzaghi as "*secondary permeability.*" In most practical situations rock-mass permeability is primarily a function of joint conditions including spacing, aperture width, and the nature of the filling or of cavity size and distribution. Because of the great irregularity of these features, including their normally discontinuous nature in most rock masses, an estimation of rock-mass permeability can be subject to substantial error. Where rock is essentially intact, values are extremely low for most rock types, except for porous rocks such as some sandstones (argillaceous or highly cemented sandstones may have low permeabilities). Some typical values are given in Table 3.12.

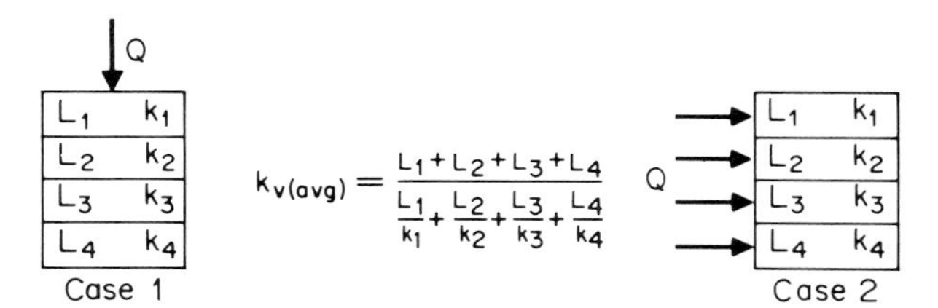


$$k_{h(avg)} = \frac{k_1 L_1 + k_2 L_2 + k_3 L_3 + k_4 L_4}{L_1 + L_2 + L_3 + L_4}$$

**where $Q$ = quantity of flow**
**$L$ = flow path length**
**$k$ = coefficient of permeability**

NOTE: **The electrical analogy**

**If $L_i = L_2 = L_3 = L_4 = 1$, and $k_1 = 1$, $k_2 = 2$, $k_3 = 3$, $k_4 = 4$, then in case 1, $k_{v(avg)} = 1.9$ and in case 2 $k_{h(avg)} = 2.5$**

**FIG. 8.23 An evaluation of the effect of stratification on permeability.** [*From Salzman (1974).*[11]]

## Pore-Water Pressures

### *General*

*Pore-water pressure* $u$ or $u_w$ is the pressure existing in the water in the pores, or void spaces, of a saturated soil element.

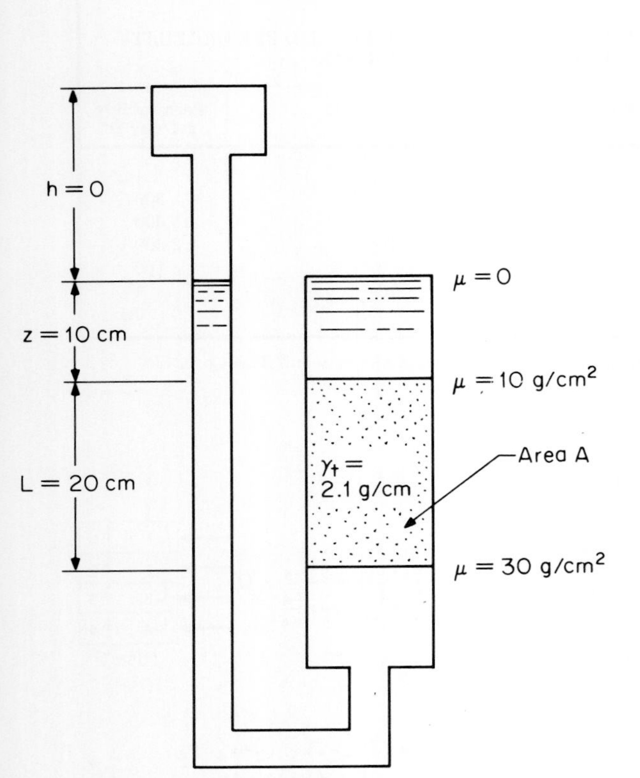


**FIG. 8.24 Pore-water pressures for the no flow condition and buoyancy water pressures.** [*After Lambe and Whitman (1969).*[12] *Adapted with permission of John Wiley & Sons, Inc.*]

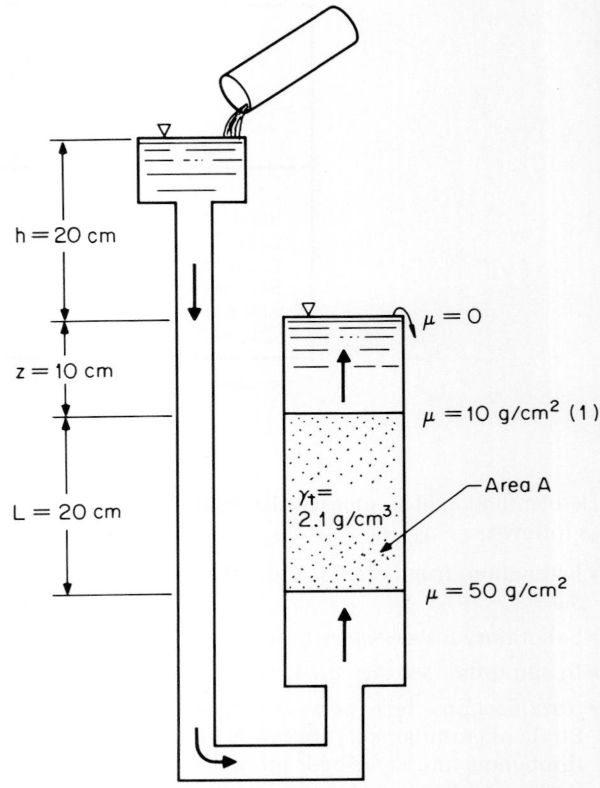


**FIG. 8.25 Boundary water pressure during upward flow. At (1) there is 20 g/cm² lost in seepage.** [*After Lambe and Whitman (1969).*[12] *Adapted with permission of John Wiley & sons, Inc.*]

*Cleft-water pressure* refers to the pressures existing on the water in saturated joints or other fractures in rock masses.

*Excess hydrostatic pressure* is that pressure capable of causing the flow of water ($h\gamma_w$ in Fig. 8.27).

### *No-Flow and No-Applied-Stress Condition*

For the condition of no flow and no applied stress, pore-water pressures are equal to the unit weight of water times the depth below the free-water surface as shown on Fig. 8.24, expressed as:

$$u = \gamma_w z_w \tag{8.4}$$

*Buoyancy pressures* refer to the vertical pressures acting on each end of the soil column; on the specimen bottom the buoyancy force equals 30 g/cm².

### *Upward Flow Condition*

In Fig. 8.25, a head of 20 cm (seepage force) causes an increase in pore pressure, at the base of the soil column supported on a screen, to u = 50 g/cm². The tail water is barely overflowing and the 20-cm head has been dissipated in viscous friction loss in the soil specimen.

*Boundary water pressures* act on the specimen; on the bottom they are equal to the *buoyancy force* (30 g/cm²) plus the *seepage force* (20 g/

cm²), and on the top they are equal to the water pressure (10 g/cm²).

*Effective Stresses*

The effective stresses ($\bar{\sigma}_v = p - u$; see *Art. 3.4.2*) may be found either from boundary forces considering total soil weight, or seepage forces considering submerged weights. At the bottom of the specimen on Fig. 8.25, $\sigma_v$ equals either:

1. Total specimen weight $LA\gamma_t$ plus the overlying water weight $zA\gamma_w$ minus the pore-water pressure $(h + L + z)A\gamma_w$, or
2. Submerged specimen weight $LA\gamma_b$ minus the seepage force $hA\gamma_w$. For $\bar{\sigma}_v = 0$, $LA\gamma_b - LA\gamma_w = 0$ and a "quick" condition exists.

It is the development of high pore pressures as water levels rise in slopes that causes the reduction in shear strength and slope failure. High pore pressures also develop in earth dams when the phreatic level in the dam arises (Fig. 8.29) and require consideration during design. They are not of concern in properly designed and constructed dams.

*Pore-Pressure Ratio $r_u$*

Defined as a ratio between the pore pressure and the total overburden pressure or between the total upward force due to water pressure and the total downward force due to the weight of overburden pressure. It is used frequently in computer program analysis of slope stability problems.

$$r_u = \frac{\text{volume of sliding mass under water} \times \text{unit weight of water}}{\text{volume of sliding mass} \times \text{unit weight of soil}}$$

or, since the unit weight of water is approximately equal to one-half the unit soil weight, it can be expressed approximately by:

$$r_u = \frac{\text{cross-sectional area of sliding mass under water}}{2 \times \text{total cross-sectional area of sliding mass}}$$

*Applied Stresses*

Applied stresses cause an increase in pore pressures. Loading a clayey soil causes the process of consolidation (see *Art. 3.5.4*). The pore water first carries the load, then, as the water drains from the soil, pore pressures dissipate, the voids become smaller, and the load is transferred to the soil skeleton. If the load is applied rapidly, however, the soil has no time to drain, friction is not mobilized, and the substantially lower undrained shear strength prevails. Even if drainage occurs, the frictional component of strength will be reduced by the amount of pore pressure.

The pressure in the pore water is referred to as the *neutral stress* because it does not contribute either to compression or to an increase in shearing resistance.

*Piezometers*

These devices, described in Art. 4.4.2, are installed in the field to measure water tables and pore pressures. Several conditions are illustrated on Fig. 8.26.

## Seepage

*Velocity*

The average seepage velocity $v_s$ of water flowing through the pores of a saturated soil mass is equal to the discharge velocity ($v_d = ki$) times the ratio $(1 + e)/e$, where $e$ is the void ratio, or the discharge velocity divided by the effective porosity ($n_e$), expressed as:

$$v_s = \frac{ki}{n_e} \quad \text{ft/day or cm/s} \tag{8.5}$$

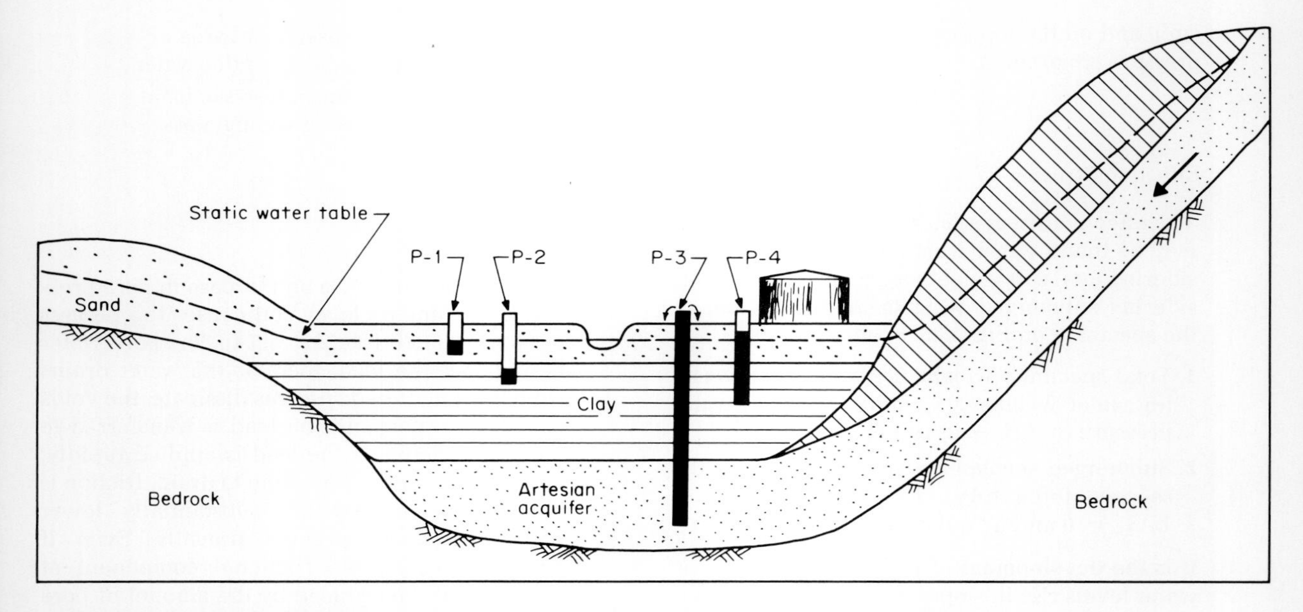


**FIG. 8.26 Various possible conditions of pore-water pressures as measured by piezometers:**

- **P-1—Installed in a sand stratum, measures the static water table**
- **P-2—Installed in a desiccated, overconsolidated, and still only partially saturated clay, reflects negative pore pressures (pressures lower than the static water level)**
- **P-3—Installed in an artesian aquifer, shows excess hydrostatic pressure (pressure higher than the static water level)**
- **P-4—Installed in a clay below the depth of desiccation beneath a newly constructed storage tank, shows excess hydrostatic pressure because the clay is still consolidating under the applied load**

**(See also Figs. 8.27 and 8.28 for flow causing excess pressures.)**

The practical significance of seepage velocity lies in the field evaluation of $k$, using dye tracers and measuring the time for the dye to travel the distance between two holes and in estimating the rate of movement in pollution-control studies. Some relationships between permeability, hydraulic gradient, and the rate of groundwater flow are given on Table 8.2.

*Pressures and Liquefaction*

Seepage pressures $j$ are stresses in the soil caused by the flow of water. They are equal to the hydraulic gradient $i$ times the unit weight of water, expressed as:

$$j = h/L\gamma_w = i\gamma_w \tag{8.6}$$

and act in the direction of the flow.

Where the flow tends to be upward, as at the toe of a dam, along slopes, or in the bottoms of excavations, the pressure is resisted by the weight of the overlying soil column.

*"Quick conditions" (boiling or liquefaction)* occur when the upward gradient increases until the seepage pressure exceeds the submerged weight of the soil, and the soil column is uplifted. The result is complete loss of intergranular friction and the supporting capacity of the soil. (*"Cyclic" liquefaction* occurs under dynamic loadings; see *Art. 11.3.3*).

*Critical gradient* is the hydraulic gradient required to produce liquefaction and equals the ratio of the submerged unit weight of the soil to the unit weight of water:

$$i_{cr} = \frac{\gamma_t - \gamma_w}{\gamma_w} = \frac{\gamma_b}{\gamma_w} \quad (8.7)$$

In coarse to fine sand, $i_{cr}$ is about 0.9 to 1.0, and in layman's terms results in "quicksand." The example given in Fig. 8.25 is barely stable.

The *factor of safety* against liquefaction, usually taken as 3 or more because of the disastrous nature of such a failure, is defined as

$$FS = i_{cr}/i \quad (8.8)$$

Because seepage pressure is directly proportional to the hydraulic gradient, the most dangerous areas for liquefaction to occur are those where the upward gradient is large and the counterbalancing weight is small. The analysis of flow through soils and seepage pressures may be performed with *flow nets* (*see Art. 8.3.3*).

### *Rock Masses*

Water pressures that develop in rock-mass fractures often are termed *cleft-water pressures.* When too high, they can result in instability of rock foundations for concrete dams, and they are the common cause of slope failures. Seepage can result in the softening of joint fillings, and the development of high pore pressures in the filling material can reduce strength.

*Stress changes* can significantly affect seepage and permeability in rock masses. Compressive stresses cause closure of joints even under relatively low stress levels, reducing seepage flow, although sufficient closure of other voids to reduce permeability occurs in most rocks only under relatively high stress levels. Tensile stresses can increase permeability and flow, with the increase commonly occurring as a rock slope begins to deflect. The failure of the Malpasset Dam (*see Art. 8.3.4*) is considered to be the result of tensile stresses increasing under the toe of the concrete arch dam in the foundation gneiss. It has been estimated that the gneiss had a permeability 1000 times greater under the tensile stresses than when in compression. The greater permeability permitted an increase in uplift pressures beneath the foundations, resulting in excessive deflections of the dam.

**TABLE 8.2**
**PERMEABILITY, HYDRAULIC GRADIENT, AND GROUNDWATER FLOW-RATE RELATIONSHIPS FOR VARIOUS SOIL GRADATIONS**

| Soil type | Permeability, cm/s | Gradient $i$ | Time to move 30 cm | $n_e$ |
|---|---|---|---|---|
| Clean sand | $1.0 \times 10^{-2}$ | 0.10<br>0.01 | 2.5 hr<br>25.0 hr | 0.30 |
| Silty sand | $1.0 \times 10^{-3}$ | 0.10<br>0.01 | 1.4 days<br>14.0 days | 0.40 |
| Silt | $1.0 \times 10^{-4}$ | 0.10<br>0.01 | 14.0 days<br>140.0 days | 0.40 |
| Clayey sand | $1.0 \times 10^{-5}$ | 0.10<br>0.01 | 174 days<br>4.8 years | 0.50 |
| Silty clay | $1.0 \times 10^{-6}$ | 0.10<br>0.01 | 4.8 years<br>48.0 years | 0.50 |
| Clay (intact) | $1.0 \times 10^{-7}$ | 0.10<br>0.01 | 48.0 years<br>480 years | 0.50 |

*Leakage*

Leakage occurs through natural slopes; through the embankment, foundation, or abutments of dams; and beneath sheeted excavations. Sloughing of the downstream face of a dam embankment or a natural slope is a fairly common phenomenon. It usually occurs where the phreatic level intersects the slope. Seepage forces in the zone of emergence cause a loosening of the surface materials and raveling, and local failures occur.

Leakage through dam foundation materials is more common than through the embankment, since foundation soils are generally less dense and more erratic than the structure which results from manufacturing an embankment. Uncontrolled seepage beneath an embankment manifests itself as springs near the toe; and, as fine soil particles are carried along (i.e., piping occurs) they are deposited on the surface around the springs as "*sand boils*." These can be found also at the toe of a cut slope, on the surface after earthquakes in areas of fine-grained cohesionless soils, or behind levees during flood stages.

Underseepage also can cause the development of excess pore pressures under the embankment toe. The loss of stability of the foundation materials can result in a deep downstream slide. Since failure does not relieve the pore pressures, sliding will continue, and if not immediately corrected, failure of the dam may occur.

### Piping

*Piping* is the progressive erosion of soil particles along flow paths. Fine soil particles near the point of emergence can be removed by flow, and as they wash away, flow and erosion increase in the soil mass, in time developing channels which result in greater flows and erosion, and finally catastrophic failure (*see Art. 8.3.4*, discussion of dams). The term piping is used also to refer to the phenomenon of boiling described previously.

Piping through an embankment occurs in finer soils along layers of free-draining coarse materials, through cracks in embankment soils, or adjacent to rock masses where fractures are in contact with fine-grained embankment soils. Embankment cracks can result either from shrinkage of clay soils or from differential settlement of the embankment or its foundation. Embankment settlement caused by compression of the foundation materials can result in breaking of outlet pipes, which permits piping of the embankment soils into the pipes. Except for overtopping, piping has caused a far greater number of earth dam failures than any other activity.

Soils susceptible to piping in the natural state, including "dispersive clays," are described in Art. 10.5.5.

## 8.3.3 FLOW SYSTEMS AND ANALYSIS

### General

*Flow Systems*

All flow systems extend physically in three directions. The flow of water through saturated soil is a form of streamlined flow (the tangent of any point on a flow line is in the direction of the velocity at that point) and can be represented by the Laplace equation for three-dimensional flow through porous media [DeWeist (1965)[13]]. The equation in effect states that the change in gradient in the $x$ direction plus the change in gradient in the $y$ direction plus the change in gradient in the $z$ direction equals zero.

In practice, seepage problems can be two- or three-dimensional. Fortunately, most engineering problems can be resolved by assuming two-dimensional flow. Most three-dimensional seepage problems are extremely complex in their solutions.

*Analytical Methods*

A number of analytical methods are available for solving the Laplace equations including:

- Electrical analog [Karplus (1968),[14] Meehan and Morgenstern (1968)[15]]
- Relaxation method
- Finite element method [Zienkiewicz et al. (1966)[16]]
- Conforming mapping configurations
- Flow nets (this article)
- Well formulas (this article)

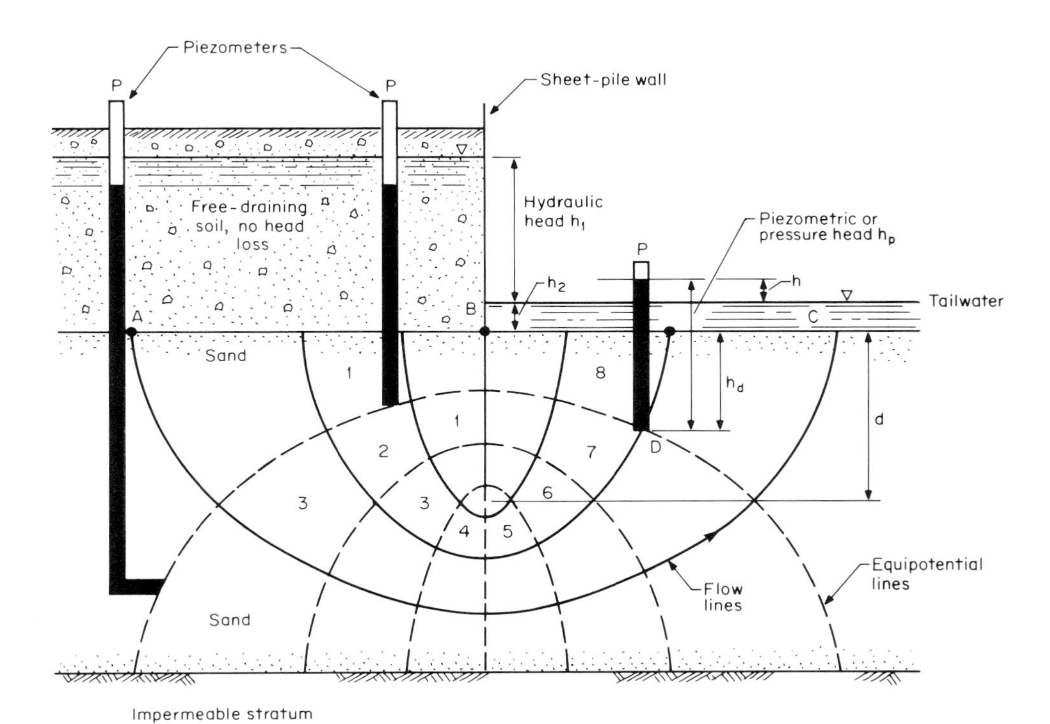


**FIG. 8.27 An example of a confined flow net beneath sheet piling.**

*Solutions:*

1. **Find *seepage quantity beneath sheet piling.***
   **Given: $k = 0.001$ m/min, $h_1 = 8$ m, $h_2 = 2$ m, $h_d = 6$ m**
   **$L$ = length of sheet pile wall along excavation = 30 m**
   **Solution: From the flow net, $N_f = 4$, $N_e = 8$. Quantity**

$$q = k(N_f/N_e)h_1 \qquad \text{(Eq. 8.9)}$$

$$q \text{ (per meter of wall)} = 0.001 \times 4/8(8) = 0.004 \text{ m}^3/\text{min}$$

$$Q \text{ (per total wall length)} = 0.004 \times 30 = 0.12 \text{ m}^3/\text{min}$$

2. **The *pore pressure at point D* = $u_w = (h_1 + h_2 + h_d - 7/8h_1)\gamma_w = 9$ T/m². That part, $(h_1 - 7/8h_1)\gamma_w = h\gamma_w$, due only to the flow of water, is the *excess hydrostatic pressure.***
3. ***Liquefaction* occurs when the exit gradient $i$ is approximately 1; $i$ can be expected to be highest on the outside of the wall at point *B*. $i = \Delta h/l$, where $\Delta h = h_1/N_e$, and $l$ = the length of the side of the square where exit occurs, or 3 m. Therefore, $i = 8/8/3 = 0.33$, which is a marginal value in terms of desired values for FS (3 to 5). Terzaghi demonstrated that heave from boiling occurs within a distance from the wall of about $d/2$.**

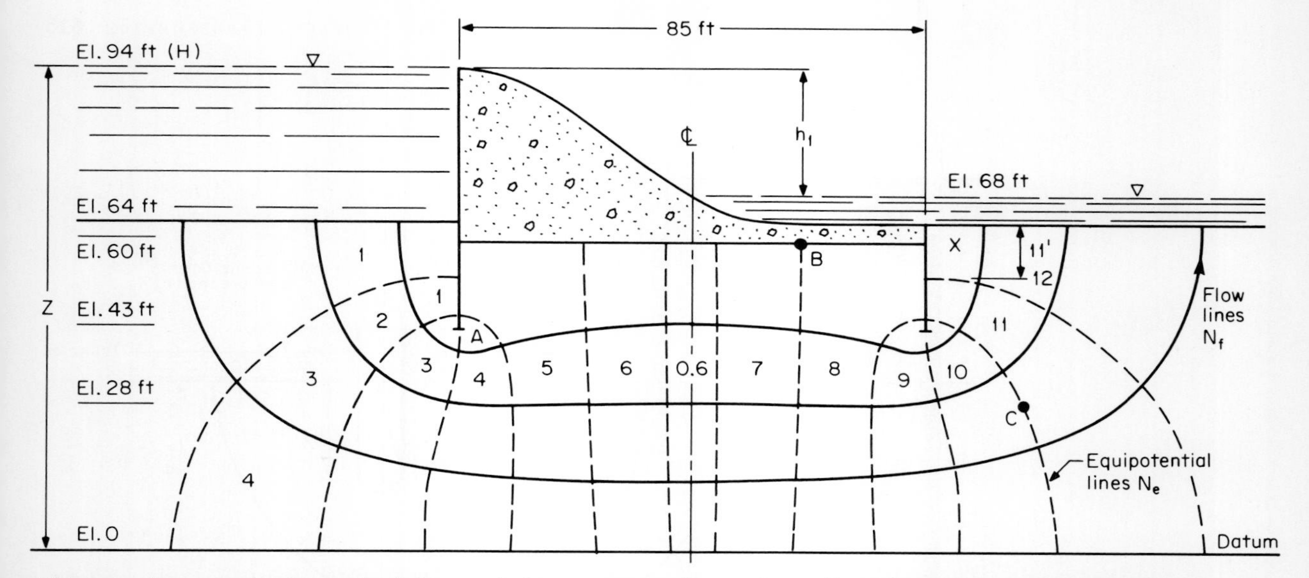


**FIG. 8.28 An example of a confined flow net below a concrete dam with cutoffs to illustrate seepage uplift.** [*From Lambe and Whitman (1969).*[12] *Adapted with permission of John Wiley & Sons, Inc.*]

**Example:**

**For the flow net shown find the pressure head and water pressure $p$ at points $A$, $B$, and $C$; the quantity of seepage; and the gradient in *zone* x. $k$ = 0.1 ft/min.**

***Solution:***

**1. From the figure: $N_f = 4$, $N_e = 12.6$, $S = N_f/N_e = 0.32$.**

**2. Total head at any point in the flow domain $h = p/\gamma_w + Z$.**

**(Selection of the datum determines sign of Z; i.e., if tail water were datum 0, then Z would be negative downward.)**

**3.**

| Point | (A) Elevation head Z, ft | (B) Total head h, ft $(H - N_f)\left(\frac{h_i}{N_e}\right) = h_t$ | (C) Pressure head $p/\gamma_w$, ft (col. B − col. A) | (D) Water pressure p, lb/ft² (col. C × 62.4) |
|---|---|---|---|---|
| A | 43 | $94 - 3\left(\frac{26}{12.6}\right) = 87.8$ | 44.8 | 2800 |
| B | 60 | $94 - 7.6\left(\frac{26}{12.6}\right) = 78.3$ | 18.3 | 1142 |
| C | 28 | $94 - 10.6\left(\frac{26}{12.6}\right) = 72.1$ | 44.1 | 2750 |

**4. Inserting a piezometer in the soil at point C will show the water level to rise 44.1 feet above point C.**

**5. The quantity of seepage is: $q = Skh = 0.32(0.1)(26)$**

$$q = 0.83 \text{ ft}^3/\text{min/ft of dam}$$

**6. The gradient in region x is: $i_x = \Delta h/l = 26/12.6/11 = 0.19$.**

**7. Seepage pressure in zone x = 0.19 × 62.4 = 11.9 pcf.**

## Flow Nets

### *Description*

A flow net is a two-dimensional graphical presentation of flow consisting of a net of flow lines and equipotential lines, the latter connecting all points of equal piezometric level along the flow lines.

### *Applications*

Flow nets are used to evaluate:

1. Seepage quantities exiting through or beneath a dam or other retaining structure
2. Inflow quantities into wells or other openings in the ground

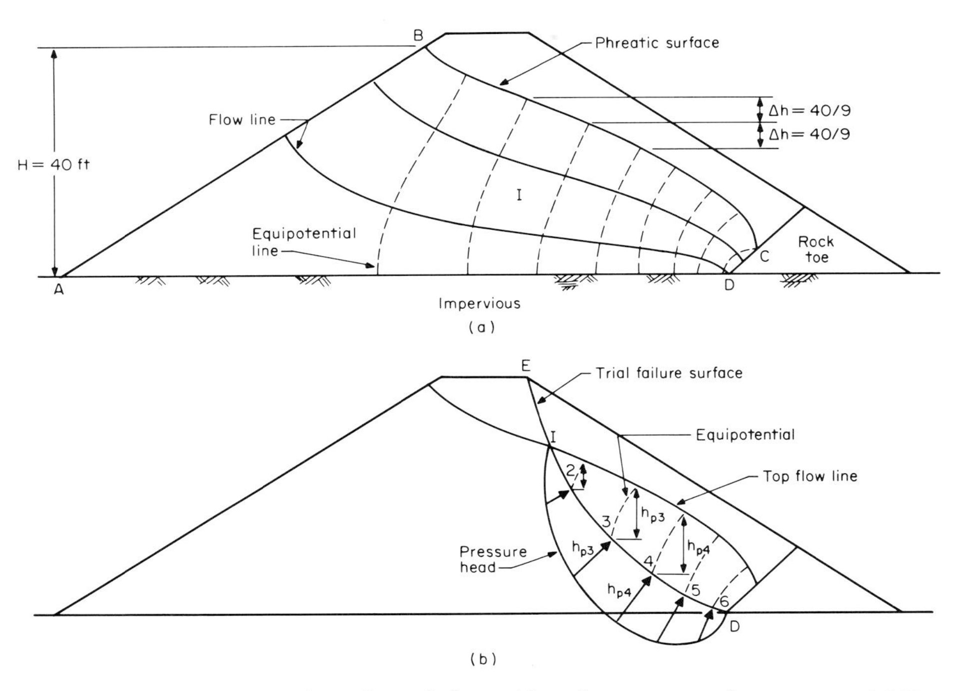


**FIG. 8.29 Unconfined flow through earth dam with rock toe to control toe seepage: (*a*) Use of flow net to find seepage quantity and gradient. Given: $k$ = 0.005 ft/s, $H$ = 40 ft, $\mathcal{S} = N_f/N_e$ = 2.65/9 = 0.294. Seepage through dam = $Q/L = kH\mathcal{S}$ = 0.0005 × 40 × 0.294 = 59 × $10^{-4}$ ft³/min/ft. Gradient in square $I = i_I = \Delta h/l_1$ = 40/9/11.2 = 0.40. (*b*) Use of flow net to find pore-water pressures on a failure surface: (1) layoff trial failure surface on flow net, (2) measure pressure heads as elevation differences between phreatic line and trial failure surface at equipotential lines, and (3) pore-water pressure = $h_p \times \gamma_w$.** [*From Lambe and Whitman (1969).*[12] *Adapted with permission of John Wiley & Sons, Inc.*]

3. Seepage pressures that result in uplift below dewatered excavations or at the toe of dams
4. Exit gradients and the potential for liquefaction in dams, slopes, or excavations
5. Pore pressures along potential failure surfaces in slopes

*Flow Conditions for Analysis*

*Confined flow* refers to the case where the phreatic surface is known; it commonly occurs beneath cutoff walls. Examples are given on Figs. 8.27 and 8.28, graphical presentations of a flow net about an impervious cutoff wall penetrating a pervious flow medium and of flow beneath a concrete dam with a cutoff.

*Unconfined flow* refers to the case where the location of the phreatic line is not known; it commonly occurs in earth dams (Fig. 8.29) and slopes (Fig. 8.30). In the earth dam illustrated, the rock toe is provided to ensure that the phreatic surface does not emerge along the slope face, since this could result in high exit seepage forces, erosion, and slope instability.

*Flow Net Construction*

Flow net construction is a graphical procedure accomplished by trial and error, subdividing the flow zone of a scaled drawing of the problem as nearly as possible into equidimensional quadrilaterals bounded by *flow lines* and *equipotential*

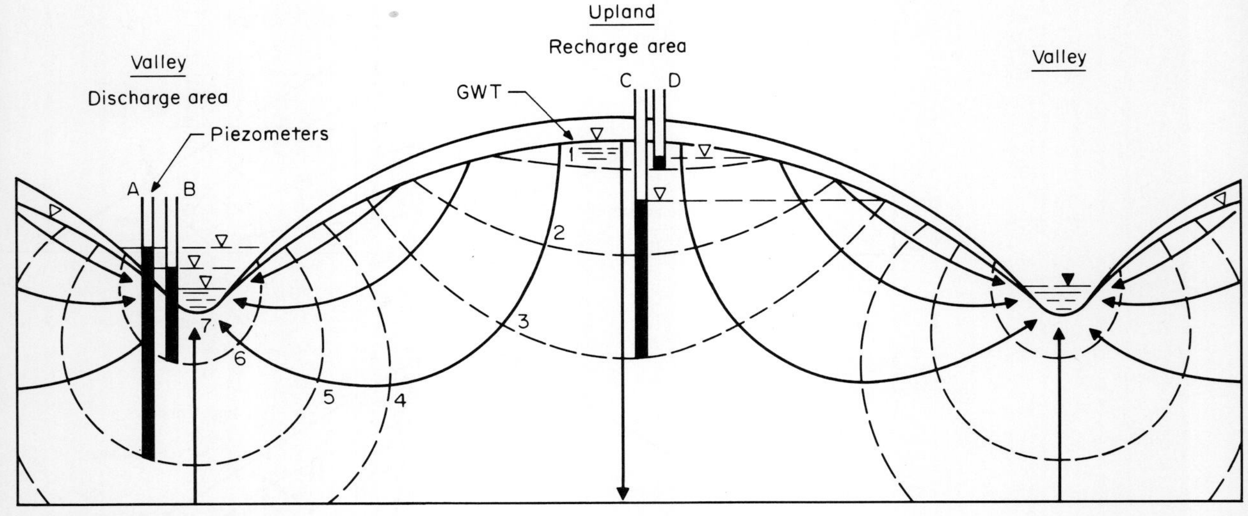


**FIG. 8.30 Simplified regional flow system in uniformly permeable materials.** [*After Hubbert (1940)*,[21] *from Patton and Hendron (1974)*.[20]]

*lines* crossing at right angles as shown on Fig. 8.27. [For details see Taylor (1948),[17] Cedergren (1967).[18]]

Assumptions are that Darcy's law is valid, and the soil formation is homogeneous and isotropic.

A *flow line* is represented on Fig. 8.27 as the path along which a particle of water flows on its course from point *A* to point *C* through the saturated sand mass. Each flow line starts at some point along *AB* where it has a pressure head $h_p$; thereafter the flowing particle gradually dissipates this head in viscous friction until it reaches line *BC*. (In this case the soil above line *AB* is a free-draining gravel in which there is assumed to be no head loss.) Along each flow line there is a point where the water has dissipated any specific portion of its potential.

*Equipotential lines* connect all such points of equal piezometric level on the flow lines. The level may be determined in the field by piezometers.

*Anisotropic conditions* resulting from stratification cause horizontal permeability to be greater than vertical. This is accounted for in flow net construction by shrinking the dimensions of the cross section in the direction of the greater permeability. For example, if $k_h > k_v$, the horizontal scale is reduced by multiplying the true distance by $\sqrt{k_v/k_h}$, producing a transformed section, and the flow net is constructed in the ordinary manner. Anisotropic effects are discussed by Harr (1962)[19] and DeWeist (1965).[13] The effect of anisotropic conditions on equipotential lines in rock masses is illustrated on Fig. 8.32.

### *Analysis*

*Seepage quantity* can be calculated, once the flow net is drawn, from the expression:

$$q = \frac{N_f}{N_e} kh \qquad (8.9)$$

where $N_f$ = number of flow channels (space between any adjacent pair of flow lines)
$N_e$ = number of equipotential drops along each flow channel
$k$ = coefficient of permeability
$h$ = total head loss ($h_1$ on Fig. 8.27)
$N_f/N_e$ = $\mathcal{S}$, the shape factor

$q$ = discharge or quantity of flow per foot (or meter), commonly given in $ft^3/s$ per running foot in the English system, where $k$ is given in ft/s, or $m^3/s$ per meter with $k$ given in m/s in the metric or SI system

Examples of computations of seepage quantities for confined flow conditions are given on Figs. 8.27 and 8.28, and for unconfined flow conditions on Fig. 8.29*a*.

*Seepage pressure* is equal to the hydraulic gradient times the unit weight of water ($p_s = i\gamma_w$) and acts in a direction at right angles to the equipotential lines and parallel to the flow lines. In the example of Fig. 8.28, $p_s$ in zone $x = 0.19 \times 62.4 = 11.9$ pcf. This seepage is resisted by the submerged weight of the overlying soil column, for the average sand about 60 pcf. Therefore, liquefaction or boiling is not imminent since $p_s$ is much less than the weight of the overlying soil. Other examples are given on Figs. 8.27 and 8.29.

*Pore-water pressures* may be determined from flow nets as illustrated on Fig. 8.29*b*. The application to stability analysis of slopes is illustrated on Fig. 9.77.

### *Conclusions*

Flow nets are useful tools, since even a crude flow net will permit fairly accurate determinations of seepage quantities and pressures in soils. They are somewhat time-consuming to construct, and each time the dimensions are changed (for example, when the depth of a cutoff wall is increased), new flow nets are constructed.

In critical problems where high potential for seepage uplift and pore-water pressures exist, the values obtained from flow-net analysis should be verified by measurements with instruments, such as piezometers, to monitor the development of actual pore pressures in or beneath an embankment, at the toe, or in a slope.

## Natural Flow Systems in Slopes

### *Simplified Regional Flow Systems*

Classical descriptions usually consider groundwater flow systems to be hydrostatic, whereas, in actuality, nonhydrostatic distributions are common in the vicinity of slopes [Patton and Hendron (1974)[20]]. The general flow system in hilly terrain proposed by Hubbert (1940)[21] is shown on Fig. 8.30. In the upland recharge area the flow tends to be downward, and in the valley lowlands in the discharge area the flow tends to be upward. The conditions given in the figure are for a relatively uniform material; the nonhydrostatic distributions along the slope, as illustrated by the equipotential lines, are apparent. If low permeabilities are present the differences between the actual and the hydrostatic distribution will be accentuated.

### *Slope Seepage*

Flow systems within the slope and adjacent to its face are important in slope-stability problems. Slope seepage is often shown in the geotechnical literature with the flow lines parallel to the water table (Fig. 8.31*a*). This case actually exists normally only during other than wet periods. During the wet season when failures are likely, conditions are those illustrated on Fig. 8.31*b*. There is a downward pore-pressure gradient in the upper portion of the slope, and an *upward gradient* in the lower portions. It is the upward gradient that results in instability during heavy rains.

The major difference between the two conditions given in Fig. 8.31 lies in the effect on the discharge area. In the case of parallel flow (*a*) there would be no adverse effects to placing an impervious fill at the slope toe, since the flow would not be impeded. In the case of (*b*), however, there would be a buildup of pore pressures at the toe and within the slope, possibly leading to failure. Deposits of colluvium, resulting from natural slope failures, also block the discharge area and consequently are usually unstable.

### *Rock Masses*

Variations in equipotential distributions for different permeability configurations in rock masses are shown on Fig. 8.32. The significance with respect to flow of bedding that dips parallel to the slope is readily apparent in Fig. 8.32*c*.

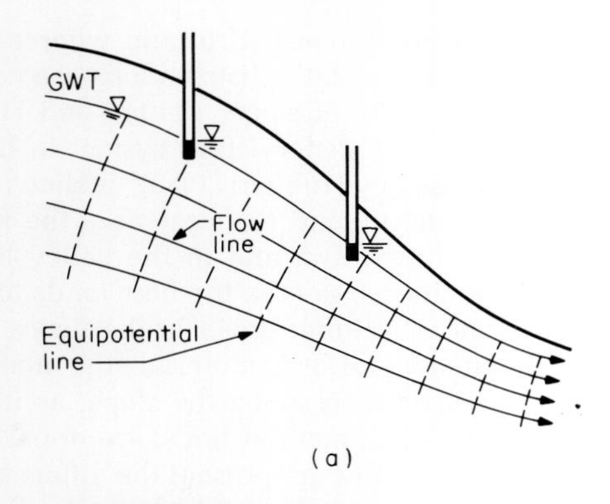


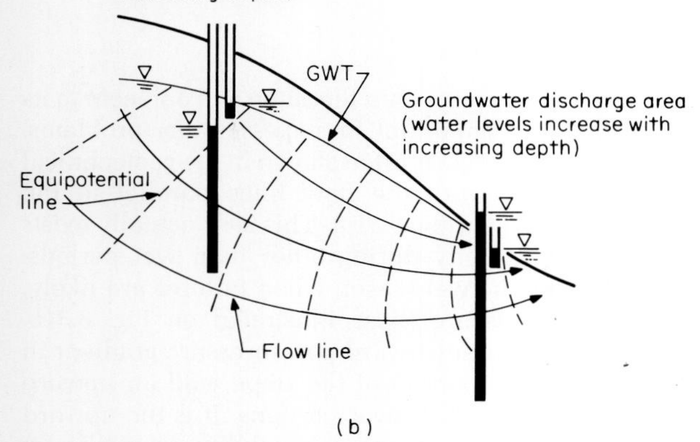


**FIG. 8.31 Comparison of the normal concept of groundwater flow in slopes with the more typical occurrence causing instability: (*a*) groundwater flow assumed parallel to groundwater table (common in geotechnical literature, but seldom found in practice) and (*b*) typical groundwater flow in slopes.** [*From Patton and Hendron (1974).*[20]]

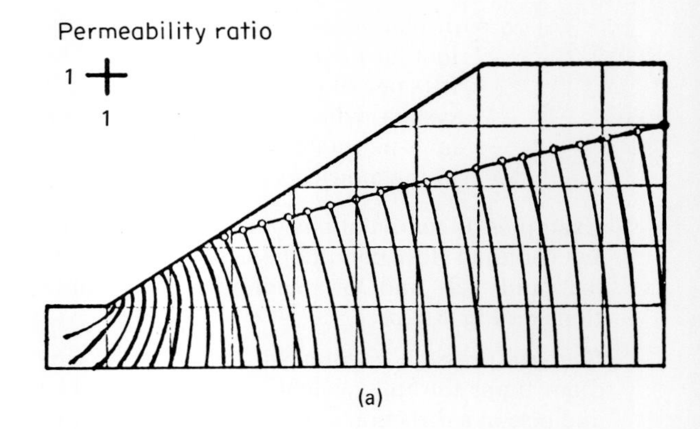


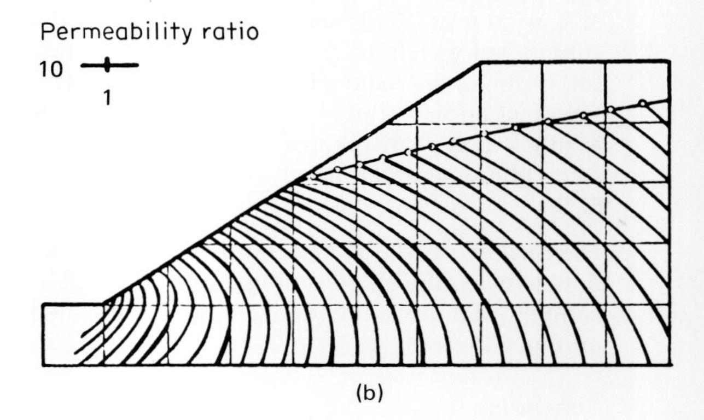


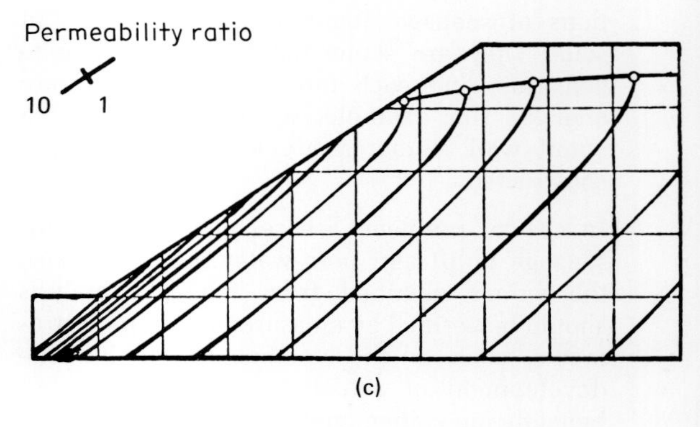


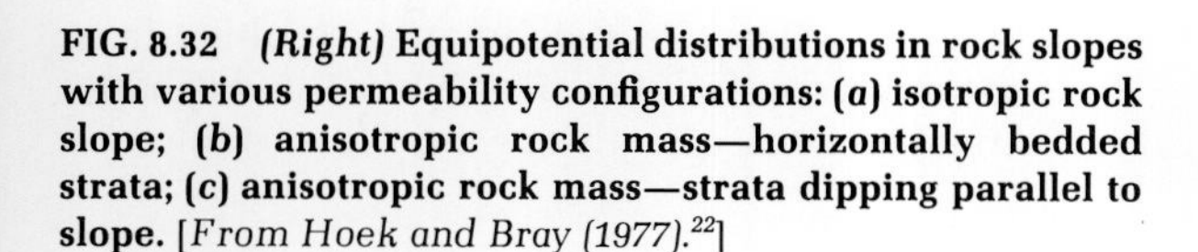

**FIG. 8.32 *(Right)* Equipotential distributions in rock slopes with various permeability configurations: (*a*) isotropic rock slope; (*b*) anisotropic rock mass—horizontally bedded strata; (*c*) anisotropic rock mass—strata dipping parallel to slope.** [*From Hoek and Bray (1977).*[22]]

## Flow to Wells

### General

A water well is a vertical excavation constructed for the purpose of extracting groundwater for water supply, or for the purpose of dewatering or controlling water during construction or other operations.

There are two general cases of flow to wells:

1. *Single wells*, or a pattern of wells, affecting a zone that is essentially circular or elliptical in area
2. *Slots*, considered as a continuous line drain such as a stone-filled trench, or a line of closely spaced wells such as a row of well points

The two general types of wells are *gravity* or water-table wells, which penetrate unconfined aquifers and *artesian wells* which penetrate confined strata.

### Gravity Wells: Characteristics

Two cases of gravity wells are illustrated on Fig. 8.33; in (*a*) the well fully penetrates an unconfined aquifer and in (*b*) the well only partially penetrates the aquifer. The various relationships pertaining to case *a* are described below; case b, the partially penetrating well, is described in Mansur and Kaufman (1962).[23]

A *cone of depression* is produced in the water table surrounding the well as pumping lowers the water level in the well and extracts water from the surrounding water-bearing strata. *Drawdown* is the vertical distance between the original water table and the bottom of the cone of depression. It can extend to horizontal distances from the well as much as 10 times the well depth, or greater, and under certain conditions it can result in ground subsidence (*see Art. 10.2*).

During pumping a number of hydraulic observations are possible, on the assumption that the cone of depression is not influenced by other wells:

1. *Yield*, the quantity of water pumped per unit of time (gal/min, liters per minute, etc.) can be measured to provide a quantitative amount for the well yield without respect to drawdown or well size.
2. *Drawdown* during a selected period of pumping can be measured. If the pumping level becomes stationary after a period of pumping, the natural groundwater supply to the cone of depression is *equal* to the quantity pumped, thereby providing information on the supply available. The drawdown necessary to produce the water pumped is a direct function of the permeability of the water-producing stratum, and the permeability of the aquifer supplying a number of wells can be determined from the drawdown of the respective wells, provided that frictional resistances caused by well screens and filters into the various wells are equal.

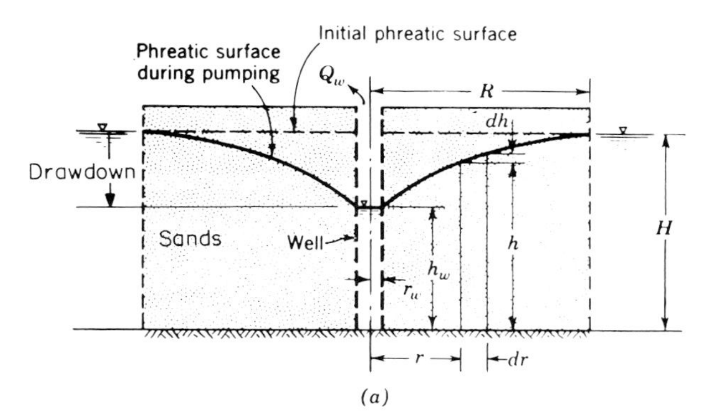


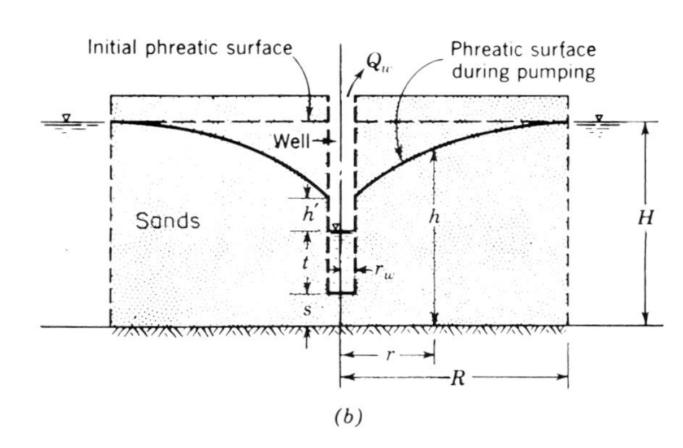


**FIG. 8.33 Flow to a gravity wall from an unconfined aquifer providing circular seepage source: (*a*) fully penetrating well and (*b*) partially penetrating well (also gives the height of free discharge $h'$ to be accounted for in computations in either case).** [*From Mansur and Kaufman (1962).*[23] *Reprinted with permission of McGraw-Hill Book Company.*]

3. *Specific capacity* (gallons per foot of drawdown) can be calculated. It provides the best measure for comparison of yield between two or more wells, and depends on the permeability and the thickness of the aquifer (transmissibility), and the frictional resistance at the well entrance.
4. *Overpumping* of an aquifer is indicated if the water level does not rise to its original level after pumping is stopped.
5. *Safe yield* is indicated if the recovery between pumping periods is complete.
6. *Rate of recovery* depends on, and subsequently indicates, the permeability of the surrounding aquifer. It can be determined by means of an electric probe lowered into the well to measure the rise in the water level as a function of time *(see Art. 2.3.7)*.
7. *Shape of the cone of depression* during pumping is determined from measuring the depth to the water table in a number of observation wells or piezometers distributed about the well. (At least two should be located along each of several sets of perpendicular lines extending from the well.) From the shape of the cone and the distance $R$ it extends from the well, the hydraulic behavior of the well and the permeability of the aquifer penetrated by the well can be calculated.

### Gravity Wells: Analysis

Because of the heterogeneity of formations, calculations of quantities before data are available from pumping tests are only rough approximations. Well yields are best determined from pumping tests, and observation wells provide important additional data on drawdown. Because of the low velocities of groundwater flow, true equilibrium conditions usually occur only after some time interval of pumping.

*Quantity* of flow to a fully penetrating gravity well (Fig. 8.33*a*) at equilibrium may be expressed in terms of permeability ($k_{mean}$ because of stratification effects) and the depression cone characteristics as:

$$Q_w = \frac{\pi k(H^2 - h_w^2)}{\log_e (R/r_w)} \tag{8.10}$$

where $H$ is the height to the original groundwater table.

*Permeability* can be computed by rearrangement of Eq. 8.10 from:

$$k_{mean} = \frac{Q \log_e \dfrac{r_2}{r_1}}{\pi(h_2^2 - h_1^2)} \tag{8.11}$$

where $k_{mean}$ represents the overall stratum permeability, $h_1$ and $h_2$ are the heights of the phreatic surface referenced to an impermeable stratum, and $r_1$ and $r_2$ are the distances from the well to monitoring piezometers where $h_1$ and $h_2$ were measured.

*Drawdown*, $H - h$, is used in dewatering problems to evaluate system effectiveness, as well as to estimate the possible effects of overextraction and ground subsidence. It may be found by calculating the head $h$ at a distance $r$ from the well with the expression:

$$h = \sqrt{\frac{Q_w}{\pi k} \log_e \frac{r}{r_w} - h_w^2} \tag{8.12}$$

At distances from the well exceeding approximately 1.0 to 1.5 times the height $H$ to the original groundwater table, the drawdown will be equal to that computed from Eq. 8.12. At closer distances to the well the drawdown will be less than computed, with the difference increasing in magnitude with decreasing distance. It is significant that in a frictionless gravity well the water level in the well will be lower than the piezometric surface at the periphery of the well as shown on Fig. 8.33*b*. The difference $h'$ in the two water levels is the height of free discharge. Equation 8.10 provides an accurate estimate for $Q$ if the height of water $(t + s)$ is used for $h_w$.

### Artesian Wells

Flow from a confined aquifer to a fully penetrating artesian well is illustrated on Fig. 8.34.

*Quantity* and *permeability* are related by the expression:

$$Q_w = \frac{2\pi kD(H - h_w)}{\log_e (R/r_w)} \tag{8.13}$$

*Drawdown*, $H - h$, at any distance $r$ from the

well, may be computed from the following expression for the head at distance $r$:

$$h = \frac{Q_w}{2\pi kD} \log_e \frac{r}{r_w} + h_w \qquad (8.14)$$

Equations 8.13 and 8.14 *are valid* only when there is no head loss in the well; i.e., the head at the well $h_w$ is equal to the water level in the well. Since some head is required to force water through the filter and well screen, there will be some head loss. The above equations are valid, therefore, provided that $h_w$ is considered as the head at the periphery of the well, and not the water level in the well. Relationships are available for estimating head loss [see p. 314 in Mansur and Kaufman (1962),[23] for example].

*Combined Artesian-Gravity Flow*

In the artesian case above, the water level remains in the impervious stratum. It is possible at high pumping rates to lower the water table to below the top of the aquifer, or pervious stratum. Under these conditions the flow pattern close to the well is similar to that of a gravity well, whereas at distances farther from the well the flow is artesian.

*Overlapping Cones of Depression*

When several wells are close together, the cones of depression overlap, causing interference, and the water table becomes depressed over a large area. At any point where the cones overlap, the drawdowns are the *sum* of the drawdowns caused by the individual wells. When wells are too closely spaced, flow to each is impaired and the drawdowns are increased.

*Slots*

A line of wells, such as well points, or a dewatering trench may be simulated by a *slot*. Solutions may be found in Mansur and Kaufman (1962),[23] for fully or partially penetrating slots, from single- or two-line sources, in gravity, artesian gravity, and artesian conditions.

## 8.3.4 PRACTICAL ASPECTS OF GROUNDWATER

### General

The practical aspects of groundwater problems

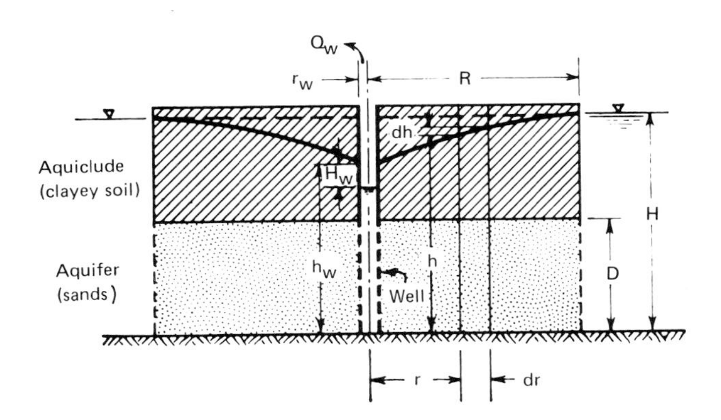


**FIG. 8.34 Flow from confined aquifer to a fully penetrating artesian well from a circular seepage source.** [*From Mansur and Kaufman (1962).*[23] *Reprinted with permission of McGraw-Hill Book Company.*]

can be grouped into three major categories: flow quantities, stability problems, and water quality.

*Quantity of flow* is of concern for water supply and for excavations made for structures, mines, and tunnels and through, beneath, and around dams.

*Stability* in soil formations is related to porewater pressures occurring in slopes, excavation bottoms, and beneath embankments and pavements; in rock masses it is related to softening or removal of fillings in fractures, and by the development of high pore- or cleft-water pressures in slopes or beneath structures.

*Water quality* is of concern in water supply from wells and with respect to deterioration of concrete and corrosion of other materials. Groundwater requires protection against pollution, not only from the aspects of water supply, but also from the aspects of contamination of adjacent water bodies and the effect on aquatic life.

Groundwater and seepage control are discussed in Art. 8.4, pollution control in Art. 8.5.2.

### Water Supply

*Soil formations* normally providing suitable quantities for water supply are clean sands or sand and gravel strata. Overwithdrawal on a

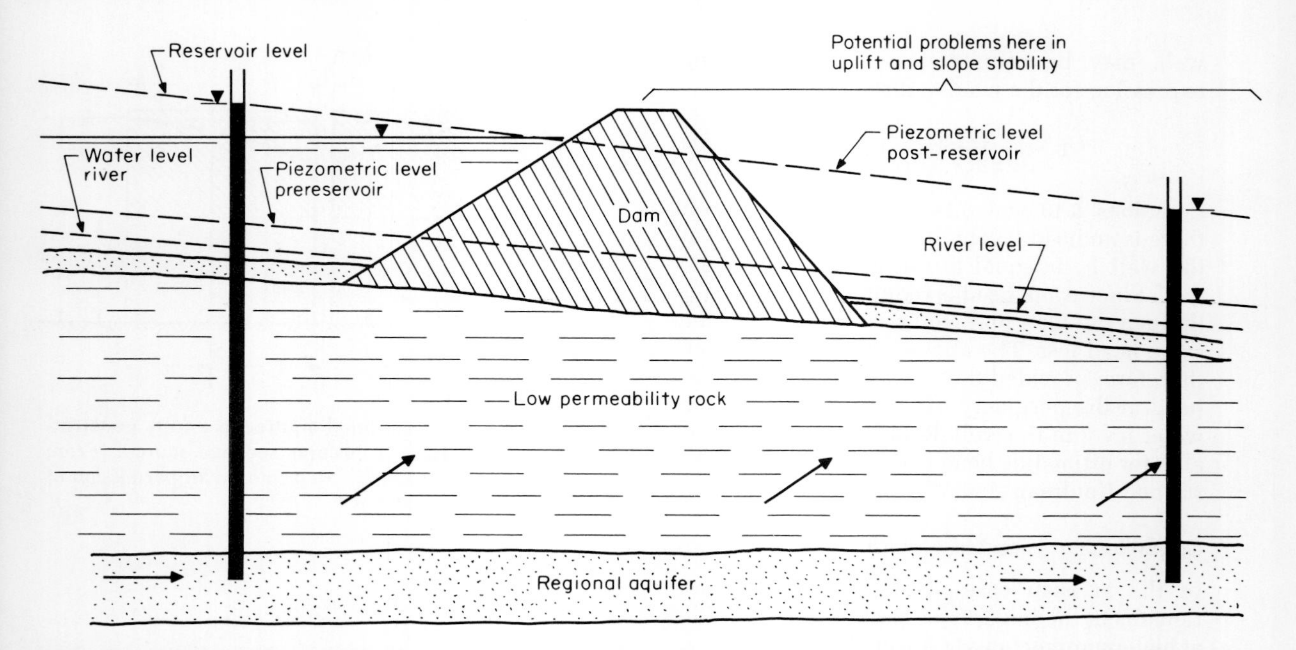


**FIG. 8.35 Possible stability problems caused by reservoir changing the regional piezometric system.** [*From Patton and Hendron (1974).*[20]]

long-term basis can result in surface subsidence *(see Art. 10.2)*.

*Rock masses* constituting the most significant aquifers are some sandstones, intensely fractured or vesicular rocks, or cavernous limestone. In igneous and metamorphic rocks, groundwater sources are most likely to be found in major fault zones or concentrations of joints. Groundwater extraction from soluble rocks can cause rapid cavern growth and ground collapse *(see Art. 10.4)*.

### Slopes

As discussed in Chap. 9, the increase in pore pressures in soil slopes, or the increase in cleft water pressures in rock slopes, is the major cause of slope failures.

Flows from rock faces are common but can become dangerous when increased by rainwater infiltration or blocked by freezing. A dry slope face does not necessarily indicate a lack of seepage forces in the mass. In slow-draining masses, such as tightly jointed rock, the face may dry because the evaporation rate often exceeds the seepage rate, but seepage forces remain in the mass at some depth and their increase can result in failure.

### Reservoirs

Reservoirs have a regional affect on groundwater conditions as rising reservoir levels cause a regional rise in piezometric levels.

In rock masses, more or less horizontally bedded alternating pervious and impervious strata can be the cause of high pore pressures in the foundation rock. Within the reservoir area the effect of rising piezometric levels is partially offset by the weight of the reservoir acting on the valley floor and sides, but downstream piezometric changes can have substantial effect, causing heaving of the downstream valley floor, boils (Fig. 8.35), and slope failures.

Slope instability in the reservoir area can present a possible hazard when the slopes are resting in a barely stable condition. Steep slopes with colluvium, or sedimentary rocks dipping in the slope direction with the beds daylighting along the slope, are particularly susceptible to slides when rising pool elevations change the hydrostatic conditions in the slope. The slope failure at Vaiont Dam *(see Art. 9.2.3)* appears to have been at least partially triggered by rising water levels.

**Dams**

*Foundation Stability*

Stability analysis of dam foundations is performed as a matter of standard procedure, but the case of the failure of the Malpasset Dam, near Fréjus, France, bears some discussion because conditions were unusual and the results catastrophic. The buildup of seepage pressures in the rock mass beneath the thin-arch concrete dam is considered to be the cause of the failure on December 2, 1959, that took 400 lives [Jaeger (1972)[24]].

The rock foundation of Malpasset Dam was a tightly jointed, finely fissured gneiss considered as competent, with more than adequate strength to support the compressive stresses and arch thrust. Failure studies hypothesize that as the water load behind the dam rose to 58 m, the dam deflected slightly and placed the rock under the heel in tension, causing the normally tight fissures in the gneiss to open slightly, increasing its permeability. As the water level rose, uplift pressures increased and a progressive displacement of the foot of the dam and rotation of the shell began. The shell transferred a tremendous thrust to the left abutment which failed, resulting in the collapse of the dam.

Seepage beneath dams is normally controlled by grouting or constructing a cutoff trench, or installing a pressure relief system (*see Art. 8.4.7*), none of which was provided at Malpasset because of the apparent tightness of the rock. Since Malpasset, it has become standard practice to drain the foundation rock beneath concrete-arch dams.

*Seepage Losses and Piping Failures*

Seepage can account for large losses of storage through abutment and foundation rock that is porous, heavily fractured, or cavernous, and, in the case of earth dams, can cause piping of embankment materials which can lead to failure. During the inspection of earth dams, all seepage points are noted including flow quantities and water condition. Normally the inspector becomes concerned when flows are muddy, since the indication is that piping and erosion are occurring. In the case of the Teton Dam failure of June 5, 1970, cited below, however, seepage was reported to have remained clear, almost to the time of failure.

The collapse of the Teton Dam in eastern Idaho, which resulted in 11 deaths and $400 million in damages, was caused by uncontrolled seepage [Penman (1977),[25] *Civil Engineering* (1977),[26] Fecker (1980)[27]]. Cross sections of the foundation, cutoff trench, and embankment are given on Fig. 8.36.

Figure 8.36*a* shows a cross section across the valley floor, showing the grout curtain. In addition to the curtain grouting (long, solid lines), in places blanket grouting was done (dotted lines). Below the valley floor were alluvial materials, so a cutoff trench was excavated down to rock. In both abutments above El. 5100 the rock was badly fractured near the surface, so keytrenches 70 ft deep were blasted.

Figure 8.36*b* shows a cross section of a keytrench excavated into the left and right abutments. At the bottom of the keytrench, one or three rows of holes were drilled, and grout pumped into them. Note the steep sides of the keytrench (0.5:1). Some think the steep sides may have led to differential settlement of the earth placed here, and that this may have led to an open pipe across the keytrench, which grew and led to failure. (It has been suggested that arching across the keytrench reduced stresses and favored the differential movement and fractures.)

Figure 8.36*c* shows the maximum cross section of the dam. Classes of earth materials used in making the embankment are as follows: (1) Selected clay, silt, sand, gravel, and cobbles compacted by tamping rollers to 6-in layers; (2) selected sand, gravel, and cobbles compacted by crawler-type tractors to 12-in layers; (3) miscellaneous material compacted by rubber-tired

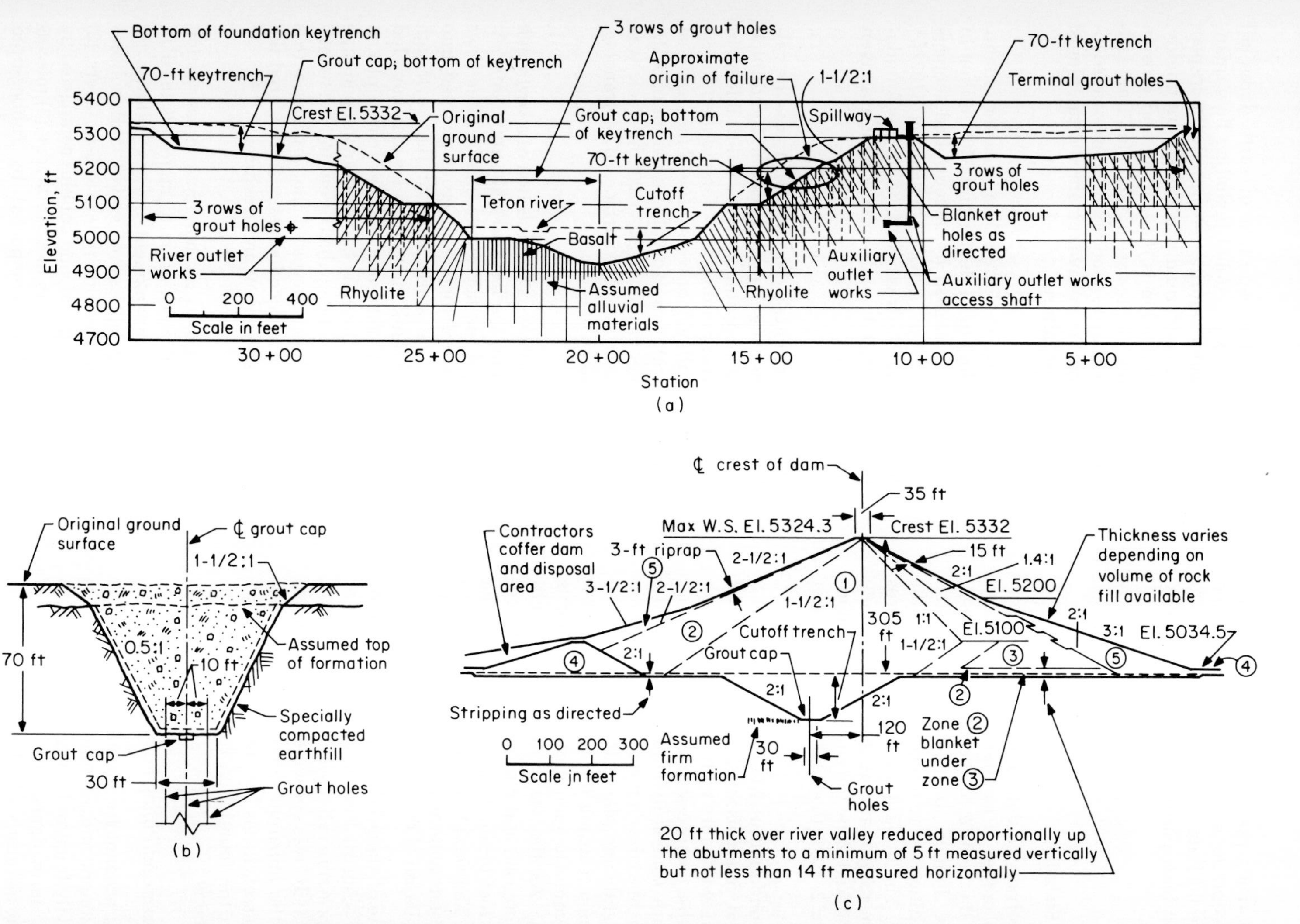


**FIG. 8.36 Elements of the Teton Dam. Failure originated along the right abutment as shown in part (*a*).** [*From Civil Engineering (1977).*[26] *Reprinted with the permission of the American Society of Civil Engineers.*]

rollers to 12-in layers; (4) selected silt, sand, gravel, and cobbles compacted by rubber-tired rollers to 12-in layers; and (5) rockfill placed in 3-ft layers.

Foundation rock was a badly fractured and porous rhyolite, described as a welded tuff, that was grouted beneath the entire length of the dam and well into the abutments. A core trench was backfilled with loessial soils compacted slightly on the dry side of optimum moisture, placed in direct contact with untreated, fractured rock.

During reservoir filling, when the water level was about 3 m below sill level, minor seepage was observed on the right bank downstream from the spillway on June 3. The seepage was clear and flowing at about 20 gal/min, continuing at this rate until the morning of June 5 when it increased markedly to 50 to 60 $ft^3/s$, although still remaining clear. Within 1½ hr the seepage increased to about 1000 $ft^3/s$, a whirlpool formed in the reservoir, and within an hour the embankment ruptured, releasing a wall of water downstream. Two panels of experts decided that the cause of failure lay in the erodible, pipable silt (loess) placed in the core trench in direct contact with the untreated fractured rock. Water flowing through the core trench, but over the grout curtain, opened up a passage. One panel concluded that the water flowed through cracks in the core material which were caused either by differential settlement or by hydraulic fracturing from water pressures. The water flowing through the fractures enlarged the openings by piping until failure occurred. It subsequently has been thought that sealing the core trench walls with blanket grouting and designing a filter layer between the core material and the trench walls might have prevented the failure.

## Open Excavations

Most open excavations made below the water table require control of groundwater to permit construction to proceed in the dry and to reduce lateral pressures on the retaining system. The problems of bottom heave, boiling, and piping should also be considered. Backslope subsidence and differential settlement of adjacent structures may be caused by groundwater lowering, or by piping and raveling of soils through the retaining structure if it contains holes.

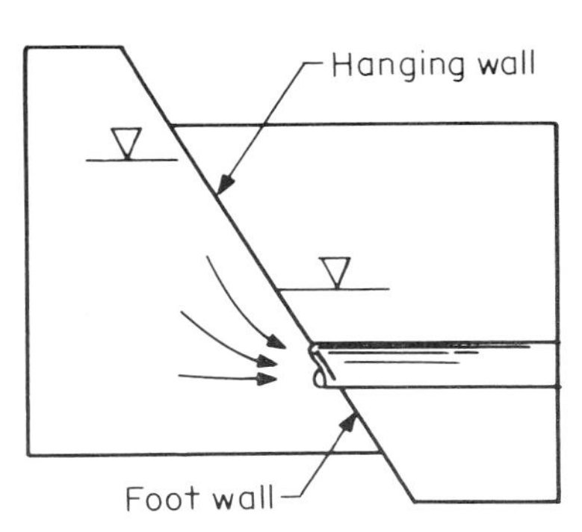


**FIG. 8.37 Differential water pressures along the sides of a normal or dip-slip fault can be extremely hazardous to tunnels approaching from the hanging-wall side.**

Structures built below the water table must be protected against uplift when the water table is permitted to rise to its original level and also against water infiltration and dampness, as in basements.

## Tunnels in Rock

*High flows* under *high pressures*, occurring suddenly, are the most serious problem encountered during tunnel construction in rock. Water pressures equal to full hydrostatic head can cause bursting of the roof, floor, or heading, even in hard but jointed rock. High flows, but not necessarily high pressures, can be encountered in porous rocks such as vesiculated lavas or cavernous limestones.

*Squeezing ground* refers to the relatively weak plastic material that moves into a tunnel opening under pressure from surrounding rocks immediately upon exposure and can be aggravated by seepage forces.

*Running ground*, the sudden inrush of slurry and debris under pressure, occurs in crushed rock zones, shear zones, and fault zones. The materials in fault zones can be highly fragmented and pervious, saturated and lacking in cohesive binder. Such materials are usually associated with the foot wall or hanging wall of a fault, as illustrated on Fig. 8.37.

When encountered in excavation, the saturated debris flows as a slurry into the tunnel, often under high initial pressures and quantities. Sharp et al. (1973)[28] cite a case where tunneling

**TABLE 8.3**
**INDICATORS OF CORROSIVE AND INCRUSTING WATERS***

| Corrosive water | Incrusting water |
|---|---|
| A pH less than 7 | A pH greater than 7 |
| Dissolved oxygen in excess of 2 ppm | Total iron (Fe) in excess of 2 ppm |
| Hydrogen sulfide ($H_2S$) in excess of 1 ppm, detected by a rotten egg odor | Total manganese (Mn) in excess of 1 ppm in conjunction with a high pH and the presence of oxygen |
| Total dissolved solids in excess of 1000 ppm, indicating an ability to conduct electric current great enough to cause serious electrolytic corrosion | Total carbonate hardness in excess of 300 ppm |
| Carbon dioxide ($CO_2$) in excess of 50 ppm | |
| Chlorides in excess of 500 ppm | |

*From Cording et al. (1975)[29] after Johnson (1963).[30]

for the San Jacinto Tunnel in California encountered maximum flows of 16,000 gal/min (60,800 liters/minute) with water pressures as high as 600 psi (40 kg/cm$^2$). As is often the case, the flows were associated with a fault zone filled with clay gouge. On the hanging wall side of the fault, piezometric pressures were relatively low and controllable by draining, whereas on the footwall side the piezometric level was high and perched (Fig. 8.37). When the tunnel penetrated the clay gouge, 3,000 m$^3$ of crushed material and water were suddenly released into the tunnel.

*Corrosion* of concrete tunnel linings can be high where water flows over gypsum or other sulfates and can contain solutions of sulfuric salts, as described below.

### Pavements

High groundwater tables and capillarity provide the moisture resulting in "pumping" or frost-heaving of pavements in fine-grained granular soils, or heaving from swelling clays.

### Water Quality

#### *Importance*

Two basic aspects of water quality are:

- The desired quality of *water supply* depends upon its intended use, e.g., potable water, industrial (pure for pharmaceutical use, much less pure for cooling), or irrigation.
- *Corrosive* effects on construction materials, primarily concrete, are influenced by water quality.

#### *Soluble Salts*

Normally salts are found in solution in all groundwater. Concentrations of soluble salts are dependent on water movement, temperature, and origin. As groundwater passes through rocks and soils, their minerals are subject to dissolution; sedimentary rocks generally are more soluble than are the igneous and metamorphic types. The salts occur in the form of carbonates, bicarbonates, and sulfates, principally of Ca, Na, and Mg. In calcareous rocks the water is enriched with carbonates; in ferruginous rocks, with iron oxides. The principal agent causing solution is $CO_2$, resulting from rainwater infiltrating through surface organic materials.

Some definitions are as follows:

- Hard water contains a high level of carbonates, in the order of 50 g of $CaCO_3$ for 1000 liters or water. Pipes become encrusted and soap does not foam.
- Soft water contains a low level of carbonates.
- Mineral water is groundwater with a minimum

of 1 g of dissolved salt per liter, but the salt cannot be $CaCO_3$ or $MgCO_3$.

*Effects*

*Corrosive and incrusting water* causes deterioration of underground metal piping and geotechnical instrumentation and the clogging of drains and piezometers. Indicators of a high potential for corrosion and incrusting are given on Table 8.3.

*Aggressive salts* attack concrete, causing its deterioration below the water table. The chemical agents normally aggressive to concrete are $CO_2$, chlorides, magnesium, sulfates, and ammonia, with calcium and magnesium sulfates being encountered less frequently. The value of the pH must also be considered since it also affects corrosion. Corrosion and destruction of concrete are affected by the type of cement and aggregate used, the water-cement ratio, the age of the concrete, and the geologic conditions to which it is exposed. The action of sulfates in contact with concrete is summarized on Table 8.4, where it is shown that concentrations greater than 1000 mg/liter of sulfate in water can cause considerable deterioration. The degree of aggressiveness of pH and aggressive $CO_2$ is given in Table 8.5. A case where moisture migrating upward through pyritic shales produced sulfuric acid which leached cement from a concrete floor, turning it into a "mush," is described in Art. 10.6.3.

*Improvement of concrete resistance* is accomplished by mixing the concrete with chemical additives or protecting the concrete from contact with aggressive waters by epoxy coatings or some other method of impermeabilization. Sulfate-resistant cements are available.

**TABLE 8.4**
**EFFECT OF SULFATE SALTS ON CONCRETE***

| Degree of attack | Sulfate in water sample, mg/L |
|---|---|
| Negligible | 0–150 |
| Positive | 150–1000 |
| Considerable | 1000–2000 |
| Severe | >2000 |

*After USBR (1973b).[31]

**TABLE 8.5**
**EFFECT OF AGGRESSIVE $CO_2$ ON CONCRETE***

| Degree of aggressiveness | pH | Aggressive $CO_2$, mg/L |
|---|---|---|
| Neutral | >6.5 | <15 |
| Weak | 6.5–5.5 | 15–30 |
| Strong | 5.5–4.5 | 30–60 |
| Very strong | <4.5 | >60 |

*From Chiossi (1975).[32]

## 8.4 GROUNDWATER AND SEEPAGE CONTROL

### 8.4.1 INTRODUCTION

#### General

Groundwater and seepage control is a most important consideration in the stability of natural slopes, dams, and levees; excavations for structures, cut slopes, open-pit mines, tunnels, and shafts; buried structures; pavements; and side-hill fills.

Investigation must consider the control of groundwater and seepage for conditions both during and after construction, in the recognition of the probability that conditions are likely to be different from those encountered during explorations as a result of either natural causes or causes brought about by construction itself.

#### Control Necessity

*During construction*, groundwater and seepage control are required to:

- Provide a dry excavation and permit construction to proceed efficiently
- Reduce lateral loads on sheeting and bracing in excavations
- Stabilize "quick" bottom conditions, and prevent bottom heave and piping
- Improve supporting characteristics of foundation materials
- Increase stability of excavation slopes and sidehill fills
- Reduce air pressure in tunneling operations, and
- Cut off capillary rise and prevent pumping and frost heaving of pavements

*After construction* groundwater and seepage control is required to:

- Reduce or eliminate uplift pressures on bottom slabs and permit economies from the reduction of slab thicknesses for basements, buried structures, canal linings, spillways, drydocks, etc.
- Provide for dry basements
- Reduce lateral pressures on retaining structures
- Control embankment seepage in earth- and rock-fill dams
- Control foundation and abutment seepage in all dams
- Control seepage and pore pressures beneath pavements, side-hill fills, and cut slopes
- Prevent surface and groundwater contamination from pollutants.

**Uncontrolled Seepage**

Many investigations concentrate on keeping an excavation dry, but the control of forces due to seepage is equally important in the prevention of failures, which in terms of uncontrolled seepage forces have been divided into two categories by Cedergren (1967)[18]:

*Category I: Failure caused by migration of particles to free exits or into coarse openings*

1. Piping failures of dams and levees caused by:
   a. Lack of filter protection
   b. Poor compaction along conduits, along foundation trenches, etc.
   c. Holes in embankments from animals, decomposed wood, etc.
   d. Filters or drains with openings too large
   e. Open seams or joints in rock
   f. Gravel or other coarse strata in foundations or abutments
   g. Cracks in rigid drains, reservoir linings, dam cores, etc., caused by mass deformation
   h. Any other natural or human-made imperfection in the embankment or foundation
2. Clogging of drains and filter systems

*Category II: Failures caused by uncontrolled saturation and seepage forces*

1. Most slope failures, including highway and other cut slopes, open pit mines, and reservoir slopes caused by seepage forces.
2. Deterioration and failure of roadbeds caused by insufficient structural drainage.
3. Earth embankment and foundation failures caused by excess pore pressure.
4. Retaining wall failures caused by unrelieved hydrostatic pressures.
5. Canal linings, and slabs for basements, spillways, dry docks, and other buried structures (such as partially buried tanks for sewage treatment plants which are often located adjacent to streams) uplifted by unrelieved pressures.
6. Most liquefaction failures of dams and slopes caused by earthquake forces.

**Investigation *(See Also Art. 2.3.7)***

*Prior to Construction*

*Reconnaissance* using imagery interpretation and site visits provides an overview of water-table conditions. All springs and other seepage conditions from slopes and cuts should be noted. Precipitation data should be gathered to provide the basis for formulating judgments regarding existing groundwater conditions, i.e., higher than normal, normal, or lower than normal *(see Art. 9.3.4)*. On important projects such data should go back for at least 25 years to determine if the region is in a long wet or dry period.

*Explorations* should be extended to depths significantly below any excavations to define all groundwater conditions including the depth to the existing static water table and perched and artesian conditions. Control methods may vary with the conditions. Artesian conditions and large quantities flowing in open formations will be costly to control, whereas a perched water table may simply require a gravity drainage system which permits the water to drain to a lower, open stratum with a lower piezometric level. Major tunnel projects and other underground construction in rock at substantial depths are often best explored with pilot tunnels.

*In situ tests* are performed to provide data on permeabilities and drawdown and to evaluate the potential for seepage loss through abutments and foundations for dams, especially in rock masses (see Art. 3.3.4).

*Piezometers* (see Art. 4.4.2) are installed to monitor pore pressures and changes in groundwater conditions.

*Conclusions* should be reached regarding the worst conditions likely to be encountered during the life of the project, and to foresee possible changes in groundwater conditions brought about naturally or by construction.

*During Construction*

Changes in groundwater conditions occurring during construction are monitored with the piezometers.

During tunnel construction, pilot holes drilled in advance of the heading provide data to forewarn of hazardous conditions.

*After Construction*

Where the possible development of seepage forces may endanger the performance of the structure, conditions are monitored with piezometers.

Where the problem of excessive seepage through a dam embankment, abutments, or foundation exists, attempts to locate the seepage paths are advisable to aid in designing treatment. Various dyes and tracers have been injected into boreholes and observations made of their exit point, but positive location of seepage channels is usually extremely difficult (see Art. 2.3.7). The acoustical emissions device (see Art. 4.3.5), which measures microseisms, may provide useful information. Testing indicates that minimum flow rates required for detection with acoustical emissions devices are in the order of 45 ml/s for clear water seepage and 10 ml/s for turbid water, such as that in a dam undergoing piping erosion [Koerner et al. (1981)[32a]].

**Control Methods Summarized**

Control methods may be placed in three main categories:

1. *Cutoffs and barriers,* which, when constructed or installed properly, have the potential to seal off flow
2. *Dewatering systems,* which serve to lower the water table and reduce pore-water pressures, or in some cases only to reduce pore pressures
3. *Drains,* which serve to control flow, in some cases lowering the water table, in others reducing pore-water pressures and seepage forces

Other factors:

- *Filters* are provided between zones with significant differences in permeability to control flow velocities and prevent migration of fines and piping. They are most important adjacent to drains where it is desirable to prevent clogging or piping.
- *Surface treatments* are provided to deter or prevent infiltration of water on slopes.
- Groundwater and seepage control methods and their common applications are summarized on Table 8.6. A general comparison of many of the various methods in terms of soil gradation characteristics is given in Fig. 8.38. Not included are liners, walls, and drains which are normally an integral part of a structure.

## 8.4.2 CUTOFFS AND BARRIERS

**Liners, Blankets and Membranes**

*Clay blankets* are placed on materials with moderate permeability, extending upstream from a dam embankment to increase the horizontal length of flow paths and decrease seepage quantities and pressures at the downstream toe, as discussed also in Art. 8.4.7.

*Clay or plastic liners* are used to seal off leakage from off-channel reservoirs, waste-disposal ponds, and sanitary landfills.

*Asphaltic or coal-tar membranes* are used to waterproof the exterior of basement walls below the water table. Where basement dampness is to be minimized, as in rooms with delicate instruments, a layered system is used, alternating asphalt compounds with plastics or epoxies.

*Asphaltic concrete, concrete,* or *welded steel plates* have been used as impervious membranes on the upstream slopes of dam embankments constructed of rock fill or gravel. At times

**TABLE 8.6**
**GROUNDWATER CONTROL METHODS AND COMMON APPLICATIONS**

| Method | Engineering problem: Earth and rock-fill embankments | Dam foundations | Slopes | Retaining structures (slopes) | Open excavations | Closed excavations (tunnels, etc.) | Building basements | Site surcharging | Pavements | Pollution-control systems |
|---|---|---|---|---|---|---|---|---|---|---|
| **Cutoff:** | | | | | | | | | | |
| Liners* | X | | | | | | | | | X |
| Earth walls* | X | | | | | | | | | X |
| Slurry walls*†‡ | | X | | | X | X | X | | | X |
| Ice walls† | | | | | X | X | | | | |
| Concrete walls* | | X | | | X | X | X | | | X |
| Sheet pile walls† | | | | | X | | | | | |
| Grout walls*‡ | | X | | | X | X | | | | |
| **Dewater:** | | | | | | | | | | |
| Sump pumps† | | | | | X | | | | | |
| Wellpoints† | | X | | | X | | | | | X |
| Deep wells†‡ | | X | X | | X | X | | | | X |
| Electroosmosis† | | | X | | X | X | | | | |
| **Drain:** | | | | | | | | | | |
| Blanket drains* | X | | | X | | | X | | X | |
| Trench drains* | | X | | | | | X | | X | |
| Triangular drains* | X | | | X | | | | | | |
| Circular vertical drains* | | | X | | | | | | | |
| Circular subhorizontal drains*‡ | | | X | X | | X | | | | |
| Drainage galleries† | | | X | | | X | | | | |
| Relief or bleeder wells†‡ | | X | X | | | | | X | | |

*Permanent installation.
†Construction control.
‡Corrective measure after construction.

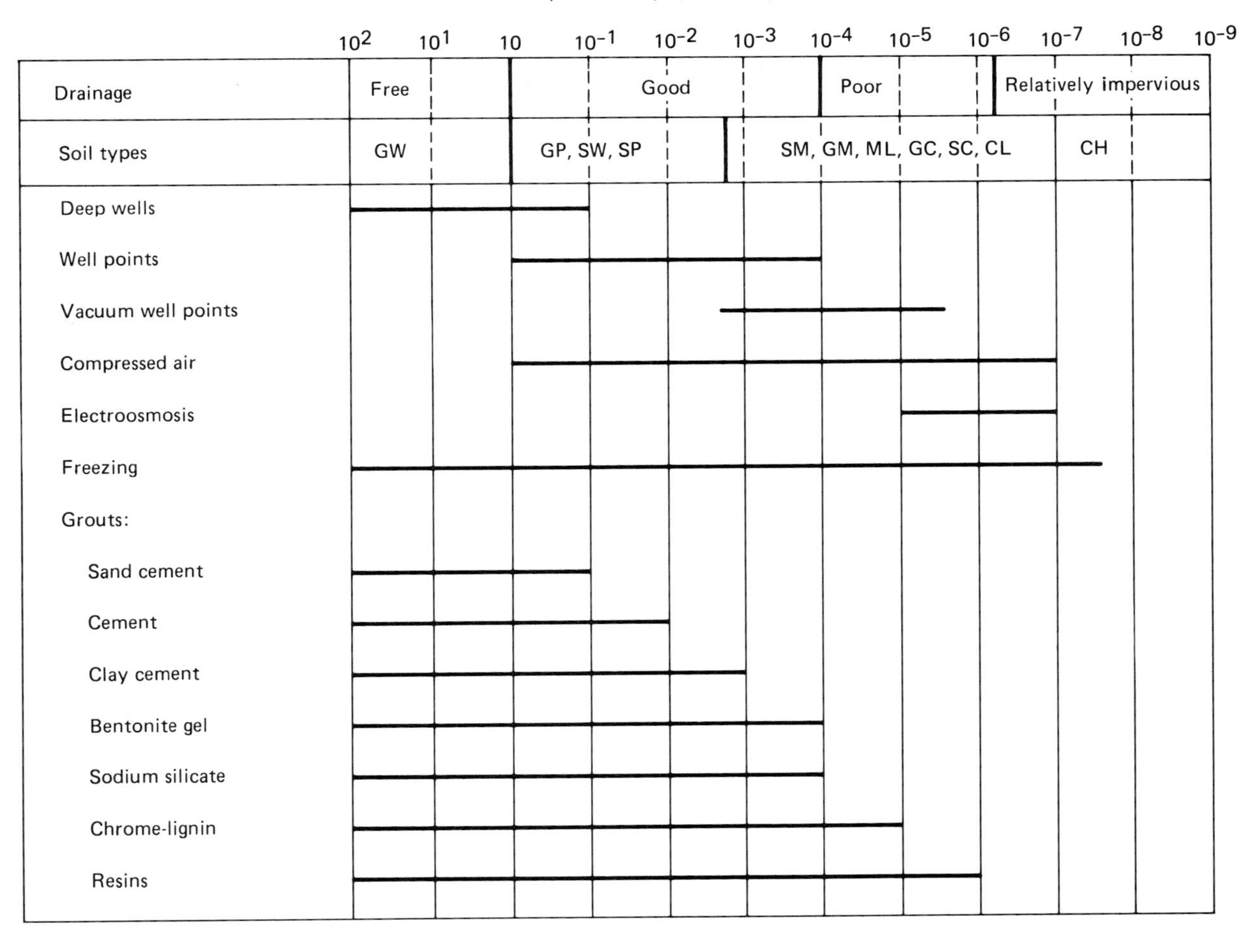


**FIG. 8.38 General applicability of some methods for controlling groundwater and seepage in soils as a function of grain-size characteristics. (Other methods not shown include liners, walls, and drains.)**

they are placed in the body of the dam [Sherard et al. (1963)[4]].

## Walls

### *Compacted Earth Walls*

Walls of compacted earth constructed to form a cutoff can include a homogeneous embankment for a dam, a clay core within a dam, or a clay-filled core trench in a dam foundation, as discussed in Art. 8.4.7.

### *Concrete Walls*

Walls of concrete are commonly used in foundation excavations and as tunnel linings where a permanent water barrier is required. In addition, they provide high supporting capacity to retain the excavation. Vertical concrete walls cast in braced excavations extending down to impervious material effect a positive cutoff beneath dams (Fig. 8.51e), but their use in the United States has decreased in recent years because of their relatively high cost compared with other methods.

### *Sheet Pile Walls*

*Sheet pile walls* provide significant efficiency in seepage control only when the interlocks are extremely tight, a difficult condition to ensure, especially when sheets encounter cobbles, boul-

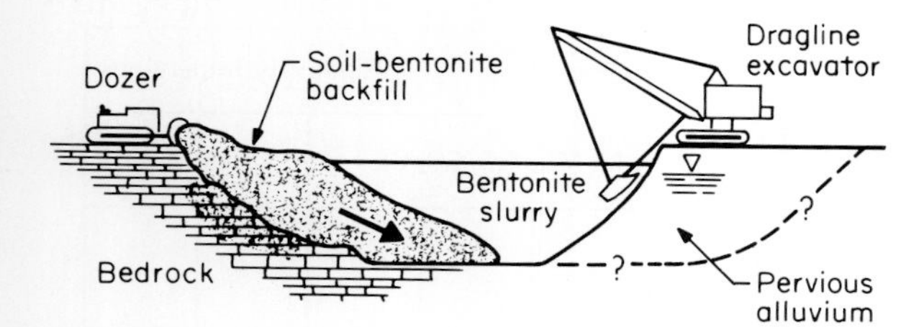


**FIG. 8.39 Dragline excavates trench as slurry is pumped in, then displaced with soil-bentonite backfill.** [*From Miller and Salzman (1980).*[34] *Reprinted with the permission of the American Society of Civil Engineers.*]

**FIG. 8.40 Slurry wall excavation with clamshell. Concrete guide walls extend to a depth of 1 m.**

ders, broken rock, or other obstructions. Long sheets, in particular, tend to bend and deflect. At best they provide only a partial cutoff. Commonly used to retain excavations, they are normally combined in free-draining soils with a dewatering system, which serves also to reduce lateral earth pressures. Thin "curtain" walls of sheeting have been used for years as a cutoff for foundation seepage beneath dams, but their use is now much less frequent, not only because of the difficulties in obtaining a tight barrier, but also because of relatively high costs and their susceptibility to corrosion.

*Slurry Walls*

*Slurry walls and the slurry trench method* provide a positive cutoff for seepage beneath dams, in open excavations, in tunnel construction, and in pollution control systems, and are becoming extremely popular. The general procedure (except for tunnels) involves excavating a trench while keeping it filled with a bentonite slurry to retain the trench walls, then displacing the slurry with some material to form a permanent and relatively incompressible impervious wall. The *bentonite slurry*, when pure, will have a density in the order of 1.06 g/cm$^3$, but density can be increased by adding silt or sand. The most important property controlling slurry characteristics is viscosity, which must have the correct value to allow proper displacement by the backfill and to assure trench stability and good filter cake formation on the trench walls. A minimum viscosity of 40-s Marsh is normally required, but some fluidity should be maintained as evidenced by the slurry's ability to pass through a Marsh funnel [D'Appolonia (1980)[33]]. There are two general approaches to installation procedures:

1. The trench is excavated to an impervious stratum by either a dragline (Fig. 8.39) or a clamshell bucket with a narrow grab (Fig. 8.40). In the latter case, shallow concrete walls 60 to 100 cm apart are formed as guides. As the excavation proceeds, the bentonite slurry is pumped into the trench to retain the walls. A backfill consisting of a silty sand suspended in a thick slurry is then pushed into the trench (Fig. 8.39) where it displaces the thinner bentonite slurry. The backfill is formed by mixing a minimum bentonite content of 1% by dry weight [D'Appolonia (1980)[33]].
2. The trench is excavated to an impervious stratum using the clamshell as shown in Fig. 8.40, or a clamshell mounted on a rigid Kelly bar, in a series of panels as shown on Fig. 8.41. As each panel is completed the bentonite is displaced with a cement-bentonite slurry. The Kelly-bar clamshell can excavate coarse sands and gravels and soft rock with greater confidence than can a dragline or cable-suspended "grab." The bentonite can be displaced also with tremie concrete to form a diaphragm wall.

*Concrete*

*Diaphragm walls of concrete* provide permanent building walls constructed by the slurry trench technique. The general procedure is illustrated on Fig. 8.42. The trench excavation is made in panels up to 7 m in length, extending to the bottom of the foundation level. A guide tube is placed at the earth end of a panel when the panel is completely excavated and a prefabricated cage is lowered into the excavation. The bentonite slurry is then replaced by tremie concrete and the guide tubes withdrawn before the panel concrete takes its final set. Construction then proceeds to the next panel.

*Ice Walls*

*Ice walls* provide a temporary expedient for controlling seepage during the construction of open excavations, tunnels, and shafts. They are most useful in thick deposits of "running sands" and saturated silts, or where grout materials may contaminate water supply. Ice walls have been used for shaft construction to depths of over 300 m in the mining industry for many years.

The procedure for deep shaft construction is as follows [Lancaster-Jones (1969)[36]]:

1. A freezing plant is installed with adequate capacity to ensure that the wall will be sufficiently thick and continuous.
2. One or two rings of cased boreholes are drilled outside the shaft perimeter to depths sufficient to fully penetrate the saturated, potentially troublesome zone, and brine pipes are installed.

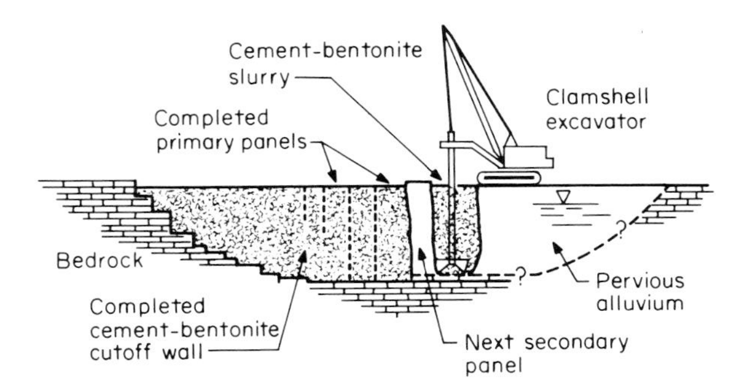


**FIG. 8.41 Rigid Kelly bar clamshell excavating panels while placing soil-bentonite slurry which is displaced by cement bentonite slurry or tremie concrete.** [*From Miller and Salzman (1980).*[34] *Reprinted with the permission of the American Society of Civil Engineers.*]

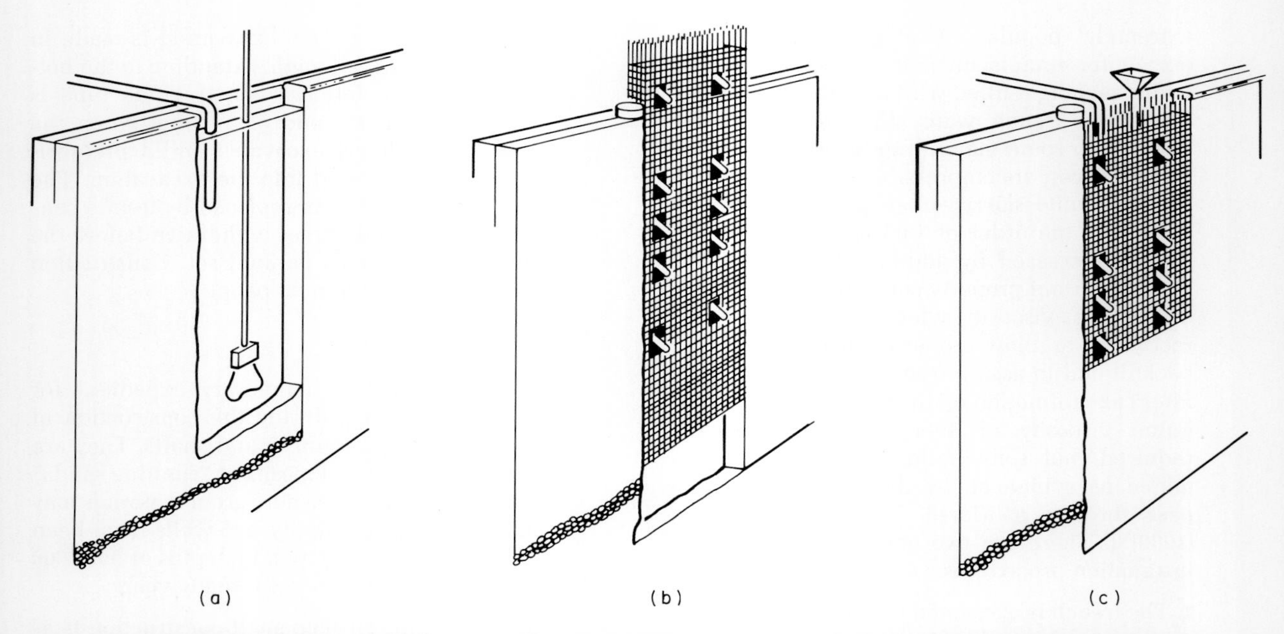


**FIG. 8.42 Sequence of slurry and diaphragm wall construction (World Trade Center, New York City). Slurry trench excavations over 20 m in height through thick deposits of organic silt were stable.** [*From Kapp (1969).*[35]]

3. The ice cylinder is formed by pumping brine solution of calcium chloride injected at −20°C, or even liquid propane at −44°C.
4. The cooling process usually takes from 2 to 4 months, and when complete, the shaft sinking begins. Care is required not to damage the brine pipes, which must continue to function during the excavation.
5. Concrete is used normally for lining but the temperature changes during hydration and when the ice thaws can cause substantial cracking, requiring repairs with cement or epoxy grouts.

The method is time-consuming and costly, generally causing delays of about 6 months or more. Careful installation is required to ensure a thick and continuous wall, since even a small flow through an opening can be disastrous during excavation. Substantial ground heave may occur from freezing operations near the surface, followed by ground collapse when thawing occurs. The impact on nearby structures must be evaluated carefully to guard against distress. The amount of heave and subsequent collapse depends on the types of materials near the surface.

## Grouting

### *Applications*

Grout injection into pervious soils and rock formations is a common, permanent solution to contain flows, but often provides an imperfect wall. Grouts are used also to strengthen soil and rock formations. During construction, grouting is used beneath dam foundations, in tunneling, and in excavations. In the last case, grout is injected behind pervious sheeting to provide flow control. Postconstruction installations are usually corrective measures to control flows through or beneath embankments.

### *Types*

Grouts can consist of soil-cement mixtures, cement, or chemicals. Common cement and chemical grouts, their composition, and application in terms of characteristics of the materials to be treated are summarized on Table 8.7; some are also shown on Fig. 8.38.

**TABLE 8.7**
**COMMON CEMENT AND CHEMICAL GROUTS***

| Grout type | Composition | Application |
|---|---|---|
| Sand-cement | Loose volume sand-cement ratio varies from about 2:1 to 10:1. Bentonite or fly ash additives reduce segregation and increase pumpability. Water-cement ratios from about 2:1 to 5:1 by volume. | Grouting large foundation voids, mud-jacking, and contact grouting around structure periphery. Slush grouting without pressure to fill surface irregularities in rock foundation. Depending upon sand gradation in mix, will penetrate gravels with $D_{10}$ about ¾ in. For usual mix, strengths vary with water content from 100 to 700 psi. |
| Clay-cement | Loose volume clay-cement ratio varies from about 3:1 to 8:1. Water-clay ratios from about ¾:1 to 2:1 by volume. Water-cement ratios from about 3:1 to 10:1. | Comparatively large voids and fractures where clay is added for economy. Penetrates coarse sand with $D_{10}$ as small as 1 mm; fissures in the range of 0.06 to 0.01 mm depending upon pressures, water-cement ratio, and cement types. Not appropriate for large voids with vigorous groundwater movement. Strength depends on water-cement ratio and averages about 100 psi for typical mix. |
| Portland cement | Water-cement ratios generally between 1:1 and 1:4. To reduce bleeding and segregation, add bentonite, silica gels, pozzolans. To increase pumpability, add ligno-sulphonates. To accelerate set, add calcium chloride. | Penetrates coarse sand with $D_{10}$ as small as 1 mm; fissures in the range of 0.06 to 0.01 mm depending upon pressures, water-cement ratio, and cement type. Not appropriate for grouting large voids with vigorous groundwater movement. Most common material for decreasing permeability and increasing strength. Strength depends on water-cement ratio and averages about 100 psi for average mix. |
| Bentonite gel | Dispersed clay slurry with flocculating agent such as aluminum sulfate to cause suspension to coagulate after injection. | Materials down to fine sand with $D_{10}$ between 0.2 and 1.0 mm, depending on grain size of clay. Relatively low cost, but slurry may be removed by vigorous flow. No significant strength increase. |
| Sodium silicate (single stage) | Sodium silicate with a setting agent such as sodium aluminate in water solution. May be combined in cement or soil-cement mixes. | Sands with $D_{10}$ as small as about 0.08 mm. Setting agent used to obtain set time ranging from few minutes to several hours. Compressive strength about 100 psi. (*Two stage:* successive injection of sodium silicate and calcium chloride produces strength ranges from 500 to 1000 psi.) |
| Epoxy resins | See text. | |

*After NAVFAC (1971).[37]

*Selection* of the grout type depends on the porosity of the geologic formation to be treated, the rate of groundwater flow, and the desired compressive strength of the grouted formation. In general, sand-cement grouts are used to seal large cavities and fractures, and clay and portland cement grouts are used to seal relatively small fractures and coarse-grained soils. Bentonite gels are relatively low-cost and seal sands, but there is no strength increase. Chemical grouts are used in fine-grained soils with effective particle size smaller than 1 mm (the limit of penetration of cement particles), but without a significant clay portion. Although substantially more costly than cement grouts, additives allow very short set times, in the order of minutes, if necessary. Since silica gels have a viscosity approaching that of water and contain no solid particles, penetration is usually excellent.

*Epoxy resins* are relatively new, and are finding many applications such as to seal piping channels or small flow channels in soils and rocks, as well as basement walls, etc. They are relatively expensive but offer a number of advantages. They react normally under adverse conditions, whether the medium injected is corrosive or not, in acid or akaline environments, organic soils, etc. Viscosity can be controlled to permit adaptation to any soil or rock type that will pass water, and the hardening time can be varied from slow to quick by the use of additives. Injected with simple equipment, epoxy resins can be dissolved in inexpensive solvents, such as alcohol. The finished product is of high quality including resistance to physical changes such as compression and failure by rupture or bending, and has almost no appreciable shrinkage. Since it is strong, but not completely rigid (as are cement grouts), it permits deformation when set in a soil or rock medium, without fracturing, which is especially important along contact zones of differing materials.

*Installation*

A *grout curtain* injected in a single line of holes installed in a soil or rock formation often will suffice to control seepage. The curtain is thickened by adding rows of grout holes if necessary. Occasionally grouting will achieve a complete cutoff, but generally it is difficult to obtain 100% penetration, especially in rock fractures, or where conditions of high groundwater flows persist. Cedergren (1967)[18] states that to achieve 90% reduction in permeability in the grouted zone, 99% of the cracks must be grouted; therefore, grouting often results in only a partial control of seepage.

In *rock formations* the dip and bearing of the injection holes should be chosen as normal to the joint set as possible to provide for maximum intersection of the joints, otherwise many will remain ungrouted. Spacing of drill holes depends on the joint concentrations and the effectiveness of the curtain desired.

*Injection pressures* vary with conditions. Care is required in rock masses to obtain a pressure that is greater than water pressure but less than that required to cause new fractures, although at times fracturing is desired, such as during stage grouting, to increase penetrations. When grouting close to the surface of an embankment or slope, or beneath foundations, care is required that pressures do not exceed lateral or vertical overburden pressures or ground heave will result. In rock masses, pressures of 7 kg/cm$^2$ are usual although pressures as high as 40 kg/cm$^2$ have been used in some cases [Jaeger (1972)[24]]. In fact, in recent years pressures as high as 100 kg/cm$^2$ have been used to *compress* rock masses.

*Checking grout-curtain effectiveness* is necessary and it is accomplished in rock masses with water-pressure tests (WPT) *(see Art. 3.3.4)*. The limits of water losses given in Table 8.8 may be used as a standard for grout curtain effectiveness [Fecker (1980)[27]]. Since the Lugeon criterion of 1 liter/min/m at 10 kg/cm$^2$ pressure for dams is a criterion seldom achieved, the criteria of Terzaghi and the Water Reservoir Commission represent an upper limit.

## 8.4.3 DEWATERING

### General

*Applications*

Dewatering systems are used primarily to lower the water table to permit construction in the dry, but they serve also to reduce seepage pressures in slopes and the bottom of excavations. Dewatering can be achieved with sump pumps, wellpoints, deep wells, and occasionally electroosmosis. The procedure selected depends on the

**TABLE 8.8**
**RECOMMENDED SATISFACTORY WATER LOSS IN WATER-PRESSURE TESTING FOR GROUT CURTAIN EFFECTIVENESS***

| Reference | Satisfactory water loss, L/min/m at 10 kg/cm² pressure |
|---|---|
| Lugeon (1933) | |
| $H > 30$ m | 1† |
| $H < 30$ m | 3 |
| Heitfeld (1964) | |
| $H = 100$ m | 3 |
| $H = 50$ m | 4 |
| $H = 20$ m | 4, 8 |
| Terzaghi (1929) | 5‡ |
| Water Res. Comm. Std. | 7‡ |
| Int. Teton Dam Rev. Group (1977) | 18 |

*After Fecker (1980).[27]
†Criteria seldom achieved.
‡Suggested upper limit of loss.

material to be dewatered and the depth below the water table to which dewatering is desired. Systems generally applicable for various soil gradations are given on Fig. 8.38.

*Selection*

In the *selection* of the dewatering system to be employed a number of factors require consideration, as follows:

- Depth to the water table during construction, the nature of the water (static, perched, or artesian), and the estimated flow quantities into the excavation from the sides and bottom (and in the case of tunnels, from the top and heading).
- Flow quantity into the dewatering system, which depends on the permeability and thickness of the water-bearing formations, the depth to which the water table must be lowered to provide a dry excavation and prevent piping, and the corresponding hydraulic gradients producing flow toward the excavation.
- The coefficient of permeability of the various strata is the most important factor. It can be estimated from gradation curves *(see Art. 3.3.2)*, or it can be measured by full-scale pumping tests using piezometers to measure drawdown *(see Art. 8.3.3)*. The probable inflow rates are estimated from well formulas, slot formulas, or flow nets.

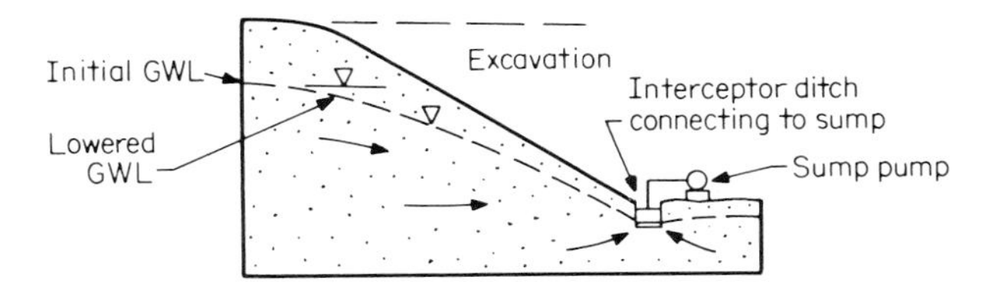


**FIG. 8.43 Dewatering with sump pumps where neither seepage nor the depth below the water table is too great.**

- The magnitude of surface subsidence to be anticipated during dewatering and its effect on nearby structures must be evaluated, and methods selected for its control *(see Art. 10.2.5)*.

### Sump Pumping

The simplest procedure for controlling water in open excavations is to provide interceptor ditches at the slope toe, or in the bottom of a sheeted excavation, and connect them to sumps from which water is pumped (Fig. 8.43). The procedure usually is suitable where flows are not too large, as in silty or clayey sands where the depth below the water table is not great and heads are low. The ditch and sump bottom

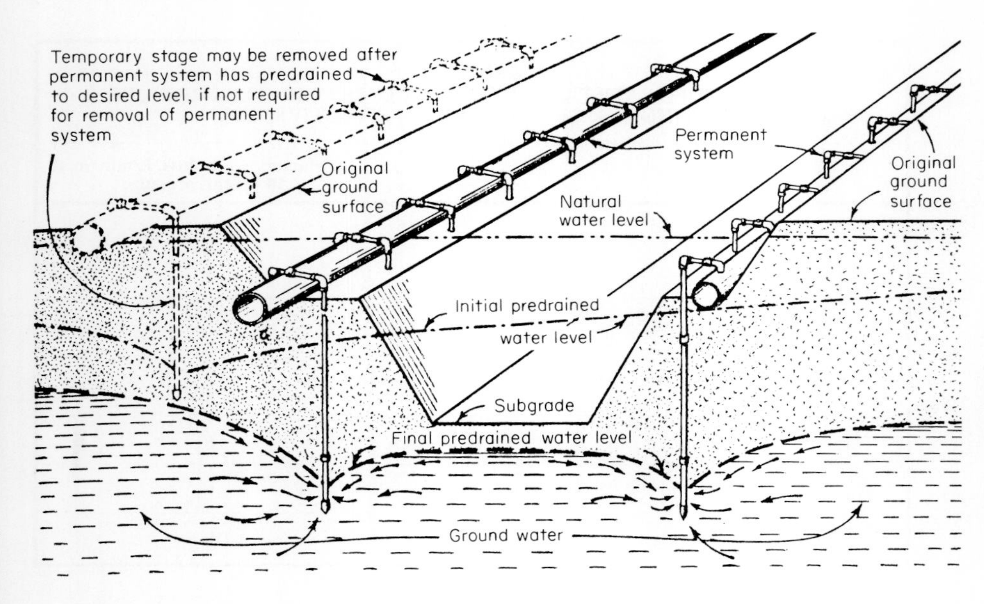


**FIG. 8.44 Typical wellpoint system installation.** [*From Mansur and Kaufman (1962)*[23] *and Griffin Wellpoint Corp. Reprinted with permission of McGraw-Hill Book Company.*]

should be lined with a gravel or crushed-stone blanket to contain the migration of fines.

### Wellpoints

#### *Single and Multistage Systems*

Standard wellpoints are commonly used in sandy soils with $k$ values ranging from 10 to $10^{-4}$ cm/s. They consist of small well screens or porous points 2 to 3 in in diameter and from 1 to 3½ ft in length. Installation, shown on Fig. 8.44, is usually between 3 to 12 ft on centers, depending on soil permeability. The wellpoints are attached to 6- to 12-in-diameter header pipes, which are connected to a combined vacuum and centrifugal pump.

Application of a *single-stage* system is limited by the width and depth of the excavation. The effective lift of suction pumps is about 15 to 18 ft (4.5 to 5.4 m). A *multistage* system consists of rows of wellpoints set on benches at elevation intervals less than 15 ft (4.5 m). They are required for excavations that cannot be adequately dewatered by a single stage as determined by either the excavation depth or width, or both. The width is significant since the peak of the drawdown curve must be far enough below the excavation bottom at all points to prevent uplift and boiling.

Nomographs for estimating wellpoint spacing required to lower the water table to various depths for various soil types in uniform conditions are given in Fig. 8.45, and for stratified conditions in Fig. 8.46.

#### *Vacuum Wellpoints*

In silts in which gravity methods cannot drain pore water held by capillary forces, vacuum wellpoints are required.

To cause silt to change in an excavation bottom from a soft "quick" condition to a strong, firm condition usually requires only a small change in pore pressure. This is achieved by installing the wellpoint in the silt stratum and sealing the upper portion of the well with clay (such as bentonite balls), as shown on Fig. 8.47. If a sand layer is affected by the system, the vacuum may not be applied effectively to the silt.

#### *Jet-Eductor Wellpoint System*

Noncohesive soils to depths of 50 to 100 ft (15 to 30 m) can be dewatered with a jet-eductor system, useful for controlling uplift pressures in deep excavations. It is generally limited to small yields from each wellpoint (less than 10 to 15 gal/min or 38 to 57 liters/min).

The system consists of a wellpoint attached to

the bottom of a jet-eductor pump with one pressure pipe and one slightly larger pipe as shown on Fig. 8.48.

## Deep Wells

*Applications*

Deep wells are necessary for deep, wide excavations, where the soil below the excavation becomes more pervious with depth and the water table cannot be lowered adequately with well points. There must be sufficient depth of pervious materials below the level to which the water table is to be lowered to permit adequate submergence of the well screen and pump. Deep wells are particularly effective in highly stratified formations containing free-draining gravels. They are used also to dewater tunnels and to stabilize deep-seated slide masses, but they are costly and continuous pumping is required to maintain stability.

*Method*

The wells, containing submersible or turbine pumps, usually are installed outside the work area at spacings of 20 to 200 ft (6 to 60 m). The well diameter can vary from 6 to 18 in (15 to 45

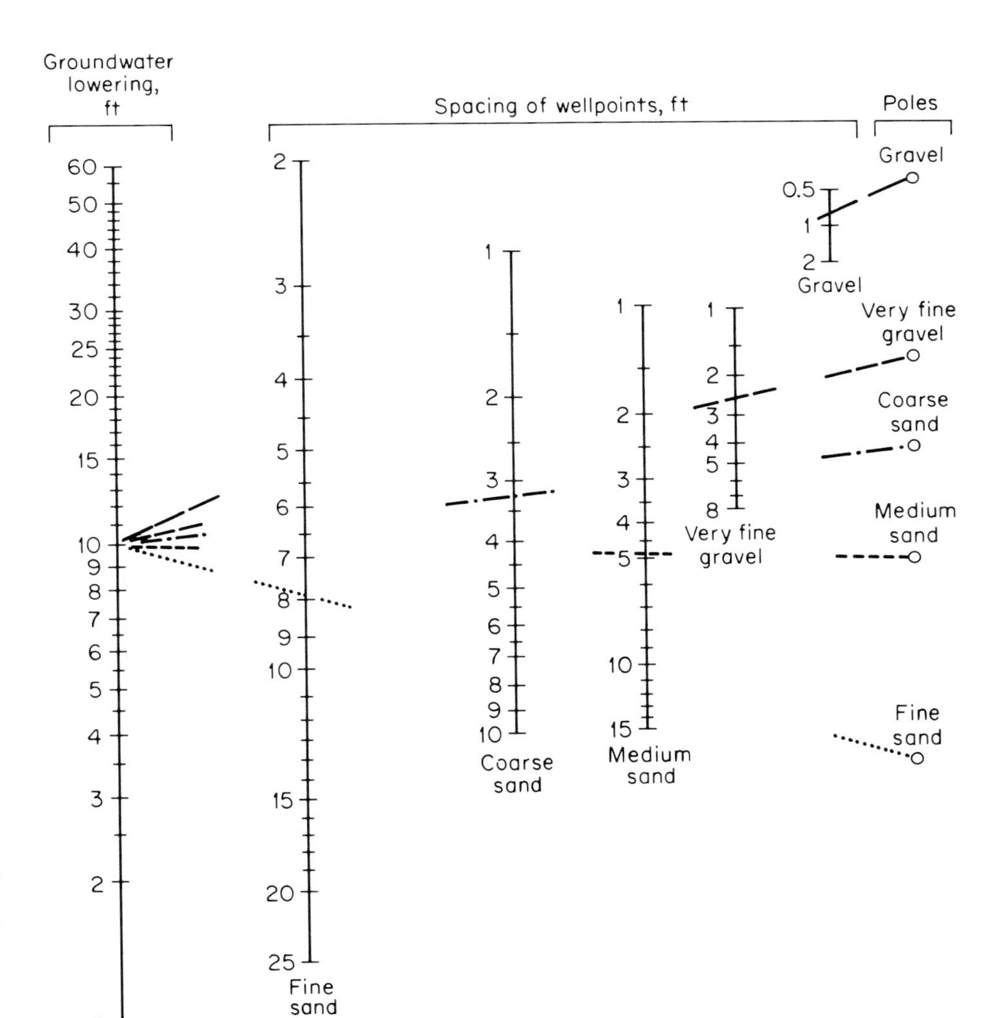


**FIG. 8.45 Wellpoint spacing for uniform clean sands and gravels.** [*From Mansur and Kaufman (1962)*[23] *and Moretrench Corp. Reprinted with permission of McGraw-Hill Book Company.*]

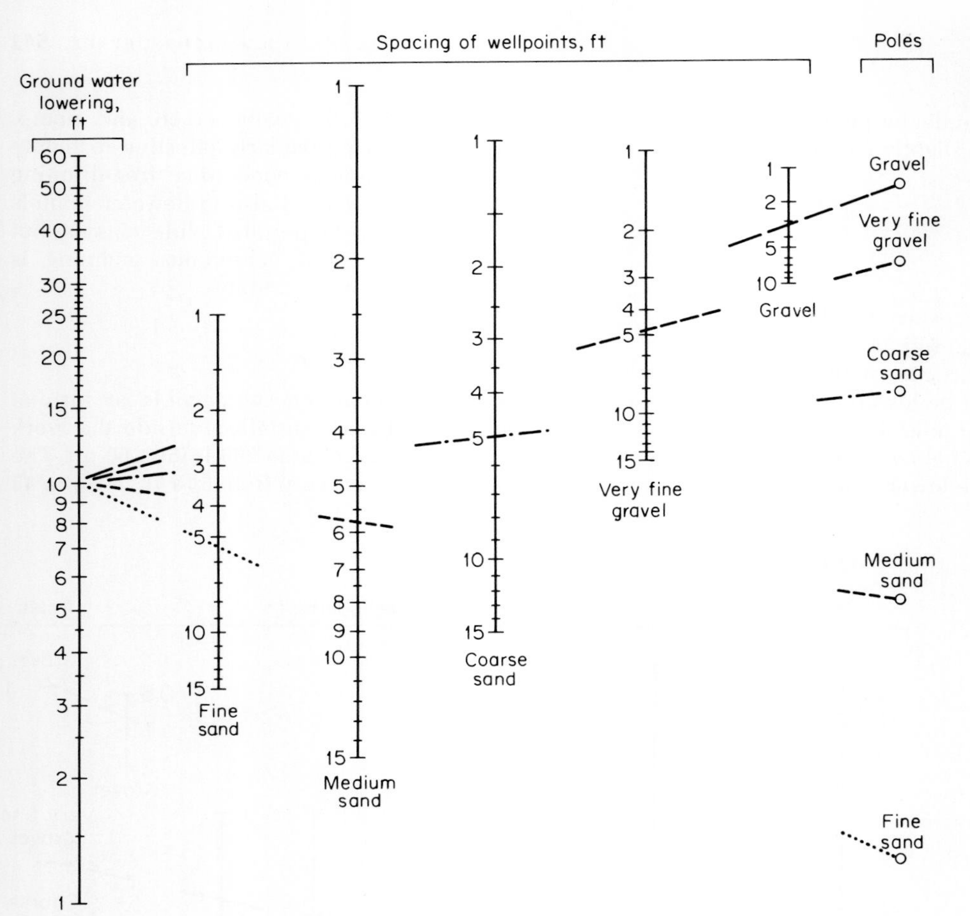


**FIG. 8.46 Wellpoint spacing for stratified sands and gravels.** [*From Mansur and Kaufman (1962)*[23] *and Moretrench Corp. Reprinted with permission of McGraw-Hill Book Company.*]

cm), and the screen is commonly of lengths of from 20 to 75 ft (6 to 22.5 m). Often used in combination with wellpoints as shown in Fig. 8.54, the deep wells are installed around the perimeter of the excavation and the wellpoints within the excavation.

### Electroosmosis

*Application*

Electroosmosis is used to increase the strength of *silts* when encountered as thick deposits in open excavations, slopes, or tunnels. It has been used infrequently in the United States because of its relatively high cost, and perhaps because of the sparseness of suitable conditions.

*Method*

Two electrodes are installed to the depth of dewatering in saturated soil and a direct current applied. The induced current causes the pore water to flow from the anode to the cathode from which it is removed by pumps. The electrodes are arranged so that the seepage pressures are directed away from the exposed face of an excavation and hence add to the stability.

## 8.4.4 DRAINS

### General

*Purpose*

Drains provide a controlled path along which flow can occur, thereby reducing seepage pressures in earthen dams, slopes, retaining structures, pavements, side-hill fills, tunnels, and spillways. In combination with pumps, drains reduce uplift forces on structures below the water table such as basements, concrete tanks for sewage treatment plants, and dry docks. Some types of drains are incorporated into

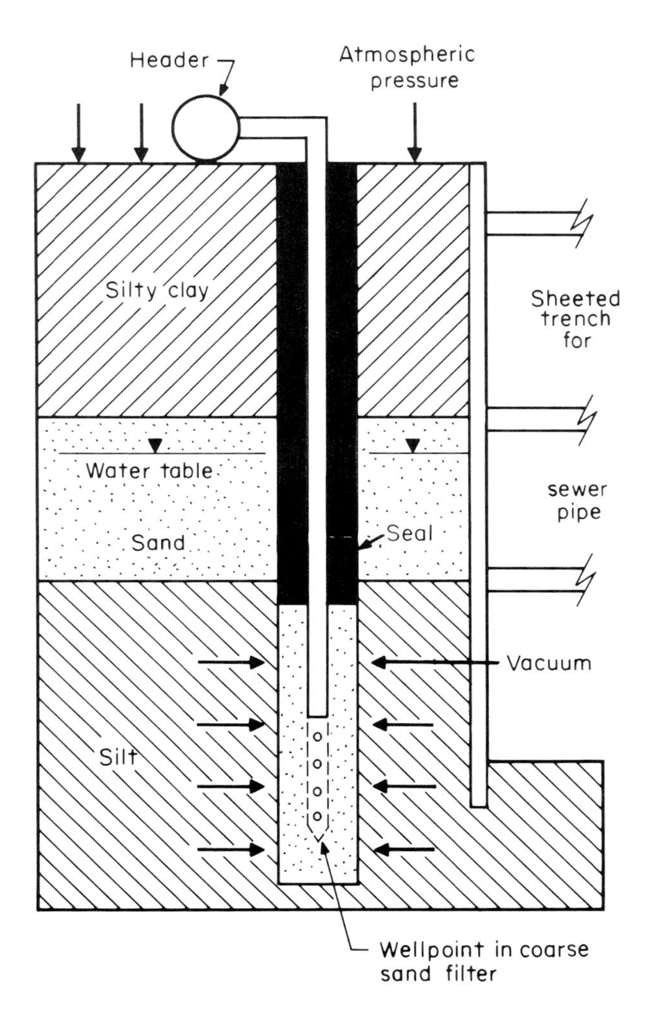


**FIG. 8.47** ***(Left)*** **Installation of a vacuum wellpoint to reduce pore pressures in a silt stratum to provide firm trench bottom for sewer pipe foundations. Separate system is used for the sand layer.**

**FIG. 8.48** ***(Below)*** **Operation of a Griffin jet-eductor wellpoint.** [*From Tschebotarioff (1973).*[38] *Reprinted with permission of McGraw-Hill Book Company.*]

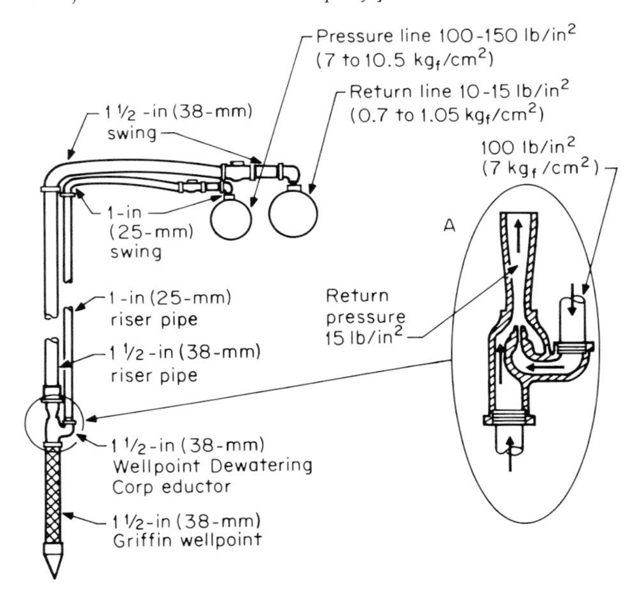


design, whereas others are used as corrective measures.

*Description*

Drains can be described by their shape and form as either blanket, trench, triangular or trapezoidal, or circular. Some types are constructed only of soil materials, some combine soils with a pipe to collect flow, and a few are open. Drains of large dimensions, such as toe drains for dams, are usually constructed of free-draining materials such as gravel or even stone.

The thickness of drainage layers depends on the flow quantities to be controlled and the permeability of the material available. A drain can be constructed of sand, but the thickness required may be several times greater than if a pea gravel were used. Most drains are provided with *filters* (*see Art. 8.4.5*) to protect against clogging and to prevent migration of finer materials from the adjacent soils. Descriptions and applications are described also in Art. 8.4.6.

*Clogging*

Drain performance is reduced or eliminated by clogging which results from the infiltration of fines, or from the incrustation of water. Drain holes have been plugged by calcite leached from grout curtains [Christensen (1974)[39]].

*Design*

Drain design is based on Darcy's law or on flow nets and is described in Cedergren (1967),[18] for example.

**Blankets**

Blanket drains extend longitudinally, with or without pipes. Horizontal blankets are used in pavements, building underdrains, and earth dams.

Inclined or vertical blankets are used in earth dams (chimney drains) or behind retaining walls.

**Trench Drains**

Trench drains are normally vertical, extending longitudinally to intercept flow into pavements or beneath side-hill fills, or transversely to provide a drainage way downslope. Along the toe of dams or slopes they provide for pressure relief.

They contain pipes (perforated, jointed, slotted, or porous) which during installation are placed near the bottom of the trench and surrounded with pea gravel. The trench backfill is of carefully selected pervious filter material designed to prevent both piping and the infiltration of the adjacent soil, and compacted to prevent surface settlement. The surface is sealed with clay or some other impervious material to prevent rainwater infiltration.

**Triangular or Trapezoidal Drains**

Triangular or trapezoidal drains extend longitudinally along the toe of earth dams, in which case they are normally very large; or in the body of the dam embankment; or behind retaining walls where they are connected to other drains.

**Circular Drains**

*Vertical circular drains* of relatively small diameters (10 to 20 cm usually) are used to relieve water pressures (1) in slopes where perched conditions overlie an open material of lower piezometric level and drainage can occur under gravity forces *(see Art. 9.4.4)* and (2) along the toe of earth embankments or slopes (relief or bleeder wells). Larger-diameter drains are used to relieve excess hydrostatic pressure, allowing consolidation during surcharging (sand drains), or to relieve pore pressures which may lead to liquefaction during earthquakes *(see Art. 11.3.3)*.

*Horizontal or subhorizontal* drains include large open *galleries* excavated in rock masses to improve gravity flow and decrease seepage pressures, and subhorizontal gravity drains of slotted pipe, which are one of the most effective means of stabilizing moving slopes, especially where large masses are involved *(see Art. 9.4.4)*.

## 8.4.5 FILTERS

**Purpose**

Filters are used to reduce flow velocities and prevent the migration of fines, clogging of drains, and piping of adjacent soils where flow passes across zones with significant differences in permeability. When water flows across two strata of widely differing gradation, such as silt to a gravel, the silt will wash into the gravel and a pipe or cavity will be created which could lead to structural collapse. In addition, the silt may clog the gravel, stopping flow and causing an increase in water pressures.

Filters are used commonly with blanket and trapezoidal or triangular drains, and often with trench drains. Circular drains including wellpoints or slotted pipe are installed with filter materials, except in the cases of subhorizontal drains and galleries.

**Design Criteria**

*Objectives*

A filter is intended to permit water to pass freely across the interface of adjacent layers, without substantial head loss, and still prevent the migration of fines. See Bennett (1952),[40] Cedergren (1967)[18] and (1975),[41] Lambe and Whitman (1969),[12] Terzaghi and Peck (1967),[42] and USBR (1974).[43]

*Design Basis*

Empirical relationships have been developed to satisfy the above criteria based on gradation characteristics as follows:

- Piping criterion: The 15% size of filter material

must not be more than 4 to 5 times the 85% size of the protected soil

- Permeability criterion: The 15% size of the filter material must be at least 4 to 5 times the 15% size of the protected soil, but not greater than 20 to 40 times, expressed as:

$$\frac{D_{15}\text{ (filter)}}{D_{15}\text{ (protected soil)}} \text{ must be} >4 \text{ or } 5, \text{ but} <20 \text{ to } 40$$
$$\frac{D_{15}\text{ (filter)}}{D_{85}\text{ (protected soil)}} \text{ must be} <4 \text{ or } 5 \qquad (8.15)$$

The filter soil designed may be too fine-grained to convey enough water, provide a good working surface, or pass water freely without loss of fines to a subdrain pipe. Under these circumstances a *second filter* layer is placed on the first filter layer and the first layer is then considered as the soil to be protected, and the second layer as the soil to be designed.

### Discharge through Drain Pipes

Water flowing through filters often is carried away through subdrainage pipe, of which there are many forms including:

- Plastic pipe (corrugated or smooth) with holes or slots
- Asbestos cement pipe with holes or slots
- Corrugated metal pipe (bituminous-coated) with holes or slots
- Clay tile with open joints
- Porous concrete pipe

The pipes should be surrounded with a filter soil designed so as not to enter the holes or slots. Cleanout points often are provided.

## 8.4.6 SURFACE TREATMENTS

### Purpose

Surface treatments are used to deter or prevent water infiltration so as to improve slope stability where a natural slope appears potentially unstable or a cut slope is being excavated, and to prevent increased saturation of potentially collapsible or swelling soils.

### Techniques

Slope stabilization is improved *(see Art. 9.4.3)* by preventing or minimizing the surface infiltration of water from upslope by:

- Planting vegetation which reduces the occurrence of shrinkage cracks which permit water to enter, deters erosion, and binds the surface materials
- Grading all depressions to prevent water from ponding and to allow runoff
- Sealing all surface cracks with a deformable material
- Installing surface drains to collect runoff and direct it away from the potentially unstable area

Areas adjacent to structures founded on potentially collapsible or swelling soils are sometimes paved to prevent direct infiltration from rainfall.

## 8.4.7 TYPICAL SOLUTIONS TO ENGINEERING PROBLEMS

### Earth and Rock-Fill Dams

*Embankment Control*

Some schemes for the control of seepage through earth and rock-fill dams are illustrated on Fig. 8.49. The objective of design is to retain the reservoir and prevent flow from emerging from the downstream face (Fig. 8.49*a*) where piping erosion may result.

A homogeneous embankment can be constructed from some natural materials such as glacial tills or residual soils containing a wide range of grain sizes which, when compacted, provide high strengths and adequately low permeabilities. Low dams, in particular, are constructed of these materials. Most dams, however, are zoned to use a range of site materials and require an impervious core to control flow within the shell, which provides the embankment strength (Fig. 8.49*e*).

Internal drainage control of the phreatic surface is provided in all dams. Because earth dams are constructed by compacting borrowed soil materials in layers, there is always the possibility for the embankment to contain relatively pervious horizontal zones permitting lateral drainage

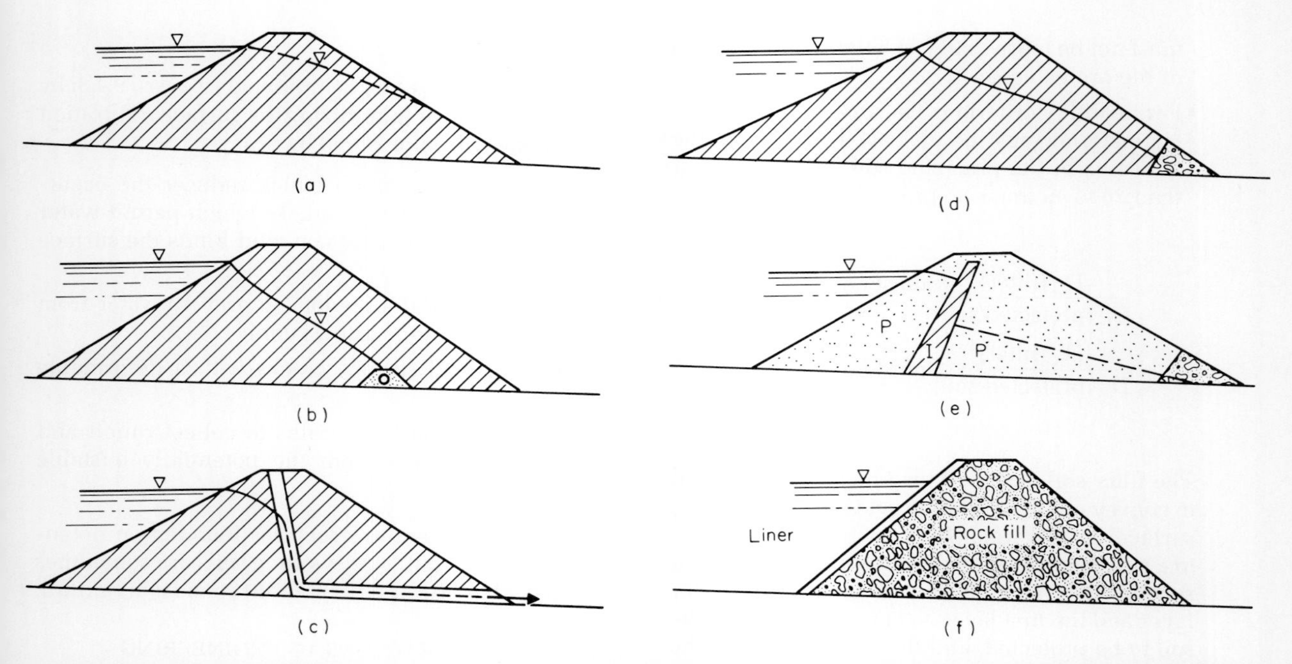


**FIG. 8.49 Control of seepage through earth and rock-fill dams: (*a*) Embankment without seepage control. Seepage exits on downstream slope. (*b*) Longitudinal drain in homogeneous embankment. (*c*) Chimney and blanket drain in homogeneous embankment. (*d*) Toe drain in homogeneous embankment. (*e*) Zoned dam to control seepage. There are any number of combinations of pervious and impervious zones. (*f*) Rock-fill dam with upstream liner.**

which must be prevented from intersecting the downstream face. Various types of drains are used, including longitudinal drains (Fig. 8.49*b*), chimney and blanket drains (Fig. 8.49*c*), and toe drains (Fig. 8.49*d*). Details of two types of toe drains are given on Fig. 8.50.

Rock-fill dams may be provided with an impervious core section, or the upstream face may be covered with a membrane of concrete, asphaltic concrete, or welded steel (Fig. 8.49*f*).

*Foundation and Abutment Control*

Various methods of controlling seepage in pervious materials beneath dams are illustrated on Fig. 8.51. If the dam foundation and abutments consist of pervious materials, the reservoir will suffer large losses and high seepage pressures may build up at the toe (Fig. 8.51*a*). Some form of cutoff may then be required, depending upon the purpose of the dam. (Flood control dams, for example, are not a concern regarding water loss, as long as seepage pressures are not excessive.)

A *core trench,* formed by extending the embankment core zone of compacted impervious soils through the pervious foundation, is often the most economical cutoff solution (Fig. 8.51*b*). Ideally, the core trench should extend into impervious materials. Sherard et al. (1963)[4] state that compacted fill cutoffs are commonly constructed to depths of about 75 ft (22.5 m) when necessary, but below this depth become increasingly expensive. Because the trench is excavated in granular soils or other free-draining materials in a stream valley, the primary problem to overcome is maintenance of a dry excavation during core construction.

A *blanket* formed of impervious materials, extending some distance upstream from the embankment (Fig. 8.51*c*), reduces toe-seepage forces in materials with moderate permeability by increasing the length of flow paths. Usually the same material used for the embankment core zone is used for the blanket. Thickness and length are dependent upon the permeability of the blanket material; permeabilities, stratification, and thickness of the pervious stratum; and reservoir depth [Bennett (1952)[40]]. The thickness usually ranges from 2 to 10 ft (0.7 to 3 m).

A *grout curtain*, extending from a core trench or the core wall, can be installed to almost any practical depth, and has the added advantage of reducing the amount of excavation and dewatering necessary. The controlling factors to be evaluated in the determination of the necessity of grouting a rock foundation (Fig. 8.51*d*) are the rock-mass characteristics and the height of water contained by the dam. From a historical viewpoint, dams with a height of less than 50 ft (15 m) on rock that is not excessively fractured have not been grouted, whereas dams with a pool depth of 100 ft (30 m) or more, have had foundations grouted [Sherard et al. (1963)[4]]. Conditions normally requiring grouting include badly fractured zones such as those from faulting, pervious sandstones, and vesicular and cavernous rocks. The grouting of limestones is common but often unsuccessful practice; a difficult problem to assess is the depth of grouting required when limestone extends for substantial depths. In pervious rock masses, grout curtains usually extend for some distance beyond the abutments to contain seepage losses.

Either a *concrete wall or a slurry wall* (Fig. 8.51e) provides an effective cutoff. Slurry walls have been used with success even in highly pervious gravel and sand formations to depths of 80 ft (24 m) [Sherard et al. (1963)[4]]. Consideration is required regarding there being a possible source of embankment settlement.

*Relief wells or trenches*, installed at the downstream toe (Fig. 8.51*f*), are relatively inexpensive and usually highly effective in relieving seepage pressures. They have the advantage that they can be installed after construction, if necessary. Well spacing depends on the amount of seepage to be controlled and often 50 to 100 ft (15 to 30 m) on centers is adequate to reduce seepage pressures to acceptable limits in soils. It is advantageous that the wells penetrate to the full depth of the pervious stratum. Inside pipe diameter should be 6 in or larger if heavy flows are anticipated. The wells usually consist of perforated pipe of metal or wood, surrounded by a gravel pack, although in recent years pipe of plastic, concrete, or asphalt-coated galvanized metal has been used. Installation should be done with the same care as with water wells; i.e., the wells should be drilled by a method that does not seal the pervious stratum with fines, then surged with a rubber piston to remove muddy drilling fluid. Seepage water from wells is discharged at ground surface in a horizontal pipe connected to a lined drainage ditch running along the embankment toe. Relief wells have the disadvantages of decreasing the average seepage path and, therefore, increasing the

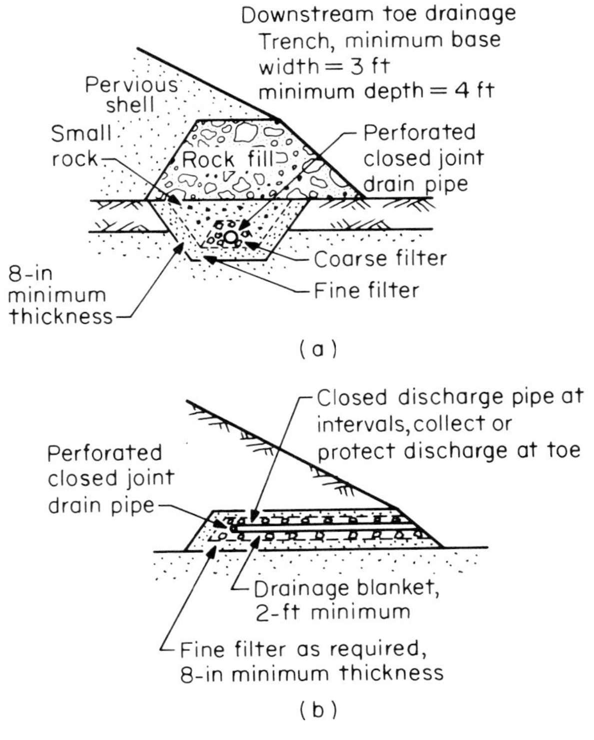


**FIG. 8.50 Examples of toe drains for earth dams: (*a*) impervious surface layer and (*b*) homogeneous pervious foundation.** [*From NAVFAC (1971).*[37]]

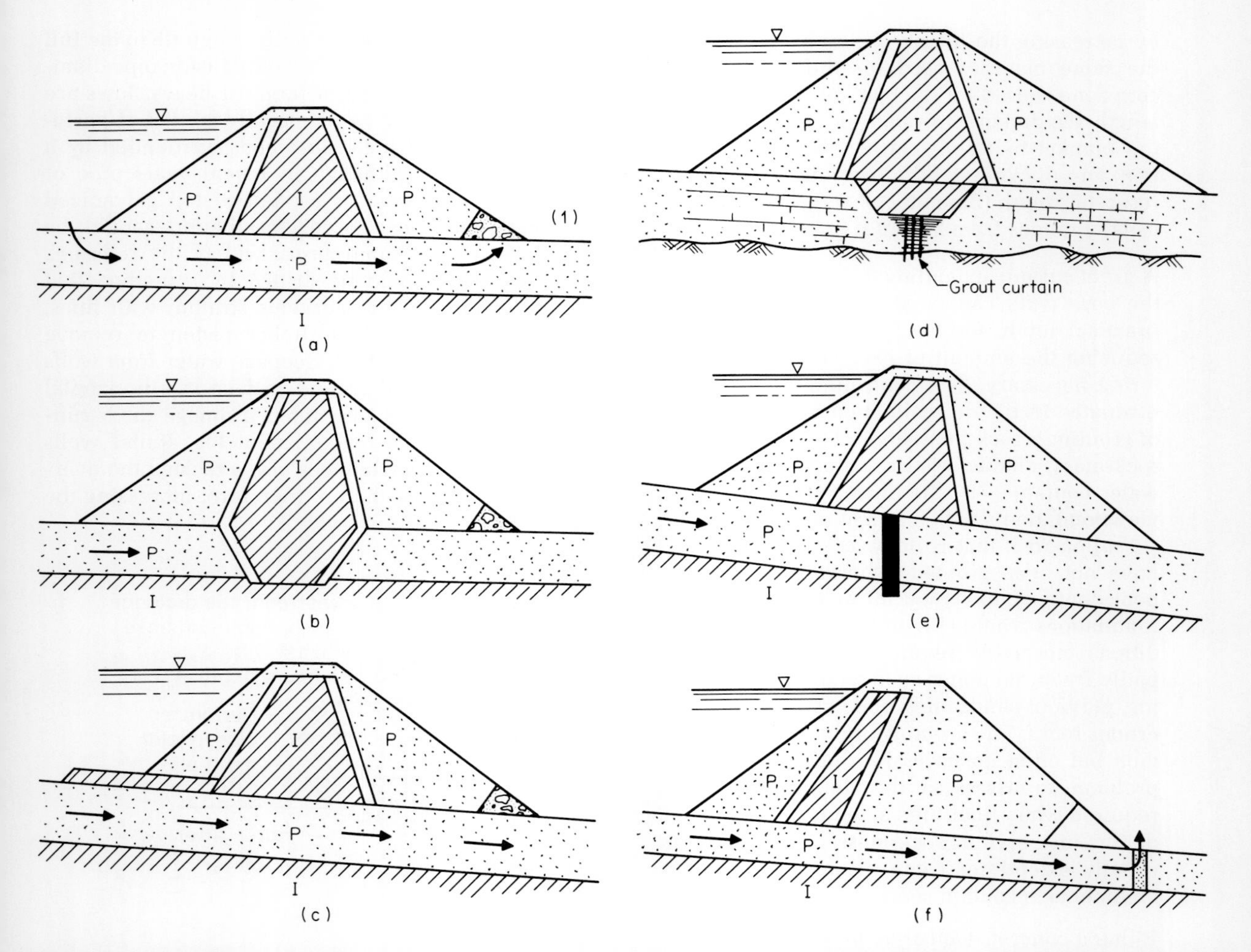


**FIG. 8.51 Controlling seepage in pervious materials beneath dams: (*a*) uncontrolled foundation seepage causes water loss and uplift at (1); (*b*) rolled earth cutoff; (*c*) upstream blanket to reduce head; (*d*) grout curtain in fractured rock; (*e*) concrete wall or slurry-trench cutoff; (*f*) downstream relief wells or trench. (Sections are schematic only; upstream slopes are normally about 2½ to 3½:1 and downstream 2 to 3:1.)**

underseepage quantity, and they are subject to deterioration and require periodic inspection and maintenance. Unless protected, metal pipes corrode and wooden pipes rot and are attacked by organisms.

### Concrete Dams

*Grout curtains* are sometimes used to reduce uplift pressures beneath foundations and reduce pore pressures in the abutments of concrete dams founded in rock masses.

*Drain holes* should be installed as common practice in rock masses to assure that seepage pressures are relieved since complete cutoff with a grout curtain is seldom achieved. A typical configuration of grout curtains and a relief-well system beneath a thin-arch concrete dam is shown on Fig. 8.52.

*Drain spacing* depends on rock quality but often

ranges from 3 to 10 m on centers. The depth extends to perhaps 75% of that of the grout curtain. Drains should be inspected periodically to assure that plugging is not occurring, either from the deposition of minerals carried in groundwater or minerals from the grout itself.

*Piezometers* should be installed to monitor pore- or cleft-water pressures. An indication of high pressures requires consideration of the installation of additional drains.

### Slopes

*See also Arts. 9.4.3 and 9.4.4. Cut slopes in soils* are often provided with vegetation to prevent erosion and deter infiltration, longitudinal surface drains upslope of the cut and along benches to carry away runoff, and transverse surface drains downslope to direct runoff on long cuts. In some cases subhorizontal drains are installed to relieve pore pressures along potential failure surfaces. In silts, temporary relief may be obtained by electroosmosis.

*Cut slopes in rock masses* may require subhorizontal drains to relieve cleft-water pressures and improve stability.

*Potentially unstable or actively unstable slopes* in many cases can be stabilized by the installation of drains to decrease pore pressures along failure surfaces. There are a number of possibilities as follows:

- Subhorizontal drains often are the most practical and economical solution. They consist of 2-in-diameter or larger pipe, forced into a drill hole, made at a slight inclination upslope, which extends beneath the phreatic surface for some distance. The length and depth of the drains depend on the amount of groundwater lowering in the slope that is desired. Drains will be severed in moving slopes and require reinstallation.
- Drainage galleries are sometimes used in rock masses where ground collapse is unlikely.
- Vertical wells, which require continuous pumping, have been used to stabilize large moving masses. Vertical gravity drains are effective in relieving seepage pressures caused by perched water tables where impervious strata are underlain by a free-draining material with a lower piezometric level.
- Relief wells and trenches have been used to relieve pressures where seepage emerges at the toe of slope, at times the beginning point of instability.

### Concrete Retaining Walls

Drainage is required to prevent the buildup of hydrostatic pressures behind concrete walls retaining slopes. Design depends on flow quantities anticipated and can range from buried blankets to merely weep holes. Several schemes are shown on Fig. 8.53. Flow is collected in longitudinal drains and carried beyond the wall, or discharged through openings in the wall.

### Open Excavations

*Support by walls* is normally not required if relatively steep slopes remain stable and there is adequate space within the property for the increased excavation dimension. In any event, groundwater control usually is required to:

- Provide for a dry excavation
- Relieve pressures along the slope face, or against walls
- Relieve bottom pressures to control piping and heave

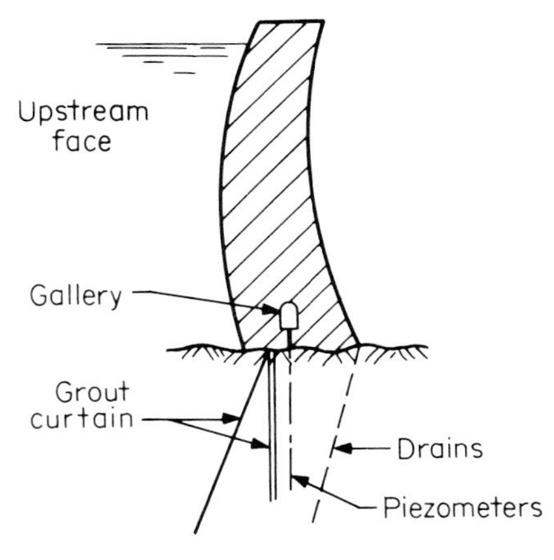


**FIG. 8.52 Schematic of grout curtain and drains beneath thin-arch concrete dam.**

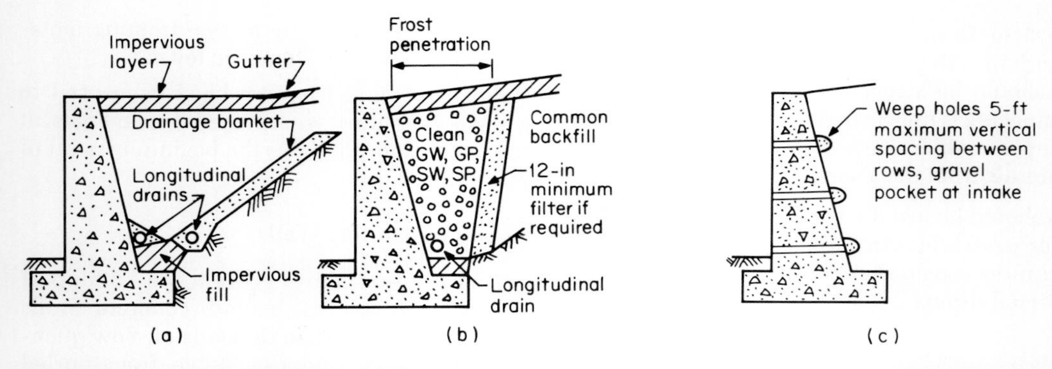


**FIG. 8.53 Examples of retaining wall drainage: (*a*) complete drainage; (*b*) prevention of frost thrust; (*c*) minimum drainage.** [*From NAVFAC (1971).*[37]]

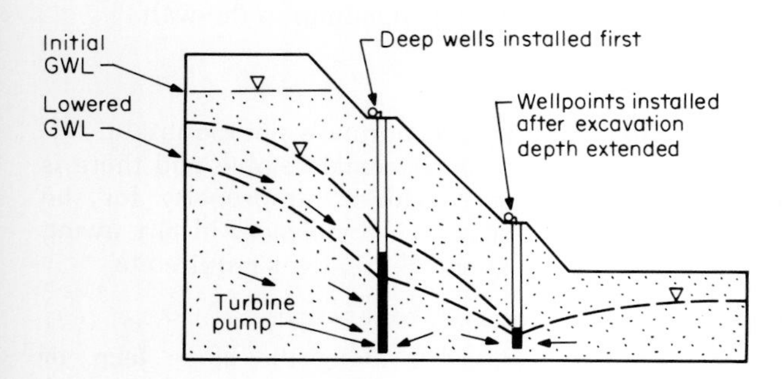


**FIG. 8.54 Dewatering with deep wells and wellpoints in free-draining materials.**

**FIG. 8.55 Anchored sheet-pile wall retaining clayey silts interbedded with sands in excavation adjacent to the ocean behind dike in upper right of photo. Wellpoint header pipe in lower left and wellpoint discharge in lower right.**

*Sloped, unsupported excavation dewatering systems* may include:

- Sump pumping in relatively shallow excavations in soils with moderate flows
- Single-stage wellpoints to depths of 15 ft in sandy soils
- Multistage wellpoints at 15-ft depth intervals in sandy soils
- Deep wells for free-draining soils at substantial depths, used in conjunction with wellpoints placed at shallower depths as illustrated on Fig. 8.54.

*Supported Excavations*

Sheet piling (Fig. 8.55) or soldier piles and lagging are sometimes used in conjunction with wellpoints. Driving the sheet piling to penetrate impervious soils provides a partial seepage cutoff, as does driving the sheeting into sands below excavation bottom to a depth equal to at least half the head difference between the water table and the excavation bottom.

Slurry walls backfilled with concrete (diaphragm walls) provide an impervious barrier and may become an integral part of the final structure, as illustrated on Fig. 8.56. Care is required to ensure that no openings occur in the wall through which groundwater can penetrate. If they occur, they can usually be corrected with injection grouting.

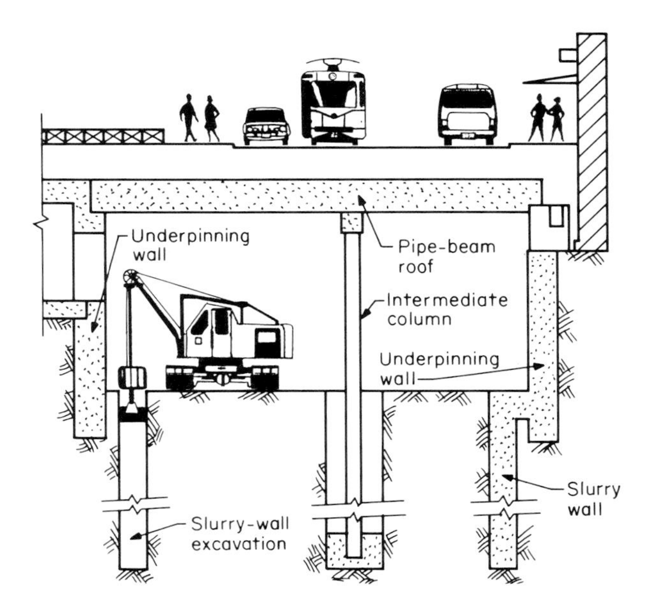


**FIG. 8.56 Slurry wall construction for five-story-high metro station (Antwerp, Belgium). After pipe-beam roof is installed and underpinned, and partial excavation underneath is completed, deeper parts of the station structure are built starting with slurry trench foundation walls. Intermediate columns are also founded on deep slurry walls. Floor slabs are cast in place.** [*From Musso (1979).*[44] *Adapted with the permission of the American Society of Civil Engineers.*]

## Building Basements

*Uplift protection* for building basements may be provided by

- Adequate basement slab thickness
- Slab tied down with anchors
- Gravity wells to relieve a perched water condition
- Underdrain system

*Underdrain systems* must be designed for the life of the structure, removing the maximum flow of water without significant soil loss or clogging. The system consists of several filter layers directly beneath the floor to intercept the water and carry it to subdrainage pipes and finally to an outlet as illustrated on Fig. 8.57. The outlet can be drained by gravity if the topographic configuration of the adjacent land permits, but more

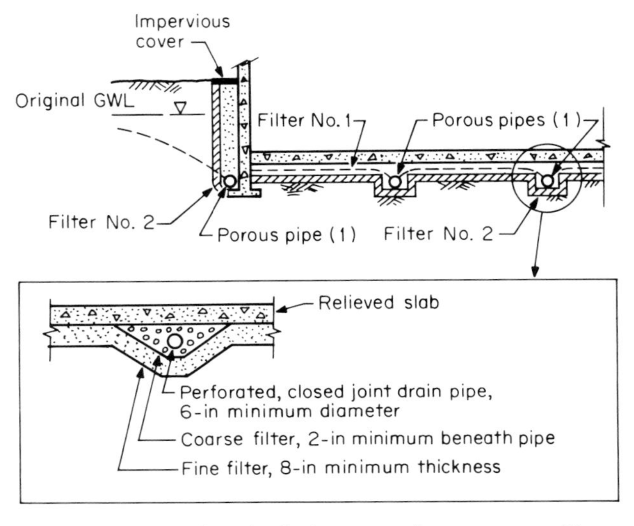


**FIG. 8.57 Typical underdrain system for structures. Pipes (1) lead either to gravity discharge or to a storage basin for removal by pumping.**

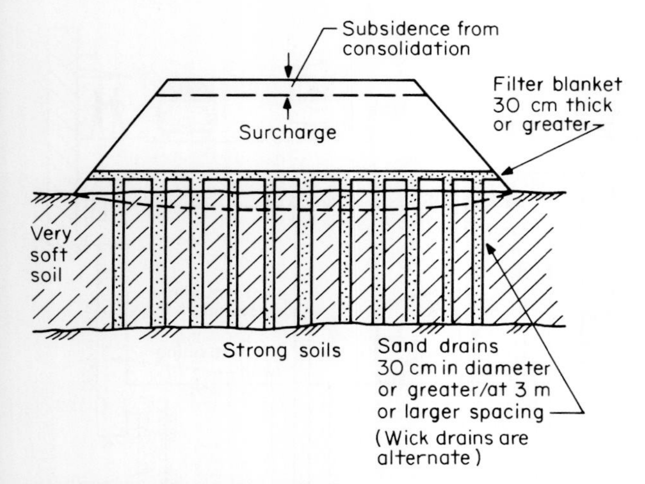


**FIG. 8.58 Vertical sand drains to accelerate drainage and consolidation of weak soils beneath a surcharge.**

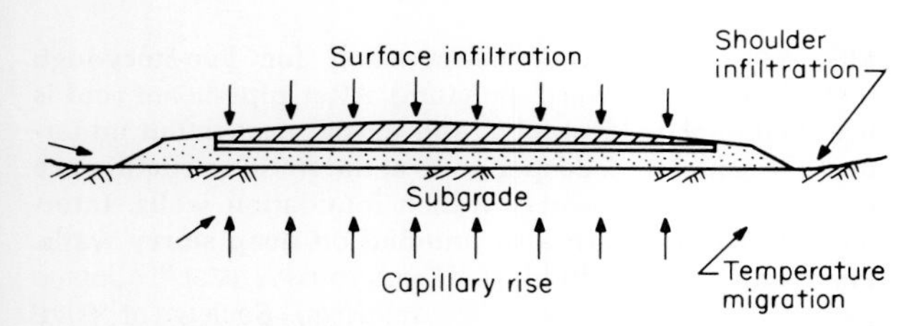


**FIG. 8.59 Sources of moisture infiltrating pavement and subgrade.**

often the discharge is collected in a tank or sump and removed by pumping. High water levels should be assumed in design since only a slight change in the underdrain system design will be required for conservative design insignificantly affecting costs.

*Damp-proofing* of basements is assisted by the underdrain system, but protective covering should always be applied on the outside of the walls during construction. After construction, leakage may often be corrected with grout injection, for which epoxy resins may be extremely effective.

### Site Surcharging

*Purpose* of surcharging is to preconsolidate soft soils and improve their supporting ability.

*Sand or wick drains* are relief wells installed to provide flow paths for the soil water, thereby reducing the time required for consolidation and the resulting prestress. Flow is upward through the drains to a free-draining blanket placed on the surface beneath the surcharge, from which the flow is discharged. The general system configuration is illustrated on Fig. 8.58.

### Pavements

#### *Failure Causes*

Pavement failures are usually related to groundwater, but water entering pavement cracks is equally important. Failure causes are saturation and softening of the subgrade and subbase, pumping from traffic, frost heaving, and swelling soils. Water can enter a pavement system from the surface through cracks and construction joints when they are not kept sealed by maintenance programs; lateral inflow occurs from the shoulders and median strips; and upward seepage occurs from a rising water table, capillary action, or the accumulation of condensation from temperature variations, as illustrated on Fig. 8.59. Provision for adequate drainage usually is the single most important aspect of pavement design.

#### *Control Measures*

*Watertight surfaces,* which require a strong maintenance program, prevent water infiltration through the pavement.

*Underdrainage* is always required, even in arid climates because saturation can occur from capillary rise and temperature migration. Prudent design assumes that water will enter from the surface and a layer of free-draining material is provided. Its thickness depends on its permeability. The roadway is crowned and sloped to allow percolation in the drainage layer to exit beyond the pavement and enter drainage ditches.

*Blanket drains* are optional considerations to control high hydrostatic pressures beneath the roadway and are barriers to capillary moisture. They are connected to trench drains as shown on Fig. 8.60*a*.

*Longitudinal trench drains* are installed along

the roadway, even in level or near-level terrain, as well as in side-hill cuts (Fig. 8.60*b*), to intercept groundwater and permit base-course drainage. Perforated, jointed, slotted or porous pipe is placed near the trench bottom and surrounded with pea gravel. Trench backfill is of carefully selected pervious filter material designed to prevent piping and infiltration of adjacent soil. The surface is sealed with clay or some other impervious material to prevent rainwater infiltration. Rainwater runoff is controlled by surface ditches.

*Transverse interceptor drains* or a *drainage blanket*, as shown on Fig. 8.60*c*, are necessary to control flows beneath side-hill fills and provide stability.

*All drains* must flow freely under gravity and discharge at locations protected against erosion.

### Tunnels

*Predrainage* of high-pressure water trapped by geologic structure in rock masses such as a fault zone may be accomplished by pilot "feeler" holes drilled in front of the tunnel heading. At the very least, the pilot holes help to disclose severe groundwater conditions. The holes are spaced to explore at angles from the heading, as well as directly in front, since in rock masses a water-bearing zone can be present at any orientation with the tunnel heading.

*Grouting* is used in soil or rock. Very difficult conditions are caused by saturated crushed rock zones associated with faulting and folding. They have been combatted in some cases by sealing the tunnel heading with a concrete bulkhead and injecting grout to stabilize the mass. Lancaster-Jones (1969)[36] describes a case where four phases of treatment were carried out to stabilize a wet, crushed quartzite sand before tunneling could proceed. A preliminary injection was made with sodium silicate to assist penetration of the subsequent cement grout which was pumped under a pressure of up to 1400 psi (100 kg/cm$^2$) to fissure the ground and compress the sand. This formed a strong network of cement in the fractured ground and sealed the main water passage. A third treatment was made with chemical grouts injected under pressures of 140 psi (10 kg/cm$^2$), designed to set in 15 minutes. This grout tended to bind the sand grains together, increasing the shear strength. A final injection of cement was made to further compress the mass. The final result was so satisfactory that explosives were required to advance the excavation.

*Slurry moles* are finding increasing application in lieu of drainage or air pressure in saturated granular soils. Cutting is accomplished with a large open-head cutting wheel which operates behind a bentonite slurry, pumped directly to the excavation face [NCE (1979)[45]].

*Shield tunneling* is a method commonly used in soft ground, often in conjunction with predrainage by deep wells.

*Deep wells* may be used to dewater soil or rock masses along a tunnel alignment.

*Freezing* during tunnel construction has been used occasionally [Jones and Brown (1978)[46]].

### Pollution Control

*See Art. 8.5.2.*

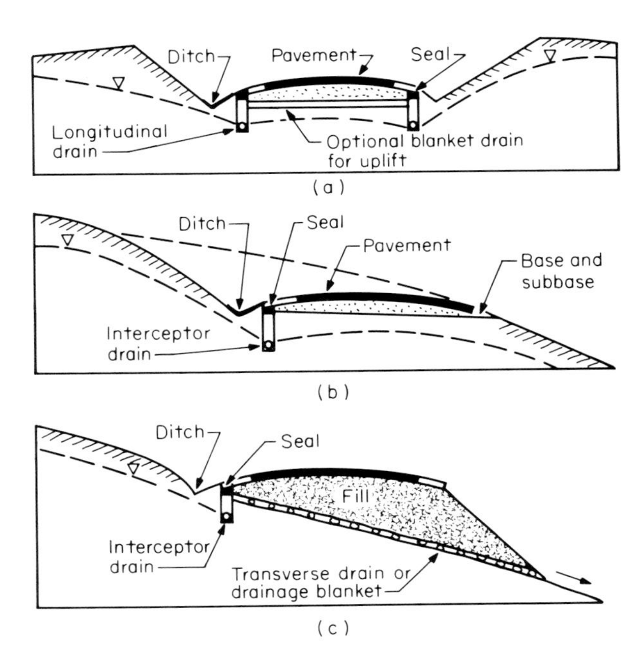


**FIG. 8.60 Some methods of pavement drainage control: (*a*) shallow cut in level ground; (*b*) side-hill cut; (*c*) side-hill fill.**

## 8.5 ENVIRONMENTAL CONSERVATION

### 8.5.1 WATER CONSERVATION AND FLOOD CONTROL

#### Resources and Hazards

*Aspects*

Groundwater is an important natural resource, and, especially in areas of low rainfall, its availability often is critical to the growth and even survival of cities. *Overextraction* results in a number of detrimental effects:

- The water table declines and the cost of extraction increases, and often the quality decreases as well so that treatment is required. In some cases, it is conceivable that groundwater could cease to be a viable source of water supply.
- In areas with marginal rainfalls, a desert environment could be created where one did not previously exist, thereby degrading the ecosystem in general, and increasing the incidence of erosion.
- Land subsidence has occurred in many areas *(see Art. 10.2)* which results in increased flooding incidence, as well as surface faulting.
- Along seacoasts, groundwater pollution is caused by saltwater intrusion.

Other aspects of conservation related to water are:

- Pollution resulting from surface and subsurface disposal of liquid and solid wastes
- Increased erosion and siltation resulting from uncontrolled land development
- Increased flooding incidence resulting from overdevelopment of floodplains
- Salinization and destruction of croplands resulting from irrigation

*Environmental Planning*

Conservation of water and protection against floods and other undesirable effects such as erosion require comprehensive regional environmental planning.

#### Case Study: Pima County, Arizona

*Introduction*

A study of the impact of urbanization on the natural environment in the semiarid climate of Tucson, Arizona, and the adjacent area of Cañada del Oro was made to determine the measures that could be taken to minimize the impact and protect the public [Hunt (1974)[47]].

*Background*

The city of Tucson is situated in a basin at elevations about 700 m above sea level, bounded on three sides by mountains rising to as high as 3500 m above sea level. In the valley, up to 700 m of sediments have filled over the basement rock.

One hundred years ago, according to reports, water was plentiful in the basin, lush grass covered the valley floor, and water flowed all year in the Santa Cruz River. Even today average precipitation in the basin is 300 mm and as much as 750 mm in the nearby mountains, but the Santa Cruz winds its way through the city as a dry riverbed (Fig. 8.15), except for short periods when it carries runoff from flash floods, and the basin is semidesert with sparse vegetation.

Extraction of large quantities of groundwater from the basin for farming, the mining industry and the growing population has caused the water table to drop as much as 40 m since 1947 alone. Human activities have disturbed nature's balance and created desert conditions and other associated problems, including increased erosion, flooding, land subsidence, and surface faulting.

*Study Purpose and Scope*

The Pima County Association of Governments recognized the need to establish guidelines for water conservation and flood protection in the valley, and for the planning and development of the essentially virgin area of 340 km$^2$ known as Cañada del Oro, adjacent to the city on the lower slopes of the mountains (see high-altitude stereo photos, Fig. 2.11). The association engaged a consultant to perform an environmental protection study.

The study scope included evaluations of the following aspects:

- Groundwater and surface water resources, and the principal groundwater recharge areas and their maintenance and protection
- Minimization of flooding and erosion

- Suitability of the soil-water regime for liquid and solid waste disposal
- Potential for severe areal subsidence from increasing groundwater withdrawal
- Maintenance of the vegetation and wildlife regimes
- Foundation conditions for structures as affected by subsidence faulting
- Extraction of construction materials and their consequences, especially sand and gravel operations in the washes which were increasing the erosion hazard along the banks and bridge foundations

*Master Planning Concepts Developed for Cañada del Oro*

A *geologic constraint map,* given as Fig. 8.61, was prepared of the Cañada del Oro area based on information obtained from the interpretation of stereo pairs of both high- and low-altitude aerial photos, ground reconnaissance, and a review of the existing literature including the soil association map for the area prepared by the Soil Conservation Service. With the map as the data base, a number of master planning concepts were developed.

For the Cañada del Oro area it was recommended that the major washes and floodplains be zoned to prohibit any development except that accessory to recreational uses. In addition to being flood-prone areas, they are the major source of recharge for much of the basin. (In some semiarid regions, recharge is being accomplished by deep-well injection of water stored temporarily in flood-holding reservoirs to avoid high losses to evaporation [ENR (1980)[49]].

Water conservation requires a reduction in ostentatious uses such as watering lawns and swimming pools by discouraging their use with substantial tax premiums. Landscaping with desert vegetation, for example, will help to conserve water.

Residential development should be located to avoid blocking the larger of the natural drainage ways that traverse the area as sheet wash gullies *(see Fig. 2.11)* as shown on the *tentative land-use map* given as Fig. 8.62. Roadways should be planned to parallel the natural drainage ways with a minimum of crossovers (Fig. 8.14) or blocking. Housing constructed on a cluster or high-density basis will minimize site grading, roadways, and other disturbances to the surface, and reduce the costs of waste disposal. Extensive site grading removes valuable vegetation such as the saguaro cactus (which requires years to grow), contributes to degrading the entire ecosystem, and can result in dust-bowl conditions.

A dendritic drainage pattern should be maintained to direct flows away from developments, and care should be taken to control floods and to minimize erosion. Solutions to similar conditions in Albuquerque, New Mexico, are described by Bishop (1978).[50]

**Salinization**

Salinization is associated with a desert or low-rainfall environment. It is caused by irrigation, or by natural runoff migrating downslope, carrying large amounts of salts which are deposited in the valley trough to accumulate and result in nonfertile areas.

Surface drainage to eliminate the ponding of water affords some control. Subsurface collector drains in irrigated valleys carry away the saline waters to larger drains and canals for discharge in some location environmentally acceptable.

### 8.5.2 GROUNDWATER POLLUTION CONTROL

**Pollution**

*Sources*

Forms of pollutants include liquid wastes moving directly into the groundwater system or leachates from liquid or solid wastes flowing or percolating into the groundwater system.

*Liquid waste* pollutes by percolation or direct contact, and originates from:

- Domestic sources disposed into septic tanks, with or without leaching fields, which produce biological contaminants;
- Industrial sources disposed into shallow unlined pits or reservoirs, or by deep-well injection, which produce chemical contaminants; and
- Spills from chemical plants or other industrial sources.

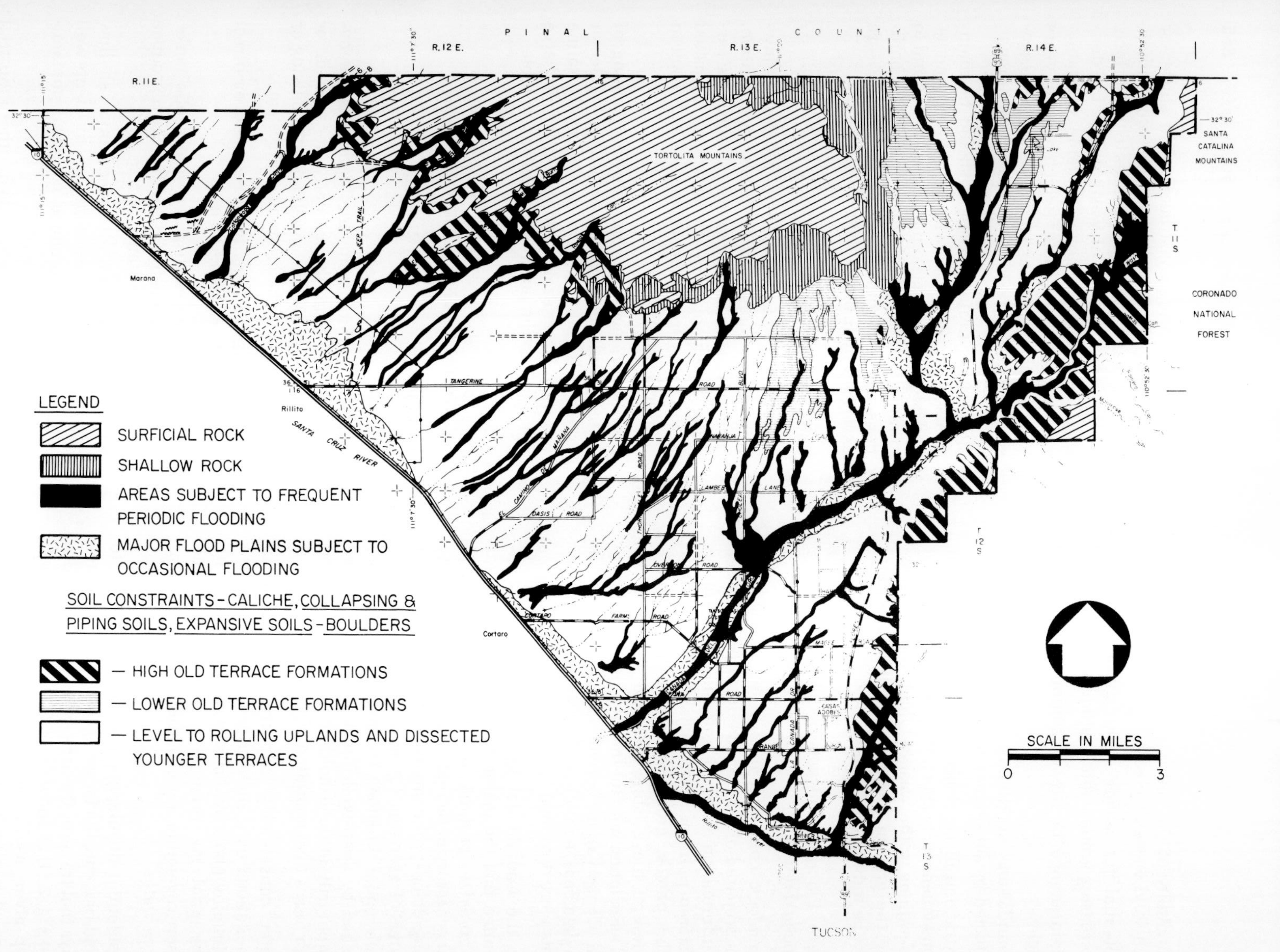


**FIG. 8.61 Geologic constraint map (Cañada del Oro area, Tucson, Arizona). (See Fig. 2.11 for high-altitude stereo pair of aerial photos of the area.)** [*From Ward (1975).*[48]]

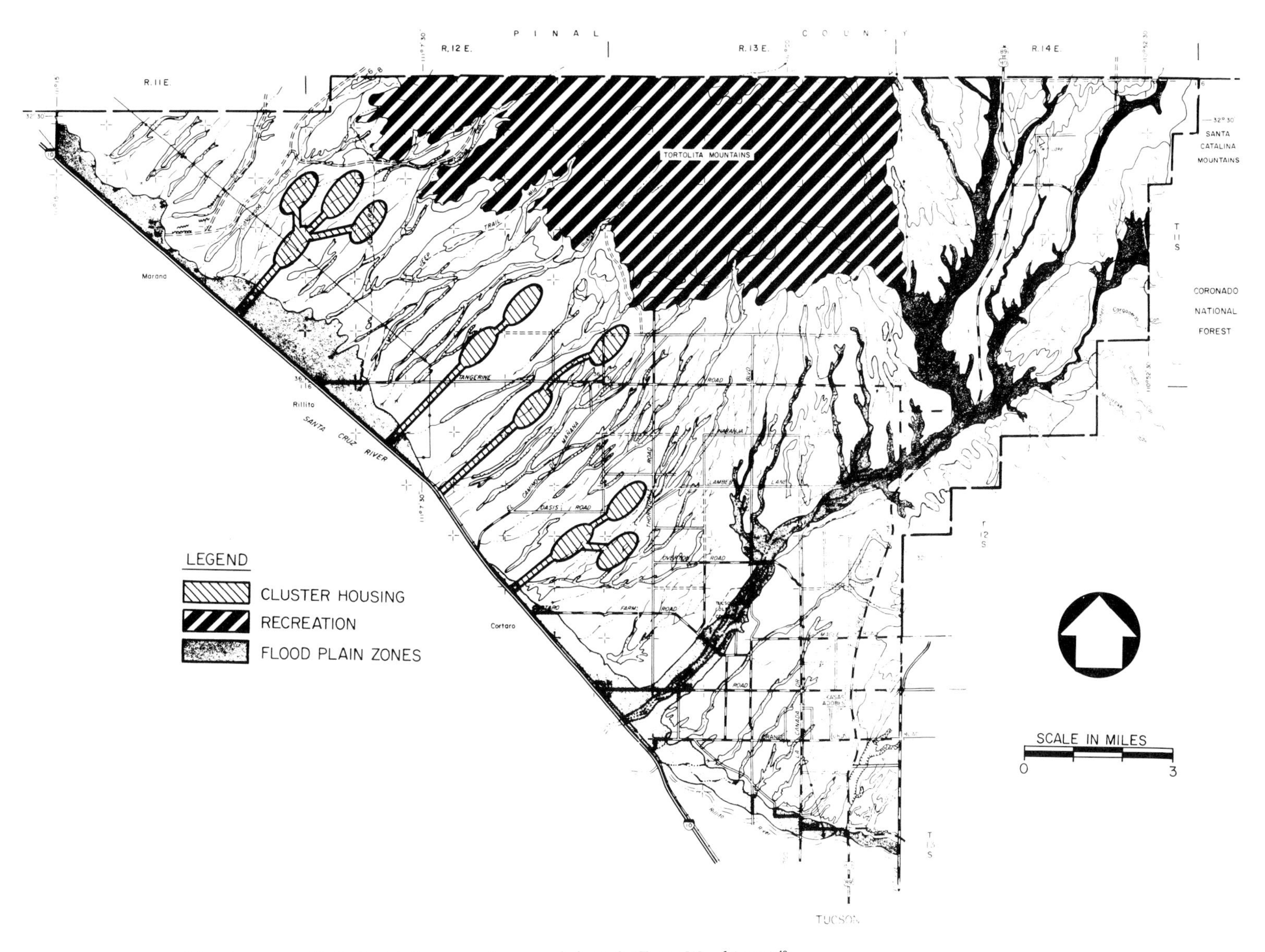


FIG. 8.62 **Suggested land-use map (Cañada del Oro area, Tucson, Arizona).** [*From Ward (1975).*[48]]

*Solid waste* produces leachates from groundwater or rainwater percolating through:

- Domestic sources disposed of in garbage or rubbish dumps or sanitary land fills, producing biological and chemical contaminants.
- Industrial waste dumps which produce chemical contaminants. Solid wastes result from such industries as coal and phosphate mining, power generation (fly ash and nuclear wastes), pulp and paper manufacturing, etc.

*Occurrence*

Pollution occurs when the liquid waste or leachate moves away from the disposal area. Pollution potential depends on the mobility of the contaminant, its accessibility to the groundwater system, the reservoir characteristics, and climate.

Permeable soils permit relatively rapid movement, but depending upon the rate of movement, biological contaminants may be partially or effectively filtered by movement. Chemical constituents, however, are generally free to move rapidly when they enter the groundwater flow system and can travel relatively large distances. Cavernous or highly fractured rock will directly and rapidly transmit all pollutants for great distances where running water is present. Impervious materials retard movement, or restrict leachates to the local vicinity of the waste disposal site, and pollution of underlying aquifers is negligible as long as the impervious stratum is adequately thick.

Climatic conditions are also significant. In areas of high rainfall, the pollution potential is greater than in less moist areas. In semiarid regions there may be little or no pollution potential because all infiltrated water is either absorbed by the material holding the contaminants, or is held in soil moisture and eventually evaporated.

The character and strength of the contaminant are dependent, in part, on the length of time that infiltrated water is in contact with the waste and the amount of infiltrated water. The *maximum potential* for groundwater pollution occurs in areas of shallow water tables where the waste is in constant contact with the groundwater and leaching is a continual process.

**Liquid Waste Disposal**

*Sewage plant treatment* prior to discharge into the surface or groundwater system is the most effective means of preventing groundwater pollution, provided the treatment level is adequate for the particular sewage.

*Lined reservoirs for industrial waste* are constructed with plastic or clay liners to provide a barrier to infiltration, but storage capacity limitations usually require that the waste eventually be treated. Moreover, open reservoirs can affect air quality.

*Deep-well injection* into porous sandstones, limestones, and fractured rock masses can be a risky solution unless it is certain beyond doubt that an existing or potential aquifer cannot be contaminated.

*Septic Tanks and Leaching Fields*

Sanitary wastes from homes and other buildings often are disposed into septic tanks with or without leaching fields. Biological decomposition of solids by anaerobic bacteria takes place in the septic tank and part of the solids remain. A large proportion of the harmful microorganisms, however, is not removed from the waste.

Effluent from the tank may be discharged into the ground by means of a seepage bed, pit, or trench. The rate at which the soil absorbs the effluent is critical to the system operation. If it is not absorbed rapidly enough, it may back up into the drains from the building and eventually rise to the ground surface over the seepage area. If it drains too rapidly, it may travel unfiltered into wells or surface water supplies and contaminate them with various types of disease-bearing organisms. If the flow rate is intermediate, nature will act as a purification system through microbial action, adsorption, ion exchange, precipitation, and filtration.

Septic tanks and leaching fields will not work in clayey or silty soils, or below the water table, or in frozen ground, since flow is required for their effectiveness. They are suitable in lightly populated areas where land is adequate and geologic conditions are favorable, or in developments where use is intermittent, such as vacation dwellings. When used in impervious materials without leaching fields, the waste should be pumped periodically from the septic tank into trucks. Continuous use over a long period of time in densely populated areas, even with favorable geological conditions, may eventually result in pollution of surface and subsurface waters.

*Protection against Spills*

Spills from chemical plants and other industrial sources often cannot be avoided in many industrial processes, especially over long time intervals. It is best, therefore, to locate plants where spills cannot contaminate valuable aquifers.

Control can be achieved, however, with systems of impervious barriers such as a slurry-wall cutoff and a wellpoint dewatering system. In the example illustrated on Fig. 8.63, pollutants from plant spills enter the granular materials beneath the plant, mix with groundwater, and flow through the sheet-pile bulkhead into the adjacent river unless contained with an impervious barrier and dewatering system as shown.

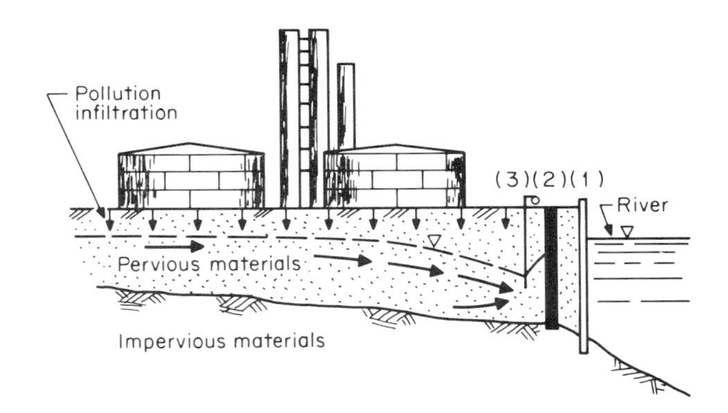


**FIG. 8.63 Pollution control system for chemical plant adjacent to river or other water body. (1) Pollutants would flow through the sheet-pile bulkhead into river. (2) A slurry wall penetrating the pervious materials into the impervious materials acts as a barrier. The wall should extend along the entire plant waterfront and then extend perpendicular to the river inland for some distance, or surround the plant, to contain spillage. (3) The artificial basin is dewatered periodically with wellpoints to prevent flooding from rainfall runoff. The effluent is pumped to storage facilities and held for treatment.**

### Solid-Waste Disposal

Disposal in open areas carries with it an inherent potential for pollution of water resources, regardless of the manner of disposal or the composition of the waste material. Industrial wastes subjected to rainfall and groundwater will produce chemical constituents; leachates from open dumps or sanitary landfills usually contain both biological and chemical constituents.

Since the pollution of an aquifer or nearby water body depends on the ability of the leachate to seep through the waste and underlying natural materials, waste dumps should be located over, or in, impervious materials, or the waste dump area should be lined with impervious materials, or totally confined with a vertical seepage barrier. Location of solid waste dumps in old sand and gravel pits or limestone quarries is common and extremely risky unless dump seepage is contained.

## 8.5.3 ENVIRONMENTAL PLANNING ASPECTS SUMMARIZED

### Water Supply

- Evaluate the water quality and water balance quantity, or the withdrawal quantity that does not exceed recharge, as a measure of the permissible withdrawal for urban populations, farming, and industrial use.
- Identify the major recharge areas and provide for their protection against blockage and pollution.
- Evaluate the potential for land subsidence from overwithdrawal and formulate contingency plans.
- Provide contingency plans to supplement water supply for the future when the normal situation occurs of continuing urban growth and groundwater depletion.
- Provide for, or plan for, wastewater treatment to maintain and improve water quality, and provide for solid-waste disposal to avoid or minimize pollution.

### Flood Control

- Avoid construction in the floodplains of rivers.
- If development is necessary, protection and control is achieved by raising grades, building dikes or other structures, and constructing holding reservoirs upstream.

### Erosion Protection

- Protect riverbanks only where structures are endangered. When protection extends over considerable distances, consider the effects downstream of increased discharges.
- During development, plant vegetation, divert runoff, slow flow velocities, and protect the surface of flow channels. These treatments serve also to minimize *siltation*.
- Minimize deforestation and removal of existing vegetation.

## REFERENCES

1. Smith, D. W. (1977) "Why Do Bridges Fail?", *Civil Engineering*, ASCE, November, pp. 59–62.
2. Gray, D. H., Leiser, A. T. and White, C. A. (1980) "Combined Vegetative-Structural Slope Stabilization," *Civil Engineering*, ASCE, January, pp. 82–85.
3. Dallaire, G. (1976) "Controlling Erosion and Sedimentation at Construction Sites," *Civil Engineering*, ASCE, October, pp. 73–77.
4. Sherard, J. L., Woodward, R. J., Gizienski, S. G. and Clevenger, W. A. (1963) *Earth and Earth Rock Dams*, John Wiley & Sons, New York.
5. USBR (1973) *Design of Small Dams*, U.S. Bureau of Reclamation, U.S. Govt. Printing Office, Washington, D.C.
6. Linsley, R. K., Kohler, M. A. and Paulhus, J. L. H. (1958) *Hydrology for Engineers*, McGraw-Hill Book Co., New York.
7. Rahn, P. H. (1975) "Lessons Learned from the June 9, 1972 Flood in Rapid City, South Dakota," *Bull. Assoc. Engrg. Geol.*, Vol. XII, No. 2.
8. Gilluly, J., Waters, A. G. and Woodford A. O. (1959) *Principles of Geology*, W. H. Freeman and Co., San Francisco.
9. Krynine, D. P. and Judd, W. R. (1957) *Principles of Engineering Geology and Geotechnics*, McGraw-Hill Book Co., New York.
10. Linsley, R. K. and Franzini, J. B. (1964) *Water Resources Engineering*, McGraw-Hill Book Co., New York.
11. Salzman, G. S. (1974) "Seepage and Groundwater," lecture notes prepared for Gilbert Assocs., Joseph S. Ward & Assocs., Caldwell, New Jersey.
12. Lambe, T. W. and Whitman, R. V. (1969) *Soil Mechanics*, John Wiley & Sons, New York.
13. DeWeist, R. J. M. (1965) *Geohydrology*, John Wiley & Sons, New York, p. 187.
14. Karplus, W. J. (1968) *Analog Simulation: Solution of Field Problems*, McGraw-Hill Book Co., New York.
15. Meehan, R. L. and Morgenstern, N. R. (1968) "The Approximate Solution of Seepage Problems by a Simple Electrical Analogue Method," *Civil Engineers and Public Works Review*, Vol. 53, pp. 65–70.
16. Zienkiewicz, O. C., Mayer, P. and Cheung, Y. K. (1966) "Solution of Anisotropic Seepage by Finite Elements," *Proc. ASCE, J. Engrg. Mech. Div.*, Vol. 92, No. EM1, pp. 111–120.
17. Taylor, D. W. (1948) *Fundamentals of Soil Mechanics*, John Wiley & Sons, New York.
18. Cedergren, H. (1967) *Seepage, Drainage and Flownets*, McGraw-Hill Book Co., New York.
19. Harr, M. E. (1962) *Groundwater and Seepage*, McGraw-Hill Book Co., New York.
20. Patton, F. D. and Hendron, A. J., Jr. (1974) "General Report on Mass Movements," *Proc. 2nd Intl. Cong.*, Intl. Assoc. Engrg. Geol., São Paulo, V-GR 1 to 57.
21. Hubbert, M. K. (1940) "The Theory of Groundwater Motion," *J. of Geology*, Vol. 48, November–December, pp. 785–944.
22. Hoek, E. and Bray, J. W. (1977) *Rock Slope Engineering*, 2d ed., Inst. of Mining and Metallurgy, London.
23. Mansur, C. I. and Kaufman, R. I. (1962) "Dewatering," *Foundation Engineering*, G. A. Leonards, McGraw-Hill Book Co., New York, Chap. 3.
24. Jaeger, C. (1972) *Rock Mechanics and Engineering*, Cambridge Univ. Press, Cambridge, England.
25. Penman, A. D. M. (1977) "The Failure of Teton Dam," *Ground Engineering*, September, pp. 18–27.

26. *Civil Engineering* (1977) "Teton Dam Failure," ASCE, August, pp. 56–61.

27. Fecker, E. (1980) "The Influence of Jointing on the Failure of Teton Dam: A Review and Commentary," *Bull. Intl. Assoc. Engrg. Geol.*, No. 21, pp. 232–238.

28. Sharp, J. C., Maini, Y. N. and Brekke, T. L. (1973) "Evaluation of the Hydraulic Properties of Rock Masses," *New Horizons in Rock Mechanics, Proc. 14th Symp. on Rock Mechs.*, ASCE, New York, pp. 481–500.

29. Cording, E. J., Hendron, A. J., Jr., Hansmire, W. H., Mahar, J. W., MacPherson, H. H., Jones, R. A. and O'Rourke, T. D. (1975) *Methods for Geotechnical Observations and Instrumentation in Tunneling*, Dept. Civ. Engrg., Univ. of Illinois, Urbana, Vol. I and II (Appendices).

30. Johnson, E. E. (1963) "Basic Principles in Water Well Design," *The Johnson National Drillers Journal*, Vol. 35, No. 5, St. Paul, Minn.

31. USBR (1973*b*) *Concrete Manual*, U.S. Bureau of Reclamation, U.S. Govt. Printing Office, Washington, D.C.

32. Chiossi, N. J. (1975) *Geologia Aplicada á Engenharia*, Gremio Politécnico, São Paulo.

32*a*. Koerner, R. M., McCabe, W. M. and Baldivieso, L. F. (1981) "Acoustic Emission Monitoring of Seepage," *Proc. ASCE., J. Geotech. Engrg. Div.*, Vol. 107, No. GT4, April, pp. 521–526.

33. D'Appolonia, D. J. (1980) "Soil-Bentonite Slurry Trench Cutoffs," *Proc. ASCE, J. Geotech. Engrg. Div.*, ASCE, Vol. 106, No. GT4, April, pp. 399–417.

34. Miller, E. A. and Salzman, G. S. (1980) "Value Engineering Saves Dam Project," *Civil Engineering*, ASCE, August, pp. 51–55.

35. Kapp, M. S. (1969) "Slurry Trench Construction for Basement Wall of World Trade Center," *Civil Engineering*, ASCE, April.

36. Lancaster-Jones, P. F. F. (1969) "Methods of Improving the Properties of Rock Masses," *Rock Mechanics in Engineering Practice*, Stagg and Zienkiewicz, eds., John Wiley & Sons, New York, Chap. 12.

37. NAVFAC (1971) *Design Manual DM 7, Soil Mechanics, Foundations and Earth Structures*, Naval Facilities Engineering Command, Alexandria, Va., March.

38. Tschebotarioff, G. P. (1973) *Foundations, Retaining and Earth Structures*, McGraw-Hill Book Co., New York.

39. Christensen, D. L. (1974) "Unusual Foundation Developments and Corrective Action Taken," *Foundations for Dams*, ASCE, New York, pp. 343–370.

40. Bennett, P. T. (1952) "Seepage Control Soil Mechanics Design," *Engineering Manual for Civil Works Construction*, Part CXIX, Chap. 1, February, U.S. Army Corps of Engineering.

41. Cedergren, H. (1975) "Drainage and Dewatering," *Foundation Engineering Handbook*, Winterkorn and Fang, eds., Van Nostrand Reinhold Co., New York, Chap. 6.

42. Terzaghi, K. And Peck, R. B. (1967) *Soil Mechanics in Engineering Practice*, John Wiley & Sons, New York.

43. USBR (1974) *Earth Manual*, U.S. Bureau of Reclamation, U.S. Govt. Printing Office, Washington, D.C.

44. Musso, G. (1979) "Jacked Pipe Provides Roof for Underground Construction in Busy Urban Area," *Civil Engineering*, ASCE, November, pp. 79–82.

45. NCE (1979) "Tunneling," *NCE International*, Inst. of Civ. Engrs., London, February, pp. 28–31.

46. Jones, J. S. and Brown, R. E. (1978) "Temporary Tunnel Support by Artificial Ground Freezing," *Proc. ASCE, J. Geotech. Engrg. Div.*, Vol. 104, No. GT10, October, pp. 1257–1276.

47. Hunt, R. E. (1974) "Engineering Geology and Urban Planning for the Cañada del Oro Area, Tucson, Arizona, USA," *Proc. 2d Intl. Cong.*, Intl. Assoc. Engrg. Geol., São Paulo, Vol. I, Theme III-1.

48. Ward (1975) "Environmental Protection Study, Tucson, Arizona," Spec. Pub., J. S. Ward & Assocs., Caldwell, New Jersey.

49. ENR (1980) "Flood Control Plan Recycles Water," *Engineering News-Record*, April 24, pp. 28–29.

50. Bishop, H. F. (1978) "Flood Control Planning in Albuquerque," *Civil Engineering*, ASCE, April, pp. 74–76.

**BIBLIOGRAPHY**

ASCE (1977) *Geotechnical Practice for Disposal of Solid Waste Materials, Proc. ASCE*, Spec. Conf., Univ. of Michigan, June.

Cyanamid (1975) "All about Cyanamid AM-9 Chemical Grout," Amer. Cyanamid Co., Wayne, New Jersey.

Giefer, G. J. and Todd, D. K. (1972) *Water Publications by State Agencies*, Water Information Center, Port Washington, New York.

HRB (1973) "Soil Erosion: Causes and Mechanisms: Prevention and Control," Spec. Report 135, Highway Research Board, Washington, D.C.

Tolman, C. F. (1937) *Groundwater*, McGraw-Hill Book Co., New York.

Walton, W. C. (1970) *Groundwater Resource Evaluation*, McGraw-Hill Book Co., New York.

Xanthakos, P. P. (1979) *Slurry Walls*, McGraw-Hill Book Co., New York.

# PART III

# THE GEOLOGIC HAZARDS

**CHAPTERS**



**PURPOSE AND SCOPE**

Part III sets forth the basis for recognizing, understanding, and treating the geologic hazards to provide for safe and economical construction. It invokes general concepts rather than rigorous mathematical analyses.

**SIGNIFICANCE**

Geologic hazards represent substantial danger to humans and their works. The hazards may exist as a consequence of natural events, but often they are the result of human activities.

*Slope failures,* such as landslides and avalanches, can occur in almost any hilly or mountainous terrain, or offshore, often with a very frequent incidence of occurrence, and can be very destructive, at times catastrophic. The potential for failure is identifiable, and therefore forewarning is possible, but the actual time of occurrence is not predictable. Most slopes can be stabilized, but under some conditions failure cannot be prevented by reasonable means.

*Ground subsidence, collapse,* and *expansion* usually are the result of human activities and range from minor to major hazards, although loss of life is seldom great as a consequence. Their potential for occurrence evaluated on the basis of geologic conditions, is for the most part readily recognizable and they are therefore preventable or their consequences are avoidable.

*Earthquakes* represent the greatest hazard in terms of potential destruction and loss of life. They are the most difficult hazard to assess in terms of their probability of occurrence and magnitude as well as their vibrational characteristics, which must be known for aseismic design of structures. Recognition of the potential on the basis of geologic conditions and historical events provides the information for aseismic design.

*Floods* (Chap. 8) have a high frequency of occurrence, and under certain conditions can be anticipated. Protection is best provided by avoiding potential flood areas, which is not always practical. Prevention is possible under most conditions, but often at substantial costs.

# CHAPTER NINE

# LANDSLIDES AND OTHER SLOPE FAILURES

## 9.1 INTRODUCTION

### 9.1.1 GENERAL

#### Origins and Consequences of Slope Failures

Gravitational forces are always acting on a mass of soil or rock beneath a slope. As long as the strength of the mass is equal to or greater than the gravitational forces, the forces are in balance, the mass is in equilibrium and movement does not occur. An imbalance of forces results in slope failure and movement in the forms of creep, falls, slides, avalanches, or flows.

Slope failures can range from being a temporary nuisance by partially closing a roadway, to destroying structures, to being catastrophic and even burying cities.

#### Failure Oddities

- Some failures can be predicted, others cannot, although most hazardous conditions are recognizable.
- Some conditions can be analyzed mathematically, others cannot.
- Some conditions cannot be treated to make them stable; they should be avoided.
- Some forms occur without warning; many other forms give warning, most commonly in the form of early surface cracks.
- Some move slowly, others progressively or retrogressively, others at great velocities.
- Some move short distances; others can move for many kilometers.
- Some involve small blocks; others involve tremendous volumes.
- Some geologic formations have characteristic failure forms; others can fail in a variety of forms, often complex.

#### Chapter Objectives

The objectives of this chapter are to provide the basis for:

- Prediction of slope failures through the recognition of the geologic and other factors that govern failure
- Treatment of slopes that are potentially unstable and pose a danger to some existing development
- Design and construction of stable cut slopes and sidehill fills
- Stabilization of failed slopes

### 9.1.2 HAZARD RECOGNITION

#### General

Slope failures occur in many forms. There is a wide range in their predictability, rapidity of occurrence and movement, and ground area affected, all of which relate directly to the consequences of failure. Recognition permits the selection of some slope treatment which will either *avoid*, *eliminate*, or *reduce* the hazard.

**TABLE 9.1**
**A CLASSIFICATION OF SLOPE FAILURES**

| Type | Form | Definition |
|---|---|---|
| Falls | Free fall | Sudden dislodgment of single or multiple blocks of soil or rock which fall in free descent. |
| | Topple | Overturning of a rock block about a pivot point located below its center of gravity. |
| Slides | Rotational or slump | Relatively slow movement of an essentially coherent block (or blocks) of soil, rock, or soil-rock mixtures along some well-defined arc-shaped failure surface. |
| | Planar or translational | Slow to rapid movement of an essentially coherent block (or blocks) of soil or rock along some well-defined planar failure surface. |
| | *Subclasses* | |
| | Block glide | A single block moving along a planar surface. |
| | Wedges | Block or blocks moving along intersecting planar surfaces. |
| | Lateral spreading | A number of intact blocks moving as separate units with differing displacements. |
| | Debris slide | Soil-rock mixtures moving along a planar rock surface. |
| Avalanches | Rock or debris | Rapid to very rapid movement of an incoherent mass of rock or soil-rock debris wherein the original structure of the formation is no longer discernible, occurring along an ill-defined surface. |
| Flows | Debris<br>Sand<br>Silt<br>Mud<br>Soil | Soil or soil-rock debris moving as a viscous fluid or slurry, usually terminating at distances far beyond the failure zone; resulting from excessive pore pressures. (Subclassed according to material type.) |
| Creep | | Slow, imperceptible downslope movement of soil or soil-rock mixtures. |
| Solifluction | *See Art. 7.7.3* | Shallow portions of the regolith moving downslope at moderate to slow rates in Arctic to sub-Arctic climates during periods of thaw over a surface usually consisting of frozen ground. |
| Complex | | Involves combinations of the above, usually occurring as a change from one form to another during failure with one form predominant. |

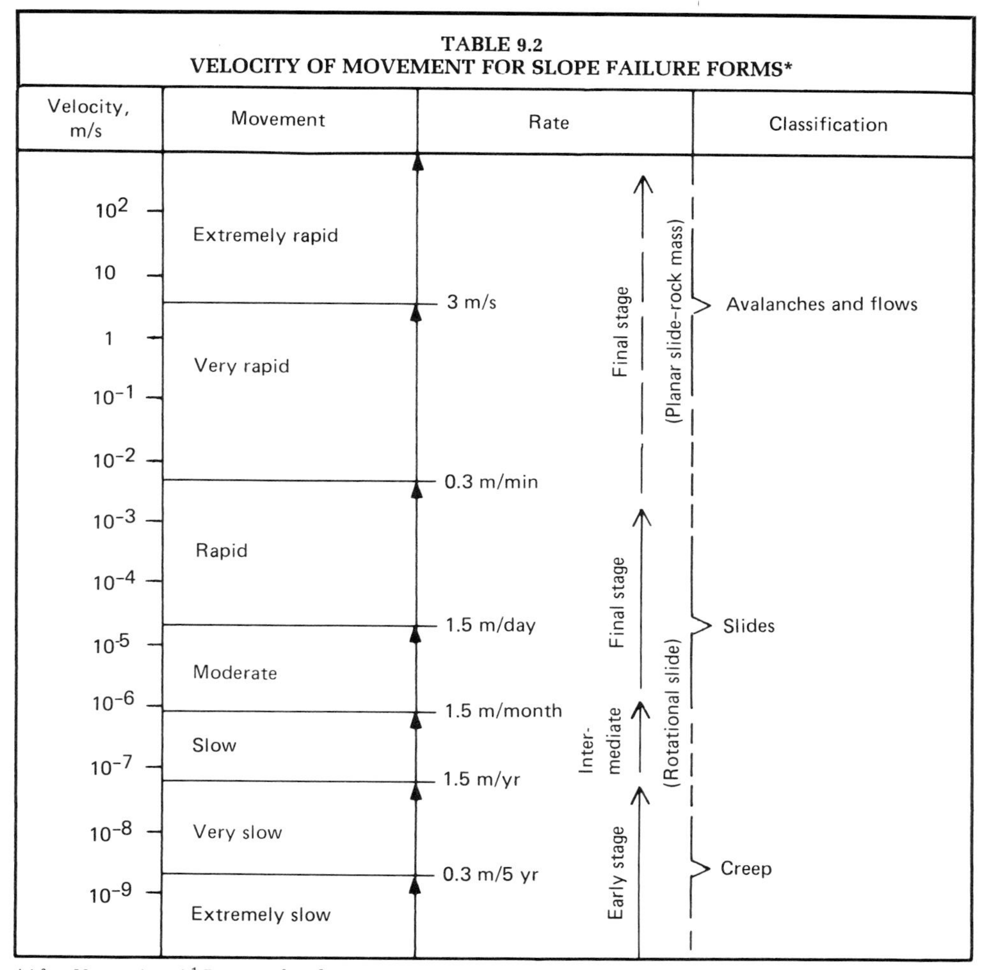

**TABLE 9.2**
**VELOCITY OF MOVEMENT FOR SLOPE FAILURE FORMS***

| Velocity, m/s | Movement | Rate | Classification |
|---|---|---|---|
| $10^2$, 10 | Extremely rapid | Final stage (Planar slide-rock mass) | |
| | | 3 m/s | Avalanches and flows |
| 1, $10^{-1}$, $10^{-2}$ | Very rapid | | |
| | | 0.3 m/min | |
| $10^{-3}$, $10^{-4}$ | Rapid | Final stage | |
| | | 1.5 m/day | Slides |
| $10^{-5}$, $10^{-6}$ | Moderate | (Rotational slide) | |
| | | 1.5 m/month | |
| $10^{-7}$ | Slow | Inter-mediate | |
| | | 1.5 m/yr | |
| $10^{-8}$ | Very slow | Early stage | |
| | | 0.3 m/5 yr | Creep |
| $10^{-9}$ | Extremely slow | | |

*After Varnes (1958).[1] Reprinted with permission of the Transportation Research Board.

Hazard recognition and successful treatment require thorough understanding of a number of factors including:

- Types and forms of slope failures (classification)
- Relationship between geologic conditions and the potential failure form
- Significance of slope activity, or amount and rate of movement
- Elements of slope stability
- Characteristics of slope failure forms *(see Art. 9.2)*
- Applicability of mathematical analysis *(see Art. 9.3)*

### Classification of Slope Failures

A classification of slope failures is given on Table 9.1.

Major factors of classification include:

- Movement form: fall, slide, slide-flow (avalanche), flow
- Failure surface form: arc-shaped, planar, irregular, ill-defined
- Mass coherency: coherent, with the original structure essentially intact although dislocated, or incoherent, with the original structure totally destroyed
- Constitution: single or multiple blocks, heterogeneous mass without blocks or a slurry

| TABLE 9.3 GEOLOGIC CONDITIONS AND TYPICAL FORMS OF SLOPE FAILURES | |
|---|---|
| **Geologic condition** | **Typical movement forms** |
| Rock masses: general | Falls and topples from support loss<br>Wedge failure along joints, or joints, shears, and bedding<br>Block glides along joints and shears<br>Planar slide along joints and shears<br>Multiplanar failure along joint sets<br>Dry rock flow |
| Metamorphic rocks | Slides along foliations |
| Sedimentary rocks | Weathering degree has strong affect |
| Horizontal beds | Rotational, or a general wedge through joints and along bedding planes |
| Dipping beds | Planar along bedding contacts; block glides on beds from joint separation |
| Marine shales, clay shales | Rotational, general wedge, or progressive through joints and along mylonite seams |
| Residual and colluvial soils | Depends on stratum thickness |
| Thick deposit | Rotational, often progressive |
| Thin deposit over rock | Debris slide, planar; debris avalanche or flow |
| Alluvial soils | Depends on soil type and structure |
| Cohesionless | Runs and flows |
| Cohesive | Rotational, or planar wedge |
| Stratified | Rotational, or wedges, becoming lateral spreading in fine-grained soils |
| Aeolian deposits | Variable |
| Sand dunes or sheets | Runs and flows |
| Loess | Block glides; flows during earthquakes |
| Glacial deposits | Variable |
| Till | Rotational |
| Stratified drift | Rotational |
| Lacustrine | Rotational becoming progressive |
| Marine | Rotational to progressive; rotational becoming lateral spreading; flows |

- Failure cause: tensile strength or shear strength exceeded along a failure surface, hydraulic excavation, or excessive pore pressures

Other factors to consider include:

- Mass displacement: amount of displacement from the failure zone, which can vary from slight to small, to very far. Blocks can move together with similar displacements, or separately with varying displacements.
- Material type: rock blocks or slabs, soil-rock mixtures (debris), sands, silts, blocks of overconsolidated clays, or mud (weak cohesive soils).
- Rate of movement during failure: varies from extremely slow and barely perceptible to extremely rapid as given on Table 9.2.

### Slope Failure Forms Related to Geologic Conditions

Various geologic conditions are associated with typical forms of slope failures as summarized on

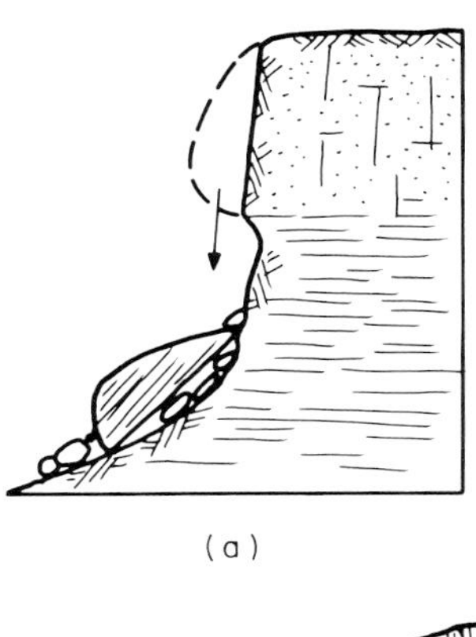

(a)

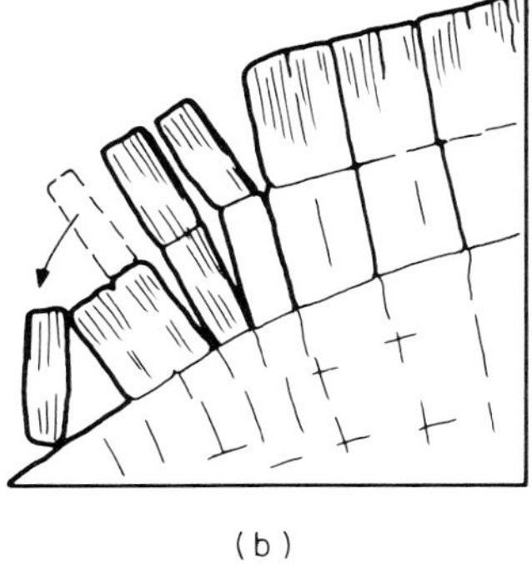

(b)

**FIG. 9.1** ***(Above)*** **Forms of falls in rock masses: (*a*) free fall and (*b*) toppling by overturning.**

**FIG. 9.2** ***(Right)*** **Slide forms in rock masses. (*a*) Rotational slide failed through joints and weak basal horizontal bed. (*b*) Translational sliding of blocks along a weak planar surface such as shale. (*c*) Planar slide failed along steeply dipping beds after cutting along lower slope. (*d*) Wedge failure scar. Failure occurred along intersecting joints and bedding planes when cut was made in obliquely dipping beds.** ***(See Arts. 9.2.3 and 9.2.4.)***

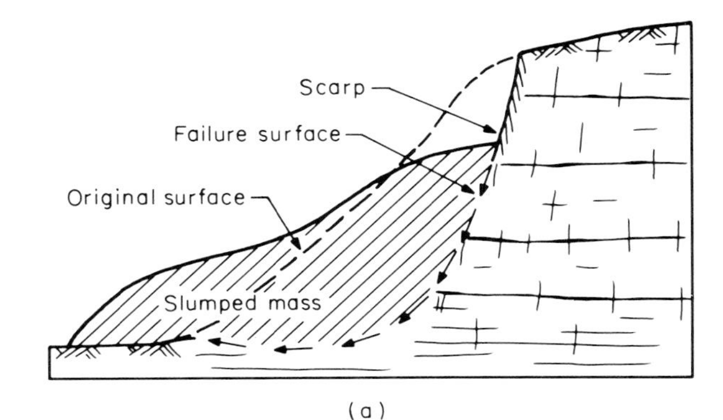


(a)

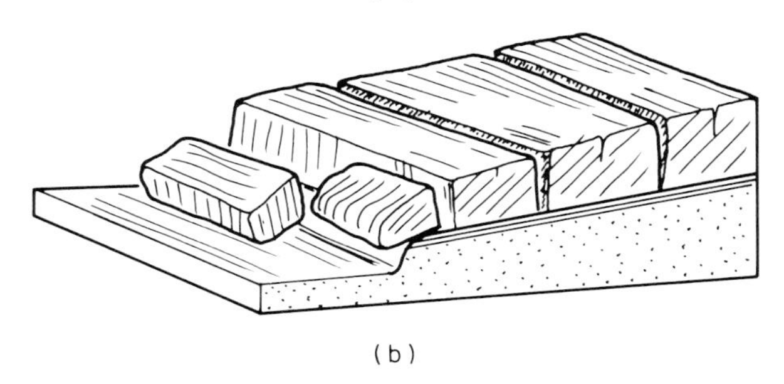

(b)

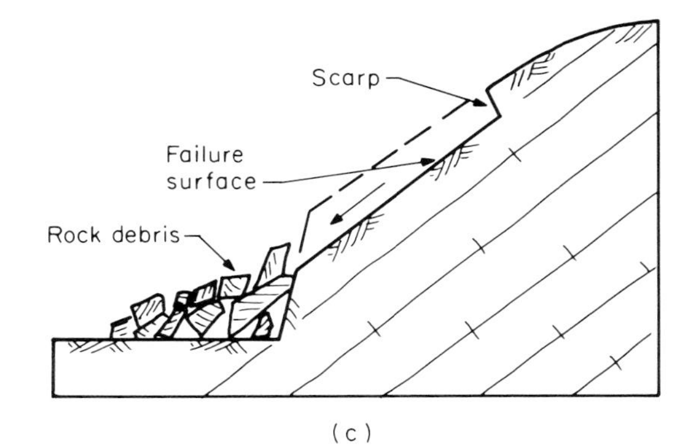


(c)

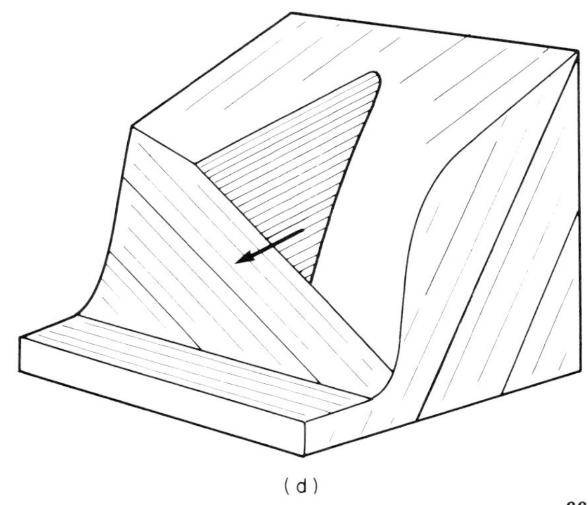

(d)

Table 9.3, which provides some bases for the anticipation of the failure hazard. Detailed descriptions of the various forms are given in Art. 9.2. Some forms of falls and slides in rock masses are illustrated on Figs. 9.1 and 9.2, slides in soil formations on Fig. 9.3, and avalanches and flows in rock, soil, and mixtures on Fig. 9.4.

### Slope Activity

Slope activity relates to the amount and rate of slope movement that occur. Some failure forms occur suddenly on stable slopes without warning, although many forms occur slowly through a number of stages. Failure implies only that movement has occurred, but not necessarily that it has terminated; therefore, it is necessary to

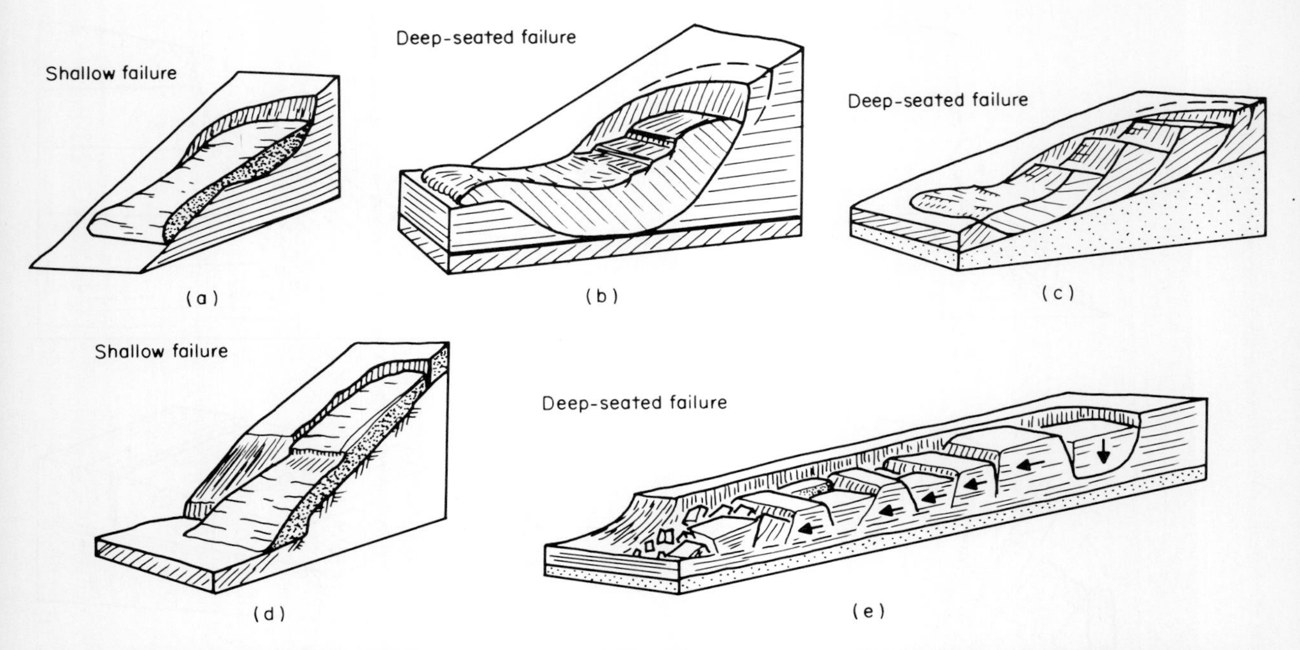


**FIG. 9.3 Slide forms in soil formations. (*a*) Single block failed along slope as a result of high groundwater level, or strength increase with depth in cohesive soils. (*b*) Single block in homogeneous cohesive soils failed below toe of slope because of either a stronger or a weaker soil boundary at base. (*c*) Failure of multiple blocks along the contact with strong material. (*d*) Planar slide or slump in thin soil layer over rock. Often called debris slides. Common in colluvium and develop readily into flows. (*e*) Failure by lateral spreading. Occurs in glaciomarine or glaciolacustrine soils. (Parts *a*, *b*, and *c* are rotational forms; parts *d* and *e* are planar or translational forms.)**

establish descriptive criteria for failure, or stability, in terms of stages. The amount and rate of movement vary with the failure stage for some failure forms.

*Slide forms of failure* may be classed by five stages of activity:

1. *Stable slope:* No movement has occurred in the past, or is occurring now.
2. *Early failure stage:* Creep occurs, with or without the development of tension cracks on the surface (*see Fig. 9.22*). Movement velocities may range from 0.1 to 1.6 m/year (*see Table 9.2*). Common to many slopes.
3. *Intermediate failure stage:* Progressive slumps and scarps begin to form during rotational slides, and blocks begin to separate during planar slides, as tension cracks grow in width and depth. Movement velocities may range up to about 5 cm/day, accelerating during rainy seasons and storms and diminishing during dry periods. Movement is affected also by flooding, high tides, and earthquake loadings. The slope is essentially intact, and may remain in this condition for many years (*see Fig. 9.88*).
4. *Partial total failure:* A major block or portion of the unstable mass has moved to a temporary location leaving a large scarp on the slope (*see Fig. 9.89*).
5. *Complete failure:* The entire unstable mass has displaced to its final location (*see Fig. 9.91*), moving rapidly at rates of about 1 m/min for the case of rotational slides. Planar slides in rock masses commonly reach velocities of 20 to 60 km/hr [Banks and Strohn (1974)[2]]. Large planar slides in rock masses can achieve tremendous velocities, at times in the order of 300 km/hr, as has been computed for the Vaiont slide (*see Art. 9.2.3*). Habib (1975)[3] considers these high velocities to be the result of movement of the rock mass over a cushion of water that negates all frictional resistance. The cushion is caused by heat, generated by shearing forces, which vaporizes the pore water. Such velocities are the major reason for the often disastrous effects of planar rock

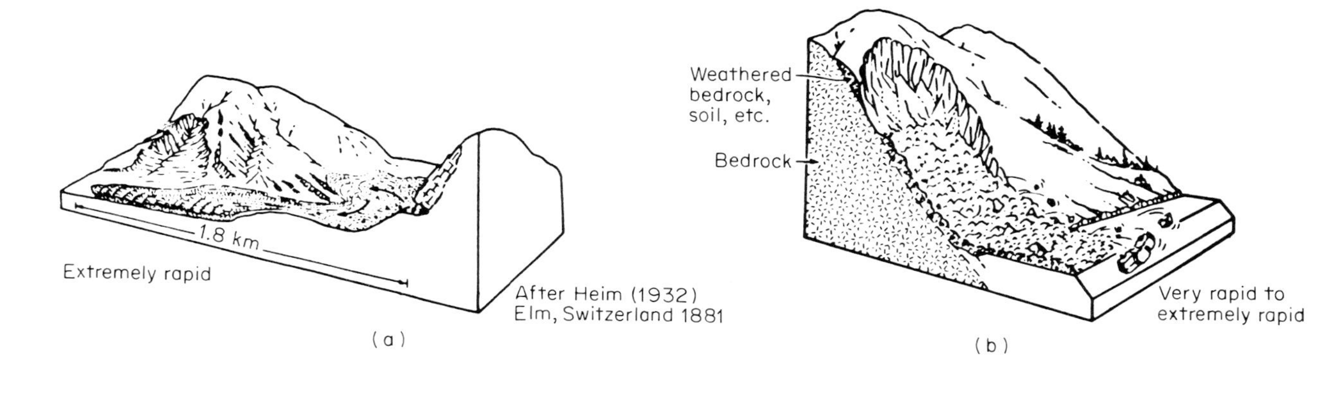


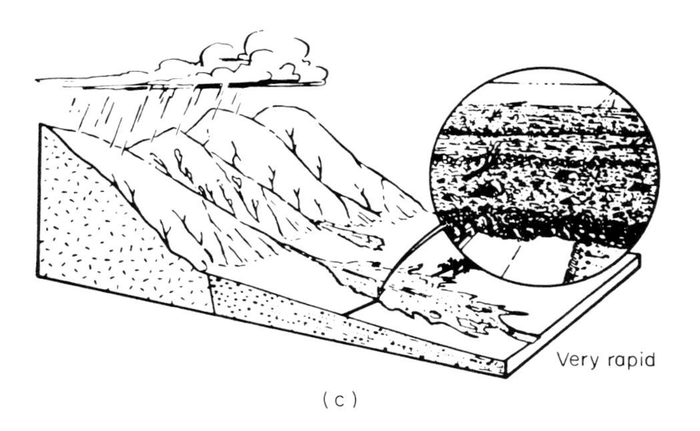


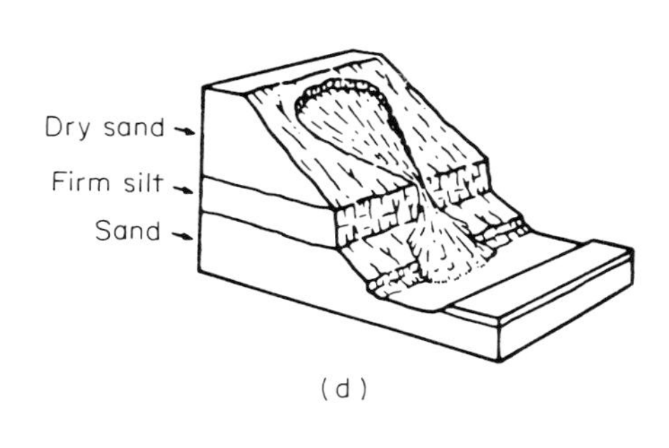


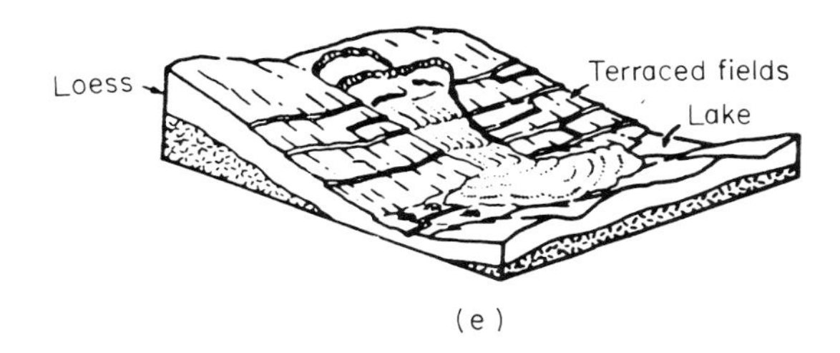


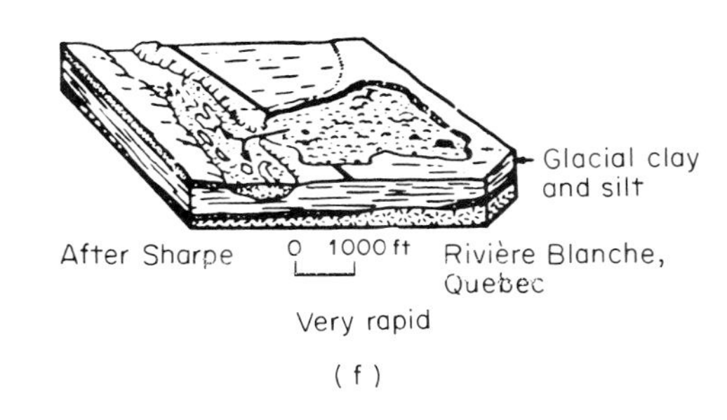


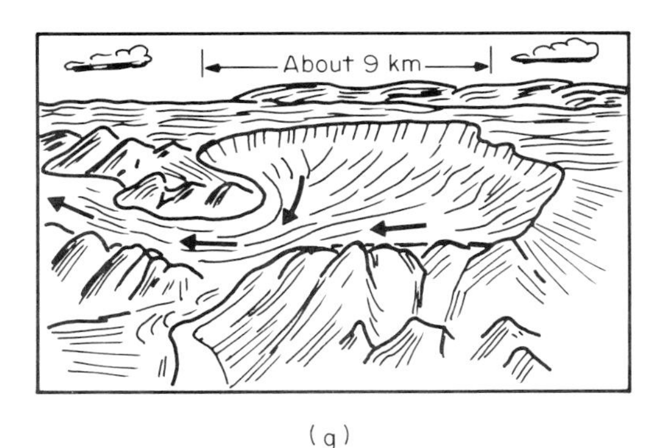


**FIG. 9.4 Avalanches and flows in rock, debris, and soil. (*a*) Rock fragment flow or rockfall avalanche. [This type of movement occurs only when large rockfalls and rockslides attain unusual velocity. Extremely rapid (more than 130 ft/s at Elm, Switzerland.)] (*b*) Debris avalanche. (*c*) Debris flow. (*d*) Sand run: rapid to very rapid. (*e*) Dry loess flow caused by earthquake (Kansu Province, China, 1920). Extremely rapid movement. (*f*) Soil or mud flow. (*g*) Achacolla mud flow (La Paz, Bolivia). Huge mass of lacustrine soils slipped off the altiplano and flowed downstream for 25 km *(see Fig. 9.54)*.** [*Parts a–f from Varnes (1958).*[1] *Reprinted with permission of the Transportation Research Board.*]

slides. Slide failures are usually progressive, and can develop into failure by lateral spreading, as well as into avalanches and flows.

*Avalanches and flows* may develop from slide forms as mentioned above, or may undergo an early stage, but total final failure often occurs suddenly without warning on a previously stable slope as the result of some major event such as a very large rainfall or an earthquake. Velocities are usually very rapid to extremely rapid as given on Table 9.2.

*Falls* may occur suddenly, but often go through an early stage evidenced by the opening of tension cracks.

**Deposition**

Colluvium is the residue of soil materials composing the soil mass, generally resulting from complete failure. Its characteristics are described in Art. 7.3.2.

9.1.3 RATING THE HAZARD AND THE RISK

**Significance**

An existing or potential slope failure must be evaluated in terms of the degree of the hazard and the risk when plans for the treatment are formulated (*see Art. 9.4*). Some conditions cannot be improved and should be avoided; in most, however, the hazard can be eliminated or reduced.

*Hazard* refers to the slope failure itself in terms of its potential magnitude and probability of occurrence.

*Risk* refers to the consequences of failure on human activities.

**Hazard Degree**

The rating basis for hazard is the potential magnitude and probability of failure.

- *Magnitude* refers to the volume of material which may fail, the velocity of movement during failure, and the land area which may be affected. It depends very much on the form of failure as related to geology, topography, and weather conditions.
- *Probability* is related in a general manner to weather, seismic activity, changes in slope inclinations, and other transient factors.

*No Hazard*

A slope is not likely to undergo failure under any foreseeable circumstances.

*Low Hazard*

A slope may undergo total failure (as compared with partial failure) under extremely adverse conditions which have a low probability of occurrence (for examples, a 1000-year storm, or a high-magnitude earthquake in an area of low seismicity), or the potential failure volume and area affected are small even though the probability of occurrence is high.

*Moderate Hazard*

A slope probably will fail under severe conditions which can be expected to occur at some future time, and a relatively large volume of material is likely to be involved. Movement will be relatively slow and the area affected will include the failure zone and a limited zone downslope (moderate displacement).

*High Hazard*

A slope is almost certain to undergo total failure in the near future under normal adverse conditions and will involve a large to very large volume of materials, or a slope may fail under severe conditions (moderate probability), but the potential volume and area affected are enormous, and the velocity of movement very high.

**Risk Degree**

The *rating basis* for risk is the type of project and the consequences of failure.

*No Risk*

The slope failure will not affect human activities.

*Low Risk*

An inconvenience easily corrected, not directly endangering lives or property, such as a single block of rock of small size causing blockage of a small portion of roadway and easily avoided and removed.

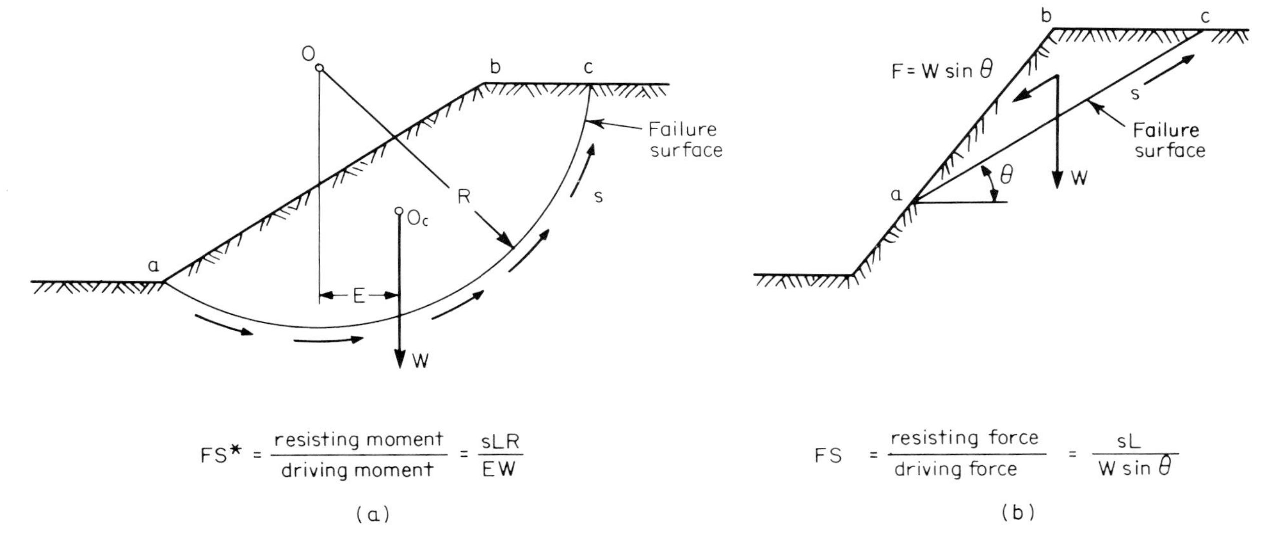


**FIG. 9.5 Forces acting on cylindrical and planar failure surfaces. (*a*) Rotational cylindrical failure surface with length *L*. Safety factor against sliding, FS. (*b*) Simple wedge failure on planar surface with length *L*. (*Note that the expression for FS is generally considered unsatisfactory—see text.)**

*Moderate Risk*

A more severe inconvenience, corrected with some effort, but not usually directly endangering lives or structures when it occurs, such as a debris slide entering one lane of a roadway and causing partial closure for a brief period until it is removed.

*High Risk*

Complete loss of a roadway or important structure, or complete closure of a roadway for some period of time, but lives are not necessarily endangered during the failure.

*Very High Risk*

Lives are endangered at the time of failure by, for example, the destruction of inhabited structures or a railroad when there is no time for a warning.

### 9.1.4 ELEMENTS OF SLOPE STABILITY

#### General

*Dependent Variables*

Stated simply, slope failures are the result of gravitational forces acting on a mass which can creep slowly, fall freely, slide along some failure surface, or flow as a slurry. Stability can depend on a number of complex variables, which can be placed into four general categories as follows:

1. Topography—in terms of slope inclination and height
2. Geology—in terms of material structure and strength
3. Weather—in terms of seepage forces and runoff quantity and velocity
4. Seismic activity—as it affects inertial and seepage forces.

It is important to note that, although topography and geology are usually constant factors, there are situations where they are transient.

*Mechanics of Sliding Masses*

Masses that fail by sliding along some well-defined surface, moving as a single unit (as opposed to progressive failure or failure by lateral spreading), are the only slope failure form that can be analyzed mathematically in the present state of the art (*see Art. 9.3*). The diagrams given as Fig. 9.5 illustrate the concept of failure that occurs when driving forces exceed resisting forces.

In the figure, the weight of mass $W$ bounded by slice *abc* [in (*a*) acted on by the lever arm $E$; in (*b*) a function of the inclination of the failure surface] causing the driving force, is resisted by the shear strength $s$ mobilized along the failure surface of length $L$ [in case (*a*) acted on at *a* by lever arm $R$). The expression for factor of safety FS given on the figure is commonly encountered but is generally considered unsatisfactory because the resisting moment and the driving moment in (a) are ambiguous. For example, the portion of the rotating mass to the left of the center of rotation could be considered as part of the resisting moment. For this reason, FS is usually defined as:

$$FS = \frac{\text{shearing strength available along sliding surface}}{\text{shearing stresses tending to produce failure along surface}} \qquad (9.1)$$

The four major factors influencing slope stability are illustrated on Fig. 9.6 and described in the following sections.

## Slope Geometry (Fig. 9.6*a*)

*Significance*

*Driving forces* and *runoff* are increased as slope inclination and height increase.

*Runoff quantity* and *velocity* are related directly to amount of erosion, and under severe conditions cause "hydraulic excavation," resulting in avalanches and flows (see discussion of runoff below.)

*Inclination*

Geologic formations often have characteristic inclinations at which they are barely stable in the natural state, for example, residual soils at 30 to 40° colluvium at 10 to 20°, clay shales at 8 to 15°, and loess, which often stands vertical to substantial heights.

Inclination is increased by:

- *Cutting* during construction, which should be controlled by analysis and judgment
- *Erosion*, as a result of undercutting at the slope toe by wave or stream activity, of seepage exiting from the slope face, or of removal of materials by downslope runoff. All these are significant natural events
- *Tectonic movements* in mountainous terrain, a very subtle and long-term activity which provides a possible explanation for the very large failures that occur from time to time and for which no other single explanation appears reasonable. An example is the disastrous rock slide at Goldau, Switzerland (*see Art. 9.2.3*)

*Slope Height*

Slope height is *increased* by filling at the top, erosion below the toe, or tectonic activity. It is *decreased* by excavation and erosion at the top, or by placing a berm at the bottom. The driving forces are affected in failure forms where the *limited* slope condition applies (*see Fig. 9.5*).

## Material Structure (Fig. 9.6*b*)

*Significance*

Material structure influences the failure form and the location and shape of the potential failure surface, and can be considered in two broad categories: uniform and nonuniform.

*Uniform Materials*

Uniform materials consist of a single type of soil or rock, essentially intact and free of discontinuities. From the aspect of slope stability, they are restricted to certain soil formations. *Rotational failure* is normal; the depth of the failure surface depends on the location of the phreatic surface and on the variation of strength with depth. *Progressive failures* are common, and falls and flows possible; flows are common in fine-grained granular soils.

*Nonuniform Materials*

Formations containing strata of various materials, and discontinuities represented by bedding, joints, shears, faults, foliations, and slickensides are considered nonuniform. The controlling factor for stability is the orientation and strength of the discontinuities, which represent surfaces of weakness in the slope.

*Planar slides* occur along the contacts of dipping beds of sedimentary rock and along joints, fault and other shear zones, slickensides, and foliations. Where a relatively thin deposit of soil overlies a sloping rock surface, progressive failure is likely and may develop into a debris ava-

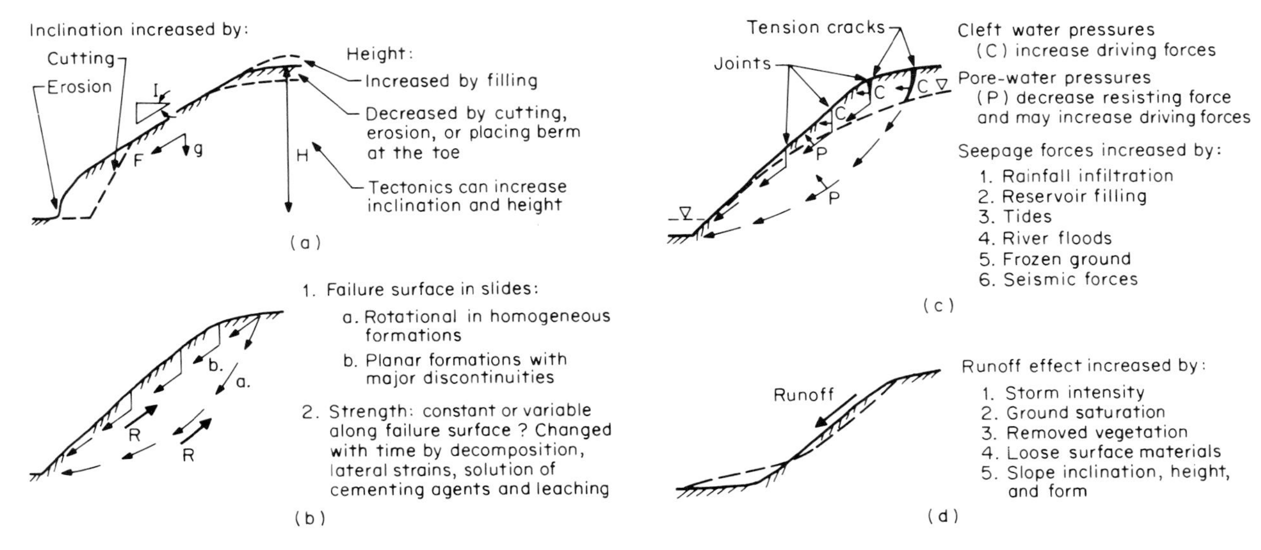


**FIG. 9.6 The major factors influencing slope stability: (*a*) increasing slope inclination and height increases the driving forces *F*; (*b*) geologic structure influences form and location of failure surface, material strength provides the resisting force *R*; (*c*) seepage forces reduce resisting forces along failure surface and increase driving forces in joints and tension cracks: (*d*) runoff quantity and velocity are major factors in erosion, avalanches, and flows.**

lanche. Along relatively flat-lying strata of weak material, failure can develop progressively in the form of lateral spreading, and can develop into a flow.

*Rotational slides* occur in horizontally bedded soil formations, and in certain rock formations such as clay shales and horizontally bedded sedimentary rocks.

*Falls* occur from lack of tensile strength across joints in overhanging or vertical rock masses.

*Changes* in the orientation of the discontinuities with respect to the slope face occur normally as a result of excavation, but can also be caused by tectonic activity. *Joint intensity* can be affected by construction blasting.

### Material Strength (Fig. 9.6*b*)

#### *Significance*

Material strength provides the resisting forces along a surface of sliding. It is often neither the value determined by testing, nor the constant value assumed in analysis.

#### *Variations along the Failure Surface*

Slopes normally fail at a range of strengths, varying from *peak* to *residual*, distributed along the failure surface as a function of the strains. Slopes that have undergone failure in the past will have strengths at or near residual, depending upon the time for restitution available since failure.

#### *Changes with Time*

*Chemical weathering* is significant in residual soils and along discontinuities in rock masses in humid climates, and provides another possible explanation for the sudden failure of rock-mass slopes that have remained stable for a very long period of time under a variety of weather and seismic conditions.

*Lateral strains* in a slope tend to reduce the peak strength toward the residual, a significant factor in the failure of slopes in clay shales and some overconsolidated clays containing recoverable strain energy [Bjerrum (1966)[4]], as well as in materials where slope movements have occurred.

**FIG. 9.7 Exposed rock surface remaining after runoff from torrential rains removed all vegetation, soil, and loose rock, depositing the debris mass at the toe of the slope (BR 116, km 56, Teresopolis, R.J., Brazil).**

*Solution* of cementing agents reduces strength.

*Leaching* of salts from marine clays increases their sensitivity, and therefore their susceptibility to liquefaction and flow [Bjerrum et al., (1969)[5]].

### Seepage Forces (Fig. 9.6*c*)

*Significance*

Seepage forces may reduce the resisting forces along the failure surface or increase the driving forces. *(See also Art. 8.3.3.)*

*Factors Causing Increased Seepage Forces*

In general, seepage forces are increased by rainfall infiltration or reservoir filling, which raises the water table or some other phreatic surface (perched water level); sudden drawdown of a flooded stream or an exceptionally high tide; melting of a frozen slope that had blocked seepage flow; and earthquake forces.

*Rising groundwater level* is a common cause. Variables affecting such a rise include: rainfall accumulation and increase in ground saturation for a given period, the intensity of a particular storm, the type and density of ground vegetation, drainage characteristics of the geologic materials, and the slope inclination and other features of topographic expression. Vegetation, geology, and topography influence the amount of infiltration that can occur, and careful evaluation of these factors often can provide the rea-

**FIG. 9.8 Creep ridges and erosion in residual soils after removal of vegetation (state of Rio de Janeiro, Brazil).**

sons for failure to occur at a particular location along a slope rather than at some other position during a given storm or weather occurrence.

*Earthquake forces (see Art. 11.3.4)* can cause an increase in pore-air pressures, as well as pore-water pressures. Such an increase is believed to be the cause of the devastating extent of the massive landslides in loess during the 1920 earthquake in Kansu, China, which left 200,000 or more dead.

### Runoff (Fig. 9.6*d*)

*Significance*

The quantity and velocity of runoff are major factors in erosion, and are a cause of avalanches and flows. Storm intensity, ground saturation, vegetation, frozen ground, the nature of the surficial geologic materials, and slope inclination and other topographic features affect runoff.

*Hydraulic Excavation*

Many avalanches and flows are caused by hydraulic excavation during intense storms, a common event in tropical and semiarid climates. Water moving downslope picks up soils loosened by seepage forces, and as the volume and velocity increase, the capacity to remove more soil and even boulders increases, eventually resulting in a heavy slurry which removes everything loose in its path as it flows violently downslope. The scar of a debris avalanche is illustrated in Fig. 9.7. Failure could hardly have been foreseen at that particular location along the slope, since conditions were relatively uniform.

## 9.2 SLOPE FAILURE FORM CHARACTERISTICS

### 9.2.1 CREEP

### General

*Creep* is the slow, imperceptible deformation of slope materials under low stress levels, which normally affects only the shallow portion of the slope, but can be deep-seated where a weak zone exists. It results from gravitational and seepage forces, and is indicative of conditions favorable for sliding.

### Recognition

Creep is characteristic of cohesive materials and soft rock masses on moderately steep to steep slopes. Its major surface features are parallel transverse slope ridges ("cow paths") as illustrated on Fig. 9.8, and tilted fence posts, poles and tree trunks as shown on Fig. 9.9. Straight tilted tree trunks indicate recent movement, whereas bent tree trunks indicate old continuing movement *(see Art. 9.5.2)*.

**FIG. 9.9 Bending and tilting tree trunks on a creeping hillside of varved clays (Tompkins Cove, New York). Straight tilted trunks indicate recent movement. Bent tree trunks indicate old, continuing movement.**

### 9.2.2 FALLS

#### General

*Falls* are the sudden failures of vertical or near-vertical slopes involving single or multiple blocks wherein the material descends essentially in free fall. *Toppling*, or overturning of rock blocks, often results in a fall.

In soils, falls are caused by the undercutting of slopes due to stream or wave erosion, usually assisted by seepage forces. In rock masses, falls result from undercutting by erosion or human excavation; increased pressures in joints from frost, water, or expanding materials; weathering along joints combined with seepage forces; and differential weathering wherein less-resistant beds remove support from stronger beds (*see Fig. 6.95*).

Their engineering significance lies normally in the occurrence of a single or a few blocks falling on a roadway, or occasionally encountering structures on slopes. At times, however, they can be massive and very destructive as shown on Fig. 9.10.

#### Recognition

Falls are characteristic of vertical to near-vertical slopes in weak to moderately strong soils and jointed rock masses as shown on Fig. 9.11. Before total failure some displacement often occurs, as indicated by tension cracks; after total failure, a fresh rock surface remains and talus debris accumulates at the toe (*see Fig. 6.70*).

### 9.2.3 PLANAR SLIDES IN ROCK MASSES

#### General

*Forms* of planar slides in rock masses include:

- Sliding as a unit, or units, downslope along one or more planar surfaces (also termed *translational slide*).

**FIG. 9.10 Rockfall destroyed a powerhouse (Niagara Falls, New York). Movement may actually have been in the form of a huge topple.** *(Photo by B. Benedict, 1956).*

**FIG. 9.11 Topple of a large block of granite. Adjacent block is breaking free along joints (Rio de Janeiro, Brazil).**

FIG. 9.12 *(Left)* Small granite block glide (Rio de Janeiro, Brazil).

FIG. 9.13 *(Below)* Exfoliation loosening granite slabs. Impact wall on right was constructed to deflect falling and sliding blocks from building on lower slopes behind trees. Damage from falls and slab slides is a serious problem in Rio de Janeiro.

- Block glide involving a single unit of relatively small size, as shown on Fig. 9.12.
- Slab glide involving a single unit of relatively small to large size as shown on Fig. 9.13.
- Massive rock slide involving multiple units, small to very large in size *(see Fig. 9.16)*.
- Wedge failures along intersecting planes involving single to multiple units, small to very large in size. A small wedge failure is illustrated on Fig. 9.14.

Block and slab slides can be destructive, but massive rock slides are often disastrous in mountainous regions and in many cases cannot be prevented, only avoided.

**Recognition**

Planar slides are characteristic of:

- Bedded formations of sedimentary rocks dipping downslope at an inclination similar to, or less than, the slope face. They result in block glides or massive rock slides (see Examples below)
- Faults, foliations, shears or joints forming long, continuous planes of weakness that intersect the face of the slope
- Intersecting joints result in wedge failures, which can be very large in open pit mines
- Jointed hard rock results in block glides
- Exfoliation in granite masses results in slab glides

*Surface features:*

- Before total failure, tension cracks often form during slight initial displacement.
- After total failure, blocks and slabs leave fresh scarps. Massive rockslides leave a long fresh surface denuded of vegetation, varying in width from narrow to wide and with a large debris mass at the toe of the slope and beyond. Since they can achieve very high velocities, they can terminate far beyond the toe.

**Examples of Major Failures**

*Goldau, Switzerland*

In September, 1806, a massive slab 1600 m long, 330 m wide, and 30 m thick broke loose and slid downslope during a heavy rainstorm, destroying a village and killing 457 persons. The slab consisted of Tertiary conglomerate with a calcareous binder resting on a 30° slope. At its interface with the underlying rock was a porous layer of weathered rock.

**FIG. 9.14 Scar of wedge failure in 40-year-old railroad cut in amphibolite gneiss (Tres Ranchos, Goias, Brazil).**

Three possible causes were offered by Terzaghi (1950)[6]:

1. The slope inclination gradually increased from tectonic movements.
2. The shearing resistance at the slab interface gradually decreased because of progressive weathering, or from removal of cementing material.
3. The piezometric head reached an unprecedented value during the rainstorm. Terzaghi was hesitant to accept this as the only cause,

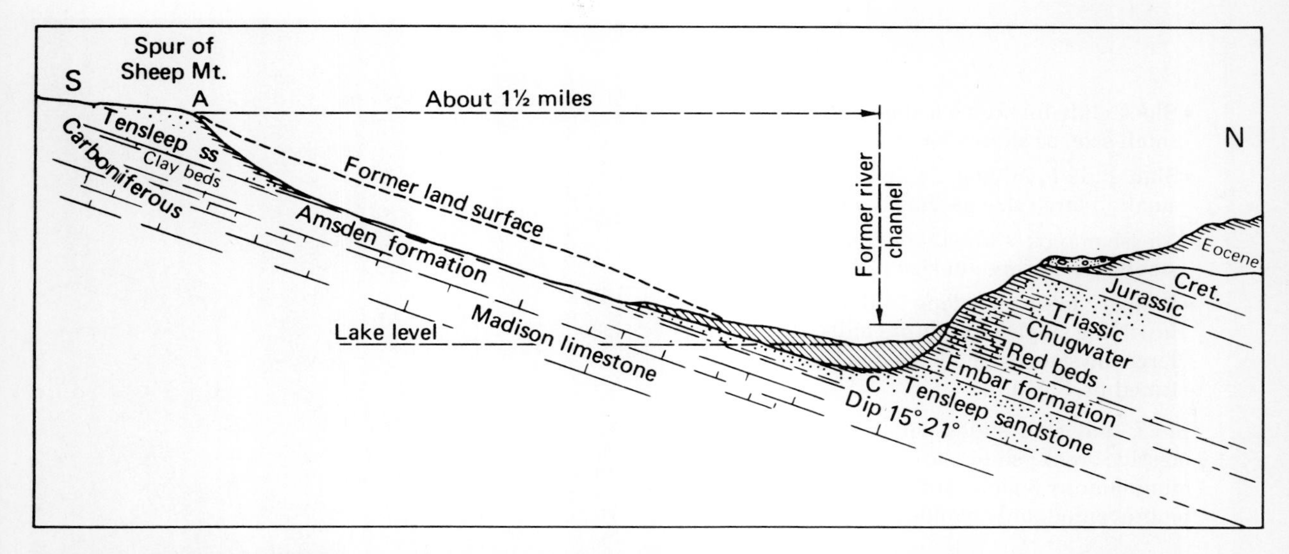


**FIG. 9.15 Geologic section after the landslide in the Gros Ventre River Valley, Wyoming.** [*From Alden (1928), as presented in Tank (1973).*[7]]

**FIG. 9.16 The scar of the Gros Ventre slide as seen from the Gros Ventre River in August 1977.**

since he considered it unlikely that in the entire geologic history of the region, there had not been a more severe storm. Therefore, he concluded that the slide resulted from two or more changing conditions.

*Gros Ventre, Wyoming*

On June 23, 1925, following heavy rains and melting snow, approximately 50 million $m^3$ slid in a few minutes down the mountainside along the Gros Ventre River near Grand Teton National Park in Wyoming. The debris formed a natural dam as high as 75 m which blocked the river, and resulted in a lake 5 km long. Almost 2 years later, in May 1927, water from heavy rains and melting snow filled the reservoir, overtopped the natural dam, eroded a large channel, and released flood waters which resulted in a number of deaths.

A geologic section is given on Fig. 9.15, and the slide scar which is still evident in 1977, 52 years later, is illustrated on Fig. 9.16. Failure occurred along clay layers in the carbonaceous Amsden formation, dipping downslope. It appears that water entered the joints and pores of the Tensleep sandstone saturated the clay seams, and reduced or eliminated the normal stresses.

*Vaiont, Italy*

On October 9, 1963, the worst dam disaster in history occurred when more than 300 million $m^3$ of rock slid into the reservoir formed by the world's highest thin-arch concrete dam causing a tremendous flood which overtopped the dam and flowed into the Piave River valley, taking some 2600 lives. The slide involved an area on the south side of the valley roughly 2.3 km in width and 1.3 km in length, as shown on Fig. 9.17. The natural slope was in the order of 20 to 30°.

A geologic section is given as Fig. 9.18. The valley had formed in the trough of a syncline, and the beds forming the limbs dipped downslope at inclinations a few degrees steeper than the slope. The south slope consisted of Jurassic sedimentary rocks, primarily limestones and marls occasionally interbedded with clay seams (bentonite clay at residual strength; Patton and Hendron, unpublished). Tectonic activity had caused regional folding, faulting, and fracturing of strata, and some of the tectonic stresses probably remained as residual stresses in the mass. Erosion of the valley caused some stress relief of the valley walls, resulting in numerous rebound joints that produced blocky masses. In addition, groundwater had attacked the limestone, leaving cavities and contributing to the generally unstable conditions [Kiersch (1965)[8]].

The slide history is given by Kiersch (1965)[8]. Large-scale slides had been common on the Vaiont valley slopes, and evidence of creep had been observed near the dam as early as 1960, when the dam was completed at its final height of 267 m. During the spring and summer of 1963, the slide area was creeping at the rate of 1 cm/week. Heavy rains occurred during August and September and movement accelerated to 1 cm/day. In mid-September movement accelerated to 20 to 30 cm/day, and on the day of failure, 3 weeks later, it was 80 cm/day. Since completion of the dam, the pool had been filled gradually and the elevation maintained at about 50 m below the crest or lower. During September, the pool rose at least 20 m higher, submerged the toe of the sliding mass, and caused the groundwater level to rise in the sliding mass. Collapse was sudden and the entire mass to a depth of 200 m broke loose and slid to the valley floor in 30 to 60 s, displacing the reservoir and causing a wave that rose as much as 140 m above reservoir level. The dam itself was only slightly damaged by the wall of water but was rendered useless.

Sliding was apparently occurring along the clay seams, but the actual collapse is believed to have been triggered by artesian pressures and the rising groundwater levels which decreased the effective weight of the sliding mass and, thereby, the resisting force at the toe.

### 9.2.4 ROTATIONAL SLIDES IN ROCK

#### General

In this slide form, a spoon-shaped mass begins failure by rotation along a cylindrical rupture surface; cracks appear at the head of the unstable area, and bulging appears at the toe as the mass slumps. At final failure the mass has displaced substantially, and a scarp remains at the head *(see Art. 9.2.5 for nomenclature)*. The major

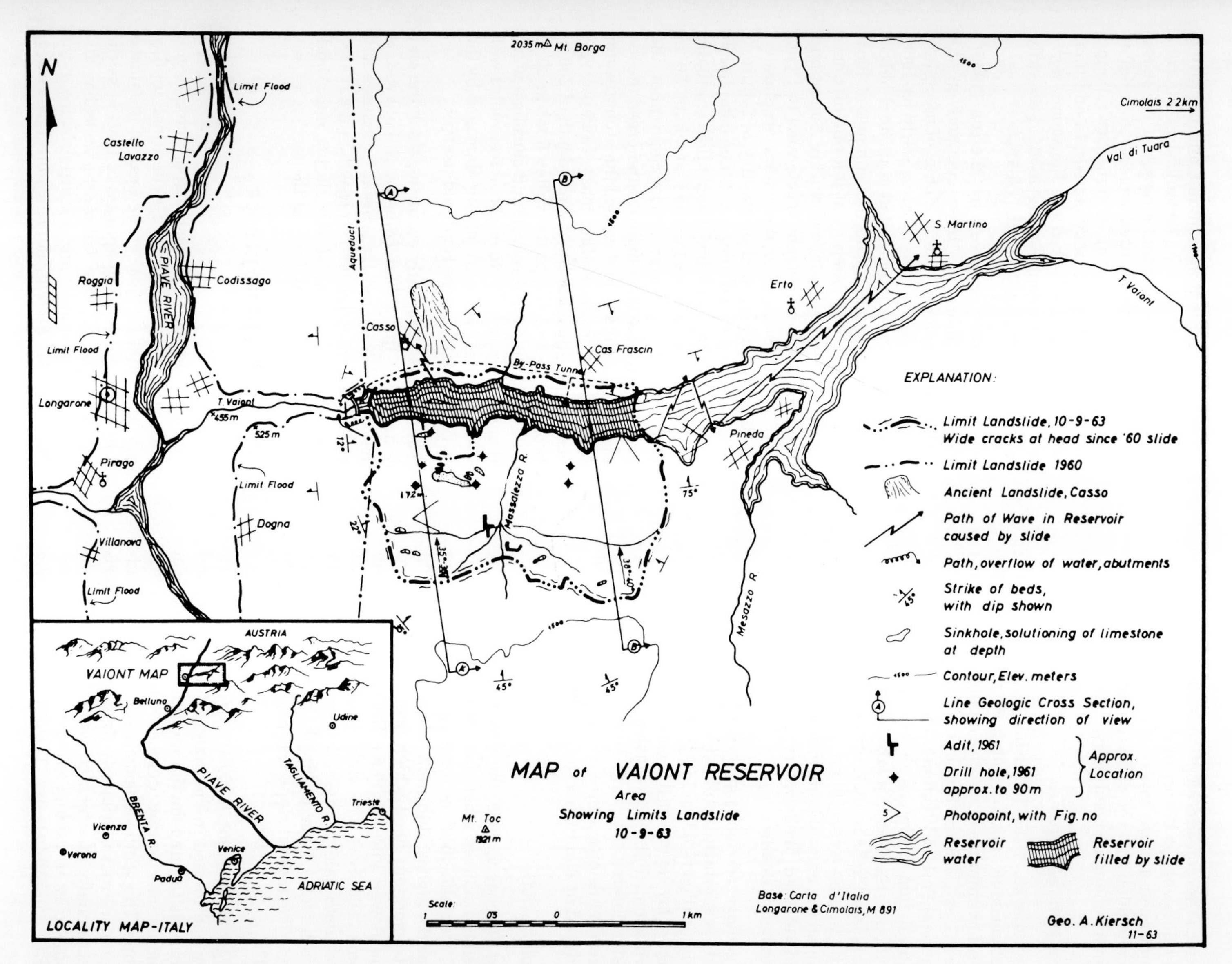


FIG. 9.17 Map of Vaiont Reservoir area and limits of slide. [From Kiersch (1965).[8]]

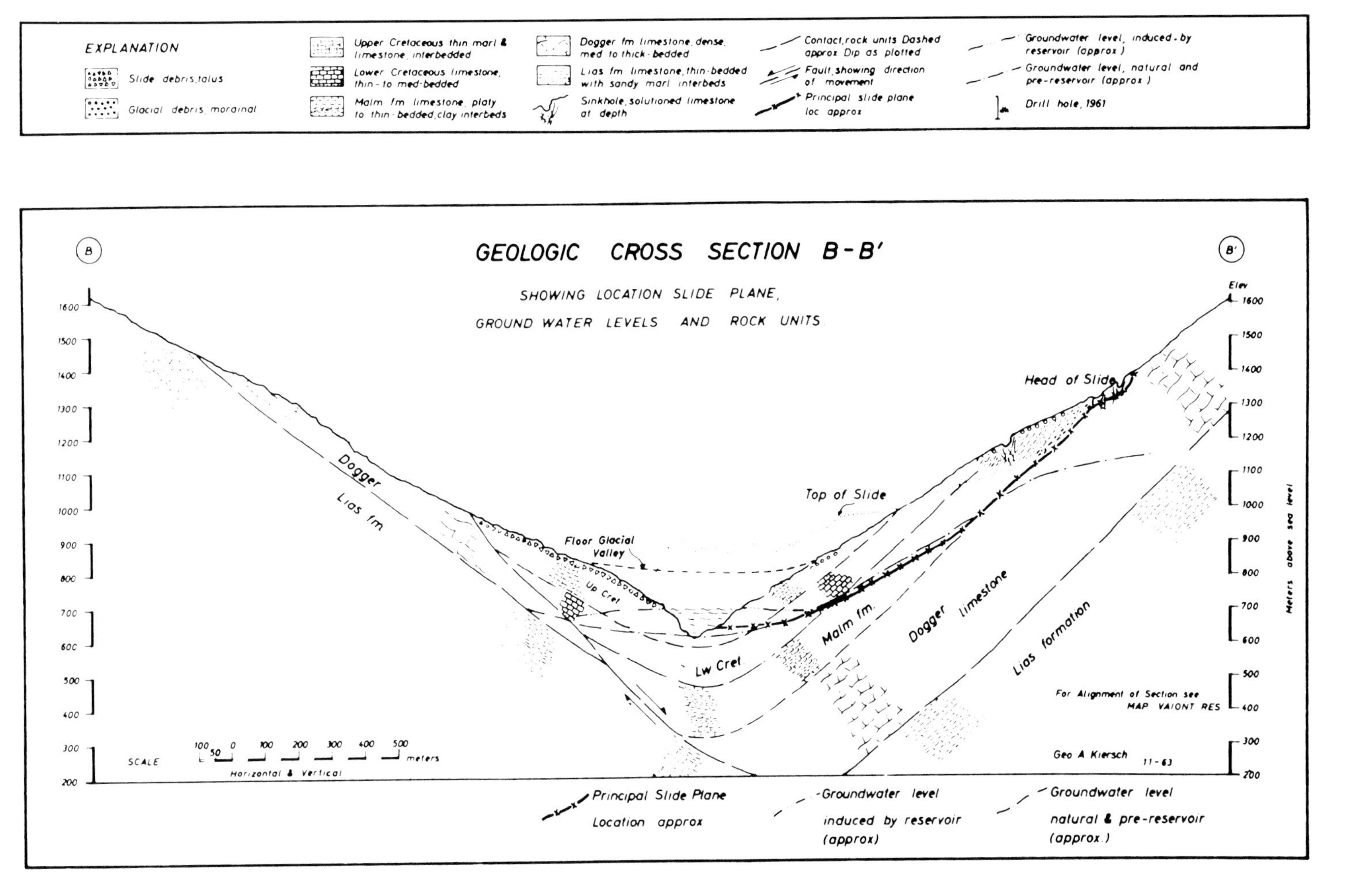


**FIG. 9.18 Geologic section through slide at Vaiont, Italy.** [From Kiersch (1965).[8]]

**FIG. 9.19 Slumping in road cut made in clay shales to heights of 10 m (Rio do Sol, Santa Catarina, Brazil).**

causes are an increase in slope inclination, weathering, and seepage forces.

**Recognition**

Rotational slides are essentially unknown in hard-rock formations, but are common in marine shales and other soft rocks, and in heavily jointed stratified sedimentary rocks with weak beds.

*Marine shales,* with their characteristic expansive properties and highly fractured structure, are very susceptible to slump failures (Fig. 9.19), and their wide geographic distribution makes such failures common *(see Art. 6.7.3).* Natural slope angles are low, about 8 to 15°, and stabilization is often difficult. Failure is often progressive and can develop into large moving masses *(see Art. 9.2.6).*

*Stratified sedimentary rocks* can on occasion result in large slides, and in humid climates slope failures can be common [Hamel (1980)[9]] (see example below).

*Surface features* before total failure are tension cracks; after total failure, a head scarp remains along with spoon-shaped slump topography *(see Art. 9.2.5).*

**Example of Major Failure**

*Event*

At the Brilliant cut, Pittsburgh, Pennsylvania on March 20, 1941, a rotational slide involving 110,000 $m^3$ of material displaced three sets of railroad tracks and caused a train to be derailed [Hamel (1972)[10]]. A plan of the slide area is given on Fig. 9.20*b*.

*Geological conditions* are illustrated on the section given as Fig. 9.20*a*. The basal stratum, zone 1, is described as "soft clay shale and indurated clay (a massive slickensided claystone)". The

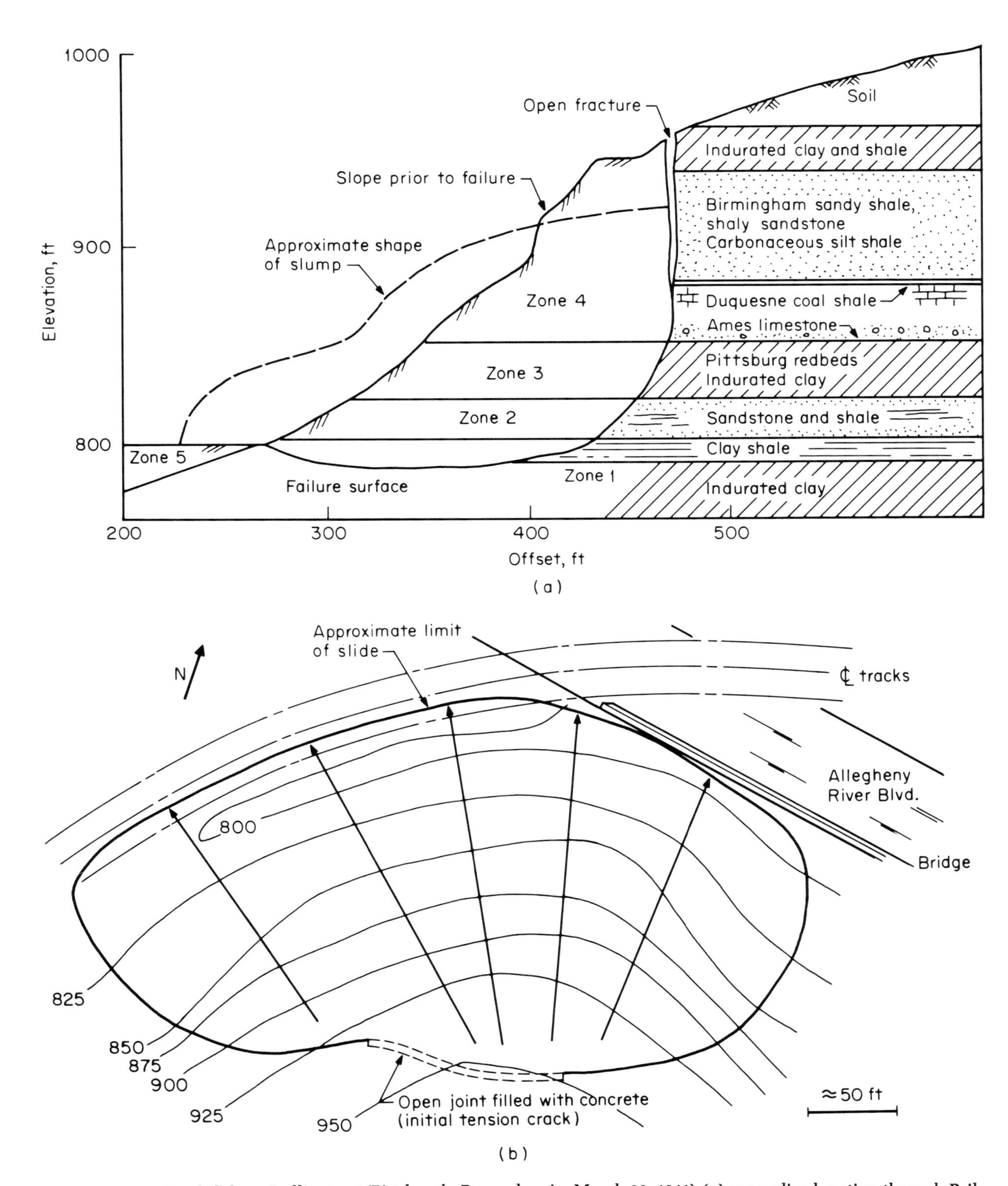


**FIG. 9.20 Rotational slide at Brilliant cut (Pittsburgh, Pennsylvania, March 20, 1941): (*a*) generalized section through Brilliant cut and (*b*) plan of slide area.** [*From Hamel (1972).*[10]]

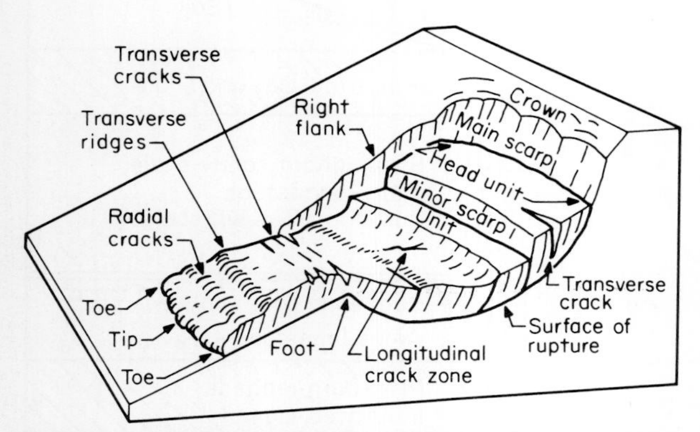


**FIG. 9.21 Characteristics and nomenclature of a rotational slide.** [*From Varnes (1958).*[1] *Reprinted with permission of the Transportation Research Board.*]

Birmingham shale of zone 4 is heavily jointed vertically.

*Slide History*

In the 1930s a large tension crack opened at the top of the slope. Sealing with concrete to prevent infiltration was unsuccessful in stopping movement and the crack continued to open over a period of several years. The rainfall that entered the slope through the vertical fractures normally drained from the slope along pervious horizontal beds. On the day of failure, which followed a week of rainfall, the horizontal passages were blocked with ice. Hamel (1972)[10] concluded that final failure was caused by water pressure in the mass, and the failure surface was largely defined by the existing crack at the top of the slope and the weak basal stratum.

### 9.2.5 ROTATIONAL SLIDES IN SOILS

#### General

A common form of sliding in soil formations is the rotation about some axis of one or more blocks bounded by a more or less cylindrical failure surface. The characteristics at total failure and descriptive nomenclature are given on Fig. 9.21.

The major causes are seepage forces and increased slope inclination, and relict structures in residual soils. Usually neither the volume of mass involved nor the distance moved is great; therefore, the consequences are seldom catastrophic although slump slides cause substantial damage to structures. If their warning signs are recognized they can usually be stabilized or corrected.

#### Recognition

*Occurrence*

Slump or rotation slides are characteristic of relatively thick deposits of cohesive soils without a major weakness plane to cause a planar failure. The depth of the failure surface varies with geology.

Deep-seated failure surfaces are common in soft to firm clays and glaciolacustrine, and glaciomarine soils. Deep to shallow failure surfaces are common in residual soils, depending on the strength increase with depth and relict rock defects. Relatively shallow failure surfaces are characteristic of colluvial soils.

*Surface Features*

During *early failure stages* tension cracks begin to form as shown on Figs. 9.22 and 9.23. After *partial failure*, in a progressive mode, the slope exists as a series of small slumps and scarps with a toe bulge as shown on Fig. 9.24, or it may rest with a single large scarp and a toe bulge as illustrated on Fig. 9.25. After *total failure*, surface features include various head scarps, concentric and deep tension cracks (Fig. 9.26), and a large mass of incoherent material at the toe (Fig. 9.27). *(See also Fig. 9.21.)*

*Slump landforms* remaining after total failures provide forewarning of generally unstable slope conditions. They include spoon-shaped irregular landforms, as seen from the air (Figs. 9.28 and 9.29) cylindrical scarps along terraces and water courses *(see Fig. 2.18)*, and hummocky and irregular surfaces, as seen from the ground (Figs. 9.30 and 9.31). In the stereo pair of aerial photos shown in Fig. 9.28, the slump failure mass has stabilized temporarily but probably will reactivate when higher-than-normal seasonal rainfall arrives. A small recent failure scar exists along the road in the center of the slide mass. The rounded features of the mass, resulting from weathering, and vegetation growth indicate that the slide is probably 10 to 15 years old, or more. In the photo, it can be seen that the steep high-

(a)

(b)

FIG. 9.22 (*a*) Small scarp along tension crack appears in photo (middle right). Small highway cut is far below to the left. Scarp appeared after soil was removed from small slump failure at the toe (BR 101, Santa Catarina, Brazil). Movement is in residual soil. If uncorrected, a very large failure will develop. (*b*) Tension crack in same slope found in another location.

**FIG. 9.23 Stereo pair of aerial photos showing tension cracks of incipient slides, such as at 1, along the California coast, a short distance east of Portuguese Bend.**

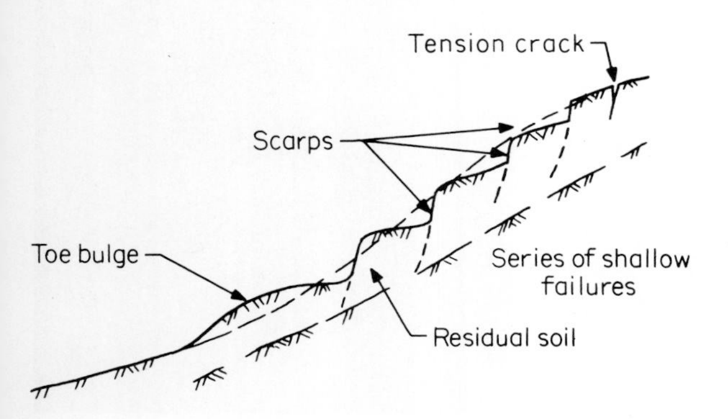


**FIG. 9.24 During the intermediate stage (during partial failure), residual soils often fail progressively, forming a series of slumps in tropical climates. Blocks move downhill during rainy periods and stabilize during dry periods.**

FIG. 9.25 Partial slump failure, showing single scarp and toe bulge, occurred after cutting in fine-grained glacial till (Mountainside, New Jersey). Cut continued to fail until stabilized by benching and installation of trench drain along the toe.

Fig. 9.26 Part of head scarp and concentric cracking in residual soil slope after total failure partially covered a roadway in Santa Catarina, Brazil (BR 101, near Joinville). Failure occurred after torrential rains in 1974.

**FIG. 9.27 Large rotational slide in residual soils at final failure (Rio Santos Highway, Brazil). Note 30-m-high head scarp (at left) and toe bulge (at right) which is partially covering the roadway. Total length from head to toe is about 200 m. History of this slide is described in Art. 9.3.3.**

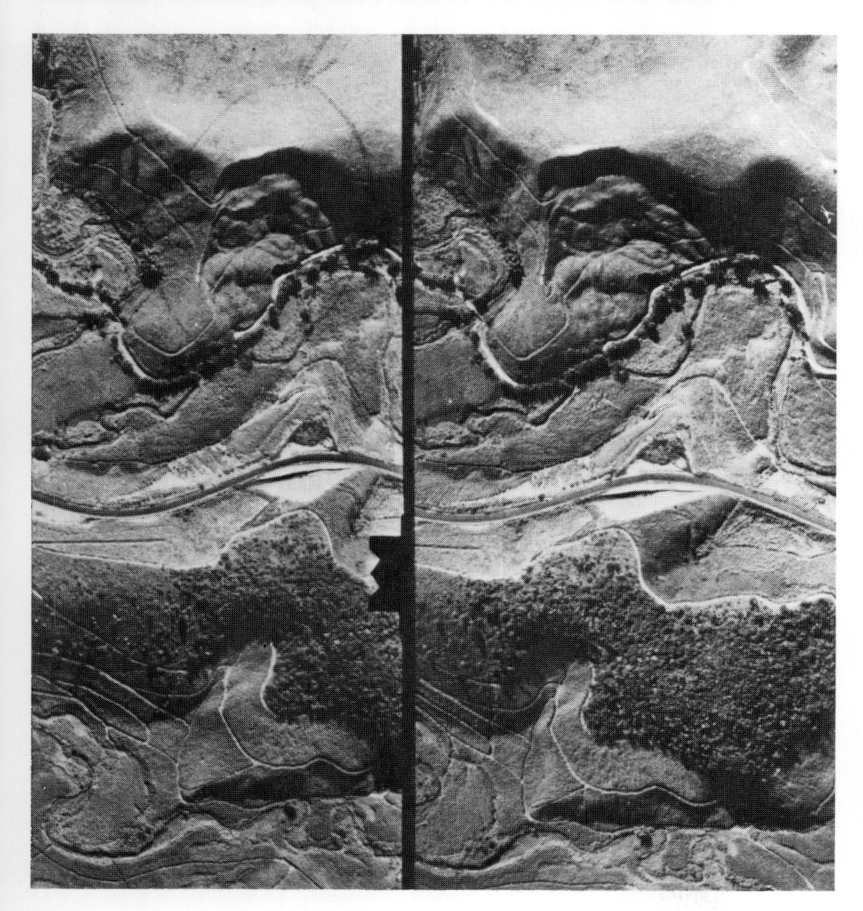

**FIG. 9.28 Stereo pair of aerial photos showing slump failure landform (Scale 1:8000). (*See also Fig. 2.12,* which shows a stereo pair of aerial photos of the general area at a scale of 1:40,000.)** [*From Hunt and Santiago (1976).*[11]]

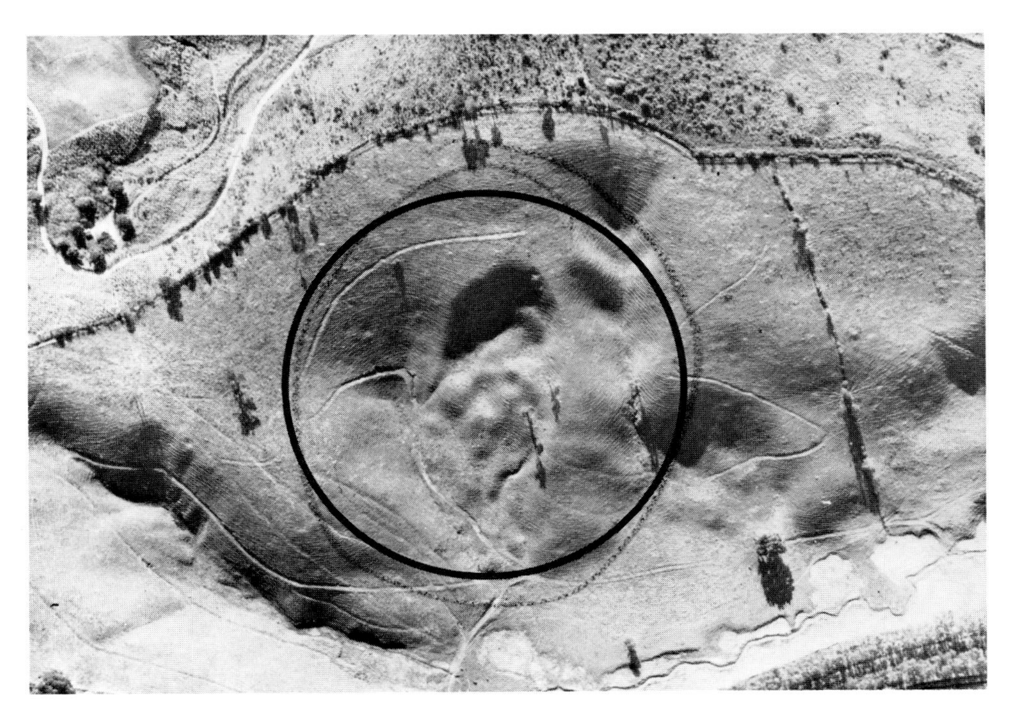

FIG. 9.29 Old slump scar in residual soils located near slide in Fig. 9.28. Photo is a portion of Fig. 2.13 enlarged to show the classical spoon shape.

FIG. 9.30 Slump landform in glaciolacustrine soils showing shallow slopes, creep ridges, and seepage (Barton River, Vermont). Trees in upper left are growing on slide area shown on the stereo pair given as Fig. 2.18. Slope failures are common in this region in the spring when the ground thaws and rains arrive.

**FIG. 9.31 Slump-slide landform (valley of the Rio Choqueyapa, La Paz, Bolivia). High center scarp in strong sands and gravels remains after failure of underlying lacustrine soils. Slopes were extremely unstable prior to channelization of the river, because of river erosion and flood stages. Grading of old slide in upper left is not arresting slope movements as evidenced by cracks in new highway retaining wall (not apparent in photo). Slope failures continue to occur from time to time throughout the valley. (Photo taken in 1973.)**

way cut on the opposite side of the valley appears stable, indicative of different geologic conditions. In general, the geology consists of residual soils derived from metamorphic rocks in a subtropical climate. In Fig. 9.29, an old slump scar in residual soils, weathering has strongly modified the features. In the photo, the tongue lobe at the intersection of the trails and the creep ridges are to be noted. The location is near the slide of Fig. 9.28, as shown on Fig. 2.12, a smaller-scale stereo pair of the area.

## 9.2.6 LATERAL SPREADING AND PROGRESSIVE FAILURE

### General

Failure by lateral spreading is a form of planar failure which occurs in both soil and rock masses. In general, the mass strains along a planar surface, such as shown on Fig. 9.3e, which represents a weak zone. Eventually, blocks progressively break free as movement retrogresses toward the head. The major causes are seepage forces and increased slope inclination and height.

Failure in this mode is essentially unpredictable by mathematical analysis, since one cannot know at what point the first tension crack will appear, forming the first block. Nevertheless, the conditions for potential instability are recognizable, since they are characteristic of certain soil and rock formations. Failure usually develops gradually, involving large volumes, but can be sudden and disastrous. Under certain conditions it is unavoidable and uncontrollable

**FIG. 9.32 Lateral spreading by block gliding on a very large scale (note road in foreground). Lacustrine colluvium in the valley of the Rio La Paz, La Paz, Bolivia. Located in central valley *(see Fig. 9.54)*.**

from the practical viewpoint, and under other conditions control is difficult at best.

### Recognition

The failure mode is particularly common in river valleys and occurs characteristically in stiff fissured clays, in clay shales, and in horizontal or slightly dipping strata with a continuous weak zone such as occur in glaciolacustrine and glaciomarine soils. Colluvium over gently sloping residual soils or rock also fails progressively in a form of lateral spreading.

*Surface features* are characterized during the *early stages* by tension cracks, although failure can be sudden under certain conditions such as earthquake loadings. During the *progressive failure* tension cracks open and scarps form, separating large blocks as shown on Fig. 9.32. The cracks can extend far beyond the slope face when a large mass goes into tension, even affecting surface structures as shown on Fig. 9.33. *Final failure* may not develop for many years, and when it occurs it may be in a form resembling a large slump slide, or it may develop into a flow with individual blocks floating in a highly disturbed mass, depending upon natural conditions as described in the examples below.

### Failure Examples

*Clay Shales: General*

Clay shales, particularly those of marine origin, are susceptible to several modes of slope failures as shown on Fig. 9.34, of which progressive failure involves the largest volumes and can be the most serious from the engineering viewpoint. Their characteristics are described in Art. 6.7.3, the most significant of which are their content of montmorillonite and their high degree of overstress. Excavation, either natural or human, results in lateral strains causing the strength along certain planes to be reduced to residual

**FIG. 9.33 One-year-old church being split in half from slope movements although located over 1 km from the slope shown on Fig. 9.31 (La Paz, Bolivia, 1972).**

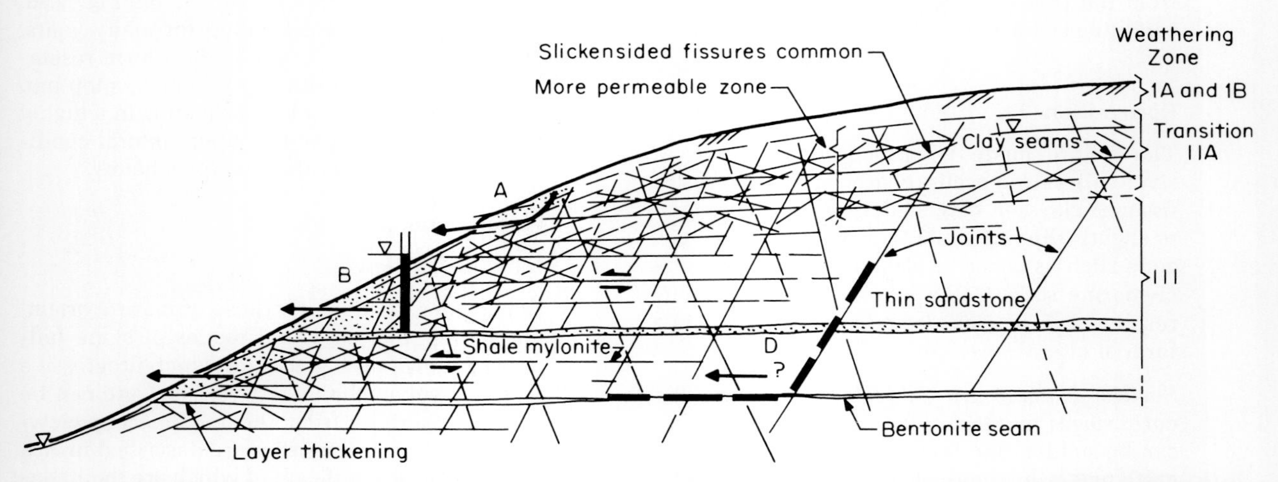


**FIG. 9.34 Failure forms in weathered clay shales. (A) Surface slump in shallow weathered zone. (B) Wedge failure along joints and sandstone seam. (C) Wedge failure along thin bentonite seam may develop into large progressive failure to (D) or beyond.** [*From Deere and Patton (1971).*[12]]

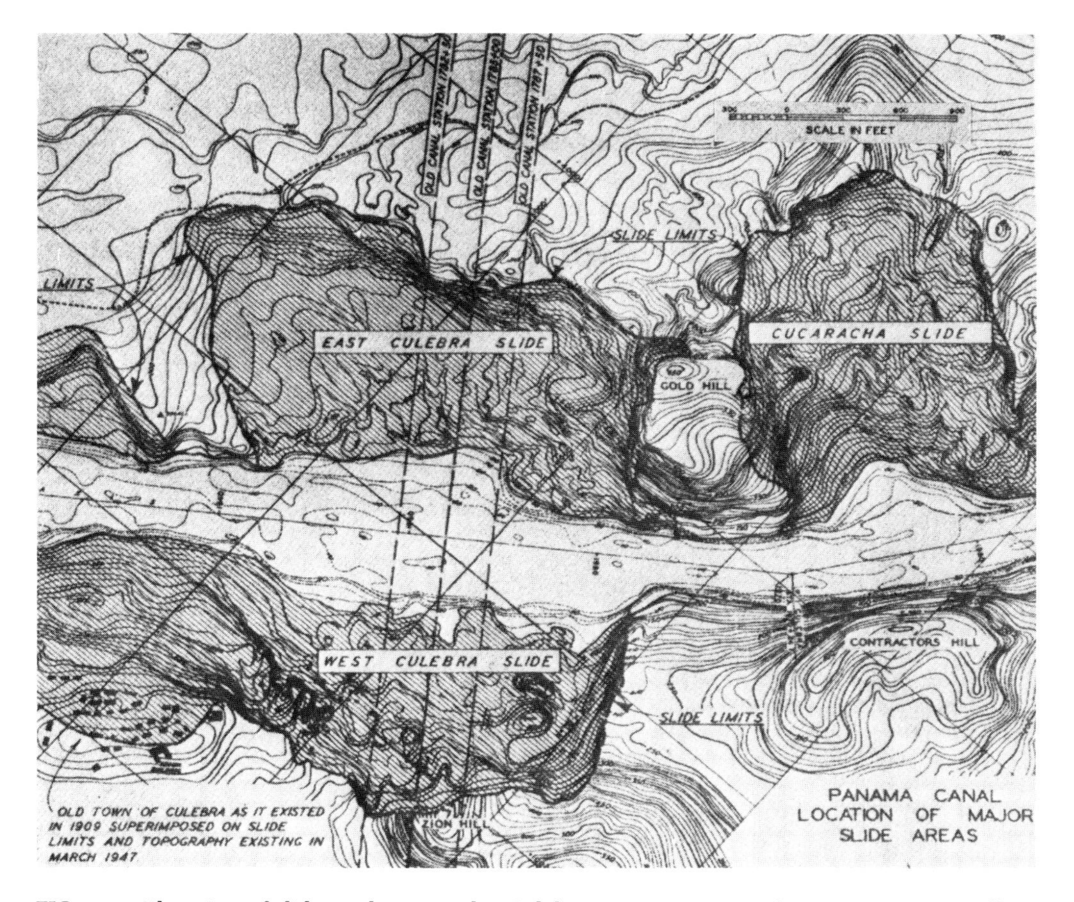


**FIG. 9.35 Plan view of slides and topography, Culebra cut, Panama Canal.** [*From Binger (1948).*[13]]

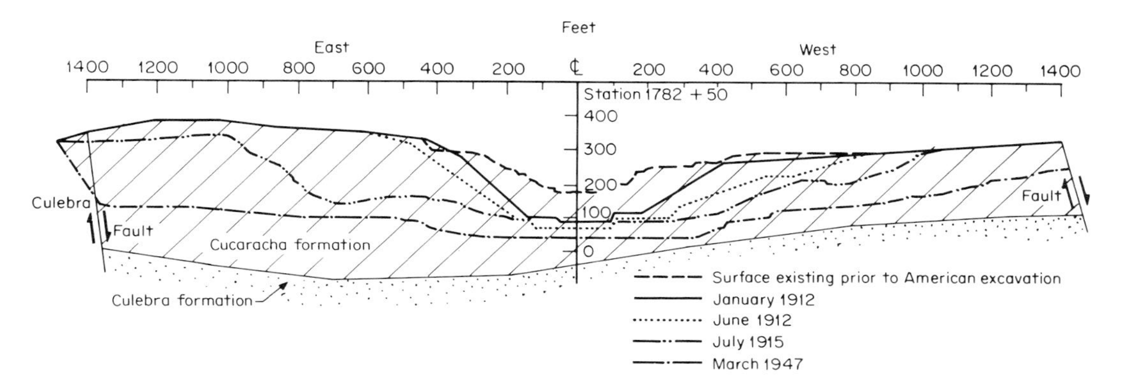


**FIG. 9.36 Sketch of east and west Culebra slides (Panama Canal) showing progress of slide movement: Cucaracha tuffaceous shale and Culebra tuffaceous shale, siltstone, and sandstone.** [*From Binger (1948).*[13]]

values. Water entering the mass through open tension cracks and fractures assists in the development of failure conditions.

*Clay Shale: Panama Canal Slides*

EVENT Massive slides occurred during 1907 and 1915 in the excavation for the Panama Canal in the Culebra cut [Binger (1948),[13] Banks (1972)[14]].

GEOLOGY On the plan view of the slide areas (Fig. 9.35) the irregular to gentle topography of the Cucaracha formation (Tertiary) is apparent. The Cucaracha is a montmorillonitic shale with minor interbedding of sandstone and siltstone more or less horizontally bedded but occasionally dipping and emerging from natural slopes. It is heavily jointed and slickensided, and some fractures show secondary mineral fillings. Natural slopes in the valley were relatively gentle, as shown on the geologic section given as Fig. 9.36, generally about 20° or less. Laboratory consolidation tests gave values for preconsolidation pressure as high as 200 kg/cm$^2$.

SLIDE HISTORY Excavations in the order of 100 m in depth were required in the Cucaracha formation. Some minor sliding occurred as the initial excavations were made on slopes of 1:1 through the upper weathered zones to depths of about 15 m. The famous slides began to occur when excavations reached about 30 m. They were characterized by a buckling and heaving of the excavation floor, at times as great as 15 m; a lowering of the adjacent ground surface upslope; and substantial slope movements. Continued excavation resulted in progressive sliding on a failure surface extending back from the cut as far as 300 m. The causes of the sliding are believed to be stress relief in the horizontal direction, followed by the expansion of the shale, and finally rupture along a shallow arc surface [Binger (1948)[13]].

ANALYSIS Banks (1972)[14] found that at initial failure conditions, the effective strength envelope yielded $\phi = 19°$ and $\bar{c} \approx 0$. For the case of an infinite slope *(see Art. 9.3.2) without* slope seepage these values would produce a stable slope angle of 19°, or for the case of seepage parallel and coincident with the slope face, $\frac{1}{2}\bar{\phi}$, or 9.5°. Since movements had occurred the value 9.5° is considered to be the residual strength.

SOLUTION The slides were finally arrested by massive excavation and cutting the slopes back to 9.5° ($\frac{1}{2}\phi_r$), which is flatter than the natural slopes. Banks reported that measurements with slope inclinometers indicated that movement was still occurring in 1969, and that the depth of sliding was at an elevation near the canal bottom.

*Coastal Plain Sediments: Portugese Bend Slide*

EVENT At Portuguese Bend, Palos Verdes Hills, California *(see Fig. 10.3 for location)*, a slide complex with a maximum width of roughly 1200 m and a head-to-toe length of about 1400 m began moving significantly in 1956 and is still moving. Coastal plain sediments are involved, primarily marine shales. This slide may be classed as progressive block glides or failure by lateral spreading. It is one of the most studied active slides in the United States [Jahns and Vonder Linden (1973)[15]].

PHYSIOGRAPHY The limit of the slide area is shown on Fig. 9.37, and the irregular hummocky topography is shown on the stereo pair of aerial photos in Fig. 9.38. In the slide area the land rises from the sea in a series of gently rolling hills and terraces to more than 200 m above sea level. The hills beyond the slide area rise to elevations above 360 m and the cliffs along the oceanfront are roughly 40 m above the sea. A panoramic view of the slide is given as Fig. 9.39.

GEOLOGY The slide zone occurs in Miocene sediments *(see Art. 7.4.4)* of heavily tuffaceous and sandy clays interbedded with relatively thin strata of bentonitic clays. When undisturbed, the beds dip seaward at about 10 to 20°, which more or less conforms with the land surface as illustrated on the section (Fig. 9.40). A badly crushed zone of indurated clayey silt forming a soil "breccia" (Fig. 9.41) is found in the lower portions of the slide area. Present movement of the slide appears to be seated at a depth of about 30 m below the surface in the "Portuguese tuff," originally deposited as a marine ash flow.

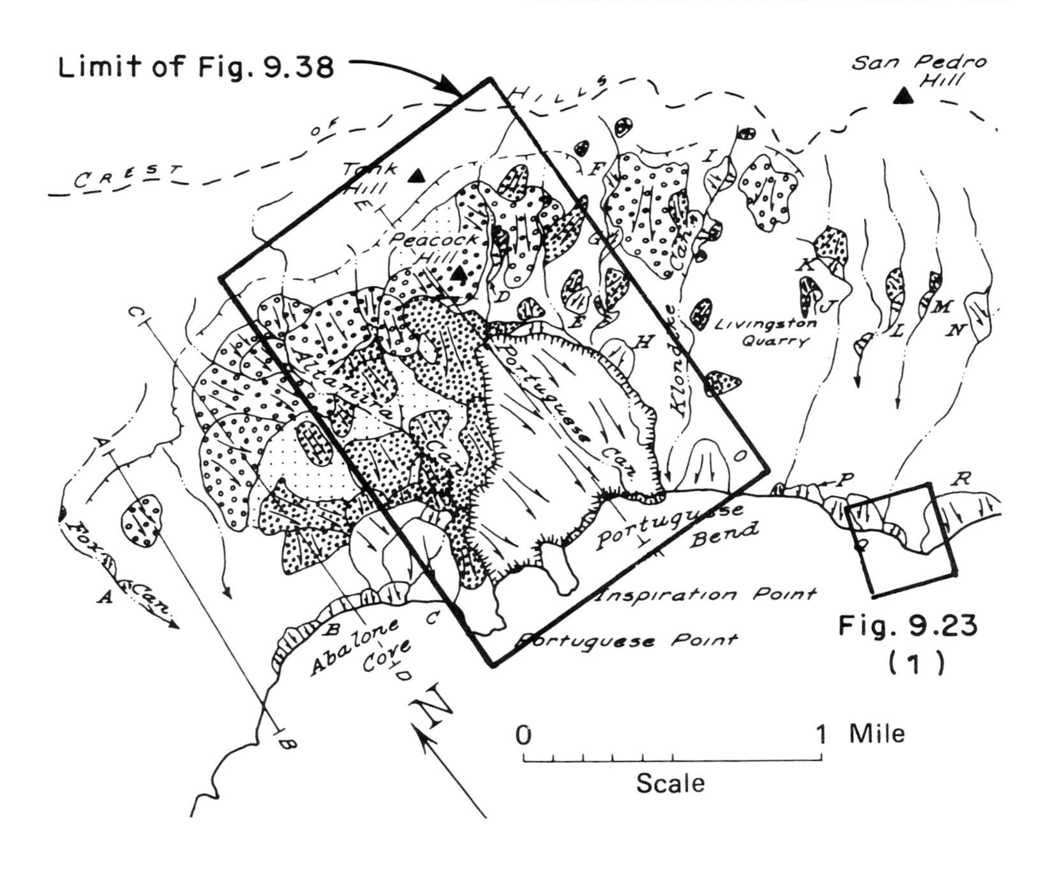


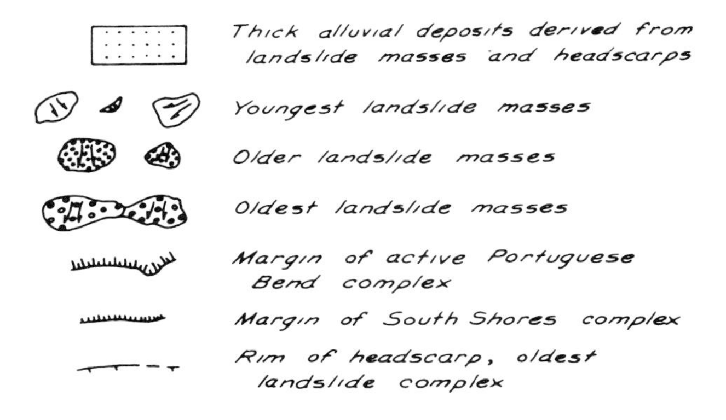


**FIG. 9.37 Distribution of principal landslides and landslide complexes (Palos Verdes Hills).** [*From Johns and Vonder Linden (1973).*[15]]

SLIDE HISTORY The area has been identified as one with ancient slide activity *(see Fig. 9.37)*. Using radiometric techniques, colluvium older than 250,000 years has been dated, and intermediate activity dated at 95,000 years ago. In recent times some block movement was noted in 1929, but during the 1950s, when housing development began on the present slide surface, the slide was considered as *inactive*. Significant modern movement began in 1956, apparently triggered by loading the headward area of the slide with construction fill for a roadway.

**FIG. 9.38 Stereo pair of aerial photos of the Portuguese Bend area. (January 14, 1973. Scale 1:24,000.)**

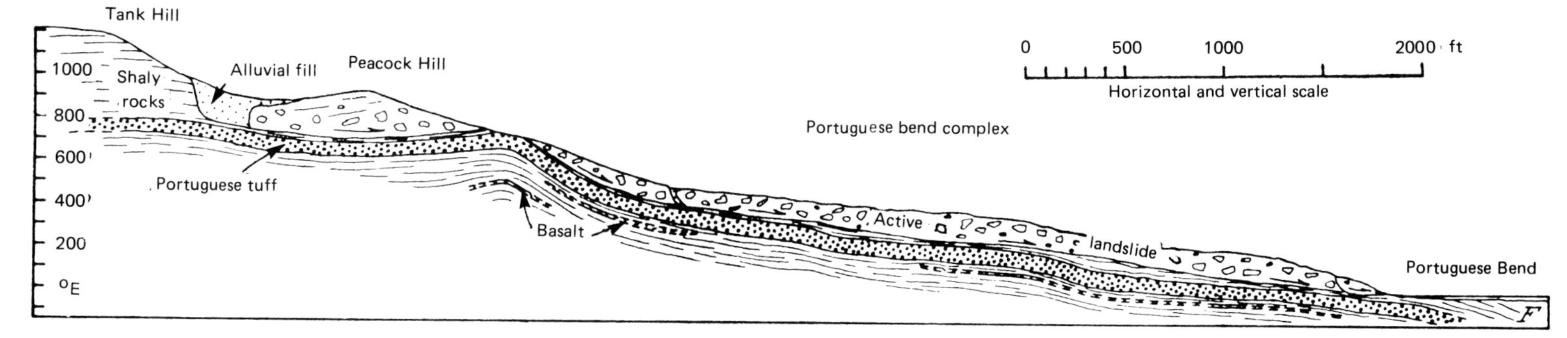


**FIG. 9.40 Geologic section through the Portuguese Bend slide. For location see Fig. 9.37, section along line e–f.** [*From Johns and Vonder Linden (1973).*[15]]

**FIG. 9.39 Panorama of the Portuguese Bend landslide looking south. The highway on the left is continually moving, and the old abandoned road appears in the photo center. The broken ground on the right is the head scarp of rotational slides in the frontal lobe of the unstable mass. (Photo taken in 1973.)**

**FIG. 9.41 Soil "breccia" of fragments of indurated clayey silt in a crushed uplifted zone at Portuguese Bend.**

SLIDE MOVEMENTS The mass began moving initially during 1956 and 1957, at rates of 5 to 12 cm/year, continuing at rates varying from 15 to 60 cm/year during 1958, then 1 to 3 m/year during 1961 to 1968. After 1968 a dramatic increase in movement occurred. Eventually 120 houses were destroyed over a 120-hectare area. Studies have correlated acceleration in rate of advance with earthquake activity, abnormally high tides, and rainfall [Easton (1973)[16]]. The average movement in 1973 ranged from about 8 cm/day during the dry season, to 10 cm/day during the rainy season, to peaks of 15 cm/day during heavy rains. Rainfall penetrating deeply into the mass through the many large tension cracks builds up considerable hydrostatic head to act as a driving force on the unstable blocks supported by material undoubtedly at residual strength. The maximum horizontal displacement between 1966 and 1970 was about 40 m and the maximum vertical displacement about 12 m. An interesting feature of the slide is the gradual and continuous movement without the event of total collapse.

STABILIZATION Because of the large area involved and the geologic and other natural conditions, there appears to be no practical method of arresting slide movements. The cracks on the surface are too extensive to consider sealing to prevent rainwater infiltration, and the strength of the tuff layer is now inadequate to restrain gravity movement even during the dry season. A possible solution to provide stability might be to increase the shearing resistance of the tuff by chemical injection. Since this would be extremely costly, it appears prudent to leave the unstable area as open space although continuous maintenance of the roadway in Fig. 9.39 will be necessary.

### *Glaciolacustrine Soils*

Glaciolacustrine soils composing slopes above river valleys normally are heavily overconsolidated *(see Art. 7.6.4)*. Shear strengths, as measured in the laboratory, are often high, with cohesion ranging from 1 to 4 kg/cm$^2$. Therefore, these soils would not usually be expected to be slide-prone on moderately shallow slopes, and normal stability analysis would yield an adequate factor of safety against sliding [Bjerrum (1966)[4]]. Sliding is common, however, and often large in scale, even on shallow slopes.

In the Seattle Freeway slides (Fig. 9.42), failure occurred along old bedding plane shears associated with lateral expansion of the mass toward the slopes when the glacial ice in the valley against the slopes disappeared [Palladino and Peck (1972)[17]]. Similar conditions probably existed at the site of the slide occurring at Kingston, New York, in the Hudson River valley in August 1915 [Terzaghi (1950)[6]]. The Kingston slide was preceded by a period of unusually heavy rainfall. Factors contributing to failure as postulated by Terzaghi were the accumulation of stockpiles of crushed rock along the upper edge of the slope and perhaps the deforestation

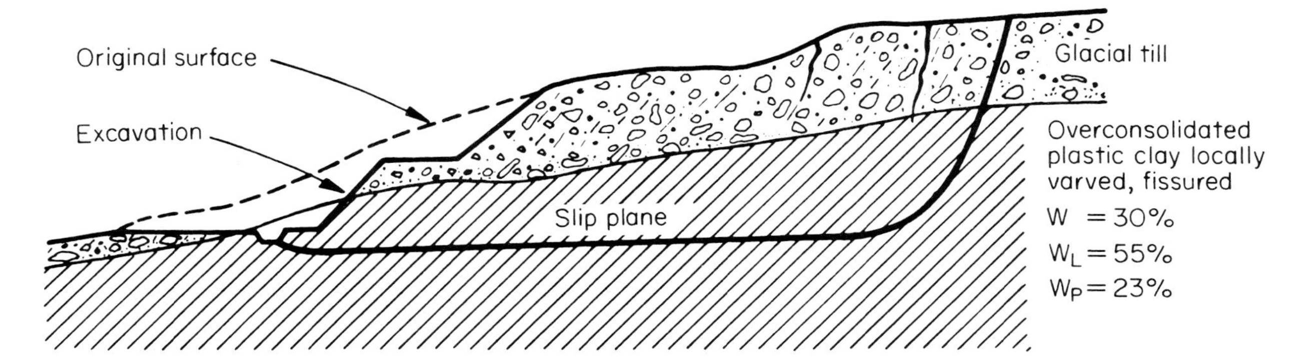


**FIG. 9.42 Failure surface in overconsolidated, fissured clays, undergoing progressive failure as determined by slope inclinometer measurements (Seattle Freeway).** [*From Bjerrum (1966).*[4]]

of outcrops of the aquifer underlying the varved clays which permitted an increase in pore-water pressures along the failure surface.

### *Glaciomarine Soils: South Nation River Slide*

EVENT The South Nation River slide in Casselman, Ontario, of May 16, 1971, is typical of many slides occurring in the sensitive Champlain clays of glaciomarine origin, in Quebec province, Canada [Eden et al. (1971)[18]]. These clays are distributed in a broad belt along the St. Lawrence River and up the reaches of the Saguenay River. Most of the slides occur along riverbanks, commencing as either a slump or block glide and retrogressing through either slumping or lateral spreading. At times the frontal lobes of the slides liquefy and become flows (Fig. 9.4*f*). *(See Art. 9.2.11.)*

GEOLOGY Glaciomarine soils in general are described in Art. 7.6.5. The stratigraphy at the South Nation River slide prior to failure consisted of 2 to 7 m of stratified silty fine sands overlying the Champlain clay (Leda clay) as shown on Fig. 9.43. The undrained strength of the clay was about 0.5 kg/cm$^2$, its sensitivity ranged from 10 to 100, the average plastic limit was 30% and liquid limit 70%, and the natural water content was at the liquid limit.

SLIDE HISTORY An all-time record snowfall of 170 in (432 cm) occurred during the 1970–71 winter and gradual melting resulted in saturation of

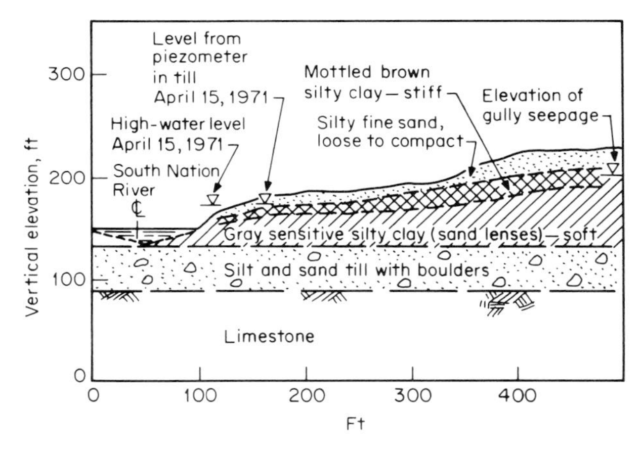


**FIG. 9.43 Stratigraphy prior to the South Nation River slide. An overconsolidated crust over soft, sensitive clays.** [*From Eden et al. (1971).*[18] *Reprinted with permission of the National Research Council of Canada.*]

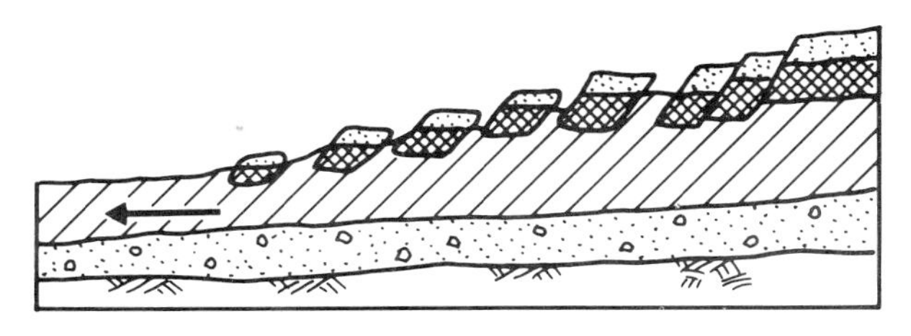

**FIG. 9.44 Stratigraphy after slide at South Nation River. Blocks broke loose and moved by lateral spreading.** [*After Mollard (1977).*[19]]

the upper clays. The slide occurred at the end of the snow-melting season during a heavy rainstorm. A contributing factor was the river level at the slide toe area. It had risen as much as 9 m during spring floods, remained at that level for a week, then dropped back rapidly to preflood levels. At the time of the slide, groundwater at the lower part of the slope was observed to be nearly coincident with the surface. From the appearance of the ground after failure, it appears that the slide retrogressed as a series of slumps as shown on Fig. 9.44.

*Glaciomarine Soils: Turnagain Heights Slide*

EVENT Much of the damage to the Anchorage, Alaska, area from the March 1964 earthquake was caused by landslides induced by seismic forces. The slides occurred in the city in the Ship Creek area and along the waterfront formed by the Knik Arm of Cook Inlet. The largest slide occurred at Turnagain Heights, a bluff some 20 m high overlooking Knik Arm. Many homes were destroyed in the slide area of 50 hectares as illustrated on Fig. 9.45. The slide at Turnagain Heights is an example of sliding along horizontal strata. It was planar and evolved by block gliding or slump failure at the bluff, followed by lateral spreading of the mass for a width of 2500 m and extending as much as 270 m inland.

GEOLOGY Anchorage and the surrounding area are underlain by the Bootlegger Cove clay of glaciomarine origin. Soil stratigraphy at the bluff consisted of a thin layer of sand and gravel overlying a clay stratum over 30 m thick as shown on Fig. 9.46*a*. The consistency of the upper portions of the clay was stiff to medium, becoming soft at a depth of about 14 m. The soft zone extended to a depth of about 7 m below sea level. Layers of silt and fine sand were present at depths of a meter or so above sea level.

SLIDE HISTORY Seed and Wilson (1967)[20] postulated that cyclic loading induced by the earthquake caused liquefaction of the silt and fine sand lenses resulting in instability and block gliding along the bluff. Blocks continued to break loose and glide retrogressively, resulting in lateral spreading of the mass which came to rest with a profile more or less as illustrated in Fig. 9.46*b*. The movement continued for the duration of the earthquake (more than 3 min), but essentially stopped once strong ground motion ceased.

CONCLUSIONS The magnitude of the 1964 event was 8.5 (Richter) with an epicenter 130 km east of Anchorage. Previous earthquakes of slightly lower magnitudes but closer epicenters had occurred, but the Turnagain Heights area had not been affected *(see Art. 11.3.4)*. Seed and Wilson (1967)[20] concluded that, in light of previous earthquake history, the slide was the result of a continuous increase in pore pressures caused by the long duration of the 1964 event, and that it is extremely unlikely that any analysis would have anticipated the extent of inland transgression of the failure. Considering the local stratigraphy and seismic activity, however, the area certainly should be considered as one with a high slope failure hazard.

### 9.2.7 DEBRIS SLIDES

#### General

Debris slides involve a mass of soil, or soil and rock fragments, moving as a unit or a number of units along a steeply dipping planar surface. They often occur progressively and can develop into avalanches or flows. Major causes are increased seepage forces and slope inclination, and the incidence is increased substantially by stripping vegetation. Very large masses can be involved, with gradually developing progressive movements, but at times total failure of a single block can occur suddenly.

#### Recognition

Occurrence is common in colluvial or residual soils overlying a relatively shallow, dipping rock surface.

During the *initial stages* of development, tension cracks are commonly formed. After *partial failure* the tension cracks widen and the complete dislodgement of one or more blocks may occur, often leaving a clean rock surface and an elliptical failure scar as shown on Fig. 9.47. *Total failure* can be said to have occurred when the failure surface reaches to the crest of the hill. If uncorrected, failure often progresses upslope as blocks break loose.

**FIG. 9.45 Failure by block gliding and lateral spreading resulted from the 1964 earthquake, Turnagain Heights, Anchorage, Alaska.** *(Photo courtesy of U.S. Geological Survey, Anchorage.)*

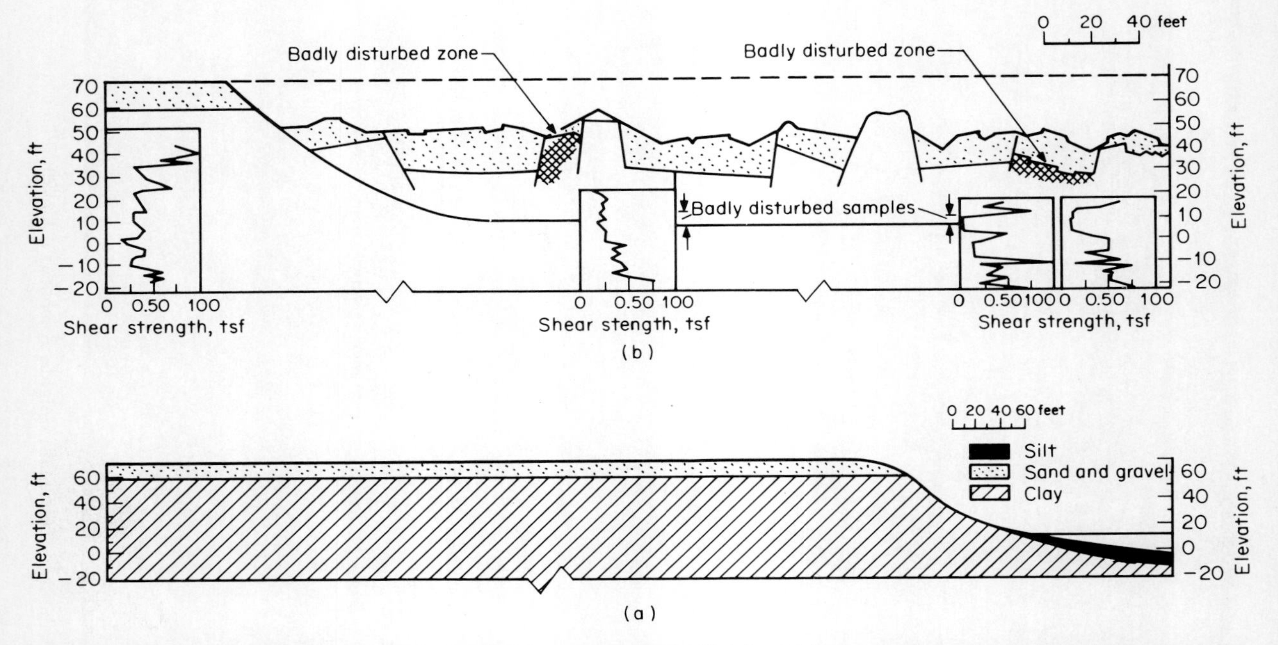


**FIG. 9.46 Soil profiles through east end of slide area at Turnagain Heights (a) before and (b) after failure.** [*From Seed and Wilson (1967).*[20]]

### Examples of Major Failures

*Pipe Organ Slide*

DESCRIPTION The Pipe Organ slide in Montana [Noble (1973)[21]], triggered by a railroad cut made at the toe, involved about 9 million m$^3$ of earth. The failure surface developed at depths below the surface of 40 to 50 m in a Tertiary colluvium of stiff clay containing rock fragments, which overlay a porous limestone formation.

SLIDE MOVEMENTS During sliding, movement continued for a year at an average rate of 5 cm/week, developing in a progressive mode. Total movement was about 4 m, and the length of the sliding mass was 600 m along the flatter portions of a mountain slope.

STABILIZATION Movement was arrested by the installation of pumped wells, which were drilled into the porous limestone and later converted to gravity drains. The water perched on the sliding surface near the interface between the colluvium and the limestone drained readily into the limestone and the hydrostatic pressures were relieved.

*Golden Slide*

DESCRIPTION A railroad cut into colluvium of about 15 m thick caused a large slide near Golden, Colorado [Noble (1973)[21]]. The colluvium is an overconsolidated clay with fragments of clay shale and basalt overlying a very hard blue-gray clayey siltstone. The water table was midway between the ground surface and the failure surface, and there was evidence of artesian pressure at the head of the slide.

SLIDE MOVEMENTS Movement apparently began with a heavy rainfall and was about 2.5 cm/day. Tension cracks developed in the surface and progressed upslope with time. About 500,000 m$^3$ of material were moving within a length of about 300 m.

STABILIZATION Cutting material from the head of the slide and placing approximately 100,000 m³ against a retaining wall at the toe which penetrated into underlying sound rock was unsuccessful. Movement was finally arrested by the installation of horizontal drains as long as 120 m, and vertical wells which were being pumped daily at the time that Noble (1973)[21] prepared his article.

*Colluvium on Shale Slopes*

The Pennington shale of the Cumberland plateau in Tennessee and the sedimentary strata in the Appalachian plateau of western Pennsylvania develop thick colluvial overburden which is the source of many slide problems in cuts and side-hill fills. The geology, nature of slope problems, and solutions are described in detail by Royster (1973)[22], (1979),[23] and Hamel (1980).[9]

### 9.2.8 DEBRIS AVALANCHES

#### General

Debris avalanches are very rapid movements of soil and rock debris which may, or may not, begin with rupture along a failure surface. All vegetation and loose soil and rock material may be scoured from a rock surface as shown on Fig. 9.7. Major causes are high seepage forces, heavy rains, snowmelts, snowslides, earthquakes, and the creep and gradual yielding of rock strata.

Failure is sudden and without warning, and essentially unpredictable except for the recognition that the hazard exists. Effects can be disastrous in built-up areas at the toes of high steep slopes in suitable geologic conditions (see examples below).

**FIG. 9.47 Small debris-slide scar along the Rio Santos Highway, Brazil. Note seepage along the rock surface, the sliding plane. Failure involved colluvium and part of the underlying rock. The area was subsequently stabilized by a concrete wall.**

### Recognition

Debris avalanches are characteristic of mountainous terrain with steep slopes of residual soils where topography causes runoff concentration (*see Fig. 9.85*) or badly fractured rock such as illustrated in Fig. 9.48.

There is usually no initial stage, although occasionally tension cracks may be apparent under some conditions. Total failure occurs suddenly either by a rock mass breaking loose or by "*hydraulic excavation*" which erodes deep gullies in soil slopes during torrential rains as shown on Fig. 9.49. All debris may be scoured from the rock surface and deposited as a terminal lobe at a substantial distance from the slope. As shown on Fig. 9.50, the force is adequate to move large boulders, and erosion can cause the failure area to progress laterally to affect substantial areas as shown on Fig. 9.51.

### Examples of Major Failures

*Rio de Janeiro, Brazil*

EVENT A debris avalanche occurred during torrential rains in 1967 in the Laranjeiras section of Rio de Janeiro which destroyed houses and two apartment buildings, causing the death of more than 130 persons. The avalanche scar and a new retaining wall are shown on Fig. 9.52.

CLIMATIC CONDITIONS Hundreds of avalanches and slides occurred in Rio de Janeiro and the nearby mountains during the unusually heavy rains of 1966 and 1967 when intensities as high as 100 mm/hr were recorded [Jones (1973)[24]].

**FIG. 9.48 Scarred surface remaining after a rock and debris avalanche in a limestone quarry. The rock is heavily jointed with sets oriented more or less parallel to the slope and across the bedding plane. Failure was induced by wedging from water and ice pressures, and occurred in the early spring.**

**FIG. 9.49 The force of hydraulic excavation is evident in this photo taken in the typical V-shaped scarred zone of a debris avalanche. Location is near the crest of the hill in Fig. 9.50. The bedrock surface is exposed.**

During a 3-day storm beginning on January 10, 1966, a gaging station at Alto da Boa Vista, in the mountains a few kilometers from the city, recorded 675 mm (26.2 in) of rainfall. It was an unprecedented amount. Although heavy rains occur each year during the summer months of January and February, with rainfall averaging 171 mm during January, most of the rain falls during intense storms. The potential for slope failures is very much dependent upon the accumulated rainfall and associated water-table conditions for a given rainy season (*see Art. 9.3.4*).

LOCAL GEOLOGY Typical profiles in the residual soils (*see Figs. 6.85 and 9.69*) along the coastal mountains of Brazil show that these soils are most impervious near the surface and that permeability increases downward through the soil profile into the underlying decomposed and fractured crystalline rocks. Fissures in the outer portions of the residual and colluvial soils close during rainfall; they thereby block drainage and cause a rapid increase in pore pressures, resulting in sudden failure, which combined with high runoff develops into an avalanche or even a flow. To minimize the slope-failure hazard, the city of Rio has zoned some areas of high steep slopes to prohibit construction, and has undertaken the construction of numerous stabilization works throughout the city.

*Ranrahirca and Yungay, Peru*

EVENT NO. 1 One of the most disastrous debris avalanches in modern history occurred in the Andes Mountains of Peru on January 10, 1962, when in a period of 7 min, 3500 lives were lost and seven towns, including Ranrahirca, were buried under a mass of ice, water, and debris

**FIG. 9.50 Debris avalanche that covered BR 101 near Tubarão, Santa Catarina, Brazil, during the torrential rains of 1974. Note the minibus for scale. The debris lobe crossed the highway and continued for a distance of about 200 m and carried boulders several meters in diameter. Debris has been removed from the roadway. It is unlikely that a failure at this particular location could have been foreseen.**

[McDonald and Fletcher (1962)[25]]. The avalanche began with the collapse of Glacier 511 from the 7300-m-high peak, Nevada Huascaran. Triggered by a thaw, 3 million tons of ice fell and flowed down a narrow canyon picking up debris and spilling out onto the fertile valley at an elevation 4000 m lower than the glacier and 15 km distant. The debris remaining in the towns ranged from 10 to 20 m thick.

EVENT NO. 2 The catastrophe was almost duplicated on May 31, 1970, when the big Peruvian earthquake caused another avalanche from Nevada Huascaran which buried Yungay, adjacent to Ranrahirca, as well as Ranrahirca again, taking at least 18,000 lives [Youd (1978)[26]]. During the 1962 event, Yungay had been spared. The average velocity of the avalanche has been given as 320 km/hr [Varnes (1978)[27]] and the debris flowed upstream along the Rio Santa for a distance of approximately 2.5 km. As with the 1962 failure, the avalanche originated when a portion of a glacier on the mountain peak broke loose.

### 9.2.9 DEBRIS FLOWS

#### General

Debris flows are similar to debris avalanches except that the quantity of water in the debris-flow mass causes it to flow as a slurry; in fact, differentiation between the two forms can be dif-

**FIG. 9.51 Very large debris avalanche scar and numerous smaller scars after the torrential rains of January 1974 (near Tubarão, Santa Catarina, Brazil). Unpredictable, such events are common in wet climates around the world in mountainous terrain.**

ficult. The major causes are very heavy rains, high runoff, and loose surface materials.

### Recognition

Occurrence is similar to debris avalanches, but debris flows are more common in steep gullies in arid climates during cloudbursts, and the failing mass can move far from its source *(see Fig. 9.4c)*.

## 9.2.10 ROCK-FRAGMENT FLOWS

### General

A rock mass can suddenly break loose and move downslope at high velocities as a result of the sudden failure of a weak bed or zone on the lower slopes causing loss of support to the upper mass. Weakening can be from weathering, frost wedging, or excavation. Failure is sudden, unpredictable, and can be disastrous.

### Recognition

High, steep slopes in jointed rock masses offer the most susceptible conditions. The avalanche illustrated in Fig. 9.48 could also be classed as a dry rock flow because of its velocity and lack of water. In the initial stages tension cracks may develop; after the final stage a scarred surface remains over a large area, and a mass of failed debris may extend far from the toe of the slope.

**FIG. 9.52 Scar of debris avalanche (February 18, 1967) that destroyed two apartment buildings and took 132 lives in the Laranjeiras section of Rio de Janeiro. The buttressed wall was constructed afterward. Debris avalanches need not be large to be destructive.**

**Example of Major Failure**

*Event*

The Turtle Mountain slide of the spring of 1903 destroyed part of the town of Frank, Alberta, Canada. More than 30 million m$^3$ of rock debris moved downslope and out onto the valley floor for a distance of over 1 km in less than 2 min.

*Geology*

The mountain is the limb of an anticline composed of limestone and shales as shown in the section (Fig. 9.53). Failure was sudden, apparently beginning in bedding planes in the lower shales [Krahn and Morgenstern (1976)[28]] which are steeply inclined.

*Cause*

Terzaghi (1950) postulated that the flow was caused by joint weathering and creeping of the soft shales, probably accelerated by coal mining operations along the lower slopes.

### 9.2.11 SOIL AND MUD FLOWS

**General**

Soil and mud flows generally involve a saturated soil mass moving as a viscous fluid, but at times can consist of a dry mass. Major causes include earthquakes causing high pore-air pressures (loess) or high pore-water pressures; the leaching of salts from marine clays increasing their sensitivity, followed by severe weather conditions; lateral spreading followed by a sudden collapse of soil structure; and heavy rainfall on a thawing mass or the sudden drawdown of a flooded water course.

Flows occur suddenly, without warning, and can affect large areas with disastrous consequences.

### Recognition

Occurrence is common in saturated or nearly saturated fine-grained soils, particularly sensitive clays, and occasional in dry loess or sands (sand runs).

During initial stages flows may begin by slump failure followed by lateral spreading in the case of sensitive clays. During final failure, a tongue-shaped lobe of low profile extends back to a bottleneck-shaped source area with a small opening at the toe of the flow as shown on Fig. 9.4*f*, and a distinct scarp remains at the head. Flowing masses can extend for great distances, at times measured in kilometers.

### Examples of Major Failures

*Achocallo Mudflow, La Paz, Bolivia*

Believed triggered by an earthquake some thousands of years ago, an enormous portion of the rim of the Bolivian altiplano (elevation 4000 m), roughly 9 km across, slipped loose and flowed down the valley of the Rio Achocallo into the Rio La Paz at an elevation approximately 1500 m lower. The flow remnants extend downstream today for a distance of about 25 km, part of which are shown on the aerial oblique panorama given as Fig. 9.54.

The altiplano is the remains of an ancient lake bed, probably an extension of Lake Titicaca, underlain by a thick stratum of sand and gravel beneath which are at least several hundred meters of lacustrine clays and silts, interbedded with clays of volcanic origin. The head scarp of the mudflow is shown in Fig. 9.55; the city of La Paz is situated beyond the far edge of the rim. The photo *(see Fig. 10.34)* of a large piping tunnel was taken in the bowl-shaped valley about 3 km downslope from the rim, and the photo *(see Fig. 9.32)* of lateral spreading was taken in the valley of the Rio La Paz.

*Province of Quebec, Canada*

EVENT In Saint Jean Vianney, on May 4, 1971, a mass of glaciomarine clays completely liquefied, destroying numerous homes and taking 31 lives [Tavenas et al. (1971)[29]].

DESCRIPTION The flow began in the crater of a much larger 500-year-old failure (determined by carbon-14 dating). Soil stratigraphy consisted of about 30 m of disturbed clays with sand pockets from the ancient slide debris, overlying a deep layer of undisturbed glaciomarine clay. Occurring just after the first heavy rains following the spring thaw, the flow apparently began as a series of slumps from the bank of a small creek, which formed a temporary dam. Pressure built up behind the dam, causing it to fail, and 9 million m$^3$ of completely liquefied soils flowed downstream with a wavefront 20 m in height and a velocity estimated at 26 km/hr. The flow finally discharged into the valley of the Saguenay River, 3 km from its source.

CAUSES Slope failures in the marine clays of Quebec are concentrated in areas that seem to be associated with a groundwater flow regime resulting from the existence of valleys in the underlying rock surface [Tavenas et al. (1971)[29]]. The valleys cause an upward flow gradient and an artesian pressure at the slope toes. The upper part of the soil profile is subjected to a down-

**FIG. 9.53 Geologic section at Turtle Mountain.** [*From Krahn and Morgenstern (1976).*[28]]

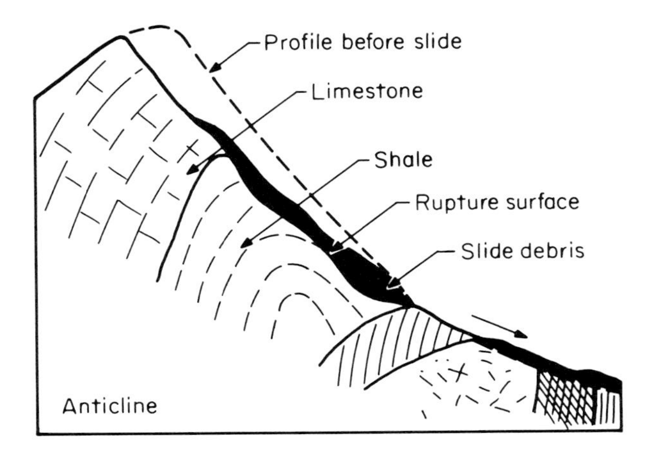

**FIG. 9.54 Believed triggered by an earthquake some thousands of years ago, an enormous portion of the rim of the Bolivian altiplano, roughly 9 km across (photo right), slipped loose and flowed down the valley of the Rio Achocallo into the Rio La Paz, for a total distance of over 25 km (photo left). The flow is apparent in the photo as the light-colored area in the valley.**

**FIG. 9.55 The head scarp of the Achocallo mudflow (La Paz, Bolivia).** ***(See Fig. 9.54.)***

ward percolation of surface water because of the existence of sand strata. The downward percolation and upward flow produce an intense leaching of the clay, resulting in a decrease of the undrained shear strength and an increase in sensitivity *(see Art. 7.6.5)*. The evidence tends to indicate that the leaching is a function of the gradient. Field studies have shown a close relationship between the configuration of the bedrock and the properties of the underlying clay deposit.

### *Norwegian "Quick" Clays*

REGIONAL GEOLOGY Approximately 40,000 km$^2$ of Norway has deposits of glaciomarine clays which overlie an irregular surface of granite gneiss, similar to conditions in Quebec. During postglacial times, the area has been uplifted to place the present surface about 180 m above sea level. Typical stratigraphy includes 5 to 7 m of a stiff, fissured clay overlying normally consolidated soft marine clay which extends to depths greater than 70 m in some locations *(see Art. 7.6.5)*. Rock varies from outcropping at the surface in stream valleys to over 70 m in depth. The quick clays are formed by leaching of salts, but the leaching is believed to be caused by artesian pressure in the rock fractures from below, rather than downward percolation of water [Bjerrum et al. (1969)[5]]. Sensitivity values are directly related to the amount of leaching and the salt content, and are greatest where rock is relatively shallow, about 15 to 35 m.

*Slope failures* are common events. The natural slopes are stable at about 20° where there is a stiff clay crust. Seepage parallel to the slopes occurs in fissures in the clay, and the stiff clay acts as a cohesionless material with slopes at $i = \frac{1}{2}\bar{\phi}$, and $\phi = 38°$ (*see the infinite slope problem in Art. 9.3.2*). Stream erosion causes small slides in the weathered stiff clay; the sliding mass moves into the soft clay, which upon deformation becomes quick and flows. In one case cited by Bjerrum et al. (1969),[5] 200,000 $m^3$ flowed away from the source in a few minutes.

SOLUTION Since stream degrading appears to be a major cause of the flows, Bjerrum et al. (1969)[5] proposed the construction of small weirs to impede erosion in streams where failures pose hazards.

### 9.2.12 SEAFLOOR INSTABILITY

#### General

Various forms of slope failures have been recognized offshore, including deep rotational slides (Fig. 9.56) and shallow slumps, flows, and collapsed depressions (Fig. 9.57).

Major causes are earthquakes, storm waves inducing bottom pressures, depositional loads accumulating rapidly and differentially over weak sediments, and biochemical degradation of organic materials forming large quantities of gases in situ which weaken the seafloor soils.

Offshore failures can occur suddenly and unpredictably, destroying oil production platforms, undersea cables, and pipelines. Large flows, termed "turbidity currents," can move tremendous distances.

#### Recognition

Occurrence is most common in areas subjected to earthquakes of significant magnitude and on gently sloping seafloors with loose or weak sediments, especially in rapidly accreting deltaic zones.

After failure the seafloor is distorted and scarred with cracks, scarps, and flow lobes similar to those features which appear on land as illustrated on Fig. 9.58. Active areas are explored with side-scan sonar (Fig. 9.59) and high-resolution geophysical surveys (*see Fig. 9.56*).

#### Examples of Major Failures

*Gulf of Alaska*

The major slide illustrated on the high-resolution seismic profile given as Fig. 9.56 apparently occurred during an earthquake, and covers an area about 15 km in length. Movement occurred on a 1° slope and is considered to be extremely young [Molnia et al. (1977)[30]]. As shown on the figure, the slide has a well-defined head scarp, disrupted bedding, and a hummocky surface.

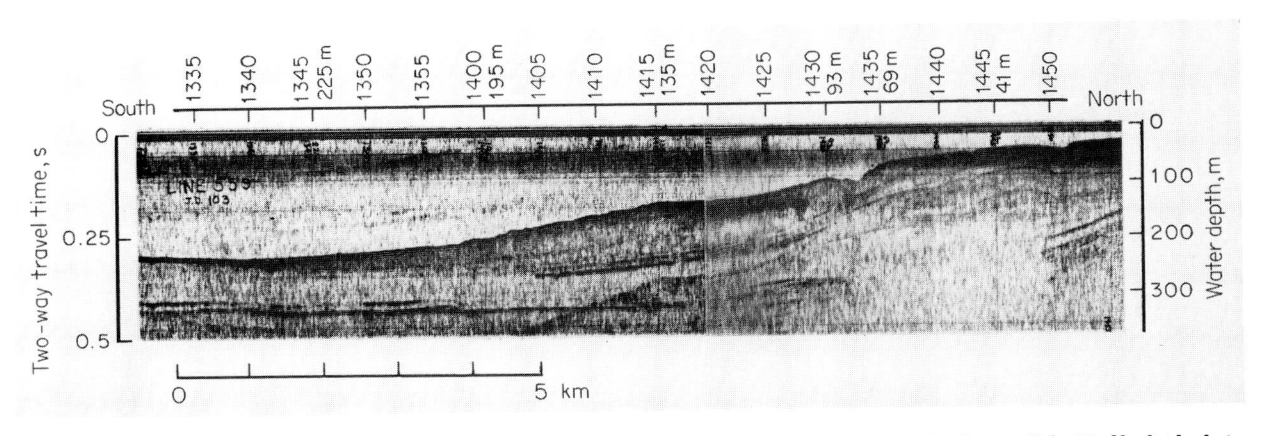


**FIG. 9.56 High-resolution seismic reflection profile showing a portion of the Kayak Trough slump slide (Gulf of Alaska).** [*From Molnia et al. (1977).*[30]]

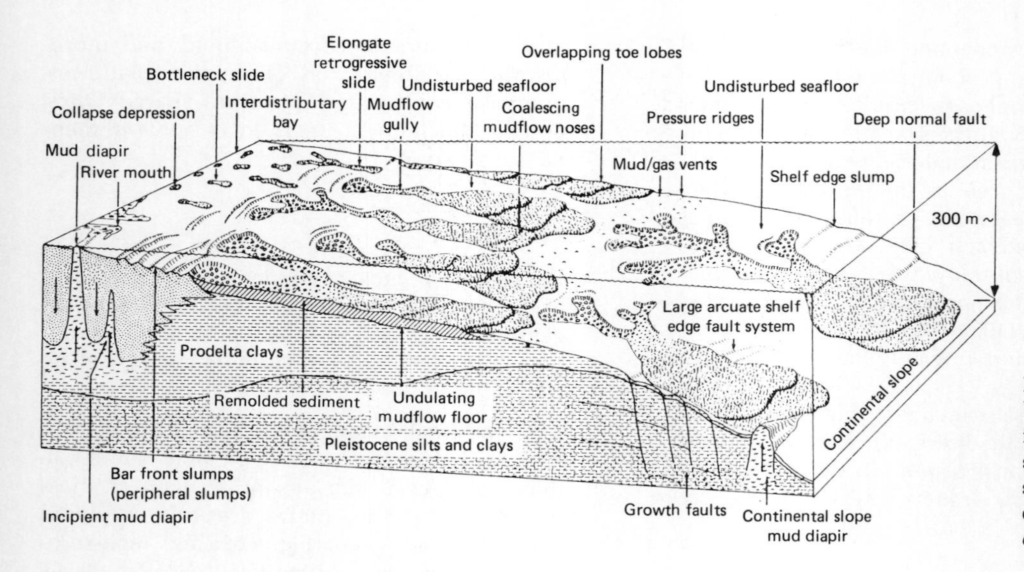


**FIG. 9.57 Schematic block diagram illustrating the various forms of slope failure in the offshore Mississippi River delta.** [*From Coleman et al. (1980).*[31]]

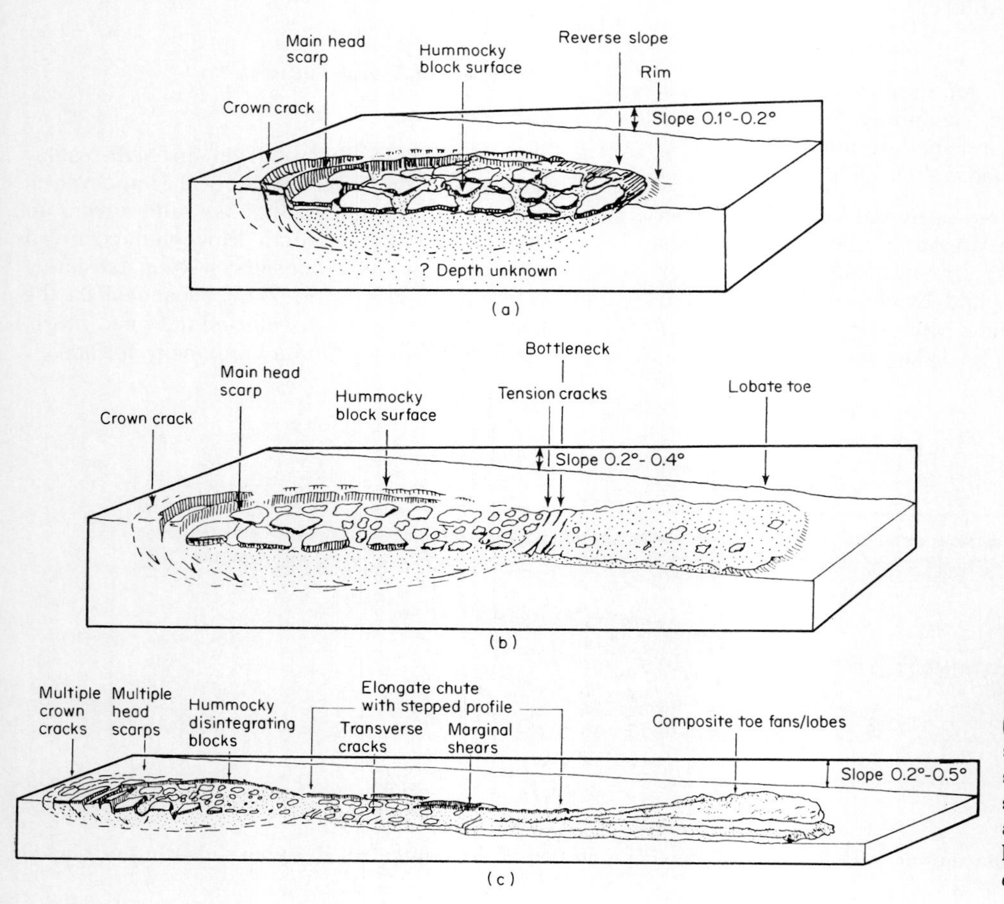


**FIG. 9.58 Morphology of several forms of slope failures in the offshore Mississippi River delta: (*a*) a collapsed depression; (*b*) a bottleneck slide; (*c*) an elongate slide, mudflow gullies, and depositional flow lobe.** [*From Coleman et al. (1980).*[31]]

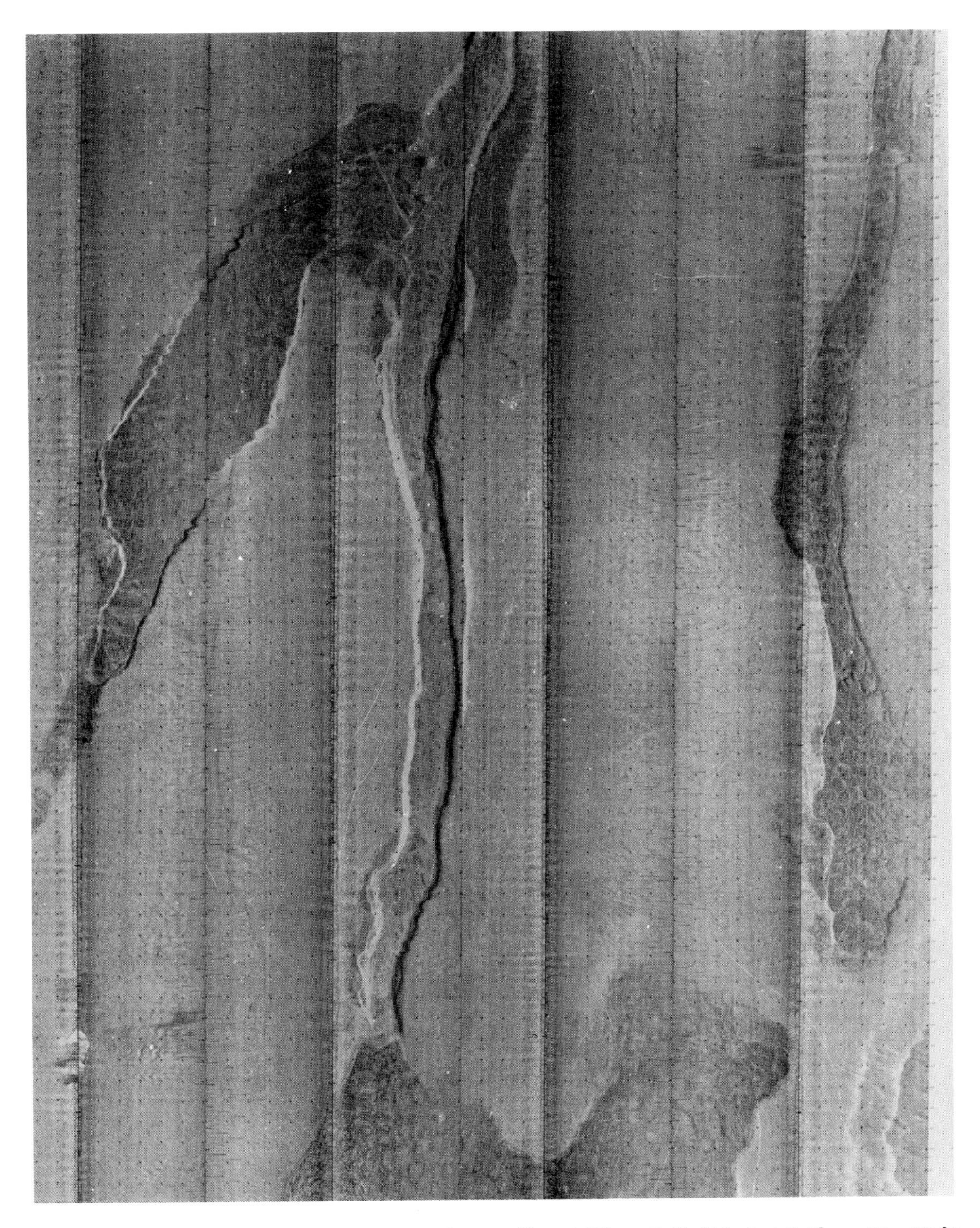

**FIG. 9.59 Side-scan sonar mosaic illustrating seafloor mudflows (offshore Gulf of Mexico). Grids are 25 m (82 ft) apart and the mosaic covers an area approximately 1.5 km in length.** *(Mosaic courtesy of Dr. J. M. Coleman, Coastal Studies Institute, Louisiana State University.)*

*Gulf of Mexico*

Movements are continually occurring offshore of the Mississippi River delta. During hurricane Camille in August 1969, wave-induced bottom pressures caused massive seafloor movements that destroyed two offshore platforms and caused a third to be displaced over a meter on a bottom slope that was very flat, less than 0.5% [Focht and Kraft (1977)[32]].

*Grand Banks, Newfoundland*

The earthquake of November 1929 *(see Art. 11.3.4)* caused a section of the continental shelf to break loose and subsequently mix with seawater. It moved offshore for a distance of about 925 km and broke a dozen submarine cables. Geologists called this flow a "turbidity current" [Richter (1958)[33]].

## 9.3 ASSESSMENT OF SLOPES

### 9.3.1 GENERAL

#### Objectives

The assessment of an existing unstable or potentially unstable slope, or of a slope to be cut, provides the basis for the selection of slope treatments. Treatment selection requires forecasting the form of failure, the volume of material involved, and the degree of the hazard and risk.

Assessment can be based on quantitative analysis in certain situations, but in many cases must be based on qualitative evaluation of the slope characteristics and environmental factors including weather and seismic activity.

#### Key Factors to be Assessed

- *History* of local slope failure activity as the result of construction, weather conditions, seismic activity *(see Art. 11.3.4)*, or other factors, in terms of failure forms and magnitudes
- *Geologic conditions* including related potential failure forms and their suitability for mathematical analysis, material shear strength factors (constant, variable, or subject to change or liquefaction, Art. 9.3.2), and groundwater conditions
- *Slope geometry* in terms of the influence of inclination, height, and shape on potential seepage forces, runoff, and failure volume
- *Surficial indications of instability* such as creep, scars, seepage points, and tension cracks
- *Degree of existing slope activity (see Art. 9.1.2)*
- *Weather factors* (rainfall and temperature) in terms of the relationship between recent weather history and long-term conditions (less severe, average, more severe) in view of present slope activity, stability of existing cut slopes, groundwater levels, and slope seepage

### 9.3.2 STABILITY ANALYSIS: A BRIEF REVIEW

#### General Principles

*Basic Relationships*

Stability analysis of slopes by mathematical procedures is applicable only to the evaluation of failure by *sliding* along some definable surface. Avalanches, flows, falls, and progressive failure cannot be assessed mathematically in the present state of the art.

*Slide failure* occurs when the shearing resistance available along some failure surface in a slope is exceeded by shearing stresses imposed on the failure surface. *Static analysis* of sliding requires knowledge of the location and shape of the potential failure surface, the shear strength along the failure surface, and the magnitude of the driving forces. Statically determinate failure forms may be classified as:

- *Infinite slope*—translation on a plane parallel to the ground surface whose length is large compared to its depth below the surface (end effects can be neglected) [Morgenstern and Sangrey (1978)[42]].
- *Finite slope, planar surface*—displacement of one or more blocks, or wedge-shaped bodies, along planar surfaces with finite lengths.
- *Finite slope, curved surface*—rotation along a curved surface approximated by a circular arc, log-spiral, or other definable cylindrical shape.

FAILURE ORIGIN As stresses are usually highest at the toe of the slope, failure often begins there and progresses upslope, as illustrated on Fig. 9.60, which shows the distribution of active and passive stresses in a slope where failure is just

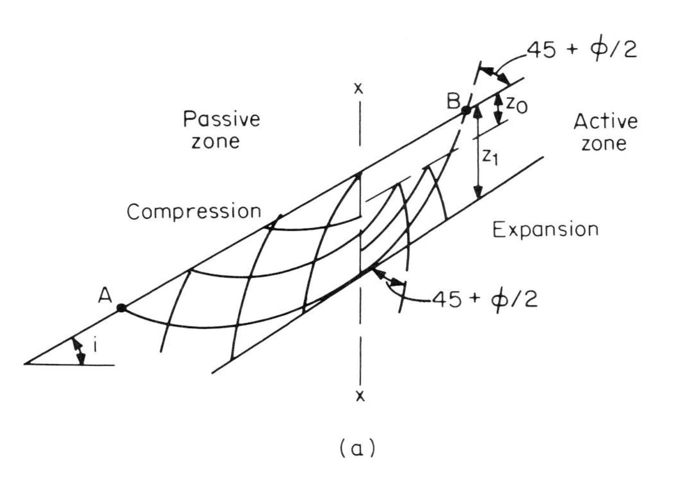


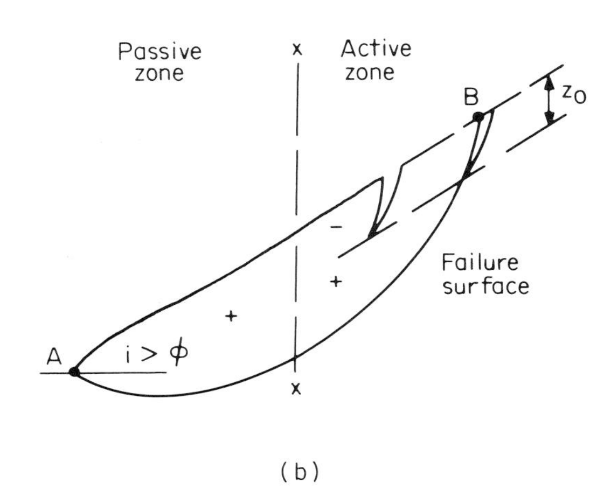


**FIG. 9.60 Active and passive stresses acting on a slope. The passive resistance at the lower portions is most significant. (*a*) Distribution of stresses on slope with $i > \phi$ (failure beginning). (*b*) A case of rupture where + indicates zones of passive stress and − indicates zones of active stress, subjected to tension cracks.**

beginning. Failures can begin at any point along a failure surface, however, where the stresses exceed the peak strength. Because failure often is progressive, it usually occurs at some average shear strength which can be considerably less than the peak strength measured by testing techniques.

*Limit Equilibrium Analysis*

Most analytical methods applied to evaluate slope stability are based on limiting equilibrium, i.e., on equating the driving or shearing forces due to water and gravity to the resisting forces due to cohesion and friction.

*Shearing forces* result from gravity forces and internal pressures acting on a mass bounded by a failure surface. Gravity forces are a function of the weight of the materials, slope angle, depth to the failure surface, and in some cases, slope height. Pressures develop in joints in rock masses from water, freezing, swelling materials, or hydration of minerals and, in soils, from water in tension cracks and pores.

*Resisting forces*, provided by the shear strength along the failure surface, are decreased by an increase in pore pressures along the failure surface, by lateral strains in overconsolidated clays in clay shales, by dissolution of cementing agents and leaching, or by the development of tension cracks (which serve to reduce the length of the resisting surface).

*Safety factor* against rupture, as given in Eq. 9.1 is:

$$FS = \frac{\text{shearing strength available along the sliding surface}}{\text{shearing stresses tending to produce failure along the surface}}$$

### Shear Strength Factors *(See Also Art. 3.4)*

*Strength Parameters*

The basic strength parameters are the angle of internal friction $\phi$ and cohesion $c$. Frictional resistance $\phi$ is a function of the normal stress, and the maximum frictional shear strength is expressed as:

$$S_{max} = N \tan \phi \qquad (9.2)$$

Cohesion $c$ is independent of the normal stress and acts over the area of the failure surface.

*Total and Effective Stresses*

In the *total stress condition*, the measured stress includes both pore-water pressures and stresses from grain-to-grain contact. In the *effective stress condition*, stresses from grain-to-grain contact are measured, which increase as pore pressures dissipate. Effective stress equals total stress minus pore pressure.

*Pore-water pressures* ($U$ for total, $u$ for unit pressures) are induced either by a load applied to a saturated specimen, or by the existence of a phreatic surface above the sliding surface. They directly reduce the normal force component $N$, and shearing resistance is then expressed as:

$$S_{max} = (N - U) \tan \phi \qquad (9.3)$$

In Fig. 9.66, therefore, if pore pressures become equal to the normal component of the weight of the block, there will be *no* shearing resistance.

*Failure Criteria*

*The Mohr-Coulomb* criterion defines failure in terms of unit shear strength and total stresses, as:

$$s = c + \sigma_n \tan \phi \qquad (9.4)$$

*The Coulomb-Terzaghi* criterion accounts for pore-water pressures by defining failure in terms of effective stresses as:

$$s = \bar{c} + \bar{\sigma}_n \tan \bar{\phi} \qquad (9.5)$$

where $\bar{c}$ = effective cohesion

$\bar{\sigma}_n$ = effective normal stress = $p - u$, where $p$ = total normal stress. In a slope the total pressure $p$ per unit of area at a point on the sliding surface equals $h_z \gamma_t / \cos^2 \theta$, where $h_z$ = the vertical distance from the point on surface of sliding to top of slope; $\gamma_t$ = slope unit weight of soil plus water; $\theta$ = inclination of surface of sliding at point with respect to horizontal

$u$ = pore-water pressure; in a slope $u = h_w \gamma_w$, the piezometric head times the unit weight of water

$\bar{\phi}$ = the effective friction angle

The strength parameters representing shearing resistance in the field are a function of the kind of materials, slope history, drainage conditions, and time. Most soils (except purely granular materials and some normally consolidated clays) are represented by both parameters $\phi$ and $c$, but whether both will act during failure depends primarily on drainage conditions and the stress history of the slope.

*Undrained vs. Drained Strength*

*Undrained conditions* exist when a fully saturated slope is sheared to failure so rapidly that no drainage can occur, as when an embankment is placed rapidly over soft soils. Such conditions are rare except in relatively impervious soils such as clays. Soil behavior may then be regarded as purely cohesive *(see Fig. 3.30)* and $\phi$ as equal to 0. Results are interpreted in terms of total stresses, and $s_u$, the undrained strength, applies. The case of sudden drawdown of an adjacent water body is an undrained condition, but analysis is based on the consolidated undrained (CU) strength of the soil before the drawdown. This strength is usually expressed in terms of the consolidated-undrained friction angle.

*Drained or long-term conditions* exist in most natural slopes, or some time after a cut is made and drainage permitted. Analysis is based on effective stresses, and the parameters $\bar{\phi}$ and $\bar{c}$ will be applicable.

*Peak and Residual Strength*

The foregoing discussion, in general, pertains to peak strengths. When materials continue to strain beyond their peak strengths, however, resistance decreases until a minimum strength, referred to as the ultimate or residual strength, is attained. The residual strength, or some value between residual and peak strengths, normally applies to a portion of the failure surface for most soils; therefore, the peak strength is seldom developed over the entire failure surface.

*Progressive failure*, when anticipated, has been approximately evaluated by using the residual strength along the upper portion of the failure surface, and the peak strength at maximum normal stress along the lower zone [Conlon as reported in Peck (1967)[34] and Barton (1972)[35]].

*Stiff fissured clays and clay shales* seldom fail in natural slopes at peak strength, but rather at some intermediate level between peak and residual. Strength is controlled by their secondary structure. The magnitude of peak strength

**FIG. 9.61 Creep deformation in varved clays (Roseton, New York).**

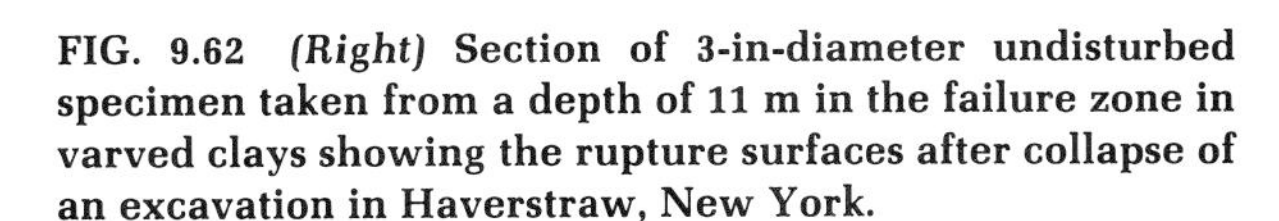

**FIG. 9.62 *(Right)* Section of 3-in-diameter undisturbed specimen taken from a depth of 11 m in the failure zone in varved clays showing the rupture surfaces after collapse of an excavation in Haverstraw, New York.**

varies with the magnitude of normal stress, and the strain at which peak stress occurs also depends on the normal stress [Peck (1967)[34]]. Because the normal stress varies along a failure surface in the field, the peak strength cannot be mobilized simultaneously everywhere along the failure surface.

*Residual strength* applies in the field to the entire failure surface where movement has occurred or is occurring. Deere and Patton (1971)[12] suggest using $\phi_r$ (the residual friction angle) where preexisting failure surfaces are present.

*Other Strength Factors*

*Stress levels* affect strength. Creep deformation occurs at stress levels somewhat lower than those required to produce failure by sudden rupture. A steady, constant force may cause plastic deformation of a stratum that can result in intense folding, as illustrated on Fig. 9.61. Shear failure by rupture occurs at higher strain rates and stress levels and distinct failure surfaces are developed as shown on Fig. 9.62. The materials are genetically the same, i.e., varved clays from the same general area.

The strength of *partially saturated* materials *(see Art. 3.4.2)* cannot be directly evaluated by effective stress analysis since both pore-air and pore-water pressures prevail. *Residual soils,* for example, are often partially saturated when sampled. In Brazil, effective stress analysis has sometimes been based on parameters measured from direct shear tests performed on saturated

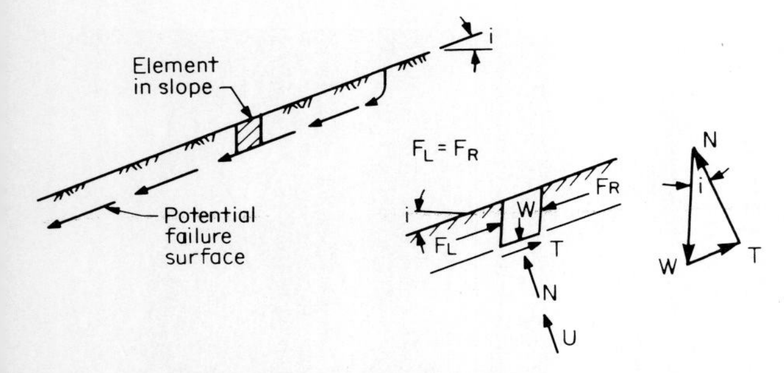


**FIG. 9.63 The infinite slope and forces on an element. (Total shear resistance = $T_{max}$ = $N \tan \phi$: $N = W \cos i$.)**

specimens to approximate the most unfavorable field conditions [Vargas and Pichler (1957)[36]]. Depending upon the degree of field saturation, the saturated strengths may be as little as 50% of the strength at field moisture. *Apparent cohesion* results from capillary forces in partially saturated fine-grained soils such as fine sands and silts; it constitutes a temporary strength which is lost upon saturation and, in many instances, on drying.

*Spontaneous liquefaction* occurs and the mass becomes fluid in fine-grained, essentially cohesionless soils when the pore pressure is sufficiently high to cause a minimum of grain-to-grain contact *(see Art. 8.3.2)*. After failure, as the mass drains and pore pressures dissipate, the mass can achieve a strength higher than before failure. High pressures can develop in pore air or pore water.

*Changes with time* occur from chemical weathering, lateral strains, solution of cementing agents, or leaching of salts *(see Art. 9.1.4)*.

*In Situ Rock Strength*

Effective stress analysis normally is applicable because the permeability of the rock mass is usually high. In clay shales and slopes with preexisting failure surfaces, the residual friction angle $\phi_r$ is often applicable, with pore pressures corresponding to groundwater conditions.

Two aspects that require consideration regarding strength are that strength is either governed by (1) planes of weakness that divide the mass into blocks (*see Art. 6.4.4,* discussion of block behavior) or (2) the degree of weathering controls, and soil strength parameters apply.

Seepage or cleft-water pressures affect the frictional resistance of the rock mass in the same manner that pore pressures affect the strength of a soil mass.

### Failure Surface Modes and Stability Relationships

*General: Two Broad Modes of Failure*

*Infinite slope* mode involves translation on a planar surface whose length is large compared with its depth. This mode is generally applicable to cohesionless sands, some colluvial and residual soil slopes underlain by a shallow rock surface, and some cases of clay shale slopes.

*Finite or limited slope* mode involves movement along a surface limited in extent. The movement can be along a straight line, a circular arc, a log-spiral arc, or combinations of these. There are two general forms of finite slope failures: wedges and circular failures. *Wedge analysis* forms are generally applicable to jointed or layered rock, intact clays on steep slopes, stratified soil deposits containing interbedded strong and weak layers, and clay shale slopes. *Cylindrical* failure surfaces are typical of normally consolidated to slightly overconsolidated clays and common to other cohesive materials including residual, colluvial, and glacial soils where the deposit is homogeneous.

*Infinite-Slope Analysis*

The infinite slope and forces acting on an element in the slope are illustrated on Fig. 9.63. In the infinite-slope problem, neither the slope height nor the length of the failure surface is considered when the material is cohesionless.

Relationships at equilibrium between friction $\phi$ and the slope angle $i$ for various conditions in a cohesionless material are given on Fig. 9.64, in which $T$ = total shearing resistance, summarized as follows:

- Dry slope: $i = \phi$ (angle of repose for sands), $T = N \tan \phi$.

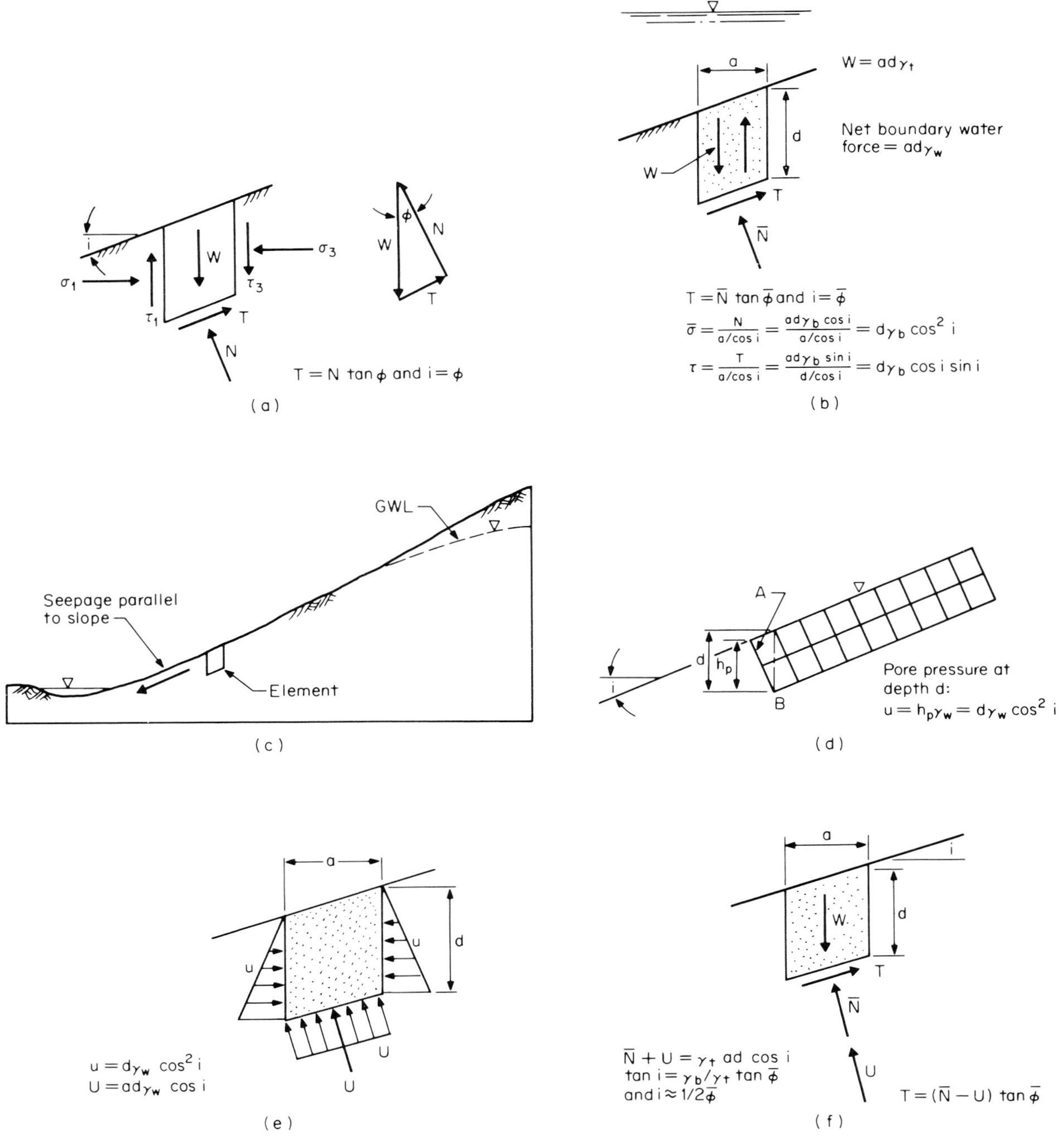


**FIG. 9.64 Equilibrium of an infinite slope in sand under dry, submerged, and slope seepage conditions: (*a*) slope in dry sand; (*b*) submerged slope in sand; (*c*) seepage in a natural slope; (*d*) flow net of seepage parallel to slope; (*e*) boundary pore pressures; (*f*) force equilibrium for element with seepage pressures.** [*After Lambe and Whitman (1969).*[37] *Adapted with permission of John Wiley & Sons, Inc.*]

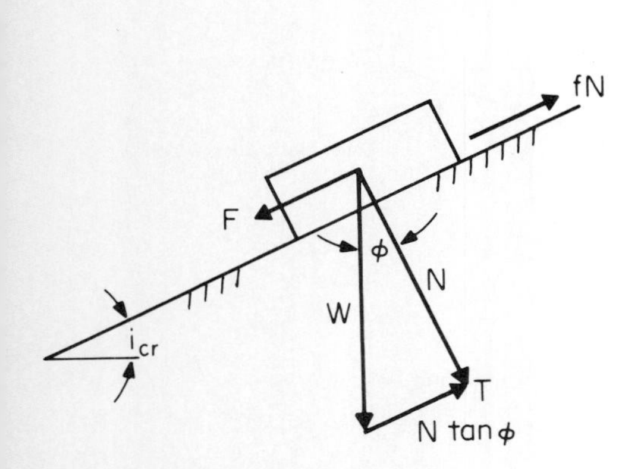


FIG. 9.65 Simple sliding block.

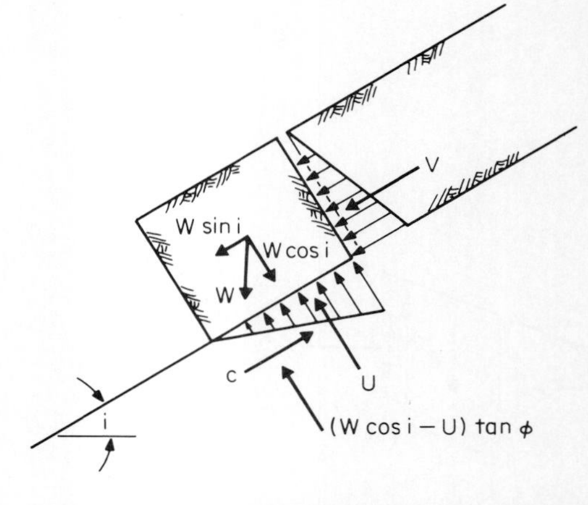


FIG. 9.66 Block with cleft-water pressures and cohesion.

- Submerged slope: $i = \phi$, $T = \overline{N} \tan \overline{\phi}$, and

$$FS = \frac{(W \cos i) \tan \overline{\phi}}{W \sin i} \qquad (9.6)$$

- Seepage parallel to slope with free water surface coincident with the ground surface (Fig. 9.64 c, d, e, and f): $i = \frac{1}{2} \overline{\phi}$, and $T = (\overline{N} - U) \tan \overline{\phi}$.
- Infinite-slope conditions can exist in soils with cohesion which serves to increase the stable slope angle $i$. These conditions generally occur where the thickness of the stratum, and therefore the position of the failure surface that can develop, are limited by a lower boundary of stronger material. Many colluvial and clay shale slopes are found in nature at $i = \frac{1}{2} \phi_r$, the case of seepage parallel to the slope with the free water surface coincident with the ground surface.

*Finite Slope: Planar Failure Surface*

CASE 1 Single planar failure surface with location assumed, involving a single block and no water pressures (Fig. 9.65). Driving force $F$ = block weight component $= W \sin i$. Resisting force $T = N \tan \phi = (W \cos i) \tan \phi$

$$FS = \frac{(W \cos i) \tan \phi}{W \sin i} \qquad (9.7)$$

where $i_{cr} = \phi$.

CASE 2 Single block with cleft-water pressures and cohesion along the failure surface with location assumed (Fig. 9.66).

$$FS = \frac{cA + (W \cos i - U) \tan \phi}{W \sin i + V} \qquad (9.8)$$

where $A$ = block base area
$V$ = total joint water pressure on upstream face of block
$U$ = total water pressure acting on the base area (boundary water pressures)
$c$ = cohesion, independent of normal stress, acting over the base area
$W$ = total weight of block, based on $\gamma_t$

CASE 3 Simple wedge acting along one continuous failure surface with cohesion and water pressure; failure surface location known (Fig. 9.67).

$$FS = \frac{cL + (W \cos \theta - U) \tan \phi}{W \sin \theta} \qquad (9.9)$$

where $L$ = length of failure surface.

CASE 4 Simple wedge with tension crack and cleft-water pressures $V$ and $U$. Failure surface location known; tension crack beyond slope crest (Fig. 9.68*a*); tension crack along slope (Fig. 9.68*b*). Fig. 9.69 gives an example of a simple wedge developing in residual soils. In Fig. 9.68:

$$FS = \frac{cL + (W \cos \theta - U - V \sin \theta) \tan \phi}{W \sin \theta + V \cos \theta} \qquad (9.10)$$

where $L = (H - z) \operatorname{cosec} \theta$
$U = \frac{1}{2} \gamma_w z_w (H - z) \operatorname{cosec} \theta$
$V = \frac{1}{2} \gamma_w z_w^2$
$W = \frac{1}{2} \gamma_t H^2 \{[1 - (z/H)^2] \cot \theta - \cot i\}$ (Fig. 9.68*a*)
or $W = \frac{1}{2} \gamma_t H^2 [(1 - z/H)^2 \cot \theta (\cot \theta \tan i - 1)]$ (Fig. 9.68*b*)

CASE 5 Single planar failure surface in clay: *location unknown; Culmann's simple wedge* (Fig. 9.70). Assumptions are that the failure surface is planar and passes through the slope toe, shear strength is constant along the failure surface in a homogeneous section, and there are no seepage forces. In practice, seepage forces are applied as in Eq. 9.9. The solution is generally

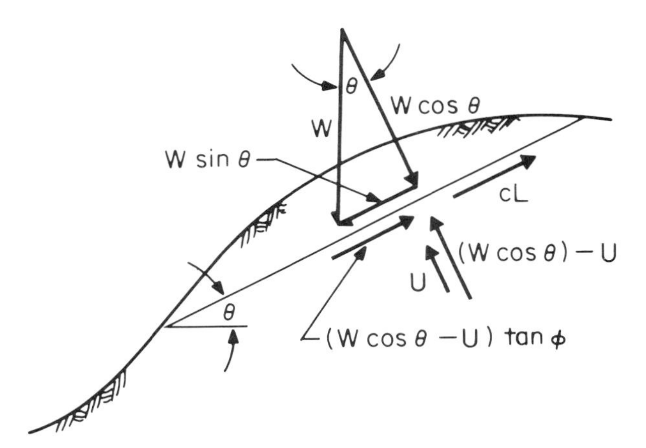


**FIG. 9.67** ***(Above)*** **Simple wedge acting along continuous surface with cohesion.**

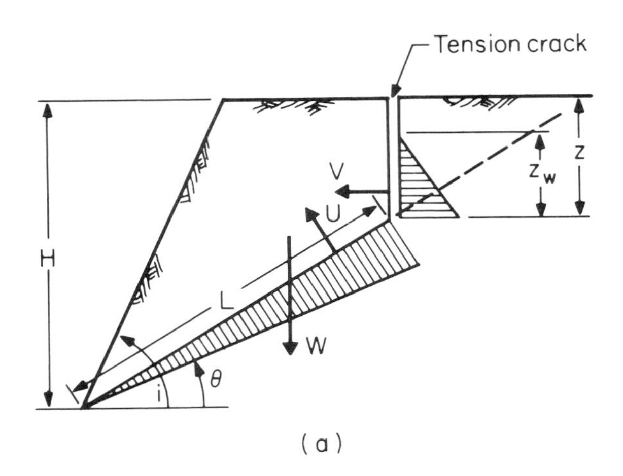


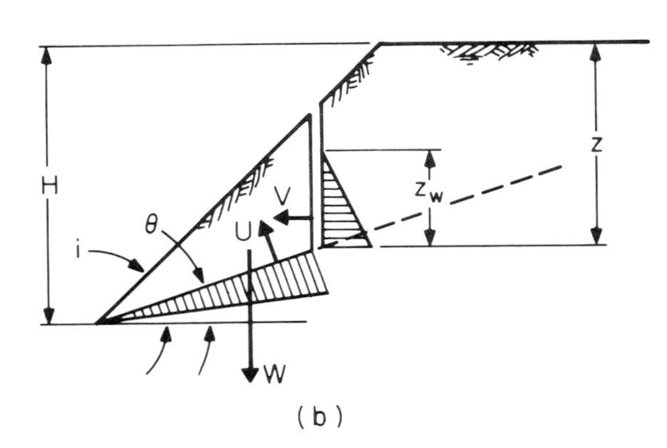


**FIG. 9.68** ***(Right)*** **Plane failure analysis of a rock slope with a tension crack: (*a*) tension crack in upper slope surface and (*b*) tension crack in slope face.** [*From Hoek and Bray (1977).*[38]]

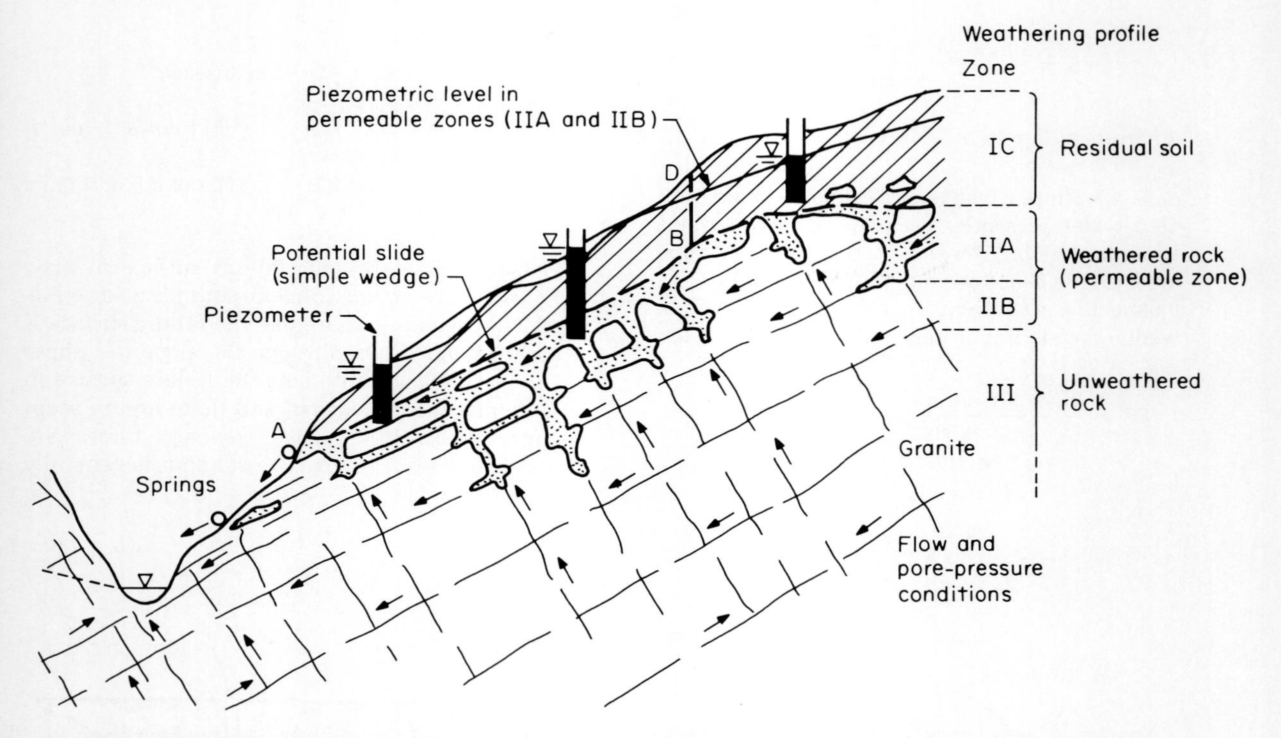


**FIG. 9.69 Development of simple wedge failure in residual soil over rock.** [*After Patton and Hendron (1974).*[39]]

considered to yield reasonable results in slopes that are vertical or nearly so, and is used commonly in Brazil to analyze forces to be resisted by anchored curtain walls *(see Art. 9.4.6)*. The solution requires finding the critical failure surface given by:

$$\theta_{cr} = \frac{i + \phi}{2} \tag{9.11}$$

$$FS = \frac{cL + (W \cos \theta) \tan \phi}{W \sin \theta} \tag{9.12}$$

where $W = \frac{1}{2} \gamma_t LH \operatorname{cosec} i \sin (i - \theta)$.

CASE 6 Critical height and tension crack in clay (Fig. 9.71). The *critical height* $H_{cr}$ is defined as the maximum height at which a slope can stand before the state of tension, which develops as the slope yields, is relieved by *tension cracks* [Terzaghi (1943)[40]]. Terzaghi gave the critical height in terms of total soil weight, having concluded that the tension crack would reach to one-half the critical height, as:

$$H'_{cr} = \frac{2.67c}{\gamma_t} \tan (45° + \phi/2) \tag{9.13}$$

or for the case of $\phi = 0$,

$$H'_{cr} = \frac{2.67c}{\gamma_t} \tag{9.14}$$

Field observations indicate that the tension crack depth $z_c$ ranges from $\frac{1}{3}H$ to $\frac{1}{2}H$. In practice $z_c$ is often taken as $\frac{1}{2}H_{cr}$ of an unsupported vertical cut, or as:

$$z_c = 2c/\gamma \tag{9.15}$$

which is considered conservative [Tschebotarioff (1973)[41]].

CASE 7 Multiple planar failure surfaces are illustrated as follows, relationships not included:

- Active and passive wedge force system applicable to rock or soil and rock slopes—Fig. 9.72
- General wedge or sliding block method appli-

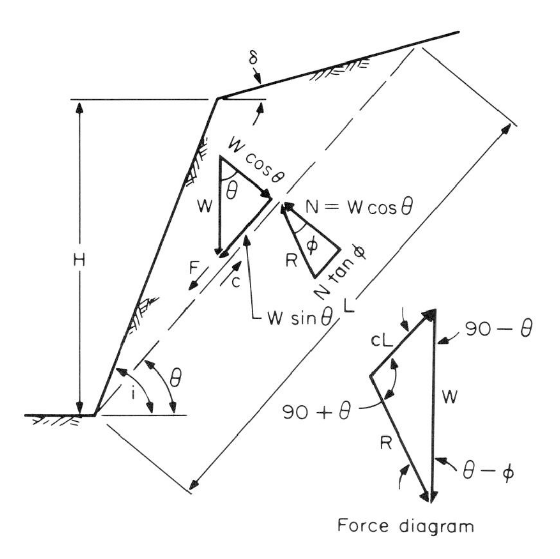


**FIG. 9.70 Culmann's simple wedge in clay.**

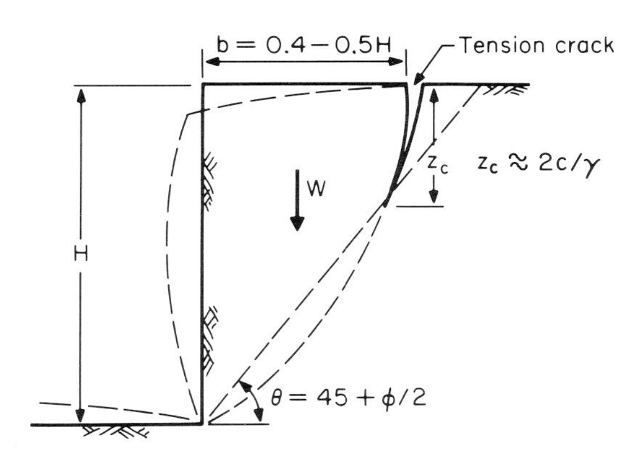


**FIG. 9.71 Critical height of a vertical slope in clay and the tension crack.**

cable to soil formations and earth dams—Fig. 9.73

- Intersecting joints along a common vertical plane—Fig. 9.74
- Triangular wedge failure, applicable to rock slopes—Fig. 9.75

*Finite Slope: Cylindrical Failure Surface*

In rotational slide failures, methods are available to analyze a circular or log-spiral failure surface, or a surface of any general shape. In all cases, the location of the critical failure surface is found by trial and error, by determining the safety factor for various trial positions of the surface until the lowest value of safety factor is reached. The forces acting on a free body taken from a slope are given on Fig. 9.76.

FRICTION CIRCLE METHOD See Taylor (1948).[44] Charts based on total stresses are used to find FS in terms of slope height and angle and of soil parameters $c$, $\phi$, and unit weight. The direction of the resultant normal stress for the entire free body is slightly in error because the resultant is not really tangent to the friction circle, but the

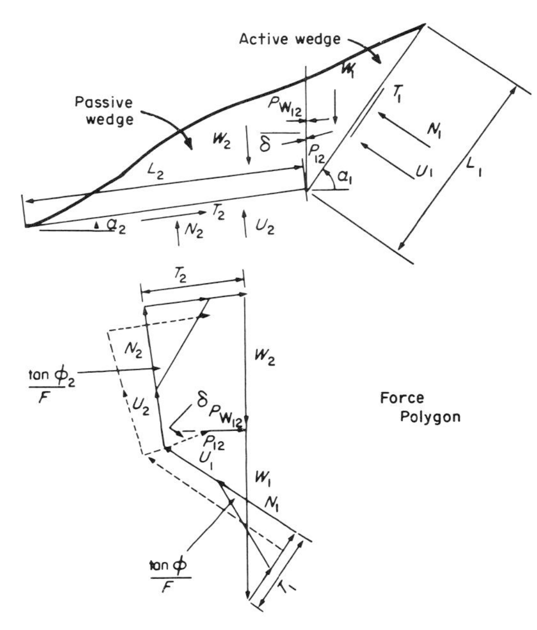


$W_1$, $W_2$ weight of a wedge
$U_1$, $U_2$ resultant water pressure acting on the base of the wedge
$N_1$, $N_2$ effective force normal to the base
$T_1$, $T_2$ shear force acting along the base of the wedge
$L_1$, $L_2$ length of the base
$\alpha_1$, $\alpha_2$ inclination of the base to the horizontal
$P_{W12}$ resultant water pressure at the interface
$P_{12}$ effective force at the interface
$\delta$ inclination of $P_{12}$ to the horizontal

**FIG. 9.72 Forces acting on two wedges: one active, one passive.** [*From Morgenstern and Sangrey (1978).*[42] *Reprinted with permission of the National Academy of Sciences.*]

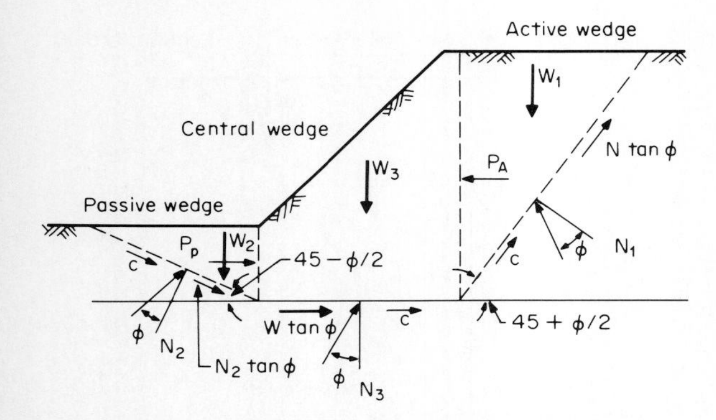


**FIG. 9.73 The general wedge or sliding block concept.** [*From NAVFAC (1971).*[43]]

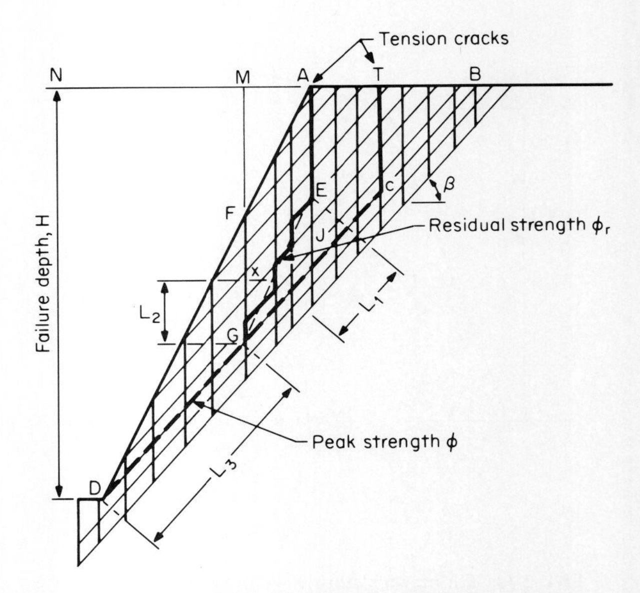


**FIG. 9.74 Geometry of two intersecting joint sets failing progressively (the multilinear failure surface).** [*From Barton (1972).*[35]]

analysis provides a lower bound for safety and is therefore conservative. The Taylor charts are strictly valid only for homogeneous slopes with no seepage. They consider that shear strength is mobilized simultaneously along the entire failure surface and that there is no tension crack. They are used for rough approximations and preliminary solutions of more complex cases. If the strength values vary along the failure surface they are averaged to obtain working values. This must be done with judgment and caution. For the foregoing conditions of validity, solutions using the charts are in close agreement with the method of slices described below.

ORDINARY METHOD OF SLICES (SWEDISH CIRCLE OR FELLENIUS METHOD) As illustrated on Fig. 9.77, the mass above a potential failure surface is drawn to scale and divided into a number of slices with each slice having a normal force resulting from its weight. A flow net is drawn on the slope section (Fig. 9.77*a*) and the pore pressures determined as shown on Fig. 9.77*b*. The equilibrium of each slice is determined and FS found by summing the resisting forces and dividing by the driving forces as shown on Fig. 9.77*c*. The operation is repeated for other circles until the lowest safety factor is found. The method does not consider all of the forces acting on a slice, as it omits the shear and normal stresses and porewater pressures acting on the sides of the slice, but usually (although not always) it yields conservative results. However, the conservatism may be high.

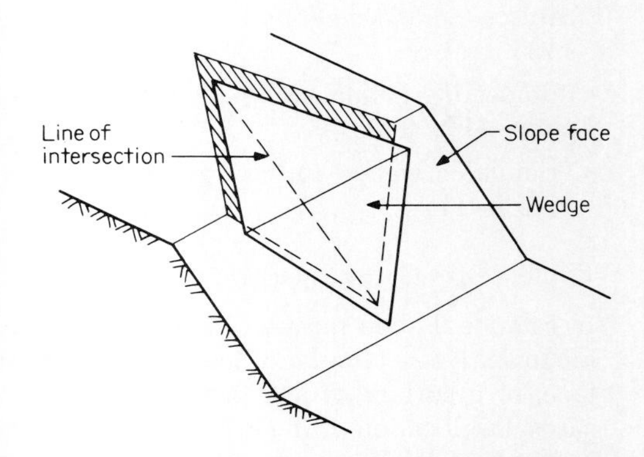


**FIG. 9.75 Geometry of a triangular wedge failure.** [*From Hoek and Bray (1977).*[38]]

BISHOP'S METHOD OF SLICES This method considers the complete force system, but is complex and requires a computer for solution. The results, however, are substantially more accurate than either the ordinary method or the modified Bishop method [Bishop (1955)[45]].

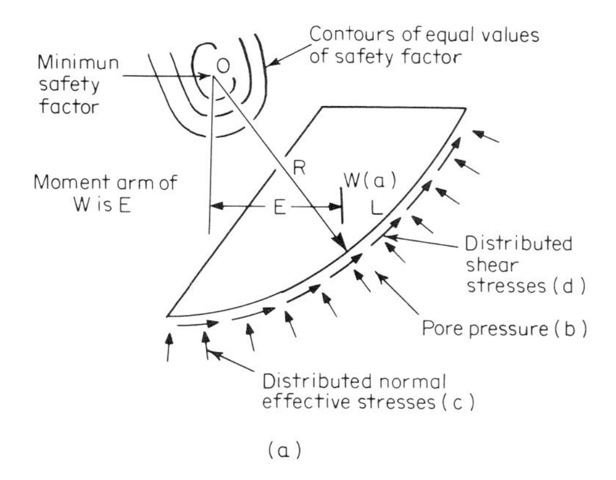


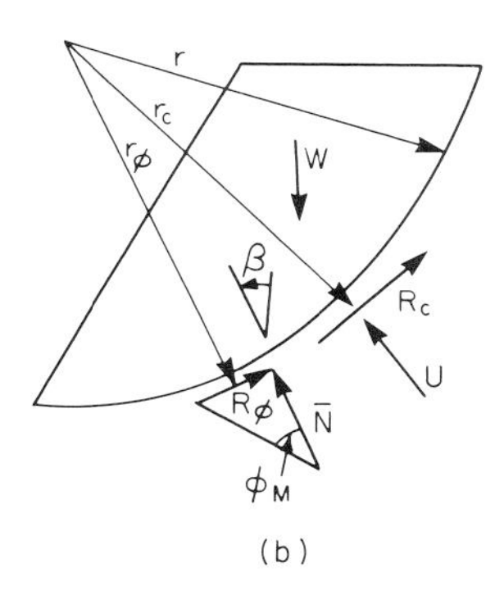


**FIG. 9.76** *(Above)* **Forces acting on a free body with circular failure: (*a*) distributed stresses and (*b*) resultant forces.**

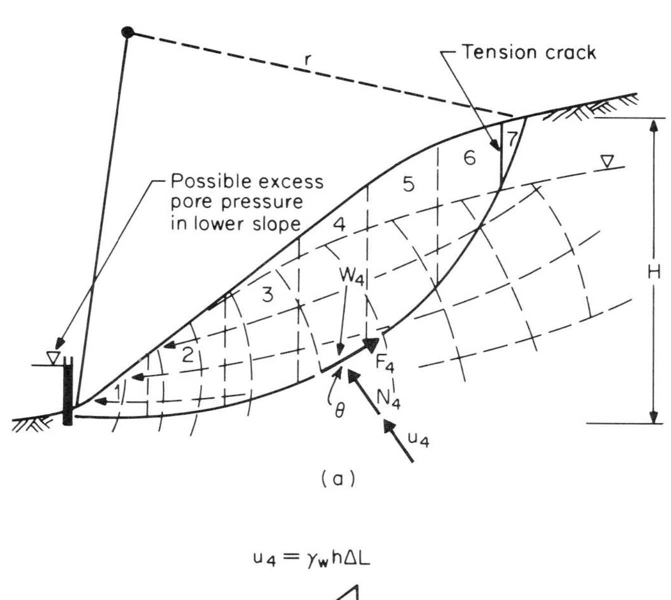


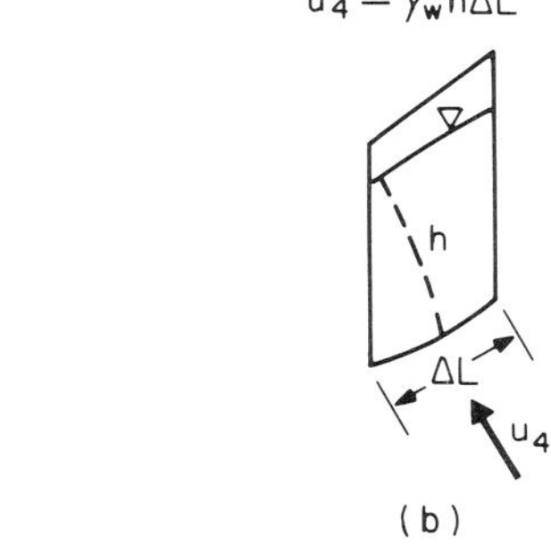


**FIG. 9.77** *(Right)* **The ordinary method of slices. (*a*) Draw the slope and flow net to scale. Select failure circle and divide into equal slices of similar conditions. (*b*) Pore pressure for slice $u_4$. (*c*) Determine forces and safety factor.**

$$\overline{N}_i = W_i \cos\theta - u_i L_i \quad \text{and} \quad F = W_i \sin\theta$$

$$FS = \frac{\Sigma \bar{c}\, \Delta L_i + \Sigma \overline{N}_i \tan \bar{\phi}}{\Sigma W_i \sin \theta_i}$$

**Repeat for other r values to find $FS_{min}$**

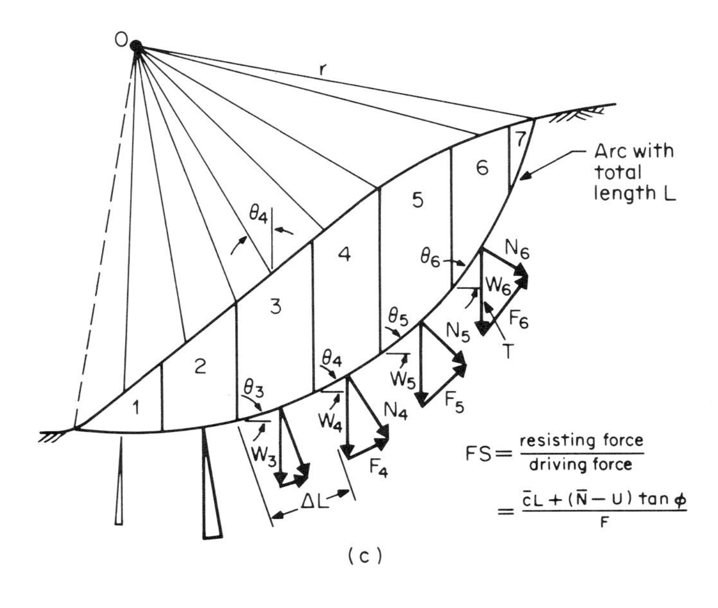

**TABLE 9.4**
**COMPARISON OF ELEMENTS AND CLASSIFICATION OF GEOLOGICAL AND ENGINEERING FAILURE FORMS**

| | Elements of slope failures* | | | | | | Engineering failure forms† | | | | |
|---|---|---|---|---|---|---|---|---|---|---|---|
| Geologic failure forms | Slope inclination | Slope height | Material structure | Material strength | Seepage forces | Runoff | Infinite slope | Single planar failure surface (simple wedge and sliding block) | Multiple planar failure surfaces (oblique surfaces intersecting parallel surfaces) | Multiple planar failure surfaces intersecting obliquely (wedge) | Cylindrical failure surface |
| Falls | P | N | P | P | P | N | N | N | N | P | N |
| Planar slides (translational, block glides) | P | S | P | P | P | M | A | A | A | A | N |
| Rotational slides in rock | P | P | P | P | P | M | N | N | N | N | A |
| Rotational slides in soil | P | P | P | P | P | M | N | N | N | N | A |
| Lateral spreading and progressive failure | S | M | P | P | P | N | N | N | N | N | N |
| Debris slides | P | M | P | P | P | N | S | S | S | S | N |
| Debris avalanches | P | S | S | S | P | P | N | N | N | N | N |
| Debris flows | P | S | S | S | P | P | N | N | N | N | N |
| Rock fragment flow | P | S | P | P | P | N | N | N | N | N | N |
| Soil and mud flows | S | S | S | P | P | M | N | N | N | N | N |
| Submarine slides | S | S | P | P | P | N | N | N | N | N | S |

*P—primary cause; S—secondary cause; M—minor effect; N—little or no effect.
†A—application; S—some application; P—poor application; N—no application.

MODIFIED BISHOP METHOD This method is a simplified Bishop method [Janbu et al. (1956)[46]], widely used for hand calculations since it gives reasonably accurate solutions for circular failure surfaces.

MORGENSTERN AND PRICE METHOD See Morgenstern and Price (1965).[47] This method can be used to analyze any shape of failure surface and satisfies all equilibrium conditions. It is based on the Bishop method and requires a computer for solution. There are several theoretically possible positions for the line of action of the resultant forces between slices, and the line of action must be checked to determine if it is a possible one.

JANBU'S METHOD This method is an approximate method applicable to noncircular as well as circular failure surfaces and suitable for hand calculations. It is sufficiently accurate for many practical cases [Janbu et al. (1956),[46] Janbu (1973)[48]].

### Earthquake Forces

*Pseudostatic methods* have been the conventional approach in the past [Terzaghi (1950)[6]]. The stability of a potential sliding mass is determined for static loading conditions, and the effects of earthquake forces are accounted for by including equivalent horizontal forces acting on the mass. The horizontal force is expressed as the product of the weight of the sliding mass and a *seismic coefficient* which is expressed as a fraction of the acceleration of gravity *(see Art. 11.3.4).*

*Dynamic analysis techniques* provide for much more realistic results but also have limited validity. These techniques are described by Newmark (1965)[49] and Seed (1966).[50]

### Summary

The applicability of mathematical analysis to various slope failure forms and the elements affecting slope failures are summarized on Table 9.4.

General methods of stability analysis for sliding masses and the applicable geologic conditions are summarized on Table 9.5.

Strength parameters acting at failure under various field conditions are summarized on Table 9.6.

## 9.3.3 SLOPE CHARACTERISTICS

### General

Qualitative assessment of slopes provides the basis for predicting the potential for failure and selecting practical methods for treatment, and for evaluating the applicability of mathematical solutions.

The two major elements of qualitative assessment are slope characteristics (geology, geometry, surface conditions, and activity) and the environment (weather conditions of rainfall and temperature, and earthquake activity). The discussion in Art. 9.2 presents relationships between the mode of slope failure and geologic conditions, as well as other slope characteristics, giving a basis for recognizing potential slope stability problems.

**TABLE 9.5**
**GENERAL METHOD OF STABILITY ANALYSIS AND APPLICABLE GEOLOGIC CONDITION FOR SLIDES**

| General method of analysis | Geologic conditions |
|---|---|
| Infinite slope—(depth small compared with length of failure surface) | Cohesionless sands. Residual or colluvial soils over shallow rock. Stiff fissured clays and marine shales in the highly weathered zone. |
| Limited slope<br>Simple wedge (single planar failure surface) | Sliding block.<br>Interbedded dipping rock or soil.<br>Faulted or slickensided material.<br>Stiff to hard cohesive soil, intact, on steep slope. |
| General wedge (multiple planar failure surfaces) | Sliding blocks in rock masses.<br>Closely jointed rock with several sets.<br>Weathered interbedded sedimentary rock.<br>Clay shales and stiff fissured clays.<br>Stratified soils.<br>Side-hill fills over colluvium. |
| Cylindrical arc | Thick residual or colluvial soil.<br>Soft marine or clay shales.<br>Soft to firm cohesive soils. |

**TABLE 9.6**
**STRENGTH PARAMETERS ACTING AT FAILURE UNDER VARIOUS FIELD CONDITIONS**

| Material | Field conditions | Strength parameters |
|---|---|---|
| (a) Cohesionless sands | Dry | $\phi$ |
| (b) Cohesionless sands | Submerged slope | $\bar{\phi}$ |
| (c) Cohesionless sands | Slope seepage with top flow line coincident with and parallel to slope surface | $\bar{\phi}$ |
| (d) Cohesive materials | Saturated slope, short-term or undrained conditions ($\phi = 0$) | $s_u$ |
| (e) Cohesive materials (except for stiff fissured clays and clay shales) | Long-term stability | $\bar{\phi}$, $\bar{c}$ |
| (f) Stiff fissured clays and clay shales | Part of failure surface | $\bar{\phi}_r$ |
| (g) Soil or rock | Part of failure surface<br>Existing failure surfaces | $\bar{\phi}$, $\bar{c}$<br>$\bar{\phi}_r$ |
| (h) Clay shales or existing failure surfaces | Seepage parallel and top flow line coincident to slope surface | $\bar{\phi}_r$ |
| (i) Pore-water pressures | Reduce $\phi$ in e, f, and g in accordance with seepage forces, $\gamma$ applicable; or boundary water forces, $\gamma_t$ applicable *(see Art. 8.3.2)*<br>In c and h pore pressures reduce effectiveness by 50% | $(p - u) \tan \bar{\phi}$ |

## Geologic Conditions

*Significant Factors*

- *Materials* forming the slope (for rock, the type and the degree of weathering; for soil, the type as classed by origin, mode of occurrence, and composition) as well as their *engineering properties*.
- *Discontinuities* in the formations, which for rock slopes include joints, shears, bedding, foliations, faults, slickensides, etc., and for soils include layering, slickensides, and the bedrock surface.
- *Groundwater* conditions; static, perched or artesian, and seepage forces.

*Conditions With a High Failure Incidence*

- Jointed rock masses on steep slopes can result in falls, slides, avalanches, and flows varying from a single block to many blocks.
- Weakness planes dipping down and out of the slope can result in planar failures with volumes ranging from very large to small.
- Clay shales and stiff fissured clays are frequently unstable in the natural state where they normally fail by shallow sloughing, but cuts can result in large rotational or planar slides.
- Residual soils on moderate to steep slopes in wet climates may fail progressively, generally

involving small to moderate volumes, although heavy runoff can result in debris avalanches and flows, particularly where bedrock is shallow.

- Colluvium is generally unstable on any slope in wet climates and when cut can fail in large volumes, usually progressively.
- Glaciolacustrine soils normally fail as shallow sloughing during spring rains, although failures can be large and progressive.
- Glaciomarine and other fine-grained soils with significant granular components can involve large volumes in which failure may start by slumping, may spread laterally, and under certain conditions may become a flow.
- Any slope exposed to erosion at the toe, or cut too steeply, or experiencing deformation.

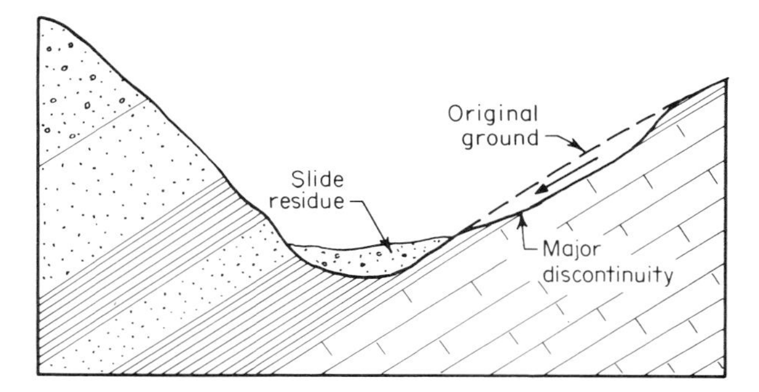


**FIG. 9.78 Dipping beds of sedimentary rock in mountainous terrain. Failures on the right side are large dimension and occur along bedding plane weaknesses; those on the left are smaller in scale and occur along joints.**

**FIG. 9.79 Near-vertical slope in 40-year-old railroad cut standing stable in amphibolite gneiss because of vertical jointing (Tres Ranchos, Goias, Brazil). Compare with Fig. 9.80.**

**FIG. 9.80 Same cut as in Fig. 9.79, but different station. Wedge failure along dipping joints.**

### Examples of the Influence of Geologic Conditions

A general summary of typical forms of slope failures as related to geologic conditions is given on Table 9.3.

Dipping beds of sedimentary rocks in mountainous terrain (Fig. 9.78) are often the source of disastrous slides or avalanches. Very large planar slides failing along a major discontinuity may occur as shown on the right-hand side of the figure where the beds incline in the slope direction. On the left-hand side the slope is steeper and usually more stable because of the bedding orientation. Failures will generally be small, evolving under joint sets, although disastrous avalanches have occurred under these conditions, such as the one at Turtle Mountain, Alberta.

Orientation of joints with respect to the slope face controls stability and the form of failure. The near-vertical slope in the 40-year-old railroad cut illustrated in Fig. 9.79 is stable in decomposed amphibolite gneiss because of the vertical jointing. The cut shown in Fig. 9.80 is near that of Fig. 9.79 but at a different station and on the opposite side of the tracks. Here the slope is much flatter, roughly 1:1, but after 40 years is still experiencing failures such as that of the wedge shown in the photo which broke loose along the upper joints and slid along a slickensided surface. These examples illustrate how joint orientation controls slope stability, even in "soft" rock. The cuts were examined as part of a geologic study for 30 km of new railroad to be constructed in the same formation but some distance away.

Sea erosion undercutting jointed limestone illustrated in Fig. 9.81 is causing concern over the possible loss of the roadway, which is the only link between the town of Tapaktuan, Sumatra, and its airport. A fault zone may be seen on the right-hand side of the photo. For the most part the joints are vertical and perpendicular to the cliff face, shown as plane *a* in Fig. 9.82, and the conditions are consequently stable. Where the joints are parallel to the face and inclined into it, as shown by plane *b* on the figure, a potentially unstable condition exists. This condition was judged to prevail along a short stretch of road beginning to the right of the photo, illustrated on Fig. 9.83. The recommendation was to cut into the landward slope and relocate the roadway away from the sea cliff along this short stretch.

The major cause of instability in colluvial soil slopes is illustrated on Fig. 9.84. The slide debris impedes drainage at the toe and causes an increase in pore-water pressures in (*b*) over those in (*a*). The sketch also illustrates the importance of placing piezometers at different depths because of pressure variations. Conditions in (*b*) apply also to the case of a side-hill embankment for which a free-draining blanket beneath the fill would be necessary to provide stability.

**FIG. 9.81 Sea erosion undercutting limestone and causing rockfalls (Tapaktuan, Sumatra).**

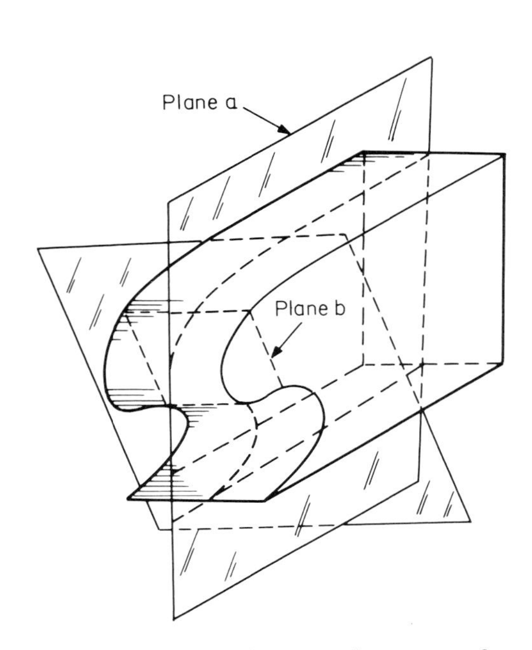


**FIG. 9.82 Orientation of fracture planes controls rock-mass stability. Plane *a* represents stable conditions; plane *b*, unstable.**

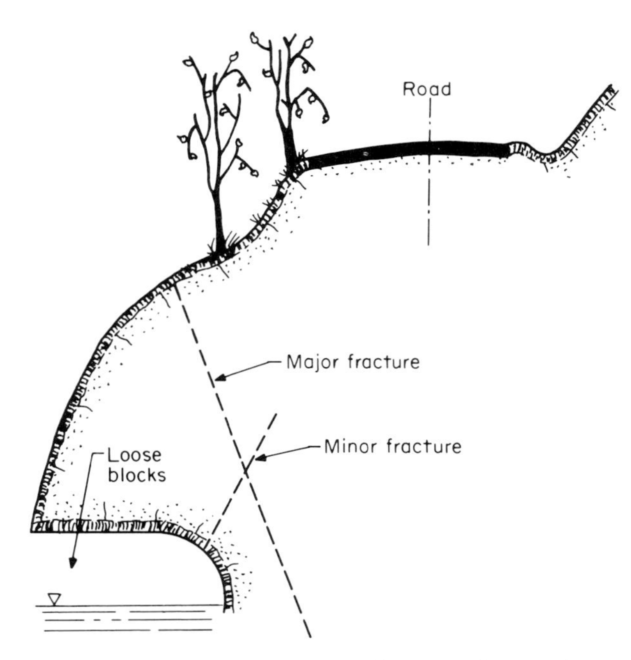


**FIG. 9.83 Possible orientation of fracture planes at km 10 + 750 which might lead to a very large failure (Tapaktuan, Sumatra).**

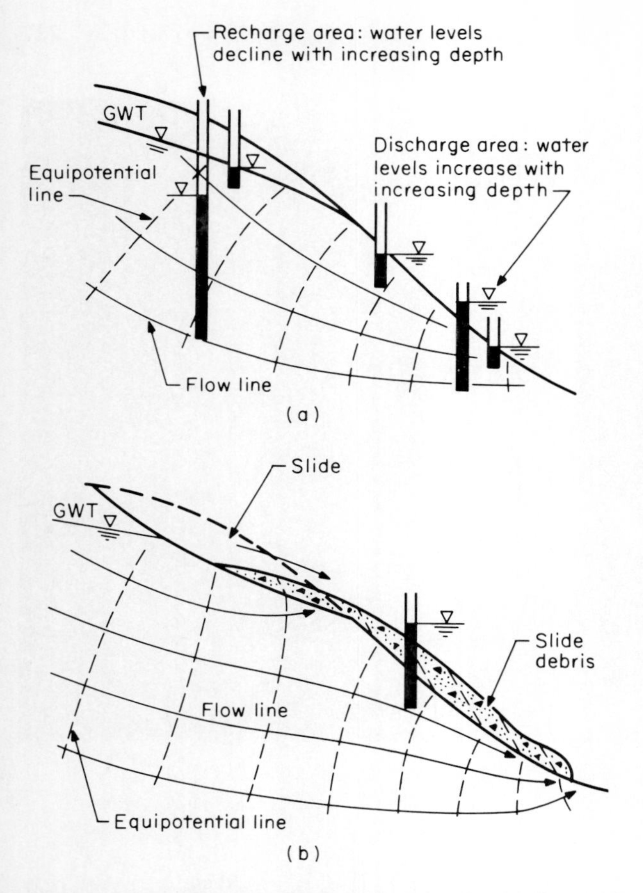


**FIG. 9.84 The effect of colluvium on groundwater flow in a slope: (*a*) groundwater flow in slope before slide and (*b*) groundwater flow in slope with mantle of slide debris.** [*From Patton and Hendron (1974).*[39]]

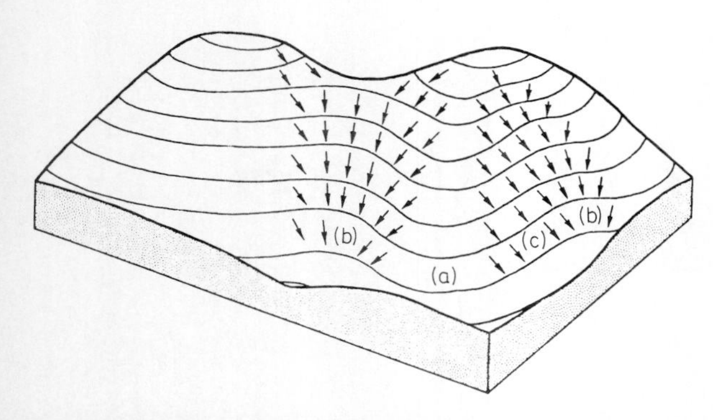


**FIG. 9.85 The influence of topography on runoff and seepage forces. Cuts and natural slopes at *a* are relatively stable compared with those located at *c* or *b*.**

### Slope Geometry

*General*

The significant elements of slope geometry are inclination, height, and form. Aspects of inclination and height, as they relate to a particular point along a slope, are described in Art. 9.1.4. This section is more concerned with the form and other characteristics of an entire slope as they affect seepage and runoff, which can be dispersed by the geometrical configuration of the slope or can be concentrated. The difference influences slope stability.

The examples given are intended to illustrate the importance of considering the topography of an entire slope during roadway planning and design, not only the immediate cut or fill area.

*Topographic Expression*

In both natural and cut slopes the topographic expression has a strong influence on where failure may occur since landform provides the natural control over rainfall infiltration and runoff when geologic factors are constant. In Fig. 9.85 runoff is directed away from the nose-form at *a* and a cut made there will be stable at a much steeper angle than at *b*, where runoff is concentrated in the swale. Runoff and seepage at *c* are less severe than at *b* but still a problem to be considered. Natural slides, avalanches, and flows usually will not occur at *a*, but rather at *b* and *c*, with the highest incidence at *b*.

*Location of Cut on Slope*

Cuts in level ground or bisecting a ridge perpendicular to its strike will be stable at much steeper inclinations than cuts made along a slope, parallel to the strike (side-hill cuts.). The side-hill cut in Fig. 9.86 intercepts seepage and runoff from upslope and will be much less stable on its upslope side than on its downslope side where seepage is directed away from the cut. The treatment to provide stability, therefore, will be more extensive on the upslope side than on the opposite side.

The significance of cut locations along a steeply inclined slope in mountainous terrain in a tropical climate is illustrated on Fig. 9.87. A cut made at location 1 will be much less stable than at location 3, and treatment will be far more costly because of differences in runoff and seepage quantities. River erosion protection or retention of the cut slope at 1 can be more costly than the roadway itself. Retention would not be required at 3 if a stable cut angle were selected, but might be required at 2 together with positive seepage control.

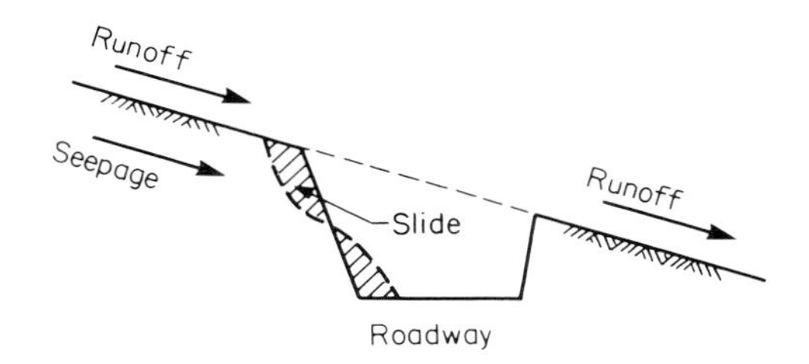


**FIG. 9.86 Upslope side of hillside cut tends to be much less stable than downslope side because of runoff and seepage.**

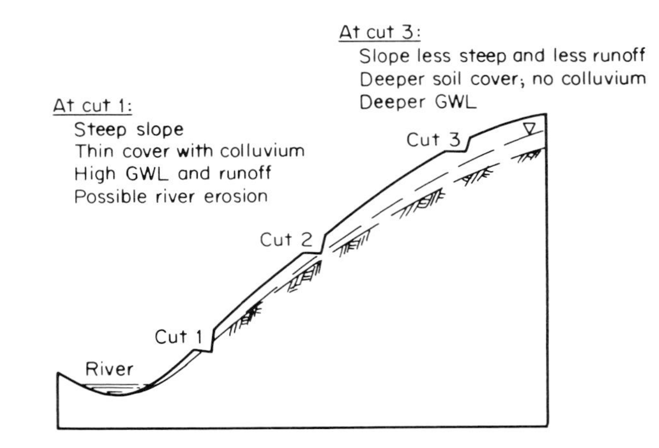


**FIG. 9.87 Stability problems may be very much related to the cut location along a steeply inclined slope in mountainous terrain in a tropical climate, and generally decrease in the upslope direction.**

### Surface Conditions

*Seepage Points*

Observations of seepage points should be made in consideration of the weather conditions prevailing during the weeks preceding the visit, as well as the season of the year, and regional climatic history.

No slope seepage during a rainy period may be considered as very favorable for stability, if there is no blockage from ice, colluvium, etc. On the other hand, seepage during a dry period signifies that a substantial increase in seepage will occur during wet periods.

Toe seepage indicates a particularly dangerous condition, especially during dry periods.

*Vegetation*

Density of vegetation is an important factor in slope stability. Recently cleared upslope areas for logging, farming, or grazing are very likely to be locations where failures will occur in freshly made cuts, or in old cuts during severe weather conditions. Removal of vegetation permits an increase in erosion, a reduction in strength in the shallow portions of the slope from the loss of root structure, an increase in infiltration during rainy periods, and an increase in evaporation during dry spells resulting in surface desiccation and cracking.

Certain types of vegetation may be indicators of potential instability. For example, in tropical climates, such as in Brazil and Indonesia, banana plants seem to favor colluvial soil slopes, probably because colluvium has a higher moisture content than the residual soils in the same area.

*Indications of Instability*

Surface features indicating instability include tilted or bending tree trunks, tilted poles and fence posts, tension cracks along the slope and beyond the crest, and slump and hummocky topography, as described and illustrated in Arts. 9.1.2 and 9.2.

### Slope Activity

*Degrees of Activity*

Slopes reside at various degrees of activity, as discussed in Art. 9.1.2, ranging through stable slopes with no movement, early failure stages with creep and tension cracks, intermediate failure stages with significant movement, partial

**FIG. 9.88 Slump movement caused by a cut made in residual soils for the Rio Santos Highway (Itaorna, Brazil) has remained more or less as shown for a period of at least 4 years. (Photo taken in 1978.)**

total failures with substantial displacement, to complete failure with total displacement.

Evidence that a slope is unstable requires an assessment of the imminence of total collapse; and, if movement is occurring, how much time is available for treatment and stabilization. Tension cracks, in particular, serve as an early warning of impending failure and are commonly associated with the early stages of many failures. Their appearance, even together with scarps, does not necessarily mean, however, that failure is imminent. The slope shown in Fig. 9.88 appears precipitous but the slide has moved very little over a 4-year period, an interval of lower than normal rainfall *(see Art. 9.3.4).* Note the "nose" location; a massive slide has already occurred along the slope at the far left in the photo which closed the highway for a brief time.

The most significant factors indicating approaching total failure for many geologic conditions are velocities of movement and accelerations.

*Movement Velocity vs. Failure*

Methods of monitoring slope movements, both surficially and internally, are described in Art. 4.5.6. During surficial monitoring, both vertical and horizontal movements should be measured, and evaluated in terms of velocity and acceleration. There is a lack in the literature of definitive observational data that relate velocity and acceleration to final or total failure in a manner suitable for formulating judgments as to when failure is imminent. Movement velocities before total failure and the stabilization treatment applied are summarized on Table 9.7 for a few cases from the literature. From the author's experience and literature review, it appears that, as a rule of thumb, if a slope of residual or colluvial soils is moving at a rate in the order of 2 to 5 cm/day (0.8 to 2.0 in/day) during a rainy season with the probability of storms, and if the velocity is increasing, final failure is imminent and may occur during the first heavy rain, or at some time during the rainy season. The following two case histories relate slope movements to total failure.

CASE 1 A roadway cut made in residual soils in the coastal mountains of Brazil continued to show instability through creep, tension cracks, small slumps, and periodic encroachment on the roadway for a period of several years. An intermediate failure stage occurred on November 29, 1977, after a weekend of moderately heavy rain and a period during which the highway department had been removing material from the slope toe. Figure 9.89, a photo taken from a helicopter about 10 days after the failure, illustrates the general conditions. Tension cracks have opened at the base of the forwardmost transmission tower, a large scarp has formed at midslope, and the small gabion wall at the toe has failed.

Figure 9.90 illustrates the tension crack and the distortion in the transmission tower shown in Fig. 9.89. The maximum crack width was about 30 cm and the scarp was as high as 50 cm. Slope-movement measurements were begun immediately by optical survey and the transmission lines were quickly transferred to a newly constructed tower situated farther upslope.

For the first 2 weeks after the initial movement, a period of little rainfall, the vertical drop along the scarp was about 2 to 3.5 cm/day (0.8 to 1.4 in/day). In 5 weeks, with occasional rainy periods, the scarp had increased to 3 m (10 ft), and finally after a weekend of heavy rains the slide failed totally in its final stage within a few hours, leaving a scarp about 30 m (100 ft) in height as shown on Fig. 9.91, and partially block-

**TABLE 9.7**
**VELOCITIES OF SLIDE MOVEMENTS BEFORE TOTAL FAILURE AND SOLUTIONS***

| Location | Material | Movement velocity | Solution | Reference |
|---|---|---|---|---|
| Philippine Islands | Weathered rock (open-pit mine) | 2 cm/day | Horizontal adits | Brawner (1975)[51] |
| Santos, Brazil | Colluvium (cut) | 2.5 cm/day | Trenches, galleries | Fox (1964)[52] |
| Rio de Janeiro state, Brazil | Residuum (cut) | 0.4–2.2 cm/day | None applied, no total failure in 20 years | Garga and DeCampos (1977)[53] |
| Rio Santos Highway, Brazil | Residuum (cut) | 2–3.5 cm/day for first 2 weeks, 30 cm/day during 6th week, then failure | Removal of failure mass | Hunt (1978) *(see Art. 9.3.3)* |
| Golden slide | Debris (cut) | 2.5 cm/day | Horizontal drains, vertical wells | Noble (1973)[21] |
| Pipe Organ slide | Debris (cut) | 5 cm/week | Gravity drains | Noble (1973)[21] |
| Vaiont, Italy | Rock-mass translation | 1 cm/week, then 1 cm/day, then 20–30 cm/day and after 3 weeks, 80 cm/day and failure | | Kiersch (1965)[8] |
| Portuguese Bend | Lateral spreading | 1956–1957, 5–12 cm/year<br>1958, 15–60 cm/year<br>1961–1968, 1–3m/year<br>1968, dramatic increase, houses destroyed<br>1973, 8 cm/day during dry season<br>1973, 10 cm/day during wet season<br>1973, 15 cm/day during heavy rains<br>No total mass failure | None | Easton (1973)[16] |

*From examples given in Chap. 9.

FIG. 9.89 *(Above)* Four-year-old rotational slide at Muriqui, Rio de Janeiro, Brazil, initiated by road cut, began major movements after weekend of heavy rains, endangering transmission towers.

FIG. 9.90 *(Left)* Tension crack and distortion of transmission tower shown in Fig. 9.89.

ing the roadway. Excavation removed the slide debris, and 2 years afterward the high scarp still remained. The tower in the photo was again relocated farther upslope. Although future failures will occur, they will be too far from the roadway to cover the pavement. The relocation of the slope by permitting failure is an example of one treatment method.

CASE 2 Another situation is similar to that described in case 1 although the volume of the failing mass is greater. A 25° slope of residual soil has been moving for 20 years since it was activated by a road cut [Garga and DeCampos (1977)[53]]. Each year during the rainy season movement occurs at a velocity measured by slope inclinometer ranging from 0.4 to 2.2 cm/day (0.2 to 0.9 in/day). The movement causes slide debris to enter the roadway, from which it is removed. During the dry season movement ceases, and as of the time of the report (1977), total failure had not occurred.

### 9.3.4 WEATHER FACTORS

#### Correlations Between Rainfall and Slope Failures

*Significance*

Ground saturation and rainfall are the major factors in slope failures and influence their incidence, form, and magnitude. Three aspects are important:

1. Climatic cycles over a period of years, i.e., high annual precipitation vs. low annual precipitation
2. Rainfall accumulation in a given year in relationship to normal accumulation
3. Intensities of given storms

**FIG. 9.91 Total collapse of slope shown in Fig. 9.89 occurred 6 weeks later during the rainy season after several days of heavy rains. *See also Fig. 9.27.***

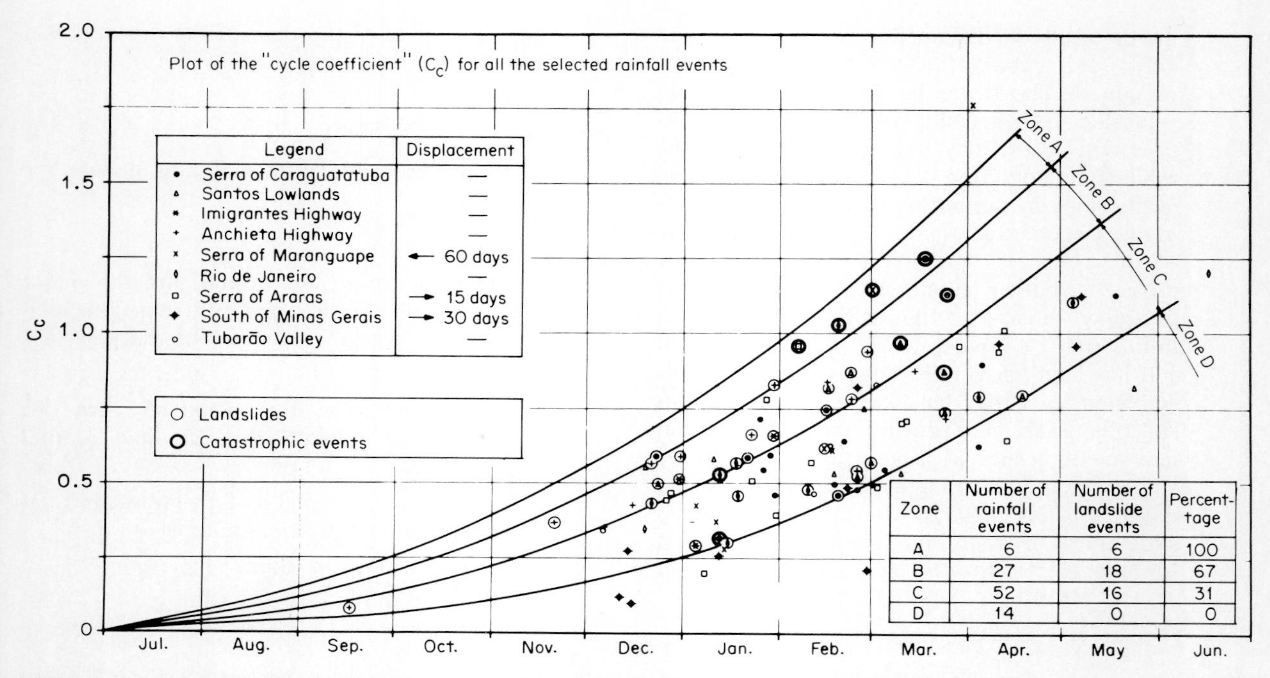


**FIG. 9.92 Comparison of landslide events and cycle coefficient $C_c$ for coastal mountains of Brazil.** [*From Guidicini and Iwasa (1977).*[54]]

*Cumulative Precipitation vs. Mean Annual Precipitation*

A study of the occurrence of landslides relative to the cumulative precipitation record up to the date of failure as a percentage of the mean annual precipitation (termed the *cycle coefficient* $C_c$) was made by Guidicini and Iwasa (1977).[54] The study covered nine areas of the mountainous coastal region of Brazil, which has a tropical climate characterized by a wet season from January through March and a dry season, June through August.

Cumulative precipitation causes ground saturation to increase and the water table to rise. A rainstorm occurring during the dry season or at the beginning of the wet season will have a lesser effect on slope stability than a storm of the same intensity occurring near the end of the wet season. A plot by month of the occurrence of failures as a function of the cycle coefficient is given as Fig. 9.92. It is seen that the most catastrophic events occur toward the end of the rainy season, when the cumulative precipitation is higher than the mean annual.

Regarding rainfall intensity, Guidicini and Iwasa concluded that:

- Extremely intense rainfalls, about 12% greater than the mean annual rainfall (300 mm in 24 to 72 hr) or more, can cause natural slope failures in their area, regardless of the previous rainfall history.
- Intense rainfalls, up to 12% of the mean annual, where the precipitation cycle is normal or higher, will cause failures, but if the preceding precipitation level is lower than the mean annual, failures are not likely even with intensities to 12%.
- Rainfalls of 8% or less of annual precipitation will generally not cause failures, regardless of the preceding precipitation, because the gradual increase in the saturation level never reaches a critical magnitude.
- A *danger level chart* (Fig. 9.93) was prepared by Guidicini and Iwasa for each study area,

intended to serve communities as a guide for assessing failure hazard in terms of the mean cumulative precipitation for a given year.

## Evaluating Existing Cut Slope Stability

### *General*

It is often necessary to evaluate a cut slope that appears stable, to formulate judgments as to whether it will remain so. If a cut slope has been subjected during its lifetime to conditions drier than normal and there have been no major storms, it can be stated that the slope has not been tested under severe weather conditions, and it may be concluded that it is not necessarily a potentially stable slope. If a cut is failing under conditions of normal rainfall, it can be concluded that it will certainly undergo total failure at some future date during more severe conditions.

### *Case 1*

In a study of slope failures on the island of Sumatra, examination was made of several high, steep slopes cut in colluvium which were subjected to debris and slump slides during construction. Failure occurred during a normal rainy season of 500 mm (20 in) for the month of occurrence. The cuts were reshaped with some benching and flatter inclinations and have remained stable for a year of near-normal rainfall of about 2500 mm (98 in) with monthly variations from 80 to 673 mm (3 to 26 in).

Rainfall records were available during the study for only a 5-year period, but during the year before the cuts were made, 1685 mm (66 in) were reported for the one month of December during monsoon storms. The cuts cannot be considered stable until subjected to a rainfall of this magnitude, unless there is an error in the data or the storm was a very unusual occurrence. Neither condition appears to be the case for the geographic location.

### *Case 2*

A number of examples have been given of slope failures along the Rio Santos Highway which passes through the coastal mountains of Brazil. Numerous cuts were made in the years 1974 and 1975 without retention, and a large number of relatively small slides and other failures have occurred. The solution to the problem adapted by the highway department, in most cases, is to allow the failures and subsequently clean up the roadway. As of 1980, except for short periods, during the slide illustrated in Fig. 9.91, the road has remained in service.

A review of the rainfall records for the region during the past 40 years reveals that the last decade has been a relatively dry period with rainfall averaging about 1500 to 2000 mm (59 to 79 in), but during the previous 30 years the annual rainfall averaged 2500 to 3500 mm (98 to 138 in). One storm in the period dropped 678 mm (27 in) in 3 days *(see Art. 9.2.8)*. In view of the already unstable conditions along the roadway, if the weather cycle changes from the currently dry epoch to the wetter cycle of the previous epoch, a marked increase in incidence and magnitude of slope failures can be anticipated.

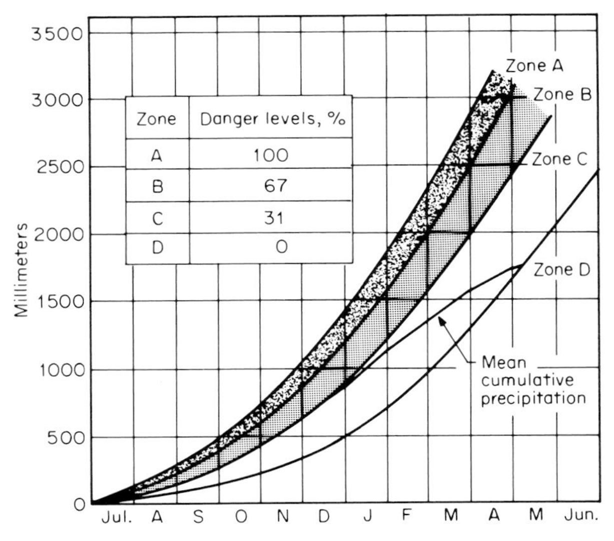


| Zone | Danger levels, % |
|---|---|
| A | 100 |
| B | 67 |
| C | 31 |
| D | 0 |

**FIG. 9.93 Correlation between rainfall and landslides (Serra de Caraguatatuba, São Paulo, Brazil). Chart based on rainfall record station No. E2-65-DAEE, installed in May 6, 1928, in the town of Caraguatatuba. Mean annual precipitation of 1905 mm, based on 46-year record. Approximate station elevation of 10 m.** [*From Guidicini and Iwasa (1977).*[54]]

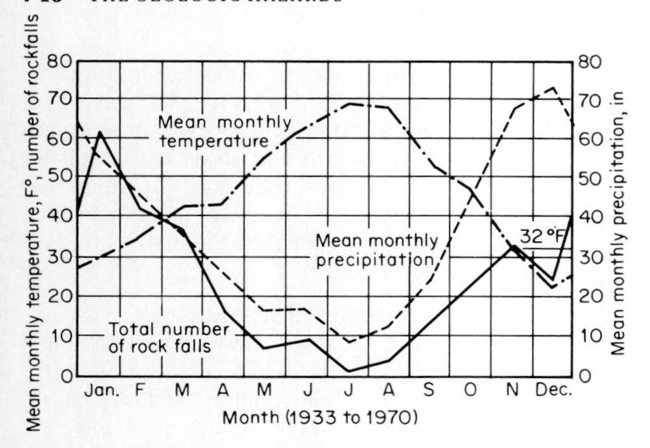


**FIG. 9.94 Number of rockfalls, mean monthly temperature, and mean monthly precipitation in the Fraser Canyon of British Columbia for 1933–1970.** [*Peckover (1975); from Piteau (1977).*[55]]

### Temperatures

Freezing temperatures and the occurrence of frost in soil or rock slopes are highly significant. Ground frost can wedge loose rock blocks and cause falls, or in the spring months can block normal seepage, resulting in high water pressures which cause falls, debris avalanches, slides, and flows. A relationship between the number of rock falls, mean monthly temperature, and mean monthly precipitation is given on Fig. 9.94.

## 9.3.5 HAZARD MAPS

### Purpose

Degrees of slope-failure hazards along a proposed or existing roadway or other development can be illustrated on *slope hazard maps*. Such maps provide the basis not only for establishing the form of treatment required, but also for establishing the degree of urgency for such treatment in the case of existing works, or the programming of treatment for future works. They represent the product of a regional assessment.

### Example

*The Problem*

A 7-km stretch of existing mountain roadway with a 20-year history of slope failures including rotational slides, debris slides, avalanches, and rock falls was mapped in detail with respect to slope stability, to provide the basis for the selection of treatments and the establishment of treatment priorities. A panoramic photo of the slopes in the higher elevations along the roadway is given as Fig. 9.95.

**FIG. 9.95 Panorama of mountain road presented on the hazard map given as Fig. 9.96. This section, between km 52 and 53.5, is characterized by debris slides of residual soils over inclined rock surfaces. A debris avalanche that occurred at km 56 is illustrated on Fig. 9.7.**

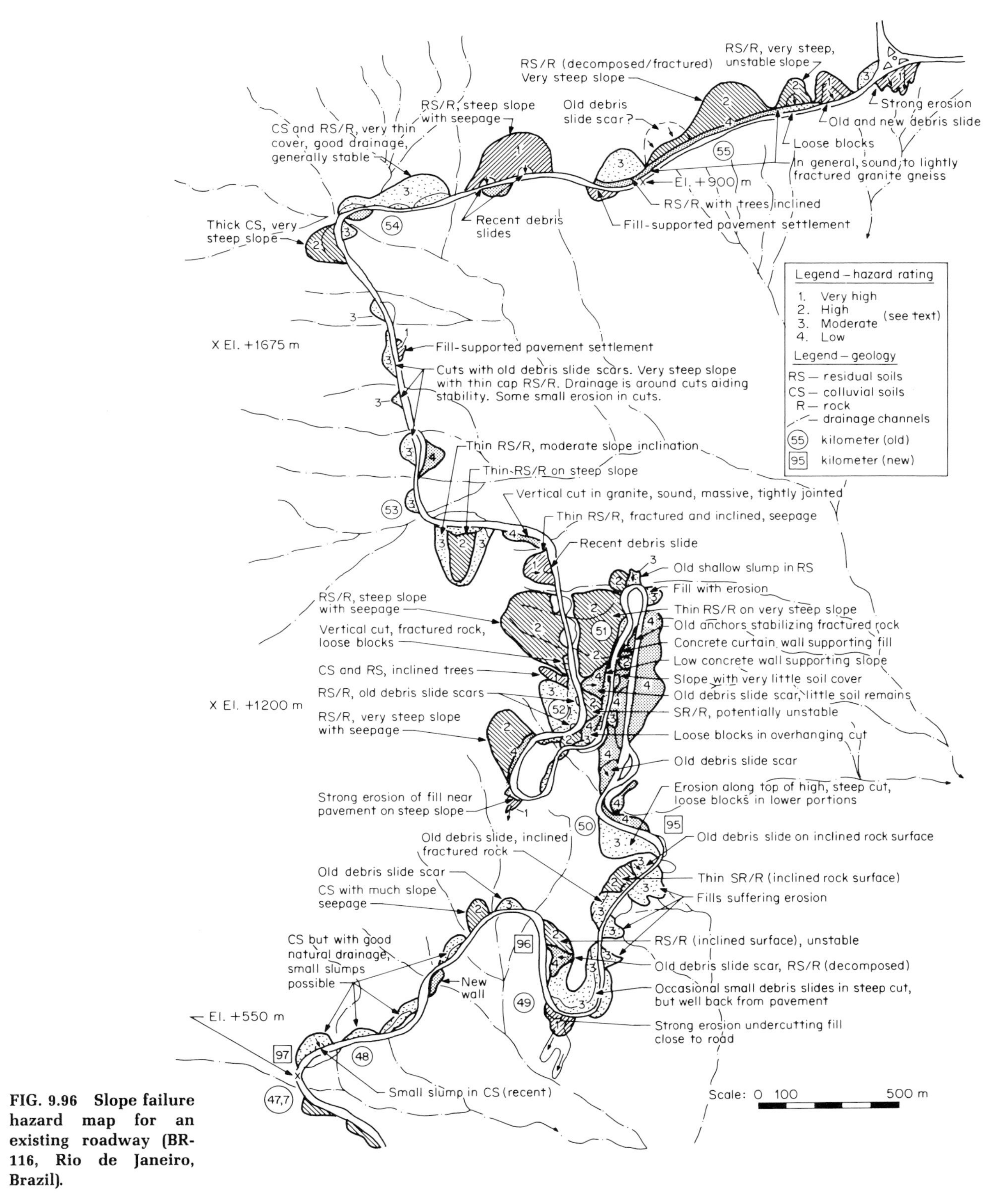


**FIG. 9.96 Slope failure hazard map for an existing roadway (BR-116, Rio de Janeiro, Brazil).**

*The Slope Failure Hazard Map*

The map (Fig. 9.96) illustrates the location of cuts and fills, drainage, and the degree of hazard. Maps accompanying the report gave geologic conditions and proposed solutions. The maps were prepared by enlarging relatively recent aerial photographs to a scale of 1:10,000 to serve as a base map for plotting, since more accurate maps illustrating the topography and locations of cuts and fills were not available.

Five degrees of hazard were used to describe slope conditions:

1. *Very high:* Relatively large failures will close the roadway. Slopes are very steep with a thin cover of residual or colluvial soils over rock, and substantial water penetrates the mass. Fills are unstable and have suffered failures.
2. *High:* Relatively large failures probably will close the road. Failures have occurred already in residual or colluvial soils and in soils over rock on moderately steep slopes. Fills are unstable.
3. *Moderate:* In general, failures will not close the road completely. Relatively small failures have occurred in residual soils on steep slopes, in colluvial soils on moderate slopes with seepage, in vertical slopes with loose rock blocks, and in slopes with severe erosion.
4. *Low:* Low cuts, cuts in strong soils or stable rock slopes, and fills. Some erosion is to be expected, but in general slopes are without serious problems.
5. *No Hazard:* Level ground, or sound rock, or low cuts in strong soils.

## 9.4 TREATMENT OF SLOPES

### 9.4.1 GENERAL CONCEPTS

#### Selection Basis

*Basic Factors*

The first factor to consider in the selection of a slope treatment is its purpose, which can be placed in one of two broad categories:

- Preventive treatments which are applied to stable, but potentially unstable natural slopes or to slopes to be cut or to side-hill fills to be placed
- *Remedial* or *corrective* treatments which are applied to existing unstable, moving slopes or to failed slopes

Assessment is then made of other factors, including the degree of the failure hazard and risk *(see Art. 9.1.3)* and the slope condition, which can be considered in four general groupings and which itself is related to the hazard and risk.

*Slope Conditions*

*Potentially unstable natural slopes* range from those subject to falls or slides where development along the slope or at its base can be protected with reasonable treatments, to those where failures may be unpreventable and will have disastrous consequences. The potential for the latter failures may be recognizable, but since it cannot be known when the necessary conditions may occur, the failures are essentially unpredictable. Some examples of slope failures that could not have been prevented with any reasonable amount of expenditure include those at Nevada Huascaran in the Andes which destroyed several towns *(see Art. 9.2.8)*, the Achocallo mudflow near La Paz *(see Art. 9.2.11)*, and the thousands of debris avalanches and flows that have occurred in a given area during heavy storms in mountainous regions in tropical climates *(see Art. 9.2.8)*.

*Unstable natural slopes* undergoing failure may or may not require treatment depending upon the degree of hazard and risk, and in some instances, such as Portuguese Bend *(see Art. 9.2.6)*, stabilization may not be practical because of the expenditures required.

*Unstable cut slopes* in the process of failure need treatment, but stabilization may not be economically practical.

*New slopes* formed by cutting or filling may require treatment by some form of stabilization.

*Initial Assessment*

An initial assessment is made of the slope conditions, the degree of the hazard, and the risk. There are then three possible options to consider for slope treatment: avoid the high-risk hazard; accept the failure hazard; or stabilize the slope to eliminate or reduce the hazard.

### Treatment Options

#### *Avoid the High-Risk Hazard*

CONDITIONS Where failure is essentially unpredictable and unpreventable by reasonable means and the consquences are potentially disastrous, as in mountainous terrain subject to massive planar slides or avalanches, or slopes in tropical climates subject to debris avalanches, or slopes subject to liquefaction and flows, the hazard should be avoided.

SOLUTIONS Avoid development along the slope or near its base and relocate roadways or railroads to areas of lower hazard where stabilization is feasible, or avoid the hazard by tunneling.

#### *Accept the Failure Hazard*

CONDITIONS Low to moderate hazards, such as partial temporary closure of a roadway, or a failure in an open-pit mine where failure is predictable but prevention is considered uneconomical, may be accepted.

OPEN-PIT MINES Economics dictates excavating the steepest slope possible to minimize quantities to be removed, and most forms of treatment are not feasible; therefore, the hazard is accepted. Slope movements are monitored to provide for early warning and evacuation of personnel and equipment. In some instances measures may be used to reduce the hazard where large masses are involved, but normally failures are simply removed with the equipment available.

ROADWAYS Three options exist besides avoiding the hazard, i.e., accept the hazard, reduce the hazard, or eliminate it. Acceptance is based on an evaluation of the degree of hazard and the economics of prevention. In many cases involving relatively small volumes, failure is self-correcting and most, if not all, of the unstable material is removed from the slope by the failure; it only remains to clean up the roadway. These nuisance failures commonly occur during or shortly after construction when the first adverse weather arrives. The true economics of this approach, however, depends on a knowledgeable assessment of the form and magnitude of the potential failure, and assurance that the risk is low to moderate. Conditions may be such that small failures will evolve into very large ones *(see Fig. 9.22)* or that a continuous and costly maintenance program may be required. Public opinion regarding small but frequent failures of the nuisance type also must be considered.

#### *Eliminate or Reduce the Hazard*

Where failure is essentially predictable and preventable, or is occurring or has occurred and is suitable for treatment, slope stabilization methods are applied. For low- to moderate-risk conditions, the approach can be either to eliminate or to reduce the hazard, depending on comparative economics. For high-risk conditions the hazard should be eliminated.

### Slope Stabilization

#### *Methods*

Slope stabilization methods may be placed in four general categories:

1. Change slope geometry to decrease the driving forces or increase the resisting forces.
2. Control surface water infiltration to reduce seepage forces.
3. Control internal seepage to reduce the driving forces and increase material strengths.
4. Provide retention to increase the resisting forces.

Stabilization methods are illustrated generally on Fig. 9.97 and summarized on Table 9.8 with respect to conditions and general purpose. "General Purpose" indicates whether the aim is to prevent failure, or to treat the slope by some remedial measure.

#### *Selection*

In the selection of the stabilization method or methods, consideration is given to a number of factors including:

- Material types composing the slope and intensity and orientation of the discontinuities

**TABLE 9.8**
**SUMMARY OF SLOPE TREATMENT METHODS FOR STABILIZATION**

| Treatment | Conditions | General purpose (preventive or remedial) |
|---|---|---|
| **CHANGE SLOPE GEOMETRY** | | |
| Reduce height | Rotational slides | Prevent/treat during early stages |
| Reduce inclination | All soil/rock | Prevent/treat during early stages |
| Add weight to toe | Soils | Treat during early stages |
| **CONTROL SURFACE WATER** | | |
| Vegetation | Soils | Prevent |
| Seal cracks | Soil/rock | Prevent/treat during early stages |
| Drainage system | Soil/decomposing rock | Prevent/treat during early stages |
| **CONTROL INTERNAL SEEPAGE** | | |
| Deep wells | Rock masses | Temporary treatment |
| Vertical gravity drains | Soil/rock | Prevent/treat during early stages |
| Subhorizontal drains | Soil/rock | Prevent/treat—early to intermediate stages |
| Galleries | Rock/strong soils | Prevent/treat during early stages |
| Relief wells or toe trenches | Soils | Treat during early stages |
| Interceptor trench drains | Soils (cuts/fills) | Prevent/treat during early stages |
| Blanket drains | Soils (fills) | Prevent |
| Electroosmosis* | Soils (silts) | Prevent/treat during early stages: temporarily |
| Chemicals* | Soils (clays) | Prevent/treat during early stages |

| Treatment | Conditions | General purpose (preventive or remedial) |
|---|---|---|
| **RETENTION** | | |
| Concrete pedestals | Rock overhang | Prevent |
| Rock bolts | Jointed or sheared rock | Prevent/treat sliding slabs |
| Concrete straps and bolts | Heavily jointed or soft rock | Prevent |
| Cable anchors | Dipping rock beds | Prevent/treat early stages |
| Wire meshes | Steep rock slopes | Contain falls |
| Concrete impact walls | Moderate slopes | Contain sliding or rolling blocks |
| Shotcrete | Soft or jointed rock | Prevent |
| Rock-filled buttress | Strong soils/soft rock | Prevent/treat during early stages |
| Gabion wall | Strong soils/soft rock | Prevent/treat during early stages |
| Crib wall | Moderately strong soils | Prevent |
| Reinforced earth wall | Soils/decomposing rock | Prevent |
| Concrete gravity walls | Soils to rock | Prevent |
| Anchored concrete curtain walls | Soils/decomposing rock | Prevent/treat—early to intermediate stages |
| Bored or root piles | Soils/decomposing rock | Prevent/treat—early stages |

*Provides strength increase.

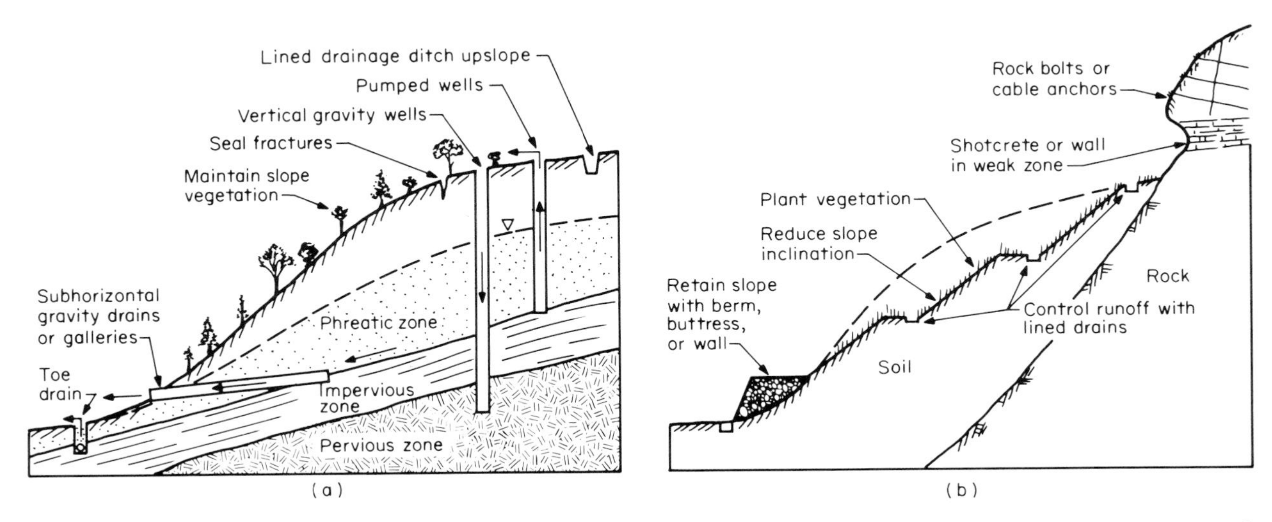


**FIG. 9.97 The general methods of slope stabilization: (*a*) control of seepage forces and (*b*) reducing the driving forces and increasing the resisting forces.**

- Slope activity
- Proposed construction, whether cut or side-hill fill
- Form and magnitude of potential or recurring failure (summary of preventive and remedial measures for the various failure forms is given on Table 9.9)
- Time available for remedial work on failed slopes, judged on the basis of slope activity, movement velocity and acceleration, and existing and near-future weather conditions
- Degree of hazard and risk
- Necessity to reduce or eliminate the hazard

*Hazard Elimination*

The hazard can be eliminated by sufficient reduction of the slope height and inclination combined with an adequate surface drainage system or by retention.

Retention of rock slopes is accomplished with pedestals, rock bolts, bolts and straps, or cable anchors; retention of soil slopes is accomplished with the addition of adequate material at the toe of the slope, or with properly designed and constructed walls.

*Hazard Reduction*

The hazard can be decreased by partially reducing the height and inclination or adding material at the toe; by planting vegetation, sealing cracks, installing surface drains, and shotcreting rock slopes; and by controlling internal seepage. In the last case, one can never be certain that drains will not clog, break off during movements, or be overwhelmed by extreme weather conditions.

*Time Factor*

Where slopes are in the process of failing, the time factor must be considered. Time may not be available for carrying out measures that will eliminate the hazard; therefore, the hazard should be reduced and perhaps eliminated at a later date. The objective is to arrest the immediate movement. To the extent possible, treatments should be performed during the dry season when movements will not close trenches, break off drains, or result in even larger failures when cuttings are made.

In general the time required for various treatment measures are as follows:

- Sealing surface cracks and constructing interceptor ditches upslope are performed within several days, at most.
- Excavation at the head of a slide or the removal of loose blocks may require 1 to 2 weeks.
- Relief of internal water pressures may require 1 to 4 weeks for toe drains and trenches and 1

**TABLE 9.9**
**FAILURE FORMS: TYPICAL PREVENTIVE AND REMEDIAL MEASURES**

| Failure form | Prevention during construction | Remedial measures |
|---|---|---|
| Rock fall | Base erosion protection.<br>Controlled blasting excavation.<br>Rock bolts and straps, or cables.<br>Concrete supports, large masses.<br>Remove loose blocks.<br>Shotcrete weak strata. | Permit fall, clean roadway<br>Rock bolts and straps.<br>Concrete supports.<br>Remove loose blocks.<br>Impact walls. |
| Soil fall | Base erosion protection. | Retention. |
| Planar rock slide | Small volume: remove or bolt.<br>Moderate volume: provide stable inclination or bolt to retain.<br>Large volume: install internal drainage or relocate to avoid. | Permit slide, clean roadway.<br>Remove to stable inclination or bolt.<br>Install internal drainage or relocate to avoid. |
| Rotational rock slide | Provide stable inclination and surface drainage system.<br>Install internal drainage. | Remove to stable inclination.<br>Provide surface drainage.<br>Install internal drains. |
| Planar (debris) slides | Provide stable inclination and surface drainage control.<br>Retention for small to moderate volumes.<br>Large volumes: relocate. | Allow failure and clean roadway.<br>Use preventive measures. |
| Rotational soil slides | Provide stable inclination and surface drainage control, or retain. | Permit failure, clean roadway.<br>Remove to stable inclination, provide surface drainage, or retain.<br>Subhorizontal drains for large volumes. |
| Failure by lateral spreading | Small scale: retain.<br>Large scale: avoid and relocate, prevention difficult. | Small scale: retain.<br>Large scale: avoid. |
| Debris avalanche | Prediction and prevention difficult. Treat as debris slide.<br>Avoid high-hazard areas. | Permit failure, clean roadway; eventually self-correcting.<br>Otherwise relocate.<br>Small scale: retain or remove. |
| Flows | Prediction and prevention difficult. Avoid susceptible areas. | Small scale: remove.<br>Large scale: relocate. |

to 2 months for the installation of horizontal or vertical drains.

- Counterberms and buttresses at the toe require space, but can be constructed within 1 to 2 weeks.
- Retention with concrete walls can require 6 months or longer.

### 9.4.2 CHANGING SLOPE GEOMETRY

#### Natural Slope Inclinations

*Significance*

In many cases the natural slope represents the maximum long-term inclination, but in other cases the slope is not stable. The inclination of existing slopes should be noted during field reconnaissance, since an increase in inclination by cutting may result in failure.

*Some Examples of Natural Slope Inclinations*

- Hard, massive rocks: Maximum slope angle and height is controlled by the concentration and orientation of joints and by seepage. The critical angle for high slopes of hard, massive rock with random joint patterns and no seepage acting along the joints is about 70° [Terzaghi (1962)[56]].
- Interbedded sedimentary rocks: Extremely variable, depending upon rock type, climate, and bedding thickness as well as joint orientations and seepage conditions. Along river valleys, natural excavation may have reduced stresses sufficiently to permit lateral movement along bedding planes and produce bedding-plane mylonite shear zones. On major projects such shears should be assumed to exist until proven otherwise. The slickensides illustrated in Fig. 6.56 are probably shale mylonite.
- Clay shales: 8 to 15°, but often unstable. When interbedded with sandstones, 20 to 45° *(see Art. 6.7.3)*.
- Residual soils: 30 to 40°, depending upon parent rock type and seepage.
- Colluvium: 10 to 20°, and often unstable.
- Loess: Often stands vertical to substantial heights.
- Sands: dry and "clean," are stable at the angle of repose ($i = \phi$).
- Clays: Depends upon consistency, whether intact or fissured, and the slope height.
- Sand-clay mixtures: Often stable at angles greater than repose as long as seepage forces are not excessive.

#### Cut Slopes in Rock

*Excavation*

The objective of any cut slope is to form a stable inclination without retention. Careful blasting procedures are required to avoid excessive rock breakage resulting in numerous blocks. Line drilling and presplitting during blasting operations minimize disturbance of the rock face.

*Typical Cut Inclinations*

Hard masses of igneous or metamorphic rocks, widely jointed, are commonly cut to $1H{:}4V$ (76°) as shown on Fig. 9.98.

Hard rock masses with joints, shears, or bedding representing major discontinuities and dipping downslope are excavated along the dip of the discontinuity as shown in Fig. 9.99, although all material should be removed until the original slope is intercepted. If the dip is too shallow for economical excavation, slabs can be retained with rock bolts *(see Art. 9.4.6)*.

Hard sedimentary rocks with bedding dipping vertically and perpendicularly to the face as in Fig. 9.100, or dipping into the face; or horizontally interbedded hard sandstones and shales *(see Fig. 6.95)* are often cut to $1H{:}4V$, but in this case, the shales should be protected from weathering with shotcrete or gunite if they have expansive properties.

Clay shales, unless interbedded with sandstones, are often excavated to $6H{:}1V$ (9.5°).

Weathered or closely jointed masses (except clay shales and dipping major discontinuities) require a reduction in inclination to between $1H{:}2V$ to $1H{:}1V$ (63 to 45°) depending on conditions, or require some form of retention.

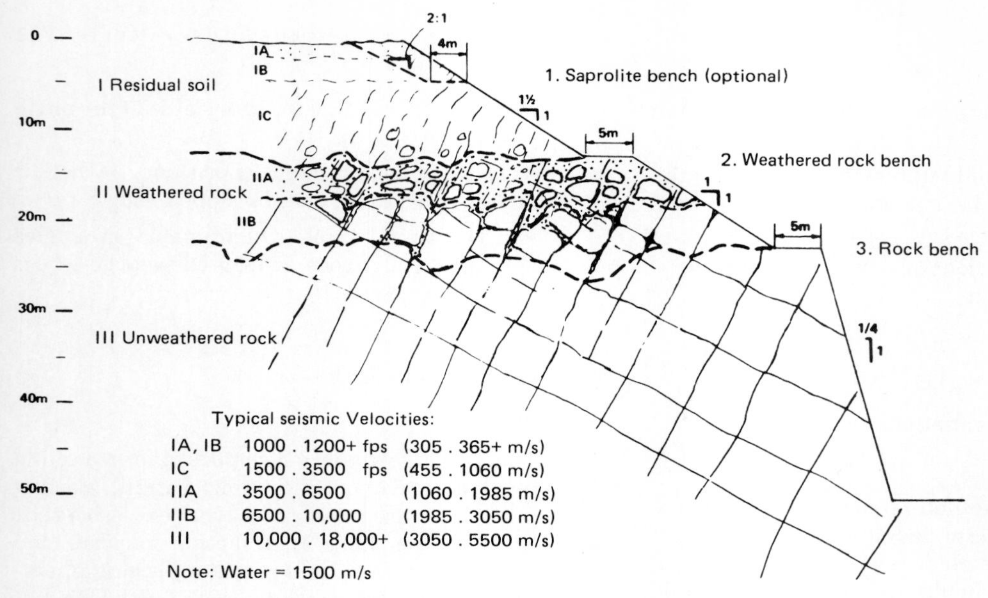


**FIG. 9.98 Typical cut slope angles for various rock and soil conditions.** [*From Deere and Patton (1971).*[12]]

**FIG. 9.99 The face of the major joint surface in siltstone is stable but the overhanging blocks will fail along the same surface unless removed or retained (Sidikalang, Sumatra).**

**FIG. 9.100 Vertical beds in highly fractured, hard arenaceous shale, moderately folded, are stable in near-vertical cuts (Sidikalang, Sumatra). Decreasing inclination would increase susceptibility to erosion.**

*Benching*

Benching is common practice in high cuts in rock slopes but there is disagreement among practitioners as to its value. Some consider benches as undesirable because they provide takeoff points for falling blocks [Chassie and Goughnor (1975)[57]]. To provide for storage they must be of adequate width. Block storage space should always be provided at the slope toe to protect the roadway from falls and topples.

FIG. 9.101 Cuts in colluvium over rock are potentially highly unstable and require either removal of the soil or retention (Rio de Janeiro).

FIG. 9.102 Cut slopes at the beginning of a 300-m-deep excavation in highly decomposed igneous and metamorphic rocks for a uranium mine. Bench width is 20 m, height is 16 m, and inclination is $1H$:$1.6V$ (57°). Small wedge failure at right occurred along kaolinite-filled vertical joint. Part of similar failure shows in lower left.

## Cut Slopes in Soils

*Typical Inclinations*

Thin soil cover over rock (Fig. 9.101): The soil should be removed or retained as the condition is unstable.

Soil-rock transition (strong residual soils to weathered rock) such as in Fig. 9.102 are often excavated to between $1H{:}2V$ to $1H{:}1V$ (63 to 45°) although potential failure along relict discontinuities must be considered.

Most soil formations are commonly cut to an average inclination of $2H{:}1V$ (26°) but consideration must be given to seepage forces and other physical and environmental factors to determine if retention is required. Slopes between benches are usually steeper.

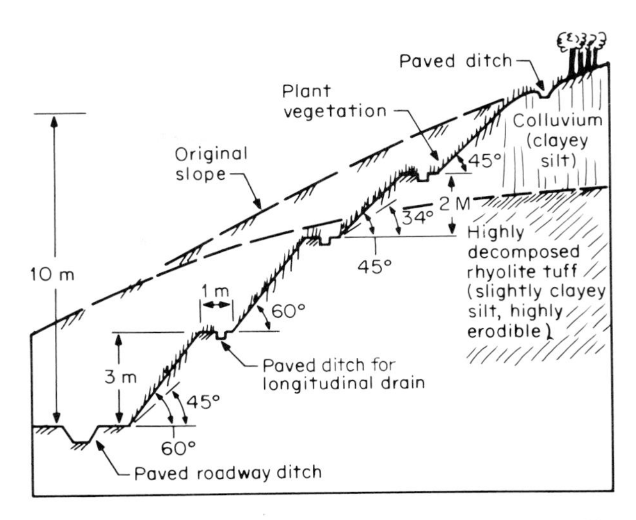


**FIG. 9.103 Benching scheme for cut in highly erodable soils in a tropical climate. Low benches permit maximum inclination to reduce the effect of runoff erosion. Cut before benching is shown in Fig. 8.6.**

*Benching and Surface Drainage*

Soil cuts are normally designed with benches, especially for cuts over 8 to 10 m high. Benches reduce the amount of excavation necessary to achieve lower inclinations because the slope angle between benches may be increased.

Drains are installed as standard practice along the slopes and the benches to control runoff as illustrated on Figs. 9.103 and 9.104.

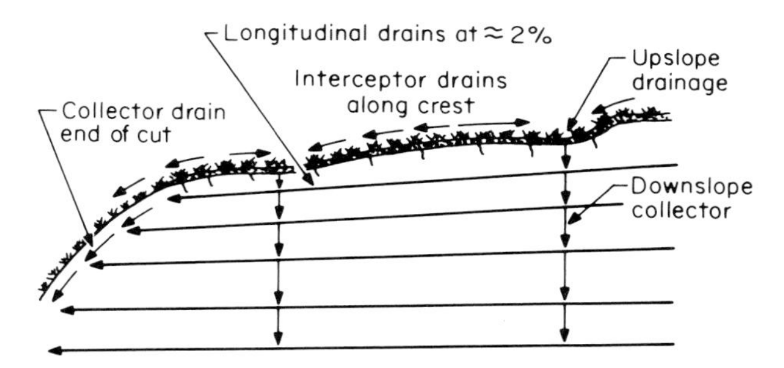


**FIG. 9.104 Sketch of slope face in Fig. 9.103 showing system of longitudinal and downslope drains to control erosion. All drains should be paved with cement-soaked burlap bags *(see Fig. 8.8)* or some other permanent protection.**

## Failing Slopes

If a slope is failing and undergoing substantial movement, the removal of material from the head to reduce the driving forces can be the quickest method of arresting movement of relatively small failures. Placing material at the toe to form a counterberm increases the resisting forces. Benching may be effective in the early stages, but it did not fully stabilize the slope illustrated in Fig. 9.31, even though a large amount of material was removed. An alternative is to permit movement to occur and remove debris from the toe; eventually the mass may naturally attain a stable inclination.

Changing slope geometry to achieve stability once failure has begun usually requires either the removal of very large volumes or the implementation of other methods. Space is seldom available in critical situations to permit placement of material at the toe, since very large volumes normally are required.

### 9.4.3 SURFACE WATER CONTROL

**Purpose**

Surface water is controlled to eliminate or reduce infiltration and to provide erosion protection. External measures are generally effective, however, only if the slope is stable and there is no internal source of water to cause excessive seepage forces.

### Infiltration Protection

*Planting* the slope with thick, fast-growing native vegetation not only strengthens the shallow soils with root systems, but discourages desiccation which causes fissuring. Not all vegetation works equally well, and selection requires experience. In the Los Angeles area of California, for example, Algerian ivy has been found to be quite effective in stabilizing steep slopes [*Sunset* (1978)[58]]. Newly cut slopes should be immediately planted and seeded. Burlap bags or sprayed mulch helps to increase growth rate and provide protection against erosion during early growth stages.

*Sealing* cracks and fissures with asphalt or soil cement will reduce infiltration but will not stabilize a moving slope since the cracks will continue to open. *Grading* a moving area results in filling cracks with soil, which helps to reduce infiltration.

### Surface Drainage Systems

*Cut Slopes* should be protected with interceptor drains installed along the crest of the cut, along benches, and along the toe (Fig. 9.103). On long cuts the interceptors are connected to downslope collectors (Fig. 9.104). All drains should be lined with nonerodable materials, free of cracks or other openings, and designed to direct all concentrated runoff to discharge offslope.

*With failing slopes,* installation of an interceptor along the crest beyond the head of the slide area will reduce runoff into the slide. But the interceptor is a temporary expedient, since in time it may break up and cease to function as the slide disturbance progresses upslope.

### Erosion Protection

In addition to plantings, erosion protection along the slope can be achieved with wattling bundles as discussed in Art. 8.2.2.

## 9.4.4 INTERNAL SEEPAGE CONTROL

### General

*Purpose*

Internal drainage systems are installed to lower the piezometric level below the potential or existing sliding surface.

*System Selection*

Selection of the drainage method is based on consideration of the geologic materials, structure, and groundwater conditions (static, perched, or artesian), and the location of the phreatic surface.

*Monitoring*

As the drains are installed, the piezometric head is monitored by piezometers and the efficiency of the drains is evaluated. The season of the year and the potential for increased flow during wet seasons must be considered, and if piezometric levels are observed to rise to dangerous values (as determined by stability analysis, or from monitoring slope movements), the installation of additional drains is required.

*Cut Slopes*

Systems to relieve seepage forces in cut slopes are seldom installed in practice, but they should be considered more frequently, since there are many conditions where they would aid significantly in maintaining stability.

*Failing Slopes*

The relief of seepage pressures is often the most expedient means of stabilizing a moving mass. The primary problem is that, as mass movement continues, the drains will be cut off and cease to function; therefore, it is often necessary to install the drains in stages over a period of time. Installation must be planned and performed with care, since the use of water during drilling could possibly trigger a total failure.

### Methods (See Fig. 9.97*a*)

*Deep wells* have been used to stabilize many deep-seated slide masses, but they are costly since continuous or frequent pumping is required. Check valves normally are installed so that when the water level rises, pumping begins. Deep wells are most effective if installed in relatively free-draining material below the failing mass.

*Vertical gravity drains* are useful in perched water-table conditions where an impervious stratum overlies an open, free-draining stratum with a lower piezometric level. The drains permit seepage by gravity through the confining stratum and thus relieve hydrostatic pressures (see Art. 9.2.7, discussion of the Pipe Organ slide). Clay strata over granular soils, or clays or shales over open-jointed rock, offer favorable conditions for gravity drains where a perched water table exists.

*Subhorizontal drains* are one of the most effective methods to improve stability of a cut slope, or to stabilize a failing slope. Installed at a slight angle upslope to penetrate the phreatic zone and permit gravity flow, they usually consist of perforated pipe, of 2-in diameter or larger, forced into a predrilled hole of slightly larger diameter than the pipe. Horizontal drains have been installed to lengths of more than 100 m. Spacing depends on the type of material being drained; fine-grained soils may require spacing as close as 3 to 8 m, whereas, for more permeable materials, 8 to 15 m may suffice.

*Drainage galleries* are very effective for draining large moving masses but their installation is difficult and costly. They are used mostly in rock masses where roof support is less of a problem than in soils. Installed below the failure zone to be effective, they are often backfilled with stone. Vertical holes drilled into the galleries from above provide for drainage from the failure zone into the galleries.

*Interceptor trench drains* can be installed upslope to intercept groundwater flowing into a cut or sliding mass, but they must be sufficiently deep. Perforated pipe is laid in the trench bottom, embedded in sand, and covered with free-draining material, then sealed at the surface *(see Art. 8.4.4)*. Interceptor trench drains are generally not practical on steep, heavily vegetated slopes because installation of the drains and access roads requires stripping the vegetation, which will further decrease stability.

*Relief trenches* relieve pore pressures at the slope toe (*see Art. 8.4.4*, discussion of earth dams). They are relatively simple to install. Excavation should be made in sections and quickly backfilled with stone so as not to reduce the slope stability and possibly cause a total failure. Generally, relief trenches are most effective for small slump slides (Fig. 9.25) where high toe seepage forces are the major cause of instability.

*Electroosmosis (see Art. 8.4.3)* has been used occasionally to stabilize silts and clayey silts, but the method is relatively costly, and not a permanent solution unless operation is maintained.

*Chemicals* have sometimes been injected to increase soil strength. In a number of instances the injection of a quicklime slurry into predrilled holes has arrested slope movements as a result of the strength increase from chemical reaction with clays [Handy and Williams (1967),[59] Broms and Bowman (1979)[60]]. Strength increase in saltwater clays, however, was found to be low.

### Examples

#### *Case 1: Open-Pit Mines [Brawner (1975)[51]]*

GENERAL Problems encountered in open-pit mines in soft rock (coal, uranium, copper, asbestos) during mining operations include both bottom heave of deep excavations (in the order of several hundred meters in depth) and slides, often involving millions of tons.

SOLUTIONS Deep vertical wells, which have relieved artesian pressures below mine floors where heave was occurring, have arrested both the heave and the associated slope instability. Horizontal drainage in the form of galleries and boreholes as long as 150 m installed in the toe zone of slowly moving masses arrested movement even when large failures were occurring. In some cases vacuum pumps were installed to place the galleries under negative pressures. Horizontal drains, consisting of slotted pipe installed in boreholes, relieve cleft-water pressures in jointed rock masses.

#### *Case 2: Failure of a Cut Slope [Fox (1964)[52]]*

GEOLOGICAL CONDITIONS The slope consists of colluvial soils of boulders and clay overlying schist interbedded with gneiss as shown on the section given on Fig. 9.105. Between the colluvium and the relatively sound rock is a zone of highly decomposed rock.

SLIDE HISTORY An excavation was made to a depth of 40 m into a slope with an inclination of

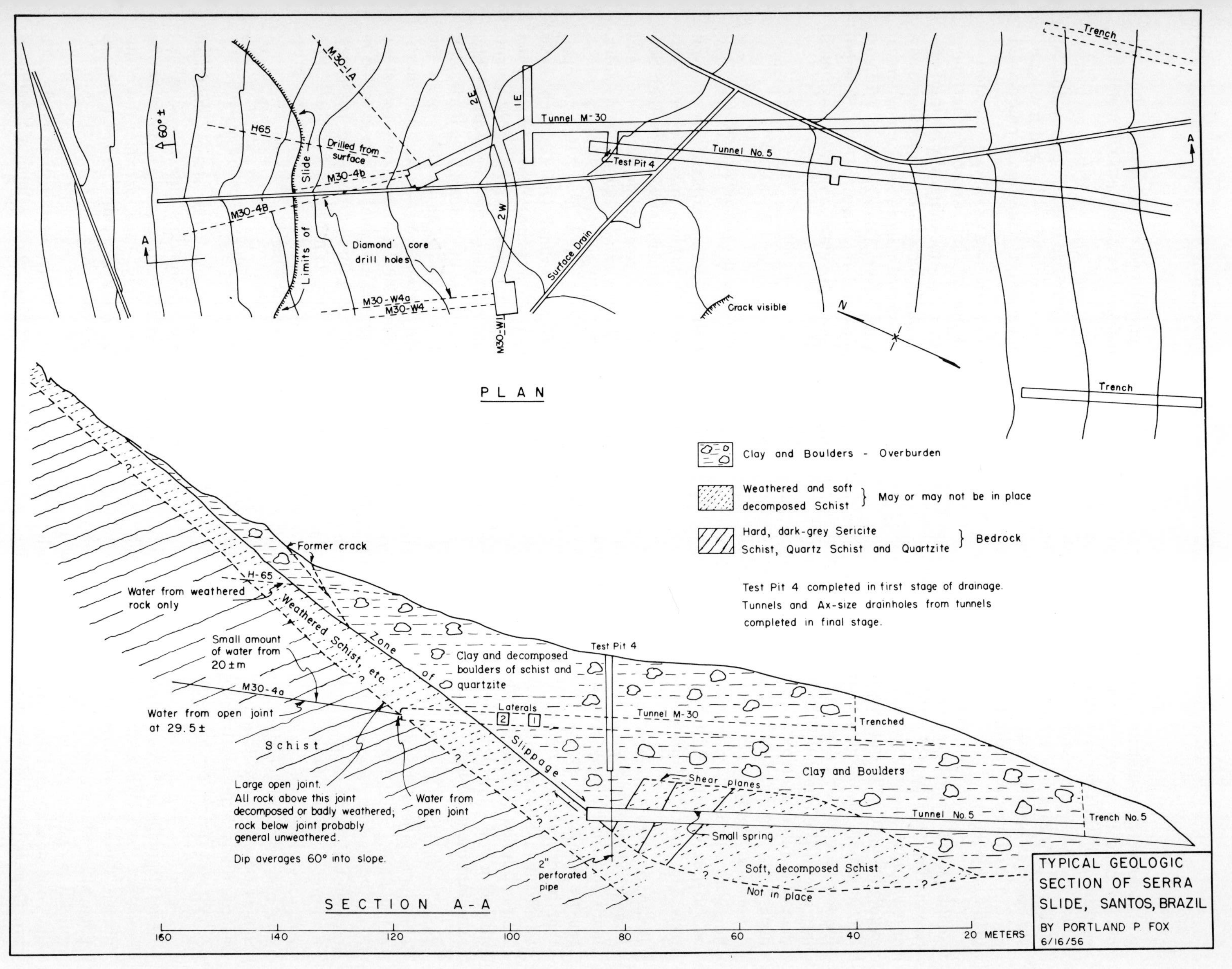


**FIG. 9.105 Stabilization of a failure in colluvial soil slope using lateral drains and galleries (Santos, Brazil).** [*From Fox (1964).*[52]]

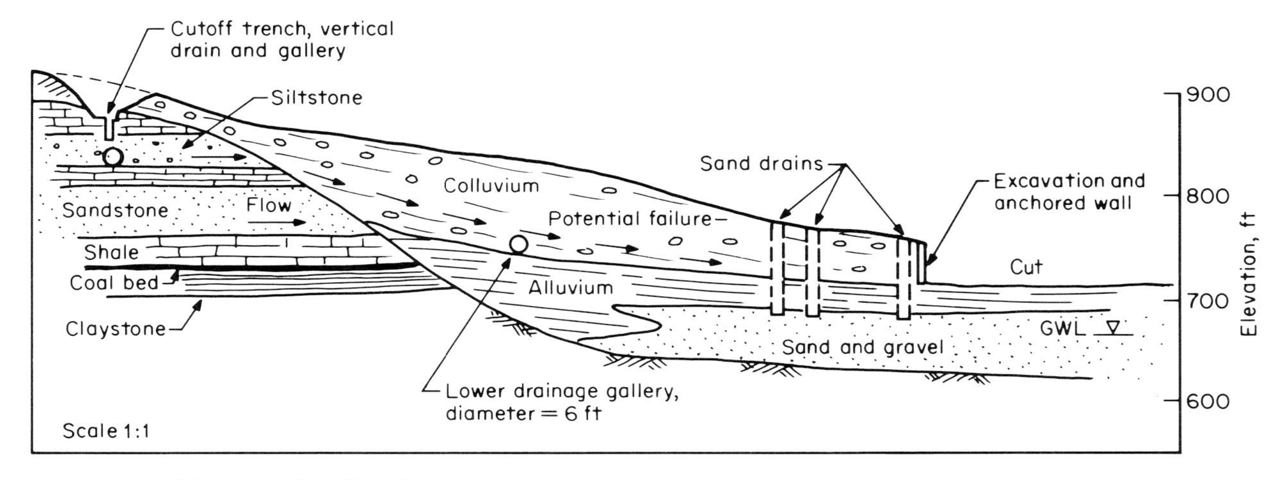


**FIG. 9.106 Stabilization of a colluvial slope (Weirton, West Virginia) with vertical drains and galleries.** [*From D'Appolonia et al. (1964).*[61]]

about 28°. Upon its completion cracks opened, movement began, and springs appeared on the surface. The excavation was backfilled and the ground surface was graded to a uniform slope and covered with pitch. Monuments were installed to permit observations of movements. Even after the remedial measures were invoked, movement continued to endanger nearby structures. The greatest movement was about 2.5 cm/day. Failure had reduced the preexisting strengths to the extent that the original slope inclination was unstable. Piezometers installed as part of an investigation revealed that the highest pore pressures were in the fractured rock zone, under the colluvium.

REMEDIAL MEASURES To correct the slide a number of horizontal drill holes and galleries were extended into the fractured rock as shown on Fig. 9.105. The holes drained at rates of 10 to 100 liters/minute, and the water level in the piezometers continued to fall as work progressed. The slide was arrested and subsequent movements were reported to be minor.

*Case 3: Construction of a Large Cut Slope*

See D'Appolonia et al. (1967).[61]

GEOLOGIC CONDITIONS As illustrated on the section (Fig. 9.106), conditions were characterized by colluvium derived from sandstones and shales, overlying rock, and granular alluvium. Explorations were thorough and included test pits which revealed the overburden to be slickensided, indicating a high potential for instability.

TREATMENT Construction plans required a cut varying from 6 to 18 m in height in the colluvium along the slope toe. To prevent any movement, a system of trenches, drains, and galleries was installed. A cutoff trench, vertical drain, and gallery were constructed upslope, where the colluvium was relatively thin, to intercept surface water and water entering the colluvium from a pervious siltstone layer. A 2-m-diameter drainage gallery was excavated in the colluvium at about midslope to intercept flow from a pervious sandstone stratum, and to drain the colluvium. Sand drains were installed downslope, near the proposed excavation, to enable the colluvium to drain by gravity into the underlying sands and gravels lying above the static water level, thereby reducing pore pressures in the colluvium. An anchored sheet-pile wall was constructed to retain the cut face; the other systems were installed to maintain the stability of the entire slope and reduce pressures on the wall.

## 9.4.5 SIDE-HILL FILLS

### Failures

Construction of a side-hill embankment using slow-draining materials can be expected to

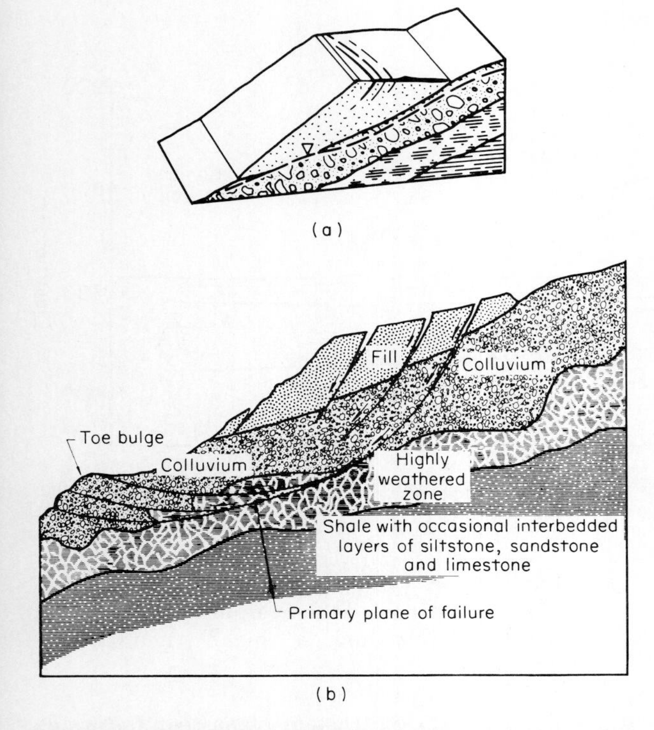


**FIG. 9.107 Development of rotational failure in side-hill fill with inadequate subsurface drainage. (*a*) Early failure stage: concentric cracks show in pavement. (*b*) Rotational failure of side-hill fill over thick colluvium.** [*Part b from Royster (1973).*[22]]

block natural drainage and evaporation. As seepage pressures increase, particularly at the toe as shown on Fig. 9.107*a*, the embankment strains and concentric tension cracks form. The movements develop finally into a rotational failure as shown on Fig. 9.107*b*, a case of deep colluvium. Figure 9.108 illustrates a case of shallow residual soils.

Side-hill fills placed on moderately steep to steep slopes of residual or colluvial soils, in particular, are prone to be unstable unless seepage is properly controlled, or the embankment is supported by a retaining structure.

## Stabilization

### *Preventive*

*Interceptor trench drains* should be installed along the upslope side of all side-hill fills as standard practice to intercept flow as shown on Fig. 9.109. Perforated pipe is laid in the trench bottom, embedded in sand, covered by free-draining materials, and then sealed at the surface. Surface flow is collected in open drains and all discharge, including that from the trench drains, is directed away from the fill area.

*A free-draining blanket* should be installed between the fill and the natural slope materials

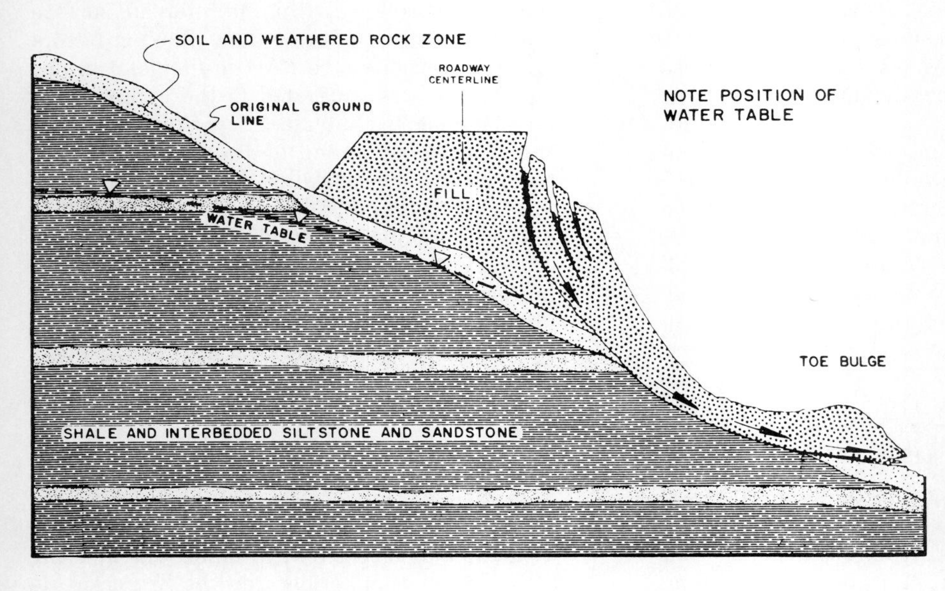


**FIG. 9.108 Development of rotational failure in side-hill fill underlain by thin formation of residual soils and inadequate subsurface drainage.** [*From Royster (1973).*[22]]

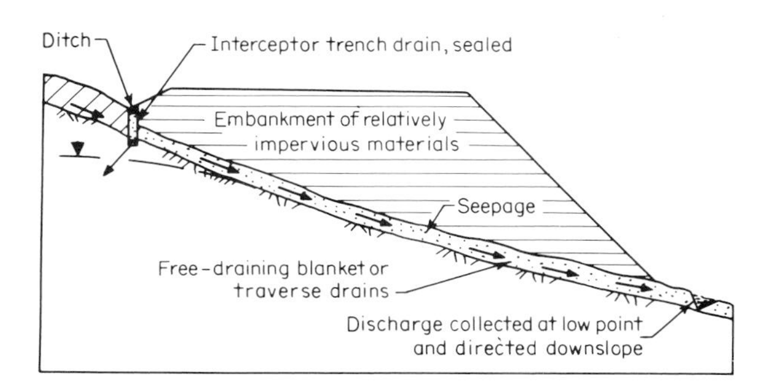


**FIG. 9.109 Proper drainage provisions for a side-hill fill.**

to relieve seepage pressures from shallow groundwater conditions wherever either the fill or the natural soils are slow-draining, as shown on Fig. 9.109. It is prudent to *strip* potentially unstable upper soils, which are often creeping on moderately steep to steep slopes, to a depth where stronger soils are encountered, and to place the free-draining blanket over the entire area to be covered by the embankment. Discharge should be collected at the low point of the fill and drained downslope in a manner that will provide erosion protection.

*Transverse drains* extending downslope and connecting with the interceptor ditches upslope, parallel to the roadway, may provide adequate subfill drainage where anticipated flows are low to moderate.

*Retaining structures* may be economical on steep slopes that continue for some distance beyond the fill, if stability is uncertain *(see Fig. 9.121).*

*Corrective*

*After the initial failure stage,* subhorizontal drains may be adequate to stabilize the embankment if closely spaced, but they should be installed during the dry season since the use of water to drill holes during the wet season may accelerate total failure. An alternative is to retain the fill with an anchored curtain wall *(see Fig. 9.121).*

*After total failure,* the most practical solutions are either reconstruction of the embankment with proper drainage, or retention with a wall.

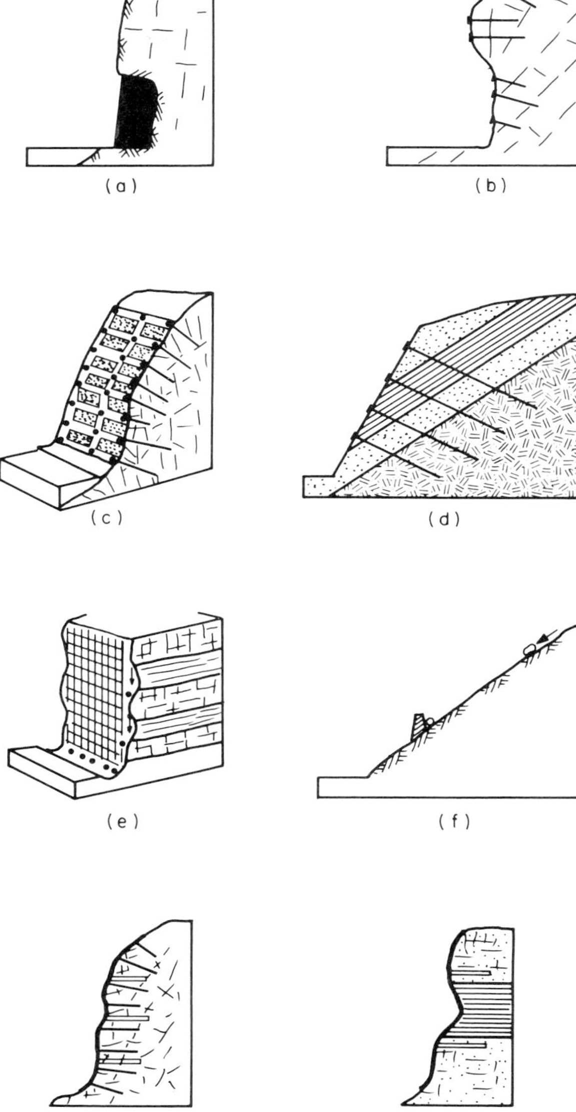


**FIG. 9.110 Various methods of retaining hard rock slopes: (*a*) concrete pedestals for overhangs; (*b*) rock bolts for jointed masses; (*c*) bolts and concrete straps for intensely jointed masses: (*d*) cable anchors to increase support depth; (*e*) wire mesh to constrain falls; (*f*) impact walls to deflect or contain rolling blocks; (*g*) shotcrete to reinforce loose rock, with bolts and drains; (*h*) shotcrete to retard weathering and slaking of shales.**

(a)

(b)

**FIG. 9.111 Support of granite overhang with pedestals (Rio de Janeiro). Ancient slide mass of colluvium appears on lower slopes. (*a*) Side view and (*b*) face view.**

9.4.6 RETENTION

## Rock Slopes

### *Methods Summarized*

The various methods of retaining hard rock slopes are illustrated on Fig. 9.110 and described briefly below.

- *Concrete pedestals* are used to support overhangs, where their removal is not practical because of danger to existing construction downslope, as illustrated on Figs. 9.111*a* and *b*.
- *Rock bolts* are used to reinforce jointed rock masses or slabs on a sloping surface.
- *Concrete straps* and rock bolts are used to support loose or soft rock zones or to reduce the number of bolts as shown on Fig. 9.112.
- *Cable anchors* are used to reinforce thick rock masses.
- *Wire meshes*, hung on a slope, restrict falling blocks to movement along the face.
- *Concrete impact walls* are constructed along lower slopes to contain falling or sliding blocks or deflect them away from structures *(see Fig. 9.13)*.
- *Shotcrete* (Fig. 9.113) is used to reinforce loose fractured rock, or to prevent weathering or slaking of shales or other soft rocks, especially where interbedded with more resistant rocks *(see Fig. 6.95)*.
- *Gunite* is similar to shotcrete except that the aggregate is smaller.

### *The More Common Methods*

*Rock bolts* are tensile units used to place the mass in compression, and should be installed as near to perpendicular to the joint as practical. The ordinary types consist of rods installed in drill holes either by driving and wedging, driving and expanding, or by grouting with mortar or resins as illustrated on Fig. 9.114. Bolt heads are then attached to the rod and torqued against a metal plate to impose the compressive force on the mass. Weathering of rock around the bolt head may cause a loss in tension; therefore, heads are usually protected with concrete or other means, or used in conjunction with concrete straps in high-risk conditions.

**FIG. 9.112 Stabilization of exfoliating granite with rock bolts and concrete straps (Rio de Janeiro).**

*Fully grouted rock bolts*, illustrated in Fig. 9.115, provide a more permanent bolt than those shown in Fig. 9.114. The ordinary bolt is subject to loss in tension with time from several possible sources including corrosion from attack by aggressive water, anchorage slip or rock spalling around and under the bearing plate, and block

**FIG. 9.113 Shotcrete applied to retain loose blocks of granite gneiss in cut. Unprotected rock exposed in lower right (Rio de Janeiro).**

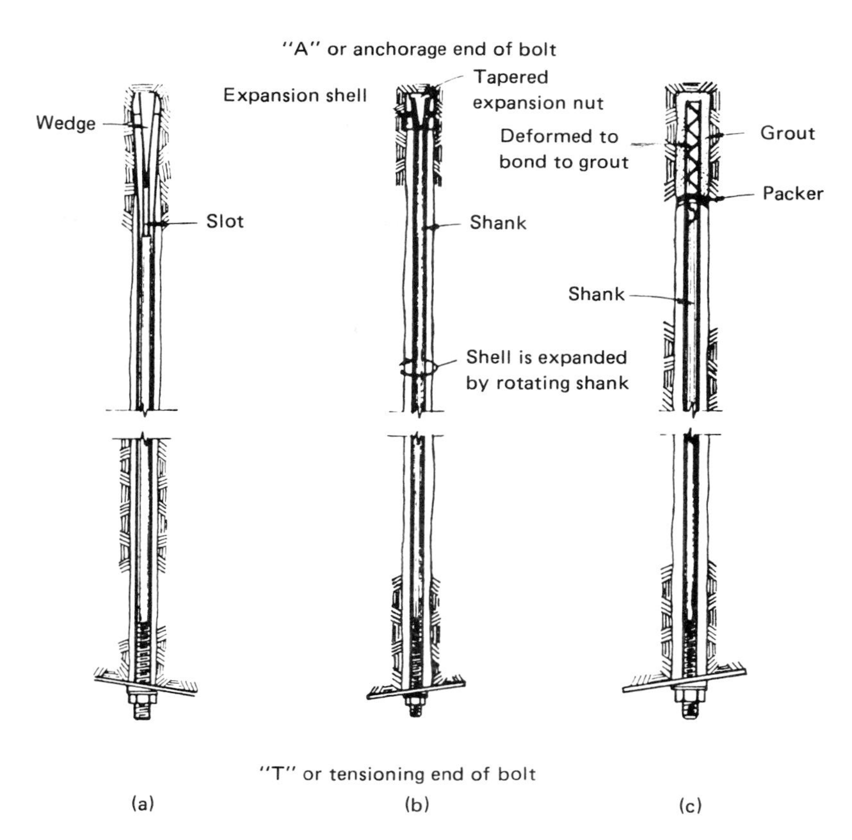


**FIG. 9.114 Types of ordinary rock bolts: (*a*) drive-set or slot and wedge bolt; (*b*) torque-set or expansion bolt; (*c*) grouted bolt.** [*From Lang (1972).*[62]]

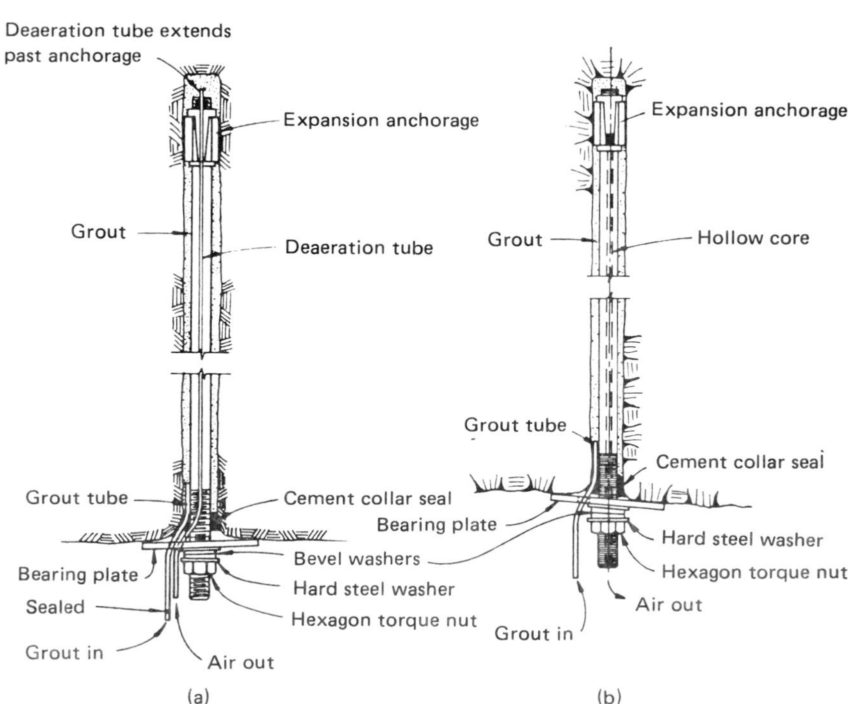


**FIG. 9.115 Fully grouted rock bolts: (*a*) grouted solid expansion anchorage bolt and (*b*) hollow-core grouted rock bolt.** [*From Lang (1972).*[62]]

**FIG. 9.116 Loose granite gneiss blocks retained with rock bolts, and blocks of gneiss and a soft zone of schist retained with bolts and concrete straps on natural slope (João Monlevade, M.G., Brazil).** *(Courtesy of Tecnosolo, S.A.)*

movement along joints pinching the shaft. Grouting with resins is becoming more and more common because of easy installation and the rapid attainment of capacity within minutes of installation. Care is required during grouting to minimize grout spread, which results in decreasing mass drainage, especially where bolts are closely spaced. Drain holes may be required.

A major installation of bolts and straps is illustrated on Fig. 9.116, part of a 60-m-high rock slope (Fig. 9.117), at the base of which is to be constructed a steel mill. The consultant selected the support system rather than shaving and blasting loose large blocks for fear of leaving a weaker slope in a high-risk situation.

*Shotcrete,* when applied to rock slopes, usually consists of a wet-mix mortar with aggregate as large as 2 cm (¾ in) which is projected by air jet directly onto the slope face. The force of the jet compacts the mortar in place, bonding it to the rock, which first must be cleaned of loose particles and loose blocks. Application is in 8- to 10-cm (3- to 4-in) layers, each of which is permitted to set before application of subsequent layers. Weep holes are installed to relieve seepage pressures behind the face. Since shotcrete acts as reinforcing and not as support, it is used often in conjunction with rock bolts. The tensile strength can be increased significantly by adding 25-mm-long wire fibers to the concrete mix.

### Soil Slopes

*Purpose*

Walls are used to retain earth slopes where space is not available for a flat enough slope or excessive volumes of excavation are required, or to obtain more positive stability under certain conditions. Except for anchored concrete curtain walls, other types of walls which require cutting into the slope for construction are seldom suitable for retention of a failing slope.

*Classes*

The various types of walls are illustrated on Fig. 9.118. They may be divided into four general classes, with some wall types included in more than one class: gravity walls, nongravity walls, rigid walls, and flexible walls.

*Gravity walls* provide slope retention by either

**FIG. 9.117 The 60-m-high slope of Fig. 9.116 before treatment. The lower portions have been excavated into relatively sound gneiss. The workers (in circle) give the scale. Several large blocks weighing many tons broke loose during early phases when some scaling of loose blocks was undertaken.**

their weight alone, or their weight combined with the weight of a soil mass acting on a portion of their base or by the weight of a composite system. They are free to move at the top thereby mobilizing active earth pressure. Included are rock-filled buttresses, gabion walls, crib walls, reinforced earth walls, concrete gravity walls, cantilever walls, and counterfort walls. A somewhat complex system of gravity walls installed to correct a debris slide is illustrated on Fig. 9.119. The reconstructed roadway was supported on a reinforced earth wall in turn supported by an embankment and a buttress at the toe of the 360-ft-high slope. Horizontal and longitudinal drains were installed to relieve hydrostatic pressures in that part of the slide debris which was not totally removed.

*Nongravity walls* are restrained at the top and not free to move. They include basement walls, some bridge abutments, and anchored concrete curtain walls. Anchored concrete curtain walls, such as the one illustrated in Fig. 9.120, can be constructed to substantial heights and have a very high retention capacity. They are constructed from the top down by excavation of a series of benches into the slope and formation of a section of wall, retained by anchors, in each bench along the slope. Since the slope is thus retained completely during the wall construc-

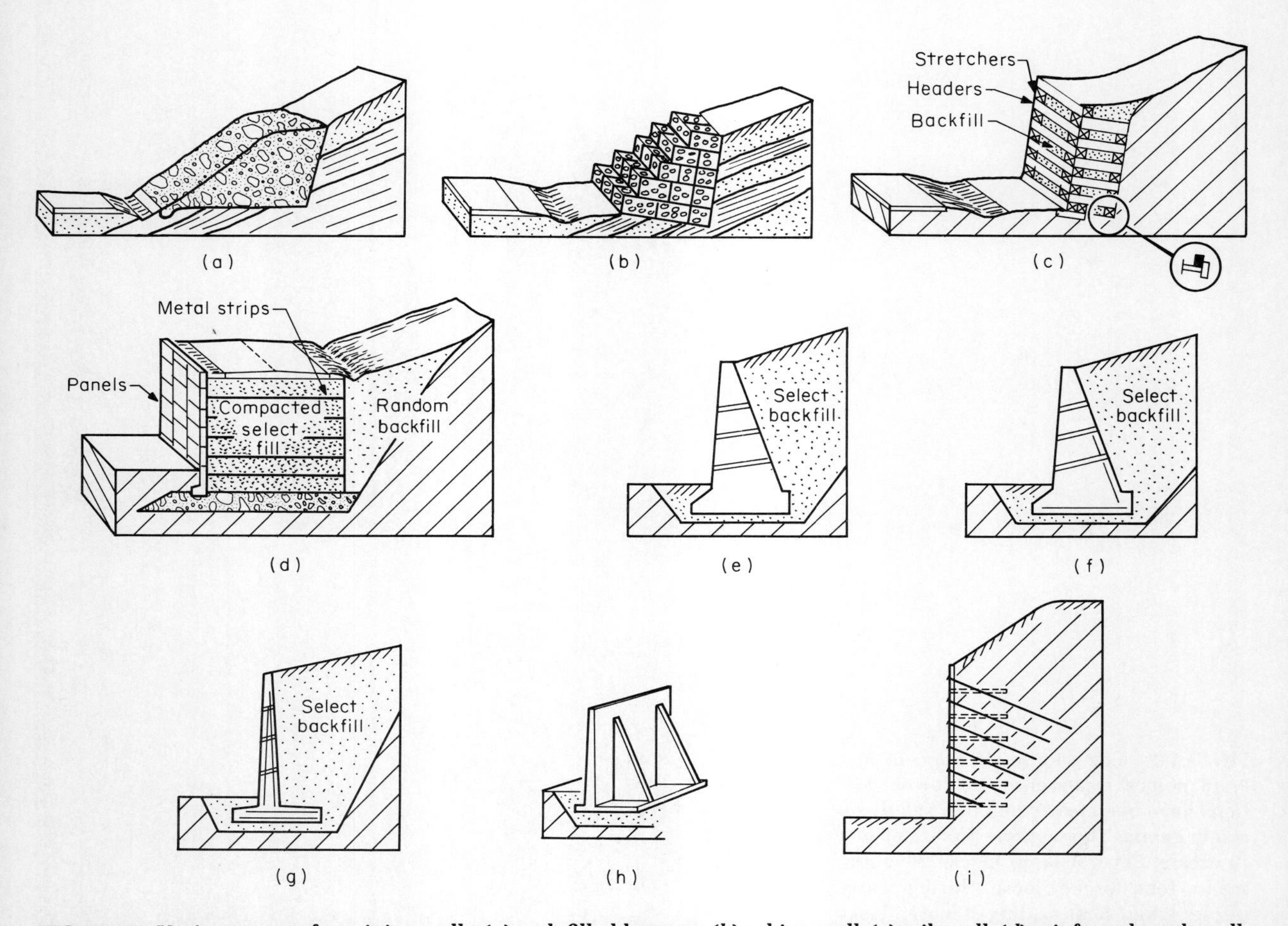


**FIG. 9.118 Various types of retaining walls: (*a*) rock-filled buttress; (*b*) gabion wall; (*c*) crib wall; (*d*) reinforced earth wall; (*e*) concrete gravity wall; (*f*) concrete-reinforced semigravity wall; (*g*) cantilever wall; (*h*) counterfort wall; (*i*) anchored curtain wall.**

**FIG. 9.120 Anchored curtain wall being constructed to a height of 25 m and length of 150 m, completed to 15-m height (João Monlevade, M. G., Brazil). The wall, of maximum thickness of 50 cm, is constructed in sections 1.5 m high from the top down, with each section anchored to provide continuous slope support. Geology is residual soils and weathered schist and gneiss.** [*From Hunt and Costa Nunes (1978).*[64]] ▶

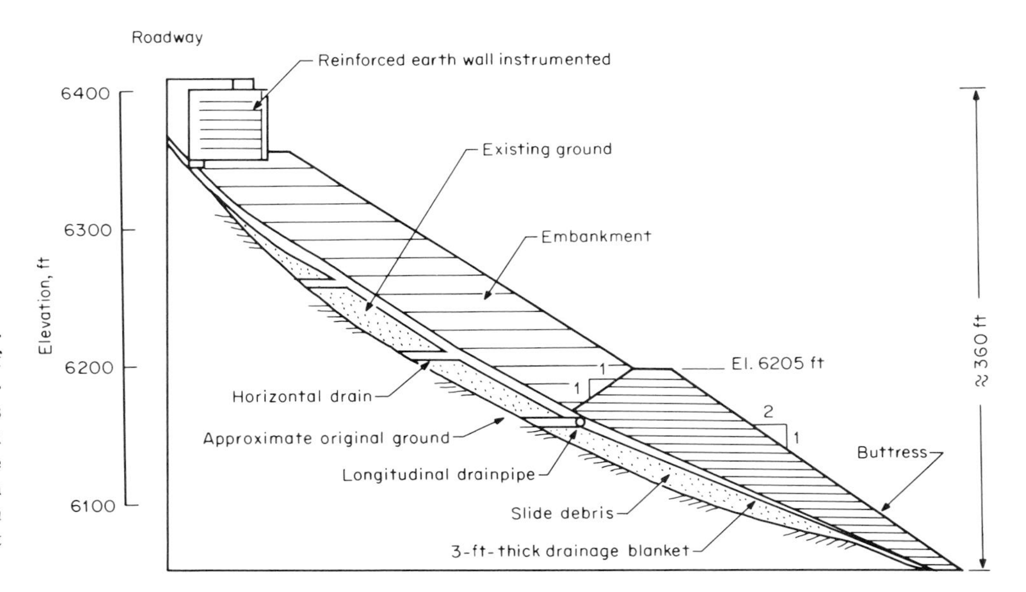


**FIG. 9.119 Section illustrating method of reconstructing Highway 39, Los Angeles County, California, on slide debris. The slide and its correction occurred in 1969.** [*From Gebney and McKittrick (1975).*[63]]

**FIG. 9.121 Anchored concrete curtain wall supports fill placed over colluvium. Wall is pile-supported to rock. Slope movements are occurring downslope to the right but the wall is stable (Highway BR 277, Parana, Brazil).**

**FIG. 9.122 Slope stabilization with anchored premolded concrete panels which conform readily to the shape of the slope. New headquarters building for Banco Nacional Desenvolvimento Economico (BNDE), Rio de Janeiro. Designed and constructed by Tecnosolo S.A.**

tion, the system is particularly suited to potentially unstable or unstable slopes. An example of an anchored curtain wall retaining a side-hill fill is shown in Fig. 9.121. A variation of the anchored curtain wall consists of anchored premolded concrete panels. As shown on Fig. 9.122, the advantage of the system is that the wall conforms readily to the slope configuration.

*Rigid walls* include concrete walls: gravity and semigravity walls, cantilever walls, and counterfort walls. Anchored concrete curtain walls are considered as semirigid.

*Flexible walls* include rock-filled buttresses, gabion walls, crib walls, reinforced earth walls, and anchored sheet-pile walls.

### *Wall Characteristics*

The general characteristics of retaining walls are summarized on Table 9.10. Also included are bored piles and root piles, not shown on Fig. 9.118.

*Wall Selection and Design Elements*

The wall type is tentatively selected on the basis of an evaluation of the cut height, materials to be supported, wall purpose, and a preliminary economic study.

Earth pressures are determined (magnitude, location, and direction), as influenced by the slope inclination and height; location and magnitude of surcharge loads; wall type, configuration and dimensions; depth of embedment; magnitude and direction of wall movement; soil parameters for natural materials and borrowed backfill; and seepage forces.

Wall stability is evaluated with respect to adequacy against overturning, sliding along the base, foundation bearing failure, and settlement. The slope must be evaluated with respect to formation of a possible failure surface beneath the wall.

Structural design proceeds when all of the forces acting on the wall have been determined.

Beyond the foregoing discussion, the design of retaining walls is not within the scope of this volume.

## 9.5 INVESTIGATION: A REVIEW

### 9.5.1 GENERAL

**Study Scopes and Objectives**

*Regional Planning*

Regional studies are performed to provide the basis for planning urban expansion, transportation networks, large area developments, etc. The objectives are to identify areas prone to slope failures, and the type, magnitude, and probability of occurrence. Hazard maps pictorialize the findings.

*Individual Slopes*

Individual slopes are studied when signs of instability are noted and development is endangered, or when new cuts and fills are required for development. Studies should be performed in two phases: phase 1, to establish the overall stability, is a study of the entire slope from toe to crest to identify potential or existing failure forms and their failure surfaces, and phase 2 is a detailed study of the immediate area affected by the proposed cut or fill.

**Considerations**

*Failure Forms and Hazard Degrees*

Engineers and geologists must be aware of which natural slope conditions are hazardous, which can be analyzed mathematically with some degree of confidence, which are very sensitive to human activities on a potentially catastrophic scale, which can be feasibly controlled, and which are to be avoided. They should also be aware that in the present state of the art there are many limitations in our abilities to predict, analyze, prevent, and contain slope failures.

*Rotational slides* are the forms most commonly anticipated, whereas the occurrence of other forms is often neglected during slope studies. They are generally the least catastrophic of all forms, normally involve a relatively small area, give substantial warning in the form of surface cracking, and usually result in gradual downslope movement during the initial development stages. Several potential failure forms can exist in a given slope, however.

*Planar slides* in mountainous terrain, which usually give warning and develop slowly, can undergo sudden total failure, involving huge volumes and high velocities with disastrous consequences.

*Falls, avalanches,* and *flows* often occur suddenly without warning, move with great velocities, and can have disastrous consequences.

*Stability Factors*

Slope geometry and geology, weather conditions, and seismic activity are the factors influencing slope stability, but conditions are frequently transient. Erosion, increased seepage forces, strength deterioration, seismic forces, tectonic activity, as well as human activity, all undergo changes with time and work to decrease slope stability.

**TABLE 9.10**
**RETAINING WALL CHARACTERISTICS**

| Wall type | Description | Comments |
|---|---|---|
| Rock-filled buttress | Constructed of nondegradable, equidimensional rock fragments with at least 50% between 30 to 100 cm and not more than 10% passing 2-in sieve [Royster (1979)[23]]. | Gradation is important to maintain free-draining characteristics and high friction angle, which combined with weight provides retention. Capacity limited by $\phi$ of approximately = 40° and space available for construction. |
| Gabion wall | Wire baskets, about 50 cm each side, are filled with broken stone about 10 to 15 cm across. Baskets are then stacked in rows. | Free-draining. Retention is obtained from the stone weight and its interlocking and frictional strength. Typical wall heights are about 5 to 6 m, but capacity is limited by $\phi$. |
| Crib wall | Constructed by forming interconnected boxes from timber, precast concrete, or metal members and then filling the boxes with crushed stone or other coarse granular material. Members are usually 2 m in length. | Free-draining. Height of single wall is limited to an amount twice the member length. Heights are increased by doubling box sections in depth. High walls are very sensitive to transverse differential settlements, and the weakness of cross members precludes support of high surcharge loads. |
| Reinforced earth walls | A compacted backfill of select fill is placed as metal strips, called ties, are embedded in the fill to resist tensile forces. The strips are attached to a thin outer skin of precast concrete panels to retain the face. | Free-draining and tolerant of different settlements, they can have high capacity and have been constructed to heights of at least 18 m. Relatively large space is required for the wall. (*See also Fig. 9.119.*) |
| Concrete gravity wall | A mass of plain concrete. | Requires weep holes, free-draining draining backfill, large excavation. Can take no tensile stresses and is uneconomical for high walls. |
| Semigravity concrete wall | Small amount of reinforcing steel is used to reduce concrete volume and provide capacity for greater heights. | Requires weep holes, free-draining backfill, and large excavation. Has been constructed to heights of 32 m [Kulhawy (1974).[65]] |
| Cantilever wall | Reinforced concrete with a stem connected to the base. The weight of earth acting on the heel is added to the weight of the concrete to provide resistance. | Requires weep holes and free-draining backfill; smaller excavation than gravity walls but limited to heights of about 8 m because of inherent weakness of the stem-base connection. |

**TABLE 9.10**
**RETAINING WALL CHARACTERISTICS (*Continued*)**

| Wall type | Description | Comments |
|---|---|---|
| Counterfort wall | A cantilever wall strengthened by the addition of counterforts. | Used for wall heights over 6 to 8 m. |
| Buttress wall | Similar to counterfort walls except that the vertical braces are placed on the face of the wall rather than on the backfill side. | As per cantilever and counterfort walls. |
| Anchored reinforced-concrete curtain wall | A thin wall of reinforced concrete is tied back with anchors to cause the slope and wall to act as a retaining system.<br>A variation by Tecnosolo S. A. uses precast panels as shown in Fig. 9.122. | Constructed in the slope from the top down in sections to provide continuous retention of the slope during construction. (All other walls require an excavation which remains open while the wall is erected.) Retention capacity is high and they have been used to support cuts in residual soils over 25 m in height. Drains are installed through the wall into the slope. *See also Figs. 9.120 and 9.121.* |
| Anchored steel sheet-pile wall | Sheet piles driven or placed in an excavated slope and tied back with anchors to form a flexible wall. | Seldom used to retain slopes because of its tendency to deflect and corrode and its costs, although it has been used successfully to retain a slope toe in conjunction with other stabilization methods (*see Fig. 9.106*). |
| Bored piles | Bored piles have been used on occasion to stabilize failed slopes during initial stages and cut slopes. | Height is limited by pile capacity in bending. Site access required for large drill rig unless holes are hand-excavated. |
| Root piles | Three-dimensional lattice of small-diameter, cast-in-place, reinforced-concrete piles, closely spaced to reinforce the earth mass. | Trade name "Fondedile." A retaining structure installed without excavation. Site access for large equipment required. |

*Selection of Slope Treatments*

Slope treatments are selected primarily on the basis of judgment and experience, and normally a combination of methods is chosen.

*For active slides of large dimensions,* consideration should be given chiefly to external and internal drainage; retaining structures are seldom feasible.

*Active slides of small dimensions* can be stabilized by changing their geometry, improving drainage, and when a permanent solution is desired, containing them by walls. An alternative, which is often economically attractive, is to permit the slide to occur and to remove material continuously from the toe until a stable slope has been achieved naturally. The risk of total failure, however, must be recognized.

*Cut slopes* are first approached by determining the maximum stable slope angle; if too much excavation is required, or if space does not permit a large cut, alternative methods employing retention are considered. It must be noted that side-hill cuts are potentially far more dangerous than cuts made in level ground, even with the same cut inclination, depth, and geologic conditions. The significant difference is likely to be seepage conditions.

*Side-hill fills* must always be provided with proper drainage, and on steep slopes retention usually is prudent.

*Slope Activity Monitoring*

Where potentially dangerous conditions exist, monitoring of slope activity with instrumentation is necessary to provide early warning of impending failures.

*Hazard Zoning*

In cities and areas where potentially dangerous conditions exist and failures would result in disastrous consequences, such as on or near high, steep slopes or on sensitive soils near water bodies or courses, development should be prohibited by zoning regulations. Pertinent in this respect is a recent slide in Goteburg, Sweden [ENR (1977)[66]], a country with a long history of slope failures in glaciolacustrine and glaciomarine deposits. Shortly after heavy rains in early December 1977, a slide occurred taking at least eight lives and carrying 67 single-family and row houses into a shallow ravine. Damage was over $7 million. The concluding statement in the article: "Last week's slide is expected to spark tighter controls of construction in questionable areas."

## 9.5.2 REGIONAL AND TOTAL SLOPE STUDIES

### Preliminary Phases

*Objectives and Scope*

The objectives of the preliminary phases of investigation, for either regional studies or for the study of a particular area, are to anticipate forms, magnitudes, and incidences of slope failures.

The study scope includes collection of existing data, generation of new data through terrain analysis, field reconnaissance, and evaluation.

*Existing Data Collection*

Regional data to be collected include: slope failure histories, climatic conditions of precipitation and temperature, seismicity, topography (scales of 1:50,000 and 1:10,000), and remote-sensing imagery (scales 1:250,000 to 1:50,000).

At the project location, data to be collected include topography (scales of 1:10,000 to 1:2,000, depending upon the area to be covered by the project, and contour intervals of 2 to 4 m), and remote-sensing imagery (scale of 1:20,000 to 1:6,000). Slope sections are prepared at a 1:1 scale showing the proposed cut or fill in its position relative to the entire slope.

*Terrain Analysis*

On a regional basis, terrain analysis is performed to identify unstable and potentially unstable areas, and to establish preliminary conclusions regarding possible failure forms, magnitudes, and incidence of occurrence. A preliminary map is prepared, showing topography, drainage, active and ancient failures, and geology. The preliminary map is developed into a hazard map after field reconnaissance.

At the project location, more detailed maps are prepared illustrating the items given above, and including points of slope seepage.

*Field Reconnaissance*

The region or site location is visited and notations are made regarding seepage points, vegetation, creep indications, tension cracks, failure scars, hummocky ground, natural slope inclinations, and exposed geology. The data collected during terrain analysis provide a guide as to the more significant areas to be examined.

*Preliminary Evaluations*

From the data collected, preliminary evaluations are made regarding slope conditions in the region or project study area, the preliminary engineering geology and hazard maps are modified, and an exploration program is planned for areas of particular interest.

**Explorations**

*Geophysical Surveys*

*Seismic refraction profiling* is performed to determine the depth to sound rock and the probable groundwater table, and is most useful in differentiating between colluvial or residual soils and the fractured-rock zone. Typical seismic velocities from the weathering profile that develops in igneous and metamorphic rocks in warm, humid climates are given on Table 2.10 and Fig. 9.98. Surveys are made both longitudinal and transverse to the slope. They are particularly valuable on steep slopes with a deep weathering profile where test borings are time-consuming and costly.

*Resistivity profiling* is performed to determine the depth to groundwater and to rock. Profiling is generally only applicable to depths of about 5 to 10 m, but very useful in areas of difficult access. In the soft, sensitive clays of Sweden, the failure surface or potential failure surface is often located by resistivity measurements since the salt content, and therefore the resistivity, often changes suddenly at the slip surface [Broms (1975)[67]].

*Test Boring Program*

Test borings are made to confirm the stratigraphy determined by the geophysical explorations, to recover samples of the various materials, and to provide holes for the installation of instrumentation.

The depth and number of borings depend on the stratigraphy and uniformity of conditions, but where the slope consists of colluvial or residual soils, borings should penetrate to rock. In other conditions, the borings should extend below the depth of any potential failure surface, and always below the depth of cut for an adequate distance.

Sampling should be continuous through the potential or existing rupture zone, and in residual soils and rock masses care should be taken to identify slickensided surfaces.

Groundwater conditions must be defined carefully, although the conditions existing at the time of investigation are not likely to be those during failure.

*In Situ Measurements*

*Piezometers* yield particularly useful information if in place during the wet season. In clayey residual profiles, confined water-table conditions can be expected in the weathered or fractured rock zone near the interface with the residual soils, or beneath colluvium. A piezometer set into fractured rock in these conditions may disclose artesian pressures exceeding the hydraulic head given by piezometers set into the overlying soils, even when they are saturated (Fig. 9.123).

*Instrumentation* is installed to monitor surface deformations, to measure movement rates, and to detect the rupture zone if the slope is considered to be potentially unstable or is undergoing movement.

*Nuclear probes* lowered into boreholes measure density and water content, and have been used to locate a failure surface by monitoring changes in these properties resulting from material rupture. In a relatively uniform material, the moisture and density logs will show an abrupt change in the failure zone from the average values [Cotecchia (1978)[68]].

*Dating Relict Slide Movements*

Radiometric dating of secondary minerals in a ruptured zone or on slickensided surfaces, or of organic strata buried beneath colluvium, provides a basis for estimating the age of previous major movements.

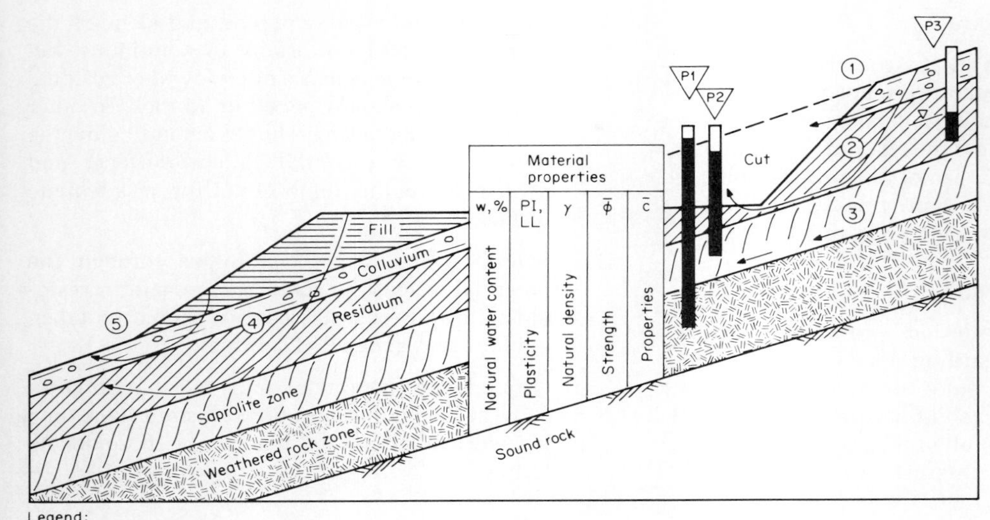


Legend:

1. Shallow slump in colluvium
2. Total failure of cut in residuum caused by very high-seepage pressures in saprolite
3. Large-scale failure of slope caused by very high-seepage pressures in the fractured rock zone
4. Large rotational failure of fill through residuum
5. Slump failure at fill toe through the colluvium

**FIG. 9.123 Schematic of typical section prepared at scale of 1:1 showing tentative cut and fill imposed on 30° in residual soil profile as basis for analysis. Piezometers P1 and P2 show excess water pressures in saprolite and fractured rock compared with saturation zone in residuum at P3. Several possible failure conditions requiring evaluation are shown.**

Growth ring counts in trees which are inclined in their lower portion and vertical above also provide data for estimating the age of previous major slope movements. The date of the last major movement can be inferred from the younger, vertical-growing segments [Cotecchia (1978)[68]]. Slope failures cause stresses in the tree wood which result in particular tissues (reaction or compressed wood) which are darker and more opaque than normal unstressed wood. On the side toward which the tree leans there is an abrupt change from the growth rings of normal wood to those of compression wood. By taking small cores from the tree trunk it is possible to count the rings and estimate when the growth changes occurred, and thus to date approximately the last major slope movement.

## Slope Assessment

### *Data Presentation*

A *plan* of the slope area is prepared showing contours, drainage paths, seepage emerging from the slope, outcrops, tension cracks and other failure scars, and other significant information. *Sections* are prepared at a 1:1 scale illustrating the stratigraphy and groundwater conditions as determined from the explorations, as well as any relict failure surfaces.

### *Evaluations and Analyses*

Possible failure forms are predicted and existing failures are delineated as falls, slides, avalanches, or flows, and the degree of the hazard is judged. Depending upon the degree of risk, the decision is made to avoid the hazard or to eliminate or reduce it. For the cases of falls, avalanches, flows, and failures by lateral spreading, the decision is based on experience and judgment. Slides may be evaluated by mathematical analysis, but in recognition that movements may develop progressively.

Preliminary analysis of existing or potential failures by sliding includes the selection of potential failure surfaces by geometry in the case of planar slides, or analytically in the case of rotational slides, or by observation in the case of an existing slide. An evaluation is made of the safety factor against *total failure* on the basis of existing topographic conditions, then under conditions of the imposed cut or fill. For preliminary studies, shear strengths may be estimated from published data, or be measured by laboratory or in situ testing. In the selection of the strength parameters, consideration is given to field conditions (Table 9.6) as well as to changes that may occur with time (reduction from weathering, leaching, solution). Other transient conditions also require consideration, especially if the safety factor for the entire slope is low and could go below unity with some environmental change.

### 9.5.3 DETAILED STUDY OF CUT, FILL, OR FAILURE AREA

#### General

Detailed study of the area of the proposed cut or fill, or of the failure, is undertaken after the stability of the entire slope is assessed. The entire slope is often erroneously neglected in studies of cuts and side-hill fills, and is particularly important in mountainous terrain.

#### Explorations

*Seismic refraction surveys* are most useful if rock is anticipated within the cut, and there are boulders in the soils which make the delineation of bedrock difficult with test and core borings.

*Test and core borings,* and *test pits* are made to recover samples, including undisturbed samples, for laboratory testing. In colluvium, residuum, and saprolite, the best samples are often recovered from test pits, but these are usually limited to depths of 3 to 5 m because of practical excavation considerations.

In situ testing is performed in materials from which undisturbed samples are difficult or impossible to procure.

#### Laboratory Testing

Laboratory strength testing should duplicate the field conditions of pore-water pressures, drainage, load duration, and strain rate that are likely to exist as a consequence of construction operations, and samples should usually be tested in a saturated condition. It must be considered that conditions during and at the end of construction (short-term) will be different than long-term stability conditions. In this regard, the natural ability of the slope to drain during cutting plays a significant role.

#### Evaluation and Analysis

*Sections* illustrating the proposed cut, fill, or failure imposed on the slope are prepared at a 1:1 scale. The selection of cut slope inclination is based on the engineer's judgment of stability and is shown on the section together with the stratigraphy, groundwater conditions measured, and the soil properties on Fig. 9.123.

*Analyses* are performed to evaluate stable cut angles and side-hill fill stability, and the necessity for drainage and retention. Consideration must be given to the possibility of a number of failure forms and locations as shown on Fig. 9.123, as well as to changing groundwater and other environmental conditions.

### 9.5.4 INSTRUMENTATION AND MONITORING

#### Purpose

Where movement is occurring, where safety factors against sliding are low, or where a major work would become endangered by a slope failure, instrumentation is required to monitor changing conditions which may lead to total failure.

Slope-stability analysis is far from an exact science, regardless of the adequacy of the data available, and sometimes the provision for an absolutely safe slope is prohibitively costly. In this case, the engineer may wish to have contingency plans available such as the installation of internal drainage systems or the removal of material from upslope, etc., if the slope shows signs of becoming unstable.

In *unstable* or *moving slopes,* instrumentation is installed to locate the failure surface and determine pore-water pressures for analysis, and to measure surface and subsurface movements, velocities, and accelerations which provide indications of impending failure. In *cut slopes,* instrumentation monitors movements and changing stress conditions to provide early warning and permit invoking remedial measures when low safety factors are accepted in design.

### Instrumentation Methods Summarized

Instrumentation is discussed in detail in Chap. 4, and for slopes is illustrated on Figs. 4.40 and 4.41.

*Surface movements* are monitored by survey nets, tiltmeters (on benches), convergence meters, surface extensometers, and terrestrial photography. Accuracy ranges from 0.5 to 1.0 mm for extensometers, to 30 mm for the geodimeter, to 300 mm for the theodolite [Blackwell et al. (1975)[69]].

*Subsurface deformations* are monitored with inclinometers, deflectometers, shear-strip indicators, steel wire and weights in boreholes, and the accoustical emissions device. Accuracy for extensometers and inclinometers usually ranges from 0.5 to 1.0 mm, but the accuracy depends considerably on the deformation pattern and in many instances cannot be considered better than 5 to 10 mm.

*Pore-water pressures* are monitored with piezometers.

*All instruments* should be monitored periodically and the data plotted *as it is obtained* to show changing conditions. Movement accelerations are most significant.

### Example of an Instrumentation Network

*Project*

A large slide had occurred in a rock mass during excavation of a near-vertical cut for a highway at the Libby Dam site in Montana [Heinz (1975)[70]].

*Geologic Conditions*

The significant material involved was argillite, a metamorphosed sedimentary rock, with strong jointing along and intersecting the bedding planes.

*Instrumentation*

Instruments were installed to measure slope movements (multipositional extensometers, inclinometers, and the acoustical emission device) and pore-water pressures (electrical piezometers for pervious joints and bedding plane contacts, vibrating wire piezometers for uplift pressures in the mass, and pneumatic transducer piezometers as backup for the electrical instruments).

## REFERENCES


1. Varnes, D. J. (1958) "Landslide Types and Processes," *Landslides and Engineering Practice*, E. B. Eckel, ed., Highway Research Board Spec. Report No. 29, Washington, D.C.
2. Banks, D. C. and Strohn, W. E. (1974) "Calculation of Rock Slide Velocities," *Proc. 3d Intl. Cong. for Rock Mechs.*, Intl. Soc. for Rock Mechs., Denver, Vol. 118, pp. 839–847.
3. Habib, P. (1975) "Production of Gaseous Pore Pressure During Rock Slides," *J. Intl. Soc. Rock Mechs.*, Vol. 7, No. 4, November, p. 193.
4. Bjerrum, L. (1967) "Progressive Failures in Slopes of Overconsolidated Plastic Clay and Clay Shales," *Terzaghi Lectures 1963–1972*, ASCE (1974), pp. 139–189.
5. Bjerrum, L, Loken, T., Heiberg, S. and Foster, H. (1969) "A Field Study of Factors Responsible for Quick Clay Slides," *Proc. 7th Intl. Conf. Soil Mechs. and Found. Engrg.*, Mexico City, pp. 531–540, Vol. 2.
6. Terzaghi, K. (1950) "Mechanism of Landslides," *Engineering Geology* (Berkey volume), The Geologic Society of America, pp. 83–123.
7. Alden, W. C. (1928) "Landslide and Flood at Gros Ventre, Wyoming," Art. 16 in *Focus on Environmental Geology*, R. Tank, ed., Oxford University Press, New York (1973), pp. 146–153.
8. Kiersch, G. A. (1965) "The Vaiont Reservoir Disaster," *Focus on Environmental Geology*, R. Tank, ed., The Oxford University Press (1973), Art. 17, pp. 153–164.
9. Hamel, J. V. (1980) "Geology and Slope Stability in Western Pennsylvania," *Bull. AEG*, Vol. XVII, No. 1, Winter, pp. 1–26.
10. Hamel, J. V. (1972) "The Slide at Brilliant Cut,"

*Stability of Rock Slopes, Proc. ASCE, 13th Symp. on Rock Mechs.*, Urbana, Ill. (1971), pp. 487–572.

11. Hunt, R. E. and Santiago, W. B. (1976) "A funcão critica do engenheiro geólogo em estudos de implantacão de ferrovias," *Proc. 1° Cong. Brasileiro de Geologia de Engenharia*, Rio de Janeiro, Aug., Vol. 1, pp. 79–98.
12. Deere, D. U. and Patton, F. D. (1971) "Slope Stability in Residual Soils," *Proc. ASCE, 4th Pan American Conf. Soil Mechs. Found. Engrg.*, San Juan, P. R., pp. 87–170.
13. Binger, W. V. (1948) "Analytical Studies of Panama Canal Slides," *Proc. 2d Intl. Conf. Soil Mechs. and Found. Engrg.*, Rotterdam, Vol. 2, pp. 54–60.
14. Banks, D. C. (1972) "Study of Clay Shale Slopes," *Stability of Rock Slopes, Proc. ASCE, 13th Symp. on Rock Mechs.*, Urbana, Ill. (1971), pp. 303–328.
15. Jahns, R. H. and Vonder Linden, C. (1973) "Space-Time Relationships of Landsliding on the Southerly Side of Palos Verdes Hills, California," *Geology, Seismicity and Environmental Impact*, Spec. Pub. Assoc. Engrg. Geol., Los Angeles, pp. 123–138.
16. Easton, W. H. (1973) "Earthquakes, Rain and Tides of Portuguese Bend Landslide, California," *Bull. Assoc. of Engrg. Geol.*, Vol. 8, No. 3.
17. Palladino, D. J. and Peck, R. B. (1972) "Slope Failures in an Overconsolidated Clay, Seattle, Washington," *Geotechnique*, Vol. 22, No. 4, pp. 563–595.
18. Eden, W. J., Fletcher, E. B. and Mitchell, R. J. (1971) "South Nation River Landslide, 16 May 1971," *Canadian Geotechnical J.*, Vol. 8, No. 3, August.
19. Mollard, J. D. (1977) "Regional landslide types in Canada," *Reviews in Engineering Geology*, Vol. III, *Landslides*, Geological Society of America, pp. 29–56.
20. Seed, H. B. and Wilson, S. D. (1967) "The Turnagain Heights Landslide, Anchorage Alaska," *Proc. ASCE, J. Soil Mechs. Found. Engrg. Div.*, Vol. 93, No. SM4, July, pp. 325–353.
21. Noble, H. L. (1973) "Residual Strengths and Landslides in Clay and Shale," *Proc. ASCE, J. Soil Mechs. Found. Engrg. Div.*, Vol. 99, No. SM9, September.
22. Royster, D. L. (1973) "Highway Landslide Problems Along the Cumberland Plateau, Tennessee," *Bull. Assoc. Engrg. Geol.*, Vol. X, No. 4.
23. Royster, D. L. (1979) "Landslide Remedial Measures," *Bull. Assoc. of Engrg. Geologists*, Spring, Vol. XVI, No. 2, pp. 301–352.
24. Jones F. O. (1973) "Landslides of Rio de Janeiro and the Serra das Araras Escarpment, Brasil," *U.S. Geological Survey Paper No. 697*, U.S. Govt. Printing Office, Washington, D.C.
25. McDonald, B. and Fletcher, J. E. (1962) "Avalanche—3500 Peruvians Perish in Seven Minutes," *National Geographic Magazine*, June.
26. Youd, T. L. (1978) "Major Cause of Earthquake Damage Is Ground Failure," *Civil Engineering*, ASCE, April, pp. 47–51.
27. Varnes, D. J. (1978) "Slope Movement Types and Processes," Chap. 2, *Landslides: Analysis and Control*, Special Report No. 176, National Academy of Sciences, Washington, D.C., pp. 11–35.
28. Krahn, J. and Morgenstern, N. R. (1976) "Mechanics of the Frank Slide," *Proc. ASCE, Rock Engineering for Foundations and Slopes*, Vol. 1, pp. 309–332.
29. Tavenas, F., Chagnon, J. Y. and La Rochelle, P. (1971) "The Saint-Jean-Vianney Landslide: Observations and Eyewitness Accounts," *Canadian Geotechnical J.*, Vol. 8, No. 3.
30. Molnia, B. F., Carlson, P. R. and Bruns, T. R. (1977) "Large Submarine Slide in Kayak Trough, Gulf of Alaska," *Reviews in Engineering Geology*, Vol. VII, *Landslides*, D. R. Coates, ed., Geologic Society of America, pp. 137–148.
31. Coleman, J. M., Prior, D. B. and Garrison, L. E. (1980) "Subaqueous Sediment Instabilities in the Offshore Mississippi River Delta," Open File Report No. 80.01, Bureau of Land Management, U.S. Dept. of Interior.
32. Focht, J. A., Jr. and Kraft, L. M., Jr. (1977) "Progress in Marine Geotechnical Engineering," *Proc. ASCE, J. Geotech. Engrg. Div.*, Vol. 103, No. GT10, October, pp. 1097–1118.
33. Richter, C. F. (1958) *Elementary Seismology*, W. H. Freeman and Co., San Francisco, p. 125.
34. Peck, R. B. (1967) "Stability of Natural Slopes," *Proc. ASCE, J. Soil Mechs. Found. Engrg. Div.*, Vol. 93, No. SM4, July.
35. Barton, N. (1972) "Progressive Failure of Excavated Rock Slopes," *Stability of Rock Slopes, Proc. ASCE, 13th Symp. on Rock Mechs.*, Urbana, Ill., (1971), pp. 139–170.
36. Vargas, M. and Pichler, E. (1957) "Residual Soils and Rock Slides in Santos, Brazil," *Proc. 4th Intl. Conf. Soil Mechs. Found. Engrg.*, Vol. 2, pp. 394–398.
37. Lambe, T. W. and Whitman, R. V. (1969) *Soil Mechanics*, John Wiley & Sons, New York.
38. Hoek, E. and Bray, J. W. (1977) *Rock Slope Engi-*

*neering*, 2d ed., The Institute of Mining and Metallurgy, London.

39. Patton, F. D. and Hendron, A. J., Jr. (1974) "General Report on Mass Movements," *Procs 2d Intl. Cong. Intl. Assoc. Engrg. Geol.*, São Paulo, p. V-GR 1.
40. Terzaghi, K. (1943) *Theoretical Soil Mechanics*, John Wiley & Sons, New York.
41. Tschebotarioff, G. P. (1973) *Foundations, Retaining and Earth Structures*, McGraw-Hill Book Co., New York.
42. Morgenstern, N. R. and Sangrey, D. A. (1978) "Methods of Stability Analysis," Chap. 7, *Landslides: Analysis and Control*, Schuster and Krizek, eds., Spec. Report 176, National Academy of Sciences, Washington, D.C., pp. 155–172.
43. NAVFAC (1971) *Design Manual, Soil Mechanics, Foundations and Earth Structures*, DM-7 Naval Facilities Engineering Command, Alexandria, Va., March.
44. Taylor, D. W. (1948) *Fundamentals of Soil Mechanics*, John Wiley & Sons, New York.
45. Bishop, A. W. (1955) "The Use of the Slip Circle in the Stability Analysis of Earth Slopes," *Geotechnique*, Vol. 5, No. 1, pp. 7–17.
46. Janbu, N., Bjerrum, L. and Kjaernsli, B. (1956) *Soil Mechanics Applied to Some Engineering Problems*, Norwegian Geotechnical Institute, Oslo, Publ. 16, pp. 5–26.
47. Morgenstern, N. R. and Price, V. E. (1965) "The Analysis of the Stability of General Slip Surfaces," *Geotechnique*, Vol. 15, No. 1, pp. 79–93.
48. Janbu, N. (1973) "Slope Stability Computations," in *Embankment Dam Engineering*, Hirschfield and Poulos, eds., John Wiley & Sons, New York, pp. 47–86.
49. Newmark, N. M. (1965) "Effects of Earthquakes on Dams and Embankments," *Geotechnique*, Vol. 15, No. 2, June, London.
50. Seed, H. B. (1966) "A Method for Earthquake Resistant Design of Earth Dams," *Proc. ASCE, J. Soil Mechs. Found. Engrg. Div.*, Vol. 92, No. SM1, pp. 13–41.
51. Brawner, C. O. (1975) "Case Examples of Instability of Rock Slopes," *J. Assoc. of Prof. Engrs. of British Columbia*, February, Vol. 26, No. 2.
52. Fox, P. P. (1964) "Geology, Exploration, and Drainage of the Serra Slide, Santos, Brasil," *Engineering Geology Case Histories Numbers 1–5*, The Geological Society of America, Engrg. Geol. Div., pp. 17–24.
53. Garga, V. K. and De Campos, T. M. P. (1977) "A Study of Two Slope Failures in Residual Soils," *Proc. 5th Southeast Asian Conf. on Soil Engrg.*, Bangkok, pp. 189–200.
54. Guidicini, G. and Iwasa, O. Y. (1977) "Tentative Correlation Between Rainfall and Landslides in a Humid Tropical Environment," *Bull. Intl. Assoc. of Engrg. Geol.*, No. 16, December pp. 13–20.
55. Piteau, D. R. (1977) "Regional Slope-Stability Controls and Engineering Geology of the Fraser Canyon, British Columbia," *Reviews in Engineering Geology*, Vol. III, D. R. Coates, ed., Geologic Society of America, Boulder, Colo. pp. 85–112.
56. Terzaghi, K. (1962) "Stability of Steep Slopes on Hard, Unweathered Rock," *Geotechnique*, No. 12, pp. 251–270.
57. Chassie, R. G. and Goughnor, R. D. (1976) "States Intensifying Efforts to Reduce Highway Landslides," *Civil Engineering*, ASCE, New York, April, p. 65.
58. *Sunset*, (1978) "If hillside slides threaten in Southern California, November is planting action month," *Sunset Magazine*, November, pp. 122–126.
59. Handy, R. L. and Williams, W. W. (1967) "Chemical Stabilization of an Active Landslide," *Civil Engineering*, ASCE, August, pp. 62–65.
60. Broms, B. B. and Boman, P. (1979) "Lime Columns—A New Foundation Method," *Proc. ASCE, J. Geotech. Engrg. Div.*, Vol. 105, No. GT4, April, pp. 539–556.
61. D'Appolonia, E. D., Alperstein, R. A., and D'Appolonia, D. J. (1967) "Behavior of a Colluvial Soil Slope" *Proc. ASCE, J. Soil Mechs. and Found. Engrg. Div.*, Vol. 93, No. SM4 July, pp. 447–473.
62. Lang, T. A. (1972) "Rock Reinforcement," *Bull. Assoc. Engrg. Geol.*, Vol. IX, No. 3, Summer, pp. 215–239.
63. Gebney, D. S. and McKittrick, D. P. (1975) "Reinforced Earth: A New Alternative for Earth-Retention Structures," *Civil Engineering*, ASCE, October.
64. Hunt, R. E. and Costa Nunes, A. J. da (1978) "Retaining Walls: Taking It from the Top," *Civil Engineering*, ASCE, May, pp. 73–75.
65. Kulhawy, F. H. (1974) "Analysis of a High Gravity Retaining Wall," *Proc. ASCE, Conf. Analysis and Design in Geotech. Engrg.*, Univ. of Texas, Austin, Vol. I, pp. 159–171.
66. ENR (1977) "Landslide kills eight, destroys 76 homes," *Engineering News-Record*, Dec. 8, p. 11.
67. Broms, B. B. (1975) "Landslides," *Foundation Engineering Handbook*, Winterkorn and Fang,

eds., Van Nostrand Reinhold Co., N.Y., Chap. 11, pp. 373–401.

68. Cotecchia, V. (1978) "Systematic Reconnaissance Mapping and Registration of Slope Movements," *Intl. Assoc. of Engrg. Geol.*, Bull. No. 17, June pp. 5–37.
69. Blackwell, G., Pow, D. and Klast, L. (1975) "Slope Monitoring at Brenda Mine," *Proc. 10th Canadian Rock Mechs. Symp.*, Kingston, Ontario, September, pp. 45–79.
70. Heinz, R. A. (1975) "Insitu Soils Measuring Devices" *Civil Engineering*, ASCE, October, pp. 62–65.

## BIBLIOGRAPHY

Barata, F. E. (1969) "Landslides in the Tropical Region of Rio de Janeiro," *Proc. 7th Intl. Conf. Soil Mechs. and Found. Engrg.*, Mexico City, Vol. No. 2, pp. 507–516.

Coates D. R., ed., (1977) *Landslides, Reviews in Engineering Geology*, Vol. III, The Geologic Society of America, Boulder, Colo.

Costa Nunes, A. J. da (1966) "Slope Stabilization—Improvements in the Prestressed Anchorages in Rocks and Soils," *Proc. 1st Intl. Conf. on Rock Mechs.*, Lisbon.

Costa Nunes, A. J. da (1969) "Landslides in Soils of Decomposed Rock Due to Intense Rainstorms," *Proc. 7th Intl. Conf. Soil Mechs. and Found. Engrg.*, Mexico City.

Costa Nunes, A. J. da, Fonseca, A.M.C.C., and Hunt, R. E. (1980) "A Broad View of Landslides in Brasil," *Geology and Mechanics of Rockslides and Avalanches*, B. Voight, ed., Elsevier Scientific Pub. Co., Amsterdam.

Dunn, J. R. and Banino, G. M. (1977) "Problems with Lake Albany Clays," *Reviews in Engineering Geology*, Vol. III, *Landslides*, the Geologic Society of America, Boulder, Colo., pp. 133–136.

ENR (1977b) "Italian pile system supports slide in its first major U.S. application," *Engineering News-Record*, Nov. 24, p. 16.

Guidicini, G. and Nieble, C. M. (1976) *Estabilidade de Taludes Naturais e de Escavacão*, Editora Edgard Blucher Ltda., São Paulo, 170p.

Hunt. R. E. and Shea, G. P. (1980) "Avoiding Landslides on Highways in Mountainous Terrain," *Proc. International Road Federation, IV IRF African Highway Conf.*, Nairobi, January.

Hutchinson, J. N. (1977) "Assessment of the Effectiveness of Corrective Measures in Relation to Geological Conditions and Types of Slope Movement," *Bull. Intl. Assoc. Engrg. Geol.*, Landslides and Other Mass Movements, No. 16, December, pp. 131–155.

Kalkani, E. C. and Piteau, D. R. (1976) "Finite Element Analysis of Toppling Failure at Hell's Gate Bluffs, British Columbia," *Bull. Assoc. Engrg. Geol.*, Vol. XIII, No. 4, Fall.

Kennedy, B. A. (1972) "Methods of Monitoring Open Pit Slopes," *Stability of Rock Slopes, Proc. ASCE, 13th Symp. on Rock Mechs.*, Urbana, Ill. (1971), pp. 537–572.

Kenney, T. C., Pazin, M., and Choi, W. S. (1977) "Design of Horizontal Drains for Soil Slopes," *Proc., ASCE, J. Geotech. Engrg. Div.*, Vol. 103, No. GT11, November, pp. 1311–1323.

Koerner, R. M., Lord, A. E., Jr., and McCabe, W. M. (1978) "Acoustic Emission Monitoring of Soil Stability," *Proc., ASCE, J. Geotech. Engrg. Div.*, Vol. 104, No. GT 5, May, pp. 571–582.

La Rochelle, P. (1970) "Regional Geology and Landslides in the Marine Clay Deposits of Eastern Canada," *Canadian Geotechnical J.*, Vol. 8.

Lo, K. Y. and Lee, C. F. (1973) "Analysis of Progressive Failure in Clay Slopes," *Proc. 8th Intl. Conf. Soil Mechs. Found. Engrg.*, Moscow, Vol. 1, pp. 251–258.

Pinckney, C. J., Streiff, D., and Artim, E. (1979) "The Influence of Bedding-Plane Faults in Sedimentary Formations on Landslide Occurrence Western San Diego County, California," *Bull. Assoc. of Engrg. Geologists*, Spring, Vol. XVI, No. 2, pp. 289–300.

Piteau, D. R. and Peckover, F. L. (1978) "Engineering of Rock Slopes," Chap. 9 in *Landslides: Analysis and Control*, Schuster and Krizek, eds., Spec. Report No. 176, National Academy of Sciences, Washington, D.C., pp. 192–227.

Quigley, R. M. et al. (1971) "Swelling Clay in Two Slope Failures at Toronto, Canada," *Canadian Geotechnical J.*, Vol. 8.

Skempton, A. W. and Hutchinson, J. (1969) "Stability of Natural Slopes and Embankment Foundations," State-of-the-Art Paper, *Proc. 7th Intl. Conf. Soil Mechs Found. Engrg.*, Mexico City, pp. 291–340.

Taylor, D. W. (1940) "Stability of Earth Slopes," *Contributions to Soil Mechanics 1925 to 1940*, Boston Society of Civil Engineers.

Terzaghi, K. (1956) "Varieties of Submarine Slope Failures," *Proc. 8th Texas Conf. Soil Mechs. Found. Engrg.*, Chap. 3.

Terzaghi, K. (1960) "Selected Professional Reports—Concerning a Landslide on a Slope Adjacent to a Power Plant in South America," reprinted in *From Theory to Practice on Soil Mechanics*, John Wiley & Sons, New York.

Terzaghi, K. and Peck, R. B. (1967) *Soil Mechanics in Engineering Practice*, John Wiley & Sons, New York.

Vanmarcke, E. H. (1977) "Reliability of Earth Slopes," *Proc. ASCE, J. Geotech. Engrg. Div.*, Vol. 103, No. GT 11, November, pp. 1247–1265.

Záruba, Q. and Mencl, V. (1969) *Landslides and Their Control*, Elsevier Scientific Pub. Co., Amsterdam.

# CHAPTER TEN

# GROUND SUBSIDENCE, COLLAPSE, AND HEAVE

## 10.1 INTRODUCTION

### 10.1.1 GENERAL

#### Origins

The hazardous vertical ground movements of subsidence, collapse, and heave, for the most part, are the result of human activities that change an environmental condition. Natural occurrences, such as earthquakes and tectonic movements, also affect the surface from time to time.

#### Significance

Subsidence, collapse, and heave are less disastrous than are either slope failures or earthquakes in terms of lives lost, but the total property damage that results each year probably exceeds that of the other hazards. A positive prediction of their occurrence is usually very difficult, and uncertainties always exist, although the conditions favorable to their development are readily recognizable.

### 10.1.2 THE HAZARDS

A summary of hazardous vertical ground movements, causes and important effects is given on Table 10.1.

### 10.1.3 CHAPTER SCOPE AND OBJECTIVES

#### Scope

Ground movements considered in this chapter are caused by some internal change within the subsurface such as extraction of fluids or solids, solution of rock or a cementing agent in soils, erosion, or physicochemical changes, rather than movements brought about by the application of surface loadings from construction activity (i.e., ground settlements resulting from embankments, buildings, etc.).

#### Objectives

The objectives are to provide the basis for recognizing the potential for surface movements, and for preventing or controlling the effects.

## 10.2 GROUNDWATER AND OIL EXTRACTION

### 10.2.1 SUBSURFACE EFFECTS

#### Groundwater Withdrawal

*Aquifer Compaction*

Lowering the groundwater level reduces the buoyant effect of water, thereby increasing the effective weight of the soil within the depth through which the groundwater has been lowered. For example, for a fully saturated soil the buoyant force of water is 62.4 pcf (1 t/m$^3$) and if the water table is lowered 30 m, the increase in effective stress on the underlying soils will be 3.0 tsf (30 t/m$^3$), a significant amount. If the prestress in the soils is exceeded, compression occurs and the surface subsides. In an evaluation of the effect on layered strata of sands and clays, the change in piezometric level in each

**TABLE 10.1**
**SUMMARY OF HAZARDOUS VERTICAL GROUND MOVEMENTS**

| Movement | Description | Causes | Important effects |
|---|---|---|---|
| Regional subsidence | Downward movement of ground surface over large area | Seismic activity,* groundwater extraction, oil and gas extraction | Flooding, growth faults, structure distortion |
| Ground collapse | Sudden downward movement of ground surface over limited area | Subsurface mining, limestone cavity growth, piping cavities in soils, leaching of cementing agents | Structure destruction, structure distortion |
| Soil subsidence | Downward movement of ground surface over limited area | Construction dewatering, compression under load applied externally, desiccation and shrinkage | Structure distortion |
| Ground heave | Upward movement of ground surface | Expansion of clays and rocks, release of residual stresses,† tectonic activity,‡ ground freezing§ | Structure distortion, weakening of clay shale slopes |

*See Art. 11.3.3.
†See Art. 6.6.
‡See Art. 11.3.3, Appendix A.
§See Art. 7.7.3.

compressible stratum is assessed to permit a determination of the change in effective stress in the stratum. Compression in sands is essentially immediate; cohesive soils exhibit a time delay as they drain slowly during consolidation *(see Art. 3.5.4)*. Settlements are computed for the change in effective stress in each clay stratum from laboratory consolidation test data.

The amount of subsidence, therefore, is a function of the decrease in the piezometric level, which determines the increase in overburden pressures, and the compressibility of the strata. For clay soils the subsidence is a function of time.

*Construction Dewatering*

Lowering the groundwater for construction projects has the same effect as "aquifer compaction," i.e., it compresses soil strata because of an increase in effective overburden stress.

### Oil Extraction

Oil extraction differs from groundwater extraction mainly because much greater depths are involved, and therefore much greater pressures. Oil (or gas) extraction results in a reduction of pore-fluid pressures, which permits a transfer of overburden pressures to the intergranular skeleton of the strata.

In the Wilmington oil field, Long Beach, California, Allen (1973)[1] cites compaction as taking place primarily by sand grain arrangement, plastic flow of soft materials such as micas and clays, and the breaking and sharding of grains at stressed points. Overall, about two-thirds of the total compaction at the Wilmington field is attributed to the reservoir sands and about one-third to the interbedded shales [Allen and Mayuga (1969)[2]]. During a period of maximum subsidence in 1951–1952, faulting apparently occurred at depths of 450 to 520 m, shearing or damaging hundreds of oil wells.

### 10.2.2 SURFACE EFFECTS

#### Regional Subsidence

*General*

Surface subsidence from fluid extraction is a common phenomenon and probably occurs to some degree in any location where large quantities of water, oil, or gas are removed. Short-term detection is difficult because surface movements are usually small, are distributed over large areas in the shape of a dish, and increase gradually over a span of many years.

*Some Geographic Locations*

LONDON, ENGLAND A drop in the water table by as much as 60 m has resulted in a little more than 2 cm of subsidence, apparently without any detrimental effects because of the stiffness of the clays.

SAVANNAH, GEORGIA The city has undergone as much as 10 cm of subsidence since 1933 because of water being pumped from the Ocala limestone, also apparently without detrimental effects [Davis et al. (1962)[3]].

MEXICO CITY In the hundred years or so between the mid-1800s and 1955, the city experienced as much as 6 m of subsidence from compression of the underlying soft soils because of groundwater extraction. By 1949 the rate was 35 cm/year. Surface effects have been serious.

HOUSTON, TEXAS A decline in the water table of 90 m since 1890 has caused as much as 2.7 m of subsidence with serious surface effects.

LONG BEACH, CALIFORNIA The city has suffered as much as 9 m of subsidence from oil extraction between 1928 and 1970 with serious effects.

LAKE MARICAIBO, VENEZUELA The area underwent as much as 3.3 m of subsidence between 1926 and 1954 due to oil extraction.

PO DELTA OF ITALY, NIIGATA, JAPAN The areas have been affected by gas withdrawal.

#### Flooding, Faulting, and Other Effects

*Flooding* results from grade lowering and has been a serious problem in coastal cities such as Houston and Long Beach in the United States, and Venice in Italy. In Venice, although subsidence from groundwater withdrawal has been reported to be only 5.5 cm/year, the total amount combined with abnormally high tides has been enough to cause the city to be inundated periodically. The art works and architecture of the city have been damaged as a result. Flood incidence also increases in interior basins where stream gradients are affected by subsidence.

*Faulting*, or *growth faults*, occur around the periphery of subsided areas. Although displacements are relatively small, they can be sufficient to cause distress in structures and underground utilities, and sudden drops in roadways. Oil extraction can cause movement along existing major faults.

*Differential movement* over large distances affects canal flows, such as in the San Joaquin Valley of California and over short distances causes distortion of structures, as in Mexico City.

*Grade lowering* can also result in the loss of head room under bridges in coastal cities and affect boat traffic, as in Houston [ENR (1977a)[4]].

#### Local Subsidence from Construction Dewatering

*Drawdown* of the water table during construction can cause surface subsidence for some distance from the dewatering system. Differential settlements reflect the cone of depression. The differential settlements can be quite large, especially when peat or other organic soils are present, and the effect on adjacent structures can be damaging.

During the construction of the Rotterdam Tunnel, wellpoints were installed to relieve uplift pressures in a sand stratum which underlay soft clay and peat. The groundwater level in observation wells, penetrating into the sand, at times showed a drop in water level of 12.8 m. Settlements were greatest next to the line of wellpoints: 51 cm at a distance of 9 m and 8 cm at a

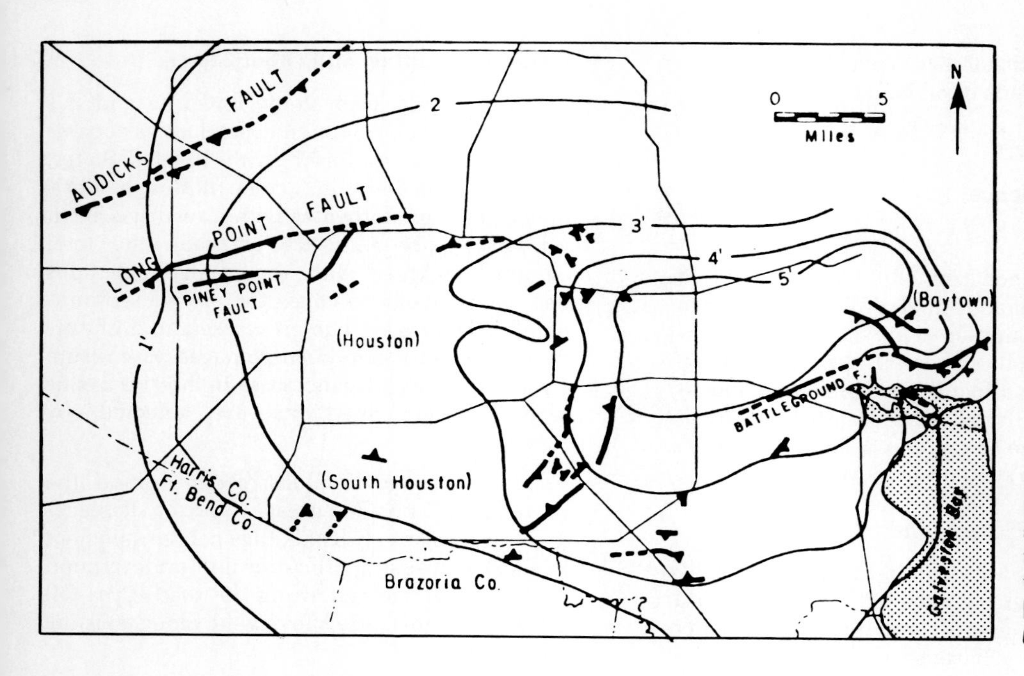


**FIG. 10.1** **Surface faulting and cumulative subsidence (in feet) in the Houston area between 1906 and 1964.** [*From Castle and Youd (1972).*[12]]

distance of 110 m. The water level was lowered for about 2.9 years and caused an effective stress increase as high as 1.3 $kg/cm^2$ [Tschebotarioff (1973)[5]].

### 10.2.3 PHYSIOGRAPHIC OCCURRENCE

#### General

Although subsidence can occur in any location where large quantities of fluids are extracted, its effects are felt most severely in coastal areas and inland basins.

#### Coastal Areas

Many examples of coastal cities subsiding and suffering flooding can be found in the literature, and any withdrawal from beneath coastal cities with low elevations in reference to sea level must be performed with caution.

#### Interior Basins

In the semiarid to arid regions of the western United States, the basins often are filled with hundreds of meters of sediments which serve as natural underground reservoirs for the periodic rainfall and runoff from surrounding mountains. When groundwater depletion substantially exceeds recharge, the water table drops and subsidence occurs (see Tucson, Arizona, case study, *Art. 8.5.1*). Subsidence can reach significant amounts: for example, several meters in the San Joaquin and Santa Clara valleys in California and in Elroy, Arizona, and 0.9 m in Las Vegas, Nevada [ENR (1977b)[6]]. Around the Tucson basin, where the water level has dropped as much as 40 m since 1947, it has been suggested that minor faulting is occurring and may be the reason for distress in some home foundations [Davidson (1970),[7] Peirce (1972)[8]]. Other effects will be increased flooding due to changes in stream gradients and the loss of canal capacity due to general basin lowering.

The basin in which Mexico City is situated is filled with thick lacustrine sediments of volcanic origin, and groundwater withdrawal has resulted in serious consequences.

### 10.2.4 SIGNIFICANT EXAMPLES

#### Houston, Texas (Water Extraction: Flooding and Faulting)

Between 1906 and 1964, 1.5 m of subsidence occurred, and recent reports place the subsidence at 2.7 m [*Civil Engineering* (1977)[9]]. The cost of the subsidence, including flood damage, since 1954 has been estimated to be $110 million and is growing at the annual rate of $30 million [Spencer (1977)[10]].

The problem of growth faults in the Houston metropolitan area is severe. Activity has been recognized on more than 40 normal faults, which according to Van Siclen (1967),[11] are prehistoric. Major surface faults and the cumulative subsidence between 1906 and 1964 are given on Fig. 10.1, and a profile of subsidence and groundwater decline for a distance of about 22 km is given on Fig. 10.2. The drop in water level of almost 90 m causes an increase in overburden pressure of about 9 kg/cm$^2$ which is believed to be causing downward movement along the old faults as clay beds interbedded with sand aquifers consolidate [Castle and Youd (1972)[12]]. The faults cause distress in structures, large deflections of roadways, and rupture of utility lines.

### Mexico City (Water Extraction: Subsidence and Foundation Problems)

*Geologic Conditions*

The basin of the valley of Mexico City, 2240 m above sea level, has been filled with 60 to 80 m of Pleistocene soils including interbedded sands, sands and gravels, and lacustrine volcanic clay, which overlie a thick deposit of compact sand and gravel. As shown on the section given as Fig. 7.60, the soils to depths of about 33 m are very soft to medium-stiff clays. Void ratios as high as 13 and water contents as high as 400% indicate the very high compressibility of the soils (*see Fig. 3.87*). The clays are interbedded with thin sand layers, and thick sand strata are found at depths of 33, 45, and 73 m.

*Ground Subsidence*

In the 100 years prior to 1955 a large number of water wells were installed between depths of 50 and 500 m in the sand and sand and gravel layers. The wells caused a large reduction in piezometric head, especially below 28 m, but the surface water table remained unaltered because of the impervious shallow clay formation. A downward hydraulic gradient is induced because of the difference in piezometric levels between the ground surface and the water-bearing layers at greater depths. The flow of the descending water across the highly compressible silty clay deposits increases the effective stresses, produces consolidation of the weak clays, and thus causes the surface subsidence [Zeevaert (1972)[13]].

In some places, as much as 7.6 m of subsidence has occurred in 90 years, with about 5 m occurring between 1940 and 1970. The maximum rate, with respect to a reference sand stratum at a depth of 48 m, was 35 cm/year, and was reached in 1949. About 80 to 85% of the subsidence is attributed to the soils above the 50-m depth. In 1955 the mayor of the city passed a decree prohibiting all pumping from beneath the city, and since then the rate of lowering of the piezometric levels and the corresponding rate of subsidence have decreased considerably. As of 1970, however, subsidence was still continuing at rates of about 4 cm/year, although the piezometric levels have remained practically unchanged since 1957. This indicates that the compressible soils are still consolidating under the increased effective stresses.

*Foundation Problems*

Before the well shutdown program, many of the wells that were poorly sealed through the upper thin sand strata drew water from these strata, causing dish-shaped depressions around the wells. The result was severe differential settlements causing tilting of adjacent buildings and breakage of underground utilities.

A different problem occurs around the perimeter of the basin. As the water table drops, the weak bentonitic clays shrink by desiccation,

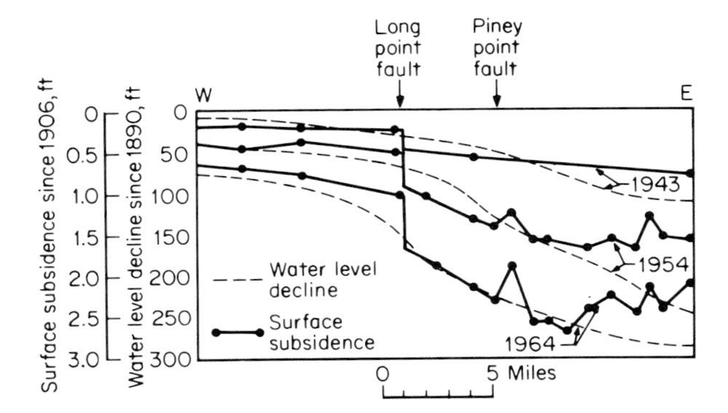


**FIG. 10.2 Profiles of subsidence and groundwater decline along a section trending due west from Houston.** [*From Castle and Youd (1972).*[12]]

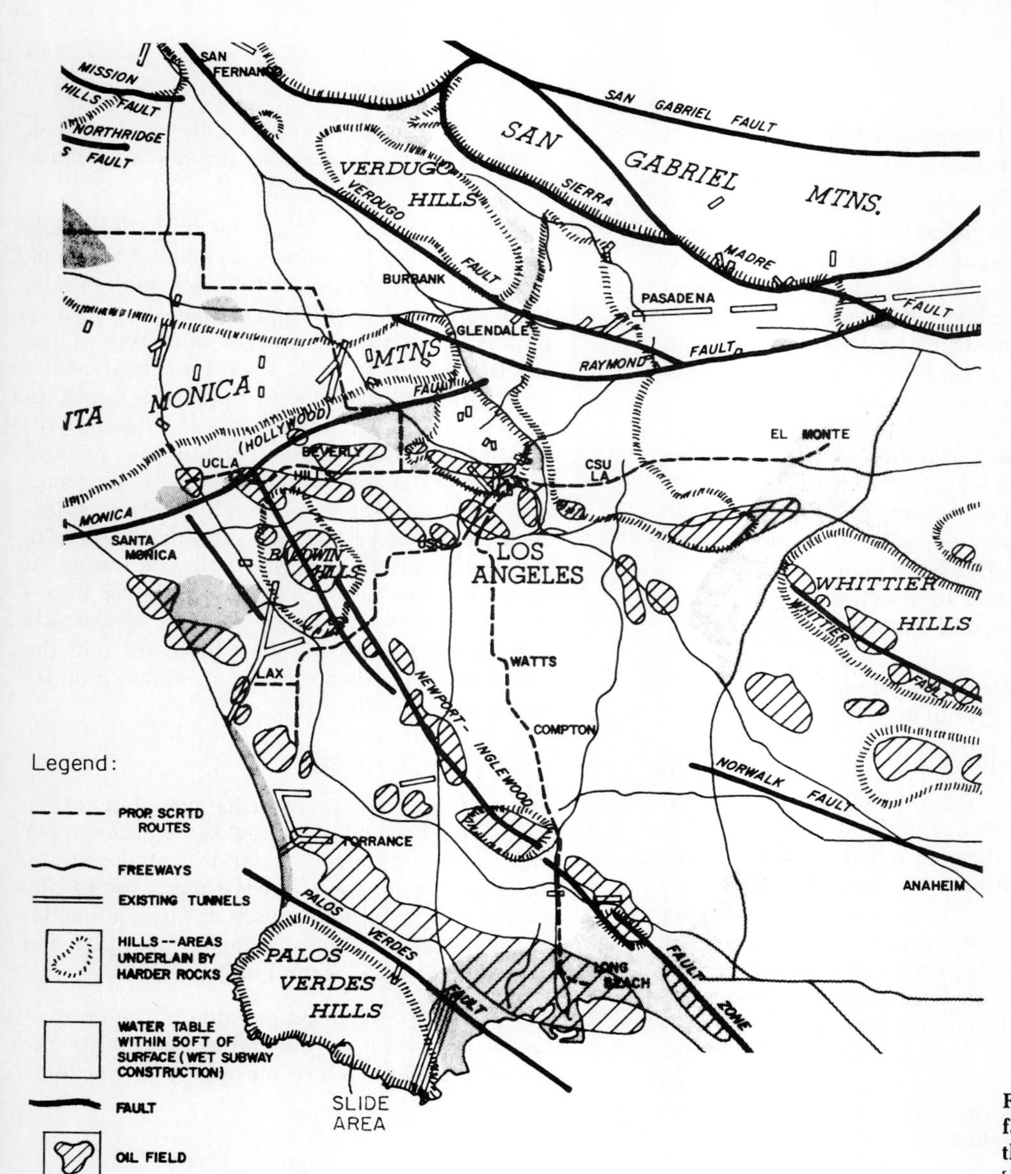


**FIG. 10.3 Water table, faults, and oil fields in the Los Angeles Basin.** [*From Proctor (1973).*[15]]

resulting in surface cracks opening to as much as a meter in width and 15 m in depth. When the cracks open beneath structures, serious damage results.

The major problems, however, have been encountered within the city. Buildings supported on piles can be particularly troublesome. End-bearing piles are usually driven into the sand stratum at 33 m for support. As the ground surface tends to settle away from the building because of the subsidence, the load of the overburden soils is transferred to the piles through "negative skin friction" or "downdrag." If the piles do not have sufficient capacity to support both the building load and the downdrag load, settlements of the structure result. On the other hand, if the pile capacities are adequate, the structure will not settle but the subsiding adjacent ground will settle away from the building. When this is anticipated, utilities are installed with flexible connections and allowances are made in the first-floor design to permit the sidewalks and roadways to move downward with respect to the building.

Modern design attempts to provide foundations that enable a structure to settle at about the same rate as the ground subsides [Zeevaert (1972)[13]]. The "friction-pile compensated foundation" is designed such that downdrag and consolidation will cause the building to settle at the same rate as the ground subsidence. The Tower Latino Americano, 43 stories high, is supported on a combination of end-bearing piles and a compensated raft foundation [Zeevaert (1957)[14]]. The piles were driven into the sand stratum at 33 m, and a 13-m-deep excavation was made for the raft which had the effect of removing a substantial overburden load, subsequently replaced by the building load. The building was completed in 1951 and the settlements as of 1957 occurring from consolidation of the clay strata were as predicted, or about 10 cm/year.

### Long Beach, California (Oil Extraction: Subsidence and Flooding)

*Geologic Conditions*

The area is underlain by 600 m of "unconsolidated" sediments of late Pliocene, Pleistocene, and Holocene age, beneath which are 1400 m of oil-producing Pliocene and Miocene formations including sandstones, siltstones, and shales.

*Ground subsidence* began to attract attention in Long Beach during 1938–1939 when the extraction of oil began from the Wilmington field located primarily in the city (see site location map, Fig. 10.3). A peak subsidence rate of over 50 cm/year was reached in 1951–1952, and by 1973 subsidence in the center of a large bowl-shaped area had reached 9 m vertically, with horizontal movements as great as 4 m.

*Flood protection* for the city, which is now below sea level in many areas, is provided by extensive diking and concrete retaining walls.

### Baldwin Hills Reservoir Failure (Oil Extraction: Faulting)

*Event*

On December 14, 1963, the Baldwin Hills Reservoir, a pumped storage reservoir located in Los Angeles (Fig. 10.3), failed and released a disastrous flood [Jansen et al. (1967)[16]].

*Background*

The dam was located close to the Inglewood oil field and the Inglewood fault passed within 150 m of the west rim of the reservoir, as shown on Fig. 10.4. The Inglewood fault is part of the major Newport-Inglewood fault system (Fig. 10.3). During construction excavation in 1948, two minor faults were found to pass through the reservoir area, but a board of consultants judged that further movement along the faults was unlikely. The dam was well instrumented with two strong-motion seismographs, tiltmeters, settlement measurement devices, and observation wells.

*Surface Movements*

Between 1925 and 1962, at a point about 1 km west of the dam, about 3 m of subsidence and about 15 cm of horizontal movement occurred. In 1957 cracks began to appear in the area around the dam. Six years later, failure occurred suddenly and the narrow breach in the dam was found to be directly over a small fault. It was judged that 15 cm of movement had occurred along the fault, which ruptured the lining of the reservoir and permitted the sudden release of water. It appears likely that subsidence from oil extraction caused the fault displacement, since there was no significant seismic activity in the area for at least the prior month.

## 10.2.5 SUBSIDENCE PREVENTION AND CONTROL

### Groundwater Extraction

*General*

Subsidence from groundwater extraction cannot be avoided if withdrawal exceeds recharge, resulting in significant lowering of the water table or a reduction in piezometric levels at depth, and if the subsurface strata are compressible.

*Prediction*

Prior to the development of a groundwater resource, studies should be made to determine the water-balance relationship and estimate

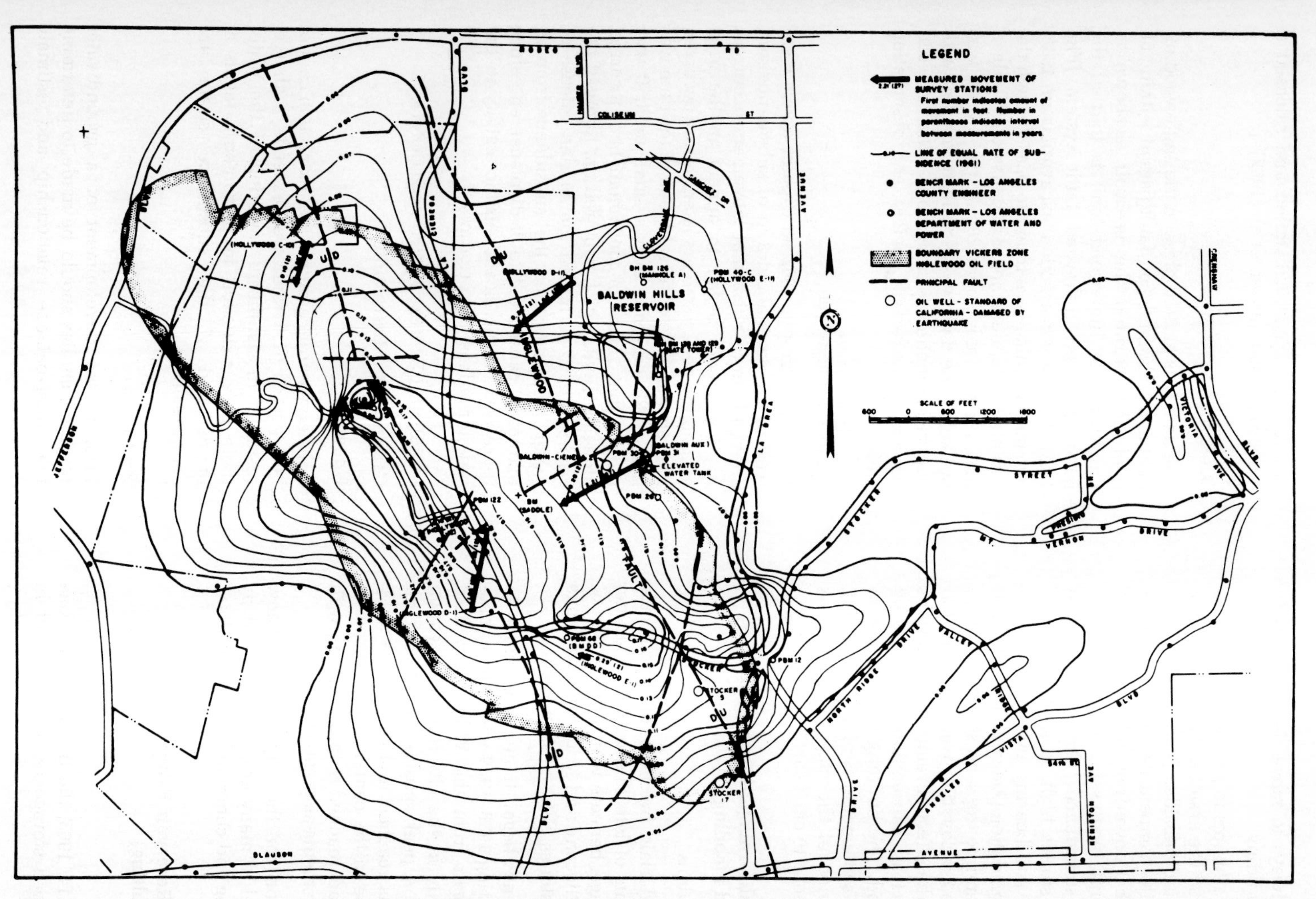


FIG. 10.4 **Subsidence rates and locations of faults (Baldwin Hills, California).** [*From Jansen et al.* (1967).[16]]

magnitudes of subsidence. The water-balance relationship is the rate of natural recharge compared with the anticipated maximum rate of withdrawal. If recharge equals withdrawal, the water table will not drop and subsidence will not occur. If withdrawal significantly exceeds recharge, the water table will be lowered.

Estimates of the subsidence to be anticipated for various water-level drops are made to determine the maximum overdraft possible before surface settlement begins to be troublesome and cause flooding and faulting. By using concepts of soil mechanics it is possible to compute the surface settlements for various well-field layouts, including differential deflections and both immediate and long-term time rates of settlements.

The increase in the flood hazard is a function of runoff and drainage into a basin or the proximity to large water bodies and their relative elevations, including tidal effects. The growth of faults is difficult to assess both in location and magnitude of displacement. In general, locations will be controlled by the locations of relict faults. New faults associated with subsidence normally are concentrated in a concentric pattern around the periphery of the subsiding area. The center of the area can be expected to occur over the location of wells where withdrawal is heaviest.

*Prevention*

Only control of overextraction prevents subsidence. Approximate predictions as to when the water table will drop to the danger level can be based on withdrawal, precipitation, and recharge data. By this time the municipality must have provisions for an alternate water supply to avoid the consequences of overdraft.

*Control*

Where subsidence from withdrawal is already troublesome, the obvious solution is to stop withdrawal. As shown in the case of Mexico City, however, underlying soft clays continue to consolidate for many years, even after withdrawal ceases, although at a much reduced rate. Artificial recharging will aid the water balance ratio.

*Recharging* by pumping into an aquifer requires temporary surface storage. The Santa Clara Valley Water District (San Jose, California) has been storing storm water for recharge pumping for many years [*Civil Engineering* (1980)[17]]. In west Texas and New Mexico, dams are to be built to impound flood waters to be used as pumped-groundwater recharge [ENR (1980)[18]].

Where the locale lacks terrain suitable for water storage, deep-well recharging is not a viable scheme, and recharge is permitted to occur naturally. Venice considered a number of recharge schemes, in addition to a program to cap the city wells begun in 1965. Apparently natural recharge is occurring and measurements indicate that the city appears to be rising at the rate of about 1 mm every 5 years [*Civil Engineering* (1975)[19]].

### Oil and Gas Extraction

*Prediction* of subsidence from oil and gas extraction is difficult with respect to both magnitude and time. Therefore, it is prudent to monitor surface movements and to have contingency plans for the time when subsidence approaches troublesome amounts.

*Control* by deep-well recharging appears to be the most practical solution for oil and gas fields. At Long Beach, water injection into the oil reservoirs was begun in 1958; subsidence has essentially halted and about 20 km$^2$ of land area has rebounded, in some areas as much as 0.3 m [Allen (1973)[1]].

### Construction Dewatering

*Prediction and Control*

Before the installation of a construction dewatering system in an area where adjacent structures may be affected, a study should be made of the anticipated drop in water level as a function of distance, and settlements to be anticipated should be computed considering building foundations and soil conditions. In many cases, condition surveys are made of structures and all signs of existing distress recorded as a precaution against future damage claims. Before dewatering, a monitoring system is installed to permit observations of water level and building movements during construction operations (see *Fig.*

4.36). The predicted settlements may indicate that preventive measures are required.

*Prevention*

Prevention of subsidence and the subsequent settlement of a structure is best achieved by placing an impervious barrier between the dewatering system and the structure, such as a slurry wall *(see Art. 8.4.2)*. Groundwater recharge to maintain water levels in the area of settlement-sensitive structures is considered to be less reliable.

*Surcharging*

Surcharging of weak compressible layers is a positive application of construction dewatering. If a clay stratum, for example, lies beneath a thickness of sands adequate to apply significant load when dewatered, substantial prestress can be achieved if the water table remains lowered for a long enough time. Placing a preload on the surface adds to the system effectiveness.

## 10.3 SUBSURFACE MINING

### 10.3.1 SUBSIDENCE OCCURRENCE

#### General

Extraction of materials such as coal, salt, sulfur and gypsum from "soft" rocks often results in ground subsidence during the mining operation, or at times many years after operations have ceased. Subsidence can also occur during tunneling operations.

In the United States ground subsidence from mining operations has occurred in about 30 states, with the major areas located in Pennsylvania, Kansas, Missouri, Oklahoma, Montana, New Jersey, and Washington [*Civil Engineering* (1978)[20]]. Especially troublesome in terms of damage to surface structures are the Scranton–Wilkes-Barre and Pittsburgh areas of Pennsylvania and the midlands of England where coal has been, and is still being, mined. Paris, France, and surrounding towns have suffered surface collapse, at times swallowing houses,

**FIG. 10.5 Building damaged by mine subsidence (Pittsburgh).** *(Photo courtesy of Richard E. Gray.)*

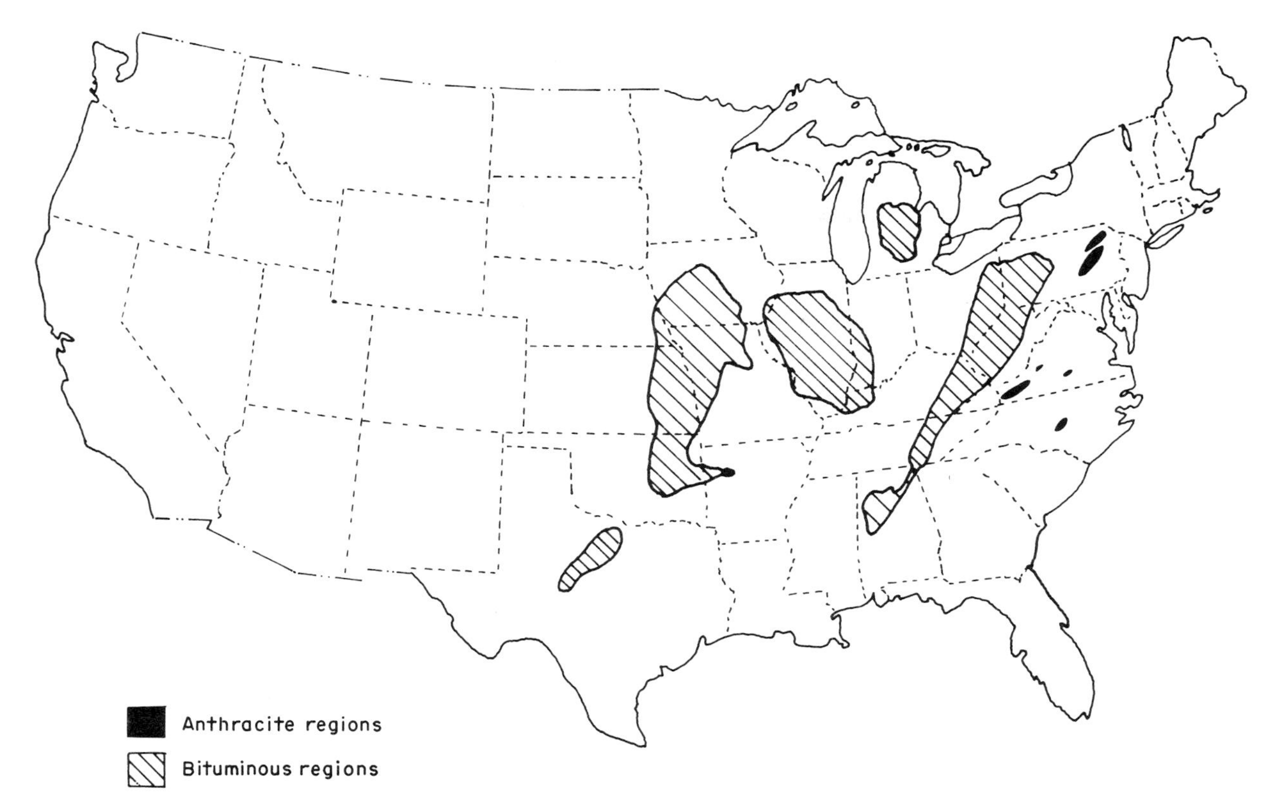


**FIG. 10.6 Approximate extent of coal fields of the Pennsylvanian formations in the eastern United States.** [*From Averitt (1967).*[22]]

over the old underground limestone and gypsum quarries that were the source of building stone for the city in the eighteenth century [Arnould (1970)[21]]. In the Paris area, collapse has been intensified by groundwater pumping *(see Art. 10.4.2).*

### Coal Mining

From the aspects of frequency of occurrence and the effects on surface structures, coal mining appears to be the most important subsurface mining operation. Mine collapse results in irregular vertical displacement, tilting, and horizontal strains at the surface, all resulting in the distortion of structures such as illustrated on Fig. 10.5. The incidence and severity of subsidence are a function of the coal bed depth, its thickness, percent of material extracted, tensile strength of the overburden, and the strength of pillars and other roof supports.

There are two general methods of extraction:

- *Room and pillar method* is used in the United States. Pillars are left in place to support the roof, but subsequent operations rob the pillars, weakening support. Collapse is often long-term.
- *Longwall panel method* is used in Europe. It involves complete removal of the coal. Where the mine is relatively shallow and the overlying materials weak, collapse of the mine and surface subsidence progress with the mining operation.

The approximate extent of coal fields in the eastern United States in the Pennsylvania formations are given on Fig. 10.6.

#### 10.3.2 ROOM AND PILLAR METHOD (ALSO "BREAST AND HEADING" METHOD)

### Extraction

*Early Operations*

In the anthracite mines of the Scranton–Wilkes-Barre and Pittsburgh areas of Pennsylvania, "first mining" proceeded historically by driving openings, called breasts, in the up-dip direction

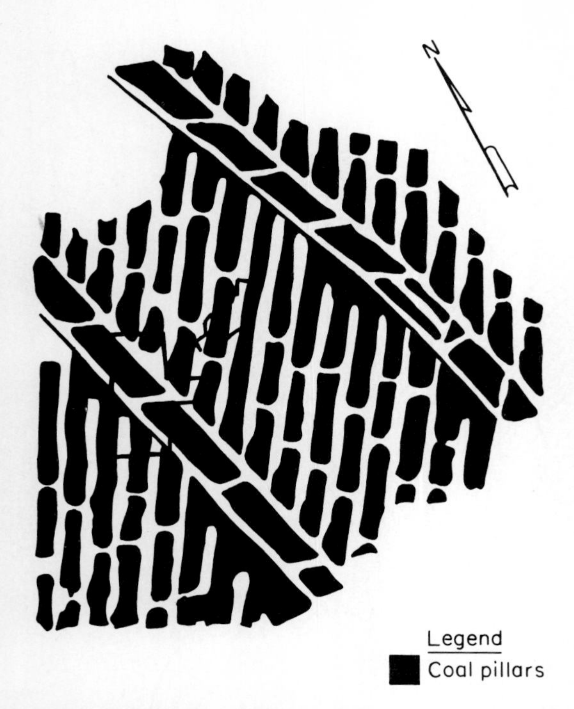


**FIG. 10.7 Plan of coal mine room and pillar layout (Westmoreland County, Pennsylvania).** [*From Gray and Meyers (1970).*[23]]

within each vein, which were then connected at frequent intervals with heading openings. The usual width of the mine openings ranged from 5 to 8 m and the distance between center lines of adjacent breasts varied generally from 15 to 24 m with the shorter distances in the near-surface veins and the greater distances in the deeper veins. Pillars of coal were left in place to support the roof and the room-pillar configurations were extremely variable. One example is given on Fig. 10.7.

"Robbing" occurred subsequent to first mining and consisted of removing the top and bottom benches of coal in thick veins and trimming coal from pillar sides.

Collapse occurs with time after mine closure. The coal often is associated with beds of clay shale which soften and lose strength under sealed, humid mine conditions. Eventually failure of the pillars occurs, the roof collapses, and the load transfer to adjacent pillars causes them to collapse. If the mined area is large enough and the roof thin enough, subsidence of the surface results.

Old mines may contain pillars of adequate size and condition to provide roof support or robbed pillars, weakened and in danger of collapse. On the other hand, collapse may have already occurred.

*Modern Operations*

Two major coal seams of the Pennsylvanian underlie the Pittsburgh area, each having an average thickness of 2 m: the Pittsburgh coal and the Upper Freeport coal. The Pittsburgh seam is shallow, generally within 60 m of the surface, and has essentially been worked out. In 1970, the Upper Freeport seam was being worked to the north and east of the city where it lies at depths of 100 to 200 m [Gray and Meyers (1970)[23]].

In the new mines complete extraction is normally achieved. A system of entries and cross-entries is driven initially to the farthest reaches of the mine before extensive mining. Rooms are driven off the entries to the end of the mine, and when this is reached, a second, or retreat phase, is undertaken. Starting at the end of the mine, pillars are removed and the roof is permitted to fall.

In active mines where surface development is desired, some companies are guaranteeing safety against subsidence if approximately 50% of the coal in the zone beneath the structure or development is purchased by the developers and not extracted by the mining company. The area of limited extraction is determined by taking an area 5 m wide around the proposed structure and projecting it downward at an angle of 15° from the vertical to the level of the mine as shown on Fig. 10.8.

## Mine Collapse Mechanisms

*General*

Three possible mechanisms which cause mine collapse are roof failure, pillar failure, or pillar foundation failure.

*Roof Failure*

Roof stability depends upon the development of an arch in the roof stratum, which in turn depends on the competency of the rock in relation to span width. In weak, fractured sedimentary rocks, this often is a very difficult problem to assess, since a detailed knowledge of the engineering properties and the structural defects of the rock is required, and complete information on these conditions is difficult and costly to obtain. If the roof does have defects affecting its capability, it is likely that it will fail during mining operations, not at some later date, as is the case with pillars.

Roof support does become important when the pillars weaken or collapse, causing the span length to increase, which in turn increases the loads on the pillars. Either roof or pillars may then collapse.

*Pillar Failure*

The capability of a pillar to support the roof is a function of the compressive strength of the coal, the cross-sectional area of the pillar, the roof load, and the strength of the floor and roof. The pillar cross-sectional area may be reduced in time by weathering and spalling of its walls, as shown on Fig. 10.9, to the point where it cannot support the roof and failure occurs.

*Pillar Punching*

A common cause of mine collapse appears to be punching of the pillar into either the roof or the floor stratum. Associated with coal beds, clay shale strata are often left exposed in the mine roof or floor. Under conditions of high humidity or a flooded floor in a closed mine, the clay shales soften and lose their supporting capacity. The pillar fails by punching into the weakened shales, the roof load is transferred to adjacent pillars which in turn fail, resulting in a lateral progression of failures. If the progression involves a sufficiently large area, surface subsidence can result, depending upon the type and thickness of the overlying materials.

*Earthquake Forces*

In January 1966, during the construction of a large single-story building in Belleville, Illinois, settlements began to occur under a section of the building, causing cracking. It was determined that the settlements may have started in late October or early November 1965. An earthquake was reported in Belleville on October 20,

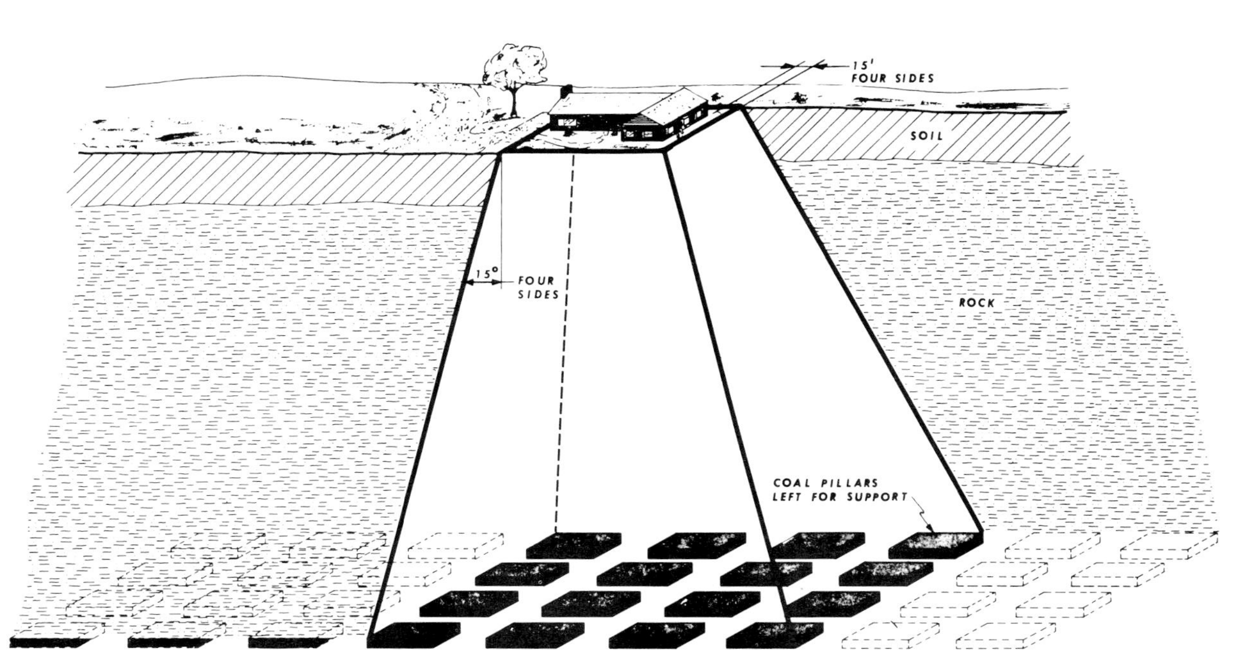


**FIG. 10.8 Sketch of pillar arrangement used to support the surface during present-day mining operations in the Pittsburgh area.** [*From Gray and Meyers (1970).*[23]]

**FIG. 10.9 Spalling of a coal pillar in a mine room (Pittsburgh).** *(Photo courtesy of Richard E. Gray.)*

1965. The site was located over an old mine in a coal seam 2 to 2.5 m thick at a depth of 40 m, which was closed initially in about 1935, then reworked from 1940 to 1943. Mansur and Skouby (1970)[24] considered that building settlements were the result of pillar collapse and mine closure initiated by the earthquake. Some investigators, however, consider that the collapsing mine was the shock recorded in Belleville.

### 10.3.3 LONGWALL PANEL EXTRACTION

#### NCB Studies

Based on surveys of 157 collieries, the National Coal Board of Great Britain [NCB (1963),[25] (1966)[26]] developed a number of empirical relationships for the prediction of the vertical component of surface displacement $s$ and the horizontal component of surface strain $e$, associated with the trough-shaped excavation of the longwall panel method of coal extraction, illustrated on Fig. 10.10.

#### Mine Conditions

Surveys were made over coal seams that were inclined up to 25° from the horizontal, were about 1 to 5 m in thickness $m$, and ranged in depth $h$ from 30 to 800 m. Face or panel width $w$ varied from 30 to 450 m, and the panel width to depth ratio $w/h$ varied from 0.05 to 4.0. Panel widths were averaged if the panel sides were nonparallel. The foregoing conditions assume that there is no zone of special support within the panel areas. The physical relationships are illustrated on Fig. 10.10.

### Subsidence Characteristics

Subsidence was found to occur within a day or two of extraction in a dish-shaped pattern over an area bounded by lines projected upward from the limits of the collapsed area at the angle of the draw $\alpha$ as shown on Fig. 10.10. The angle of the draw was found to be in the range of 25 to 35° for beds dipping up to 25°.

*Greatest maximum subsidence* $S_{max}$ possible was found to be approximately equal to 90% of the seam thickness, occurring at values of $w/h$ greater than about 1.2. At values of $w/h$ less than 0.2, the maximum subsidence was less than 10% of the seam thickness.

*Critical width* is the panel width required to effect maximum subsidence (Fig. 10.10); as the panel width is extended into the zone of supercritical width, additional vertical subsidence does not occur, but the width of the subsiding area increases accordingly. Detailed discussion and an extensive reference list are given in Voight and Pariseau (1970).[28]

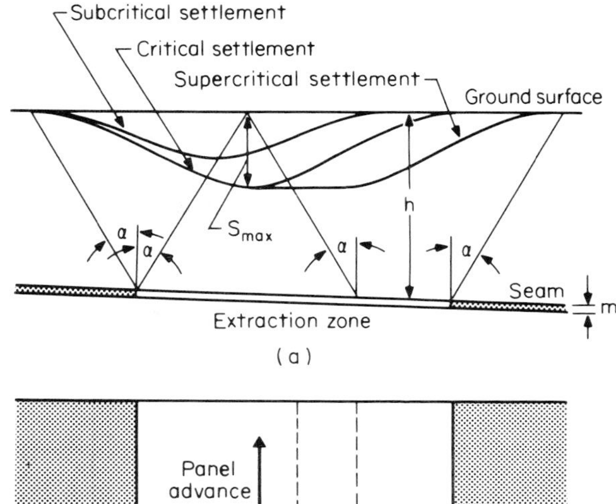


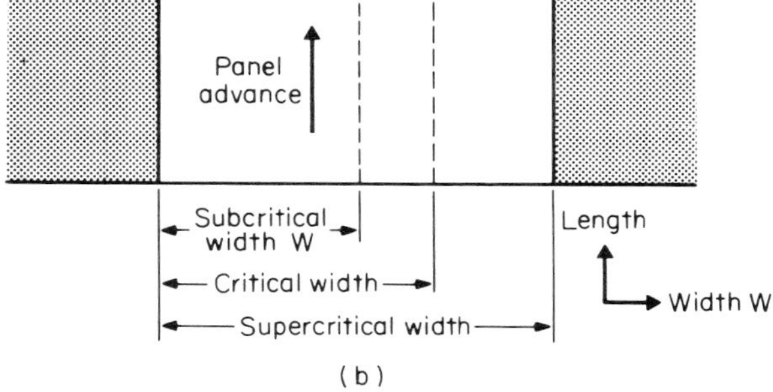


**FIG. 10.10 Mine subsidence vs. critical depth concept for longwall panel extraction developed by the National Coal Board of Great Britain: (a) section and (b) plan.** [*After Mabry (1973).*[27]]

## 10.3.4 STRENGTH PROPERTIES OF COAL

### General

Pillar capacity analysis requires data on the strength properties of coal. A wide range of values has been obtained by investigators either by testing specimens in the laboratory, or by back-analysis in which the strength required to support an existing roof is calculated for conditions where failure has not occurred. For the determination of stability of a working mine, the strength of fresh rock specimens governs, whereas for problems involving stability after a lapse of many years, the strength of weathered specimens of the pillar and its roof and floor support pertain.

Typical coals contain a system of orthogonal discontinuities consisting of horizontal bedding planes and two sets of vertical cracks called "cleats" which are roughly perpendicular to one another as shown on Fig. 10.11. This pattern makes the recovery of undisturbed specimens difficult.

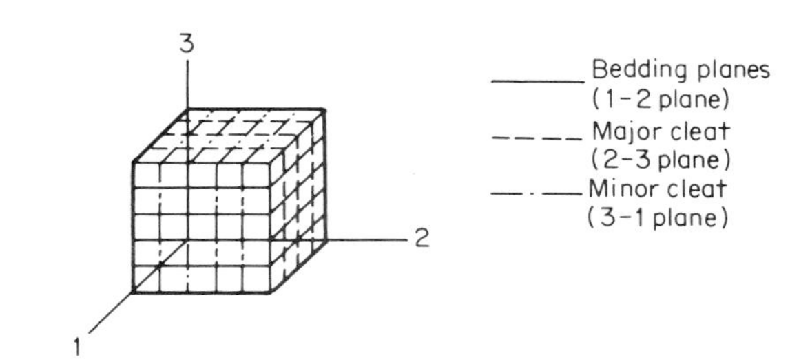


**FIG. 10.11 Bedding and cleat orientation of coal and the principal material axis.** [*From Ko and Gerstle (1973).*[29]]

Some unconfined compression-strength values for coals from various locations are given on Table 10.2.

### Triaxial Compression Tests

A series of triaxial compression tests was performed in the laboratory on specimens from Eire, Colorado; Sesser, Illinois; and Bruceton,

**TABLE 10.2**
**UNCONFINED COMPRESSION STRENGTH VALUES FOR COAL**

| Specimen source | Description | Strength, kg/cm² | Comments | Reference |
|---|---|---|---|---|
| South Africa | 2-ft cube | 19 | Failure sudden | Voight and Pariseau (1970)[28] |
| Not given | Not given | 50–500 | None | Farmer (1968)[30] |
| Pittsburgh coal | Sound pillar | ~57 | Pillar with firm bearing, $H = 3$ m, sides, 4.8 × 4.8 m | Greenwald et al. (1941)[31] |
| Anthracite from Pennsylvania | 1-in cubes | 200*<br>404* | First crack appears<br>Crushing strength | Mabry (1973)[27] from Griffith and Conner (1912)[32] |

*Values selected by Mabry (1973)[27] from those obtained during a comprehensive study of the strength of 116 cubic specimens reported on by Griffith and Conner (1912).[32]

Pennsylvania, by Ko and Gerstle (1973).[29] Confining pressures of 50, 100, 250, and 600 psi were used, with the maximum pressures considered as the overburden pressure on a coal seam at a depth of 180 m (600 ft). While confining pressure was held constant, each specimen was loaded axially to failure while strains and stresses were recorded. The test specimens were oriented on three perpendicular axes ($\alpha$, $\beta$, and $\gamma$), and the failure load was plotted as a function of the confining pressure $P_c$ to obtain the family of curves presented on Fig. 10.12.

It was believed by the investigators that the *proportional limit* represents the load level at which microcracking commences in the coal. The *safe limit* is an arbitrary limit on the applied load set by the investigators to avoid internal microcracking.

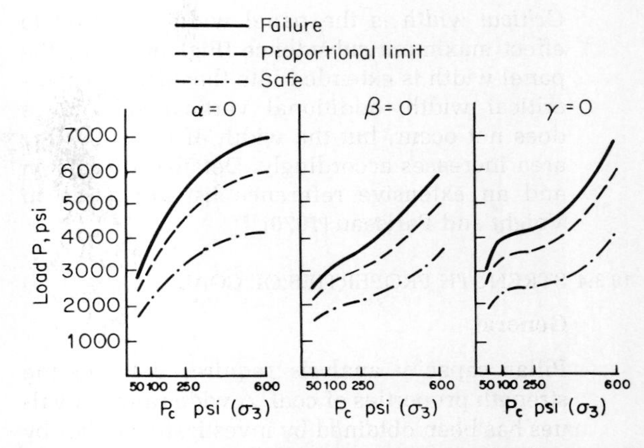


**FIG. 10.12 Triaxial compression test results from coal specimens (14.2 psi = 1 kg/cm²).** [*From Ko and Gerstle (1973).*[29]]

## 10.3.5 INVESTIGATION

### Data Collection

Collection of existing data is a very important phase of investigation for projects to be constructed over mines. The data should include information on local geology, local subsidence history, and mining operations beneath the site.

Data on mining operations beneath the site are obtained from the mining company that performed the extraction. The data should include information on the mine limits, percent extraction, depth or depths of seams, pillar dimensions, and the closure date. Other important data that may be available include pillar conditions, roof and floor conditions, flooding incidence, amount of collapse that has already occurred, and accessibility to the mine.

### Explorations

*Exploration scopes* will vary depending upon the comprehensiveness of the existing data and the accessibility of the mine for examination. Actual inspection of mine conditions is extremely important, but often not possible in old mines.

*Preliminary explorations* where mine locations and collapse conditions are unknown or uncertain may include the use of:

- Gravimeters to detect anomalies indicating openings.
- Rotary probes, if closely spaced, to detect cavities and indicate collapse conditions.
- Borehole cameras to photograph conditions, and borehole TV cameras, some of which are equipped with a zoom lens with an attached high-intensity light, to inspect mines remotely.
- Acoustical emissions devices, where mines are in an active collapse state, to locate the collapse area and monitor its growth.

*Detailed study* of conditions requires core borings to obtain cores of the roof, floor, and pillars, permitting an evaluation of their condition by examination and laboratory testing.

### Collapse Probability Assessment

#### *General*

ROOF CONDITIONS In the experience of Gray and Meyers (1970),[23] damage to single-story construction has not occurred over old, inactive mines excavated by room and pillar methods, where at least 90% of the cover over the mine is relatively competent rock and exceeds about 30 m in thickness. The material type, its resistance to weathering, and its strength in both the mine roof and floor are factors to consider, as well as percent extraction.

PERCENT EXTRACTION AND PILLAR CONDITIONS In general if extraction does not exceed 50%, and the pillar conditions and their support are sound, there is a low probability of collapse, as long as the roof span is not excessive.

#### *Examples of Mine Conditions vs. Subsidence*

In the Pittsburgh area there is a wide range in the percentage of coal extracted from old mines in any given location. A number of cases have been cited by Gray and Meyers (1970).[23]

CASE 1 Building was damaged severely in 1959 by subsidence under the end of the building. Mine closed in 1943; extraction was 75% under damaged end of building, none under the undamaged end. Pillars remaining were approximately 3 × 3 m; mine depth was 33 m.

CASE 2 New school site. Mine depth was 15 m; mine was closed by 1922, and extraction was approximately 70%. Some subsidence had occurred but voids still remained in the mine. School was moved to a nearby location where extraction was only 30% and the mine depth 20 m.

CASE 3 Prestressed-concrete water-tank site. Mine depth was 27 m. Mine was closed in about 1918, and extraction was about 52%. The mine was unflooded, and pillar, roof, and floor conditions were sound. Tank was supported on the surface.

CASE 4 New building site. Mine depth was 80 m; mine was closed in 1918, and had complete extraction. Surface subsidence of 1.2 m was complete by 1919. Building was supported on normal shallow foundations.

CASE 5 New building site. Mine depth 11 to 23 m, mine closed in 1930, extraction about 85%. Surface subsidence of about 1 m occurred shortly after mine closure. Building was founded on doubly reinforced continuous footings.

CASE 6 General. When test borings encountered voids and it was not possible to determine with confidence the percent extraction, structures were either relocated or supported on deep foundations (*see Art. 10.3.6*).

### Finite Element Method Application

#### *General*

The prediction of surface distortions beneath a proposed building site caused by the possible collapse of old mines by use of the finite element method is described by Mabry (1973).[27] The site is in the northern anthracite region of Pennsylvania near Wilkes-Barre, and is underlain by four coal beds at depths ranging from 80 to 170 m, with various percentages of extraction $R$, as illustrated on Fig. 10.13.

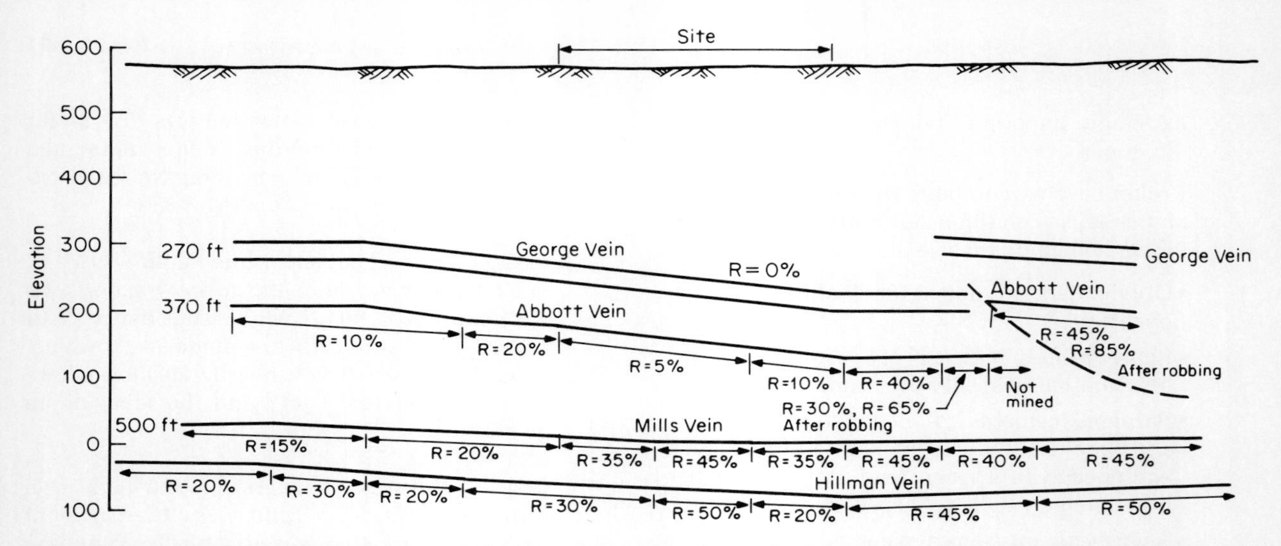


**FIG. 10.13 Section illustrating coal mines and percent extraction *R* beneath a proposed construction site near Wilkes-Barre, Pennsylvania.** [*From Mabry (1973).*[27]]

*Finite Element Model*

Pillar analysis revealed low safety factors against crushing and the distinct possibility of subsidence. To evaluate the potential subsidence magnitude, a finite element model was prepared incorporating the geometry of the rock strata and mines to a depth of 220 m, and engineering properties including density, strength, and deformation moduli of the intervening rock and coal strata. The finite element mesh is given on Fig. 10.14.

*Analysis and Conclusions*

Gravity stresses were imposed and the ground surface subsidence due to the initial mining in the veins was determined. Subsequent analysis was made of the future surface distortions, such as would be generated by pillar weathering and eventual crushing. Pillar weathering was simulated by reducing the joint stiffness (*see Art. 6.4.4*), and pillar collapse or yield was simulated by setting the joint stiffness in the appropriate intervals of the coal seams to zero. After pillar weathering in the coal seams was evaluated by changing joint stiffness values, the intervals of the veins were "collapsed" in ascending order of computed safety factors for several extraction ratios. The results are summarized on Table 10.3.

After an evaluation of all of the available information, it was the judgment of the investigator that the more realistic case for plant design was case 2 (Table 10.3), and that the probability of cases 3 and 4 developing during the life of the structure was very low.

### 10.3.6 SUBSIDENCE PREVENTION AND CONTROL AND FOUNDATION SUPPORT

#### New Mines

In general, new mines should be excavated on the basis of either total extraction, permitting collapse to occur during mining operations (if not detrimental to existing overlying structures), or partial extraction leaving sufficient pillar sections to prevent collapse and resulting subsidence at some future date. Legget (1972)[33] cites the case where the harbor area of the city of Duisburg, West Germany, purposely was lowered 1.75 m by careful, progressive longwall mining of coal seams beneath the city, without damage to overlying structures.

#### Old Mines

Solutions are based on predicted distortions and their probability of occurrence.

*Case 1*

No or small surface distortions are anticipated

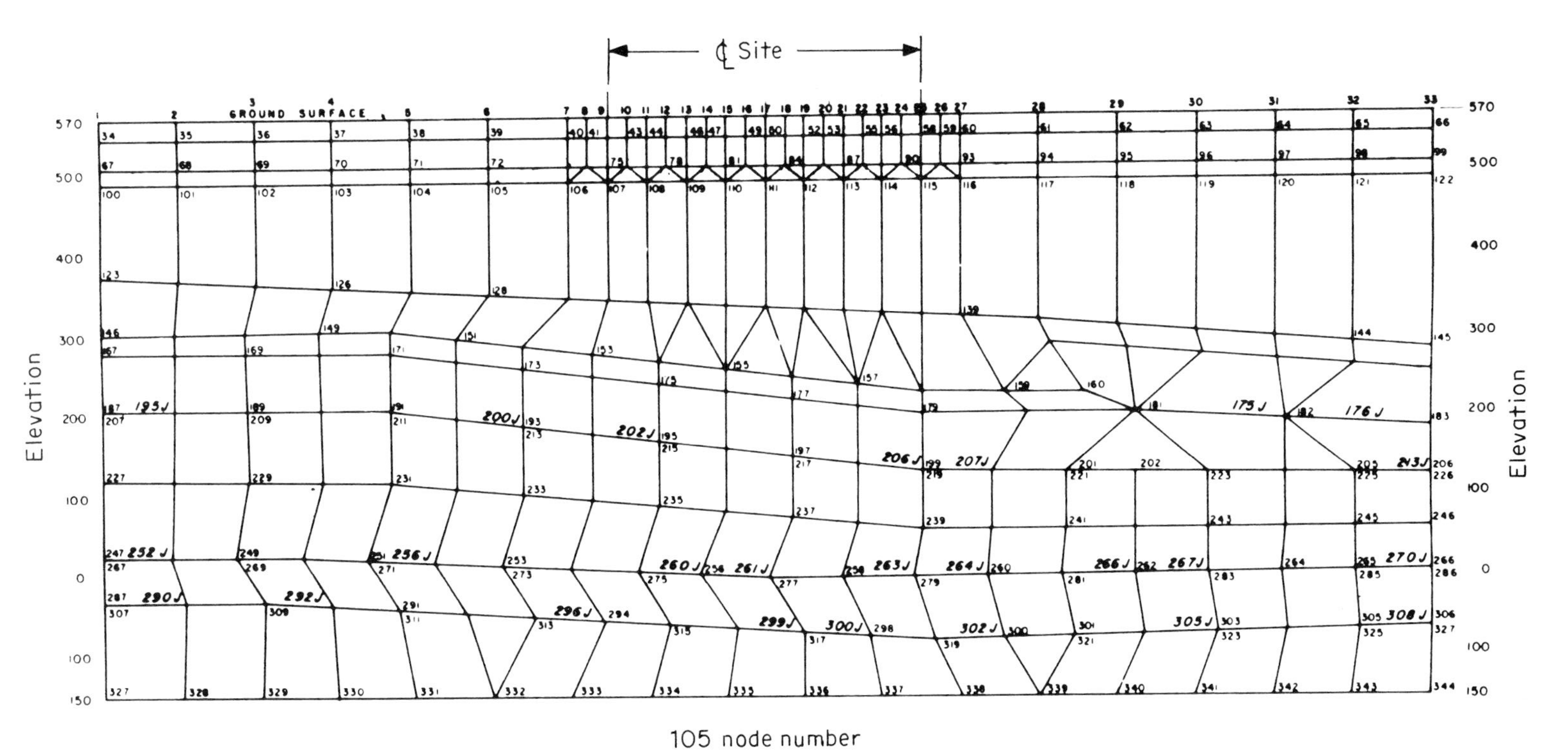


**FIG. 10.14 Finite element mesh for conditions beneath site near Wilkes-Barre, Pennsylvania.** [*From Mabry (1973).*[27]]

| TABLE 10.3 SUMMARY OF FINITE ELEMENT ANALYSIS OF COAL MINE STUDY* | | | | |
|---|---|---|---|---|
| | | | Maximum distortion within plant site | |
| Condition | Safety factor against pillar failure | Cumulative settlement, cm | Angular rotation, radians | Horizontal strain (+ = tension) |
| Weathering in all veins | | 1.0 | $20.0 \times 10^{-6}$ | $\pm 12.0 \times 10^{-6}$ |
| Collapse in Hillman, $R$ = 50% | 0.80 | 4.0 | $3.6 \times 10^{-4}$ | $+1.7 \times 10^{-4}$ |
| Collapse in Mills and Hillman, $R$ = 40% | 0.89 to 0.99 | 29.0 | $3.0 \times 10^{-3}$ | $+10.1 \times 10^{-4}$ |
| Collapse in Mills, $R$ = 30% | 1.1 | 63.2 | $4.8 \times 10^{-3}$ | $+19.2 \times 10^{-4}$ |

*From Mabry (1973).[27]

when conditions include adequate pillar support, or complete collapse has occurred, or the mined coal seam is at substantial depth overlain by competent rock. Foundations may include mats, doubly reinforced continuous footings, or articulated or flexible design to allow compensation for some differential movements of structures.

*Case 2*

Large distortions are anticipated, or small distortions cannot be tolerated, when pillar support is questionable, collapse has not occurred, and the mine is at relatively shallow depths. Solutions may include:

- Relocate project to a troublefree area.
- Provide mine roof support with construction of piers in the mine or installation of grout columns, or completely grout all mine openings from the surface within the confines of a grout curtain installed around the site periphery.
- Install drilled piers from the surface to beneath the mine floor.

## 10.4 SOLUTION OF ROCK

### 10.4.1 GENERAL

#### Significance

*Ground subsidence* and *collapse* in soluble rock masses can result from nature's activities, at times aided by humans, or from human-induced fluid or solid extraction. Calcareous rocks, such as limestone, dolomite, gypsum, halite, and anhydrite are subject to solution by water, which causes the formation of cavities of many shapes and sizes. Under certain conditions, the ground surface over these cavities subsides or even collapses, in the latter case forming sinkholes *(see Art. 6.2.3)*.

#### The Hazard

*Geographic distribution* is widespread, and there are many examples in the literature of damage to structures and even deaths caused by ground collapse over soluble rocks. Examples are the destruction of homes in central Florida [Sowers (1975)[34]] (Fig. 10.15); the sudden settlement of a seven-story garage in Knoxville, Tennessee [ENR (1978)[35]]; and a foundation and structural failure in an Akron, Ohio, department store that resulted in one dead and 10 injured [ENR (1969)[36]]. Collapses resulting in substantial damage and in some cases deaths have also been reported for locations near Johannesburg and Paris *(see Art. 10.4.3)*. Subsidence and sinkholes associated with the removal of halite have been reported for areas around Detroit; Windsor, Ontario; and Hutchinson, Kansas.

*Collapse incidence* is much less than for slope failures, but nevertheless the recognition of its potential is important, especially since the potential may be increasing in a given area. Collapse does occur as a natural phenomenon, but the incidence increases substantially in any given area with an increase in groundwater withdrawal.

## 10.4.2 SOLUTION PHENOMENON AND DEVELOPMENT

### Characteristics of Limestone Formations

*General*

Limestone, the most common rock experiencing cavity development, is widely distributed throughout the world, and is exposed in large areas of the United States, as shown on Fig. 10.16. The occurrence, structure, and geomorphology of carbonate rocks is described in detail in Art. 6.2.3 and will only be summarized briefly in this section.

*Rock Purity and Cavity Growth*

*Purer limestones,* normally found as thick beds of dense, well-indurated rock, are the most susceptible to cavity growth. At least 60% of the rock must be carbonate materials for karst development, and a purity of 90% or more is required for full development [Corbel (1959)[38]].

*Impure limestones* are characteristically thinly bedded and interbedded with shales and are resistant to solution.

**FIG. 10.15 Collapse of two houses into a funnel-shaped sink in Bartow, Florida, in 1967. Cause was ravelling of medium to fine sand into chimneylike cavities in limestone at depths of 15 to 25 m below the surface.** *(Photo courtesy of George F. Sowers.)*

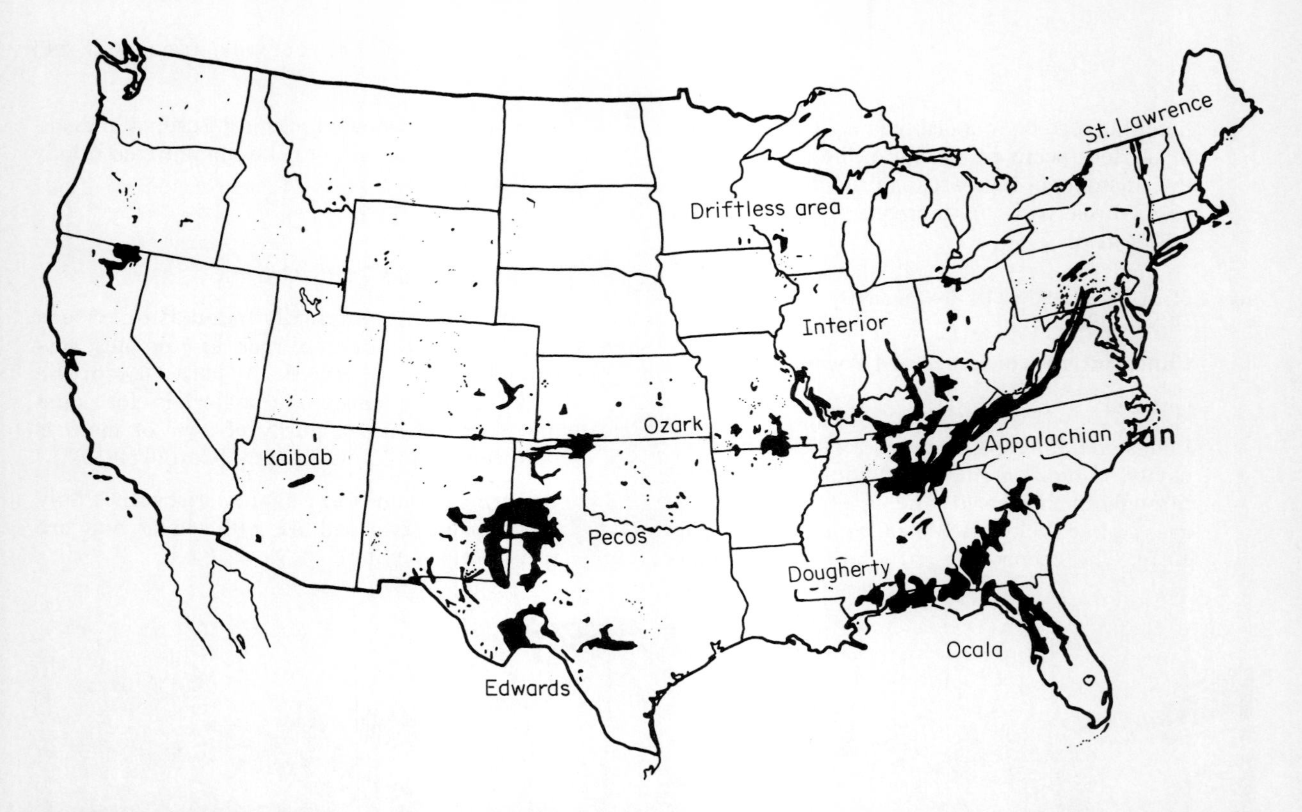


**FIG. 10.16 Distribution of karst regions in the United States.** [*Compiled by William E. Davies; from White (1968).*[37]]

*Jointing*

Groundwater moves in the rock along the joints, which are usually the result of strain energy release (residual from early compression) that occurs during uplift and rebound subsequent to unloading by erosion. This dominant origin causes most joints to be normal to the bedding planes. Major joints, cutting several beds, usually occur in parallel sets and frequently two sets intersect, commonly at about 60°, forming a conjugate joint system.

**Cavity Growth, Subsidence, and Collapse**

*Solution*

Groundwater moving through the joint system at depth and rainfall entering the joint system from above result in solution of the rock. As rainwater passes through the surface organic layer, it becomes a weak acid which readily attacks the limestone. Solution activity is much greater, therefore, in humid climates with heavy vegetation than in dry climates with thin vegetation.

*Geologic Conditions and Cavity Growth Form*

*Horizontal beds* develop cavities vertically and horizontally along the joints, which grow to caverns as the solution progresses. Cavern growth is usually upward; surface subsidence occurs when the roof begins to deflect, or when broken rock in the cavern provides partial support, preventing a total collapse. When a cavern roof lacks adequate arch to support overburden pressures, collapse occurs *(see Fig. 6.24)* and a sinkhole is formed.

*Dipping beds* develop cavities along joint dips as shown on Figs. 6.89 and 10.17, creating a very irregular rock surface. As the cavity grows, the overburden moves into the void, forming a soil arch. With further growth the arch collapses and a sinkhole results. In granular soils the soil may suddenly enter a cavity by *ravelling,* wherein the arch migrates rapidly to the surface, finally collapsing.

*Horizontal beds overlain by thick granular overburden* are also subject to sudden ravelling into cavities developing along joints. An example of a large sink developing under these conditions is given as Fig. 10.15. Sowers (1975)[34] states, "Ravelling failures are the most widespread and probably the most dangerous of all the subsidence phenomena that are associated with limestone." The author considers this statement to apply to conditions like those in central Florida, i.e., relatively thick deposits of granular alluvium overlying limestone undergoing cavity development from its surface.

*Natural Rate of Cavity Growth*

It has been estimated that the rainfall in the sinkhole region of Kentucky will dissolve a layer of limestone 1 cm thick in 66 years [Flint et al. (1969)[40]]. Terzaghi (1913),[41] reporting on a geologic study that he made in the Gačka region of Yugoslavia, observed that solution proceeded much more rapidly in heavily forested areas than in areas covered lightly by grass or barren of vegetation. In an analysis he assumed that 60% of the annual rainfall, or 700 mm/year, entered the topsoil and that the entire amount of carbon dioxide developed in the topsoil was used up in the process of solution, which removed the limestone at the rate of 0.5 mm/year, or 1 cm in 20 years.

*Collapse Causes Summarized*

Collapse of limestone cavities can result from:

- Increase in arch span from cavity growth until the strength is insufficient to support the overburden weight
- Increase in overburden weight over the arch by increased saturation from rainfall or other sources, or from groundwater lowering, which removes the buoyant force of water
- Entry of granular soils by ravelling into a cavity near the rock surface
- Applications of load to the surface from structures, fills, etc.

### Geomorphic Features of Karst

*Karst* refers in general to the characteristic, readily recognizable terrain features that

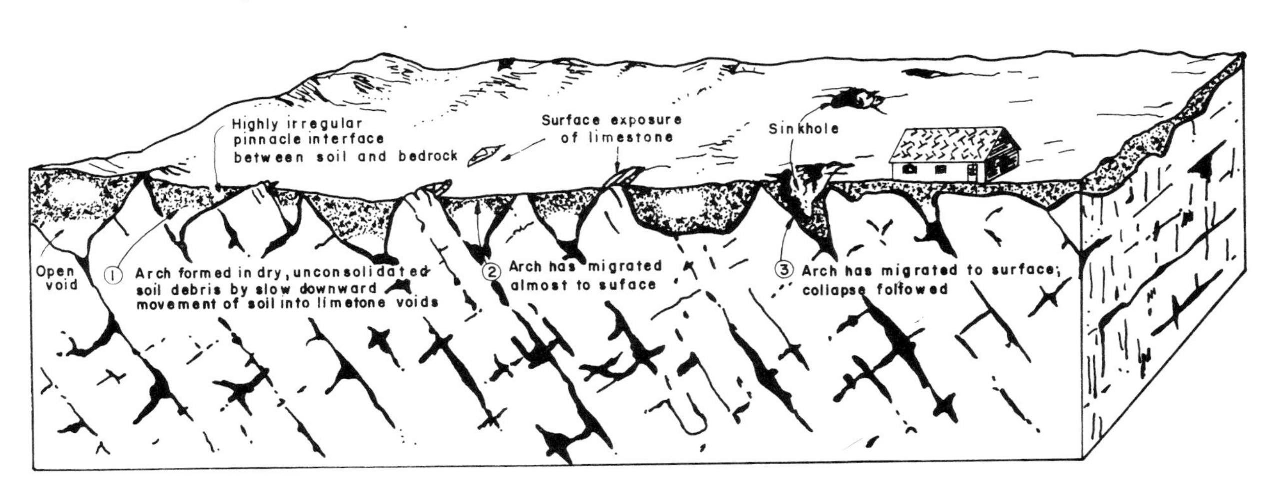


**FIG. 10.17 Hypothetical section through a carbonate valley showing stages of sinkhole development.** [*From Pennsylvania Geological Survey (1972).*[39]]

**FIG. 10.18 Sink 20 m in diameter and 13 m deep formed suddenly in December 1973 in Pierson, Florida, after 3 days of continuous pumping from nearby irrigation wells. The limestone is about 30 m in depth.** *(Photo courtesy of George F. Sowers.)*

develop in the purer limestones. The important characteristics of karst topography are its predominantly vertical and underground drainage, lack of surface drainage systems, and the development of circular depressions and sinks. At times streams flow a short distance and suddenly disappear into the ground.

*Youthful karst* is characterized by numerous sinkholes and depressions, as well as deranged and intermittent drainage (*see Table 6.2*) as shown on Fig. 6.25, a USGS quad sheet of an area near Versailles, Kentucky.

*Mature karst* is characteristic of humid tropical climates. The landform consists of numerous rounded, steep-sided hills ("haystacks" or "pepinos") as illustrated on Fig. 6.26, a portion of the USGS quad sheet for Manati, Puerto Rico.

*Buried karst* is illustrated by the ERTS image of Florida (*see Fig. 6.27*), showing numerous lakes which have filled subsidence depressions. In the Orlando region the limestone is often buried under 20 to 30 m of alluvium.

## Groundwater Pumping Effects

### *Significance*

Groundwater withdrawal greatly accelerates cavity growth in soluble rocks, and lowering of the water table increases overburden pressures. The latter activity, which substantially increases the load on a naturally formed arch, is probably the major cause of ground subsidence and collapse in limestone regions [Prokopovich (1976)[42]]. Even if groundwater withdrawal is controlled with the objective of maintaining a water balance and preserving the natural water table, the water table drops during severe and extensive droughts and collapse activity increases significantly, as occurred in central Florida during the spring of 1981.

*Examples*

PIERSON, FLORIDA The sink illustrated on Fig. 10.18, 20 m in diameter and 3 m deep, formed suddenly in December 1973, after 3 days of continuous pumping from nearby irrigation wells. The limestone is about 30 m in depth.

JOHANNESBURG, SOUTH AFRICA A large pumping program was begun in 1960 to dewater an area, underlain by up to 1000 m of the Transvaal dolomite and dolomitic limestone, for gold mining operations near Johannesburg. In December 1962, a large sinkhole developed suddenly under the crushing plant adjacent to one of the mining shafts, swallowed the entire plant, and took 29 lives. In 1964 the lives of five persons were lost when their home suddenly fell into a rapidly developing sinkhole. Between 1962 and 1966 eight sinkholes larger than 50 m in diameter and 30 m in depth had formed in the mine area [Jennings (1966)[43]].

PARIS, FRANCE Groundwater withdrawal has increased the solution rate and cavern growth in old gypsum quarries beneath the city and some suburban towns [Arnould (1970)[21]]. Ground collapse has occurred, causing homes to be lost and industrial buildings to be damaged. In one case it was estimated that pumping water from gypsum at the rate of about 85 $ft^3$/min removed 136 lb of solids per hour.

HERSHEY, PENNSYLVANIA: Increased dewatering for a quarry operation caused groundwater levels to drop over an area of 2600 hectares and soon resulted in the appearance of over 100 sinkholes [Foose (1953)[44]]. The original groundwater levels were essentially restored after the quarry company sealed their quarry area by grouting.

ROUND ROCK, TEXAS The Edwards limestone outcrops in the area *(see Fig. 2.20)* and is known to be cavernous in some locations. During a study for new development, interpretation of stereo pairs of air photos disclosed several sinkholes and depressions in one area of the site as illustrated on Fig. 10.19 [Hunt (1973)[45]]. During reconnaissance of the site in 1972, discovery of a sink 7 m wide and 1 m deep (Fig. 10.20) that did not appear on the photos, dated 1969, indicated recent collapse activity. Since the Edwards overlies the major aquifer in the area and groundwater withdrawals have increased along with area development, it may be that the incidence of collapse activity is increasing. An engineering geology map of the site, on which three general conditions are zoned, including areas judged to have collapse potential, is given as Fig. 2.19. Regional topography and a site location map are given on Fig. 2.21.

### 10.4.3 INVESTIGATION

#### Preliminary Phases

*Data Collection*

Existing data are gathered to provide information on regional rock types and their solubility, bedding orientation and jointing, overburden types and thickness, and local aquifers and groundwater withdrawal.

*Terrain Analysis*

Topographic maps and remote-sensing imagery are interpreted by terrain analysis techniques. Some indicators of cavernous rock are:

- Surface drainage: lack of second- and third-order streams; intermittent streams and deranged drainage; streams ending suddenly.
- Landform: sinks and depressions or numerous dome-shaped, steep-sided hills with sinks in between *(see Art. 10.4.2)*.
- Photo tone: Soils in slight depressions formed over cavity development will have slightly higher moisture contents than those in adjacent areas and will show as slightly darker tones on black-and-white aerial photos such as Fig. 10.19. Infrared also is useful in detecting karstic features because of differences in soil moisture.

*Preliminary Evaluation*

From a preliminary evaluation of the data, judgments are formulated regarding the potential for

**FIG. 10.19 Stereo pair of aerial photos of area near Round Rock, Texas, showing sinks and depressed areas from partial collapse of shallow limestone.** [*From Hunt (1973).*[45]]

**FIG. 10.20 Recent sink, not present on air photos dated 1969** ***(see Fig. 10.19).*** **Photo taken in 1972.** [*From Hunt (1973).*[45]]

or the existence of cavity development from natural or human causes, and the possible locations, type, and size of ground subsidence and collapse. From this data base explorations are programmed.

## Explorations

### *General*

The location of all important cavities and the determination of their size and extent is a very difficult and usually impossible task. Explorations should never proceed without completion of data collection and terrain analysis.

### *Geophysics (See Art. 2.3.2)*

Explorations with geophysical methods can provide useful information, but the degrees of reliability vary.

Seismic refraction surveys may result in little more than "averaging" the depth to the limestone surface if it is highly irregular.

Cross-hole surveys may indicate the presence of cavities if they are large.

Electrical resistivity, at times, has indicated shallow cavity development.

Gravimeter surveys were found to be more useful than other geophysical methods (seismic cross-hole surveys or electrical resistivity) in a study for a nuclear power plant in northwestern Ohio [Millet and Moorhouse (1973)[46]].

Ground-probing radar, still undergoing development, shows promise in applications to cavity identification.

### *Test Borings*

Test and core borings are programmed to explore anomalies detected by geophysical explorations and terrain analysis, as well as to accomplish their normal purposes. Voids are disclosed by the sudden drop of the drilling tools and loss of drilling fluid. The material in a sinkhole usually is very loose compared with the surrounding materials, and may overlie highly fractured rock where a roof has collapsed.

*Test pits* are useful to allow examination of the bedrock surface. Although the normal backhoe reach is limited to 3 to 5 m, on important projects, such as for dams or construction with heavy foundation loadings, deeper excavations, perhaps requiring dewatering, may be warranted.

*Rotary Probes*

If cavity presence has been confirmed, it is usually prudent to make either core borings, rotary probes, or pneumatic percussion drill holes at the location of each footing before final design, or before construction. The objective is to confirm that an adequate thickness of competent rock is present beneath each foundation.

*Proof testing* with air drills is an alternate to rotary probes at each footing location, particularly for drilled piers where installations are relatively deep. Proof testing the bottom of each pier founded on rock with air drills is much less costly than rotary probes or core borings.

### Evaluation and Analysis

*Basic Elements*

The following elements should be considered:

- Overburden thickness and properties and thickness variations, which are likely to be substantial
- Bedrock surface characteristics, i.e., weathered, sound, relatively sound and smooth with cavities following joint patterns from the rock surface *(see Fig. 6.22)*, or highly irregular in configuration and soundness *(see Fig. 10.17)*
- Cavities within the rock mass—location, size, and shape
- Arch characteristics—thickness, span, soundness, and joint characteristics and properties
- Groundwater depth and withdrawal conditions—present vs. future potential

*Analysis*

Analysis proceeds in accordance with a normal foundation study (except where foundations may overlie a rock arch, then rock-mechanics principles are applied to evaluate the minimum roof thickness required to provide adequate support to foundations). A generous safety factor is applied to allow for unknown rock properties.

## 10.4.4 SUPPORT OF SURFACE STRUCTURES

### Avoid the High Hazard Condition

*Project relocation* should be considered where cavities are large and at relatively shallow depths, or where soluble rock is deep but overlain by soils subject to ravelling. The decision is based on the degree of hazard presented, which is related directly to the occurrence of groundwater withdrawal or its likelihood in the future. Groundwater withdrawal represents very high hazard conditions. In fact, the probability and effects of groundwater withdrawal are the *most important* considerations in evaluating sites underlain by soluble rock.

### Foundation Treatments

*Dental Concrete*

Cavities that can be exposed by excavation can be cleaned of soil and filled with lean concrete *(see Fig. 6.22)*, which provides suitable support for shallow foundations.

*Grouting*

Deep cavities that cannot be reached by excavation often are filled by grout injection, but the uncertainty will exist that not all cavities and fractures have been filled, even if check explorations are made subsequent to the grouting operations. Grouting has the important advantage of impeding groundwater movement and therefore cavity growth, even where pumping is anticipated.

*Deep Foundations*

Deep, heavily loaded foundations, or those supporting settlement-sensitive structures, when founded on soluble rock, should be proof-tested, whether grouted or not (see discussion of explorations in Art. 10.4.3).

## 10.5 SOIL SUBSIDENCE AND COLLAPSE

### 10.5.1 GENERAL

**Causes**

Subsidence in soils results from two general categories of causes:

1. *Compression* refers to the volume reduction occurring under applied stress from grain rearrangement in cohesionless soil, or consolidation in a cohesive soil (*see Art. 3.5.4*). The phenomenon is very common and always occurs to some degree under foundation loadings.
2. *Collapse* is the consequence of a sudden closure of voids, or a void, and is the subject of this article. Collapsible or metastable soils undergo a sudden decrease in volume when internal structural support is lost; piping soils are susceptible to the formation of large cavities, which are subject to collapse.

**The Hazard**

*Subsidence* from compression or soil collapse is a relatively minor hazard, resulting in structural distortions from differential settlements.

*Piping erosion* forms seepage channels in earth dams and slopes and in severe cases results in collapse of the piping tunnel which can affect the stability of an earth dam or a natural or cut slope.

### 10.5.2 COLLAPSIBLE OR METASTABLE SOILS

**Collapse Mechanisms**

*Temporary Internal Soil Support*

Internal soil support, which is considered to provide a temporary strength, is derived from a number of sources including *capillary tension* which provides a temporary strength in partially saturated fine-grained cohesionless soils; *cementing agents*, which may include iron oxide, calcium carbonate, or clay in the clay-welding of grains; and *other agents* which include silt bonds, clay bonds, and clay bridges, as illustrated on Fig. 10.21.

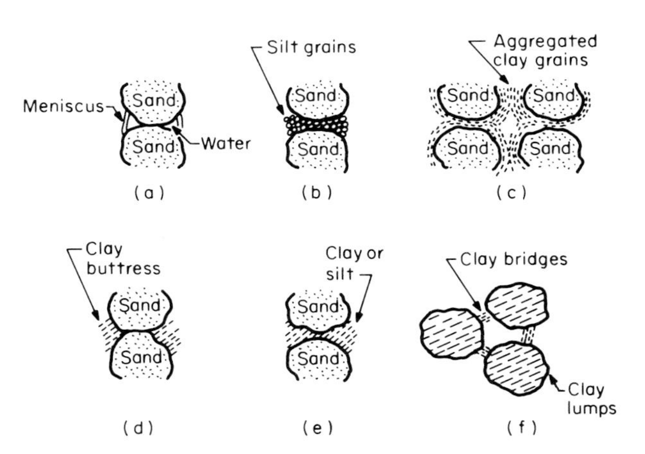


**FIG. 10.21 Typical collapsible soil structures: (*a*) capillary tension; (*b*) silt bond; (*c*) aggregated clay bond; (*d*) flocculated clay bond; (*e*) mudflow type of separation; (*f*) clay bridge structure.** [*From Clemence and Finbarr (1980)*[47]*; adapted from Barden et al. (1973).*[48]]

*Collapse Causes*

*Wetting* destroys capillary bonds, leaches out cementing agents, or softens clay bonds and bridges in an open structure. Local shallow wetting occurs from surface flooding or broken pipelines, and subsidence can be substantial and nonuniform. Intense, deep local wetting from the discharge of industrial effluents or irrigation also can result in substantial and nonuniform subsidence. Slow and relatively uniform rise in the groundwater level usually results in uniform and gradual subsidence.

*Increased saturation* under an applied load can result in gradual settlement, or in a sudden collapse as the soil bonds are weakened.

*Applied load of critical magnitude* can cause a sudden collapse of the soil structure when the bonds break in a brittle type of failure, even at natural moisture content.

**Susceptible Soils**

*Loess*

Distribution, characteristics, and engineering properties of loess and other fine-grained aeolian soils are described in Art. 7.5.3.

*Valley Alluvium: Semiarid to Arid Climate*

*Origin* and some characteristics of valley alluvium susceptible to collapse are described in Art. 7.4.1.

**FIG. 10.22 Subsidence after 3 months caused by ground saturation around test plot in San Joaquin Valley, California.** [*From Curtin (1973).*[49] *Photo courtesy of California Dept. of Water Resources.*]

*Subsidence studies in the San Joaquin valley* of California were undertaken by the U.S. Bureau of Reclamation to evaluate problems and solutions for the construction of the California aqueduct [Curtin (1973)[49]]. Test sites were selected after extensive ground and aerial surveys of the 4500-km$^2$ study area. Test procedures to investigate subsidence potential involved either inundating the ground surface by ponding or filling bottomless tanks with water as shown on Fig. 10.22.

One of the large ponds overlays 75 m (250 ft) of collapsible soils. Water was applied to the pond for 484 days, during which an average settlement of 4.1 m occurred. Benchmarks had been set at the surface and at 7.6-m (25-ft) intervals to a depth of 45 m (150 ft). A plot of the subsidence and compaction between benchmarks as a function of time is given on Fig. 10.23. It is seen that the effects of the test influenced the soils to a depth of at least 150 ft. The subsidence appears to be the summation of soil collapse plus compression from increasing overburden pressure due to saturation. The test curves show an almost immediate subsidence of 1 ft upon saturation within the upper 25 ft; thereafter, the shape of the curves follows the curve expected from normal consolidation.

*San Joaquin valley soils* are described by Curtin (1973)[49] as having a texture similar to that of loess, characterized by voids between grains held in place by clay bonds, with bubble cavities formed by entrapped air, interlaminar openings in thinly laminated sediments, and unfilled polygonal cracks and voids left by disintegration of entrapped vegetation. The classification ranges from a poorly graded silty sand to a clay, in general with more than 50% passing the 220 sieve. Dry density ranges from 57 to 110 pcf (0.9 to 1.8 g/cm$^2$) with a porosity range of 43 to 85%. The predominant clay mineral is montmorillonite, and the clay content of collapsing soils was reported to be from 3 to 30%. The observation was made that soils with a high clay content tended initially to swell rather than collapse. During field testing, the benchmarks on the sur-

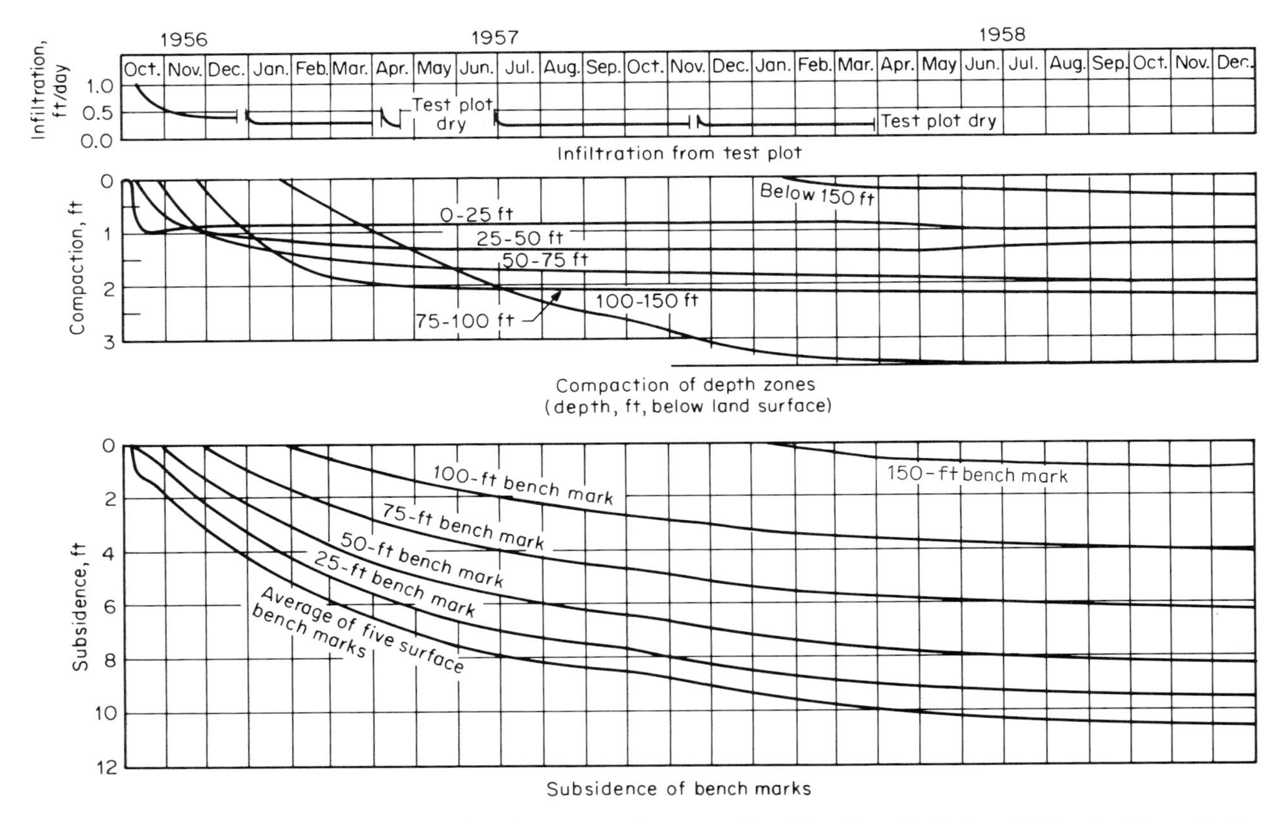


**FIG. 10.23 Subsidence measured by benchmarks, test pond B, west side of San Joaquin Valley, California.** [*From Curtin (1973).*[49]]

face first rose upon inundation as the montmorillonite swelled; soon subsidence overcame the swelling and the benchmarks moved downward. A laboratory consolidation test curve showing soil collapse upon the addition of water is given as Fig. 10.24.

In describing the soils in western Fresno County, Bull (1964)[50] notes that maximum subsidence occurs where the clay amounts to about 12% of solids; below 5% there is little subsidence, and above 30% the clay swells.

*Tucson, Arizona* also suffers from the collapsing soil problem, and damage to structures has been reported [Sultan (1969)[51]].

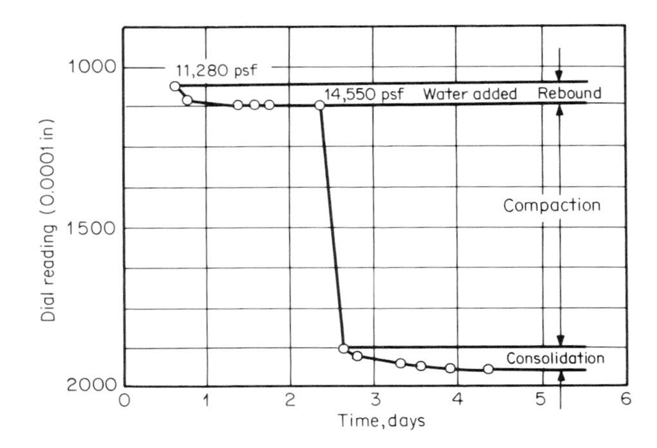


**FIG. 10.24 Laboratory consolidation test curve of compression vs. time for a collapsible soil from the San Joaquin Valley.** [*From Curtin (1973).*[49]]

### *Residual Soils*

GEOGRAPHIC OCCURRENCE Collapse has been reported occurring in residual soils derived from

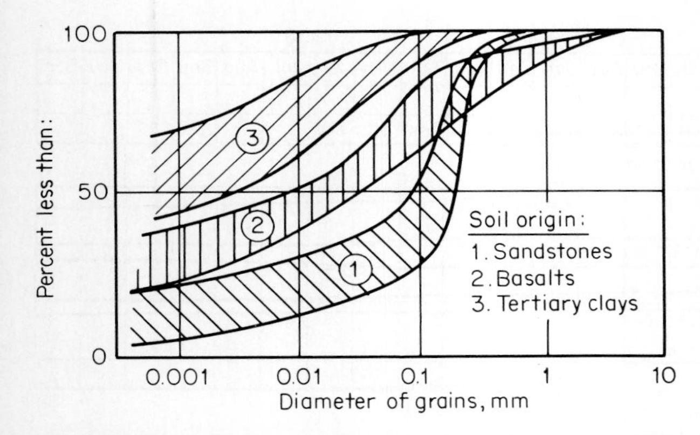


**FIG. 10.25 Gradation curves for typical porous clays of Brazil.** [*From Vargas (1972).*[53]]

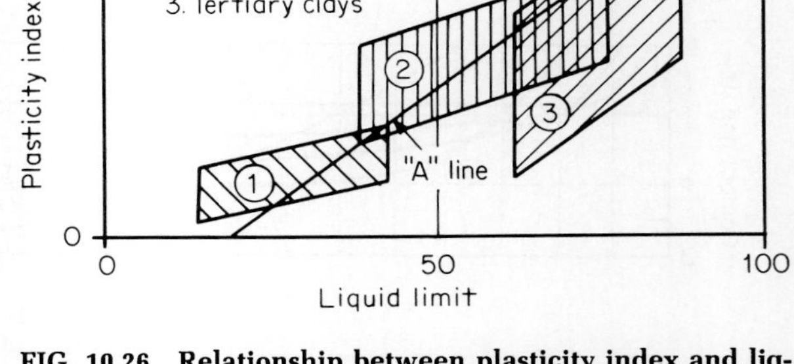


**FIG. 10.26 Relationship between plasticity index and liquid limit, porous clays of Brazil.** [*From Vargas (1972).*[53]]

granite in South Africa and northern Rhodesia [Brink and Kantey (1961)[52]], and from sandstones and basalts in Brazil [Vargas (1972)[53]].

*Porous clays of Brazil* (argila porosa) occur intermittently in an area of hundreds of square kilometers ranging through the central portions of the states of São Paulo, Paraná, and Santa Catarina. The terrain consists of rolling savannah and the annual rainfall is about 1200 mm, distributed primarily during the months of December, January, and February. The remaining 9 months are relatively dry. Derived from Permian sandstones, Triassic basalts, and Tertiary sediments, the soils are generally clayey. Typical gradation curves are given on Fig. 10.25, and plasticity index ranges on Fig. 10.26.

The upper zone of these formations, to depths of 4 to 8 m, typically yields low SPT values, ranging from 0 to 4 blows/ft; void ratios of 1.3 to 2.0 are common. Below the "soft" (dry) upper zone, a hard crust is often found. A typical boring log is given as Fig. 10.27; it is to be noted that groundwater was not encountered, which is the normal condition. Some laboratory test data are also included on the figure. A profile giving percent water content, plasticity index, liquid limit, gradation, and porosity is also given on Fig. 7.7.

Although the soil is essentially a clay, its open, porous structure provides for high permeability and the rapid compression characteristics of a sand, hence the term "porous clay." The porous characteristics are the result of leaching out of iron and other minerals which are carried by migrating water to some depth where they precipitate to form the aforementioned hard zone, which often contains limonite nodules. The effect of saturation on a consolidation test specimen is given on Fig. 10.28, a plot of void ratio vs. pressure. The curve is typical of collapsing soils. In the dry condition strengths are high, and excavation walls will stand vertical for heights greater than 4 or 5 m without support in the same manner as loess.

*Recognition of porous clays* often can be accomplished from terrain analysis. Three factors appear to govern the development of the weak, open structure: a long relatively dry period followed by heavy summer rains, a relatively high ground elevation in rolling, hilly terrain with a moderately deep water table, and readily leachable materials. Examination of aerial photos, such as Fig. 10.29, reveals unusual features for a clay soil: characteristically thin vegetation and lack of any surface drainage system, both indicative of the open porous structure. Here and there, where terrain is relatively level, bowl-shaped areas, often 3 to 4 m deep and 20 m across, with no apparent existing drainage, seem to indicate areas of possible natural collapse (Fig. 10.29). These may have occurred during periods of very heavy rains which either created ponds or fell on zones that had been very much weakened by leaching.

| Depth, m | "N" | Soil type | Laboratory test data | | | | |
|---|---|---|---|---|---|---|---|
| | | | LL | PI | w,% | γ | e |
| 0 | | | | | | | |
| 1 ■ | 3 | Very soft to soft | 45 | 21 | 32 | 1.25 | 1.94 |
| 2 ■ | 2 | red silty clay | 46 | 19 | 31 | 1.19 | 2.01 |
| 3 ■ | 5 | ( CL ) | 48 | 24 | 34 | 1.51 | 1.53 |
| 4 ■ | 5 | | 43 | 16 | 32 | 1.39 | 1.39 |
| 5 | 10 | Stiff tan to brown | Notes: | | | | |
| 6 | 12 | very silty clay | ■ — block sample | | | | |
| 7 | 16 | | $\gamma_t$ — gm/cm$^3$ | | | | |
| 8 | 14 | | N — blows per foot (SPT) | | | | |
| 9 | 11 | | GWL — groundwater table not encountered | | | | |
| 10 | 14 | | | | | | |
| 11 | 14 | | | | | | |
| 12 | 36 | becoming hard | | | | | |
| 13 | 50 | | | | | | |
| 14 | 31 | | | | | | |
| 15 | 5/1 cm (refusal) | | | | | | |

**FIG. 10.27 Test boring log and laboratory test data for a porous clay derived from basalt (Araras, São Paulo, Brazil).**

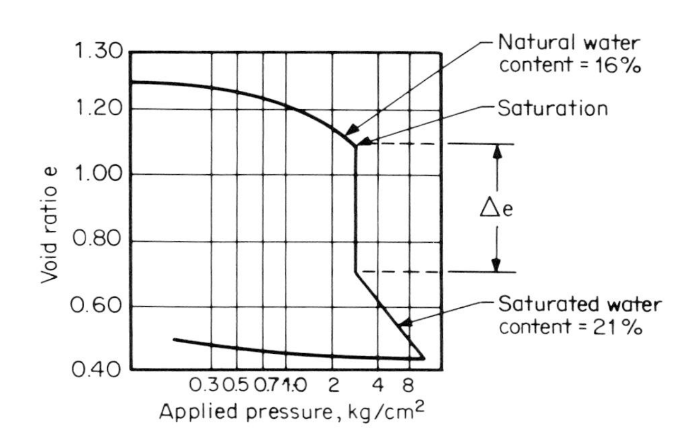


**FIG. 10.28 Effect of saturation on the pressure vs. void ratio curve of a porous clay from Brazil.** [*From Vargas (1972).*[53]]

**FIG. 10.29 Stereo pair of aerial photos of area of porous clays showing possible ancient collapse zones (state of São Paulo, Brazil). Although soils are clays, the lack of drainage patterns indicates high infiltration.** [*From Hunt and Santiago (1976).*[54]]

10.5.3 PREDICTING COLLAPSE POTENTIAL

### Preliminary Phases

#### *Data Collection*

Preliminary knowledge of the local geology from a literature review aids in anticipating soils with a collapse potential, since they are commonly associated with loess and other fine-grained aeolian soils, and fine-grained valley alluvium in dry climates. The susceptibility of residual soil is difficult to determine from a normal literature review. Rolling terrain with a moderately deep water table in a climate with a short wet season and a long dry season should be suspect.

*Terrain Analysis*

Loess and valley alluvium are identified by their characteristic features described in Chap. 7. Residual soils with collapse potential may show a lack of surface drainage channels indicating rainfall infiltration rather than runoff, especially where the soils are known to be clayey. Unexplained collapse depressions may be present.

**Explorations**

*Test Borings and Sampling*

Drilling using continuous-flight augers (no drilling fluid) will yield substantially higher SPT values in collapsible soils than will drilling with water, which tends to soften the soils. A comparison of the results from both methods on a given project provides an indication of collapse potential. Undisturbed sampling is often difficult in these materials.

In residual soils, a "soft" upper zone with low SPT values will be encountered when drilling fluids are used, often underlain by a hard zone or crust which may contain limonite nodules or concretions.

*Test pits* are useful for close examination of the soils and description in the undisturbed state, in situ natural density tests, and the recovery of block samples for laboratory testing.

*Simple Hand Test*

A hand-size block of the soil is broken into two pieces and each is trimmed until the volumes are equal. One is wetted and molded in the hand and the two volumes then compared. If the wetted volume is obviously smaller, then collapsibility may be suspected [Clemence and Finbarr (1980)[47]].

**Field Load Tests**

*Ground saturation* by ponding or using bottomless tanks is useful for evaluating collapse and subsidence where very large leakage may occur, as through canal linings.

*Full-scale or plate load tests*, for the evaluation of foundation settlements, should be performed under three conditions to provide comparative data at founding level:

1. Soils at natural water content when loads are applied.
2. Loads applied while soil is at the natural water content until the anticipated foundation pressure is reached. The ground around and beneath the footing then is wetted by pouring water into auger holes. (In very dry climates and clayey soils, several days of treatment may be required to achieve an adequate level of saturation. In all cases natural soil densities and water contents should be measured before and after wetting.)
3. Soils at the test plot are wetted prior to any loading, and then loads are applied.

**Laboratory Testing**

*Natural Density*

When the natural density of fine-grained soils at the natural moisture content is lower than normal, about 1.4 g/cm$^3$ or less, collapse susceptibility should be suspected *(see Table 7.10)*.

*Density vs. liquid limit* as a collapse criterion is given by Zur and Wiseman (1973)[55] as follows:

$$\frac{D_0}{D_{LL}} < 1.1, \text{ soil prone to collapse}$$

$$\frac{D_0}{D_{LL}} > 1.3, \text{ soil prone to swell}$$

where $D_0$ = in situ dry density
$D_{LL}$ = dry density of soil at full saturation and at moisture content equal to the liquid limit

*Normal Consolidation Test*

Loads are applied to the specimen maintained at its natural water content: the specimen is wetted at the proposed foundation stress, and compression is measured as shown on Fig. 10.28 *(see also Art. 3.5.4)*.

*Double Consolidation Test*

The double consolidation test is used to provide data for the calculation of estimates of collapse magnitude [Jennings and Knight (1975),[56] Clemence and Finbarr (1980)[47]].

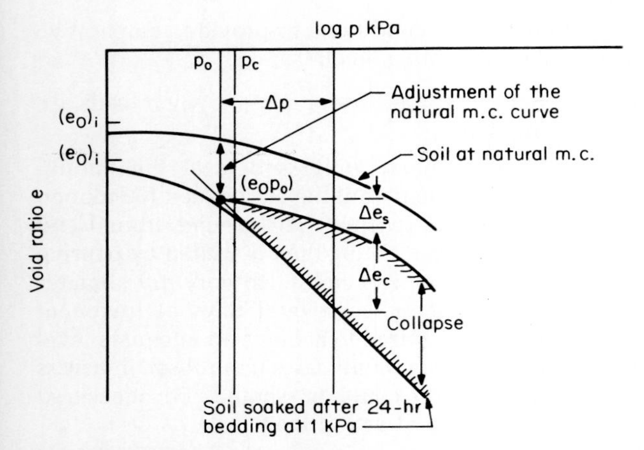


**FIG. 10.30 Double consolidation test *e-p* curves and adjustments for a normally consolidated soil.** [*From Clemence and Finbarr (1980).*[47]]

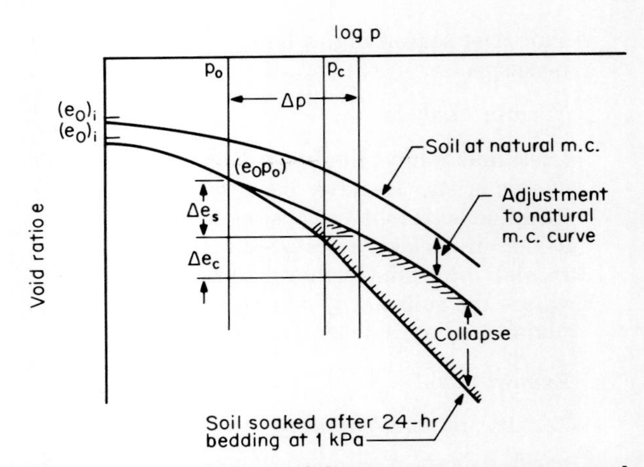


**FIG. 10.31 Double consolidation test *e-p* curves and adjustments for an overconsolidated soil.** [*From Clemence and Finbarr (1980).*[47]]

Two specimens of similar materials (preferably from block samples) are trimmed into consolidometer rings and placed in the apparatus under a light 1-kPa (0.01-kg/cm$^2$) seating load for 24 hr, after which one specimen is submerged and the other kept at its natural water content for an additional 24 hr. Load applications on each specimen are then carried out in the normal manner.

The *e*-log-*p* curve for each test is plotted on the same graph, along with the overburden pressure $p_o$ and the preconsolidation pressure $p_c$ for the saturated specimen. The curves for a normally consolidated soil ($p_c/p_o = 0.8$ to 1.5) are given on Fig. 10.30, and for an overconsolidated soil ($p_c/p_o > 1.5$) on Fig. 10.31. The curve for the natural water content condition is relocated to the point ($e_0$, $p_o$) given on the figures.

For an increase in foundation stress $\Delta P$, and wetted soil, the settlement $\rho$ is estimated from the expression:

$$\rho = \frac{\Delta e_s}{1 + e_0} + \frac{\Delta e_c}{1 + e_0} \qquad (10.1)$$

10.5.4 TREATMENT AND SUPPORT OF STRUCTURES

### Evaluate the Degree of Hazard and Risk

*Degree of hazard* is basically a function of the probability of significant ground wetting and of the magnitude of the potential collapse, if the critical pressure that will cause collapse at the natural water content is not approached. Sources of ground wetting have been given under Art. 10.5.2 in the discussion of collapse causes.

- *Low-hazard* conditions exist where potential collapse magnitudes are small and tolerable, or the probability for significant ground wetting is low.
- *Moderate-hazard* conditions exist where the potential collapse magnitudes are undesirable but the probability of substantial ground wetting is low.
- *High-hazard* conditions exist where the potential collapse magnitudes are undesirable and the probability of occurrence is high.

*Degree of risk* relates to the sensitivity of the structure to settlement and to the importance of the structure.

### Reduce the Hazard

*Prevent ground wetting* and support structures on shallow foundations designed for an allowable bearing value sufficiently below the critical pressure to avoid collapse at natural water content. The critical pressure is best determined by in situ plate-load tests, and the allowable soil pressure is based on FS = 2 to 3, depending upon the settlement tolerances of the structure.

In some cases, such as large grain elevators, where the load and required size of a mat foundation impose bearing pressures in the order of the critical pressure, the structures have been permitted to settle as much as 1 ft (30 cm), provided that tilting is avoided. This solution has been applied to foundations on relatively uniform deposits of loess with natural water contents in the order of 13% in eastern Colorado and western Kansas. In collapsible soils derived from residual soils, however, such solutions may not be applicable because of the likelihood that variation in properties will result in large differential settlements. Adequate site drainage should be provided to prevent ponding and all runoff should be collected and directed away from the structure. Avoid locating septic tanks and leaching fields near the structure, and construct all utilities and storm drains carefully to ensure tightness. Underground water lines near the structures have been double-piped or encased in concrete to assure protection against exfiltration. In a desert environment, watering lawns, which is commonly accomplished by flooding, should be avoided, and areas should be landscaped with natural desert vegetation.

*Lime stabilization* has been used in Tucson, Arizona, to treat collapsible soils which have caused detrimental settlements in a housing development [Sultan (1969)[51]]. A water-lime mixture was pumped under high pressure into 2-in-diameter holes to depths of about 5 ft, and significant movements were arrested.

*Hydrocompaction* to preconsolidate the collapsible soils was the solution used by the California Department of Water Resources for the construction of the California aqueduct [Curtin (1973)[49]]. Dikes and unlined ditches were constructed and flooded along the canal route to precompact the soils at locations where collapse potential was considered high. A section of the canal being constructed over areas both precompacted and not precompacted is shown on Fig. 10.32; the subsidence effects on the canal sides are evident on the photo. When hydrocompaction is used, however, the possibility of long-term settlements from consolidation of clay soils under increased overburden pressures should be considered.

*Vibroflotation* was experimented with before construction of the California aqueduct, but adequate compaction was not obtained in the fine-grained soils along the alignment.

*Dynamic compaction* involves dropping 8- to 10-ton tamping blocks from heights of 30 to 120 ft (9 to 36 m). The drops, made by a crane in carefully regulated patterns, produce high-energy shock waves that have compacted soils to depths as great as 60 ft (18 m) [ENR (1980)[57]].

### Avoid the Hazard

Settlement-sensitive structures may be supported on deep foundations that extend beyond the zone of potential collapse or, if the collapsible soils extend to limited depths, shallow foundations may be established on controlled compacted earth fill after the collapsible soils are excavated.

In Brazil, buildings constructed over collapsible residual soils are normally supported on piers or piles penetrating to the hard stratum at depths of 3 to 5 m. Floors for large industrial buildings are often supported on the porous clays or small amounts of fill, and protection against ground wetting is provided. The risk of subsidence is accepted with the understanding that some relevelling by "mudjacking" may be required in the future.

## 10.5.5 PIPING SOILS AND DISPERSIVE CLAYS

### General

Soils susceptible to piping erosion and dispersion are not a cause of large-scale subsidence. Ground collapse can occur, however, when the channels resulting from piping and dispersion grow to significant dimensions.

### Piping Phenomena

*Piping* refers to the erosion of soils caused by groundwater flow when the flow emerges on a free face and carries particles of soil with it *(see Art. 8.3.2)*.

*Occurrence* in natural deposits results from water entering from the surface, flowing through the soil mass along fractures or other openings, finally to exit through the face of a stream bank or other steep slope as illustrated on Fig. 10.33. As the water flows, the opening increases in size,

**FIG. 10.32 Mendoza test plot showing prototype canal section along the California aqueduct. Crest width is 108 ft and length is 1400 ft. Note that both the lined and unlined sections of the canal are subsiding where the land was not precompacted. Concentric subsidence cracks indicate former locations of large test ponds.** [*From Curtin (1973).*[49] *Photo courtesy of California Dept. of Water Resources.*]

at times reaching very large dimensions as shown on Fig. 10.34, forming in the remains of the enormous ancient mudflow shown on Fig. 9.54. The massive movement of the flow destroyed the original structure of the formation and it came to rest in a loose, remolded condition in which fissures subsequently developed. Rainwater entering surface cracks passes along the relict fissures and erodes their sides.

### Dispersive Clays

*Occurrence*

Erosion tunnels from piping in earth dams constructed with certain clay soils are a relatively common occurrence that can seriously affect the stability of the embankment [Sherard et al. (1972)[58]]. It was originally believed that the clay soils susceptible to dispersion erosion were lim-

**FIG. 10.33** *(Above)* **Piping erosion in road cut, Tucson, Arizona.** *(Photo courtesy of Robert S. Woolworth, 1972.)*

**FIG. 10.34 Tunnel about 7 m high formed from piping in colluvial-lacustrine clayey silts, near La Paz, Bolivia. The vertical slopes are about 15 m in height.**

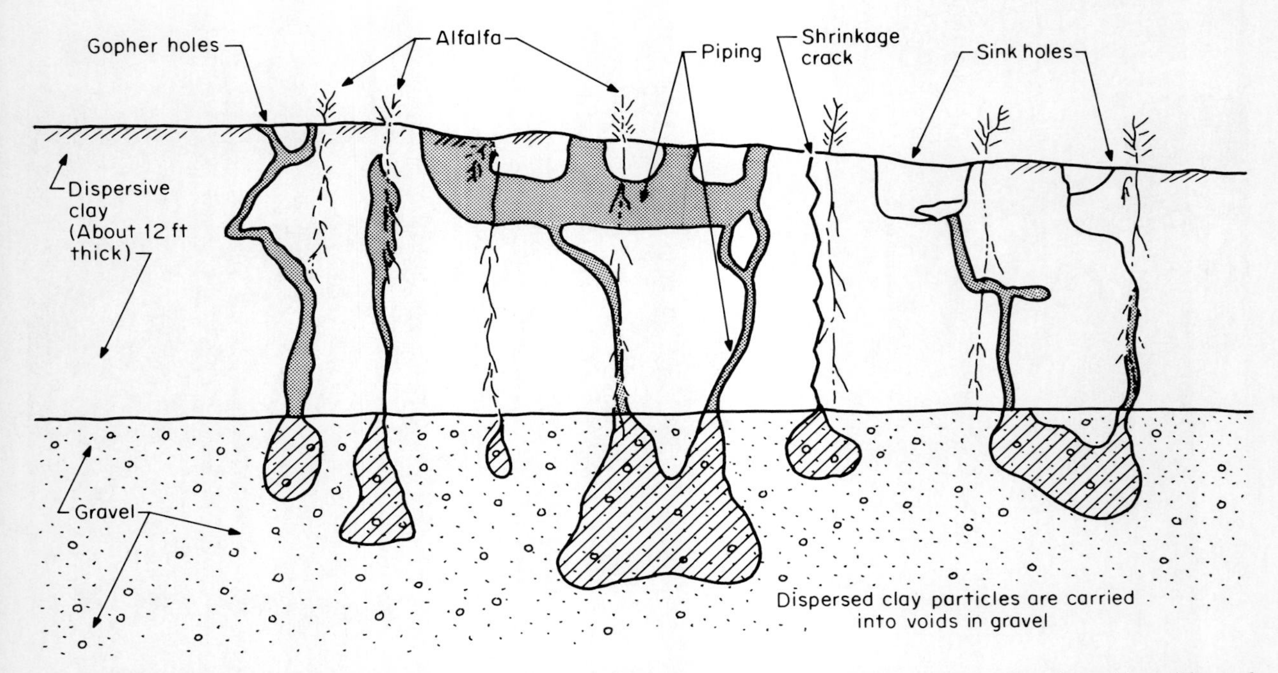


**FIG. 10.35 Damage to agricultural fields in Arizona from piping and sinkhole formation in dispersive clays.** [*From Sherard et al. (1972)*[58]; *after Carroll (1949).*]

ited to dry climates, but in recent years these soils and their related problems have been found to exist in humid climates. Sherard et al. (1972)[58] cite examples in the United States from Oklahoma and Mississippi and from western Venezuela.

*The Phenomenon*

Dispersive clays erode in the presence of water by dispersion or deflocculation. In certain clay soils in which the electrochemical bond is weak, contact with water causes individual particles to detach or disperse. Flowing fresh water readily transports the dispersed particles in the form of piping erosion, in time creating voids or tunnels in the clay mass as illustrated on Figs. 10.35 and 10.36. Any fissure or crack from desiccation or settlement can provide the initial flow channel. In expansive clays, the cracks will close by swelling and dispersion will not occur. The failure of many small dams and dikes has been attributed to dispersive clays.

*Soil Susceptibility*

The main property governing susceptibility to dispersion appears to be the quantity of dissolved sodium cations in the pore water relative to the quantities of other main cations (calcium and magnesium); i.e., the higher the percentage of sodium cation, the higher the susceptibility to dispersion. Soil scientists refer to this relationship as "exchangeable sodium percentages," or ESP (*see also Art. 5.3.3*).

As of 1976, there appeared to be no good relationship between the ESP and the index tests used by the geotechnical engineer to classify soils, except for the identification of expansive clays, which are nonsusceptible. Highly dispersive clays frequently have the same Atterberg limits, gradation, and compaction characteristics as nondispersive clays. These clays plot above the *A* line on the plasticity chart (see *Fig. 3.12*) and are generally of low to medium plasticity (LL = 30 to 50%; CL-CH classifications), although cases of dispersion have been reported in clay with a high liquid limit.

*The pinhole test* is a simple laboratory test developed to identify dispersive clays [Sherard et al. (1976)[59]].

### Prevention of Piping and Dispersion

*Piping in natural deposits* is difficult to prevent. The formation of large tunnels is rare and limited to loose, slightly cohesive silty soils where they are exposed in banks and steep slopes and seepage can exit from the slope. A practical preventive measure is to place a filter at the erosion tunnel outlet to reduce flow exit velocities, but in some cases such as shown on Fig. 10.36, the area should be avoided, since tunnels are likely to open at other locations.

*Piping in dispersive clays* used in embankment construction is prevented by the proper design of filters to control internal seepage *(see Art. 8.4.5)* and by the use of materials that are not susceptible to the phenomenon. Where piping is already occurring it may be necessary to reconstruct the embankment using proper design and materials, if placing a filter at the outlet is not effective.

## 10.6 HEAVE IN SOIL AND ROCK

### 10.6.1 GENERAL

### Origins of Ground Heave

Heave on a regional basis occurs from tectonic activity *(see Art. 11.3.1 and Appendix A)*. On a local basis heave occurs from stress release in rock excavations *(see Art. 6.6)*, expansion from freezing *(see Art. 7.7.3)*, and expansion from swelling in soil or rock *(this article)*.

### The Swelling Hazard

*Swelling in Geologic Materials*

Clay soils and certain minerals readily undergo volume change, shrinking when dried *(see Fig. 5.11)*, or swelling when wet. When in the dry state, or when less than fully saturated, some clays have a tremendous affinity for moisture, and in some cases may swell to increase their volume by 30% or more. Pressures in excess of 8 kg/cm$^2$ can be generated by a swelling material when it is confined. Once the material is permitted to swell, however, the pressures reduce. There is a time delay in the swelling phenomenon. Noticeable swell may not occur for over a year after completion of construction, depending on the soil's access to moisture, but may continue for 5 years or longer.

*Damage to Structures*

Ground heave is a serious problem for structures supported on shallow foundations, or deep foundations if they are not isolated from the swelling soils. Heave results in uplift and cracking of floors and walls, and in severe cases, in the rupture of columns. It also has a detrimental effect on pavements. Shrinkage of soils also causes damage to structures. Damages to structures in the United States each year from swelling soils alone has been reported to amount to $700 million [ENR (1976)[60]].

### Geographic Distribution

*Swelling Soils*

Swelling soils are generally associated with dry climates such as exist in Australia, India, Israel, the United States, and many countries of Africa *(see Art. 7.2.2,* Black Cotton Soils*)*. In the United States, foundation problems are particularly prevalent in Texas, Colorado, and California, and in many areas of residual soils, as shown on the map given as Fig. 10.37.

*Swelling in Rocks*

Swelling in rocks is associated primarily with clay shales and marine shales *(see Art. 6.7.3)*. In

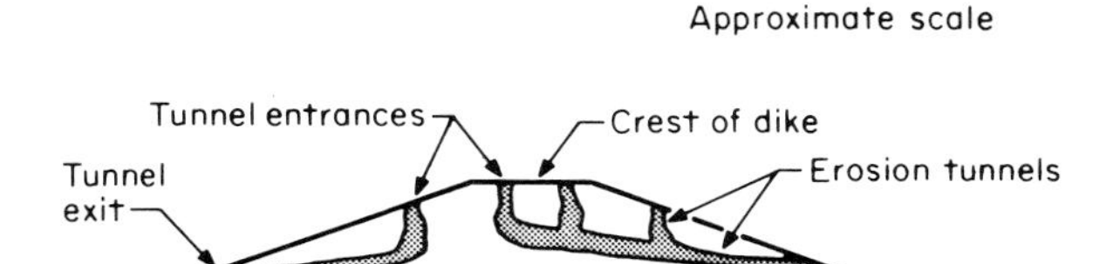


**FIG. 10.36 Schematic of typical rainfall erosion tunnels in clay flood-control dike in badly damaged section.** [*From Sherard et al. (1972).*[58]]

the United States, it is particularly prevalent in Texas, California, Montana, and North and South Dakota. The distribution of materials susceptible to swelling in Texas is given on Fig. 7.54.

## 10.6.2 SWELLING IN SOILS

### Determining Swell Potential

*Basic Relationships*

The phenomenon of adsorption and swell is complex and not well understood, but it appears to be basically physicochemical in origin. Swell potential is related to the percentage of the material in the clay fraction (defined as less than 2 $\mu$, 0.002 mm), the fineness of the clay fraction, the clay structure, and the type of clay mineral *(see Art. 5.3.3)*. Montmorillonite has the highest potential for swell, followed by illite, with kaolinite being the least active. Thus, mineral identification *(see Art. 5.3.3)* is one means of investigating swell potential.

*Clay Activity*

Swell potential has been given by Skempton (1953)[62] in terms of *activity* defined as the ratio of the plasticity index to the percent finer by weight than 2 $\mu$. On the basis of activity, soils have been classed as inactive, normal, and active *(see Table 5.28)*. The activity of various types of clay minerals as a function of the plasticity index and the clay fraction is given on Fig. 10.38. It is seen that the activity of sodium montmorillonite is many times higher than that of illite or kaolinite.

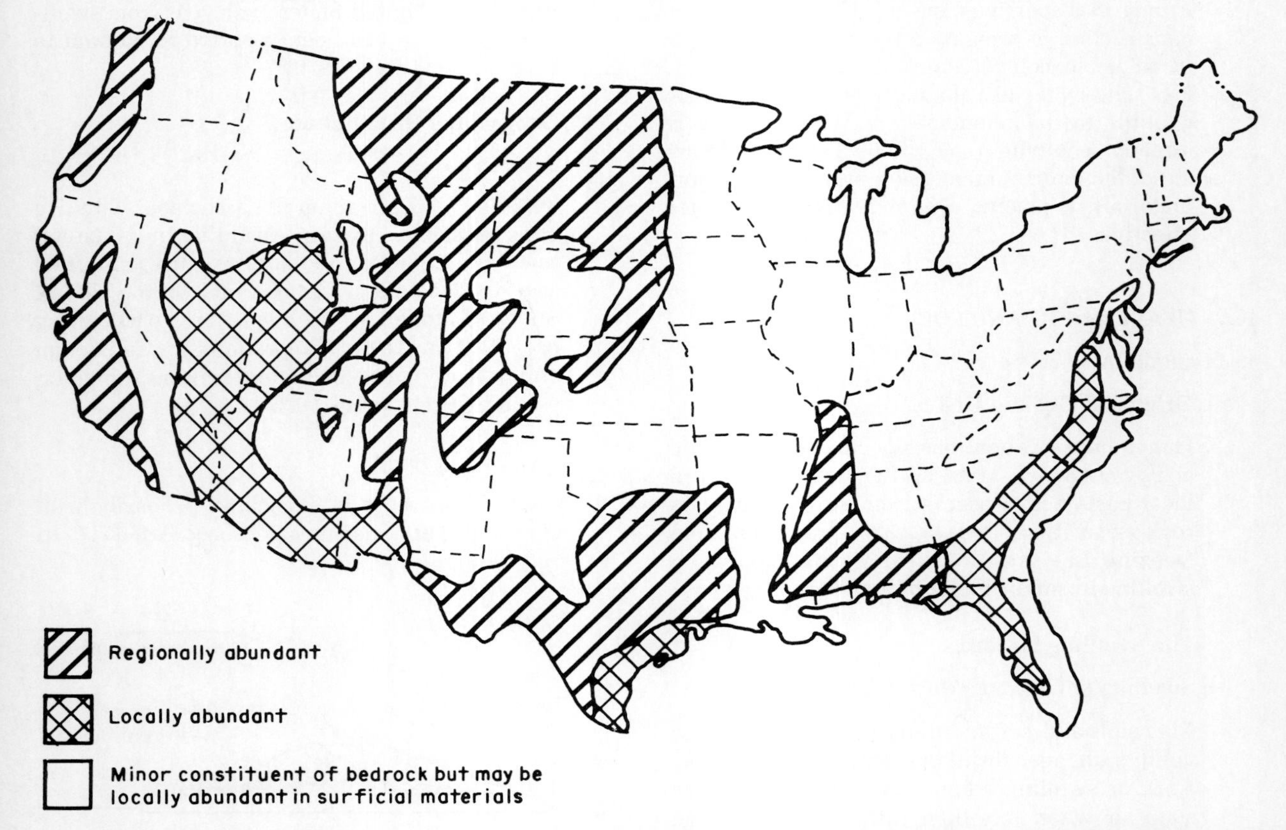


**FIG. 10.37 Distribution of expansive soils in the United States. They are most widespread in areas labeled "regionally abundant," but many locations in these areas will have no expansive soils, while in some unshaded portions of the map, some expansive soils may be found.** [*From Godfrey (1978)*.[61]]

**TABLE 10.4**
**RELATION OF SOIL INDEX PROPERTIES TO PROBABLE VOLUME CHANGES FOR HIGHLY PLASTIC SOILS***

| Data from index tests† | | | Estimation of probable expansion,‡ percent probable total volume change (dry to saturated condition) | Degree of expansion |
|---|---|---|---|---|
| Colloid content, % < 0.001 mm | Plasticity index | Shrinkage limit, % | | |
| >28 | >35 | <11 | >30 | Very high |
| 20–31 | 25–41 | 7–12 | 20–30 | High |
| 13–23 | 15–28 | 10–16 | 10–20 | Medium |
| <15 | <18 | >15 | <10 | Low |

*From USBR (1974).[64]

†All three index tests should be considered in estimating expansive properties.

‡Based on a vertical loading of 1.0 psi as for concrete canal lining. For higher loadings the amount of expansion is reduced, depending on the load and clay characteristics.

*Prediction from Index Tests*

Any surface clay with plasticity index greater than about 25 (CH clays) and a relatively low natural moisture content approaching the plastic limit must be considered as having swell potential. The colloid content (percent minus 0.001 mm), the plasticity index, and shrinkage limit are used by the USBR (1974)[64] as criteria for estimating the probable total volume change from the dry to the saturated condition, as given on Table 10.4.

In the method developed by Seed et al. (1962),[65] expansion was measured as percent swell by placing samples at 100% maximum density and optimum water content in a Standard AASHO compaction mold under a surcharge of 1 psi (0.07 kg/cm²) and then soaking them. A family of curves given as Fig. 10.39 was developed to describe the percent swell potential for various clay types in terms of activity and clay fraction present.

The chart given as Fig. 10.40 provides another basis for estimating potential expansiveness.

Index tests have limitations. They do not always identify the swell potential for all natural deposits, nor provide information on the true amount of heave or pressures that may develop. Because the tests are made on remolded specimens, the natural structure of the material is destroyed and other environmental factors are ignored.

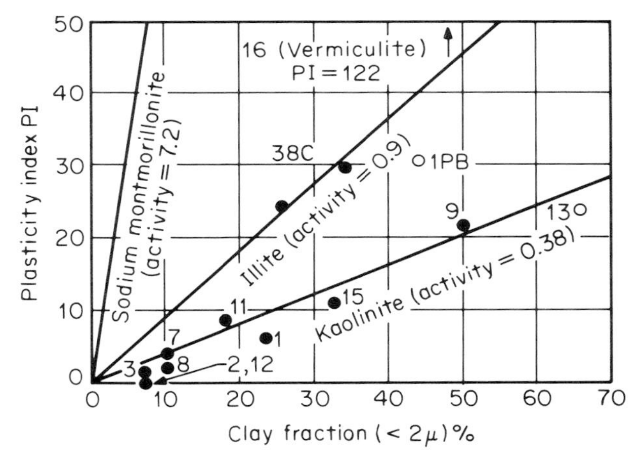


**FIG. 10.38 Clay fraction and plasticity index of natural soils in relation to activity chart of Skempton (1953).[62]** [*After Basu and Arulanandan (1973).[63]*]

*Laboratory Tests*

Tests in the laboratory on undisturbed samples trimmed into the consolidometer will provide an indication of potential heave and the swelling pressures that may develop *(see Art. 3.5.4)*.

## Environmental Factors

*Basic Factors*

WATER-TABLE DEPTH For a soil to develop substantial heave it must lie above the static water

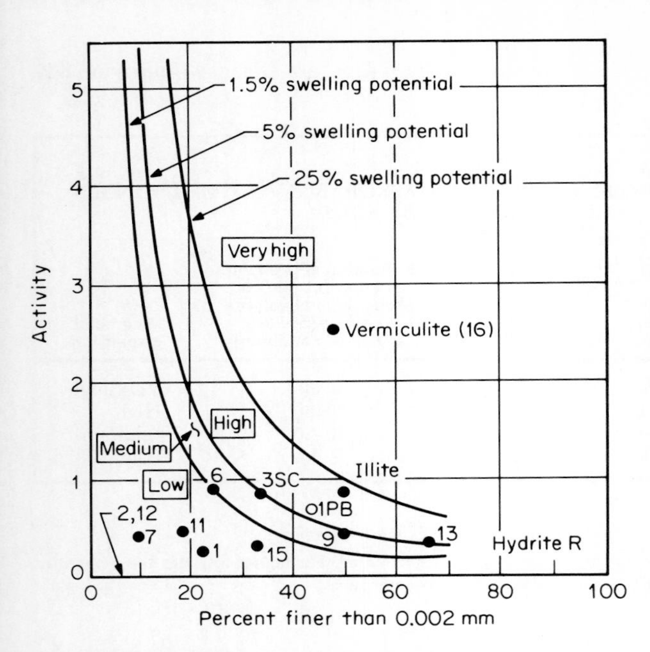


**FIG. 10.39 Classification chart for swelling potential.** [*After Seed et al. (1962)*[65]*; from Basu and Arulanandan (1973).*[63]]

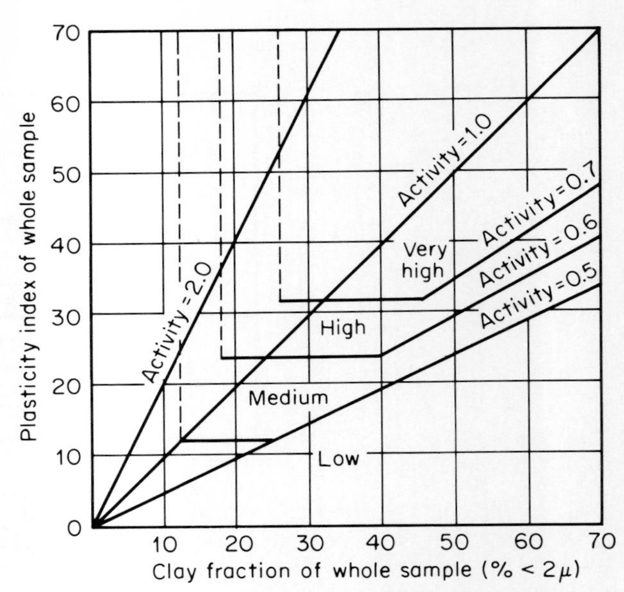


| Potential expansiveness | Inch per foot of soil* |
|---|---|
| Very high | 1.0 |
| High | 0.5 |
| Medium | 0.25 |
| Low | 0 |

*After Van der Merwe (1975).[65b]

**FIG. 10.40 Proposed modified chart for determining expansiveness of soils.** [*From Williams and Donaldson (1980)*[65a]*; after Van der Merwe (1975)*[65b].]

table in a less than saturated state and have moisture available to it. Moisture can originate from capillary action or condensation as well as in the form of free water.

CLIMATE Highly expansive soils are found in climates from hot to cold. The controlling factor is long periods with little or no rainfall, permitting the water table to drop and the soils to decrease in moisture. The soils dry and shrink, large cracks open on the surface, and fissures develop throughout the mass, substantially increasing its permeability. In Texas during the dry season these cracks can extend to depths of 6 m. Rainfall or other moisture then has easy access for infiltration to cause swelling.

*Topography* affects runoff and infiltration. Poorly drained sites have a higher potential for ground heave than slopes.

### *Environmental Changes Cause Surface Movements*

*Decrease in moisture,* causing shrinkage and fissuring, results from:

- Prolonged dry spell or groundwater pumping producing a drop in the water table
- Growth of trees and other vegetation producing moisture loss by transpiration
- Heat from structures such as furnaces, boiler plants, etc., producing drying.

*Increase in moisture,* causing swelling and heave, results from:

- Rainfall and a rise in the water table
- Drilling holes, such as for pile foundations, through a perched water table which permits permeation into a lower clay stratum [ENR (1969)[66]]
- Retarding evaporation by covering the ground with a structure or a pavement
- Thermoosmosis, or the phenomenon by which moisture migrates from a warm zone outside a building area to the cool zone beneath the building

- Condensation from water lines, sewers, storm drains, and canals
- Removal of vegetation which increases susceptibility to fissuring and provides access for water

*Time Factor*

Usually a year or more passes after construction is completed before the effects of heave are apparent, although heave can occur within a few hours when the soil has sudden access to free water as from a broken water main or a clogged drain.

A plot of yearly rainfall and heave as a function of time for a house in the Orange Free State in South Africa is given as Fig. 10.41. It shows almost no heave for the first year, then a heave of 11.6 cm occurring over a period of 4 years, after which movement essentially stops. The long-term effect is the result of the slow increase, due to natural events, of moisture content beneath a covered area.

### 10.6.3 SWELLING IN ROCK MASSES

#### Marine and Clay Shales

*Characteristics*

As described in Art. 6.7.3, montmorillonite is a common constituent of marine and clay shales; therefore, these shales have a high swell potential.

They are found commonly disintegrated and badly broken and microfissured from weathering and expansion of the clay minerals. As a result, their shallower portions consist of a mass of hard fragments in a soil matrix. The hard fragments can vary from a medium-hard rock (see *Fig. 6.91*, Permian shales from Goias, Brazil)

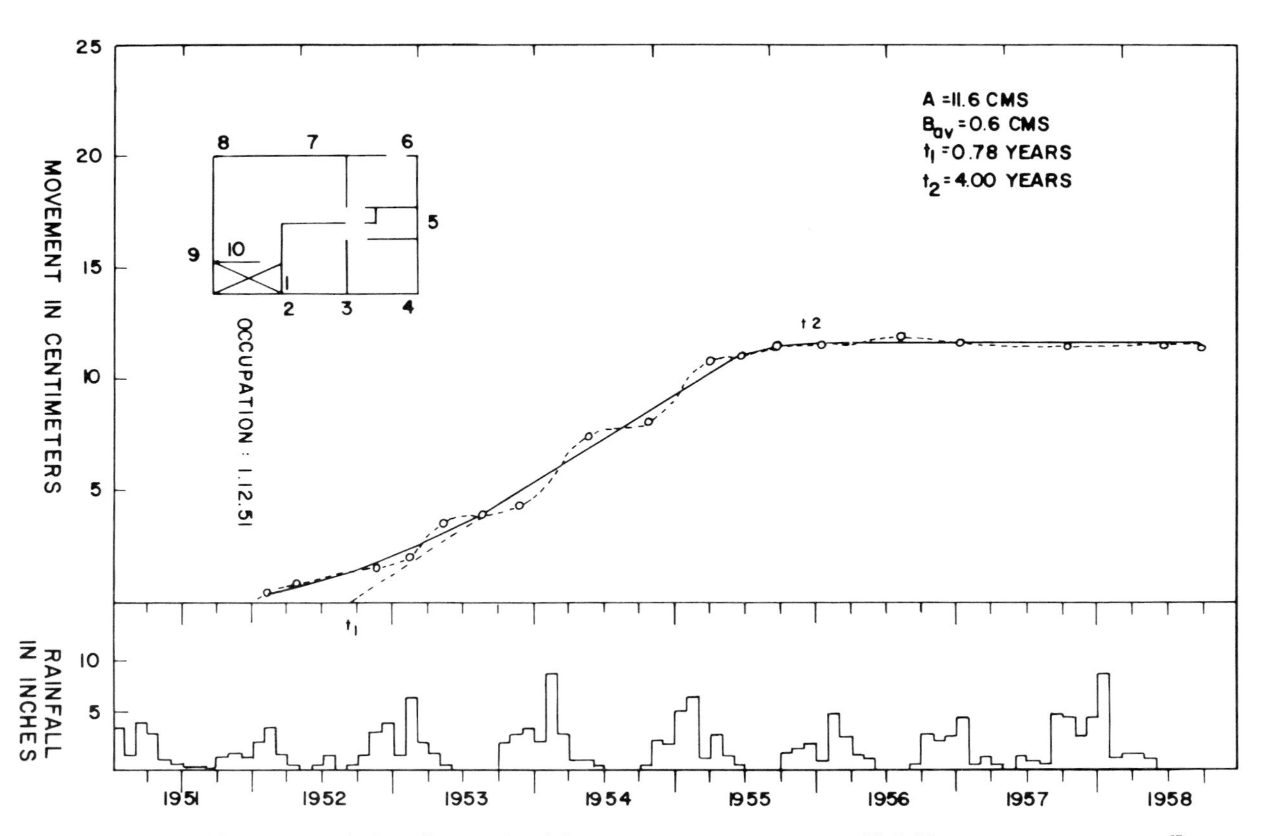


**FIG. 10.41 Typical heave record of single-story brick house in Orange Free State gold fields.** [*From Jennings (1969).*[67]]

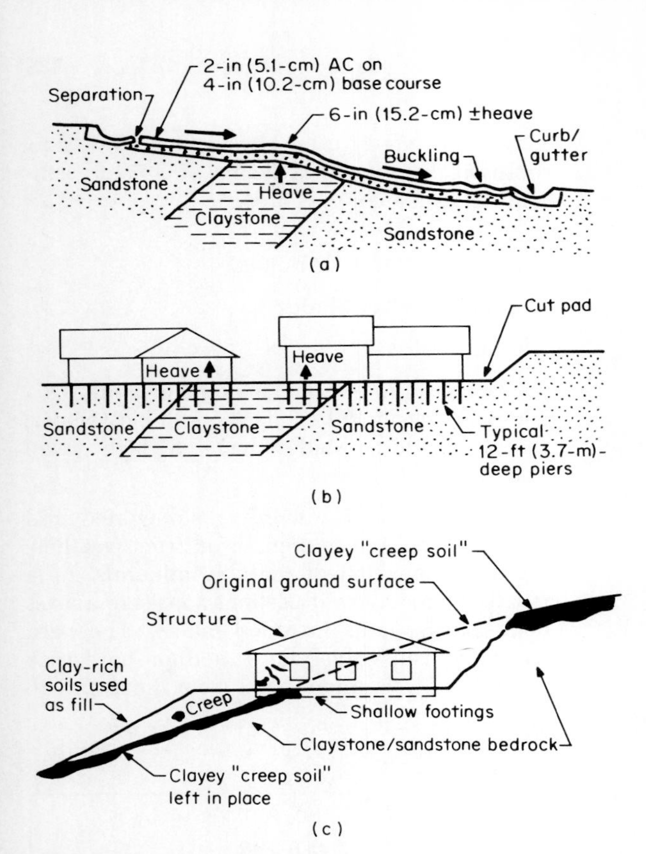


**FIG. 10.42 Problems of heave and creep in the interbedded claystones and sandstones at Menlo Park, California: (*a*) typical pavement damage; (*b*) damage to houses on shallow piers; (*c*) conditions resulting in creep damage.** [*From Meehan et al. (1975).*[68]]

to a very hard indurated clay (*see Plate 5.18*, a clay shale from Santa Catarina, Brazil). Where the weathering is primarily mechanical from the swelling of the clay minerals the disintegrated shale can be found extending downward nearly from the surface, or where chemical weathering has dominated it can be found under a residue of highly expansive clay.

Classification of these materials as either a residual soil (*see Art. 7.2.3*) or a weathered rock (*see Art. 6.7.3*) is difficult because of their properties and characteristics. Clay and marine shales are often described as soils in engineering articles and as rock in most geologic publications. They truly are a transitional material, even referred to at times as claystone.

*Example: Menlo Park, California*

PROJECT A large urban development suffered substantial and costly damage from the expansion of clay shales [Meehan et al. (1975)[68]].

GEOLOGIC CONDITIONS Prior to development, the area was virgin hilly terrain underlain by interbedded sandstones and claystones dipping typically at 40 to 60°. In the lower elevations the claystones are overlain by 1 to 3 m of black clayey residual or colluvial soils. The weathered portions of the claystones are yellow to olive brown in color, becoming olive gray in fresh rock. The fresh claystone (soft rock, with a hardness of 2) has two major joint systems, the master system spaced at about 1 m and the secondary system from 2 to 5 cm. Groundwater is generally about 10 m below the surface. Typical claystone properties are:

- *Index properties:* LL = 70%, PI = 20, $G_s$ = 2.86, $\gamma_t$ = 135 pcf (2.2 g/cm$^3$)
- *Clay mineral:* Montmorillonite
- *Swell pressures:* Measured in the consolidometer, 4 to 9 kg/cm$^2$

THE PROBLEMS Damage from heave to houses and street pavements has been severe, with movements in the order of 10 to 15 cm. The major problem is that of differential movements because of the alternating surface exposure of the dipping beds of sandstone and claystone as illustrated on Fig. 10.42*a* and *b*. Claystone swelling where structures have been placed in cuts typically occurs over several years following construction. It may not be observed for 1 or 2 years, but generally continues to be active for 5 to 7 years after construction. Slopes are unstable, even where gentle. In dry weather the heavy black clays exhibit shrinkage cracks a meter or so deep. During the winter rains, the cracks close and the clay becomes subject to downhill creep, at about ½ in (1.2 cm) per year. Downhill movement of fills has occurred, causing severe distress in structures located partly in cut and partly on fill as shown on Fig. 10.42*c*. Attempted and successful solutions are described in Art. 10.6.4.

### Other Rocks

*Pyritic Shales*

In Kansas City, the Missouri limestone has been quarried underground and the space left from the operations has been used since 1955 for warehousing, manufacturing, office, and laboratory facilities. Underlying the limestone and exposed in the floors of many facilities is a black, pyritic shale of Upper Pennsylvanian age. Sulfide alteration of the pyrite results in swelling, and in recent years as much as 3 to 4 in (8 to 11 cm) of floor heave has occurred, causing severe floor cracking as well as cracking of the mine pillars left to provide roof support [Coveney and Parizek (1977)[69]]. Possible solutions are discussed in Art. 10.6.4.

Heaving from the swelling of a black, pyritic carbonaceous shale is reported to have caused damage to structures in Ottawa, Canada [Grattan-Bellew and Eden (1970)[70]]. In some instances, the concrete of floors placed in direct contact with the shales has turned to "mush" over a period of years. Apparently the pyrite oxidizes to produce sulfuric acid which reacts with calcite in the shale to produce gypsum, the growth of which results in the heave. The acid builds up in the shale to lower the pH to an observed value of 3, leaching the cement from the concrete. The phenomenon does not appear to affect the more deeply embedded footing foundations.

*Gneiss and other metamorphic rocks* may contain seams of montmorillonite which can be troublesome to deep foundations, tunnels, and slopes.

## 10.6.4 TREATMENTS TO PREVENT OR MINIMIZE SWELLING AND HEAVE

### Foundations

*Excavations*

Sections as small as practical should be opened in shales, and water infiltration prevented. The opening should be covered immediately with foundation concrete, cyclopic concrete, or compacted earth. The objective is to minimize the exposure of the shales to weathering, which occurs very rapidly, and is especially important for dams and other large excavations.

*Deep foundations,* which generally are drilled piers, extending below the permanent zone of saturation eliminate the heave hazard. The piers, grade beams, and floors must be protected against uplift from swelling forces.

*Shallow rigid mat* or *"rigid" interconnected continuous footings* may undergo heave as a unit, but they provide protection against differential movements when adequately stiff.

*Other methods, often unsatisfactory,* include:

- Preflooding to permit expansion, then designing to contend with settlements of the softened material and attempting to maintain a balance between swell and consolidation.
- Injection with lime has met with some success, but in highly active materials it may aggravate swelling, since the lime is added with water.
- Excavation of the upper portion of the swelling clays, mixing that portion with lime, which substantially reduces their activity, and then replacing the soil-lime mixture as a compacted fill is a costly, and not always satisfactory, procedure.

### Floors

Floors should consist of a structural slab not in contact with the expansive materials or supported on a free-draining gravel bed which permits breathing. Infiltration of water must be prevented.

### Pavements

Foundations for pavements are prepared by excavating soil to some depth, which depends on clay activity and environmental conditions, and replacing the materials with clean granular soils, or replacing with the same soil compacted on the wet side of optimum, or replacing with the same soil mixed with lime. Protection against surface-water infiltration is provided at the pavement edges, adequate underdrainage is provided, and pavement cracks are sealed.

### Solutions at Menlo Park

*Site Grading*

The procedures cited in Meehan et al. (1975)[68] to correct the problems described in Art. 10.6.3

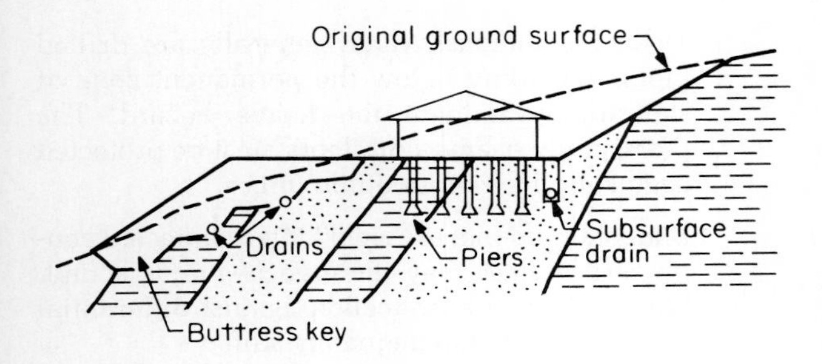


**FIG. 10.43 Site grading and foundation solutions at Menlo Park, California.** [*From Meehan et al. (1975).*[68]]

called for stripping of the surficial soils and keying fills into slopes in a series of steps. Subsurface drains were being provided to prevent water migration into the expansive weathered rock as illustrated on Fig. 10.43.

*Foundation support* was being provided by drilled piers taken to depths of 6 to 10 m, or to the sandstones if shallower—a satisfactory but costly solution for homes.

*Lime injected* into closely spaced holes was not successful in reducing heave, and at times the situation was aggravated because the lime was added with water.

*Pavements* were treated by ripping the subgrade to a depth of about 5 ft, injecting a lime slurry into the loosened claystone, then recompacting the surface before paving or repaving or constructing new streets with "full-depth" asphaltic concrete with typical thicknesses of 5 in laid directly on the subgrade. The success history of the treatments for pavements was about 2 years, when Meehan et al (1975)[68] wrote their article.

### Possible Solutions for Kansas City

For the problems in the pyritic shales in Kansas City *(see Art. 10.6.3)*, several possible solutions appear feasible:

- Immediate coating of the shale after excavation with bitumen or a comparable airtight substance
- Removal of the shales beneath the floor areas and replacement with concrete to some fairly substantial depth, but with consideration given to the phenomenon reported for the shales in Ottawa *(see Art. 10.6.3)*
- Bypass the old mine pillars as roof support with concrete supports founded beyond the active zone of the shale

## REFERENCES


1. Allen, D. R. (1973) "Subsidence, Rebound and Surface Strain Associated with Oil Producing Operations, Long Beach, California," *Geology, Seismicity and Environmental Impact,* Spec. Pub., Assoc. of Engrg. Geol., University Publishers, Los Angeles.
2. Allen, D. R. and Mayuga, M. N. (1969) "The Mechanics of Compaction and Rebound, Wilmington Oil Field, Long Beach, California," *Land Subsidence,* Vol. II, Pub. No. 89, IASH-UNESCO-WHO, pp. 410–423.
3. Davis, G. H., Small, J. B. and Counts, H. B. (1964) "Land Subsidence Related to Decline of Artesian Pressure in the Ocala Limestone at Savannah, Georgia," *Engineering Geology Case Histories No. 4,* The Geological Society of America, Engineering Geology Div., New York, pp. 185–192.
4. ENR (1977*a*) "Houston Land Subsidence Sinks 1970 Bridge Specs," *Engineering News-Record,* June 9, p. 11.
5. Tschebotarioff, G. P. (1973) *Foundations, Retaining and Earth Structures,* 2d ed., McGraw-Hill Book Co., New York.
6. ENR (1977*b*) "USGS Bets on Las Vegas: Sinking Is No Emergency," *Engineering News-Record,* Dec. 1, p. 17.
7. Davidson, E. S. (1970) "Geohydrology and Water

Resources of the Tucson Basin, Arizona," *Open File Report*, Geological Survey, U.S. Dept. Of Interior, Tucson, Arizona, May.

8. Pierce, H. W. (1972) "Geologic Hazards and Land-Use Planning," *Earth Science and Mineral Resources in Arizona*, Vol. 2, No. 3, September, Ariz. Bur. Mines.
9. *Civil Engineering* (1977), "Houston Area Tackles Subsidence," ASCE, March, p. 65.
10. Spencer, G. W. (1977) "The Fight to Keep Houston from Sinking," *Civil Engineering*, ASCE, September, pp. 69–71.
11. Van Siclen, D. C. (1967) "The Houston Fault Problem," Amer. Inst. Prof. Geol., *Proc. 3d Annual Mtg. Texas Section*, Dallas.
12. Castle, R. O. and Youd, T. L. (1972) "Discussion: The Houston Fault Problem," *Bull. Assoc. Engrg. Geol.*, Vol. IX, No. 1.
13. Zeevaert, L. (1972) *Foundation Engineering for Difficult Subsoil Conditions*, Van Nostrand Reinhold Co., New York, pp. 245–281.
14. Zeevaert, L. (1957) "Foundation Design and Behavior of Tower Latino Americano in Mexico City," *Geotechnique*, Vol. VII, No. 3, September.
15. Proctor, R. J. (1973) "Geology and Urban Tunnels—Including a Case History of Los Angeles," *Geology, Seismicity and Environmental Impact*, Spec. Pub. Assoc. Engrg. Geol., University Publishers, Los Angeles, pp. 187–193.
16. Jansen, R. B., Dukleth, G. W., Gordon, B. B., James, L. B. and Shields, C. E. (1967) "Earth Movement at Baldwin Hills Reservoir," *Proc. ASCE, J. Soil Mechs. Found. Engrg. Div.*, Vol. 93, No. SM4, July.
17. *Civil Engineering* (1980) "Subsidence—A Geological Problem with a Political Solution," ASCE, May, pp. 60–63.
18. ENR (1980) "Flood Control Plan Recycles Water," *Engineering News-Record*, April 24, pp. 28–29.
19. *Civil Engineering* (1975) "Venice Is Actually Rising?", ASCE, November, p. 81.
20. *Civil Engineering* (1978) "Ground Surface Subsidence Is a World-Wide Problem," ASCE, June, p. 18.
21. Arnould, M. (1970) "Problems Associated with Underground Caverns in the Paris Region," *Proc. Annual Mtg., Assoc. Engrg. Geol.*
22. Averitt, P. (1967) "Coal Resources of the United States, January, 1967," *USGS Bull. 1275*, U.S. Govt. Printing Office, Washington, D.C.
23. Gray, R. E. And Meyers, J. F. (1970) "Mine Subsidence and Support Methods in the Pittsburgh Area," *Proc. ASCE, J. Soil Mechs. Found. Engrg. Div.*, Vol. 96, No. SM4, July.
24. Mansur, C. I. and Skouby, M. C. (1970) "Mine Grouting to Control Building Settlement," *Proc. ASCE, J. Soil Mechs. Found. Engrg. Div.*, Vol. 96, No. SM2, March.
25. NCB (1963) "Principles of Subsidence Engineering," *Info. Bull. 63/240*, National Coal Board of the United Kingdom, Production Dept.
26. NCB (1966) "Subsidence Engineers Handbook," National Coal Board of the United Kingdom, Production Dept.
27. Mabry, R. E. (1973) "An Evaluation of Mine Subsidence Potential," *New Horizons in Rock Mechanics, Proc. ASCE, 14th Symp. Rock Mechs.*, Univ. Park, Pa., June (1972).
28. Voight, B. and Pariseau, W. (1970) "State of Predictive Art in Subsidence Engineering," *Proc. ASCE, J. Soil Mechs. Found. Engrg. Div.*, Vol. 96.
29. Ko, H. Y. and Gerstle, K. H. (1973) "Constitutive Relations of Coal," *New Horizons in Rock Mechanics, Proc. ASCE, 14th Symp. Rock Mechs.*, Univ. Park, Pa., June (1972), pp. 157–188.
30. Farmer, I. W. (1968) *Engineering Properties of Rocks*, E. & F. N. Spon, London.
31. Greewald, H. P., Howarth, H. C. and Hartmann, I. (1941) "Experiments on Strength of Small Pillars of Coal in the Pittsburgh Bed," U.S. Bureau of Mines, *Report of Investigations 3575*, June.
32. Griffith, W. and Conner, E. T. (1912) "Mining Conditions Under the City of Scranton, Pennsylvania," *Bull. 25, Bureau of Mines*, U.S. Dept. of Interior, Washington, D.C.
33. Legget, R. F. (1972) "Duisburg Harbour Lowered by Coal Mining," *Canadian Geotechnical J.*, No. 9, November.
34. Sowers, G. F. (1975) "Failures in Limestones in Humid Subtropics," *Proc. ASCE, J. Geotech. Eng. Div.*, Vol. 101, No. GT8, August.
35. ENR (1978) "Limestone Cavity Sinks Garage," *Engineering News-Record*, Feb. 9, p. 11.
36. ENR (1969) "Sinkhole Causes Roof Failure," *Engineering News-Record*, Dec. 11, p. 23.
37. White, W. B. (1968) "Speleology," *Encyclopedia of Geomorphology*, Dowden, Hutchinson & Ross Publ., Stroudsburg, Pa., pp. 1036–1039.
38. Corbel, J. (1959) "Erosion en terrain calcaire," *Ann. Geog. 68*, pp. 97–120.
39. Penn. Geol. Survey (1972) *Engineering Charac-*

*teristics of the Rocks of Pennsylvania*, Pennsylvania Geological Survey.

40. Flint, R. F., Longwell, C. R. and Sanders, J. E. (1969) *Physical Geology*, John Wiley & Sons, New York.
41. Terzaghi, K. (1913) "Landforms and Subsurface Drainage in the Gačka Region in Yugoslavia," reprinted in *From Theory to Practice in Soil Mechanics*, John Wiley & Sons, New York (1960).
42. Prokopovich, N. P. (1976) "Some Geologic Factors Determining Land Subsidence," *Bull. Assoc. Engrg. Geol.*, No. 14, December, pp. 75–81.
43. Jennings, J. E. (1966) "Building on Dolomites in the Transvaal," *Trans. South African Inst. of Civ. Engrs.*, January.
44. Foose, R. M. (1953) "Groundwater Behavior in the Hershey Valley," *Bull. Geol. Soc. Amer.*, No. 64.
45. Hunt, R. E. (1973) "Round Rock, Texas New Town: Geologic Problems and Engineering Solutions," *Bull. Assoc. Engrg. Geol.*, Vol. X, No. 3, Summer.
46. Millet, R. A. and Moorhouse, D. C. (1973) "Bedrock Verification Program for Davis-Besse Nuclear Power Station," *Proc. ASCE, Spec. Conf. Structural Design Nuclear Power Plant Facilities*, Chicago, December, pp. 89–113.
47. Clemence, S. P. and Finbarr, A. O. (1980) "Design Considerations and Evaluation Methods for Collapsible Soils," *Proc. ASCE*, Preprint 80-116, April, 22p.
48. Barden, L., McGown, A. and Collins, K. (1973) "The Collapse Mechanism in Partially Saturated Soil," *Engineering Geology*, pp. 49–60.
49. Curtin, G. (1973) "Collapsing Soil and Subsidence," *Geology, Seismicity and Environmental Impact*, Spec. Pub., Assoc. Engrg. Geol., University Publishers, Los Angeles.
50. Bull, W. B. (1964) "Alluvial Fans and Near-surface Subsidence in Western Fresno County, California," *Geological Survey Prof. Paper 437-A*, Washington, D.C.
51. Sultan, H. A. (1969) "Foundation Failures on Collapsing Soils in the Tucson, Arizona Area," *Proc. 2d Intl. Res. and Engrg. Conf. on Expansive Clay Soils*, Texas A & M Univ., College Station, Texas.
52. Brink, A. B. A. and Kantey, B. A. (1961) "Collapsible Grain Structure in Residual Granite Soils in Southern Africa," *Proc. 5th Intl. Conf. Soil Mechs. Found. Engrg.*, Paris, Vol. 1, pp. 611–614.
53. Vargas, M. (1972) "Fundação de Barragems de Terra Sobre Solos Porosos," VIII *Seminario Nacional de Grandes Barragems*, São Paulo, Nov. 27.
54. Hunt, R. E. and Santiago, W. B. (1976) "A função critica do engenheiro geólogo em estudos de implatação de ferrovias," *Proc. $l_o$ Cong. Brasileiro de Geologia de Engenharia*, Rio de Janeiro, August, Vol. 1, pp. 79–98.
55. Zur, A. and Wiseman, G. (1973) "A Study of Collapse Potential of an Undisturbed Loess," *Proc. Intl. Conf. Soil Mechs. and Found. Engrg.*, Moscow, Vol. 2.2, p. 265.
56. Jennings, J. E. and Knight, K. (1975) "A Guide to Construction on or with Materials Exhibiting Additional Settlement Due to 'Collapse' of Grain Structure," *6th Regional Conf. for Africa on Soil Mechs. Found. Engrg.*, September, pp. 99–105.
57. ENR (1980) "Reclaimed Site Pays Off," *Engineering News-Record*, Jan. 3, p. 14.
58. Sherard, J. L., Decker, R. S. and Ryker, N. L. (1972) "Piping in Earth Dams of Dispersive Clay," *Performance of Earth and Earth Supported Structures, Proc. ASCE*, Purdue Univ., Vol. I, Part 1.
59. Sherard, J. L., Dunnigan, L. P., Decker, R. S. and Steele, E. F. (1976) "Pinhole Test for Identifying Dispersive Soils," *Proc. ASCE, J. Geotech. Engrg. Div.*, Vol. 102, No. GT1, January, pp. 69–85.
60. ENR (1976) "Expansive Soils Culprit in Cracking," *Engineering News-Record*, Nov. 4, p. 21.
61. Godfrey, K. A. (1978) "Expansive and Shrinking Soils—Building Design Problems Being Attacked," *Civil Engineering*, ASCE, October, pp. 87–91.
62. Skempton, A. W. (1953) "The Colloidal Activity of Clays," *Proc. 3rd Intl. Conf. Soil Mechs. Found. Engrg.*, Zurich, Vol. I, pp. 57–61.
63. Basu, R. and Arulanandan, K. (1973) "A New Approach to the Identification of Swell Potential for Soils," *Proc. 3rd Intl. Conf. on Expansive Soils*, Haifa, Israel.
64. USBR (1974) *Earth Manual*, U.S. Bureau of Reclamation, Federal Center, Denver, Colo.
65. Seed, H. B., Woodward, R. J. and Lundgren, R. (1962) "Prediction of Swelling Potential for Compacted Clays," *Proc. ASCE, J. Soil Mechs. Found. Engrg. Div.*, Vol. 88, No. SM3, June.

65*a*. Williams, A. A. B. and Donaldson, G. W. (1980) "Building on Expansive Soils in South Africa: 1973–1980," *Proc. 4th Intl. Conf. Expansive Soils*, Denver, Colo., Vol. II, pp. 834–844.

65*b*. Van der Merwe, D. H. (1975) Contribution to Specialty Session B, "Current Theory and Practice for Building on Expansive Clays," *Proc. 6th Reg. Conf. for Africa on Soil Mechs. and Found. Engrg.*, Durban, Vol. 2, pp. 166–167.

66. ENR (1979) "Groundwater Bottled Up to Check Moving Piles," *Engineering News-Record*, Oct. 18, p. 26.

67. Jennings, J. E. (1969) "The Engineering Problems of Expansive Soils," *Proc. 2d Intl. Res. and Engrg. Conf. on Expansive Soils*, Texas A&M Univ., College Station, Texas.

68. Meehan, R. L., Dukes, M. T. and Shires, P. O. (1975) "A Case History of Expansive Clay Stone Damage," *Proc. ASCE, J. Geotech. Engrg. Div.*, Vol. 101, No. GT9, September.

69. Coveney, R. M. and Parizek, E. J. (1977) "Deformation of Mine Floors by Sulfide Alteration," *Bull. Assoc. Engrg. Geol.*, Vol. XIV, No. 3, Summer, pp. 131–156.

70. Grattan-Bellew, P. E. and Eden, W. J. (1975) "Concrete Deterioration and Floor Heave Due to Biogeochemical Weathering of Underlying Shale," *Canadian Geotechnical J.*, Vol. 12, No. 3, August, pp. 372–378, National Research Council of Canada, Ottawa.

## BIBLIOGRAPHY

Braun, W. (1973) "Aquifer Recharge to Lift Venice," *Underground Services*.

Christie, T. L. (1973) "Is Venice Sinking?", *Focus on Environmental Geology*, R. Tank, ed., Oxford University Press, New York, pp. 121–123.

Dahl, H. D. and Choi, D. S. (1975) "Some Case Studies of Mine Subsidence and Its Mathematical Modeling," *Applications of Rock Mechanics, Proc. ASCE, 15th Symp. on Rock Mechs.*, Custer State Park, North Dakota, pp. 1–22.

Davies, W. E. (1958) "Caverns of West Virginia," *West Vir. Geol. Econ. Surv. Pub. 19A*.

Dudley, J. H. (1970) "Review of Collapsing Soils," *Proc. ASCE, J. Soil Mechs. Found. Engrg. Div.*, Vol. 96, No. SM3, May.

ENR (1974) "Texas Puts Land Subsidence Cost at $110 million," *Engineering News-Record*, Oct. 31, p. 18.

Fuqua, W. D. and Richter, R. C. (1960) "Photographic Interpretation as an Aid in Delineating Areas of Shallow Land Subsidence in California," *Manual of Photographic Interpretation*, Amer. Soc. of Photogrammetry, Washington, D.C.

Goodman, R., Korbay, S. and Buchignani, A. (1980) "Evaluation of Collapse Potential over Abandoned Room and Pillar Mines," *Bull. Assoc. Engrg. Geol.*, Vol. XVII, No. 1, Winter, pp. 27–38.

Green, J. P. (1973) "An Approach to Analyzing Multiple Causes of Subsidence," *Geology, Seismicity and Environmental Impact, Spec Pub. Assoc. Engrg. Geol.*, University Publishers, Los Angeles.

Greenfield, R. J. (1979) "Review of the Geophysical Approaches to the Detection of Karst," *Bull. Assoc. Engrg. Geol.*, Vol. XVI, No. 3, Summer, pp. 393–408.

Herak, M. and Stringfield, V. T. (1972) *Karst*, Elsevier Publ. Co., New York.

Holtz, W. G. and Gibbs, H. (1956) "Engineering Properties of Expansive Clays," *Trans. ASCE*, Vol. 120.

Lamoreaux, P. E. (1979) "Remote Sensing Techniques and the Detection of Karst," *Bull. Assoc. Engrg. Geol.*, Vol. XVI, No. 3, Summer, pp. 383–392.

Legget, R. F. (1973) *Cities and Geology*, McGraw-Hill Book Co., New York.

Livneh, M. and Greenstein, J. (1977) "The Use of Index Properties in the Design of Pavements on Loess and Silty Clays," Transportation Research Inst. Pub. No. 77-6, Technion-Israel University, Haifa.

Liszkowski, J. (1975) "The Influence of Karst on Geological Environment in Regional and Urban Planning," *Bull. Assoc. Engrg. Geol.*, December.

Mathewson, C.C., Dobson, B.M., Dyke, L.D. and Lytton, R.L. (1980) "System Interaction of Expansive Soils with Light Foundations," *Bull. Assoc. Engrg. Geol.*, Vol. XVII, No. 2, Spring, pp. 55–94.

Mayuga, M. N. and Allen, D. R. (1973) "Long Beach Subsidence," *Focus on Environmental Geology*, R. Tank, ed., Oxford University Press, New York, p. 347.

Meyer, K. T. and Lytton, A. M. (1966) "Foundation Design in Swelling Clays," paper presented to Texas Section, ASCE, October.

Mitchell, J. K. (1973) "Influences of Mineralogy and Pore Solution Chemistry on the Swelling and Stability of Clays," *Proc. Intl. Conf. on Expansive Soils*, Haifa, Israel.

Poland, J. F. (1973) "Land Subsidence in the Western United States," *Focus on Environmental Geology*, R.

Tank, ed., Oxford University Press, New York, p. 335.

Price, D. G., Malkin, A. B. and Knill, J. L. (1969) "Foundations of Multi-story Blocks on the Coal Measures with Special Reference to Old Mine Workings," *Quarterly, J. Engrg. Geol.*, London, June.

Prokopovich, N. P. (1975) "Past and Future Subsidence Along San Luis Drain, San Joaquin Valley, California," *Bull. Assoc. Engrg. Geol.*, Vol. XII, No. 1.

Soderberg, A. D. (1979) "Expect the Unexpected: Foundations for Dams in Karst," *Bull. Assoc. Engrg. Geol.*, Vol. XVI, No. 3, Summer, pp. 409–426.

Sowers, G. F. (1962) "Shallow Foundations," *Foundation Engineering*, G. A. Leonards, ed., McGraw-Hill Book Co., New York, Chap. 6.

Spanovich, M. (1968) "Construction over Shallow Mines: Two Case Histories," *ASCE Annual and National Mtg. on Structural Engrg.*, Pittsburgh, September, *Preprint 703*.

Turnbull, W. J. (1969) "Expansive Soils—Are We Meeting the Challenge?", *Proc. 2d Intl. Conf. Expansive Clay Soils*, Texas A&M Univ., College Station, Texas.

Van Siclen, D. C. and DeWitt, C. (1972) "Reply: The Houston Fault Problem," *Bull. Assoc. Engrg. Geol.*, Vol. IX, No. 1.

Zeevaert, L. (1953) "Pore Pressure Measurements to Investigate the Main Source of Subsidence in Mexico City," *Proc. 3d Intl. Cong. Soil Mechs. Found. Engrg.*, Zurich, Vol. II, p. 299.

Zeitlen, J. G. (1969) "Some Approaches to Foundation Design for Structures in Expansive Soil Areas," *Proc. 2d Intl. Res. and Engrg. Conf. on Expansive Clay Soils*, Texas A&M Univ., College Station, Texas.

# CHAPTER ELEVEN

# EARTHQUAKES

## 11.1 INTRODUCTION

### 11.1.1 GENERAL

**The Hazard**

*Earthquakes* are the detectable shaking of the earth's surface resulting from seismic waves generated by a sudden release of energy from within the earth. Surface effects can include damage to or destruction of structures; faults and crustal warping, subsidence and liquefaction, and slope failures offshore or onshore; and tsunamis and seiches in water bodies.

*Seismology* is the science of earthquakes and related phenomena [Richter (1958)[1]]. The Chinese began keeping records of earthquakes about 3000 years ago, and the Japanese have kept records from about 1600 A.D. Scientific data, however, were lacking until the first seismographs were built in the late 1800s. Strong-motion data, the modern basis for aseismic design, did not become available until the advent of the accelerograph, the first of which was installed in Long Beach, California, in 1933.

**Engineering Aspects**

*Objectives* of engineering studies are to design structures to resist earthquake forces, which may have a wide variety of characteristics.

*Important elements of earthquake studies* include:

- Geographic distribution and recurrence of events
- Positions as determined by focus and epicenter
- Force as measured by intensity or magnitude
- Attenuation of the force with distance from the focus
- Duration of the force
- Characteristics of the force as measured by (1) amplitude of displacement in terms of the horizontal and vertical acceleration of gravity and (2) its frequency component, i.e., ground motion
- Response characteristics of engineered structures and the ground

*Earthquake damage factors* to be considered include:

- Magnitude, frequency content, and duration of the event
- Proximity to populated areas
- Local geologic conditions
- Local construction practices

### 11.1.2 GEOGRAPHIC DISTRIBUTION

**Worldwide**

*General Distribution*

The relationship between earthquake zones and the tectonic plates *(see Appendix A.3)* is given on Fig. 11.1. It is seen that concentrations are along the boundaries of subducting plates and zones of seafloor spreading, the concept of which is illustrated on Fig. 11.2. The great Precambrian shields of Brazil, Canada, Africa, India, Siberia,

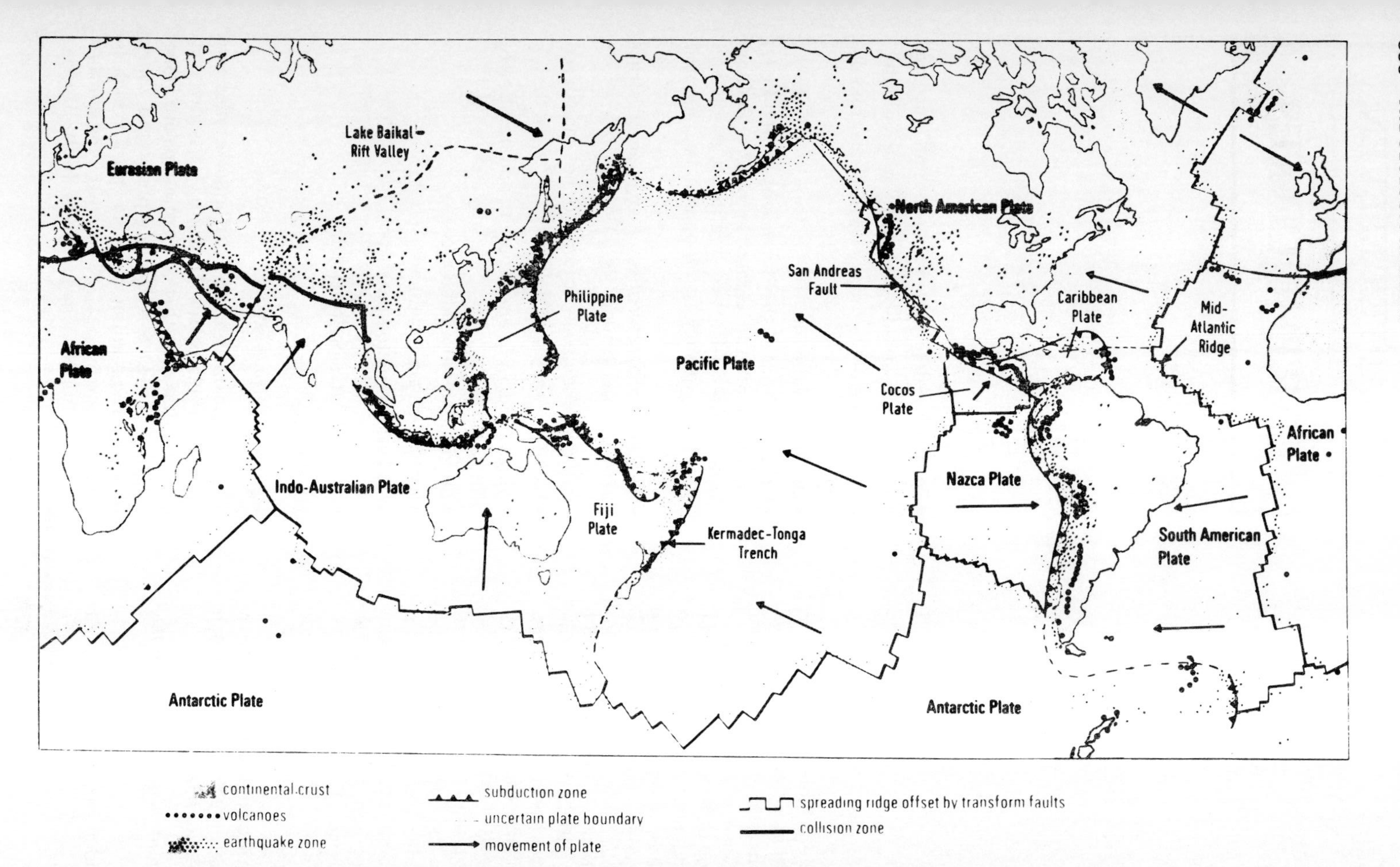


**FIG. 11.1 Worldwide distribution of earthquakes and volcanos in relation to the major tectonic plates.** [*From Bolt et al. (1975).*[2]]

and Australia are generally aseismic although their margins are subjected to earthquake activity.

*Earthquake occurrence* predominates in three major belts: island chains and land masses forming the Pacific Ocean; the mid-Atlantic ridge; and an east-west zone extending from China through northern India, Turkey, Greece, Italy, and western North Africa to Portugal. Countries with a high incidence of damaging earthquakes include Chile, China, Greece, India, Indonesia, Italy, Iran, Japan, Mexico, Morocco, Peru, the Philippines, Rumania, Spain, Turkey, Yugoslavia, the East Indies, and California in the United States.

*Important worldwide events* are summarized on Table 11.1. Selection is based on consideration of the number of deaths; the effect on the land surface in terms of faulting, subsidence, or landslides; and the contribution to the knowledge of seismology and earthquake engineering. The table also serves to illustrate the partial distribution of events by city and country.

*Two General Classes*

PLATE-EDGE EARTHQUAKES The boundaries of the lithospheric plates are defined by the principal global seismic zones in which about 90% of the world's earthquakes occur.

INTRAPLATE EARTHQUAKES Areas far from the plate edges are characterized by fewer and smaller events, but large destructive earthquakes occur from time to time such as those of New Madrid, Charleston, and northern China (Table 11.1). These events and others indicate that the lithospheric plates are not rigid and free of rupture.

### Continental United States

The distribution of the more damaging earthquakes in the continental United States through 1966 is given on Fig. 11.3. The largest events recorded include those of 1811 and 1812 in New Madrid, Missouri, and that of 1886 in Charleston, South Carolina, which had magnitudes estimated to be greater than 8. The New Madrid quakes had tremendous effects on the central lowlands, causing as much as 5 to 7 m of subsidence in the Mississippi valley, forming many large lakes, and were felt from New Orleans to Boston [Guttenberg and Richter (1954)[4]].

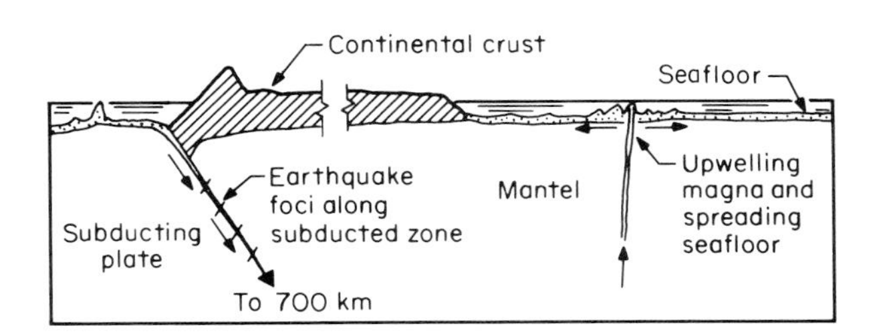


**FIG. 11.2 The concept of a subducting plate and the spreading seafloor.** [*From Deitz and Holden (1970).*[3]]

11.1.3 CHAPTER OBJECTIVES AND SCOPE

### Objectives

The objectives are to summarize and interrelate all of the aspects of earthquakes including causes, characteristics, and surface effects to provide a basis for recognizing the hazard potential, for investigating quakes comprehensively, and for minimizing their consequences.

There is no field in geotechnical engineering in which the state of the art is changing more rapidly, and it is expected that some of the concepts and methodology presented may quickly become obsolete.

### Scope

The earthquake phenomenon is described in terms of its geographic distribution, its location as determined by focus and epicenter, its force as measured by intensity and magnitude, attenuation of the force with distance, its causes and predictability, and ground and structural response to its forces, including effects on the geological environment such as faults and crustal warping, liquefaction and subsidence, slope failures, tsunamis, and seiches. Structural response is treated only briefly to provide a background for the understanding of those elements of earthquake forces that require determination for the analysis and design of structures during investigation. Sources of information pertaining to earthquake occurrence and characteristics are given in Appendix B.

**TABLE 11.1**
**IMPORTANT EARTHQUAKES OF THE WORLD**

| Location | Date | Magnitude* | Importance |
|---|---|---|---|
| Corinth, Greece | 856 | | 45,000 dead. |
| Chihli, China | 1290 | | 100,000 dead. |
| Shensi, China | 1556 | | 830,000 dead. |
| Three Rivers, Quebec | Feb. 5, 1663 | | Strongest quake, NE North America (est. $I$ = X). |
| Calcutta, India | 1737 | | 300,000 dead. |
| Lisbon, Portugal | Nov. 1, 1755 | ≈8.7 | 60,000 dead; caused large tsunamis, and seiches in lakes to distances of 3500 km. |
| Calabria, Italy | Feb. 5, 1788 | | 50,000 dead. |
| New Madrid, Missouri | Dec. 16, 1811 and Feb. 7, 1812 | ≈8 | Three great quakes occurred in this period that affected an area of over 8000 km$^2$ and caused large areas to subside as much as 6 m. |
| Charleston, South Carolina | Aug. 31, 1886 | ≈8 | Large-magnitude shock in low-seismicity area. |
| San Francisco, California | Apr. 18, 1906 | ≈8.3 | About 450 dead, great destruction, especially from fire. San Andreas fault offset for 430 km or more; one of the longest surface ruptures on record. First "microzonation map" prepared by Wood. |
| Colombia-Ecuador border | Jan. 31, 1906 | 8.9 | Largest magnitude known, in addition to one in Japan, 1933. |
| Messina, Italy | Dec. 28, 1908 | 7.5 | 120,000 dead. |
| Kansu, China | Dec. 16, 1920 | 8.5 | 200,000 dead or more. Perhaps the most destructive earthquake. Entire cities destroyed by flows occurring in loess. |
| Tokyo, Japan | Sept. 1, 1923 | 8.2 | The Kwanto earthquake; 143,000 dead, primarily from fires. |
| Attica, New York | Aug. 12, 1929 | 7.0 | Largest recent event in NE United States and eastern Canada. |
| Grand Banks, Newfoundland | Nov. 18, 1929 | ≈7.5 | Undersea quake caused turbidity currents which broke 12 undersea cables in 28 locations. |
| Long Beach, California | Mar. 10, 1933 | 6.3 | Small shock caused much destruction to poorly designed and constructed buildings. Resulted in improved building code legislation regarding schools. First shock recorded on an accelerograph. |
| Quetta, India | May 30, 1935 | 7.5 | 30,000 dead. |
| Western Turkey | Dec. 26, 1939 | 8.0 | 20,000 to 30,000 dead. |
| El Centro, California | May 18, 1940 | 6.5 | The Imperial valley event. The first quake to provide good strong motion data from an accelerograph. Provided the basis for seismic design for many years. |
| Kern Co., California | July 21, 1952 | 7.7 | First major shock in California after earthquake-resistant construction began; showed the value of resistant design. |
| Churchill Co., Nevada | Dec. 16, 1954 | 7.1 | Caused much surface faulting in an area 32 by 96 km, with as much as 6 m vertical and 36 m horizontal displacement. |

*Magnitude of 4.0 is usually given as the threshold of damage.

**TABLE 11.1**
**IMPORTANT EARTHQUAKES OF THE WORLD (*Continued*)**

| Location | Date | Magnitude* | Importance |
|---|---|---|---|
| Mexico City, Mexico | July 28, 1957 | 7.5 | Maximum acceleration only 0.05 to 0.1g in the city, but caused the collapse of multistory buildings because of weak soils. |
| Hebgen Lake, Montana | Aug. 17, 1959 | 7.1 | Triggered large landslide in mountainous region which took 14 lives. |
| Agadir, Morocco | Feb. 29, 1960 | 5.8 | Small shallow-focus event destroyed the poorly constructed city and caused 12,000 deaths of 33,000 population. Previous heavy shock was in 1751. |
| Central Chile | May 21, 1960 | 8.4 | Strong, deep-focus quake was felt over large area, and generated one of the largest tsunamis on record. Much damage in Hilo, Hawaii, from a 10-m-high wave and in Japan from a 4-m-high wave. |
| Skopje, Yugoslavia | July 26, 1963 | 6.0 | 2,000 dead, city 85% destroyed by relatively small shock in poorly constructed area. |
| Anchorage, Alaska | Mar. 27, 1964 | 8.6 | 3-min-long acceleration caused much damage in Anchorage, Valdez, and Seward, particularly from landsliding. |
| Niigata, Japan | June 16, 1964 | 7.5 | City founded on saturated sands suffered much damage from subsidence and liquefaction. Apartment houses overturned. |
| Parkfield, California | June 27, 1966 | 5.6 | Low magnitude, short duration, but high acceleration shock (0.5g) caused little damage. The San Andreas broke along a 37-km length and displacement continued for months after the shock. An accelerograph located virtually on the fault obtained good strong-motion data. |
| Caracas, Venezuela | July 29, 1967 | 6.3 | 277 dead, much damage but occurred selectively. |
| Near Chimbote, Peru | May 31, 1970 | 7.8 | 50,000 dead, including 18,000 from avalanche triggered by the quake. |
| San Fernando, California | Feb. 9, 1971 | 6.5 | Strongest damaging shock in Los Angeles area in 50 years resulted in 65 deaths from collapsing buildings, caused major damage to modern freeway structures. Gave highest accelerations yet recorded: 0.5 to 0.75g with peaks over 1.0g. *(See Art. 11.2.4.)* |
| Managua, Nicaragua | Dec. 23, 1972 | 6.2 | 6000 dead; shallow focal depth of 8 km beneath the city. |
| Guatemala City, Guatemala | Feb. 3, 1976 | 7.9 | 22,000 dead, great damage over 125-km radius. |
| Tangshan, China | July 28, 1976 | 7.8 | 655,000 dead (*Civil Engineering*, November 1977). |
| Mindanao, Philippines | Aug. 18, 1976 | 7.8 | Over 4000 dead in northern provinces. |
| Vrancea, Rumania | Mar. 4, 1977 | 7.2 | 2000 dead; much damage to Bucharest. |
| El Asnam, Algeria | Oct. 10, 1980 | 7.2 | Over 3000 dead. |
| Eboli, Italy | Nov. 30, 1980 | 6.8 | Over 10,000 dead. |

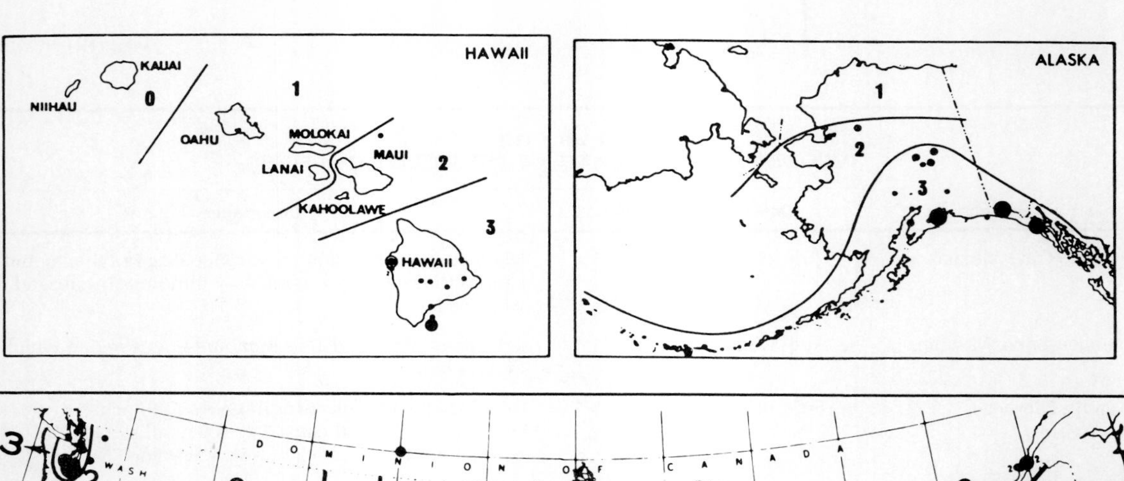


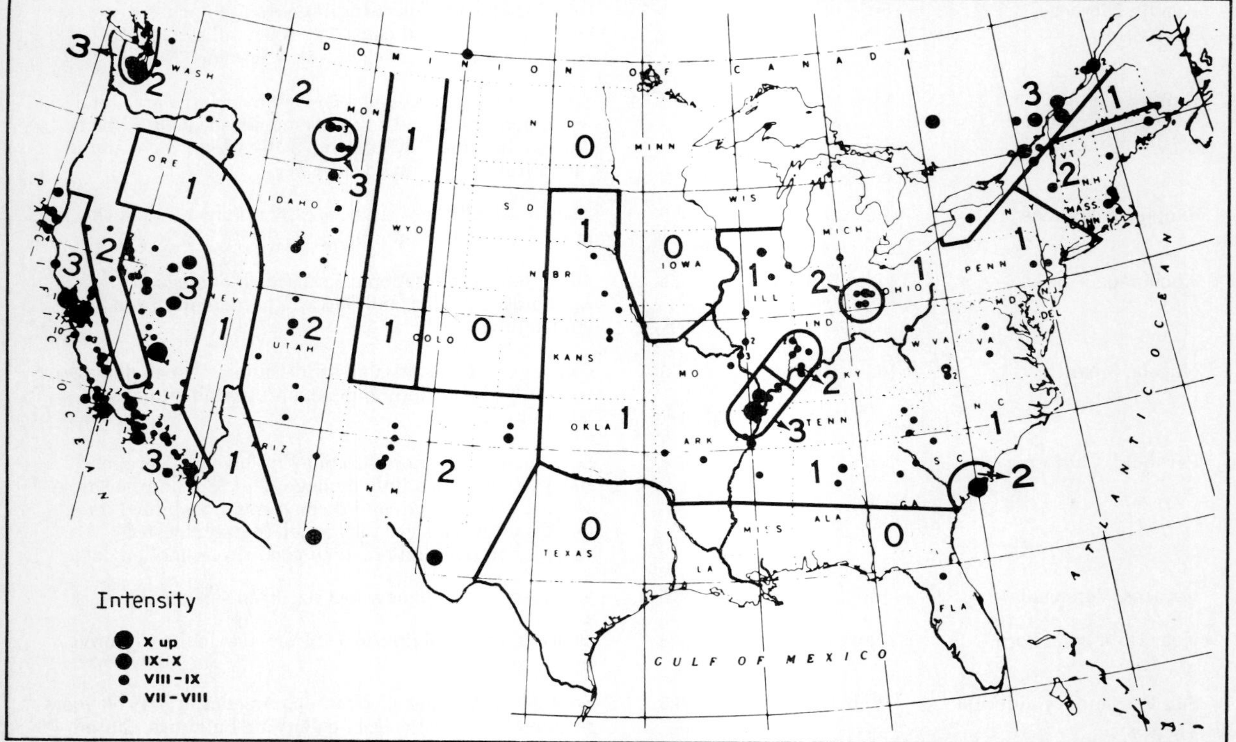


| Zone | Maximum acceleration g, % | M |
|---|---|---|
| 3 (near a great fault) | 50 | 8.5 |
| 3 (not near a great fault) | 33 | 7.0 |
| 2 | 16 | 5.75 |
| 1 | 8 | 4.75 |
| 0 | 4 | 4.25 |

SOURCE: From Housner (1965).[5]

**FIG. 11.3 Seismic risk zones and location of damaging earthquakes in the United States through 1966. See table above for significance of zones. Compare with Fig. 11.15.** *(Map prepared by the U.S. Geological Society.)*

## 11.2 EARTHQUAKE ELEMENTS

### 11.2.1 THE SOURCE

#### Tectonic Earthquakes

*General*

Tectonic earthquakes are those associated with the natural overstress existing in the crust as described in Appendix A. This overstress is evidenced by crustal warping, faults, and residual stresses in rock masses, as well as earthquakes. It is generally accepted that large earthquakes are caused by a rupture in or near the earth's crust that is usually associated with a fault or series of faults, but primarily along one dominant fault termed the *causative fault (see Art. 6.5).*

*Elastic rebound theory* is described by Richter (1958)[1]: "The energy source for tectonic earthquakes is potential energy stored in the crustal rocks during a long growth of strain. When the accompanying elastic stress accumulates beyond the competency of the rocks, there is fracture; the distorted blocks then snap back toward equilibrium, and this produces an earthquake." Earthquakes at very shallow focus may be explained by the elastic rebound theory, but the theory does not explain deep-focus events.

*Plastic Yielding*

At a depth of about 5 km or so, the lithostatic pressure is about equal to the strength of massive rock at the temperature (500°C) and pressure present. Rock deformation under stress, therefore, would be expected to be plastic yielding rather than the brittle rupture needed for a large release of energy. The cause of earthquakes that originate with deeper foci is not clearly understood. The dilantancy theory has been used to explain rupture at substantial depths (Art. 11.2.7).

*Deep-focus earthquakes* appear to be generally associated with tectonic plates and spreading seafloor movements.

#### Volcanic Activity

*Earthquake Relationships*

The worldwide distribution of volcanic activity is shown on Fig. 11.1, where it is seen that volcanos are generally located near plate edges. Large earthquakes were at one time attributed to volcanic activity but there is usually a separation of about 200 km or more between belts of active volcanos and major tectonic activity. The seismic shocks occurring before, during, and after eruptions are referred to by Richter (1958)[1] as *volcanic tremors.*

*The Volcano Hazard*

Eruptions, laval flows, and particles thrown into the atmosphere present the hazardous aspects of volcanic activity. In the last 2000 years, there have been relatively few tremendous and disastrous eruptions. Mt. Vesuvius erupted in 79 AD and destroyed the ancient city of Pompeii. Mont Pelée in Martinique erupted violently in 1902, destroying the city of St. Pierre and leaving but two survivors. Krakatoa in Indonesia literally "blew up" in 1883 in what was probably the largest natural explosion in recorded history.

The most recent notable event occurred with *Mt. St. Helens,* 70 km northeast of Portland, Oregon, which started ejecting steam, ash and gas on March 27, 1980, after a week of intermittent earth shaking. Finally on May 18, 1980, the mountaintop exploded, sending ash and debris some 20 km into the air. Avalanches, debris flows, and huge mudflows, in addition to the blast forces, caused widespread devastation and flooding, and about 60 deaths. A layer of ash was deposited over thousands of square kilometers *(see Art. 7.5.4).*

*Flowing lava* is perhaps the most common cause of destruction. An example is Iceland's first geothermal power plant [ENR (1976)[6]]. As work on the $45 million plant was nearing completion the site was shaken for about a year with tremors, and then fissures opened about 1.6 km away and lava was spewed, threatening the installation. Apparently the lava was contained by dikes erected around the plant area. Many of the 11 deep wells drilled to capture the geothermal energy were severely damaged by subsurface movements.

*Volcanic Hazard Manual* is being prepared by UNESCO (1980).

#### Other Natural Causes

Minor earth shaking over a relatively small surface area occasionally can be attributed to the

| TABLE 11.2<br>ARTIFICIAL RESERVOIRS WITH INDUCED SEISMICITY* | | | | | | |
|---|---|---|---|---|---|---|
| **Location (dam, country)** | **Dam height, m** | **Capacity, $m^3 \times 10^9$** | **Basement geology** | **Date impounded** | **Date of first earthquake** | **Seismic effect** |
| L'Oued Fodda, Algeria | 101 | 0.0002 | Dolomitic marl | 1932 | 1/33 | Felt |
| Hoover, United States | 221 | 38.3 | Granites and Precambrian shales | 1935 | 9/36 | Noticeable ($M$ = 5) |
| Talbingo, Australia | 176 | 0.92 | | | | Seismic ($M$ < 3.5) |
| Hsinfengkiang, China | 105 | 11.5 | Granites | 1959 | | High activity ($M$ = 6.1) |
| Grandval, France | 78 | 0.29 | | 1959–60 | 1961 | MM intensity V in 1963 |
| Monteynard, France | 130 | 0.27 | Limestone | 1962 | 4/63 | $M$ = 4.9 |
| Kariba, Zimbabwe | 128 | 160 | Archean gneiss and Karoo sediments | 1958 | 7/61 | Seismic ($M$ < 6) |
| Vogorno, Switzerland | 230 | 0.08 | | 8/64 | 5/65 | |
| Koyna, India | 103 | 2.78 | Basalt flows of Deccan trap | 1962 | 1963 | Strong ($M$ = 6.5); 177 people killed |
| Benmore, New Zealand | 110 | 2.04 | Greywackes and argillites | 12/64 | 2/65 | Significant ($M$ = 5.0) |
| Kremasta, Greece | 160 | 4.75 | Flysch | 1965 | 12/65 | Strong ($M$ = 6.2); 1 death, 60 injuries |
| Nuzek Tadzik, USSR | 300 | 10.5 | | 1972 (to 100 m) | | Increased activity ($M$ = 4.5) |
| Kurobe, Japan | 186 | | | 1960–69 | | Seismic ($M$ = 4.9) |

*From Bolt et al. (1975).[2]

collapse of mines or caverns, to large slope failures such as avalanches, or to meteorites striking the earth.

## Human-Induced Causes

### *Reservoirs*

Filling reservoirs behind dams, forming lakes in the order of 100 m or more in depth, creates stress changes in the crust which may be of sufficient magnitude over a large area to induce earthquakes, especially where faults are near, or within, the reservoir area. The cause of reservoir-induced earthquakes is not clearly understood but seems to be more closely associated with an increase in pore- and cleft-water pressures in the underlying rocks than from the reservoir weight. Artificial reservoirs associated with seismic activity and some of their characteristics are given on Table 11.2. Of the 52 reservoirs over 100 m in height in the United States (1973), only about 20% caused seismic activity from water impounding [Bolt et al. (1975)[2]]. Over 10,000 shocks have been recorded in the area of Lake Mead behind Hoover Dam ($H$ = 221 m) since its impoundment in 1935, with the largest having an intensity of MM = IV occurring in 1939. The 236-m-high Oroville Dam in California had not caused detectable seismicity within 10 km from the date of its impoundment in 1968 through early 1975. Following a series of small shocks, an event of magnitude $M$ of 5.7 occurred on Aug. 1, 1975.

At Nurek Dam, Tadzhikistan, U.S.S.R., under construction in a seismically active area, the number of yearly events has increased significantly since 1972 when the water level reached 116 m [ENR (1975)[7]]. When completed, Nurek Dam will be the world's highest at 315 m. At the Hsinfengkaing Dam ($H$ = 105 m) 160 km from Canton, China, constructed in an area that had no record of damaging earthquakes, a shock of $M$ = 6.1 occurred 7 months after the reservoir was filled, causing a crack 82 m long in the upper dam structure.

*Accelerographs* (Art. 11.2.3) are used to instrument large dams to monitor reservoir-induced seismicity. The practice is being applied also to dams lower than 100 m where there is substantial risk to the public if failure should occur.

*Deep-Well Withdrawal and Injection*

Faulting and minor tremors occurred in the Wilmington oil field, Long Beach, California, associated with the extraction of oil *(see Art. 10.2.2).*

Pumping waste fluids down a borehole to depths of 4 km below the surface near Denver caused a series of shocks as described in Art. 11.2.7.

*Nuclear Explosions*

Underground nuclear blasts cause readily detectable seismic tremors.

### Focus and Epicenter

*Nomenclature*

The position of the earthquake source is described by the *focus*, or *hypocenter*, which is the location of the source within the earth, as shown on Fig. 11.2, and the *epicenter*, which is the location on the surface directly above the focus. Depending upon geologic conditions, the epicenter may or may not be the location where surface effects are most severe.

Earthquakes are classed on the basis of depth of focus as follows:

- *Normal or shallow:* 0 to 70 km (generally within the earth's crust)
- *Intermediate:* 70 to 300 km
- *Deep:* Greater than 300 km. (None has been recorded greater than about 720 km, and no magnitude greater than 8.6 has been recorded below 300 km.)

*Some Depth Relationships*

Southern California events generally occur at about 5 km depth, whereas Japanese events generally occur at less than 60 km with more than half less than 30 km.

Foci depths for a number of events apparently define the edge of a subducting plate in some locations, as shown on Fig. 11.2.

Focus depth can be significantly related to surface damage. The Agadir event (1960) of magnitude 5.8 had a very shallow focal depth of about 3 km, but since it was essentially beneath the city, its effects were disastrous. A magnitude 5.8 event usually is considered to be moderate, but the shallow focus combined with the very weak construction of the city resulted in extensive damage, although the total area of influence was small. The Chilean event of 1960 with a magnitude of 8.4, however, had a focal depth of about 65 km, and although it was felt over a very large area, there was no extreme damage.

## 11.2.2 SEISMIC WAVES

### Origin

Earthquake occurrence causes an energy release which moves as a *shock front* or *strain pulse* through the earth, which is considered as an elastic medium. The pulse becomes an oscillatory wave in which particles along the travel paths are "excited" and move in orbits repeating cyclically. In a simple two-dimensional diagram, the oscillation is shown as a wave shape with a crest and a trough as given on Fig. 11.4.

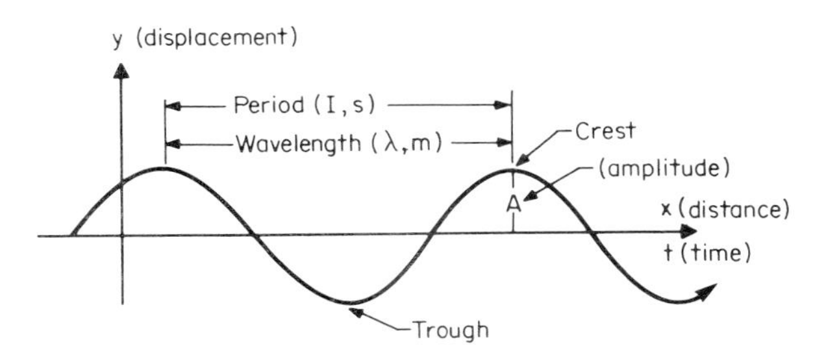


**FIG. 11.4 Characteristics of elastic waves.**

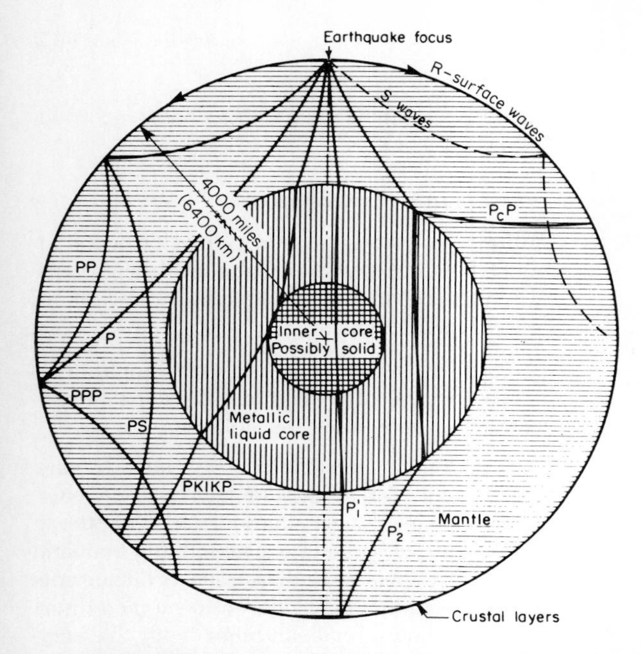


- **PP—longitudinal waves reflected 1 time at earth's surface**
- **P—direct longitudinal wave passed 1 time below continental layers**
- **PPP—longitudinal waves reflected 2 times at earth's surface**
- **PS—longitudinal waves transformed on reflection from the earth's surface**
- **PKIKP—longitudinal waves reflected 1 time from inner surface of core, passing twice through core**
- **$P_cP$—longitudinal waves reflected from outer surface of core**
- **$P_1P_2$—direct longitudinal waves that have traversed the core**
- **S—shear waves**
- **R—surface waves: Rayleigh and Love waves**

**FIG. 11.5 Cross section of the earth showing the paths of the more common earthquake waves.** [*After Hodgson (1964).*[8] *Adapted with permission of Prentice-Hall Inc.*]

If the wave energy produces displacements in the geologic materials within their elastic limits, the materials return to their original volume and shape after the energy wave has passed. The waves are then termed *elastic waves.* They *propagate* through the earth and along its surface as various types of *seismic waves* as shown on Fig. 11.5.

## Wave Types

### *Body Waves*

PRIMARY OR COMPRESSION WAVES (P WAVES) The initial shock applies a compressive force to the materials, causing a wave motion in which the particles move back and forth in the direction of propagation. They are termed as longitudinal, compressional, or primary waves. Normally they are identified as *primary waves (P waves)* because they travel faster than any elastic waves and are the first to arrive at a distant point.

An indication of the compressive effect that seismic waves can have, even at great distances, has been reported by Rainer (1974)[9]. He made a comparison between the incidence of rock bursts occurring in a mine in Bleiberg, Germany, with earthquakes on a worldwide basis and found a strong correlation between rock bursts at Bleiberg and the large shocks that occurred in Agadir (1960), Skopje (1969), San Fernando (1971), and Nicaragua (1972).

*Shear, transverse, or secondary waves (S waves)* are generated where the initial pressure pulse, or the *P* wave which it generates, strikes a free surface or a change in material in a direction other than normal. The shape of the transmitting material is then changed by shear rather than compression. *S* waves can travel only in a solid because their existence depends on the ability of the transmitting medium to resist changes in shape (the shear modulus). (*P* waves can travel in any matter that resists compression or volume change to solid, liquid, or gas.) The *S* waves move at slower velocities than *P* waves and arrive later at a distant point even though they are both generated at the same instant. Both *P* and *S* waves travel through the earth in direct, refracted, or reflected paths, depending upon the material through which they are traveling.

### *Surface Waves or Long Waves (L waves)*

Long waves travel along a free surface of an elastic solid bounded by air or water. They are defined by the motion through which a particle in its path moves as the wave passes.

*Rayleigh (R) waves* cause the particles to move vertically in an elliptical orbit, or to "push up, pull down" in the direction of propagation.

*Love (Q) waves* cause the particles to vibrate transverse to the direction of wave advance, with no vertical displacement.

Both Rayleigh and Love waves move at slower velocities than *P* or *S* waves, and as they travel they disperse into rather long wave trains (long periods). (The comparative arrivals of *P*, *S*, and *L* waves are shown on the seismogram given as Fig. 11.7.)

### Propagation Velocity

The velocity with which the seismic waves travel through the earth is termed the *propagation velocity*, which can be expressed in terms of elastic moduli and material density for *P* waves ($V_p$) and *S* waves ($V_s$) as follows:

$$V_p = \sqrt{\frac{K + (4/3)G}{\rho}} \quad \text{m/s} \qquad (3.20)$$

$$V_s = \sqrt{\frac{G}{\rho}} \quad \text{m/s} \qquad (3.21)$$

where $K$ = dynamic bulk modulus
$G$ = dynamic shear modulus
$\rho$ = material bulk density

NOTE: The elastic moduli are defined on Table 3.34 and Fig. 3.68 and expressions for the dynamic elastic parameters $E_d$, $G_d$, and $K$ and Poisson's ratio given in terms of $V_p$ and $V_s$ are found on Table 3.35. When these velocities are propagated synthetically in the field (*see Art. 2.3.2*) or in the laboratory (*see Art. 3.5.5*), values for $E_d$, $G_d$, and $K$ can be computed.

### Characteristics

Seismic waves (Fig. 11.4) may be described by the quantities of vibratory motion, i.e., amplitude, wavelength, period, and frequency. Amplitude and frequency are the two parameters commonly used to define vibratory motion in earthquake studies.

*Amplitude A* is the displacement from the mean position or one-half the maximum displacement.

*Wavelength* $\lambda$ is the distance between crests.

*Period T* is the time of a complete vibration, or the time a wave travels distance $\lambda$, expressed as:

$$T = \frac{1}{f} = \frac{2\pi}{\omega} \qquad (11.1)$$

where $f$ = frequency and $\omega$ = circular frequency.

*Frequency f* is the number of vibrations per second (or oscillation in terms of cycles per unit of time), given normally in *hertz* (Hz) with units of cycles per second, expressed as:

$$f = \omega/2\pi \qquad (11.2)$$

Ground shaking is felt generally in the ranges from 20 Hz (high frequency) to less than 1 Hz (low frequency of long waves).

*Circular frequency* $\omega$ defines the rate of oscillation in terms of radians per unit of time; $2\pi$ rad is equal to one complete cycle of oscillations.

$$\omega = 2\pi f = 2\pi/T \qquad (11.3)$$

## 11.2.3 GROUND MOTION

### Elements

Ground motion occurs as the seismic waves reach the surface and is described in terms of several elements which are derived from the characteristics of seismic waves, and include displacement, velocity, and acceleration.

*Displacement* y at a given time *t* is a function of position x and time *t* in Fig. 11.4, expressed as:

$$y = A \sin \frac{2\pi}{\lambda}(x + vt) \qquad (11.4)$$

*Velocity* v, termed the *particle* or *vibrational velocity*, is expressed as:

$$v = \lambda/T = f\lambda, \text{ or } v = dy/dt = \dot{y} \qquad (11.5)$$

*Acceleration a* is

$$a = d^2y/dt^2 = \ddot{y} \qquad (11.6)$$

If Eq. 11.5 is substituted for v and $f$ is expressed in terms of the circular frequency $\omega$ from Eq. 11.3, y, v, and $a$ can be expressed as:

*Ground displacement:* $y = A \sin \omega t$ (11.7)

*Ground velocity:* $v = \dot{y} = \omega A \cos \omega t$ (11.8)

*Ground acceleration:* $a = y = -\omega^2 A \sin \omega t = -\omega^2 y$ (11.9)

Acceleration is given in terms of the *acceleration of gravity* g:

$$1g = 32 \text{ ft/s}^2 \text{ or } 980 \text{ cm/s}^2$$
$$\text{or } 1g = 980 \text{ gal}$$

where 1 gal = 1 cm/s$^2$.

### Strong Ground Motion

Strong ground motion refers to the degree of ground shaking produced as the seismic waves reach the surface. It has an effect on structures, and is applied in both the horizontal and vertical modes.

*Characteristics* of ground motion are those of the waveform plus the duration of shaking.

- *Amplitude* is differential along the wavelength and causes differential displacement of structures.
- *Wavelength*, when much larger (long period) than the length of the structure, will cause tall structures to sway, but differential displacement will be negligible.
- *Frequency* causes shaking of the structure as the wave crests pass beneath, and contributes to the acceleration magnitude.
- *Acceleration* is a measure of the force applied to the structure.
- *Duration* is the time of effective strong ground motion, and induces fatigue in structures and pore pressures in soils.

*Examples* illustrate the variability of duration.

- The San Francisco event (1906) started with a relatively small ground motion which increased to a maximum amplitude at the end of about 40 s, stopped for 10 s, then began again more violently for another 25 s.
- Agadir, Morroco (1960) was essentially destroyed in 15 s and 12,000 of a population of 33,000 were killed.
- Guatemala City (1976) was first struck by an event of magnitude 6.5 which lasted for 20 s. The first quake was followed by a number of smaller shocks *(aftershocks)*. Two days later, two more shocks occurred, one with a magnitude of 7.5, and in the following week more than 500 shocks were registered.
- Anchorage, Alaska (1964) experienced a duration in the order of 3 min which resulted in widespread slope failures where they had not occurred before, even though the area had been subjected to strong ground motion on a number of occasions (*see Arts. 9.2.6* and *11.3.4*).

### Detecting and Recording

#### *Seismographs*

*Seismic-wave amplitudes* are detected and recorded on *seismographs*. A seismometer, the detection portion of the instrument, is founded on rock and includes a "steady mass" in the form of a pendulum, which is damped. Seismic waves cause movement of the instrument relative to the pendulum, which remains stationary. In modern instruments the movements are recorded electromechanically and stored on magnetic tape. Operation and recording are continuous.

*Recording stations* have "sets" of instruments, each set having three seismographs. Complete description of ground motion amplitude requires measurements of three components at right angles: the vertical component, and the north-south and east-west components. Instruments designed for different period ranges are also necessary, since no one instrument can cover all of the sensitivity ranges required. In North America it is common to have one set sensitive to periods of 0.2 to 2.0 s, and another set sensitive to periods of 15 to 100 s. The worldwide network of seismograph stations as of March 1977 is shown on Fig. 11.6.

*Seismograms* are the records obtained of ground motion amplitudes; an example is given as Fig. 11.7. The *Richter magnitude* (*see Art. 11.2.4*) is assigned from the maximum amplitude recorded. The distance between the epicenter and the recording seismograph is determined from the arrival times of the *P*, *S*, and *L* waves. By comparing records from several stations, the source of the waves can be located in terms of direction and distance. Epicenters are calculated by NOAA from information received from the worldwide network.

Seismographs are too sensitive to provide information of direct use in seismic design, and a strong earthquake near the normal seismograph will displace the reading off scale or even dam-

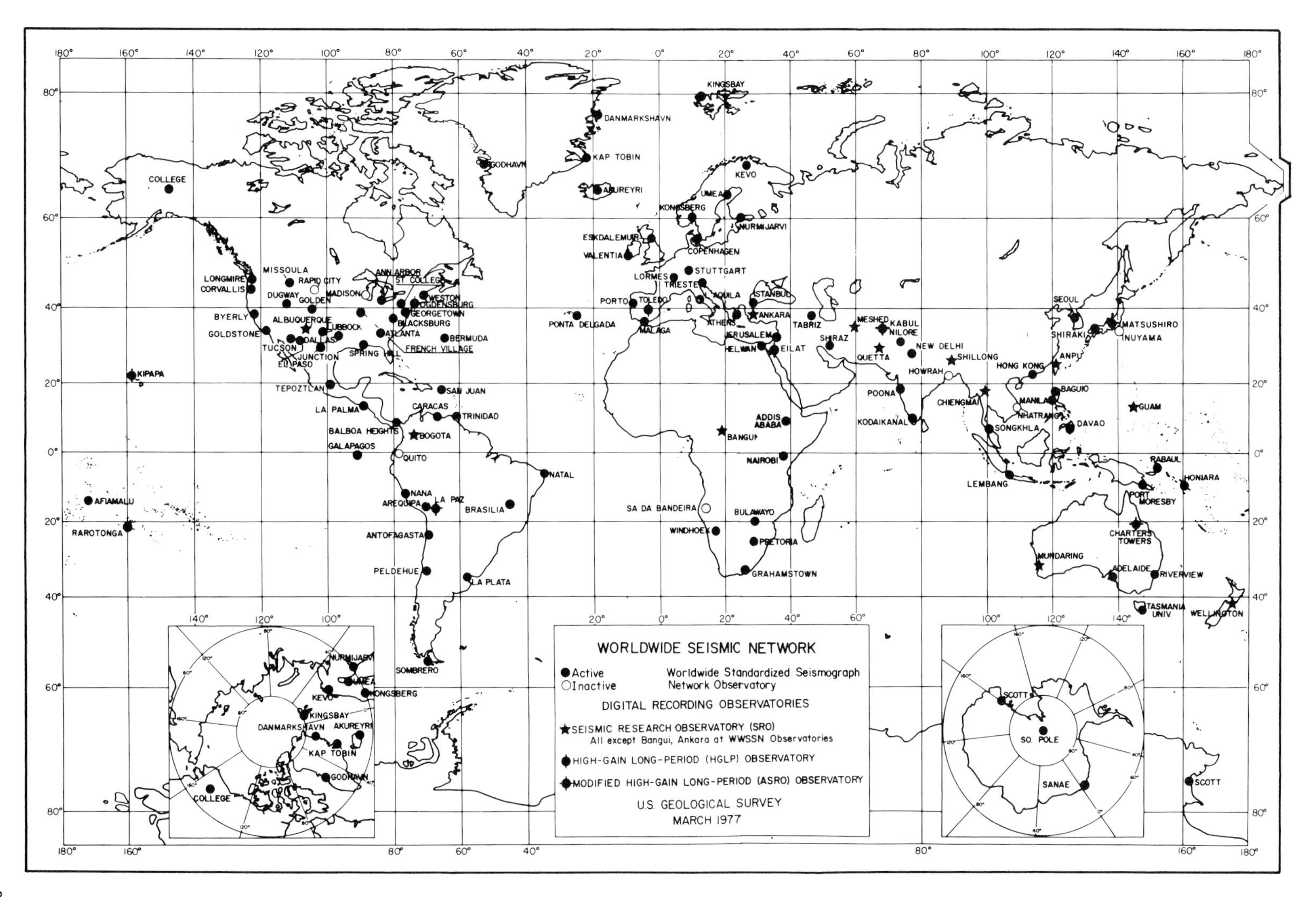


FIG. 11.6 Locations of the Worldwide Standardized Seismograph Network stations and the Global Digital Seismograph Network.

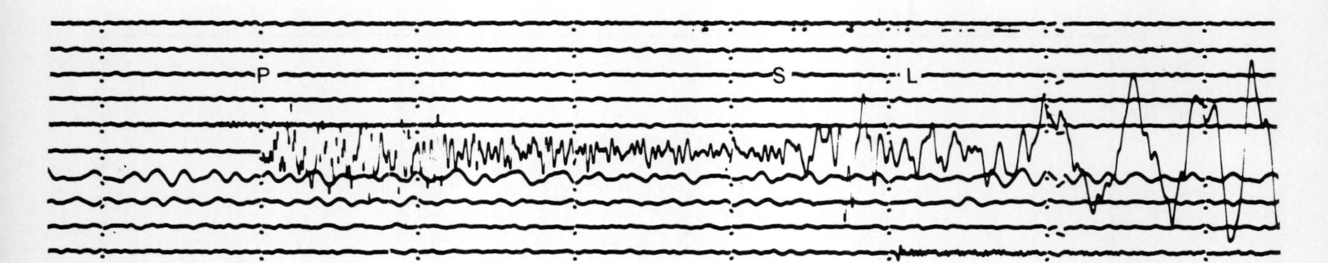


**FIG. 11.7 Seismograph from a long-period, vertical instrument for an event of September 26, 1959, located off the coast of Oregon and recorded in Tucson, Arizona. The long-period waves are well-represented on the record and the *P*, *S*, and *L* waves are noted.** *(Compare with the synthetic records obtained from a refraction seismograph, Fig. 2.25.)* **The *P* and *S* waves have been damped (attenuated) as they traveled through the earth.** [*From Neuman (1966).*[10]]

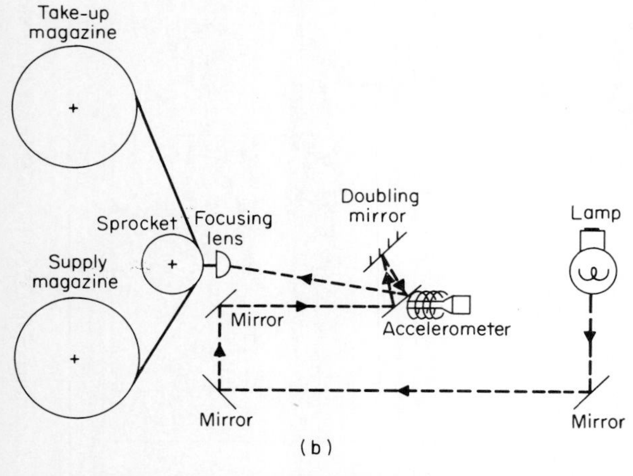


**FIG. 11.8 Photograph of the SMA-1 strong motion accelograph and sketch of the optical system.** *(Courtesy of Kinemetrics, Inc.)*

age the instrument. Instruments are normally located on sound bedrock to eliminate local effects of soils or weakened rock structure, and therefore do not provide information on these materials.

*Accelerographs (Strong-Motion Seismographs) (Accelerometers)*

The *purpose* of the strong-motion seismograph is to provide ground-response criteria in an area of interest for the dynamic design of structures. It measures and records the three components of *absolute ground acceleration* over a period range of 0.1 to 3 or 4 s, or even 10 s.

The *instrument* is illustrated on the photo and sketch given as Fig. 11.8. It does not operate continuously, but rather is designed to begin operating and recording when affected by a small horizontal movement.

*Location* of most accelerographs is on the ground surface, and not necessarily on rock; therefore, data correlations between sites are difficult unless subsurface conditions are known for each.

*Accelerograms* are the records obtained of ground accelerations g as illustrated on Fig. 11.9. Ground motion displacements and velocities are then computed from the acceleration records by integration of Eqs. 11.7 and 11.8.

*Network installations* in the United States as of 1975 are given on Fig. 11.10. Installation on a worldwide basis has been proceeding slowly because the instruments are relatively expensive. To provide adequate coverage of a given

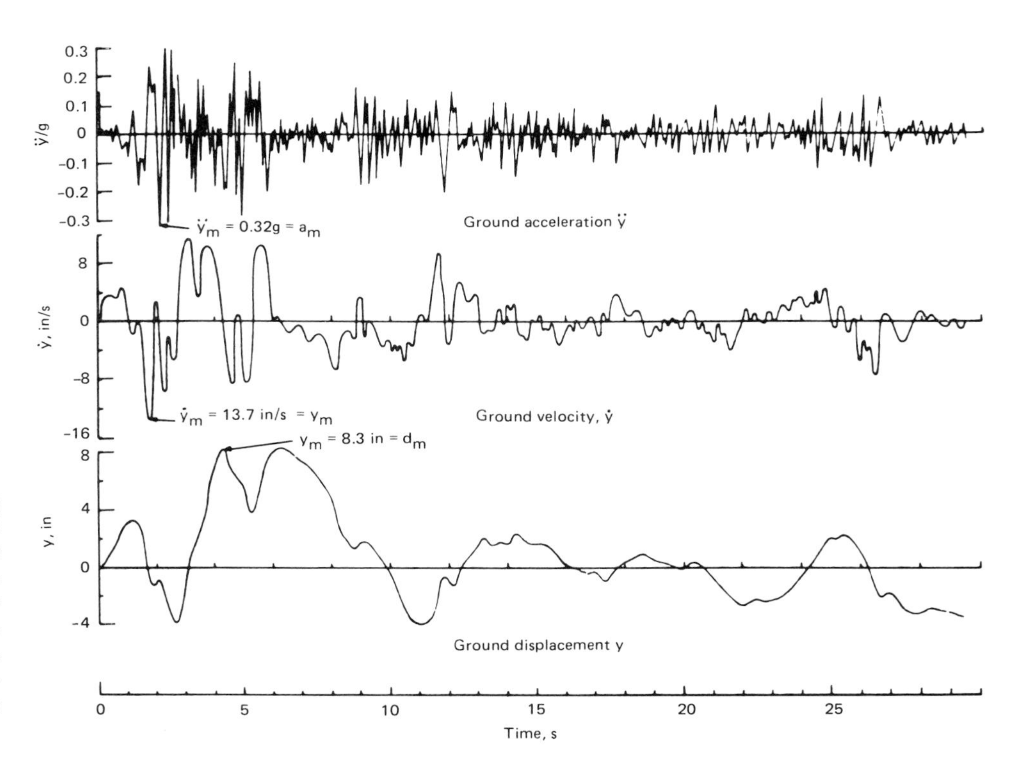


**FIG. 11.9 Strong ground-motion record of the N-S component from El Centro, California, earthquake of May 18, 1940. Ground acceleration is integrated to obtain velocity and displacement.** [*From USAEC (1972).*[11]]

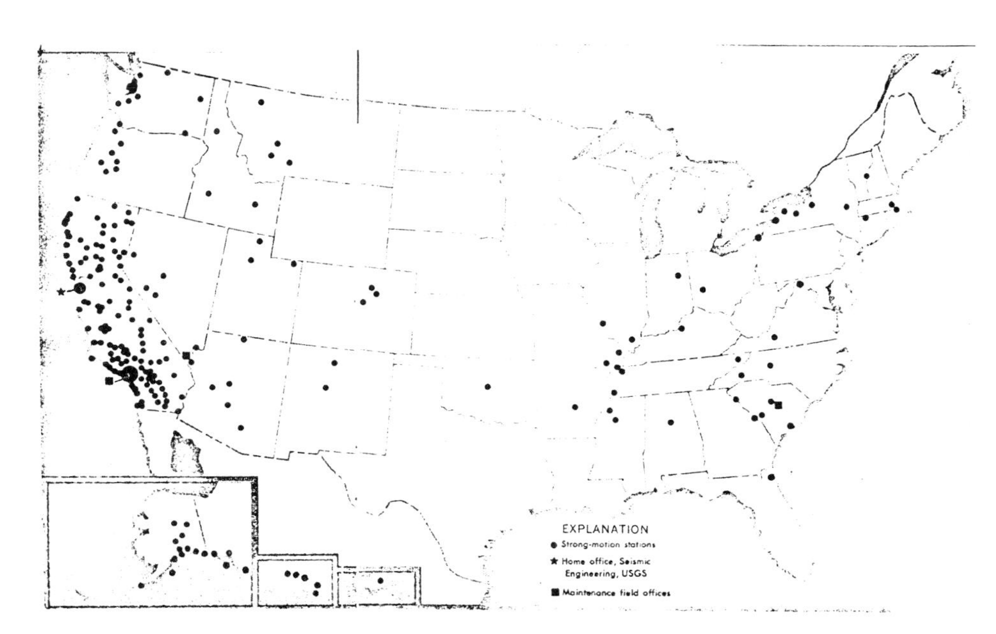


**FIG. 11.10 Strong-motion accelerograph network in the United States as of 1975. Between 1963 and 1975 the network expanded from 70 to 1300 instruments, most of which are in California, where many are located in buildings.** [*From USGS (1975).*[12]]

**TABLE 11.3**
**EARTHQUAKES WITH STRONG RECORDED GROUND ACCELERATION***

| Recording station | Horizontal distance to epicenter (E) or fault (F), km | Component | Maximum acceleration, % gravity | Remarks |
|---|---|---|---|---|
| *May 16, 1968, Japan. Magnitude = 7.9.* | | | | |
| Hachinohe | ca. 200 (E) | N–S<br>E–W | 24<br>24 | Port area. Small shed.<br>Soft soil. |
| *July 21, 1952, Kern County, California. Magnitude = 7.7.* | | | | |
| Taft | 40 (E) | N21°E<br>S69°E | 15<br>18 | In service tunnel between buildings. Alluvium. |
| *October 17, 1966, Peru. Magnitude = 7.5.* | | | | |
| Lima | 200 (E) | N08°E<br>N82°W | 42<br>27 | Small building. Coarse dense gravel and boulders. |
| *April 13, 1949, Puget Sound, Washington. Magnitude = 7.1.* | | | | |
| Olympia | 16 (E) | S04°E<br>S86°W | 16<br>27 | Small building. Filled land at edge of Sound. Focal depth $h$ = 50 km. |
| *December 11, 1967, India. Magnitude = 6.5.* | | | | |
| Koyna Dam | 8 (E) | Along dam axis.<br>Normal dam axis. | 63<br>49 | Dam gallery. |
| *January 21, 1970, Japan. Magnitude = 6.8.* | | | | |
| Hiroo | 18 (E) | E–W<br>N–S | 44<br>≈30 | Focal depth $h$ = 60 km. |
| *December 21, 1964, Eureka, California. Magnitude = 6.6.* | | | | |
| Eureka | 24 (E) | N79°E<br>N11°W | 27<br>17 | Two-story building.<br>Alluvium. |
| *August 6, 1968, Japan. Magnitude = 6.6.* | | | | |
| Uwajima | 11 (E) | Transverse<br>Longitudinal | 44<br>36 | Itashima Bridge site.<br>Soft alluvium.<br>Focal depth $h$ = ≈40 km. |
| *May 18, 1940, El Centro, California. Magnitude = 6.5.* | | | | |
| El Centro | 6 (F) | N–S<br>E–W | 32<br>21 | Two-story heavy reinforced concrete building with massive concrete engine pier. Alluvium. |

*From Bolt et al. (1975).[2]

†Maximum acceleration ≧ 0.15g on 31 records within 42 km of the faulted zone.

## TABLE 11.3
## EARTHQUAKES WITH STRONG RECORDED GROUND ACCELERATION* (*Continued*)

| Recording station | Horizontal distance to epicenter (E) or fault (F), km | Component | Maximum acceleration, % gravity | Remarks |
|---|---|---|---|---|
| *February 9, 1971, San Fernando, California. Magnitude = 6.5.†* | | | | |
| Pacoima Dam Abutment | 3 (F) | S14°W<br>N76°W | 115<br>105 | Small building on rocky spine adjacent to dam abutment. Highly jointed diorite gneiss. |
| Lake Hughes Station No. 12 | 25 (E) | N21°E<br>N69°W | 37<br>28 | Small building. 3-m layer of alluvium over sandstone. |
| Castaic Dam Abutment | 29 (E) | N21°E<br>N69°W | 39<br>32 | Small building. Sandstone. |
| *March 10, 1933, Long Beach, California. Magnitude = 6.3.* | | | | |
| Vernon | 16 (E) | N08°E<br>S82°E | 13<br>15 | Basement of six-story building. Alluvium. |
| *December 23, 1972, Nicaragua. Magnitude = 6.2.* | | | | |
| Managua | 5 (F) | E–W<br>N–S | 39<br>34 | Esso Refinery. Alluvium |
| *June 30, 1941, Santa Barbara, California. Magnitude = 5.9.* | | | | |
| Santa Barbara | 16 km (E) | N45°E<br>S45°E | 24<br>23 | Two-story building. Alluvium. |
| *June 27, 1966, Parkfield, California. Magnitude = 5.6.* | | | | |
| C–H No. 2 | 0.08 (F) | N65°E<br>N25°W | 48<br>Failed | Small building. Alluvium. |
| *September 4, 1972, Bear Valley, California. Magnitude = 4.7.* | | | | |
| Melendy Ranch | 8.5 km (E) | N29°W<br>N61°E | 69<br>47 | ≈19 m from San Andreas fault. Small building. Alluvium. (No damage.) |
| *June 21, 1972, Italy. Magnitude = 4.5.* | | | | |
| Ancona | Ca. 5 (E) | N–S<br>E–W | 61<br>45 | Rock |

area requires locations on various types of geological conditions in a network layout, since the instruments are generally effective only within a radius of about 50 km from an epicenter.

*Existing data* are not extensive in terms of earthquake characteristics or ground conditions. Earthquakes with strong ground motions recorded as of 1975 are summarized on Table 11.3. In the United States, El Centro (1940) remains the severest ground motion recorded (0.32g). The recording station was located 6 km from the fault break. During the Parkfield event (1966) of $M = 5.6$, accelerations of 0.5g were recorded, but the acceleration was a single strong pulse. During the San Fernando quake (1971, $M = 6.5$), an accelerograph registered the highest acceleration ever recorded: in the 0.5 to 0.75g range with peaks over 1.0. The instrument is located on the abutment of the Pacoima Dam, about 8 km south of the epicenter, which was undamaged. Local topography, and the location, are believed to have influenced the very high peak acceleration [Seed et al. (1975)[13]].

*Seismoscopes*

*Purpose* of the seismoscope is to simulate the effects of ground motion on a structure in terms of displacement for a given period, rather than to obtain measurements of ground-motion components. This permits the production of an instrument that is much less costly than an accelerograph.

The *instrument* is set for a particular period and structural damping, and the motion in a single horizontal plane is simulated during an earthquake and recorded on smoked glass. Instruments have been developed that contain a number of elements having various periods and damping to enable recording a number of points on the response spectrum (*see Art. 11.4.4*).

11.2.4 INTENSITY AND MAGNITUDE

**Earthquake Strength Measurements**

Two different scales are commonly used to provide a measure of earthquake strength as related to ground-motion forces at the surface: intensity and magnitude.

- *Intensity* is a qualitative value based on the response of people and objects on the earth's surface. Given as "felt" reports, values and their geographic distribution very much reflect population density.
- *Magnitude* is a quantitative value computed from seismograms.

*Seismic moment* is a parameter suggested in recent years to rate the strength of an earthquake. It includes the rigidity of the rock in which the rupture occurs, times the length of fault face which moves, times the amount of slip. The San Fernando event has been computed to have a seismic moment of nearly $10^{26}$ ergs.

**Intensity (I or MM)**

*Modified Mercalli Scale of Intensity (MM)*

Intensity scales were developed as a basis for cataloging the force of an event for comparison with others, and the change in force with distance (attenuation) from the epicenter. The first intensity scale was developed by DeRossi of Italy and Forel of Switzerland in 1883 (in the literature referred to as RF, followed by a Roman numeral representing the intensity).

The DeRossi-Forel scale was improved by Mercalli in 1902 and modified further in 1931. In 1956 Richter produced the version given on Table 11.4. It correlates ground motion with damage to structures having various degrees of structural quality. (The author has added the columns for approximate comparative peak ground velocity, acceleration, and magnitude.)

*Data Presentation*

*Isoseismal maps* are prepared for affected areas showing zones of equal intensities. The intensity distribution for the Kern County shock of July 21, 1952, has been overlaid on a physiographic diagram of southern California in an attempt to show some relationship with geologic conditions in Fig. 11.11.

*Regional seismicity maps* also are prepared on the basis of intensities (Figs. 11.3 and 11.12) and in some cases developed into *seismic risk* or *seismic hazard maps* (Fig. 11.3). In recent years these maps have been prepared in terms of either the Richter magnitude or effective peak acceleration (Fig. 11.15).

## TABLE 11.4
## MODIFIED MERCALLI SCALE, 1956 VERSION*

| M§ | Intensity | Effects | v,† cm/s | g‡ |
|---|---|---|---|---|
| | I. | Not felt. Marginal and long-period effects of large earthquakes (for details see text). | | |
| 3 | II. | Felt by persons at rest, on upper floors, or favorably placed. | | |
| | III. | Felt indoors. Hanging objects swing. Vibration like passing of light trucks. Duration estimated. May not be recognized as an earthquake. | | 0.0035–0.007 |
| 4 | IV. | Hanging objects swing. Vibration like passing of heavy trucks; or sensation of a jolt like a heavy ball striking the walls. Standing motor cars rock. Windows, dishes, doors rattle. Glasses clink. Crockery clashes. In the upper range of IV wooden walls and frame creak. | | 0.007–0.015 |
| | V. | Felt outdoors; direction estimated. Sleepers wakened. Liquids disturbed, some spilled. Small unstable objects displaced or upset. Doors swing, close, open. Shutters, pictures move. Pendulum clocks stop, start, change rate. | 1–3 | 0.015–0.035 |
| 5 | VI. | Felt by all. Many frightened and run outdoors. Persons walk unsteadily. Windows, dishes, glassware broken. Knickknacks, books, etc., off shelves. Pictures off walls. Furniture moved or overturned. Weak plaster and masonry D cracked. Small bells ring (church, school). Trees, bushes shaken (visibly, or heard to rustle—CFR). | 3–7 | 0.035–0.07 |
| 6 | VII. | Difficult to stand. Noticed by drivers of motor cars. Hanging objects quiver. Furniture broken. Damage to masonry D, including cracks. Weak chimneys broken at roof line. Fall of plaster, loose bricks, stones, tiles, cornices (also unbraced parapets and architectural ornaments—CFR). Some cracks in masonry C. Waves on ponds; water turbid with mud. Small slides and caving in along sand or gravel banks. Large bells ring. Concrete irrigation ditches damaged. | 7–20 | 0.07–0.15 |
| | VIII. | Steering of motor cars affected. Damage to masonry C; partial collapse. Some damage to masonry B; none to masonry A. Fall of stucco and some masonry walls. Twisting, fall of chimneys, factory stacks, monuments, towers, elevated tanks. Frame houses moved on foundations if not bolted down; loose panel walls thrown out. Decayed piling broken off. Branches broken from trees. Changes in flow or temperature of springs and wells. Cracks in wet ground and on steep slopes. | 20–60 | 0.15–0.35 |
| 7 | IX. | General panic. Masonry D destroyed; masonry C heavily damaged, sometimes with complete collapse; masonry B seriously damaged. (General damage to foundations—CFR.) Frame structures, if not bolted, shifted off foundations. Frames racked. Serious damage to reservoirs. Underground pipes broken. Conspicuous cracks in ground. In alluviated areas sand and mud ejected, earthquake fountains, sand craters. | 60–200 | 0.35–0.7 |
| 8 | X. | Most masonry and frame structures destroyed with their foundations. Some well-built wooden structures and bridges destroyed. Serious damage to dams, dikes, embankments. Large landslides. Water thrown on banks of canals, rivers, lakes, etc. Sand and mud shifted horizontally on beaches and flat land. Rails bent slightly. | 200–500 | 0.7–1.2 |
| | XI. | Rails bent greatly. Underground pipelines completely out of service. | | >1.2 |
| | XII. | Damage nearly total. Large rock masses displaced. Lines of sight and level distorted. Objects thrown into the air. | From Fig. 11.14 | |

NOTE: *Masonry A, B, C, D.* To avoid ambiguity of language, the quality of masonry, brick or otherwise, is specified by the following lettering (which has no connection with the conventional Class A, B, C construction).

- *Masonry A:* Good workmanship, mortar, and design; reinforced, especially laterally, and bound together by using steel, concrete, etc.; designed to resist lateral forces.
- *Masonry B:* Good workmanship and mortar; reinforced, but not designed to resist lateral forces.
- *Masonry C:* Ordinary workmanship and mortar; no extreme weaknesses such as non-tied-in corners, but masonry is neither reinforced nor designed against horizontal forces.
- *Masonry D:* Weak materials, such as adobe; poor mortar; low standards of workmanship; weak horizontally.

*From Richter (1958).[1] Adapted with permission of W. H. Freeman and Company.

†Average peak ground velocity, cm/s.

‡Average peak acceleration (away from source).

§Magnitude correlation.

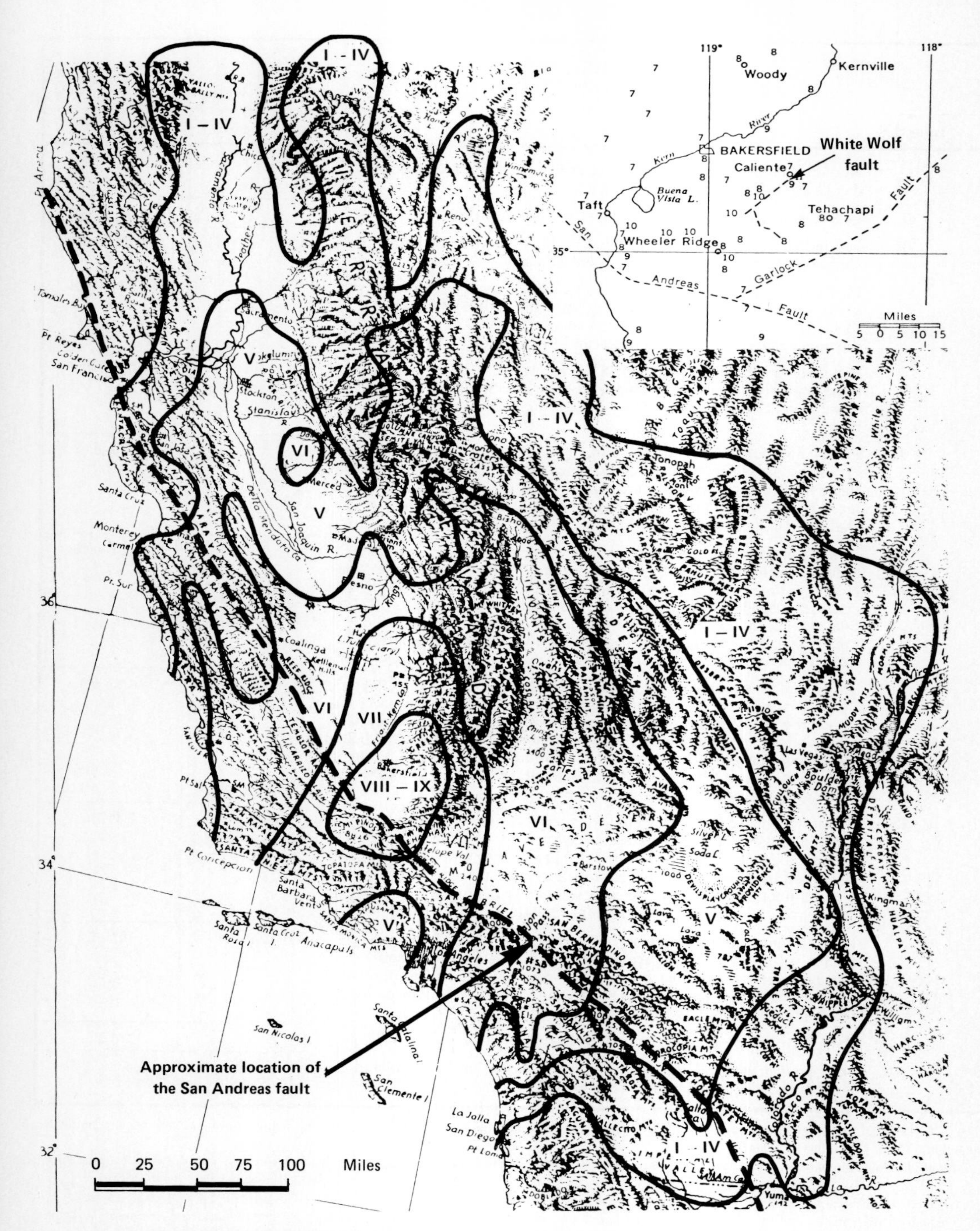


**FIG. 11.11 Intensity distribution of the July 21, 1952, earthquake in Kern County, California, which occurred along the "inactive" White Wolf fault.** [*After Murphy and Cloud (1952).*[14]] **The map has been overlaid onto the physiographic diagram of southern California for comparison with geologic conditions as revealed by physiography.** [*Physiographic diagram from Raisz (1946).*[14a]]

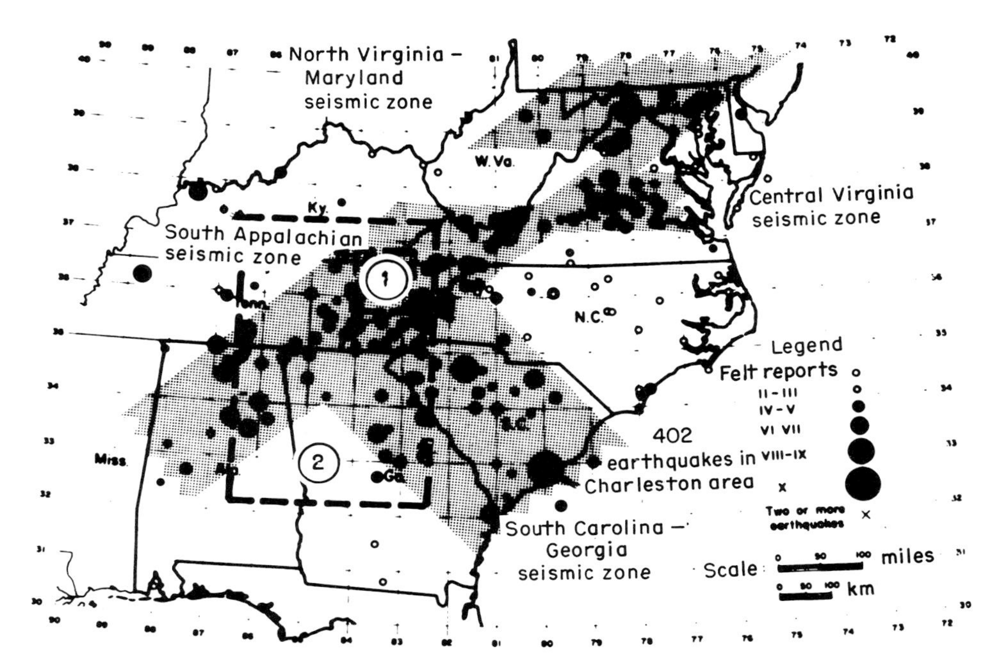


**FIG. 11.12 Seismicity (1754–1971) and earthquake zone map for the southeastern United States.** [*From Bollinger (1973).*[15]] **Epicenters shown by open and solid circles; zones shown by stippling.** ***(For area 1, see the LANDSAT image in Fig. 6.39; for area 2, see the geology map coverage in Fig. 2.2.)***

## Magnitude (M)

### *The Richter Scale*

The concept of magnitude was developed in 1935 by C. F. Richter for defining the *total energy* of seismic waves radiated from the focus based on instrumental data for shallow earthquakes in southern California. He defined the magnitude of local earthquakes $M_L$ as "the logarithm to the base 10 of the maximum seismic wave *amplitude* (in thousands of a millimeter) recorded with a standard seismograph at a distance of 100 km from the earthquake epicenter." Relationships have been developed subsequently to account for distance variations between the recording instrument and the focus, as well as the difference in wave trains between deep and shallow-focus shocks [see, for example, Bolt (1978)[16]].

Amplitudes vary enormously among different earthquakes. An increase in one magnitude step has been found to correlate with an increase of 30 times the energy released as seismic waves [Bolt et al. (1975)[2]]. An earthquake of magnitude 8.0, for example, releases almost 1 million times the energy of one of magnitude 4.0, hence the necessity for a logarithmic scale.

### *Significance of Magnitude*

The largest quakes have had a magnitude of 8.9 (Table 11.1). In general, magnitudes greater than 5.0 generate ground motions sufficiently severe to cause significant damage to poorly designed and constructed structures, and $M = 4$ is generally considered as the damage threshold.

Magnitude is not, however, a measure of the damage that may be caused by an earthquake, the effect of which is influenced by many variables. These include natural conditions of geology (soil type, rock depth and structure, water-table depth), focal depth, epicentral distance, shaking duration, population density, and construction quality.

### *Correlations with Magnitude*

Empirical correlations between $M$, $I_o$, and g are given on Fig. 11.13, where $I_o$ signifies epicentral intensity. Near the source there is no strong correlation with g [Bolt (1978)[16]]. Another correlation between $I_o$ and g is given on Fig. 11.14.

Acceleration also has been related to magnitude by the expression given by Esteva and Rosenblueth (1969)[18] as follows:

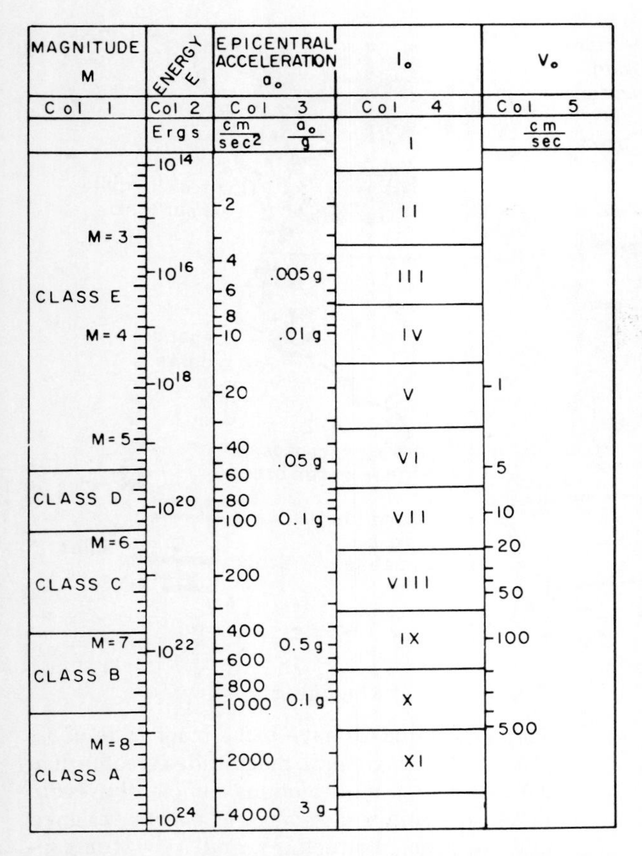


**FIG. 11.13 A summary of rough relationships between magnitude, energy, and epicentral acceleration and between acceleration, intensity, and ground velocity. Approximations are for an order of magnitude.** [*From Lomnitz and Rosenblueth (1976).*[17] *Reprinted with permission of Elsevier Scientific Publishing Company.*]

$$g = 2000e^{0.8M}R^{-2} \qquad (11.10)$$

where $R$ = focal distance and g is in cm/s² per 1000 cm/s².

### Seismic Risk Maps

An early seismic risk map is given on Fig. 11.3 with correlations between zones, maximum accelerations, and magnitude $M$. Zone 3 was considered as having high risk of damaging earthquakes; zone 2, moderate risk; zone 1, low risk; and zone 0, essentially no risk. It was subsequently updated by Algermissen and others in 1969 and the boundary lines changed to incorporate new data.

A recent seismic risk map for the United States is given as Fig. 11.15. It presents contours of effective peak rock acceleration in terms of probability of occurrence in a 50-year period.

## 11.2.5 ATTENUATION

### Description

Attenuation is the decay or dissipation of energy or intensity of shaking with distance from the source, occurring as the seismic waves travel through the earth, and results in the *site intensity* of *rock excitation*.

The epicentral area extends for some distance about the epicenter, in which there is no attenuation, then with increasing distance there is wide regional variation in intensity distribution. It is affected by geology, topography, and length of fault rupture. Variations are illustrated by isoseismal maps such as Figs. 11.11, 11.16, or 11.28; or by seismicity maps such as Fig. 11.12, which are used to develop attenuation relationships. Figure 11.16 presents a comparison of intensity distributions from two earthquakes of different magnitudes but with fairly close epicenters.

### Estimations

*Relationships*

General attenuation law is given in terms of intensity by [Christian et al. (1978)[21]]:

$$I_s = C_1 + C_2 I_o - C_3 \ln (R + C) \qquad (11.11)$$

where $I_s$ = site intensity
$I_o$ = epicenter intensity
$R$ = focal distance, km
$C, C_1, C_2, C_3$ = empirical constants
$\ln (R + C)$ = an error term appended in statistical analysis because of the uncertainty of attenuation laws and to account for standard deviation within and outside the epicentral area

Relationships have been developed for a number of regions relating focal distance $R$ in km to the magnitude $M$ and site intensity $I_s$. Esteva and Rosenblueth (1969)[18] proposed expressions

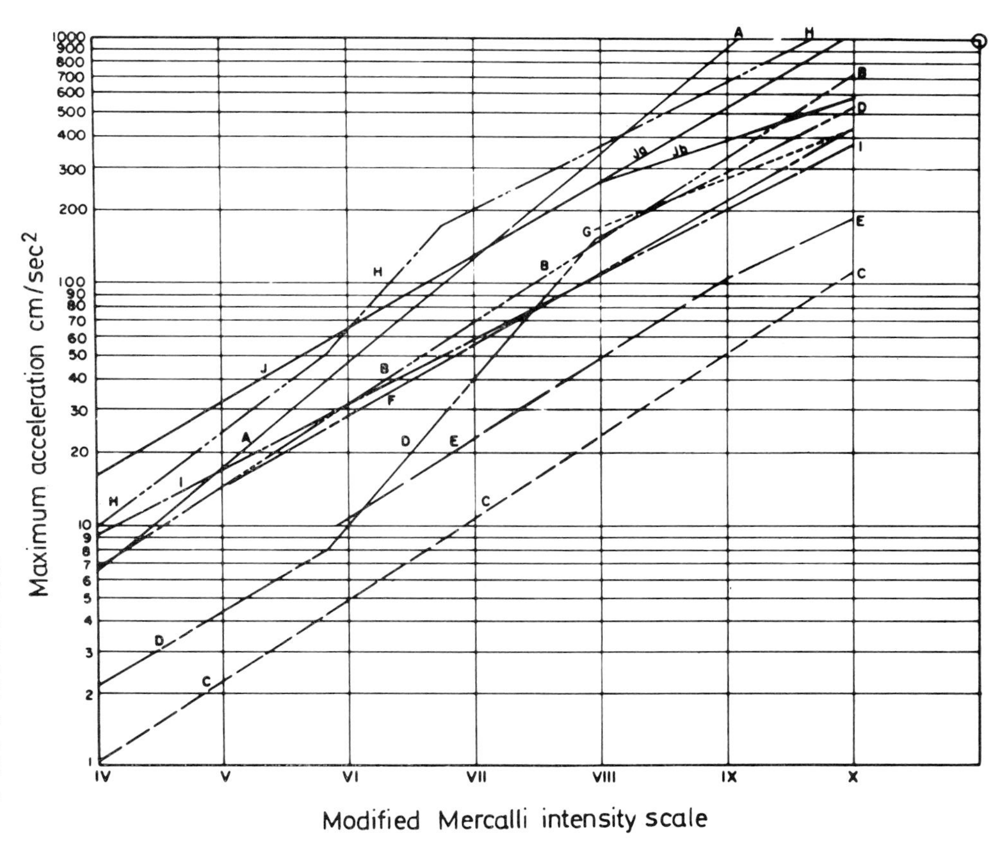


**FIG. 11.14 Empirical correlations of MM intensity with peak ground acceleration measured in earthquakes as determined by various investigators. ○ represents the San Fernando event of February 1971.** [*From Bolt et al. (1975).*[2]]

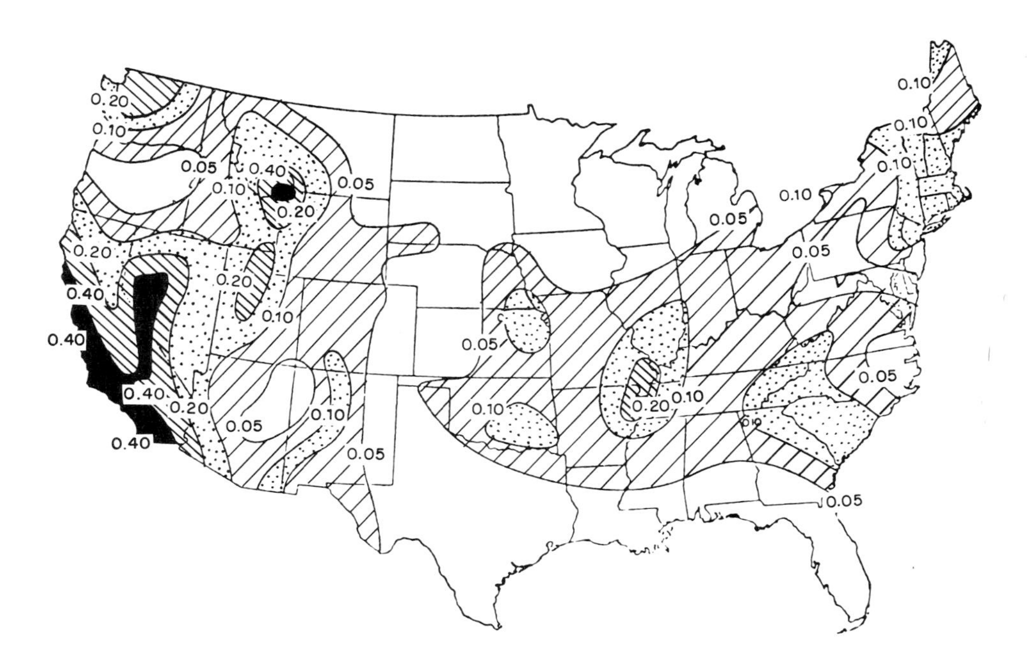


**FIG. 11.15 Seismic risk map of the continental United States prepared in 1976. Contours, given in decimal fractions of gravity, indicate effective peak acceleration in rock with a 90% probability of not being exceeded in 50 years.** [*From Applied Technology Council (1977),*[19] *as presented by Shah and Benjamin (1977).*[20]]

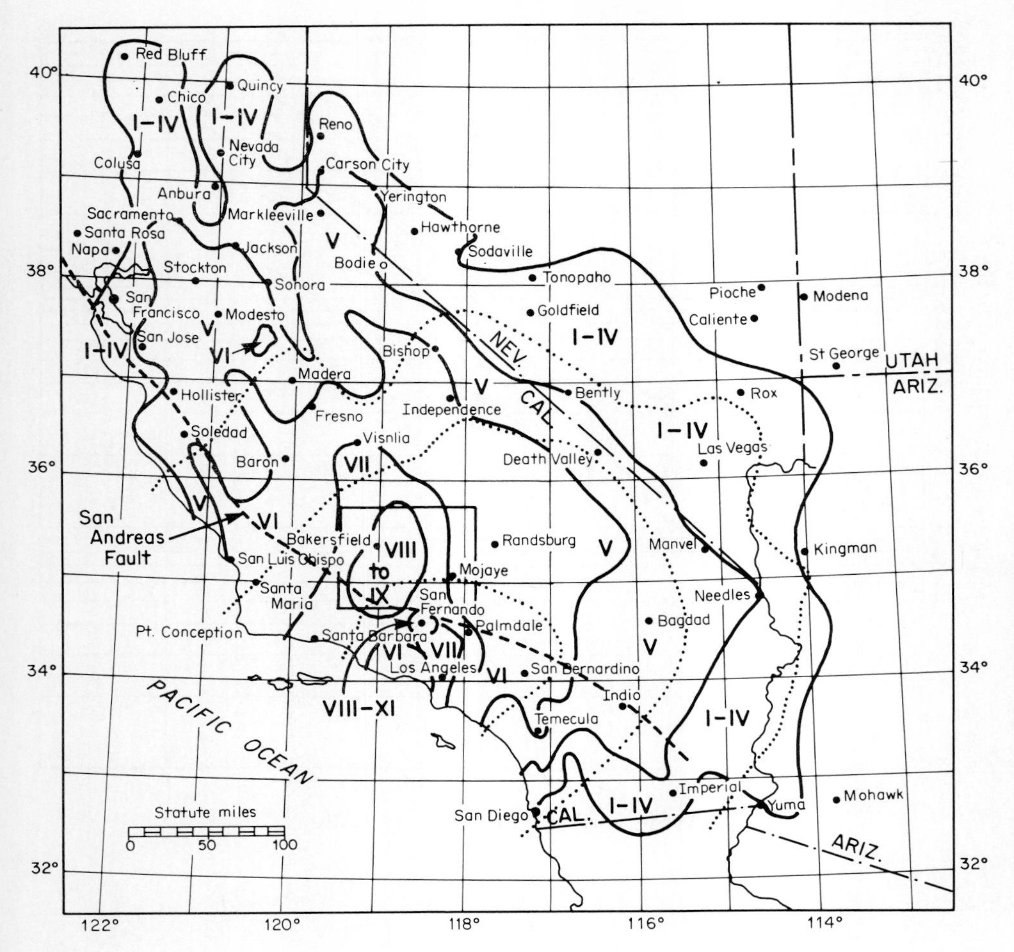


**FIG. 11.16** Isoseismal map showing the intensity distribution of the Kern County event of 1952 ($M$ = 7.7) and that of the San Fernando event of 1971 ($M$ = 6.6) (dashed lines). The differences in energy attenuation are clearly related to the differences in magnitude between the two earthquakes. *(From United States Geological Survey.)*

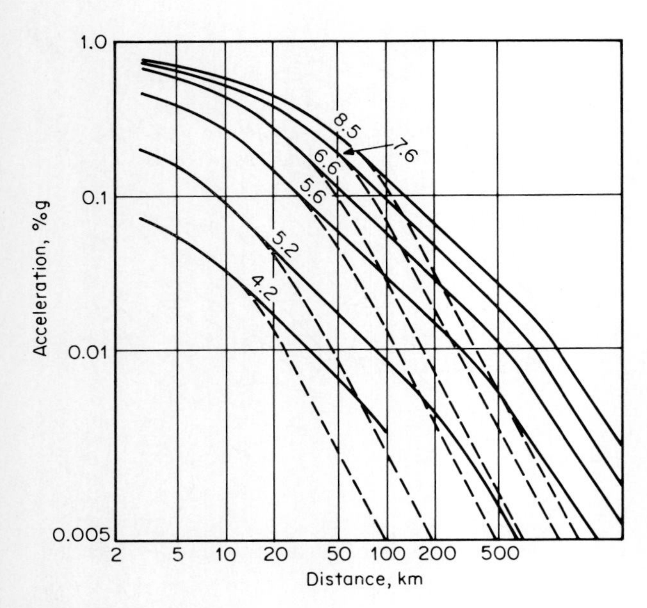


**FIG. 11.17** Acceleration attenuation curves for the United States. The solid lines are curves for the eastern region (east of longitude 105°). The dashed lines together with solid lines at close distances are the attenuation curves used for the western United States and are taken from Schnable and Seed (1973). It is to be noted that under certain conditions the area of shaking in the eastern United States is very much larger than in the western regions under similar earthquake conditions (see Fig. 11.18). [*From Algermissen and Perkins (1976).*[23]]

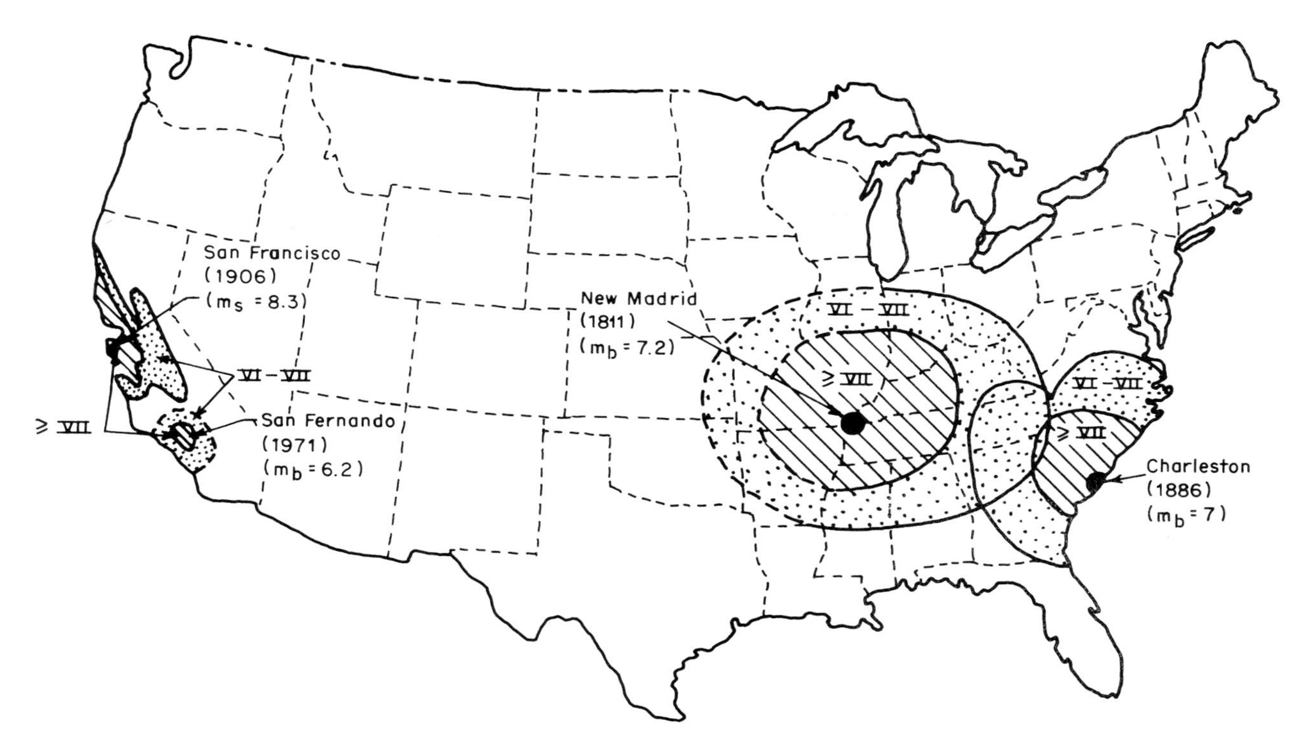


**FIG. 11.18 Comparison of areas of minor ($I$ = VI to VII) and major ($I$ > VII) damage for four major U.S. earthquakes. The damage area for the western half of the New Madrid event is inferred because there were no settlements in the area at that time ($m_b$ = body wave magnitude; $m_s$ = surface wave magnitude).** [*After Nuttli (1979).*[24]]

generally applicable to North America as follows:

$$I_s = 8.16 + 1.45M - 2.46 \log_{10} R \quad (11.12)$$

Where the focal distance is less than 100 km, focal depth becomes important. Ergin (1969)[22] proposed the following:

$$I_s = I_o - 3 \log_{10} (R/h) \quad (11.13)$$

where $h$ = depth of focus in km.

*Graphs and Charts*

A family of attenuation curves giving magnitude and acceleration as a function of distance from the source, prepared by Algermissen and Perkins (1976),[23] is given as Fig. 11.17. Attenuation as a function of distance from the rupture of a causative fault is given on Figs. 11.25 and 11.26.

**Comments**

Attenuation does relate in many instances to focal depth; very shallow focus events will be felt over relatively small areas; however, geology and topography are significant factors. In a comparison of earthquakes of similar intensities in the eastern United States with those in the western states, it can be shown that those in the east often affect areas 100 times greater than those in the west as illustrated on Fig. 11.18. In addition, those in the east often are not associated with evidence of surface faulting [Nuttli (1979)[24]].

11.2.6 AMPLIFICATION

**Description**

*Ground Amplification Factor*

Site intensity is often amplified by soil conditions. An increase in ground acceleration with respect to base rock excitation is termed the *ground amplification factor.*

*Stable Soil Conditions*

Under conditions where the soils are stable (nonliquefiable), the influence of local soil con-

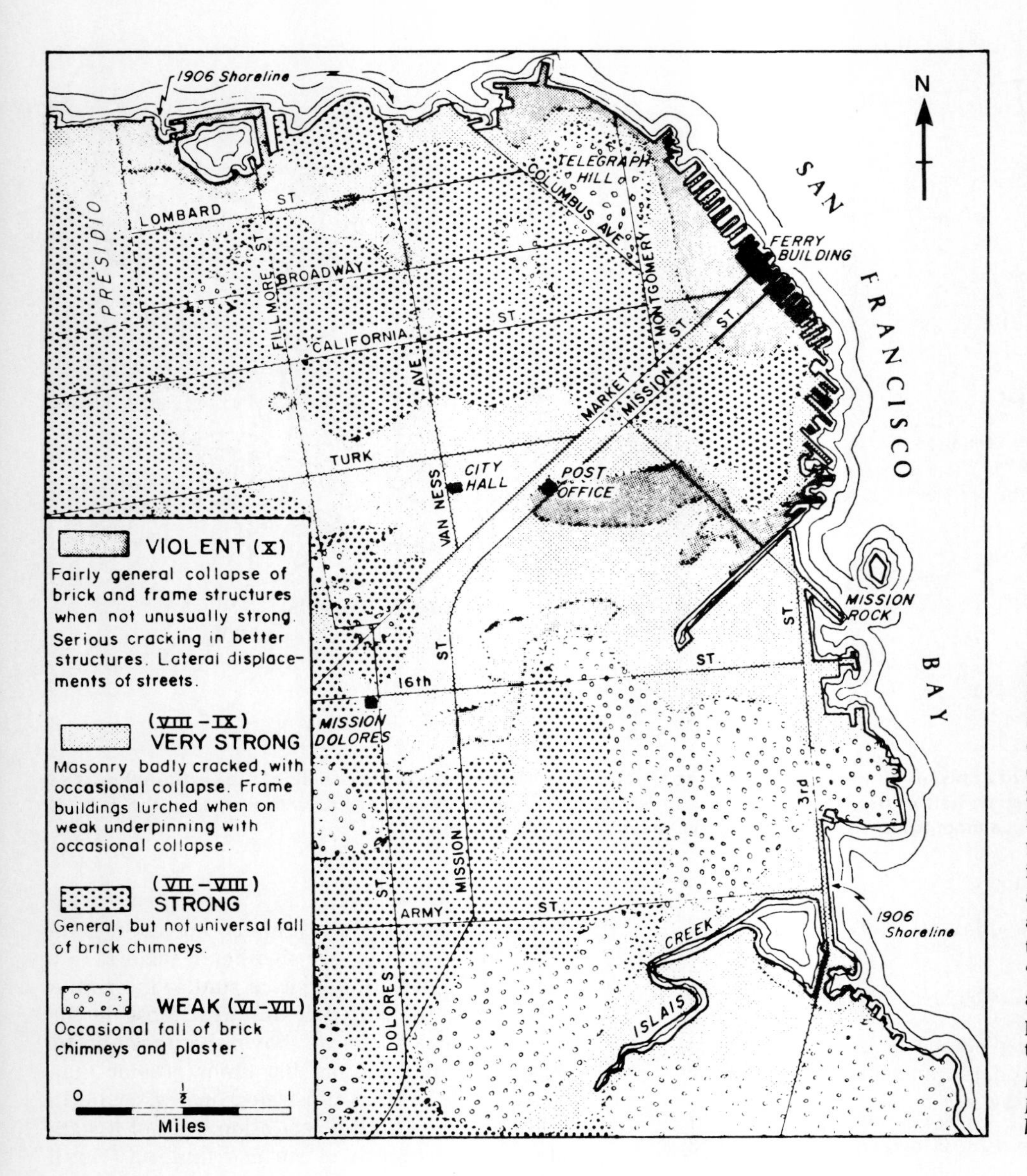


**FIG. 11.19 Earthquake intensity distribution in a portion of San Francisco from the 1906 event ($M$ = 8.3). The San Andreas fault is located about 10 mi west of the Ferry Building. Note the large variations for intensity for areas of equivalent distance from the fault. "Violent" areas are associated with the poorest ground conditions.** [*From USGS, prepared by Wood, 1906, presented by Steinbrugge (1968).*[27a]]

ditions on ground motions can take the form of dynamic amplification, which can result in an increase in peak amplitudes at the surface or within a specific layer. The shaking duration may also be increased. The factors influencing the occurrence are not well understood, although, in general, amplification is a function of soil type (densities and dynamic properties are most significant), depth, and extent. Attenuation may occur under certain conditions.

*Unstable (liquefiable) soils* are described in Art. 11.3.3

**Influencing Factors**

*Soil Type*

Zeevaert (1972)[25] concluded that in the valley of Mexico City the acceleration in the lacustrine soils is approximately 2 times larger than in the compact sands and gravels surrounding the valley. This difference would probably result in intensities of VII in the lacustrine soils and VI in the sands and gravels.

Ground-motion records from a peat layer in Seattle showed that motion was amplified for a

**TABLE 11.5**
**GROUND FOUNDATION FACTOR (FOR SOIL AMPLIFICATION)***

| Factor *F* | Soil type | Depth of soil |
|---|---|---|
| 1.0 | Rock<br>Dense to very dense coarse-grained soils<br>Very stiff to hard fine-grained soils | All depths |
| | Compact coarse-grained soils<br>Firm to stiff fine-grained soils | 0 to 50 ft |
| 1.3 | Compact coarse-grained soils<br>Firm to stiff fine-grained soils | >50 ft |
| | Very loose and loose coarse-grained soils<br>Very soft to soft fine-grained soils | 0 to 50 ft |
| 1.5 | Very loose to loose coarse-grained soils<br>Very soft to soft fine-grained soils | >50 ft |

*After CGS (1978).[30]

distant event (270 km) but attenuated for one nearby [Seed (1970*a*)[26]]. A possible explanation may be that the seismic waves from the distant event pass parallel to the layering in the peat, whereas those from nearby areas pass upward through the peat (the effect of different period characteristics).

*Soil Thickness*

The Caracas event of 1967 provides an example of the effect of thickness variation on intensity in a formation essentially similar throughout its lateral extent [Seed (1970*a*),[26] Espinosa and Algermissen (1972)[27]]. During the earthquake far more damage occurred in the east end where soil depths range from 90 to 210 m, than in the west end where depths range from 18 to 60 m. Four multistory buildings collapsed in the east end while none collapsed in the west.

A damage map of the San Francisco earthquake of 1906, prepared by Wood in 1906, is given as Fig. 11.19. In relating damage to ground conditions Wood showed that in areas of shallow rock only minor damage occurred to structures, whereas in areas of uncontrolled fill over deep soft soils, such as former stream valleys or along the waterfront, the effect was "violent."

*Foundation Depth*

Ground motions are usually given for the surface but design requires ground motion at the foundation level. Seed et al. (1975)[28] present data showing that during the Tokyo-Higashi-Matsuyana earthquake of July 1, 1968, accelerations recorded for structures near the ground surface were on the average about 4 times larger than for buildings founded at a depth of about 24 m.

*Source Distance*

The amplifications at *close distances* from the source appear to be more influenced by topographic expression and geologic structure than by local soil conditions [Faccioli and Reséndiz (1976)[29]].

For *large epicentral distances* from the source, however, local amplification can be considerable, as it can be for motions of smaller intensities, and stratigraphies characterized by sharp contrasts of seismic impedance [Faccioli and Reséndiz (1976)[29]].

### Ground Amplification Factors

Amplification factors in general may range from 1 to 2 for large earthquakes to 10 or more for microtremors, and may be higher for soft soils than for firm soils [Seed (1970*a*)[26]]. Factors vs. soil type and depth as given in the National Building Code of Canada are provided on Table 11.5. A relationship among acceleration, intensity, and ground conditions is also given on Fig. 11.20. In using such approximations consideration must be given to the various aforementioned influencing factors.

Ground amplification factors are most reliably obtained from accelerograms from the site or

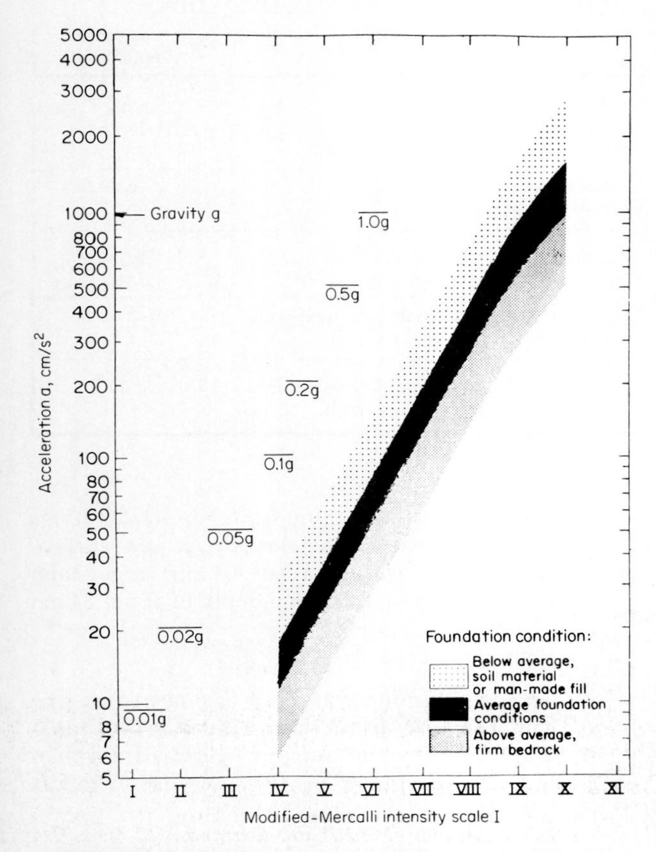


**FIG. 11.20 Comparison of earthquake acceleration and intensity and the relationship to foundation conditions.** [*From Coulter et al. (1973).*[31]]

from sites having similar conditions, and even these must be evaluated in consideration of source distance and foundation depths.

## 11.2.7 RECURRENCE AND FORECASTING

### General

*Prediction Basis*

Forecasting the location, magnitude, and time of occurrence of an earthquake is the role of the seismologist, and is necessary for seismic design and for the early warning of an impending event.

A number of factors are considered in forecasting events:

- Statistical analysis of historical data (recurrence analysis)
- Measurements of fault movements, crustal warping, and stress increases
- Changes in seismic-wave velocities (dilatancy theory)
- Changes in the earth's magnetic field and other geophysical properties

*Seismic Risk Analysis*

Seismic risk analysis is based on probabilistic and statistical procedures to assess the probable location, magnitude, occurrence, and frequency of earthquake occurrence. Procedures require evaluation of historical records and of the regional and local geology, particularly with respect to faults and their activity. The recurrence of events of various magnitudes is examined, and then the attenuation relationships are evaluated to allow the development of the probability of ground motion at the site for various magnitudes in terms of the geologic conditions [Donovan and Bornstein (1978)[32]].

### Statistical Analysis and Recurrence Equations

*Limiting Factors*

Prediction of an event for a given location during investigations is usually based on statistical analysis of recorded historical events, but the limitations in the accuracy of such predictions must be recognized. It is known where earthquakes are likely to occur from recorded history, but it must be considered that major events can occur in areas where they would be totally unexpected. New Madrid and Charleston, for example, are essentially singular events.

Comparing the span of modern history and its recorded events with the span of even recent geologic history results in the realization that data are meager as a basis for accurate prediction. For example, activity can be cyclic. A region apparently can go through several centuries without seismic activity and then enter a period with numerous events. The Anatolian zone of Turkey, with 2000 years of recorded events, now is an active seismic area, although it has had periods of inactivity for as long as 250 years [Bollinger (1976)[33]]. By comparison, the history of significant population in the United States is scarcely 250 years.

On a historical basis, records of events are very

much related to population density and area development, especially for events of moderate to low magnitudes felt over limited areas.

*General Recurrence Relationships*

*Occurrence frequency* of shocks of any given magnitude for the world in general and most of the limited areas that have been studied is roughly about 8 to 10 times that for shocks about one magnitude higher. The relationship can be represented by:

$$\log_{10} N = a - bM \qquad (11.14)$$

where $N$ = number of shocks of magnitude $M$ or greater per unit of time
$a, b$ = constants for a given area based on statistical analysis of recorded data
$a$ = $\log_{10} N(0)$, or the logarithm of the number of earthquakes greater than $M = 0$ for a given time period, given in units of earthquakes per year
$b$ = $\log_{10} \{[1 - F(M)]/M\}$ where $F(M)$ is the cumulative probability distribution of earthquake magnitudes

References for the above are Richter (1958)[1] and Lomnitz (1974).[34]

*Recurrence relation* is found also expressed in terms of $I_o$ as

$$\log N = \alpha - \beta I_o \qquad (11.15)$$

where $N$ = annual number of earthquakes with epicentral intensities equal to or greater than $I_o$
$\alpha, \beta$ = empirical constants which describe the decay rate of occurrence with increasing epicentral intensity in a manner similar to $a$ and $b$ in Eq. 11.14 [Christian et al. (1978)[21]]

*Some Regional Relationships*

Recurrence equations have been developed for various regions expressing the number of earthquakes per year $N$ in terms of the maximum magnitude $M$ or MM intensity $I_o$ for the magnitude range of interest. The data are taken from seismicity maps such as Fig. 11.12. Statistically the computed number of events per year for a given magnitude usually is presented as a *return time* once in so many years. The general relationships vary from region to region with geologic conditions, stress level, and perhaps magnitude. The variation in stress level is important in that both short-term and secular changes in earthquake frequency are thereby permitted [Bollinger (1976)[33]].

In *southeastern United States* [Bollinger (1976)[33]], from Fig. 11.12:

$$\log N = 3.01 - 0.59 I_o \text{ (for V} \leq I_o \leq \text{VIII)} \qquad (11.16)$$

The frequency of occurrence for events of various intensities expressed by Eq. 11.16 is given on Table 11.6. Several interpretations are possi-

**TABLE 11.6**
**FREQUENCY OF EARTHQUAKE OCCURRENCE IN THE SOUTHEASTERN UNITED STATES***

| $I_o$ | Return period, years | Years since last occurrence | Number expected per 100 years |
|---|---|---|---|
| V | 0.9 | 0 | 115 |
| VI | 3.4 | 1 | 30 |
| VII | 13.0 | 48 | 8 |
| VIII | 51.0 | 63 | 2 |
| (IX, X) | (200,780)† | (90) | (0.5, 0.1) |

*From Bollinger (1976).[33]
†Extrapolated values in parentheses.

ble from these data: the region is overdue for the occurrence of a damaging shock ($I_o$ = VII or VIII), or there is a change toward a lower level of activity, or the maximum intensity in some of the historical data has been overestimated.

For the *Ramapo fault in New Jersey* (period 1937–1977) [From Aggarwal and Sykes (1978)[35]]:

$$\log N = 1.70 \pm 0.13 - 0.73M \quad (11.17)$$

Equation 11.17 gives a recurrence of shocks of $M = 7.0$ of once every 97 years.

In *southern California* (period 1934–1943) [Richter (1958)[1]]:

$$\log N = 4.77 - 0.85 \text{ (for } M \text{ below 7.0)} \quad (11.18)$$

For *the World* (period 1918–1956) [Richter (1958)[1]]:

$$\log N = 7.81 - 0.58M \; (M < 7.1) \quad (11.19)$$

$$\log N = 9.1 - 1.1M \; (M \geq 7.1) \quad (11.20)$$

## Early Warning Indicators

### *General*

At the time of this writing there is no certain way of predicting where or when an earthquake may occur, although a number of tentative methods are under long-term study, most of which are related to subtle geologic changes with time.

*Geologic changes* occurring with time include:

- Fault displacement, or tilting or warping of the surface.
- Stress increase in fault zone or in surface rocks.
- Fluctuation of gravitational or magnetic fields above normal levels. (Before the Hollister, California event of November 1974, of $M = 5.2$, the magnetic field rose above the normal level.)
- Change in arrival times of transient *P* waves (dilatancy theory).
- Change in radon emissions from soils and subsurface waters.

*Animal reactions* are considered significant to the Chinese. It appears that domestic animals can sense microseisms, and shortly before an earthquake they become highly nervous. The Chinese have even evacuated cities in recent years on the basis of animal reactions preceding an event.

### *Dilatancy Theory*

*The dilatancy theory* (or $V_p/V_s$ anomaly, or seismic velocity ratio method) is based on the observation that the arrival times of transient *P* waves traveling through the earth's crust undergo a gradual decrease when compared with the arrival times of *S* waves, until just before an earthquake. Then the arrival time difference returns to normal relatively quickly and is followed by the shock [Scholtz et al. (1973),[36] Whitcomb et al. (1973)[37]]. The *time period* from return to normal of the *P* wave velocity until the actual event has been found by examination of preearthquake records to be roughly one-tenth the time interval during which the decreasing velocities occurred. If the time of the initial decrease and the return to normal are known, predictions can be made as to when the earthquake is likely to occur and the anticipated magnitude.

The magnitude has been found to be proportional to the time interval of the decreasing velocities and the return to normal. If the decrease and the return to normal have taken place in a few days or weeks, the event can be expected to be relatively small; whereas if the velocity decrease is stretched out for many months or years, the magnitude will probably be quite high. Current estimates are: *P* wave changes occurring for as long as 10 years will precede an $M = 8$ event, a 1-year interval will precede an $M = 7$ event, and a few months will precede an $M = 6$ event [Whitcomb et al. (1973)[37]].

*Rock-mass behavior* under stress provides an explanation for the dilatancy theory. As the crustal pressures preceding a quake approach the failure point in rock masses under high stress, a myriad of tiny cracks open. This causes the decrease in velocity of the *P* waves, since their velocity is reduced when they travel across air-filled openings. As groundwater seeps into the cracks, the velocity increases until all of the cracks are filled and velocity returns to normal. The presence of the water "lubricates" the cracks, reducing rock strength, permitting failure, and producing the earthquake.

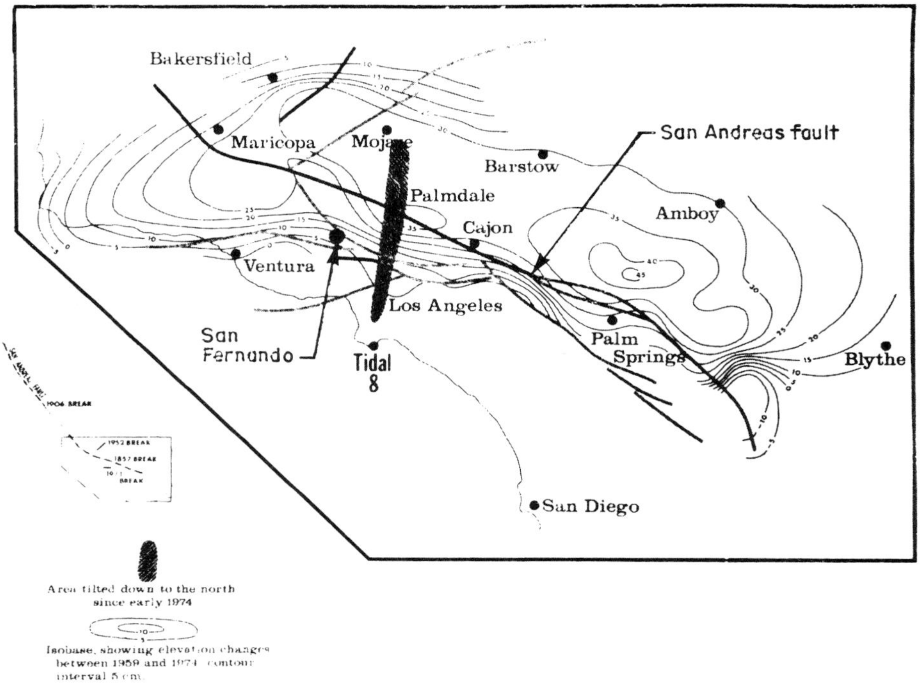

FIG. 11.21 Contours of surface warping (centimeters), area of Palmdale, California. Area uplifted 45 cm between 1959 and 1974, but between 1974 and 1977 an area at Palmdale had dropped 18 cm. Recent surveys show that the uplifted area is larger than previously thought, and that the shape and size of uplift change with time. [*From Hamilton (1978).*[39]]

*Deep well injection* apparently has verified the "lubrication" effect (which is in reality probably a pore-pressure effect). A series of shocks occurred between 1962 and 1967 near Denver, Colorado, following the pumping of liquid wastes down a borehole into rock at a depth of 4 km below the Rocky Mountain arsenal, a region where earthquakes were almost unknown. After the waste pumping was suspended, the number of events declined sharply. A similar experiment was carried out by the USGS at Chevron Oil Company's oil field in Rangely, Colorado, in 1972. Water was forced under high pressure into a number of deep wells and a series of minor earthquakes occurred. Activity ended immediately when the water was pumped from the wells [Raleigh et al. (1972)[38]].

*Surface Warping*

Overstresses in the earth's crust cause surface warping which may predate an earthquake. Records from the literature appear meager regarding the phenomenon. Data are available on ground surface elevation changes for many locations in the United States from the *Vertical Division Network, National Geodetic Survey, Rockville, Maryland.*

NIIGATA, JAPAN Japanese geologists reported that a land area near Niigata had risen 13 cm in 10 years before the 1964 event ($M = 7.5$).

PALMDALE, CALIFORNIA Measurements by the USGS have determined that an area of about 4500 km$^2$ around Palmdale rose as much as 45 cm between 1959 and 1974 as shown on Fig. 11.21. The area, now known as the Palmdale bulge, is centered on the San Andreas fault. Recent data indicate that between 1974 and 1977, Palmdale has dropped 18 cm [Hamilton (1978)[39]]. Two earthquakes of $M = 5.7$ and 5.2 which centered on the bulge, occurred on March 15, 1979 [ENR (1979)[40]].

*Research and Monitoring Networks*

*National Center for Earthquake Research* (USGS), Menlo Park, California, was established in 1966. Measurements are being made of changes in crustal strains and elevations over long distances, electrical resistivity, magnetic fields, and radon emission from soils and sub-

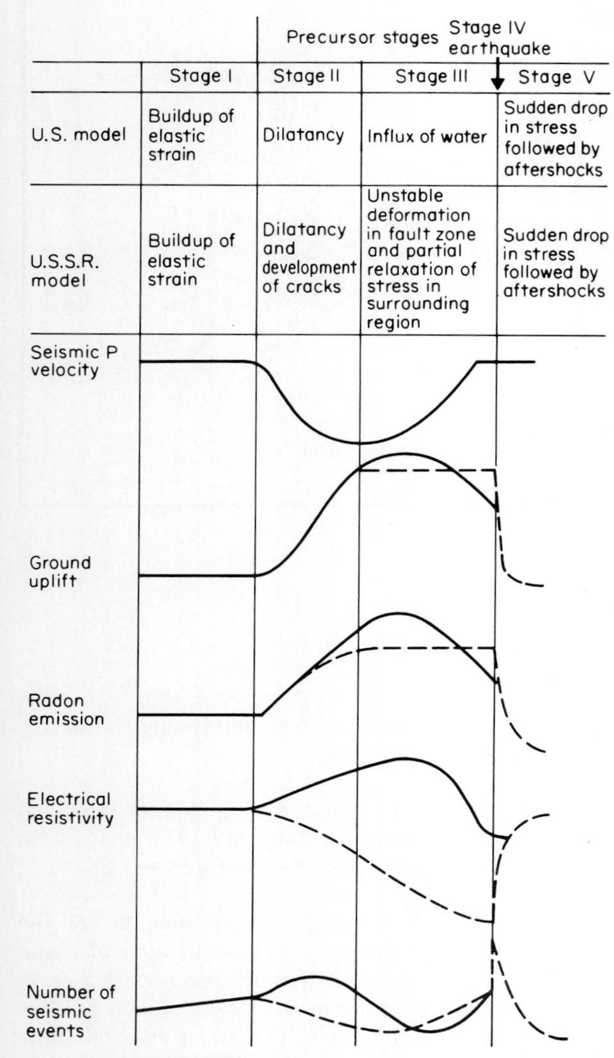


**FIG. 11.22 Schematic diagrams of expected changes in some physical parameters as a function of time before, during, and immediately after an earthquake according to two models developed in the United States and the Soviet Union. The solid lines represent the "dilatancy-instability" model developed by the Soviet Union; the dashed line is the "dilatancy-fluid flow" model of the United States. Radon emission may be a function of both water flow and rate of creation of new surface area by the growth of cracks or by rock-mass fracturing. The expected behavior of electrical resistivity in the ground has not yet been measured with sufficient accuracy between the models.** [*From National Research Council-National Academy of Sciences (1976).*[41] *Reprinted with permission of the National Academy of Sciences.*]

surface waters. The objective is to develop models of precursors to earthquakes such as shown on Fig. 11.22. Some instrumentation used to monitor ground changes is described in *Art. 4.5.7.*

*Monitoring networks* were being installed around Los Angeles, northern New York state, and Charleston, South Carolina as of December 1975. These networks will provide for the detection and evaluation of *P* wave velocity changes and other dilatancy effects, with the objective of providing an early warning system for earthquakes.

## 11.3 SURFACE EFFECTS ON THE GEOLOGIC ENVIRONMENT

### 11.3.1 FAULTING

#### General

*Geologic Aspects*

The terminology, characteristics, and identification of faults and a summary of investigation methodology are given in Art. 6.5.

*Importance in Earthquake Engineering*

Shallow-focus earthquakes, usually the most destructive, are frequently associated with faulting which can consist of a single main fracture, or of a system including subsidiary fractures. Fault identification is an important element in studies to evaluate the probability of earthquake occurrence and magnitude.

Correlations have been made from earthquake data in some geographic regions (principally in the United States) to develop a number of relationships:

- Length of fault rupture vs. earthquake magnitude
- Distance from the causative fault vs. the acceleration of gravity
- Fault displacement vs. magnitude

*Fault Study Elements*

During engineering studies for seismic design, the following aspects related to faulting are considered:

| TABLE 11.7 CRITERIA USED FOR RECOGNIZING AN ACTIVE FAULT* | |
|---|---|
| **General criteria** | **Specific criteria** |
| Geological | Active fault indicated by following features: *Young geomorphic features:* Fault scarps, triangular facets, fault scarplets, fault rifts, fault slice ridges, shutter ridges, offset streams, enclosed depressions, fault valleys, fault troughs, sidehill ridges, fault saddles. *Ground features:* Open fissures, "mole tracks" and furrows, rejuvenated streams. *Subsurface features:* Stratigraphic offset of Quaternary deposits, folding or warping of young deposits, en echelon faults in alluvium, groundwater barriers in recent alluvium. |
| Historical | Description of past earthquakes, surface faulting, landsliding, fissuring, and other phenomena from historical manuscripts, news accounts, and other publications. Indications of fault creep or geodetic monument movements may be indicated in recent reports. |
| Seismological | High-magnitude earthquakes and microearthquakes, when instrumentally well-located, may indicate an active fault. A lack of known earthquakes cannot be used to indicate that a fault is inactive. |

*After Cluff et al. (1972).[42]

| TABLE 11.8 SOME METHODS OF DATING THE MINIMUM AGE OF LAST DISPLACEMENTS ON FAULTS* |
|---|
| Determining the age of undisplaced strata overlying the fault through the use of fossils, radiometric dating, or paleomagnetic studies. |
| Determining the age of cross-cutting undisturbed dikes, sills, or other intrusions. |
| Determining the rate of development of undisturbed soil profiles across a fault. |
| Radiometric dating of minerals caused by the fault movement or of undeformed minerals in the fault zone. |
| Dating of geomorphic features along or across the fault. |
| Dating techniques in fault investigations—*see Appendix A.4.* |

*From Adair (1979).[43]

- Positive identification that a fault (or faults) is present
- Fault activity: establish the "*capable fault*" by judging if it is potentially active, or inactive
- Displacement amount and form (dip-slip, strike-slip, etc.) that might be expected
- Earthquake magnitude that might be generated by rupture (related generally to length)
- Estimated site acceleration after attenuation from the capable fault

### Fault Activity (The Capable Fault)

#### *Significance*

In recognition that shallow-focus events are associated with faulting, but that many ancient faults are not under stress and therefore are "dead" or inactive and not likely to be the source of a shock, it becomes necessary, in order to predict a possible earthquake, to identify a fault as active, potentially active, or inactive (dead). Seismic design criteria often are based

**TABLE 11.9**
**SYSTEM FOR CLASSIFICATION OF FAULT ACTIVITY BASED ON AVAILABLE DATA***

| Activity classification and definition | Criteria | | | Studies to further define activity |
|---|---|---|---|---|
| | Historical | Geological | Seismological | |
| Active—a tectonic fault which has a history of strong earthquakes and surface rupture, or a fault which can be demonstrated to have an interval of recurrence short enough to be significant during the life of the particular project. The recurrence time period considered significant for individual projects will vary with the consequence of activity. | 1. Surface faulting and associated strong earthquakes, 2. Tectonic fault creep, or geodetic indications of movement. | 1. Geologically young† deposits have been displaced or cut by faulting, 2. Fresh geomorphic features characteristic of active fault zones present along fault trace, 3. Physical groundwater barriers produced in geologically young† deposits. | Earthquake epicenters are assigned to individual faults with a high degree of confidence. | Additional investigations and explorations are needed to define: 1. The exact location of individual fault traces 2. The recurrence interval 3. The projected magnitude of future events 4. The type of surface deformation associated with the surface faulting 5. The probable source of energy release in respect to the site |
| Potentially active—a tectonic fault which has not ruptured in historic time, but available evidence indicates that rupture has occurred in the past and the recurrence period could be short enough to be of significance to particular projects. | No reliable report of historic surface faulting. | 1. Geomorphic features characteristic of active fault zones subdued, eroded, and discontinuous. 2. Faults are not known to cut or displace the most recent alluvial deposits, but may be found in older alluvial deposits. 3. Water barrier may be found in older materials. 4. Geological setting in which the geometric relationship to active or potentially active faults suggests similar levels of activity. | Alignment of some earthquake epicenters along fault trace, but locations are assigned with a low degree of confidence. | Additional investigations are needed to resolve: 1. The time interval of past activity 2. The recurrence of activity 3. The possible locations of the individual fault traces The classification becomes less important if the fault does not cross the project site and a known active fault which is capable of producing frequent high-magnitude earthquakes is located closer to the structure or project under study, and would therefore be the more significant fault. |

*From Cluff et al. (1972).[42]

†The exact age of the deposits will vary with each project and depends upon the acceptable level of risk and the time interval which is considered significant for that project.

on the identification of the active or potentially active faults *(capable faults)* and their characteristics.

### The Capable Fault

*U.S. Nuclear Regulatory Commission* (NRC) considers a fault capable if it is undergoing creep, or has undergone movement in the past 35,000 years (during the Holocene epoch), or has undergone more than one movement during the past 500,000 years, or is connected to a capable fault, or is in the locale of instrumentally recorded and located events (in general, two or more recorded events of MM intensity III or greater).

*International Atomic Energy Commission* (IAEC) considers a fault capable if it has undergone movement in late Quaternary, if there is topographic evidence of surface rupture, if there are instrumentally recorded and located events, if there is creep along the fault, or if it is connected to a capable fault.

*Japan* grades active faults according to the amount of displacement per unit of time, creep along the fault, movement during the Quaternary, and expected future movement.

### Identification and Classification

The external and internal evidences of faulting are described in Art. 6.5.3. The general criteria for recognition of an active fault are given on Table 11.7, the methods of dating the minimum age of the last fault displacement are listed on

## TABLE 11.9
## SYSTEM FOR CLASSIFICATION OF FAULT ACTIVITY BASED ON AVAILABLE DATA* (*Continued*)

| Activity classification and definition | Criteria | | | Studies to further define activity |
|---|---|---|---|---|
| | Historical | Geological | Seismological | |
| Activity uncertain—a reported fault for which insufficient evidence is available to define its past activity or its recurrent interval. The following classifications can be used until the results of additional studies provide definitive evidence. | Available information is insufficient to provide criteria that are definitive enough to establish fault activity. This lack of information may be due to the inactivity of the fault or due to a lack of investigations needed to provide definitive criteria. | | | This classification indicates that additional studies are necessary if the fault is found to be critical to the project. The importance of a fault with this classification depends upon the type of structure involved, the location of the fault in respect to the structure, and the consequences of movement. |
| *Tentatively active*—predominant evidence suggests that the fault may be active even though its recurrence interval is very long or poorly defined. | Available information suggests evidence of fault activity, but evidence is insufficient to be definitive. | | | |
| *Tentatively inactive*—predominant evidence suggests that fault is not active. | Available information suggests evidence of fault inactivity, but evidence is insufficient to be definitive. | | | |
| Inactive—a fault along which it can be demonstrated that surface faulting has not occurred in the recent past, and that the recurrence interval is long enough not to be of significance to the particular project. | No historic activity. | Geomorphic features characteristic of active fault zones are not present and geological evidence is available to indicate that the fault has not moved in the recent past and recurrence is not likely during a time period considered significant to the site. Should indicate age of last movement: Holocene, Pleistocene, Quaternary, Tertiary, etc. | Not recognized as a source of earthquakes. | No additional investigations are necessary to define activity. |

Table 11.8, and a system for classification of fault activity based on available data is given on Table 11.9.

*Instrumentation of fault movements* is illustrated on Fig. 4.42.

### *Limitations in Identification*

STRATIGRAPHY In glaciated areas it may be difficult to find a well-stratified Holocene section older than 35,000 years; therefore, the favored approach of trenching to permit examination for bedding ruptures in Holocene strata may not be applicable. In addition, fault displacement does not always extend to the surface; weathered rock and soil near the surface sometimes can adsorb the slip. After the Alaska event (1964), locations were found where 2 m of displacement were adsorbed by 20 m of weathered rock [Bolt et al. (1975)[2]]. In such cases, trenches may prove inconclusive.

PRESENT "DEAD" FAULTS Many faults that have not been carefully studied may be considered to be dead or inactive because they have not been the locus of recorded events or activity, but may be potentially active. The Kern County event (1952) occurred along the White Wolf fault (Fig. 11.11), which was little known and considered as a dead fault, although approximately 64 km in length. It may be connected to either the San Andreas or the Garlock fault, the two largest in California, which are located only about 24 km apart. In August 1975 an earthquake of $M = 5.7$ had its epicenter near the Oroville Dam on one of the faults of the Foothills System of the Sierra

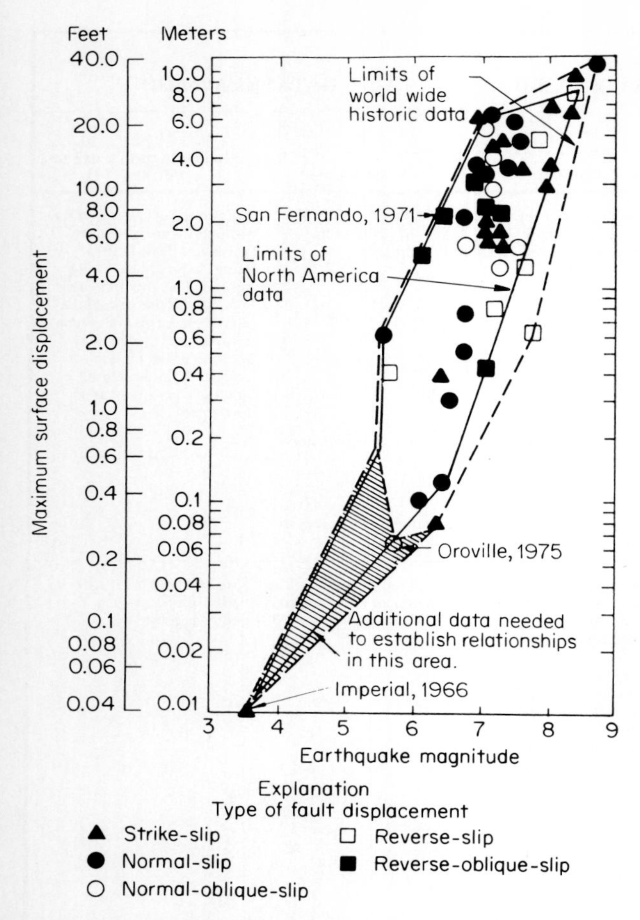


**FIG. 11.23 Relationship between maximum surface displacement and earthquake magnitude reported for historic events of surface faulting throughout the world.** [*From Taylor and Cluff (1977).*[45]]

Nevada range of California, which was considered as a dead fault. In 1952 the Ramapo fault in northeastern New Jersey *(see Fig. 6.61)* was considered to be long dead. In the intervening years the installation of a seismograph station at Lamont plus increased area development and habitation have revealed that there is a substantial amount of activity along the fault and today it even has its own recurrence equation *(see Eq. 11.17)*.

## Fault Displacements

### *Importance*

Correlations have been made among amount of displacement, fault length along which displacement occurs, and magnitude of the event. Displacement can vary substantially and does not occur in uniform amounts along the fault, and some sections may not displace at all. In general, displacement is related to magnitude of event.

### *Creep*

Before or after an earthquake, slow movement can occur along a fault (tectonic creep) which can range from a few millimeters to a centimeter or more every year. This fault slippage apparently occurs in faults filled with gouge from previous rupture as strain energy accumulates in the rock below the gouge zone.

Strike-slip is occurring along the San Andreas in Hollister, California. Records show that the two sides of the fault element passing beneath the town and a winery on the outskirts are moving past each other at the rate of about 1 cm/year. In this region creep is not occurring along an extensive line, but rather is limited to certain areas.

### *Strike-Slip Displacement*

One of the largest movements on record is the 6 m of horizontal displacement which occurred during the San Francisco quake of 1906; vertical movement did not exceed about 1 m. Horizontal movement of the Imperial Valley event (El Centro) of 1940 reached 3 m. East of El Centro, at a location along Highway 40, displacement across the roadway was 46 cm. By 1966 displacement was 64 cm because of fault creep and slips over the 26-year interval.

### *Dip-Slip Displacement*

Surface tilting and warping often result in dip-slip displacement. California events normally have only a meter or so of displacement. The largest recorded vertical displacement appears to be the 10 m that may have occurred in Assam in 1897, or possibly the 14 m that may have occurred in the Yakutat Bay, Alaska, quake of 1899. The problem in determining displacements on old scarps is that of erosional changes as discussed by Wallace (1980).[44]

### *Displacement vs. Magnitude*

Relationships between maximum surface displacement for various types of faults and earthquake magnitude based on recorded events is given on Fig. 11.23.

## Rupture Length

*General*

Surface rupture length along a fault varies greatly; in California earthquakes it has been generally in the range of 1 to 60 km. The crustal deformation that occurred during the Alaskan event of 1964 was the most extensive yet studied in a single earthquake [Bolt et al. (1975)[2]]. Vertical displacements occurred along the Alaskan coastline for a distance of almost 1000 km, including a broad zone of subsidence of as much as 2 m along the Kodiak-Kenai-Chugach mountain ranges, and a major zone of uplift of as much as 11 m along the coastline. The large extent is perhaps the reason for the unusual duration of 3 min. During the 1906 San Francisco event, the San Andreas is estimated to have ruptured for a length of 430 km.

*Rupture Length vs. Magnitude*

The energy released by a shallow-focus earthquake has been related to surface length of fault rupture as shown on Fig. 11.24. The majority of events plotted are from the western United States, Alaska, and northern Mexico and may not apply elsewhere, although good agreement has been found at the higher magnitudes with quakes in Turkey and Chile. Such relationships as given on the figure have been used to estimate the potential magnitude of an earthquake by assuming that a fault will rupture along its entire identified length, or perhaps only one-half to one-third of its length, depending on its activity and the degree of risk involved.

A rough relationship between surface fault rupture length $L$ in kilometers and surface wave magnitude $M_s$, based on worldwide data, is given [from Bolt (1978)[16]] as:

$$M_s = 6.03 + 0.76 \log L \qquad (11.21)$$

*Duration vs. Length*

Shaking duration in large earthquakes depends very much on the length of faulting. The longer the length of fault rupture, the greater is the duration of the time in which the seismic waves reach a given site. A fault can rupture progressively, however, providing a long duration such as in Alaska (1964), or it can rupture in a sequence of breaks providing a duration such as in San Francisco in 1906 (see *Art. 11.2.3*).

## Attenuation from the Fault

*Close Proximity to Rupture*

Peak intensities or magnitudes are not necessarily located at the surficial expression of the fault; the depth of focus and fault inclination will affect surface response. Housner (1970*a*)[46] suggests that the rate of decrease in magnitude is relatively small over a distance from the fault equal to the vertical distance to the focus, and beyond this point the drop-off increases rapidly. Many investigators report the lack of markedly greater shaking damage to structures adjacent to

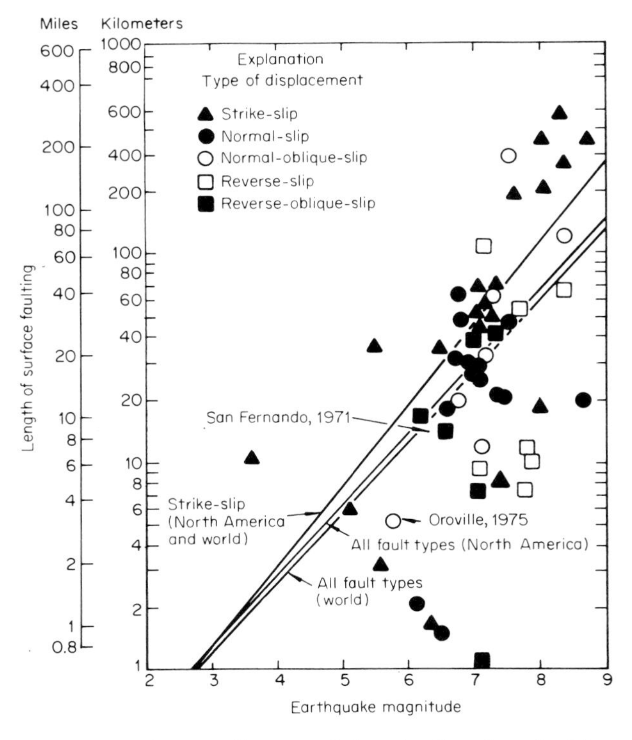


**FIG. 11.24 Scatter diagram of *length* of surface faulting related to earthquake magnitude from historical events of surface faulting throughout the world. Lines are least-squares fits.** [*From Taylor and Cluff (1977).*[45]]

the fault as compared with those some distance beyond [Bonilla (1970)[47]].

*Moderate Distances and Beyond*

In most geological environments high accelerations are severely reduced at even moderate distances. Wave attenuation from frictional resistance occurs exponentially, and if shaking is caused by shear waves in the surficial part of the crust, then even at short distances the exponential decay becomes very effective.

A relationship between peak bedrock acceleration, magnitude, and distance from the causative fault for focal depths of 0 to 20 km is given on Fig. 11.25.

Attenuation of maximum acceleration with distance from fault rupture for California earthquakes, prepared from strong motion records, is given on Fig. 11.26. Curve I applies to high-intensity sources, associated with effective modes of dislocation, rock types, rupture depths, etc., as, for example, San Fernando, 1971. Peak acceleration near a high-intensity fault would appear to be close to 0.6g on the average. The expectation ranges indicate that 90% of the time the peak acceleration would be less than 0.4g. Curve II applies to medium-intensity sources, produced by fault dislocations of a less efficient kind, such as El Centro, 1940. In either case, for distances over 100 km from the source, peak accelerations on rock are unlikely to exceed 0.1g.

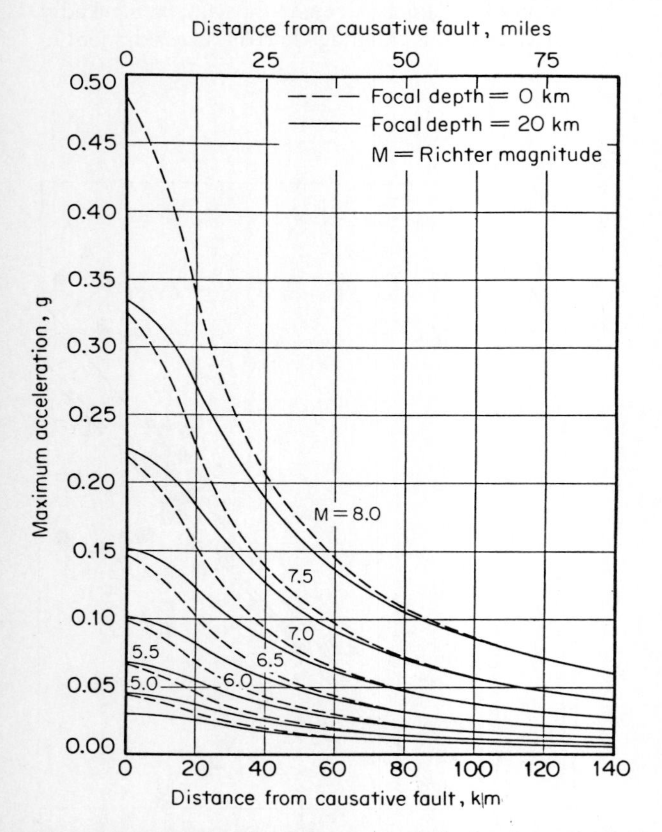


**FIG. 11.25 Relationship between peak bedrock acceleration, earthquake magnitude, and distance from the causative fault for focal depths of 0 and 20 km.** [*From Leeds (1973).*[48]]

Relationships among *predominant periods* of maximum acceleration, magnitude, and distance from the causative fault are given in Fig. 11.27.

*Isoseismal Maps*

The strong relationship between intensity distribution and fault rupture, where rupture length is long, is illustrated on Fig. 11.28, the intensity distribution for the San Francisco 1906 earthquake. Intensities generally of VIII to IX are given along the fault, but intensities of IX are given also in isolated areas as far as 64 km from the fault. Two branches of the San Andreas also are shown, the Hayward and the Calaveras faults. The map suggests that movement also may have occurred along these faults during the 1906 quake.

## Seismicity Maps and Tectonic Structures

*Importance*

In studies for seismic design, seismicity maps are overlain with geologic maps to obtain correlations with tectonic structures. For example, a comparison of the seismicity map of the southeastern United States (*see Fig. 11.12*) with the geologic map of the area (*see Fig. 2.2*) shows a very strong correlation between the epicenters and faults in the zone trending SW–NE through Tennessee, Alabama, and Georgia. The ERTS image (*see Fig. 6.39*) of the area where Tennessee, Kentucky, and West Virginia join shows a very evident major cross-fault trending in the direction of Charleston, South Carolina. The

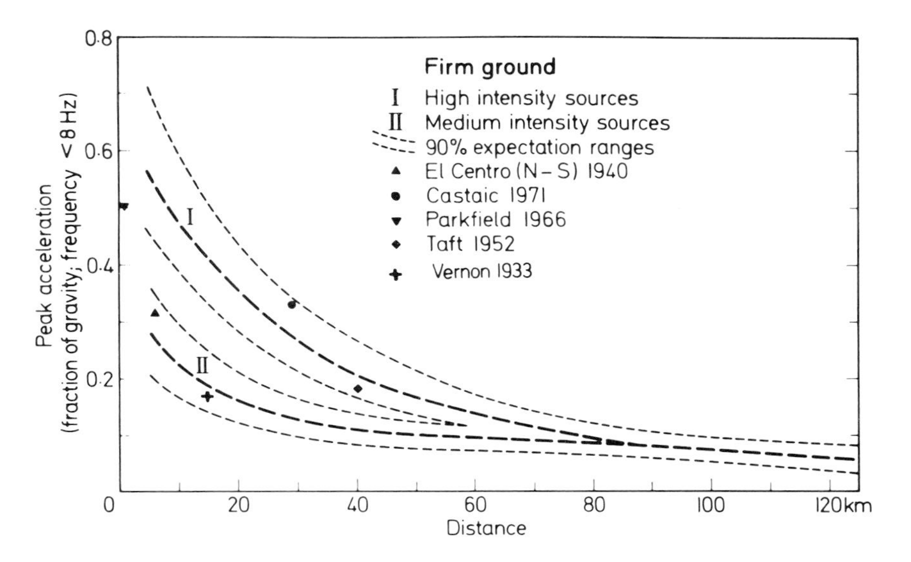


**FIG. 11.26 Attenuation of maximum acceleration with distance from fault rupture for California earthquakes prepared from strong-motion records. Frequencies are less than 8 Hz, and generally in the low to intermediate range. Acceleration is for rock or strong soils.** [*After Bolt (1973).*[49]]

recurrence equation for the area is given as Eq. 11.16.

*Preliminary Investigation Case Study*

*Study purpose* was to determine the seismological characteristics of a region of the U.S. east coast by relating tectonic structures with epicenter locations. The following discussion demonstrates the value of a detailed search of existing literature.

*Study scope* included:

- Preparation of a seismicity map from catalog data covering an area within 200 mi (320 km) of the proposed project site
- Location of known tectonic structures (faults and fold structures) already identified on geologic maps *(see Fig. 6.49)* plotted on the seismicity map
- Interpretation of remote-sensing imagery to identify lineaments, which are then plotted on the map
- Research of the literature for data from previous studies, including gravimetric and magnetic surveys
- Postulation of the lineament most likely to be responsible for the majority of events in the study region

*Investigation* results for the study area showing

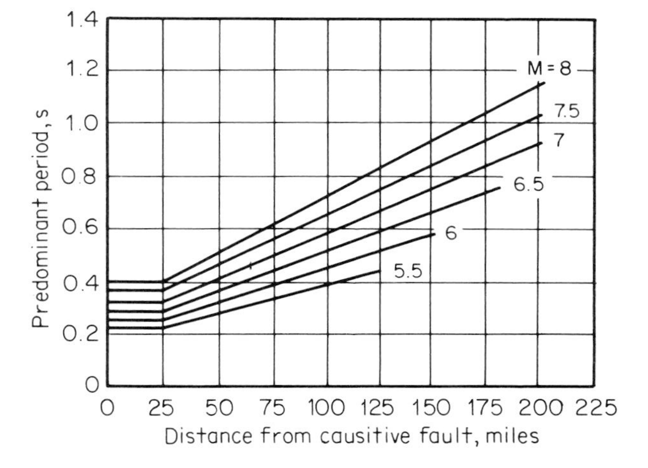


**FIG. 11.27 Relationships between predominant periods of maximum acceleration, magnitude, and distance from causative fault.** [*From Seed et al.*[50]*; as presented in Atomic Energy Commission (1972).*[11]]

major faults, other geologic structures, and epicenter locations of $I$ = V or greater are given on Fig. 11.29.

Literature review disclosed that previous investigators had encountered evidence that a major fault zone may exist striking NE-SW, slightly off-parallel with the coast from Connecticut to Virginia, but the structure is not located on any

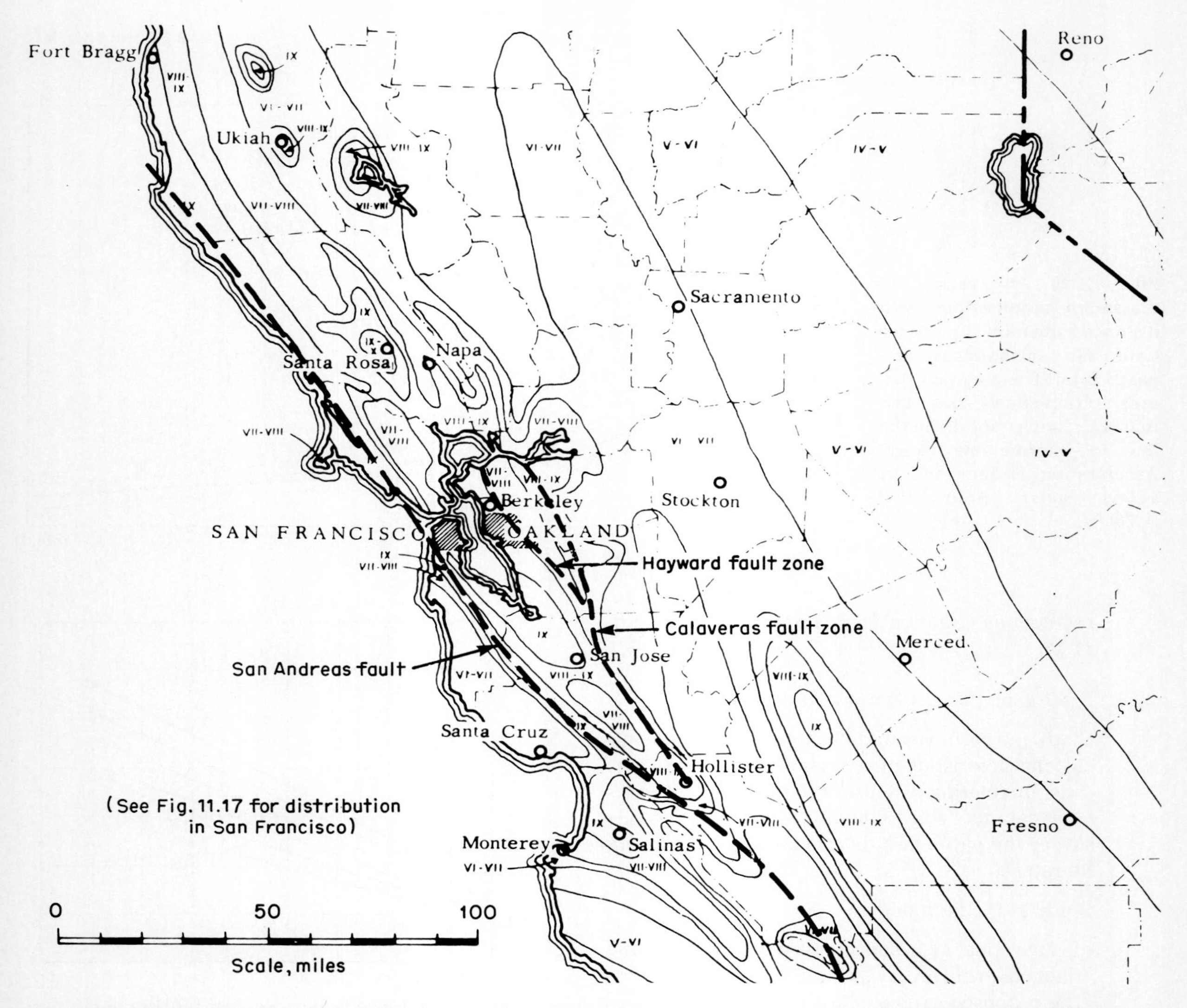


**FIG. 11.28 Isoseismal map of the 1906 San Francisco earthquake given in the Rossi-Forel scale which roughly parallels the modified Mercalli through $I$ = X. *(See also Figs. 6.66 and 6.67.)*** [*After Environmental Science Services Administration (1969)*[51] *and Lawson et al. (1908).*[51a]]

published geologic maps (as of 1974). Studies of aeromagnetic data suggested a major fault in the basement rocks underlying the Chesapeake Bay area [Higgins et al. (1974)[53]], and studies of the fall zone in Delaware revealed a lateral fault and vertical faults in the basement complex between Newark and Delaware [Spoljaric (1972)[54]].

Examination of LANDSAT imagery led to postulating the location of a major lineament as shown on Fig. 11.29. The Landsat mosaic *(see Fig. 7.36)* shows the Delaware and Potomac Rivers and Chesapeake Bay to change direction suddenly from flowing southeast to flowing southwest. This offset in river direction is typical of strike-slip faulting. In addition, a lineament is apparent on the ERTS image given as Fig. 2.8, extending approximately 42 km between Trenton on the Delaware River, northeast to New Brunswick. The lineament follows several streams, trending N40°E over Cretaceous clays.

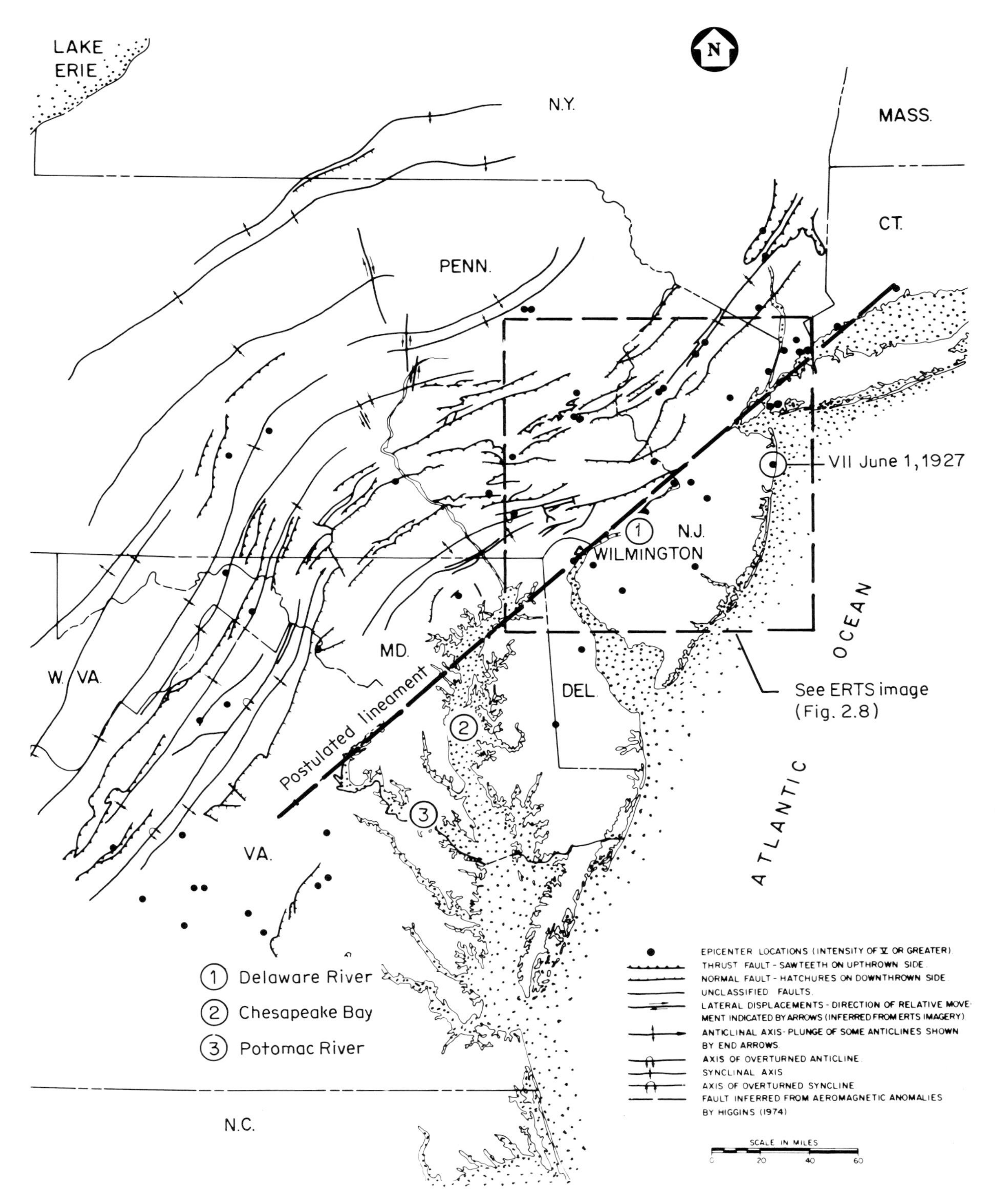


**FIG. 11.29 Map of earthquake intensities, epicenters, major geologic structures, and postulated fall line lineament, Middle Atlantic region, United States.** [*From Hunt and Sabatino (1974).*[52]]

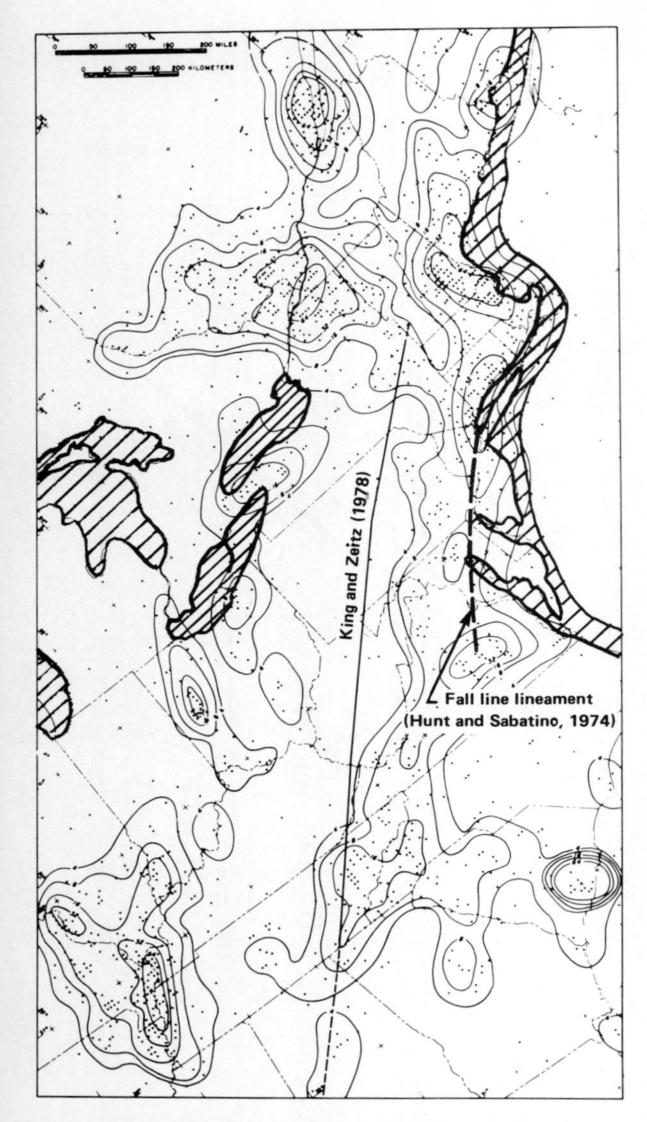


**FIG. 11.30 Seismotectonic map of the eastern United States** [*from Hadley and Devine (1974)*[56]] **showing earthquake epicenters of modified Mercalli III or greater, recorded from 1800 to 1972. Contours show number of epicenters per 10,000 km². Location of lineament is based on magnetic and gravity data.** [*From King and Zeitz (1978).*[55]]

It is possible, using LANDSAT imagery, to trace the lineament for several hundred kilometers between New York City and southwestern Virginia, and perhaps as far south as Alabama. The data from the studies of Higgins et al. and Spoljaric fall on the proposed lineament.

The Atlantic coastal plain, in the general area of the fall zone, has been the location of a number of seismic events, and a large number of epicenters are located generally along the postulated fault zone. Spoljaric concluded from earthquake data that the faults in the Newark-Wilmington area may still be active. A compilation of the seismic events within 320 km of Wilmington of MM = V or greater revealed that, of 57 events, 30 fall within 64 km of the suggested fault zone, and 17 fall within 32 km.

A parallel lineament 1400 km in length, located approximately 200 km inland from the fall line lineament, has been postulated by King and Zeitz (1978)[55] on the basis of magnetic and gravity data. This lineament also extends into the active seismic zone of the southeastern United States. Both postulated lineaments are shown on Fig. 11.30, a *seismotectonic map* of the eastern United States. With the limited data available, it is possible only to speculate on the influence of these lineaments on the potential seismicity levels in the eastern United States continental margin in consideration of their possible very large lengths.

### 11.3.2 SOIL BEHAVIOR

#### General

Excitation emanating as a stress wave from an underlying bedrock surface applies a cyclic shearing stress to soils. On the basis of response to bedrock motions, soils are divided into two general classes:

- *Stable soils* undergo elastic and plastic deformations but serve to dampen seismic motion and still maintain some strength level.
- *Unstable soils* are subject to sudden compaction, or a complete loss of strength by cyclic liquefaction (*see Art. 11.3.3*).

#### Characteristic Properties of Soils Under Cyclic Strain

*Shear modulus G (see Table 3.34 and Fig. 3.34)* is the relation between shear stress and shear strain which occurs under small amplitudes, such as earthquake loadings.

*Internal damping ratio D* or λ *(see Fig. 3.34)* pertains to the dissipation of energy during cyclic loading. Shear modulus and damping are the most important characteristics needed for analysis of most situations.

*Strength* and *stress-strain relationships* in general must be considered for large deformations such as are produced by sea-wave forces on pile-supported structures.

*Poisson's ratio* $\nu$ is required for the description of dynamic soil response, but varies within relatively close limits and affects seismic response only slightly. It is independent of frequency in the range of interest in earthquake engineering, and in contrast to $E$ and $G$, is insensitive to thixotropic effects. General ranges are $\nu = 0.25$ to 0.35 for cohesionless soils and $\nu = 0.4$ to 0.5 for cohesive soils.

### Soil Reaction to Dynamic Loads

*Initially*, cyclic loading causes partially irreversible deformations, irrespective of strain amplitude, and load-unload stress-strain curves do not coincide.

*Subsequently*, after a few cycles of similar small-strain amplitudes, differences between successive reloading curves tend to disappear and the stress-strain curve becomes a closed loop (*see Fig. 3.34*) which can be described by two parameters: *shear modulus*, defined by the average slope, and *damping ratio*, defined by the ratio of the specific enclosed areas as shown on the figure. It reflects the energy that must be fed into the soil to maintain a steady state of free vibration.

### Cyclic Shear Related to Earthquake Characteristics

*Simple shear stress-strain characteristics* at low strains are important in site response analysis because the significant earthquake strain amplitudes normally do not exceed $10^{-4}$ or $10^{-5}$, and are usually in the range of $10^{-1}$ to $10^{-3}$. Higher strains might occur during site response to a large earthquake, but the number of cycles at high strain amplitude are likely to be few.

The effect of *number of cycles* at low strain amplitudes is not great.

The effect of *loading frequency* is negligible within the range encountered in most earthquakes, i.e., 0.1 to 20 Hz.

*Strain amplitude* is the most significant characteristic. Shear modulus decreases markedly with an increase in strain amplitude as shown on Fig. 3.34 [Taylor and Larkin (1978)[57]].

### Shear Modulus and Damping Ratio

#### *Factors Affecting Values*

*Main factors* affecting values for shear modulus and damping ratio in all soils are shear strain amplitude, initial effective mean principal stress, void ratio, shear stress level, and number of loading cycles.

*Cohesive soil* values are affected also by stress history (OCR), saturation degree, effective strength parameters, thixotropy, and temperature.

*Shear strain amplitude* affects the shear modulus as follows:

- *Cohesionless soils:* $G$ decreases appreciably for amplitudes greater than $10^{-4}$, below which $G$ is nearly constant
- *Cohesive soils:* $G$ decreases with increase in amplitude at all levels [Faccioli and Reséndiz (1976)[29]]

#### *Measurement of Values*

*Laboratory tests* are used to measure the variation in shear modulus and damping as a function of stress-strain amplitude up to levels of strong-motion interest (*see Arts. 3.4.4 and 3.5.5*).

*In situ or field tests* (*see Art. 3.5.5*) take the form of direct-wave seismic surveys (*see Art. 2.3.2*), which provide compression and shear-wave velocities from which $G$ and other dynamic properties are computed. Because the moduli are obtained at lower amplitudes than those imposed by earthquakes, they are likely to be somewhat higher than reality. Values obtained from in situ testing are scaled down by comparing the results with those obtained for the same soils from the laboratory testing. Approximate strain ranges for earthquake laboratory and field tests are compared on Fig. 11.31.

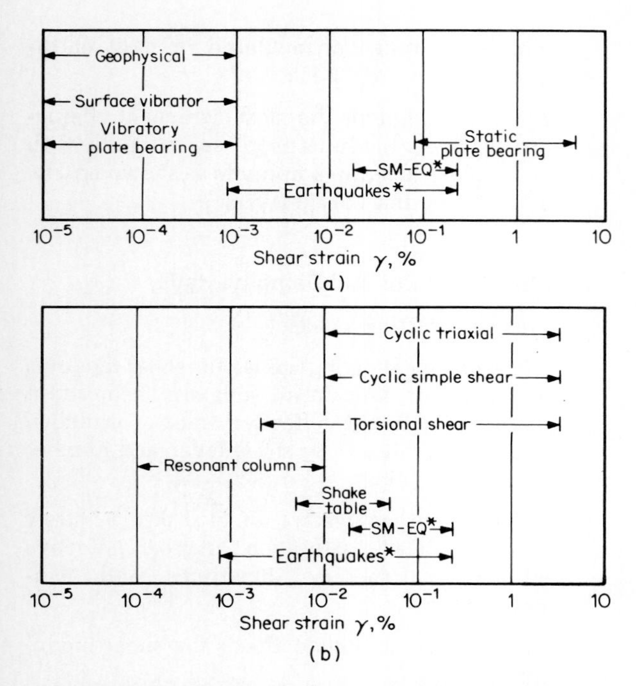


***Range of shear strain denoted as "earthquake" represents an extreme range for most earthquakes. "SM-EQ" denotes strains induced by strong-motion earthquakes.**

**FIG. 11.31 Comparison of approximate strain ranges for earthquakes, field and laboratory testing: (*a*) field tests and (*b*) laboratory tests.** [*From USAEC (1972).*[11]]

*Evaluation of Data*

The evaluation of shear modulus and damping ratio data is described in USAEC (1972)[11] and Hardin and Drenvich (1972*a*),[58] (1972*b*)[59].

*Applications* to soil-structure interaction problems are discussed in Art. 11.4.5.

### 11.3.3 SUBSIDENCE AND LIQUEFACTION

#### General

Earthquake-induced vibrations can be the cause of several significant phenomena in soil deposits, including:

- *Compaction* of granular soils resulting in surface subsidence, which at times occurs over very large areas
- *Liquefaction* of fine sands and silty sands, which results in a complete loss of strength and causes structures to settle or even overturn and slopes to fail
- *Reduction in strength* in soft, cohesive soils (strain softening), which results in settlement of structures that can continue for years and also results from a form of liquefaction

#### Subsidence from Compaction

*Causes*

*Cyclic shear strains* densify granular soils, resulting in subsidence. *Horizontal motions* induced by shocks cause compaction as long as the cycles are relatively close together, even if the cyclic shear strains are relatively small. *Vertical accelerations* in excess of 1g are required to cause significant densification of sands, which is far greater than most surface accelerations during earthquakes. This has been demonstrated by laboratory tests [Whitman and DePablo (1969)[60]].

*Susceptibility Factors*

As noted in the discussion of *liquefaction* below, the susceptibility of soils to compaction during ground shaking depends on soil gradation, relative density or void ratio, confining pressure, amplitude of cyclic shear stress or shear strain, and number of stress cycles or duration.

Compaction subsidence and liquefaction are closely related; the major difference in occurrence is the ability of the material to drain during cyclic loading. Compaction occurs with good soil drainage.

*Occurrence*

NEW MADRID EVENTS OF 1811 AND 1812 Ground subsidence extended over enormous areas, and was reported to be as great as 5 to 7 m in the Mississippi valley.

HOMER, ALASKA (1964) A deposit of alluvium 140 m in original thickness subsided 1.3 m [Seed (1970*a*),[26] (1975)[61]].

NIIGATA, JAPAN (1964) Many structures underlain

by sand settled more than 1 m [Seed (1970*a*),[26] (1975)[61]].

## The Liquefaction Phenomenon

### *Cyclic Liquefaction in Granular Soils*

DEFINED Cyclic liquefaction refers to the response of a soil, subjected to dynamic loads or excitation by transient shear waves, which terminates in a complete loss of strength and entry into a liquefied state. (*Cyclic* liquefaction differs from the liquefaction that occurs during the upward flow of water under static conditions.)

DESCRIBED If a saturated sand is subjected to ground vibrations it tends to compact and decrease in volume; if the sand cannot drain rapidly enough, the decrease in volume results in an increase in pore pressure. When the pore pressure increases until it is equal to the overburden confining pressures, the effective stress between soil particles becomes zero, the sand completely loses its shear strength and enters a liquefied state.

ORIGIN Wylie and Streeter (1976)[62] hypothesized that the shearing motion of the soil causes a slippage or sliding of soil grains which weakens the soil skeleton temporarily and causes the constrained modulus to be reduced. At the time of shear reversal, the particles do not slide, so the skeleton recovers much of its original strength, but in a slightly consolidated form. The consolidation reduces the pore volume, thereby tending to increase pore pressure and to reduce the effective stress in the soil skeleton, Since shear modulus and maximum shear stress depend on effective stress, the horizontal shaking causes a trend towards zero effective stress, and hence liquefaction. Drainage by percolation tends to reduce pore-pressure rise and cause stabilization.

GROUND RESPONSE The phenomenon can occur in a surficial deposit or in a buried stratum. If it develops at depth, the excess hydrostatic pressures in the liquefied zone will dissipate by upward water flow. A sufficiently large hydraulic gradient will induce a "quick" or liquefied condition in the upper layers of the deposit. The result is manifested on the surface by the formation of boils and mudspouts and the development of "quicksand" conditions. As the ground surface liquefies and settles in an area with a high groundwater table, the water often will flow from the fissures of the boils and flood the surface. Even if surface liquefaction does not occur, subsurface liquefaction can result in a substantial reduction in the bearing capacity of the overlying layers.

*Surface effects* can be dramatic as shown by occurrences in Alaska and Japan in 1964 and Chile in 1960. Buildings settled and tilted (Fig. 11.32), islands submerged, dry land became large lakes, roads and other filled areas settled, differential movement occurred between bridges and their approach fills, and trucks and other vehicles even sank into the ground.

### *Soft Cohesive Soils*

*Partial liquefaction* can be said to occur in soft cohesive soils. Longitudinal waves, because of their characteristics of compression and dilation, induce pore-water pressures in saturated clays. The seismically induced pore pressures reduce the shear strength of the soil, and subsequently the bearing capacity, resulting in partial or total failure. Deformation in soft to medium consistency clays from the horizontal excitation will be essentially pure shear [Zeevaert (1972)[25]].

*Increased settlements* of existing structures may result. An example of partial liquefaction is a building in Mexico City founded on soft silty clays that was not undergoing significant settlements until the July 28, 1957, earthquake. During the quake the building settled 5 cm and continued to settle for years afterward at rates of 3 to 5 cm/year. The shear forces from the earthquake-induced seismic waves reduced the shear strength of the clay and significantly increased compressibility by the phenomenon of "strain softening."

*Rupture of foundation members* may result as the seismic shear forces cause buildings to translate horizontally, imposing high earth pressures on walls and bending forces on piles and piers, especially where soft clays are penetrated. Soft-

**FIG. 11.32 Apartment houses at Niigata, Japan, overturned when supporting soils liquefied during the 1964 earthquake.** *(Photograph courtesy of H. Bolton Seed.)*

ening of the clay due to the cyclic strains may be a factor in inducing rupture.

## Occurrence of Liquefaction

*Geographic Distribution*

Incidence of occurrence is not great in comparison with the large number of earthquakes that occur annually. Studies of earthquake records have produced relatively few cases where liquefaction was reported, even though the records extended back to 1802 [Seed (1975),[61] Christian and Swiger (1975)[63]]. Known cases of liquefaction were reported for 13 locations of which two were earth dams. Magnitudes were generally greater than 6.3.

- *Japan:* Mino Qwari (1891), Tohnankai (1944), Fukui (1948), Niigata (1964), and Tokachioki (1968)
- *United States:* Santa Barbara (1925, the Sheffield Dam), El Centro (1940), San Francisco (1957), San Fernando (1971, the Van Norman Reservoir Dam), San Francisco (1971)
- *Others:* Chile (1960), Alaska (1964), and Caracas (1967)

*Geologic Factors and Susceptibility*

*Geologic factors* influencing the susceptibility to liquefaction include sedimentation processes, age of deposition, geologic history, water-table depth, gradation, burial depth, ground slope, and the nearness of a free face.

The potential *susceptibility* for soils of various geologic origins *(see Chap. 7)* in terms of age are summarized on Table 11.10. Susceptibility is seen to decrease as the age of the deposit increases, which reflects prestressing by

**TABLE 11.10**
**ESTIMATED SUSCEPTIBILITY OF SEDIMENTARY DEPOSITS TO LIQUEFACTION DURING STRONG SEISMIC SHAKING***

| Type of deposit | General distribution of cohesionless sediments in deposits | Likelihood that cohesionless sediments, when saturated, would be susceptible to liquefaction (by age of deposit) | | | |
|---|---|---|---|---|---|
| | | <500 year | Holocene | Pleistocene | Prepleistocene |
| **Continental Deposits** | | | | | |
| River channel | Locally variable | Very high | High | Low | Very Low |
| Floodplain | Locally variable | High | Moderate | Low | Very low |
| Alluvial fan and plain | Widespread | Moderate | Low | Low | Very low |
| Marine terraces and plains | Widespread | | Low | Very low | Very low |
| Delta and fan-delta | Widespread | High | Moderate | Low | Very low |
| Lacustrine and playa | Variable | High | Moderate | Low | Very low |
| Colluvium | Variable | High | Moderate | Low | Very low |
| Talus | Widespread | Low | Low | Very low | Very low |
| Dunes | Widespread | High | Moderate | Low | Very low |
| Loess | Variable | High | High | High | Unknown |
| Glacial till | Variable | Low | Low | Very low | Very low |
| Tuff | Rare | Low | Low | Very low | Very low |
| Tephra† | Widespread | High | High | ? | ? |
| Residual soils | Rare | Low | Low | Very low | Very low |
| Sebka‡ | Locally variable | High | Moderate | Low | Very low |
| **Coastal Zone** | | | | | |
| Delta | Widespread | Very high | High | Low | Very low |
| Estuarine | Locally variable | High | Moderate | Low | Very low |
| Beach | | | | | |
| High wave energy | Widespread | Moderate | Low | Very low | Very low |
| Low wave energy | Widespread | High | Moderate | Low | Very low |
| Lagoonal | Locally variable | High | Moderate | Low | Very low |
| Fore shore | Locally variable | High | Moderate | Low | Very low |
| **Artificial** | | | | | |
| Uncompacted fill | Variable | Very high | | | |
| Compacted fill | Variable | Low | | | |

*From Youd and Perkins (1978).[64]

†Tephra—coastlines where slopes consist of unconsolidated volcanic ash or bombs.

‡Sebkha—flat depression, close to water table, covered with salt crust, subject to periodic flooding and evaporation. Inland or coastal.

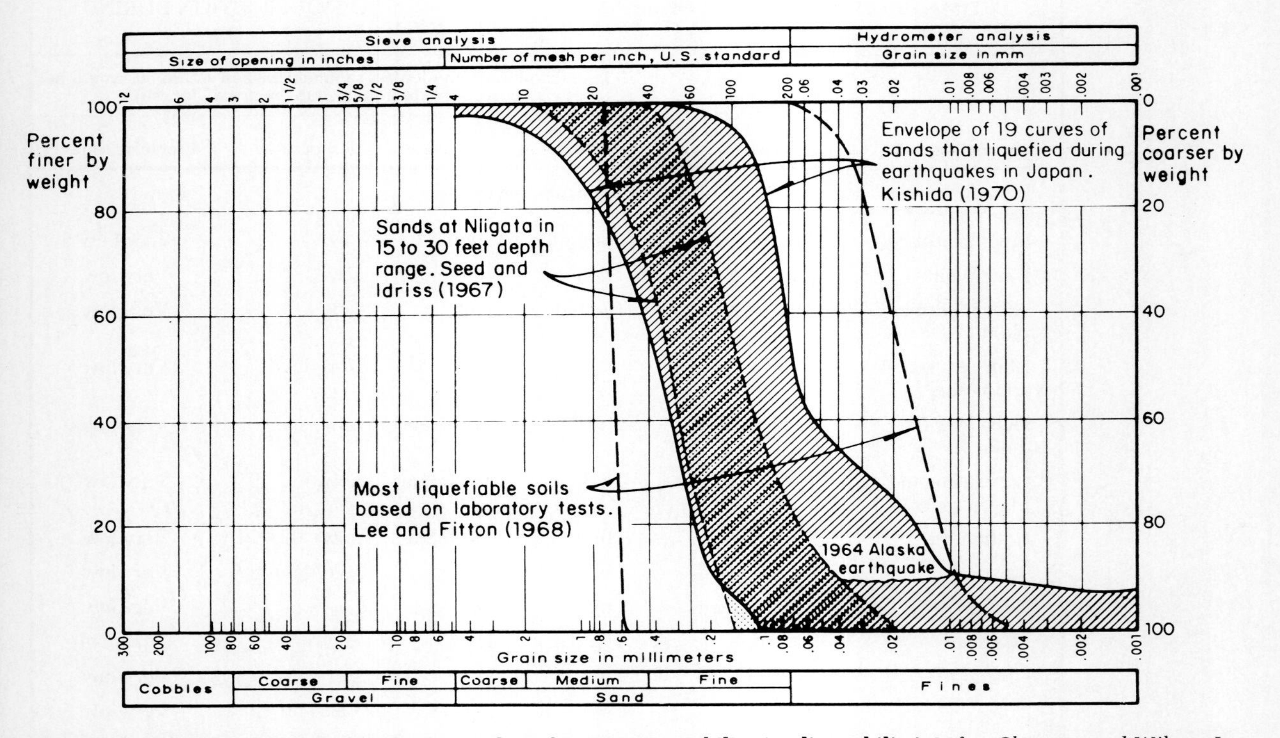


**FIG. 11.33 Effect of grain size distribution on liquefaction susceptibility (cyclic mobility).** [*After Shannon and Wilson, Inc. and Agbabian-Jacobsen Associates (1971)*[65]*; from Finn (1972).*[66]]

removal of overburden or densification by ancient earthquakes. The *greatest* susceptibility is in coastal areas where saturated fine-grained granular alluvium predominates, often with limited confinement, and recent alluvium appears more susceptible than older alluvia. *Offshore* liquefaction must be considered since the seafloor can become unstable from earthquakes or wave forces during large storms (*see Art. 11.3.4*).

## Factors of Liquefaction Potential

### General

GRADATION As shown on Fig. 11.33, fine sands and silty sands are most susceptible, especially when they are poorly graded. Permeability is relatively low and drainage slow.

GROUNDWATER CONDITIONS To be susceptible the stratum must be below the groundwater level and saturated, or nearly so, without the capacity to drain freely.

RELATIVE DENSITY D'Appolonia (1970)[67] suggested that liquefaction might occur where $D_R$ values were as high as 50% during ground accelerations in the order of 0.1g, but for sands with $D_R$ in the range of 75% or greater, liquefaction was unlikely.

BOUNDARY DRAINAGE CONDITIONS AND SOIL STRATIGRAPHY These factors affect the rate of pore-pressure increase.

INITIAL EFFECTIVE OVERBURDEN PRESSURE Also known as depth effect, this pressure influences susceptibility.

DURATION, AMPLITUDE, AND PERIOD OF INDUCED VIBRATIONS These factors influence liquefaction potential.

### Soil Conditions

SUSCEPTIBLE SOILS Seed (1975),[61] from a study of four case histories, concluded that it appears that liquefaction occurs in relatively uniform, cohesionless soils for which the 10% size is between 0.01 and 0.25 mm, and the uniformity coefficient *(see Art. 3.2.3)* is between 2 and 10. In general, the liquefiable soils had SPT *N* values less than 25. Zeevaert (1972)[25] considers that, under certain conditions, soft to medium clays undergo a partial liquefaction, although they do not become fluid in level ground. Many types of clays and clayey silts are susceptible to liquefaction on slopes as discussed in Art. 11.3.4.

NONSUSCEPTIBLE SOILS Gravels and sandy gravels, regardless of *N* values [Seed (1975)[61]], and stiff to hard clays or compact sands, regardless of being situated on level ground or in slopes, appear to be nonsusceptible to liquefaction.

### Foundation Damage Susceptibility

*Case Study: Niigata, Japan [Seed (1975)[61]]*

GENERAL During the earthquake in Niigata, Japan, in 1964, liquefaction caused a great amount of damage but distribution was random. The city is underlain by sands up to depths of 30 m. Gradation is characterized generally by a 10% size ranging from about 0.07 to 0.25, with a uniformity coefficient between 2 and 5 (uniformly or poorly graded).

DAMAGE ZONING Three zones were established relating damage to soil conditions:

- *Zone A:* Coastal dune area with dense granular soils and a relatively deep water table experienced very little damage to structures.
- *Zone B:* Relatively old alluvium of medium-compact to loose sands, with a high groundwater table experienced relatively light damage to structures.
- *Zone C:* Recent alluvium of loose sands with high water table experienced heavy damage. The primary difference between zones B and C appears to be in soil density as revealed by SPT *N* values. Heavy damage implies that buildings suffered large settlements or tilted, such as illustrated on Fig. 11.32, a building supported on spread footings. Reinforced concrete buildings in zone C were supported either on shallow spread footings or on piles.

DAMAGE VS. SPT VALUES Japanese engineers found that buildings supported on shallow foundations suffered heavy damage where *N* values were less than 15, and light to no damage where *N* waves were between 20 and 25. For buildings supported on pile foundations (lengths ranged from 5 to 18 m), damage was heavy if the *N* value at the tip was less than 15. A relationship among foundation depth, SPT value, and extent of damage, developed from Seed's study (1975),[61] is given on Table 11.11. Seed concluded that two important factors vary with depth: the SPT value as affected by soil confinement *(see Art. 3.4.5)*, and ground acceleration, which is usually considered to be larger at the ground surface and to decrease with depth.

**TABLE 11.11**
**RELATIONSHIP BETWEEN FOUNDATION DEPTH, SPT, AND DAMAGE AT NIIGATA***

| Foundation depth range, m | *N* value at foundation base | Damage relationship |
|---|---|---|
| 0–5 | 14 | Apparently adequate to prevent damage by settlement or overturning |
| 5–8 | 14–28 | Required to prevent heavy damage |
| 8–16 | 28 | Required to prevent heavy damage |

*After Seed (1975).[61]

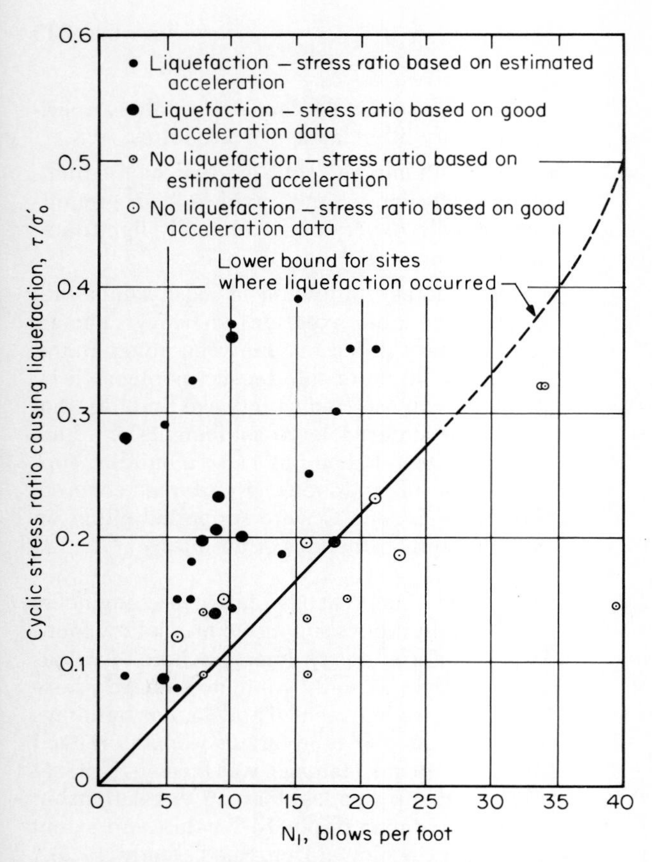


**FIG. 11.34 Correlation between the cyclic stress ratio causing liquefaction and the penetration resistance of sand.** [*From Seed (1976).*[70]]

## Predicting the Liquefaction Potential

*General*

Various investigators have studied locations where liquefaction did not occur in attempts to obtain correlations for predicting liquefaction potential on the basis of material density, initial effective overburden stress, and the earthquake-induced cyclic horizontal shear stress [Castro (1975),[68] Christian and Swiger (1975),[63] Seed et al. (1975),[69] Seed (1976)[70]].

*Cyclic Stress Ratio*

The cyclic stress ratio has been proposed as a basis for anticipating liquefaction potential [Seed et al. (1975),[69] Seed (1976)[70]]. It is defined as the ratio of the average horizontal shear stress ($\tau_h$) induced by an earthquake, to the initial effective overburden pressure ($\bar{\sigma}_o$).

For sites where liquefaction occurred, a lower bound was plotted in terms of the cyclic stress ratio vs. corrected SPT values $N_1$, where $N_1$ is the measured $N$ value corrected to an effective overburden pressure of 1 tsf (*see Art. 3.4.5*). The correlation is given on Fig. 11.34.

The cyclic stress ratio at any depth in the ground causing liquefaction can be calculated with reasonable accuracy from the relationship given by Seed (1976)[70]:

$$\frac{\tau_{hav}}{\bar{\sigma}_o} = 0.65 \frac{a_{max}}{g} \frac{\sigma_o}{\bar{\sigma}_o} r_d \qquad (11.22)$$

where $a_{max}$ = maximum acceleration at the ground surface

$\sigma_o$ = total overburden pressure on the stratum under consideration

$\bar{\sigma}_o$ = effective overburden pressure on the stratum under consideration

$r_d$ = a stress reduction factor varying from a value of 1 at the ground surface to 0.9 at a depth of 30 ft (9 m)

The cyclic stress ratio and $N_1$ are then entered into Fig. 11.34 to determine the liquefaction potential. (The important factor of *duration* is not considered in the evaluation.)

### Treatment for Liquefaction Prevention

*Avoid construction* in seismically active areas underlain by loose fine-grained granular soils where the water table may rise to within about 10 m of the surface, especially adjacent to water bodies.

*Relatively shallow deposits* may be treated by excavation and replacement of the susceptible soils with engineered compacted fill, or by the support of structures on foundations bearing on nonsusceptible soils.

*Moderately deep deposits* may be treated by densification with vibroflotation or dynamic compaction, or by strengthening with pressure grouting, or by improvement of internal drainage. The latter may be accomplished with cylindrical, vertical gravel or rock drains. A series of charts is presented by Seed and Booker (1977)[71] which provides a basis for the design and selection of a suitable drain system for effective stabilization of potentially liquefiable sand deposits by relieving pore pressures generated by cyclic loading as rapidly as they are generated.

### 11.3.4 SLOPE FAILURES

#### Natural Slopes

*General Occurrence*

Seismic forces cause numerous slope failures during earthquakes, often as a result of the development of high pore pressures. Such pressures are most likely to be induced in heavily jointed or steeply dipping stratified rock on steep slopes, and in saturated fine-grained soils even on shallow slopes. Loess, or other deposits of fine sands and silts, and clays with seams and lenses of fine sand or silt are all highly susceptible.

*Debris Slides and Avalanches*

Shallow debris slides are probably the most common form of slope failure during earthquakes and can be extremely numerous in hilly or mountainous terrain. Very large mass movements occur on high, steep slopes such as in the avalanche that buried most of the cities of Yungay and Ranrahirca during the 1970 Peruvian event *(see Art. 9.2.8)*.

An earthquake-induced debris avalanche in relatively strong materials occurred at Hebgen Lake, Montana, during the August 17, 1959, event ($M = 7.1$). Approximately 43 million $m^3$ of rock and soil debris broke loose and slid down the mountainside, attaining speeds estimated at 150 km/hr when it crossed the valley. Its momentum carried it 120 m up the opposite side of the valley, and the material remaining in the valley formed a natural dam and new lake.

*Lateral Spreading*

Common in lowlands along water bodies, lateral spreading results in considerable damage, especially to bridges and pipelines. During the Alaska quake of 1964, 266 bridges were severely damaged as a consequence of lateral spreading of floodplain deposits toward stream channels. During the San Francisco event of 1906, every major pipeline break occurred where fills overlay the soft bay muds [Youd (1978)[72]].

The Turnagain Heights failure that occurred during the 1964 Alaska quake is described in detail in Art. 9.2.6 *(see Fig. 9.45)*. Of interest is the previous earthquake history for the area without the incidence of major sliding including: $M = 7.3$ (1943) with an epicentral distance of 60 km, $M = 6.3$ (1951) with an epicentral distance of 80 km, and $M = 7$ (1954) with an epicentral distance of 100 km. The 1964 event had $M = 8.3$ with an epicentral distance 120 km, but a duration of about 3 min, which appears to have been the cause of many large slope failures.

*Flows*

Flows can be enormous in extent under certain conditions. During the 1920 earthquake in Kansu, China, formations of loess failed, burying entire cities. The cause apparently was the development of high pore-air pressures. The flow debris that extends for a distance of 25 km down the valley of the Rio La Paz near La Paz, Bolivia, is considered to be the result of an ancient earthquake *(see Art. 9.2.11)*.

*Offshore*

*Flows or "turbidity currents"* offshore can also reach tremendous proportions. An earthquake during November 1929 is considered the cause of the enormous "turbidity current" off the coast of Newfoundland. It is speculated that a section of the Continental Shelf broke loose from the Grand Banks, mixed with seawater and formed a flow that moved downslope along the continental rise to the lower ocean floor for a distance of about 925 km. Its movements were plotted from the sequential breaking of a dozen marine cables in about 13 hr, which yielded an average velocity of 70 km/hr [Hodgson (1964)[8]].

*Large submarine flows* occurred during the Alaskan event of 1964, carrying away much of the port facilities of Seward, Whittier, and Valdez. At Valdez, 75 million $m^3$ of deltaic sediments moved by lateral spreading, resulting in displacements in the city behind the port as great as 6 m [Youd (1978)[72]].

*Large submarine slides* may also occur, such as the one depicted on the high-resolution seismic profile given as Fig. 9.56, presumably caused by

**FIG. 11.35 The Lower San Fernando Dam after the San Fernando earthquake of February 9, 1971. A major slide occurred along the upstream face caused by liquefaction of hydraulic fill in the dam body.** *(Photograph courtesy of the U.S. Geological Survey.)*

an earthquake affecting the Kayak Gulf of Alaska.

### Earth Dams and Embankments

*Occurrence*

If *earth embankments,* such as those for roadway support, fail during earthquakes, it is usually by lateral spreading due to foundation failure such as occurred in San Francisco and the coastal cities of Alaska as described in the previous article.

*Earth dams,* when *well built,* can withstand moderate shaking, in the order of 0.2g or more, with no detrimental effects. Dams constructed of clay soils on clay or rock foundations have withstood extremely strong shaking ranging from 0.35 to 0.8g (from an $M = 8.5$ event) with no apparent damage. The greatest risk of damage or failure lies with dams constructed of saturated cohesionless materials that may be subjected to strong shaking. A review of the performance of a large number of earth dams during a number of earthquakes is presented by Seed et al. (1978)[73]. They list six dams in Alaska, California, Mexico, and Nevada that are known to have failed, three dams in California and Nevada known to have suffered heavy damage, and numerous dams in Japan that suffered embankment slides.

*Foundation failure* appears to have caused the collapse and total failure of the Sheffield Dam, near Santa Barbara, California, during the 1925 quake [Seed et al. (1969)[74]].

*Case Study: The Upper and Lower San Fernando Dams [Seed et al. (1975)[75]]*

EVENT During the 1971 San Fernando earthquake ($M$ = 6.6), the two dams on the lower Van Norman Reservoir complex, located about 14 km from the epicenter, suffered partial failures. If either dam had failed completely, a major disaster would have occurred, since some 80,000 persons were living downstream.

DESCRIPTIONS The Upper San Fernando Dam was 24 m high at its maximum section. During the earthquake the crest moved downstream about 1.5 m and settled about 0.9 m. Severe longitudinal cracking occurred on the upstream slope, but there was no overtopping or breaching. The Lower San Fernando Dam was 43 m high at its maximum section, and it suffered a major slide in the upstream slope and part of the downstream slope, leaving about 1.5 m of freeboard in a very precarious position as shown on the photo (Fig. 11.35).

Both dams were constructed by a combination of compacted fill and semihydraulic fill during times when little was known about engineered compacted fill. The Lower Dam was completed in 1915, and raised twice, in 1924 and 1930. The Upper Dam was completed in 1922. The dams had withstood the not-so-distant Kern County event of 1952 ($M$ = 7.7). The locations of the events of 1971 and 1952 are given on the isoseismal maps (*see Fig. 11.16*).

*Instrumentation* had been installed in recent years, including piezometers in the Upper Dam and two seismoscopes on the Lower Dam, one on the crest and one on the rock of the east abutment.

*Pseudostatic analysis* performed before the earthquake had evaluated the stabilities against strong ground motion and found the dams to be safe (*see Art. 9.3.2*, discussion of earthquake forces).

*Dynamic analysis* of the response of the dams to earthquake loadings appeared to provide a satisfactory basis for assessing the stability and deformations of the embankments [Seed et al. (1975)[75]].

- *Lower Dam:* Dynamic analysis indicated the development of a zone of liquefaction along the base of the upstream shell, which led to failure. Evidence of liquefaction was provided by the seismoscopes, which indicated that the slide had developed after the earthquake had continued for some time, when ground motions had almost ceased following the period of strong ground shaking. The investigators concluded that since the slide did not occur when the induced stresses were high, but rather under essentially static load conditions, there was a major loss of strength of some of the soil in the embankment during ground shaking.
- *Upper Dam:* Dynamic analysis indicated that the dam would not undergo complete failure, but that the development of large shear strains would lead to substantial deformations in the embankment.

### Analytical Methods

*Dynamic Analysis*

The procedure is essentially similar to that performed for structures as described in Art. 11.4. Characteristics of the motion developed in the rock underlying the embankment and its soil foundation during the earthquake are estimated, the response of the foundation to the base rock excitation is evaluated, and the dynamic stresses induced in representative elements of the embankment are computed.

Representative samples of the embankment and foundation soils are subjected to laboratory testing under combinations of preearthquake stress conditions and superimposed dynamic stresses to permit assessment of the influence of earthquake-induced stresses on the potential for liquefaction and deformations. From the data the overall deformations and stability of the dam sections are analyzed.

*Pseudostatic Analysis* is discussed briefly in Art. 9.3.2.

*Earthquake Behavior Analysis of Earth Dams*

Pseudostatic analysis is now generally recognized as being inadequate to predict behavior during earthquakes. The Committee on Earthquakes of the International Commission on Large Dams [ICOLD (1975)[76]] has recommended the dynamic analysis approach for high embankment dams whose failure may cause loss of life or major damage.

Design proceeds first by conventional methods followed by dynamic analysis to investigate any deficiencies which may exist in the pseudostatic design. A simplified procedure for estimating dam and embankment earthquake-induced deformations is presented by Makdisi and Seed (1978)[77].

### 11.3.5 TSUNAMIS AND SEICHES: RESPONSE OF LARGE WATER BODIES

#### Tsunamis

*General*

*Tsunamis* are long sea waves which can reach great heights when they encounter shorelines, where they represent a very substantial hazard.

*Causes* are not precisely known, although they are associated with earthquakes. Seismologists generally agree that they reflect some sudden change in seafloor topography such as upthrusting or downdropping along faults, or less frequently the sliding of unconsolidated material down continental shelves.

*Occurrence* is infrequent, but tsunamis are potentially very damaging.

*Geographic Occurrence*

PACIFIC OCEAN REGIONS The regions of most frequent occurrence include the Celebes Sea, Java Sea, Sea of Japan, and the South China Sea, but in recent years tsunamis have struck and caused damage in Crescent City, California, and Alaska from the 1964 event; and in Hawaii (Hilo), Japan, and Chile from the 1960 Chilean quake. Japan probably has the greatest incidence of tsunamis of any country, and has been subjected to 15 destructive tsunamis since 1956, eight of them disastrous [Leggett (1973)[78]]. One of the worst tsunamis in history occurred along the northeast coast in June 1896 when a wave 23 to 30 m above sea level rushed inland, destroying entire villages and killing more than 27,000 people. The cause was considered to be a nearby earthquake. The Hawaiian Islands are subjected to a serious tsunami about once every 25 years [Leggett (1973)[78]].

ATLANTIC OCEAN Occurrence is very infrequent. Following the Lisbon earthquake of 1755, the sea level was reported to have risen to 6 m at many points along the Portuguese coast, and in some locations to 15 m. The Grand Banks event of 1929, with the epicenter located about 400 km offshore, caused a great tsunami that was very destructive along the Newfoundland coast and took 27 lives. It may have been caused by the turbidity current described in Art. 11.3.4.

INDIAN OCEAN (BAY OF BENGAL) Relatively frequent occurrence.

MEDITERRANEAN SEA Occasional occurrence.

*Characteristics*

AT SEA Tsunamis can be caused by nearby earthquakes, or as often occurs, by earthquakes with epicenters thousands of kilometers distant from the land areas they finally affect. They are never observed by ships at sea because in the open sea, *wave amplitudes* are only a meter or so. They travel at great *velocities*, in the order of 700 km/hr or more. The tsunami caused by the 1960 Chilean quake ($M = 8.4$) reached Hawaii, a distance of 10,500 km, in a little less than 15 hr, and Japan, 17,000 km distant, in 22 hr. (The velocity of water waves is given approximately by the relationship $v = \sqrt{gD}$, where $g$ is the acceleration of gravity and $D$ is water depth. Wavelength $\lambda$ is given by the relationship $\lambda = vT$, where $T$ is the period. Tide gages around the Pacific showed the Chilean tsunami to have a period of about 1 hr; therefore, its *wavelength* was about 700 km.)

COASTAL AREAS The magnitude of a tsunami at its source is related to the earthquake magnitude. When it arrives at a coastline, the effect is influenced by offshore seafloor conditions, wave direction, and coastline configuration. Wavelengths are accentuated in bays, particularly where they have relatively shallow depths and topographic restrictions. The wave funnels into the bay and builds to great heights. Containing tremendous energy, the wavefront runs up onto the shore, at times reaching several kilometers inland. The crest is followed by the trough during which there is a substantial drawdown of sea level, exposing the seafloor well below the low-tide level. After an interval of 30 min to an hour, depending upon the wave period, the water rises and the second wave crest, often higher than the first, strikes the beach. This sequence may continue for several hours, and the third or fourth wave may sometimes be the highest. At Hilo, Hawaii, after the Chilean event of 1960, the first wave reached 1.2 m above mean sea level, the second 2.7 m, and the third, 10 m. On Honshu and Hokkaido, Japan, the water rose 4 m along the coast during the Chilean tsunami.

*Early Warning Services*

After the very damaging 1946 tsunami, an early warning service was established by the USGS and centered in Hawaii. When seismograph stations in Hawaii show a Pacific Ocean focus earthquake, radio messages are sent to other Pacific seismograph stations requesting data from which to determine the epicenter. Adequate time is available to compute when tsunami waves might arrive and to so warn the public in coastal areas.

In general, a "watch" is initiated for magnitudes of 7.5 or greater. "Warnings" are issued if tide gages detect a tsunami [Kerr (1978)[79]]. Unless waves strike the shores near the epicenter, however, there is no way for people on distant shorelines to know if a tsunami has been generated. Even though the Chilean earthquake caused tsunamis along the Chilean coastline, many people in Hawaii chose to ignore the warning and not to move to higher ground. Japanese officials similarly ignored the warning, since a Chilean earthquake had never before caused a tsunami in Japan. The tsunami reached Hawaii within 1 min of the predicted arrival time.

*Hazard Prediction*

When important structures or new communities are located along shorelines, the potential for the occurrence of tsunamis should be evaluated. There is no quantitative method available for such an evaluation.

Qualitatively there are several high-hazard conditions to evaluate:

- Regional tsunami history and recurrence.
- Near onshore earthquakes: recurrence and magnitude.
- Coastline configuration: Irregular coastlines with long and narrow bays and relatively shallow waters appear to be more susceptible than regular coastal plains when exposed to tsunami waves generated by distant earthquakes of large magnitudes.

### Seiches

*Description*

Seiches are caused when ground motion starts water oscillating from one side to the other of a closed or partly closed water body, such as a lake, bay, or channel.

*Occurrence*

Large seiches are formed when the period of the arrivals of various shocks coincides with the natural period of the water body, which is a function of its depth, and sets up resonance.

During the 1959 earthquake at Hebgen Lake, Montana, a witness standing on Hebgen Dam saw the water in the reservoir disappear from sight in the darkness, then return with a roar to flow over the dam. The fluctuation continued appreciably for 11 hr with a period of about 17 min. The first four oscillations poured water over the dam.

The Lisbon event of 1977 ($M = 8.7$) set up seiches all over Europe with the most distant ones reported from Scandinavia, 3500 km distant.

## 11.4 EARTHQUAKE-RESISTANT DESIGN: AN OVERVIEW

### 11.4.1 INTRODUCTION

### Ground Motion

*Dynamic Forces*

The large amounts of energy released during earthquakes travel through the earth as various types of seismic waves with varying oscillation

**FIG. 11.36 Modern freeway structures at the interchange of Highways 5 and 210, San Fernando, California, damaged by the earthquake of February 9, 1971.** *(Photograph courtesy of the U.S. Geological Survey.)*

frequencies and amplitudes or displacements. The oscillating particles in the wave possess velocity and exert a force due to the acceleration of gravity (*see Art. 11.2.3*).

In *rock* and *other nearly elastic geologic materials,* these dynamic forces result in transient deformations which are recovered under the low strains of the seismic waves. Interest in the properties of these materials lies primarily in their ability to transmit the seismic waves.

In *weaker deposits,* such as alluvium, colluvium, and aeolian soils, however, the base-rock excitation transmitted to the soils is usually amplified. In addition, some materials respond to the cyclic shear forces by densifying, liquefying, or reducing in shear strength (unstable soils).

*Significance*

The effect of the dynamic forces, therefore, can be considered in two broad categories:

1. The effect on structures subjected to the forces transmitted through the ground which result in *ground shaking*
2. The effect on the geologic material itself, primarily in the form of *response to cyclic shear forces*

*Field Measurements*

Ground displacement (amplitude), used in Richter's relationship to compute magnitude, is measured by seismographs.

The force imposed on structures, in terms of the acceleration of gravity, which has both horizontal and vertical components, is measured by accelerographs.

**Surface Damage Relationships**

*Factors*

Earthquake destruction is related to a number of factors including the magnitude, proximity to populated areas, and duration of the event, the local geologic and topographic conditions, and the local construction practices.

*Example*

The effects of some of these factors are illustrated by a comparison of damage to two cities during earthquakes of similar magnitude: Managua (1972), $M = 6.2$, and San Fernando (1971), $M = 6.6$; both events affected an area with about 400,000 inhabitants. The Managua quake resulted in 6000 deaths compared with 60 in San Fernando; the difference is reflected in soil conditions and local building practices. Managua is located over relatively weak lacustrine soils, and relatively few structures have been constructed with consideration for seismic forces. The San Fernando valley is filled with relatively compact soils and most major structures have been constructed according to modern practices. Even with "modern" practices, however, a number of new structures were severely damaged as shown on Figs. 11.36 and 11.37.

## 11.4.2 STRUCTURAL RESPONSE

### General Characteristics

*Reaction to Strong Ground Motion*

Dynamic forces are imposed on structures by strong ground motion. Structural response is related to the interaction between the characteristics of the structure in terms of its mass, stiffness, and damping capability, and the characteristics of the ground motion in terms of the combined influence of the amplitude of ground accelerations, their frequency components, and the duration.

The elements of ground motion are amplitude $A$, displacement $y$, frequency $f$, and period $T$, and the derivatives velocity $v$ and acceleration $a$.

**FIG. 11.37 Damage suffered by the Olive View Hospital during the San Fernando earthquake of February 9, 1971 ($M = 6.6$; $I_{max}$ = VIII − XI). Damage to the newly constructed reinforced buildings included the collapse and "pancaking" of the two-story structure in the upper left (1), the collapse of the garage and other structures in the foreground (2), and the toppling of 3 four-story stairwell wings (3). Vertical accelerations combined with horizontal accelerations were a significant cause in the collapse.** *(Photograph courtesy of the U.S. Geological Survey.)*

*Energy Transmitted to Structures*

*Maximum vibration velocity* imposed by the elastic wave as it passes beneath the structure in terms of circular frequency $\omega$ and amplitude is

$$v_{max} = 2\pi f A \qquad (11.23)$$

The *greatest acceleration* to which the structure is subjected is

$$a_{max} = 4\pi^2 f^2 A \qquad (11.24)$$

The *greatest force* applied to the structure is found when acceleration is expressed in terms of the mass. This allows the development of an expression relating force to the acceleration of gravity, frequency, and amplitude as follows:

$$F_{max} = m\,a_{max} = \frac{W}{g}\,a_{max} = \frac{W}{g}(4\pi^2 f^2 A) \qquad (11.25)$$

where $W$ = weight. It is seen that the force varies as the square of the frequency.

*Kinetic energy* is possessed by a body by virtue of its velocity; the energy of a body is the amount of work it can do against the force applied to it, and *work* is the product of the force required to displace a mass and the distance through which the mass is displaced. Kinetic energy is expressed by

$$KE_{max} = \frac{1}{2}\,m\,v^2_{max} = \frac{1}{2}\frac{W}{g}\,v^2_{max} \qquad (11.26)$$

where $v$ = the velocity with which a structure moves back and forth as the seismic wave passes [Leet (1960)[80]].

*Energy transmitted* to the structure may be represented in several ways:

- Motion amplitude (displacement), or frequency, or acceleration (results from combining amplitude and frequency)
- Force with which the energy moves the structure
- Energy itself, defined in terms of the motion which it produces (kinetic energy)

*Ground Shaking and Analysis*

Base excitation of the structure from ground shaking results in horizontal and vertical deflections for some interval of time and imposes strains, stresses, and internal forces on the structural elements. The shaking intensity depends on the maximum ground acceleration, frequency characteristics, and duration.

Analysis requires definition of the system motion in terms of time-dependent functions, and determination of the forces imposed on the structural members.

### Response Modes

Structures exhibit various modes of response to ground motion depending on their characteristics.

*Peak horizontal ground acceleration* relates closely to the lateral forces imposed on a structure and is the value used normally for approximating earthquake effects.

*Vertical acceleration* of ground motion can cause crushing of columns. During a downward acceleration of a structure, the stress in the columns is less than static. When the movement reverses and becomes upward, an acceleration is produced which causes an additional downward force which adds loads to the columns. This effect was a major factor in the collapse of the Olive View Hospital during the San Fernando event (see *Fig. 11.37*). Vertical accelerations are also important input to the design of certain structures such as massive dams and surface structures such as pipelines.

*Differential displacements* beneath structures can cause distortion and failure of longitudinal members.

*Frequency*, in relation to the natural frequency of the structural elements, governs the response of the structure. Low frequencies (long periods) cause tall structures to sway.

*Duration* of shaking or *repeated application* of forces causes fatigue in structural members (and a continuous increase in soil pore pressures).

### Dynamic Reaction of Structures

*Source*

The dynamic reaction of a structure to ground motion is governed by its characteristic period $T$ or structural frequency of vibration $\omega$ which is related to structural mass, stiffness, and damping capability.

### *Characteristic Periods*

- Very rigid structure: 0.1 s (very high frequency, $f = 10$ Hz)
- Relatively stiff structures of 5 to 6 stories: 0.4–0.5 s ($f = 2$ Hz)
- Relatively flexible structures of 20 to 30 stories: 1.5 to 2.5 s ($f = 0.5$ Hz)
- Very flexible structures where deformation rather than strength governs design, and wind loads become important: 3 to 4 s (very low, $f = 0.25$ Hz)

### *Forces on Structural Members*

*Static loads* result in stresses and deflections.

*Dynamic loads* result in time-varying deflections which involve accelerations. These engender inertia forces resisting the motion, which must be determined for the solution of structural dynamics problems. The complete system of inertia forces acting in a structure is determined by evaluating accelerations, and therefore displacements, acting at every point in the structure.

### *Deflected Shape of Structure*

The deflected shape of a structure may be described in terms of either a lumped-mass idealization or generalized displacement coordinates.

In the *lumped-mass idealization*, it is assumed that the entire mass of the structure is concentrated at a number of discrete points, located judiciously to represent the characteristics of the structure, at which accelerations are evaluated to define the internal forces developing in the system.

*Generalized displacement coordinates* are provided by Fourier series representation.

In either case, the number of displacement components of coordinates required to specify the position of all significant mass particles is called the *number of degrees of freedom* of the structure [Clough (1970)[81]].

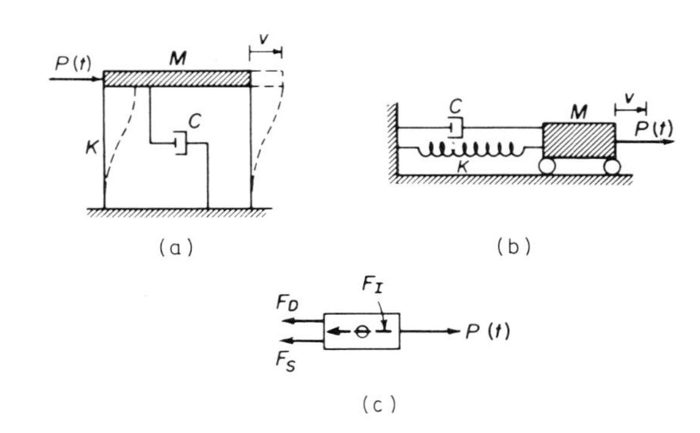


**Inertia force $F_I = M\ddot{V}$**
**Damping force $F_D = C\dot{v}$ $\quad P(t) = F_I + F_D + F_S$**
**Elastic force $F_S = Kv$**
**where $M$ = mass; $C$ = damping coefficient; $K$ = total spring constant; $v$ = displacement; $P(t)$ = external load [period $T = 2\pi\,(M/K)^{1/2}$]**

**FIG. 11.38 Single-degree-of-freedom systems: (*a*) simple frame; (*b*) spring-mass system; (*c*) forces acting on mass.**

### *Single-Degree-of-Freedom System*

Two types of single-degree-of-freedom systems are shown on Fig. 11.38. In both cases the system consists of a single rigid (lumped) mass $M$ so constrained that it can move with only one component of simple translation. The dynamic forces acting on a simple building frame founded on the surface may be represented by a simple mass-spring-dashpot system as shown on Fig. 11.38*a*. (A dashpot is an energy adsorber.)

*Translation motion* is resisted by weightless elements having a total spring constant $K$ (stiffness) and a damping device which adsorbs energy from the system. The damping force $C$ is proportional to the velocity of the mass. The fundamental period $T$ is expressed by:

$$T = 2\pi(M/K)^{1/2} \tag{11.27}$$

During earthquakes the motion is excited by an external load $P(t)$ which is resisted by an inertia force $F_I$, a damping force $F_D$ and an elastic force $F_S$. The resisting forces are proportional to the acceleration, velocity, and displacement of the

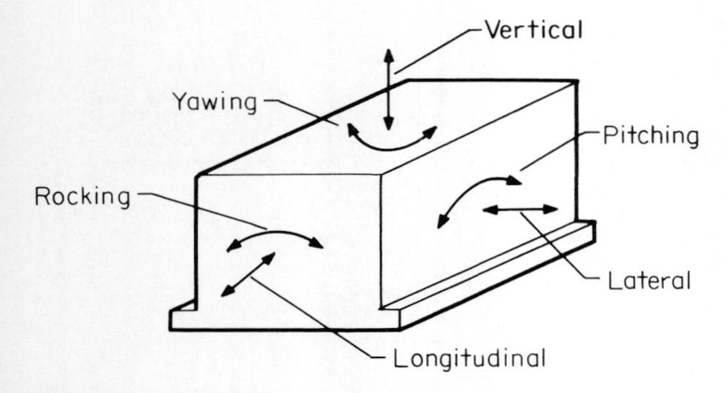


**FIG. 11.39 Six modes, or degrees of freedom, of foundation vibration. Translation modes: vertical, longitudinal, lateral. Rotational modes: rocking, pitching, yawing.**

mass given in terms of the differential of displacement $v$ with respect to time as follows: $F_I = M\ddot{v}$, $F_D = C\dot{v}$, and $F_S = Kv$. Equilibrium requires that:

$$P(t) = F_I + F_D + F_S$$

where the force $P(t)$ is the ground acceleration input, equal to the product of the mass and the acceleration, or

$$P(t) = Ma_{max} \tag{11.28}$$

*Multi-Degree-of-Freedom Systems*

In *multistory buildings,* each story can be considered a single-degree-of-freedom element with its own mass concentrated at floor levels and its own equations of equilibrium similar to the single-story building. One can proceed with analysis by assuming that displacements are of a specific form, for example, that they increase linearly with height.

*Free-foundation bases,* such as those for mechanical equipment, can move in a number of directions as illustrated on Fig. 11.39.

*Dynamic Response*

The dynamic response of a structure is defined by its displacement history, i.e., by the time variation of the coordinates which represent its degrees of freedom, and by its period of vibration $T$ or frequency $\omega$. The displacements are determined from equations of motion, which are expressions of the dynamic equilibrium of all forces acting on the structure.

**External Forces**

*General*

The external forces acting on the structure can be assigned on the basis of the conventional or simple approach or of the comprehensive approach.

*Conventional or Simple Approach*

The conventional or simple approach considers only values for acceleration g, which are obtained from provisions in the Uniform Building Code in the United States, or from some local building code. These codes often refer to seismic probability maps such as Fig. 11.15, or simply give values for acceleration in terms of g for various locations. This approach does not take into account all of the significant dynamic properties of either structures or earthquakes.

*Comprehensive Approach*

All ground response factors (see *Art. 11.4.3*) are considered in the comprehensive approach. The design earthquake (see *Art. 11.4.6*) is considered in terms of peak acceleration, frequency content, and duration by statistical analysis of recorded events; information obtained by strong motion seismographs; and geologic conditions. In terms of the structure, the design spectrum, damping, and allowable design stresses are specified. This approach is becoming standard practice for all high-risk structures such as nuclear power plants, 50-story buildings, large dams, long suspension bridges, and offshore drilling platforms.

## 11.4.3 SITE GROUND-RESPONSE FACTORS

**Design Bases**

Seismic design criteria are based on ground motion characteristics, including acceleration, frequency content, and shaking duration, which are normally given for excitation of rock or strong soils. Soil conditions have an effect on these values.

### Maximum Acceleration

*Peak horizontal acceleration* is considered to be closely related to the lateral forces imposed on a structure, and is the value normally used for approximating earthquake forces.

*Peak vertical acceleration* is generally accepted as about one-half of the mean horizontal acceleration, but close to dip-slip faulting the fraction may be substantially higher. In Managua (1972), for example, recorded near the epicenter were values of $a_{h(peak)} = 0.35g$ and $a_{v(peak)} = 0.28g$. Hall and Newmark (1977)[82] recommended taking design motions in the vertical direction as two-thirds those in the horizontal direction.

*"Effective" peak acceleration* is often selected for design because very high peaks frequently are of short duration and have little effect on a structure. Low-magnitude events, in the order of $M = 4.5$, can have peaks of 0.6g, but these large accelerations usually occur only as two or three high-frequency peaks, which are probably *S* wave arrivals, and carry little energy.

*Selection of values* is described in Art. 11.4.6.

### Frequency Content

*Significance*

Maximum accelerations occur when the ground-motion period (frequency) approaches or equals the period of the structure, and resonance occurs. Ground-motion amplitude decreases with distance by geometrical spreading and frictional dissipation (attenuation). The high frequencies (shorter periods) are close to the source; at distances in the order of 100 km it is the longer vibrational periods of the Rayleigh waves (1 to 3 s) that cause ground shaking.

Building periods are usually in the low to intermediate frequency ranges, and buildings are therefore subject to resonance from the long-period waves. Tall buildings are caused to sway, and, when in close proximity, to beat against each other as in Los Angeles during the Kern County event of 1952, where the focus was 125 km distant, and in Mexico City in 1957, where the focus was 300 km distant. In both cases, old and weak, but smaller structures, did not suffer damage.

*Design Approach [Hudson (1972)[83]]*

In *high-frequency systems* (>5 Hz), the relationship between horizontal ground accelerations and lateral forces on a structure governs design.

*Intermediate* (~1 Hz) and *low frequency* (<0.2 Hz, long periods) systems include most buildings and engineering works. Ground accelerations alone are not considered a good approximation of the actual lateral forces. In addition to $a_h$, the maximum ground velocity for intermediate frequencies and the maximum ground displacement for low frequencies should be specified.

### Duration

Duration, a measure of the number of cycles, is associated with fatigue in structures and has a major effect on the amount and degree of damage.

*Length of faulting* is considered to have a strong effect.

*Bracketed durations*, prepared from strong-ground-motion records, are often specified for design (*see Art. 11.4.6*).

### Comparisons of Acceleration, Frequency, and Duration

- Parkfield (1966, $M = 5.5$): High peak $a$ (0.5g) but high frequency and very short duration caused little damage.
- Mexico City (1957, $M = 7.5$): Lower peak $a$ (0.01 to 0.1g) had lower frequencies and a longer duration, causing complete collapse of multistory buildings in a geologic basin with weak soils.
- Anchorage (1964, $M = 8.6$): High-magnitude event was similar to many historical events except for unusual duration of 3 min which resulted in the liquefaction failure of many natural, previously stable, slopes.

### Soil Condition Effects

*Ground Amplification Factor*

Bedrock excitation accelerations generally increase in magnitude as soil thickness increases and soil stiffness decreases. The ground amplification factor ranges generally between 1 and 2 for strong motions, and in excess of 10 for microtremors (see *Art. 11.2.6*). Cases have been reported, however, of factors as low as 0.5 for deep deposits of soft to stiff clay [Seed (1975)[61]].

*Effects on Frequency*

Local soil conditions filter the motion so as to amplify those frequencies that are at or near the fundamental frequency of the soil profile [Whitman and Protonotarios (1977)[84]], but frequencies are diminished by attenuation.

Various earthquakes may have the same peak acceleration but if it occurs with differing periods, the ground response will differ and structural damage may be selective. For example, in San Francisco (1957) in *stiff soils*, acceleration peaks occurred at low values of the fundamental period (0.4 to 0.5 s); therefore, maximum accelerations would tend to be induced in relatively stiff structures 5 to 6 stories in height, rather than in high-rise buildings. In *deep deposits* of *soft soils*, however, peak acceleration occurred at intermediate values of the fundamental period (1.5 to 2.6 s), which would induce maximum acceleration in multistory buildings of 20 to 30 stories, leaving lower, stiffer buildings unaffected [Seed (1975)[61]].

*Other Factors*

- *Depth effects:* Accelerations at foundation level can be substantially lower than at the surface, as discussed in Art. 11.2.6.
- *Subsidence* and *liquefaction* are discussed in Art. 11.3.3.
- *Slope failures* are discussed in Art. 11.3.4.

*Microzonation Maps*

Geologic conditions and ground response factors presented as microzonation maps have been prepared for a few urban locations. They are useful for planning and preliminary design, and emphasize hazardous areas.

## 11.4.4 RESPONSE SPECTRA

### General

*Description*

If a given ground acceleration is applied to the base of a single-degree-of-freedom system, the behavior of the system as measured by its maximum displacement will depend upon the exciting force and upon the natural period and damping of the system. For a given excitation and a particular value of damping, the maximum displacement of the system could be plotted vs. the natural period of the system. A family of such curves, for various values of damping, would then form a *response spectrum* [Hudson (1970)[85]].

*Applications*

Response spectra are used as input in dynamic analysis of linear elastic systems; i.e., they are a convenient means of evaluating the maximum lateral forces developed in structures subjected to a given base motion. If the structure behaves as a single-degree-of-freedom system, the maximum acceleration and thus the maximum inertia force may be determined directly from the acceleration response spectrum if the fundamental period of the structure is known.

They are useful also for comparing the response of structures in a given earthquake where soil conditions vary, or for comparing a number of earthquakes in the same area, as a function of soil conditions and natural periods.

### Spectra Development

*Description*

The *response function* is a complex integral expressing the response of a damped structure (Fig. 11.38) in which are related a number of factors including acceleration, a time function, a damping factor, and the circular frequency $\omega$ equated to displacement at time $t$. *Spectral velocity* $S_v$ is the maximum value of the response function (a pseudovelocity, not equal to the maximum velocity in a damped system). The *maximum displacement* is $S_d = S_v/\omega$, and the *spectral acceleration* is $S_a = S_d\omega^2$, where $\omega = 2\pi/T$ (Eq. 11.3). The pseudovelocity spectrum is defined by [Housner (1970*a*)[46]]:

$$\left(\frac{T}{2\pi}\right) S_a = S_v = \left(\frac{2\pi}{T}\right) S_d$$

For any given earthquake acceleration at time $t$ (taken from a strong-motion record such as Fig.

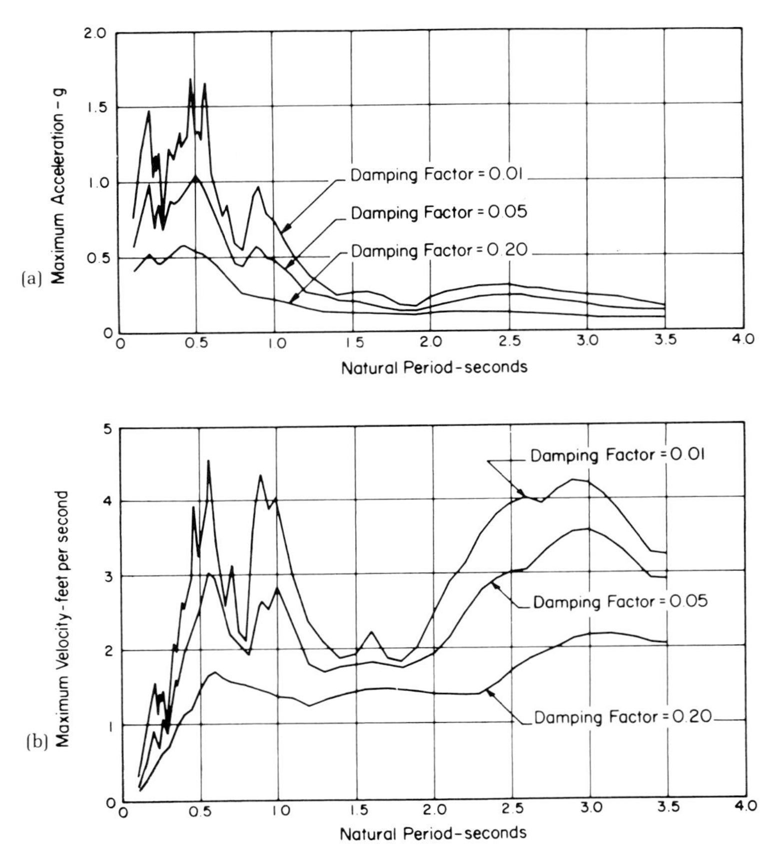


**FIG. 11.40 Acceleration and velocity response spectra for El Centro (1940), developed from the accelerogram records given on Fig. 11.9. If the computed period for the structure illustrated on Fig. 11.38 is found to be 0.2 s, then the maximum acceleration is found from (*a*) to be 1.0g and the maximum velocity from (*b*) is found to be 1 ft/s. Displacement would be found from the displacement spectrum (not shown).** [*After Seed (1975).*[61]]

11.9), and for any specified damping ratio, the spectral velocity is computed as a function of structural frequency or period. The single points obtained are connected to form the graph. Velocity and acceleration response spectra for various damping ratios for the El Centro event (1940) are given on Fig. 11.40.

### *Combined Ground-Motion Records*

To allow application on a regional basis, the response spectra of a number of events are normalized to a standard intensity level and the irregularities smoothed to obtain the velocity, acceleration, and displacement spectra given on

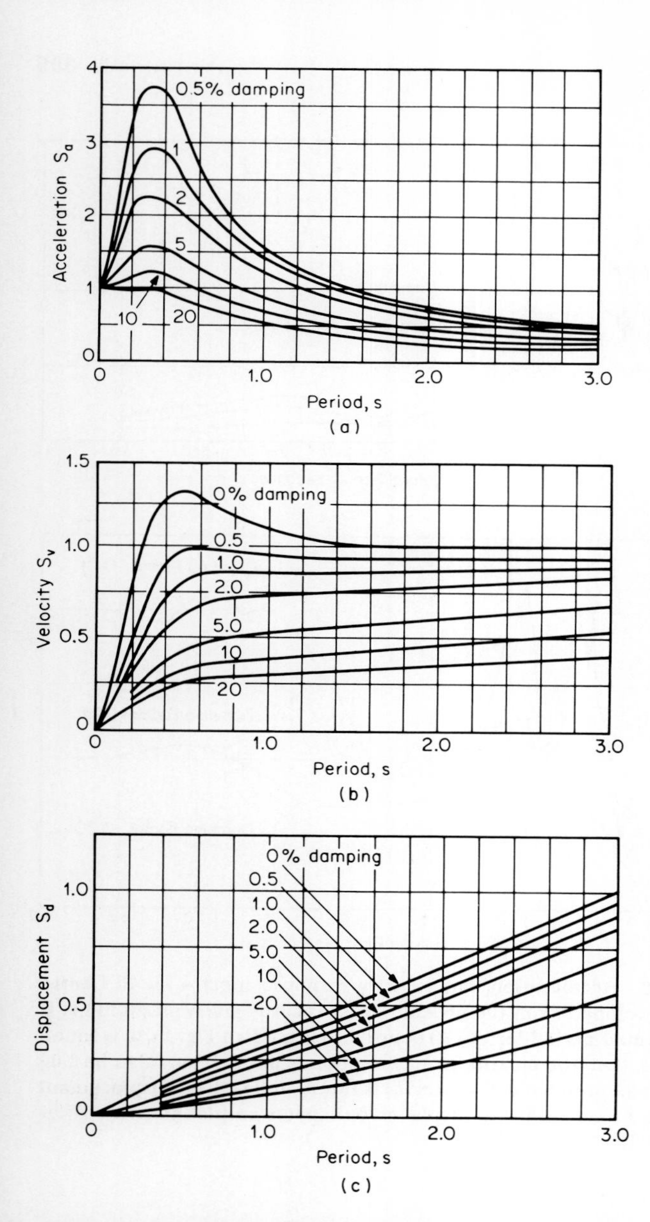


**FIG. 11.41 Design spectra for $S_a$, $S_v$, and $S_d$ as functions of period and damping. Ordinate scale is arbitrary: (*a*) spectral acceleration $S_a$; (*b*) spectral velocity $S_v$; (*c*) spectral displacement $S_d$.** [*From USAEC (1963).*[86]]

Fig. 11.41. These families of curves represent the averages of response spectra for two components of ground motion of four strong-motion records experienced in California, plotted for various damping percentages. The values for $S_a$, $S_v$, and $S_d$, as a function of period and scaled at some value for acceleration as a percentage of g, can be presented as a combined plot such as in Fig. 11.42.

*Spectral envelopes*, providing upper bounds, are also developed for design. The combined plot given as Fig. 11.43, developed in 1973 for nuclear power plant studies, is considered generally applicable to rock or strong soils when the response frequency will be greater than 5 Hz [Newmark et al. (1973)[88]].

## Limitations in Applications

### *Strong-Motion Data*

Available information from strong-motion records is inadequate for the preparation of average response spectra or spectral envelopes computed with a high degree of confidence for all important geologic conditions. They are most accurate where the data are available for a particular region in which construction is to be situated [Bolt et al. (1975)[2]].

### *Peak Ground Velocity Basis*

From a study of ground-motion parameter relationships, McGuire (1978)[89] concluded that the use of design spectra of fixed shape, representing a fixed frequency content and scaled to a peak ground acceleration, is adequate only as a first approximation of ground motion. For sites affected by large, distant earthquakes, the procedure will underestimate spectral velocities except at high frequencies.

The strong-motion data studies by McGuire indicated that the mean ratio of spectral velocity at intermediate frequencies (~1 Hz) to peak ground acceleration increases with earthquake magnitude and distance more than does the ratio of peak ground velocity. The effects of epicentral distances on frequency characteristics are shown on Fig. 11.44 for two strong shocks and a weak shock. Consequently, McGuire considers it most appropriate to estimate magnitudes directly, rather than scaling them from the estimates of peak ground velocity.

### *Elastic Response of Structures Basis*

Large structures often are not designed to remain rigid or elastic during large earthquakes because of economics. Elastic response spectra do not necessarily indicate how local soil conditions will affect inelastic response of structures [Whitman and Protonotarios (1977)[84]].

## 11.4.5 DYNAMIC ANALYSIS

### Method Selection

*General*

The method selected for dynamic analysis varies from the relatively simple conventional approach to comprehensive analytical procedures. The selection depends upon the degree of risk and hazard.

*Structure Purpose and Type*

Pertains to failure consequences (the *risk*).

- *Conventional structures,* moderate risk, include industrial plants, moderate-height buildings, fossil-fuel power plants, moderate height dams, etc.
- *Lifeline structures,* moderate to high risk, include roadways, railways, tunnels, canals, pipelines (gas and liquid fuel, water, and sewage), electric power, etc.
- *Critical public structures,* high risk, include schools and hospitals.
- *Unconventional structures,* high to very high risk, include nuclear power plants, 50-story buildings, long suspension bridges, large dams, offshore drilling platforms, etc.

*Earthquake Occurrence and Magnitude*

Pertains to the *hazard.*

- *High hazard:* Frequent occurrence of moderate to high magnitude events ($I > \text{V}$), or occasional occurrence of high magnitude events ($I > \text{VIII}$).
- *Low hazard:* Generally areas of no activity, or events seldom exceed $I = \text{V}$, and activity is normally low.

*Approach*

*Conventional design* is applied for conventional structures in low-hazard areas, and considers only g forces.

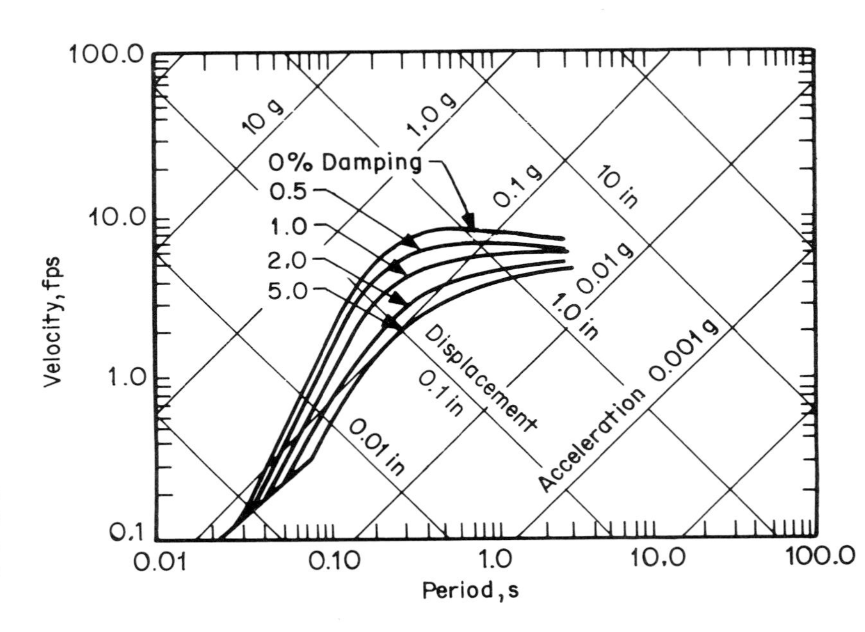


**FIG. 11.42 Combined plot of design spectrum giving $S_a$, $S_v$, and $S_d$ as a function of period and damping, scaled linearly for a peroid of 0 s.** [*From USAEC (1972).*[11]].

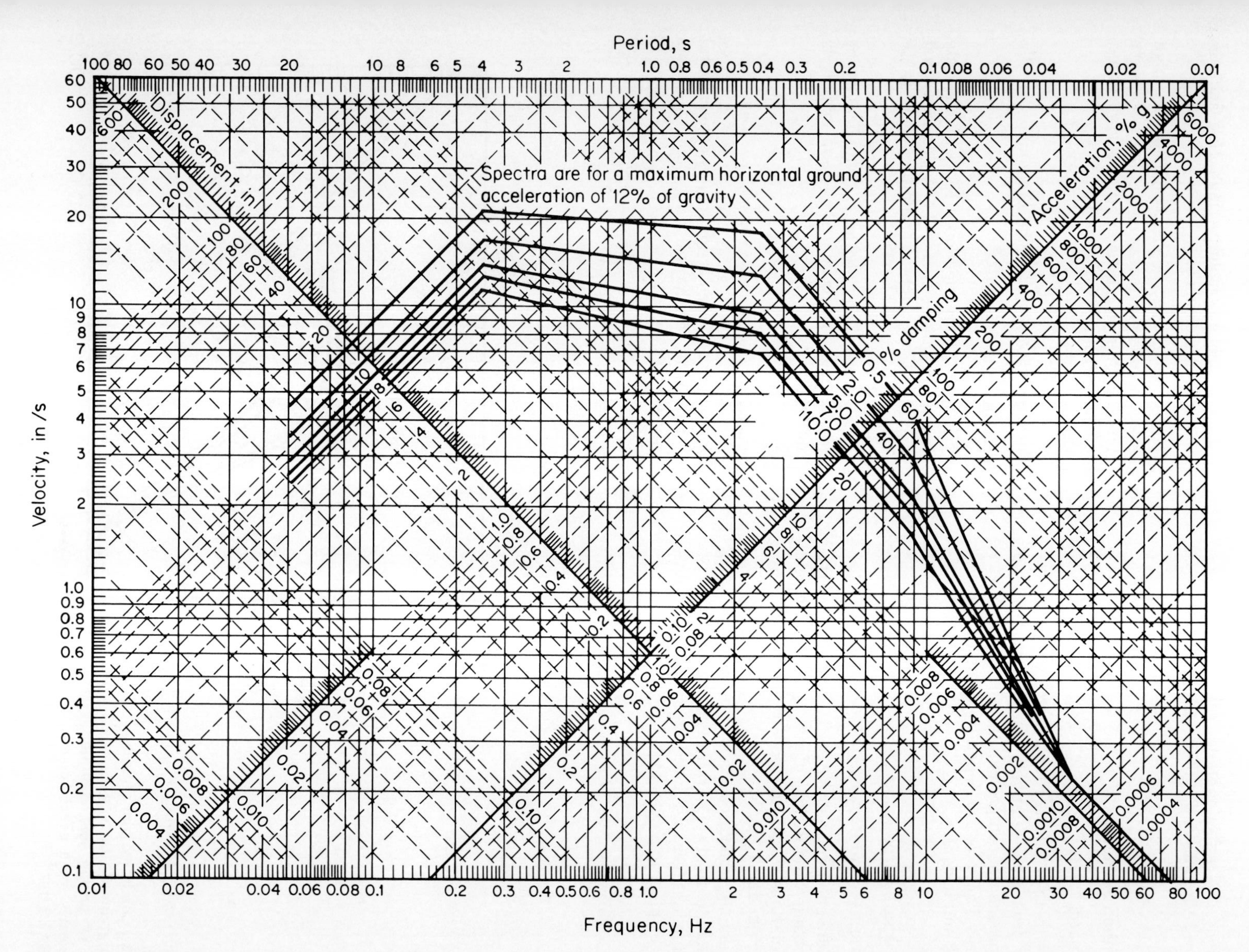


**FIG. 11.43 Design response spectra for evaluation of nuclear power plants.** *[From Newmark et al. (1973)[88] as presented in Leeds (1973).[48]]*

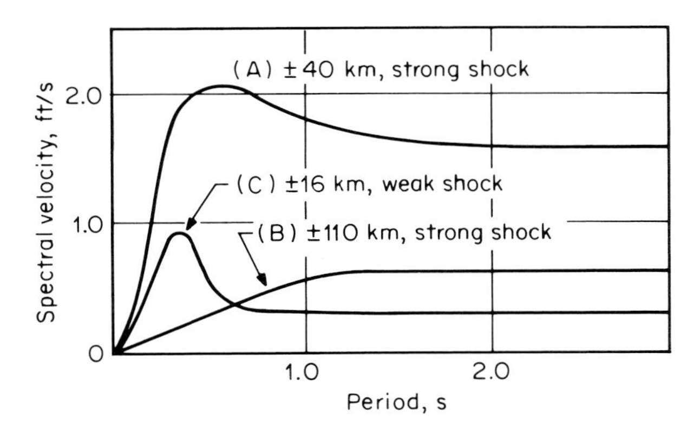


**FIG. 11.44 Effect of epicentral distance on frequency characteristics.** [*After Gama (1980).*[90]]

*Comprehensive design* is applied to unconventional structures in all areas and to conventional structures, critical public structures, and lifelines in high-hazard areas. The approach considers all site response factors in the development of the design earthquake (*see Art. 11.4.6*) which are employed in more comprehensive analysis.

## Conventional Approach

Considers only g forces, such as in Eq. 11.28, which are given in building codes or estimated from risk maps or by other procedures (*see Art. 11.4.6*).

## Comprehensive Analytical Methods

### *General*

The combined influence of ground accelerations, their frequency contents, and, to some extent, ground shaking duration, in relation to the period and damping of the structure, are considered in comprehensive analytical methods. Input can be from actual strong-motion records, from simulated earthquake motion, or from response spectra.

Analysis considers the structure as being linear-elastic and having single or multiple degrees of freedom. Ground shaking causes base excitation, which is established by equations expressing the time-dependent (dynamic) force in terms of the structure's characteristics (mass, stiffness, and damping factors) as related to time-dependent acceleration, velocity, and displacement. A dynamic mathematical model is developed representing the structure and all of its elements, closely simulating the interaction effect of components on each other and the response of each to the dynamic forces.

### *Time-Motion Analysis*

Time-motion analysis is a sophisticated method of evaluating seismic response of structures and requires substantial computer time. The basic data input may be ground motion from an actual strong-motion record, or simulated earthquake motion. The corresponding response in each configuration of the vibrating system (mode) is calculated as a function of time. The total response, obtained by summing all significant modes, can be evaluated for any desired instant.

### *Response Spectrum Analysis*

Response spectra are developed synthetically or from accelerograms to reveal directly the effects of a given ground acceleration, frequency content, and to some extent duration, on a structure for a range of periods and various levels of structural damping. Analysis yields only the maximum response for each mode rather than the time-dependent motion obtained from an actual ground-motion record.

## Soil-Structure Interaction Analysis (SSI)

### *General*

The *purpose* of a soil-structure interaction analysis is to evaluate a coupled bedrock-soil-structure system including resonance and feedback effects, for foundations on or below the surface.

FEEDBACK FROM STRUCTURAL OSCILLATIONS For relatively light structures founded on strong rock, the influence of the structure is minimal and the structural model excitation is essentially the same as for the prescribed ground motion. The situation is different for massive structures on strong soils and conventional structures on deep, weak deposits. Feedback of structural oscillations to the underlying soils, in these cases, may significantly affect the motion at the soil-structure interface which in turn may result in amplification or reduction of the structural response. Deformation of the soil formation also

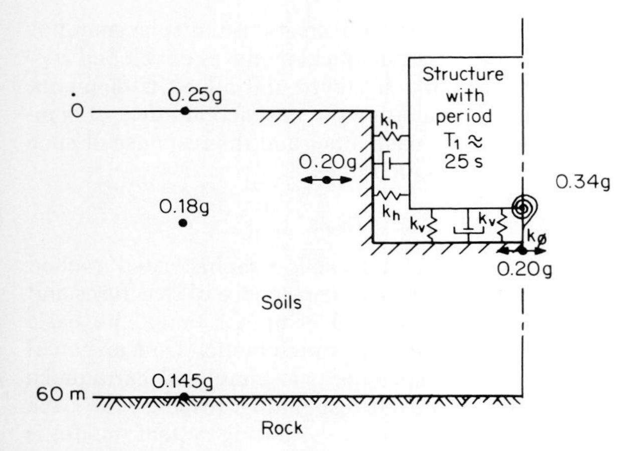


**FIG. 11.45 Soil-structure-interaction model for half-space analysis.** [*From Seed et al. (1975).*[28]]

may be caused by the feedback from the horizontal, vertical, or rotational oscillatory motion of the structure. The problem is complicated by founding below ground level, the usual procedure for most heavy structures.

SOIL CLASSES IN SEISMIC LOADINGS *Stable soils* undergo plastic and elastic deformations but will dampen seismic motion, will maintain some characteristic strength level, and are amenable to SSI analysis. Both the dynamic input to the soil from the excitation of the underlying rock (which basically applies a shearing stress to the soils), as well as the feedback from structural oscillation, should be considered in dynamic analysis, for which dynamic soil properties should be used (see *Art. 11.3.2*).

*Unstable soils* are subject to a sudden and essentially complete strength loss by liquefaction, or sudden compression resulting in subsidence, and are not readily considered in SSI analysis.

*Half-Space Analysis* [*Seed et al. (1975)*[28]]

THE MODEL The soil effects on structural response are represented by a series of springs and dashpots (energy adsorbers) in a theoretical half space surrounding the structure as shown in Fig. 11.45. The approach has limitations when applied to buried structures and is best used to analyze surface structures.

GROUND MOTIONS For the problem illustrated, the horizontal earthquake motion was specified at the ground surface in the free field (a location where interaction between soil and structure is not occurring). It was assumed that the ground surface motions for the relatively thin soil deposit were the result of vertically propagating shear waves and the maximum accelerations and corresponding time histories were found for other depths down to the bedrock surface. A maximum acceleration at the foundation base level was selected. [The procedure is described in Schnabel et al. (1971)[91]].

ANALYSIS Representative values for the spring constants were computed and an analysis of structural response made for damping ratios of 7 and 15%, which led to values of maximum acceleration at the base of the structure of 0.38 and 0.32g respectively.

LIMITATIONS As of 1974, the approach does not consider material damping, can only be applied to one- or two-layer soil systems, and provides no means for determining strains induced in the soils. These and other limiting factors are summarized on Table 11.12. The strains induced in the soils very much affect soil deformation moduli *G* used in the determination of the spring constant.

### Finite Element Method of Analysis

The model idealizes the soil continuum as a system of finite elements interconnected at a finite number of nodal points. Either triangular or rectangular elements can be used, depending upon the geometry of the conditions being modeled.

*Soil Profile Presentation*

In most cases the soils are considered to be equivalent linear-elastic materials. Soil response is described by formulating stiffness and mass matrices, and a nodal solution or a time-marching integration is effected, depending upon the capability of the particular computer program employed. Response-time histories of displacement, velocity, and acceleration can be computed for each nodal point. Soil characteristics required include shear modulus, Poisson's ratio, soil unit weight, and damping coefficients.

**TABLE 11.12**
**COMPARISON OF HALF-SPACE ANALYSIS WITH FINITE ELEMENT ANALYSIS***

| Consideration | Half-space theory or interaction springs | Finite element analysis |
|---|---|---|
| Deformability of soil profile and variation of accelerations with depth | Usually assumes accelerations are constant with depth | Can take account of deformations of profile and variability of accelerations with depths |
| Characteristics of motion below the base of structure | Usually assumes that the motions below the base of the structure (and usually around structure) are the same as those in the free field | Can readily take into account the influence of interaction on the characteristics of the motions below the base of the structure |
| Determination of soil motions adjacent to structure | Provides no means for determining motions adjacent to structure | Provides means for determining motions adjacent to the structure |
| Determination of soil deformation characteristics | Characteristics can only be approximated | Characteristics can be determined on rational basis |
| Determination of damping effects | Damping can only be estimated | Damping can be appropriately characterized and considered in analysis |
| Effects of adjacent structures | Effects cannot be considered | Effects can readily be evaluated |
| Inclusion of high-frequency effects | Effects are appropriately included | Effects may well be masked by computational errors, including: (1) the use of a coarse mesh, (2) the use of Rayleigh damping, and (3) the use of too few modes |
| Lateral extent of model | Not a factor in analysis | Must be large enough to provide required boundary conditions |
| Three-dimensional configuration | Can be considered in analysis | Must be represented by two-dimensional model |

*From Seed et al. (1975).[28]

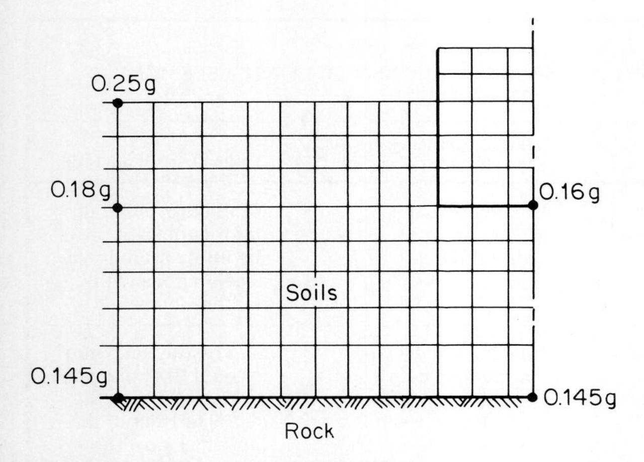


**FIG. 11.46 Finite element model of soil-structure system of Fig. 11.45.** [*From Seed et al. (1975).*[28]]

The variations of the shear moduli and damping coefficients with strain are considered.

*Input or control motion* is based on actual earthquake data, or on synthetic records, to obtain vertical and horizontal excitation in terms of g, and can be specified as located in the free field at the surface, or at foundation level.

*Comparison with Half-Space Analysis*

In the finite element method (FEM) model illustrated on Fig. 11.46 of the problem illustrated on Fig. 11.45, Seed et al. (1975)[28] found that the maximum acceleration at the base of the structure would be only 0.16g, or roughly half that found by the half-space analysis.

Damping was appropriately characterized by soil modulus values compatible with the strains that developed in the different elements representing the soil deposit. A comparison between the FEM and the half-space approach for buried structures is summarized on Table 11.12.

*General Procedure*

The procedure is illustrated by a sample analysis of major structures for a nuclear power plant presented by Idriss and Sadigh (1976),[92] as shown on Fig. 11.47. Ground motion was specified at foundation level.

STEP 1 Because the control motion is specified typically at some point below the surface in the free field, a deconvolution analysis is necessary to determine compatible base-rock motions. These are the motions that must develop in an underlying rock formation to produce the specified motions at the control point. *SHAKE* is a computer program used in the deconvolution analysis, in which the most important assumptions are

- The site response is dominated by shear shaking from below, and all other modes of seismic energy are neglected.
- The shear shaking is undirectional and the site responds with a state of plain strain.
- The stress-strain trajectories are cyclic.
- There are no residual displacements.
- There is no soil liquefaction.

STEP 2 Base-rock motion from step 1 is then used for a two-dimensional analysis (plain strain) of the soil-structure system, leading to an evaluation of the motions at any selected points such as the base of the structure, operating floor, etc. *LUSH* is a computer program used for the two-dimensional analysis. It is possible to use different damping properties for each element of the finite element model, and to account for the high-frequency ranges necessary in the study of soil-structure interaction for nuclear power plants.

In the LUSH analysis, the selection of the vertical element size considers the wavelength of the shortest wave, the shear velocity in the element, and the highest frequency of interest. Typically, maximum frequency values of 15 to 25 Hz are used in SSI analysis involving the massive structures of nuclear power plants. It is, however, generally believed that most interaction effects between the structure and the underlying soil would involve frequencies well below 20 Hz (*see Art. 11.4.2*).

Acceleration spectra are computed for various depths beneath, and various distances from, the structures. The acceleration spectra are plotted as the magnification ratio vs. the period for various accelerations in terms of gravity (expressed for convenience as g values). The variations of

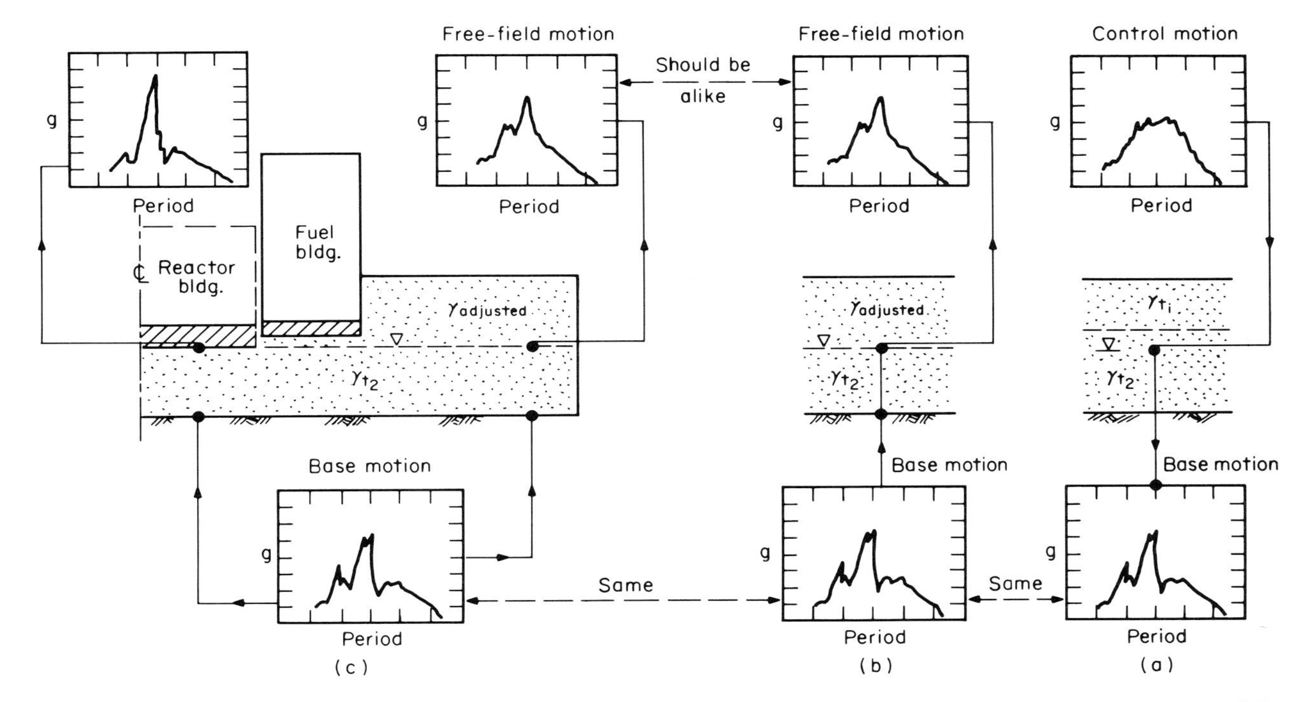


**FIG. 11.47 Representation of general analysis procedure used in soil-structure-interaction analysis: (*a*) soil deposit model used in deconvolution analysis to obtain compatible motion of base rock; (*b*) soil deposit representing free field with modified soil properties (resulting from turbine building); (*c*) finite element model of soil-structure system.** [*After Idriss and Sadign (1976).*[92]]

maximum horizontal acceleration with distance along horizontal planes at various depths are then plotted.

### Conclusions

From the study comparing half-space analysis with the FEM, Seed et al. (1975)[28] concluded that there is still much to be learned about soil-structure interaction, that even sophisticated analyses do not have the capability to incorporate all aspects of reality, and that considerable judgment is required in evaluating the results obtained in any analysis.

In addition to the limitations of the analytical procedures, there are uncertainties in measuring soil properties and ground-motion characteristics. Thus the problem is not only extremely complicated, but it is not well-defined. The result is necessary conservatism in the design of important structures.

## 11.4.6 THE DESIGN EARTHQUAKE

### Definitions

*General*

The design earthquake normally is defined as the specification of the ground motion as a basis for design criteria to provide resistance to a moderate earthquake without damage, and to provide resistance to a major event without collapse.

*Nuclear Power Plants*

Two loading levels are given by the U.S. Nuclear Regulatory Commission (USNRC):

- *Safe-shutdown earthquake (SSE)* is the largest vibratory motion that could conceivably occur at any time in the future. The operator must be able to shut the plant down safely after such an event, even if some of the plant components are damaged.

- *Operating-basis earthquake (OBE)* is the vibratory ground motion through which the safety features of the plant must remain functional while the plant remains in operation. Some portions of the plant needed for power generation but not required for safe shutdown or radiation protection, however, need not be designed to resist the OBE and conceivably the plant could cease to produce power if subjected to an event of OBE magnitude. *Category 1 structures* are those critical to safe operation and shutdown and include the reactor containment; the auxiliary, fuel handling, radioactive waste, and control buildings; and the intake screen house at the cooling water source. The OBE is often substantially lower than the SSE.

### Approaches to Selection

*Conventional Approach*

Conventional structures in areas of low seismic hazard are based on selection of values for g, and normally such values are obtained from existing publications such as national or local building codes, or seismic risk and probability maps such as Fig. 11.15.

Alternatively, if there are no published criteria, as may be the situation in many countries, g may be estimated from records of intensity or magnitude by conversion, using relationships such as Fig. 11.13, Table 11.13, or Eq. 11.10. In unfamiliar areas without building codes, if the hazard degree is in doubt, then the comprehensive approach is undertaken.

*Comprehensive Approach*

Design of unconventional structures or structures in high hazard zones (*see Art. 11.4.5*) is based on the selection of values for peak effective horizontal and vertical g, frequency content, and duration.

Existing data are collected and reviewed within a radius of about 320 km from the proposed structure, including catalogs of intensity reports and event magnitudes, strong-motion records, geologic maps, known faults, and other information, and a site response study is performed.

### Site Response Study

*Recurrence Analysis (Art. 11.2.7)*

The first step in a site response study is to estimate *site intensity* $I_s$ from analysis of earthquake history data. All known events of $I$ = IV or greater, occurring within 320 km of the site (NRC requirement), are located and zoned by intensity as a seismicity map, such as given on Fig. 11.12. Statistical analysis is performed to determine the probable return of events of various magnitudes to locations where they have occurred in the past (the source, usually given as the epicenter).

Events of significant magnitudes with return periods usually of 50 or 100 years are selected, depending on the importance of the structure. Events of the highest magnitude and those with a recurrence interval closest to the economic life of the structure are of major interest. Judgment is required to evaluate the results of the recurrence analysis. If the maximum intensities of some of the historical data are overestimated, the results might indicate that either the region is overdue for the occurrence of a damaging shock, or that there is a regional change toward a lower level of intensity.

Site intensity $I_s$ is estimated by the application of attenuation laws and relationships (*see Art. 11.2.5*) such as Fig. 11.17 or Eq. 11.11 or 11.12, or it is imposed on a "capable" fault and then attenuated to the site.

*Geologic Study*

Geologic study is performed to locate fault structures and to identify capable faults. Fault systems are correlated with intensity distributions, and faults greater than 1 mi in length require detailed study to determine their capability (*see Art. 11.3.1*). The potential magnitude may be estimated from the length of the capable fault (or faults) as given on Fig. 11.24 or Eq. 11.21. Assumptions regarding rupture length have varied by practitioners from one-half to one-third of the total [Adair (1979)[43]].

It is assumed that the focus, whether the intensity is selected from the recurrence analysis or is based on possible rupture length, will be located

**TABLE 11.13**
**DESIGN SEISMIC HORIZONTAL GROUND MOTIONS***

| | Design acceleration, gravities g | | Design velocity, in/s | |
|---|---|---|---|---|
| **Magnitude†** | **Ground motion‡** | **Structures§** | **Ground motion‡** | **Structures§** |
| 8.0 | 0.60 | 0.33 | 29 | 16 |
| 7.5 | 0.45 | 0.22 | 22 | 11 |
| 7.0 | 0.30 | 0.15 | 14 | 7 |
| 5.5 | 0.12 | 0.10 | 6 | 5 |

*From Hall and Newmark (1977).[82]

†*Magnitudes* are considered as the design maximum earthquake (were developed for four seismic zones along the trans-Alaska pipeline).

‡*Ground Motion:* Peak values which may affect slope stability, liquefaction of cohesionless materials, or apply strains to underground piping.

§*Structural Design:* Peak values used for design of structures or other facilities. Since they account for structural features and response, and soil-structure interaction, they are generally less than those used to define soil response.

at the closest point on the capable fault. This is considered a reasonable assumption for shallow-focus events. Peak acceleration for rock excitation at the site, in terms of g, is estimated by attenuation from the fault as given on Fig. 11.25.

The earthquake considered in planning the ultimate design is usually the largest which might occur at the closest approach to the site of any capable fault. Judgment is used in the selection of the design event; main considerations are recurrence probability and magnitude.

Site intensity, magnitudes, or estimates of g are thus found for rock excitation. Evaluations obtained by these foregoing procedures are, at best, only approximations applicable to shallow-focus shocks.

### *Selection of Site Ground Motion*

*Strong-motion records* for an earthquake of the design magnitude for various ground conditions in the site area would provide the most reliable data on acceleration and frequency content, but as yet such data are relatively meager. In many cases it is necessary to estimate site ground motion from correlations.

*Horizontal acceleration for rock excitation* may be estimated roughly from intensity or magnitude values such as given on Fig. 11.13 or Eq. 11.10, or from Table 11.13, recently developed from an evaluation of strong-motion records.

*Vertical acceleration for rock excitation* has been estimated as $\frac{1}{2}a_h$, or higher if the site is close to a capable fault. Hall and Newmark (1977)[82] recommend using two-thirds of the values given on Table 11.13.

*Frequency content* is estimated from Fig. 11.27, giving the predominant period vs. $M$ vs. distance from the causative fault, or estimated from Table 11.13, giving design velocity vs. $M$. For long-distance earthquakes, in the order of 100 km or more, the possibility is considered of sway in high-rise structures due to long-period waves.

*Duration of shaking* is estimated on the basis of the *bracketed duration* in terms of distance and magnitude from Table 11.14, which provides these data for an acceleration >0.05g and a frequency >2 Hz, or is developed from strong-motion records. Consideration is given to the fact that the duration of large earthquakes depends largely upon the length of faulting.

*Soil conditions* usually influence acceleration by amplification. Maximum accelerations from the same earthquakes occur at different periods for different soil types and thicknesses. Damage is selective and varies with building height and period and other factors (*see Art. 11.4.3*). *Soil amplification factor* may be estimated from Fig. 11.20, which gives general relationships

**TABLE 11.14**
**BRACKETED DURATION IN SECONDS***

| Distance, km | Magnitude 5.5 | 6.0 | 6.5 | 7.0 | 7.5 | 8.0 | 8.5 |
|---|---|---|---|---|---|---|---|
| 10 | 8 | 12 | 19 | 26 | 31 | 34 | 35 |
| 25 | 4 | 9 | 15 | 24 | 28 | 30 | 32 |
| 50 | 2 | 3 | 10 | 22 | 26 | 28 | 29 |
| 75 | 1 | 1 | 5 | 10 | 14 | 16 | 17 |
| 100 | 0 | 0 | 1 | 4 | 5 | 6 | 7 |
| 125 | 0 | 0 | 1 | 2 | 2 | 3 | 3 |
| 150 | 0 | 0 | 0 | 1 | 2 | 2 | 3 |
| 175 | 0 | 0 | 0 | 0 | 1 | 2 | 2 |
| 200 | 0 | 0 | 0 | 0 | 0 | 1 | 2 |

*From Bolt (1973).[49]

NOTES.: 1. *Bracketed duration:* The elasped time, for a particular frequency range, between the first and last acceleration excursions on an accelerogram record greater than a given amplitude level.

2. Acceleration > 0.05g, frequency >2 Hz.

between *I*, acceleration in terms of g, and foundation conditions, or Table 11.5 which gives the factor *F* in terms of soil type and depth. The alternate procedure for evaluating soil response is by SSI analysis employing dynamic soil properties (*see Art. 11.4.5*).

### Synopsis

Site acceleration can be given for design in several ways:

- Maximum peak or effective peak acceleration
- Acceleration at a given period
- Real or synthetic time motion which provides for structural periods and damping
- Continuous spectrum of time motion based on actual recorded events

## 11.5 INVESTIGATION: IMPORTANT STRUCTURES IN HIGH-HAZARD AREAS

### 11.5.1 INTRODUCTION

#### Purpose

This article tends to follow procedures and guidelines employed in studies for nuclear power plants in the United States, but is intended as a conservative guide for important structures located in any high-hazard seismic region.

#### Objectives and Scope

*Investigation objective* is basically safe and economical construction which requires:

- Identification and treatment of geologic hazards (avoid, reduce, or eliminate)
- Estimation of the design earthquake
- Establishment of foundation design criteria
- Evaluation of structural response to dynamic forces

*Study scopes* range from simple to complex, depending upon several major factors, including:

- Importance of structure and the degree of risk (*see Art. 11.4.5*)
- Regional seismicity (the degree of hazard) and adequacy of available data
- Physiographic conditions (mountains, coastline, plains, etc.)
- Regional and local geology (hazards, rock types and structure, soil types and characteristics, and groundwater conditions)

An *investigative team* to accomplish the objectives includes geologists, seismologists, geophysicists, geotechnical engineers, and structural engineers.

#### The PSAR (Preliminary Safety Analysis Report)

A preliminary safety analysis report (PSAR) covers the general requirements for investigations for nuclear power plants in the U.S. *Regulatory Guide 1.70* (1975), "Standard Format and Content of Safety Analysis Reports for Nuclear Power Plants." This guide, issued by the U.S. Nuclear Regulatory Commission, lists the information to be provided to the USNRC in the PSAR by the applicant for nuclear power plant construction permits.

### 11.5.2 PRELIMINARY PHASE

#### Purpose

During the preliminary phase, existing data on regional and local seismicity and the natural environment are collected and reviewed either to provide a data base for a site selection or a feasibility review of a previously selected site.

### Seismicity

*Study scope* depends upon the current regional seismic activity, the historical activity, and the completeness and type of historical records available.

*Hazard degree* is evaluated from an existing data review. *World seismicity maps* are useful for overseas projects for an overview to establish the general site location in relation to plate edges. *Seismic risk maps* or *microzonation maps* provide data on the hazard degree and ground response. *National* and *local building codes* provide information on design criteria.

*Data adequacy* is evaluated. Codes, strong-ground-motion records, response spectra, recurrence studies, microzonation maps, etc. may provide information adequate for design, or a suitable design earthquake or other design basis may have already been established for the area.

*New data are generated* if existing seismicity data are inadequate. A review is made of published catalogs giving values for $I$ and $M$, and a seismicity map is prepared showing hazard zones and intensities. Recurrence analysis is performed for the region, and events of highest $M$ and recurrence closest to the life of the structure (40 years for nuclear power plants) are selected. Intensities or magnitudes are attenuated to the site, or are imposed on the capable fault at its closest proximity (see *Art. 11.5.4*). It may be necessary to consider the possible impact of induced seismicity from a nearby reservoir (see *Art. 11.2.1*).

### The Natural Environment

The *purpose* of a review of existing data is the identification of important natural factors affecting the suitability of the site location from the aspect of hazard degree and from the anticipation of potential foundation problems.

#### *Scope*

Items of interest are physiography, climate, and geology. In addition to the collection and review of existing data, new data may be generated during the preliminary phase by terrain analysis techniques and field reconnaissance. Data are collected also for other environmental aspects including flora and fauna, but are not included in the scope of this discussion.

#### *Physiography*

Information on physiography is obtained from topographic maps and remote-sensing imagery. The importance of physiography lies in its direct relationship in many cases to the geologic hazards, and with respect to seismicity it influences attenuation (see *Art. 11.2.6*), although the relationships are not well-defined. Some very general associations may be established:

- *Mountainous regions:* Characteristic features are slope failures, variable regional geology, intense surface fault systems, and, in some locations, volcanos. Intensity distribution may be reduced and highly modified, as in the western United States.
- *Great Plains and other large areas of reduced relief:* Regional geology is likely to be more uniform, but limestone and potential collapse conditions more prevalent, fault systems less pronounced and identifiable, and intensity distribution over very large areas relatively uniform, as in the central United States (see *Fig. 11.18*).
- *River valleys* are possible flood areas with potentially liquefiable soils in the floodway, pastoral, and estuarine zones; relatively poor foundation conditions; and fault systems less pronounced and identifiable in broad, mature valleys.
- *Coastal areas* exhibit great geological variations, but may contain liquefiable or otherwise unstable soils, and, depending upon location and configuration, may be subjected to the tsunami hazard.

#### *Climate*

Climate relates to hazards and geologic conditions, as an aid or as a requirement for predictions. Some typical associations are the evaluation of the flood hazard, the type and depth of residual soil development, and the potential for slope failures, collapsing soils, and expansive soils.

**TABLE 11.15**
**MINIMUM FAULT LENGTH TO BE CONSIDERED IN ESTABLISHING SAFE-SHUTDOWN EARTHQUAKE***

| Distance from site, miles | Minimum length, miles |
|---|---|
| 0 to 20 | 1 |
| Greater than 20 to 50 | 5 |
| Greater than 50 to 100 | 10 |
| Greater than 100 to 150 | 20 |
| Greater than 150 to 200 | 40 |

*From NRC (1975).[93]

*Regional Geologic Conditions*

Regional conditions include information on rock types and structures (faults, floods, etc.), and the hazards of slope failures, ground subsidence or collapse from fluid or solid extraction or from natural causes, regional warping and tilting, and volcanos.

*Local Site Conditions*

Local conditions include information on physiography and geology. Evaluation of conditions may result in recommendations to abandon the site and select another location if the constraints are judged to be too severe. The most severe constraints are local active faulting and warping, high liquefaction potential, large-scale unstable or potentially unstable slopes, volcanism, ground collapse, and tsunami potential.

### 11.5.3 DETAILED STUDY OF REGIONAL AND LOCAL GEOLOGIC CONDITIONS

**Fault Studies**

*Prepare Geologic Structure Map*

Data from the literature and terrain analysis techniques are used to prepare a map showing all tectonic structures, including lineaments, within 320 km of the site, with the primary objective of locating all faults over 1 mi (1.6 km) in length. If a known active or capable fault is within the 320-km radius, and is a major fault, such as the San Andreas, then it is considered the limit of the study area. The geologic structure map is overlaid with the seismicity map and events correlated with lineaments to identify potentially capable faults.

The NRC [U.S. Nuclear Regulatory Commission (1975)[93]] provides guidance in determining which faults may be of significance in evaluating the safe-shutdown earthquake. In general, either capable or noncapable faults with lengths less than those indicated on Table 11.15 need not be considered.

*Investigate for Capable Faults*

Investigations start near the site and extend outward to locate the design capable fault. Judgment is required to evaluate the significance of lineament or fault length, site proximity, faulting evidence, etc. in regard to the necessity of a detailed investigation to evaluate the capability of a particular fault. Fault characteristics and identification are described in Arts. 6.5 and 11.3.1.

*Detailed reconnaissance* and *imagery interpretation* are performed for each lineament of interest to identify faulting evidence, fault type, past displacements, and apparent activity on the basis of observations.

*Explorations* are made of candidate faults with geophysical surveys, vertical and angle borings (including coring, sampling, and sensing with nuclear probes), and trenches to determine fault existence, zone width, and geometric attitude.

*Radiometric dating* (Appendix A.4 and Table 11.8) is performed to evaluate fault activity.

*Fault Zone Extent and Control Width*

Fault traces near the plant site are mapped along the fault for a distance of 10 miles in both directions from the point of its nearest approach to the site. The NRC [U.S. Nuclear Regulatory Commission (1975)[93]] defines the *control width* as the maximum width of the mapped fault traces. Because surface faulting may have occurred beyond the limit of the mapped fault traces, detailed faulting investigation of a zone beyond the control width is required. The width of this zone depends upon the largest potential earthquake related to the fault as given on Table 11.16.

### Investigate Other Major Hazards

- Flood potential *(see Art. 8.2.3)*
- Slope stability *(see Arts. 9.5 and 11.3.4)*
- Ground subsidence and collapse, faulting, and induced seismic activity from fluid extraction *(see Art. 10.2)*
- Ground subsidence and collapse from subsurface mining *(see Art. 10.3.5)*
- Ground collapse from failure of cavities in soluble rock *(see Art. 10.4.3)*
- Liquefaction and subsidence potential *(see Art. 11.3.3)*

### Site Soil and Foundation Studies

*Objectives*

SOIL FORMATIONS Determine stratigraphy and soil types, identify the potential for ground compression and heave, and measure static and dynamic strength and deformation properties and permeability.

ROCK FORMATIONS Determine stratigraphy and rock types, and identify degree and extent of weathering and distribution and nature of discontinuities. Measure pertinent engineering properties.

GROUNDWATER CONDITIONS Locate the static, perched, and artesian conditions, and determine water chemistry. Evaluate susceptibility to changes with time and weather, or other transient conditions.

*Explorations*

Terrain analysis and field reconnaissance are performed to provide data for a detailed geologic map of surficial conditions.

Stratigraphy is investigated with geophysics, test borings, trenches, pits, etc.

Samples for identification and laboratory testing are obtained from borings, trenches, and pits.

*Property Measurements*

Static and dynamic properties are measured in situ and in the laboratory. Of particular interest are cyclic shear moduli and damping ratios for SSI studies. The very low strains of in situ direct seismic tests should be correlated with laboratory results to obtain simulations with earthquake strains.

*Instrumentation*

Piezometers are installed as standard procedure to monitor groundwater fluctuations. Other instrumentation may be installed to monitor natural slopes, areas of ground collapse or subsidence, or active or potentially active faults, depending upon the time available for the study.

### SAR (Safety Analysis Report)

A *safety analysis report* (SAR) is a requirement of the USNRC from *The Standard Format* (Section 2.5.1.1) to include the following information:

Discuss all geologic, seismic, and man-made hazards within the region of the site and relate them to the regional physiography, tectonic structures and tectonic provinces, geomorphology, stratigraphy, lithology, geologic and structural history, and geochronology. The above information should be discussed, documented by appropriate references, and illustrated by a regional physiographic map, surface and subsurface geologic maps, isopach maps, regional gravity and magnetic maps, stratigraphic sections, tectonic and structure maps, fault maps, a site topo-

**TABLE 11.16**
**DETERMINATION OF ZONE REQUIRING DETAILED FAULTING INVESTIGATION***

| Magnitude of earthquake | Width of zone of detailed study |
|---|---|
| Less than 5.5 | 1× control width |
| 5.5–6.4 | 2× control width |
| 6.5–7.5 | 3× control width |
| Greater than 7.5 | 4× control width |

*From NRC (1975).[93]

graphic map, a map showing areas of mineral and hydrocarbon extraction, boring logs, aerial photographs, and any maps needed to illustrate such hazards as subsidence, cavernous or karst terrain, irregular weathering conditions, and landslide potential.

### 11.5.4 EVALUATION AND ANALYSIS

#### Ground-Motion Prediction

*Select Design Earthquake Magnitude for Rock Excitation (Art. 11.4.6)*

From the recurrence analysis, events of the highest *M* and recurrence closest to the life of the structure are selected. They are either attenuated to the site directly, or are imposed on capable faults. A number of capable faults may be identified, requiring evaluation of intensity and magnitude from recorded events and fault lengths, to arrive at the design capable fault.

The earthquake used in ultimate design is considered to be the largest which might occur at the closest approach to the site of any capable fault.

*Evaluate Other Factors (Art. 11.4.6)*

EARTHQUAKE CHARACTERISTICS Evaluate horizontal acceleration for rock excitation, vertical acceleration for rock excitation (higher near capable fault), frequency content and the effect of long-distance events, and shaking duration.

SOIL PROFILE EFFECTS (ART. 11.4.3) The use of values for *I* or *M* and, in most cases, of strong-motion records, results in estimates of the acceleration of rock. Rock motion must be converted to soil motion either at the surface or at foundation level. This normally is accomplished either by applying amplification factors, or by soil-structure interaction analysis based on measured soil dynamic properties. Both frequency and duration should be considered. Ground amplification factors vary with soil type, thickness, and rigid-boundary conditions (i.e., bedrock surface configuration). Soil conditions usually lead to amplification, but at times attenuation occurs; maximum accelerations occur at different periods for different soil types and thicknesses during the same earthquake. Damage can be selective and varies with building rigidity, height, and period, among other factors, such as construction type and quality.

*Ideal Solution*

The ideal solution is to match the design-basis earthquake with strong ground-motion data for a similar event. Accelerographs measure the three components of acceleration (two horizontal and one vertical) for various period ranges (usually 0.1 to 3 or 4 s, or to 10 s). Selection at present is restricted by the lack of strong-motion data for many geologic conditions, but in time application should become more of a standard practice.

#### Foundation Design Criteria and Structural Response

*Foundation Evaluation and Selection*

Evaluations are made of potentially unstable soils to determine the possibility of subsidence, liquefaction, or permanent reduction in strength when they are subjected to dynamic shear forces. The evaluation includes the possibility of developing high shearing stresses in soft clays, which could lead to rupture of deep foundations, and the possibility of embankment failures. (This evaluation should be made as early as feasible during the investigation, since it may show that the site should be abandoned.)

Suitable foundation types are selected for the structure, or structures, from evaluations of the mechanical properties of the soils and other factors.

*Seismic Input for Design Response*

A seismic design analysis can vary from a relatively simple one in which only a horizontal acceleration force is applied to the structure, to a complex one in which all elements of ground motion are considered.

*Complex analysis* considers the combined influence of the amplitude of ground motions, their frequency contents, and to some extent duration of shaking in terms of period and damping. A design response spectrum may be applied that represents an average of several appropriate spectra developed into a design envelope such as that given as Fig. 11.43. The influence of the frequency content is important, and some investigators recommend that instead of scaling the entire response spectra to ground accelerations, a better overall picture of ground response is obtained by specifying in addition the maximum ground velocity for intermediate frequencies and the maximum ground displacement for the lower (long-period) frequencies. Magnitudes for design spectra can be determined from equations that estimate amplitudes directly, instead of scaling them from the estimates of peak ground velocity. Alternatively, complete input motion can be obtained from actual accelerograph records, or produced synthetically from given base-rock motion for a particular site.

*Dynamic Analysis (Art. 11.4.5)*

For foundations supported on rock or strong soils, response is evaluated by time-motion methods or response-spectrum analysis. For structures founded on relatively deep and weak soils, the feedback from structural oscillations is evaluated by SSI analysis. The base-rock motion, given in terms of g varying with period, is used to evaluate the motions at any point, such as the structure base, employing values for the dynamic shear modulus and damping ratios of the soils. An alternate procedure is to specify a control or input motion at some point in the "free field" from which comparable rock motions are determined.

## 11.5.5 LIMITATIONS IN THE PRESENT STATE OF THE ART

### General

Limitations of knowledge regarding earthquakes as pertaining to engineering are severe and require the application of considerable judgment based on experience, as well as generous safety factors on any project. The limitations are well-recognized, and substantial effort is being applied by various disciplines involved with earthquake engineering to improve the various ignorance factors, some of the more significant of which are given below. Unfortunately, capabilities in rigorous mathematical analysis appear to be far in advance of the capabilities of generating accurate and representative data.

### Earthquake Characteristics

*Focal-Depth Effects*

Procedures to define the design earthquake are considered approximately valid for the continental western United States, where modern earthquakes are generally shallow and fault-related. The applicability to intermediate and deep-focus events, which usually are not associated with surface faulting, is not known or tested.

*Occurrence Prediction*

Seismicity data in many locations are meager, usually based on "felt" reports, and cover only a relatively brief historical period (200 years or so) in comparison with "recent" geologic time of say, about 10,000 years from the last glacial age. Seismic activity has been found to be cyclic in many areas, with cycle lengths longer than the time interval of data in some areas. (How can it be known when essentially singular events such as New Madrid or Charleston may return, or may occur in some other region with "historically" low activity?)

Felt reports are based on the response of people and structures; therefore, they depend on development and demography for registry. As the world population increases, so does the incidence of earthquakes. Seismograph stations have only been in operation for about 100 years, and only since World War II has there been a substantial number of installations in many areas.

Relationships between intensity and magnitude are very general; therefore, conversion of $I$ from felt reports to $M$ or g is uncertain. Recurrence predictions of $M$ and the time interval must be considered as broad approximations for most locations.

Other methods of earthquake prediction, used for early warning rather than design, such as those described in Art. 11.2.7, have only been studied in very recent years.

*Attenuation*

Attenuation, an important element in evaluating data from an existing event, depends on many variables such as topography, geology, fault rupture length, and focal depth, and can undergo extreme variation in a given area, as related to both distribution and areal extent. Relationships have not been well-defined and estimates must be considered only as broad approximations.

## Ground Conditions and Response

### *Fault Identification and Capability Evaluation*

Faulting does not always extend to the surface; therefore, positive identification can be difficult. Estimation of *capability* has many uncertainties, and the knowledge of causes and the basis for anticipation of activity are not well-established. Faults, considered as dead, and not studied thoroughly, can suddenly become active.

Relationships between fault rupture length and event magnitude are based on relatively few data and restricted geographic areas. Relationships between percent of rupture length and $M$, or between magnitude propagation (stress drop) and duration, have barely been addressed. Relationships between fault displacement and $M$, especially for low-intensity events, need study. It is not unusual to find structures in a risky condition with foundations bearing on rock located over a fault in areas of low seismicity.

### *Ground Response*

Relatively few events have been recorded on strong ground-motion instruments and the greater majority of these are from the western United States. Foundation and topographic conditions for instrument locations vary and require consideration when accelerograms are evaluated. Relationships among peak and effective horizontal acceleration, vertical acceleration, frequency content, and duration are not established for many conditions of geology and topography, including effects of distance and foundation depth.

Effects of soil conditions have not been well-defined; although soil is generally considered to amplify rock excitation, cases of attenuation have been reported. Relationships among soil type, depth, layering effects, bedrock boundary configuration, and response characteristics of acceleration and frequency content are not well-established.

### *Dynamic Soil Properties*

Dynamic shear moduli and damping ratios can be measured with acceptable accuracy, primarily in cyclic simple-shear testing, but basically only for those soils in which high-quality undisturbed sampling is possible. Evaluations of other soils must rely upon estimates of properties from various correlation procedures.

## REFERENCES


1. Richter, C. F. (1958) *Elementary Seismology*, W. H. Freeman & Co., San Francisco.
2. Bolt, B. A., Horn, W. L., Macdonald, G. A. and Scott, R. F. (1975) *Geological Hazards*, Springer-Verlag, New York.
3. Dietz, R. S. and Holden, J. C. (1970) "Reconstruction of Pangaea; breakup and dispersion of the continents; Permian to the present," *J. Geophysical Research*, Vol. 75, No. 26, pp. 4939–4956.
4. Guttenberg, B. and Richter, C. F. (1954) "Seismicity of the Earth and Associated Phenomena," Princeton Univ. Press, Princeton, N.J.
5. Housner, G. W. (1965) "Intensity of Ground Shaking Near the Causative Fault," *Proc. 3rd World Conf. on Earthquake Engrg.*, New Zealand, Vol. I.
6. ENR (1976) "Geothermal Plant Threatened by Volcanoes," *Engineering News-Record*, December 9, p. 11.
7. ENR (1975) "Reservoir Filling Linked to Quake," *Engineering News-Record*, December 18.
8. Hodgson, J. H. (1964) *Earthquakes and Earth Structures*, Prentice-Hall Inc., Englewood Cliffs, N.J.
9. Rainer, H. (1974) "Are there connections between earthquakes and the frequency of rock bursts in the mine at Blieburg?", *J. Intl. Soc. of Rock Mechanics*, Vol. 6, No. 2, August.
10. Neumann, F. (1966) "Principles Underlying the Interpretation of Seismograms," Spec. Pub. No. 254 (revised edition), ESSA, Coast and Geodetic Survey, U.S. Govt. Printing Office, Washington, D.C., p. 50.
11. USAEC (1972) *Soil Behavior Under Earthquake Loading Conditions*, National Technical Information Service TID-25953, U.S. Dept. of Commerce, Oak Ridge National Laboratory, Oak Ridge, Tenn., January.

12. USGS (1975) "The National Strong-Motion Instrumentation Network," *Earthquake Information Bulletin*, United States Geological Survey, January–February, Vol. 7, No. 1, pp. 16–19.

13. Seed. H. B., Lee, K. L., Idriss, I. M. and Makdisi, F. I. (1975) "The Slides in the San Fernando Dams During the Earthquake of February 9, 1971," *Proc. ASCE, J. Geotech. Engrg. Div.*, Vol. 101, No. GT7, July.

14. Murphy, L. M. and Cloud, W. K. (1954) "United States Earthquakes, 1952," U.S. Dept. of Commerce, *Coast and Geodetic Survey, Serial No. 773*, U.S. Govt. Printing Office.

14*a*. Raisz, Erwin (1946) "Map of the Landforms of the United States," 4th ed., Inst. of Geographical Exploration, Harvard Univ., Cambridge, Mass.

15. Bollinger, G. A. (1973) "Seismicity and Crustal Uplift in the Southeastern United States," *Amer. J. of Sci.*, Vol. 273 A, pp. 396–408.

16. Bolt, B. A. (1978) *Earthquakes: A Primer*, W. H. Freeman & Co., San Francisco.

17. Lomnitz, C. and Singh, S. K. (1976) "Earthquakes and Earthquake Prediction," *Seismic Risk and Engineering Decisions*, Lomnitz and Rosenblueth, eds., Elsevier Scientific Pub. Co., New York, Chap. 2.

18. Esteva, L. and Rosenblueth, E. (1969) "Espectos de temblores a distancias moderadas y grandes," *Bol. Soc. Mexicano Ing. Sismica*, 2, pp. 1–18.

19. ATC (1977) "Recommended Comprehensive Seismic Design Provisions for Buildings," Report ATC-3-05, January, Applied Technology Council, Palo Alto, California.

20. Shah, H. C. and Benjamin, J. R. (1977) "Lifeline Seismic Criteria and Risk—A State of the Art Report," *The Current State of Knowledge of Lifeline Earthquake Engineering, Proc. ASCE*, August 30–31, pp. 384–393.

21. Christian, J. T., Borjeson, R. W. and Tringale, P. T. (1978) "Probabalistic Evaluation of OBE for Nuclear Power Plant," *Proc. ASCE, J. Geotech. Engrg. Div.*, Vol. 104, No. GT7, July, pp. 907–919.

22. Ergin, K. (1969) "Observed intensity-epicentral distance in earthquakes," *Bull. Seismol. Soc. Amer.*, No. 59, pp. 1227–1238.

23. Algermissen, S. T. and Perkins, D. M. (1976) "A Probabalistic Estimate of Maximum Acceleration in Rock in the Contiguous United States," U.S. Geological Survey, Open File Report 76-416.

24. Nuttli, O. W. (1979) "Seismicity of the Central United States," *Reviews in Engineering Geology*, Vol. IV, *Geology in the Siting of Nuclear Power Plants*, Geological Society of America, Boulder, Colo., pp. 67–93.

25. Zeevaert, L. (1972) *Foundation Engineering for Difficult Soil Conditions*, Van Nostrand Reinhold Book Co., New York.

26. Seed, H. B. (1970*a*) "Soil Problems and Soil Behavior," *Earthquake Engineering*, Chap. 10, Prentice-Hall Inc., Englewood Cliffs, N.J.

27. Espinosa, A. F. and Algermissen, S. T. (1972) "Soil Amplification Studies in Areas Damaged by the Caracas Earthquake of July 29, 1967," *Proc. Intl. Conf. on Microzonation*, Seattle, November, Vol. II, pp. 455–464.

27*a*. Steinbrugge, K. V. (1968) "Earthquake hazard in the San Francisco Bay area: a continuing problem in public policy," Univ. of California, Berkeley, Calif.

28. Seed, H. B., Lysmer, J. and Hwang, R. (1975) "Soil-Structure Interaction Analysis for Seismic Response," *Proc. ASCE, J. Geotech. Engrg. Div.*, Vol. 101, No. GT5, May, pp. 439–458.

29. Faccioli, E. and Reséndiz, D. (1976) "Soil Dynamics: Behavior Including Liquefaction," *Seismic Risk and Engineering Decisions*, Lomnitz and Rosenblueth, eds., Chap. 4, Elsevier Scientific Publishing Co., New York, pp. 71–140.

30. CGS (1978) *Canadian Foundation Engineering Manual*, Canadian Geotech. Soc., Montreal, Part 1.

31. Coulter, H. W., Waldron, H. H. and Devine, J. F. (1973) "Seismic and Geologic Siting Considerations for Nuclear Facilities," *Proc. 5th World Conf. Earthquake Engrg.*, Paper No. 302, Rome.

32. Donovan, N. C. and Bornstein, A. E. (1978) "Uncertainties in Seismic Risk Procedures," *Proc. ASCE, J. Geotech. Engrg. Div.*, Vol, 104, No. GT7, July, pp. 869–887.

33. Bollinger, G. A. (1976) "The Seismic Regime in a Minor Earthquake Zone," *Proc. ASCE, Numerical Methods in Geomechanics*, Vol. II, pp. 917–937.

34. Lomnitz, C. (1974) *Global Tectonics and Earthquake Risk*, Elsevier Scientific Pub. Co., Amsterdam.

35. Aggarwal, Y. P. and Sykes, L. R. (1978) "Earthquakes, Faults and Nuclear Power Plants in Southern New York and Northern New Jersey," *Science*, Vol. 200, No. 4340, April 28, Amer. Assoc. for Advancement of Sci., pp. 425–429.

36. Scholz, C. H., Sykes, L. R. and Aggarwal, N. (1973) "Earthquake Prediction: A Physical Basis," *Science*, 181.

37. Whitcomb, J. H., Garmany, J. D. and Anderson, R. (1973) "Earthquake Prediction: Variation of Seismic Velocities Before the San Fernando Earthquake," *Science*, 180.

38. Raleigh, C. B., Healy, J. H. and Bredehoeft, J. D. (1972) "Faulting and Crustal Stress at Rangely, Colorado," Geophysical Monogram No. 16, Amer. Geophys. Union, Washington, D.C.

39. Hamilton, R. M. (1978) "Earthquake Hazards Reduction Program—Fiscal Year 1978 Studies Supported by the U.S. Geological Survey," Geological Survey Circular 780, U.S. Dept of the Interior.

40. ENR (1979) "Quakes Nudge Palmdale Bulge," *Engineering News-Record*, March 22, p. 3.

41. NRC-NAS (1976) "Predicting Earthquakes: A Scientific and Technical Evaluation with Implications for Society," Report on the Panel on Earthquake Prediction of the Comm. on Seismology, National Research Council, National Academy of Sciences, Washington, D.C., Appendix A.

42. Cluff, L. S., Hansen, W. R., Taylor, C. L., Weaver, K. D. et al. (1972) "Site Evaluation in Seismically Active Regions—An Interdisciplinary Team Approach," *Proc. Intl. Conf. on Microzonation*, Seattle, October, Vol. II, pp. 957–987.

43. Adair, M. J. (1979) "Geologic evaluation of a site for a nuclear power plant," *Reviews in Engineering Geology*, Vol. IV, *Geology in the Siting of Nuclear Power Plants*, Geological Society of America, pp. 27–39.

44. Wallace, R. E. (1980) "Discussion-Nomograms for Estimating Components of Fault Displacements from Measured Height of Fault Scarp," *Bull. Assoc. Engrg. Geol.*, Vol. XVII, No. 1, Winter, pp. 39–45.

45. Taylor, C. L. and Cluff, L. S. (1977) "Fault Displacement and Ground Deformation Associated with Surface Faulting," *Proc. ASCE, The Current State of Knowledge of Lifeline Earthquake Engineering*, Specialty Conf., Univ. of California, Los Angeles, pp. 338–353.

46. Housner, G. W. (1970*a*) "Strong Ground Motion," *Earthquake Engineering*, Chap. 4, Prentice-Hall Inc., Englewood Cliffs, N.J.

47. Bonilla, M. G. (1970) "Surface Faulting and Related Effects," *Earthquake Engineering*, R. L. Weigel, ed., Prentice-Hall Inc., Englewood Cliffs, N.J., Chap. 3.

48. Leeds, D. J. (1973) "The Design Earthquake," *Geology, Seismicity and Environmental Impact*, Spec. Pub., Assoc. of Engrg. Geol., Los Angeles, Calif.

49. Bolt, B. A. (1973) "Duration of Strong Ground Motion," *5th World Conf. Earthquake Engrg.*, Rome.

50. Seed, H. B., Idriss, I. M. and Kiefer, F. W. (1969) "Characteristics of Rock Motion During Earthquakes," *Proc. ASCE, J. Soil Mechs. Found. Engrg. Div.*, Vol. 95, No. SM5, September, pp. 1199–1218.

51. ESSA (1969) "Studies in Seismicity and Earthquake Damage Statistics, Appendix B," U.S. Dept of Commerce, Environmental Science Services Administration, Coast and Geodetic Survey.

51*a*. Lawson, A. C. et al. (1908) "The California Earthquake of April 18, 1906," Carnegie Inst. of Washington, 2 vols. and atlas.

52. Hunt, R. E. and Sabatino, A. J. (1974) "ERTS Imagery and Earthquake Epicenter Data Support the Possible Existence of a Major Zone of Faulting in the Central Atlantic Seaboard Region," unpublished.

53. Higgins, M. W., Zeitz, I., and Fisher, G. W. (1974) "Interpretation of Aeromagnetic Anomalies Bearing on the Origin of Upper Chesapeake Bay and River Course Changes in the Central Atlantic Seaboard Region: Speculations," *Geology*, Vol. II, No. 2, Geologic Society of America.

54. Spoljaric, N. (1972) "Geology of the Fall Zone in Delaware," R. L. 19, Delaware Geologic Society.

55. King, E. R. and Zeitz, L. (1978) "The New York-Alabama Lineament: Geophysical Evidence for a Major Crustal Break in the Basement Beneath the Appalachian Basin," *Geology*, GSA Bull. Vol. 6, No. 5, May, pp. 312–318.

56. Hadley, J. B. and Devine, J. F. (1974) "Seismotectonic Map of the Eastern United States," U.S. Geol. Survey Misc. Field Studies Map MF 620, scale 1:5,000,000.

57. Taylor, P. W. and Larkin, T. J. (1978) "Seismic Site Response of Nonlinear Soil Media," *Proc. ASCE, J. Geotech. Engrg. Div.*, Vol. 104, No. GT3, March, pp. 369–383.

58. Hardin, B. O. and Drnevich, V. P. (1972*a*) "Shear Modulus and Damping in Soils: Measurement and Parameter Effects," *Proc. ASCE, J. Soil Mechs. Found. Engrg. Div.*, Vol. 98, No. SM6, June, pp. 603–624.

59. Hardin, B. O. and Drnevich, V. P. (1972*b*) "Shear Modulus and Damping in Soils: Design Equations and Curves," *Proc. ASCE, J. Soil Mechs. Found. Engrg. Div.*, Vol. 98, No. SM6, July, pp. 667–692.

60. Whitman, R. V. and DePablo, P. O. (1969) "Den-

sification of Sand by Vertical Vibrations," *Proc. 4th World Conf. on Earthquake Engineering*, Santiago, Chile.

61. Seed, H. B. (1975) "Earthquake Effects on Soil-Foundation Systems," *Foundation Engineering Handbook*, Chap. 25, Winterkorn and Fang, eds., Van Nostrand Reinhold Book Co., New York.
62. Wylie, E. B. and Streeter, V. L. (1976) "Characteristics Method for Liquefaction of Soils," *Proc. ASCE, Numerical Methods in Geomechanics*, ASCE, New York, Vol. II, pp. 938–954.
63. Christian, J. T. and Swiger, W. F. (1975) "Statistics of Liquefaction and SPT Results," *Proc. ASCE, J. Geotech. Engrg. Div.*, Vol. 101, No. GT11, November.
64. Youd, T. L. and Perkins, D. M. (1978) "Mapping Liquefaction-Induced Ground Failure Potential," *Proc. ASCE, J. Geotech. Engrg. Div.*, Vol. 104, No. GT4, April, pp. 433–446.
65. Shannon & Wilson Inc. and Agbabian-Jacobsen Associates (1971) "Soil Behavior Under Earthquake Loading Conditions," Report prepared for USAEC, Contract W-7405-eng.-26.
66. Finn, W. D. (1972) "Soil Dynamics—Liquefaction of Sands," *Proc. Intl. Conf. on Microzonation*, Seattle, November, Vol. I, pp. 87–112.
67. D'Appolonia, E. (1970) "Dynamic Loadings," *Proc. ASCE, J. Soil Mechs. and Found. Engrg. Div.*, Vol. 95, No. SM1, January, p. 49.
68. Castro, G. (1975) "Liquefaction and Cyclic Mobility of Saturated Sands," *Proc. ASCE, J. Geotech. Engrg. Div.*, Vol. 101, No. GT6, June.
69. Seed, H. B., Mori, K. and Chan, C. K. (1975) "Influence of Seismic History on the Liquefaction Characteristics of Sands," Report No. EERC 75-25, Earthquake Engineering Center, Univ. of California, Berkeley, August.
70. Seed, H. B. (1976) "Evaluation of Soil Liquefaction Effects on Level Ground During Earthquakes," State-of-the-Art-Paper, *Liquefaction Problems in Geotechnical Engineering*, ASCE Preprint 2752, New York, pp. 1–104.
71. Seed, H. B. and Booker, J. R. (1977) "Stabilization of Potentially Liquefiable Sand Deposits Using Gravel Drains," *Proc. ASCE, J. Geotech. Engrg. Div.*, Vol. 103, No. GT7, July, pp. 757–768.
72. Youd, T. L. (1978) "Major Cause of Earthquake Damage is Ground Failure," *Civil Engineering*, ASCE, April, pp. 47–51.
73. Seed, H. B., Makdisi, F. I. and DeAlba, P. (1978) "Performance of Earth Dams during Earthquakes," *Proc. ASCE, J. Geotech. Engrg. Div.*, Vol. 101, No. GT7, July, pp. 967–994.
74. Seed, H. B., Lee, K. L. and Idriss, I. M. (1969) "Analysis of the Sheffield Dam Failure," *Proc. ASCE, J. Soil Mechs. and Found. Engrg. Div.*, Vol. 95, No. SM6, November.
75. Seed, H. B., Idriss, I. M., Lee. K. L. and Makdisi, F. I. (1975) "Dynamic Analysis of the Slide in the Lower San Fernando Dam during the Earthquake of February 9, 1971," *Proc. ASCE, J. Geotech. Engrg. Div.*, Vol. 101, No. GT9, September, pp. 889–911.
76. ICOLD (1975) "A Review of Earthquake Resistant Design of Dams," Bull. 27, Intl. Comm. on Large Dams, March.
77. Makdisi, F. I. and Seed, H. B. (1978) "Simplified Procedure for Estimating Dam and Embankment Earthquake-Induced Deformations," *Proc. ASCE, J. Geotech. Engrg. Div.*, Vol. 104, No. GT7, July, pp. 849–868.
78. Leggett, R. F. (1973) *Cities and Geology*, McGraw-Hill Book Co., New York.
79. Kerr, R. A. (1978) "Tidal Waves: A New Method Suggested to Improve Prediction," *Science*, Vol. 200, No. 4341, May, pp. 521–522.
80. Leet, L. D. (1960) *Vibrations from Blasting Rock*, Harvard Univ. Press, Cambridge, Mass.
81. Clough, R. W. (1970) "Earthquake Response of Structures," *Earthquake Engineering*, R. L. Weigel, ed., Prentice-Hall, Inc., Englewood, N.J., Chap. 12.
82. Hall, W. J. and Newmark, N. M. (1977) "Seismic Design Criteria for Pipelines and Facilities," *The Current State of Knowledge of Lifeline Earthquake Engineering, Proc. ASCE*, New York, pp. 18–34.
83. Hudson, D. E. (1972) "Strong Motion Seismology," *Proc. Intl. Conf. on Microzonation*, Seattle, October, Vol. I, pp. 29–60.
84. Whitman, R. V. and Protonotarios, J. N. (1977) "Inelastic Response to Site-Modified Ground Motions," *Proc. ASCE, J. Geotech. Engrg. Div.*, Vol. 103, No. GT10, October, pp. 1037–1053.
85. Hudson, D. E. (1970) "Ground Motion Measurements," *Earthquake Engineering*, Chap. 6, Prentice-Hall Inc., Englewood Cliffs, N.J.
86. USAEC (1963) "Nuclear Reactors and Earthquakes," U.S. Atomic Energy Commission Report TID-7024, August, Lockheed Aircraft Corp., Sunnyvale, Calif.
87. Housner, G. W. (1970*b*) "Design Spectrum,"

*Earthquake Engineering*, Chap. 5, Prentice-Hall Inc., Englewood, N.J.

88. Newmark, N. M., Blume, J. A. and Kapur, K. K. (1973) "Design Response Spectrum for Nuclear Power Plants," Annual Meeting, ASCE, San Francisco, Calif.
89. McGuire, R. K. (1978) "Seismic Ground Motion Parameter Relations," *Proc. ASCE, J. Geotech. Engrg. Div.*, Vol. 104, No. GT4, April, pp. 481–490.
90. Gama, C. D. da (1980) "Desenvolvimento de mapas de risco sésmico," *I Mesa Redonda Sobre Risco Sísmico*, Assoc. Brasileiria de Geologia de Engenharia, São Paulo, 25 June, pp. 40–53.
91. Schnabel, P., Seed, H. B. and Lysmer, J. (1971) "Modifications of Seismograph Records for Effects of Local Soil Conditions," *Report No. EERC 71-8*, Earthquake Engineering Research Center, Univ. of California, Berkeley, December.
92. Idriss, I. M. and Sadigh, K. (1976) "Seismic SSI of Nuclear Power Plant Structures," *Proc. ASCE, J. Geotech. Engrg. Div.*, Vol. 102, No. GT7, July, pp. 663–682.
93. NRC (1975) U.S. Nuclear Regulatory Commission, Title 10, Chap. 1, Code of Federal Regulations—Energy, Reactor Site Criteria; 10 CFR, Part 100, Appendix A, Seismic and Geologic Siting Criteria for Nuclear Power Plants.

## BIBLIOGRAPHY

Ambraseys, N. N. (1960) "On the Seismic Behavior of Earth Dams," *Proc 2nd Intl. Conf. Earthquake Engrg.*, Tokyo, July.

Arango, I. and Dietrich, R. J. (1972) "Soil and Earthquake Uncertainties on Site Response Studies," *Intl. Conf. on Microzonation*, Seattle, November.

Bergstrom, R. N., Chu, S. L. and Small R. J. (1969) "Dynamic Analysis of Nuclear Power Plants for Seismic Loadings," Presentation reprint, ASCE Annual Meeting, Chicago, October.

Blázquez, R., Krizek, R. J. and Bažant, Z. P. (1980) "Site Factors Controlling Liquefaction," *Proc. ASCE, J. Geotech. Engrg. Div.*, Vol. 106, No. GT7, July, pp. 785–802.

Bolt, B. A. (1970*a*) "Elastic Waves in the Vicinity of the Earthquake Source," *Earthquake Engineering*, R. L. Weigel, ed., Prentice-Hall, Inc., Englewood Cliffs, N.J., Chap. 1.

Bolt, B. A. (1970*b*) "Causes of Earthquakes," *Earthquake Engineering*, R. L. Weigel, ed., Prentice-Hall Inc., Englewood Cliffs, N.J., Chap. 2.

Bolt, B. A. (1972) "Seismicity," *Proc. Intl. Conf. on Microzonation*, Seattle, October, Vol. I, pp. 13–28.

Bolt, B. A. and Hudson, D. E. (1975) "Seismic Instrumentation of Dams," *Proc. ASCE, J. Geotech. Engrg. Div.*, Vol. 101, No. GT11, November.

Bonilla, M. G. and Buchanan, J. M. (1970) "Interim Report on Worldwide Surface Faulting," U.S. Geological Survey, Open-File Report.

Cluff, L. S. and Brogan, G. E. (1974) "Investigation and Evaluation of Fault Activity in the USA," *Proc. 2d Intl. Cong.*, Intl. Assoc. Engrg. Geol., São Paulo, Vol. I.

Donovan, N. C., Bolt, B. A. and Whitman, R. V. (1976) "Development of Expectancy Maps and Risk Analysis," Preprint 2805, ASCE Annual Convention and Exposition, Philadelphia, Pa., September.

Epply, R. A. (1965) "Earthquake History of the United States," Part I, Strong Earthquakes of the United States (Exclusive of California, Nevada), U.S. Govt. Printing Office, Washington, D.C.

Esteva, L. (1976) "Seismicity," *Seismic Risk and Engineering Decisions*, Lomnitz and Rosenblueth, eds., Chap. 6, Elsevier Scientific Publishing Co., New York.

Fischer, J. A., North, E. D. and Singh, H. (1972) "Selection of Seismic Design Parameters for a Nuclear Facility," *Proc. Intl. Conf. on Microzonation*, Seattle, October, Vol. II, pp. 755–770.

Haimson, B. C. (1973) "Earthquake Related Stresses at Rangely, Colorado," *New Horizons in Rock Mechanics, Proc. ASCE, 14th Symp. on Rock Mechs.*, Univ. Park, Pa., June (1972).

Hudson, D. E. (1965) "Ground Motion Measurements in Earthquake Engineering," *Proc. Symp. on Earthquake Engineering*, The Univ. of British Columbia, Vancouver, B.C.

Lamar, D. L., Merifield, P. M. and Proctor, R. J. (1973) "Earthquake Recurrence Intervals on Major Faults in Southern California," *Geology, Seismicity and Environmental Impact*, Spec. Pub., Assoc. of Eng. Geol., Los Angeles, Calif.

Lomnitz, C. and Rosenblueth, E. (1976) *Seismic Risk and Engineering Decisions*, Elsevier Scientific Pub. Co., Amsterdam.

Martin, G. M., Finn, W. D. and Seed, H. B. (1975) "Fundamentals of Liquefaction under Cyclic Loading," *Proc. ASCE, J. Geotech. Engrg. Div.*, Vol. 101, No. GT5, May.

Newmark, N. M. and Hall, W. J. (1973) "Seismic Design Spectra for Trans-Alaska Pipeline," *Proc. 5th*

*World Conf. on Earthquake Engineering*, Rome, Paper No. 60.

Panovko, Y. (1971) *Elements of the Applied Theory of Elastic Vibration*, MIR Publishers, Moscow.

Park, T. K. and Silver, M. L. (1975) "Dynamic Triaxial and Simple Shear Behavior of Sand," *Proc. ASCE, J. Geotech. Engrg. Div.*, Vol. 101, No. GT6, June.

Pensien, J. (1970) "Soil-Pile Interaction," *Earthquake Engineering*, Prentice-Hall Inc., Englewood Cliffs, N.J.

Pyke, R., Seed, H. and Chan, C. K. (1975) "Settlement of Sands Under Multidirectional Shaking," *Proc. ASCE, J. Geotech. Engrg. Div.*, Vol. 101, No. GT4, April.

Scholz, C. H. (1972) "Crustal Movement in Tectonic Areas," *Tectonophysics* 14 (3/4).

Seed, H. B. (1970*b*) "Earth Slope Stability During Earthquakes," *Earthquake Engineering*, Chap. 15, Prentice-Hall Inc., Englewood Cliffs, N.J.

Seed H. B. and Schnable, P. B. (1972) "Soil and Geologic Effects on Site Response During Earthquakes," *Intl. Conf. on Microzonation*, Seattle, Wash., November.

Sherif, M. A., Bostrom, R. C. and Ishibashi, I. (1974) "Microzonation in Relation to Predominant Ground Frequency, Amplification and Other Engineering Considerations," *Proc. 2d Intl. Cong. Intl. Assoc. of Engrg. Geol.*, São Paulo, Vol. 1, pp. 11-2.1 to 2.11.

Sherard, J. L., Cluff, L. S. and Allen, C. R. (1974) "Potentially Active Faults in Dam Foundations," *Geotechnique*, Vol. XXIV, No. 3, September.

Steinbrugge, K. V. (1970) "Earthquake Damage and Structural Performance in the U.S.," *Earthquake Engineering*, Chap. 9, Prentice-Hall Inc., Englewood Cliffs, N.J.

Wallace R. E. (1970) "Earthquake Recurrence Intervals on the San Andreas Fault," *Bull. Geologic Society of America*, Vol. 81, pp. 2875–2890.

Wiegel, R. L. (1970) "Tsunamis," *Earthquake Engineering*, Chap. 11, Prentice-Hall Inc., Englewood Cliffs, N.J.

Wong, R. T., Seed, H. B. and Chan, C. K. (1975) "Cyclic Loading Liquefaction of Gravelly Soils," *Proc. ASCE, J. Geotech. Engrg. Div.*, Vol. 101, No. GT6, June.

Zaslawsky, M. and Wight, L. H. (1976) "Comparison of Bedrock and Surface Seismic Input for Nuclear Power Plants," *Proc. ASCE, Numerical Methods in Geomechanics*, Vol. II, ASCE, New York, pp. 991–1000.

# APPENDIX A

# The Earth and Geologic History

## A.1 SIGNIFICANCE TO THE ENGINEER

To the engineer, the significance of geologic history lies in the fact that although surficial conditions of the earth appear to be constants, they are not truly so, but rather are transient. Continuous, albeit generally barely perceptible, changes are occurring because of warping, uplift, faulting, decomposition, erosion and deposition, and the melting of glaciers and ice caps. The melting contributes to crustal uplift and sea level changes. Climatic conditions are also transient and the direction of change is reversible.

It is important to be aware of these transient factors which can invoke significant changes within relatively short time spans, such as a few years or several decades. They can impact significantly on conclusions drawn from statistical analysis for flood-control or seismic-design studies based on data that extend back 50, 100, or 200 years, as well as other geotechnical studies.

To provide a general perspective, the earth, global tectonics, and a brief history of North America are presented.

## A.2 THE EARTH

### A.2.1 GENERAL

*Age* has been determined to be 4½ billion years.

*Origin* is thought to be a molten mass that subsequently began a cooling process that created a crust over a central core. Whether the cooling process is continuing is not known.

### A.2.2 CROSS SECTION

From seismological data, the earth is considered to consist of four major zones as illustrated on Fig. 11.5: crust, mantle, and outer and inner cores.

*Crust* is a thin shell of rock averaging 30 to 40 km in thickness beneath the continents, but only 5 km thickness beneath the seafloors. The lower portions are a heavy basalt ($\gamma = 3$ t/m$^3$) surrounding the entire globe, overlain by lighter masses of granite ($\gamma = 2.7$ t/m$^3$) on the continents.

*Mantle* underlies the crust separated from it by the Moho (Mohorovičić discontinuity). Roughly 3000 km thick, the nature of the material is not known, but it is much denser than the crust and is believed to consist of molten iron and other heavy elements.

*Outer core* lacks rigidity and is probably fluid.

*Inner core* begins at 5000 km and is possibly solid ($\gamma \sim 12$ t/m$^3$), but conditions are not truly known. The center is at 6400 km.

## A.3 GLOBAL TECTONICS

### A.3.1 GENERAL

During geologic time the earth's surface has been undergoing constant change. Fractures occur from faulting which is hundreds of kilometers in length in places. Mountains are pushed up, then eroded away, and their detritus deposited in vast seas. The detritus is compressed, formed into rock and pushed up again to form new mountains, and the cycle is repeated. From time to time masses of molten rock well up from the mantle to form huge flows that cover the crust.

*Tectonics* refers to the broad geologic features of the continents and ocean basins, as well as the forces responsible for their occurrence. The origins of these forces are not well understood, although it is apparent that the earth's crust is in a state of overstress as evidenced by folding, faulting, and other mountain-building processes. Four general hypotheses have been developed to describe the sources of global tectonics [Hodgson (1964),[1] Zumberge and Nelson (1972)[2]].

### A.3.2 THE HYPOTHESES

*Contraction hypothesis* assumes that the earth is cooling, and because earthquakes do not occur below 700 km, the earth is considered static below this depth, and is still hot and not cooling. The upper layer of the active zone, to a depth of about 100 km, has stopped cooling and shrinking. As the lower layer cools and contracts, it causes the upper layer to conform by buckling, which is the source of the surface stresses. This hypothesis is counter to the spreading seafloor or continental drift theory.

*Convection-current hypothesis* assumes that heat is being generated within the earth by radioactive disintegration and that this heat causes convection currents that rise to the surface under the midocean rifts, causing tension to create the rifts, then moves toward the continents with the thrust necessary to push up mountains, and finally descend again beneath the continents.

*Expanding earth hypothesis,* the latest theory, holds that the earth is expanding because of a decrease in the force of gravity, which is causing the original shell of granite to break up and spread apart, giving the appearance of continental drift.

*Continental drift theory* is currently the most popular, but is not new, and is supported by substantial evidence. Seismology has demonstrated that the continents are blocks of light granitic rocks "floating" on heavier basaltic rocks. It has been proposed that all of the continents were originally connected as one or two great land masses and at the *end of the Paleozoic era* (Permian period) they broke up and began to drift apart as illustrated in sequence on Fig. A.1. The proponents of the theory have divided the earth into "plates," with each plate bounded by an earthquake zone as shown on Fig. 11.1. Wherever plates move against each other, or a plate plunges into a deep ocean trench, such as exists off the west coast of South America or the east coast of Japan, so that it slides beneath an adjacent plate *(see Fig. 11.2),* there is high seismic activity. This concept is known as "plate tectonics" and appears to be compatible with the relatively new concept of *seafloor spreading,* as shown on Fig. 11.2.

## A.4 GEOLOGIC HISTORY

### A.4.1 NORTH AMERICA: PROVIDES A GENERAL ILLUSTRATION

The geologic time scale for North America is given on Table A.1, relating periods to typical formations. A brief geologic history of North America is described in Table A.2. These relationships apply in a general manner to many other parts of the world. Most of the periods are separated by major crustal disturbances *(orogenies).* Age determination is based on fossil identification (paleontology) and radiometric dating.

The classical concepts of the history of North America have been modified somewhat to conform with the modern concept of the continental drift hypothesis. The most significant modification is the consideration that until the end of the Permian period the east coast of the United States was connected to the northwest coast of Africa as shown on Fig. A.1.

### A.4.2 RADIOMETRIC DATING

Radiometric dating determines the age of a formation by using the decay rate of a radioactive element.

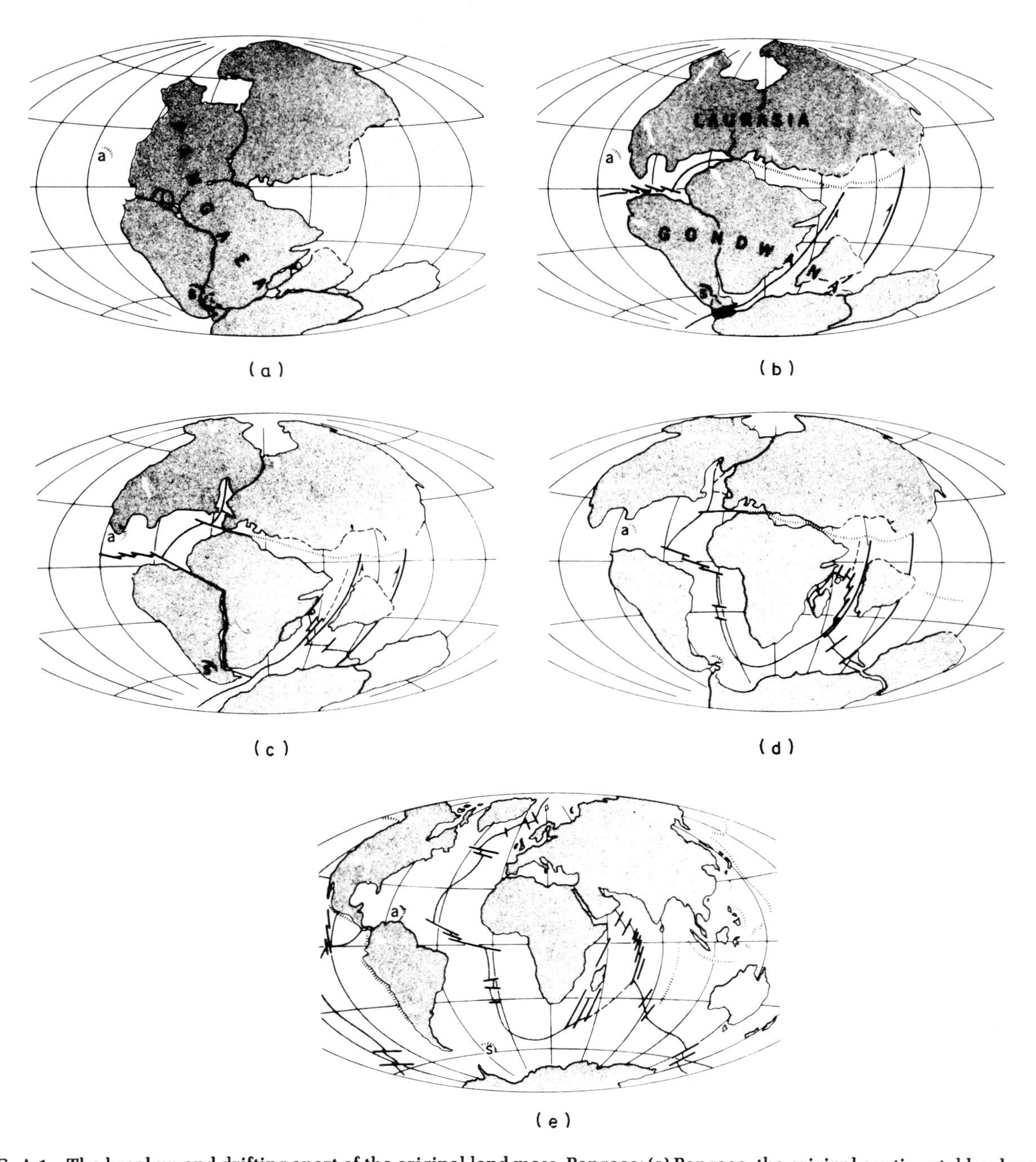


**FIG. A.1 The breakup and drifting apart of the original land mass, Pangaea: (a) Pangaea, the original continental land mass at the end of the Permian, 225 million years ago; (b) Laurasia and Gondwana at the end of the Triassic, 180 million years ago; (c) positions at the end of the Jurassic, 135 million years ago (North and South America beginning to break away); (d) positions at the end of the Cretaceous, 65 million years ago; (e) positions of continents and the plate boundaries at the present.** [From Dietz and Holden (1970).[3]]

**TABLE A.1**
**GEOLOGIC TIME SCALE AND THE DOMINANT ROCK TYPES IN NORTH AMERICA**

| Era | Period | | Epoch | Dominant formations | Age, millions of years | |
|---|---|---|---|---|---|---|
| Cenozoic | Quaternary | | Holocene | Modern soils | 0.01 | |
| | | | Pleistocene | North American glaciation | 2.5–3 | |
| | Neogene | Tertiary | Pliocene | "Unconsolidated" coastal-plain sediments | 7 | |
| | | | Miocene | | 26 | |
| | Paleogene | | Oligocene | | 37 | |
| | | | Eocene | | 54 | |
| | | | Paleocene | | 65 | 65 |
| Mesozoic | Cretaceous | | | Overconsolidated clays and clay shales | | 135 |
| | Jurassic | | | Various sedimentary rocks | | 180 |
| | Triassic | | | Clastic sedimentary rocks with diabase intrusions | | 225 |
| Paleozoic | Permian | | | Fine-grained clastics, chemical precipitates, and evaporites. Continental glaciation in southern hemisphere. | | 280 |
| | Pennsylvanian | | Carboniferous | Shales and coal beds | | 310 |
| | Mississippian | | Carboniferous | Limestones in central United States. Sandstones and shales in east. | | 345 |
| | Devonian | | | Red sandstones and shales | | 400 |
| | Silurian | | | Limestone, dolomite and evaporites, shales | | 435 |
| | Ordovician | | | Limestone and dolomite, shales | | 500 |
| | Cambrian | | | Limestone and dolomite in late Cambrian, sandstones and shales in early Cambrian | | 600 |
| Precambrian | Precambrian | | | Igneous and metamorphic rocks | | about 4.5 billion years |

**TABLE A.2**
**A BRIEF GEOLOGIC HISTORY OF NORTH AMERICA***

| Period | Activity |
| --- | --- |
| Precambrian | Period of hundreds of millions of years during which the crust was formed and the continental land masses appeared. |
| Cambrian | Two great troughs in the east and west filled with sediments ranging from detritus at the bottom, upward to limestones and dolomites, which later formed the Appalachians, the Rockies, and other mountain ranges. |
| Ordovician | About 70% of North America was covered by shallow seas and great thicknesses of limestone and dolomite were deposited.<br>There was some volcanic activity and the eastern landmass, including the mountains of New England, started to rise (Taconic orogeny). |
| Silurian | Much of the east was inundated by a salty inland sea; the deposits ranged from detritus to limestone and dolomite, and in the northeast large deposits of evaporites accumulated in landlocked arms of the seas.<br>Volcanos were active in New Brunswick and Maine. |
| Devonian | Eastern North America, from Canada to North Carolina, rose from the sea (Arcadian orogeny). The northern part of the Appalachian geosyncline received great thicknesses of detritus that eventually formed the Catskill Mountains.<br>In the West, the stable interior was inundated by marine waters and calcareous deposits accumulated.<br>In the east, limestone was metamorphosed to marble. |
| Carboniferous | Large areas of the east became a great swamp which was repeatedly submerged by shallow seas. Forests grew, died, and were buried to become coal during the Pennsylvanian portion of the period. |
| Permian | A period of violent geologic and climatic disturbances. Great wind-blown deserts covered much of the continent. Deposits in the west included evaporites and limestones.<br>The Appalachian Mountains were built in the east to reach as high as the modern Alps (Alleghanian orogeny).<br>The continental drift theory (Art. A.3.2) considers that it was toward the end of the Permian that the continents began to drift apart. |
| Triassic | The Appalachians began to erode and their sediments were deposited in the adjacent nonmarine seas.<br>The land began to emerge toward the end of the period and volcanic activity resulted in sills and lava flows; faulting occurred during the Palisades orogeny. |
| Jurassic | The Sierra Nevada Mountains, stretching from southern California to Alaska, were thrust up during the Nevadian disturbance. |

*(Continued)*

**TABLE A.2**
**A BRIEF GEOLOGIC HISTORY OF NORTH AMERICA* (Continued)**

| Period | Activity |
|---|---|
| Cretaceous | The Rocky Mountains from Alaska to Central America rose out of a sediment-filled trough. |
| | For the last time the sea inundated much of the continent and thick formations of clays were deposited along the east coast. |
| Tertiary | The Columbia plateau and the Cascade Range rose, and the Rockies reached their present height. |
| | Clays were deposited and shales formed along the continental coastal margins, reaching thicknesses of some 12 km in a modern syncline in the northern Gulf of Mexico that has been subsiding since the end of the Appalachian orogeny. |
| | Extensive volcanic activity occurred in the northwest. |
| Quaternary | During the Pleistocene epoch, four ice ages sent glaciers across the continent, which had a shape much like the present. |
| | In the Holocene epoch (most recent), from 18,000 to 6000 years ago, the last of the great ice sheets covering the continent melted and sea level rose almost 100 m. |
| | Since then, sea level has remained almost constant, but the land continues to rebound from adjustment from the tremendous ice load. In the center of the uplifted region in northern Canada, the ground has risen 138 m in the last 6000 years and is currently rising at the rate of about 2 cm/year [Walcott (1972)[4]]. |
| | Evidence of ancient postglacial sea levels is given by raised beaches and marine deposits of late Quaternary found around the world. In Brazil, for example, Pleistocene sands and gravels are found along the coastline as high as 20 m above present sea level. |

*The geologic history presented here contains the general concepts accepted for many decades, and still generally accepted. The major variances, as postulated by the continental drift concept, are that until the end of the Permian, Appalachia (a land mass along the U.S. east coast region) may have been part of the northwest coast of Africa (Fig. A.1*a*), and that the west coast of the present United States may have been an archipelago of volcanic islands known as Cascadia. [Zumberge and Nelson (1972).[2]]

In radioactive elements, such as uranium, the number of atoms which decay during a given unit of time to form new stable elements is directly proportional to the number of atoms of the radiometric element of the sample. This decay rate is constant for the various radioactive elements and is given by the half life of the element—that is, the time required for any initial number of atoms to be reduced by one-half. For example, when once-living organic matter is carbon-dated, the amount of radioactive carbon (carbon 14) remaining and the amount of ordinary carbon present are measured, and the age of a specimen is computed from a simple mathematical relationship. A general discussion on dating techniques can be found in Murphy et al. (1979)[5]. The various isotopes, effective dating range, and minerals and other materials that can be dated are given on Table A.3.

In engineering problems the most significant use is for the dating of materials from fault zones to determine the age of most recent activity (*see Art. 11.3.1*). The technique is useful also in dating soil formations underlying colluvial deposits as an indication as to when the slope failure occurred.

**TABLE A.3**
**SOME OF THE PRINCIPAL ISOTOPES USED IN RADIOMETRIC DATING**

| Isotope | | | | |
|---|---|---|---|---|
| **Parent** | **Offspring** | **Parent half life, years** | **Effective dating range, years** | **Material that can be dated** |
| Uranium 238 | Lead 206 | 4.5 billion | 10 million to 4.6 billion | Zircon, uraninite, pitchblende |
| Uranium 235 | Lead 207 | 710 million | | |
| Potassium 40* | Argon 40<br>Calcium 40 | 1.3 billion | 100,000 to 4.6 billion | Muscovite, biotite hornblende, intact volcanic rock |
| Rubidium 87 | Strontium 87 | 47 billion | 10 million to 4.6 billion | Muscovite, biotite, microcline, intact metamorphic rock |
| Carbon 14* | Nitrogen 14 | 5,730 ± 30 | 100 to 50,000 | Plant material: wood, peat charcoal, grain.<br>Animal material: bone, tissue.<br>Cloth, shell, stalactites, groundwater and seawater. |

*Most commonly applied to fault studies; Carbon 14 for carbonaceous matter, or K-Ar for noncarbonaceous matter such as fault gouge.

## REFERENCES

1. Hodgson, J. H. (1964) *Earthquakes and Earth Structure*, Prentice-Hall Inc., Englewood Cliffs, N.J.
2. Zumberge, J. H. and Nelson, C. A. (1972) *Elements of Geology*, 3d ed., John Wiley & Sons, New York.
3. Dietz, R. S. and Holden, J. C. (1970) "Reconstruction of Pangaea; breakup and dispersion of continents, Permian to present," *J. Geophysical Research*, Vol. 75, No. 26, pp. 4939–4956.
4. Walcott. R. L. (1972) "Late Quaternary Vertical Movements in Eastern North America: Quantitative Evidence of Glacio-Isostatic Rebound," *Reviews of Geophysics and Space Physics*, Vol. 10, No. 4, November, pp. 849–884.
5. Murphy, P. J., Briedis, J. and Peck, J. H. (1979) "Dating techniques in fault investigations," *Geology in the Siting of Nuclear Power Plants, Reviews in Engineering Geology IV*, The Geological Society of America, Boulder, Colo., pp. 153–168.

## BIBLIOGRAPHY

Dunbar, C. O. and Waage, K. M. (1969) *Historical Geology*, 3d ed., John Wiley & Sons, New York.

Guttenberg, B. and Richter, C. F. (1954) *Seismicity of the Earth and Related Phenomenon*, Princeton Univ. Press, Princeton, N.J.

# APPENDIX B

# Procurement of Geologic Publications, Maps and Remote-Sensing Imagery, and Earthquake Information

**TABLE B.1**
**GENERAL PROCUREMENT OF PUBLICATIONS FROM THE U.S. GEOLOGICAL SURVEY, U.S. DEPARTMENT OF THE INTERIOR***

| Source | Publications |
|---|---|
| U.S. Government Printing Office | Professional papers, bulletins, and water-supply papers. |
| Public Inquiries Offices (PIOs) | Professional papers, bulletins, and water-supply papers sold over the counter. |
| Distribution Offices (USGS) | Catalogs and lists of publications |
| Eastern Region—east of Mississippi River | Lists by state of geologic and water-supply reports and maps |
| Western Region—west of Mississippi River | Circulars and open-file reports |
| Regional basis—PIOs | Geologic quadrangle maps (GQ) |
| | Geophysical Investigations Maps (GP) (aeromagnetic) |
| | Hydrological investigations atlases |
| | Geologic map indexes by state |
| | Topographic maps including orthophoto maps |
| | River survey maps |
| EROS Data Center | ERTS and LANDSAT Imagery |
| | USGS aerial photography |
| | NASA aircraft data |
| Regional Engineering Offices | Aerial photographs |

*For addresses and details see Tables B.2, B.3, and B.4.

## TABLE B.2
## INFORMATION SOURCES: U.S. GEOLOGICAL SURVEY AND OTHERS

### PUBLIC INQUIRIES OFFICES (PIOs)

ALASKA: Anchorage (99501)
108 Skyline Building, 508 2nd Avenue
Phone: (907) 277-0577

CALIFORNIA: Los Angeles (90012)
7638 Federal Building
300 North Los Angeles Street
Phone: (213) 688-2850
_________, San Francisco (94111)
504 Custom House, 555 Battery Street
Phone: (415) 556-5627

COLORADO: Denver (80202)
1012 Federal Building
1961 Stout Street
Phone: (303) 837-4160

TEXAS: Dallas (75242)
Room 1 C45, 1100 Commerce Street
Phone: (214) 749-3230

UTAH: Salt Lake City (84138)
8102 Federal Office Building
125 South State Street
Phone: (801) 524-5652

WASHINGTON: Spokane (99201)
678 U.S. Court House Building
West 920 Riverside Avenue
Phone: (509) 456-2524

DISTRICT OF COLUMBIA: Washington (20244)
1028 GSA Building
19th and F Streets NW
Phone: (202) 343-8073

### DISTRIBUTION OFFICES

U.S. Geological Survey
Branch of Distribution
Eastern Region
1200 South Eads Street
Arlington, VA 22202

U.S. Geological Survey
Branch of Distribution
Western Region
Building 41, Federal Center
Denver, CO 80225

### STATE GEOLOGICAL AGENCIES

ALABAMA
Geological Survey of Alabama
P.O. Drawer 0
University, AL 35486

ALASKA
Division of Geological and Geophysical Surveys
3001 Porcupine Drive
Anchorage, AK 99501

ARIZONA
Arizona Bureau of Mines
University of Arizona
Tucson, AZ 85721

ARKANSAS
Arkansas Geological Commission
Vardelle Parham Geological Center
Little Rock, AR 72201

CALIFORNIA
Division of Mines and Geology
California Department of Conservation
1416 9th Street—13th Floor
Sacramento, CA 95814

COLORADO
Colorado Geological Survey
1845 Sherman Street
Denver, CO 80203

CONNECTICUT
Connecticut Geological and Natural History Survey
Box 128, Wesleyan Station
Middletown, CT 06457

DELAWARE
Delaware Geological Survey
University of Delaware
Newark, DE 19711

FLORIDA
Bureau of Geology
P.O. Drawer 631
Tallahassee, FL 32302

GEORGIA
Department of Natural Resources
19 Hunter Street, S.W.
Atlanta, GA 30334

HAWAII
Division of Water and Land Development
Department of Land and Natural Resources
P.O. Box 373
Honolulu, HI 96809

IDAHO
Idaho Bureau of Mines and Geology
Moscow, ID 83843

**TABLE B.2**
**INFORMATION SOURCES: U.S. GEOLOGICAL SURVEY AND OTHERS (*Continued*)**

STATE GEOLOGICAL AGENCIES (*Continued*)

ILLINOIS
Illinois State Geological Survey
124 Natural Resources Building
Urbana, IL 61801

INDIANA
Department of Natural Resources
Indiana Geological Survey
611 North Walnut Grove
Bloomington, IN 47401

IOWA
Iowa Geological Survey
Geological Survey Building
16 West Jefferson Street
Iowa City, IA 52240

KANSAS
State Geological Survey of Kansas
The University of Kansas
Lawrence, KS 66044

KENTUCKY
Kentucky Geological Survey
University of Kentucky
307 Mineral Industries Building
Irvington, KY 40506

LOUISIANA
Louisiana Geological Survey
Box G, University Station
Baton Rouge, LA 70803

MAINE
Maine Geological Survey
Room 211, State Office Building
Augusta, ME 04330

MARYLAND
Maryland Geological Survey
214 Latrobe Hall
Johns Hopkins University
Baltimore, MD 21218

MASSACHUSETTS
Department of Public Works
93 Worcester Street
Wellesley Hills, MA 02181

MICHIGAN
Michigan Department of Conservation
Geological Survey Division
Stevens T. Mason Building
Lansing, MI 48926

MINNESOTA
Minnesota Geological Survey
1633 Eustis Street
St. Paul, MN 55108

MISSISSIPPI
Mississippi Geological, Economic, and Topographical Survey
P.O. Box 4915
Jackson, MS 39216

MISSOURI
Division of Geological Survey and Water Resources
P.O. Box 250
Rolla, MO 65401

MONTANA
Montana Bureau of Mines and Geology
Montana College of Mineral Science and Technology
Butte, MT 59701

NEBRASKA
Conservation and Survey Division
University of Nebraska
Lincoln, NB 68508

NEVADA
Nevada Bureau of Mines
University of Nevada
Reno, NV 89507

NEW HAMPSHIRE
Department of Resources and Economic Development
Office of State Geologist
University of New Hampshire
Durham, NH 03824

NEW JERSEY
New Jersey Bureau of Geology and Topography
John Fitch Plaza—Room 709
P.O. Box 1889
Trenton, NJ 08625

NEW MEXICO
State Bureau of Mines and Mineral Resources
Campus Station
Socorro, NM 87801

NEW YORK
New York State Museum and Science Service, Geological Survey
New York State Education Building
Albany, NY 12224

NORTH CAROLINA
Office of Earth Resources
Department of Natural and Economic Resources
P.O. Box 27687
Raleigh, NC 27611

NORTH DAKOTA
North Dakota Geological Survey
University Station
Grand Forks, ND 58201

OHIO
Ohio Division of Geological Survey
Ohio Department of Natural Resources
Fountain Square
Columbus, OH 43224

OKLAHOMA
Oklahoma Geological Survey
The University of Oklahoma
Norman, OK 73069

(*continued*)

**TABLE B.2**
**INFORMATION SOURCES: U.S. GEOLOGICAL SURVEY AND OTHERS (*Continued*)**

STATE GEOLOGICAL AGENCIES (*Continued*)

OREGON
State Department of Geology and Mineral Industries
1069 State Office Building
1400 S.W. Fifth Avenue
Portland, OR 97201

PENNSYLVANIA
Bureau of Topographic and Geological Survey
Department of Environmental Resources
Harrisburg, PA 17120

PUERTO RICO
Mineralogy and Geology Section
Economic Development Administration of Puerto Rico
Industrial Laboratory
P.O. Box 38
Roosevelt, PR 00929

SOUTH CAROLINA
Division of Geology
P.O. Box 927
Columbia, SC 29202

SOUTH DAKOTA
South Dakota State Geological Survey Science Center
University of South Dakota
Vermillion, SD 57069

TENNESSEE
Department of Conservation
Division of Geology
G-5 State Office Building
Nashville, TN 37219

TEXAS
Bureau of Economic Geology
The University of Texas
University Station, Box X
Austin, TX 78712

UTAH
Utah Geological and Mineralogical Survey
103 Utah Geological Survey Building
University of Utah
Salt Lake City, UT 84112

VERMONT
Vermont Geological Survey
University of Vermont
Burlington, VT 05401

VIRGINIA
Virginia Division of Mineral Resources
P.O. Box 3667
Charlottesville, VA 22903

WASHINGTON
Washington Division of Mines and Geology
335 General Administration Building
P.O. Box 168
Olympia, WA 98501

WEST VIRGINIA
West Virginia Geological and Economic Survey
P.O. Box 879
Morgantown, WV 26505

WISCONSIN
Wisconsin Geological and Natural History Survey
University of Wisconsin
1815 University Avenue
Madison, WI 53706

WYOMING
Geological Survey of Wyoming
Box 3008, University Station
University of Wyoming
Laramie, WY 82071

**TABLE B.2**
**INFORMATION SOURCES: U.S. GEOLOGICAL SURVEY AND OTHERS (*Continued*)**

| U.S. GEOLOGICAL SURVEY | |
|---|---|
| National Center<br>Reston, VA 22092 | |
| NATIONAL TECHNICAL INFORMATION SERVICE (NTIS) | |
| U.S. Department of Commerce<br>Springfield, VA 22151 | |
| SUPERINTENDENT OF DOCUMENTS | |
| U.S. Government Printing Office<br>Washington, D.C. 20402 | |
| REGIONAL ENGINEERING OFFICES (USGS) | |
| Atlantic Region Engineer<br>U.S. Geological Survey<br>1109 N. Highland Street<br>Arlington, VA 22210 | Rocky Mountain Region Engineer<br>U.S. Geological Survey<br>Building 25, Federal Center<br>Denver, CO 80225 |
| Central Region Engineer<br>U.S. Geological Survey<br>Box 133<br>Rolla, MO 65401 | Pacific Region Engineer<br>U.S. Geological Survey<br>345 Middlefield Road<br>Menlo Park, CA 94025 |
| CARTOGRAPHIC DIVISION<br>SOIL CONSERVATION SERVICE | |
| Federal Center Building<br>Hyattsville, MD 20782 | |
| NATIONAL OCEAN SURVEY | |
| Department of Commerce<br>Washington Science Center<br>Rockville, MD 20852 | |

**TABLE B.3**
**PROCUREMENT OF REMOTE-SENSING IMAGERY**

(*A*) ERTS-1, LANDSAT, SKYLAB, HIGH-ALTITUDE NASA STEREO PHOTOS

EROS Data Center, USGS
Sioux Falls, SD 57198
(605) 594-6511

(*B*) LANDSAT FROM FOREIGN COUNTRIES

Integrated Satellite Information
Box 1630
Prince Albert, Saskatchewan S6U 5T2
Canada

Instituto de Pesquisas Especiais (INPE)
Attn. Divisão de Banco de Dados
Av. dos Astronautas 1758
Caixa Postal 515
12.200 São Jose dos Campos, SP
Brazil

Department of Forestry and Agriculture
Building 810
Pleasantville
St. Johns, Newfoundland A1A 1P9
Canada

Telespazio—SES
Corso D'Italia
L-3 Roma
Italy

(*C*) AERIAL PHOTOGRAPHS (GENERALLY AT 1:20,000) OF THE UNITED STATES AND POSSESSIONS

1. USGS: National Cartographic Information Center (NCIC) provides catalogs.

NCIC-Reston (Headquarters)
USGS, 507 National Center
Reston, VA 22092
(703) 860-6045

NCIC-Mid-Continent
USGS, 1400 Independence Road
Rolla, MO 65401
(314) 364-3680, ext. 107

NCIC-Rocky Mountain
USGS Topographic Div.
Stop 510, Box 25046
Denver Federal Center
Denver, CO 80225
(303) 232-2326

NCIC-Western
USGS, 345 Middlefield Road
Menlo Park, CA 94025
(415) 323-2427

2. SCS provides a catalog, "Status of Aerial Photography"

U.S. Department of Agriculture
Soil Conservation Service
Cartographic Division
6505 Belcrest Road
Hyattsville, MD 20782
(301) 436-8756

3. Agricultural Stabilization Conservation Service (ASCS)

Aerial Photography Division
ASCS-USDA
2505 Parley's Way
Salt Lake City, UT 84109
(801) 524-5856

**TABLE B.3**
**PROCUREMENT OF REMOTE-SENSING IMAGERY (*Continued*)**

(*D*) AERIAL PHOTOS WITH SOME COLOR AND INFRARED COVERAGE IN THE UNITED STATES

1. U.S. Department of Agriculture
   Forest Service
   P.O. Box 2417
   Washington, DC 20013
   (707) 235-8638

2. NOAA, National Ocean Survey
   6001 Executive Blvd.
   Rockville, MD 20852
   Attn. Coastal Mapping Div.
   (301) 443-8601

(*E*) OLD AERIAL PHOTOS

National Archives (GSA)
Cartographic Records Division
Washington, DC 20408
(202) 962-0173

(*F*) OUTSIDE THE UNITED STATES, PHOTOS MAY BE AVAILABLE FROM THE USAF FLIGHTS (COUNTRY RELEASES REQUIRED FROM DIPLOMATIC REPRESENTATIVES)

1. Photographic Records and Services Div.
   Aeronautical Chart and Information Center
   Washington, DC

2. Defense Mapping Agency
   Building 56
   U.S. Naval Observatory
   Washington, DC 20305
   (202) 254-4406

(*G*) SLAR IMAGERY

1. Repository for Air Force radar imageries from certain programs:
   Goodyear Aerospace Corporation
   Litchfield Park, AZ 85340
   (Attn. SLAR Imagery Depository)

2. National Cartographic Information Center
   Reston, VA

3. Westinghouse Electric Corp.
   Philadelphia, PA

**TABLE B.4**
**PROCUREMENT OF EARTHQUAKE INFORMATION**

| WORLDWIDE EARTHQUAKE DISTRIBUTION, SEISMICITY MAPS |
|---|
| National Geophysical and Solar Terrestrial Data Center<br>Code D 62<br>NOAA*/EDS<br>Boulder, CO 80302 |
| Detailed data for a given area are provided in a general catalog of "felt" reports available, titled "*United States Earthquakes*": |
| ESSA (Environmental Science Service Administration)<br>USGS, Dept. of the Interior |
| Catalogs on worldwide earthquakes are also available from the International Seismological Center in Edinburgh, Scotland. |
| EARTHQUAKE INFORMATION BULLETIN (BIMONTHLY) |
| U.S. Geological Survey<br>U.S. Govt. Printing Office<br>Washington, DC 20402 |

*NOAA is National Oceanic and Atmospheric Administration.

# APPENDIX C

# Conversion Tables

| ENGLISH TO METRIC TO THE INTERNATIONAL SYSTEM* | | | | |
|---|---|---|---|---|
| **Units†** | **English** | **Metric** | | **SI** |
| Length | 1 mi = 1760 yds = 5280 ft | 1.609 km | or | 1609 m |
| | 1 ft = 12 in | 0.3048 m | or | 30.48 cm |
| | 1 in | 2.54 cm | or | 25.4 mm |
| Area | 1 $mi^2$ = 640 acres | 2.59 $km^2$ | or | $2.59 \times 10^6$ $m^2$ |
| | 1 acre = 43,560 sq $ft^2$ | 0.4047 ha | or | 4047 $m^2$ |
| | 1 $ft^2$ = 144 $in^2$ | 0.0929 $m^2$ | or | 929.0 $cm^2$ |
| | 1 $in^2$ | 6.452 $cm^2$ | | 6.452 $cm^2$ |
| Volume | 1 acre·ft = 43,560 $ft^3$ | 1233.49 $m^3$ | | 1233.49 $m^3$ |
| | 1 $yd^3$ = 27 $ft^3$ = 1728 $in^3$ | 0.7646 $m^3$ | | 0.7646 $m^3$ |
| | 1 $ft^3$ = 7.48 gal | 0.0283 cu m | or | 28.32 liters |
| | 1 U.K. gal | 4546 $cm^3$ | or | 4.546 liters |
| | 1 U.S. gal = 231 $in^3$ | 3785 $cm^3$ | or | 3.785 liters |
| | 1 $in^3$ | 16.387 $cm^3$ | | 16.387 $cm^3$ |
| Mass | 1 ton (short) = 2000 lb | 0.9072 tonne | or | 907.18 kg |
| | 1 ton (long) = 2240 lb | 1.016 tonnes | or | 1016 kg |
| | 1 lb = 16 oz | 0.4536 kg | or | 453.6 g |
| | 1 oz | 28.352 k | | 28.352 g |
| Density | 1 pcf | 16.019 $kg/m^3$ | | 0.157 $kN/m^3$ |
| Force | 1 tonf (short) | 0.9072 tonnef | | 8.897 kN |
| | 1 tonf (long) | 1.016 tonnesf | | 9.964 kN |
| | 1 lbf | 0.4536 kgf | | 4.448 N |
| Pressure or stress | 1 tsf (short) = 2000 psf | 9.764 $t/m^2$ | | 95.76 kPa |
| | 1 tsf (short) = 13.89 psi | 0.9764 $kg/cm^2$ | | 95.76 kPa |
| | 1 tsf (long) = 2240 psf | 10.936 $t/m^2$ | | 107.3 kPa |
| | 1 psf = 0.00694 psi | 0.000488 $kgf/cm^2$ | | 0.04788 kPa |
| | 1 psi = | 0.0703 $kgf/cm^2$ | | 6.895 kPa |
| | 1 bar = | 1.02 $kg/cm^2$ | | $10^5$ Pa |
| | 1 atm = 33.9 ft of water (39.2°F) | 76 mm mercury at 0°C | | |
| | 1 atm = 1.058 tsf = 14.7 psi | 1.033 $kgf/cm^2$ | | 1.0133 bars |

(*continued*)

| ENGLISH TO METRIC TO THE INTERNATIONAL SYSTEM* (*Continued*) | | | |
|---|---|---|---|
| **Units†** | **English** | **Metric** | **SI** |
| Velocity | 1 ft/year = $1.9025 \times 10^{-6}$ ft/min | $0.9659 \times 10^{-6}$ cm/s | $0.9659 \times 10^{-8}$ m/s |
| Acceleration of gravity | 1 $ft/s^2$<br>32 $ft/s^2$ | 0.3048 $m/s^2$<br>980 $cm/s^2$ | 0.3048 $m/s^2$<br>980 gals |
| Flow | 1 gal/min = 192 $ft^3$/day | 0.0038 $m^3$/min | 3.8 liters/minute |
| Temperature | t°F = 1.8 t°C + 32 | t°C = (t°F − 32)/1.8 | |
| Water, unit weight | 62.4 pcf | 1 $g/cm^3$ = 1 $t/m^3$ | 9.81 $kN/m^3$ |

| METRIC TO ENGLISH TO SI | | | |
|---|---|---|---|
| **Units** | **Metric** | **English** | **SI** |
| Length | 1 km = 1000 m<br>1 m = 100 cm<br>1 cm = 10 mm | 0.6214 mi<br>3.28 ft = 39.37 in<br>0.3937 in | 1 km<br>1 m<br>1 cm |
| Area | 1 $km^2$ = 100 ha<br>1 ha = 10,000 $m^2$<br>1 $m^2$ = 10,000 $cm^2$ | 0.386 $mi^2$<br>2.47 acres<br>10.764 $ft^2$ = 1550 $in^2$ | 1 $km^2$<br>1 ha<br>1 $m^2$ |
| Volume | 1 $m^3$ = 1000 liter<br>1 liter = 1000 $cm^3$ | 1.31 $yd^3$ = 35.315 $ft^3$<br>0.264 gal = 61 $in^3$ | 1 $m^3$<br>1 liter |
| Mass | 1 tonne = 1000 kg<br>1 kg = 1000 g | 2204.6 lb<br>2.205 lb | 1 tonne<br>1 kg |
| Density | 1 $t/m^3$ = 1 $g/cm^3$ | 62.428 pcf | 9.807 $kN/m^3$ |
| Force | 1 tonnef = 1000 kgf<br>1 kgf = 1 kilopond = 1000 g | 1.10 tonf (short)<br>2.205 lbf | 9.807 kN<br>9.807 N |
| Unit weight | 1 $t/m^3$ = 1 $g/cm^3$ | 62.428 pcf | 9.807 $kN/m^3$ |
| Pressure or stress | 1 $t/m^2$ = 0.1 $kg/cm^2$<br>1 $kg/cm^2$<br>1 $kg/cm^2$<br>1.019 $kg/cm^2$ | 0.1024 tsf (short)<br>1.024 tsf = 14.223 psi<br>0.9678 atm<br>14.495 psi | 9.807 kPa<br>98.07 kPa<br><br>1 bar (100 kPa) |
| Velocity | 1 cm/s | 1.9685 ft/min | 1 cm/s |
| Acceleration of gravity | 1 $cm/s^2$ = 0.01 $m/s^2$ | 0.3048 $ft/s^2$ | 1 gal |
| Flow | 1 $m^3$/s = 1000 liters/second<br>1 liter/second | 15,800 gal/min<br>15.8 gal/min = 0.263 gal/s | 1 $m^3$/s |

*Metric and SI units are generally the same except for force and pressure; therefore under the SI column some additional useful metric conversions are given.

†For abbreviations see *Symbols* at end of table.

| SYMBOLS | | | |
|---|---|---|---|
| **Time** | **English** | **Metric** | **SI** |
| mo—month<br>hr—hour<br>min—minute<br>s—second | in—inch<br>ft—foot<br>yd—yard<br>mi—mile<br>oz—ounce<br>lb—pound<br>t—ton<br>gal—gallon<br>psi—$lb/in^2$<br>tsf—$tons/ft^2$<br>pcf—$lbs/ft^3$ | mm—millimeters<br>cm—centimeters<br>m—meters<br>km—kilometers<br>ha—hectares<br>g—grams<br>kg—kilograms<br>t—tonnes<br>f—force | N—newton (force)<br>Pa—pascal = $N/m^2$ (pressure)<br>1 N = 0.2248 lb<br>= 0.102 kg = 102 g<br>1 kPa = 0.0104 tsf (short)<br>= 0.145 psi<br>= 0.102 $kgf/cm^2$<br>1 bar = 100 kPa |
| Pressure | | | |
| atm—atmosphere | | | |

NOTE: M = mega ($10^6$), k = kilo ($10^3$), h = hecto $(10)^2$, da = deka = (10), d = deci ($10^{-1}$), c = centi ($10^{-2}$), m = milli ($10^{-3}$), $\mu$ = micro ($10^{-6}$).

| SUMMARY TABLE OF CONVERSIONS | | | | | | | |
|---|---|---|---|---|---|---|---|
| | | System of units | | | Mutual proportion | | |
| **Name** | **Symbol** | **SI** | **Metric** | **Imperial** | **SI** | **Metric** | **Imperial** |
| Length | $l$ | mm<br>m | mm<br>m | in<br>ft | 1<br>1 | 1<br>1 | 0.03937<br>3.281 |
| Area | $A$ | $mm^2$<br>$m^2$ | $mm^2$<br>$m^2$ | $in^2$<br>$ft^2$ | 1<br>1 | 1<br>1 | $1.55 \times 10^3$<br>10.764 |
| Volume | $V$ | milliliter<br>$m^3$ | $cm^3$<br>$m^3$ | $in^3$<br>$ft^3$ | 1<br>1 | 1<br>1 | 0.061<br>35.315 |
| Velocity | $v$ | mm/s<br>m/s | mm/s<br>m/s | in/s<br>ft/s | 1<br>1 | 1<br>1 | 0.03937<br>3.281 |
| Rate of flow | $Q$ | milliliter/second<br>$m^3/s$ | $cm^3/s$<br>$m^3/s$ | $ft^3/s$<br>$ft^3/s$ | 1<br>1 | 1<br>1 | $3.531 \times 10^{-3}$<br>35.315 |
| Mass | $m$ | g<br>t | g<br>t | lb<br>lb | 1<br>1 | 1<br>1 | $2.205 \times 10^{-3}$<br>2205 |
| Density | $\rho$ | $kg/m^3$<br>$t/m^3$ | $kg/m^3$<br>$t/m^3$ | $lb/ft^3$<br>$lb/ft^3$ | 1<br>1 | 1<br>1 | 0.06243<br>62.4 |
| Force | $P, F$ | N<br>kN<br>MN | kgf<br>kgf<br>tf | lbf<br>lbf<br>kips | 1<br>1<br>1 | 0.101971<br>101.971<br>101.971 | 0.22481<br>224.81<br>224.81 |
| Stress pressure | $\sigma, p$ | $kN/m^2$<br>$MN/m^2$ | $kgf/cm^2$<br>$tf/m^2$ | $lbf/in^2$<br>$tf/ft^2$ | 1<br>1 | 0.0101971<br>101.971 | 14.504<br>9.3238 |
| Unit weight | $\gamma$ | $kN/m^3$ | $gf/cm^3$ | $lbf/ft^3$ | 1 | 0.10197 | 6.3657 |

# APPENDIX D

# Symbols

**SYMBOLS**

| Symbol | Represents | Art. |
|---|---|---|
| $A$ | Activity of clay | 5.3.3 |
| $A$ | Amplitude | 11.2.2 |
| $A$ | Area | 3.3.1 |
| $A$ | Pore-pressure parameter | 3.4.2 |
| Å | Angstrom | 5.3.3 |
| $a$ | Acceleration | 11.2.2 |
| $a_v$ | Compressibility coefficient | 3.5.4 |
| $B$ | Bulk modulus | 3.5.2 |
| $B$ | Pore-pressure parameter | 3.4.2 |
| $C$ | Compressibility | 3.5.4 |
| $C$ | Pore-pressure parameter | 3.4.2 |
| $C_c$ | Compression index | 3.5.4 |
| $C_N$ | Depth correction factor | 3.4.5 |
| $C_u$ | Uniformity coefficient | 3.2.3 |
| $C_R$ | Gradation range | 3.2.3 |
| CD | Consolidated drained triaxial test | 3.4.4 |
| CU | Consolidated undrained triaxial test | 3.4.4 |
| CBR | California bearing ratio | 3.4.5 |
| CPT | Cone penetrometer test | 3.4.5 |
| $C_\alpha$ | Secondary-compression coefficient | 3.5.4 |
| $c$ | Cohesion | 3.4.2 |
| $\bar{c}$, $c'$ | Cohesion based on effective stresses | 3.4.2 |
| $c_a$ | Apparent cohesion | 3.4.2 |
| $c_p$ | Compression-wave velocity | 3.5.5 |
| $c_s$ | Shear-wave velocity | 3.5.5 |
| $c_v$ | Consolidation coefficient | 3.5.4 |
| $D$ | Constrained modulus | 3.5.2 |
| $D$ | Damping ratio | 3.4.2 |
| $D$ | Diameter | 3.4.3 |
| $D_R$ | Relative density | 3.2.3 |
| $D_{10}$ | Diameter at which 10% of soil is finer | 3.3.2 |
| $D_{15}$ | Diameter at which 15% of soil is finer | 8.4.5 |
| $D_{85}$ | Diameter at which 85% of soil is finer | 8.4.5 |
| $d$ | Pipe diameter | 3.3.4 |
| $E$ | Moment arm | 9.1.4 |
| $E$ | Young's modulus | 3.5.2 |
| $E_c$ | Compression modulus | 3.5.4 |
| $E_d$ | Dynamic Young's modulus | 3.5.2 |
| $E_i$ | Initial tangent modulus | 3.5.2 |
| $E_r$ | In situ static rock modulus | 3.5.3 |
| $E_s$ | In situ static soil modulus | 3.5.2 |
| $E_{se}$ | Secant modulus | 3.5.2 |
| $E_{sr}$ | Static recovery modulus | 3.5.2 |
| $E_t$ | Tangent modulus | 3.5.2 |
| $E_{seis}$ | Dynamic field modulus from $V_p$ | 3.5.3 |
| $e$ | Void ratio | 3.2.3 |
| $e_0$ | Initial void ratio | 3.5.4 |
| $e_f$ | Final void ratio | 3.5.4 |
| $F$ | Force | 3.4.2 |
| $F$ | Ground amplification factor | 11.2.6 |

(continued)

## SYMBOLS (*Continued*)

| Symbol | Represents | Art. |
|---|---|---|
| FR | Friction ratio | 2.3.4 |
| FS | Factor of safety | 9.1.4 |
| $f$ | Coefficient of friction | 3.4.1 |
| $f$ | Frequency | 11.2.2 |
| $f_s$ | Shaft friction, CPT | 2.3.4 |
| $G$ | Shear modulus | 3.5.2 |
| $G_d$ | Dynamic shear modulus | 3.5.2 |
| $G_s$ | Specific gravity of solids | 3.2.1 |
| $G_w$ | Specific gravity of water | 3.2.1 |
| GWL | Groundwater level, static | 2.3.7 |
| $g$ | Acceleration of gravity | 11.2.3 |
| $H$ | Hydraulic head | 3.3.4 |
| $H$ | Height | 9.1.4 |
| $H$ | Stratum thickness | 3.5.2 |
| $H_c$ | Capillary rise | 3.3.1 |
| $H_A$ | Abrasion test hardness | 3.2.1 |
| $H_R$ | Schmidt hardness | 3.2.1 |
| $H_T$ | Total hardness | 3.2.1 |
| $h$ | Hydraulic head | 3.3.3 |
| $h_e$ | Elevation head | 8.3.2 |
| $h_p$ | Pressure head | 8.3.2 |
| $h_t$ | Total hydraulic head | 8.3.2 |
| $I$, $I_o$ | Epicentral intensity | 11.2.4 |
| $I_s$ | Site intensity | 11.2.5 |
| $I_s$ | Point-load index | 3.4.3 |
| $i$ | Angle of slope with horizontal | 9.3.2 |
| $i$ | Angle of joint asperities | 6.4.4 |
| $i$ | Hydraulic gradient | 3.3.1 |
| $i_{cr}$ | Critical hydraulic gradient | 8.3.2 |
| JRC | Joint roughness coefficient | 6.4.4 |
| $j$ | Seepage force per unit volume | 8.3.2 |
| $K$ | Permeability coefficient (geology) | 8.3.2 |
| $K$ | Dynamic bulk modulus | 3.5.2 |
| $K$ | Pressure meter constant | 3.5.4 |
| $K$ | Stiffness or spring constant | 11.4.2 |
| $K_a$ | Active stress coefficient | 3.4.2 |

| Symbol | Represents | Art. |
|---|---|---|
| $K_o$ | At-rest earth pressure coefficient | 3.4.2 |
| $K_p$ | Passive stress coefficient | 3.4.2 |
| $k$ | Permeability coefficient | 3.3.1 |
| $k_h$ | Horizontal $k$ | 3.3.4 |
| $k_{mean}$ | Average $k$ | 3.3.4 |
| $k_v$ | Vertical $k$ | 3.3.4 |
| $k$, $k_s$ | Joint stiffness | 6.4.4 |
| $k_n$ | Normal joint stiffness | 6.4.4 |
| $k_s$ | Subgrade reaction modulus | 3.5.4 |
| $k_{sh}$ | Horizontal $k_s$ | 3.5.4 |
| $k_{sv}$ | Vertical $k_s$ | 3.5.4 |
| $k_t$ | $k_s$ from test | 3.5.4 |
| $L$ | Long waves | 11.2.2 |
| $L$ | Length | 3.3.3 |
| LI | Liquidity index | 3.2.3 |
| LL | Liquid limit | 3.2.3 |
| $M$ | Magnitude | 11.2.4 |
| $M$ | Mass | 11.4.2 |
| MM | Modified Mercalli scale | 11.2.4 |
| $m$ | Mass | 11.4.2 |
| $m_v$ | Volume-change coefficient | 3.5.4 |
| $N$ | Normal force component | 3.4.1 |
| $N$ | Standard penetration test (SPT) resistance | 3.5.4 |
| $N'$, $N_1$ | $N$ (SPT) corrected for $\bar{\sigma}_{vo}$ | 3.5.4 |
| $N_c$ | Bearing capacity factor | 3.4.5 |
| $N_f$ | Number of flow channels | 8.3.3 |
| $N_e$ | Number of equipotential drops | 8.3.3 |
| $\overline{N}$ | Effective normal force | 9.3.2 |
| $n$ | Porosity | 3.2.1 |
| $n_e$ | Effective porosity | 8.3.2 |
| $O$ | Origin | 9.1.4 |
| $O_c$ | Center of gravity | 9.1.4 |
| OCR | Overconsolidation ratio | 3.5.4 |
| $P$ | Force, pressure | 3.4.1 |
| $P$ | Primary waves | 2.3.2 |

## SYMBOLS (*Continued*)

| Symbol | Represents | Art. |
|---|---|---|
| $P'$ | Effective overburden pressure | 3.4.4 |
| $P_a$ | Atmospheric pressure | 8.3.2 |
| $P_a$ | Active force | 3.4.2 |
| $P_p$ | Passive force | 3.4.2 |
| PI | Plasticity index | 3.2.3 |
| PL | Plasticity limit | 3.2.3 |
| $P_L$ | Limiting pressure | 3.5.4 |
| $p$ | Unit pressure | 3.5.2 |
| $p_c$ | Preconsolidation pressure | 3.5.4 |
| $p_o$ | Overburden pressure | 3.5.4 |
| $p_s$ | Seepage pressure | 8.3.3 |
| Q | Load | 3.5.1 |
| Q | Rate of flow | 3.3.3 |
| Q | Quick, or CU, UU triaxial test | 3.4.4 |
| $Q_{all}$ | Allowable bearing value | 5.2.7 |
| $q$ | Rate of flow per unit area | 3.3.1 |
| $q_c$ | Cone point resistance (CPT) | 2.3.4 |
| $R$ | Radius | 8.3.3 |
| $R$ | CU triaxial test | 3.4.4 |
| $R$ | Resultant force | 9.1.4 |
| $R$ | Rayleigh waves | 2.3.2 |
| $R$ | Distance from the source | 11.2.5 |
| RQD | Rock quality designation | 2.4.5 |
| $r$ | Radius | 3.3.4 |
| $r_w$ | Well radius | 8.3.3 |
| $S$ | Saturation percent | 3.2.3 |
| $S$ | Settlement | 3.5.2 |
| $S$ | Shear waves | 2.3.2 |
| $S$ | Shearing resistance, strength | 3.4.2 |
| $S$ | Slow or CD triaxial test | 3.4.4 |
| $\mathcal{S}$ | Shape factor | 8.3.3 |
| $S_{max}$ | Maximum shearing resistance | 9.3.2 |
| SL | Shrinkage limit | 3.2.3 |
| SS | Split-barrel sampler | 2.4.2 |
| SPT | Standard penetration test | 3.5.4 |
| $S_t$ | Sensitivity | 3.4.2 |
| $s_r$ | Remolded undrained shear strength | 3.4.2 |
| $s_u$ | Undrained shear strength | 3.4.2 |
| $s_v$ | Spectral velocity | 11.4.4 |
| $s_a$ | Spectral acceleration | 11.4.4 |
| $s_d$ | Spectral displacement | 11.4.4 |
| $T$ | Shear force | 3.4.1 |
| $T$ | Period | 11.2.2 |
| $T$ | Transmissibility | 8.3.2 |
| $T_v$ | Theoretical time factor | 3.5.4 |
| $t$ | Time | 3.5.4 |
| $U$ | Average consolidation ratio | 3.5.4 |
| $U$ | Total pore pressure | 9.3.2 |
| $U_c$ | Unconfined or uniaxial compressive strength | 3.4.3 |
| $U_{ult}$ | Ultimate unconfined strength | 3.4.3 |
| UD | Undisturbed sample | 2.4.2 |
| UU | Unconsolidated undrained triaxial test | 3.4.4 |
| $u$ | Unit pore pressure | 3.4.2 |
| $u_w$ | Unit pore-water pressure | 3.4.2 |
| $V$ | Total volume of sample | 3.2.3 |
| $V_a$ | Volume of air or gas | 3.2.3 |
| $V_s$ | Volume of solids | 3.2.3 |
| $V_v$ | Volume of voids | 3.2.3 |
| $V_w$ | Volume of water | 3.2.3 |
| $V_t$ | Total flow volume | |
| $V$ | Joint water pressure | 9.3.2 |
| $V$ | Velocity | 2.3.2 |
| $V_p$ | Compression-wave velocity | 2.3.2 |
| $V_s$ | Shear-wave velocity | 2.3.2 |
| $V_r$ | Rayleigh-wave velocity | 3.5.5 |
| $V_F$ | Field $V_p$ | 3.5.3 |
| $V_L$ | Laboratory $V_p$ | 3.5.3 |
| $V_{Fs}$ | Field $V_s$ | 5.2.7 |
| $V_{Ls}$ | Laboratory $V_s$ | 5.2.7 |
| $v$ | Flow velocity | 3.3.4 |

## SYMBOLS (*Continued*)

| Symbol | Represents | Art. |
|---|---|---|
| $v$ | Particle or vibration velocity | 11.2.3 |
| $v_d$ | Discharge velocity | 8.3.2 |
| $v_s$ | Seepage velocity | 8.3.2 |
| $W$ | Total weight | 9.1.4 |
| $W_s$ | Weight of solids | 3.2.3 |
| $W_w$ | Weight of water | 3.2.3 |
| $W_t$ | Total sample weight | 3.2.3 |
| $W_D$ | Sample weight, dense | 3.2.3 |
| $W_L$ | Sample weight, loose | 3.2.3 |
| $W_N$ | Sample weight, natural | 3.2.3 |
| $w$ | Water content | 3.2.3 |
| $w_n$ | Natural water content | 3.2.3 |
| x | Distance along x axis | 11.2.2 |
| y | Distance along y axis | 11.2.2 |
| y | Displacement | 11.2.2 |
| y | Unit deflection | 3.5.4 |
| Z, z | Depth | 3.5.1 |
| $z_c$ | Depth of tension crack | 9.3.2 |
| $z_w$ | Water depth | 8.3.2 |
| $\alpha$ | Angle of obliquity | 3.4.1 |
| $\alpha$ | Compression modulus factor | 3.5.4 |
| $\alpha$ | Inclination of force angle | 3.4.1 |
| $\beta$ | Modulus reduction factor | 3.5.3 |
| $\gamma_D$ | Maximum density | 3.2.3 |
| $\gamma_L$ | Loose density | 3.2.3 |
| $\gamma_N$ | Natural density | 3.2.3 |
| $\gamma_t$ | Total unit weight | 3.2.3 |
| $\gamma_d$ | Dry unit weight | 3.2.3 |
| $\gamma_b$ | Buoyant unit weight | 3.2.3 |
| $\gamma_s$ | Saturated unit weight | 3.2.3 |
| $\gamma_w$ | Unit weight of water | 3.2.3 |
| $\Delta, \delta$ | Change or increment | 3.4.2 |
| $\epsilon$ | Strain | 3.5.2 |
| $\zeta$ | Unit shear strain | 3.5.2 |
| $\theta$ | Angle between failure surface and horizontal | 9.3.2 |
| $\theta$ | Angle between normal stress and major principal stress | 3.4.1 |
| $\theta_{cr}$ | Slope of failure surface | 3.4.1 |
| $\lambda$ | Damping ratio | 3.4.2 |
| $\lambda$ | Wavelength | 11.2.2 |
| $\mu$ | Microns | 10.6.2 |
| $\mu_r$ | Loading time-rate correction factor, vane shear test | 3.4.2 |
| $\nu$ | Poisson's ratio | 3.4.1 |
| $\rho$ | Mass or bulk density | 3.2.1 |
| $\rho$ | Consolidation settlement | 3.5.4 |
| $\Sigma$ | Sum | 3.5.2 |
| $\sigma, \sigma_n$ | Normal stress | 3.4.1 |
| $\bar{\sigma}_1, \bar{\sigma}_n$ | Effective normal stress | 3.4.2 |
| $\sigma_{nf}$ | Normal stress at failure | 3.4.1 |
| $\sigma_1, \sigma_2, \sigma_3$ | Principal stresses | 3.4.1 |
| $\sigma_d$ | Deviator stress | 3.4.2 |
| $\sigma_h$ | Horizontal stress | 3.4.1 |
| $\sigma_v$ | Vertical stress | 3.4.1 |
| $\bar{\sigma}_o, \bar{\sigma}_{vo}$ | Effective vertical overburden stress | 3.5.4 |
| $\sigma_j$ | Joint uniaxial compressive strength | 6.4.4 |
| $\tau$ | Shear stress | 3.4.1 |
| $\tau$ | Joint shear strength | 6.4.4 |
| $\tau_{max}$ | Maximum shear stress | 3.4.2 |
| $\phi$ | Friction angle | 3.4.2 |
| $\bar{\phi}, \phi'$ | Effective friction angle | 3.4.2 |
| $\phi_r$ | Residual strength friction angle | 3.4.2 |
| $\omega$ | Circular frequency | 11.2.2 |

NOTE: Not included are symbols for minerals or chemical elements (Chap, 5), classification systems (Chap 5), or dimensions (Appendix C).

# APPENDIX E

# Engineering Properties of Geologic Materials: Data and Correlations

| Materials | Properties |  | Reference |
|---|---|---|---|
|  | Class* | Details |  |
| ROCK |  |  |  |
| General | I-B | Range in total hardness | Fig. 3.1 |
|  | P, C, S | General engineering properties of common rocks | Table 5.16 |
|  | P | Typical values for $k$ | Table 3.12 |
|  | P | $k$, $n_{avg}$, specific yield | Table 8.1 |
|  | C | RQD vs. $E_r$ (rock mass) | Fig. 3.79 |
|  | C | $Q_{all}$ for various rock types (NYC Bldg. Code) | Table 5.22 |
|  | S | $U_c$ vs. consistency | Fig. 3.39 |
|  | S | $U_c$ vs. Schmidt hardness | Fig. 3.40 |
|  | S | $U_c$ and rock classification | Fig. 5.7 |
| Fresh, intact | I-B, S | $\gamma_d$, $n$, $U_c$, $\phi$, $\phi_r$ for various rock types | Table 3.26 |
|  | I-B, C, S | E, $\nu$, for various rock types | Table 3.26 |
| Decomposed | C | $Q_{all}$ for various rocks (NYC Bldg. Code) | Table 5.22 |
|  | S | $c$, $c'$, $\phi$, $\phi'$, $\phi_r$ for various rock types | Table 3.27 |
| Marine and clay shales | I-B, S | $\gamma_d$, $w$, LL, PI, $\bar{c}$, $\bar{\phi}$, $\bar{\phi}_r$ | Table 3.30 |
|  | I-B, S | LL vs. $\phi_r$; marine shales of NW United States | Fig. 6.94 |
|  | I-B | $\gamma_d$, $w_n$, LL, PI, minus 0.074 mm, activity (LA area) | Table 7.8 |
| Coal | S | Triaxial compression tests: load vs. $\sigma_3$ | Fig. 10.12 |
|  | S | $U_c$ for various conditions | Table 10.2 |
| Joints | S | $\phi$, $\phi_r$, various joint types in hard rock | Table 6.8 |
|  | S | $\phi$, $\phi_r$ for joint fillings | Table 3.27 |

(continued)

| Materials | Class* | Details | Reference |
|---|---|---|---|
| | | **Properties** | |
| | | ROCK | |
| Gouge | S | $\phi$, $\phi_r$ for foliation shear, mylonite seams, faults | Table 3.27 |
| | S | $\phi_r$ vs. PI | Fig. 3.33 |
| Minerals | I-B | $G_s$, hardness (Moh's scale) | Table 5.4 |
| | I | Characteristics | Table 5.5 |
| | | SOILS: BY GRADATIONS | |
| Various | P, C, E, S | General engineering properties | Table 5.31 |
| | P | $k$ ranges for general formations | Table 3.12 |
| | P | $k$ values for various soil types | Table 3.14 |
| | P | $k$ ranges; soil formations classed by origin | Table 3.13 |
| | P | $k$, $n$, and specific yield | Table 8.1 |
| | P | $k$, $i$ vs. flow rate | Table 8.2 |
| | P | $kv$ ranges and methods of measurement | Table 3.11 |
| | P | $k$ ranges vs. $D_{10}$ vs. $H_c$ and frost heave susceptibility | Table 3.10 |
| | C | $E$, $\nu$ | Table 3.33 |
| | C, S | $E_c$, $P_L$ (from pressuremeter) | Table 3.37 |
| | S | Ratio of $q_c/N$ (from CPT) | Table 3.24 |
| | S | Pore-pressure parameter $A$ at failure | Table 3.17 |
| | S | CBR vs. soil class vs. rating as subgrade, etc. | Fig. 3.62 |
| Compacted fills | L-B, P, C, S | $\gamma_d$, optimum moisture, % compression, $c$, $\phi$, $k$, CBR and $k_s$ | Table 3.31 |
| Cohesionless (see also Various) | | | |
| | I-B | General gradation characteristics | Fig. 3.11 |
| | I-B, S | Gradation and $D_R$ vs. $N$, $\gamma_d$, $e$, $\phi$ | Table 3.28 |
| | P | $k$ vs. $D_{10}$ | Fig. 3.14 |
| | P | $k$ vs. $D_R$ vs. gradation | Fig. 3.15 |
| | C | Compactness vs. $D_R$ and $N$ | Table 3.23 |
| | C | $k_t$ | Table 3.41 |
| | C | Strain modulus relations ($E$ and $G$) | Fig. 3.97 |
| | S | $q_c$ vs. $D_R$ vs. $p'$ | Fig. 3.60 |
| | S | $q_c$ vs. $p'$ vs. $\phi$ | Fig. 3.59 |
| | S | $\phi$ vs. $D_R$ vs. gradation | Fig. 3.63 |
| | S | Susceptibility to liquefaction during strong seismic shaking | Table 11.10 |
| | S | Gradation vs. liquefaction susceptibility | Fig. 11.33 |
| | S | Cyclic stress ratio vs. $N_1$ and liquefaction | Fig. 11.34 |
| Cohesive (see also Various and soils classed by origin) | | | |
| | I-B | Plasticity chart (LL vs. PI vs. classification) | Fig. 3.12 |
| | C, I-B | $e$-log-$p$ curves for various clays, $\gamma_d$, w, LL, PI | Fig. 3.87 |

| Materials | Properties: Class* | Properties: Details | Reference |
|---|---|---|---|
| | | ROCK | |
| | C | $k_t$ | Table 3.42 |
| | E | Clay fraction and PI vs. activity | Fig. 10.38 |
| | E | %—0.001 mm, PI, SL vs. % volume change for $p = 1$ psi | Table 10.4 |
| | E | %—0.002 vs. activity and % swell for $p = 1$ psi | Fig. 10.39 |
| | S | Sensitivity classification ($s_u/s_r$) | Table 3.18 |
| | S | Normalized ($s_u/p'$) from $q_c$ for estimating OCR | Fig. 3.71 |
| | I-B, S | Consistency vs. $N$, $\gamma_s$, $U_c$ | Table 3.29 |
| | S | $N$ vs. $U_c$ vs. plasticity classes | Fig. 3.64 |
| | I-B, S | $\gamma_d$, $w_n$, LL, PI, $s_u$, $\bar{\phi}$, $\bar{c}$, $\phi_r$ various materials | Table 3.30 |
| | | SOILS: CLASSED BY ORIGIN | |
| Residual | I-B, C | Gneiss, humid climate: $N$, $e$, general character | Table 7.5 |
| | I-B | Gneiss, humid climate, $w$, PL, LL, $p_c$ vs. depth | Fig. 7.4 |
| | I-B | Igneous and metamorphics: $N$, $w$, PI vs. depth | Fig. 7.5 |
| | I-B | Porous clays, general: gradations | Fig. 10.25 |
| | I-B | Porous clays, general: PI vs. LL | Fig. 10.26 |
| | I-B | Porous clays, basalt: $\gamma_t$, LL, PI, $w$, $e$ vs. depth | Fig. 10.27 |
| | C | Porous clays, basalt: saturation effect, $e$-log-$p$ | Fig. 10.28 |
| | I-B | Porous clays, clayey sandstone: $w$, LL, PL, $n$ vs. Z | Fig. 7.7 |
| | I-B, S | $\gamma_d$, $w$, LL, PI, $\bar{c}$, $\bar{\phi}$, (natural $w$ and soaked) | Table 3.30 |
| | I-B | Sedimentary rocks: $N$, $w$, PI, LL vs. Z | Fig. 7.9 |
| | C, I-B | Gneiss, basalt $e$-log-$p$ curves, $\gamma_d$, $w$, LL, PI | Fig. 3.87 |
| | I-B, S | Basalt and gneiss N, LL, PI, $e$, $\phi$, $c$; | Table 7.4 |
| Colluvium | I-B, S | $\gamma_d$, $w$, LL, PI, $\bar{c}$, $\bar{\phi}$, $\phi_r$ | Table 3.30 |
| Alluvial | I-B, C | Fluvial: $\gamma_d$, $w_n$, LL, PI, $e$-log-$p$ curves | Fig. 3.87 |
| | I-B, C | Fluvial (backswamp): $\gamma_d$, $w_n$, LL, PI, $e$-log-$p$ curve | Fig. 3.87 |
| | I-B, S | Fluvial (backswamp): $\gamma_d$, $w_n$, LL, PI, $s_u$ | Table 3.30 |
| | I-B, S | Fluvial (backswamp): $w_n$, LL, PI, $s_u$ | Fig. 7.27 |
| | I-B | Estuarine: $N$, LL, PI, $w_n$ | Fig. 7.39 |
| | I-B, S | Estuarine: $\gamma_d$, $w_n$, LL, PI, $s_u$ (various) | Table 3.30 |
| | I-B, S | Estuarine: $w_n$, LL, PL, $s_u$, $p'$ (Maine) | Fig. 7.40 |
| | I-B, S | Estuarine: $\gamma$, $w_n$, LL, PL, $p_c$, $s_u$, $S_t$ (Thames River) | Fig. 7.41 |
| | I-B | Coastal: $N$, $w_n$, LL, PI | Fig. 7.47 |
| | I-B | Coastal plain: $N$, $w_n$, LL, PI (Atlantic) | Fig. 7.53 |
| | I-B, S | Coastal plain: $w_n$, LL, PL, $s_u$, $m_v$ (London) | Fig. 7.57 |
| | I-B, C | Coastal plain: $\gamma_d$, $w_n$, LL, PI, $e$-log-$p$ (Texas) | Fig. 3.87 |
| | I-B, S | Coastal plain: $\gamma_d$, $w_n$, LL, PI, $s_u$, $\bar{c}$, $\bar{\phi}$, $\phi_r$, $e$ | Table 3.30 |
| | I-B, S, E | Coastal plain: $w_n$, PI, SL, $U_c$, $P_{swell}$, $\Delta V$ (Texas) | Art. 7.4.4 |

(continued)

| Materials | Properties: Class* | Properties: Details | Reference |
|---|---|---|---|
| | | ROCK | |
| | I-B, C | Lacustrine, Mexico City: $\gamma_d$, $w_n$, LL, PI, $e$-log-$p$ | Fig. 3.87 |
| | I-B, S | Lacustrine, Mexico City: $\gamma_d$, $w_n$, LL, PI, $s_u$, $S_t$ | Table 3.30 |
| | I-B, S | Lacustrine, Mexico City: $N$, $w_n$, $U_c$, $p'$ | Fig. 7.60 |
| | I-B, S | Marine, various: $\gamma_d$, $w_n$, LL, PI, $s_u$ | Table 3.30 |
| | C | Valley alluvium (collapsing soil): $e$-log-$p$ (sat.) | Fig. 10.24 |
| Loess | I-B | Gradation and plasticity: Kansas-Nebraska | Fig. 7.71 |
| | C | $\gamma_d$ vs. $P$ compression curves; prewet, wet | Fig. 7.72 |
| | C | Settlement upon saturation vs. $\gamma$ | Table 7.10 |
| | I-B, S | $\gamma_d$, $w_n$, LL, PI, $\bar{c}$, $\bar{\phi}$: Kansas-Nebraska | Table 3.30 |
| | S | $\tau$ vs. $\sigma_n$ vs. density; wet | Fig. 7.74 |
| Glacial | I-D | Till: typical $N$ values, various locations | Fig. 7.84 |
| | C | Till: $E_c$ (pressuremeter) | Table 3.37 |
| | I-B, S | Till: $\gamma_d$, $w_n$, LL, PI, $s_u(U_c)$ | Table 3.30 |
| | I | Stratified drift, $N$ values | Fig. 7.88 |
| | I-B | Lacustrine, various locations: $N$, $w_n$, LL, PI vs. $Z$ | Fig. 7.97 |
| | I-B | Lacustrine, East Rutherford, N.J.: $N$, $w_n$, LL, PI vs. $Z$ | Fig. 7.98 |
| | I-B, C | Lacustrine, New York City: $w_n$, LL, PI, $e$-log-$p$ | Fig. 3.87 |
| | I-B, S | Lacustrine, various: $\gamma_d$, $w_n$, LL, PI, $s_u$, $\bar{c}$, $\bar{\phi}$ | Table 3.30 |
| | I | Lacustrine, New York City: plasticity chart | Fig. 7.100 |
| | S | Lacustrine, New York, New Jersey, Connecticut: OCR vs. $s_u/p'$ | Fig. 7.101 |
| | S | Lacustrine, Chicago: $w_n$, LL, PL, $U_c$ vs. $Z$ | Fig. 7.96 |
| | I-B, C, S | Lacustrine, New York City: $w_n$, LL, PI, $D_{10}$, $e$, $p_c$, $C_v$, OCR, $C_\alpha$, $s_u$ (clay and silt varves) | Table 7.12 |
| | I-B, S | Marine, various: $\gamma_d$, $w_n$, LL, PI, $s_u$, $S_t$ | Table 3.30 |
| | I-B, S | Marine, Norway: $\gamma$, $w_n$, LL, PL, $c/p$ $S_t$ | Fig. 7.102 |
| | I-B, S | Marine, Norway: $\gamma$, $w_n$, LL, PL, $s_u$, $c/p$, $S_t$ | Fig. 7.103 |
| | I-B, S | Marine, Canada: $w_n$, $p'$, $p_c$, $S_t$, $\gamma$ | Fig. 7.104 |
| | I-B, C, S | Marine, Boston: $w_n$, LL, PL, $C_c$, $p_c$, $p'$, $s_u$ | Fig. 7.105 |
| | | GROUNDWATER | |
| | | Indicators of corrosive and incrusting waters | Table 8.3 |
| | | Effect of sulfate salts on concrete | Table 8.4 |
| | | Effect of aggressive $CO_2$ on concrete | Table 8.5 |

*I-B—Index and basic properties
P—Permeability
C—Compressibility
S—Strength

# SUBJECT INDEX

# LOCALITY INDEX

**AUSTRALIA–NEW ZEALAND**



**CENTRAL AMERICA**



**EUROPE**



**MIDDLE EAST**



**NORTH AMERICA**

**WORLD**

# ABOUT THE AUTHOR

Roy E. Hunt received his B.S. in Geology from Upsala College in 1952, and his M.A. in Soil Mechanics and Foundation Engineering from Columbia University in 1956. He is a fellow of both the American Society of Civil Engineers and the American Institute of Consulting Engineers.

Since 1952 he has worked for a number of consulting firms specializing in geotechnical engineering. He has been a project geologist; chief of a soil mechanics laboratory; a project engineer; and the principal of a firm. His experience ranges in many directions, including work on such diverse projects as earth and concrete dams; highways, airfields; foundations for structures; tunnels; development of large land areas; environmental conservation; retaining structures; and slope failures. These projects have taken him across the United States and around the world, to France, Israel, Brazil, Bolivia, Indonesia, and several other countries.

Mr. Hunt has published widely, with articles on soil and foundation engineering, slope failures, retaining structures, and applications of remote-sensing interpretation and terrain analysis.